Advances in Science, Technology & Innovation

IEREK Interdisciplinary Series for Sustainable Development

Advances in Science, Technology & Innovation (ASTI) is a series of peer-reviewed books based on important emerging research that redefines the current disciplinary boundaries in science, technology and innovation (STI) in order to develop integrated concepts for sustainable development. It not only discusses the progress made towards securing more resources, allocating smarter solutions, and rebalancing the relationship between nature and people, but also provides in-depth insights from comprehensive research that addresses the **17 sustainable development goals (SDGs)** as set out by the UN for 2030.

The series draws on the best research papers from various IEREK and other international conferences to promote the creation and development of viable solutions for a **sustainable future and a positive societal** transformation with the help of integrated and innovative science-based approaches. Including interdisciplinary contributions, it presents innovative approaches and highlights how they can best support both economic and sustainable development, through better use of data, more effective institutions, and global, local and individual action, for the welfare of all societies.

The series particularly features conceptual and empirical contributions from various interrelated fields of science, technology and innovation, with an emphasis on digital transformation, that focus on providing practical solutions to **ensure food, water and energy security to achieve the SDGs.** It also presents new case studies offering concrete examples of how to resolve sustainable urbanization and environmental issues in different regions of the world.

The series is intended for professionals in research and teaching, consultancies and industry, and government and international organizations. Published in collaboration with IEREK, the Springer ASTI series will acquaint readers with essential new studies in STI for sustainable development.

ASTI series has now been accepted for Scopus (September 2020). All content published in this series will start appearing on the Scopus site in early 2021.

More information about this series at http://www.springer.com/series/15883

Ferdinando Trapani · Nabil Mohareb ·
Federica Rosso · Denia Kolokotsa ·
Sreetheran Maruthaveeran ·
Mahmoud Ghoneem
Editors

Advanced Studies in Efficient Environmental Design and City Planning

Springer

Editors
Ferdinando Trapani
Department of Architecture
University of Palermo
Palermo, Italy

Federica Rosso
Department of Civil, Construction and Environmental Engineering
Sapienza University of Rome
Rome, Italy

Sreetheran Maruthaveeran
Department of Landscape Architecture
Faculty of Design and Architecture
Universiti Putra Malaysia
Serdang, Malaysia

Nabil Mohareb
Faculty of Architectural Engineering
Beirut Arab University
Beirut, Lebanon

Denia Kolokotsa
Department of Environmental Engineering
Technical University of Crete
Chania, Greece

Mahmoud Ghoneem
Department of Architecture
Faculty of Fine Arts
Helwan University
Helwan, Egypt

ISSN 2522-8714 ISSN 2522-8722 (electronic)
Advances in Science, Technology & Innovation
IEREK Interdisciplinary Series for Sustainable Development
ISBN 978-3-030-65180-0 ISBN 978-3-030-65181-7 (eBook)
https://doi.org/10.1007/978-3-030-65181-7

This Springer imprint is published by the registered company Springer Nature Switzerland AG
The registered company address is: Gewerbestrasse 11, 6330 Cham, Switzerland

Series Editor's Foreword

The world is in the midst of a disturbance of growth, consuming populations, and environmental degradation which produces many environmental challenges such as global climate change, excessive fossil fuel dependency, and the growing demand for energy. All of these are likely to be major challenges and one of the greatest problems facing humanity, nowadays. The majority of the world cities are confronted with similar and comparable challenges, including—for instance—increases in resource consumption, mobility demand, pollution, urban sprawl, social inequities, water degradation, recycling issues, erosion of fertile topsoil, and depletion of ecosystems. Since its early days, the discourse around "smart cities" has included environmental sustainability as one of its core principles.

In this context, environmental design is the fundamental principle of how to shape our cities. Almost half of the energy consumed is used in cities and urban built-up areas, that's why it is necessary to avoid mistakes in urban design at early stages in order to retain more sustainable cities. City planning within the approach of Green Urbanism reflects upon practical strategies focusing on increasing sustainability beyond and within the scope of individual buildings.

This book on *Advanced Studies in Efficient Environmental Design and City Planning* discusses how urbanism is affected by the paradigms of ecology and how smart and green concepts, systems, and interventions could truly contribute to the construction, management, operation, adaptation and maintenance of sustainable and resilient cities, in the framework of both new urban developments and the transformation of existing urban environments.

The significance of this book lies within the pressing need for the integration of sustainability principles in the urban design process of cities and the general need for a sustainable city development. It will be of particular relevance to the rapid urban growth of developing cities. As within the book's theme, the metaphor of Green Urbanism should inspire our imagination and reflection toward sustainable and resilient urbanism. It is a reference for those interested in taking an action toward the environmental issues, helping them to develop their proposals in city planning and its revolutions to be more sustainable.

Cairo, Egypt — Mourad Amer

Scientific Committee

Acknowledgements

We would like to thank the authors of the research papers that were selected for addition in this book. We would also like to thank the reviewers who contributed with their knowledge and constructive feedback in hopes of ensuring the manuscript is of the best quality possible. A special thanks goes to the Editors of this book for their foresight in organizing this volume and diligence in doing a professional job in editing it. Finally, we would like to express our appreciation to the IEREK team for supporting the publication of the best research papers submitted to the conference.

Introduction

It is safe to acknowledge that the field of sustainability, a newly emergent and multidisciplinary field, is rapidly evolving. Having gained momentum and proved its standing over the years, it is a pivotal response to rapid urbanization, specifically in cities of the developing world, in addition to utter dependence on finite energy supply with an ever-increasing demand, combined with potential damage to the world's environmental health through effluent gas emissions and climate change. Consequently, the world is left with an urgent need to address energy efficiency related issues by presenting the combined strategies and management frameworks of urban planning, architecture and construction to overcome challenges of mobility demand, pollution, urban sprawl, social inequities, water degradation, recycling issues, fertile topsoil erosion, and ecosystem exhaustion.

Thus, to deem a successful city as such, it must respond to the needs of the economy, environment, and society by offering resilient infrastructure, security, and efficiency. In this publication, the authors explore the relationship between city planning and energy consumption, identifying, as a result, the factors affecting energy efficiency at, both, the micro- and macro-levels. As main components of energy-efficient cities, energy-efficient buildings and systems represent a key area for climate change mitigation and reduction of global energy consumption. In part one of this book on *Energy and Smart Systems*, the authors present comprehensive and significant research on passive and active design technologies, smart automation technologies, and shading device systems; all of which lay a solid foundation for green building and environmental efficiency. The chapters titled "Multivariable Coupling Influence on Energy-Efficient Skylight Design", "Performance of *L*-shaped Mini Louvers in Tropical Cities in Cooling Energy Reduction: Case Study of Mumbai, Mexico City, and Lagos" and "The Effect of Shading Devices with Integrated Photovoltaics on Energy Efficiency of Buildings" demonstrate multiple and diverse energy efficiency measures, coupled with renewable energy sources, such as photovoltaic integrated shading devices, simple L-shaped mini louvers, design of skylights with exterior shades, which could significantly improve energy performance and reduce electricity consumption and carbon emissions in residential and office buildings.

This book further delves into strategies and measures taken to advance into efficient environmental design, planning and sustainable development at a city-scale level. In its chapters, authors set conditions for the governance of megaprojects to ensure their sustainability as they focus on environmental, socio-economic, institutional sustainable strategies. That being said, and with the existence of various urban sustainability measurements and indicators, questions relating to the accuracy of these methods and their limitations cannot be overlooked and may deliver useful insight into the assessment of city performance. Thus, analyses of informal settlements in Africa and Asia are undertaken with the aim to develop a more comprehensive understanding of informal settlements, establish an understanding of the community's needs and how addressing those needs can eventually contribute to the achievement of the Sustainable Development Goals and advise future development strategies in developing countries.

Through international case studies from Libya, India, New York, London, Paris, Tokyo, Algiers, Amman, and more, the authors develop plans and propose strategies to contribute to the achievement of Sustainable Development Goals. In the chapter titled "The Sustainable Strategy to Create Decentralized Economic Hubs for Regional Development", the decentralization strategy is put forward which is based on providing economic hubs to support and integrate with the medium- and small-scale cities by providing the missing core services for the region, while the chapter on "Innovative Sustainable Strategies to Mobility and Urban Transport Problems in Benghazi", the ASI (Avoid, Shift and Improve) and TOD (Transit-Oriented Development), aims to help cut down transport costs, reduce pollution, and achieve high quality and sustainable mobility and urban transport systems.

In an endeavor to respond to major urban environmental and societal challenges, through a reduction of the ecological footprint, adaptation to climate change and promotion of health, Urban Green Infrastructure (GI) has been developed as a strategic planning approach. In *Urban Landscapes and Green Infrastructure,* the authors examine current inequalities in the distribution and access to green spaces, innovate processes of transforming green waste, and attempt to integrate informal settlements through the use of landscape. The aforementioned, in realizing current pressures on green spaces due to urban sprawl, derelict land, and infill development, concluded a major contribution to the conservation and enhancement of biodiversity and environmental quality, promotion of social cohesion and adaptation of cities to climate change—an inevitable and global phenomenon with impact on urban life.

In exploring the role of green infrastructures and the environment in adapting cities for climate change, the importance and functionality of urban green spaces, often overlooked and undervalued, are highlighted. The impact of climate change has already become more noticeable over the years. Endured by the people and the built infrastructure, the damage of heat waves, flooding, physical illnesses, wildfires and more, also disrupted the built environment. *The Climate and the Environment* explores the potential of greening, urban climate modeling and building thermal performance enhancement and efficiency in supporting climate change adaptation of urban environments. In this section, the need to work at local levels to develop Climate Action Plans and take community-wide action is discussed. Not only a contributor to waste heat, CO_2 emissions and climate change, the vast expansion of the global urban population accompanied by the dynamic growth of cities has resulted in implications and environmental degradation, in part, due to the unbalanced urban integration of urban infrastructure into the urban fabric. Furthermore, these impacts may be further aggravated by poorly ventilated urban fabrics and dense morphologies making it a necessity to address current environmental implications, explore new and innovative mitigation strategies, and alter the way cities are evolving.

Existing theories of sustainable development, in their journeys toward the sustainable city, have failed to address human-oriented transportation systems which consists of healthy neighborhoods, integrated public-soft transportation networks, and transit-oriented development. For this reason, researchers in this volume come up with branding strategies for cities in rural areas and develop models for integrating human-oriented transportation system frameworks as regeneration tools for healthy neighborhoods and communities.

Cairo, Egypt Mourad Amer

Contents

Reviving Cities' Urban Fabric

Dialogues of Sustainable Urbanization: Housing, Transport, and Well-Being

Energy and Smart Systems

Multivariable Coupling Influence on Energy-Efficient Skylight Design

Fei Yu and Jiawei Leng

Abstract

Despite the great achievements made in energy-efficient architectural design over the past decade, the majority of studies have used single-variable or bivariate-orthogonal methods, which treat the variables separately. Thus, the existing literature largely neglects multivariable interactions, which are often encountered in design practice. To address this insufficiency, the response surface method is introduced and applied to a roof glazing system (skylights with exterior shades) of an office atrium in Nanjing, China. The method is used to analyze multivariable interactions and optimization processes where architectural features are parameterized and formulated in mathematical models. Model generation, effect visualization, and optimization all contribute to general conclusions and design strategies for architects. The study reveals several findings, including that up to 72% of total energy demand could be reduced by changing the value of design factors. As for energy-saving design strategies, in regions like Nanjing that have hot summer–cold winter climates, small slats are recommended. The window-to-floor ratio, the number of strips, blind reflectance, and the U-value of the skylights are suggested to reach their minimum, while glazing transmittance should be at its maximum. In addition, substitute relations of factors are proposed, indicating that architects have more choices to reach similarly low levels of total energy demand when multivariable interactions are considered.

Keywords

Multivariable • Coupling influence • Energy-efficient • Skylights • Optimization • Simulation

Symbols and abbreviations

WFR	Window–floor ratio
d	The width of a strip skylight
Δd	The distance between two strip skylights
$\Delta d'$	The distance from the skylight which is at the edge of the atrium's roof to the roof edge
n	The amount of the strip skylights
l	The length of the atrium's roof
U	The thermal conductivity coefficient
$SHGC$	Solar heat gain coefficient
$Tvis$	Visible light transmittance
Rb	Reflectance of blinds
E_c	Cooling energy requirement/GJ
E_h	Heating energy requirement/GJ
E_l	Lighting energy requirement/GJ
E_t	Total energy requirement/GJ

1 Introduction

The widely recognized need to cut down energy consumption has driven discussion on energy-efficient architectural design in recent years. The extant literature includes studies on many variables contributing to energy-efficient building design, including architectural design elements and control systems (HVAC, lighting system, etc.), for which the evaluation criteria mainly focus on energy consumption or the energy demand index (Harish & Kumar, 2016). Despite advances made in studies dealing with the influence of variable parameters on energy performance, the effect of multivariable interactions—the ordinary state in architectural design—has been largely neglected. This is mainly because they employ the single-variable method of discussing one variable at a time or the bivariate-orthogonal method of analyzing interactions two by two as cross-comparison.

F. Yu · J. Leng (✉)
School of Architecture, Southeast University, Sipailou, Nanjing, 210018, People's Republic of China
e-mail: jw_leng@seu.edu.cn

F. Trapani et al. (eds.), *Advanced Studies in Efficient Environmental Design and City Planning*, Advances in Science, Technology & Innovation, https://doi.org/10.1007/978-3-030-65181-7_1

The bivariate-orthogonal method is one of the most popular methods when facing multiple variables. The method simplifies the entire work into several bivariable discussions by means of the control variate method and to discuss the interactions between variables in pairs (Jaber & Ajib, 2011). For example, Goia et al. (2013) discussed the influence of WWR (5 values) combined with different orientations (4 values), building scales (3 values), and HVAC systems (4 values) on total energy demand in Pairs. As the variables of the research are discussed two by two and shown in line charts to make comparisons, the discussion tends to become unclear and the multiple trends hard to discover.

Another way to deal with the discussion of multivariable combinations is to combine different variables together as a nominal unit and make comparison without analyzing the quantified tendency of each parameter. The nominal unit may vary from structures such as complex glazing systems (Buratti et al., 2013) to building cases (Boyano et al., 2013). By combining the variables together, less simulation work was required. In this way, the researchers could easily choose the cases with the best/worst performance and compare different design strategies (Grynning et al., 2014). However, the levels of each variable were rather limited—sometimes even less than three (Wang et al., 2013). As a result, the conclusions of the trend analysis for a general or a specific variable could hardly be clearly defined, let alone predictability and scalability.

In addition, with advances in computer science, there are various intelligent algorithms that could be applied for simulation, prediction, and optimization (Amasyali & El-Gohary, 2018), e.g., genetic algorithms (GA) and particle swarm optimization (PSO). These algorithms have been proved beneficial in optimizing different building components where multiple variables are involved. By integrating these algorithms, the simulation time cost could be dropped, and more energy could be saved in the optimal results (Magnier & Haghighat, 2010). However, these methods are black box models with particular outputs as the optimal results (Machairas et al., 2014). The interaction relationship between multiple variables was still hard to be described (Coakley et al., 2014).

Admittedly, statistic equations with multiple variables could be formed to correlate data. There are studies working on this with the aim of energy consumption prediction and effect estimation (Foucquier et al., 2013). Thus, they pay more attention to the precision of the prediction and the coupling influence was less concerned. For example, Korolija et al. (2013) developed regression models to predict office building annual energy consumption. Although various factors were involved, including built forms, orientations, and glazings, the effect of building parameter coupling influence was not investigated.

To address the limitations in current studies relating to multivariable interaction analysis, the response surface methodology (RSM) is applied in this research. RSM analyzes the recorded value of response metrics with process factors changing by design so as to make the conclusion more objective and effective (Teglia et al., 2015). It was tested by Zhang et al. (2018) and proved to work effectively in the nearly/net zero building energy prediction. All the attempts involving the RSM demonstrate its ability to describe multivariable interactions in the form of fitting formulas and graphs after regression analysis is performed on groups of designed experiments. In addition, for optimization where the suggested optimal solutions should not just be selected based on limited experiments but rather a range of settings, this approach has been found to be effective (Baş & Boyacı, 2007).

To present how RSM can cope with building energy simulation for multivariable research, a sample research is proposed in this paper focused on strip skylights with exterior shades on the roof of an atrium in Nanjing, China. The framework of an RSM-based study is given, after which factor parameterization, model prediction, interaction analysis, and the optimization process are presented step by step. A predictive model between multiple factors and energy demand is proposed to measure the effect and interrelationship of factors, while optimal solutions are given for energy-saving design strategies. Stressing the substitute relations between factors and response, this paper not only shows how quantified multivariable research in the building simulation field could be carried out with the application of RSM, but also indicates that architects could reach similar energy-efficient goals with more design choices. This may, to some extent, change traditional attitudes toward the goal of energy saving, which is often regarded as merely a design restriction.

Based on preliminary results, in Sect. 2, the theory and the application of RSM are introduced. In Sect. 3, simulation input settings for a sample study are constructed. Section 4 demonstrates the quantified simulation results, including the predictive model and optimal solutions. Section 5 discusses the implications and limitations of the research and the method. Finally, Sect. 6 ends with our main conclusions.

2 Methodology

2.1 Response Surface Methodology Working Steps

RSM is a mathematic quantification approach with graphics technology to form the statistic relation between factors and responses (Baş & Boyacı, 2007). Using DesignExpert, a commonly used RSM software (Kumar et al., 2007), the steps of RSM can be described as follows:

(a) Variable definition
This step primarily defines the type and value range of research factors. They can be either numerical or nominal, with the ranges set by the researchers. The names of the responses are also listed in this stage.

(b) Representative point selection
There are several sampling methods suitable for multiple response optimization that have a different required number of experiments. For higher precision and fewer trials, D-optimality is suggested to work efficiently for multiple factors (Fang et al. 2005).

(c) Mathematical model generation
Mathematical models in the form of polynomial equations, describing the relation between factors and responses, could be formed after a series of tests (Fig. 1) to check the models' degree of fitting, reliability, and predictability (Baş & Boyacı, 2007). First, the fitting degrees of four classical models—linear, 2FI, full quadratic, and cubic—are analyzed after regression analysis. The one which is statistically significant (at 0.05 level) with the highest coefficients of determination (R^2) is detected and noted as "suggested." Based on the suggested model, stepwise regression analysis is made for each response, where unsignificant terms (p 0.05) can be eliminated. The fit and the statistical adequacy of the model could be tested after ANOVA and graphical diagnostic analysis. The ones pass the statistical examinations could be identified as "predictive models" for further trend analysis and optimization. If the statistical adequacy test failed, the fitting formula would be rejected. Then, the form of the factors needs to be changed and the rest procedure would be circulated.

(d) Trend analysis
With the predictive models, the general disciplines of multiple variables can be analyzed. Contour graph, a two-dimensional display of the surface plots with lines of constant response, was used for visualizing a two-factor coupling. Thus, the influence of the factor couplings could be investigated.

(e) Optimization
By setting an optimization goal and conditions, optimization points for each independent variable could be obtained based on the predictive models. The numerical optimization in DesignExpert uses a desirability function, from 0 to 1, to measure the proximity of the solution to the optimal goal. The goal-seeking process was from several random starting points to find the "best" maximum of the desirability function.

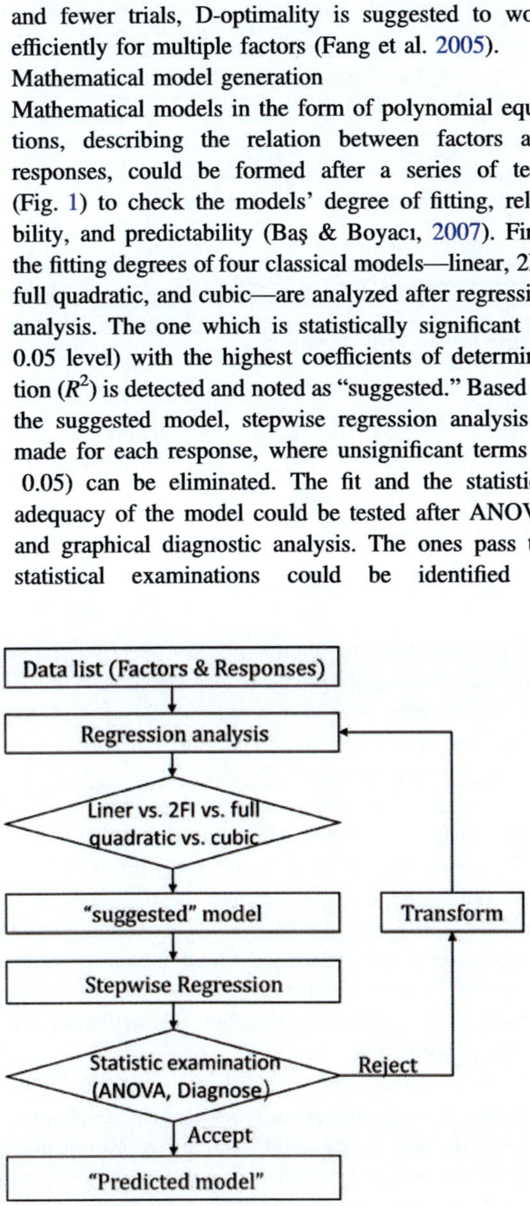

Fig. 1 RSM model generation procedure

2.2 RSM for Energy-Saving Building Design

The concept of the RSM used in studies of building energy simulation can be written as

$$R_j = f(X_{j1}, X_{j2}, \ldots X_{ji}) \tag{1}$$

X_{ji} is the element in design states that architects can adjust; R_j is any independent response that can be measured by all factors; and f is the "predictive models" between one response and several factors.

To find the relationship f, groups of corresponding X_{ji} and R_j values are needed. The value groups of X_{ji} are chosen within their own ranges set by the researchers. Each response R_j is an evaluation index of the research, the values of which are measured by simulation software after imputing the values of factors. It should be declared that the mapping relationship f is a mathematic result of regression analysis with probably no physical meaning. Even so, the fitting formulas in the form of multinomial equations are useful in gaining an overall understanding of tendency, quick simulations, prediction, and optimization.

2.3 Framework of This Study

The research procedure based on RSM is shown in Fig. 2. First, the fixed simulation settings were decided according to the literature review. Second, factors of the research objects

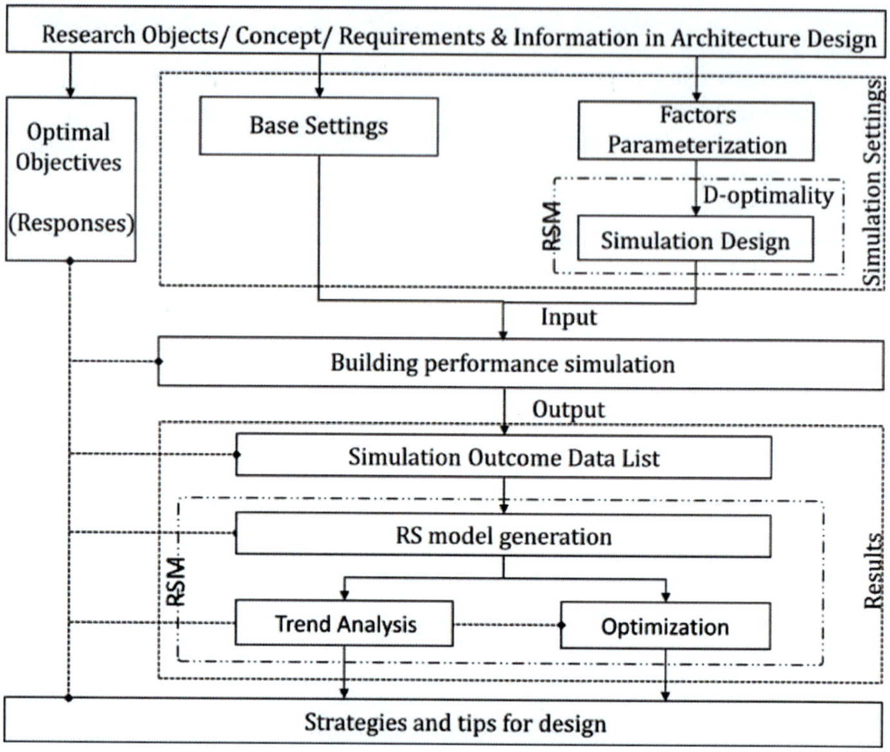

Fig. 2 Research procedure based on RSM

were parameterized with the decision of their value ranges considering the need in architecture design. Using the *D*-optimality method in RSM, an input list was defined. The third step was creating the building simulation for each simulation and recording the value of responses together with the factors as a data list. Afterward, by applying RSM, mathematic models were generated for each response, and trend analysis and optimization were then carried out based on the models. Finally came the stage where some strategies and tips could be concluded.

During this process, the software DesignExpert V8.0 was used for the application of RSM (Kumar et al., 2007), while EnergyPlus was chosen as the building energy simulation software, considering its widely accepted accuracy (Zhao & Magoulès, 2012). Discussions on descriptive statistics, correlation analysis and factor elimination, trend analysis, and optimal solutions led to our conclusions about energy-saving design strategies.

3 Simulation Settings

Widely recognized for its accuracy, the energy consumption simulation software EnergyPlus (Version 8.6) was chosen as the computer simulation approach in this research. It was supplemented by the OpenStudio (Version 8.6) modeling plug-in for SketchUp (Version 8.0) to reduce the difficulties in model creation. The base model settings as well as research factor parameterization were decided based on the use of the software and its database.

3.1 Base Model Settings

The location was in Nanjing (N32.83°, E118.8°), Jiangsu Province, a hot summer–cold winter climate city in China. The atrium space was 20 m × 20 m × 25 m (Tao & Bin, 2004) oriented north–south. It was around with a 6-layer, 12-m-deep functional space, in which the height of the first layer was 5 m while the rest were 4 m (see Fig. 3). The façade construction (e.g., roofs, floors, and walls) was based on the EnergyPlus material library (see Table 1).

According to the specifications of the local standard (GB50034-2013), the illumination requirement of the atrium should be set as 200 lx, while the unified glare value should be 22. The artificial lighting control type was continuous dimming with 4 reference points (see Fig. 4) for daylighting and glare calculation.

For each thermal zone, there was a calculation point in the center. The occupant density and activity levels were the standard settings in EnergyPlus library according to *ASHRAE Standard 189.1*. The cooling and heating thermostat setpoint values were 28 and 18 °C according to the local standard (DGJ32J96-2010). To simulate energy requirement,

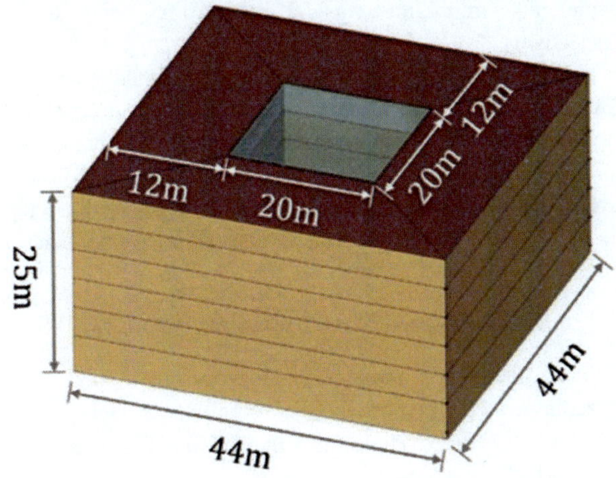

Fig. 3 Scale of the base model

the ideal HVAC system was applied in this research with suggested parameter inputs according to the EnergyPlus reference guide. The schedule for occupancy, light, equipment, and HVAC was on Monday to Friday from 08.00 to 18.00 h local time.

3.2 Factor Parameterization

For this research, skylights with exterior shades were discussed. The design factors relating to it were specified and parameterized as below.

3.2.1 Geometry Factors

For this research, we only discuss the strip-shaped skylights that were evenly arranged and long enough to cover the entire roof.

The main characteristics of the geometry of the strip-shaped skylights are presented in Fig. 5. The geometry involves several factors: window–floor ratio (WFR), the width of a strip skylight d, the distance between two strip skylights Δd, the distance from the skylight which is at the edge of the atrium's roof to the roof edge $\Delta d'$, the number of the strip skylights on the roof (n), and the categoric factor *Direction*. Considering the model was fixed in orientation, the strip-shaped skylights had two *Directions*: east–west (*E–W*) and south–north (*S–N*).

There were transformation relations between WFR, d, Δd, $\Delta d'$, and n. The expressions were as follows:

$$n \cdot d = \mathrm{WFR} \cdot l \tag{2}$$

$$(n-1) \cdot \Delta d + 2\Delta d' = (1 - WFR) \cdot l \tag{3}$$

where l is the length of the atrium's roof (a fixed number defined in Sect. 3.1). From expressions, it could be illustrated that at least three factors among WFR, d, Δd, $\Delta d'$, and n were needed to define the form. To simplify the problem, the relationship between Δd and $\Delta d'$ was set as $\Delta d = 2\Delta d'$; thus, Eq. (2) could be transferred into

$$n \cdot \Delta d = (1 - \mathrm{WFR}) \cdot l \tag{4}$$

Thus, the value of d, Δd, and $\Delta d'$ could be decided based on WFR and n

$$d = \mathrm{WFR} \cdot l/n \tag{5}$$

$$\Delta d = (1 - \mathrm{WFR}) \cdot l/n \tag{6}$$

$$\Delta d' = (1 - \mathrm{WFR}) \cdot l/2n \tag{7}$$

Considering scope definition, WFR and n were chosen as research factors. To exclude the cases in which there was no skylight on the roof or the roof was entirely made of glass, the scope of WFR in this research was [0.1, 0.9]. The n of skylight was an integer from 1 to 20 so that the width of each skylight would not be too small.

3.2.2 Material Settings of Skylights

In this study, glazing material characteristics were parameterized to investigate the general trend with fewer required

Table 1 Base model settings from EnergyPlus database

Parameter	Details	
Material	Roof	Roof membrane + Roof insulation + Metal decking[a]
	Façade	1In Stucco + 8In Concrete HW + Wall insulation + 1/2In gypsum[a]
	Ground touched basis	MAT-CC054 HW concrete + CP02 carpet pad[a]
Infiltration		0.00030226 m^3/s-m^2[a]
Ventilation		0.009438948864 m^3/s-person[a]
Internal gains	Occupant	0.107639104 person/m^2[a]
	Lighting	12.5937752 w/m^2[a]
	Equipment	0.753473729 w/m^2[a]

[a]Represents the settings were chosen from EnergyPlus database according to ASHRAE/ISO standard

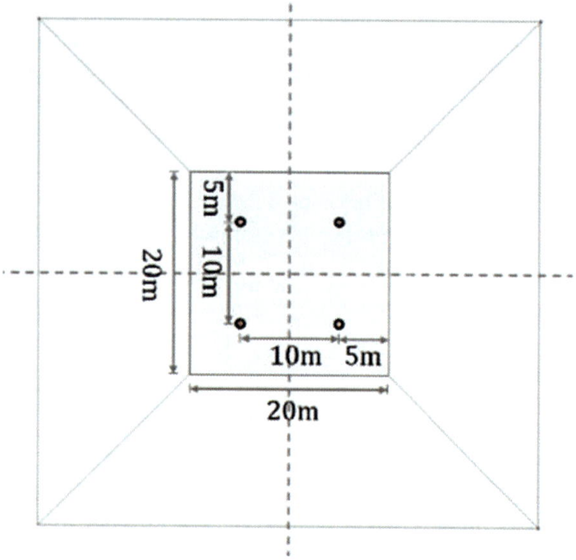

Fig. 4 Daylighting reference points

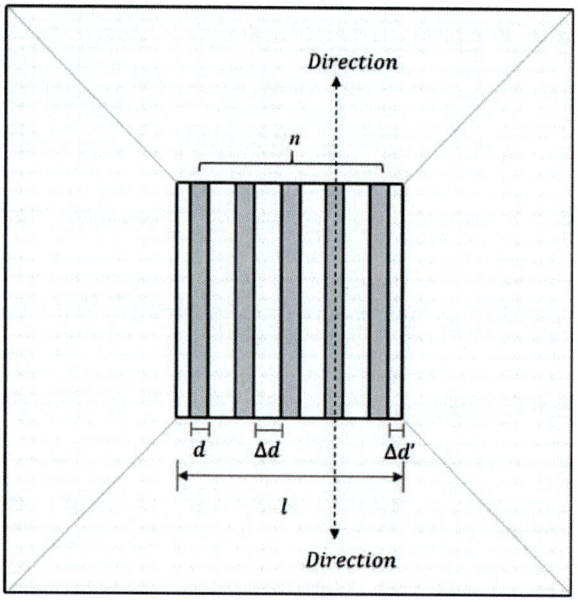

Fig. 5 Skylight strip geometry factors

simulations. According to the EnergyPlus material category settings, the material property of a glazing system could also be described as by three parameters: the thermal conductivity coefficient (U), solar heat gain coefficient (SHGC), and visible light transmittance ($Tvis$). To identify their ranges, 206 groups of common glazing material from EnergyPlus dataset file were analyzed (Table 2). For U, $Tvis$, and SHGC, the maximum was 5.89 W, 0.91, and 0.91, respectively, while the minimum was 0.79 W, 0.05, and 0.11, respectively.

For simulation designs, factors need to be independent. Thus, correlation tests were conducted with consideration of technology limitations in existing materials. Among the three factors, $Tvis$ and U were independent to each other. SHGC had significant correlation with both $Tvis$ and U (at the 0.01 level). With the step-in regression method, a predictive model was formed:

$$\text{SHGC} = 0.0082 + 0.0419 \cdot U + 0.7079 \cdot Tvis \quad (8)$$

The model was proved credible and had good predictability, as it passed the significance test with $R^2 = 0.9303 > 0.9$. Thus, the value of SHGC can be calculated by U and $Tvis$.

In general, the skylight glazing material factors were U and $Tvis$ in this study. They were both continuous variables within the limit: $U \in [0.79, 5.89]$, $Tvis \in [0.05, 0.91]$.

3.2.3 Exterior Blind Settings

For blind settings, two slat types (*Big* and *Small*) were investigated with different values of solar energy and visible light reflectance. For *Small* and *Big* type, the slat widths were 0.025 m and 0.2 m, respectively, while the distance between every two slats was 0.018 and 0.2 m. The distance between blind and glass was 0.05 for *Small* and 0.1 m for *Big* type. For parameters relating to the material of the blind, we assumed in this research that the solar reflectance and visible reflectance of blinds were the same, named as *Rb*, ranging from 0.2 to 0.8. The slats were horizontal with 0.001-m thickness. They were controlled automatically in this research with the schedule called *On-Night-If-Heating-And-On-Day-If-Cooling*. The slat angle could be adjusted by solar incident angle from 0° to 180°.

3.3 Experiment Summary

Applying the *D*-optimal statistical algorithm, the simulation plan was formed including a data list for simulation input settings. The required number of simulations was determined by the polynomial order—in this research the cubic model—and the levels of the numerical and categorical factors involved (Table 3). To test lack of fit, 5 additional runs were added. In total, 109 simulations were required for the strip-skylight-with-exterior-blind model. The responses of this research were annual cooling, heating, and lighting energy requirement (named as E_c, E_h, and E_l, respectively) as well as the total of the three (E_t). The input settings of the sampling points are derived in the whole design space by DesignExpert (see Fig. 6b–d).

Table 2 Statistical analysis on 206 groups of glazing materials from EnergyPlus database

	No.	Minimum	Maximum	Mean	Mode	Standard deviation
U	206	0.785	5.894	2.72194	5.778	1.28666
SHGC	206	0.107	0.905	0.42456	0.426	0.216745
Tvis	206	0.045	0.913	0.42716	0.073a	0.285975

[a]Represents there is more than one mode and the value shown in the table is the minimum mode

Table 3 Summary of simulation design

Factors					Response	Experiment number
Variables	Name	Class	Minimum	Maximum	Name	
A	WFR	Numerical	0.10	0.90	Cooling energy demand E_c/GJ	109 (104 + 5)
B	*n*	Numerical	1	20		
C	*U*	Numerical	0.79	5.89	Heating energy demand E_h/GJ	
D	*Tvis*	Numerical	0.04	0.91		
E	*Rb*	Numerical	0.20	0.80	Lighting energy demand E_l/GJ	
F	*Direction*	Categorical	*E–W*	*S–N*		
G	*SlatType*	Categorical	*Small*	*Big*	Total energy demand E_t/GJ	

(a) Output of E_c, E_h, and E_l in descending order of E_t; (b) input of WFR and *n* in descending order of E_t; (c) input of *U* and *Tvis* in descending order of E_t; (d) input of *Rb*, *Direction*, and *SlatType* in descending order of E_t.

4 Results

4.1 Experiment Outputs of E_c, E_h, E_l, and E_t

Simulation values obtained for E_c, E_h, and E_l of all 109 runs as well as their input settings are presented in Fig. 6a, in descending order of the total amount (E_t). From over 220 GJ to less than 60 GJ, 72% of the yearly total energy demand could be cut down by changing the settings of the research factors. The E_h reduced dramatically from nearly 100 GJ to around 10 GJ while that of lighting fluctuated around 40 GJ. With E_t decreased, the percentage of E_c, E_h, and E_l varied greatly. E_h, for example, ranked the highest among the three when the E_t was high, but was the smallest part of E_t at the end. All these indicate that large energy-saving potentials are involved in strip skylight design with exterior shades.

Despite the energy-saving potential shown in Fig. 6, the relations between factors and responses were still unclear. There were some tendencies where, for example, the values of WFR and *U* were at their upper boundary (0.9 and 5.59, respectively) when the E_t reaches its maximum, but the general tendency was hard to conclude. It even appeared that, for each factor, no matter what the value was within its range, E_t could be saved a lot by adjusting the settings of other factors. This may have been due to the coupling influence between factors.

4.2 Mathematical Model Generation for E_t

Table 4 shows a fit summary output comparing the appliance of polynomial models with different orders. From the table, the quadratic model was suggested to fit the total energy demand, as it has statistical significance ($p < 0.0001$) and highest predicted R^2 (0.9819).

Afterward, multiple regression coefficients for coded variables (Table 5) were obtained employing a step-in regression analysis to predict the specified polynomial model for the total energy demand. Only terms with $p \leq 0.05$ were included. Table 5 presents all the significant model terms in reducing the E_t. No factor was completely ignored. The terms with smaller *p*-values were considered to have much more significant impact. Thus, in this case, factors *n* and Direction have much less significant impact on E_t than others.

The regression models were found to be highly significant with $R^2 = 0.990$. Moreover, to test the degree of fitting, in-sample data (Fig. 7a) and out-of-sample data (Fig. 7b) were both investigated. The predicted versus actual response plots show the values were scattered around the 45° line randomly, indicating that the models were able to predict the E_t in the design space.

(a) The residuals versus predicted plots of in-sample data and (b) the residuals versus predicted plots of out-of-sample data.

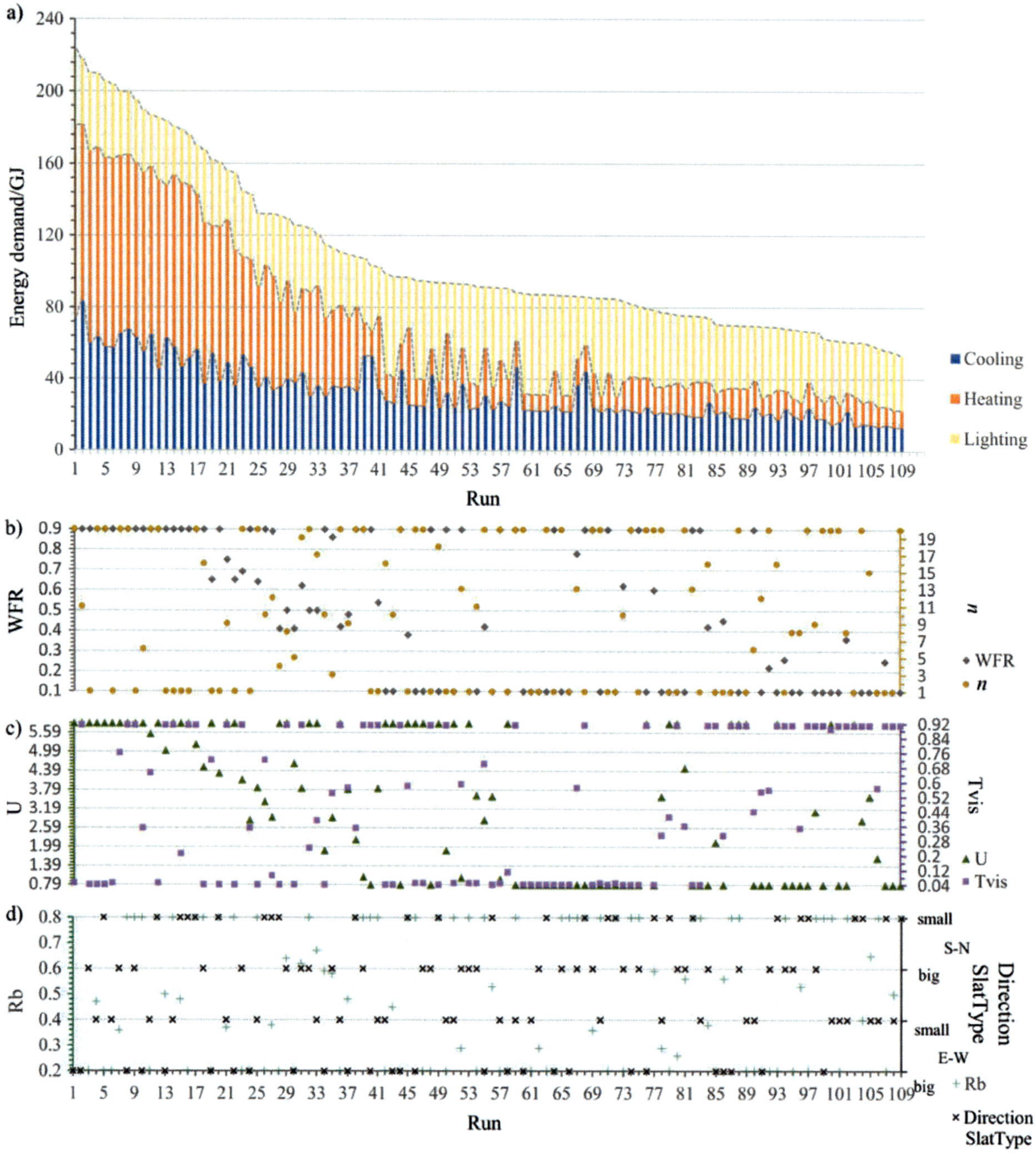

Fig. 6 Simulation data of 109 runs

With the coded model term coefficients (Table 5), the predictive model with uncoded (actual) factors for E_t could be expressed as follows.

When Direction = *E–W*, SlatType = *Small*,

$$\begin{aligned} E_t = {} & 92.18 - 59.29 \cdot WFR + 0.14 \cdot n + 1.51 \cdot U \\ & - 82.85 \cdot Tvis + 0.90 \cdot Rb + 25.28 \cdot WFR \cdot U \\ & + 33.65 \cdot WFR \cdot Tvis - 1.67 \cdot U \cdot Tvis - 3.31 \cdot U \cdot Rb \\ & + 10.53 \cdot Tvis \cdot Rb + 33.73 \cdot WFR^2 + 44.64 \cdot Tvis^2 \end{aligned} \quad (9)$$

Table 4 Suggestions on the choice of fitting model (E_t/GJ)

Source	Sequential p-value	Adjusted R^2	Predicted R^2	
Linear	<0.0001	0.7339	0.7061	Suggested
2FI	<0.0001	0.9786	0.9701	
Quadratic	<0.0001	0.9883	0.9819	
Cubic	0.5140	0.9894	0.3248	

Table 5 ANOVA test for the predictive model of E_t

Source	F-value	p-value Prob > F	Coefficient estimate	Standard error	95% CI low	95% CI high
Model	539.07	<0.0001				
A-WFR	3709.89	<0.0001	32.1681	0.5281	31.1190	33.2172
B-n	6.21	0.0145	1.3378	0.5370	0.2712	2.4044
C-U	3115.01	<0.0001	29.8802	0.5354	28.8167	30.9436
D-Tvis	162.44	<0.0001	−6.7616	0.5305	−7.8154	−5.7078
E-Rb	23.85	<0.0001	−2.5897	0.5302	−3.6430	−1.5365
F-Direction	6.71	0.0112	−1.2304	0.4751	−2.1740	−0.2867
G-*SlatType*	90.34	<0.0001	4.5108	0.4746	3.5681	5.4534
AC	1904.25	<0.0001	25.8314	0.5920	24.6555	27.0072
AD	100.66	<0.0001	5.8425	0.5823	4.6857	6.9992
AG	16.87	<0.0001	2.1743	0.5294	1.1227	3.2258
CD	9.91	0.0022	−1.8476	0.5868	−3.0133	−0.6819
CE	18.84	<0.0001	−2.5341	0.5839	−3.6940	−1.3743
DE	5.56	0.0205	1.3710	0.5813	0.2163	2.5257
DG	42.34	<0.0001	3.4637	0.5323	2.4064	4.5211
EG	4.02	0.0480	−1.0608	0.5292	−2.1120	−0.0096
A^2	17.05	<0.0001	5.3968	1.3071	2.8004	7.9932
E^2	41.72	<0.0001	8.4074	1.3016	5.8220	10.9928
Intercept			96.3790	1.4321	93.5344	99.2236

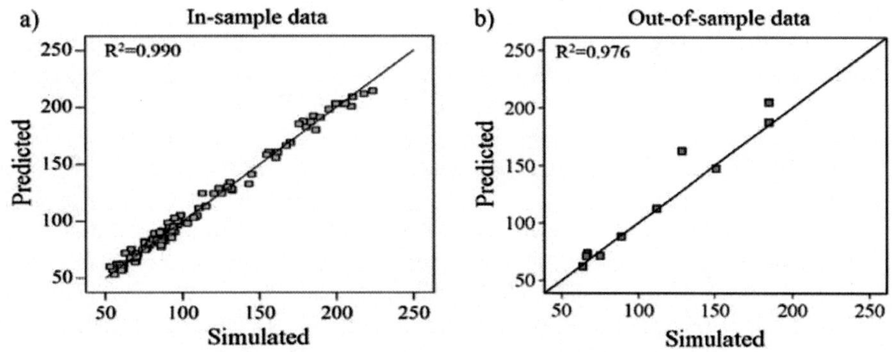

Fig. 7 Graphic diagnostic test of the predictive model

When Direction = *S–N*, SlatType = *Small*,

$$
\begin{aligned}
E_t = {} & 89.72 - 59.29 \cdot \mathrm{WFR} + 0.14 \cdot n + 1.51 \cdot U \\
& - 82.85 \cdot Tvis + 0.90 \cdot Rb + 25.28 \cdot \mathrm{WFR} \cdot U \\
& + 33.65 \cdot \mathrm{WFR} \cdot Tvis - 1.67 \cdot U \cdot Tvis - 3.31 \cdot U \cdot Rb \\
& + 10.53 \cdot Tvis \cdot Rb + 33.73 \cdot \mathrm{WFR}^2 + 44.64 \cdot Tvis^2
\end{aligned} \tag{10}
$$

When Direction = *E–W*, SlatType = *Big*,

$$
\begin{aligned}
E_t = {} & 91.65 - 48.42 \cdot \mathrm{WFR} + 0.14 \cdot n + 1.51 \cdot U \\
& - 66.89 \cdot Tvis - 6.17 \cdot Rb + 25.28 \cdot \mathrm{WFR} \cdot U \\
& + 33.65 \cdot \mathrm{WFR} \cdot Tvis - 1.67 \cdot U \cdot Tvis - 3.31 \cdot U \cdot Rb \\
& + 10.53 \cdot Tvis \cdot Rb + 33.73 \cdot \mathrm{WFR}^2 + 44.64 \cdot Tvis^2
\end{aligned} \tag{11}
$$

When Direction = *S–N*, SlatType = *Big*,

$$
\begin{aligned}
E_t = {} & 89.19 - 48.42 \cdot \mathrm{WFR} + 0.14 \cdot n + 1.51 \cdot U \\
& - 66.89 \cdot Tvis - 6.17 \cdot Rb + 25.28 \cdot \mathrm{WFR} \cdot U \\
& + 33.65 \cdot \mathrm{WFR} \cdot Tvis - 1.67 \cdot U \cdot Tvis - 3.31 \cdot U \cdot Rb \\
& + 10.53 \cdot Tvis \cdot Rb + 33.73 \cdot \mathrm{WFR}^2 + 44.64 \cdot Tvis^2
\end{aligned} \tag{12}
$$

4.3 Trend Between Total Energy Demand and Multifactors

According to the predictive model generated in Sect. 4.2, comparisons between the equations could be made to understand the impact of the categorical factors *Direction* and *SlatType*. For factor *Direction*, the only difference between the expressions is the constant term. When the value of factors WFR, $n, U, Tvis, Rb$ remained unchanged, there is Eq. (9) – Eq. (10) = Eq. (11) – Eq. (12) = 2.46 > 0. This means the total energy demand of *E–W* direction stripes would be always larger that of *S–N* direction stripes when under the same circumstance. On the other hand, since the interactions between categorical factor *SlatType* and numerical ones WFR, *Tvis*, and *Rb* occurred, the tendency was hard to conclude.

To understand the impact of interactions, contour plots were drawn in Fig. 8 illustrating the most significant interactions in the predictive model with p-values less than 0.0001: WFR-*U* (Fig. 8a, b), WFR-*Tvis* (Fig. 8c, d), and *Rb*-*U* (Fig. 8e, f). The impact of the categorical factor *SlatType* was also compared for the discussion of each interaction. For each contour graph, the factors not in the discussion were set constant to the average value in the simulation region while the *Direction* was set as *S–N*. The different colors in each graph represent different response values of total energy demand from 53 GJ (blue) to 224 GJ (red).

According to the graphs in Fig. 8, the values of total energy demand varied greatly in the simulation region and could reach a relatively small value of less than 60GJ. Moreover, the responses corresponding to the contour plots indicated that differences occurred between categories. When the *SlatType* changed, the trend of the response varied. In addition, the value of total energy demand tended to be smaller when the category *Small* was chosen for *SlatType*.

As for numerical factors, the changing tendency of the response was affected by interactions. For WFR-*U* (Fig. 8a, b), the trend was unidirectional. The total energy demand decreased when the WFR or *U* became small, regardless of the categories. For WFR-*Tvis* (Fig. 8c, d), when a small *SlatType* was chosen, the trend was unidirectional with a negative relation between the response and the factor *Tvis* (Fig. 8c). When the *SlatType* was big, however, the trend became symmetrical (Fig. 8d). At high *Tvis*, the relation between total energy demand and *Tvis* was positive. But at low *Tvis*, the relation became negative. The minimum of the contour could be achieved when *Tvis* was around 0.53, while the value of WFR was the minimum. As for *Rb*-*U* (Fig. 8e, f), it can be observed that the factor *Tvis* was less effective than the factor *U*, since the contour lines were more parallel to the *y*-axis.

4.4 Optimization for Energy-Saving Design

The variety of factors together with interactions involved in the predictive model made it hard to determine the settings in which minimal total energy response could be obtained. Thus, optimization was carried out based on the predicted formulas. The objective was the minimum of the response, while the factors were remained in the simulation region. The searches began with 100 random starting points and ended with 50 optimal solutions (desirability from 0.8 to 1). The optimal results with the highest desirability obtained by DesignExpert's in-built optimizer are listed in Table 6.

According to the table, ten optimal solutions were offered with their desirability equal to 1. Although the predicted value of the optimal response varied, they all fluctuated around 53 GJ. Considering the deviation of the predictive model was inevitable, all the nine optimal solutions were regarded as substitutes for each other. As for the setting of factors, the categorical factors were the same among the nine optimal solutions, while the values of numerical factors differed. The Direction was *S–N* (south–north) for every solution, and a small *SlatType* was favored.

To visualize the tendency of the numerical factors' value settings, a transformed radio map was drawn (Fig. 9). The polylines represent different solutions, while the round

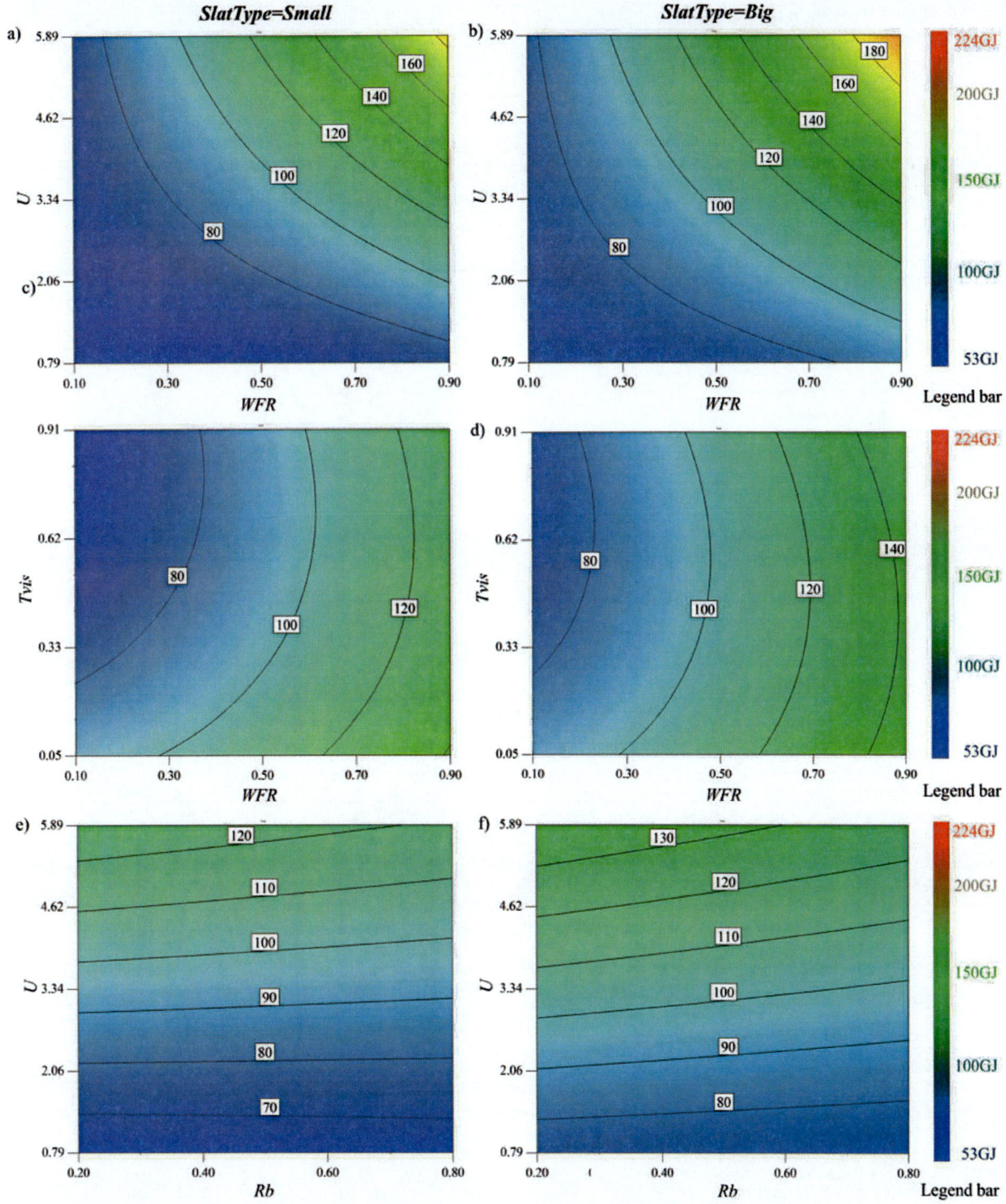

Fig. 8 Contour plots of total energy demand depending on interactions. **a** WFR-*U* interaction, when *SlatType* = *Small*, *Rb* = 0.5, *n* = 11, *Tvis* = 0.48, *Direction* = *E–W*; **b** WFR-*U* interaction, when *SlatType* = *Big*, *Rb* = 0.5, *n* = 11, *Tvis* = 0.48, *Direction* = *E–W*; **c** WFR-*Tvis* interaction, when *SlatType* = *Small*, *Rb* = 0.5, *n* = 11, *U* = 3.34, *Direction* = *E–W*; **d** WFR-*Tvis* interaction, when *SlatType* = *Big*, *Rb* = 0.5, *n* = 11, *U* = 3.34, *Direction* = *E–W*; **e** *Rb–U* interaction, when *SlatType* = *Small*, WFR = 0.5, *n* = 11, *Tvis* = 0.48, *Direction* = *E–W*; **f** *Rb–U* interaction, when *SlatType* = *Big*, WFR = 0.5, *n* = 11, *Tvis* = 0.48, *Direction* = *E–W*.

Table 6 Optimal solutions targeted at reducing total energy demand

No.	WFR	*Rb*	*n*	*U*	*Tvis*	*Direction*	*SlatType*	E_t	Desirability
1	0.18	0.28	1	0.79	0.78	*S–N*	Small	53.0975	1
2	0.11	0.23	1	0.83	0.79	*S–N*	Small	53.0687	1
3	0.13	0.21	4	0.84	0.90	*S–N*	Small	53.0555	1
4	0.14	0.26	1	0.87	0.90	*S–N*	Small	53.0980	1
5	0.13	0.21	5	0.81	0.90	*S–N*	Small	53.1032	1
6	0.13	0.20	6	0.79	0.85	*S–N*	Small	53.0722	1
7	0.20	0.22	1	0.79	0.74	*S–N*	Small	52.9602	1
8	0.13	0.30	1	0.79	0.85	*S–N*	Small	53.1012	1
9	0.12	0.21	3	0.85	0.89	*S–N*	Small	53.9175	1
10	0.13	0.22	3	0.84	0.82	*S–N*	Small	53.0811	1

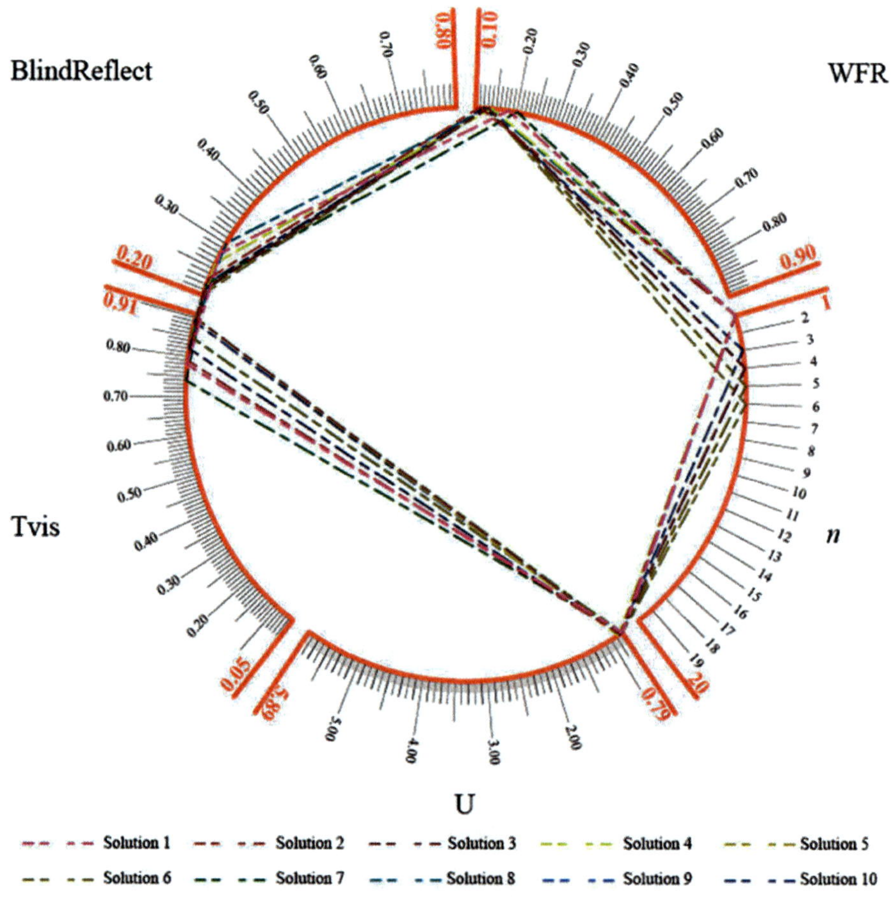

Fig. 9 Factor settings of optimal solutions within their simulation range

boundary shows the simulation region of the factors. Clearly from the graph, in general, the ranges covering the solution settings were all relatively small and close to the low boundary of their own experiment region except for *Tvis*, which fluctuated from 0.83 to 0.91 with a tendency to reach its maximum. Among all the numerical factors, the *U*-value seemed to be the most strictly limited, while the values of *Tvis* and number were more likely to fluctuate.

5 Discussion

Aimed at saving energy, the statistical relationship between total energy demand and design elements of strip skylights with exterior shades on the roof of an atrium was generated and optimized based on applying RSM to a simulation design. The simulation output results showed great potential lies in energy demand reduction by considering multivariable couplings. The predictive models generated by RSM, in addition, analyzed the relationship between multiple factors and a response. Moreover, 10 optimal solutions with different settings but similar desirabilities were found, implying that substitute relations accrued in multivariable relationships. Thus, designers could have more choices to reach similarly low level of total energy demand. Although a design would have multiple criteria and limitations, the substitute relation proposed still indicates that the goal of saving energy would be more beneficial to stimulate design thinking rather than only offer restrictions.

The focus of this study was to offer general disciplines and suggestions for the energy-saving process in the design stage, while understanding the effects and interrelationships between multiple design factors and energy demands, rather than to offer a global solution. The skylight optimal design in this research is an example showing how architects could get optimal suggestions with a certain goal by using the predictive models.

Admittedly, uncertainties do exist as errors are inevitable in the formula-fitting process. Moreover, since the simulation statistics were calculated by the simulation software EnergyPlus instead of being measured directly, the formulas might not be precise enough to predict the exact value of the energy demand and would be further improved. They could, though, be used to make quick comparison between different design proposals due to the less stringent requirement of their precision. Furthermore, complex situations (e.g., more design parameters) would lead to different optimum results and require further studies. Also, the research in this paper was based on the weather database of a certain climate in China. Therefore, more locations with different climates should be analyzed in the future. Despite these limitations, this study can be taken as a starting point to achieve a grasp of the disciplines when multivariable coupling influence occurs and to support decision-making in pre-design stages.

6 Conclusion

A multivariable research approach is proposed to understand how interactions may affect the energy-saving design of skylights with exterior shades on the roof of an atrium. The total energy demand (cooling + heating + lighting) could be reduced up to 72% by changing the value of factors. Predictive models were formed by fitting the statistical relationship between total energy demand and some design elements. Ten optimum solutions were found to get the smallest total energy demand, all of which prefer a small type of slat and south–north direction. Although the settings of other factors were different, there was a general tendency that window–floor ratio, number of strips, blind reflectance, and the *U*-value of the skylights should reach their minimum, while *Tvis* should be at its maximum.

During the whole process, RSM was applied to generate a statistical model reflecting the relationship between factors and responses. From the design element parameterization to optimization, an example was presented showing the procedure to study multivariable coupling influence and to find the possible optimum solutions based on predictive models. In addition, substitute relations were proposed in the optimal results indicating that designers could have various solutions to reach a similarly low level of total energy demand if multivariable interactions are taken into consideration. As a result, the research shows that, to some extent, the goal of saving energy could stimulate thinking in the green building design, instead of simply being considered as a restriction for architects.

Acknowledgements This research was supported/partially supported by National Key R&D Program of China (Grant No. 2016YFC0700200, No. 2016YFC0700203).

References

Amasyali, K., & El-Gohary, N. M. (2018). A review of data-driven building energy consumption prediction studies. *Renewable and Sustainable Energy Reviews, 81*, 1192–1205. https://doi.org/10.1016/j.rser.2017.04.095

Baş, D., & Boyacı, İH. (2007). Modeling and optimization I: usability of response surface methodology. *Journal of Food Engineering, 78* (3), 836–845. https://doi.org/10.1016/j.jfoodeng.2005.11.024

Boyano, A., Hernandez, P., & Wolf, O. (2013). Energy demands and potential savings in European office buildings: case studies based on EnergyPlus simulations. *Energy and Buildings, 65*, 19–28. https://doi.org/10.1016/j.enbuild.2013.05.039

Buratti, C., Moretti, E., Belloni, E., & Cotana, F. (2013). Unsteady simulation of energy performance and thermal comfort in non-residential buildings. *Building and Environment, 59*, 482–491. https://doi.org/10.1016/j.buildenv.2012.09.015

Coakley, D., Raftery, P., & Keane, M. (2014). A review of methods to match building energy simulation models to measured data. *Renewable and Sustainable Energy Reviews, 37*, 123–141. https://doi.org/10.1016/j.rser.2014.05.007

Fang, H., Rais-Rohani, M., Liu, Z., & Horstemeyer, M. F. (2005). A comparative study of metamodeling methods for multiobjective crashworthiness optimization. *Computers and Structures, 83* (25/26), 2121–2136. https://doi.org/10.1016/j.compstruc.2005.02.025

Foucquier, A., Robert, S., Suard, F., Stéphan, L., & Jay, A. (2013). State of the art in building modelling and energy performances prediction: a review. *Renewable and Sustainable Energy Reviews, 23*, 272–288. https://doi.org/10.1016/j.rser.2013.03.004

Goia, F., Haase, M., & Perino, M. (2013). Optimizing the configuration of a façade module for office buildings by means of integrated thermal and lighting simulations in a total energy perspective. *Applied Energy, 108*, 515–527. https://doi.org/10.1016/j.apenergy.2013.02.063

Grynning, S., Time, B., & Matusiak, B. (2014). Solar shading control strategies in cold climates—heating, cooling demand and daylight availability in office spaces. *Solar Energy, 107*, 182–194. https://doi.org/10.1016/j.solener.2014.06.007

Harish, V. S. K. V., & Kumar, A. (2016). A review on modelling and simulation of building energy systems. *Renewable and Sustainable Energy Reviews, 56*, 1272–1292. https://doi.org/10.1016/j.rser.2015.12.040

Jaber, S., & Ajib, S. (2011). Thermal and economic windows design for different climate zones. *Energy and Buildings, 43*(11), 3208–3215. https://doi.org/10.1016/j.enbuild.2011.08.019

Korolija, I., Zhang, Y., Marjanovic-Halburd, L., & Hanby, V. I. (2013). Regression models for predicting UK office building energy consumption from heating and cooling demands. *Energy and Buildings, 59*, 214–227. https://doi.org/10.1016/j.enbuild.2012.12.005

Kumar, T. A., Kumar, A., & Raheman, H. (2007). Biodiesel production from jatropha oil (*Jatropha curcas*) with high free fatty acids: an optimized process. *Biomass and Bioenergy, 31*(8), 569–575. https://doi.org/10.1016/j.biombioe.2007.03.003

Machairas, V., Tsangrassoulis, A., & Axarli, K. (2014). Algorithms for optimization of building design: a review. *Renewable and Sustainable Energy Reviews, 31*, 101–112. https://doi.org/10.1016/j.rser.2013.11.036

Magnier, L., & Haghighat, F. (2010). Multiobjective optimization of building design using TRNSYS simulations, genetic algorithm, and artificial neural network. *Building and Environment, 45*(3), 739–746. https://doi.org/10.1016/j.buildenv.2009.08.016

Tao, L., & Bin, Y. (2004). Approaches to the atrium space design in ecological buildings. *Architectural Journal*, 68–69 (in Chinese).

Teglia, C. M., Montemurro, M., De Zan, M. M., & Cámara, M. S. (2015). Multiple responses optimization in the development of a headspace gas chromatography method for the determination of residual solvents in pharmaceuticals. *Journal of Pharmaceutical Analysis, 5*, 296–306. https://doi.org/10.1016/j.jpha.2015.02.004

Wang, W., Tian, Z., & Ding, Y. (2013). Investigation on the influencing factors of energy consumption and thermal comfort for a passive solar house withwater thermal storage wall. *Energy Build, 64*, 218–223. https://doi.org/10.1016/j.enbuild.2013.05.007

Zhang, S., Sun, Y., Cheng, Y., Huang, P., Oladokun, M. O., & Lin, Z. (2018). Response-surface-model-based system sizing for Nearly/Net zero energy buildings under uncertainty. *Applied Energy, 228*, 1020–1031. https://doi.org/10.1016/j.apenergy.2018.06.156

Zhao, H., & Magoulès, F. (2012). A review on the prediction of building energy consumption. *Renewable and Sustainable Energy Reviews, 16*(6), 3586–3592. https://doi.org/10.1016/j.rser.2012.02.049

Performance of *L*-shaped Mini Louvers in Tropical Cities in Cooling Energy Reduction: Case Study of Mumbai, Mexico City, and Lagos

Alexander Rani Suryandono, Agus Hariyadi, and Hiroatsu Fukuda

Abstract

Sustainability is one of the biggest issues in green buildings. Many researches are done for finding optimization of building that can be implemented from the design phase to the building operation. One of the strategies to reduce energy is using sun shading devices. This paper shows one possibility of using small *L*-shaped mini louvers to reduce cooling energy. Rather than using custom devices and advance technologies, this research focuses on simple and widely available materials that can be applied without the needs of specific skills and maintenance. Several *L*-shaped mini louvers designs as shading device are simulated using computer software to show their performance of reducing cooling energy. The use of a double glass panel window without shading device will be the base case for cooling energy comparison. The simulation will be done in Grasshopper using ladybug and honeybee plugins which have widely used and validated energy simulation engines: EnergyPlus. These shading devices will be simulated in three different cities: Mumbai, Mexico City, and Lagos. The simulation will be done in one-year cycle for eight different orientations. In total, 168 simulations will be analyzed to show the results and draw conclusion of these shading devices. The conclusion shows that *L*-shaped mini louvers can be used as one of the strategies to achieve more energy-efficient buildings, notably reducing annual cooling energy by 30–32%.

A. R. Suryandono (✉) · H. Fukuda
The University of Kitakyushu, Kitakyushu, Japan
e-mail: alexanderrs@ugm.ac.id

A. Hariyadi
Universitas Gadjah Mada, Yogyakarta, Indonesia

Keywords

Mini louvers • Sun shading • Energy-efficient building • Cooling energy • Computer simulation

1 Introduction

Around fourteen years ago, Al Gore (2006) presented a warning about climate change, especially global warming, in a documentary movie and book: an Inconvenient Truth. Various researches were done to deal with this issue, including researches that were related to building. Three years ago, Al Gore published his book: an Inconvenient Sequel (2017) as a proof of his prediction on his previous book: an Inconvenient Truth. Related to this paper topic, there is a shift in temperature changes of the earth. Hansen et al. (2012) showed changes of earth summer temperature. Extremely hot days in annual average temperature in 1951–1980 only cover 0.1% of the earth. However, in the annual average temperature in 2005–2015, extremely hot days cover 14.5% of the earth. It means that, even though many researches, along with their implementations in real projects, were done, however, the effects of climate change still happen. More sustainable researches and projects must be done to confront the negative impacts of global warming. According to the data from International Energy Agency in the World Energy Outlook 2016 (n.d.), electricity demand for cooling and air conditioning take a big portion of total world energy consumption. China, India, and Southeast Asian countries showed strong growth demand of energy for space cooling. In India, 8% average annual increase in electricity comes from cooling in the residential sector. They have projections that using today's growth by 2040 developing countries in Asia account for 46% of worldwide power consumptions. International Energy Agency Market Report 2018 (n.d.) showed that cooling energy use in buildings increase more than double since 2000. Cooling

F. Trapani et al. (eds.), *Advanced Studies in Efficient Environmental Design and City Planning*,
Advances in Science, Technology & Innovation, https://doi.org/10.1007/978-3-030-65181-7_2

energy is the fastest growing end-use energy in buildings. This condition is driven by warmer temperatures, increased human's activities, population and economic development. Since many researches focus on large-scale and more business-oriented buildings, this paper aim to open the possibilities of introducing sustainable issues in household level. It means that everyone can be more aware on energy efficiency in the built environment, even such at a small space. Using a simple building element, such as *L*-shaped mini louvers that are studied in this paper, people can participate to make more optimized and energy-efficient buildings.

2 Previous Researches

Openings in buildings such as windows and door design must be taken into consideration. Shading device is one of the strategies to optimize the functions of building opening while maintaining sustainability issues. Shading devices are used for controlling the sun exposure to the buildings. Hariyadi et al. (2017) noted that *sudare* blinds reduce 5% Overall Thermal Transfer Value (OTTV) in Jakarta, Indonesia. These blinds also reduce 6% of office buildings thermal energy in the same location. *Sudare* is Japanese traditional outdoor blinds that are similar with small external louvers. Kohler et al. (2017) showed that awning as an external shading device can reduce more than 50% of annual solar heat gains. They conduct the simulation test using EnergyPlus and Berkeley Lab Window 2016 in Delhi. They also conducted simulations in Denver, USA, to show their awning's performance in a different location. Dutta et al (2017) concluded that moveable external window shading in tropical climate of northern hemisphere can reduce annual cooling energy. The south-oriented window has maximum energy consumption followed by east, west, and north in Kolkata, India. The simulations using TRNSYS software show that 9.8% annual cooling energy can be saved by putting their shading into a building. The maximum energy savings occur in June by cutting 14.9%. Not only for new buildings, El-Darwish and Gomaa (2017) showed that retrofitting old building can also save energy. They compared many design strategies to save energy by applying solar shading, adding air tightness, and changing glazing on windows. The solar shading device was the best by cutting up to 23% energy, followed by 8% energy savings of glazing strategies, and 2% lowering energy consumption using air tightness. Their experiments were done in a building in Northern Egypt. Idchabani et al. (2017) compared two shading device types: overhangs and fins in Morocco. Their research showed that overhangs work better than fins in that specific area. Ghosh and Neogi (2018) simulated strategies using EnergyPlus for energy-efficient buildings such as window-to-wall ratio, and window positioning to propose a new shading design. They tested their new shading design in Kolkata, India, and evaluated in Naples and Hanoi. Suryandono et al. (2018a) simulated the comparison of *L*-shaped mini louvers of solar radiation reduction in Singapore and Sydney. The simulation results showed that the proposed model can reduce up to 75% and 85% annual total direct and indirect solar radiation in Singapore and Sydney accordingly in North orientation. Suryandono et al. (2018b) also conducted simulation of solar radiation and cooling energy reduction using these mini louvers in three different cities in Japan: Sapporo, Tokyo, and Naha from July 1 to August 31, the hottest period in Japan. These louvers work best for south orientation by reducing around 83% solar radiation in Naha and 85% in Sapporo using their proposed model. It leads to annual cooling energy reduction 13% in Naha and 43% in Sapporo.

3 Research Methods

This study used simulation to analyze several shading device designs for dealing with the problem of cooling energy consumption in buildings. The idea of proposing sun shading devices comes from previous researches as well as awareness of available building materials in the market that might be used. The shading devices were simulated in the computer. Rhinoceros version 5 is three-dimensional software that is used for the main environment for this study. It has Grasshopper as graphical algorithm plug-in that contains ladybug and honeybee. Ladybug and Honeybee are two plug-ins that act as EnergyPlus interface simulation engine software.

To do simulation, building models, completed with its material properties and proposed shading design, were made in Rhinoceros using parametric Grasshopper scripts. Weather data for three selected locations were directly downloaded from onebuilding.org in using Grasshopper script. Energy simulations were done using Honeybee, with its EnergyPlus simulation engines. The results are analyzed to show the performance of each model in all locations and orientations. The research workflow in this paper is presented in Fig. 1.

4 Simulations

All the building models were made parametrically in Rhinoceros using Grasshopper scripts. Simulations were also done in Rhinoceros software, along with its plug-ins Grasshopper to run EnergyPlus-based simulation using Ladybug and Honeybee. First, model 1 serves as base case the study in this paper. The size of the models is 3000 mm

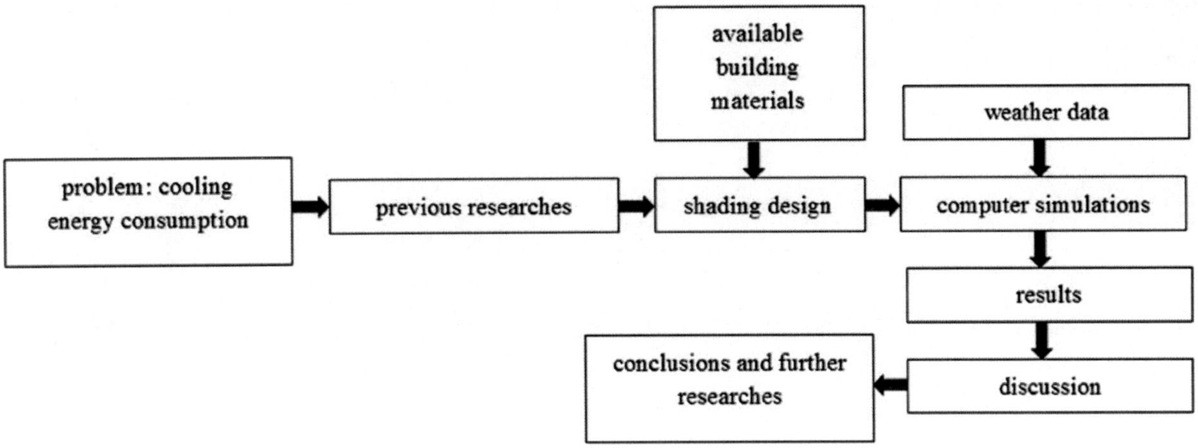

Fig. 1 Research workflow

by 3000 mm with one 2000 mm by 2000 mm single glass window in the middle of one side. All other three sides are walls, with flat roof, and flat floor. All the constructions are default properties from EnergyPlus materials. The details of these constructions are as follows: Roof constructions consist of 100 mm lightweight concrete, ceiling air space resistance, and acoustic tile. Exterior wall constructions consist of 100 mm brick, 200 mm heavyweight concrete, 50 mm insulation board, wall air space resistance, and 19 mm gypsum board. Window constructions consist of double 3 mm clear glass with 13 mm air space in between. Floor constructions consist of 50 mm insulation board and 200 mm heavyweight concrete. Grasshopper plug-in in Rhinoceros allows parametric design; so, using modifications of the same script, six other designs are made. Model 1, which is used for base case in this study, consists of one single pane glass window without any shading device. *L*-shaped mini louvers with 12 mm size are placed in exterior surface of the window in model 2a. Model 2b has 25 mm *L*-shaped mini louvers with 25 mm gap among the louvers. The length of these *L*-shaped mini louvers is 2000 mm, the same as window's width. Instead of louvers, model 2c has additional horizontal shading devices in front of glass window with the same length with the window (2000 mm) and 1000 mm depth. The gap among the horizontal shading is 1000 mm. In the three-dimensional models, all the shading devices are placed 1 mm outside exterior surface window. This 1 mm gap outside glass window surface is needed to avoid overlapping surface that might affect simulation. Moreover, the 1 mm gap mimics the presence of adhesive or bonding agent to attach the *L*-shaped mini louvers at the exterior window surface. Model 3a, 3b, and 3c has the same properties and design with model 2a, 2b, and 2c. However, instead of placed 1 mm outside exterior surface, all the model 3s have 50 mm distance from the window to find the effect of shading placement in their performance of cooling energy consumption. This 50 mm distance from exterior glass window surface similar to the condition when the exterior *L*-shaped mini louvers is attached to window frame rather than directly to the glass. The entire three-dimensional model and the simulation environment setting can be seen in Fig. 2.

All the models use midrise apartment zone programs schedule on zone usage for one-year circle simulation from EnergyPlus. The weather data is opened using Ladybug. The Honeybee software is used to run EnergyPlus simulation in Grasshopper using Rhinoceros interface. The simulation output is four-zone energy uses: lights electric, equipment electric, total heating, and total cooling. This paper shows cooling energy reductions in three selected locations for all eight orientations so only annual cooling energy are presented in tables. However, Mexico City is chosen to show sun shading's performance in subtropical area which is not only effecting the cooling but also heating energy consumption. For Mexico City, annual heating energy are also presented in table, complete with total annual energy, which are the combined annual cooling and heating energy consumption altogether. Total simulations for all models and orientations in this study are 168 simulations. One simulation generates energy consumptions every hour in one year. There are 8760 hourly based simulations. These annual energy are analyzed and compared among them to explore shading device performance in every locations for all eight building orientations.

5 Results

The first location in this paper case study is Lagos, Nigeria. The weather files exact location is at Murtala Muhammed International Airport (Onebuilding, n.d.-a). Model 1

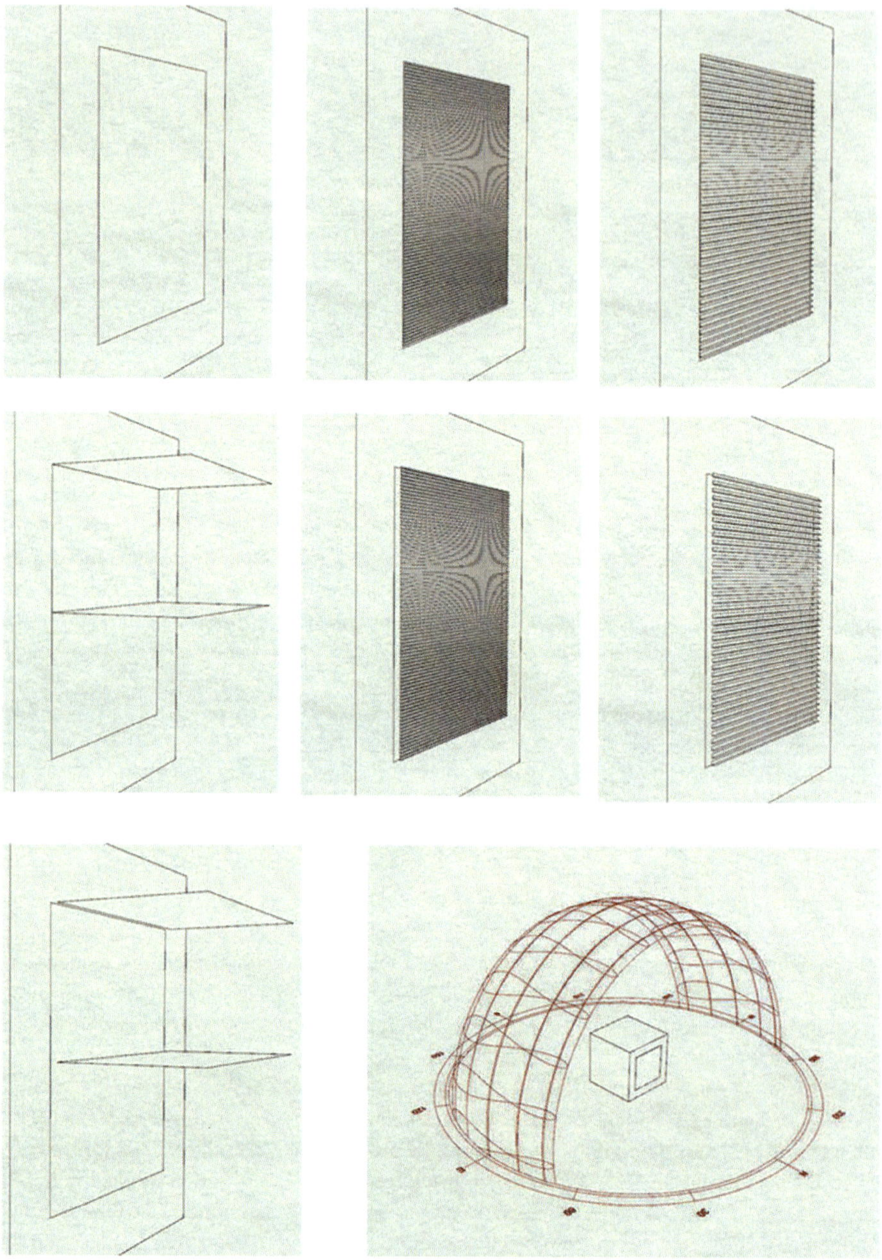

Fig. 2 Model 1 (top left), model 2a (top middle), model 2b (top right), model 2c (middle left), model 3a (center), model 3b (middle right), model 3c (bottom left), simulation environment setting (bottom right)

simulation results of annual cooling energy for eight orientations or the base case are shown at Table 1. In all eight orientations, west facade opening leads the highest cooling energy for the simulated zone. In total, 15,479,254.68 kJ is consumed. North orientation building consumes lowest cooling energy 12,848,523.51 kJ in a year. It is 17% lower than cooling energy for opening at the west side of the building. The results of all simulated models are presented in Table 2. Table 2 is presented in percentage in radar Chart 1.

When the shading devices are placed at 1 mm outside the exterior surface of the glass window, model 2b work best in the north, northwest, southwest, south, southeast, and northeast façade. Model 2b reduces annual cooling energy consumption 19% in the north façade, 20% in the northwest facade, 13% in the southwest facade, 13% in the south façade, 12% in the southeast facade, and 20% in the northeast facade. Highest amount annual cooling energy saving by model 2b is in the southwest orientation by reducing 3413.82 mJ. Model 2c performs best to lower annual cooling energy consumption in the west, southwest, and east facade by 14, 13 and 13% accordingly. Model 2c can save highest annual cooling energy of 3663.09 mJ in the west

Table 1 Model 1 simulation result of annual cooling energy in Lagos

	North	Northwest	West	Southwest	South	Southeast	East	Northeast
Model 1/base case cooling energy (mJ)	12,848.52	14,076.05	15,479.25	14,939.54	13,814.69	14,710.28	15,367.83	14,248.35
Cooling energy percentage in comparison with west façade energy (%)	83	91	100	97	89	95	99	92

Table 2 Simulation results of annual cooling energy in Lagos

	Annual cooling energy (mJ)	North	Northwest	West	Southwest	South	Southeast	East	Northeast
1	Model 1/base case	12,848.52	14,076.06	15,479.25	14,939.55	13,814.69	14,710.29	15,367.83	14,248.36
2	Model 2a	11,455.55	12,201.72	13,054.33	12,540.06	11,616.45	12,448.68	13,058.23	12,342.77
3	Model 2b	10,457.99	11,207.77	12,056.83	11,525.73	10,587.26	11,426.37	12,053.53	11,345.52
4	Model 2c	11,069.40	11,439.15	11,816.15	11,569.15	11,185.36	11,576.89	11,874.65	11,564.98
5	Model 3a	9986.69	10,138.84	10,314.54	10,186.05	10,023.31	10,252.45	10,406.66	10,186.09
6	Model 3b	9638.16	9794.94	9969.05	9834.51	9657.58	9898.85	10,058.66	9839.03
7	Model 3c	11,201.10	11,554.85	11,926.08	11,729.41	11,359.91	11,735.12	12,013.56	11,702.84

Chart 1 Simulation results of annual cooling energy in percentage comparison with base case in Lagos

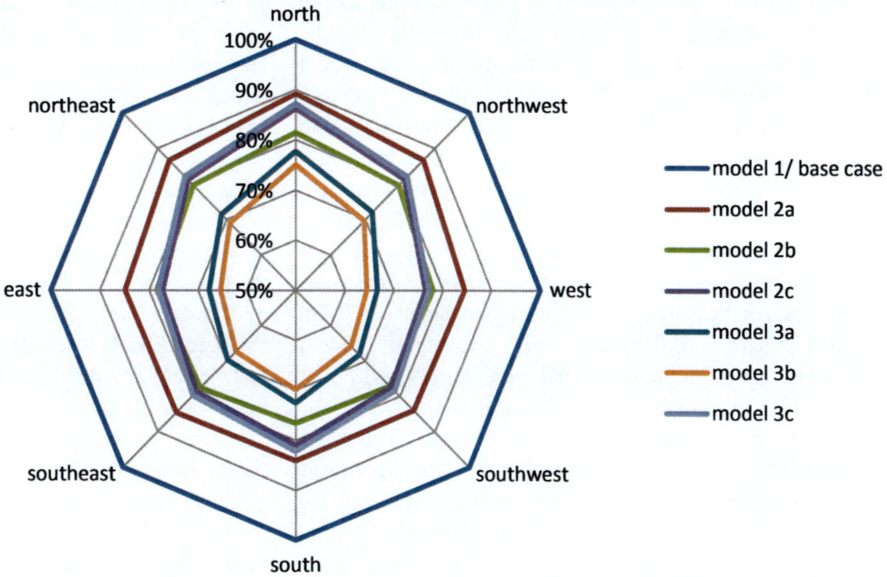

façade. Model 2a cannot perform better than the two previous models to diminish cooling energy in all eight orientations. However, in comparison with model 1 as base case, they significantly reduce 16% of annual cooling energy in the west, southwest, and south façade. Even in the worse orientations in north orientation building, model 2a still can cut 11% of annual cooling energy which is around 1392.97 mJ.

The situations are significantly different when all the shading devices are placed 50 mm in front of the exterior surface of the glass window. *L*-shaped mini louvers show performances which are better than common horizontal shading device in all building orientations. Model 3b shows best performance in comparison with other model. By utilizing model 3b, annual cooling energy consumption can be lowered to 75% (north orientation), 70% (northwest and south orientation), 66% (southwest orientation), 67% (southeast orientation), 65% (east orientation), and 69% (northeast orientation). They work best on west façade by cutting 36% of annual building cooling energy which is around 5510.20 mJ. Model 3a also shows better performance than model 2a. This shading design works best for west orientation by cutting 33% of annual cooling energy or around 5164.71 mJ in comparison with the base case. Model 3c performance is worse than model 2c in all building orientations.

Mumbai, India, is the second chosen location for this paper case study. The weather data at Chhatrapati Shivaji Maharaj International Airport, Mumbai, is used for running the simulations (Onebuilding, n.d.-b). The model 1 as base case simulation results for eight orientations is presented in Table 3. The southwest orientation has the highest building annual cooling energy consumption by 17,790.91 mJ. North façade opening uses the lowest annual cooling energy in comparison with the other 7 facades by 12,881.26 mJ or only 72% of south west's highest. The results of all simulated models are presented in Table 4. Table 4 is presented in percentage in radar Chart 2.

Similar with simulation results in Lagos, Nigeria, model 2b and 2c showing best performance in several façade, meanwhile model 2a cannot perform similarly with model 2b and 2c. Model 2b cuts 13% and 18% of annual cooling energy in the north and northwest façade. Highest amount annual cooling energy saving by model 2b is in the south orientation by reducing 4579.41 mJ. In the south, southeast, east, and northeast façade, 73, 75, 78, 83% of annual cooling energy is consumed in comparison with model 1 base case. Model 2c performs best among other models 2 in the west, southwest, southeast, and east faced by consuming 77, 74, 75, and 78% of annual cooling energy accordingly. Model 2c can save highest annual cooling energy of 4544,641.32 kJ in the southwest façade.

Moving the shading device away from glass window exterior surface makes the *L*-shaped mini louvers performs significantly better. Similar with the condition in Lagos, the model 2c and 3c shows no significant changes of cutting annual cooling energy. Model 3b achieves the best results in all eight orientations. In comparison with base case, model 3b consumes 83% (north orientation), 75% (northwest orientation), 66% (west orientation), 64% (south and southeast orientation), 67% (east orientation), and 76% (northeast orientation). Notably, model 3b reduces 37% of annual cooling energy on southwest façade, which has the highest consumption without additional shading devices. It means that around 6547.92 mJ of annual cooling energy can be saved.

Mexico City is chosen to explore the performance of proposed shading devices in subtropical climate. Shading device blocks sunlight which may affect heating energy during colder condition. Specific weather data at Ciudad Juarez International Airport, Mexico City, is used for simulation (n.d.). The base case model 1 simulation results are given in Table 5. The highest cooling energy, 2536.23 mJ, occurs in the southwest façade. Meanwhile, the north façade is the lowest one with its only 38% in comparison with the southwest facade. However, north façade has the highest heating energy with 928.27 mJ annual heating energy. The southeast and east façade similarly consumes 72% of heating energy in comparison with the highest one. The southwest façade consumes the highest 3296.32 mJ, while north is the lowest by consuming 1903.25 mJ or only around 58% of the highest annual cooling and heating energy altogether. The results of all simulated models are presented in Table 6. Table 6 is presented in percentage at radar Charts 3 for annual cooling energy, annual heating energy, and total annual cooling plus heating energy.

Model 3b performs best in eight building orientations among other models, followed by model 3a. The *L*-shaped mini louvers of model 3b cut 65% in the north façade, 74% in the northwest façade, 81% in the west façade, 82% in the southwest façade, 83% in the south façade, 79% in the southeast façade, 76% in the east façade, and 70% in the northeast façade of annual cooling energy. Model 3b works best by saving around 2089.44 mJ in the southwest façade. Best cooling energy saving for model 3a is also in the southwest façade by around 2003.44 mJ of annual cooling energy reduction.

However, reverse condition happen for annual heating energy. Model 3b performs worst among other models. Model 1 as base case, which is a glass window without shading devices performs best by allowing sunlight to help heating room space during colder condition. Model 3b

Table 3 Model 1 simulation result of annual cooling energy in Mumbai

	North	Northwest	West	Southwest	South	Southeast	East	Northeast
Model 1/base case cooling energy (mJ)	12,881.26	14,518.50	17,167.58	17,790.91	17,096.86	17,486.64	16,779.03	14,339.76
Cooling energy percentage in comparison with southwest façade (%)	72	82	96	100	96	98	94	81

Table 4 Simulation results of annual cooling energy in Mumbai

	Annual cooling energy (mJ)	North	Northwest	West	Southwest	South	Southeast	East	Northeast
1	Model 1/base case	12,881.26	14,518.51	17,167.58	17,790.91	17,096.87	17,486.65	16,779.03	14,339.76
2	Model 2a	12,243.40	13,012.83	14,549.14	14,554.39	13,740.15	14,374.95	14,299.31	12,936.25
3	Model 2b	11,173.59	11,924.48	13,414.21	13,355.82	12,517.45	13,175.08	13,166.18	11,852.75
4	Model 2c	11,932.13	12,483.67	13,226.83	13,246.27	12,728.69	13,111.40	13,110.82	12,413.68
5	Model 3a	11,045.89	11,258.64	11,656.61	11,641.29	11,291.43	11,566.03	11,571.93	11,233.06
6	Model 3b	10,692.50	10,898.24	11,280.75	11,242.98	10,885.95	11,167.71	11,198.73	10,874.45
7	Model 3c	11,998.65	12,565.24	13,420.13	13,435.26	12,964.02	13,294.06	13,192.81	12,520.85

Chart 2 Simulation results of annual cooling energy in percentage comparison with base case in Mumbai

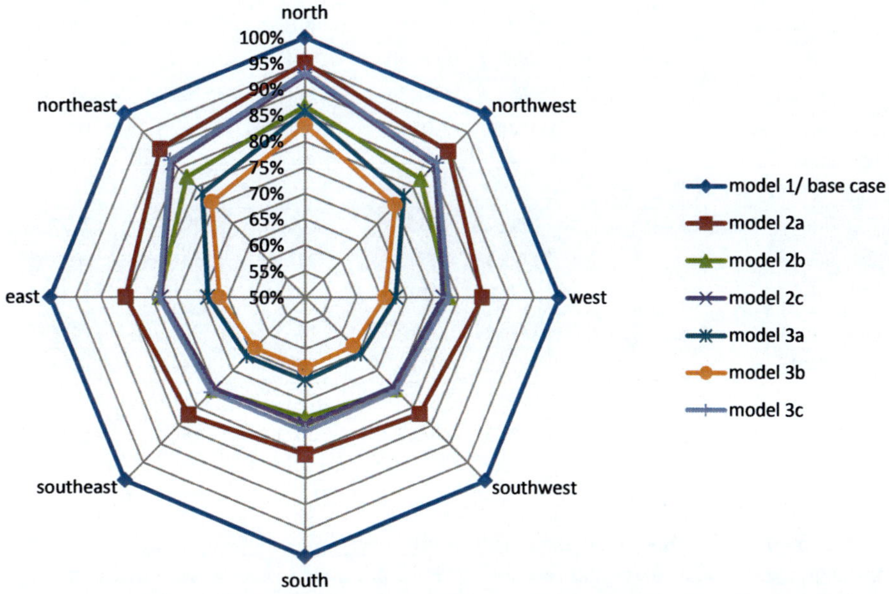

Table 5 Model 1 simulation result of annual cooling, heating, and total cooling plus heating energy in Mexico City

	North	Northwest	West	Southwest	South	Southeast	East	Northeast
Model 1/base case cooling energy (mJ)	974.97	1420.88	2267.94	2536.23	2140.88	1810.69	1532.90	1197.11
Cooling energy percentage in comparison with southwest façade (%)	38	56	89	100	84	71	60	47
Model 1/base case heating energy (mJ)	928.27	883.87	770.02	760.09	726.92	666.93	664.60	799.06
Heating energy percentage in comparison with north façade (%)	100	95	83	82	78	72	72	86
Model 1/base case total cooling and heating energy (mJ)	1903.25	2304.75	3037.97	3296.32	2867.81	2477.62	2197.51	1996.18
Total cooling and heating energy percentage in comparison with north façade (%)	58	70	92	100	87	75	67	61

Table 6 Simulation results of annual cooling, heating, and total cooling plus heating energy in Mexico City

	Annual cooling energy (mJ)	North	Northwest	West	Southwest	South	Southeast	East	Northeast
1	Model 1/Base case	974.98	1420.88	2267.95	2536.23	2140.89	1810.69	1532.91	1197.12
2	Model 2a	761.34	891.22	1292.90	1354.47	1081.02	1033.85	968.79	854
3	Model 2b	499.82	633.79	981.71	1016.98	745.69	704.42	677.92	581.48
4	Model 2c	635.76	732.39	906.21	937.71	823.47	790.16	759.16	707.24
5	Model 3a	407.26	442.59	508.67	532.79	450.19	455.62	448.91	430.72
6	Model 3b	337.92	371.61	430.21	446.79	369.68	375.56	374.03	360.91
7	Model 3c	656.16	768.83	953.44	987.08	888.96	831.21	806.52	735.53
	Annual heating energy (mJ)	North	Northwest	West	Southwest	South	Southeast	East	Northeast
1	Model 1/base case	928.27	883.88	770.03	760.1	726.93	666.94	664.6	799.07
2	Model 2a	1048.02	1020.92	928.59	950.06	922.77	799	765.32	930.79
3	Model 2b	1253.06	1205.43	1093.71	1130.31	1092.21	912.33	859.74	1093.25
4	Model 2c	1134.64	1122.95	1067.77	1085.99	1032.49	910.28	879.36	1006.79
5	Model 3a	1429.79	1432.40	1390.93	1411.76	1393.97	1252.02	1212.56	1358.61
6	Model 3b	1546.20	1543.85	1499.98	1526.67	1513.50	1357.92	1316.00	1472.92
7	Model 3c	1118.04	1105.65	1050.01	1067.15	1002.93	892.44	858.4	986.64
	Total annual cooling and heating energy (mJ)	North	Northwest	West	Southwest	South	Southeast	East	Northeast
1	Model 1/base case	1903.25	2304.76	3037.98	3296.33	2867.82	2477.63	2197.51	1996.19
2	Model 2a	1809.36	1912.15	2221.49	2304.54	2003.79	1832.84	1734.11	1784.79
3	Model 2b	1752.88	1839.22	2075.42	2147.29	1837.90	1616.75	1537.66	1674.72
4	Model 2c	1770.40	1855.34	1973.99	2023.70	1855.96	1700.45	1638.51	1714.03
5	Model 3a	1837.05	1874.99	1899.60	1944.54	1844.17	1707.64	1661.46	1789.34
6	Model 3b	1884.12	1915.47	1930.19	1973.46	1883.18	1733.48	1690.03	1833.83
7	Model 3c	1774.21	1874.48	2003.45	2054.24	1891.89	1723.65	1664.92	1722.17

makes the room consumes more than twice heating energy in the southwest, south, and southeast by 201%, 208%, and 204% accordingly. The worst condition occurs in the southwest orientation. Building with model 3b shading device attached consumes 766.57 mJ higher than building without shading device as base case. The smallest annual heating energy addition is in the north orientation with model 2a attached by around 119.74 mJ.

The total combination of annual cooling and heating energy is needed to show the performance of shading devices during one-year cycle. However, even though annual heating energy of the glass window without shading is the lowest for all building orientations, it still consumes the highest total combination of annual cooling and heating energy. Model 2b performs best for the north, northwest, south, southeast, east and northeast orientation by only consumes 92, 80, 64, 65, 70, and 84% in comparison with the base case. Model 3a shows the best results for the west, southwest, and south orientation by consuming 63, 59, and 64% of total annual cooling and heating energy in comparison with base case building. All of the annual cooling and heating energy results from the simulation in Mexico City can be seen in Chart 3.

6 Discussion

From the three selected locations, all simulation results show that shading devices have impact to reduce buildings' energy consumptions, especially in lowering cooling energy. In tropical cities: Lagos and Mumbai, the average percentage of annual cooling energy consumption for all building orientations are presented in Table 7. It is clearly seen that in both locations model 3b performs best and followed by model 3a. Model 3b has 25 mm *L*-shaped mini louvers with 25 mm gap among louvers. The mini louvers are attached 50 mm outside exterior surface of glass window. Model 3a has significantly smaller 12 mm *L*-shaped mini louvers with 12 mm gap among them. Like model 3b, these mini louvers are also attached 50 mm outside exterior surface of glass window.

Chart 3 Simulation results of annual energy in Mexico City: annual cooling energy (top), annual heating energy (middle), and total cooling plus heating energy (bottom)

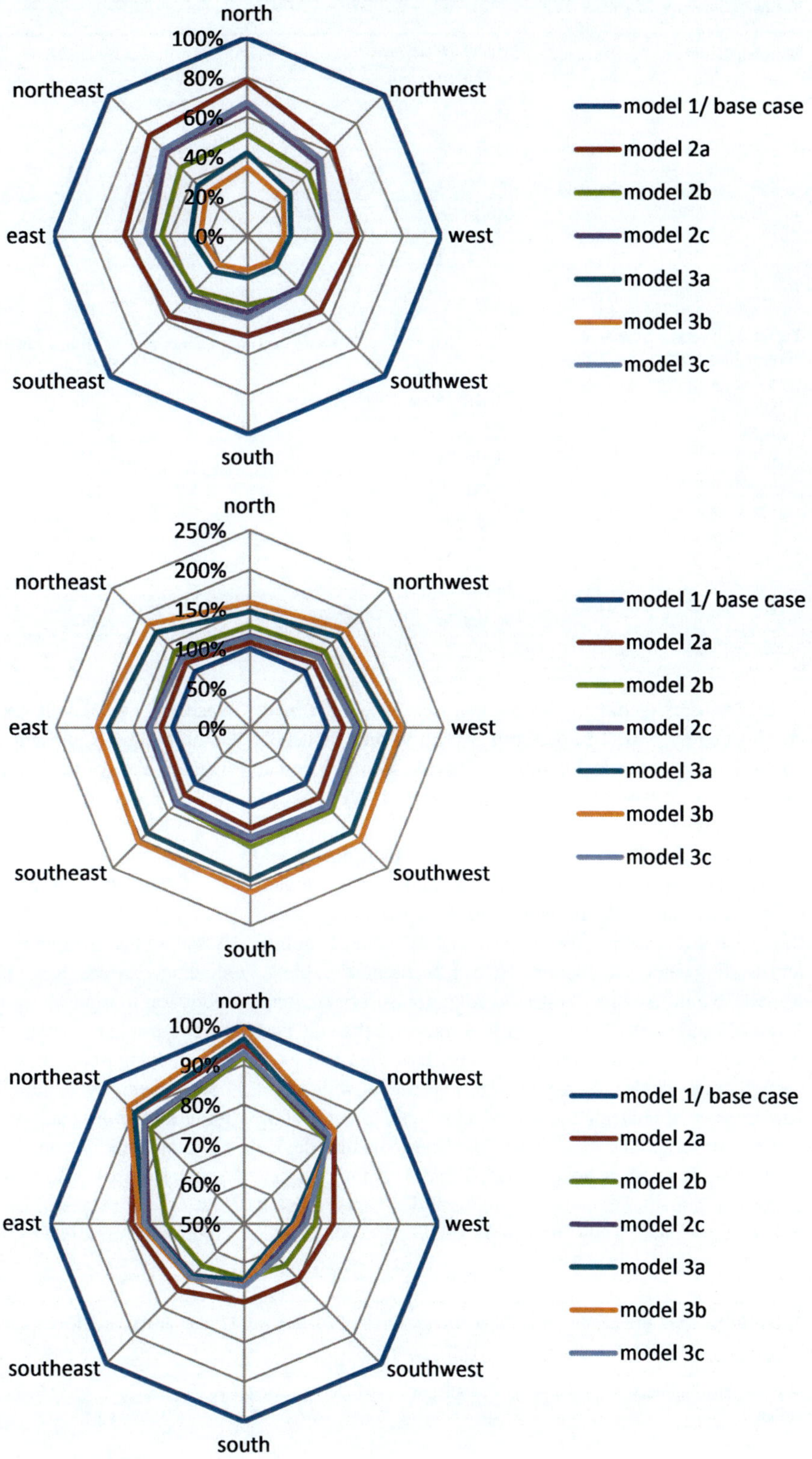

Table 7 Average of annual cooling energy in Lagos and Mumbai

	Lagos (%)	Mumbai (%)
Model 1/base case	100	100
Model 2a	86	86
Model 2b	79	79
Model 2c	80	81
Model 3a	71	72
Model 3b	68	70
Model 3c	81	81

Table 8 Average of annual cooling, heating, and total cooling plus heating energy in Mexico City

	Annual cooling energy (%)	Annual heating energy (%)	Total annual energy (%)
Model 1/base case	100	100	100
Model 2a	62	118	79
Model 2b	43	139	74
Model 2c	48	133	74
Model 3a	28	176	75
Model 3b	24	190	76
Model 3c	50	130	75

In Lagos, Nigeria, model 3b can cut 32% of annual cooling energy. Model 3a performs 3% worse than model 3b by cutting 29% of annual cooling energy. In Mumbai, India, model 3b can save annual cooling energy to 70% while model 3a save 72%.

Different conditions occur in locations which have cooling and heating period throughout the whole year. Mexico City is an example in this paper shows that shading device design strategy has positive impact of total annual cooling and heating energy reductions. It can be seen in Table 8. Model 2b and 2c works best in average for all orientations. Model 3a and model 3c performs slightly worse than the best options by only around 1% differences. It is seen that even though small in size, *L*-shaped mini louvers showing good performance of reducing the annual total cooling and heating energy consumptions. However, if the right shading device is chosen the result is better. From Table 9, if the designer choose combination among six proposed designs, the result will be better than using just single solution for all façade. Based on simulation results, the optimized shading device combination is presented in Table 6. It is seen that the average 28% of total annual cooling and heating energy can be cut in comparison with model 1 as base case.

7 Conclusions and Further Researches

Based on the three locations chosen for the simulation study, shading devices have positive impact on cooling energy reduction in tropical countries. It also shows positive impact in sub-tropical country. Even though the annual heating energy increased, but overall cooling and heating energy can be reduced. Even though the size is small, *L*-shaped aluminum profile can be used as mini louvers to reduce building energy consumptions, notably cooling energy in tropical climate. *L*-shaped mini louvers performance is the best among proposed design, even in comparison with more commonly used horizontal overhang as shading device.

Table 9 Selected shading devices to optimize total annual cooling and heating energy in Mexico City

Orientation	North	Northwest	West	Southwest	South	Southeast	East	Northeast	Average
Chosen shading device model	Model 2b	Model 2b	Model 3a	Model 3a	Model 2b/model 3a	Model 2b	Model 2b	Model 2b	
Total annual cooling and heating energy (%)	92	80	63	59	64	65	70	84	72

In Lagos, Nigeria, model 3b works best in average by reducing 32% (36,793.77 mJ) of annual cooling energy followed by model 3a by 29% (33,989.92 mJ). In Mumbai, India, model 3b also shows better performance among other models by cutting 30% (39,819.26 mJ) of annual cooling energy followed by model 3a by 28% (36,795.68 mJ). In Mexico City, Mexico, using one shading device model for all building orientations, 26% annual total cooling and heating energy can be saved. Both model 2b and 2c can be used to achieve it. Model 2b saves 5599.61 mJ while model 2c cuts 5549.08 mJ of total annual cooling and heating energy. However, using different types of shading device in this paper, total annual cooling and heating energy saving can be up to 28% (5978.17 mJ). Combination of model 2b and 3a is used to achieve this condition.

This research shows that even though *L*-shaped mini louvers have relatively small sizes, their application as shading device can save energy in buildings. They even perform better than more common shading devices, such as horizontal overhangs that are used in this paper simulation.

This research offers *L*-shaped aluminum profiles as basic elements to create simple mini louvers rather than uses custom sun shading or complicated technologies to reduce cooling energy consumptions. *L*-shaped aluminum profile has many advantages. They are easy to be obtained. *L*-shape profile is available in many sizes, from as small as 5 mm by 5 mm to 30 mm by 300 mm. It is easy to be attached in new or retrofitting old projects. The aluminum itself is self-cleaning and requires low or almost no maintenance. Aluminum is also lightweight. It means that once they are used as louvers, they do not need additional support and even can be put directly at the exterior surface of glass window using outdoor weather resistance double-faced tape. Other than gravitational load, the small size of the louvers also reduces the building load cause by wind in comparison with larger shading devices.

This paper shows only the beginning of *L*-shaped mini louvers more comprehensive research. Further research can be done to show *L*-shaped mini louvers performance in different locations. Other possible architecture-related research topics are visibility, daylight study, thermal comfort, and user preferences of the design and application. Construction research can also be accomplished regarding louvers additional load to the building, louvers window or frame attachment, as well as their durability. Economical approach research can also be executed to show the possibility of applying *L*-shaped mini louvers to save energy and users' money. More researches on this object will give broader and holistic performance of possible application in the real world.

Acknowledgements First author study at The University of Kitakyushu is financially funded by Indonesia Endowment Fund for Education (LPDP) scholarship. Participation on 3rd GU conference is funded by the University of Kitakyushu, Japan, and Fukuda Laboratory.

References

Dutta, A., Samanta, A., & Neogi, S. (2017). Influence of orientation and the impact of external window shading on building thermal performance in tropical climate. *Energy and Buildings, 139*, 680–689.

El-Darwish, I., & Gomaa, M. (2017). Retrofitting strategy for building envelopes to achieve energy efficiency. *Alexandria Engineering Journal, 56*(4), 579–589. https://doi.org/10.1016/j.aej.2017.05.011

Energyplus. (n.d.). *Mexico weather data*. Ciudad Juarez International Airport, Mexico City. https://www.energyplus.net/weather-download/north_and_central_america_wmo_region_4/MEX//MEX_Mexico.City.766790_IWEC/all

Ghosh, A., & Neogi, S. (2018). Effect of fenestration geometrical factors on building energy consumption and performance evaluation of a new external solar shading device in warm and humid climatic condition. *Solar Energy, 169*, 94–104. https://doi.org/10.1016/j.solener.2018.04.025

Gore, A. (2006). *An inconvenient truth: The planetary emergency of global warming and what we can do about it.* Rodale.

Gore, A. (2017). *An inconvenient sequel: Truth to power.* Rodale.

Hansen, J., Sato, M., & Ruedy, R. (2012). Perception of climate change. *Proceedings of the National Academy of Sciences, 109*, 37. https://doi.org/10.1073/pnas.1205276109

Hariyadi, A., Qingsong, M., & Hiroatsu, F. (2017). The effectiveness of the parametric design 'Sudare' blind as external shading for energy efficiency and visibility quality in Jakarta. *Architectural Engineering and Design Management, 13*(5), 384–403.

Idchabani, R., Ganaoui, M. E., & Sick, F. (2017). Analysis of exterior shading by overhangs and fins in hot climate. *Energy Procedia, 139*, 379–384

International Energy Agency. (n.d.). *Market report series: Energy efficiency 2018.* https://webstore.iea.org/download/direct/2369?fileName=Market_Report_Series_Energy_Efficiency_2018.pdf

International Energy Agency. (n.d.). *World energy outlook 2016.* https://webstore.iea.org/download/direct/202?fileName=WEO2016.pdf

Kohler, C., Shukla, Y., & Rawal, R. (2017). *Calculating the effect of external shading on the solar heat gain coefficient of windows.* Lawrence Berkeley National Laboratory, CEPT University. Energy Technologies Area.

Onebuilding. (n.d.-a). *Nigeria weather data.* Muhammed Murtala International Airport, Lagos. https://climate.onebuilding.org/WMO_Region_1_Africa/NGA_Nigeria/NGA_LA_Lagos-Muhammed.Intl.AP.652010_TMYx.2003-2017.zip

Onebuilding. (n.d.-b). *India weather data*. Chhatrapati Shivaji Maharaj International Airport, Mumbai. https://climate.onebuilding.org/WMO_Region_2_Asia/IND_India/MH_Maharashtra/IND_MH_Mumbai-Shivaji.Intl.AP.430030_TMYx.2003-2017.zip

Suryandono, A. R., Hariyadi, A., & Fukuda, H. (2018a). *L*-shaped mini louvers performance of solar radiation reduction in glass windows. In *The 12th Seatuc Symposium: Engineering Education for Sustainable Development*. OS02-10.

Suryandono, A. R., Hariyadi, A., & Fukuda, H. (2018b). Mini louvers performance of building cooling load energy reduction. In *Grand Renewable Energy 2018: How to Accelerate Renewable Energy Integration,* P-At-16.

User Perception of Energy Efficiency in School of Architecture and Built Environment (SABE) Building, Kigali, RWANDA

M. Michieletto and O. F. Adedayo

Abstract

The faculty building of School of Architecture and Built Environment (SABE) at University of Rwanda was designed based on volcano form and volcano stones was used as part of its external finish. This was done for aesthetic purposes and also to ensure that the building fits into the environment. The design was based on the principle of energy efficiency in terms of lighting and thermal considerations. The observable problem is the effectiveness of the design provision in meeting the users' needs and aspirations within the building. The aim of this paper is to examine the perception of the users of SABE building regarding the passive design considerations. The study made use of direct observation of the design features of the building using an observation checklist, while the opinion of the users was obtained using an interview guide and questionnaire. The results were analysed using descriptive statistics and presented as tables, charts and figures. The study concluded that the design features and concept adopted for the building were adequate in terms of lighting for the offices and inadequate in terms of thermal considerations. The paper concluded that there is need to introduce additional openings that would allow for controlled lighting in the classrooms, and there is need to make fixed lighting windows operable. There is also need to examine the issues of privacy when considering lighting of the interior spaces as users preferred to have opaque glass and window blinds at openings.

Keywords

Education buildings • Design • Efficiency • Energy • Users

M. Michieletto (✉) · O. F. Adedayo
Department of Architecture, School of Architecture and Built Environment (SABE), University of Rwanda, Kigali, Rwanda

1 Introduction

The issue of building design in the twenty-first century has taken a different approach with specific reference and consideration for sustainability. The issue of sustainability is a recurring point in every building design as it relates to environmental sustainability, with many architects striving to ensure that their designs fit into the environment as much as possible (Schluter & Thesseling, 2009). A key design consideration in the new buildings particularly as it relates to public buildings is that of energy efficiency, with the architects doing everything possible to achieve maximum passive energy control within the buildings. The other issue of sustainability is the management of resources available to man and this cuts across various facets of man's existence, be it built environment, consumer industry or production (Carlan, 2015; Khan, 2008; Manning, 2013; Morelli, 2011). It is therefore common to find many establishments and government authorities giving serious attention to environmental concerns as it affects the infrastructural development of their country. There is growing competition for the limited resources (finance) for the running and management of government institutions, and this is often attributed to the growing population and the competing demand for infrastructure. In an attempt to reduce the running costs of these institutions, issues related to environmental sustainability have become the focus with specific reference to energy cost savings. In existing buildings, care is taken to undertake energy audits with view of reducing the cost, while in the case of new buildings, such costs cuts are examined and built-in at the design stage (Jansson et al., 2013; Sonetti, 2013; Sozer, 2010).

Education is believed to be the bedrock of any nation's development, hence the huge investments in education by many African countries; these investments range from the construction of buildings, provision of infrastructure and development of the teachers. The design and construction of educational buildings have shifted towards energy efficiency

F. Trapani et al. (eds.), *Advanced Studies in Efficient Environmental Design and City Planning*, Advances in Science, Technology & Innovation, https://doi.org/10.1007/978-3-030-65181-7_3

in order to fit into the environmental sustainability requirements and also mitigate the challenges of climate change (Adeleke, 2011; Nisha & Jayasudha, 2016). It is now common to find institutions demanding building designs that would reduce the cost of energy consumption so that they could invest the savings in other aspects of the institutions. The SABE building was designed and built as a sustainable building by offering energy savings in terms of lighting and cooling through the use of passive design elements and choice of building forms. Observing how the building has been put to use by the users showed some conflict between intention of the architects and adaptation of the users. The motivation for this study was the adaptation strategies by the users that negated the passive sustainability design considerations. The objective of this study is to determine the users' perception regarding the passive design consideration for energy efficiency and the basis for the adaptive strategies they evolved in the course of using the SABE building.

The selling point of many architectural designs aside from the aesthetic value is the amount of energy savings the design would yield (Naboni et al., 2015). The most common feature is the application of the courtyard, the atrium, orientation, the glazing and the response to climatic factors such as temperature regulation. The success of any building design is the satisfaction of the users achieving the stated goals at the conception stage, in this case the energy efficiency. It implies that the users are key towards achieving the set objectives of the architects and the client, if the user fails to achieve the set objectives or even use the building in a different way from the manner intended then the design could be considered a failure. According to Jason et al. (2013), buildings are often designed and constructed to meet the needs and desires of the users, clients and the society at large, and these are expressed in terms of either regulations or codes which should be adhered to. It implies that any energy efficient building design of which educational buildings are inclusive should also meet the needs of the users (staff and students). The SABE building at the University of Rwanda is an example of educational building in Rwanda that sought to achieve energy efficiency from the design and construction perspective. The design sought to reduce energy consumption in terms of thermal and light design considerations; through the use of several design elements, this was according to the design concept proposed by the architect.

Post-occupancy evaluation (POE) is believed as the mechanism for determining if the objectives set out in the building design is achieved when the building is put to use in terms of judging aspirations of the users (Vischer, 2002). POE equally assists in determining the building quality from the users' perspective which would assist in documenting the success of the building and also aspects for improvement according to Watson (2003). It is against this background that originality of this study was drawn, and the study showed that the passive design considerations implemented in the building were negated because they did not meet the aspirations of the users' given the strategies adopted by the users.

1.1 Sustainability in Building Designs and Users

It had been established that POE gives the designer a mechanism for evaluating the building from the users' perspective; it is therefore imperative to examine issues of building sustainability and the user' as proposed in this section. The importance of sustainability in the building industry is no longer a new development as there has been a steady movement towards achieving efficient energy reduction in operation and maintenance of buildings (Khosla & Singh, 2014). In achieving the sustainability during the use of the building, the users become key to the efficient running of the energy saving considerations, hence their importance from the design stage. According to Harputlugil (2017), the belief that the buildings in the world account for 35–40% of the energy consumption is seen as a basis for ensuring sustainability right from the building design stage and its subsequent usage. The approach is to ensure that the building design and its subsequent construction fit into the climatic conditions of the place where it is situated.

The success of the design can only be measured when the users begin to utilise the building and try to adapt to the provisions within the building. In order to achieve sustainable energy building, Chwieduk (2003) stated that the most important step towards this goal was to strive for energy efficiency in the buildings. There is the need to ensure that the user of such buildings understands and makes use of the building appropriately. This includes using artificial lighting and cooling systems only when necessary.

It is well known that users' comfort is an integral part of the architectural designs, and when buildings are constructed, the users will always try to ensure they are comfortable within the building. This assumption is not different from those of the users of the SABE building. The improper use of any building that has been designed to be sustainable defeats the sustainability provision of the buildings (Chenari et al., 2016). Examples of activities by users that could affect the sustainability intent in a building include but not limited to;

1. The use of artificial lighting during the day.
2. The use of artificial cooling system to ventilate a space.
3. Lighting up multiple lighting units than required within a space.
4. Closing of windows simply to run the air conditioners.
5. The use of window blinds or coverings.

These activities oftentimes cannot be completely envisaged by the architects, particularly as the attitude of students and staff often varied and difficult to generalise. Their actions however often give an inkling of how to respond to some of their basic actions when designing similar buildings in future.

2 Research Method

This study comes under descriptive survey which can be achieved using the post-occupancy evaluation (POE) methods which is considered adequate for such studies according to Adedeji and Fadamiro (2012) and Zimmerman and Martin (2001). Direct observation was conducted on the building and the results are presented in tables. The design elements applied in the building design to ensure the energy efficiency were examined. A total of 96 respondents (16 Staff and 80 Students) were selected for the study from a total population of 850 users (47 Staff and 803 Students). The students were selected across the four levels of study, equally giving a distribution of 20 students per level of study. The implication of some of these design considerations is discussed based on interviews conducted with some selected users of the building which include student and staff. The analysis of their opinion is presented under specific sub-headings which specifically examines the aspect related to energy efficiency in terms of lighting and thermal considerations within the building. The building plans and images are presented as figures to further illustrate the findings. The elements used in providing a passive energy efficient building were identified and discussed. The selection of the respondents was based on purposive sampling method to ensure that all categories of users were represented as their use and need varied.

3 Findings and Discussion

3.1 Passive Design Considerations

There is a need to examine some specific elements and design considerations that were used in the design of the SABE building. The average temperature in Kigali is shown in Fig. 1; this temperature can be affected by some specific site features such as the total built-up area surrounding the SABE building and the ground finishes which aid in retaining heat energy around the site. The design of the SABE building is one that made use of courtyards and extensive open corridors to connect the different spaces within it. The building is made up of two floors and a semi-basement as can be observed in Figs. 2 and 3. The building orientation is placed along the East–West axis as the overall width of the building placed along this axis. This was however defeated because of the further sub-division of the building into smaller individual units of which some spaces have a long part of the building along the East–West axis. This sub-division affected the respondents as they could only respond to their experiences within these individual spaces rather than the overall building. The building envelope is made up of concrete roofing and sandcrete hollow block infill that is finished with volcano stones used as a local building material. The walls were rendered using cement and sand mix internally and externally in some areas of the building.

The building is also divided into two separate wings connected by an open bridge which was done to reduce the overall mass of the building by breaking the building into two and creating several corridors that link the different bridges thereby ensuring good flow of air around the building. The planning was such that the office spaces were located at the same area on the ground floor and the design concept adopted for the offices was open plan system which affected the number of facilities and fittings required. The offices are connected using corridors and courtyards with an office located in the centre of the courtyard which serves as the boardroom for the school. All the offices had windows on one side of the wall while the other openings by the corridors had fixed glass; hence, they could only admit either direct sunlight or reflected light depending on the location within the building. Two of the three offices located within the courtyard had their windows opening only to the courtyard while the boardroom had windows on both sides of the walls. The classroom areas on the ground floor which are at the same wing as the offices are connected by a large lobby with a small courtyard. Three of the classrooms within this particular area had windows on adjacent walls, while the fourth classroom had windows on only one side of the wall. The use of iron grill gates as doors ensured that that there

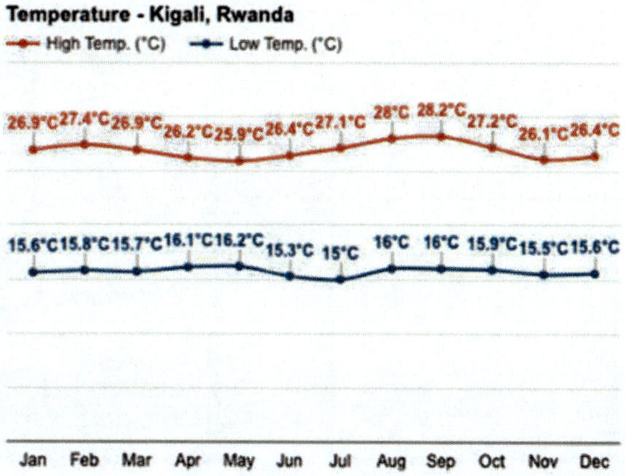

Fig. 1 Average temperature in Kigali, Rwanda. *Source* https://www.weather-atlas.com/en/rwanda/kigali-climate#temperature

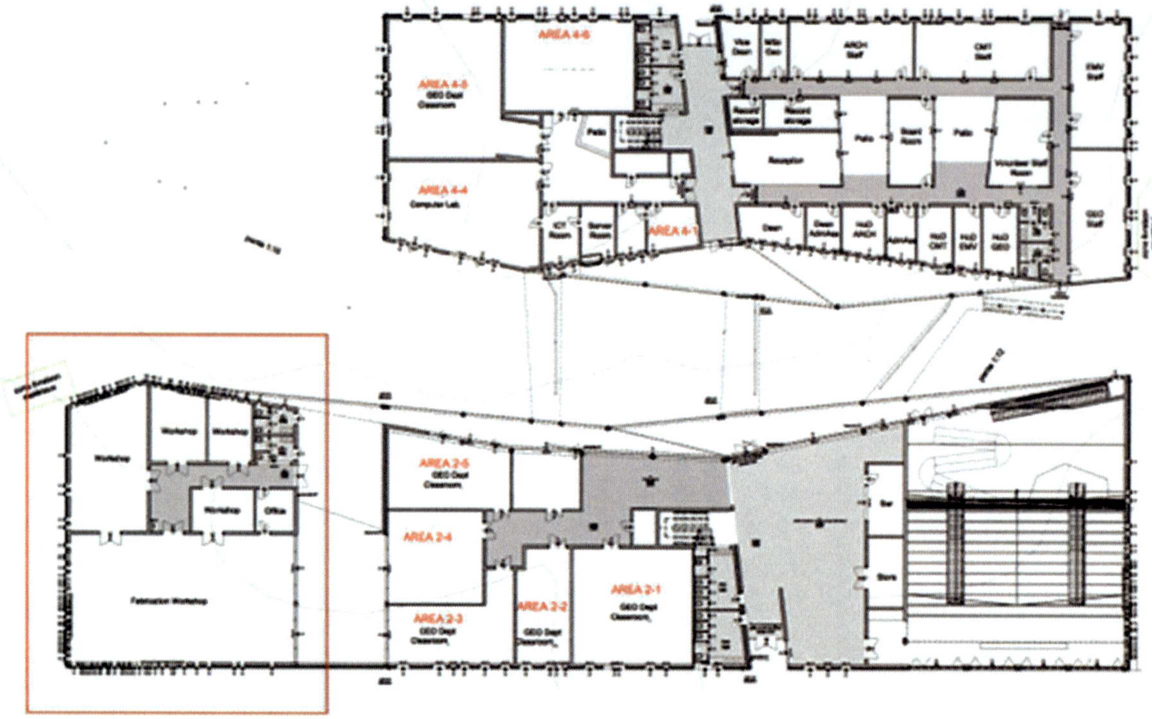

Fig. 2 Ground floor plan of School of Architecture and Built Environment

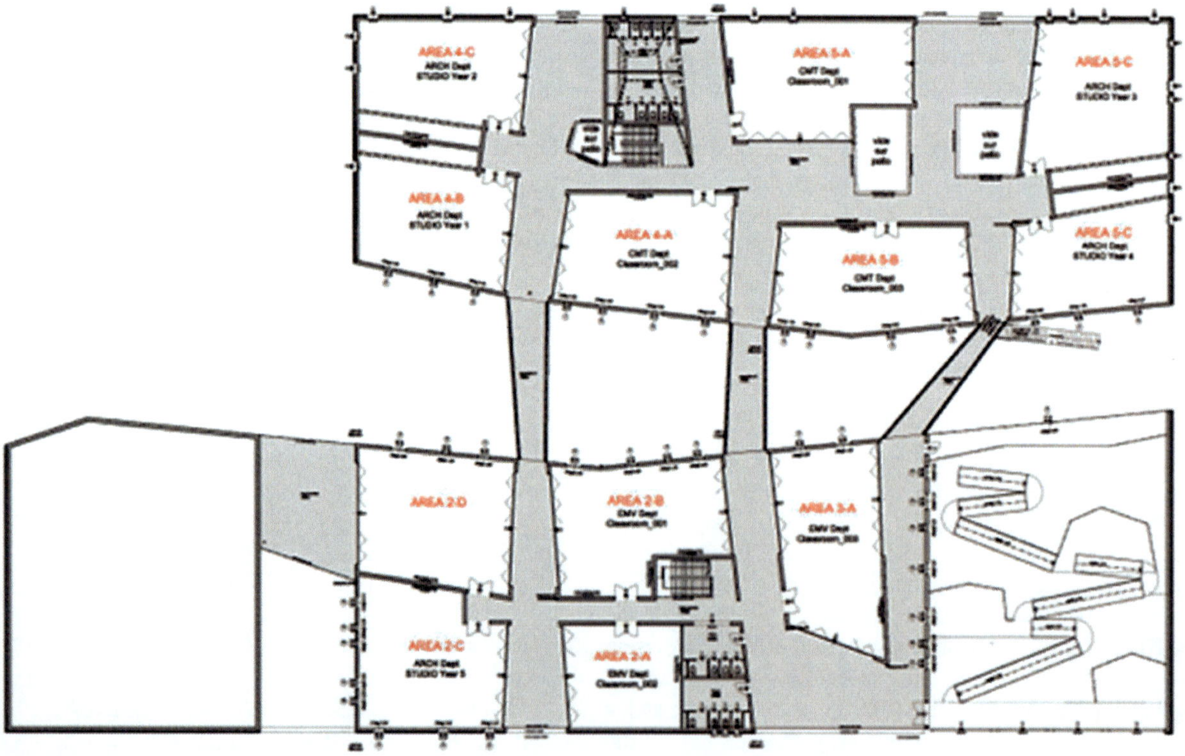

Fig. 3 First floor plan of School of Architecture and Built Environment

was free flowing air around and within the corridors which was expected to cool the building.

The ground floor of the second building is made up of an exhibition hall that is lighted using fixed glazing at one section of the wall and rectangular-shaped openings for the only source of air into the large space. This same concept was used for the auditorium which extended into the basement; the access into this hall is from a large opening expected to serve as the dining area for the kitchen. The dining area also connects to a large lobby that links the corridor to the classrooms; these classrooms have windows open on one side of the wall with glazing on another side solely for the purpose of lighting. The workshop is connected using an open courtyard accessible from outside; the workshop has a glazed side visible to the courtyard while the rectangular-shaped openings are located within two adjacent walls as shown in Fig. 4.

The first floor of the school building as shown in Fig. 3 is such that it accommodates only classroom and studio spaces. The basic design for the spaces was adjacent walls to the external part of the building with provision of windows and provision for one foldable glazed wall for the admittance of light whenever the need arose. In cases where the classrooms were adjoined by a corridor, the foldable glazed wall was provided in opposite sides of the wall with windows in only one side of the wall. The use of high-level windows made of glass louvers was adopted in each of the classrooms and studios; this was achievable given the nature of the roofing concept adopted for the building which was done on the volcano idea. The volcano concept is based on the aesthetic design concept of the mountains believed to have resulted from the eruptions of volcanoes in Rwanda, and it allowed the users to be able to relate such buildings to the local environment. The smoke is usually seen as rising through the top of the volcano eruptions, and this idea was incorporated in the design of the roof with the introduction of the louver windows at the roof top where hot air is expected to rise through.

The design considerations for the lighting in the building were hinged on the use of extensive glazing in terms of its provision of lighting for interior spaces, which was expected to reduce the need for artificial lighting. The location of these glazing within areas adjoining the courtyard was expected to reduce the direct effect of glare which was achieved in areas with overhangs on the ground floor. The low number of windows and the use of large wall areas on the exteriors of the building were expected to assist in regulating thermal balance of the building given the infill material of sandcrete hollow blocks. The use of courtyards and the separation of the unit spaces (classrooms, studios and offices) was to ensure that air could circulate around the building adequately thereby reducing the overall massing of the building. The use of the high-level windows in the classrooms and studios was to improve the ventilation within such spaces. The provided flowerbeds was expected to provide some form of cooling during hot periods and also improve the ventilation in terms of quality of air around the building (see Fig. 5). It is expected that stack effect would be achieved when the foldable walls are left open in combination with the roof windows. These are some of the perceived objectives of the building design based on the observation of the users. These design features employed by the architect were expected to ensure the sustainability of the building passively; however, with the introduction of the key element (users) in a successful building development, the objectives of the building considerations were affected.

3.2 User Perception on Thermal Considerations

The average temperature range in Kigali is at between 16 and 28 °C; this is considerably cool when compared with other hot regions within Africa such as West Africa (Nigeria) 28 °C and 40 °C and North Africa (Algeria) 23 and 45 °C; hence, it is understandable that the majority of the buildings in Rwanda do not make artificial cooling system. The design of the SABE building is equally based on such consideration with no provision for fans or air conditioners; the building is expected to be self-regulating. The stone finishing is believed to have contributed to the heat gain

Fig. 4 Cube-like openings located around the workshop space of building

Fig. 5 Pockets of flowerbed and separation of unit spaces using courtyard

within the interior of the spaces because it absorbed heat during the day and let it in at evening periods. The stones used for parts of the external finishes fall under the category of felsic rocks (light coloured rocks) which is believed to be rich in elements such as AI_2O and TiO_2, which contribute to heat emission in rocks. The use of laterite-based bricks or blocks as finishes would have reduced the heat gain as they contain far less of the elements found in the felsic rocks and also have low conductivity. This position is supported by Ramakrishnan et al. (2013) where they examined the thermal properties of rocks and laterite. The perception of the users during periods when the temperature becomes high is that the building requires some form of artificial cooling or better ventilation system. The staff offices that had only three windows which were operable were located at one side of the wall in the office space which made it difficult to cool the office whenever it got hot. In order to achieve ventilation in the space, some of the respondents stated that they had to open their doors.

> Whenever the office becomes hot and stuffy I have to open the door to expel the hot air and this is not the best option as the fixed windows on the corridor could have been made to open.

A staff in the office with windows adjoining only the courtyard was of the view that there was need to have additional opening.

> Once the office becomes hot I have to take a walk and when that is not possible, I will have to open the door to ease the effect even though the door opens to the corridor, the glazing allows my office to heat more than others.

When the weather is cold, the staff seem not to have significant challenge as they seem to have a better way of handling this challenge which does not require heating the building.

> During cold periods all I need to do is to simply wear thicker clothing and ensure that the windows are closed particularly if am the only user in the office. In cases where there are other users in the office we simply open the window a little to allow for ventilation.

It is obvious from the perception of the staff that the passive design features suit them when the temperature is low as they are able to cope conveniently within the building; however, the challenge arises when the need for cooling through ventilation arises. The suggestion for solving this challenge was stated by some respondents.

> The provision of louvers above the doors would have aided in the cooling of the office as this can easily be left open hence ventilation could be improved.
>
> All the glazed window that were made for fancy should have been made to open hence improve the ventilation of the office without us having to always leave the doors open.

In examining the perception of the staff regarding the effectiveness of the windows for passive thermal considerations in the building, the results in Table 1 show that the respondents consider the provision of window ineffective and this is believed to be largely affected by inoperable nature of the windows during the hot periods of the season. The effective perception of the windows as it relates to the cold periods is due to the individual coping activities of the respondents and the need for the space to be warmer. Using Likert scale to rank the responses of the respondents regarding, some variables of effectiveness of the windows provide within the offices. The ranking ranged from very effective (VE) to very ineffective (VI); the range of value for each option of crime frequency is as follows:

Very effective (VE)	1.0–1.49
Effective (E)	1.5–2.49
Ineffective (I)	2.5–3.49
Very ineffective (VI)	3.5–4.49.

The results in case of the classrooms/studios of which the students are the major users are quite different from those of the staff offices; this can be understood as majority of the spaces have some form of cross ventilation. The use of the foldable walls was a major contributing element towards the thermal control of the classroom/studio spaces from respondents. The opinions are shared in some statements such as;

Table 1 Respondents opinion on the effectiveness of windows for thermal control in the offices

Measured Variable		Number of respondents	Sum	Mean	Interpretation
1	The provision of windows on one side of office	16	44	2.75	Ineffective
2	Number of windows as it relates to hot periods	16	55	3.44	Ineffective
3	Number of windows as it relates to cold periods	16	30	1.88	Effective

Student 1
I feel comfortable even when the weather is hot because I can easily move my table towards the open wall and the air comes in fast.

Student 2
The windows in combination with the sliding glass wall allows us to ventilate the classroom quickly before it gets hot.

Student 3
The louvers in the roof allows the hot air to escape, especially when the rains are falling and the direction of wind is away from the roof windows, but when the rain is towards the roof openings the class can become hot because the windows are not enough especially during lectures.

Student 4
The classrooms on the ground floor are not as luck as those on the first floor because they do not have sliding walls, those closer to the windows on the ground floor enjoy better cooling than those away from it, that's why I like sitting close to the window.

Fig. 6 Example of foldable wall in the classroom/studio

The perception of the students is presented in Table 2 using similar measurement as those of Table 1. The result shows that the provision of windows on one side of the wall in a classroom/studio was considered as ineffective, however the introduction of the foldable walls and windows on two sides of the walls that they are considered effective in thermal design consideration, however the number of such windows during hot periods was found to be ineffective. The ineffectiveness perception could be due to the population of students using the space. Examples of these design features such as windows and foldable walls are shown in Figs. 6 and 7.

3.3 User Perception on Lighting Considerations

Lighting is a major contributor to the energy consumption of buildings and the SABE building is not different either. The extensive use of glazing in the building was expected to reduce the energy consumption during the day as they were expected to bring in adequate sunlight into the interior spaces. The use of fixed windows in the exterior of the office buildings improved the lighting of the offices; however, there are cases where the users considered the need for privacy above that of light hence the use of cardboards or curtains to seal the glazing on the window (see Fig. 8). This

Table 2 Respondents opinion on the effectiveness of windows and folding walls for thermal control in the classroom/studio

Measured variable		Number of respondents	Sum	Mean	Interpretation
1	The provision of windows on one side of classroom/studio	73	208	2.85	Ineffective
2	The provision of windows on two sides of classroom/studio	73	169	2.32	Effective
3	The provision of foldable walls in classroom/studio	73	112	1.53	Effective
4	Number of windows as it relates to hot periods	73	218	2.99	Ineffective
5	Number of windows as it relates to cold periods	73	153	2.10	Effective

Fig. 7 External view of windows on adjoining walls of classroom/studio

particular action affected the quantity and quality of light reaching the interior spaces as the colour of the cardboard differed. A solution to this challenge would be the use of reflective glass or tinted glass that would have given some form of privacy, because the effect of their actions mean that the need to have the artificial lighting on during day is common. One of the staff stated during interview responded by saying;

> I would rather have my privacy and use artificial lighting than to have an exposed office and not use the lighting, it is even worse at night when the light is on and outside is dark.

Natural lighting comes with its challenges such as glare and reflection, and this is considered a major challenge by a few staff because they seem to have a fixed sitting position in the shared offices. The staff at the direct path of the sunlight usually preferred to have the window blinds down all day so as to regulate the volume of sunlight and supplement it with artificial lighting. This action defeats the whole aim of the passive design consideration for lighting in the building. One of the respondents stated that;

> The sunlight is always reflecting on my computer and I can change my location within the office, hence I have to always put down the curtains to reduce the amount of light entering the office.

The artificial lighting system installed in the building is one that sought to reduce the amount of energy consumed in the building. This action is visible given the nature and number of security lighting points provided on and around the building; there is no security lighting on the external walls. The few external lightings are restricted to the bridge connecting the two buildings and the garden lightings which use low energy fittings. The external lightings and the corridor lights are fitted with motion sensors that allow for the lights to be regulated independently and remove the issue of human error of forgetting to switch off during when not in need. While this is method is commendable the same cannot be said of the interior spaces of offices, classrooms and studios are manually operated. The major problem with the

Fig. 8 Examples of cardboards and curtains used to shade window and glazing

office lighting at night is that it was considered too bright and sometimes wasteful as stated by one of the respondents. A single switch controlled the entire lighting points in the office, regardless of the number of users in the office once the switch is flipped on the entire lighting units come on (see Fig. 9).

> I do not understand why individual luminaire did not have its control switch as it is wasteful to have all the luminaire on when I'm alone in the office, the office now becomes too bright and this is uncomfortable.

The import of this opinion of the respondent is more pronounced when you consider that the office for the staff has a minimum of six luminaires working when in actual fact a single luminaire would have served. It implies that the energy consumption at that point in time could be reduced by 83.33% which is quite significant. The respondent satisfied with the lighting system installed along the corridors and the garden as it is not the same with those of the office, in the words of one of the respondents;

Fig. 9 Example of six luminaire lighting points in typical office controlled by a single switch

> The lighting along the corridors have two luminaires to one sensor so why was similar thing not done in the office as this is a good method to manage the lighting points.

In examining their satisfaction with the lighting efficiency of the building, respondents gave their opinion on a few variables which is presented in Table 3. It can be observed from Table 3 that the respondents were not satisfied with the natural lighting getting into their offices and this could be understood as a result of individual preferences; it is not enough to conclude that the natural lighting design considerations have failed, and it only implies that there is need to examine specific user needs and find a common point of solving the preferences. In the case of the artificial lighting, the major issue has to do with the control system installed for the lighting as the respondents would have loved to control the lighting points individually.

The ranking ranged from very satisfied (VS) to very unsatisfied (VU), and the range of value for each option of crime frequency is as follows:

Very satisfied (VS)	1.0–1.49
Satisfied (S)	1.5–2.49
Unsatisfied (U)	2.5–3.49
Very unsatisfied (VU)	3.5–4.49.

In the case of the students, their response to the issues of lighting had to do with night periods because they had no natural light coming into the classroom based on the area exposed within the walls. When the sun direction was through the foldable walls, they simply had to move their seats away from the path and work within the shadow which was a better option. The roof lighting was also a point of concern for respondents depending on the arrangement of seats in the class as this was often centrally located and they could not shift their seats. The challenge with the artificial lighting is the fact that it had only one control switch, when put to use all the lighting point become lit. The students shared the same opinion as the staff that there was no need to have all lighting points lit if the aim was to save energy. The responses are presented in Table 4 where it can be observed that it showed similar results to the staff as it pertained to artificial lighting, while they were satisfied with the natural

Table 3 Respondents opinion on the satisfaction with lighting design consideration in the office

Measured variable		Number of respondents	Sum	Mean	Interpretation
1	Quantity of natural lighting entering the office	16	42	2.63	Unsatisfied
2	Effectiveness of natural lighting in the office	16	48	3.00	Unsatisfied
3	Quantity of artificial lighting in office	16	39	2.44	Satisfied
4	Effectiveness of artificial lighting in the office	16	38	2.38	Satisfied
5	Control of artificial lighting in the office	16	54	3.38	Unsatisfied
6	Control of artificial lighting outside the office	16	29	1.81	Satisfied

Table 4 Respondents opinion on the satisfaction with lighting design consideration in the classroom/studio

	Measured variable	Number of respondents	Sum	Mean	Interpretation
1	Quantity of natural lighting entering the classroom/studio	73	166	2.27	Satisfied
2	Effectiveness of natural lighting in the office	73	157	2.15	Satisfied
3	Quantity of artificial lighting in office	73	150	2.05	Satisfied
4	Effectiveness of artificial lighting in the office	73	161	2.21	Satisfied
5	Control of artificial lighting in the office	73	216	2.96	Unsatisfied
6	Control of artificial lighting outside the office	73	163	2.23	Satisfied

lighting in the classrooms. It is observed that the design considerations for lighting within the classrooms and studios met the desires of the users based on the level of satisfaction.

4 Conclusion

The respondents have shown that the objective of ensuring that the building is sustainable in terms of energy efficiency could be improved upon given the adaptive methods of the users. While it appears that the classrooms met the objective in terms of the outcome of natural lighting the same cannot be said about the thermal conditions, however the users of such spaces appeared to rate it better than the office users. In the case of the staff who made use of the office, it can be concluded that there are other variables that needed to have be considered while striving to meet the energy efficiency. These issues bother specifically on reference to privacy; hence, the adaptive measures employed by the users diminished the planned objective. The common issue of artificial lighting control within the spaces (offices, classrooms and studio) was a major common ground which could be addressed by changing the lighting control. In the case of thermal cooling in the offices, the conversion of the fixed glazed windows to operable windows would greatly improve the thermal balance within the offices. A change of the glazing material used in the windows of the offices to one that allows light inwards and also provides some form of visual privacy that would help regulate the quantity of sunlight that enters the interior and remove the need to use cardboards.

In conclusion, the benefits of a good building design can only be felt when the users of such building are satisfied with the provisions within the building and do not take adaptive decisions that negates the objectives of the building which in turn can render the building unsustainable. It is not just enough to claim that the is building energy efficient and the users are not satisfied with the functionality of the building in terms of comfort.

References

Adedeji, J. A., & Fadamiro J. A. (2012). Workplace and productivity: A post occupancy evaluation of LAUTECH senate building, Ogbomosho, Nigeria. *Architecture Research, 2*(2), 14–19. https://doi.org/10.5923/j.arch.20120202.03

Adeleke, O. O. (2011). *Mitigating the effects of climate change through sustainable building environment Agenda IV.* Seminar Proceedings Organized By Architects Registration Council of Nigeria, April, 2011.

Carlan, Z. (2015). *Reducing environmental impact with solid works sustainability analysis.* Retrieved from www.javelin-tech.com on April 24, 2017.

Chenari, B., Carrilho, J. D., Silva, & M. G. (2016). Towards sustainable, energy-efficient and healthy ventilation strategies in buildings: A review. *Renewable and Sustainable Energy Reviews, 59*, 1426–1447. Retrieved from https://www.sciencedirect.com/science/article/abs/pii/S1364032116001040s

Chwieduk, D. (2003). Towards sustainable-energy buildings. *Applied Energy, 76*(1), 211–217.

Harputlugil, T. (2017). Energy efficient building design development: A retrospective approach. In *Proceedings of International Symposium on Energy Efficiency in Buildings* (pp. 296–303). Ankara, Turkey, February 13–14, 2017.

Jansson, G., Schade, J., & Olofsson, T. (2013). Requirements management for the design of energy efficient buildings. *Journal of Information Technology in Construction ITcon, 18*, 321–337.

Khan, Z. (2008). Cleaner production: An economical option for ISO certification in developing countries. *Journal of Cleaner Production, 16*, 22–27.

Khosla, S., & Singh, S. K. (2014). Energy efficient buildings. *International Journal of Civil Engineering Research, 5*(4), 361–366.

Manning, L. (2013). Corporate and consumer social responsibility in the food supply chain. *British Food Journal, 115*(1), 9–29.

Morelli, J. (2011). Environmental sustainability: A definition for environmental professionals. *Journal of Environmental Sustainability, 1*(1–2), 1–9.

Naboni, E., Malcangi, A., Zhang, Y., & Barzon, F. (2015). Defining the energy saving potential of architectural design. *Energy Procedia, 83*, 140–146. Retrieved from https://doi.org/10.1016/j.egypro.2015.12.204

Nisha, T. M., & Jayasudha, P. (2016). Sustainable and energy efficient design principles in vernacular architecture. *International Journal of Advanced Research in Biology Engineering Science and Technology (IJARBEST), 2*(11), 31–35.

Ramakrishnan, D., Bharti, R., Singh, K. D., & Nithya, M. (2013). Thermal inertia mapping and its application in mineral exploration: Results from Mamandur polymetal prospect, India. *Geophysical Journal International, 195*(1), 357–368.

Schlueter, A., & Thesseling, F. (2009). Building information model based energy/exergy performance assessment in early design stages. *Automation in Construction, 18*(2), 153–163.

Sonetti, G. (2013). *Energy and architecture—An overview*. Retrieved from https://www.epj-conferences.org or https://doi.org/10.1051/epjconf/20135402004

Sozer, H. (2010). Improving energy efficiency through the design of building envelope. *Building and Environment, 45*(12), 2581–2593.

Vischer, J. (2002). *Post occupancy evaluation: A multifaceted tool for building improvement* (pp. 23–34) Federal Facilities Council, The National Academy Press (Chapter 3).

Watson, C. (2003). Review of building quality using post occupancy evaluation. *Journal of Programme Education Building, 35*, 1–5.

Zimmerman, A., & Martin, M. (2001). Post-occupancy evaluation: Benefits and barriers. *Building Research and Information, 29*(4), 429–450.

Research on Small Sewage Treatment Technology

Xiaoxuan Wang and Weijun Gao

Abstract

In China, with the rapid development of urbanization and the gradual improvement of industrialization level, people's requirements for living environment and ecological environment are also increasing. Therefore, the development, construction, and operation of small and dispersed sewage treatment facilities are increasingly valued by the state and governments at all levels. This paper investigates and studies the construction and operation of rural sewage treatment facilities in Jilin Province. The results shows that the improved A^2O (VFL) process, SBR reactor, and MBR take over about 80% of the total rural sewage treatment facilities in Jilin Province due to their mature design and operation experience, better treatment efficiency, and stable qualified rate of effluent. In addition, this paper focuses on the reactor structure and treatment principle to analyze the improved A^2O (Vertical Flow Labyrinth) process and provides the feasibility and superiority of VFL process for rural domestic sewage treatment in cold regions in combination with the evaluation on its effect in practical application in northern cold regions.

Keywords

A^2O • VFL • Sewage treatment facilities • SBR • BR

1 Overview

With the process of society and economic development, China is facing with the problems of water shortage, low utilization rate of water resources, unbalanced development and utilization, excessive exploitation of groundwater, and low utilization rate of water resources (Liu, 2016). According to the relevant annual report of environmental statistics, the amount of domestic sewage discharged from cities and towns in China was 48.51 billion tons in 2015, and most of villages and towns lacked sewage collection and treatment facilities. On the basis of actual situation, the corresponding countermeasures are put forward, such as strengthening infrastructure construction, developing sewage treatment technology and improving water resources recycling utilization rate (Wang et al., 2008).

The sewage is various and complex in Jilin Province. The water quality and quantity are unstable in cold winter, so that the conventional treatment fails to meet discharge standards. Jilin Province is economically underdeveloped. Therefore, economic factors should also be taken into account while considering sewage treatment technology. How to treat village sewage with unstable water quality and quantity in winter is an urgent and realistic problem to be solved.

Compared with the high investment, complex operation, and different water quality characteristics of sewage treatment in big cities, the village sewage treatment has just started in Jilin Province. These technologies can be used for reference but not copied. At present, A^2O process, SBR, and MBR are commonly used in village sewage treatment in Jilin Province (Shen, 1999).

2 Rural Sewage Treatment Technology in Jilin Province

The core of each sewage treatment is the process, which is closely related to influent water quality, effluent requirements, treatment capacity, investment size, and other factors (Qin, 1989). Because the main pollutant of municipal wastewater is organic matter, most of the municipal wastewater is treated by biological treatment technology at home and abroad, which can be divided into activated sludge

X. Wang (✉) · W. Gao
Department of Architecture, The University of Kitakyushu, Kitakyushu, Japan
e-mail: wangxx_1977@126.com

F. Trapani et al. (eds.), *Advanced Studies in Efficient Environmental Design and City Planning*, Advances in Science, Technology & Innovation, https://doi.org/10.1007/978-3-030-65181-7_4

method and biofilm method (Qian, 1993; Qin, 1989; Zhou, 2003). Activated sludge wastewater treatment plants account for the vast majority because of the low efficiency and poor sanitary conditions of biofilm treatment.

There are three types of processes which widely used in Jilin Province: (1) traditional A^2O process, (2) SBR process, and (3) MBR process. This paper will briefly introduce the three wastewater treatment processes and also a new improved A^2O treatment technology (VFL).

2.1 Traditional A^2O Process

Traditional-activated sludge process is the earliest sewage treatment process in the word. It has a high efficiency of removing organic matter and suspended matter. For urban sewage, BOD_5 and SS of effluent should be below 30 mg/L. Traditional A^2O process has the effect of nitrogen and phosphorus removal at the same time. The principle is that phosphorus is released in anaerobic zone and absorbed in aerobic zone to achieve the purpose of phosphorus removal. Pollutants are oxidized and degraded in aerobic zone to remove COD and BOD_5. At the same time, under the action of nitrifying bacteria, ammonia nitrogen converted by organic nitrogen continues to be converted into nitrite nitrogen and nitrate nitrogen, and a large amount of mixed liquor with nitric acid nitrogen was refluxed to anoxic zone for denitrification (Zhou, 2003). The flow chart is shown in Fig. 1.

The main advantages of this process are high removal rate of COD, BOD_5, and SS, high removal efficiency of nitrogen and phosphorus, low operation cost, less land occupation, and good effluent quality. The disadvantages are high requirements for operation management and high investment (Bai, 2003; Li et al., 2001). For the small-scale processing program in Jilin Province, the advantages and disadvantages of A^2O are summarized as follows:

Advantages

1. Microorganisms with different populations and different environmental conditions cooperate (anoxic, anaerobic, and aerobic) to remove organic matter and also to remove nitrogen and phosphorus simultaneously.
2. The process is simple, with short cycle period and high treatment efficiency.
3. Filamentous bacteria do not multiply in large quantities under the alternating operation of three oxygen environments. SVI is generally less than 100, and sludge bulking does not occur.
4. Sludge Sedimentation is good (Zhu et al., 2002).

Disadvantages

1. The phosphorus content in sludge is high, generally more than 2.5%, so phosphorus removal is mainly achieved through sludge discharge.
2. The effect of denitrification is restricted (Gao, 1989).
3. Dissolved oxygen is strictly controlled.
4. The effect of nitrogen removal in winter is not good.
5. Small-scale equipment cannot satisfy sludge discharge and storage.
6. Poor impact resistance.

2.2 Traditional and Improved SBR Process

2.2.1 Traditional SBR Process

SBR process is also known as sequencing batch activated sludge process. It is distinctively featured by the intermittent operation, periodic influent, and effluent after aeration and precipitation for cyclic processing (Gao, 1989; Yu et al., 2004). The process flow chart is shown in Fig. 2. The core technology of SBR is the main reaction tank, which integrates primary sedimentation tank, biochemical tank and

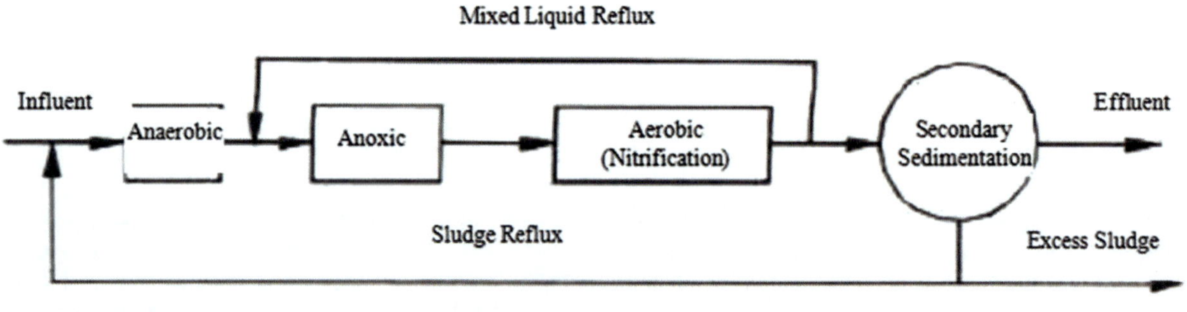

Fig. 1 A^2O process flow chart

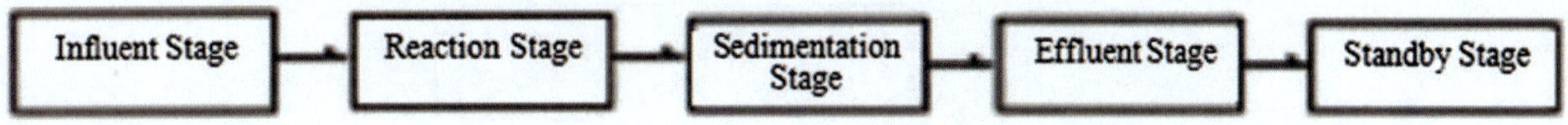

Fig. 2 SBR process flow chart

sedimentation tank, without sludge reflux. The process has been adopted by Crundy (3000 m^3/d) wastewater treatment plant in the USA and waste water treatment plant in Australia (Bai, 2003).

The process has the following advantages: stable and good effluent quality, simple operation and management, high level of automatic control; small area, low cost, flexible operation.

However, the shortcomings, such as blockage of aeration system, high failure rate, and high requirement of monitoring means, also affect the use of SBR in winter.

2.2.2 CAST Process

CAST is an improved SBR process, with the best efficiency of nitrogen and phosphorus removal among all the improved SBR process. The biggest improvement of this process to SBR is the selective zone at the front of the reactor. The sewage first enters the selective zone and mixes with the reflux mixture of the autonomous reaction zone. Under anaerobic conditions, the selective zone is equivalent to a pre-anaerobic tank which could increase the phosphorus removal efficiency (Gao, 1989). Another characteristic of this process is that it uses the principle of simultaneous nitrification and denitrification to denitrify. In the main reaction zone, the dissolved oxygen is controlled no more than 15 mg/L in the early stage of the reaction period which is in anoxic state. The original nitrate nitrogen and the nitrate nitrogen produced by simultaneous nitrification are used for denitrification. At the later stage of the reaction, the oxygen supply is increased to a aerobic state, so as to complete the phosphorus removal reaction and maintain sufficient dissolved oxygen in effluent (Qin, 1989; Yu et al., 2004). Due to simple design and operation management, this process has been applied to sewage treatment by many small sewage plants in Jilin Province.

However, shortcomings are also obvious, such as higher requirements for self-control, high power consumption, low volume utilization rate, sludge stability inferior to anaerobic nitrification, and poor low-temperature adaptation.

2.2.3 MSBR Process

MSBR is a modified SBR in combination of A^2O and SBR processes, incorporating advantages of the both. Therefore, the effluent is stable (Yu et al., 2004). MSBR is a continuous and efficient sewage treatment process, for which automatic computer control is available due to easy operation, small volume, and single tank. It can effectively treat wastewater containing high concentration of organic matter, nitrogen, and phosphorus. However, MSBR process needs further optimization and improvement due to the complex structure, varied equipment, high cost, and inconvenient management (Zhou, 2003; Yu et al., 2004; Luo, 1999).

2.3 MBR Process

Membrane bioreactor (MBR) is a water treatment technology that uses membrane separation process to replace secondary sedimentation tank in traditional activated sludge process. There are many kinds of membranes, which are classified into reaction membranes, ion exchange membranes, and osmotic membranes by separation mechanism.

MBR uses membrane separation equipment to intercept-activated sludge and macromolecule pollutants in biochemical reaction tank, omitting the secondary sedimentation tank. In traditional wastewater biological treatment technology, sludge-water separation is accomplished by gravity in secondary sedimentation tank. Its separation efficiency depends on the sedimentation performance of activated sludge. The sludge-water separation efficiency is higher when the sedimentation is better, while the sludge sedimentation depends on the operating condition of aeration tank. To improve the sedimentation performance, the operating conditions of aeration tank must be strictly controlled, which limits the application scope of this method. Membrane-bioreactor technology greatly strengthens the function of bioreactor through membrane separation technology, which greatly increases the concentration of activated sludge. In view of interdependence of hydraulic retention time (HRT) and sludge age (SRT), MBR achieves the separation of sludge retention time and hydraulic retention time, which greatly improves the efficiency of solid–liquid separation. The increase of activated sludge concentration in the aeration tank and the emergence of specific bacteria (dominant microbial population) in the sludge also accelerate the biochemical reaction. The flow chart is shown in Fig. 3.

At the same time, there are some shortcomings in membrane bioreactor, mainly in the following respects: (1) high cost of membrane, which makes higher investment of membrane-bioreactor infrastructure than that of traditional sewage treatment process; (2) membrane prone to fouling,

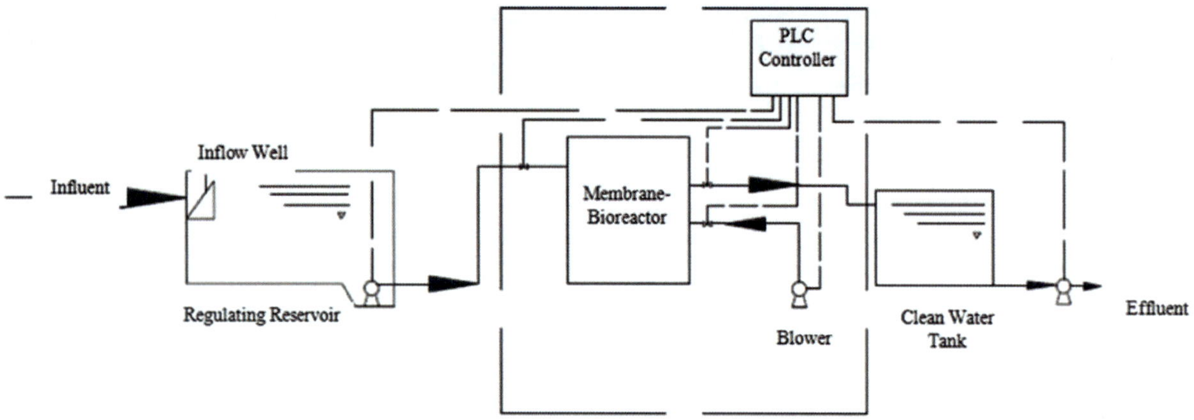

Fig. 3 MBR process flow chart

which brings inconvenience to operation and management; (3) high energy consumption: firstly, the sludge-water separation process of MBR must maintain a certain membrane driving pressure; secondly, in order to maintain sufficient oxygen transfer rate, aeration intensity must be increased in view of very high concentration of MLSS in MBR pool; additionally, in order to increase membrane flux and reduce membrane fouling, flow rate must be increased to scour the membrane surface, resulting a higher energy consumption than traditional biological treatment process.

2.4 VFL Process

Vertical flow labyrinth (VLF) technology originated in Slovakia in 1990. In 1992, it was used to construct the first small sewage treatment station. Mr. Juraj, the inventor of VFL technology, initially used it to treat domestic sewage in single dwelling houses, adapting to the domestic drainage featured by stable and qualified effluent at a large fluctuation of discharge. At the same time, the treatment equipment should meet the requirements of EU countries for sludge discharge. For example, small sewage treatment plants are allowed to discharge sludge only once a year in France and twice a year in Germany. Through 30 years of development, the technology has been widely applied in more than 35 countries around the world. The size of treatment ranges from 0.9 T/d per household to 50000 T/d for small municipal wastewater, in the form of underground integrated equipment to large reinforced concrete composite pools. This technology adapts to the characteristics of scattered households and point drainage in Europe, i.e., dispersed sewage discharge and great fluctuation. It technically solves the problem of impact resistance and stable operation of small sewage treatment facilities, and can be fully automated and unattended. The technical characteristics of VFL determine that it is the best solution for domestic sewage treatment in the process of urbanization and new rural construction in China.

2.4.1 Principle and Layout of VFL Process

VFL process adopts vertical flow labyrinth structure in anaerobic and anoxic zones, which greatly extends the flow of anaerobic and anoxic zones in structure, eliminates the adverse effects of reflux activated sludge on anaerobic and anoxic zones, and greatly improves its nitrogen removal efficiency. It is also helpful for phosphorus removal, control, and adaptation of carbon sources utilization in anaerobic and anoxic zones. The process flow chart is shown in Fig. 4.

This process not only has good effect of nitrogen and phosphorus removal, but also has flexibility of adjusting operation parameters to meet the changes of water quality and strong anti-impact load capacity. It is also prospective in coping with the improvement of effluent standards and the limitation of sludge discharge.

The effluent from VFL process precipitation tank enters the rotary drum filter, which has strong impact resistance, good filtration effect, small head loss, and all-day automatic operation. It is a sewage treatment equipment matching to VFL process, ensuring the qualified effluent. The system is equipped with mixed liquor and sludge reflux different from traditional A^2O process. The reaction process is shown in Fig. 5.

A part of activated sludge in sludge bucket of sedimentation zone returns to the front section of anoxic zone, which contains dissolved oxygen. Similarly, due to the characteristics of vertical flow structure, water flows to the second and third compartments of anoxic zone. The concentration of dissolved oxygen decreases rapidly, allowing the thorough denitrification in a longer anoxic process. Moreover, making fully use of carbon source (BOD_5) in sewage, the denitrification rate is much higher than that of denitrification by

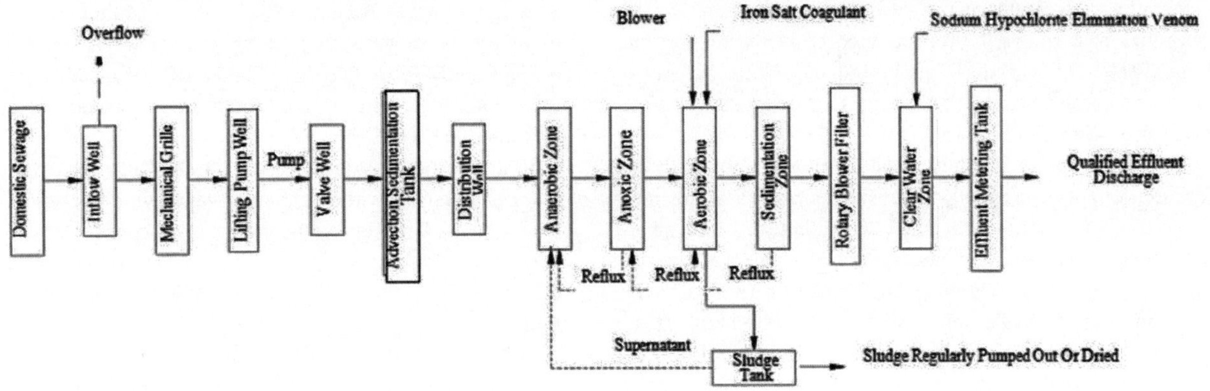

Fig. 4 VFL process flow chart

Fig. 5 VFL reaction flow chart

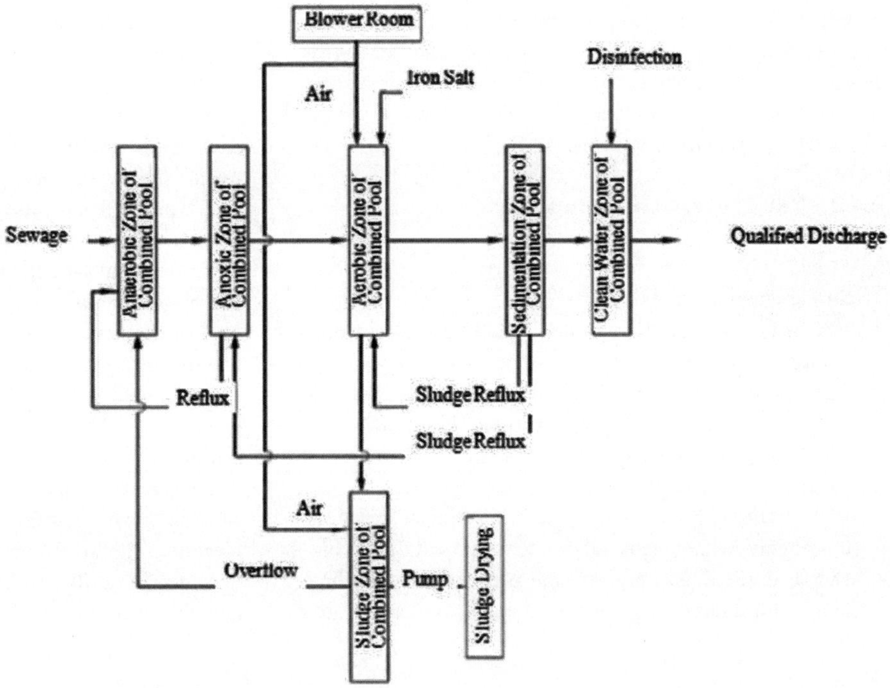

endogenous respiration. Therefore, it requires only a short denitrification residence time and small volume.

The VFL process has a unique sludge recycling route. In addition to the half of the activated sludge used for denitrification, the other half returns to the aerobic zone, and the mixed liquid in the aerobic zone is discharged into the sludge zone and then returned to the anaerobic zone. This flow allows sludge to remain active. In the aerobic zone, basically a low-load complete mixed reaction zone, the microorganism is in a low-activity state. After entering the sludge pool, the microorganism gradually turns to the endogenous respiratory state, where the sludge is partially digested. The remaining naturally screened microorganisms then enter the high-load anaerobic zone, and their activity is re-activated. The activity of adsorbing and degrading organic substances is gradually enhanced, which improves the reaction efficiency and treatment effect. At the same time, sludge can remain active for a long time and realize sludge reduction in the process of self-metabolism.

A new and efficient biological treatment device based on VFL technology combines anaerobic, anoxic, and aerobic to create the most favorable physiological environment for the microorganisms needed for sewage treatment, fully realizes the functions of organic matter degradation, nitrification and denitrification, biological phosphorus accumulation, and determines the effective volume of the structure areas according to the water quality of influent and effluent water. Aeration and sludge reflux are controlled by monitoring redox potential in aerobic zone, and the whole system is managed accordingly. The schematic diagram of VFL reactor is shown in Fig. 6.

The VFL reactor is symmetrical in left and right, with the middle long and narrow part as sludge tank. Starting unilaterally or bilaterally is decided depending on water quantity. Both anaerobic and anoxic zones of the reactor are vertical flow maze structures. The reactor is divided by its built-in vertical baffle into several reaction chambers in series, each of which is a relatively independent upflow and downflow sludge bed system, where the sludge exists in the form of granulation or flocculation. The water flow is guided by the baffle to flow upward and downward, passing the sludge bed in the reaction chambers one by one. Substrates in the influent are degraded and removed by fully contacting the microorganisms. The front section of the reactor is anaerobic zone. Sewage enters anaerobic zone, anoxic zone, aerobic zone, and sedimentation zone successively along the baffle. The plane layout is shown in Fig. 7.

2.4.2 VFL Process Characteristics

1. Strong shock load resistance
   VFL is optimized in term of pool structure and operation management. The whole system operates at high sludge concentration (sludge concentration 7–8 g/L in anaerobic and anoxic zones, 3–4 g/L in aerobic zone), with full capacity against impact load.
2. Good and stable effluent quality
   VFL technology can carry out biochemical reactions thoroughly without any advanced treatment (after biochemical reactions only through precipitation clarification and filtration) and maintain stability to meet the first A standard of "Discharge standard of pollutants for municipal wastewater treatment plant" (GB18918-2002) or even better water quality. Therefore, advanced treatment equipment and machine room are not needed.
3. Low sludge production
   VFL technology takes into account both sludge reduction and harmlessness while treating sewage, with a very low sludge yield. The sewage treatment plant constructed by VFL technology may discharge sludge for the first time after normal operation of 2–3 years and then every three months.
4. Adaptive to stringent environmental requirements
   By optimizing the operation management of system, VFL technology eliminates the odor of whole system, including anaerobic zone and sludge zone. This process gets to the root of odor in common treatment process which obsesses most of wastewater treatment plants. Therefore, without any deodorization device, the wastewater treatment station is suitable for any area even with strict environmental requirements.
5. Simple system, low energy consumption, and small daily maintenance workload
   The types and quantities of VFL technology equipment have been reduced to the minimum. Except the lifting equipment and sand removal facilities at the sewage inlet, biochemical combined pool is the core, supported by only blower and dispensing device. In this way, the daily maintenance workload of equipment has been greatly reduced, saving the energy consumption to the greatest extent.
6. No further investment
   There is no filler in the combined pool, avoiding blocking problems, and requiring no regular replacement and cleaning. The long-term use of contact oxidation process and biological filter process fillers will cause blockage, and the material will also appear aging and falling off

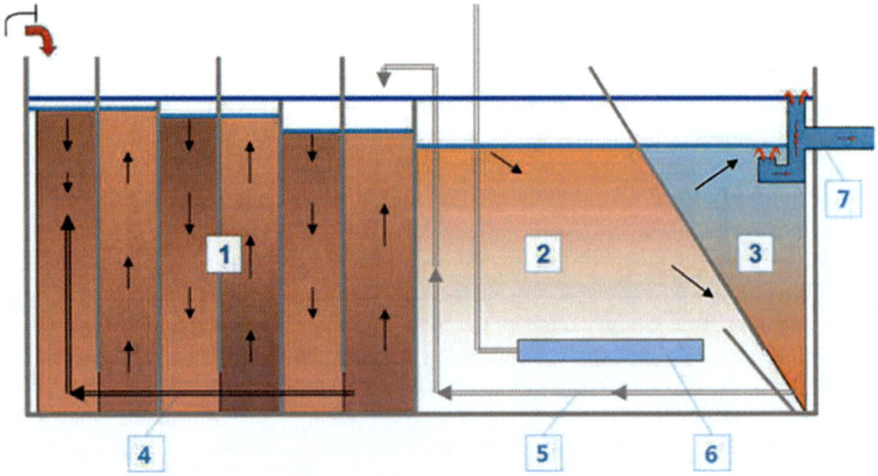

Fig. 6 VFL schematic diagram (1-anaerobic and anoxic zones-vertical flow maze structure, 2-aerobic zone, 3-sedimentation zone, 4-internal circulation, 5-sludge reflux, 6-tubular microporous aerator, and 7-drainage tank)

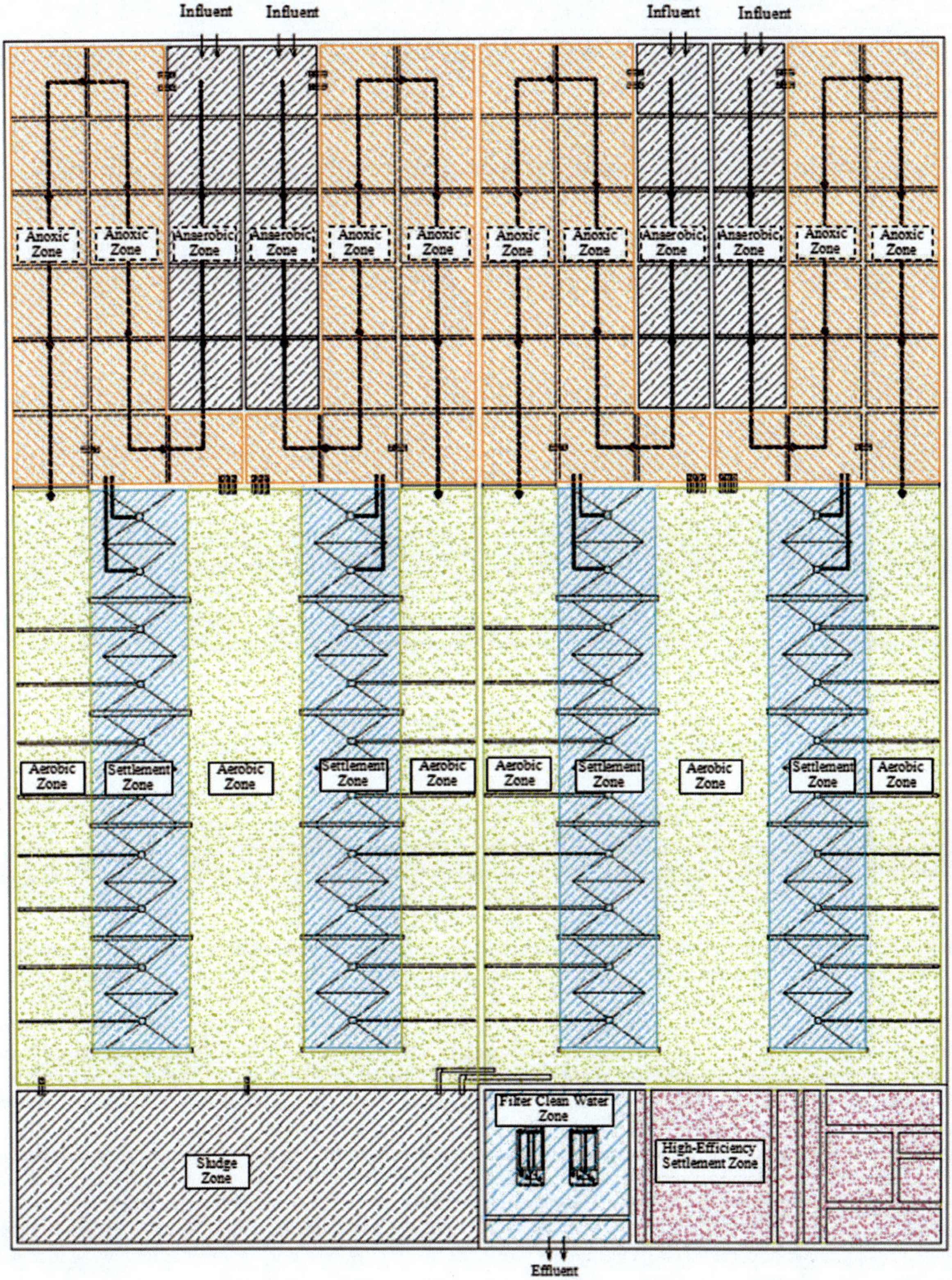

Fig. 7 VFL plane layout

Table 1 Designed water quality of Mingcheng sewage treatment plant

Index	COD	BOD_5	SS	TN	NH_3–N	TP
Date	250	120	200	40	25	5

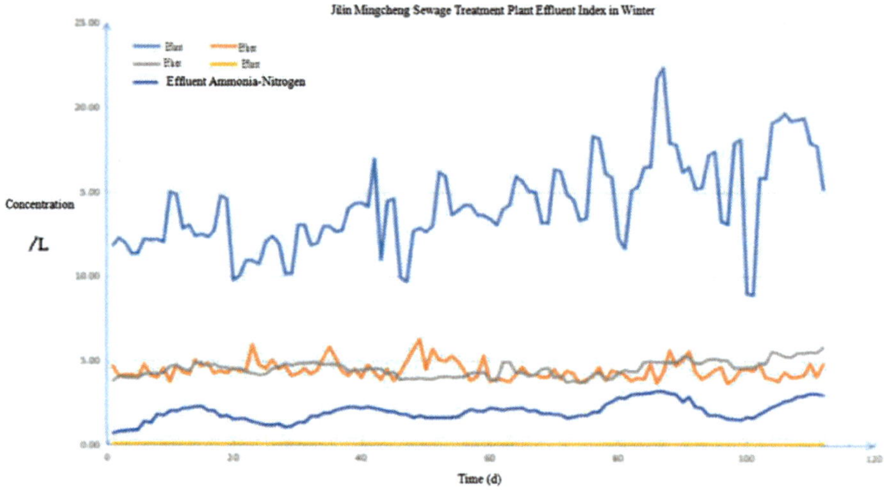

Fig. 8 Winter operation data of Mingcheng sewage treatment plant

after 5 years, so it must be replaced regularly. At present, MBR process membrane components generally have a service life of 3–5 years. Using domestic membrane, the total investment of equipment is about 40% higher than that of ordinary process, and the imported membrane is 80% higher. Moreover, the daily maintenance of back-washing and chemical cleaning requires supporting machine rooms and equipment, introducing large work-load and higher requirements for managers.

2.4.3 Practical Cases in Jilin Province

Mingcheng Economic Development Sewage Treatment Station is located in Panshi City, Jilin Province, with a designed capacity of 2500 tons per day. The main sewage collected by the Development Zone sewage treatment plant comes from old town, new district in the east city, and hardware industrial park. Old town and new district in east city are commercial and living areas. The main industry of hardware industrial park is building materials processing, which relying on iron and steel industrial park. The sewage discharge from building materials processing industry is less, accounting for only about 12% of the total sewage.

According to the actual situation of the town and considering the development trend of Mingcheng Economic Development Zone, the designed water quality of the sewage treatment plant of this project is determined as shown in Table 1.

After a long period of operation, the VFL process operates well in Mingcheng Sewage Treatment Plant. In the coldest winter, the effluent can meet the Grade 1 Class A standard of "Discharge standard of pollutants for municipal wastewater treatment plant" (GB18918-2002). The winter operation data are shown in Fig. 8.

3 Conclusion

Compared with the traditional A2O, SBR, and MBR processes, VFL process shows the following superiorities: (1) This process possesses a strong shock load resistance, and it can cope with fluctuations in water quality and quantity, (2) stable effluent, (3) low sludge production. It can greatly save the sludge quantitation cost. (4) Simple system and small daily maintenance workload greatly reduce the operating costs. (5) Adapted to cold water intake in winter in Northern China. (6) The system only occupies a small area, which significantly lowers the cost. (7) Good sludge quality and strong processing capacity under VFL hydraulic conditions. The operations of sewage treatment facilities in Mingcheng in summer and winter showed that the VFL process can satisfy Jilin Province's requirements for small-scale sewage treatment equipment. It can withstand fluctuations in water quality and quantity, and possesses strong resistance to cold. Besides, it has an extremely low sludge yield and a low operating cost. Hence, VFL process shall be the priority selection for small-scale sewage treatment equipment in Jilin Province.

References

Bai, T. (2003). Urban sewage treatment technology and development. *Mechanical and Electrical Equipment, 20*(3), 38–42.

Gao, Y. (1989). *Control project of water pollution* (Vol. II).

Li, Y., Wang, Z., & Wang, L. (2001). Discussion on urban sewage treatment technology. *Sichuan Environment, 20*(1), 40–42.

Liu, T. (2016). Theory of sewage treatment and technical overview. *Scientific and Technological Innovation, 30*, 114–115.

Luo, W. (1999). New type of sewage treatment process-MSBR. *China Water and Waste Water, 15*(6), 22–24

Qian, Y., & Mi, X. (1993). *New technology of modern waste water treatment.*

Qin, L. (1989). *Biological treatment of waste water*. Tongji University Press.

Shen, G. (1999). Several matters on the design of urban sewage treatment plant. *Chinese Environmental Protection Industry, 16*(6), 18–20.

Wang, Y., Sheng, L., Li, K., et al. (2008). Analysis on the current situation of Chinese water resources and research on the measures of sustainable development. *Academic Journal of Water Resources and Water Engineering, 19*(3), 10–14.

Yu, Z., Wang, D., & Zhang, X. (2004). (n.d.). Retrospect and prospect of urban sewage treatment technology development. *Academic Journal of Guangxi Normal University (Natural Science Edition).*

Zhou, B. (2003). Overview of domestic and international urban sewage treatment technologies. *Tianjin Municipal Engineering Design.*

Zhu, F., Yu, J., & Huo, Y. (2002). Determination of process parameters of AA/O for Guangzhou Da Tan sewage treatment plant. *Engineering Construction, 34*(4), 26–29.

Smart Energy Infrastructure and Technology: Assessing Community Preparedness for Residential Solar Uptake

Parisa Izadpanahi, Reena Tiwari, and Declan Hayward

Abstract

The new age of the modernity has brought about many changes to the urban fabric of our cities. Fossil fuels have drastically contributed to the problems of climate change, pollution, excessive carbon emissions, and countless health issues. The consequences of our past energy choices have instigated the search for alternative energy solutions as populations continue to grow and natural resources continually deplete. One of the strategies is the use of solar energy. India has planned to draw by 2022 around 100 GW of solar energy, utilising its abundant solar energy resources to meet its increased energy demands; adopting these new technologies challenges India both practically and socially. Therefore, this study looked at the community preparedness for the uptake of solar energy in residential buildings in two localities in New Delhi, Vaishali Nagar, and Vivek Vihar. The study also looked at the building typologies, overshadowing, building orientation, microclimate, and pollution to find the correlation of these environmental factors with residents' responses. Data was collected through questionnaires around residents' perceptions and opinion on climate change, solar technology, and the future of India's energy policies and infrastructure for the country. Overall, the questions were designed to shed light on the current condition of residential solar energy in India, and how decentralised energy and blockchain technology can be realised as a part of the energy growth. The findings demonstrate that the cost of solar panels and lack of information about available subsidies and technicalities are the two prime reasons for a poor solar uptake in the residential sector. Households that were hesitant to install solar, due to financial costs, also assumed there were no subsidies available. There appears to be a lack of information among residents, even among those who have solar panels installed. In regard to the peer-to-peer energy trading, 80.1% of the residents preferred to purchase power from their neighbours at a cheaper rate than from large companies at a more expensive rate. In summary, this study suggests that social and educational factors play an important role in the uptake of solar and blockchain technology, and therefore, access to clear and easily understood information is essential for residents in smart cities.

Keywords

Solar energy • Community preparedness • Peer-to-peer energy trading

1 Introduction

This research project addresses the assessment of physical, technological, policy, and financial infrastructure in New Delhi to understand suitability of introducing the cutting-edge technologies to facilitate rapid uptake of rooftop solar in India under the smart cities initiative.

The scope of this project is limited to rooftop solar and includes:

National/state/city level.

- Documenting the national policy agenda within India in regards to the energy sector
- Focussing on rooftop solar;
- Documenting the current energy supply framework within India's smart cities focussing on Delhi (supply, load, customer needs, likely growth profile, regulation, government subsidies);
- Identifying potential barriers to meeting policy targets for rooftop solar;
- Documenting the recent international trends with regards to transactive technologies and transactive energy applications, and assessing its applicability in the selected case studies.

P. Izadpanahi (✉) · R. Tiwari · D. Hayward
Curtin University, Perth, WA, Australia
e-mail: parisa.izadpanahi@curtin.edu.au

F. Trapani et al. (eds.), *Advanced Studies in Efficient Environmental Design and City Planning*,
Advances in Science, Technology & Innovation, https://doi.org/10.1007/978-3-030-65181-7_5

Neighbourhood/unit level.

- Number of houses with solar
- Built form (neighbourhood and unit)—height, roof design, orientation
- Customer energy profile (needs, affordability, usage, outages)
- Household profile (socioeconomic, number of people, ownership, vacancy rates
- Metering infrastructure
- Use of solar + blockchain (affordability, regulations around net-metering/open access arrangement, wheeling of energy and peer-to-peer transactions.

2 Project Background

India's economy and population is one of the world's fastest growing, and therefore, its energy needs are on the rise. India's energy consumption is forecasted to grow by 4.2% a year for the next twenty-five years and predicted to overtake China as the largest growth market for energy. With an approach to unlock the potential of renewable energy options, India has planned to draw by 2022 around 175 GW of energy from renewable sources out of which reliance is on solar energy for 100 GW (Mission Innovation, 2018).

Being a tropical country, India has rich solar energy resources with an average intensity of solar radiation received being 200 MW/km^2 (India Energy Portal, 2018). There is wide ranging solar application starting at the higher end from solar thermal power stations to the lower end applications in the hand of the end-users.

The government's focus on utilising renewables and energy efficient technologies is reflected on its mission of constructing 100 smart cities. In 2015, the Government of India announced Smart Cities Mission with multiple objectives that include:

- Reducing congestion, air pollution, and resource depletion;
- Promoting a clean and sustainable environment and application of smart solutions;
- Applying smart solutions to existing infrastructure;
- Making governance citizen-friendly and cost effective, increasingly relying on technology and online services to bring about accountability and transparency, reducing cost of services and providing services without having to go to municipal offices; and
- Amendments in building bye-laws for mandatory provision of rooftop solar for new construction.

This research studies Delhi as one of the 100 identified for developing as smart cities. Delhi is also on the list of 48 approved solar cities committed to renewable energy installations and energy efficiency measures. With Prime Minister Modi's target of 100 GW solar energy by 2020, a solar revolution is on its way in India. However, the question is whether the residential sector is ready to join in this revolution. Are Delhi residents ready for rooftop solar for their energy requirements? What are the challenges that they are facing in going solar? Are they ready to plunge into the future where house owners control their electricity generation, consumption, and sell their excess energy to neighbours in need?

3 India's Energy Context: Exploring Policy, Regulation, and Guidelines and Their Influence on India's Solar Energy Market

This study will review existing regulations and policies regarding rooftop solar photovoltaics SPV and observe the programs and incentives that have been implemented to aid the expansion of renewable energy into the urbanscape. Furthermore, it will attempt to understand the current condition of solar energy in India and how decentralised energy and blockchain technology can be realised as a part of this growth.

The challenges in physical transmission of electricity and also unaffordability of the service are among the key reasons as to why many do not have access to this essential utility in India. Since its introduction to the Indian renewable energy paradigm in the 1970s, the solar photovoltaics (SPV) has emerged as a frontrunner (with 31% market share, Kumar et al., 2018) along with wind (49% market share, Kumar et al., 2018) and biomass (13% market share) (Kumar et al., 2018) to empower those in rural areas and residents of dense urban residential complexes to generate and store electricity when necessary.

Government authorities, namely the Ministry of New and Renewable Energy (MNRE), have the capacity to permit people to access to electricity through means such as rooftop SPV and control major regulations and policies that promote this rapid growth industry. The MNRE plays a key role in guiding overall policy and coordinating with rooftop SPV development partners in achieving their targets. Regulatory bodies like the MNRE also make recommendations on a national and state level as to how architects can incorporate and facilitate rooftop SPV within the urban fabric both efficiently and with cost-effective approaches.

Prior to the implementation of an overarching solar target, the MNRE formed the "Working Group on Energy Policy" in 1979 identifying the need to utilise "non-conventional energy" (Yenneti, 2016). This policy was further solidified by the "Department of Non-Conventional Energy Sources" (DNES) in 1982, though the oil shock of the early 1970s had been overcome and results of alternative energy technologies were non-tangible at this stage. Following the splitting of the Ministry of Energy (MoE) in the 1990s, the Ministry of Non-Conventional Energy Sources (MNES) emerged, which became the MNRE as of 2006. Development of renewable energy around this time was primarily centred around wind, which had more than 10,000 MW installed between 1990 and 2008.

4 Indian Building Codes

Indian built environment, industry, and building codes also outline the requirements for incorporation of the rooftop SPV. "Electrical and Allied Installations" is contained within Part 3 Sect. 21 of the IBC (Bureau of Indian Standards, 2016, 58) and sets guidelines for solar energy utilisation for a number of differing building typologies. According to the code, any property or lot "having size 500 m^2 and above shall install solar PV power generation system"(Bureau of Indian Standards, 2016, 58). In addition, the code encourages that even plots smaller than 500 m^2 implement a rooftop SPV power system and outlines that the appropriate "authority shall have required provisions in the building bye-laws and mechanism for required clearances and approvals… (and) also specify minimum generation requirements" (Bureau of Indian Standards, 2016, 58). As demonstrated in Table 1, each typology in the urbanscape requires differing levels of power generation on a national level. The contents of the table are sourced from National Building Code of India (Bureau of Indian Standards, 2016).

These guidelines are essential to understanding how power generation should be influenced by the footprint of the structure, though due to the small plot allocations (most underneath 100 m^2) in dense, urbanscapes have consequently led to an underutilisation of rooftop SPV, as the panels are not specifically required by code if they are not a newly built structure (as opposed to similar Australian policies).

5 To Address the Challenge…

Background studies show that decentralised energy sourcing stands as an attractive solution for Indian households and companies alike that cannot actively access power or can only access it with severe grid unreliability due to peak demand and poor distribution methodology. Approximately, 400 million people in India, including 47.5% of the rural population, still have no access to electricity, and their remote location means that renewable energy can offer a fiscally proportionate alternative to grid connection (MNRE, 2019). Considering this demand, and in order to understand whether the energy policies and building codes could sufficiently encourage Indian citizens to uptake rooftop solars, a group of academics and students from Curtin University, Western Australia, conducted a research in two localities of New Delhi. This research also investigated whether the blockchain-based energy trading system could become viable in the lower socioeconomic context of urban India, in an effort to reduce GHG emissions and move towards the grand solar capacity targets outlined in the JNNSM (the MNRE's National Solar Mission).

The group of researchers conducted the contextual study and also explored the Australian precedents for transactive technologies. Students participating in the study as researchers were also exposed to this innovative development through architectural technology units in their third year. One of the successful examples studied was in White Gum Valley, Fremantle, Western Australia, which introduced the decentralised and transactive technology conducted by Power Ledger. This world-class precedent is the first apartment development in Australia utilising Power Ledger's blockchain technology for solar power trading

Table 1 Generation requirements—National Building Code of India

Building type	Plot size	Generation requirement
Residential (plotted houses)	100 m^2 and above	1 kWp or 5% of connected load, whichever is higher
Residential (group housing)	All sizes	Minimum 5% of connected load
Business, educational buildings, having connected load of 30 kW and above	500 m^2 and above	5 kWp of 5% of connected load, whichever is higher
Mercantile, hotels, motels, assembly, industrial and institutional buildings	500 m^2 and above	For buildings having a connected load of: 50–1000 kW–10 kWp or 5% of connected load, whichever is higher above 1000 kW–50 kWp or 5%

between residents in a strata setting, allowing them to save on their monthly power bills (Power Ledger, 2018).

6 Methodology

The Australian Government, under the National Colombo Plan, supported this project which was in collaboration with Resilience Relations and students from Delhi University. This research project had two phases. One in Australia and the other one in India. The project commenced in Australia where students were required to conduct literature review on the recent international trends in regards to transactive energy applications around rooftop solar. Students reviewed prospects for peer-to-peer transaction-based energy systems (Jogunola et al., 2017) and optimising P2P energy models for smart homes within a smart grid (Alam et al., 2017). They were also required to review and document the current Indian government energy supply framework in smart cities and identify the possible governmental funding opportunities and subsidies for community/locals. Students were able to identify a series of references relevant to the study, analysing subsidies offered by the government through third party sellers such as the grid-connected solar rooftop program, with some subsidies being offered up to 40% (MNRE, 2019), and were also informed by articles explaining the processes of claiming solar rooftop funding (Deshmukh et al., 2010; MNRE, 2019). As part of their research, and to prepare themselves to work in Indian context, they identified the challenges for adopting the technology, infrastructure, and social challenges and reviewed and documented the related Indian council policies and building codes and explored their indication in urban design. Such documents that were reviewed looked at issues posed by additional loading upon residential rooftops and relationships between building heights and proximity to streets within the National Building Code (Bureau of Indian Standards, 2016). This pre-travel background research assisted students to prepare an inventory of infrastructure and built form requirements for implementation of solar panels that were need to be audited on site.

In India, Australian students and their lecturers organised a workshop with Indian partners to frame the questionnaire together and based on the background research that were conducted in Australia. First, Australian students were teamed with Indian students with the aim to have equal voices in each team. This was of prime importance for the project since Indian students had the local knowledge and knew what factors could possibly impact on residents' preparedness for uptaking solar panels. Australian students also contributed in terms of the technological knowledge about recent international energy trends and applications, such as blockchain and transactive energy systems. The two lecturers moderated the workshop and helped the students' teams to design clear and short questions. Therefore, the workshop resulted in designing a questionnaire, drawing from prior research and understanding policy and subsidy approaches from the MNRE, as well as the local knowledge.

To administer the survey, the students attended the two localities of Vaishali Nagar and Vivek Vihar in New Delhi, and sections of the neighbourhood were randomly selected. They then used the door-knocking method and collected the data from any household who were happy to respond. Australian and Indian students surveyed 165 residents with the aim to explore the preparedness of the residents to install rooftop solar panels and to investigate the factors affecting rooftop solar panels output, e.g. location, orientation, panel efficiency, ambient temperature and available roof area. In addition to the questions asked, participants were also shown some selected imagery so they can express their opinion regarding key topics, such as peer-to-peer trading system, solar panel technology and energy storage, smart cities and policy proposals from MNRE.

Demographic data was sourced from the 2011 Indian Census, for Vivek Vihar and Vaishali Nagar. Vivek Vihar, located within East District Delhi, had a population of 247,906 in 2011, with an overall literacy rate of 89.6% (Census India 2011). The literacy rate was a positive indicator, due to verbal communication being key to the survey collection methodology, and any non-English conversation was effectively facilitated by the Indian students. Hindi is considered the main language in this locality. The Census also recorded 50,134 families residing within Vivek Vihar, with over 85,214 people being engaged in work activities (Census India, 2011). This indicated 34.3% of people earning for their household, indicating a form of income required to implement solar energy into the household.

Census data was analysed from Ghaziabad, the city within which the locality of Vaishali Nagar is located. Ghaziabad tehsil, as of 2011, indicated a population of 2,705,849, with a 80.9% literacy rate (Census India, 2011). 75.7% of people identified as Hindu (2,048,725 total) and 21.9% identified as Muslim (592,656 total). In terms of work, 896,095 people were engaged in work activities in 2011, 33.1% of the population (Census India, 2011). The socioeconomic and demographic analysis of the 2011 Census gave students an insight into how people within these villages lived and worked within their communities.

A. Usage of solar panels			
1	Do you know about solar panels and their benefits?		Yes—— No—— Unsure——
2	Have you seen solar panels in your area?		Yes—— No——
3	Do you have solar panels installed?		Yes—— No—— Interested in installing——
4	If you have solar panels: (If no please go to Q5)	a. How many panels (or, what is the system's capacity)?	
		b. Have you seen a reduction in power bill after installation of panels?	Yes (Indicate %)—— No—— Unsure——
		c. Would you recommend this to your neighbours?	Yes—— No—— Likely—— Reason:
		d. How did you find out about solar panels?	
		e. Did you receive any subsidy or incentive?	Yes—— Which subsidy? No——
		f. Are you generating excessive solar energy?	Yes—— How much excessive? No—— Unsure——
		g. Is the solar panel accessible for cleaning?	Yes—— No——
5	If you do not have solar panels:	a. What is stopping you from installing solar panels?	Cost of installation—— Lack of information—— Satisfied with current energy use—— Negative feedback from neighbours—— Not a priority—— Other————————
		b. Did you know solar energy can	Yes—— No——

(continued)

		save you money? (Highlight monetary benefits of solar energy)	Unsure—
		c. If the above issues are resolved, would you be interested in installing the panels?	Yes—— No—— Likely——
		d. How much space can you spare for the panels? (in square metres)	
		e. Does your roof allow for sunshine?	Yes—— No—— Unsure——
B. Preparedness for installation (if participant does not have solar panels)			
1	Do you use your roof for other purposes?		Yes—— (Indicate how) No——
2	Do you have any preferences for any alternative solar methods?		Mobile solar panels (flexibility)—— Wall-mounted solar panels (building integrated PV) —— Others——
3	Do you know about any relevant Municipal Corporation of Delhi (MCD) regulations for installation?		Yes—— No—— Unsure——
C. P2P questions (to be used during panel discussion)			
Peer-to-peer trading involves the anonymous selling and buying of excess energy from your neighbours. *give some examples			
1	Do you have reliable internet access?		Yes—— No——
2	Would you rather purchase power from your neighbours at a cheaper rate, than from large companies at a more expensive rate?		Yes—— No—— Reason:
3	If there was an option for storing the solar energy through battery, will you go for it and pay the additional cost of battery?		Yes—— No—— Reason:
4	Would you be more likely to uptake this trading with a government incentive?		Yes—— No—— Unsure—— Reason:

(continued)

5	If so (Yes to Q4), would you feel confident receiving this incentive	Yes—— No—— Reason:
D. Observational template locality		
Typology of the building		Detached Housing ☐ Semi-detached ☐ Apartments ☐
How tall the building is? (How many storey)		Indicate the height with a figure on the map
Type of the roof		Flat-roof ☐ Pitched roof ☐
Condition of the roof: i.e. could it take extra load?		Yes ☐ No ☐ Unclear ☐
Overshadowing from the surrounding buildings and trees, etc. Consider orientation when assessing this		Indicate the number of the hours in shade Less than 5 h ☐ Almost 5 h ☐ More than 5 h ☐
Type of the tall trees. Only care about those trees that cast shadow on buildings		Evergreen ☐ Deciduous ☐
Microclimate (pollution): is the house on a busy polluted road?		Minor pollution ☐ Average pollution ☐ Very polluted ☐
How much vertical surface area can be used for PV?		Indicate in m^2

7 Analysis

Survey and observational results were analysed in order to understand these localities' awareness and preparedness to implement solar panel and peer-to-peer trading technologies. The results were best understood when interpreted in the following sections:

- Experiences of people who do not have solar panels
- Experiences of those who own/use solar panels
- Built environment implications upon the uptake of SPV technologies.

8 Experience Without Solar Panels

This aspect sought to understand perception of solar panels, why they were not being utilised in community and group dwellings and awareness of available government subsidies. It also looked comparatively between consumption patterns to those who have solar PV. The results also revealed socioeconomic reasoning for not uptaking solar.

Results revealed a relatively positive result, with 81.6% of those surveyed knowing about solar panels, and also aware of their benefits (i.e. not drawing energy from the grid, lower cost on monthly bills). A further 52.5% of this 81.6% had observed solar panels within their locality. This reveals a larger exposure than was hypothesised, though correlates with our preparation research into policy, which showed a lower uptake of solar in residential New Delhi than was projected. As per publically available information online for Vivek Vihar (Smart Roof, 2017), it is stated that the MNRE would provide an incentive subsidy of Rs. 2.0 kWh on solar electricity generated on any new solar system. The same source explains that approximately 8.36 m^2 of shadow-free area could support 1 kW of solar panels installed, assuming roof size of 50% of the plot area. Perhaps dispersion of information being online is the problem with this particular community, and it may need to be delivered in a different manner.

Findings produced by Pillai outline that the reduction in average production cost and price of solar panels between 2005 and 2012 "has been drive by reduction in the price of polysilicon, improvements in technology, increasing market penetration of lower cost firms from China, and increases in industry investment" (Pillai, 2015). This prompted questions similar to these conclusions, seeking to understand why solar has not been implemented in lieu of lower costs. Subjective responses often arose, with "cost of installation… lack of information… satisfied with current energy consumption… negative feedback… low priority" being the standardised responses from residents. However, there was a resounding positive response when asked if these problems were resolved would they install panels—72.3% responding yes, compared with 16.8% responding that they still would not install them (Fig. 1).

These problems of installation cost and lack of information seemed to be of major concern to those who did not uptake solar in the communities. The other challenges that seemed to arise from those who were surveyed (including those with solar panels) were the exposure to higher levels of observed pollution (Fig. 2). One resident remarked upon the changing colours of leaves, with accumulating dirt and dust changing them from green to brown.

Other factors to consider included cost of monthly power bill, frequency of power cuts, and peak use of electricity. It was evident that peoples' energy requirements vastly differ, ranging anywhere between Rs. 700 and Rs. 33,000—with time of use (during peak) influencing opinion on solar panels (with most people "being out of the house" during the day). Another possible reason for some residents to not consider alternative energy methods was the range of responses regarding power cuts—some experienced them daily (up to 5 times), and some experienced none of the time. Perhaps the opinion was that with consistent power, alternative energy sources like solar power would not be required.

9 Experience with Solar Panels

Those with solar panels were asked about their knowledge of government subsidy availability, benefits of solar panel technology, and their willingness to uptake smart technology, such as P2P trading, in future. Results showed an unwavering interest to share energy "for cheap" with their neighbours (Fig. 3).

There is a demonstrable gap in ownership of solar panels in the communities that were surveyed. Only a small fraction, 3.7%, had solar PV systems installed on their buildings,

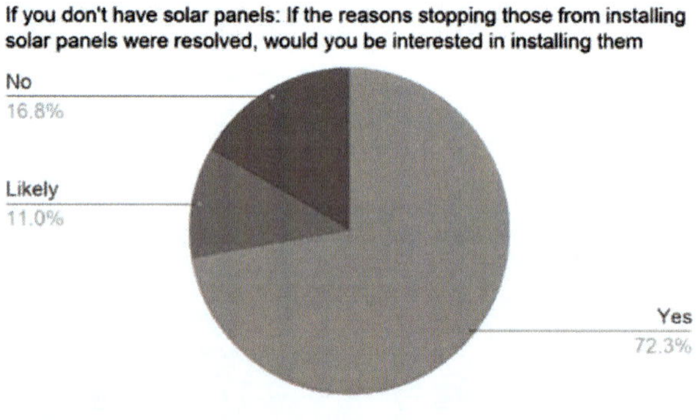

Fig. 1 Likelihood of installing solar panels upon issue resolution in Vivek Vihar and Vaishali Nagar

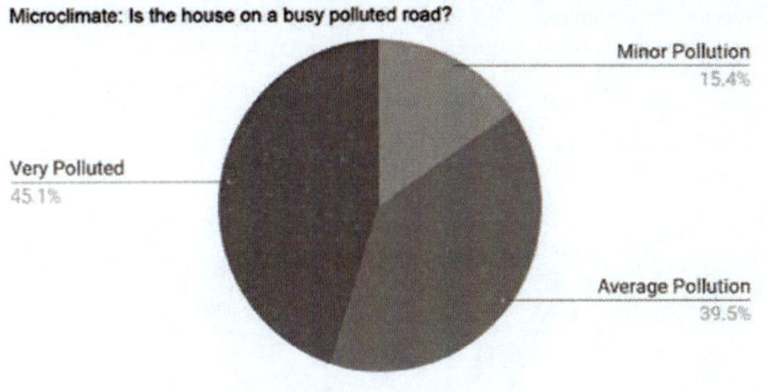

Fig. 2 Opinion response from residents on pollution in their locality in Vivek Vihar and Vaishali

whilst the majority of 71.2% stated they were disinterested in the installation of the alternative energy solution (Curtin University School of Architecture, 2019). A promising statistic was the quarter of survey participants (25.2%) who expressed interest in the installation of panels if the option were made available to them (i.e. cost, building ownership, etc.). It should also be noted that ownership and rent within buildings were subjected and worked on different stipulations, for example, the resident living and owning on the bottom floor of an apartment may not be able to use the roof space, as that may belong to the tenant just below the roof—or it may belong to an owner not living in the building at all.

In order to gain an oversight into the experiences of those who had installed solar panels, the data collected from those who did not have the panels was combined and compared to understand a proven set of benefits from the installation of solar panels. Key stakeholders of installation of panels upon residential rooftops are those who live in Delhi, who are in a position to gain from the advancements of technology. Two key findings upon analysis of data included:

- The mean monthly energy bill for those with solar panels was lower than those without solar (Fig. 4)
- The peak energy usage of those with solar was offset to the peak production time of their solar PV systems (consistent with current standard use of solar PV worldwide, Fig. 5)

However, it was interesting to find that the energy bills of 80% of solar panel users were comparatively similar to those who did not use solar panels, potentially representing a gap in knowledge regarding maintenance and upkeep, or simply the habits of those with panels differed from those without panels. A positive result was that 80% of those with panels installed stated they had received a form of government subsidy, with one participant quoting they received a tax-free subsidy, with two more benefiting from the 30% feed-in subsidy. These results are consistent with our earlier findings, demonstrating that a lack, or gap, in information is harming the uptake of rooftop solar PV systems in residential Delhi.

Another promising result from these residents, in addition to those who did not have solar panels installed, was the interest in emerging energy technologies such as decentralised P2P energy trading and alternative energy storage solutions. The form of anonymous buying and selling of excess energy seemed to be appealing to survey participants, as demonstrated by 80% saying they would be interested in purchasing power at a cheaper rate from their neighbours (Curtin University School of Architecture, 2019) (Fig. 6).

Both of these questions revealed a motivation to embrace emerging alternative energy solutions in the Delhi localities. Some people were quoted saying, "it's a cheaper source of energy", "it's more reliable", and the idea of an interesting "business prospect" was also a popular response from the participants. Those who were not interested were quoted saying "I do not like my neighbours", "I do not like to share with my neighbours", "too much hassle" and "no point". Residents were informed that the implementation of solar battery storage systems in the community would allow their

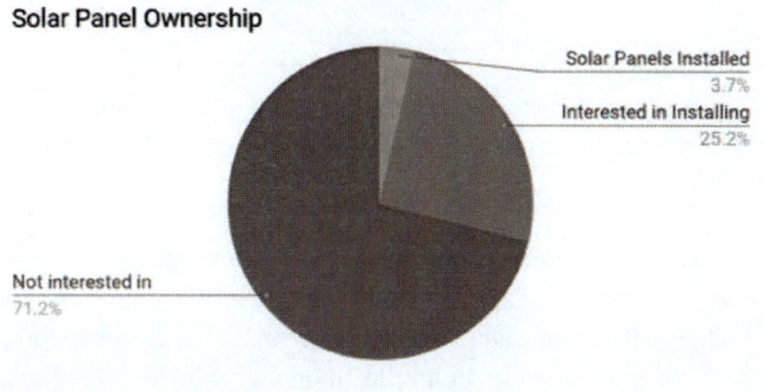

Fig. 3 Chart showing solar panel ownership of residents in Vivek Vihar and Vaishali Nagar

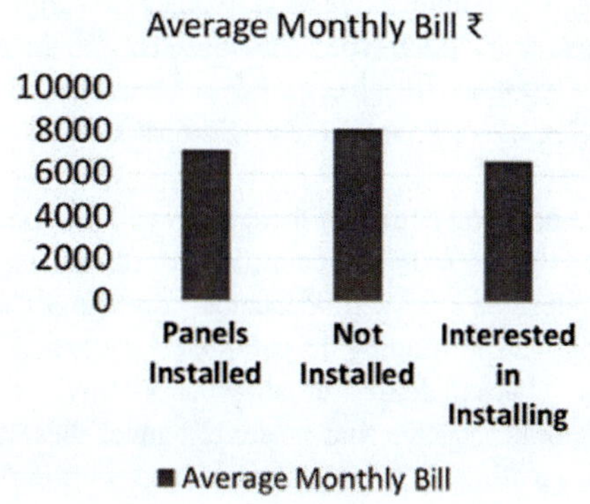

Fig. 4 Graph showing mean electricity bill across survey participants

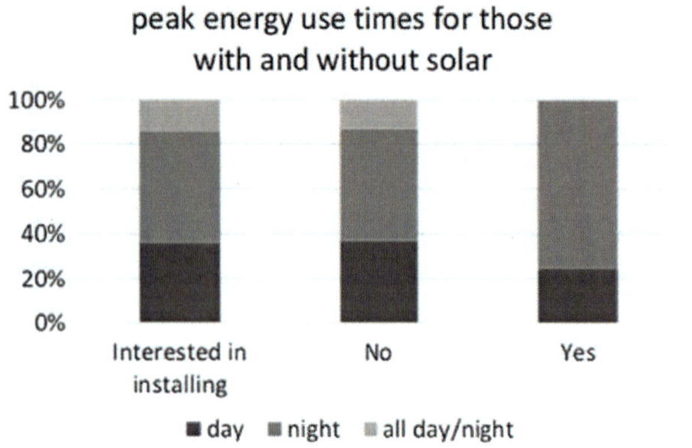

Fig. 5 Graph showing peak energy use times across survey participants

buildings to run independently from the grid, changing the offset time of power consumption and generation (Fig. 7). In addition to this, participants were asked about the inclusion of incentive from the government to implement transactive energy, to which 73.7% responded yes, 15.8% unsure, and 10.5% negative (Fig. 8). A question regarding confidence of receiving incentive seemed to correlate, with 73% responding in confidence, and 27% saying they did not feel confident in receiving incentive from the government (Fig. 9).

We believe that these results reflect positively upon the future of blockchain-based energy technology being implemented in situ with battery storage and solar PV systems in residential Delhi.

10 Built Environment Implications on Uptake of Solar Panels

The importance of New Delhi's architectural typologies and built environment context was also found to be an influential on uptake of solar PV technologies in the city. Elements such as roof conditions, overshadowing, exposure to natural canopy cover and pitch type were considered in the analysis. Building typology, height, roof type/condition, overshadowing, type of trees (and height), vertical surface area, and microclimate.

After observing homes of those surveyed in the societies, we found that 69.51% lived in apartment blocks, whilst 23.17% resided in semi-detached dwellings and the remaining 7.32% residing in completely detached housing (Fig. 10). These dense urban spaces have important socioeconomic benefits that positively affect these societies in Delhi, creating closer relationships and working towards community goals. The quantity of apartments in close proximity means that access to amenities and other services is greatly increased.

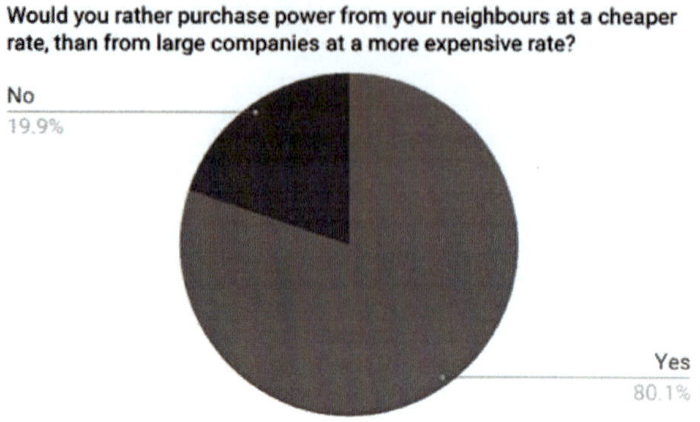

Fig. 6 Chart showing response of residents regarding the implementation of transactive energy solutions

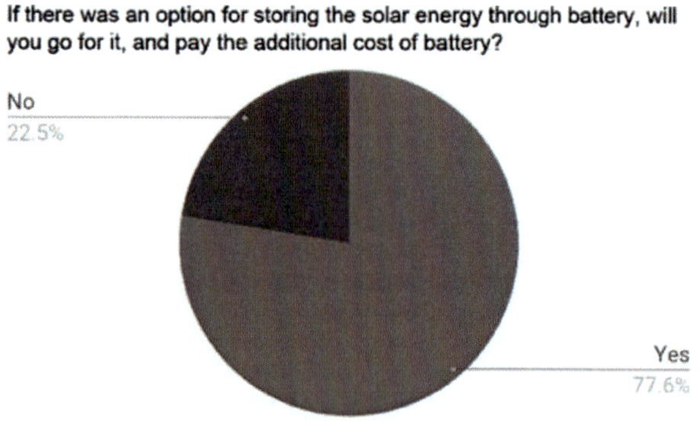

Fig. 7 Chart showing response of residents regarding installation of battery storage

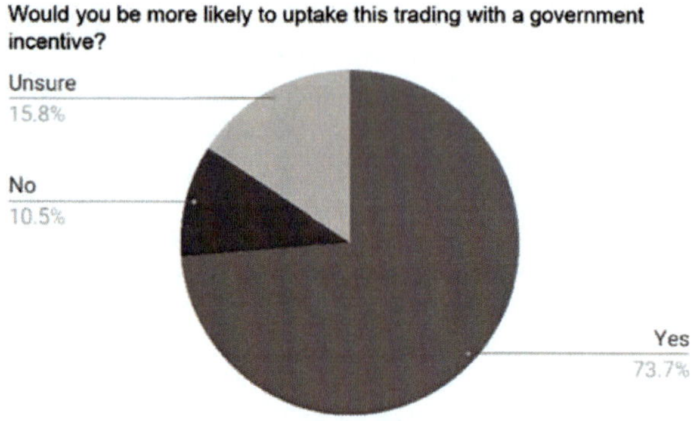

Fig. 8 Chart showing response of residents regarding government incentive

The survey gained an understanding, through observation and questioning, as to the extra load bearing capacity of participants' roof slabs. Analysis of results was surprising, with 89.6% showing potential in being able to take extra load whilst only 2.4% stated that there was no extra load capacity. Meanwhile, 8% stated that they were unsure as to the load bearing capacity of their roof. These results were

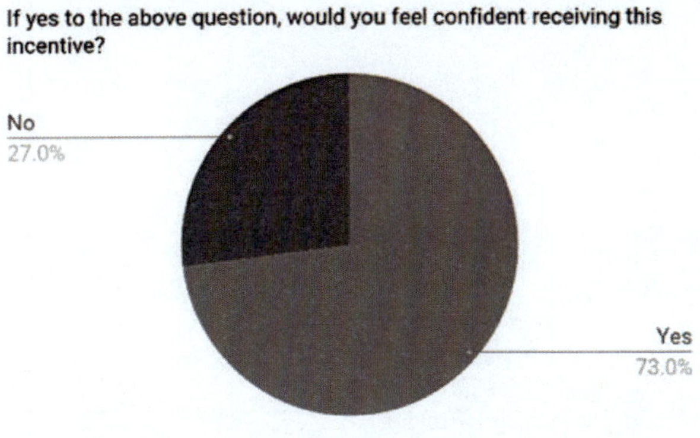

Fig. 9 Chart showing response of residents regarding confidence in receiving incentive

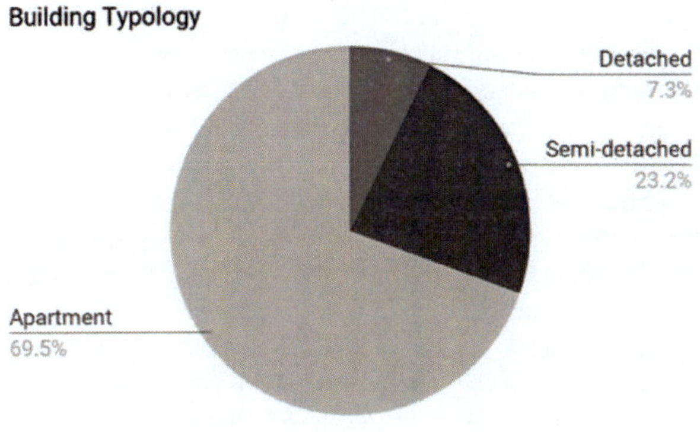

Fig. 10 Chart showing the distribution of building typologies in Vivek Vihar and Vaishali Nagar

revealing for two key reasons: many people are confident in their building's capacity to take extra imposed load, and the minority do not believe in their roof's capacity to support solar panels. It should be noted that these indicative results are prejudice to the opinion of either the observer or the survey participant at time of surveying (January 2019). With a comprehensive quantity of flat-roof residences in Vivek Vihar and Vaishali Nagar (100% flat, zero pitch), one could assume that these localities are in a prime position to implement rooftop solar PV. However, certain built implications must be addressed prior to making such assumptions. Firstly, the usage of the rooftop is integral to analysing the potentiality for solar PV in an area. For instance, the survey found that 59.3% did not use their rooftop for miscellaneous purposes, whereas 40.7% actively used their rooftop for other purposes, such as recreation, domestic use, food preparation, and external residences. Secondly, knowledge of imposed load bearing capacity on a rooftop is paramount in deciding upon a rooftop solar system (Saur Energy International, 2018).

Overshadowing awareness is also another key element, either caused by a neighbouring buildings or natural context. These buildings would have to be of equal or greater height in order to provide any significant blockage to sun access for the panels (as is the same with tree heights). The sun paths during the summer and winter months are typically high due to New Delhi's position relative to the equator. Figure 11 is a derived from a Revit model looking at the sun path over a typical four-storey block in Vivek Vihar, adjacent to a small park. It demonstrates the localities' opportunity to implement solar panel systems as a part of their approach to a consistent power source over the coming years. Of the people who had solar currently installed, none had access to clean their panels or had someone cleaning them. The heavy pollution of Delhi means that smog settles layers of fine particulates that form a blackish soot cover on any surface it settles on. This includes solar panels, meaning that cleaning is of greater priority. Access must be designed into an array and a maintenance schedule must be followed.

A final interesting statistic was the question of access internet, which came back overwhelmingly positive. 90.74% of those interviewed claimed they had reliable internet access at home. Upon further enquiry, it was found that many of those who had access were using internet via 4G on their phone, rather than a hardwired modem. This may impact the implementation of a P2P smart grid in the form of tablets/smart controls used to manage trading systems. These require a stable internet connection, and so if a home only uses 4G via phone, it may require a stay-at-home modem solution.

11 Conclusion

This study looked at the current energy infrastructure in New Delhi, addressing physical, technological, policy. It also investigated the possible opportunities for the cutting-edge smart technologies and residents' preparedness to uptake these technologies. Data was collected through questionnaires and interview with locals in two neighbourhoods in Delhi to understand the social and infrastructure challenges for up taking solar rooftops in Delhi.

The findings reflect that the cost of solar and lack of information about available subsidies and technicalities are two prime reasons for a poor solar uptake in the residential sector. Households that were hesitant to install solar, due to financial costs, also assumed there were no subsidies available. There appears to be a lack of information among residents, even among those who have solar panels installed. Access to clear and easily understood information is essential.

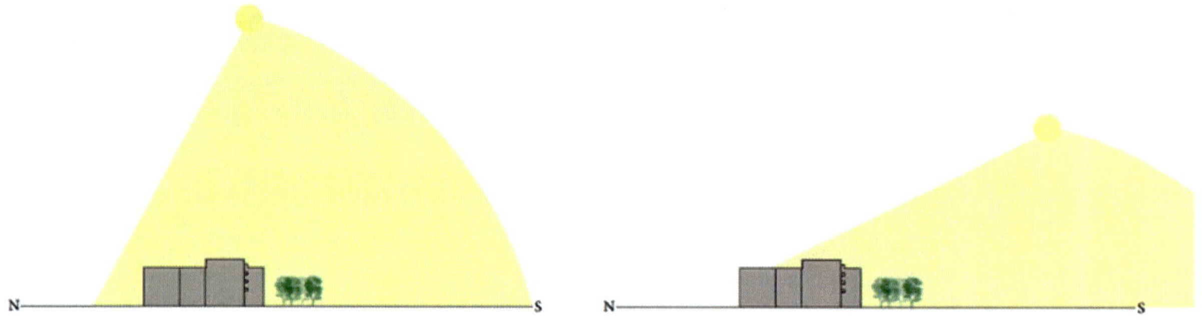

Fig. 11 Diagrams showing sun path over Vivek Vihar: summer solstice (left) and winter solstice (right)

With nearly 100% flat-roof residences in Vivek Vihar and Vaishali, one could assume that these localities are in a good position to begin implementation of rooftop solar. However, the roof spaces are often used in variety of ways. For instance, the survey found 40.7% used their rooftops for other purposes such as recreation, domestic use, food preparation, etc. Access to roof for apartment dwellers was another issue and respondents were interested in exploring options of angled solar panels on blank vertical building surfaces. A solar pergola could also be a valid option for apartment owners who do not have access or enough space on their roofs for the solar panels to be installed. This could be installed on top of the roof area, balconies, sundecks, terraces, and carports They can provide shade underneath, as well as produce renewable energy.

Overshadowing caused by either neighbouring buildings or trees was another key aspect that was explored in assessing the potential of rooftop's readiness. The study area was overrepresented with apartment typology of varying heights from one to six storeys. Overshadowing issues were noted, however towards the end of each day starting at about 3 pm. These buildings would have to be of equal or greater height in order to provide any significant blockage to sun access for the panels (as is the same with tree heights).

The survey analysis shows that over 95% of the people who had mechanical cooling and heating systems relied on the air conditioning units (AC), whereas less than 5% had built in HVAC systems. In most cases, HVAC systems are known to be more energy efficient than AC units. There needs to be a focus on ways of conserving energy, alongside with implementing renewable energy production methods.

The study found that although there were energy savings recorded by the households who used solar panels, these did not reflect the optimum use of the energy generated. Lack of awareness regarding positioning and maintenance of the panels meant limiting the efficiency of the entire system. Contributing to the dust on the solar panels is the smog in Delhi that significantly reduced the performance of the solar panels. 75% of respondents with solar PV indicated that their peak energy use was at late evening/nighttime, which meant that they would have benefited more with a battery storage system.

Around 77% of all respondents were positive about investing in battery storage if this was something that was easily available and incentivised for them and would in future help them to buy or sell excess energy to neighbours resulting in cheaper energy. 80% respondents were optimistic about peer-to-peer trading, which is a form of anonymous buying and selling excess solar energy between neighbours. When asked if respondents wanted to get the government involved in such an initiative, 73% were positive whilst only 10% were negative with a 17% being unsure. With government incentives and policy framework changes, such cutting-edge initiatives in future will make the householders confident and responsible about managing their energy needs themselves resulting in "democratisation" of energy.

References

Alam, M. R., St-Hilaire, M., & Kunz, T. (2017). An optimal P2P energy trading model for smart homes in the smart grid. *Energy Efficiency, 10*(6), 1475–1493. https://doi.org/10.1007/s12053-017-9532-5

Bureau of Indian Standards. (2016). National Building Code of India. In *Electrical and allied installations (including lightning protection of buildings and solar energy utilisation)* (Vol. 21.2.1). Bureau of Indian Standards.

Census of India. (2011). Retrieved from https://www.censusindia.co.in/ from Census India https://www.censusindia.co.in/

Curtin University School of Architecture. (2019). *Smart cities NCP research—Survey of Vivek Vihar and Vaishali Nagar.*

Deshmukh, R., Gambhir, A., & Sant, G. (2010). Need to realign India's national solar mission. *Economic and Political Weekly, 45*(12), 41–50. Retrieved from www.jstor.org/stable/25664251

India Energy Portal. (2018). *Renewable energy: Solar.* Retrieved from https://www.indiaenergyportal.org/subthemes.php?text=solar

Jogunola, O., Ikpehai, A., Anoh, K., Adebisi, B., Hammoudeh, M., Son, S.-Y., & Harris, G. (2017). State-of-the-art and prospects for peer-to-peer transaction-based energy system. *Energies, 10*(12), 2016. Retrieved from https://www.mdpi.com/1996-1073/10/12/2106

Kumar, A., Garg, V., Singla, S., Girdhar, P., & Singh, V. (2018). *Rooftop solar in India—Looking back, looking ahead*. Retrieved from Delhi.

Mission Innovation. (2018). *Innovation is vital for combating climate change and ensuring climate justice*. Retrieved from https://mission-innovation.net/participating-countries/india/

MNRE. (2019). *Solar schemes*. Retrieved from https://mnre.gov.in/solar/schemes

Pillai, U. (2015). Drivers of cost reduction in solar photovoltaics. *Energy Economics, 50*, 286–293. https://doi.org/10.1016/j.eneco.2015.05.015

Power Ledger. (2018). *Power ledger to provide peer-to-peer renewable energy trading at Yolk's evermore development at White Gum Valley*. Retrieved from https://medium.com/power-ledger/power-ledger-yolk-evermore-white-gum-valley-blockchain-energy-4ecd72c0a9c4

Saur Energy International. (2018). *Best practices for solar rooftop system*. Retrieved from https://www.saurenergy.com/solar-energy-articles/best-practices-for-solar-rooftop

Smart Roof. (2017). *Vivek Vihar*. Retrieved from https://www.gosmartroof.com/solar-panel-system/solar-at-home-Delhi-Vivek-Vihar

Yenneti, K. (2016). The grid-connected solar energy in India: Structures and challenges. *Energy Strategy Review, 11–12*, 41–51. https://doi.org/10.1016/j.esr.2016.06.002

A Comparative Study of Spanish Mid-Sized Smart Cities from the Citizen's Perspective

Amor Ariza-Álvarez, Rosa M. Arce Ruiz, Julio A. Soria-Lara, and Concepción Moreno Alonso

Abstract

The growing interest in evaluating what constitutes a Smart City has led to a proliferation of rankings based on measurement indicators. These rankings facilitate comparisons between cities and are a key instrument for benchmarking. This paper aims to contribute additional insights into the evaluation of Smart Cities by providing a set of key indicators based on six main aspects: Governance, Mobility, Environment, Economy, People, and Living. The evaluation methodology consisted mainly of 22 indicators to evaluate cities from the perspective of citizens and/or visitors as the main recipients of the smart measures. This empirical approach was applied to a total of 27 Spanish cities between 150,000 and 250,000 inhabitants. To prioritize the relevance of the set of indicators, a Delphi panel was also conducted by engaging a total of 30 experts in the field of Smart Cities, along with other internal processes. The results show differences between the Spanish cities and in the degree of development reached for the different indicators, evidencing the difficulties they are facing to achieve the Smart City goals.

Keywords

Smart City • Indicators • Sustainability • Urban development • Citizens

A. Ariza-Álvarez (✉) · R. M. Arce Ruiz · J. A. Soria-Lara · C. Moreno Alonso
Transport Research Centre (TRANSyT), Universidad Politécnica de Madrid (UPM), C/ Profesor Aranguren, 3, 28040 Madrid, Spain
e-mail: mariaamor.ariza@upm.es

J. A. Soria-Lara
Instituto de Economía, Geografía y Demografía, Consejo Superior de Investigaciones Científicas (CSIC), C/ Albasanz 26, 28037 Madrid, Spain

1 Introduction

Over half the world population currently lives in urban areas, a figure that is expected to rise to 70% by 2050 (United Nations, 2019). Cities generate development opportunities for their inhabitants, but also face major challenges regarding air pollution, traffic congestion, social inequalities, insecurity, or unemployment to name just a few (OECD, 2020). In today's ever more interconnected society, information and communication technologies (ICTs) are key for addressing these challenges. In this context, the Smart City concept can be seen as a means to achieve efficient and sustainable cities with a higher quality of life for its citizens thanks to new developments and technological innovations.

In recent years, multiple Smart City initiatives and projects have been developed worldwide (Angelidou, 2015; Lee et al., 2014; Albino et al., 2015; Zheng et al., 2020). This has led to an increasing interest to design methodological frameworks to monitor the progresses of those cities and to establish comparison between them. Various city rankings have been promoted by global companies and academic institutions, in which the Smart City concept is usually present (Giffinger et al., 2007; IDC, 2012; IMD, 2018; IESE, 2018; Moreno, 2015; PWC, 2012). However, in their assessment frameworks, much stronger emphasis has been put on ICTs and the "smartness" of the solutions, rather than on the real benefits that those innovations brought to citizens. Moreover, these Smart City rankings mostly cover large cities, and mid-sized cities are usually missed, even though majority of the urban population lives in such medium-sized cities. The role city rankings can play in identifying best smart and citizen-oriented practices for mid-sized cities is largely unexplored.

This book chapter aims to conduct a comparative evaluation of 27 Spanish mid-sized cities to determine how far they have gone in their process of transformation into Smart Cities, to identify best practices, and to show the areas in which improvement and investment are needed. For this

F. Trapani et al. (eds.), *Advanced Studies in Efficient Environmental Design and City Planning*, Advances in Science, Technology & Innovation, https://doi.org/10.1007/978-3-030-65181-7_6

purpose, a methodological framework has been developed, in which a set of 22 indicators was defined according to the six main dimensions of Smart Cities: Governance; Mobility; Environment; Economy; People; Living (Giffinger et al., 2007). Cities were evaluated for each indicator and for each Smart City dimension, arriving at six final city rankings. The evaluation methodology namely follows the citizen's perspective as the main recipients of the smart policies and strategies implemented by the cities. It is, therefore, focused on quality of service and "user experience", and it is not dependent on the technological solutions implemented by the cities, but on what these technologies provide to its inhabitants. The evaluation was carried out between May and September 2019.

The reminder of this chapter is organized as follows. Section 2 present a literature review on the Smart City concept, Sect. 3 describes the case study, Sect. 4 covers the methodological framework, while Sect. 5 summarizes the main obtained results. Finally, Sect. 6 closes the document with some concluding remarks, including reflections on future research directions.

2 The Concept of Smart City

There has been a growing number of Smart City definitions over the years (Angelidou, 2015; Baucells et al., 2016; Caragliu et al., 2011; Chourabi et al., 2012; Fernández-Añez et al., 2018; Hollands, 2008; Neirotti et al., 2014). Two mainstreams can be identified, one is technology-oriented, and the other is people-oriented. With respect to the technology-oriented approach, there is an extensive body of literature considering the role of ICT and modern technological developments (e.g. Internet of Things or big data) as key elements for Smart Cities (Al Nuaimi et al. 2015, Angelidou, 2015; González and Rossi, 2011; Hall et al., 2000; Jucevičius et al., 2014; Marsal-Llacuna and Segal, 2016; Paroutis et al., 2014; Zanella et al., 2014). For most of those authors, the Smart City concept essentially covers the use of modern technologies in urban life to improve the efficiency and effectiveness of urban processes and services (e.g. transport, infrastructures, energy). However, an increasing number of authors consider these technologies not only as enablers in addressing urban challenges, but with the aim to improve the citizens' lives and promote their economic, social, and cultural development (Bouskela et al., 2016; Caragliu et al., 2011; Giffinger et al., 2007; Hollands, 2008; Klein and Kaeger, 2008; Nam and Pardo, 2011; Neirotti et al., 2014). This has triggered a shift to a more holistic Smart City concept, in which technology is more people-oriented.

Despite there is not consensus regarding Smart City concept, achieving sustainability in Smart Cities seems to be a common desire (Ahvenniemi et al., 2017; European Commission, 2015). A smart and sustainable city can be seen as an innovative city using technology to improve the quality of life of its citizens, the efficiency of urban services, and competitiveness. It also ensures that the needs of present and future generations are met with respect to economic, social, environmental, and cultural aspects. Thus, Smart Cities must consider the UN Sustainable Developments Goals (SDGs) (United Nations, 2015) as inspirational lines to define their policies and initiatives. The assessment of Smart City performance should not only measure the efficiency of smart and innovative solutions, but also the contributions towards goals such as environmental, economic, or social sustainability (Ahvenniemi et al., 2017).

As mentioned, the Smart City concept brings together different aspects related to technology, society, and sustainability, which demands the use of different viewpoints for studying and analysing such Smart Cities. Accordingly, six main Smart City dimensions can be distinguished (Giffinger et al., 2007): Governance; Mobility; Environment; Economy; People; and Living (Fig. 1). These six dimensions constitute the main fields of action in which a Smart City should invest and act. Although some studies cover a greater number of dimensions (Nam and Pardo, 2011; Chourabi et al., 2012; Neirotti et al., 2014), most of studies define and assess the Smart City according to these six pillars (Caragliu et al., 2011; Batty et al., 2012; European Parliament, 2014, Moreno, 2015). Therefore, it is the most widely accepted classification of the Smart City concept. The six dimensions have been utilized as suitable vectors for developing this comparative study between cities. Technology is not

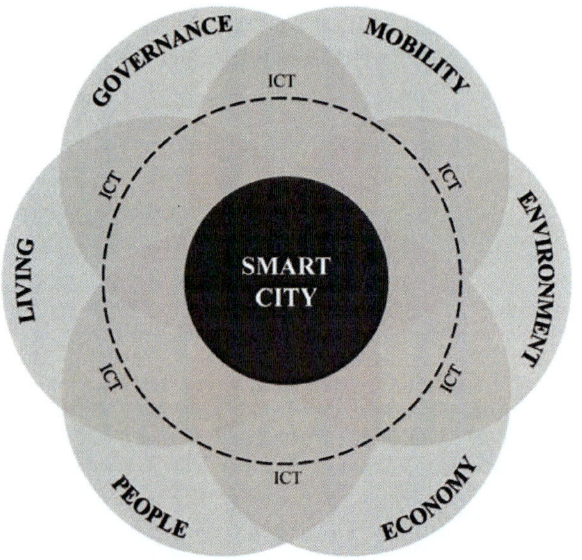

Fig. 1 Dimensions of the Smart City with ICTs as a transversal element. (ASCIMER, 2014)

considered here as dimension itself, but as a tool to improve the efficiency of smart actions and policies and as a transversal element through all the dimensions of a Smart City.

3 Case Study

The object of this evaluation is Spanish mid-sized cities. The definition of mid-sized city is context dependent. In the case of Spain, 80% of Spanish people live in city regions with more than 10,000 inhabitants, which are considered urban population. Twenty percent of this urban population lives in cities with more than 500,000 inhabitants, and around 15% lives in cities between 150,000 and 250,000 inhabitants. The latter are considered as mid-sized cities in this research. In Spain, there are 27 mid-sized cities (Spanish Statistical National Institute, 2019) that constitute the objective of this work. These cities occupy an intermediate place in the hierarchical scale of the Spanish system of cities and tend to act as sub-regional or provincial centres. This set of cities represent a homogeneous sample in relation with their population size which has allowed to draw significant conclusions from the comparison between them (Fig. 2).

4 Methodological Framework

To achieve the proposed objective of this study, the work methodology consists of five general phases: (i) selecting the indicators for each Smart City dimension, (ii) defining the evaluation scale of each indicator, (iii) obtaining the Relevance Index (RI) of each indicator, (iv) evaluating the cities in terms of each indicator, and (v) obtaining a ranking of cities for each Smart City dimension.

4.1 Definition of the Indicators

As a first step in defining the model, the set of indicators to be analysed for each of the six Smart City dimensions (Governance, Mobility, Environment, Economy, People, and Living) were identified and selected. One important consideration in the process of selecting the indicators is that they must be measurable "from outside" the municipal management itself to ensure they are independent and citizen-centred. Any indicator for which there is no information available for their assessment were discarded. Table 1 shows the indicators selected for the analysis of each Smart City dimension.

4.2 Evaluation Scales

After selecting the indicators, an evaluation scale of 0–4 was defined in order to determine the degree of maturity reached with the initiatives and strategies implemented in each city. The evaluation scales were defined according to the following generic scheme. A score of 4 corresponds to a Smart City that meets all the requirements established in the indicator, a score of 3 was given to cities that could improve in some or several of the aspects considered in the indicator, a score of 2 was given to cities that had recently begun the process of transformation and meet few of the requirements

Fig. 2 Study Case, 27 Spanish municipalities with a population between 150,000 and 250,000 inhabitants

Table 1 Selected indicators for each Smart City dimension

Governance indicators	
Smart plan/smart strategy	It considers whether cities have developed a strategic plan defined specifically for their transformation into Smart Cities and the quality of this plan
City website	It evaluates the benefits that the city website offers the citizen, as well as its accessibility and navigability
Transparency	It is intended to assess the quality of municipal transparency portals, considering the type, quantity, characteristics, level of updating, and accessibility of the information contained
Communication channels	It analyses the channels that establish communication between the administration and citizens in both directions
Geo-information	It evaluates the quality of the geolocated information of the municipality visible on its website, as well as its usefulness for citizens and visitors
Mobility indicators	
Payment modes in public transport	It evaluates the integration of information and communication technologies (ICT) to make it easier for citizens to pay for public transport modes in the city
Information for mobility	It evaluates the information that the City Council offers to citizens and visitors in relation to mobility in the city, as well as the technological innovations incorporated to improve its transmission and enhance its usefulness
Alternative transport modes	It assesses the degree of implementation of alternative mobility modes to conventional ones in the city. Specifically, measures aimed at promoting mobility by bicycle and shared vehicles have been considered
Traffic management and control	It assesses the implementation of measures for the control, management, and improvement of accessibility in relation to traffic in the city
Environment indicators	
Energy efficiency	It evaluates the deployment of energy efficiency measures with intensive use of ICTs and the installation of smart energy networks
Sustainable urban transport	It assesses measures promoted by the City Councils aimed at reducing emissions generated by urban transport, e.g. incorporation of electric or hybrid vehicles and energy efficiency measures in public transport
Efficiency in water consumption	It evaluates the implementation of measures to improve the efficiency of water consumption and management in urban processes and services
Waste management	It evaluates the implementation of smart measures for the control and management of the different parameters related to urban waste
Emission monitoring and information	It assesses the actions carried out to measure the GHG Emissions of the municipality, as well as the systems for transmitting this information to citizens
Economy indicators	
Open data	It analyses the quality of the Open Data portal that the City Council makes available to the public, mainly considering data diversity, accessibility and ease of consultation
Entrepreneurship and innovation	It analyses the effort of local public institutions to create an ecosystem of entrepreneurship and innovation in the city
People indicators	
Training in the Smart City concept	It evaluates the implementation of initiatives aimed at training citizens in the digital skills to participate and access the services of the Smart City
Municipal Wi-Fi	It analyses the existence of open and free Wi-Fi points throughout the city
Attracting and retaining talent	It assesses the measures carried out by local administrations to manage talent through its attraction, growth, and retention
Living indicators	
Social services	It evaluates the integration by the local administration of technological solutions to improve the provision of social services to its citizens
Universal accessibility	It evaluates the adoption of technological solutions to improve Universal Accessibility in the municipalities for all citizens regardless of their abilities
Technologies for enjoying the city	It analyses the measures, based on ICTs, aimed at promoting the best knowledge and enjoyment of the city among citizens and visitors

defined for the indicator, and a score of 0 was given to the cities that did not meet any of the indicator's expectations and were far from becoming Smart Cities.

This scale was particularized for each of the indicators. As an example, the evaluation scale used for the Open Data indicator was as follows. In this case, a score of 4 represents a "*Government's Open Data with a wide spectrum of data available, easy to consult, unrestricted access and possibility of re-utilization, distribution, and interoperability*". A score of 3 represents "*Open Data that requires improvements in the way the data are presented, distributed, or organized*". A score of 2 represents "*Open Data that is in process of consolidation and where there are still datasets remaining to be incorporated*". A score of 1 indicates that "*Open Data is still in process of creation, or it has no content available*". Finally, a score of 0 indicates that "*Open Data is non-existent, and it is not expected to be created*".

The evaluation scales used for each of the 22 indicators are included in Annexes 1, 2, 3, 4, 5, and 6.

4.3 Obtaining the Relevance Indexes

As there is no homogeneity in the degree to which each indicator contributes to the transformation of conventional cities into Smart Cities, it was necessary to weight the set of indicators according to their relevance for the smart transformation of the cities. To prioritize the relevance of the set of indicators, an index—here called relevance index (RI)—was determined. This index was composed of four factors.

Factor F1 incorporates the views of a group of 30 experts in the field of Smart Cities. The experts showed, according to their previous experience, their opinion about the relevance of each indicator for the transformation of conventional cities into Smart Cities. These views were obtained through an online survey where the experts signalled the relevance of each indicator according to a 5-point Likert scale. *Factor F2* considered the number of Smart City dimensions affected by each indicator: Governance, Mobility, Environment, Economy, People, and Living. This factor, that could take values from 1 to 6, was scored for each indicator by the research team following an internal deliberation process. *Factor F3* incorporates the degree to which each indicator considers information and communication technologies (ICTs) in cities' evaluation. This factor, that could take values from 0 to 3, was also scored by the research team through an internal deliberation process. Finally, *Factor F4* assessed the usefulness of each indicator to the citizens, on a scale of 0 to 3. The evaluation of the indicators for this factor was also made by the research team through an internal process. These four factors (*F1*, *F2*, *F3*, and *F4*) were then properly standardized to conform to a scale of 0–1. Finally, the RI for each indicator was obtained (1).

$$\mathrm{RI} = \frac{F1 + F2 + F3 + F4}{4} \quad (1)$$

4.4 Evaluating the Cities and Obtaining the Rankings

Once the set of indicators was defined, the 21 Spanish cities considered in this study were rated for each indicator according to its evaluation scale (from 0 to 4). The score of each city for each Smart City dimension was obtained through the weighted sum—by the RI—of its corresponding indicators. The City Councils′ websites were among the main sources consulted to evaluate the cities. These websites contain reliable and verified information about the initiatives and strategies carried out in each city on the way to its transformation into Smart Cities. Consulting these websites also allowed to carry out this evaluation from the perspective of the citizens or visitors of these cities and as potential recipients of the implemented measures. Other additional sources, such as specialized publications on the Internet (newsletter, monographs, etc.) or publications by technology companies and service providers were also analysed to conduct this evaluation. The final outputs of this study were six ranking with the 27 Spanish mid-sized cities analysed, one for each Smart City dimension. These rankings can be found in Sect. 5 of this study.

5 Results

5.1 Relevance Indexes (RIs)

Figure 3 shows the relevance indexes (RIs) considered for each indicator. The set of indicators reveals considerable variation between their RIs which range from 0.43 to 0.84 on a scale of 0 to 1 (Fig. 3).

The two indicators with the highest RI belong to the Mobility dimension: "Information for mobility" (RI 0.84) and "Payment modes in public transport" (RI 0.82). In contrast, the indicators with the lowest RI are "Transparency" (RI 0.44) and "Smart Plan" (RI 0.43), both belonging to the Governance dimension. It is precisely these two dimensions (Mobility and Governance) that show the greatest variability in the RI of their indicators. The same occurs, although to a lesser degree, with the two indicators for the Economy dimension. Thc influence of the weighting of the indicators will therefore be decisive for these three dimensions in the final result of the ranking.

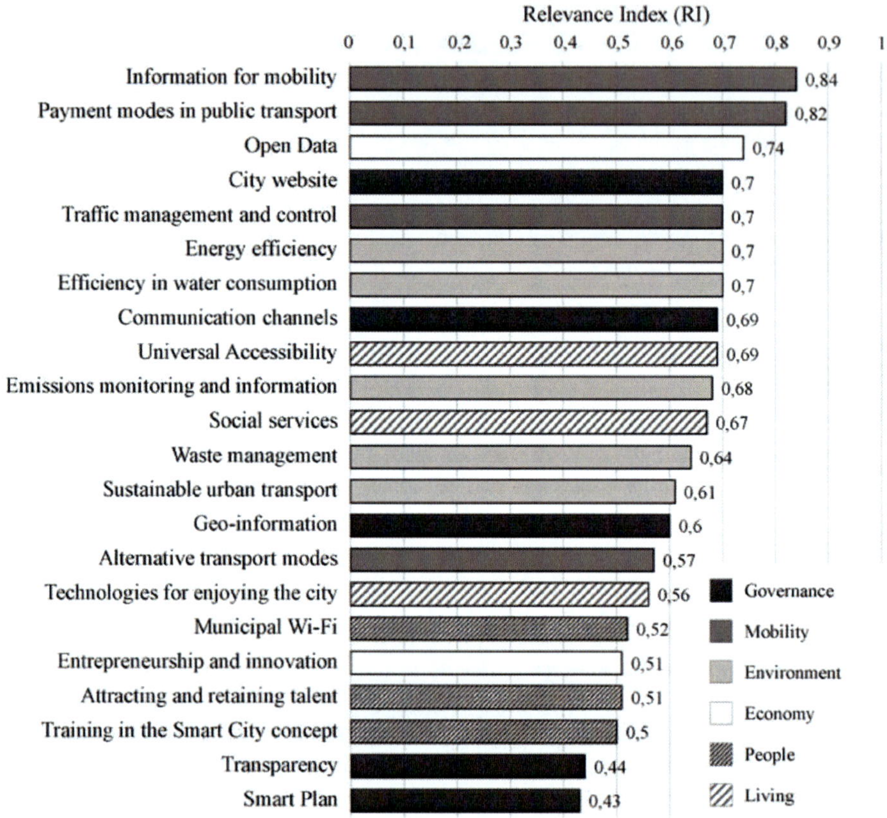

Fig. 3 Indicator's relevance indexes (RI)

There is scarcely any variability between the RI of the indicators for the Environment dimension, whose values range between 0.6 and 0.7. A similar situation occurs with the Living indicators, which range between 0.56 and 0.59. The three indicators selected for the People dimension are located in the lower part of the graph and obtained low RIs that were also very similar to each other, at around 0.5. The influence of the weighting of the indicators is not relevant for these last three Smart City dimensions.

5.2 Evaluating the Cities

As previously mentioned, the 27 cities included in this evaluation were rated on a scale of 0 to 4 for each of the 22 indicators. Figure 4 shows the evaluation of the cities for the indicators with the percentage of the total cities with each possible score (0, 1, 2, 3, 4, or 5).

In first place, two indicators can be distinguished in which most of the cities had very positive evaluations of between 3 and 4. These are: "Communication channels" and "Entrepreneurship and innovation". These indicators are the ones in which Spanish mid-sized cities have achieved the greatest degree of smartness in their relationship with the citizens. Most of the cities analysed offer their citizens a wide range of communication channels (apps, social networks, online participative processes, etc.). A substantial offer of infrastructures and services for entrepreneurs and companies with innovative projects was also identified.

In a second group of indicators, most of the cities had scores of between 2 and 3. These are: "Geo-information", "Alternative transport modes", "Efficiency in water consumption", "Municipal Wi-Fi", and "Universal accessibility". Most cities achieved a high degree of smartness for these indicators; however, room for improvement was still detected in certain aspects. Although most of the cities analysed have their own Geographic Information Systems, these are not very intuitive or feature very few information levels. In most cases, the cities have a municipal bike hire service, and there are still no municipal initiatives for shared mobility in private vehicles. Smart water management measures are deployed, but not extensively throughout the whole city or else they do not cover all the phases of the water cycle. There are free Wi-Fi points available to citizens, but these are not managed as a public service or are deployed in municipal buildings and not in outside spaces. Actions have been undertaken based on ICTs to improve the physical accessibility for people with reduced mobility or disabilities, but online accessibility to the municipal website is not guaranteed for people with disabilities.

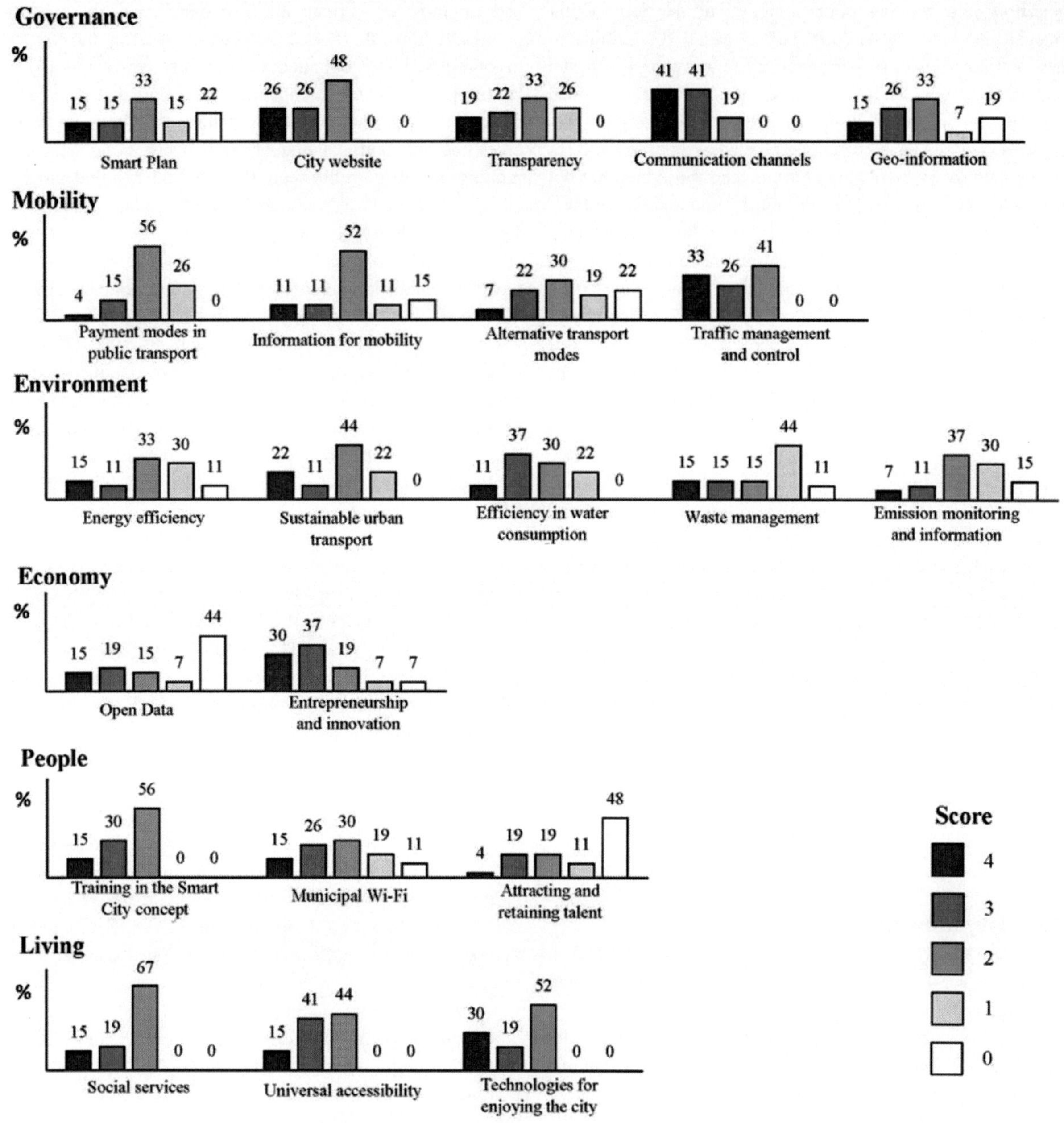

Fig. 4 Evaluation of the cities for each indicator

Most of the cities obtained a score of 2 for a large number of indicators. Most of the cities are currently in a process of transformation for the following indicators: "City website", "Payment modes in public transport", "Information for mobility", "Traffic management and control", "Sustainable urban transport", "Training in the Smart City concept", "Social services" and "Technologies for enjoying the city". Most of the city websites analysed are in the process of incorporating online procedures. Smart payment systems in public transport (payment with smartphones or bank cards) have barely been integrated. In most cases, City Councils were considered to provide citizens and visitors with uncomplete and low-quality information on the different mobility options in the city. Traffic control systems do not intensively integrate ICTs, and non-polluting vehicles have been incorporated in the public transport fleet, but charging points for electrical vehicles are very limited and little information is available on this service. Although there are activities and training courses aimed at the job-seeking population, very few actions were detected for training in new technologies or for the purpose of narrowing the digital gap in older people. In most cities, the provision of social services is conventional and there are still few online procedures. Actions have been undertaken that incorporate ICTs in the context of events, exhibitions, and museums, but not for the general enjoyment of its cultural and artistic heritage.

In another group of indicators, most of the cities obtained a score of between 2 and 1; that is, they have recently begun to implement initiatives, but still have a long way to go. These indicators are: "Smart Plan", "Transparency", "Energy efficiency", "Waste management" and "Emissions monitoring and information". Most have smart city strategies or projects but have not developed a specific plan for a fixed time horizon. Most of the transparency portals analysed are incomplete or contain out of date information of little importance to citizens. Only conventional energy efficiency measures are used in municipal buildings, equipment and urban lighting, and no smart energy networks have been incorporated. Very few smart systems have been implemented for waste collection. Information on emissions measuring and monitoring procedures in the municipality is scarce in most cases.

Finally, there are two indicators for which most of the cities obtained the worst possible score of 0 out of 1. These are "Open Data" and "Attracting and retaining talent". A significant number of the analysed cities did not have an Open Data portal at the date of this evaluation. Similarly, most of the cities did not carried out any actions for attracting, retaining, and returning talent.

5.3 City Rankings

Figure 5 shows the six city rankings for each of the Smart City dimensions.

For the Governance Ranking (Fig. 5a), only seven cities obtain a score above 3. San Sebastián tops this ranking, followed by Vitoria-Gasteiz, the most populated city of those studied, and Castellón de la Plana. In this ranking, cities with less than 200,000 inhabitants, such as San Sebastian, Castellón de la Plana or Santander, are in the top positions. Cities with more than 200,000 inhabitants have all ratings greater than 2 in this ranking. Most cities (a total of 15) have obtained global scores between 2 and 3. Finally, there have been five cities that have obtained global scores below 2, all of them with less than 200,000 inhabitants, such as Getafe or Alcorcón.

For the Mobility ranking (Fig. 5b), only three cities have obtained global scores higher than 3. San Sebastián also tops this ranking followed by Granada and Logroño, which share the second position. Again, two small cities, San Sebastián and Logroño, are among the best scored and above most populated cities. As in the eGovernance ranking, most cities (a total of 15) have obtained global scores between 2 and 3. In this group, there are both the most populated cities and others with less than 180,000 inhabitants. There are eight cities that obtain global evaluations between 1 and 2 for this ranking, all of them between 187,000 and 215,000 inhabitants. Only the city of Alcorcón obtains a global score below 1 for this Mobility ranking.

For the Environment Ranking (Fig. 5c), only five cities have obtained global scores higher than 3. Sabadell, with almost 210,000 inhabitants, tops this ranking, followed by Pamplona and San Sebastián, both with less than 200,000 inhabitants. There are six cities that have obtained global valuations between 2 and 3. Most cities (a total of 15) have obtained valuations between 1 and 2 for this ranking. Only the city of Oviedo, one of the most populated of those analysed, has obtained an overall rating below 1 for this ranking.

For the Economy Ranking (Fig. 5d), seven cities have obtained global scores higher than 3. Santander, with almost 172,000 inhabitants, tops this ranking, followed by Sabadell and Santa Cruz de Tenerife, both with more than 200,000 inhabitants. There are only four cities that have obtained global evaluations between 2 and 3, while most cities (a total of 12) have obtained scores between 1 and 2. Finally, there are four cities that have obtained a score below 1, all of them between 195,000 and 215,000 inhabitants.

For the People Ranking (Fig. 5e), only two cities have obtained a global score higher than 3. These are San Sebastián and Pamplona, both with less than 200,000 inhabitants, in first and second position, respectively. A large number of cities (a total of 10) have obtained a global valuation between 2 and 3, followed by other 13 cities with scores between 1 and 2. Alcorcón and San Cristóbal de la Laguna, both with less than 170,000 inhabitants, are in the last positions in this ranking, with global scores below 1.

For the Living Ranking (Fig. 5f), eight cities have obtained a global score of more than 3. A Coruña tops this ranking, followed by Cartagena and Elche, all of them with more than 200,000 inhabitants. The remaining 19 cities have obtained global scores between 2 and 3. Therefore, none of the 27 cities have obtained global evaluations below 2 in the Quality of Life ranking.

Globally, there are cities that stand out above the rest for being among the top positions on several occasions. The case of San Sebastián is evident, as it appears among the first five cities for four of the six rankings, and it is in the first position in Governance, Mobility and People. Santander is also among the top five cities for four rankings, leading the Economy ranking. Another outstanding example is A Coruña, which appears in the top 5 for three rankings, leading the Living ranking. Vitoria-Gasteiz, Logroño, Pamplona,

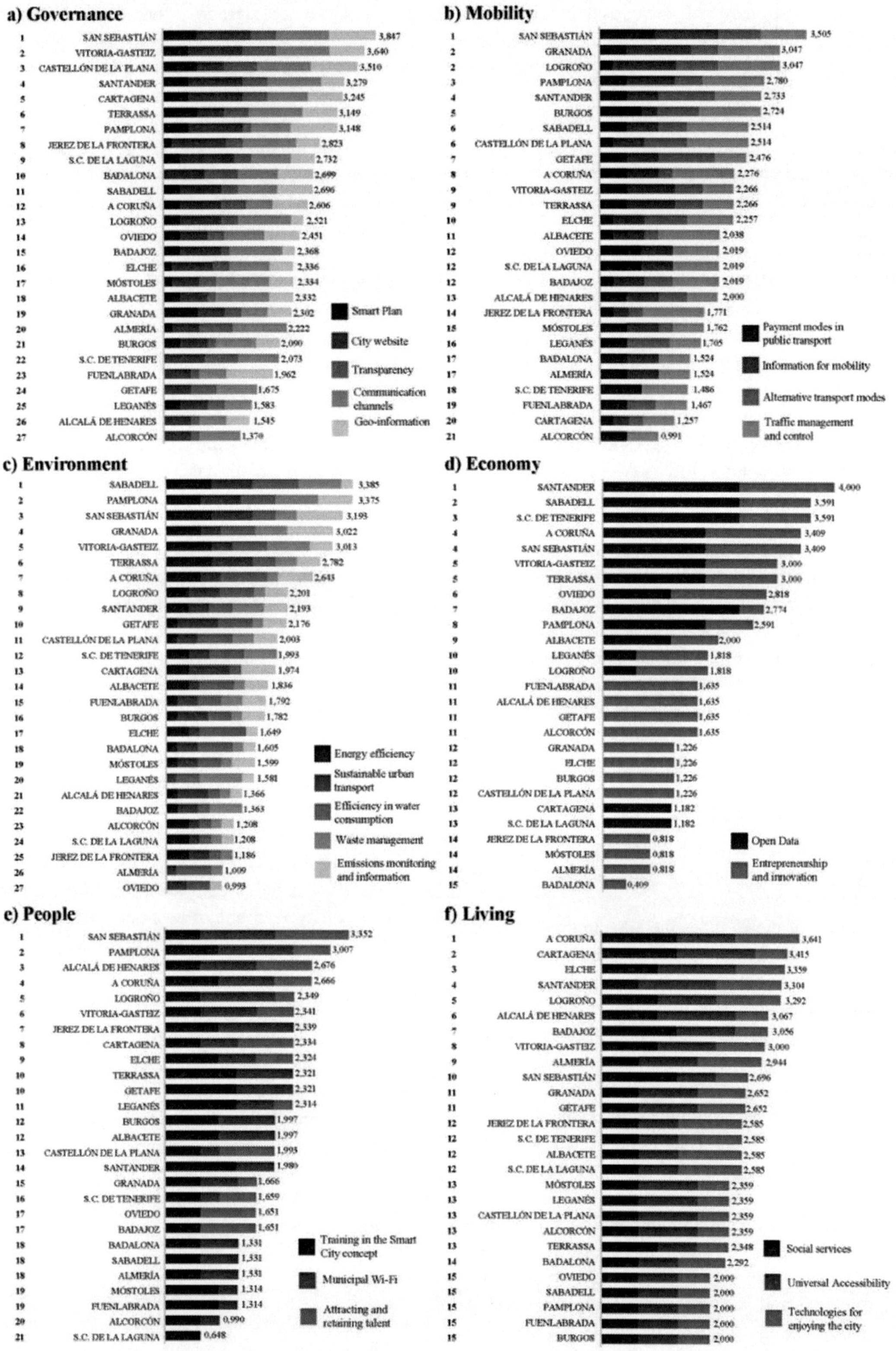

Fig. 5 City rankings for Governance (**a**), Mobility (**b**), Environment (**c**), Economy (**d**), People (**e**), and Living (**f**)

and Terrassa have also achieved very positive results and appear in the top positions on numerous occasions. All these cities have carried out a wide spectrum of measures that cover all the Smart City dimensions. However, most cities have only focused on one or two Smart City dimensions.

6 Conclusions

This comparative study provides a novel methodology for the evaluation of Smart Cities based on a detailed study of each city from the citizen's perspective as the main recipient of the measures and actions carried out. The city evaluation model designed in this work has resulted to be valid to know the degree of maturity of the transformation processes of conventional cities into Smart Cities, as well as to establish comparisons between them. Three conclusions are drawn from the results of this study:

- The set of 22 indicators used for the evaluation of cities has proved to be closely related to the concept of Smart City. However, there is considerable variation between the relevance indexes of the indicators. According to this index, the most relevant indicators are related to Mobility in cities, specifically with the information offered for the different mobility options and payment methods existing in public transport. These indicators have a great impact on the daily life of citizens and, in addition, imply a high level of incorporation of ICTs. On the contrary, the indicators with less relevance are related to Governance, specifically with the "Transparency" and the "Smart City Plans" developed. These have a lower impact on the daily life of citizens and imply a less degree of ICTs involvement.
- The assessment of the 27 cities included in this study has revealed differences in the degree of implementation of Smart City measures between indicators. In general, cities have demonstrated that they have developed a greater degree of "intelligence" in relation to the development of different channels for communication with citizens and entrepreneurship and innovation initiatives. On the contrary, a lower degree of development and "smartness" has been observed in the Open Data portals of these municipalities, as well as in initiatives for attracting and retaining talent.
- The transformation of Spanish cities into Smart Cities is being carried out, with quantifiable positive results. However, each city has focused its transformation process on different axes and with different degrees of development. In the rankings obtained in this study, there are three cities that stand out from the rest because they are among the top positions on several occasions. These cities are San Sebastián, Santander, and A Coruña. Vitoria-Gasteiz, Logroño, Pamplona, or Terrassa have also achieved very positive results and appear in the top positions on numerous occasions. However, most of the cities analysed have focused on one or several Smart City dimensions, according to their particular circumstances (number of inhabitants, municipal budget, geographical location, etc.), leaving aside the rest of dimensions. For these cities, which have obtained negative evaluations in one or several aspects, the transformation process into a Smart City must continue attending mainly to those aspects in which they have been worst valued.

The results obtained in this study open up new possible research horizons. For example, the methodology developed in this study could be applied to other sets of cities outside or within the Spanish context, which would serve to verify its effectiveness and replicability. Also, there exists the possibility of making a global ranking that brings together the six Smart City dimensions.

Annex 1. Evaluation Scales Established for the Governance Indicators

Smart Plan/Smart Strategy

- **4.** Smart City Plan with incidence on the six dimensions of the smart city, with specific strategic objectives, and with an action plan for a fixed time horizon.
- **3.** Smart City Plan with incidence on the six dimensions of the smart city, with specific strategic objectives, and with an action plan for a fixed time horizon. However, it is not available or accessible to the citizens.
- **2.** There is no action plan for a fixed time horizon.
- **1.** Smart City Plan under development.
- **0.** There is no Smart City Plan or strategy.

City Website

- **4.** City Council Website that is accessible, intuitive, with procedures organized by topic and with large number of online procedures.
- **3.** Website that is non-intuitive or poorly organized.
- **2.** Website in process of incorporation online procedures or that is confusing.
- **1.** Website under development.
- **0.** Non-existent Website.

Transparency

- **4.** Information about government organization, costs of public services, and budgets. It is updated, accessible, organized, and clear.
- **3.** The information is difficult to understand.

- **2.** The information is not updated.
- **1.** Non-relevant information.
- **0.** Non-existent transparency portal.

Communication Channels

- **4.** Large number of communication channels. The presence of the City Council in social networks, existence of institutional apps with updated information from the municipality and for the communication of incidents and suggestions, and online participatory processes.
- **3.** City Council does not make communication channels visible to citizens.
- **2.** There is no institutional apps for communication of incidents and suggestions.
- There are no online participatory processes.
- There are no communication channels between City Council and citizens.

Geo-Information

- **4.** Geolocated information of the municipality that is visible on its website and useful for citizens and visitors.
- **3.** Geolocated information that is difficult to understand for citizens.
- **2.** Geolocated information that is not useful for citizens and visitors.
- **1.** Google map or similar, without specific information for the city.
- **0.** Non-interactive maps to download.

Annex 2. Evaluation Scales Established for the Mobility Indicators

Payment Modes in Public Transport

- **4.** "Smart" public transport payment systems. NFC technology for payment with smartphones or credit card.
- **3.** Rechargeable public transport pass online or through apps.
- **2.** Online purchase of transport tickets or integration of the different modes of public transport in the payment method.
- **1.** Personal rechargeable public transport cards.
- **0.** Conventional payment per trip.

Information for Mobility

- **4.** Extensive information on the City Council website about all the mobility options in the city. Apps with real-time information have been developed integrating all the mobility options in the city.
- **3.** Apps with real-time information have been developed, but they do not integrating all the mobility options in the city.
- **2.** Incomplete information on the municipal website about mobility options in the city. They have developed public transport apps.
- **1.** Incomplete information on the municipal website about mobility options in the city. They have not developed public transport apps.
- **0.** There is no information for mobility on the municipal website.

Alternative Transport Modes

- **4.** Municipal bike rental service, with real-time information. In addition, there are municipal initiatives for shared mobility in private vehicle.
- **3.** Municipal bike rental service, with real-time information. However, there are no municipal initiatives for shared mobility in private vehicle.
- **2.** Municipal bike rental service, without real-time information.
- **1.** Traditional information for mobility by bicycle (e.g. bicycle lanes).
- **0.** No policies for alternative mobility in the city.

Traffic Management and Control

- **4.** Traffic control and management centre with intensive integration of ICTs.
- **3.** Public transport control and management systems with intensive integration of ICTs, generally for public transport services.
- **2.** Standard traffic control conditions.
- **1.** Few measures for traffic control.
- **0.** No policies for traffic management and control.

Annex 3. Evaluation Scales Established for the Environment Indicators

Energy Efficiency

- **4.** Smart energy networks in municipal buildings and urban lighting. Information about these networks is visible on the municipal website for public knowledge.
- **3.** Smart energy networks only in municipal buildings or in urban lighting.
- **2.** Conventional energy efficiency policies.
- **1.** Only awareness campaigns on energy efficiency among the population.
- **0.** No policies for energy efficiency have been implemented.

Sustainable Urban Transport

- **4.** There are charging points for electric vehicles in the city and there is information available on the municipal website. A high number of clean vehicles have been incorporated in public transport.
- **3**. There are charging points for electric vehicles in the city but there is no information available on the municipal website. Clean vehicles have been incorporated in public transport.
- **2.** There is no charging points for electric vehicles in the city or clean vehicles have not been incorporated in public transport.
- **1.** Only one pilot measure through sustainable urban transport has been implemented.
- **0.** No policies for sustainable urban transport have been implemented.

Efficiency in Water Consumption

- **4.** "Smart" management of the water cycle throughout the city. Remote management systems for water supply and irrigation of green areas.
- **3.** Remote management systems only for water supply or for irrigation of green areas throughout the city.
- **2.** Pilot projects for "smart" water management in demonstration areas.
- **1.** Policies to reduce water consumption without using ICTs (e.g. water reuse and awareness campaigns).
- **0.** No policies for efficiency in water consumption have been implemented.

Waste Management

- **4.** "Smart" systems for waste collection and management (e.g. pneumatic collection systems).
- **3.** Systems for the optimization of waste collection routes or technologies for the control of waste container filling by sensors.
- **2.** Geo-information of waste containers available for citizens (e.g. type of waste collected in each container).
- **1.** Policies for waste management without using ICTs (e.g. installation of recycling containers).
- **0.** No policies for efficient waste management have been implemented.

Emission Monitoring and Information

- **4.** Updated and real-time information on the air quality of the city is visible on the municipal website. In addition, historic municipal emissions inventories are published.
- **3.** Updated and real-time information on the air quality of the city is visible on the municipal website. However, historic municipal emissions inventories are not published.
- **2.** Air quality information is not visible enough on the municipal website, or it is not updated. Historic municipal emissions inventories are not published.
- **1.** Air quality information is not particularized for the municipality.
- **0.** Air quality information is not published on the municipal website.

Annex 4. Evaluation Scales Established for the Economy Indicators

Open Data

- **4.** Government's Open Data with a wide spectrum of data available, easy to consult, unrestricted access and possibility of re-utilization, distribution, and interoperability.
- **3.** Open Data that requires improvements in the way the data are presented, distributed, or organized.
- **2.** Open Data that is in process of consolidation and where there are still datasets remaining to be incorporated.
- **1.** Open Data is still in process of creation, or it has no content available.
- **0.** Open Data is non-existent, and it is not expected to be created.

Entrepreneurship and Innovation

- **4.** Ecosystem of entrepreneurship and innovation in the city. Municipal infrastructures for entrepreneurs and companies (e.g. co-working centres), promotion of start-ups and innovative projects, and public–private initiatives. In addition, this information is visible and available in the municipal website.
- **3.** Ecosystem of entrepreneurship and innovation in the city. However, this information is not visible or available in the municipal website.
- **2.** There is no clear strategy for encouraging entrepreneurship and innovation in the city.
- **1.** Only sporadic actions are carried out.
- **0.** No policies for entrepreneurship and innovation have been implemented.

Annex 5. Evaluation Scales Established for the People Indicators

Training in the Smart City Concept

- **4.** Regular activities promoted by the City Council for training all the citizens in the digital skills to participate and access the services of the Smart City. In addition, this information is visible in the municipal website.

- **3.** Regular activities promoted by the City Council for training all the citizens in the digital skills to participate and access the services of the Smart City. However, this information is not visible enough in the municipal website.
- **2.** Activities are aimed at a specific population group (e.g. young people).
- **1.** Only possibility of downloading basic information from the municipal website.
- **0.** No policies for training in the Smart City concept have been implemented.

Municipal Wi-Fi

- **4.** Free municipal Wi-Fi in municipal buildings, outdoor public spaces, and in public services (e.g. public transport).
- **3.** Free municipal Wi-Fi is not available in outdoor public spaces.
- **2.** Detailed information about municipal Wi-Fi is not available in the municipal website.
- **1.** There is only a few public buildings with free Wi-Fi (e.g. libraries).
- **0.** There is no municipal free Wi-Fi.

Attracting and Retaining Talent

- **4.** Strategy for the attraction, retention, and return of talent in the city. Municipal infrastructure and stay programs to host people from outside the city and policies for the return of talent to the city.
- **3.** There is no policies for attracting talent to the city or for the return of talent to the city.
- **2.** There is no clear strategy.
- **1.** Only sporadic actions are carried out.
- **0.** No policies have been implemented for attracting and retaining talent in the city.

Annex 6. Evaluation Scales Established for the Living Indicators

Social Services

- **4.** It is possible to request an online appointment for the social services. Online procedures can be started.
- **3.** It is possible to request an online appointment for the social services. Online procedures can be started.
- **2.** It is not possible to request an online appointment for the social services (this is the lowest score for this indicator).
- **1.** Not applicable.
- **0.** Not applicable.

Universal Accessibility

- **4.** Municipal website is accessible for people with visual or hearing disabilities. In addition, some action has been carried out that involves the intensive use of ICTs to improve accessibility for people with reduced mobility or with some other type of disability.
- **3.** Municipal website is accessible for people with visual or hearing disabilities. However, no action has been carried out to improve accessibility for disabled people that involve the use of ICTs.
- **2.** Architectural barriers have been removed throughout the city.
- **1.** Architectural barriers have been removed only in some parts of the city.
- **0.** No policies have been implemented for universal accessibility in the city.

Technologies for Enjoying the City

- **4.** Incorporation of ICTs throughout the city for the enjoyment of its cultural and artistic heritage (e.g. Apps, virtual reality).
- **3.** Incorporation of ICTs in various spaces or neighbourhoods in the city for the enjoyment of its cultural and artistic heritage (e.g. Apps, virtual reality).
- **2.** Some action that incorporates ICTs in the context of specific events or exhibitions in the city have been implemented.
- **1.** Some actions have been implemented, but without incorporating ICTs.
- **0.** No policies have been implemented for a better knowledge and enjoyment of the city.

References

Ahvenniemi, H., Huovila, A., Pinto-Seppä, I., & Airaksinen, M. (2017). What are the differences between sustainable and smart cities? *Cities, 60*, 234–245.

Al Nuaimi, E., Al Neyadi, H., Mohamed, N., & Al-Jaroodi, J. (2015). Applications of big data to smart cities. *Journal of Internet Services and Applications, 6*(1), 25.

Albino, V., Berardi, U., & Dangelico, R. M. (2015). Smart cities: Definitions, dimensions, performance, and initiatives. *Journal of Urban Technology, 22*(1), 3–21.

Angelidou, M. (2015). Smart cities: A conjuncture of four forces. *Cities, 47*, 95–106.

ASCIMER. (2014). *Smart cities: Concept, challenges and projects*. https://www.eiburs-ascimer.transyt-projects.com/. Accessed May 02, 2020.

Batty, M., Axhausen, K. W., Giannotti, F., Pozdnoukhov, A., Bazzani, A., Wachowicz, M., ... & Portugali, Y. (2012). Smart cities of the future. *The European Physical Journal Special Topics, 214*(1), 481–518.

Baucells, N., Moreno, C., & Arce, R. M. (2016). Mapping smart cities situation+ CITIES: The Spanish case. In *International Conference on Smart Cities* (pp. 95–105). Springer.

Bouskela, M., Casseb, M., Bassi, S., De Luca, C., & Facchina, M. (2016). *La ruta hacia las smart cities: Migrando de una gestión tradicional a la ciudad inteligente* (Vol. 454). Inter-American Development Bank.

Caragliu, A., Del Bo, C., & Nijkamp, P. (2011). Smart cities in Europe. *Journal of Urban Technology, 18*(2), 65–82.

Chourabi, H., Nam, T., Walker, S., Gil-Garcia, J. R., Mellouli, S., Nahon, K., Pardo, T. A., & Scholl, H. J. (2012, January). Understanding smart cities: An integrative framework. In *2012 45th Hawaii International Conference on System Sciences* (pp. 2289–2297). IEEE.

European Commission. (2015). *Smart cities and communities*.

European Parliament. (2014). *Mapping smart cities in the EU*. European Parliament. Directorate General for Internal Policies. Policy Department A: Economic and Scientific Policy.

Fernández-Añez, V., Fernández-Güell, J. M., & Giffinger, R. (2018). Smart City implementation and discourses: an integrated conceptual model. The case of Vienna. *Cities, 78*, 4–16.

Giffinger, R., Fertner, C., Kramar, H., & Meijers, E. (2007). *City-ranking of European medium-sized cities* (pp. 1–12). Center of Regional Science, Vienna University of Technology.

González, J. A., & Rossi, A. (2011). *New trends for smart cities, open innovation mechanism in smart cities*. European Commission with the ICT Policy Support Programme.

Hall, R. E., Bowerman, B., Braverman, J., Taylor, J., Todosow, H., & Von Wimmersperg, U. (2000). *The vision of a smart city* (No. BNL-67902; 04042). Brookhaven National Laboratory.

Hollands, R. G. (2008). Will the real smart city please stand up? Intelligent, progressive or entrepreneurial? *City, 12*(3), 303–320.

IDC. (2012). *White pater: Big data: Trends, strategies, and SAP technology* (#236135).

IESE. (2018). *Cities in motion*. Index.

IMD World Competitiveness Center's Smart City Observatory. (2019). *Smart city index*. https://www.mobilitycity.es/informe/imd-smart-city-index-2019/. Accessed May 02, 2020.

Jucevičius, R., Patašienė, I., & Patašius, M. (2014). Digital dimension of smart city: critical analysis. *Procedia-Social and Behavioral Sciences, 156*(26), 146–150.

Klein, C., & Kaefer, G. (2008, September). From smart homes to smart cities: Opportunities and challenges from an industrial perspective. In *International Conference on Next Generation Wired/Wireless Networking* (pp. 260–260). Springer.

Lee, J. H., Hancock, M. G., & Hu, M. C. (2014). Towards an effective framework for building smart cities: Lessons from Seoul and San Francisco. *Technological Forecasting and Social Change, 89*, 80–99.

Marsal-Llacuna, M. L., & Segal, M. E. (2016). The Intelligenter Method (I) for making "smarter" city projects and plans. *Cities, 55*, 127–138.

Moreno, C. (2015). *Desarrollo de un modelo de evaluación de ciudades basado en el concepto de ciudad inteligente (Smart City)* [Tesis doctoral]. https://doi.org/10.20868/UPM.thesis.39079

Nam, T., & Pardo, T. A. (2011, June). Conceptualizing smart city with dimensions of technology, people, and institutions. In *Proceedings of the 12th annual international digital government research conference: Digital government innovation in challenging times* (pp. 282–291). ACM.

Neirotti, P., De Marco, A., Cagliano, A. C., Mangano, G., & Scorrano, F. (2014). Current trends in Smart City initiatives: Some stylised facts. *Cities, 38*, 25–36.

OECD. (2020). *Cities in the World: A New Perspective on Urbanisation*, OECD Urban Studies, OECD Publishing, Paris. https://doi.org/10.1787/d0efcbda-en

Paroutis, S., Bennett, M., & Heracleous, L. (2014). A strategic view on smart city technology: The case of IBM Smarter Cities during a recession. *Technological Forecasting and Social Change, 89*, 262–272.

Price Waterhouse Coopers. (2012). *Cities of opportunity* (p. 5).

Spanish Statistical National Institute. (2019). Available online: https://www.ine.es.

United Nations. (2015). *Transforming our world: The 2030 Agenda for sustainable development*. UN Publishing. https://sustainabledevelopment.un.org/post2015/transformingourworld/publication. Accessed May 02 2020.

United Nations, Department of Economic and Social Affairs, Population Division. (2019). *World population prospects 2019: Highlights*. ST/ESA/SER.A/423.

Zanella, A., Bui, N., Castellani, A., Vangelista, L., & Zorzi, M. (2014). Internet of things for smart cities. *IEEE Internet of Things Journal, 1*(1), 22–32.

Zheng, C., Yuan, J., Zhu, L., Zhang, Y., & Shao, Q. (2020). From digital to sustainable: A scientometric review of smart city literature between 1990 and 2019. *Journal of Cleaner Production*, 120689.

Optimization of Micro-house Building Façade in Supporting the Net Zero Energy Buildings (Nze-Bs) Program for Low-Cost Economics Communities in Bandung City, Indonesia

Asep Yudi Permana, Hafiz Nurrahman,
Aathira Farah Salsabilla Permana, and Siti Rahayu

Abstract

Inaccuracies in design that result in wasteful energy will have an impact on operational costs as long as the building is used. Energy efficiency is not only able to solve any problems that become obstacles through the utilization of climate potential. The design of the facade as an opening protector should be designed to utilize the optimal bright light of the sky. The design of the facade through the use of a roof overlap is an inhibiting factor for direct solar heat entry to save energy. The house as one of the main human needs requires careful planning that takes into account the physical and sociocultural potential. Regarding the current issue of global warming, the climate is a major consideration that needs to be resolved. The design aims to create amenities for residents, through physical comfort, both spatial comfort, thermal comfort, auditory comfort, and visual comfort. However, due to the development process and the time of its use, it can cause various negative impacts on the surrounding environment. Human nature is a source of pollutants due to respiratory activities that produce CO_2 and actions that damage the environment and do not save energy. Energy waste is caused by the design that is not integrated both related to the aspects of function and climate utilization. This study complements research on the concept of designing energy-efficient micro-houses for lower economic classes. The research method uses experimental methods. In developing the micro-house design, the SketchUp v.2019, IES-VE v.5.3.1, and CFD & Ecotect software programs are used as simulation tools. While the infrared thermometer, anemometer (hot wire), lux meter/light meter, thermo-hygrometer, and environment meter as a measurement tool. The study produced a micro-house design model that considers energy saving. The results of this study can also be a reference for policymakers, academics, and the general public.

Keywords

Energy efficiency • Climate potential • Physical comfort • Energy consumption • The green concept

1 Introduction

City as a node of community life is a process of agglomeration of human diversity and other supporting objects. As a consequence of the diversity that exists, it will bring up different needs and desires in the context of the interaction between communities. The older the age of a city, the morphology of the city will contain more layers in the form of urban tissues as a place to articulate various city phenomena throughout the history of growth and development. The urban tissues will form a configuration on the physical shape of the city (Kostoff, 1977; Setioko, 2010).

Within a matter of years, there has been a marked shift and change in the patterns and order of life in several developing countries including Indonesia through the shifting of traditional agrarian life into modern industrial life (Karyono, 2014). This condition changes the behavior (Millennium & Goals, 2007; Carmona et al., 2008; Araújo et al., 2011) and way of thinking of humans to their environment. The shift of agrarian-traditional life that is in harmony with nature and does not depend on modern technology as a form of preserving nature has shifted. In its development, the shift in life services from the traditional agrarian order to modern technology often presents several problems for people's lives. This problem develops in all sectors, one of which is the housing sector. Based on data

A. Y. Permana (✉) · H. Nurrahman
Universitas Pendidikan Indonesia, Bandung, Indonesia
e-mail: yudi.permana@upi.edu

A. F. S. Permana
Institut Teknologi Nasional, Bandung, Indonesia

S. Rahayu
Athila Consultant, PT, Bandung, Indonesia

F. Trapani et al. (eds.), *Advanced Studies in Efficient Environmental Design and City Planning*,
Advances in Science, Technology & Innovation, https://doi.org/10.1007/978-3-030-65181-7_7

from the Bandung Central Statistics Agency in the last 10 (ten) years, the city of Bandung has experienced an increase of 67% (density reaches 30,000 inhabitants/km^2). This growth rate leads to the irregular physical development of the city.

One problem in the housing sector is the deterioration in environmental quality which is increasingly worrying. Meeting human needs for habitable homes is increasingly being ignored, conditions that are uncomfortable to live in, wasteful of energy, etc. (Karyono, 2014; Miller, 2012; Sukawi, 2011). Meanwhile, the house is one of the basic human needs that must be careful in its development planning, including by considering aspects of design both physical and sociocultural potential. The physical potential is related to consideration of building materials, geological conditions, and local climate. Meanwhile, the sociocultural potential is related to the local architecture and way of life of the local community.

Bandung as one of the five big cities in Indonesia is a growing city with a dense population. Along with the development of the city of Bandung where land constraints and the high price of land become a problem in the provision of decent housing for people, especially those on low incomes, resulting in poorly designed (Permana, 2014). Based on the Center for Property Studies in Indonesia, (Prianto, 2012) shows the potential for investing in property increases rapidly (UNESCAP & UNHABITAT, 2008; Sugiri et al., 2011). The community's need for housing reaches 1.2 million homes per year (Sunarti et al., 2014). This means the high public interest in needing a house greatly influences the level of housing construction in Indonesia.

The growth of houses is very significant with the increase in electricity consumption, where fuel used to obtain electricity in Indonesia is still dominated by fuel oil and coal. Thus, if it is not successful in managing fuel, it will have an impact on the high and rising basic rates of electricity prices. And, the household sector will get the impact.

The condition of Indonesia as a country whose entire territory is located in the equatorial region is generally wet tropical climate (Batubara et al., 2002; Satwiko, 2005; Soegijanto, 1999; Widayanti et al. 2010), thus making Indonesia in a position that is beneficial but also at the same time detrimental on the other hand. This is caused by high temperatures, solar radiation, rainfall, and humidity and wind characteristics (Prianto, 2012) that are different from other regions. Efforts to control this climate problem to some extent can still be done passively (Permana & Wijaya, 2018), namely through building design that considers design factors and variables such as climate, physical properties of building materials, building orientation, forms of using plants/trees as shade from sunlight, and so on. This climate control effort is not always in line with expectations that can produce the desired thermal conditions throughout the day, because building elements and the surrounding environment have limited thermal control. There are several energy use activities in buildings (houses), including air conditioning and lighting. Soegijanto (1999) explains that the higher the intensity of activities in a building/house, the higher the use of energy (Soegijanto, 1999). There are 2 (two) identifiable activities that are very related and play a direct role in energy waste, namely in air conditioning and lighting.

The concept of building with NZE-buildings is very important because if you look at global energy use the building sector is the most dominant in energy use. The biggest energy consumption in buildings both as residential and office buildings is to meet the demand for electricity used for artificial lighting, cooling, and space heating (Mintorogo, 1999; Mustika, 2010). While the biggest energy consumption in residential buildings is used to meet the electricity demand of 67.5% of the total energy. This is in line with the results of research by Hosseini and Cole (2013) which states that the building sector absorbs 40% of overall energy needs (Hosseini & Cole, 2013). The use of energy in buildings, especially for heating, cooling, and lighting, results in very high energy consumption.

Based on this background, the purpose of this study is (1) to find out which façade design is suitable for the climatic conditions of the city of Bandung; (2) to determine the thermal comfort conditions of the room in the micro-house design; (3) to find out the heat that comes into the room, especially during the day; (4) to find out the right finishing materials in the completion of the micro-house façade; and (5) to find out the concepts that remain in reducing the heat of sunlight directly into the building.

2 Literature Review

2.1 The Concept of "Efficiency-House Plus" as a Concept of Energy Efficiency and Renewable Energy

Based on the results of Fraunhofer-Gesellschaft research in 1995–1998 in Freiburg Germany, the use of photovoltaic roof systems (PV-systems) gives good results in seasonal energy storage, but when viewed from investment, costs are still very high (Erhorn, 2013; Hoffman, 1994; Invernizzi, 1998; Kaysing, 1995; Krishan, 2001). The results of this study were developed by Rolf Disch in Freiburg Germany in 2000–2006, by taking cases in 50 (fifty) houses lined up in ten rows of different sizes as shown in Fig. 1.

The results of his research resulted in a system of using public electricity networks as "seasonal energy storage" and

Fig. 1 **a** Pilot project for building photovoltaic systems (PV-systems). **b** The "Freiburg surplus-energy housing estate" building. *Source* Erhorn (2013) and Permana

can save development costs. Recommendations from this research are the energy performance of a building can be improved in several ways, for example, by choosing the appropriate building design (through the compact structure, optimal orientation), increasing thermal insulation (high-performance windows and insulation systems to protect the building envelope), optimized construction (no thermal bridging, airtight construction, and connections), and by ensuring energy-conscious behavior. The results of this study were implemented by the German government through the "the efficiency-house plus" program in 2011.

One important aspect of architectural design today that is increasingly felt important is the arrangement of energy in buildings. The crisis of non-renewable energy sources encourages architects to care more about renewable energy in designing energy-efficient buildings. User comfort will largely increase due to the action of reducing consumption through increased thermal comfort in the room (Invernizzi, 1998; Roaf et al., 2010; UNEP SBCI, 2009; Thiers & Peuportier, 2012; Permana et al., 2020; Miller, 2012). This program can be seen more clearly in Fig. 2.

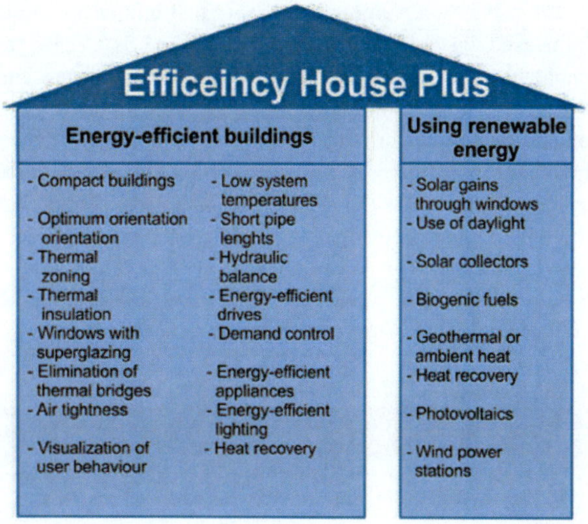

Fig. 2 Basic concept of "the efficiency-house plus." *Source* Erhorn (2013) and Permana

Based on the picture, there are at least 3 (three) main factors that greatly affect the energy savings in buildings, namely the design of the building envelope, energy management, and user awareness. The accumulation of waste in the use of energy in buildings ranges from 15–30% of the use of electricity nationally.

The design of the housing envelope cannot be separated from the consideration of climatic conditions and the surrounding environment. The shape of the imagery in buildings is an effort to anticipate tropical climate to achieve comfortable thermal conditions in the building. Completion of the design of the facade must be made unequal between those facing west, east, south, or north. Because in principle, the row of houses has different problems when viewed from the aspect of the crossing of the sun. The embodiment of architectural design for environmentally sound buildings is often referred to as green building. This is closely related to the concept of sustainable architecture (sustainable architecture) and energy-saving. The role of architects is very important in saving energy.

2.2 Between the Concept of Eco-architecture and NZE-Buildings

The eco-architecture paradigm is a building paradigm that leads to the use of the environment as a basis for consideration of the various architectural design streams. Frick and Mulyani (2006) stated that the development of ecological architecture can be traced from various approaches that began in 1920–1960 by prioritizing freedom of expression in form and function. This development continues with the birth of independent architecture, alternative architecture, and experimental architecture (Frick & Mulyani, 2006; Yuliani, 2013). The development of architecture also experienced changes when the energy crisis began to be felt so that energy-saving architecture (NZE-buildings) was born.

The complexity of the relationship between architecture and the environment is ecologically based on the input and output categories in forming a closed balance (Permana et al., 2020; Wai, 1994), diagrammatically can be seen in Fig. 3.

From Fig. 3, a building is an entity that influences and is influenced by the environment. A balance is needed to maintain a sustainable relationship so that architecture and the environment do not harm each other but improve each other's quality by forming symbiotic relationships. The act building can be interpreted as a response of its inhabitants to achieve a balance in the system of human interaction with the environment in maintaining a homeostatic state (Rachmat, 2000).

The NZE-Building program began in the 1980s since the "efficiency-house plus" program was launched by the German Federal Ministry or transport, building, and urban affairs (BMVBS). This program is the result of research through collaborative partnerships with industry in producing innovative energy-efficient building industries. The Berlin building is the first sample house in the energy efficiency development program as shown in Fig. 4.

The energy efficiency development program for residential homes undertaken by BMVBS is realized in various ways through the use of a combination of energy-efficient construction technologies with renewable energy systems. BMVBS research timeline can be seen in Fig. 5 (Kolokotsa et al. 2010; Permana et al., 2020).

Based on the curve of Fig. 5, it can be explained that the lower curve shows the target of the research project. Energy efficiency development requires 10–15 years between the application of the pilot and the establishment of renewable energy sources. This energy efficiency allows many alternative ways and does not depend on certain technologies.

2.3 Building Facade

The building facade is the most important element that cannot be removed from an architectural design product because it is this visible element that is appreciated or seen first. Through the facade, you can get an overview of the characteristics and functions of the building, besides that the facade also functions as a means of recording the history of civilization. By observing and studying, the facade design can determine the socio-cultural conditions, spiritual life, even economic and political conditions at a certain time.

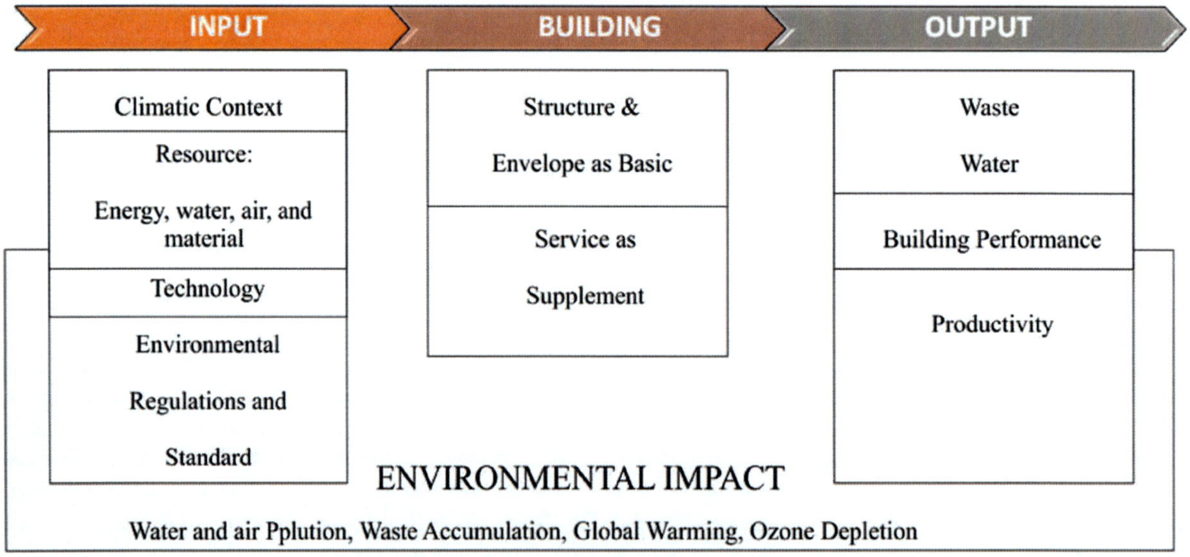

Fig. 3 Complexity of the relationship between buildings and the environment. *Source*: Wai (1994)

Fig. 4 The Berlin building. *Source* Erhorn (2013)

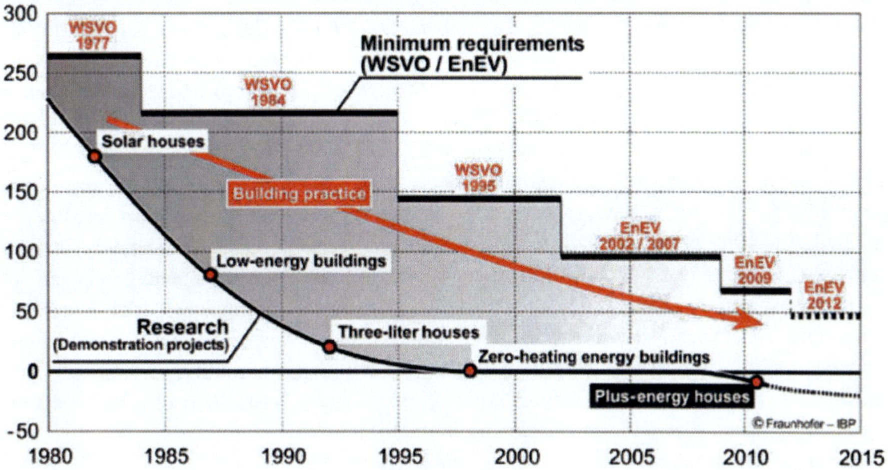

Fig. 5 BMVBS research timeline developed in 1980. *Source* Erhorn (2013)

The facade as a field of the front of the building can determine the style and characteristics of architecture. This implies that the character or characteristics of a building can be seen in how architects process the building's facade. The building facade is a composition that takes into account the functionality of windows, doors, sun visors, and roof areas (Invernizzi, 1998; Kier, 1979). The facade design is a harmonious and proportional unity of both horizontal and vertical structures, building materials, colors, and decorative elements (Kier, 1979).

The important parts of the facade are a. the door, as one of the perforations of the wall that must not be removed in the composition of the facade of the house and the shape of the door, can symbolize the character of its inhabitants; b. windows, functioning air circulation holes, entry of sunlight, and areas get a view out of the room; c. the wall, as a room divider; d. canopy is an extension of the roof area that protrudes above the wall and serves as a protection from sunlight and rainwater; e. sun-shading as a window protector for functions such as canopies.

2.3.1 Daylight Level (DiL)

In the initial design of a building, the direction of falling sunlight greatly affects the performance of the building. The potential utilization of daylight level (DiL) functions for visual comfort and energy conservation. Visual comfort can be obtained through optimizing the utilization of natural light and the design of the right aperture so that the natural light that is obtained matches the needs of visual work. While energy conservation can be achieved through the use of natural light by reducing up to 20% of the total electricity consumption needs (Latifah, 2015; Thapa & Panda, 2015).

Daylight level (DiL) has 3 (three) main obstacles, namely: (a) In the dark (cloudy/rainy), artificial lighting must still be used; (b) light sources (sky) vary in the level of illumination (when the sky is supporting the dark/large window size, it is quite optimal to receive natural light, but in a bright sky, the light becomes excessive); and (c). If natural light is received from the side of the room, close lighting levels with windows usually much higher than required, constraints b and c are conditions that will inevitably result in over-illumination (Krishan, 2001). Utilization of daylight level (DiL) can be optimized by taking into account the factors that support the basic strategy in planning light openings in buildings, including (a) building orientation and light openings; (b) space allocation and light openings; and (c) the broad aperture of light.

2.3.2 Building and Orientation

Indonesia as one of the countries located in the area around the equator, in general, window placement, must pay attention to the sun's path, north and south sides are potential places for window opening (openings), to get daylight. While the east and west positions at certain hours required protection against direct solar radiation. For buildings that are exposed to direct sunlight, especially, the openings can be processed using barriers such as utilizing trees or architectural elements such as a retaining grid or by processing space that aims to reduce the impact of direct sunlight so the room will still feel comfortable. Things that need to be considered in the design are the time where the sun's rays will emit the highest radiation around 11 a.m. noon until 3 p.m.

In tropical areas with the intensity of the sun throughout the year, the orientation of the building has a very significant effect on the lighting inside the building, ideally avoiding direct sun direction as far west and east as possible.

3 Research Methods

3.1 Method

The research approach uses the method of the experimental method (Creswell, 2007, 2010; Moleong, 2008), in which research begins with the design of a micro-house model and is followed up with the development of modeling of tropical houses. The output of this study is the optimization of the micro-house facade design for the lower economic community in the city of Bandung to contribute to energy savings. The tools used consist of 2 (two) tools, namely (1) simulation tools in the form of software SketchUp v.2019, IES-VE v.5.3.1, CFD, and Ecotech to measure the level of natural lighting and lighting on the micro-house unit model; and (2) measuring instruments in this study using an infrared thermometer, anemometer (hot wire), lux meter/light meter, thermo-hydrometer, and environment meter. The procedural simulations performed include (1) making 3D models with SketchUp v.2019 software; (2) import image files to IES-VE v.5.3.1, CFD and Ecotect software in Dxf format; (3) managing Bandung City weather and climate data with default settings to get the closest simulation of the truth; and (4) choosing time and date and sky conditions (by dividing 2 conditions, namely at the peak of the annual tropical climate, usually between June and December). As for the measuring instrument used to measure thermal comfort/room temperature, the lighting on the micro-house model is made. The object of observation in the form of modeling the micro-house with orientation direction can be rotated in 3600. The building facade is positioned in the direction the sun comes in all day (Hinrich, 2005).

The comfort measurement formulation uses 2 (two) main parameters namely human aspects (subjective factors) and environmental aspects (objective factors). From these parameters, it is translated into six indicators, namely wind speed, the average temperature of the room, air temperature, humidity, activity level, and clothing (Fanger, 1972). Equation of comfort according to Fanger (1972):

$$f(M, lcl, ta, tr, v, pa) = 0 \tag{1}$$

and, equality of body heat balance and the environment:

$$M - W = H + Ec + Cres + Eres \tag{2}$$

This equation is used to calculate the comfort based on 6 (six) parameters by calculating the balance of body heat and the environment.

3.2 Research Site

The research was done in the area of "*Taman Hewan*" (lit. Animal Park) included under the area of Balubur-Tamansari, Bandung City, Indonesia, which has uniqueness and peculiarities, both from the aspects of physical buildings and the area itself. The research location can be more clearly observed in Fig. 6.

Existing data taken from Bandung City (Permana et al., 2018) Metoblue, in August 2018, had the highest average temperature of 27 °C and the lowest in July which was 18 °C. The temperature contained in this meteoblue data is then used as initial data for the basis of designing and calculating thermal comfort.

3.2.1 Desain Micro-house

Bandung's higher population growth rate has led to the irregular physical development of the city with a density of 30,000 people/km^2. Along with that, limited land and high land prices are a problem in providing adequate housing for the community, especially for those with low incomes. Most of their houses are designed and built independently without the help of architects, so that the fulfillment of their residential needs is not designed properly (Permana, 2014). Their houses tend to build without considering the aspects of comfort, health, or structure. On average, their houses are not habitable.

The design of a micro-house (Armitage, 2017; Hildner, 2013; Mittal & Kashyap, 2015; Roaf et al., 2010; Schmeckpeper & Patterson, 2016; Scognamiglio et al., 2014) as a program in meeting the need for housing for low economic groups is one model that can be developed. In this study, it was discussed how to optimize the design of the micro-house building facade in support of the NZE-built program (Millennium & Goals, 2007; Thiers & Peuportier, 2012; Hosseini & Cole, 2013) through the utilization of both lighting and air temperature. Modeling is made by taking two types namely, type 6 m^2 and type 36 m^2. The design of 2 (two) types of micro-houses are based on the results of interviews with lower-class people as a preliminary study of meeting the needs of homes in the Balubur-Tamansari area of Bandung (Permana, 2014). The designs of these two types are shown in Fig. 7.

4 Results and Discussion

4.1 Facades and Buildings

The micro-house facade design concept responds to the influence of solar and wind radiation on the thermal comfort conditions of buildings. Wind conditions in the city of Bandung in the south-southeast have a speed of 3 m/s so that the building that leads directly to the south and southeast is resolved by adding a canopy or double-skin facade. Natural ventilation is preferred over artificial ventilation through drainage around buildings. Based on the optimization simulation of the facade of the micro house thermal conditions based on 3 (three) times (08.00–11.00; 11.00–14.00; and 14.00–17.00) with consideration of minimum and maximum lighting conditions. Then the conditions will be produced as shown in Fig. 8.

The simulation is then done by unit modeling, which produces a daylight level at each measured point so that the daylight contour is visible. Material variables that are set for input are in Table 1.

In radiance, the CIE model overcast sky was chosen to represent the Indonesian climate. Based on the simulation results (Fig. 8), it is obtained the value of natural lighting for residential spaces, minimum and maximum lighting levels, and areas above the 145 lx threshold. Meanwhile, the threshold for residential houses is set at 120 lx. And the results of the analysis of natural lighting levels, the results of the analysis show that the function of the micro house requires shade or reduction of light openings because the natural lighting that enters during the day exceeds the standard of occupancy. A more complete simulation result can be seen in Fig. 8.

Based on the simulation results using three different time conditions that have been proven to affect the results of excessive illumination of sunlight, which enters the room, it can be reduced so that it approaches the standard of daylighting for residential homes.

Whereas daylighting at night is used with two strategies that use artificial light and natural bright light using the bright moonlight that enters through a water-filled bottle system. More details can be seen in Fig. 9.

Simulation results that refer to the images for both modeling proved that the light coming from moonlight into a bottle filled with water provides illumination of bright light without using artificial light. This condition applies when the weather is not cloudy.

4.2 Material and Paint Selection

Types of facade cladding include: (1) choosing a wall covering material without plastering; (2) only the walls are plastered; (3) painted walls; and (4) walls covered with natural stone. The results of Prianto's (2012) research show that the heat levels of the exterior and interior air are not much different / relatively the same (the difference is approximately 3 °C). This means that when the outside air is hot, the indoor air is also hot, and vice versa. Prianto (2012) examined the effect of the type of wallcovering on the heat level of exterior and interior air. Based on the results of the

Fig. 6 **a** Map of the areas of the city of Bandung; **b** satellite map of the Balubur-Tamansari area; and **c** map of the Balubur-Tamansari area in Bandung City. *Source* Modification of survey results, 2013–2018

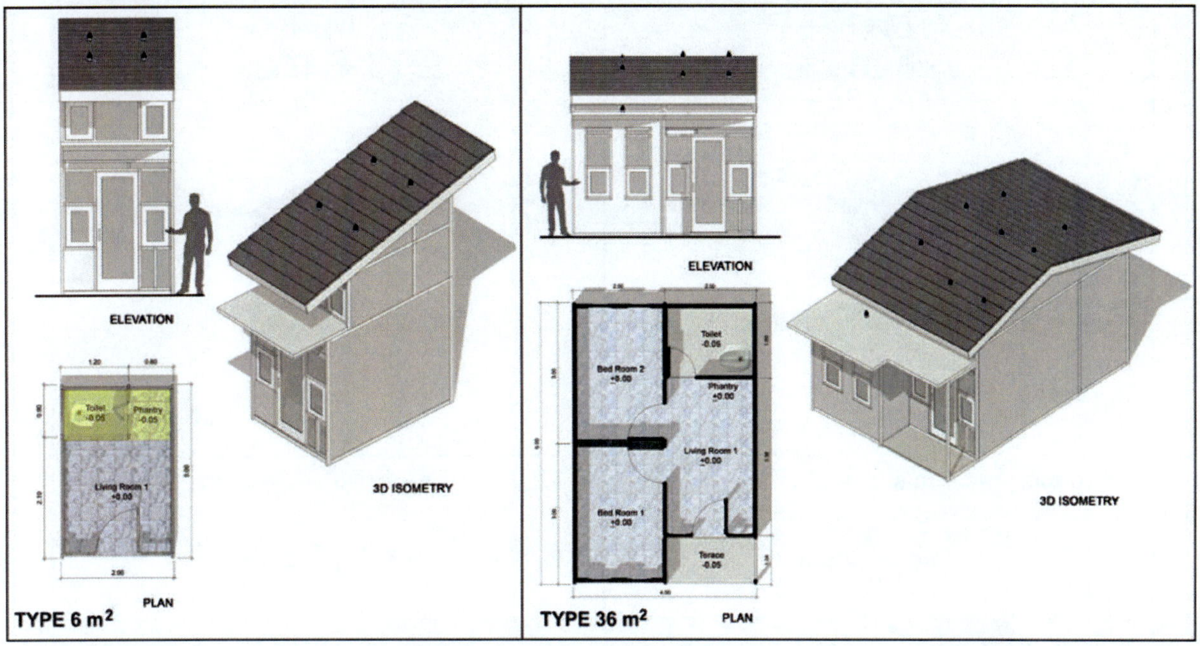

Fig. 7 Types of micro-house. *Source* Authors' document, 2018

Table 1 Material variables in the simulation unit model

Construction	Specification	Material
Wall	Reflection factor 70%	Brick finishing paint
Glass	Transmission factor 100%	Clear glass
Floor	Reflection factor 0.4	Ceramic
Ceiling	Reflection factor 0.1	Concrete floor finishing paint

Source Authors' document, 2018

comparative modeling of the difference in heat load between the finishing wall of the natural stone layer and the only plastered wall, the surface heat of the exterior walls experiences a peak that occurs when the walls are plastered.

Wall coating with wall paint, the results are seen that there is a clear difference between conditions on the surface of the outer and inner walls. During the day, the indoor temperature conditions decrease around 3 °C compared to the average temperature of the outside air. The surface temperature of the outer wall on the painted wall will still be higher than the average temperature of the outside air, but lower than the wall without paint. And the surface temperature of the interior walls also decreased compared to the surface temperature of the exterior, which is cooler 3 °C. Based on observations of the use of paint colors in the finishing of the micro-house facade, it shows that the use of blue will reduce the heat accumulation of the room up to 60% when compared to the color red. This is following the theory of color psychology proposed by Birren (1998).

While based on the modeling of the differences in heat load profiles for the four types: from the hottest conditions, the gradual opening of the outer wall is 42.25% hotter in the walls in the form of filtering, 28.75% in natural stone-coated walls, 23.75% in paint-coated walls, and 12% ceramic-coated walls. There is an interesting phenomenon where the results of the analysis show that on natural stone usage, the heat profile experiences a temperature rise again afternoon between 12.00 and 16.00. Whereas when observing the heat profile of the outside air in general, it should post daylight, and the heat load decreases.

4.3 IES-VE V.5.3.1, CFD, and Ecotect Thermal Comfort Tool

IES-VE v.5.3.1, CFD, and Ecotect thermal comfort tool is used to determine comfort, with the results of calculations that can be seen in the thermal comfort Table 2. This

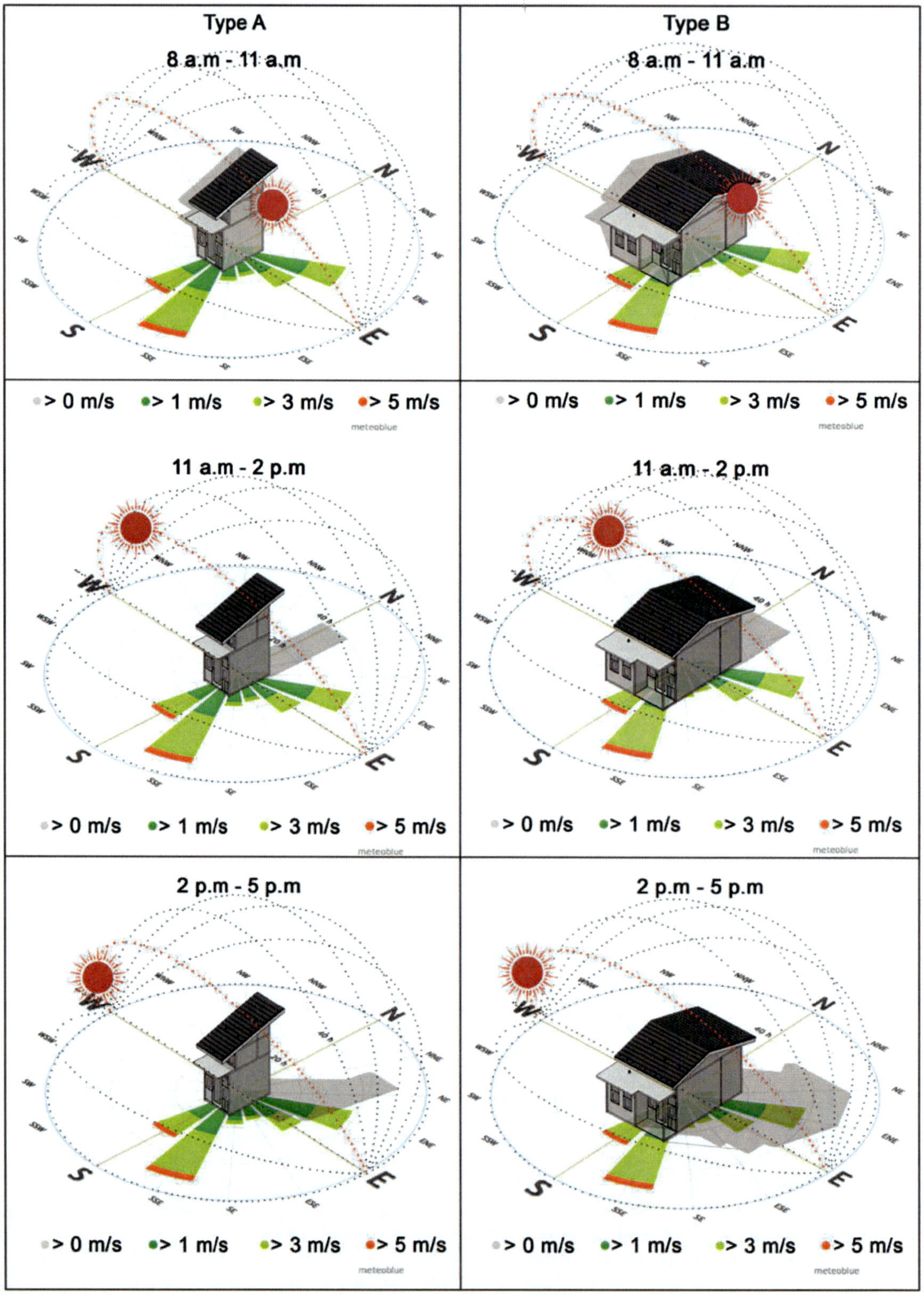

Fig. 8 Simulation on two types of micro-houses at three times. *Source* Authors' document, 2018

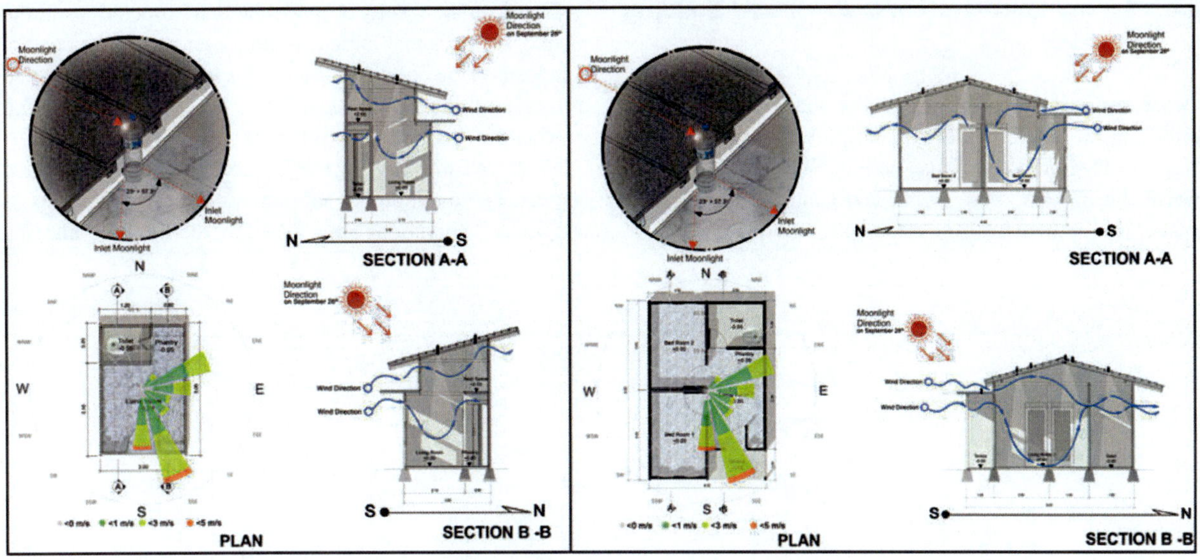

Fig. 9 Lighting simulation at night. *Source* Authors' document, 2018

Table 2 IES-VE v.5.3.1, CFD, and Ecotect thermal comfort tool

Air type	A	B
Operative temperature (°C)	29	29
Air speed (m/s)	1	3
Humidity (%)	79	79
Metabolic rate (met)	1	1
Clothing level (clo)	0.61	0.61
Result	t_m = 19.5°C rh = 4,0% W_n = 0.2 g_w/kg_{da} t_{ab} = 5.7° C t_w = -29.9° C h = 0.4 kJ/kg 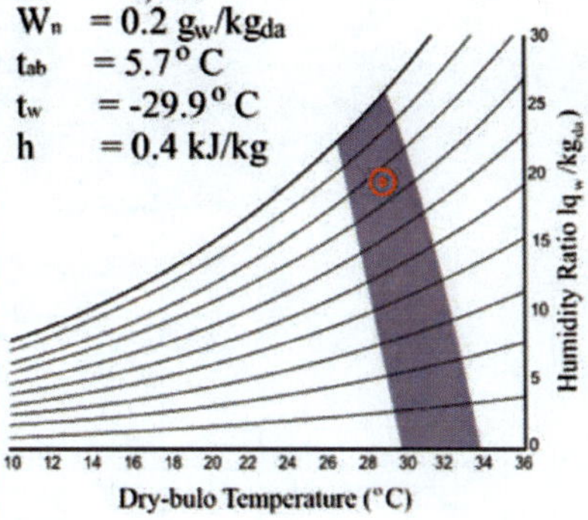 	t_m = 25.0°C rh = 4,8% W_n = 0.9 g_w/kg_{da} t_{ab} = 5.7° C t_w = -15.8° C h = 2.4 kJ/kg 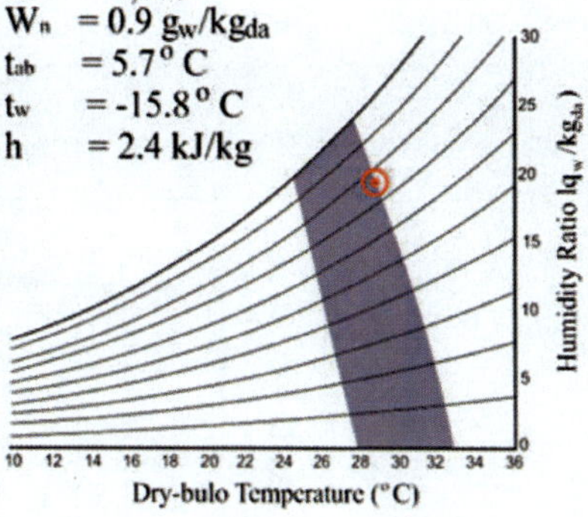

Source Authors' document, 2018

condition is simulated through two conditions, namely with 1 m / s wind and wind 3 m/s.

Based on the simulation results in Table 2 with wind conditions of 1 m/s and 3 m/s CBE Thermal Comfort Toolroom temperature in the micro-house ranges from 19.5–25 °C, the level of humidity in the space between 4 and 4.8%, and indoor water content of 0.2 ga/kg da–0.9 ga/kg da.

4.4 Energy Consumption in Processed Building Facades

Observation results for both humidity and temperature (heat profile in the micro-house modeling are grouped into four cases in the direction of the building facade, namely the facade facing east, west, north, and south. The simulation results can be seen in Figs. 10 and 11.

Based on the results of modeling and observations in the picture, it shows that humidity and air temperature affect the heat load on the building facade. This will affect comfort and energy consumption. The level of air heat on the exterior and interior walls is not much different.

Processed finishing of the facade of the building affects the surface conditions of the outer and inner walls. Room temperature conditions are affected by the finishing material

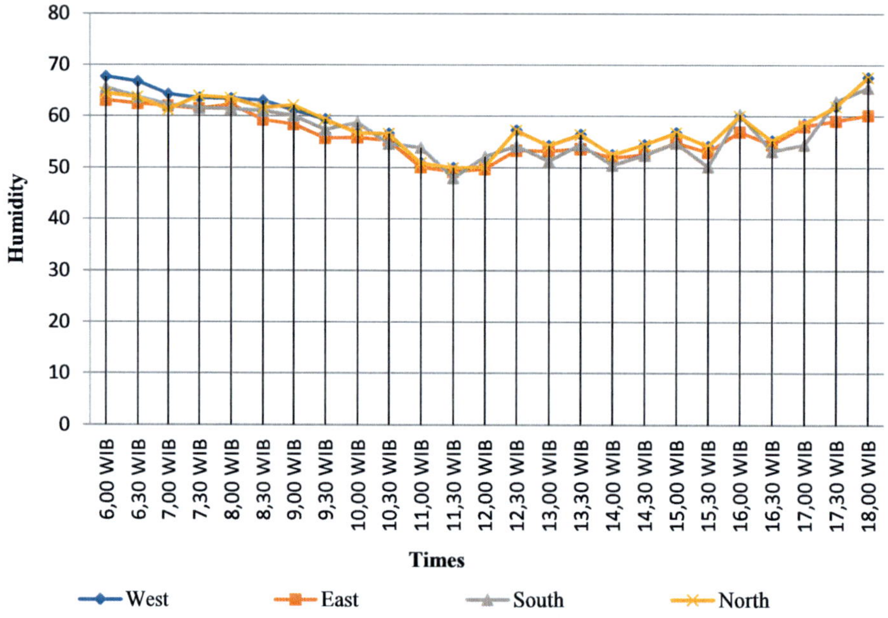

Fig. 10 Graphic observation results of humidity in the micro-house facade with a position appear east, west, south, and north. *Source* Authors' document, 2018

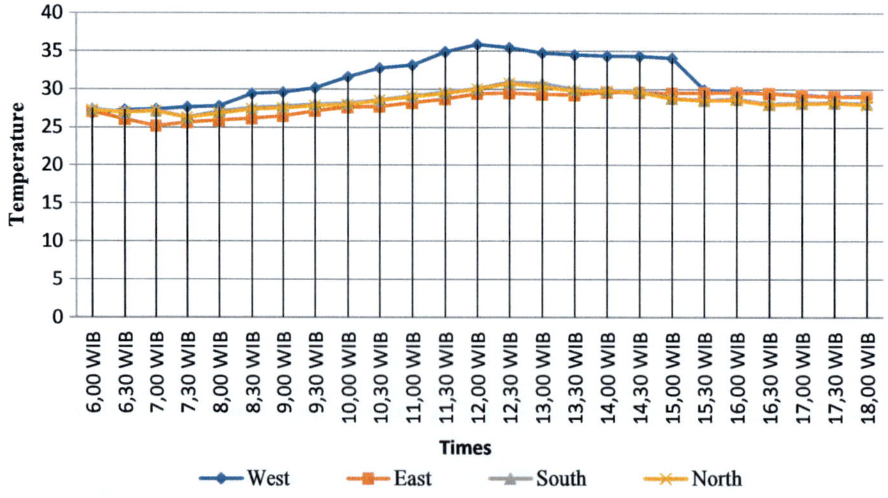

Fig. 11 Graphic observation results of temperature/heat profile on the micro-house facade with the position look east, west, south, and north. *Source* Authors' document, 2018

of the building facade. The facade of the building with finished paint has decreased to 2 °C compared to the average temperature outside.

The geographical position of the city of Bandung, which is in a tropical climate, results in direct rays of sunlight entering the room which will bring direct heat at the same heat as the heat outside the building. So it needs to be avoided direct heat sunlight. The use of canopy is one solution in overcoming direct sunlight. The existence of the canopy does not make the conditions of the atmosphere of the room dark, because the bright light of nature can still enter the building.

5 Conclusions

The optimal facade design for thermal comfort is designed taking into account the condition of the city of Bandung in a tropical climate, namely through the design of the facade by using a canopy or double skin as well as a building envelope. The tools used to measure thermal comfort use the Sun-Hour simulation and data consideration from meteoblue, as well as measurements using IES-VE v.5.3.1, CFD, and the Ecotect Thermal Comfort Tool with indoor thermal comfort results averaging 19 °C–24 °C.

The heart condition in a micro-house room is caused by the transfer of solar heat that is channeled through the design of the walls of the room, both through the openings and insulating elements. The amount of heat that enters the room during the day can reach 800–1000 $W/m^2/h$. Facade wall finishing is very significant in reducing heat load. For cold areas, it is better to use non-finished walls (exposed bricks), and vice versa for hot areas that is it is better to use plastered walls with paint finishing.

Reducing the heat of sunlight directly to the walls of buildings can be reduced by using canopies and sun shading in the shape adapted to the function and whether or not much sun is exposed to the wall. Energy efficiency is related to optimizing building facades depending on how big the role of facade wall design is in efforts to reduce the heat load of solar rays. The micro-house design model is expected to be a reference for local governments, especially the city of Bandung as policymakers, academics, and the general people. This design model needs to be developed related to the use of material, climate, and location and considers the characteristics of its users.

Acknowledgements Appreciations are addressed to (1) Kemenristek RI which has provided fund support through fundamental research scheme in 2018–2020; (2) the Head of the University of Education of Indonesia who has supported and granted the permit for using of interior design laboratory facilities during the research; and (3) the people in the education area Balubur-Tamansari, Bandung City, Indonesia.

References

Araújo, M. B., Alagador, D., Cabeza, M., Nogués-Bravo, D., & Thuiller, W. (2011). Climate change threatens European conservation areas. *Ecology Letters, 14*(5), 484–492. https://doi.org/10.1111/j.1461-0248.2011.01610.x

Armitage, L. (2017). *Micro houses: Trends and implications on the gold coast*. October 2017. Retrieved from https://www.researchgate.net/publication/320245592_Micro_Houses_Trends_and_Implications_on_the_Gold_Coast

Batubara, M., Tanimura, H., Asikhia, M. O., & Toshimori, A. (2002). *An application of the AHP to urban residential upgrading in Jakarta* (pp. 253–259).

Birren, F. (1998). Colour psychology and colour therapy. New York: University Books Inc.

Carmona, M., Heath, T., Oc, T., & Tiesdell, S. (2008). *Public spaces: The management dimension*. Taylor and Francis Group.

Creswell, J. W. (2007). *Research design qualitative, quantitative, and mixed methods approaches* (2nd ed.). Sage Publication Inc.

Creswell, J. W. (2010). *Research design: Pendekatan Kualitatif, Kuantitatif, dan Mired*. Pustaka Pelajar.

Erhorn, H. (2013). *The age of positive energy building has come*. Fraunhofer Institute of Building Physics.

Fanger, P. (1972). *Thermal confort*. McGraw-Hill.

Frick, H., & Mulyani, T. H. (2006). *Arsitektur Ekologis, Konsep arsitektur ekologis di iklim tropis, penghijauan kota dan kota ekologis, serta energi terbarukan*. Kanisius dan Soegijapranata University Press.

Hildner, C. (2013). Small houses. *Small Houses*. https://doi.org/10.1515/9783034610506

Hinrich, R. K. M. (2005). *Energy-its used and the environment*. (4th ed.). Thomson Book Cole.

Hoffman, J. (1994). *Ambiences climatises et confort thermique*. Sans lieu: Actes du GSTIC.

Hosseini, Z. S., & Cole, R. J. (2013). *Lessons from net positive energy to be applied in net positive material flows* (pp. 24–34).

Invernizzi, T. (1998). *Maion Tropical d'Asie*. Benedickt Taschen Verlag GmbH.

Karyono, T. H. (2014). *Green architecture: Pengantar Pemahaman Arsitektur Hijau di Indonesia* (edisi ke-2). PT Raja Grafindo Persada.

Kaysing, B. (1995). The microhouse: A small home you can build. *Earth*, 1–24.

Kier, R. (1979). *Urban space*. Academy Edition.

Kolokotsa, D., Rovas, D., Kosmatopoulos, E., & Kalaitzakis, K. (2010). A roadmap towards intelligent net zero and positive enerby building. *Journal Solar Energy*. Retrieved from www.elsevier.com/locate/solener

Kostoff, S. (1977). *The city assembled: The elements of urban form through history*. Thames and Hudson.

Krishan, A. (2001). *Climate responsive architecture*. Mc Graw Hill.

Latifah, N. L. (2015). *Fisika Bangunan 1*. Griya Kreasi.

Millennium, T., & Goals, D. (2007). *Cities 2006/7*.

Miller, W. (2012). Anatomy of a sub-tropical positive energy home (PEH). *Journal Building and Environment, 56*, 57–68.

Mintorogo, D. S. (1999). Strategi "Daylighting" Pada Bangunan Multi-Lantai Diatas Dan Dibawah Permukaan Tanah. *Dimensi Teknik Arsitektur, 27*(1), 64–71.

Mittal, J., & Kashyap, A. (2015). Real estate market led land development strategies for regional economic corridors—A tale of two mega projects. *Habitat International, 47*, 205–217. https://doi.org/10.1016/j.habitatint.2015.01.026

Moleong, L. J. (2008). *Qualitative research methodology*. Remaja Rosdakarya Offset.

Mustika, N. W. M. (2010). *Optimasi pencahayaan alami untuk efisiensi energi pada rumah susun dengan konfigurasi Tower Di Denpasar Studi Kasus: Rumah Susun Dinas Kepolisian Daerah Bali*. Institut Teknologi Sepuluh Nopember Surabaya.

Permana, A. Y. (2014). *Transformasi Gubahan Ruang: Pondokan Mahasiswa di Kawasan Balubur Tamansari Kota Bandung* [Program Doktor Arsitektur dan Perkotaan Fakultas Teknik Universitas Diponegoro]. Retrieved from https://eprints.undip.ac.id/62084/

Permana, A. Y., Susanti, I., & Wijaya, K. (2018). The transformation of Gegerkalong Girang area, Bandung City: amid educative and religious areas. *IOP Conference Series: Earth and Environmental Science, 213*(1), 012022. https://doi.org/10.1088/1755-1315/213/1/012022

Permana, A. Y., & Wijaya, K. (2018). Spatial change transformation of educational areas in Bandung. *IOP Conference Series: Earth and Environmental Science*, *99*(1). https://doi.org/10.1088/1755-1315/99/1/012029

Permana, A. Y., Wijaya, K., Nurrahman, H., & Permana, A. F. S. (2020). *Pengembangan desain micro house dalam program net zero eenergy buildings (NZE-Bs)* (pp. 73–81). Retrieved from https://jurnal.universitaskebangsaan.ac.id/index.php/arcade/article/view/424/279

Prianto, E. (2012). Strategi Disain Fasad Rumah Tinggal Hemat Energi. *Jurnal Riptek, Bappeda Kota Semarang, 6*(1), 54–64.

Rachmat, W. B. (2000). *Perkembangan Kota dan Beberapa Permasalahannya. Sebuah Bacaan Pelengkap untuk Sosiologi Masyarakat Kota*. Fakultas Sosial dan Politik UGM.

Roaf, S., Hyde, R., Campbell, C., & Seigert, M. (2010). Transforming markets in the built environment and adapting to climate change: An introduction. *Architectural Science Review, 53*(1), 3–11. https://doi.org/10.3763/asre.2009.0104

Satwiko, P. (2005). *Arsitektur Sadar Energi*. Andi Press.

Schmeckpeper, E., & Patterson, J. (2016). *Using a micro-house as a starting point to create an affordable house*. https://doi.org/10.18260/p.27122

Scognamiglio, A., Garde, F., & Røstvik, H. N. (2014). How net zero energy buildings and cities might look like? New challenges for passive design and renewables design. *Energy Procedia*. https://doi.org/10.1016/j.egypro.2014.11.1044

Setioko, B. (2010). *Integrasi Ruang Perkotaan di Kelurahan Meteseh Kawasan Pinggiran Kota Semarang*. Universitas Diponegoro.

Soegijanto. (1999). Pengaruh Selubung Bangunan terhadap Penggunaan Energi dalam Bangunan. *Seminar Arsitektur Hemat Energi*. Universitas Kristen Petra.

Sugiri, A., Buchori, I., & Soetomo, S. (2011). Sustainable metropolitan development: Towards an operational model for Semarang metropolitan region. *The International Journal of Environmental, Cultural. Economic and Sosial Sustainability*, *7*(5).

Sukawi. (2011). Penerapan Konsep Sadar Energi dalam Perancangan Arsitektur yang Berkelanjutan. In Fakultas Teknik Universitas Sriwijaya (Ed.), *Prosiding seminar nasional a VoER ke-3 Palembang* (pp. 136–143). Fakultas Teknik Universitas Sriwijaya.

Sunarti, Syahbana, J. A., & Manaf, A. (2014). Slum upgrading without displacement at Danukusuman Sub-District Surakarta City. *International Transaction Journal of Engineering, Management, & Applied Sciences & Technologies*, *5*(3), 213–225. Retrieved from https://tuengr.com/V05/0213M.pdf

Thapa, S., & Panda, G. K. (2015). Energy conservation in buildings—A review. *International Journal of Energy Engineering*, *5*, 95–112. https://doi.org/10.5963/IJEE0504001

Thiers, S., & Peuportier, B. (2012). Energy and environmental assessment of two high energy performance residential buildings. *Building and Environment*, *51*, 276–284. https://doi.org/10.1016/j.buildenv.2011.11.018

UNEP SBCI. (2009). Building and climate change: Summary for decision makers. In *UNEP DTIE sustainable consumption and production branch*. Paris, France.

UNESCAP, & UNHABITAT. (2008). *Perumahan bagi Kamu Miskin di Kota-kota Asia. Urbanisasi: Peran Kaum Miskin di dalam perkembangan Kota*. Bangkok Thailand: United Nations Economic and Social Commission for Asia and the Pacific (UNESCAP).

Wai, T. K. (1994). *Sustainable symbiotic systemic phenomenon*. Van Nostrand Reinhold Company.

Widayanti, R., Suparman, A., & Sekarsari, N. (2010). Kajian Aspek Pemakaian Energi Pada Sistem Bangunan Tradisional Jawa. *Seminar Teknik Arsitektur Universitas Gunadarma, 7*(06), 1–4.

Yuliani, S. (2013). *Paradigma Ekologi Arsitektur sebagai Metode Perancangan dalam Pembangunan Berkelanjutan di Indonesia*. Retrieved from https://eprints.uns.ac.id/id/eprint/12832

Commercial Development and GIS in Neighborhood Context: Case of Tirana, Albania

Andi Shameti

Abstract

This study investigated how neighborhood accessibility is reflected in the site-location by examining retail markets and shops viability of urban environmental factors supporting walking activities. Attention has been paid to urban configuration and how it supports retail activity and their site-location viability in the neighborhood. Accessibility and pedestrian traffic are the most important factors that are considered when selecting a site area. However, the hypothesis of the retail viability is linked to neighborhood's accessibility and defines a statistical relationship. The proposed methodology is demonstrated using a case study based on city of Tirana. We analyzed the socio-economic data in block area and the spatial data, supported by geographic information systems (GIS), which highlights the physical parameters of the retail site-locations in urban environment and also the employment and business activity. The analysis of the site-location was led by the use of GIS to explore where may be the next retail store could be located. As well as comparative analysis between GIS and the gravity model was done and techniques of the analysis are combined to find the most potential areas for market use. The results suggest the potential locations in urban and suburban areas for the better viability of retail to the pedestrian-friendly orientation and proximity to housing. The paper concludes in the evaluation of the neighborhood's configuration determining the business viability by various parameters and a better understanding about accessibility for new location and the areas with the next store potential stores and market enlargement.

Keywords

Commercial • Viability • Location • Density • GIS

A. Shameti (✉)
Faculty of Architecture and Urbanism, Polytechnic University of Tirana, Tirana, Albania

1 Introduction

Retails in the process of distribution to the residents are the last in the sales line, to the residents, like stores, supermarkets, hypermarkets, but they are often specialized in one direction, especially when it comes to food products. This typology is part of the daily life of every person, but the sensitivity to their quality and this philosophy has increased in recent years, as elsewhere in the world, to residents of the city of Tirana. The essential factors are the development and growth of demand, that is, the lifestyle, as a consequence of the economic and social aspect, but also of the environment in which they live. Residents of Tirana have had a strong connection with the fresh and natural product, grown in more traditional forms or otherwise called organic. On the other hand, economic and urban development has strengthened their role as neighborhoods in both aspects, of commercial real estate and the impact on lifestyle for the citizens of Tirana. In recent years, there has been an increase in the demand of many new entrepreneurs to develop businesses specialized in organic and natural products, by converting informal brands or sales into products and creating brands and sale networks.

The population directly affects the increase in demand for more services; on the other hand, the densification of areas leads to an increase in units and structures where they operate which often if not in the right ratio with the need will lead to unfair relations and for shake creates not very efficient environments. In an indirect form, it creates a territory that is too built and inefficient, as a result of which it will reduce the space and the open environment. See Fig. 1.

A GIS analysis will be done to determine the potentials for positioning these points based on demographic data and characteristics of commercial needs. Secondly, they will be evaluated with the environment constructed according to the respective areas and will determine the most important areas for the development of this trade, through gravitational models.

F. Trapani et al. (eds.), *Advanced Studies in Efficient Environmental Design and City Planning*, Advances in Science, Technology & Innovation, https://doi.org/10.1007/978-3-030-65181-7_8

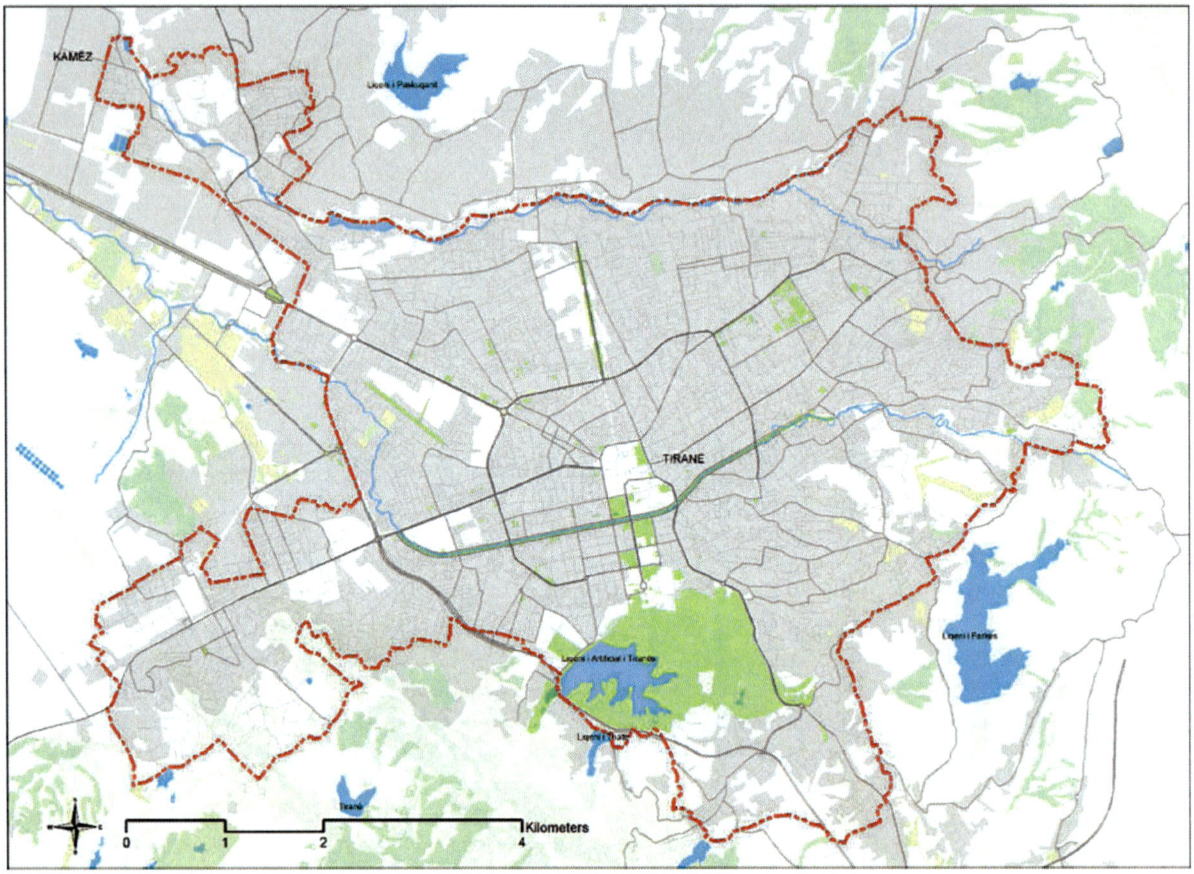

Fig. 1 Tirana urban area

2 Literature Review and Background

The city of Tirana exhibits a rapid development in the last few years, mainly due to entries deriving from the internal economy of the country and from the rapid mass flow of building constrictions. This kind of development together with high consumers leads to an increase in the number commercial area and structures.

The location analysis for them is most important than the cost due to the benefits as the income, stability of the business, and the permeability elements. Theoretically, the process mode is developed by selecting a typology of the subject and then a specific space to the store type. Factors as diversity, location, and cost have a kind of direct impact on how the type of entity or store will be determined (Baumol & Ide, 1956). Also taking into account and the consumer's relationship of positive characteristics of products such as standard, originality, marketing, cost and some socio-demographic elements. All these definitions have effects on the way the store is selected and by hypothesizing on a correlation between them, and we can look at their order according to the visual form. There is a wide theoretical field where the way of defining these subjects from the commercial aspect is debated and it is clear that the population often go to fulfill their consumption needs in retail outlets, such as markets and shops.

2.1 Importance of Location Selection

For the purposes of this research, Tirana center area and the surrounding areas have been defined as the study zone of the research. Tirana is the capital and the main city of Albania in aspects such as the population, economy, and ownership of the physical territory, often so-called metropolis. These commercial spaces operate in larger numbers around this area and the demand comes through high competition. In such areas, the densification of building and commercial structures is always expected due to demand. For most users of a specific business, often in an unconscious form, they place the location as the element that most influences their decision to select the store for their needs. Most of these cases meet their daily needs in the markets closest to them.

Such cases can be seen in areas where families live, and their decision is clear to consume and meet their daily needs at the nearest point or shop from their home or in the more frequent itineraries they make during the day. All this leads us to think that in order to have so many advantages and positive elements in a sustainable competition with similar entities, and we must use a strategy in which position is the main element. The selection of a store location generally requires an extensive a series of elements from which we have influences are mostly of a demographic type which is included in the area and around it. The characteristics they have, their size, the similarity with other subjects in the area or direct competition, the possibility of connection with public transport, parking, and other characteristics related to the entities that surround it or to the market value of the property.

2.2 Analysis Through GIS

Through new processes enabled by GIS for location zoning, there has been a significant impact on business companies in the last decade, achieving results in location analysis. The use of GIS has allowed organizations to abandon the feeling of having factual information about a place (Bennison & Hernandez, 2000). Because in addition to using analog, statistical, and gravitational methods (Rogers, 2004), since none of the methods was the best, it brought incorporation from the GIS into a single software. We also have the advantage of eliminating the unknown impacts of urban form factors. Exploring the potential of the market by the gravity model, it can also have the sense of attraction by the resident buyers (Hammond & McCullagh, 1978). Ignoring the fundamental urban design aspect such as connections accessibility and mixed uses, it can result in the creation of less sustainable and less socially equitable environment (The Urban Task Force, 1999). The significance and use of space may be different over time and the way in which they act. Mendes and Themido, they named the physical space used by the store with the phrase "store location," as well as its surrounding areas experiencing strong economic and commercial activities (Mendes & Themido, 2004). While commercial areas are a contiguous geographic area, as defined by Levy et al. (1998), the analysis of the location of the commercial spaces should be carried out for both trade and retail sites. These areas determine the perimeter of the inclusion of existing or new store areas. Also, from the retail site analyst is the evaluation of specific store alternatives that may be isolated, an unplanned or planned commercial use associated with the area characteristics, as accessibility, visibility, and traffic.

2.3 Characteristics of the Neighborhood

The traffic flow is substantial factor to impact the sales, which matter from the amount of movement of people visiting the area by vehicles or in the form of pedestrian movement. According to Levy et al. (1998), a good flow of traffic and good accessibility has a direct impact on the performance of the store, especially pedestrian traffic has a greater importance in the neighborhood then vehicular one. Not necessarily a large pedestrian traffic necessarily means a potential buyer for that product, which means that the viability of commercial spaces depends on the function and products they serve.

According to Berman and Evans (1995), the necessary or valid amount of foot traffic to make it profitable for the store must contain certain criteria as categorization by age, distribution based on realized time; survey on people who have circulated; and analysis of trade movements. Vehicle users are always consumers who avoid overcrowded commercial spaces and no parking, but above all this applies to those commercial services that have a focus on buyers from other areas. So, if these individuals who mainly perform these activities on their return from work can use a store in one area or another area, the latter should be well connected to the public transport network, thus, allows it to be tangible from a commercial and time-consuming point of view by the entire population of the city. If dedicated parking, a nearby public parking lot or other complementary space is available, it helps to operate the commercial space. The closer this service is to the customer, the more inclusive it is. Another strong link is the size and number of retail outlets in the area. The more traffic there is, the more traffic it will generate so it will also serve specific space.

Levy describes how retail entities can help each other indirectly when they are in communities with similar businesses or complement each other (Levy et al., 1998). This calculation is where the balance between potential and supply is in optimal conditions and when their goods and services are appropriate. To evaluate the accessibility of a retail site, various factors such as road patterns, crossing conditions, barriers, no lanes, and safety are evaluated together.

Specific Site: Such places or spaces should be classified according to factors of visibility, administrative costs, and use of the environment, limitations of use, impacts on the environment, and sustainability of the activity. Some other factors affecting consumer decision to shop at a supermarket are usually related to location, content, quality, service, pricing, image, and branding.

2.4 Research Problematics

These commercial spaces indicate an evolving situation. They operate in greater numbers around this area and the demands for development come through intense competition. Consumers similarly shop closer to them and near areas with similar units, but to choose a location for shop and activity, some criterias are generally needed. Finding a location is the approach that all services, government agencies, education, and public services face. In different countries, today, these are done depending on the criteria in order to protect nature reserves. In this debate of services distribution, choosing the location of the business is a very important issue that all companies face. The importance of the urban development process today is the rapid growth rate and the creation of dynamic environments with different activities that take place in urban environment, often with the multiple functional orientation and the instability of the commercial space usable. The city plans often have little impact on the creation or management of these spaces, and on the other side, the government play the role of being the main coordinator of public and private actions. Consequently, the use of space can happen to be developed by the way developers organize their private needs.

A not too clear perspective of the proposals or solutions that are taken without taking into account the relationship of different factors together has often led to not very sustainable developments in terms of viability of the activity but also of the environment in which it takes place. The need to make efficient choices that minimize these negative effects or achieve a better balance of environmental characteristics, which affect the structure or commercial activity.

2.5 Objectives of the Research

The relationship between neighborhood access and existing practices, beliefs, and processes has been explored and how much the location affects the retail sector. To support this target, it is needed a detailed analysis for determining the potential location of the different stores in Tirana, to identify the appropriate areas to develop commercial space from the built form and consumer criteria. What is the best position? Identification of the real competitors and consumer criteria, determining the potential location from the rating. Impact of neighborhood access on activity locations. The analyses will be focused in three main directions: (1) Determining retail trade boundaries for the country, based on the analysis of competitor distribution and other influencing factors such as topography. (2) Examine and compare the characteristics of the commercial area such as changing population and household, and demographic and urban characteristics. (3) Identify potential sales areas for store placements.

2.6 Research Hypothesis and Development Scenario

Recognizing the ongoing relationships and practices in the development process can acquaint us with the effects they are feeling or their tendencies. The study proves to identify the causes and phenomena that occurred and to present an analysis of how urban parameters influence the development of commercial spaces situations based on neighborhood business decision making. Accessibility and pedestrian traffic are the most important factors that are considered when selecting a site area. From these arguments, it is posed to hypothesize the evaluation of the link between the viability of the retail trade and context. However, the hypothesis of the retail viability is linked to neighborhood's accessibility and defines a statistical relationship, causes, and phenomena, such as urban parameters, affect trade development, and their efficient design. The sustainability of the proposals and solutions that are made through choices that balance and minimize the negative effects of how the territory is modeling and what impact they have on the environment and is also important for an efficient environment design. Referring to Cant (2010), the retail balance is defined by the variety of stores in an area. The most inefficient developments involve a more expensive loss of natural land, as well as a cost for all actors. If we are guided by decisions based on evaluations to minimize the risk for the sustainability of the development, we can have a more proper relationship of environment and commercials activity.

3 Methodology

3.1 Software Needs

The use of software aims to create maps and find positions through analysis through Spatial-Analyst in ArcGIS 10.1 and edit the table in Excel.

3.2 Collection and Data Editing

The background information of the urban area in the city of Tirana includes data as population distribution, real estate, building orientation, accessibility, streets, and competing services, geographic elements as: county boundaries and lakes. The data was obtained from the ASIG-geoportal, published statistics (OECD, 2008) and from data collection at the site. The demographic information was collected, processed, and customized for specialized stores, for approximate calculation of the capacity of residential and commercial premises (Fig. 2). After collecting information on the location of existing commercial spaces, the tables have been converted to the format of a database. This database will be geo-conducted through attributes using the traffic flow data and as a reference level, which will then be finalized with a shapefile of the type of point that will represent the geographical positioning and the commercial spaces. The shapefile contained information regarding the attribution of the area and the specialized stores.

27 different companies operate in the market area of Tirana, which operate as a supermarket and are organized in a network of 107 stores (Fig. 3b). Their geographical location and distribution are presented in Fig. 3a.

Colored dots on the map indicate the location of specialized shops in daily consumer goods and that are important in this field, identified in the market area of Tirana. A buffer of 400 m has been created around them, to illustrate the coverage area of each of them. As can be seen in the map, there is a market coverage concentrated mainly in the central part, where there is also an overlap in the potential market coverage.

Various materials such as aerial photographs and cartographic maps from the 2007 and 2015 orthoimages were used [2018]. They were on a scale of 1:25,000 and orthoimages with a resolution of up to 20 to 8 cm. Some auxiliary data also helped in analyzing city plans, photographs, and historical documents. Aerials, first, were georeferenced using the already rectified orthoimage, RMS was between 0.5 and 0.3, respectively, from 1980 till 2007, and secondly, they are unified in UTM (Zone 34N). Analysis is made possible through ArcGIS.

The shapefiles with their geometry data from polygon technology are created through visual digitization and attribute tablets are attached to each of the geometries. In the data that are integrated into the attribute tables, there is also the percentage quantity of permeability, access, and management. To understand the structure and how to use it, it is important to consider the elements of the environment and access properties, as well as the attributes related to management. The way and the permeability of spaces are seen as an integration factor (Forman & Alexander, 1998; Schueler, 1994).

Market area analyses are based on "Secondary Market Research", which means using information that is already available: selection of areas and their examination in detail through the analysis of "commercial areas," which includes some retail-related benefits. Include knowledge: demographic characteristics such as size, economic aspect, competitiveness, freedom of positioning of the activity, permitted uses, and possibilities of workers (Figs. 4 and 5).

Each upload of information in the shapefile contains information regarding existing data related to our objective: the accessibility of urban characteristics, neighborhood accessibility, and the analysis of existing data.

The use of spatial analysis helps us to create a range of network data from different geodemographic layers. Through it we can enter the data to be used in raster calculations. The gravity model is helping to explore the residential zone potential and the variable of these distances is obtained from the most accessible points of these distances from the average time of description through the road network.

4 Analysis

The content of the investigated context includes the physical features of "commercial space development" as well as the relationship with urban development. In the following description of the context, they are divided into two main

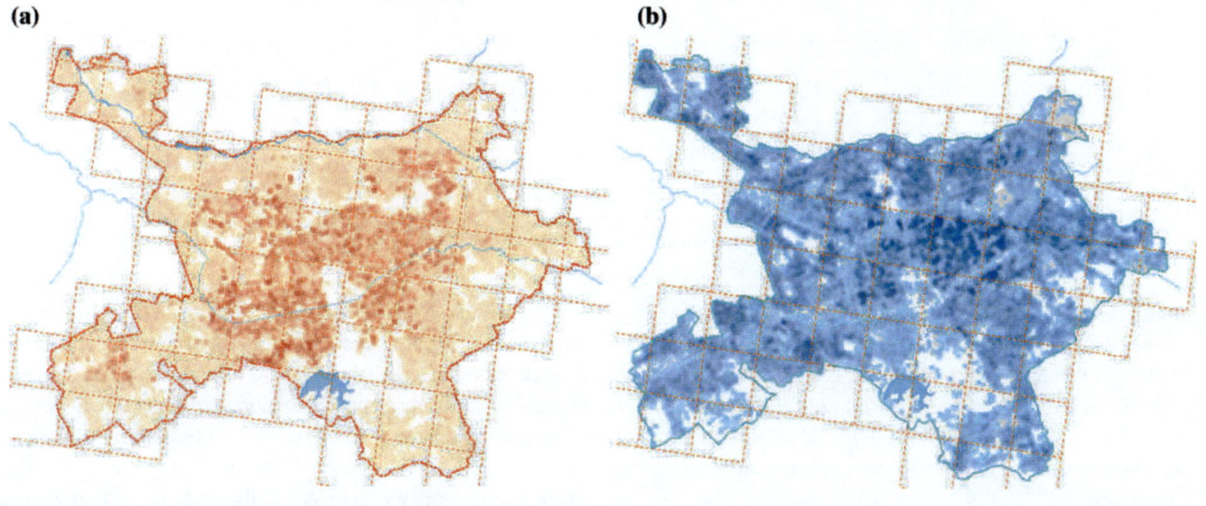

Fig. 2 a Population density. b Buildings/units density

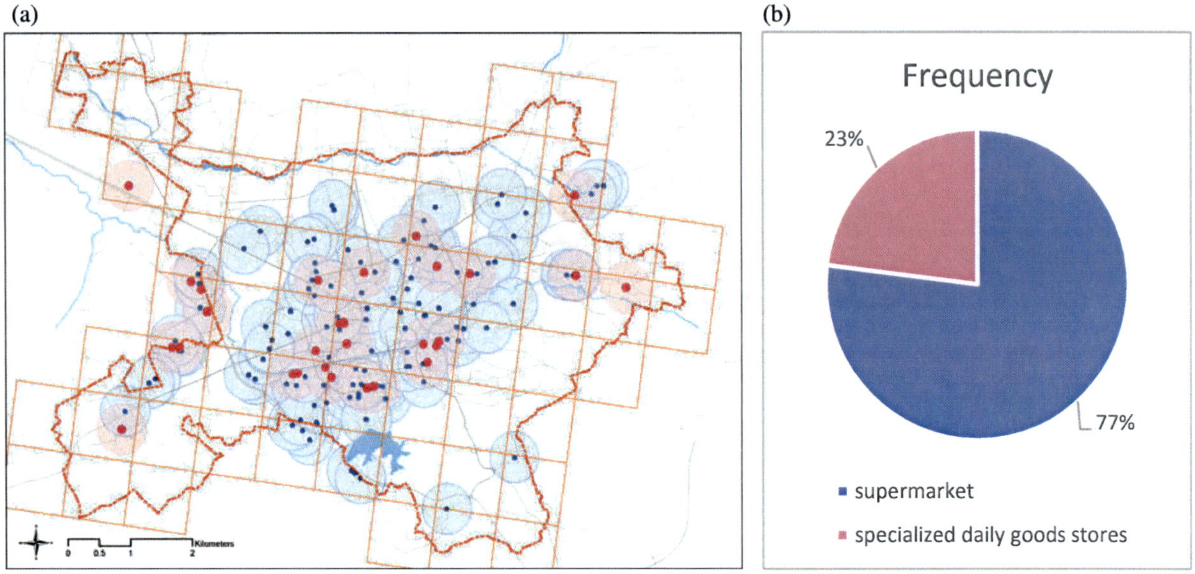

Fig. 3 **a** Distribution of competitors and directly competitors by coverage areas. **b** Frequency of competitors

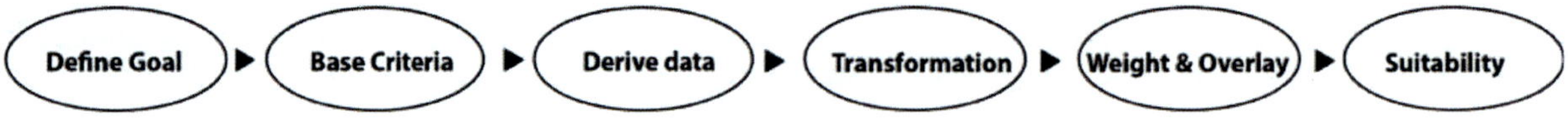

Fig. 4 Suitability from the analysis

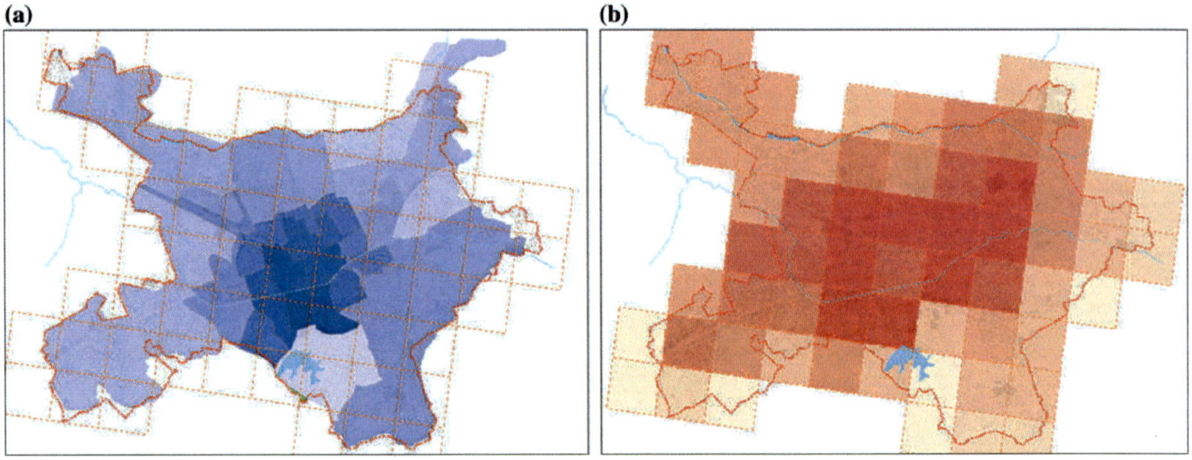

Fig. 5 **a** Market value of the immobile property. **b** The population by age between 15 and 65 years

points, in the physical expression of urban morphology, by accessing the forms and the size of spaces, function, and structures, which are clearly evidenced in the city center. After there is the possibility to define the general characteristics and also the sustainability values of their market area development from the economic, social, and population distribution.

This research is related to spatial analysis in order to identify the potential area and the important characteristics that affect retail viability from the urban form and its sustainability development. By exploring the important terminology of commercial space, according to history, economy, ecology, and social aspect. From one point of view, the examination and expression of the use of virtual-spatial

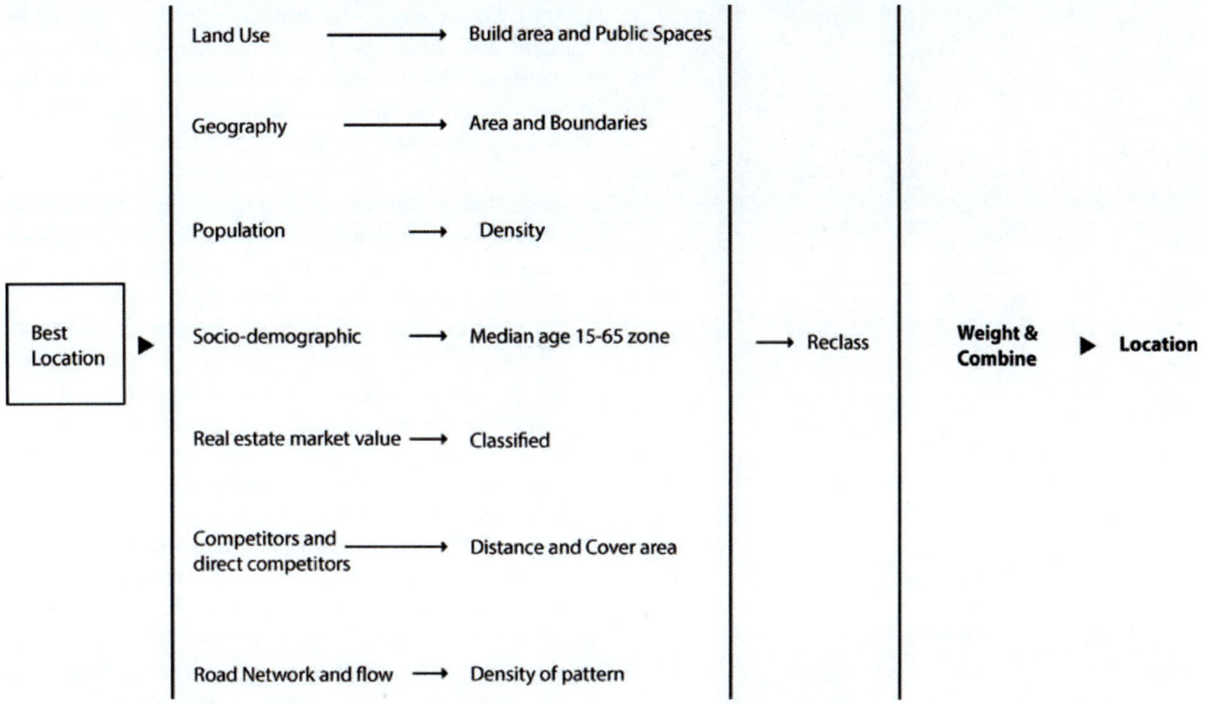

Fig. 6 Evaluating by various factors

relationships occurs through the creation of the mapping process as suggested by Ittelson et al. (1970). The spatial analysis of the function, performed through GIS, shows us a spatial relationship and the formation of spaces and the way they are used, which indicates the dimension of the space, the geometry of the space, the average per unit, and identity they occupy. Types of commercial spaces were able to be traced through field investigations and other supplementary materials, examining the material and creating a geographic information coverage. Also, the level of change was investigated using GIS technology. These spaces will be selected with the criteria based on the type of function, urban morphology, and the model of urban development (Fig. 6).

Different examples have been analyzed to recognize the impact of the intervention in different ways to the open urban space on the physical condition and socio-economic context. This will determine the transformation of typology, local, and theoretical characteristics that distinguish in central categories and function in the whole hierarchy. The transformations that have taken place in the country and other examples abroad will be part of our study, which will provide a comprehensive overview of the change process, so that we can conclude the theoretical definitions for decision making in similar situations (Figs. 7 and 8).

The final recommendations for a specific geographical location derived from all factors. There are three basic factors that influence the choice of the individual consumer: prices (own and similar products), income, and preferences. The latter variable is based on socio-demographic and domestic characteristics.

The main indicators to be used as criteria in the analysis are real estate market value to be more than 98,000 lekë/m^2; population, average age between 15 and 65 years. Other information as: road network and flows, public spaces, city zoning information, geographical information), geo-processed information of competitors (supermarkets, dairies, grocery markets, and major competitors) in the Tirana area.

"population density" > = 50 "median age" > = 15 AND "median age" <= 65 AND "SUM_HIGH_E" > 40 AND "market value" > 98,000 (Figs. 9 and 10).

The gravity model: Domestic market zone 1 km^2 = Σ (Population of each zone 1 km^2 * Market value for the zone)/And the distance of the zone x with the location of the average population density.

5 Results

5.1 Findings and Discussions

The analysis provided answers the following questions for each of the areas considered potentially suitable as the best location of stores. However, it is more prudent to use a currency for each investment, in order to benefit an activity,

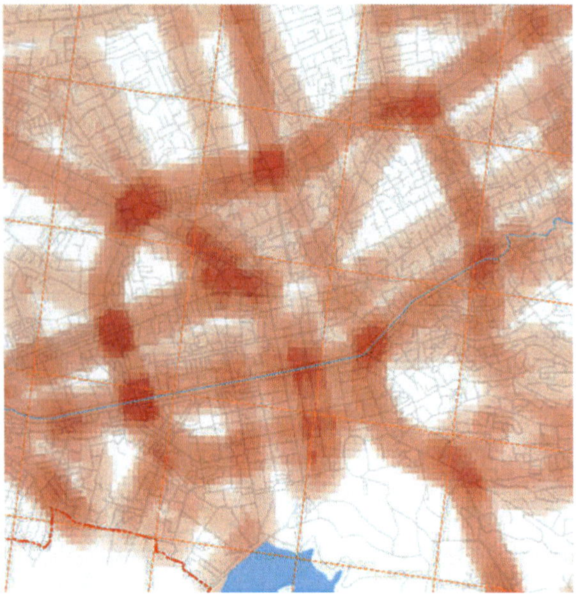

Fig. 7 Road network and flow density

Fig. 8 Network flows near the potential area in the urban form

and factors that influence the selection if it is aggravated or followed by typological meanings, both personal and personal. The evaluation of the location is based on different variables, the resources on the population, the income, and the age of the area allow us to better identify where the population and the consumer are. From an economic point of view, the act of consumption is a subject of the income process, as also considered by Samuelson and Nordhaus (2005). Consequently, the benefits affect the perception, lifestyle, and consumption trends of individuals and usually influence the choice of store. From the survey, it appears that the areas with the highest heat and in a quieter and less crowded environment seem to be more favorable.

- Accessibility through transport: Commercial spaces depend on their demographic characteristics which are generally different. Mostly, the downtown area consists of a clientele with medium to high living standards while the surrounding areas feel a slight decrease in this level. Pedestrian flows, transport, and public spaces give a positive effect to the entities close to them. Users of these spaces are non-residents, who are also car users, but who prefer such areas with parking opportunities. This means that favorite locations are commercial spaces in more suburban areas. While the central area is mostly used by pedestrian consumers which is also supported by the transport and public spaces factor best part of transportation in close proximity to these commercial spaces. This is illustrated in Figs. 11 and 12: Potential market areas for stores identified based on defined criteria (demographic and socio-economic) and the positions of identified competitors. The cherry-colored areas on the map represent the selected areas.
- The factors that affect the location are also different: According to Levy et al. (1998), people often make decisions after thinking about how to find a store faster or find one closer, focus of the same type, exchanged standards, level of service, costs, way of representing them, and the popularity of the name. Is the visibility of the place determined by the visualization from the main traffic flow? Small retail stores can take advantage of the traffic from nearby large stores, but in other cases, local businesses can harm them; therefore, it depends on the compatibility of the objective. And ranking them according to the impact they have on consumers, we have a traveling, service quality, location, convenience, price, nearness point, accessibility, visibility, and continued with the image, brand, and parking possibilities. Not all factors tell us about location, but it is important that they take into account the convenience offered by location. The demographics of the area show a population that is moderately single-family, for whom the time spent on purchases is short and the persons performing them are unable to make long trips. Most easily identifiable commercial spaces are those located on main streets or building corners, also, the spaces surrounded by public transport.
- Flows and Location: In order to study their relationship between flows and the potential location, it is possible to use observation methods to identify movement patterns at

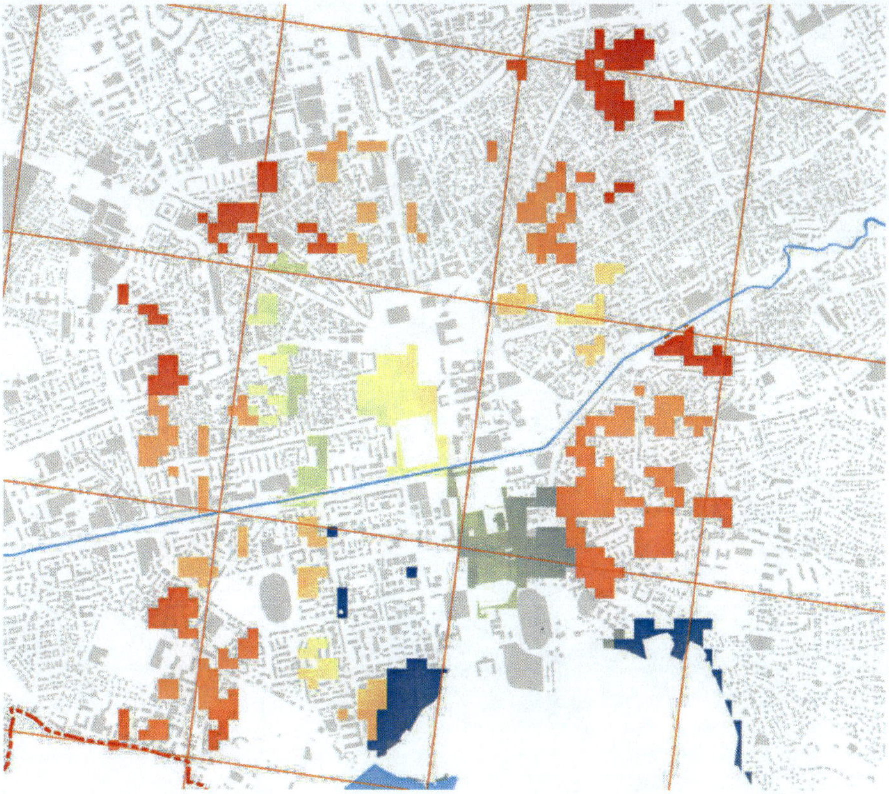

Fig. 9 Potential area by socio-demographic and density

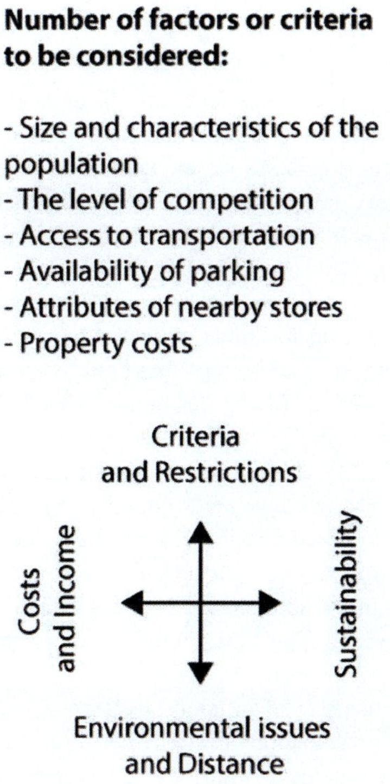

Fig. 10 Factors that affect the viability of commercial space

different times. The results are presented in the graph below. Most of the traffic is concentrated on the main roads, but close to high-density residential areas. Usually, the traffic is higher in the morning (40%) and in the afternoon (35%) after working hours. While other roads have more constant traffic during the day and a slight decrease of 25% from afternoon, this shows that the areas with better access and close to public transport have higher flows and better-quality customers than those within neighborhoods, which have consumers who prefer proximity factor for shorter time.

Through GIS analyzes that summarize geographic techniques, we obtain the results, as in Fig. 13, which shows the potential trading areas according to population density, from lowest to highest. They also feature a perimeter area that shows accessibility from 400 m users and a buffer of about 1 km. These results-based analyses reveal some site-location that may be suitable for new commercial spaces. Then, comparing the ranking findings by gravity form with the most important zone in the city, a ranking is done by interpolating the results from both models. After reviewing all the findings from the analysis of the place and their locations, we are continuing with the recommendations of the areas with the greatest demand for commercial space (see Fig. 14).

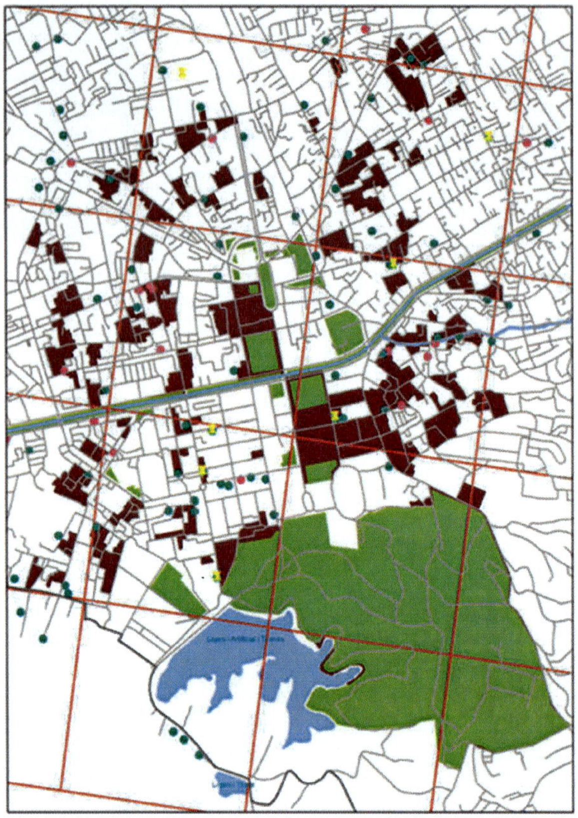

Fig. 11 Potential commercial areas according to defined criteria

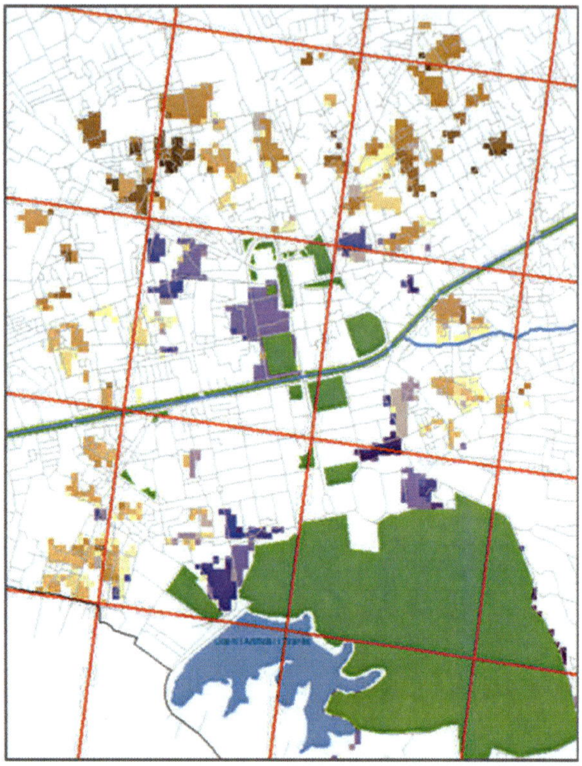

Fig. 12 Areas with the greatest potential

A discrepancy is identified when some of these commercial spaces are developed in not recommended areas, which are considered as areas with no potential or with minor importance in the retail trade for the gravity model findings. Also, areas that are directly related to residential areas with a high density certainly become places of interest and possible locations not just of the demand for consumption but often they fulfill several conditions at once.

6 Conclusions and Recommendations

As shown in this study, another very powerful and essential tool for making the best decisions about store location is the use of GIS technology. We could have avoided stores in poor locations, or finding new retail locations, together with the help of market knowledge for the final decisions. Based on the results, we realized the importance and geographic advantages that GIS and gravitational model give us. From this analysis, now we possess the knowledge for the city of Tirana of which areas of commercial development should we consider for new stores and how we can find them how. With this important geographic and correlation domain, we will have the opportunity that in future, we may have possibly fewer new store closures and we will understand how the neighborhood's configuration determining the business viability by various parameters, and with a better understanding of accessibility the leading and future location of the potential areas to the retail stores and market enlargement.

Other analyzes on a smaller scale can investigate the parameters of residential blocks, which are recommended to find some optimal locations, with the best visual communication from the main artery that bind to the traffic circulation that imposed facing the future location of the store. Not necessary, heavy traffic mean more consumers. The location of the identified stores is proposed based on the estimations of the consumer's quantity that fulfills the targeted market criteria. The selected areas are covered by public transport service. Clients and distribution vehicles can easily enter in the shop area. The areas should offer the use of public parking. The stores are visible even from the main traffic flow.

Through the use of different analysis techniques in this study, raster calculations derived the most potential development spaces in the downtown area of Tirana. Summarizes, using multiple analysis techniques and evaluating each.

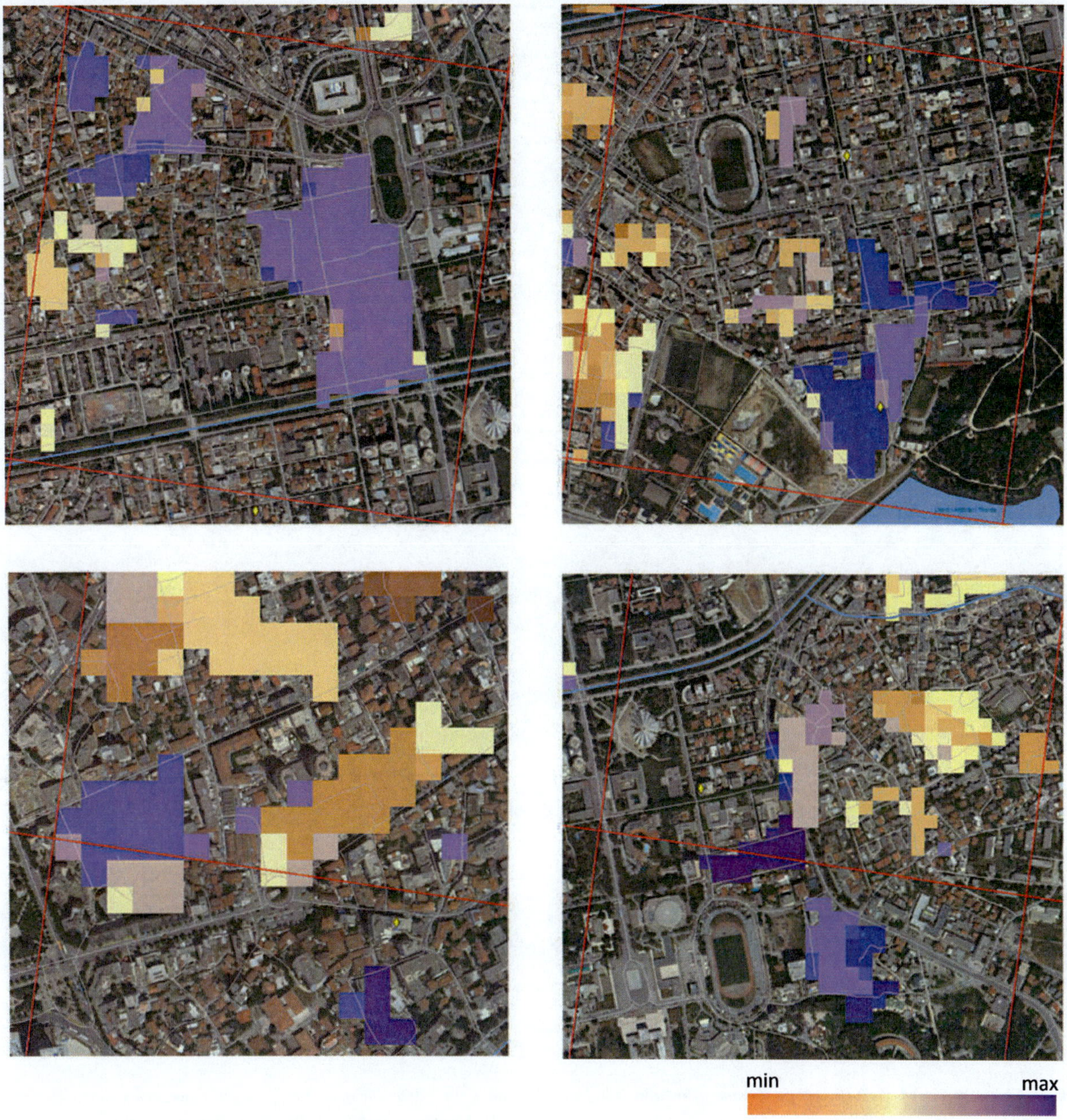

Fig. 13 **a–d** Potential locations for the new commercial activity

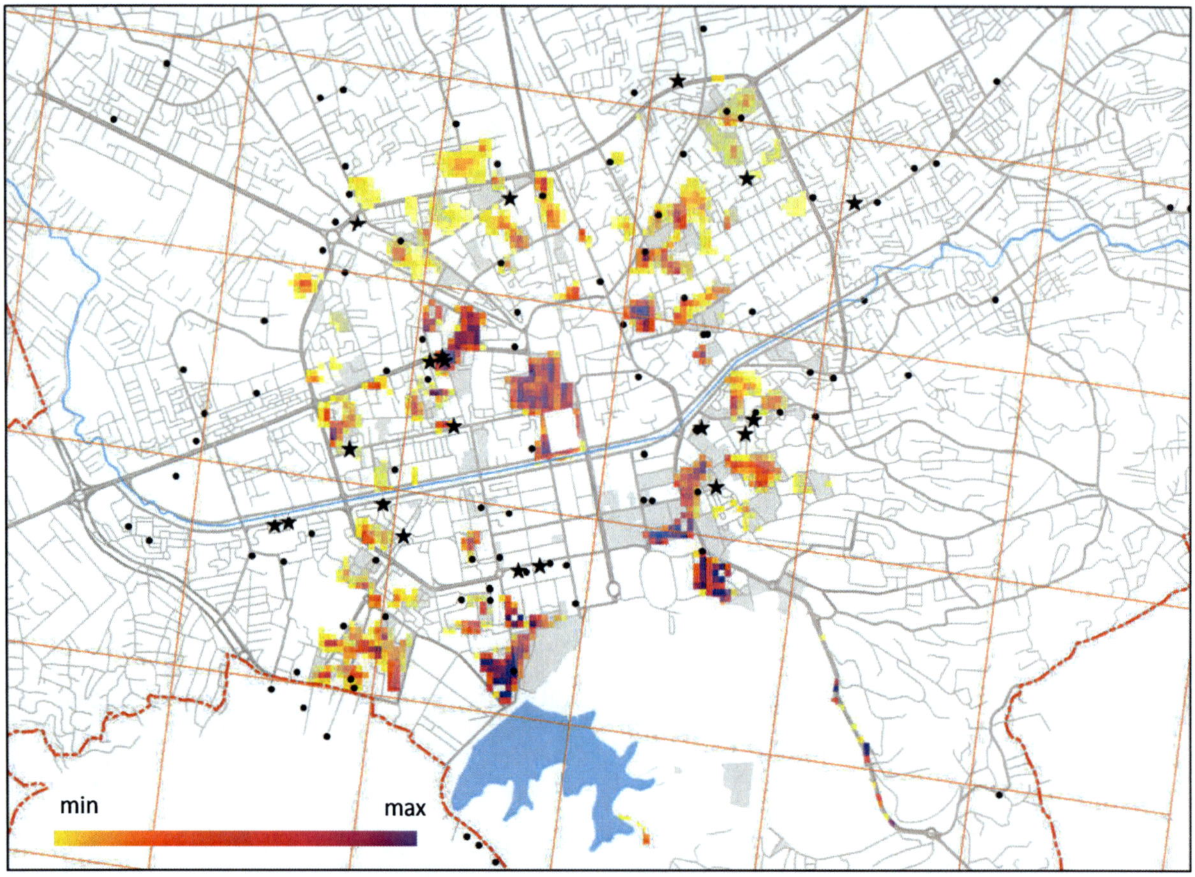

Fig. 14 Ranking of the greatest potential locations according to other competitors

References

Baumol, W. J., & Ide, E. A. (1956). Variety in retailing. *Management Science, 3,* 93–101.

Bennison, D., & Hernandez, T. (2000, February). The art and science or retail location decisions. *International Journal of Retail & Distribution Management, 28*, 357–367. Retrieved February 2, 2007.

Berman, B., & Evans, J. R. (1995). *Retail management: A strategic approach.*

Cant, M. (2010). In M. Cant (Ed.), *Introduction to retailing* (2nd ed.). Juta.

ESRI. (2013). *Environmental Science Research Institute* (The ArcGIS Book edition, ArcGIS 10.1 [computer software]). ESRI.

Forman, R. T. T., & Alexander, L. E. (1998). Roads and their major ecological. *Ecologic Systems*, 207–231.

Hammond, R., & McCullagh, P. (1978). *Quantitative techniques in geography* (2nd ed.). Oxford University Press.

Ittelson, W. H., Rivlin, L. G., & Prohansky, H. M. (1970). *The use of behavioural maps in environmental psychology*. Holt, Rinehart & Winston.

Levy, M., Weitz, B., & Grewal, D. (1998). *Retailing management* (8th ed.). McGraw Hill.

Mendes, A. B., & Themido, I. H. (2004). Multi-outlet retail site location assessment. *International Transactions in Operational Research, 11*, 1–18. Retrieved February 2, 2004.

OECD. (2008). Household behaviour and the environment: Reviewing the evidence.

Rogers, D. (2004, January 26). A status report on location analysis. *DSNRetailing Today, 14*. Retrieved February 2, 2007.

Samuelson, P. A., & Nordhaus, W. D. (2005). *Economics* (8th ed.). McGraw-Hill.

Schueler, T. (1994). The importance of imperviousness. *Watershed Protection Techniques*, 1–6.

The Urban Task Force. (1999). *Towards an urban renaissance*. Taylor and Francis.

The Effect of Shading Devices with Integrated Photovoltaics on Energy Efficiency of Buildings

Salwa El Gindi

Abstract

Due to the increase in urban population and comfort levels, the use of electricity in the building sector has rapidly grown. Recently, building energy demand has significantly increased. Buildings consume almost 40% of the world's energy. Combining energy efficiency measures with renewable energy sources in an integrated building system will improve the building performance and consequently reduce the electricity consumption and carbon emissions. Research demonstrated that it is technically possible to achieve significant reductions in energy consumption with the proper combination between energy efficiency measures and renewable energy sources. Building-integrated photovoltaics (BIPVs) have come a promising technology for energy generation and life-cycle costs reductions. In this regard, photovoltaic-integrated shading devices (PVSDs) constitute an essential part of BIPVs and play a significant role in electricity generation and energy consumption reductions. This research investigates and explores the potentials of combining PV technology in shading devices. Different types of external shading devices with integrated photovoltaic panels were examined for different orientations using DesignBuilder simulation software. The cooling loads, the electric lighting consumption, and the electric energy production related to the use of PVSD have been evaluated. The research aims at optimization of shading device configuration for energy efficiency and energy production.

Keywords

Renewable energy • Energy efficiency • Shading devices • Building-integrated photovoltaics • Photovoltaic-integrated shading devices

S. E. Gindi (✉)
Department of Architecture, Faculty of Engineering, October University for Modern Sciences and Arts, Cairo, Egypt

Nomenclature

RE	Renewable energy
BIPV	Building integrated photovoltaic
PVSD	Photo voltaic integrated shading devices

1 Introduction

Building facades have a great effect on the quality of the indoor environment and consumption of energy; therefore, they require careful design optimization (Lee et al., 2009). Shading devices are essential façade components as they can contribute significantly in reducing the heat gain into the building and providing acceptable indoor conditions (Alzoubi & Al-Zoubi, 2010). Shading devices prevent overheating by blocking unwanted energy from reaching the building. However, fixed shading devices may reduce availability of daylight and consequently increase artificial light needs. It may also block solar radiation in winter which is needed for heating (Mandalaki et al., 2012). Therefore, a balance between blocking the unwanted solar radiation in summer and providing adequate natural lighting levels and adequate amount of solar radiation in winter is needed when designing a shading device (Tzempelikos & Athienitis, 2007).

Several factors contribute to heat gains in buildings like electrical equipment, occupants, electrical lighting, etc.; however, the main factor causing heat gains is solar gains through openings. Therefore, using external shading devices to intercept the solar radiation before entering the building is an important method to reduce the heat loads (Gratia & De Herde, 2007; Kim et al., 2012).

Photovoltaic-integrated shading devices have an advantage of converting sunlight into electricity besides its function as a shading device (Sun et al., 2012). "However, the application of shading devices with integrated photovoltaics has significant challenges due to the complexity of the

F. Trapani et al. (eds.), *Advanced Studies in Efficient Environmental Design and City Planning*,
Advances in Science, Technology & Innovation, https://doi.org/10.1007/978-3-030-65181-7_9

system and the adaptability of these systems to different contextual conditions" (Lee et al., 2009). Photovoltaic-integrated shading devices PVSD has several advantages, such as clean energy generation and adding specific architectural features (Ibraheem et al., 2017). PVSDs have proven technical advantages over other types of PV installations like roof stand-alone PV systems (Mandalaki et al., 2014a, b). Studies have demonstrated that the buildings with PVSD can make significant contributions in energy savings (EIA, 2011; Fouad et al., 2019). In order to optimize the application of this technology, it is important to highlight the main parameters that influence the performance of shading devices with integrated PV such as providing optimal tilt angle with the appropriate length of the shading device and correct distance between them to prevent excessive sunlight during summer while allowing it during winter and allowing diffuse solar radiation penetration into the building (Hwang et al., 2012; DGS guide, 2008).

Due to the large glazed façades and the nature of office activities (Hwang et al., 2012), office buildings consume more energy than other types of commercial buildings (IEA, 2011). The glazed envelopes of existing office buildings result in heat losses in winter, heat gains in summer, glare, and increasing cooling and heating loads, specifically in hot arid climates. Moreover, the effective functioning hours of photovoltaics match well with the working hours of offices (Gindi et al., 2017). Hence, office buildings are suitable for PVSD application due to large energy consumption, large surfaces of glazed façades, and a schedule that matches well the building functions and electricity generation (EIA, 2011).

In Egypt, both the commercial and residential buildings consume about 47% of the electricity consumption (Abdin et al., 2006). Large amounts of electricity could have been avoided by a more considerable use and consequently reducing the carbon footprint of buildings. This suggests that office buildings need an integrated approach for implementation of energy efficiency measures and renewable energy technologies to improve their performance. Thus, the research aimed at optimizing PVSD in office buildings in Egypt to improve their performance.

1.1 Design Considerations

Several variables have to be taken into consideration when designing PVSD. Various factors such as location, orientation, and dimensions as well as the shading coefficient and daylight factor have to be considered to determine the appropriate type of shading devices which are suitable for PV integration (Mandalaki et al., 2012; Bahr, 2009; kim et al. 2010). The latitude sets several essential inputs like temperature, sky conditions, the amount of solar radiation, and other climatic parameters. The shading type and dimensions can be specified by considering latitude, and other variables (Bahr, 2009). The optimal design option for each location can be proposed (Bahr, 2013). One of the key determinants for PVSD optimization for different latitudes is building orientation (Bahr, 2009). It is found to expectedly be south (Bahr, 2013). These findings were assessed according to cooling loads reductions (Hwang et al., 2012). Youssef et al. (2015) investigated building façade envelope variations. The angle of inclination is one of the most influential parameters determining the performance of the PVSDs, which helps ensure an optimum amount for both solar gain control and electric production (Hwang et al., 2012; Bahr, 2013; Kaldellis et al., 2012). Some researchers have proposed that a tilt angle equal to the latitude of the location maximizes the harvested solar energy, especially for horizontal louvers, and have proposed that horizontal installation (0° tilt angle) decreases self-shading (Bahr, 2009; Bahr, 2014). Others researchers have proposed different solutions suggesting that a horizontal inclination angle of 60° and a vertically inclination angle that is less than 15° will have the best performance, regardless of the building's orientation (Hwang et al., 2012). Jung (2014) proposed that controlling the tilt angle was efficient concerning cooling loads reduction and visual comfort. Many studies focused on the dimension of the PV panels as one of the influential parameters (Mandalaki et al., 2014a, b; Sun & Yang 2010). They concluded that the width of the panels is more effective considering electricity production. PVSD performance differs according to their configurations (Mandalaki et al., 2012). The relationship between the height of the opening and the depth of overhangs is significant. It has been proven that the ratio of the distance between the shading device slats and the depth of the slats has a significant effect on the performance of PVSD (Hwang et al., 2012; Bahr, 2009; 2014). This ratio determines the effect of shading on the panels, and it has been utilized as a method of installation to estimate the proportion of electricity production. (Hwang et al., 2012). It is recommended that PV shading devices should be applied in such a way that it is not shaded by the panel above (Hwang et al., 2012; Sun & Yang 2010). Thus, the area, tilt angles and the spaces between shading devices are effective parameter in electricity generation. (Hwang et al. 2012) investigated how to maximize power generation in a high-rise building with BIPV systems. He examined the effect of the ratio of the distance between the shading device slats (D) and the depth of the slats (L) on the performance of such systems. He concluded that a greater D/L ratio yields a greater amount of sunlight, but this amount is unproportionate to the amount of produced power due to a decrease in the area of power production. Thus, he recommended to set the D/L ratio between 1 and 3 (Hwang et al., 2012).

1.2 Performance Evaluation

In order to evaluate the performance of PVSD in terms of visual comfort and energy efficiency, several criteria have been developed. Optimal design parameters of PVSD would be: the annual total solar insolation on the panels per meter squared of the panel area (Wh/m^2), the cooling load reduction during summer per meter squared of floor area (Wh/m^2), the geometric shading coefficient on the glass façade (%), and the average daylight factor inside the space (%) (Yoo et al., 1998). Other researchers suggest that "electricity reductions can be calculated by assessing the electricity production and the cooling load reduction per unit of PV to achieve optimum design for PV shading devices" (Sun & Yang 2010).

"For cost–benefit analysis, different design options can be assessed based on annual electricity production in correlation with electricity saving for cooling, additional electricity consumption for artificial lighting, maintenance, and cleaning cost of PV panels" (Bahr, 2014). Electricity production of PV shading devices has been examined by Hwang et al. (2012) and Kang et al. (2012). "Electricity production from PV has been a reliable indicator especially when combined with other criteria such as visual comfort" (Mandalaki et al., 2014a). In some specific cases, all the electricity production by PV can cover the artificial lighting energy (Mandalaki et al., 2014b). Yoo and Lee (2002), Yoo and Manz (2011) examined the effect of solar irradiance. Tongtuam et al. (2011) examined the annual energy yield per square meter of PVs.

The aim of this research is optimization of shading device for energy efficiency. The research aims at determining the optimum shading device configuration which will reduce the energy needs for cooling, electric lighting and produces the maximum energy by integrated photovoltaic panels.

This research analyzes the effect of applying different types of external shading devices with integrated PV panels on the energy consumption of a typical office room in an air-conditioned office building for the Egyptian climate. The energy demand of the main systems has been evaluated (cooling and lighting). The annual energy savings related to the use of different configurations have been evaluated. Also, the energy production due to the PV integration was calculated.

The main goal of this study is to provide a criteria for using shading devices with integrated PV in air-conditioned office buildings in the Egyptian climate, quantify possible energy savings and to optimize geometry and position of shading devices.

2 Methodology

In order to examine the effect of adding different types of PVSD in a typical Egyptian office building, a simulation was conducted using DesignBuilder software.

Fourteen types of shading devices mostly used in office buildings (Zhang et al., 2018) were examined for their energy consumption and energy production (Fig. 1). These types were compared to unshaded base case.

The main criteria for evaluating the efficiency of shading devices are: the cooling loads of the building, the electric lighting consumption, and the energy production of the integrated PV panels.

In order to assess the efficiency of the PVSD, four types of evaluations are used. The first evaluation is the ability of the PVSD to provide thermal comfort with least amount of energy consumption. The second is the amount of electricity generated by the PV-integrated in the SD. The third evaluation corresponds to the efficiency of the PV production in relation to the area it occupies. The fourth evaluation corresponds to the efficiency of the PV production in relation to the electricity consumption.

2.1 Location and Climate

The simulation was conducted in the climate of Cairo, Egypt, and location 30.1128° N, 31.3998° E. The weather file was obtained from the EnergyPlus website.

2.2 Case Study

In this research, the electric consumption of a room in a typical office building is examined. AC with heating set point 18 °C and cooling set point to 26°, lighting level is 300 lx and occupancy scheduling from 8:00 to 16:00 (n.b. HVAC systems only operate with building occupancy). Occupancy level is 0.1 $persons/m^2$. Thermal load related to the office electrical equipment is 14 W/m^2, cooling system (fan coils + water chiller).

Room dimensions are 3.5 m × 3.5 m × 3.5 m (Width Depth × Hight) with a window facing each orientation 1.8 m × 2 m. The area of the space is 12.25 m^2; the window/wall ratio (WWR) is 40%.

Photovoltaic panels are integrated to the shading devices to produce electricity that can supply part of the electricity needs. They follow the geometry of the shading device. DesignBuilder software is used to evaluate the energy consumption of the building and the electricity production of the PV panels integrated in the shading devices.

The invariables were the thermal loads, the geometry and dimension of the office room, window, and type of used materials. The examined variable was the type of the shading devices. A ratio D/L was used which represents the ratio of the distance between the shading device louvers to the length of the louvers (Fig. 2). The tilt angle 30° was chosen on a basis of a carried out energy optimization.

Fig. 1 Types of shading devices

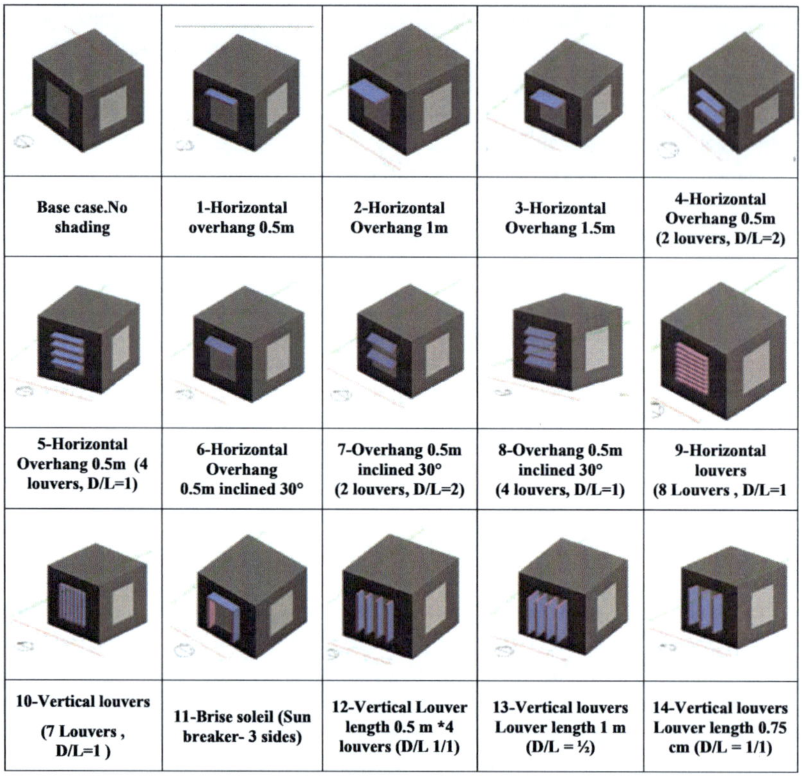

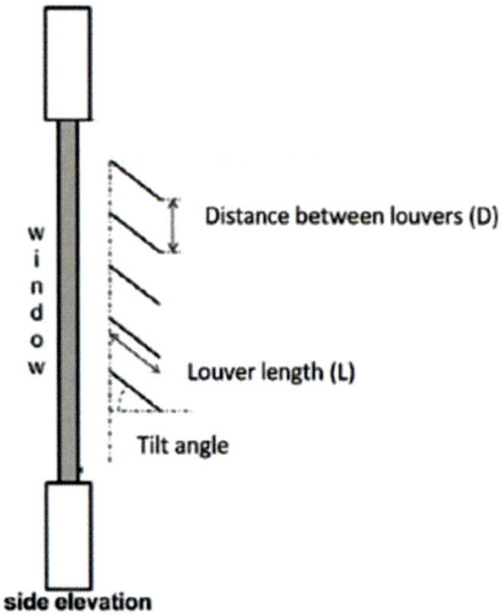

Fig. 2 Shading device

Fourteen different configurations of shading devices were assessed for East, West, South, and North orientations to determine the most appropriate type of shading device for each orientation in terms of electricity consumption reduction and PV electricity production.

2.3 Simulations

Fourteen types of shading devices are examined for each of the main four orientations to determine the most appropriate type for each orientation. These are compared to unshaded base case.

For each orientation, fourteen shading types were simulated while the other three orientations were unshaded in order to evaluate the effect of the shading and determine the most appropriate type for this particular orientation. The results were compared to unshaded base case (Figures 3, 4, 5 and 6).

DesignBuilder V4.6.015 was used for simulations for its user-friendly interface and computational fluid dynamics (CFD) simulations capability. Simulation engine of DesignBuilder is EnergyPlus which has been validated by Energy Efficiency and Renewable Energy (EERE) Program, U.S. Department of Energy (DOE). EnergyPlus provides access to mostly all the required simulation capabilities of thermal performance including natural ventilation, shading systems, glazing, thermal mass, building fabrics, and HVAC systems. It is noteworthy that shading models used by EnergyPlus have been validated by means of experimental investigation (Lomanowski, Lomanowski & Wright, 2009; Loutzenhiser et al., 2008).

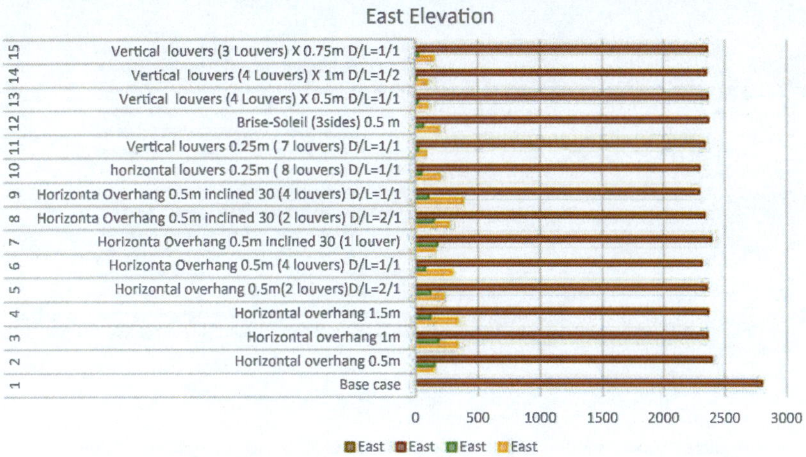

Fig. 3 Energy consumption and energy production for reference room with shading device on East elevation

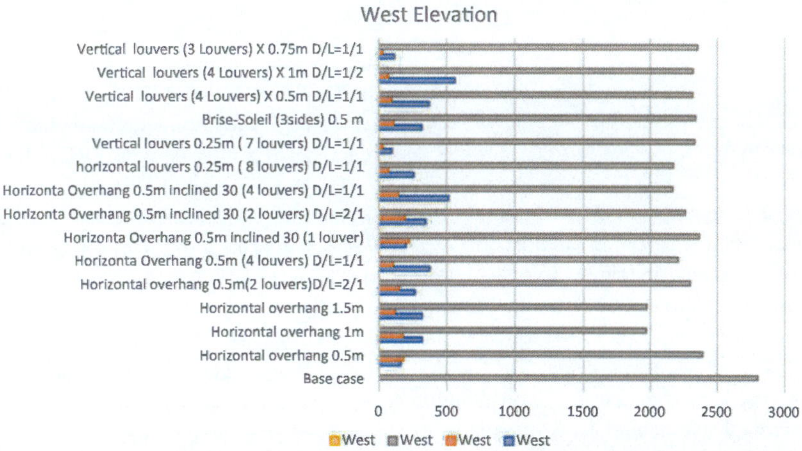

Fig. 4 Energy consumption and energy production for reference room with shading device on west elevation

Fig. 5 Energy consumption and energy production for reference room with shading dcvicc on South elevation

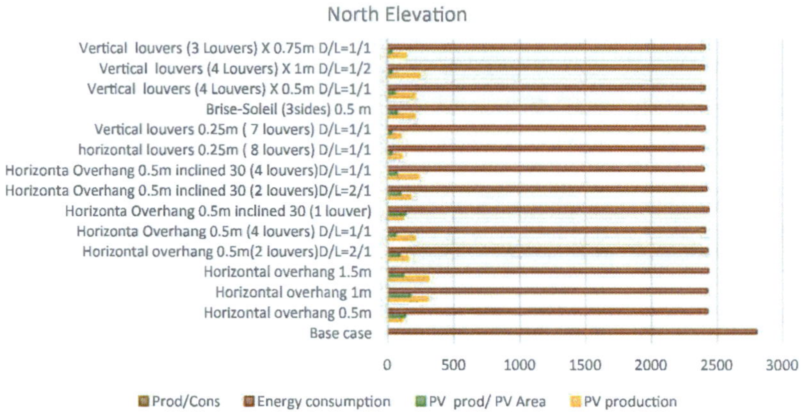

Fig. 6 Energy consumption and energy production for reference room with shading device on north elevation

3 Results

See Table 1.

3.1 Energy Consumption

For East elevation, the least annual energy consumption was achieved by horizontal overhang (0.5 m–inclined 30° * 4 louvers–$D/L = 1/1$) with savings of about 16.6% of total annual electric consumption, followed by horizontal louvers ($8L - D/L = 1$), horizontal overhang (0.5 m–4 louvers), and overhang (0.5 m–inclined 30°–2 louvers) (Fig. 7).

For West elevation, the least annual energy consumption was achieved by horizontal overhang 1 m with savings of about 29.6% of total annual electric consumption, followed by horizontal overhang (1.5 m), horizontal overhang (0.5 m–inclined 30° * 4 louvers–$D/L = 1/1$), and horizontal louvers ($8L - D/L = 1$) (Fig. 8).

For South elevation, the least annual energy consumption was achieved by horizontal louvers ($8L - D/L = 1$), with savings of about 19% of total annual electric consumption, followed by horizontal overhang (0.5 m, inclined 30° * 4 louvers $D/L = 1/1$), horizontal overhang (0.5 m, 4 louvers), and vertical louvers (Fig. 9).

For North elevation, the least annual energy consumption was achieved by horizontal louvers ($8L - D/L = 1$), with savings of about 14.3% of total annual electric consumption, followed by horizontal overhang (0.5 m–inclined 30° * 4 louvers–$D/L = 1/1$), vertical louver (length 0.5 m * 4 louvers D/L 1/1), and vertical louvers (7 louvers) (Fig. 10).

The maximum reductions in annual energy consumption were obtained by applying horizontal overhang 1 m on West elevations with savings of about 29.6% (Fig. 11).

3.2 PV Production

For East elevation, the highest annual electric energy production was achieved by horizontal overhang (0.5 m–inclined 30°–4 louvers–$D/L = 1/1$), followed by horizontal overhang 1.5 m, horizontal overhang 1 m, and horizontal overhang (0.5 m–4 louvers, $D/L = 1/1$) (Fig. 12).

For West elevation, the highest annual electric energy production was achieved by vertical louvers (4 Louvers * 1 m $D/L = 1/2$, followed by horizontal overhang (0.5 m–inclined 30°–4 louvers–$D/L = 1/1$), horizontal overhang (0.5 m–4 louvers–$D/L = 1/1$), then vertical louvers (4 Louvers * 0.5 m–$D/L = 1/1$) (Fig. 13).

For South elevation, the highest annual electric energy was achieved by horizontal overhang (0.5 m–inclined 30°–4 louvers–$D/L = 1/1$), followed by horizontal overhang (0.5 m–4 louvers–$D/L = 1/1$), horizontal overhang (0.5 m–inclined 30°–2 louvers–$D/L = 2/1$), and horizontal overhang 1.5 m (Fig. 14).

For North elevation, the highest annual electric energy production was achieved by horizontal overhang 1.5 m, followed by horizontal overhang 1 m, vertical louvers (4 louvers * 1 m–$D/L = 1/2$), and horizontal overhang (0.5 m–inclined 30°–4 louvers–$D/L = 1/1$) (Fig. 15).

The higher annual energy production was achieved by applying horizontal overhang (0.5 m–inclined 30°–4 louvers–$D/L = 1/1$) on South elevation where it covered 21.6% from the building annual energy consumption (Fig. 16).

3.3 Economics of PV Production

The ratio of the amount of electricity production from each shading device to the installed PV area is calculated to categorize the shading devices in terms of economical construction.

Table 1 PV annual energy production, annual energy consumption, PV annual electric production/installed PV area and PV annual electric production/annual electric energy consumption for different PVSD types in East, West, South, and North orientations

Type of shading	PV area m^2	East				West				South				North			
		PV prod. (kWh)	PV prod/PV area (kWh/m^2)	Annual electr. cons. (kWh)	Prod/Cons.	PV prod. (kWh)	PV prod/PV area (kWh/m^2)	Annual electr cons. (kWh)	Prod/Cons.	PV prod (kWh)	PV prod/PV area (kWh/m^2)	Annual electr. cons. (kWh)	Prod/Cons.	PV prod (kWh)	PV prod/PV Area (kWh/m^2)	Annual electr cons. (kWh)	Prod/cons.
Base case	0	0	0	2798.9	0	0	0	2798.9	0	0	0	2798.9	0	0	0	2798.9	0
Horizontal overhang 0.5 m	0.9	139.92	155.5	2390	0.06	163.24	181.4	2390	0.0683	209.88	233.2	2370.7	0.09	116.6	129.6	2425.28	0.05
Horizontal overhang 1 m	1.8	334.64	185.9	2355.32	0.14	320.65	178.1	1970.5	0.1627	419.76	233.2	2332	0.18	303.16	168.4	2425.3	0.12
Horizontal overhang 1.5 m	2.7	334.76	124	2363.72	0.14	320.88	118.8	1974.39	0.1625	424.31	157.2	2341.3	0.18	310.2	114.9	2432.28	0.13
Horizontal overhang 0.5 m (2 louvers)*D*/*L* = 2/1	1.8	221.77	123.2	2348.79	0.09	269.346	149.6	2294.45	0.1174	351.3	195.2	2319.1	0.15	157.41	87.5	2425.51	0.06
Horizontal overhang 0.5 m (4 louvers) *D*/*L* = 1/1	3.6	292.69	81.3	2308.56	0.13	375.89	104.4	2207.3	0.1703	483.85	134.4	2280.93	0.21	206.68	57.4	2410.37	0.09
Horizontal overhang 0.5 m inclined 30° (1 louver)	0.9	159.02	176.7	2385.43	0.07	201.56	224	2363.2	0.0853	239.49	266.1	2359.18	0.1	120.44	133.8	2435.86	0.05
Horizontal overhang 0.5 m inclined 30° (2 louvers)*D*/*L* = 2/1	1.8	269.36	149.6	2333.8	0.12	347.58	193.1	2258.78	0.1539	425.77	236.5	2304	0.18	175.58	97.5	2420.82	0.07
Horizontal overhang 0.5 m inclined 30° (4 louvers)*D*/*L* = 1/1	3.6	383.89	106.6	2290.32	0.17	515.76	143.3	2165.41	0.2382	603.46	167.6	2267.63	0.27	234.78	65.2	2398.86	0.1
horizontal louvers 0.25 m (8 louvers) *D*/*L* = 1/1	3.6	196.97	54.7	2292.43	0.09	260.68	72.4	2171.67	0.12	326.54	90.7	2264.23	0.14	114.63	31.8	2397.58	0.05
Vertical louvers 0.25 m (7 louvers) *D*/*L* = 1/1	3.5	88.17	25.2	2334.67	0.04	99.21	28.3	2329.6	0.0426	116.06	33.2	2287.58	0.05	100.34	28.7	2406.09	0.04
Brise Soleil (sun breaker) (3 sides) 0.5 m	2.9	191.34	66	2365.74	0.08	321.55	110.9	2335.31	0.1377	259.67	89.5	2325.32	0.11	207.58	71.6	2418.32	0.09
Vertical louvers 0.5 m (4 Louvers) *D*/*L* = 1/1	4	99.6	24.9	2360.83	0.04	373.95	93.5	2316.01	0.1615	109.42	27.4	2316.01	0.05	210.24	52.6	2407.76	0.09
Vertical louvers 1 m (4 Louvers) × 1 m *D*/*L* = 1/2	8	99.2	12.4	2346.25	0.04	562.01	70.3	2318.57	0.2424	199.2	24.9	2304.81	0.09	246.07	30.8	2399.88	0.1
Vertical louvers 0.75 m (3 Louvers) *D*/*L* = 1/1	4.5	149.4	33.2	2353.73	0.06	118.66	26.4	2353.73	0.0504	112.05	24.9	2333.45	0.05	138.88	30.9	2407.9	0.06

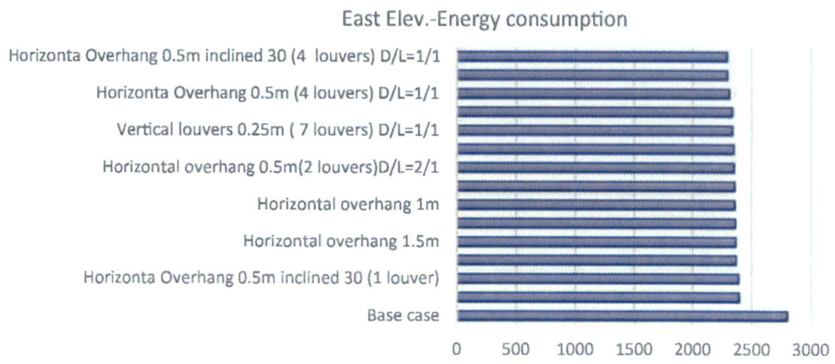

Fig. 7 East elevation annual energy consumption (kWh)

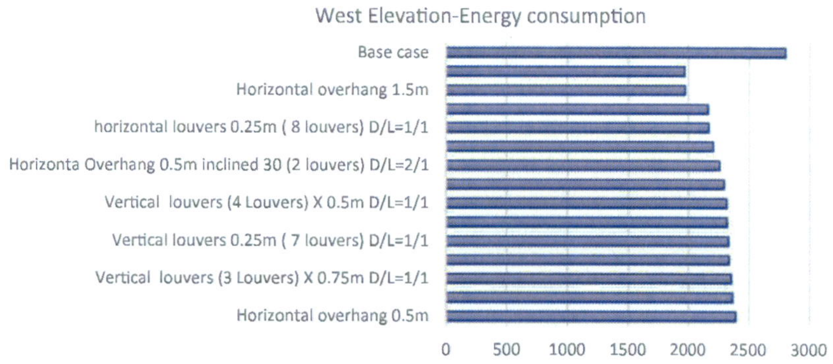

Fig. 8 West elevation annual energy consumption (kWh)

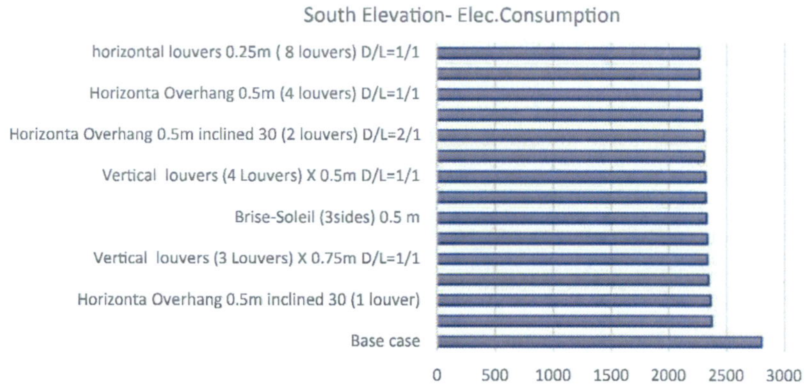

Fig. 9 South elevation annual energy consumption (kWh)

For East elevation, the highest PV production per meter square of installed PV was as follows: horizontal overhang 1 m, followed by horizontal overhang (0.5 m–inclined 30°–1 louver), horizontal overhang 0.5 m, and horizontal overhang (0.5 m–inclined 30°–2 louvers–*D*/*L* = 2/1) (Fig. 17).

For West elevation, the highest PV production per meter square of installed PV was as follows: horizontal overhang (0.5 m–inclined 30°–1 louver), horizontal overhang (0.5 m–inclined 30°–2 louvers–*D*/*L* = 2/1), horizontal overhang 0.5 m, and horizontal overhang 1 m (Fig. 18).

For South elevation, the highest PV production per meter square of installed PV was as follows: horizontal overhang 1 m, followed by horizontal overhang (0.5 m–inclined 30°–1 louver), horizontal overhang 0.5 m, and horizontal overhang 1.5 m (Fig. 19).

For North elevation, the highest PV production per meter square of installed PV was as follows: horizontal overhang 1 m, horizontal overhang (0.5 m–inclined 30°–1 louver), horizontal overhang 0.5 m, and horizontal overhang 1.5 m (Fig. 20).

Generally, it could be concluded that horizontal overhangs 1 m and horizontal overhang (0.5 m–inclined 30°–1 louver) are the most efficient from the economical point of view because they produce the larger amount of electricity with the least installed PV area (Fig. 21).

Fig. 10 North elevation annual energy consumption (kWh)

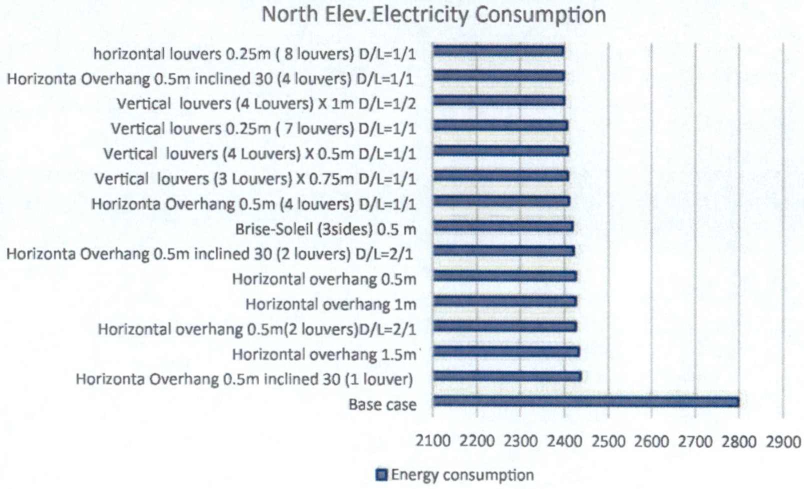

Fig. 11 Annual energy consumption (kWh)

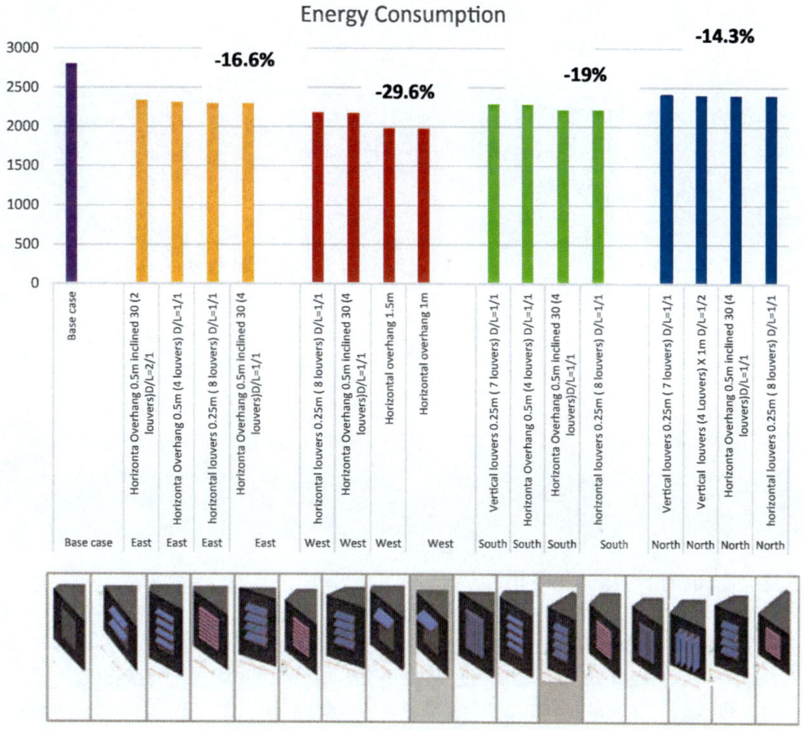

Fig. 12 East elevation PV production (kWh)

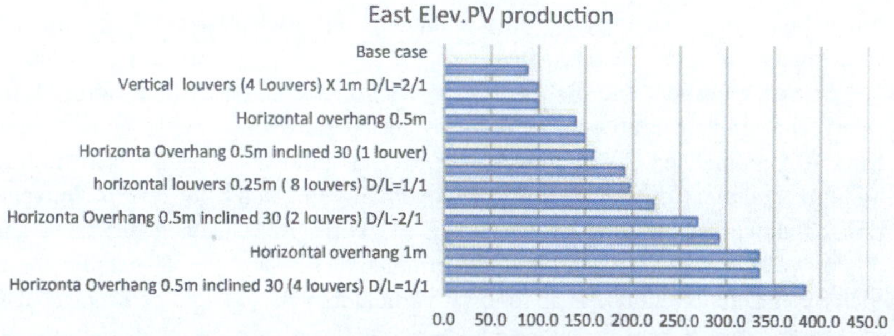

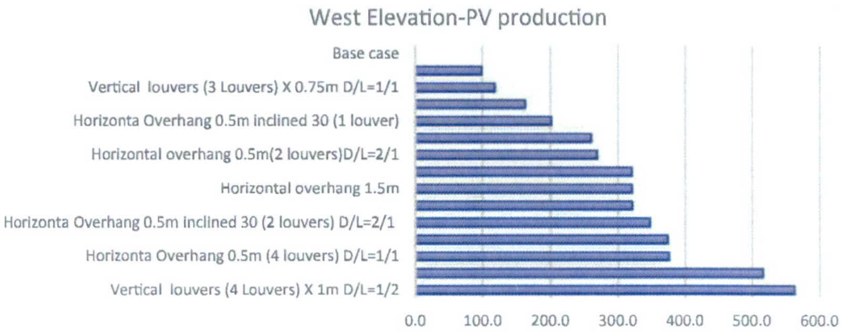

Fig. 13 West elevation PV production (kWh)

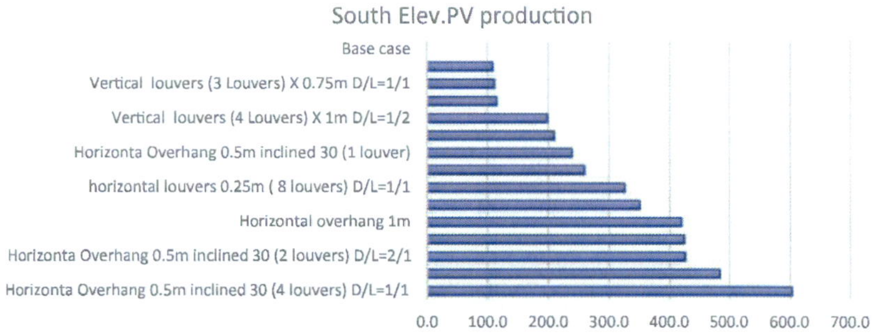

Fig. 14 South elevation PV production (kWh)

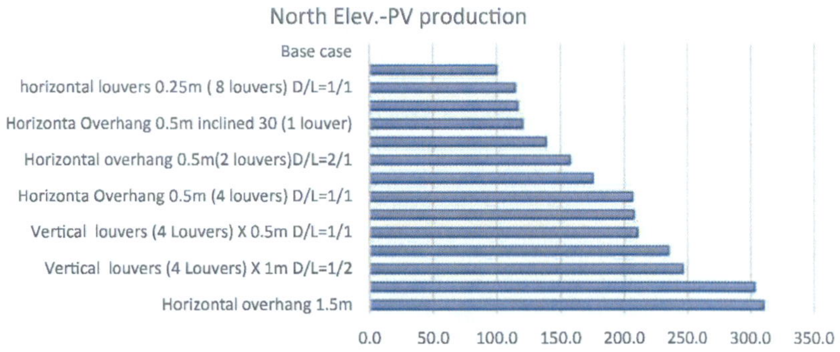

Fig. 15 North elevation PV production (kWh)

3.4 Shading Device Efficiency

3.4.1 PV Production/Electricity Consumption

In order to categorize the shading devices in terms of the more energy-efficient construction, the percentage of PV energy production (%) contribution in the annual energy consumption is used. The ratio of the amount of electricity production from each device to the amount of electricity consumed in each case is calculated.

For east elevation, the highest PV production to electricity consumption ratio was as follows: horizontal overhang (0.5 m–inclined 30°–4 louvers–*D*/*L* = 1/1), followed by horizontal overhang 1.5 m, horizontal overhang 1 m, and horizontal overhang (0.5 m–4 louvers–*D*/*L* = 1/1) (Fig. 22).

For West elevation, the highest PV production to electricity consumption ratio was as follows: vertical louvers (4 louvers * 1 m–*D*/*L* = 1/2), followed by horizontal overhang (0.5 m–inclined 30°–4 louvers–*D*/*L* = 1/1), horizontal overhang (0.5 m * 4 louvers–*D*/*L* = 1/1), and vertical louvers (0.5 m * 4 louvers–*D*/*L* = 1/1) (Fig. 23).

For South elevation, the highest PV production to electricity consumption ratio was as follows: horizontal overhang (0.5 m–inclined 30°–4 louvers–*D*/*L* = 1/1), followed by horizontal overhang 0.5 m * 4 louvers–*D*/*L* = 1/1), horizontal overhang (0.5 m inclined 30° * 2 louvers–*D*/*L* = 2/1), and horizontal overhang 1.5 m (Fig. 24).

For North elevation, the highest PV production to electricity consumption ratio was as follows: horizontal overhang 1.5 m, followed by horizontal overhang 1 m, vertical louvers (4 louvers * 1 m–*D*/*L* = 1/2) and horizontal overhang (0.5 m inclined 30° * 4 louvers–*D*/*L* = 1/1) (Fig. 25).

Generally, the most energy-efficient construction, where the percentage of PV energy production contribution in the annual energy consumption reached 27% was horizontal

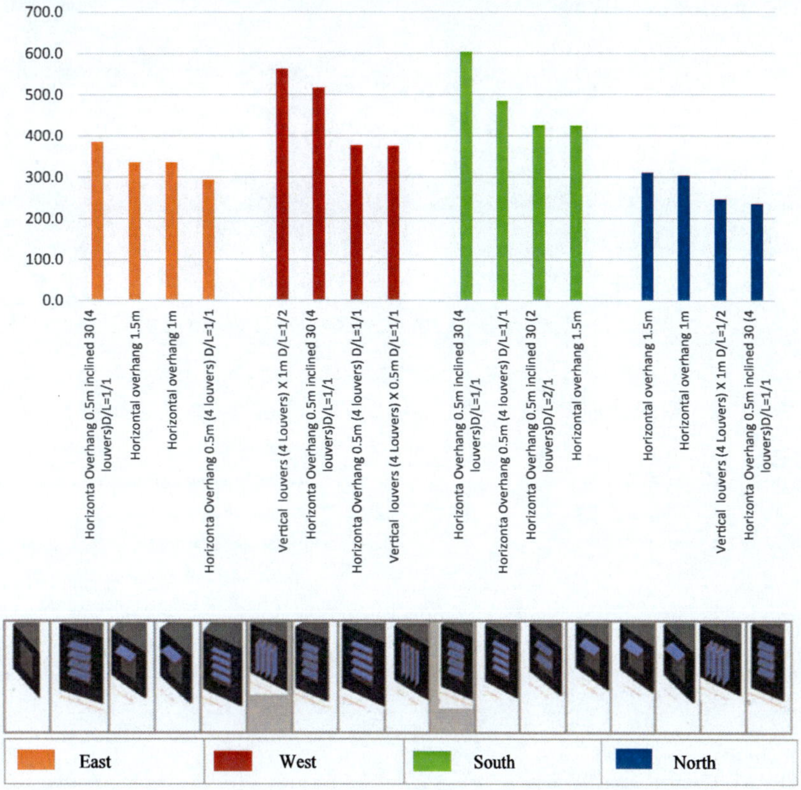

Fig. 16 PV annual electric energy production (kWh)

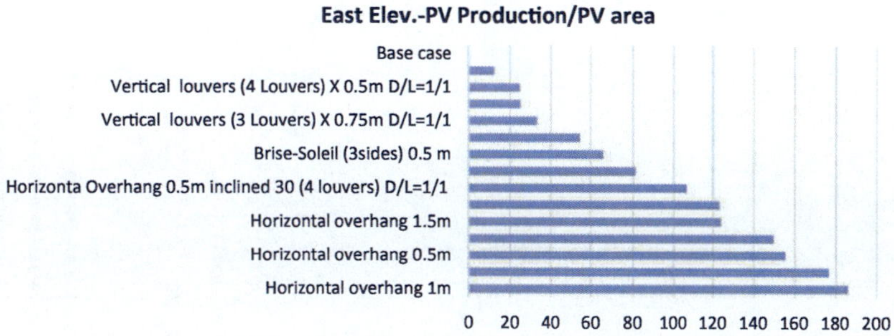

Fig. 17 East elevation PV production/PV area

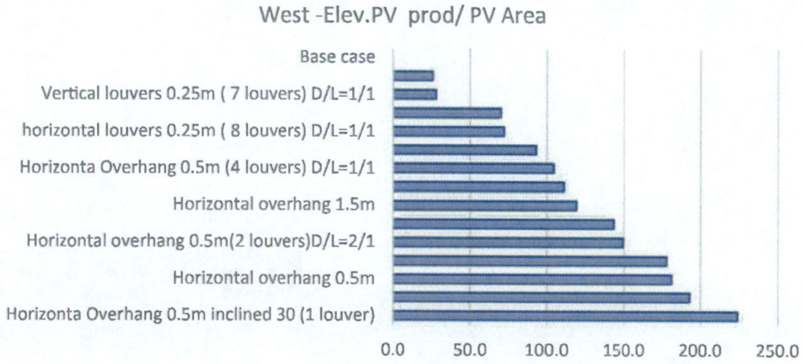

Fig. 18 West elevation PV production/PV area

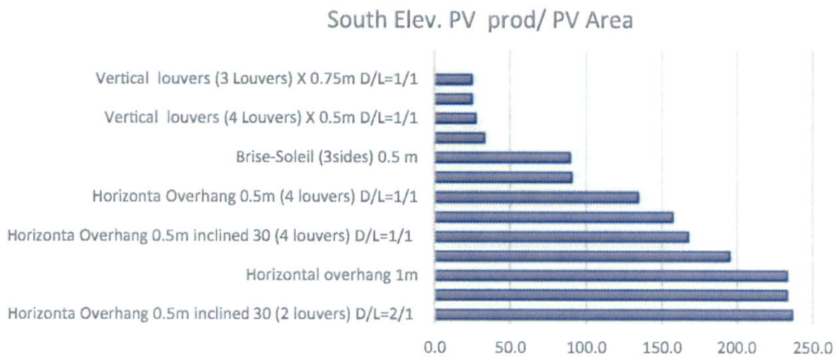

Fig. 19 South elevation PV production/PV area

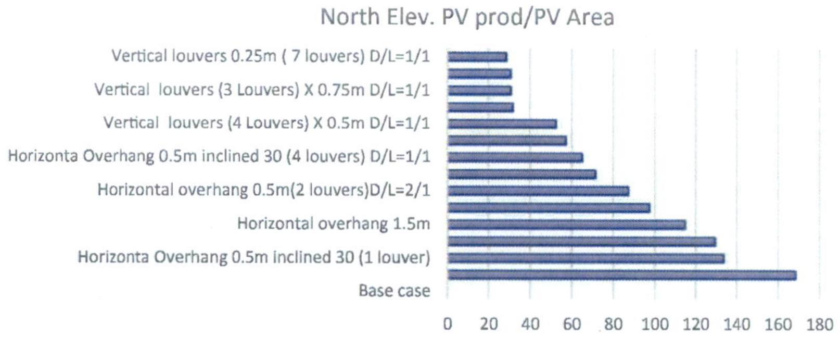

Fig. 20 North elevation PV production/PV area

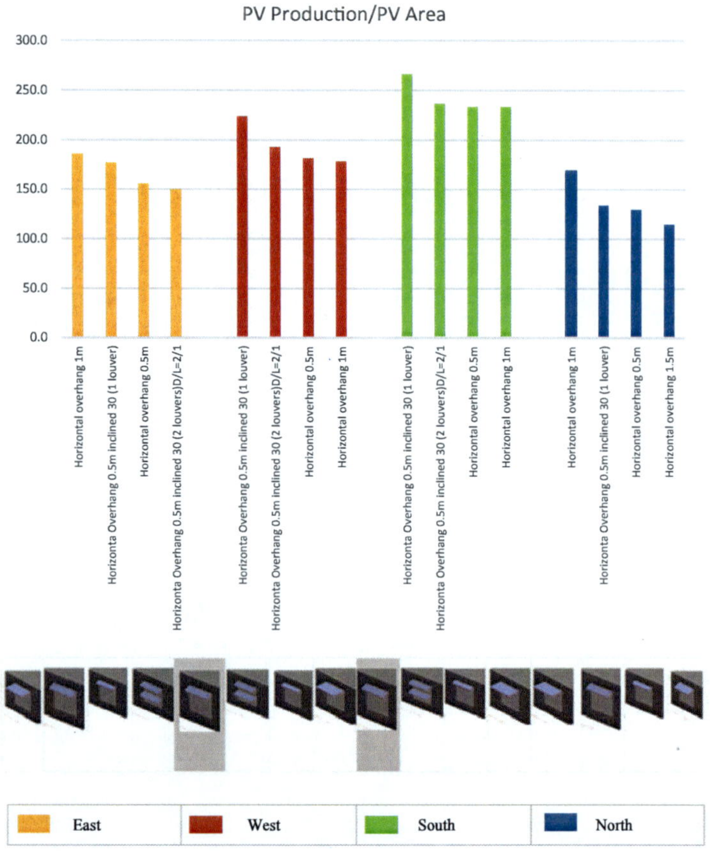

Fig. 21 PV production/PV area (kWh/m^2)

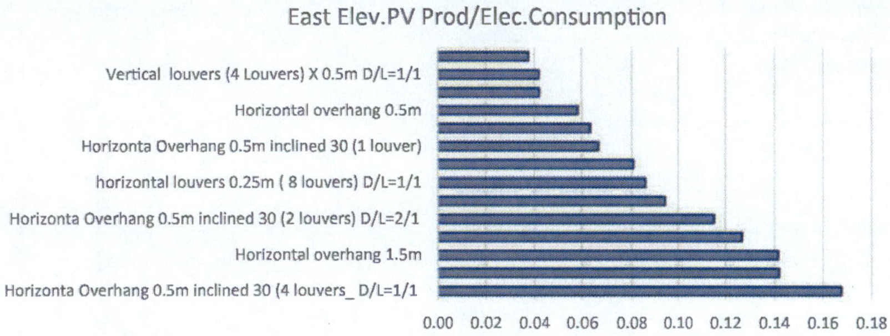

Fig. 22 East elevation PV production/electricity consumption

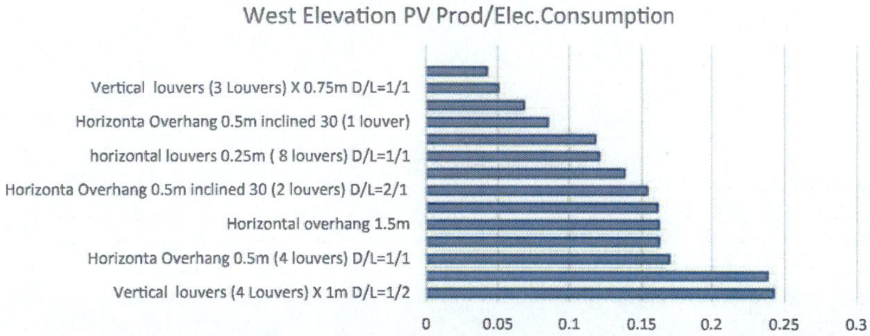

Fig. 23 West elevation PV production/electricity consumption

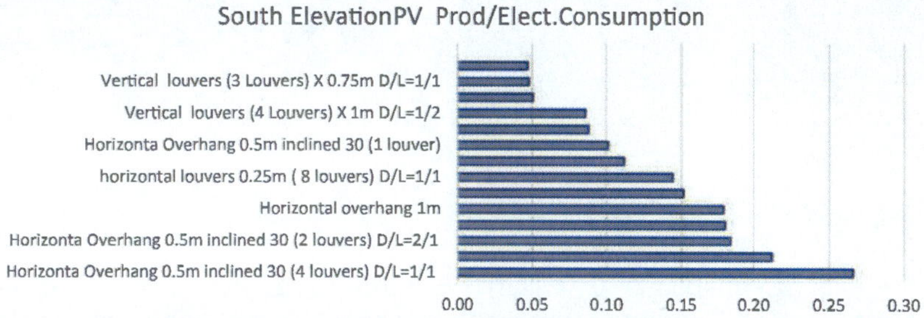

Fig. 24 South elevation PV production/electricity consumption

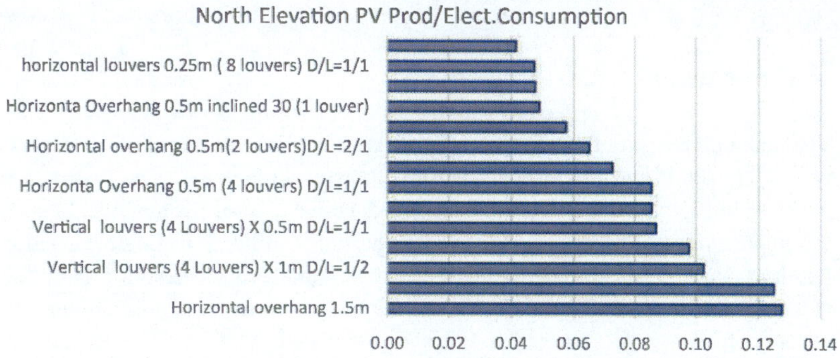

Fig. 25 North elevation PV production/electricity consumption

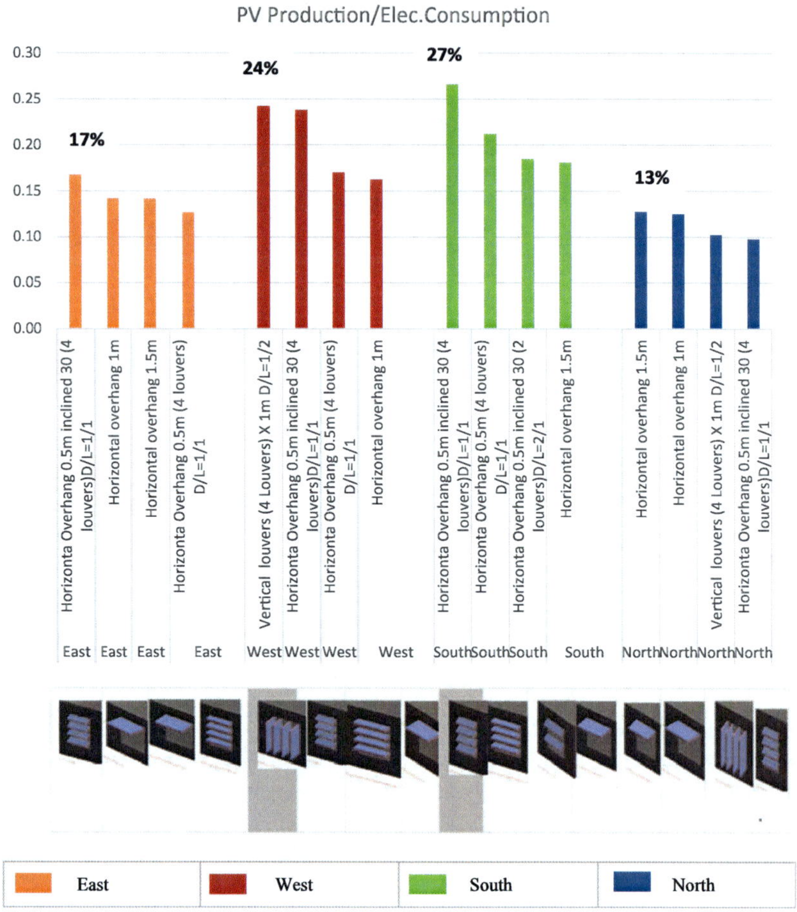

Fig. 26 Annual PV production/annual electric consumption

overhang (0.5 m inclined 30°–4 louvers and $D/L = 1/1$) in South elevation, followed by vertical louvers (1 m–4 louvers–$D/L = 1/2$) where the percent reached 24% (Fig. 26).

4 Conclusions

This research analyzes the effect of utilizing shading devices with integrated PV on energy consumption of a typical office building in Egypt. Annual electric energy reductions due to the application of shading device to air-conditioned office building were calculated by means of energy simulation software and DesignBuilder. Also, the annual electric energy production due to the integrated PV was calculated. The energy demand for cooling and lighting was analyzed as well. The research aimed at establishing criteria for applying shading devices with integrated PV, quantifying possible energy savings and optimizing geometry and position of shading devices as a function of various parameters.

By examining 14 types of shading devices with integrated PV, the following conclusions can be highlighted:

It has been proven that the shading device best performing in terms of energy consumption reduction are horizontal overhang (1 m), horizontal overhang (1.5 m) and horizontal overhangs (0.5 m–inclined 30°–4 louvers–$D/L = 1/1$). The greatest reductions in energy consumption reached 29.6% in West elevation in case of using horizontal overhangs (1 m) which highlights the importance of using shading in West elevations.

The highest PV production was achieved by using horizontal overhangs (0.5 m–inclined 30°–4 louvers–$D/L = 1/1$) in South elevation. This highlights the fact that the South elevation is the best for PV integration since it receives the higher amount of solar radiation throughout the year.

The best-performing shading device in terms of PV production according to an economic point of view is not corresponding to the best performance in terms of the maximum PV production and minimum energy consumption. Horizontal overhangs (1 m) and horizontal overhang (0.5 m–inclined 30°–1 louver) are the most efficient from the economical point of view because they produce the larger amount of electricity with the least PV area. As the number of louvers increase, overshadowing increase, thus decreasing the PV production per square meter of installed PV.

However, increasing the number of louvers is better in blocking the sun and consequently decreasing the energy consumption, but the distance between louvers should be large enough to prevent over shadowing. According to an economic point of view, the ratio of the distance between louvers to the length of the louver *D*/*L* should better be 2:1 and not less than 1:1.

The most energy-efficient constructions were: horizontal overhang (0.5 m–inclined 30°–4 louvers–*D*/*L* = 1/1) in south elevation where the amount of PV electricity production covered about 27% from the total electricity consumption followed by Vertical louvers (4 louvers * 1 m–*D*/*L* = 1/2) in West elevation.

This research highlights the significant contribution of PV-integrated shading devices in energy consumption reduction and electricity production especially in West and South orientations. Recently, due to the increasing energy demand and the great dependency on conventional energy sources in the building sector, the world is facing a major energy problem. Photovoltaic-integrated shading devices can be considered as a promising solution to energy problems since they can significantly contribute to the energy balance. Thus, there should be further development to PVSD in order to enhance their utilization. Further research is required in order to reach a more integrated approach regarding using PVSD for improvement of the quality of the interior space with reduced consumption of energy. Further research should include more quantitative measures of the PVSD, in addition to other, qualitative ones. They should also include cost–benefit analysis, assessment of glare problems, visual contact with the outside, and studies on the aesthetic effects of the PV installation as well. More studies may include also PVSD application in other building types.

References

Abdin, A., et al. (2006). Energy efficiency in the construction sector in the Mediterranean Market analysis and capacity assessment—Egypt. Study released under the MED-ENEC Project. http://medenec.com/en/TOPIC/marketstudies/Marketanalysisandcapacityassessment-Egypt.pdf.

Alzoubi, H. H., & Al-Zoubi, A. H. (2010). Assessment of building façade performance in terms of daylighting and the associated energy consumption in architectural spaces: Vertical and horizontal shading devices for southern exposure facades. *Energy Conversion and Management, 51*(8), 1592–1599.

Bahr, W. (2009, December). Integration of shading devices and photovoltaic panels into existing building facades. In *Proceedings of Green Retrofit Conference*, UAE, Dubai, 15–16.

Bahr, W. (2013, June). Optimal design parameters of the blinds integrated photovoltaic modules based on energy efficiency and visual comfort assessments. In *Proceedings of Central Europe Towards Sustainable Building Conference*, Czech Republic, Prague, 26–28.

Bahr, W. (2014). A comprehensive assessment methodology of the building integrated photovoltaic blind system. *Energy and Buildings, 82*, 703–708.

Energy Information Administration (EIA). (2011). Annual energy review, 2011. https://www.eia.gov/totalenergy/data/annual/pdf/aer.pdf.

Fouad, M., Abdalfatah, M., Kandil, H., Shehata, L. (2019). Simulation of a zero energy office building in Egypt with a photovoltaic integrated shading system. *Journal of Photonics for Energy*.

Gindi, S., Abdin, A. R., & Hassan, A. (2017). Building integrated photovoltaic retrofitting in office buildings. *Energy Procedia, 115*, 239–252. https://doi.org/10.1016/j.egypro.2017.05.022.

Gratia, E., & De Herde, A. (2007). The most efficient position of shading devices in a double-skin façade. *Energy and Buildings, 39*, 364–373.

Guide, D. G. S. (2008). *Planning and installing photovoltaic systems: A guide for installers, architects and engineers* (2nd ed.). London: Earthscan.

Hwang, T., Kang, S., & Kim, J. T. (2012). Optimization of the building integrated photovoltaic system in office buildings—Focus on the orientation, inclined angle and installed area. *Energy and Buildings, 46*, 92–104.

Ibraheem, Y., Farr, E., & Piroozfar, P. A. E. (2017). Embedding passive intelligence into building envelopes: A review of the state-of-the-art in integrated photovoltaic shading devices. *Energy Procedia, 111*, 964–973.

International Energy Agency. (2011). *World energy outlook*. OECD. https://doi.org/10.1787/weo-2011-en.

Jung, S. K. (2014). Optimal control methods for PV-integrated shading devices [PhD dissertation]. University of Michigan.

Kaldellis, J., Kavadias, K., & Zafirakis, D. (2012). Experimental validation of the optimum photovoltaic panels' tilt angle. *Renewable Energy, 46*, 179–191.

Kang, S., Hwang, T., & Kim, J. T. (2012). Theoretical analysis of the blinds integrated photovoltaic modules. *Energy and Buildings, 46*, 86–91.

Kim, J. J., Jung, S. K., Choi, Y. S., & Kim, J. T. (2010). Optimization of photovoltaic integrated shading devices. *Indoor and Built Environment, 19*(1), 114–122.

Kim, G., Lima, H. S., Limb, T. S., Schaeferc, L., Kim, J. T. (2012). Comparative advantage of an exterior shading device in thermal performance for residential buildings. *Energy and Buildings, 46*, 105–111.

Lee, E. S., Selkowitz, S. E., DiBartolomeo, D. L., Klems, J. H., Clear, R. D., Konis, K., Hitchcock, R., Yazdanian, M., Mitchell, R., & Konstantoglou, M. (2009). *High performance building façade solutions* (Report number: 4583E). Lawrence Berkeley National Laboratory.

Lomanowski, B. A., & Wright, J. L. (2009). Modeling fenestration with shading devices in building energy simulation: A practical approach. In *Proceedings of Building Simulation 2009*, Glasgow, Scotland, 27–30 (pp. 976–983).

Loutzenhiser, P. G., Manz, H., Carl, S., Simmler, H., Maxwell, G. M. (2008). Empirical validation of solar gain models for a glazing unit with exterior and interior blind assemblies. *Energy and Buildings, 40*, 330–340.

Mandalaki, M., Zervas, K., Tsoutsos, T., & Vazakas, A. (2012). Assessment of fixed shading devices with integrated PV for efficient energy use. *Solar Energy, 86*, 2561–2575.

Mandalaki, M., Papantoniou, S., & Tsoutsos, T. (2014a). Assessment of energy production from photovoltaic modules integrated in typical shading devices. *Sustainable Cities and Society, 10*, 222–223.

Mandalaki, M., Tsoutsos, T., & Papamanolis, N. (2014b). Integrated PV in shading systems for Mediterranean countries: Balance between energy production and visual comfort. *Energy and Buildings, 77,* 445–456.

Sun, L. L., & Yang, H. X. (2010). Impacts of the shading-type building-integrated photovoltaic claddings on electricity generation and cooling load component through shaded windows. *Energy and Buildings, 42*(4), 455–460.

Sun, L., Lu, L., & Yang, H. (2012). Optimum design of shading-type building-integrated photovoltaic claddings with different surface azimuth angles. *Applied Energy, 90*(1), 233–243.

Tongtuam, Y., Ketjoy, N., Vaivudh, S., & Thanarak, P. (2011). Effect of the diffuse radiation reflection from exterior wall on shading device integrated photovoltaic case of thailand building. *Energy Procedia, 9,* 104–116.

Tzempelikos, A., & Athienitis, A. (2007). The impact of shading design and control on building cooling and lighting demand. *Solar Energy, 81,* 369–382.

IEA ECBCS Annex 53. (n.d.). Annex 53. Total energy use in buildings: Analysis & evaluation methods. www.ecbcsa53.org.

Yoo, S. H., & Lee, E. T. (2002). Efficiency characteristic of building integrated photovoltaics as a shading device. *Building and Environment, 37,* 615–623.

Yoo, S. H., & Manz, H. (2011). Available remodeling simulation for a BIPV as a shading device. *Solar Energy Materials and Solar Cells, 95*(1), 394–397.

Yoo, S. H., Lee, E. T., & Lee, J. K. (1998). Building integrated photovoltaics: A Korean case study. *Solar Energy, 64,* 151–161.

Youssef, A. M. A., Zhai, Z. J., & Reffat, R. M. (2015). Design of optimal building envelopes with integrated photovoltaics. *Building Simulation—China, 8*(3), 353–366.

Zhang, X., Lau, S. K., & Lau, S. Y. (2018, August). Photovoltaic integrated shading devices (PVSDs): A review. *Solar Energy*.

The Traditional Architecture of the Heritage Villages in the Kingdom of Saudi Arabia and the Thermal Performance Efficiency

Eman Mohamed Eid Attieh and Ola Samy Ali Alhinawy

Abstract

Architecture is considered as the primary means which created by human for protecting them from external environmental conditions and for adaptation and compensation with his environment and nature which changing by various ways from place to place depending on climatic conditions and surrounding nature and available local materials. Kingdom of Saudi Arabia has many heritage villages which are unique examples of traditional architecture which are considered as the reflection of the culture, traditions, and how the community using local building materials to adapt to the warm climate of the region which has a big variation from summer to winter and also vary from extreme humidity to severe drought due to the proximity or distance from water sources. We have seen the attempts of the old Saudi architect to adapt to the climate that has emerged using many traditional smart techniques, some of which are adjoining buildings, streets, and the thickness of the walls of the building, effective attempts to improve the efficiency of the thermal performance of the internal spaces in the urban fabric of the village and buildings. The study aims are studying the environmental performance of some of the heritage villages in the Kingdom of Saudi Arabia, in the village of Al Majmaah, and the village of Hirma, in a practical way using thermal data loggers to monitor the temperature and humidity in the fabric and buildings of the heritage villages. The importance of our research is to encourage the conservation of these heritage villages, not only because they are heritage, but also because they are environmental and to encourage the use of the traditional architecture of ancient architecture for use in the current architecture. The research uses the experimental approach and the analytical approach in monitoring, analyzing, and evaluating thermal measurements. The village of Al Majmaah located in the south west of the modern city, the historic center of Majmaah includes a few heritage and archeological buildings, including the Menekh archaeological observatory, Al-Rabiah Palace, the historic Al-Sanea Academic School, the Military Palace and its annexes, the King Abdul Aziz Mosque, and its accessories. The village is located about 195 km north-west of the city of Riyadh, and the date of the establishment of the town of Hirma around 770 AH has been included as part of the state projects in the project of rehabilitation and development of the central cities of the Kingdom, which is based on the ministry of municipal and rural affairs represented by secretariats and municipalities in partnership with the general authority for tourism and national heritage.

Keywords

Traditional architecture • Heritage villages • Majmaah • Hirma • Thermal performance efficiency

1 Introduction

Architecture is the main means that people have devised to protect them from external environmental conditions, and to adapt and coexist in their environment, varying from place to place depending on the climatic conditions and the nature of the surrounding environment and local materials available and methods of construction and customs and traditions of the inhabitants of the region to integrate all these technologies together to form an architecture known as traditional architecture.

The Kingdom of Saudi Arabia has many traditional villages that represent this traditional architecture, which

E. M. E. Attieh
Professor History of Architecture, Department of Architecture, Faculty of Engineering, Menoufia University, Al Minufya, Egypt

O. S. A. Alhinawy (✉)
Lecturor in Department of Architectural Engineering, Faculty of Engineering, Menoufia University, Al Minufya, Egypt

F. Trapani et al. (eds.), *Advanced Studies in Efficient Environmental Design and City Planning*, Advances in Science, Technology & Innovation, https://doi.org/10.1007/978-3-030-65181-7_10

reflects the culture, customs, and traditions of the community, and the use of local construction is evident to adapt to the hot climate of the region, where temperatures vary between summer and winter. From extreme humidity to severe drought depending on the proximity or distance from water sources, the attempt of the ancient Saudi architect to adapt to the climate has emerged through the use of many traditional smart techniques, some of which are the buildings are close together, winding streets, and increase the thickness of the building walls, in a good attempt to improve the thermal performance efficiency of interior spaces in buildings.

2 Research Problem and Objective of Research

The research problem is the loss of Arab architectural identity, which has long been interested in finding a way to adapt to the surrounding climate, and the modern architect is not familiar with these methods but replaced by cooling and heating devices that drain energy and pollution of the environment.

The aim of the research is to study the thermal performance of some heritage villages in the Kingdom of Saudi Arabia, which are represented in Al Majmaah Heritage village and Hirma Heritage village, using instruments to measure and monitor the thermal performance of the village at the level of fabric and buildings to evaluate their thermal performance and to take advantage of those techniques developed by the old Saudi architect. In the design of the modern Arab house, which makes it better environmental efficiency, it helps to reduce and rationalize the energy consumed by refrigeration and heating devices to reach it.

3 Research Methodology

Research uses a quantitative approach by collecting, monitoring, and then analyzing data to reach results, and the research relies on monitoring thermal measurements and gathering information about the method of planning the city, and the method of construction and building materials used and study how they are correlated to thermal measurements to get results that allow us to assess the thermal performance of these villages and explain the reasons that contributed to those results.

The research also uses the comparative analysis method, through comparing the urban planning of the fabric of the two traditional villages (Majmaah, Herma), And linking the results of thermal measurements that were monitored with the urban planning of the fabric of those villages, to know the extent of the impact of urban planning on the thermal performance of the fabric and buildings of the villages

4 Temporal and Spatial Determinants

First: Time Limits: After studying the climate of the city of Majmaah in Saudi Arabia, we chose month May to apply the study, which considered the most volatile month of the year in the atmosphere, where it combines with the wind loaded with dust, rain, and high temperatures.

Secondly, the spatial determinants: It was selected from Al Majmaah Heritage Village and Hirma Heritage Village, as their location is very close to each other. they are similar in construction, building materials, finishing, and building design; while their physical layout is different, this allows us to make a sound scientific comparison in which all variables are fixed except the urban planning of the two villages, and this allows to know the impact of urban planning on the thermal performance of the urban community.

5 Practical Steps in the Search

The practical steps in the search were divided into several points:

a. **The field study of the villages**: study and survey the urban fabric, study the design, construction methods and materials of the buildings, and determine the most important places where the thermal data loggers will be installed.
b. **Monitoring thermal measurements**: Thermal measurements were monitored in different spaces of some village houses with different design and orientation, and some internal and external streets of villages with different areas and orientation, at different times of the day and different days in May 2019, and the use of meteorological archive at the station (Marqab) adjacent to the study site, to know the thermal measurements of the city, in addition to monitoring thermal measurements outside the heritage village to be a reference that facilitates comparison of thermal measurements that were monitored by the heritage village (Fig. 1).
c. **Thermal data loggers used for research**: The devices used consist of five sensors for monitoring and recording thermal measurements, and a tablet device that works to read the sensors and monitor that data and link them together and can connect a maximum of eight devices, and a diagram of them can also monitor that data by connecting them to the computer in the form Excel file (Fig. 2).

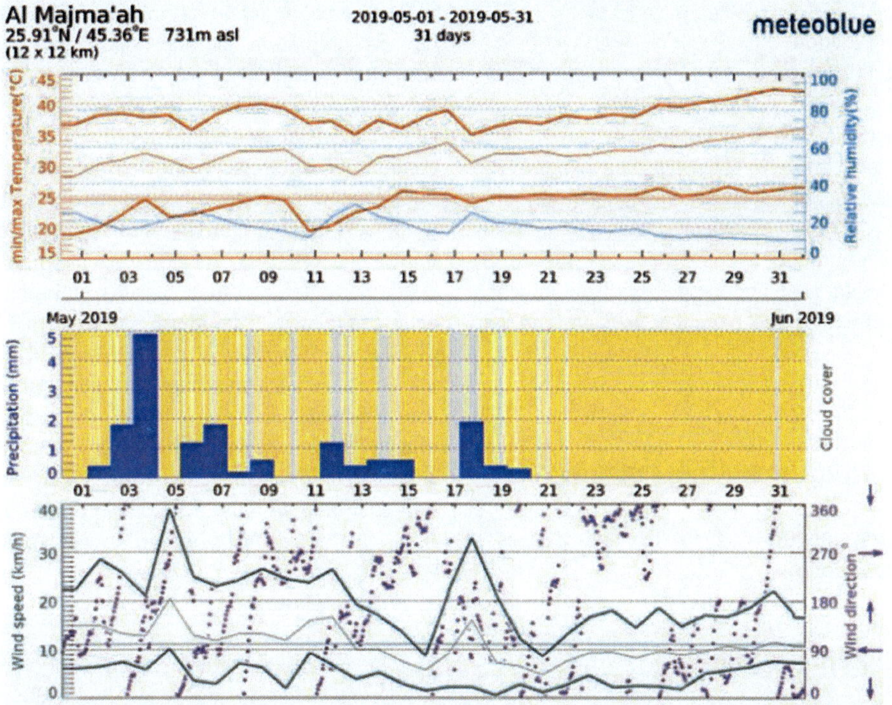

Fig. 1 Thermal measurements of Majmaah City, Saudi Arabia for May 2019. *Source* https://www.meteoblue.com/en/weather/historyclimate/weatherarchive/al-majma%27ah_saudi-arabia_109194

Fig. 2 Thermal data loggers used in research. *Source* Researchers

Brand Name: MI·SOL
Style: Standing Station
Display Size: 2.0–3.9 in.
Display Type: Digital
Model Number: A6-WS-HP3001-8MZ.

d. **Data output and analysis**: Thermal data loggers have Datalogger, and this is a feature for storing information for long periods of time, where devices monitor the temperature and humidity over long periods of time that are analyzed by the user, and the devices data are stored and then uploaded by the computer to save and read (Fig. 3).

6 Analytical Study of Heritage Villages

The research deals with the study of the two villages as mentioned above, as follows.

6.1 Majmaah Heritage Village

The historic village of Majmaah is located southwest of the modern city, and it includes a few heritage and archaeological buildings, including mount archaeological observatory, the Rabiah Palace, Ahmed Al-Sanea historical school, the Askar Palace and its annexes (Palace of the Principality), King Abdul Aziz Mosque, and the endowment of King Abdul Aziz and its accessories.

The complex has been included in the state's projects in the "Rehabilitation and Development of Mid-Cities in the Kingdom" project, which is based on the ministry of municipal affairs. The project aims to prepare a comprehensive plan for the development of the historical center of the community to preserve the heritage identity of the area improving the region's historic environment, both architecturally and urban, improving the corridors and utility network lines. Phases (I-II-III), including paving, lighting of the entire historical center, and some facade restoration work have now been completed. The fourth phase, which completes the renovation of buildings in the historical center, was also presented. A joint coordination mechanism was also established between the municipality, the commission and the tourism committee in Al Majmaah governorate on citizens' adoption of renovating, rebuilding, and rehabilitating their houses in line with the development of Al Majmaah Center (Fig. 4) (Heritage Villages and Towns, 2009, p 29).

a. **Village planning**

The layout of the village relied on conjoined buildings, winding streets and narrowing the width and shading to reduce the exposure of buildings and streets to undesirable sunlight, reduce winds loaded with dust, create a stream of gentle winds, and provide shadows on the streets pedestrians at the daylight time, also has been used wood panels to shade some of the street spaces and also drilling a well in the middle of the village to provide its people with drinking water (Figs. 5 and 6).

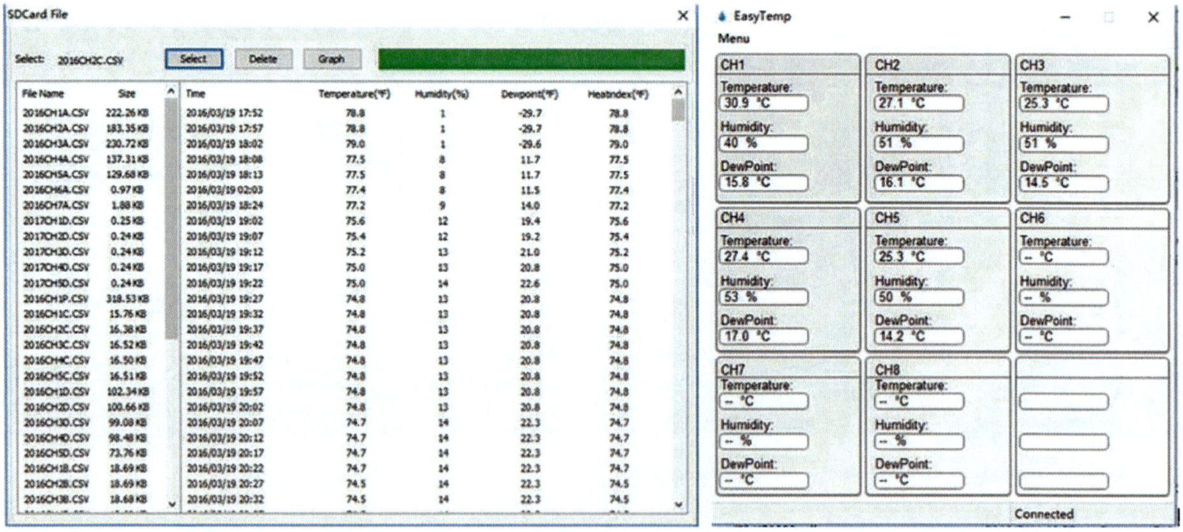

Fig. 3 How to output data from thermal data loggers. *Source* Researchers

Fig. 4 Al Majmaah Heritage Village map. *Source* Google Earth

Fig. 5 The urban fabric of Majmaah Heritage Village. *Source* Heritage villages and towns, General Authority for Tourism and Antiquities, Saudi Arabia, 2009, Pdf, p. 29

a **b** **c** **d**

Fig. 6 **a** Covering the streets, **b**, **c** narrow streets, **d** water well in the village of Majmaah Heritage. *Source* Researchers

b. **Building Materials**

Mud and straw were used in the construction of buildings, and the roof structures were made of wood, and the clay material which has a positive effect on the interior atmosphere of the building during the summer and winter because the heat energy gained from the sun during the day is stored in it and then slowly reissued into the building later in the day. In the summer, this process reduces cooling loads, meaning that the heat gained from the external or internal factors of the building is stored inside the building materials of the building itself, which keeps less amount of heat inside the house need to be disposed of during the day, which has a big effect in minimizing heat gain (Fig. 7) (General Authority for Tourism and Antiquities, 2009).

c. **Designing buildings**

The buildings were designed with a two-floor height, each floor is about three meters and the ground floor designed to contain a seating area, a storage and preparation area for food, and a barn for animals. Rooms were distributed on the first floor. The ventilation holes were made in the walls of the building in addition to the windows openings which are small spaces to reduce hot air coming out of the building during the daylight and the cold air at night (Fig. 8).

d. **Thermal monitoring of the buildings and fabric of the village**

Thermal measurements (°C, and Humidity %) were monitored for several days in May 2019 at different time of the

Fig. 7 Building materials used in Majmaah Heritage Village. *Source* Researchers

Fig. 8 Form of buildings and windows in Majmaah Heritage Village. *Source* Researchers

day with rate of reading every 3 h for different spaces for some study houses that differ in area, design and direction, and for some internal and external streets of the villages that vary in area and direction. The data were then emptied, and the temperature measurements were selected to display the hottest days (Fig. 9).

These data are in Tables 1 and 2.

6.2 Hirma Heritage Village

It is located about 195 km north-west of Riyadh city. Hirma shares with Al Majmaah in Al-Kalbi Valley as it is known, but it depends on the flow of these two valleys as they meet in this large dam which is about 400 m long. The history of the establishment of the town of his sanctuary around 770 AHS has been old (Ibrahim bin Hussein bin Medlej al-Hassani al-Anzi). The ancient historians have identified the location of Hirma, stating that it was the last border of Sudair from the northern side as well as the last border of Al-Yamamah (Fig. 10) [Heritage villages and towns, p 37].

a. **Village planning**

The village of Hirma was designed in such a way that the buildings are close together, as was the case in Al Majmaah village. However, the village Hirma did not divide the streets with narrow widths; the width of the streets in the complex was about 1.5 m; while in the sanctity increased to 3 m, the streets of the village of Hirma were not winding as in Majmaah, and no parts were covered. This has affected its thermal performance, giving a noticeable difference from the thermal performance recorded by Majmaah Village (Fig. 11).

Fig. 9 Installed thermal data loggers. *Source* Researchers

b. **Used building materials**

Mud, straw, and wood were used in construction as in the village of Majmaah, and they have a thermal performance that suits the desert climate. These materials can isolate the building during the daylight from extreme heat and store it inside for air at night inside the building to cool it. It also has good properties in providing a good amount of humidity required inside the building (Fig. 12).

c. **Design of buildings**

The design of buildings in the village of Hirma was larger than the buildings of the village of Al Majmaah. This may be because the village of Hirma was considered one of the rich villages in the Kingdom, but it did not differ in the interior design from what is in Majmaah buildings. It consisted of ground and first floors, and the ground floor contains a place to prepare and store The first floor consisted of bedrooms and living rooms, but the space of windows and openers were more wider than Majmaah buildings with small triangular openings that were made near the ceilings to be in a higher level of windows to get hot air out of them, creating a stream of cold air inside the spaces of the building (Fig. 13).

d. **Thermal performance of the buildings and fabric of the village**

The same steps were followed to monitor the thermal measurements (°C, and Humidity %) in the village of Majmaah and to monitor the thermal measurements of the village of Hirma, and the measurements were as follows in Tables 3 and 4.

7 Results

The main results of the research are as follows:

The traditional architecture of the heritage villages has been able to apply the principles of urban sustainability to adapt with the surrounding environment and climate, using local building and finishing materials, and planning methods adapted to the local climate.

Table 1 Thermal measurements of some houses of Majmaah Heritage Village

The time	The place	Outside the village		House 1	House 2	House 3	House 4
	The description	Open area outside and near the village		Northern orientation	Southern orientation	Eastern orientation	Western orientation
		Sun	Shadow				
12 am	Temp.	20.3	20.6	25.4	26.1	25.5	26.2
	Humidity	18.5	19.1	24.5	24.2	24.4	24.3
3 am	Temp.	18.2	18.5	24.8	25.4	25	25.6
	Humidity	17.1	17.4	25.3	25.3	25.6	25.2
6 am	Temp.	23.3	22.1	24.2	25.8	25.6	26
	Humidity	24.8	26.6	26.6	26.4	27	26.5
9 am	Temp.	25.4	23.4	23	23.7	23.8	23.9
	Humidity	28.4	30.5	32.5	32	33.2	32.1
12 pm	Temp.	38.8	34.8	32.4	32.9	32.6	33
	Humidity	23.5	23.7	32.1	32.6	32.5	32.7
3 pm	Temp.	37.2	34.4	32.5	33.8	32.8	34
	Humidity	23.8	24	31.6	31.3	31.9	31.4
6 pm	Temp.	31	29	28.4	31.6	28.6	31.5
	Humidity	24	24.2	28.3	28.1	28.5	28.2
9 pm	Temp.	23.8	23.5	27.1	29.4	27.3	29.1
	Humidity	22.6	23	26.9	26.5	27	26.8

Source The researchers

Table 2 Thermal measurements of some urban streets of Al Majmaah Heritage Village

The time	The place	Outside the village		Street width 1.5 M							
				Street 1		Street 2		Street 3		Street 4	
15-5-2019	The description	Open area outside and near the village		Northern orientation		Southern orientation		Eastern orientation		Western orientation	
		Sun	Shadow	Sun	Shadow	Sun	Shadow	Sun	Shadow	Sun	Shadow
12 am	Temp.	20.3	20.6	22.5	22.8	23.3	23.5	22.7	22.9	23.8	23.9
	Humidity	18.5	19.1	20.1	20.3	20.4	20.5	20.2	20.4	20.6	20.8
3 am	Temp.	18.2	18.5	20.3	20.5	20.4	20.6	20.4	20.5	20.3	20.4
	Humidity	17.1	17.4	19.2	19.4	19.6	19.5	19.5	19.3	19.4	19.6
6 am	Temp.	23.3	22.1	23.8	23.2	23.8	23.4	23.9	23.1	23.5	23.1
	Humidity	24.8	26.6	24.5	26.3	24.4	26	26.6	26.8	26.4	26.8
9 am	Temp.	25.4	23.4	24.4	24	24.6	24.2	24.8	24.2	24.3	24
	Humidity	28.4	30.5	29.8	30.2	29.7	30.1	29.7	30.3	29.5	29.9
12 pm	Temp.	38.8	34.8	36.2	33.6	36.4	33.8	36.3	33.2	36.6	33.8
	Humidity	23.5	23.7	26.5	26.8	26.7	26.9	26.7	26.9	26.8	27
3 pm	Temp.	37.2	34.4	35.8	32.4	36.1	32.6	35.7	32.4	36.4	32.8
	Humidity	23.8	24	25.5	25.7	25.5	25,8	25.4	25.7	25.6	25.9
6 pm	Temp.	31	29	29	27.2	29.4	27.9	29.2	27.6	29.6	27.8
	Humidity	24	24.2	25.3	25.5	25.2	25.6	25.3	25.5	25.6	25.8
9 pm	Temp.	23.8	23.5	25.4	25.5	25.5	25.5	25.4	25.5	25.6	25.7
	Humidity	22.6	23	24.6	24.7	24.6	24.6	24.6	24.7	24.5	24.6

Source The researchers

Fig. 10 The boundaries of the village of Hirma Heritage. *Source* Google Earth

Fig. 11 The urban fabric of Hirma Heritage Village. *Source* Heritage villages and towns, General Authority for Tourism and Antiquities, Saudi Arabia, 2009, Pdf, p. 38

Fig. 12 Building materials used in the village of Hirma heritage. *Source* The researchers

Fig. 13 Ventilation openings in the houses of the village of Hirma heritage. *Source* The researchers

7.1 Results of Thermal Performance Analysis for Both Villages

The results of the thermal analysis for each of the two villages were divided into an analysis of thermal measurements at the architectural and urban level, at the level of buildings and urban fabric of the two villages, as follows:

a. **First, the thermal performance analysis of Al Majmaah Heritage Village**

At the architectural level (buildings)

In the middle of the daylight (12 noon) temperatures dropped by more than 6 °C than outside the village, and buildings recorded higher humidity rates of up to more than 9 °C.

In the middle of the night (12 pm), the buildings maintained a good degree of temperature and humidity, with temperatures higher than outside the village by more than 5 ° C, and humidity increased by more than 6 °C.

Thus, the properties of building materials used in roofs and walls and their use with a large thickness, it is greatly contributed to isolating the interior spaces of buildings from the surrounding climate, to achieve the greatest possible thermal comfort rates for the internal spaces of the buildings, in addition to its ability to store moisture and heat during the day, to be used at night when the temperature is low.

At the urban level (the urban fabric of the village)

Planning the village so that the buildings are close together, thus reducing the walls exposed to direct sunlight and other climatic elements., Subsequently reducing the temperatures inside them, and the small size of the streets and coverage of parts of it has contributed to the work of different air pressure led to the generation of a stream of cold air, which positively affected the performance In mid-day temperatures recorded a decrease in temperature by an average (2 °C) from outside the village in the sun and shade, and humidity increased by an average (3%), which contributed to improve the thermal performance of the urban fabric and thus improve the thermal performance of buildings. Temperatures in the middle of the night recorded a rise in temperature and humidity by an average of (2%), giving a sense of warmth at night.

b. **Second: Thermal Performance Analysis of Hirma Heritage Village**

At the architectural level (buildings)

In the middle of the day (12 am), bicycles dropped by more than 4 °C than outside the village, and buildings recorded higher humidity rates of more than 6%.

In the middle of the night (12 pm), the buildings retained a good amount of temperature and humidity, recorded more than outside the village by more than (7 °C), and humidity increased by more than 7%.

Table 3 Thermal measurements of some houses of Hirma Heritage Village

The time	The place	Outside the village		House 1	House 2	House 3	House 4
	The description	Open area outside and near the village		Northern orientation	Southern orientation	Eastern orientation	Western orientation
		Sun	Shadow				
12 am	Temp.	20.3	20.6	26.6	26.9	26.4	26.9
	Humidity	18.5	19.1	23.6	23.5	23.7	23.7
3 am	Temp.	18.2	18.5	25.8	26.3	26.5	26.3
	Humidity	17.1	17.4	24.2	24.8	24.9	24.9
6 am	Temp.	23.3	22.1	25.2	25.6	25.7	26
	Humidity	24.8	26.6	26.6	26.4	27	26.5
9 am	Temp.	25.4	23.4	23.9	24.5	24.2	24.8
	Humidity	28.4	30.5	31.5	32	32.6	31.8
12 pm	Temp.	38.8	34.8	33.2	33.4	33.5	33.9
	Humidity	23.5	23.7	31.5	31.6	31.8	31.9
3 pm	Temp.	37.2	34.4	33.4	34.5	33.6	34.9
	Humidity	23.8	24	30.0	30.3	30.6	30.4
6 pm	Temp.	31	29	29.1	32.2	30.1	32.3
	Humidity	24	24.2	28.1	28.2	28.7	28.6
9 pm	Temp.	23.8	23.5	28.1	30.0	28.4	30.4
	Humidity	22.6	23	27.2	27.4	27.3	27.5

Source The researchers

Table 4 Thermal measurements of some urban streets of Hirma Heritage Village

The time	The place	Outside the village		Street Width 1.5 M							
				Street 1		Street 2		Street 3		Street 4	
15-5-2019	The description	Open area outside and near the village		Northern orientation		Southern orientation		Eastern orientation		Western orientation	
		Sun	Shadow	Sun	Shadow	Sun	Shadow	Sun	shadow	Sun	Shadow
12 am	Temp.	20.3	20.6	23.4	23.5	24.5	24.6	23.3	23.5	24.8	24.6
	Humidity	18.5	19.1	21.1	21.2	21.2	21.3	21.0	21.2	20.8	20.9
3 am	Temp.	18.2	18.5	21.0	21.0	21.3	21.5	21.8	22.0	21.9	21.7
	Humidity	17.1	17.4	20.6	20.4	20.5	20.7	20.6	20.6	20.8	20.9
6 am	Temp.	23.3	22.1	23.2	22.6	23.9	22.6	23.1	22.1	23.2	22.2
	Humidity	24.8	26.6	24.6	24.7	24.3	24.5	24.9	35.1	24.8	24.9
9 am	Temp.	25.4	23.4	25.2	23.8	25.4	23.6	25.6	23.5	25.3	23.8
	Humidity	28.4	30.5	29.9	30.2	29.8	30.0	29.9	30.1	29.8	29.9
12 pm	Temp.	38.8	34.8	37.8	33.3	38.4	33.5	38.6	33.2	38.4	33.4
	Humidity	23.5	23.7	27.1	27.2	27.3	27.4	27.3	27.5	27.4	27.3
3 pm	Temp.	37.2	34.4	36.8	32.8	36.9	32.6	36.8	32.6	37.0	32.7
	Humidity	23.8	24	26.1	26.0	26.2	26.2	26.4	26.5	26.7	26.7
6 pm	Temp.	31	29	30.1	27.4	30.8	27.9	30.2	27.5	30.6	27.5
	Humidity	24	24.2	25.8	25.6	25.9	25.9	25.7	25.8	26.0	25.9
9 pm	Temp.	23.8	23.5	26.0	26.2	26.4	26.6	26.3	26.4	26.8	26.4
	Humidity	22.6	23	24.3	24.4	24.6	24.6	24.5	24.6	24.8	24.9

Source The researchers

Consequently, the buildings performed their desired role in the thermal insulation of the building spaces during the day and night, and the storage of moisture and heat during the day to take advantage of them at night, through the large wall thickness, and building materials that have the characteristics of heat insulation, storage, and pumping in the building in the long run, but they were recorded as temperatures are higher during the day and night than recorded by Al Majmaah Heritage Village.

At the urban level (the urban fabric of the village)

The urban planning of the Hirma village heritage is different from what is in the village of Al Majmaah. The buildings were not close to the same density of buildings in Al Majmaah village. In the cold wind movement when it is in the village of Al Majmaah, temperatures did not record in the mid-day marked decrease in degrees from outside the village in the sun and shade, and humidity raised an average (2 °C). While temperatures in the middle of the night increased in temperature and humidity by an average (3 °C), an increase of 1 °C than in the village heritage of Al Majmaah gives a sense of warmth at night.

7.2 The Results of the Comparison Between the Thermal Performance of the Two Villages

Despite the similarity of the two villages in the building materials and the method of building buildings, the difference of urban planning for the fabric of the two villages has significantly affected the thermal performance of them.

8 Recommendations

- Planners should take advantage of traditional ideas and techniques in planning the traditional architecture of the local environment, which has worked to adapt to the climate, making new cities more sustainable and adaptive to the climate.
- Planners and architects must work together to reach urban communities and buildings more than the environment and with a better thermal performance by finding planning solutions and design with the choice of building materials that help to improve the thermal performance of the urban fabric and the building. As the building design and method materials for its construction integrate with planning the surrounding architectural to improve its thermal performance, therefore the architect and planner must study the local climate well and work to create an urban community that can adapt to its climate, to reduce the resort to cooling and heating devices to improve performance for them, thanks to the reduction energy consumption and environmental pollution.
- Architects must reach to design solutions that make the building more adapted to the environment and external climatic factors, to reach the greatest possible thermal comfort rates. In addition to the selection of building materials and finishing and implementation methods that make the building more sustainable and more adapted to the climate and the surrounding environment.
- Heritage village residents should be developed awareness among individuals residing in heritage villages, introducing them to the characteristics of heritage villages and their ability to adapt to the climate, and encouraging them to preserve those villages and preserve the local identity.

References

General Authority for Tourism and Antiquities. (2009). Building renovation work guide clay and stone. Saudi Arabia. Retrieved July 23, 2019, from https://scth.gov.sa/AntiquitiesMuseums/Documents/Cultural%20Village%202010%20FINAL%20small%20with%20logo.pdf

Heritage villages and towns. (2009). Building renovation work guide clay and stone (p. 29). General Authority for Tourism and Antiquities. Retrieved July 29, 2019, from https://scth.gov.sa/AntiquitiesMuseums/Documents/VILLAGE_PROGRAM_book_(law).pdf

Innovative Strategies and Measures for Achieving Sustainability

Megaprojects, Ecology and the Environment: Dimensions of Sustainability in Global Cities

Gerardo del Cerro Santamaría

Abstract

The objective of this paper is to propose a set of conditions for the governance of megaprojects in order to ensure their sustainability. We focus on several sustainable strategies: environmental, socio-economic, institutional, as well as those associated with the design, planning, and management aspects of megaprojects. As our main methodology, we use summary descriptions of case studies. Our assessment of environmental challenges and sustainability in megaprojects of four key global cities (New York, London, Paris, and Tokyo) focuses on controversies triggered by the planning and construction of megaprojects in these cities. From the viewpoint of sustainability, what specific controversies do megaprojects trigger at various socio-political scales? What are the challenges, and what are the most suitable governance arrangements that can be key for supporting sustainable cities and regions? An exploration into the conditions for sustainability in megaprojects will help to critically evaluate the urban megaproject phenomenon, and it will contribute to an understanding of the mutual relationships between globalization, infrastructure development, and sustainability. The main finding of our analysis is that environmental challenges are gradually becoming one of the key factors affecting the capacity, performance, and impact of megaprojects in urban settings. Our analysis led to highlight several sustainable strategies: environmental, socio-economic, institutional, as well as those associated with the design, planning, and management aspects of megaprojects. We conclude that suggesting that megaprojects could foster sustainability in cities and regions is a counter-intuitive idea, but the growing importance of megaprojects in the sustainable development agenda warrants a careful analysis in order to critically assess to what extent urban leaders foster sustainable urban practices, even as they face pressures from investors and promoters to keep building, and how those practices could be improved. In addition, we conclude by providing governance criteria for sustainable megaprojects development in policy discourses and practices.

Keywords

Megaprojects • Environment • Sustainability • New York city • London • Paris • Tokyo

1 Introduction

Studies about urban megaprojects and megaprojects more generally have proliferated in the past few decades, and "megaproject research" is becoming an emergent discipline with robust sets of findings and multiple open and ongoing debates. In this research, however, we do not see enough focus on the relationships between urban sustainability and megaprojects. We begin to see an interest in the specific governance arrangements that foster sustainability in megaproject design, planning, and operations. In this context, one can generally define "urban sustainability" as a set of practices allowing cities and urban areas to organize in such ways that the present needs of their inhabitants do not compromise the needs of future generations.

New regional planning approaches have not provided the kind of systematic proposals and solutions to confront and solve urban livability challenges. What is to be found, usually, is a piecemeal process of adjustment of the plans by governments that are intended to preserve the interests of the private sector. Environmental deterioration, one could argue, requires a more pro-active approach to planning, even if urban governments in the past few decades have significantly faced constraints to action due to the prevailing context of neoliberalism and globalization. As a result, multi-layer governance has almost never integrated the planning process, but rather targeted specific areas (such as

G. del Cerro Santamaría (✉)
London School of Economics, London, UK

F. Trapani et al. (eds.), *Advanced Studies in Efficient Environmental Design and City Planning*,
Advances in Science, Technology & Innovation, https://doi.org/10.1007/978-3-030-65181-7_11

disaster preparedness, river cleanups and improvements in water control infrastructure) for a concerted response. Naturally, the outstanding challenge is that of assessing to what extend such sector projects and programs provide good enough responses and coherent strategies able to address global risks to urban living and, in particular, the ecological crisis in all its magnitude (Douglass, 2010).

Thus, it is necessary to look beyond the "iron triangle" of fulfilled schedule, budget, and specifications in analyzing megaprojects.

> Big projects need to be judged for how they meet objectives over time, amid shifting societal, political, and environmental values. Measuring the success of a megaproject is not linear. There are twists and turns not only in terms of engineering and the emergence of new technology, for example, but also in the moving target of public expectations. New problems always crop up that such projects are expected to solve, long after the first blueprints were approved. The biggest plans tend get started under political leaders who are almost always gone by the time of completion (Omega Center, 2012: 35).

2 A Focus on Sustainability

Urban sustainability as a concept dates its first appearance in 1968 in an article written by Stanley A. Cain with the title "*The importance of ecological studies as a basis for land-use planning*." Cain interrogates land use and planning and its contribution to urban sustainable development. In the article, the author highlights the relevance of ecology and ecological studies for the planning process, a recurrent theme since then (Zhang & Li, 2017). Vojnovic argues that

> The three pillars of sustainable development, namely society, economy, and environment, can be equally promoted through the concepts of inter-generational and intra-generational equity. The first is concerned with maintaining the quality of natural ecological systems and their services over time, while the second is based on promoting the equitable access to resources within current generations, providing human populations with basic needs (Vojnovic, 2014: 53).

Hannan and Sutherland propose

> Six principles as a tool to assess the extent to which urban megaprojects recognize urban sustainability. These include: a) create of a place with a vibrant culture, where a diversity of social, environmental, and economic activities can take place, b) ensure social justice and contribute to intra-generational and inter-generational equity including the recognition of social, environmental, and cultural heritage, c) ensure adequate community participation and democratic governance, d) ensure urban spatial integration and promote more sustainable and efficient forms of transport, e) promote economic growth and employment creation and ensure economic viability, and f) minimize pollution and waste; maximize energy efficiency and maintain ecological integrity (2015: 42).

A megaproject can be defined as sustainable "if it is planned and executed to take into account the capacity, suitability, resistance, diversity, and balance of its urban ecosystem. We consider sustainability as an organic process that includes the environment, the economy, and the community" (del Cerro Santamaría, 2018). To be sure, suggesting that urban megaprojects could foster sustainability in cities is a counter-intuitive idea, but "the growing importance of UMPs in globalizing cities warrants a balanced analysis in order to assess and show whether and how urban leaders foster sustainable urban practices even as they face pressures from investors and promoters to keep building" (del Cerro Santamaría, 2018).

Our assessment of environmental sustainability in megaprojects of four key global cities (New York, London, Paris, and Tokyo) focuses on controversies triggered by the planning and construction of megaprojects in these cities. Our aim is to be able to provide criteria for sustainable urban governance in policy discourses and practices. In this context, we can highlight several strategies for sustainability: environmental, socio-economic, institutional, as well as those associated with the design, planning, and management aspects of megaprojects.

3 Environmental Sustainability in Tokyo, London, Paris, and New York City

3.1 Tokyo Bay

The megaproject developments around Tokyo Bay can be traced back to megastructural movements in the 1960s where it drew its inspiration and design. Kento Tange, one of the major architects and influences in the Tokyo Bay project, and also the writer of the 1960 Tokyo Bay Plan, was himself influenced by a group called the metabolists. Metabolists see the transformation of a city as an organic process, where cities can expand in (at that time) futuristic ways—towards the sea or into the sky. Tange's plan involved the conception of a city "as a gigantic structure with a hierarchical arrangement of circulations and programs." It also called for the creation of several artificial islands "to house a highway that could facilitate communication in the area" (Lin, 2007, 34).

The Tokyo Metropolitan Government is in charge of the development around Tokyo Bay. Over the years, the construction of megastructures and skyscrapers in the area proceeded without clear planning frameworks. The construction boom housed many costly ornaments,

> a glitzy hotel or world-class aquarium here, a convention center or two there—mingle with older and more familiar factories, smokestacks, and oil storage tanks (Dahlby, 2002, 113).

According to Japanese conventional wisdom, the development around Tokyo Bay has been beneficial to Japan after the devastation of World War II. It has

> worked like a powerful magnet, pulling in millions of people from around the country, providing them with new jobs and new lifestyles (Dahlby, 2002, 114).

Tokyo Bay is a gigantic conglomerate of office towers, and there have been plans to create even more megastructures, including the Shimizu megacity pyramid and a floating airport.

3.1.1 Urban Heat Island Effect

Due to megastructures that exist in the Tokyo Bay, residents of Tokyo now suffer from severe heat, especially during the summer. The Tokyo Metropolitan Government, in 2008, showed that the average temperate rise in Tokyo over the course of twentieth century has been 3 °C (Kakuchi, 2006). "The Japan Meteorological Agency also showed that the annual average temperature nationwide rose at a rate of 1.15 degree Celsius per century in 1898–2010, which is much higher than the global average temperature rise of 0.74 °C" (Hooper, 2013).

The temperature increase has been blamed on the urban heat island effect (UHI). "UHI occurred because the concrete and asphalt used in buildings and skyscrapers retain heat energy radiated from the sun and from other buildings, making the whole city "a vast absorber and then emitter of convected heat" (Hooper, 2013, 67).

This then creates "guerrilla storms," and renders the nights (always cooler than the day due to lack of sunlight) barely better than stifling days (Hooper, 2013). According to the Inter Press Service,

> The media now point to data showing how the summer nights that traditionally brought relief to Tokyoites, are not effective any more, having been blocked out by the tall structures that wall off Tokyo Bay. The much admired neon glitter of the Shinagawa seafront actually obstructs the cool breeze that once flowed into the city after dark (Kakuchi, 2006, 82).

As an expert in the UHI effect, Komichi Ikeda, deputy director of an environmental-scientists organization—the Environmental Research Institute Inc.—said that "Tokyo has turned into a nightmare caused by government policy that prioritizes the economy over environmental protection." He also stated that land owners and developers are building "at the expense of the environment and the health of residents" (Kakuchi, 2006, 85).

> As a response to the heat island effect, there are now plans to construct twin skyscrapers 246 m apart from each other and connected by a pedestrian deck. After they are completed, an existing old 12-story building that now blocks the wind from Tokyo Bay would be demolished, which is expected to create a wind path to make the area cooler than before (Edahiro, 2008).

3.1.2 Liquefaction of Reclaimed Land

The metropolitan shining life of the Tokyo Bay did attract people to reside around the area, even if not many people could afford the sky-rocketing housing prices and had to settle in the cheaper areas. However, the land there is prone to liquefaction, especially when Japan is a earthquake-prone area (Hamilton, 2011).

> Since Japan is situated at the joint of four different plates—the Pacific Plate, the Philippine Plate, the Eurasian Plate, and the North American plate—as the Pacific Plate and Philippine Plate move towards the west and go under the Eurasian plate and North American plate, a sort of "backlogging" occurs, and as the plates got kicked out, a serious earthquake and tsunami can occur. Even if the magnitude of the earthquake is small, relatively serious damage could be caused, as we have witnessed over the years (Gaulter, 2011).

Most of the Tokyo Bay Area, due to development, is constituted by reclaimed land. When an earthquake takes place, the reclaimed soil would act like liquid: the sand and water from underground rise and structures on the surface sink into the water (Hamilton, 2011). Despite the danger of liquefaction, construction and developments did not stop in the Tokyo Bay Area; in fact, numerous new reclamation projects are in the pipeline awaiting to materialize (Japan Property Central).

During the Giant Earthquake in 2011, although no megastructures fell down, the area was damaged mostly by soil liquefaction. Shinkiba, Urayasu, and Tokyo Disneyland were the most affected areas (Bhattacharya, 2011). During the liquefaction due to the earthquake,

> watery mud had covered much of the courtyards. In many places, the ground level dropped as much as a foot. Out on the street, manholes were being pushed up out of the earth. Water pipes broke. Utility poles began to lean (Hamilton, 2011, 64).

In spite of the fact that reclaimed land is highly vulnerable to liquefaction, many people continue to live in those areas because they are the only ones they can afford.

3.2 London: 2012 Olympics

The London Olympic Park, also known as the Queen Elizabeth II Olympic Park, was constructed for the 2012 Olympics. It is situated east of Stratford, adjacent to the Stratford City Development. The park occupies five boroughs: Newham, Tower Hamlets, Hackney, Waltham Forest, and Greenwich (Antebe, 2012). The park houses the Olympic Village, the Olympic Stadium, London Aquatics Center, and the Arcelor Mittal Orbit, an observation tower and a piece of public art. The motto of the London Olympics was "Toward One Planet Olympics," a vision for sustainable and low-carbon Olympic Games (Kinver, 2005). As a result, the London Olympics were assessed as the "Greenest

Olympics ever" by the Independent Commission for a Sustainable London 2012. The venues were made of at least 25% of recycled materials, and the transport arrangements were very successful (BBC, 2012).

3.2.1 Cooperation with the Rio Tinto Mine

The medals of the London Olympics were made almost entirely out of materials extracted and smelted at Kennecott Copper Mine, which is owned by Australian mining tycoon Rio Tinto Kennecott. The mine was involved in a federal lawsuit in Utah, where the owner was accused of violating the US Clean Air Act for several years (McNara, 2012). The Kennecott copper mine has produced a quarter of all US copper, as well as vast amounts of gold, silver, and molybdenum. The mine is also said to be the largest man-made mine in history. However, unregulated extraction has made the mine the largest industrial polluter in Salt Lake City. Kennecott was accused by health advocacy groups of endangering the health of Salt Lake City citizens by disregarding emission standards set by the EPA, making Salt Lake City one of the worst cities for air pollution in the US. The allegations, by doctors, lawyers, and advocates, said that the Rio Tinto mines were responsible for 30% of air pollution in Salt Lake City, inducing long-term respiratory problems on its citizens, including children (McNara, 2012).

3.2.2 Loss of Biodiversity

The construction of the Olympic Park took place in an area with immense biodiversity, which was severely affected. The five boroughs produced a biodiversity action plan (BAP) a few years before the London Olympics that detailed the local wildlife and biodiversity and focused on key habitats and species. In the report, there are more than 100 identified sites of importance for nature conservations (SINCs). There was a vast amount of woodland, standing water, reedbed, and rassland in Hackney, one of the Olympic Boroughs. However, the London Thames Gateway Development Corporation drew a picture of "fragmented ownership" and "declining industry," which depicts North East London as "an area littered with relics of industry and rusting steel planes" (Antebe, 2012, 57). Not much later, the corporation bought the land at a convenient price.

> Also affected is the lower Lea valley, known as a green lung of London with the river Lea running through its middle. It is the home to many species, including kingfishers, and it has also been for hundreds of years a popular migratory route in London for birds such as ducks, geese, warbles, and thrushes (Antebe, 2012: 54).

Lower Lea Valley is also a oasis for kestrels and herons. The reserve was ultimately removed to build a velodrome arena for track-cycling events. An official statement declared that

> extensive works were needed to remediate the land and eradicate the invasive species in the area. The redesign of the parklands in this area incorporates new wet woodland habitat [and] will encourage biodiversity as well as foster greater recreation access (Antebe, 2012, 59).

3.2.3 Mismanagement of Nuclear Waste

Documents obtained from the British Government via the freedom of information act revealed that radioactive nuclear waste was buried under the London Olympics Park, especially under the velodrome and the basketball arena (Wells, 2010).

> The Olympic Development Authority (ODA) claimed that there was "no exposure to increased levels of radiation either through direct contact or dust," yet it was revealed that more than 7000 t of waste were buried in a radioactive storage bunker under the Stratford International Station, where attendees to the Olympics travelled to. The radioactive waste included radium, polonium, thorium, and uranium that was diluted by mixing together higher and lower radioactive material, which is prohibited by the British Environment Agency. The ODA rejected a request for radiation monitoring and did not used yellow radiation signs to identify the presence of radioactive waste (Griffiths, 2010).

At the main stadium, there were significant quantities of radioactive contamination that were unknowingly excavated and spread around other areas. Public relations from the ODA claimed that

> Highly sensitive monitoring has been carried out on the site during the earthworks program, and this has detected a small amount of low level radiological that finds in the soil in the area. This soil, with mild radioactive properties present due to the past industrial activity over the last 100 years or so, is being managed so that it presents no risk to public or workers (Wells, 2010, 142).

Yet the excavation was completed before any radioactive monitoring could take place. According to public reports, "when radiation was detected in the excavations in November 2007, no action was taken, and it was not made public. Only in February 2008 were the relevant authorities alerted and remedial and protective measures started" (Wells, 2010). Paul Charman, an analyst with the Citizens Intelligence Network, expressed his concern on the management by the ODA:

> It has become clear that the real priority is the avoidance of negative PR and of any delays to the delivery schedule and the less that was known the better. The ostrich-like approach led to real injustice and left the workforce and local population vulnerable (Wells, 2010, 145).

John Large, an independent nuclear analyst, said at the time:

> The Olympic site's hurried and unplanned development may have resulted in a great deal of public harm to the local communities remaining around the site. Overall, there is some doubt about the applicability and validity of the radiological risk analysis undertaken for the future legacy use (Griffiths, 2010, 32).

3.2.4 Loss of Allotment and Greenspace

Public space was lost during the development of the Olympics. For instance, the manor garden society was right in the middle of the Olympic Park and

> consequently felt the sharp end of the compulsory purchase order driven through their 100-year-old grounds. The manor garden society claimed that their "old site will be replaced by something far inferior in the future Olympic Park […] the displacement is a betrayal of the thousands of people on allotment waiting lists in the local area, who were hoping for a genuine example of sustainability and response to local needs in the Olympic "legacy' (Antebe, 2012: 65).

Other complaints on the loss of open space include the disruption of Greenwich Park, Leabank Square, etc. In March 2012, the Save Leyton Marsh Group protested on Porters Field to oppose the construction of the Olympics practice basketball courts (Antebe, 2012).

3.3 Paris: La Défense

La Defense is a major business district of the Paris metropolitan area, located in the commune of Courbevoie, and parts of Puteaux and Nanterre, which is west of the city of Paris. It spans through a perimeter of 760 hectares and was developed at a time when the area "had seen very little development and was filled with rundown houses, was full of small factories and shanty towns" (Defacto, 2015). In 1958, the government decided to build a modern office district bordered by the historic axis, which starts from the Louvre in Central Paris and continues along Champs-Elysees, the Arc de Triomphe and ending at La Défense. MODERN EPAD (the public institution in charge of developing La Défense) was given the task to design, build, and develop the district. They brought into play the principles of "modern movement" laid down in the Athen Charter of 1933, which advocated

> the absence of roads, construction of tall buildings to favor light and the separation of pedestrian and vehicular traffic using paved pedestrian areas, bordered by a high-speed road and serviced by underpasses (Defacto, 2015, 43).

3.3.1 Disturbance of Skyline

The skyscrapers in La Defense are criticized for being disruptive to the sky view of certain monuments, including the Arc de Triomphe. In 1931, the government came up with plans to renovate the area, but with a limit to the height of the buildings along the historical Triumphal Way. Towers were only allowed at the end of the historical axis since they would affect the Étoile if allowed in the middle of the way. The first master plan of La Défense in 1964 required all office towers to observe the same rules. In this plan, all towers would be identical in size: "a base of 42 m by 24 m, a maximum height of 100 m, and a floor area of 30,000 m^2" (Defacto, 2015, 58). The plan was to build two rows of skyscrapers of 25 stories each, but this was never implemented as most companies started putting pressure on EPAD for taller office buildings.

As a result, La Défense ended up with towers of different heights. The tallest of them—the GAN tower is 589 ft. tall. The public criticized the "forest of towers," in which buildings came to disturb the view on the Arc de Triomphe as seen from the Étoile. The skyscrapers were also criticized as generic buildings—towers that are considered as a post-modernist style that "could be built anywhere" (Langmead & Christine, 2001, 31). To counter the disruption of view on the Arc de Triomphe, the government decided to build the Grande Arche. Yet, the architectural style and impact of the Grande Arche was, and still is, very controversial.

3.3.2 Criticisms Against the Grande Arche

> The Grande Arche in La Défense is a marble and glass cube measuring 110 m high by 110 m wide and 110 m deep and big enough to fit Notre Dame cathedral in its archway. The building is covered with 300,000 t of glass and Italian Carrara marble. Security studies have shown that between 2004 and 2010, one in six of its marble tiles crumbled because the porosity of the stone made it vulnerable to the weather and pollution. The area around the north side of the arch was sealed off after fears of crumbling marble falling on people below. Staff of the French ecology and housing ministries, who occupy the south side, complained of gloomy corridors and offices with oppressively low ceilings and no natural light. The roof was closed to the public after a lift accident in April 2010, even though no one was injured, and shortly, afterward, the marble slabs on the southern tower were replaced by harder wearing granite (Willsher, 2014).

3.3.3 Unequal Distribution of Natural Light

Giant skyscrapers have often been the cause of unequal distribution of natural light. After the first plan for La Défense, more and more of the built towers became increasingly high. Some of those include the FIAT Tower, which is 184 m high and has 44 floors. Employees criticized such towers for their permanent indirect lighting (Defacto, 2015).

In 2008, the government proposed to build a pyramid shaped structure designed by architecture firm Herzog and de Meuron. The building is named The Tour Triangle. Architects claimed that the shape will not block out sunlight and will also allow "optimum solar and wind power generation." Green activists demanded "shadow studies," where "new designs are computer tested in a virtual landscape to see where shadows will fall at different times of the day and year." Despite the efforts of Paris City Hall to limit heights of buildings, building shapes also affect the environment through the phenomena of overshadowing. Critics pointed out at the time that Paris did not limit the construction of buildings they way New York does it, which is by regulating and limiting the shape of the building (Rose, 2008, 55).

The Tour Triangle is also criticized for disrupting the skyline of the city center. Olivier de Rohan Chabot from the Safeguard of French Art group said that "he was not against towers in principle, but within central Paris they ruined the skyline's general harmony." Charbot said:

> Look at the Montparnasse Tower; it has crushed the Hotel des Invalides (housing Napoleon's tomb). The monument was built to be grandiose. But what has it become? A dwarf. The tower ridicules it. In this sense, it is a veritable attack on the beauty of the capital (Rose, 2008, 61).

Yves Contassot, a Paris environmental activist, indicated that "the project would cast a shadow over locals and its foundations would disturb a water table below" (Rose, 2008, 62).

3.4 New York City: New World Trade Center

3.4.1 Green Concrete

The construction of the One World Trade Center used "green concrete." When compared to traditional concrete, green concrete saved about 12 million pounds of carbon dioxide emissions, 8 million kWh of energy and 30,000 gallons of fresh water. However, the production of green concrete also involved the production of coal fly ash (Schelmetic, 2012).

Green concrete consist of

> basic traditional concrete ingredients with three extra ingredients: coal fly ash, blast furnace slag and natural pozzolan–zeolite. Coal fly ash is produced by combustion of coal. During coal combustion, fine particles are driven out of the boiler with the flue gas. Fly ash is captured by electrostatic precipitators or other particle filtration equipment, and when combined with the bottom ash (ash falls to the bottom of the boiler), the ash is called coal ash (Smrckova, 2014).

Despite the generation of green concrete is more sustainable than traditional concrete, coal fly ash possesses many environmental risks. "Depending on where the coal was mined, fly ash could contain heavy metals such as arsenic, lead, mercury, cadmium, chromium, and selenium." When fell into water, these heavy metals could poison aquatic life. The US Environmental Protection Agency also "found that people living next to coal ash disposal sites could have an increased risk of cancer and other diseases." These toxic substances can also cause substantial damage to the nervous system and cause cognitive deficits, developmental delays, and behavioral problems. Other diseases, such as heart damage, lung disease, and reproductive problems, can also occur in people who drink water contaminated by coal fly ash. Although it has been argued that the benefits of green concrete far outweigh its drawbacks, green concrete poses environmental challenges due to the composition of coal fly ash (Physicians for Social Responsibility, 2015).

3.4.2 Doubts on LEED Certification

The Freedom Tower is one of the few buildings in its scale to obtain the US Green Building Council's LEED gold certification. To secure the gold certification, the Freedom Tower needed to "achieve net zero carbon dioxide emissions (by purchasing green energy credits) and operate with at least 20% more energy efficient than New York's current building requirements" (West, 2014, 210).

> The building also features lighting that reacts with sunshine, rain harvesting, and state-of-the-art fuel cell installation. The cells convert chemical energy into electricity by an energy-efficient electrochemical process, rather than burning it. The heat and water that are generated as byproducts are also used for cooling and heating the tower. In 2008, the New York governor David Paterson praised the building for its "space-age energy technology" (West, 2014, 211).

However, superstorm Sandy "destroyed the fuel cells of the building." An internal document obtained by Climate Desk from the Port Authority of New York and New Jersey claimed that "Superstorm Sandy damaged the Freedom Tower's $10.6 million clean power fuel cells. Two hundred million gallons of water flooded and destroyed all nine fuel cells." A third of the fuel cells "went unrepaired and unreplaced, because of the inability to do so, and the pressure to honor the billion dollar deal with Conde Nast, the global publishing powerhouse and high-profile anchor tenant. Although towers three and four now have new fuel cells, the main tower's power supply has never been replaced by fuel cells due to financial pressures" (West, 2014, 215).

3.4.3 Hotter Environment

The Freedom Tower is covered with 1368 ft. of glass parapet (partly to pay homage to the fallen twin towers), hence giving the Freedom Tower a glassy exterior. The Freedom Tower has a relatively lower temperature than neighboring infrastructure, especially lower than structures built with bricks and the ground. Glassy skyscrapers are often thought to be more environmentally friendly, since the glass would reflect sunlight away from the building, hence keeping cooling and air conditioning costs down. However, the exterior environment, which is made from bricks, would become the recipient of the heat, hence making the environment hotter (Ferro, 2013).

John Frederick, a professor of atmospheric science at the University of Chicago, said that

> as we get more shiny, glassy metal buildings, reflected light will become a bigger deal [...] The high reflection is good if you are inside trying to keep intense sunlight out, but if youare outside, you will be the recipient of that reflected light." However, "the effect of one building on another building in its vicinity is something that never seems to be discussed." All in all, Frederick suspected that the effects of the Freedom Tower on urban heat island effect would be insignificant (in Ferro, 2013, 93).

4 Governance for Sustainable Megaprojects

Megaprojects exhibit and trigger significant environmental impacts in global cities. Therefore, the goal of environmental sustainability must be placed at the top of the list in any planning strategy by planners, developers, and other stakeholders. However, "the attainment of environmental sustainability does not in itself ensure megaproject sustainability, a goal that needs to be pursued holistically. One way to do it is to use the notion of *key or multiple success factors*" (Grunert & Ellegaard, 1992).

This notion

> is not new in the field of project management and, in fact, constitutes one of the topics most discussed by specialists. It is increasingly important to evaluate projects and their impacts at different times and based on multiple criteria in order to fully evaluate their performance. Success is often driven by political and/or power-related factors (Grunert & Ellegaard, 1992, 34).

According to Harris,

> due to the strongly political nature of the stakeholders throughout the supply chain and their different underlying objectives, the success factors usually considered no longer seem sufficient. This configuration requires innovative governance solutions that align the interests of the different stakeholders in a complex environment with a large number of key actors (Harris, 2017).

By using the concept of "multiple success factors," we have argued elsewhere that

> there are a number of requirements that need to be met in order to achieve sustainable megaprojects: environmental sustainability (sustainable infrastructure delivery and sustainable development zones); sustainability in design and planning; sustainability in megaproject management; institutional sustainability; and socio-economic sustainability. Thus, a megaproject can be defined as sustainable if it is planned and executed to account for the capacity, fitness, resilience, diversity, and balance of its urban ecosystem. We take the view of sustainability as an organic process including environment, economy, and community: form and efficiency (environmental factors in design, architecture, engineering, and construction) as well as policy, that is, urban plans and practices that explicitly aim at maintaining and improving the social and economic well-being of citizens (del Cerro Santamaría, 2018).

5 Sustainable Infrastructure Delivery

The next phase of socio-economic development would bring new ways of urban infrastructure delivery that place environmental concerns at the core.

> As cities and countries climb the industrial value ladder and expand their service sector to cater to growing domestic demand, environmental quality will become central to achieving sustainable economic growth. Urban residents in the more sophisticated markets are already putting a substantial price premium on high-quality urban environment. To attract the right labor pool, cities will need to raise their game further (World Bank, 2018, 33).

The backbone of the next phase of infrastructure development

> should be the "one-system' approach. Infrastructure planners need to consider the development of the entire city-wide infrastructure system, including its energy, transport, land, and water subsystems. Realizing the potential synergies between subsystems will require technology for real-time information, conservation pricing, and demand management (Ness, 2018, 74).

Foreign infrastructure providers and experts with strong track records in these areas should be poised to make a major contribution to achieving higher infrastructure performance standards in the coming years.

> Government stimulus and financing will also be critical. Central governments can redouble its commitment to environmental sustainability by continuing to pursue aggressive resource conservation and economic productivity targets, and by backing those efforts up with funding for investment in infrastructure (United Nations, 2016, 41).

6 Sustainable Development Zones

Some countries, for example, China, have implemented strategies to develop "sustainable development zones." In 2019, the Chinese Government approved three sustainable development zones, Shenzhen, Guilin, and Taiyuan, that will be in charge of implementing many of the United Nations 2030 Sustainable Development Goals:

> Shenzhen is China's innovation engine. This zone will integrate technologies in sewage treatment, waste utilization, ecological restoration, and artificial intelligence to solve issues from resource management to pollution. Guilin will focus on innovations that tackle desertification, creating solutions that can be replicated by other regions facing the threat of encroaching deserts. In Taiyuan, targeting air and water pollution, this zone will foster innovative solutions that can be replicated by regions relying on resource extraction (XinhuaNet, 2018, 22).

7 Sustainable Planning and Design

Growth and competition should not be the only focus of megaproject planning. Instead, urban leaders ought to be integrating equity and social cohesion goals.

> The financial planning of megaprojects ought to avoid strategic misrepresentation of costs and benefits, and there should exist clauses ensuring a fair distribution of benefits for the community at large. In the design process of megaprojects, contextual elements such as history and local culture should be important

factors to interpret the architecture and to assign a specific meaning (local, regional, national, global) to the architectural practices used to build megaprojects and make them visible (Del Cerro Santamaría, 2013a, b).

8 Sustainable Management

Often times planners and managers tend to believe that their projects are unique (the so-called exclusivity bias). This should be avoided as it prevents stakeholders from learning about similar projects elsewhere.

> Often there is an excessive commitment to a certain project concept at an early stage, resulting in a "block" or "capture," which makes the analysis of alternatives unlikely and leads to ad hoc commitments in later stages. The achievement and delivery of the megaproject is a high-risk stochastic activity, with overexposure to the so-called black swans, that is, extreme events with massively negative results. Managers should keep this in mind, instead of treating the projects as if they really existed in a deterministic Newtonian world of cause, effect, and control. Complexity and unplanned events need to be taken into account; budgets and time contingencies must adequately address them (Flyvbjerg, 2014, 14).

9 Institutional Sustainability

Elsewhere we have argued that

> when studying megaprojects in connection with urban growth coalitions, we can ask ourselves if the form of development represented by these large projects simply legitimizes growth machines and commercial interests or if this phenomenon can be analyzed from the prism of the role that actors and state agencies play in urban restructuring. Besides growth coalitions, the governance of megaprojects needs to take into account additional stakeholders in order to ensure institutional sustainability. There is no room in this paper to adequately develop this argument, but I will simply mention the following aspects that would need to be taken into account: (1) the role of civil society; (2) the role of local context, history and culture; (3) the importance of keeping urban variety and diversity; (4) the importance of megaproject local embeddedness; (5) the role of public space; and (6) the role of urban design professionals (del Cerro Santamaría, 2019).

10 Socio-economic Sustainability

Socio-economic sustainability usually benefits from an alignment of megaproject planning goals with regional and national socio-economic goals. This would allow the utilization of megaprojects not only as catalyst of urban development but as key elements in the formulation of national sustainability goals. As argued elsewhere,

> the embeddedness of the multiple scales of socio-economic action has paradoxically come along with a tendency toward providing megaprojects with their own strategic spatial planning tools that function independently of state and urban regulations. New institutional arrangements between administrative levels, vital for implementing strategic policies, need to emerge, due to the negative results of separating the objectives of the megaprojects from regional and national policy goals. The result can be an optimization of sustainable policy outcomes due to synergistic, multiplier effects. From this perspective, the key question for future research would be how to plan and build urban megaprojects that simultaneously foster both sustainability and competitiveness (del Cerro Santamaría, 2019).

11 Conclusion

This paper has described some of the environmental challenges encountered in four global cities (Tokyo, London, Paris, and New York) as these cities engaged in processes of redevelopment via urban megaprojects. For the sake of recapitulation, let us remind the reader of the problems that we discussed:

- **Tokyo**—Urban heat island effect, liquefaction of reclaimed land.
- **London**—Cooperation with the Rio Tinto Mine, loss of biodiversity, mismanagement of nuclear waste, loss of allotment and greenspace.
- **Paris**—Disturbance of skyline, criticisms against the Grande Arche, unequal distribution of natural light.
- **New York**—Green concrete, doubts on LEED certification, hotter environment.

Our assessment of environmental challenges and sustainability in megaprojects of these four global cities of the north focused on controversies triggered by the planning and construction of megaprojects in these cities. The main lesson of our analysis is that environmental challenges are gradually becoming one of the key factors affecting the capacity, performance, and impact of megaprojects in urban settings. Our analysis led to highlight several sustainable strategies: environmental, socio-economic, institutional, as well as those associated with the design, planning, and management aspects of megaprojects.

The growing importance of megaprojects in the sustainable development agenda warranted a careful analysis in order to assess to what extent urban leaders foster sustainable urban practices and how these practices could be improved, in a context where officials and others are pressured to expand building plans by promoters and developers. We found that any sustainable governance arrangement in megaprojects benefits from governance solutions where there is an alignment of multiple stakeholders interests and a large number of actors. Specifically, we argued that there are a number of requirements that need to be met in order to achieve an acceptable level of sustainability in megaprojects: environmental sustainability; sustainability in design, planning and management, institutional sustainability, and socio-economic sustainability. Our view is that future research on megaprojects ought to pay close attention to the interplay between sustainability and competitiveness, and how to build megaprojects that meet both goals in balanced ways. The co-variance in practices related with these two concepts ought to be at the center of the analysis in the future research.

References

Antebe, T. (2012, July 27). The real environmental impacts of holding the Olympics in East London. *Corporate Watch*. Retrieved June 15, 2015, from http://www.corporatewatch.org/news/2012/jul/26/real-environmental-impacts-holding-olympics-east-london

BBC News. (2012, August 11). London 2012: Olympic Games 'greenest ever'. *BBC News*. Retrieved June 15, 2015, from http://www.bbc.com/news/uk-19220847

Bhattacharya, S. (2011). Liquefaction of soil in the Tokyo Bay area from the 2011 Tohoku (Japan) earthquake. *Soil Dynamics and Earthquake Engineering, 31*(11), 1618–1628.

Center, O. M. E. G. A. (2012). *Mega-projects: Lessons for decision-makers. An analysis of selected international large-scale transport infrastructure projects*. Bartlett School of Planning, University College London.

Del Cerro Santamaría, G. (Ed.). (2013a). *Urban megaprojects. A worldwide view*. Emerald Publishing.

Dahlby, C. (2002, October). Tokyo Bay. *National Geographic*. Retrieved July 3, 2015, from http://ngm.nationalgeographic.com/print/features/world/asia/japan/tokyo-bay-text

Defacto. (2015). Guide to architecture: Museum de la Defense. Defacto. Retrieved July 5, 2015, from http://www.ladefense.fr/sites/default/files/kiosque/pdf/guide_architecture_uk.pdf

Del Cerro Santamaría, G. (Ed.). (2013b). Megaprojects, sustainability and competitiveness in the United Arab Emirates. Unpublished Fulbright Research Proposal.

Del Cerro Santamaría, G. (2018). Megaprojects, sustainability and competitiveness in the United Arab Emirates. Fulbright Scholar Project Proposal.

Del Cerro Santamaría, G. (Ed.). (2019). Rapid urbanization, ecology and sustainability. The need for a broad strategy, holism and transdisciplinarity. *Transdisciplinary Journal of Engineering and Science, 2019*

Douglass, M. (2010). Globalization, mega-projects and the environment: Urban form and water in Jakarta, *Environment and Urbanization Asia, 1*(1). https://doi.org/10.1177/097542530900100105

Edahiro, J. (2008, September). Efforts in Japan to mitigate the urban heat island effect. *Japan for Sustainability*. Retrieved July 3, 2015, from http://www.japanfs.org/en/news/archives/news_id027856.html

Ferro, S. (2013, September 4). Why do skyscrapers melt things? *Popular Science*. Retrieved July 6, 2015, from http://www.popsci.com/science/article/2013-09/why-do-skyscrapers-melt-things

Flyvbjerg, B. (2014). What you should know about megaprojects and why. An overview. *Project Management Journal, 45*(2), 6–19. https://doi.org/10.1002/pmj

Gaulter, S. (2011, March 11). Why Japan is prone to earthquakes. Al Jazeera. Retrieved July 3, 2015, from http://www.aljazeera.com/news/asiapacific/2011/03/2011311105046675615.html

Griffiths, I. (2010, June 20). Tons of radioactive waste casts doubt over London's Olympic stadium legacy. *The Guardian*. Retrieved June 16, 2015, from http://www.theguardian.com/business/2010/jun/20/radioactive-waste-olympic-site

Grunert, K. G., & Ellegaard, C. (1992). The concept of key success factors: theory and method. *MAPP Working Paper, 4*

Hamilton, J. (2011, April 7). In Japan, shaken soil turned soft after quake. National Public Radio. Retrieved July 3, 2015, from http://www.npr.org/2011/04/07/135181474/in-japan-shaken-soil-turned-soft-after-quake

Hannan, S., & Sutherland, C. (2015). Mega-projects and sustainability in Durban, South Africa: Convergent or divergent agendas? *Habitat International, 45*(P3), 205–212.

Harris, M. (2017). Competitive precinct projects. The five consistent criticisms of "global" mixed-use megaprojects. *Project Management Journal, 48*(6), 76–92.

Hooper, R. (2013, July 13). Effects will become more obvious as Japan's climate changes. *The Japan Times*. Retrieved July 3, 2015, from http://www.japantimes.co.jp/news/2013/07/13/national/science-health/effects-will-become-more-obvious-as-japans-climate-changes/#.VaFvQflVhHw

Japan Property Central. (2015). Reclaimed land in Japan. *Japan Property Central*. Retrieved July 3, 2015, from http://japanpropertycentral.com/real-estate-faq/reclaimed-land-in-japan/

Kakuchi, S. (2006, August 31). Environment: Japan swelters under urban heat island effect. *Inter Press Service News Agency*. Retrieved July 3, 2015, from http://www.ipsnews.net/2006/08/environment-japan-swelters-under-urban-heat-island-effect/

Kinver, M. (2005, October 24). London's "One Planet Olympics" vision. *BBC News*. Retrieved June 15, 2015, from http://news.bbc.co.uk/2/hi/science/nature/4299732.stm

Langmead, D., & Christine, G. (2001). Encyclopedia of architectural and engineering feats. Retrieved July 5, 2015, from https://books.google.com/books?id=T5J6GKvGbmMC&dq=la+defense+forest+of+towers&source=gbs_navlinks_s

Lin, Z. (2007). From megastructure to megalopolis: Formation and transformation of mega-projects in Tokyo Bay. *Journal of Urban Design, 12*(1), 73–92.

McNara, M. (2012, June 20). The great Olympic greenwash. *Aljazeera*. Retrieved June 15, 2015, from http://www.aljazeera.com/programmes/peopleandpower/2012/06/201262075757520106.html

Ness, D. (2018). Sustainable urban infrastructure in China: Towards a factor 10 improvement in resource productivity through integrated infrastructure systems. *The International Journal of Sustainable Development and World Ecology 15*(4).

Physicians for Social Responsibility. (2015). Coal ash: Hazardous to human health. *Physicians for Social Responsibility*. Retrieved July 6, 2015, from http://www.psr.org/assets/pdfs/coal-ash-hazardous-to-human-health.pdf

Rose, S. (2008, October). Paris Triangle: Beyond the shadows of doubt. The Guardian. Retrieved July 5, 2015, from http://www.theguardian.com/artanddesign/2008/oct/07/triangle.herzog.demeuron.paris

Schelmetic, T. (2012, January 3). Green facts about New York's new one world trade center. *Industry News*. Retrieved July 6, 2015, from http://news.thomasnet.com/imt/2012/01/03/green-facts-about-new-yorks-new-one-world-trade-center

Smrckova, E. (2014). Basic characteristics of green cement of CEM V/A and V/B kind. *Advanced Materials Research, 894*, 196–199.

United Nations. (2016). Executive summary of China's actions on the implementation of the 2030 agenda for sustainable development, voluntary national review 2016. https://sustainabledevelopment.un.org/memberstates/china

Vojnovic, I. (2014). Urban sustainability: Research, politics, policy and practice. *Cities, 41,* 30–44.

Wells, M. (2010, September 26). London 2012: A rubbish Olympics… *Games Monitor*. Retrieved June 19, 2015, from http://www.gamesmonitor.org.uk/node/1108

West, J. (2014, December 8). The freedom tower was supposed to be the greenest building in America. So what went wrong? *Mother Jones*. Retrieved July 3, 2015, from http://www.motherjones.com/environment/2014/12/freedom-tower-world-trade-center-green-sustainable-plan-sandy

Willsher, K. (2014, August). France pledges €200 m to restore La Défense's crumbling cube. *The Guardian*. Retrieved July 5, 2015, from http://www.theguardian.com/world/2014/aug/04/france-pledges-200m-euroes-restore-la-defense-la-grande-arche

World Bank. (2018). *China—Systematic country diagnostic: Towards a more inclusive and sustainable development (English)*. World Bank Group. http://documents.worldbank.org/curated/en/147231519162198351/China-Systematic-Country-Diagnostic-towards-a-more-inclusive-and-sustainable-development

XinhuaNet. (2018, February). China approves three demonstration zones on sustainable development. http://www.xinhuanet.com/english/2018-02/24/c_136997243.htm

Zhang, X., & Li, H. (2017). Urban resilience and urban sustainability: What we know and what do not know? *Cities, 72,* 141–148.

Elements Related to Sustainability of Urban Form: Study of an Indian City, Thane (Maharashtra, India)

Janmejoy Gupta and Jismi Varghese

Abstract

Over the last few decades, there has been a rapid increase in the rate of urbanization worldwide. The coming decades will further add changes to the size and spatial distribution of the global population. The rate of urbanization in India is also high. By 2030, it is expected that 40.76% of India's total population would reside in urban areas, as per a survey by the United Nations 'State of the World Population Report' in 2007. To cope with the enormous problems that have emerged because of swift urban growth, it has become imperative to formulate strategies for sustainable urban development. It is acknowledged that the shape, size, and density of an urban area have a large role in achieving sustainability of urban form. Urban form is the arrangement of various elements that repeat and can help to achieve sustainability of cities by optimizing itself and minimizing sprawling. This paper aims to study ways to understand the intrinsic complexity in urban forms and explore different elements related to sustainability of urban form, taking the example of a typical Indian city. It suggests interventions to achieve more sustainable urban forms.

Keywords

Spatial arrangement • Spatial distribution • Sustainable urban form • Urban sprawl • Density profile • Land-use distribution • Growth pattern

J. Gupta (✉)
School of Planning and Architecture, SPA, Vijayawada, Andhra Pradesh 520008, India

J. Varghese
Practicing Architect-Urban Planner, Navi, Mumbai, India

1 Introduction

The whole world depends on nature and healthy ecosystems to provide with good health, hygiene, and general well-being of all species. In the recent decades, humans have made large changes to ecosystems to meet the demands for food, fresh water, shelter, and energy to improve life. However, all this has weakened nature's capacity to give back services it naturally does such as purifying air and water, safeguarding from calamities, and offering natural medicines. Our actions have also increased extinction of species, further risking our own existence. Such loss of the ecosystem services increases poverty, hunger, and diseases, which is a threat to achieve overall humankind's well-being. If the human attitude and actions continue to be the same, then in the future the pressure on the already fragile ecosystem will increase tremendously. The concept of sustainable development took root from several environmental movements in earlier decades and was formally defined in 1987 by the World Commission on Environment and Development as Sustainable Development—'Development that meets the needs of the present without compromising the ability of future generations to meet their own needs.' (Brundtland Commission Report, 1987) Sustainability and sustainable development emphasize on balancing between our need toward advancement in economy and technology and the needs to safeguard the environment in which we and others live. Sustainable development involves long-term, multidimensional, and integrated perspective to evolve and accomplish a healthy society by focusing on economic, environmental, and social issues, at the same time avoiding the overutilization of important natural resources. Sustainable development of a city is closely related to its urban form. The burgeoning city form encroaching into the green periphery lands and ensuing discussions about 'sustainable development' as a common perception has brought back the debate over the forms of cities, inspiring the scholars of different fields such as planners, socialists, environmentalists, and

F. Trapani et al. (eds.), *Advanced Studies in Efficient Environmental Design and City Planning*, Advances in Science, Technology & Innovation, https://doi.org/10.1007/978-3-030-65181-7_12

geographers to explore various forms for human settlements that satisfy the criteria of sustainability in a way benefiting the functioning of the built environments (Jabareen, 2006). It is a widely accepted fact that there is a connection between the shape, size, and density of an urban area and its sustainability (Nuzzolo et al., 2014). Somehow, there is a lack of general consensus on the level and extent of this relationship. It is not at all mandatory that every urban form fulfills all the aspects of sustainability. Rather, in modern urban areas, the reverse holds true to a large extent, except some notable exceptions. Some urban forms may be sustainable in some respects but may lack in other aspects; some forms may be locally sustainable but may not respond at the large urban- or regional-level scales. Putting all together, urban form is defined as the physical characteristics that make up built-up areas, including the shape, size, density, and configuration of settlements (Williams, 2014). Sustainable urban form may be broadly termed as, at this stage, as a spatial arrangement of elements within a city, including the overall form and extent of the city, which when built on certain sustainable city-planning principles, can allow achieving sustainable cities.

2 Objectives

This paper aims to arrive at ways to understand the intrinsic complexity in urban environments better and to introduce general proposals for better urban sustainability. At a more detailed level, this paper attempts to explore different elements related to sustainability of urban form, taking the example of a typical Indian city, Thane (Maharashtra). Taking the specific case of Thane, this study attempts to assess the extent of sustainability of existing urban form of cities considering certain parameters, namely density profile, land-use distribution, and growth patterns, which have been selected as determining parameters on the basis of exhaustive literature review, and to measure its potential to develop as a sustainable one, taking the example of a typical Indian city. The selected city for study is Thane, a city in the state of Maharashtra, in the Western part of India. Thane is located very near to the financial capital and metropolis of India, Mumbai.

3 Methodology

The four basic parameters for measuring urban form have been outlined in this paper, taking the example of Thane. The four parameters being—urban spatial structure, density profile, the city's growth pattern, and land-use distribution (comparing the existing land use with the desirable values given by UDPFI Guidelines). The four parameters have been selected on the basis of literature study and urban-planning principles.

Other than study of urban spatial structure of Thane, a three-pronged methodology to study urban form sustainability of Thane has been adopted. The three factors, which contribute to the urban character and sustainability of any urban area, are as given below:

a. Density profile.
b. Land-use distribution.
c. Growth pattern.

With the help of the above three parameters, the existing form of the city, in this case, Thane, has been studied, to gather enough knowledge of the city's existing form, its expansion potential and to understand the requirements to achieve sustainability of urban form. The Holdern–Shannon's entropy coefficient, one of the more widely used tools used to measure the urban growth pattern of the city, has been used to calculate the extent of urban sprawl of Thane and thereby to calculate the extent of its compactness.

4 Literature Review

4.1 Urban Form Sustainability—Global Scenario

Today, out of the world's total population of seven billion, half of the world's population lives in cities, which are expected to rise over a billion more by 2030. Many parts of the world for the first time will have the rural population going down in comparison with the urban population. It is also expected that between 2010 and 2050, the urban population will increase by 2.5–3 billion people, making the urban share to two-thirds of the world's population. Our inherent dependence on fossil fuel-based resources will be put to acute test, if we do not change our planning and construction techniques and adopt more sustainable ways of energy use and generation.

As cities grow, it becomes important to analyze how they expand out into the surrounding geography. It has been observed that over the past centuries along with the tremendous demographic expansion the urban settlements have also sprawled outwards—spreading over valuable farmland. If this trend continues, by 2050 the area covered by the urban settlements on the planet will rise by 3 million km^2. This increase in land consumption will affect the global food security due to the shortage of farmlands. This would and already is affecting the food web and global energy patterns. In short, the continuation of the current trend of urbanization through unsustainable practices will further threaten global sustainability. Today, many American cities are becoming sustainable. The reason is that the

American cities are adopting the global sustainable development agenda that try to deal with discriminations, provide equal growth platforms for all, and ensure natural environment protection. Cities such as San Jose (CA), Baltimore (MD), and New York City (NY), among many more, are working toward a more sustainable future. Along with surveying and data monitoring systems, they are trying to achieve city-wide plans that cater to all aspects of sustainability, by defining priorities, and developing strategies through appropriate policies and investments (Prakash et al., 2017).

4.2 Urban Form Sustainability—Indian Scenario

In the last two decades, India has witnessed a tremendous change on all fronts. The speed of development in India has also accelerated. The trend of the community has shifted from the traditional rural to the modern urban which has resulted in the urban shift from the traditional rural community to the modern urban community. The rate of urbanization in India has been continuously increasing. In 1901, the urban population was 11.4%, which increased to 28.53% in 2001, and it became more than 30% in 2011. By 2030, it is expected that 40.76% of the population will be living in the urban areas as per a report published by the 'UN State of the World Population, 2007.' According to the World Bank, India, China, Indonesia, Nigeria, and the USA will contribute the major share to the world's urban population by 2050 (Business Standard, 15 June 2012). In comparison with the major countries, India's contribution toward sustainability is small and can increase substantially. The continuing challenges of rapid urbanization in India and its huge share to the global population can threaten global sustainability. Thus, it has become imperative to formulate strategies for sustainable development.

4.3 Different Types of Urban Forms

There are different types of urban forms, as well as different aspects of urbanization, each with its own concepts, which contribute toward sustainability of urban forms. The seven parameters for urban form sustainability include compactness, density, mixed land use, diversity, sustainable transport, passive solar designs, and green spaces (Jabareen, 2006). It is suggested that urban form needs to be understood at different scales, using information from a wide range of sources and addressing factors that challenge various aspects of sustainability (Clifton et al., 2008). Out of various types of urban forms, the compact urban form is the best for sustainable development. Even if the city possesses the characteristics of a compact urban form, it may disperse or sprawl in the future. To control this, it becomes important to devise policies that help to retain its compact characteristics to achieve sustainable development of the city (Kotharkar et al., 2014). To reach a sustainable urban form, for any city, it is necessary to study its existing form. Hence, it becomes important for the planners to have enough knowledge of the city's existing model and its potential extent of future expansion to achieve urban sustainable development. The Holdern–Shannon's entropy and Moran and Geary's coefficients are among the more widely used tools used to measure the urban growth pattern of the city (Mobaraki et al., 2012).

The urban spatial structure of a city is determined primarily by the movement pattern within the city. The different types of urban spatial structures existing in most urban agglomerations in the world can be broadly divided into four common types of urban spatial structures as listed below (Bertaud, 2003).

i. **Mono-centric Model**:

It is a classical model with a high-density center, which provides high number of jobs and other amenities. It attracts people from the fringe areas and the surrounding to the center through a radial movement pattern (Fig. 1).

ii. **Polycentric Model**:

There is no concentrated central zone in this model. The jobs and amenities are distributed all around the city generating random movement pattern of people across the city (Fig. 2).

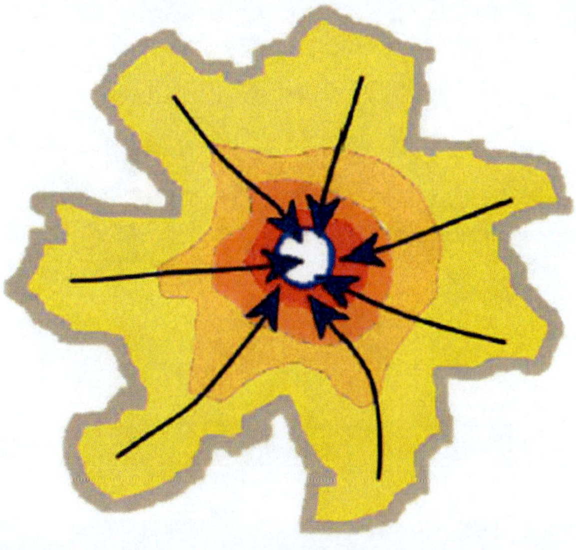

Fig. 1 Mono-centric spatial model. *Source* Bertaud (2003)

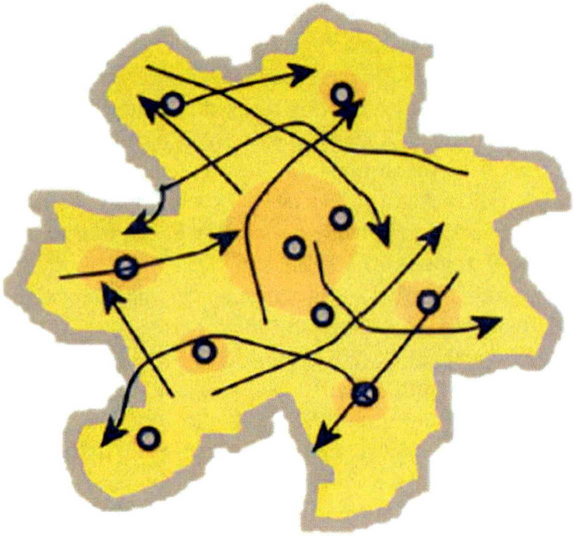

Fig. 2 Polycentric spatial model. *Source* Bertaud (2003)

iii. **Urban Village Model**:

Urban village model is a type of spatial structure, which has a lot of merits, but it is never usually occurs in real life. It has multiple centers distributed across the city. People live in proximity to their place of employment and daily amenities, which can be easily reached by walking or bicycling (Fig. 3).

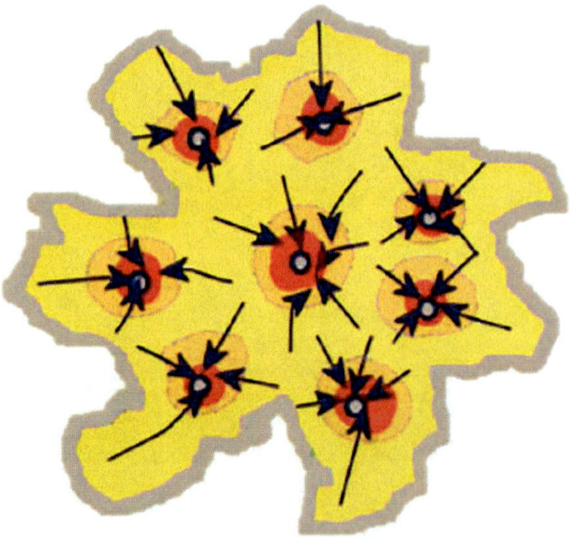

Fig. 3 Urban village spatial model. *Source* Bertaud (2003)

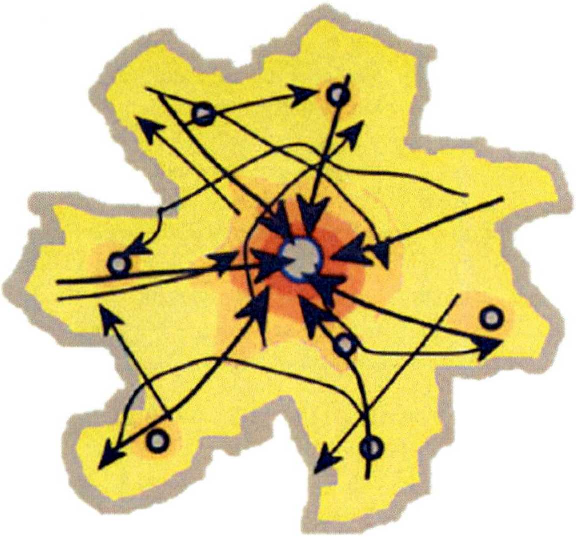

Fig. 4 Composite spatial model. *Source* Bertaud (2003)

iv. **Composite Model**:

It is a mixture of single dominant center and many other sub-centers around it. It witnesses simultaneous radial and random movement of people side by side all across the city (Fig. 4).

The relation between form, morphology, density, and energy usage is based on the spatial composition, materials, energy consumption patterns, human behavior, and their choice of transport. The form of the city indirectly affects the travel pattern and energy consumption of the city. Density and transport energy are closely related to each. Higher the density, optimum will be the energy usage but could compromise with some other factors of sustainability such as social networks and community interactions (Doherty et al., 2009). Similarly, relationship exists between the land-use and other policies that contribute to the reduction in travel energy consumption. For example, the areas that have higher residential density with higher rates of employment and higher levels of public transport are likely to show efficient energy consumption as compared with the low-density areas (Alford & Whiteman, 2008). Along with this, preservation and protection of natural ecosystems such as streams, wetlands, wildlife corridors, estuarine regions, fertile agricultural lands, alluvial soil-filled river basins, hilly regions, mountains and their associated flora and fauna are also important. It is also suggested that the use of natural drainage be encouraged and implemented, alongside creating green tributaries, controlling urban heat island effect, incorporating open parks and gardens in order to contribute to urban sustainability (Wheeler, 2010).

5 Factors That Make Indian Cities Unsustainable

Some reasons that have led to urbanization in India are the increase in government services, rural to urban migration, rapid mechanizing, providing for urbanization measures to further economic development of India in the eleventh five years plan, employment generation, and better infrastructure which caused migration to cities alongside the growth of infrastructure. Although rising urbanization brings with it the benefits of economic growth, it results in unsustainable growth as it brings many social and environmental challenges. It causes excessive resource exploitation, unplanned growth, growth of slums, environmental dilapidation, congestion, and a host of other ill-effects. India's huge population, varied as well as extensive land-cover and rural–urban migration, has caused the problems briefly outlined below, which have increased the load on the administrative bodies, policy and strategy makers, architects, and planners. A brief overview of the major problems of India making it unsustainable is as follows:

i. **Congestion**: Congestion is a situation caused due to overcrowding of people in very small areas. Population explosions in urban areas are the main cause of congestion. It is a common fact seen in Indian cities that large population is packed within a small space resulting in congestion.

ii. **Urban Sprawl**: Rapidly growing cities, both due to their exploding populations and geographical expansion, are the main cause of problems in urban areas. Huge rural to urban migration has led to burgeoning sizes of the cities. In most of the cities, the economic setup fails to deal with the rising demand due to the excessively growing rate of the population. Due to lack of planning, it gives rise to haphazard and sprawled conditions in the cities.

iii. **Housing Paucity**: When the population grows ceaselessly and in appreciable quantity, providing housing for all becomes a big challenge for the government. With the rise in demand for the houses, the real-estate costs also increase. The main sufferers of such situations are the middle- and low-income groups. Trying to meet the demand often leads to a situation of overcrowding; accompanied by poor sanitation, ventilation, solid waste disposal, power supply, and water supply.

iv. **Slums and Squatter Settlements**: Paucity of adequate quality infrastructure and dwelling spaces results in slum and squatter formation in the long run in urban areas.

v. **Environmental concerns**: Environmental vulnerability increases with each man-made and natural disaster. According to 'Disaster Management in India', a report by Ministry of Home Affairs, Government of India, 70% population of India is at risk of floods and 60% is prone to earthquakes. Due to high density and congestion the risk is higher in the urban areas. Due to excess human interference, the urban areas are gradually turning into heat islands with depleting groundwater tables and water scarcity.

vi. **Poverty**: Today, approximately one-third of the population in urban areas of India is below the poverty line. In urban areas, there are rising urban inequalities between the wealthy and the needy. As per United Nations Habitat Report, 2010, urban inequality in India rose from 34 to 38% based on consumption patterns in the period in-between 1995 to 2005.

vii. **Unemployment**: The urban unemployment problem is similar and of the same magnitude as the housing problem. Numerous urban areas across India are sprawling indefinitely, eating up green areas and farmlands on city fringes. This is followed by the subsequent migration of rural, largely uneducated, unskilled, agricultural workers to urban areas in search of employment, once their agricultural lands are sold off or taken over by real-estate agents and their subsequent unemployment in an overwhelming majority of cases.

viii. **Transport**: The number of vehicles on road has increased greatly creating traffic snarls and hold-ups, in turn affecting the effectiveness of road-based public transport. The main cause of such a situation is that an increasing number of people goes for private vehicular transport clogging the road with private vehicles. This has further led to the increase in the level of pollution. Transportation has always played a significant part in shaping the cities. The form of our cities is the reflection of how the transport networks are incorporated in the city at different stages of the city's development.

ix. **Water**: Water is one necessity to sustain life. With the growth of cities, supply of water is falling short. The rising water-pollution levels have also affected the quality and quantity of safe drinking water. Parts of urban as well as rural India have already begun to face acute water scarcity, with depleting groundwater levels at many places.

x. **Sanitation**: In the urban areas of India, prevailing sub-standard sanitation conditions is another gloomy scenario, particularly, in slums and squatters. Many slums and unauthorized settlements do not even have a proper drainage system and even if one exists, they are in very bad shape. These unsanitary conditions affect the city profile and the health of the people there.

xi. **Waste Management**: With the growth of cities, the quantity of waste generation has also increased manifold. Many Indian cities do not have proper arrangements for waste management and disposal. Due to this, a large quantity of wastes ends up in the surroundings and the dumping yards, resulting in unhygienic conditions. Existing landfills become full till the brim, afterward leaking into the ground and spilling onto their surroundings.

6 Study Area

Thane City is an industrial town in the state of Maharashtra. It is located in the northern part of the Konkan region, to the North-East of Salsette Island and to the North of Mumbai city. The population of the city as per 2011 census was put at 1,818,872. (Population Census again due in 2021, so 2011 census was the latest available census-based population record.) Thane forms a major part of the Mumbai Metropolitan Region, and it occupies a unique position in the region.

In the last forty years, the city has undergone tremendous demographic growth. The land cover of Thane City is 147 km^2 out of which around 40 ha of land consists of lakes and water bodies. The surrounding geographical attributes have directly impacted the growth and development of the urban form of the city. Its central location, easy access to infrastructure, abundant market places, civic services, and other facilities make this place a critical node in the Mumbai Metropolitan Region. The topography of the city is varied with high hills on one side and marshy lands on the other along the Thane Creek, alongside Ulhas River Bank. The availability of plain terrain is limited, which has again had its effect on the growth and development in the city (Fig. 5).

7 Measuring Urban Form of Thane

The urban form of a city comprises its physical patterns, arrangements, and structures. Urban forms are built upon a canvas influenced by our day-to-day living and our understanding about the past. Cities constantly evolve, exhibiting its unique character. It keeps changing and adapts to every new building built on it, as well as blending into itself parks, sidewalks, roads, and other amenities built on it. Each city is different and has some hidden characters that need to be studied through analysis and mapping. Monitoring and managing the spatial expansion of cities is a complex issue. So, to achieve sustainability of urban form, one needs to understand about the city past, study its present, and foresee its future growth.

Figures 6 and 7 show that the urban spatial structure of Thane is a composite model with a dominant center (Thane railway station/city CBD) and other sub-centers.

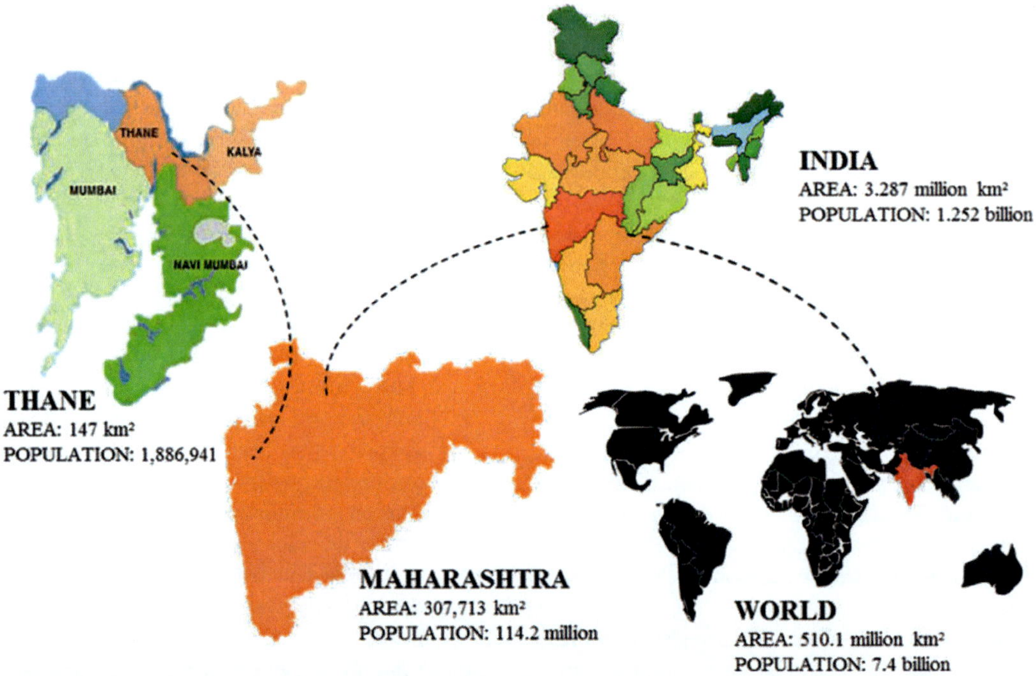

Fig. 5 Location of Thane in India. *Source* www.natmo.gov.in/National Atlas of Thematic Maps and Atlas, India

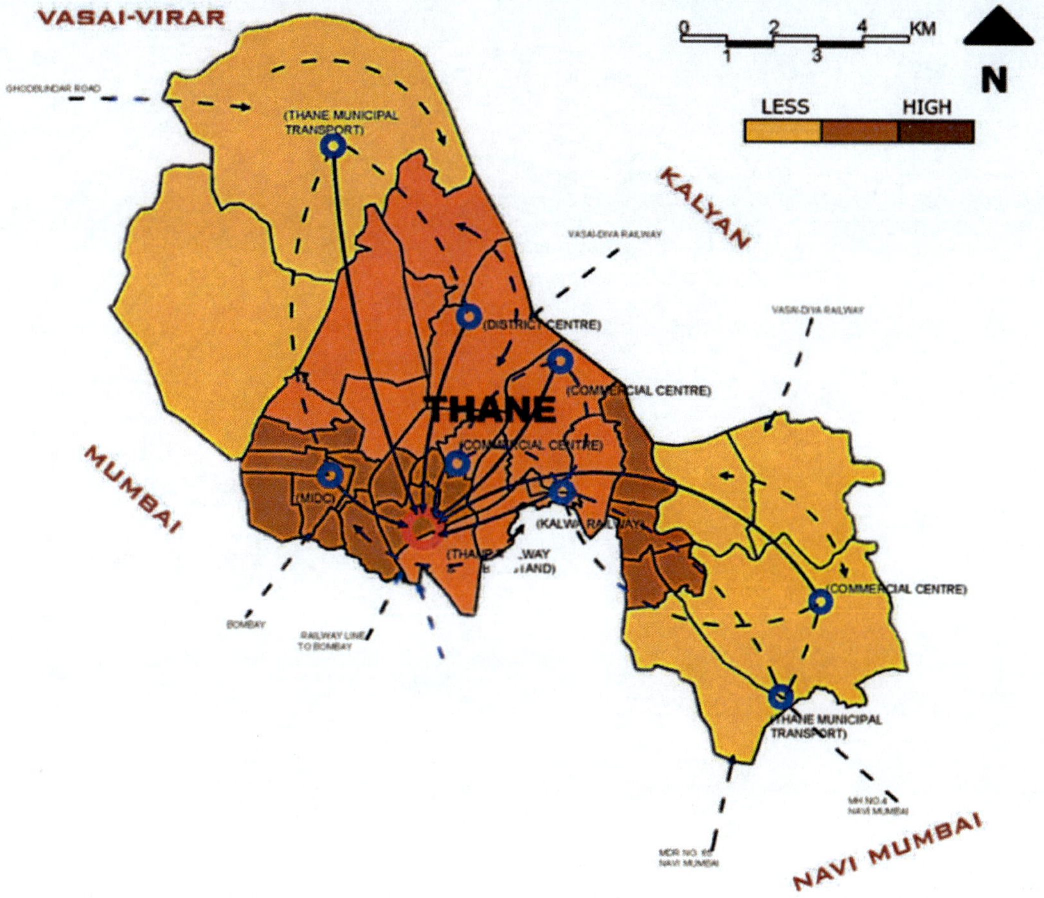

Fig. 6 Map showing spatial structure of Thane. *Source* Authors

(CBD is Central Business District.)

The movement pattern of the people across the urban area is simultaneously radial and random. Other than the urban spatial structure, three other parameters to study urban form sustainability of Thane have been adopted on the basis of previously done literature review. They are:

a. Density profile.
b. Land-use distribution.
c. Growth pattern.

With the help of the above three parameters, the existing form of the city, in this case, Thane, has been studied, to gather enough knowledge of the city's existing model and its expansion potential to achieve sustainable urban development.

a. **Density Profile**:

Urban density is a very important factor as far as the city's livability and sustainability is a concern. Urban densities are a function of the spatial structure of the city as well as the government regulations and taxation imposed. Average built-up densities not only vary across cities in the world, but also vary within the city. So, to analyze the city, it becomes important to understand its density profile. To derive the density profile of a city, the population is plotted over the *x*- and *y*-axis in which the vertical axis represents the urban densities and the horizontal axis is the distance from the CBD at an interval of 1 km. In Fig. 7, rings are plotted for the city of Thane at an interval of 1 km approximately.

This figure shows the density profile of Thane across the city area and the distribution of people around the CBD. A graph is plotted with respect to the distance from the CBD against the population (Fig. 8). It helps to understand the density distribution pattern of the city from its CBD. In the case of Thane, it is seen that the density is high at the CBD, and it goes on decreasing as it moves away from the CBD. This is the common scenario in most of the cities as the employment and other amenities are greater at the CBD, thereby attracting maximum

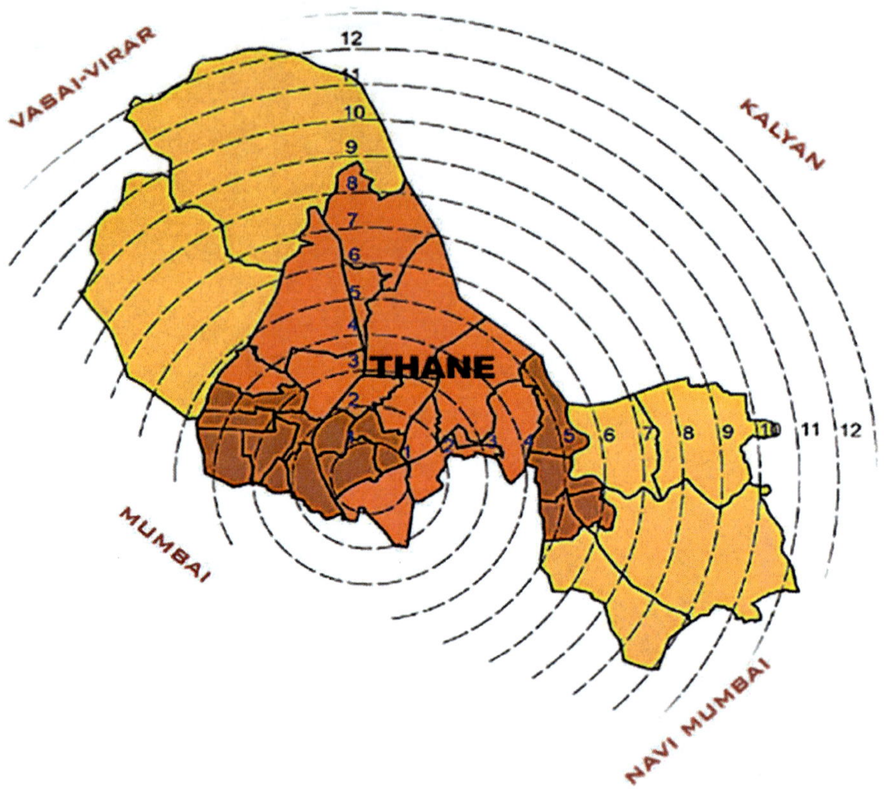

Fig. 7 Consecutive rings at a spacing of 1 km for Thane, Maharashtra, India. *Source* Authors

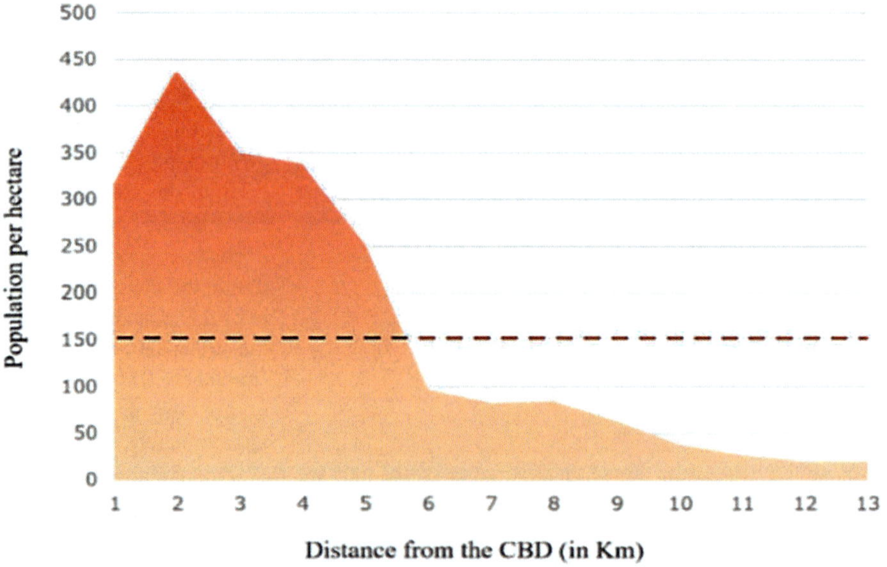

Fig. 8 Graph of distance from the CBD plotted against the population per hectare. *Source* Authors

population. The reason for this is mainly because of the availability of infrastructure facilities and employment facilities at the CBD. Specifically in the case of Thane, it is seen that the CBD is the major linking space with Mumbai and Navi Mumbai.

b. **Land Use Distribution**:

Mixed land use is one major aspect to be achieved for ensuring a sustainable urban form. It is the special integration of all kinds of places like living, work, recreation, relaxation, and entertainment. Urban Development Plans Formulation and Implementation (UDPFI) Guidelines by Ministry of Urban Development suggests a minimum land use requirement for each type of city. These values are considered to be optimum for sustainable development. Comparing a city's land use distribution with UDPFI guidelines allows assessing the gap and implementing the needful. It is observed that out of total land area in Thane, 63.56% is non-developable land which falls under natural ecosystem.

It means land available for development is less than 50%. Comparing Thane's land use with UDPFI guidelines, gaps are observed in residential, commercial, public and semi-public, recreational and transport communication sectors (Table 1).

One of the key methods for assessing the growth pattern of the city is Holdern method. It helps to study the growth pattern of the city by helping to determine its urban growth sprawl. It uses the natural logarithm (Ln) of the ratio of the population at the end of the period to the population at the beginning of a period and sums it to the logarithm of the ratio of land gross per capita at the end of the period to the land gross per capita at the beginning of the period. This equation gives the percentage of physical growth due to the increase in population and percentage of growth due to sprawling or unorganized growth. The Holdrens method is a mathematically rigorous method first described and partially developed by Harvard physicist John Holdren. A way of expressing this mathematical relationship is as follows:

ln (final population/initial population) + ln (final per capita land area/initial per capita land area) = ln (final total land area/initial total land area), i.e.,

$$\mathrm{Ln}\left(\frac{\text{population at the end of the period}}{\text{population at the beginning of the period}}\right) + \mathrm{Ln}\left(\frac{\text{land per capita at the end of period}}{\text{land per capita at the beginning of the period}}\right) = \mathrm{Ln}\left(\frac{\text{urban area at the end of the period}}{\text{urban area at the beginning of the period}}\right)$$

Population's share of growth rate = annualized average population growth rate/annualized average land development growth rate. The term 'annualized' means that the natural logarithm (ln) is applied to the rate of increase in each of the factors. Applying the same to the case of Thane, the extent of the physical growth of Thane due to population growth and due to unorganized growth on account of sprawling is calculated, as in Table 2.

$$\mathrm{Ln}\left(\frac{\text{population at the end of the period}}{\text{population at the beginning of the period}}\right) + \mathrm{Ln}\left(\frac{\text{land per capita at the end of period}}{\text{land per capita at the beginning of the period}}\right) = \mathrm{Ln}\left(\frac{\text{urban area at the end of the period}}{\text{urban area at the beginning of the period}}\right)$$

$$\mathrm{Ln}\left(\frac{1{,}818{,}872}{1{,}261{,}517}\right) + \mathrm{Ln}\left(\frac{123.732}{98.379}\right) = \mathrm{Ln}\left(\frac{14{,}700}{12{,}823}\right)$$

$$\mathrm{Ln}(1.441) + \mathrm{Ln}(1.257) = \mathrm{Ln}(1.146)$$

0.365 + 0.228 = 0.593; therefore,

$$\frac{0.365}{0.593} + \frac{0.228}{0.593} = \frac{0.593}{0.593}$$

0.61 + 0.38 = 1.

Table 1 Land use comparison—Thane & Urban Development Plans Formulation and Implementation guidelines

Land use category	UDPFI (%)	Thane City (%)	Inference
Residential	36–38	20.78	Gap of 15.22% (less)
Commercial	5–6	0.95	Gap of 4.05% (less)
Industrial	7–8	8.82	More by 0.82%
Public and semi-public	10–12	1.55	Gap of 8.45% (less)
Recreational	14–16	0.14	Gap of 13.86% (less)
Transport and communication	12–14	2.89	Gap of 9.11% (lless)
Defence purpose	Balance	1.09	
Agricultural lands		23.06	
Vacant land		2.43	
Forest		22.45	
Hilly area		0.05	
Brick Kilns		0.54	
Salt pans		0.24	
Quarries		0.18	
Sand collection areas		0.5	
Marshy land		8.06	
Tanks and Nallas		1.24	
Creeks		6.24	

Source www.mpurban.gov.in/pdf/EOI_UDPFI_Guidelines.pdf and authors

Table 2 Population for the year 2001 and 2011 of Thane City

Year	Population	Area	Land gross per capita
2001	1,261,517	12,823	98.379
2011	1,818,872	14,700	123.732

Source Census of India, 2011

Hence, it can be surmised that from 2001 to 2011 the physical growth of Thane due to population growth is 61% and due to unorganized growth on account of sprawling is 38%.

By implementing the above methods in the case of the Thane City, we can conclude that Thane has a composite urban spatial structure. The density profile study states that the growth pattern is compact but gradually dispersing. Based on the graph, as seen in Fig. 8, the population of Thane is unevenly distributed. To achieve a sustainable urban form, the population distribution needs to be even across the city. Assigning certain growth activities and reconsidering certain change in FSI will help to achieve uniform density across the city. The land use distribution study of Thane city shows that out of total land area 63.56% is non-developable land which falls under natural ecosystem. It means land available for development is less than 50%. Thus, it is important to have improved urban management, administration and firm regulations to obtain maximum benefits of the available land.

8 Conclusions

As a result of rapid urbanization, it has become a necessity to think of and implement new development ideas in urban areas. This uncontrolled expansion of the urban areas has resulted in contamination of resources and environmental degradation. On the basis of the arguments that have arose over time, many urbanists have suggested, that to achieve sustainable urban form, the compact city development model (a model with a large population on a limited amount of land) is the best form to achieve urban form sustainability. Most Indian urban centers have sprawled beyond control, and not too many cities can claim to have sustainable urban forms depending on the above study's parameters. This study has attempted to identify of how urban sprawl contributes to the unsustainability of forms of Indian cities. The study identifies that the different factors to be considered to achieve sustainable urban form include space management, optimized population density, mixed land use-based development, and diversified development. The study of growth pattern of the Thane concludes that from 2001 to 2011 the physical growth of Thane due to population growth is 61% and due to unorganized growth or sprawling is 38% which shows that the city is gradually dispersing. Appropriate time-to-time planning and legislative interventions are required to manage the urban form to help retain its compact urban form in the future.

References

Alford, G., & Whiteman, J. (2008). Macro-urban form and transport energy out-comes—Investigations for Melbourne. In *31st Australasian Transport Research-Forum*.

Bertaud, A. (2003). The spatial organization of cities: Deliberate outcome or un-foreseen consequence? *World Development Report 2003: Dynamic Development in a Sustainable World*. World-Bank.

Brundtland Report—Chapter 2: Towards Sustainable Development. (n. d.). *United Nations World Commission on Environment and Development*.

Business Standard. (2012, June 15). *Victims of urbanization: India, Indonesia and China*.

Clifton, K., Ewing, R., Knaap, G. J., & Song, Y. (2008). Quantitative analysis of urban form: A multidisciplinary review. *Journal of Urbanism: International Research on Place-Making and Urban-Sustainability*.

Doherty, M., Nakanishi, H., Bai, X., & Meyers, J. (2009). Relationships between form, morphology, density and energy in urban-environments. *GEA Background-Paper*.

Duncan & Hartman (1996). Sustainable urban transportation initiatives in Canada. In *APEC Forum on Urban Transportation*, Seoul, Korea, November 20–22, 1996.

EOI UDPFI guidelines. (2017, February 15). Urban Development and Housing Department, Government of Madhya Pradesh. www.mpurban.gov.in/pdf/EOI_UDPFI_Guidelines.pdf

Jabareen, Y. R. (2006). Sustainable urban forms: Their typologies, models, and concepts. *Journal of Planning Education and Research*.

Jenks, M. (2000). The acceptability of urban intensification. In *Achieving sustainable urban-form*.

Kotharkar, R., Bahadure, P., & Sarda, N. (2014). Measuring compact urban form: A case of Nagpur City, India. *Sustainability*. ISSN: 2071-1050.

Location of Thane in India. (2017, February 15). National Atlas & Thematic Mapping Organization, Government of India (Department of Science & Technology). www.natmo.gov.in

Maharashtra District census Handbook. (2017, January 20). Office of the Registrar General & Census Commissioner, India. Ministry of Home Affairs, Government of India. https://censusindia.gov.in/2011census/dchb/DCHB.html

Mobaraki, O., Mohammadi, J., & Zarabi, A. (2012). Urban form and sustainable development: The case of Urmia-City. *Journal of Geography and Geology, 4*(2).

Nuzzolo, A., Coppola, P., & Papa, E. (2014). Urban from and sustainability: Modeling evidences from the empirical case-study of Rome. In *ETC European Transport Conference*.

Prakash, M., Teksoz, K., Espey, J., Sachs, J., & Shank, M. (2017). Achieving a sustainable urban America. *The U.S. Cities-Sustainable Development Goals-Index.*

Wheeler, S. M. (2010). The evolution of urban form in Portland and Toronto: Implications for sustainability planning. *Local Environment—The International Journal of Justice and Sustainability*.

Williams, K. (2014). *Urban-form and infrastructure: A morphological review*. University of the West of England.

The Potential of Benghazi Lakes and Wetlands in Achieving Sustainable Master Plan

Zakeia A. Benmoussa and Eman M. Elmazek

Abstract

Lakes and wetlands are considered a nature-based and cost-effective answer that presents sustainable assets. Their management and development should be included in city sustainable plan as an essential part of the urban sustainable strategies. Benghazi is known with its wetlands and lakes that have always been a focus sector of the city landscape and urban structure through the city history. Unfortunately, the neglect, lack of good management and planning of its wetlands results in negative environmental and economic effects. In addition, lakes due to environmental and economic mishandling have become a distressing source instead of being an enjoyable recess to the city. Therefore, Benghazi residents find themselves in an isolation with their city, its nature, and landscape. This paper investigates the current situation of Benghazi major lakes and wetlands and highlights the main problems for each of them. The research surveys what was done, how it was done, current efforts, and any conventions or agreements related to the subject. From these inspections associated by opportunity analysis for each lake and wetland in Benghazi City, the research resolves sustainable strategies for these city vulnerable elements. The survey includes main issues and problems created by the wetland oversight, and their dangerous effect on the city. Some pervious calls were done to mark these questions. These efforts are taken in consideration in this research. In addition, the research lists the different types of wetlands in Benghazi and the main sustainable strategies needed to improve and profit these wetlands to the favor of people, economy, and environment.

Keywords

Wetlands • Sustainable city • Lakes • Urban sustainable strategies • Benghazi • Sustainability • Sustainable landscape

1 Introduction

Water fronts of postwar cities are the most affected areas at the city level. The coastal urban wetlands and lakes which are located in the city of Benghazi are deteriorating and having the risk of extinction as a result of prolonged periods of conflict and mismanagement. The city of Benghazi and its environs have undergone in major changes in their natural environment, which have caused a disturbance in their ecological balance and functions. The main cause of such alters is the urban projects that did not take in consideration these ecosystems. The urban planning projects in Benghazi since the mid-sixties have created a phenomenal transformation that caused undesirable infringements on Benghazi natural environment and landscape (Elbabour, 2012). The unbalance happened in these vital elements of Benghazi are affecting the balance and function of the city as whole. Throughout the history of the city, the surrounding wetlands and coastal lakes have provided many benefits in terms of economy and human wellbeing. However, nowadays due to negligence and maladministration they have become an extra pressure to the city pollution. They were abundant sources of salt which was one of Benghazi most important export products; now they are an economic tension to Benghazi master planning.

Benghazi lakes and wetlands have unique features and landscape. They mark the city landscape and nature condition. By time they were subject to many alters and deteriorations, accordingly they need continuous checking and evaluation of the projects that are targeting the urban landscape. Any projects should work on preserving the natural

Z. A. Benmoussa (✉) · E. M. Elmazek
Architecture and Urban Planning Department, Faculty of Engineering, University of Benghazi, Benghazi, Libya
e-mail: z_benmussa@yahoo.com

F. Trapani et al. (eds.), *Advanced Studies in Efficient Environmental Design and City Planning*, Advances in Science, Technology & Innovation, https://doi.org/10.1007/978-3-030-65181-7_13

elements and restore the biodiversity of these vulnerable ecosystems. In addition, there are many haphazard activities that consume the purity of Benghazi wetlands and lakes. Many legal and illegal activities have affected the natural environment of these lakes. Governmental projects that alter the nature of these lakes and destroy their wildlife such as damping them to use the land for building, putting sewage pipes into them, changing their edges destroying the wildlife, and blocking or opening channels between each other's and the sea. In addition, unaware individual activities are increasing due to the lack of law enforcement such as using them as landfills, constructing slums, and discharging sewage water into them. Despite the presence of some projects targeting the lakes to lessen their bad effects to Benghazi environs, all efforts and plans are not preserving the nature of these vulnerable ecosystems enough to obtain a sustainable master plan. These projects have developed the landscape of lakes and used them as recreation areas by applying both softscape and hardscape to the surrounding environment which alter the ecosystem and lead to wildlife lose. There is a need to a strong database of these landscape features including current problems and any possible solutions that should be taken in consideration in Benghazi sustainable plan.

1.1 Methodology and Aims

The overall aims of this paper is to scan the most prominent lakes in Benghazi landscape, establish their current situation, highlight their significance for biodiversity conservation and their values to sustain landscape and natural features of Benghazi, and recognize any environmental deterioration and factors that lead to the destruction of these sensitive natural sites. The paper also encourages environmental awareness and sustainable use of these sites through recommendations and sustainable strategies that should be implied in any development planned for these locations taking in consideration the principles of natural preservation and conservation. These strategies should also be put in action immediately to stop any trespasses that may lead to the lakes ultimate destruction. The main focus of this paper is to construct a database for research on Benghazi lakes and wetlands landscape from a sustainable urban planning point of view. Another objective of this paper is to produce an interdisciplinary debate in favor of classifying these sensitive sites as a strong city planning element.

In order to gather evidence and information about the potential of Benghazi lakes and wetlands, a comprehensive analytical review has been done to previous researches about any of these lakes including any development projects or plans. Some lakes and wetlands have geological and wildlife detailed studies, while other information are collected from governmental and non-governmentals briefs, interviews, and news. Each lake or wetland has been taken as a case by case basis study to acknowledge each site real needs, weakening circumstances, and strengthening opportunities. Five lakes and wetlands have been acknowledged in this paper by identifying the topology of each of them, history of any alters to its natural environmental either man-made or not, including structures, buildings, or edge relocating, and its potential to the biodiversity and landscape of Benghazi. Based on the collected and analyzed data, general recommendations and sustainable strategies are defined. In addition, according to the particular facts related to each lake or wetland sustainable proposals and suggestions have been based.

1.2 Wetlands and Lakes of Benghazi

The wetlands and lakes formations that abound abundantly around the historic city of Benghazi have ecological characteristics and extreme sensitivity. The contemporary urban planning projects have affected the natural landscape in Benghazi and need a review of the possibilities of its unique position and the adoption of different development methods that take into account the approaches of the "landscape ecology" in order to integrate elements of the natural environment in the urban landscape of the city neglected for some time. The debate is over the strategic importance of environmental planning and its role in maintaining the fragile landscape of Benghazi in order to achieve sustainable urban growth.

1.2.1 City of Lakes and Wetlands Between Past and Present

The city was surrounded by its northeastern, eastern, southern, and southwestern sides of a continuous strip of wetlands and lakes areas; see Fig. 1. This continuous strip of wetlands extends beyond the city boundary for tens of miles along the sea coast in the northeast and southwest directions. The lakes are fed by a subterranean spring from the canals (the Laithi River) and sometimes surface in the form of karst craters and caves (Ain Zayana and the Boudzieh/Megrine group, as well as several dry Karst craters) (Elbabour, 2012).

1.2.2 Benghazi Main Wetlands and Lakes Types

According to Ramsar Convention[1] "wetlands are defined as areas of marsh, fen, peatland, or water, whether natural or artificial, permanent or temporary, with water that is static or

[1]Ramsar convention is a convention on wetlands. It is the intergovernmental treaty that provides the framework for the conservation and wise use of wetlands and their resources.

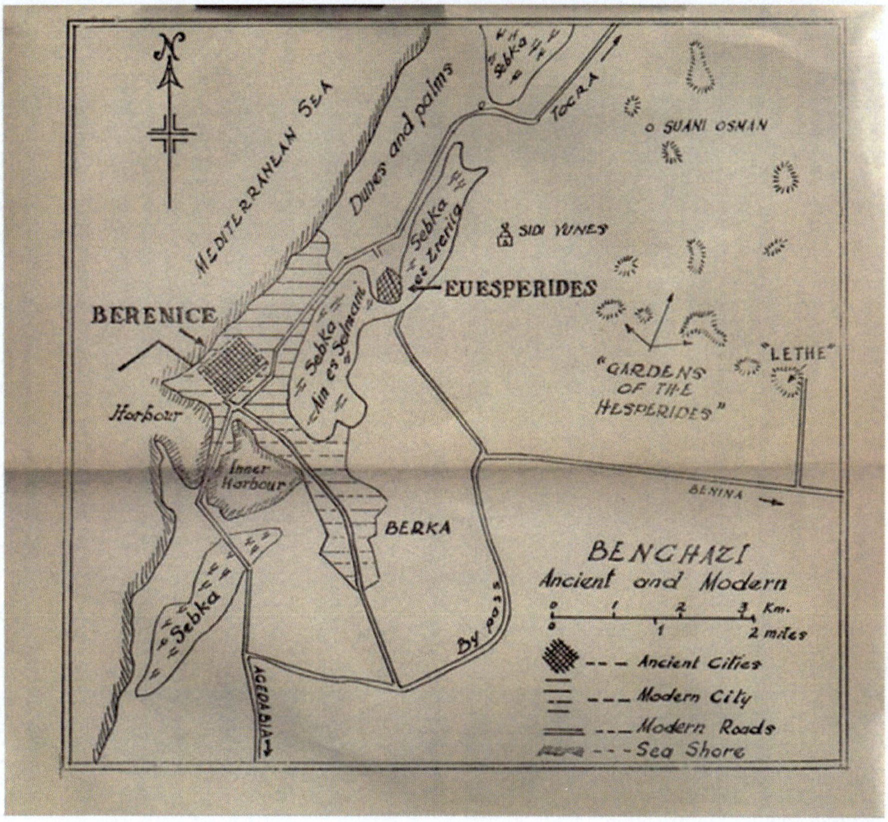

Fig. 1 Sketch of the town and its wetland landscapes which survived the of perils of time until the early decades of the twentieth century (Goodchild, 1962).

flowing, fresh, brackish or salt, including areas of marine water the depth of which at low tide does not exceed six meters." The main types of wetlands in Benghazi are swamps and saltmarshes. Examples of Benghazi wetlands are Sabkha Sabri, Sabkha Alminqar, Sabkha Ganfuda, and Sabkha Garyounis. These are the major and most sensitive wetlands in the city. As mentioned before, these wetlands have a historical, social, economic, and ecological significant to the Benghazi people. By this, the potential of investigating in these wetlands is an integral part of sustainable development for the city, as they touch the four pillars of sustainability: economy, society, culture, and environment.

In addition to wetlands, Benghazi City features coastal lakes and lagoons. These lakes have an urban development potentials not less than the wetlands. Coastal lakes and lagoons are bodies of water formed in basins (topographic depression) separated from the sea by a barrier of narrow piece of land. Coastal ponds and lakes are totally compassed by land and shielded from the sea. Water bodies that outlet to the sea are called lagoons and costal bays depending on their shapes (Oertel & Leatherman, 1985). The most significant lakes in Benghazi are Juliana Lake and Northern lakes. These two water formations gain their importance by their location in the city's western and eastern entrances. There are two main lagoons in Benghazi, 23rd July Lake and Ain Zayana Lake.

1.2.3 Case Studies of Benghazi Lakes and Wetlands

The paper analyzes five of Benghazi main lakes and wetlands. Benghazi has a numerous number of wetlands; some of which have unfortunately totally disappeared. Other are safe from any environmental hazards. The case studies here are the most prominent ones and most engaged in its urban structure. Most of them are in a haphazard condition and other have started to affect the city vision and environment in a negative way.

Sabkha Sabri

From more than 1000 years ago, the area was part of the sea. After the area drained and its population moved to Sidi Akhbarish, its territory became a marsh. After War World II the area become populated again. Nowadays Sabri suburb is one of the most populated suburbs of Benghazi. It carries a lot of social, historical, and economic city identity. In the current period, the Environmental Sanitation Office confirmed the existence of several microbes in the vicinity of Sabkha Sabri caused by the stagnation of water and this transferred many diseases, including smallpox and typhoid.

Fig. 2 Sbakha Sabri highlighted with blue lines. The aerial view shows current built area (Google Earth)

The environmental reason is that the wetland lost its ability to work as a sponge to absorb excess water. This leaded to the stagnation of water, lack of natural absorption and difficulty of natural purification, in addition to mixing with the sewage water. The undesirable infringements on the natural environment of this wetland are mainly due to land use dysfunction. Most of the wetland is covered by buildings and hardscape associated with these buildings; see Figs. 2 and 3. This happened during the phenomenal transformation of modern city of Benghazi urban planning.

West Entrance Lake (Lake Juliana)

Lake Juliana is the most prominent of the city's karst lakes and Sebkha formations.[2] It is located in Juliana coastal suburb southwest of Benghazi City center occupying an approximate area of some 1.6 km^2 (about 160 ha). It is one of the fragile urban wetlands that are shrinking by a fixed percentage and which are subjected to erosion and damage to it and its aquatic organisms. The lake is known by its significant ecological functions, biodiversity roles, and unique landscape. It is known locally by bird sanctuary as it is known to be a station to migratory birds. The lake is visited by 33 species of bird each year (EGA, 2012). This city landmark was a subject to nineteenth and twentieth century traveler's sketches, maps, photographs, and satellite images. All show a change in the size and magnitude of the lake (Abdulsamad & Elbabour, 2013). Figure 4 shows clearly the drought and water receding of the lake from 2003 to 2011. The construction on the site has contributed to the shrinking and loss of part of the biodiversity chain by nature and loss of part of its function (Elbabour, 2012).

Northern Lakes

They are a wide flat area submerged with shallow water at the northern outskirts of Benghazi. They are known by Lathama Lakes (Fig. 5). The area has been used as the city dump for many years, having all kinds of debris, old cars, oil, and dangerous waste. The site has become full of slums and stands as a dangerous environmental hazard (Fig. 6). A decision was taken to save the area by establishing The Northern Lakes Project in the 1980s. The first phase was completed by the construction of roads, bridges, and the digging of a lake. After the first phase, work stopped for a long time and environmental abuse continues on part of the lakes. In 2006, the second phase was launched to complete the project. There are aims to get rid of Sabkha in this region and work on its development. The completion rate reached 44% at that time. The project consists mainly of digging two artificial lakes that are connected to the sea by two channels surrounded by footpaths (Fig. 6; Libya 24, 2010).

Lagoon 23rd July

Lake 23rd of July is connected to Benghazi old harbor and located in the center of Benghazi (Fig. 7). The lake features Benghazi main landmarks such as Tebisti Hotel, Ozoo Hotel, the Islamic Daawa building, and Benghazi sport complex and stadium. Its area is about 100 ha with maximum depth of 5 m. It was once known as Salina della Punta. It was one of three Salinas[3] in Benghazi; Salina della Giuliana (Lake Juliana) and Salina Selmani (Sebkha Selmani) which totally disappeared due to urban planning activities. The lake was later transformed into an artificial lake and was a center of sea activities and a gathering of fishing enthusiasts. The lake features a natural park by its side that is a place of celebrating national and religious festivals. From the nineties, the lake suffered from the problem of failure in some sewage networks, which led to leakage and contamination and the death of fish except living on organic waste (Libya Today, 2008). This transfers the main recess area of Benghazi to a place with low-quality air and environment. In November 2008, a scientific symposium on the lake

[2]Karst lakes are lakes formed as the result of caves collapse in water-soluble rocks such as limestomes. Sbekha is an arabic word of coastal salts marshes. They are special type of wetland habitats that are formed in coastal areas that are relatively sheltered from harsh sea waves and where rivers or creeks deposit a special type of fine sediment.

[3]Salina means a salt pan or a salt flat.

Fig. 3 Sbakha Sabri emerged in excess water during the last war causing an environmental hazard (Alagouri Photography)

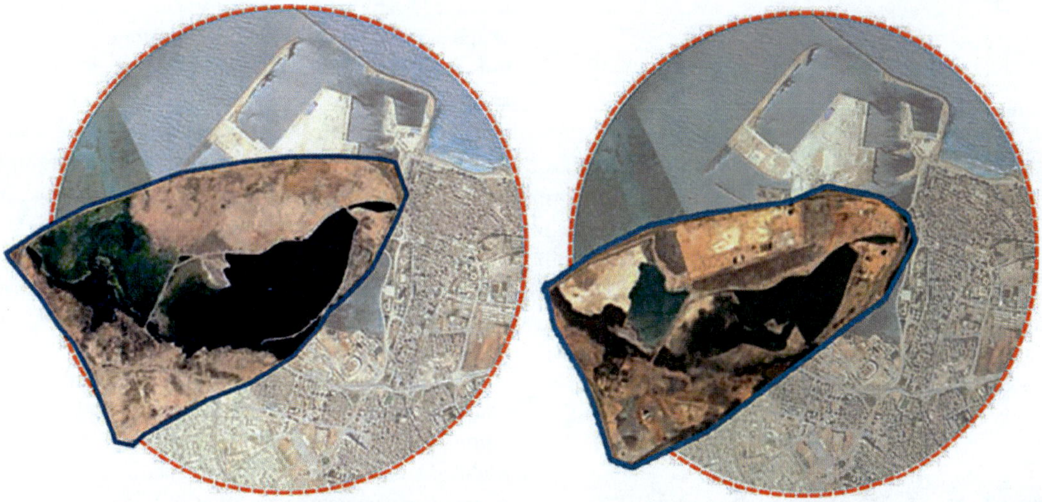

Fig. 4 Lake Juliana aerial view (2003–2011) (Google Earth)

concluded that it is necessary to pay attention to it and adopt research plans to address and correct it (Oyah Journal, 2008).

Ain Zayana Lagoon

The lake is located 15 km northeast of Benghazi. It is by a coastal shoreline. The lagoon is 1.5 km long and 3–4 km wide. It is connected by a channel at the northern end to the sea and fed with 10 underground springs from deep groundwater. The lake is surrounded by shallow swamps where it is covered with water at high tide especially east. It is considered as an important site for migratory water birds. The annual number of bird species sighted at the lagoon is 25 species (EGA, 2012). The main threats to this lagoon are hunting for rare birds and indiscriminate construction resulting in mobilization of parts of wetlands. Moreover, the area is used as a main landfill of the city. See Fig. 8 (Fishpool & Evans, 2001).

1.2.4 Summary of Results

Benghazi lakes and wetlands play a crucial role in the shaping of Benghazi master plan through history. The previous survey of the wetlands and lakes case by case has shown some main pinpoints important for the continues recommendations and strategies serving the matter sustainable plan of Benghazi. First to mention that there are different types of coastal lakes in Benghazi. Unfortunately, alterations done to some of them have transformed their typology which would cause loss in biodiversity and landscape. The lakes and wetlands in Benghazi have been a main landscape feature and losing them would cause social, environmental, and economic issues. This would stand as a barrier to get a sustainable master plan for Benghazi City. There are many trespasses from the unaware public as well as non-sustainable government plans. There would be more database and awareness about the past, the present, and current situation of

Fig. 5 Northern Lakes (Google Earth)

these sites and also the consequences due to these actions. Strategies should be constructed to sustain the wellbeing of these sites. These strategies should take in consideration using smart, and sustainable planning to preserve what still existed, restore what can be back, and finally adopt any environmental issues happened due to the deterioration of the landscape. The destruction caused different kinds of environmental pollution as in air, water, and visual pollution which is related to the urban and landscape environment.

1.3 Wetlands' Significance to Benghazi City

Through Benghazi history, its extensive wetlands have been a good source of salt. The salt had a large contribution to the city economy as it was exported aboard. This valuable source of income to the city residents gave Benghazi the name of the City of Salt back on that time. In addition, wetlands have served as effective natural defensive barriers for the city from invaders before the Italian colony built a defensive wall after they conquered the city. Wetlands also have provided many ecological and biodiversity services. To mention, they have served as natural flood control and flood hazard during rainy seasons as they worked as the sponge that protect the city from any excess water.

1.4 The Importance of Wetlands and Coastal Lakes

They are considered open spaces with high potential tourism investments. They are popular tourist destinations for locals for their unique nature, biodiversity, and important historical heritage. Many of the wetlands and lakes are locations to watch wintering water birds as mentioned before. In addition, they are home of different species of salt-tolerant plants, some of which have been used for local traditional industry and crafts. A more economic benefit is that the wetlands can be used for salt extract and export as they were in the past. Moreover, the lakes work as oasis to escape from hot summer weather for Benghazi people and enjoy a sunny day in winter weather at an abundant rich open space. A good example is the Lake Boudzira which is a prominent tourist place in the city. It is a natural lake centered by an island that is surrounded by a national park.

Environmentally, wetlands have a high ability to purify water by retaining pollutants inside the soil and the vegetation. They work as a natural filter for runoff water and polluted water from sewage and other organic and non-organic contaminants. Wetlands as well are a natural system to recharge the underground water as they are connected to the groundwater system. Lakes and wetlands are a very sensitive and effective part of the hydrology of the city. They act as a sponge or natural flood control system. Moreover, plants and animals find lakes and wetlands a

Fig. 6 Northern Lakes dangerous environmental hazards (Libya 24)

Fig. 7 23rd July Lake Figure (Google Earth) (Alagouri photography)

Fig. 8 Ain Zayana Lagoon (Google Earth) (Libyan Documentation Center)

suitable place to create the reasons of life and the most important of which are migratory water birds. In addition to be recreational areas, they can be areas for study, research, culture, and spreading environmental awareness among the community such as national, plants, birds, and natural history parks and museums.

1.5 Wetlands and Lakes Problems and Possible Fragmentations

One of the main threats that are facing wetlands in Benghazi is drying them and is establishing residential areas and highways. Moreover, they might be transferred into agricultural lands as well and their water might be displaced by other water for irrigation which is considered to be obtrusion to its nature. Lakes and wetlands are generally subject to drainage of water to irrigate other drylands. In addition, a major fragmentation to Benghazi wetlands is the disposal of pollutants and sewage water which make them unfit for wildlife. Wetlands as well as lakes could be subject to electricity generating projects which affect wintering water birds visits and trips. In addition, there is evidence of migrating bird hunting in all Benghazi wetlands and lakes. This leads to the loss of large numbers of birds and species and less future birds visits.

Benghazi lakes and lagoons become lately a major place of solid waste and trash dumping due to lack of environmental awareness. The residents found these lakes abandoned and neglected therefore the lakes has become a source of pollution and a bad view in the urban structure. Furthermore, the lakes due to unwise decisions are used for wastewater and sewage outlet. Most lakes were subject to plant removal and destruction which affected the wildlife structure and the pleasant landscape. Juliana Lake and the Northern Lakes sites are partly used as a truck depot and loading building materials such as gravel and sand. All of the above negatively render, these sensitive ecosystems as urban stains and a cause of environmental, health, and city scenery disturbance. One of the biggest threats to the lakes and wetlands is irresponsible urban planning projects and building activities that aim to change their natural status into artificial ones. Coastal lakes, for example, have projects of direct open link to the sea and/or displace their natural abundant edges with artificial hardscapes like what happened in 23rd of July Lake (Elbabour, 2018).

1.6 Previous Calls to Highlight of the Wetlands and Lakes Situation in Benghazi

There were many attempts to take advantages of the wetlands and lakes of Benghazi; however, these attempts were far from sustainable development as most of them shifts the natural status of the landscape. In 2000, Libya acceded to the Ramsar Convention. Ramsar Convention is an intergovernmental agreement that encourages the conservation and wise use of wetlands through national and regional action, and international cooperation that contributes to sustainable development. In 2007, World Wetlands Day was held in Benghazi City organized by Libya Environment General Authority and Ramsar/Medwet Project. In this day, it was pointed out the need to cooperate with EGA in announcing new Ramsar sites in and around Benghazi. In April 2012, recommendations came from the symposium on sustainable planning for the reconstruction of Libya organized by the Libyan Urban Planning Association on the importance of studying the phenomenon of aggression on nature reserves, including wetlands and the convening of a workshop dedicated to them in June of that year. Among the recommendations of this workshop are to consider, preserve, and classify wetlands in urban and rural areas as nature reserves. In addition, the workshop recommends to prevent any alteration in a method that threatens their environmental sustainability and finally to comply with international conventions on the conservation of wetlands and their natural habitats and biodiversity.

1.7 Why Wetlands and Lakes Are Essential for Sustainable Development?

Human life on earth and its wellbeing need sustainable management to the earth life resources. One of the most essential natural elements is wetlands and lakes. They affect our wellbeing, the growth of our economy, and climate mitigation and adaptation. Coastal lakes and wetlands make our cities safe by protecting the shores. They are nature-based solutions that provide low-cost protection for coastlines from water disasters. In the last two decades, water disasters are counted as 90% of all disasters. Coastal lakes and wetlands absorb excess water and reduce storm flooding. For example, during hurricane Sandy in 2012, wetlands saved the US 625 million dollars in flood damage by reduce the effects of the hurricane (MEA, 2005). In addition, they serve people and biodiversity in many ways. They are source of food and water for many habitats. It is estimated that 40% of the global species live and breed in lakes and wetlands (Costanza et al., 2014).

1.8 Sustainable Strategies for Benghazi Wetlands and Lakes Rehabilitation Plan

Benghazi wetlands and lakes have been an extensive part of the city landscape and life. Therefore, it is essential to integrate urban and environmental planning and landscaping by recognizing them as major elements to achieve a sustainable plan for Benghazi. They should be acknowledged as a key component of urban water management infrastructure. Environmental laws should ensure the preservation and restoration of the lakes and wetlands parallel to the protection of their birds, plants, and all forms of aquatic life. Citizens as well as officials should learn about the importance of Benghazi lakes and wetlands and their role in environmental, economic, and social balance. It should be clear for them how any urban changes to the nature environment will cost the city and its residents, although these projects might look upgrading.

Many studies recommend that there should be regular evaluation surveys to these ecosystems in order to check any trends to alter their nature situation. As for the samples surveyed in this paper, some recommendations are set for each case. As for Sabkha Sabri, there are many man-made features that have completely altered a big part of this sensitive wetland. The best suggestion here is to remove all this artificial elements and reclaim the area as an open space which would give it the chance to restore its original nature by years. All human activities should be stopped in Juliana Lake especially dumping and backfilling it in favor of construction. Also, bird hunting and wastewater disposal should be stopped completely. The lake should be integrated with Elkeish Lake and reclaimed to be a recreational open place. Northern lakes second phase project should be reconsidered regarding linking the lakes with the sea through two artificial channels. Also, the project includes digging the lakes and constructing roads, buildings, and sidewalks. All these are considered as a transformation from a natural to an artificial lake which threatens the sustainability of this ecosystem. Any projects that are planned should respect the nature of the wetlands including the edges of them where a lot of aquatic and terrestrial plants live. 23rd July Lake or Lagoon is an example of altering the edges and replacing them with artificial ones in favor of urban projects. The lake lost its ability to purify its water through the plants and the biolife in its edges. This lagoon renews its water by a small link to the sea; however, it is not enough with all the sewage water and solid waste disposal. It is favorable to install a second method to help purifying the water along with avoiding more polluting to the lake.

1.9 Conclusions

Wetlands and lakes are vital parts of Benghazi sustainable master plan. Establishing sustainable strategies for these sensitive ecosystems have its major influence from historical, social, economic, and environmental aspects. The current situation of these lakes and wetlands remarks them as elements that decay the city and its environment. Generally, the urban built environment is becoming dominant in Benghazi over the nature and landscape leaving these ecosystems with negligence and continuously destroying and polluting them. There are recently many calls to save the wildlife and reserve the landscape of Benghazi especially the lakes and wetlands to achieve a sustainable plan of Benghazi. However, realistic projects are far away from sustainable development regarding these ecosystems. On the opposite, many projects or most of them are targeting to divert the natural features and wildlife elements in favor of urban development. Therefore, regular and particular checks should be done to these nature systems case by case to establish references and recommendations as foundations to restore Benghazi landscape. Any alters to the nature features might lead to the failure of these hydrologic systems and loss of wildlife. Any future or current project should take in consideration the wellbeing of Benghazi lakes and wetlands. In addition, work should be done with accord to international conventions such as Rasmar. Some lakes have already lost some of their nature components partly or completely. For example, 23rd July Lake has lost its plants and edges through years, and it seems that there is no way to return it back so the solution is in install an alternative system that renews and purify the water. Also, Sebkha Sabri has lost a big area due to urban projects. The lost part can be restored by time if the area is returned back to be an open area. Each case has its own fragmentations and solutions and projects targeting, and it should analysis and acknowledge its privacy.

References

Goodchild, R. G. (1962). Benghazi: The story of a city. Department of Antiquities, Cyrene (Shahat).

MEA (Millennium Ecosystem Assessment). (2005). *Ecosystems and human wellbeing: Wetlands and water synthesis*. World Resources Institute.

Libya 24 channel. (2008). The Housing Projects and Utilities Department: Northern Lakes Project.

Libya Today Journal. (2008, June 11). The pollution in 23rd of July Lake.

Oea News. (2008, November 29). In *Environmental Scientific Seminar*, July 23, Lake.

EGA—RAC/SPA waterbird census team. (2012). Atlas of wintering waterbirds of Libya (2005–2010). Imprimerie COTIM.

Elbabour, M. M. (2012). Framework for linking urban and environmental planning—The changing landscapes of Benghazi. In *Symposium on sustainable planning and the reconstruction of Libya.*

Alagouri, I. (2013). Ibrahim Alagouri photography. https://www.pixoto.com/alagouri/recent

The Mediterranean Wetlands Initiative. (2016, February). Wetlands for a sustainable Mediterranean region. https://medwet.org/aboutwetlands/ramsarconvention/

Google Earth, Benghazi City. (n.d.). https://www.google.com/earth/

Elbabour, M. M., & Abdulsamad, E. O. (2018). Juliana Lake: A wetland in an urban environment. *Libyan Journal of Science & Technology, 7*, 2.

Abdulsamad, E. O., & Elbabour, M. M. (2013). Juliana Lake: A Ben-ghazi wetland in distress. European Geosciences Union (EGU) General Assembly, Vienna, Austria, 2013.

Fishpool, L. D., & Evans, M. I. (2001). *Important bird areas in Africa and associated islands: Priority sites for conservation*. Pisces Publications.

Costanza, R., & de Groot, R. (Eds.). (2014). Changes in the global value of ecosystem services. *Global Environmental Change, 26*.

Oertel, G. F., & Leatherman, S. P. (Eds.). (1985). *Barrier islands*. Marine Geology.

Innovative Sustainable Strategies to Mobility and Urban Transport Problems in Benghazi

Aziza Safour

Abstract

Nowadays, sustainable transport strategies and plans that involve different modes and options of travel have been considered as the suitable tools for tackling urban traffic and transport problems in cities with high population densities. In Libya, the mobility and traffic conditions in the urban areas experience number of problems, due the rapid increase in urbanization and the lack of public transportation means, which have led to an increase in the number of automobiles ownership per household. Libyan transportation authority stated that the road traffic has crucial negative impact on society's health and safety, considered that a major cause of death and injury in large Libyan cities, particularly, in Tripoli and Benghazi. In these cities, the car-based transport mode is predominated and most commuters are dependent on private travel mode to move from home to work or other destinations. While reviewing successful cities experiences that implemented comprehensive sustainable transport plans proved having access to a variety of transport modes and options can help households and road-users to reduce transport costs, improve transport safety, reduce pollution and providing easy and safe access to work, services and opportunities. Furthermore, public transport, includes buses; metro, light rail; and other modes, is one essential aspect of the overall system. This paper ascertains the remaining situation of transport and mobility in Benghazi, which is the second city in Libya, and hosts several important companies and institutions. Also, it is an important economy center and the most populous center in the Cyrenaica region. This study develops a plan for sustainable transport strategies that can be applied in or around the urban area of Benghazi to reduce transport problems, and this plan could be used as an indicator for achieving healthy and high quality of life in Benghazi community.

Keywords

Transport problems • Traffic conditions • Sustainable strategies • Variety of transport modes • Accessibility • Mobility

1 Introduction

Today's studies stated that the terms "quality of life" and "livability" are used as essential demand in environmental and land-use planning, particularly, in planning for economic development and infrastructure, where the main part of this infrastructure is the urban transport system. In Libya, the mobility and traffic conditions in the urban areas experience number of problems, and the rapid urbanization and motorization and also the increase in car ownership per household make the problems more difficult. Libyan transport authorities state that the road traffic has crucial negative impacts in social, health and safety, and they assort it as a major source of death and injury in large Libyan cities, particularly, in Tripoli and Benghazi. Where, in these cities, the car-based transport mode is predominated and most commuters are dependent on private travel mode to move from home to work or other destinations. Furthermore, in large Libyan cities, traffic congestion causes additional costs for society in terms of travel delays, longer journey times, road safety and environmental pollution (Safour, 2011).

According to the authority of Traffic and Licenses, Libya is one of the countries that have the highest percentage of deaths by accidents, and the following figure illustrates samples of the increase in the car numbers and accidents, where Fig. 1a shows the estimations of the gradually increase in car numbers in Libya from 1995 to 2025, and

A. Safour (✉)
University of Benghazi, Benghazi, Libya
e-mail: aziza.safour@uob.edu.ly

F. Trapani et al. (eds.), *Advanced Studies in Efficient Environmental Design and City Planning*, Advances in Science, Technology & Innovation, https://doi.org/10.1007/978-3-030-65181-7_14

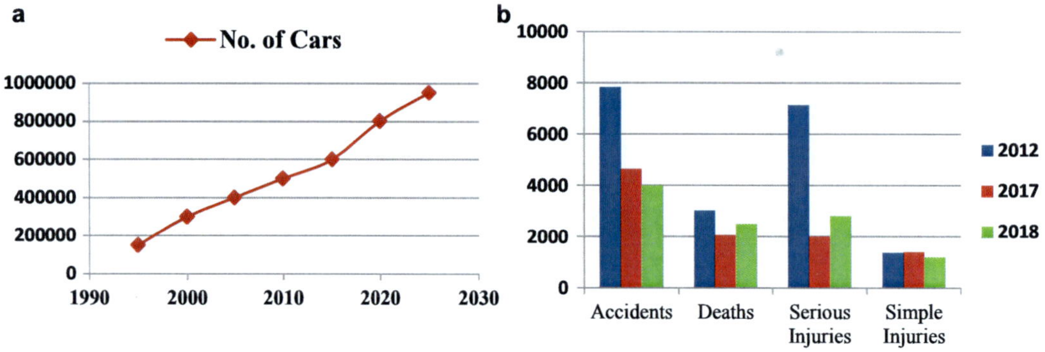

Fig. 1 **a** The increase in car numbers in Libya from 1995 to 2025 (Safour, 2011). **b** The traffic accidents and deaths in Libya in years 2012, 2017 and 2018

Fig. 1b illustrates the traffic accidents and deaths in Libya during the years 2012, 2017 and 2018.

On the other hand, the urban sustainable transport strategies have been considered a suitable tool for tackling urban transport and traffic problems. Moreover, to meet the Goals of the UN Sustainable Development agenda, in particular no. 11, sustainable transport strategies are needed to create a better livable environment for the benefit of users and transport policymakers. The study aims to set key issues that can be used as a plan for the reconstruction of the urban transport system in Benghazi through a sustainable urban development plan. This study shall be a part of comprehensive study that based on extrapolating the agenda of sustainable urban planning that identified by the United Nations. This next section explains the meaning of sustainable transport and the benefits of sustainable approaches.

2 The Urban Sustainable Transport Strategies

To meet the goal no. 11 from the United Nations Agenda (2015) "by 2030, *ensure access to safe, affordable, accessible and sustainable transport systems for all, improving road safety, notably by expanding public transport, with special attention to the needs of those in vulnerable situations, women, children, persons with disabilities and older persons*", also, to get a better quality of movement and travel, urban sustainable transport strategies that involve different modes and options of travel means have been considered as the optimal approach to solve urban traffic and transport problems in cities with high densities.

Sustainable transport is defined as an urban system that is intended to meet daily needs through a mechanism of accessibility and mobility in multiple ways, and high efficiency to achieve the social, environmental and economic goals of current generations in a manner that develops and maintains non-energy sources and to meet the needs of subsequent generations (USDOT, 2010). In simple terms, sustainable transport refers to the mobility within cities with low environmental impact and includes non-mechanical transport patterns such as walking, cycling, mass mechanical and environmentally friendly transport means such as green vehicles and car sharing (Abbas & Imarn, 2016).

The studies illustrated that transport sector has a serious environmental impacts, and it is responsible for approximately 23% of global CO_2 emission. It also has a number of other effects such as noise and pollution (IEA, 2011). Because of that *sustainability in urban transport* becomes increasingly important target.

2.1 Sustainable Transport Objectives

The main objective of the sustainable transport is *reducing the demand for transport* by decreasing the number of trips and the length of traveled distance. The urban planning and design sector can help in reducing the distances between places and people by increase the density of the spaces, and as a result, people travel less to obtain goods and services. Furthermore, the reduction in the transport demand will cut down the use of scarce resources and adverse impact on the environment and economy.

2.2 The Benefits of Sustainable Transport Strategies

Studies stated that sustainable transport strategies could be a powerful tools in reducing traffic congestion, improving the environment in the city, so they can offer benefits in terms of managing the traffic on the roads by reducing the delays and increasing traveling car speeds, encouraging road-users to use other alternatives, decreasing car numbers at peak periods, making journey times more reliable. It can offer benefits in terms of reducing pollution and emissions.

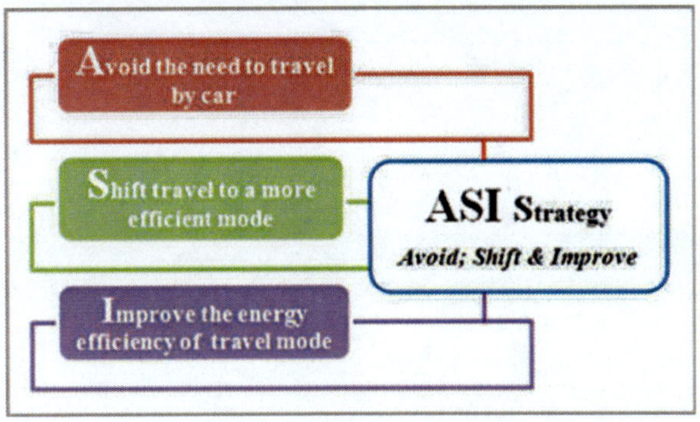

Fig. 2 Three pillars of an integrated urban transport policy

2.3 Principles of Successful Sustainable Solutions

Reviewing successful cities experiences indicated that the implementation of comprehensive sustainable transport plans and proved to be the only successful solution for urban sustainable transport strategies which is a combination of the following three principles:

- avoiding the need to travel by car; to avoid or decrease the travel by car, some issues are needed to recognized, such as integrate the better urban design and transport policies, like car park charging and road pricing.
- shifting travel to a more efficient mode; Ishii (2013) mentioned that to reduce the greenhouse gas emissions there numbers of action can be done, such as implementing strong and optimized public transport, integrating transit with efficient land use, encourage people to use mini-cars and electric two-wheelers or non-motorized transport (NMT) like walking and cycling.
- improving the energy efficiency of travel mode, where improving vehicle energy efficiency could be reduced by up to 50% by 2030.

These strategies are called the "(ASI) Avoid, Shift and Improve" strategy (Mezghzani, 2011). The successful examples pointed that having access to the variety of transport modes and options can help road-users to reduce transport costs, improve transport safety, reduce pollution and providing easy and safe access to work, services and opportunities. Figure 2 summarizes the three pillars of an integrated urban and sustainable transport policy.

3 Case Study Area, Benghazi City

Benghazi, which is the second city and the capital of East of Libya, hosts several important companies and institutions. The city had faced a very hard time during the war which started on May 2014 till July 2017. During this time, many areas of the city were massively destroyed such as the central business district (CBD) which is the main sector of Benghazi which includes the historical area and most services buildings (Safour & Elmazeg, 2018). Moreover, as a result of the war, Benghazi traffic conditions faced many problems. Another point is the rapid increase in car ownership for household in Libya makes the problems more difficult, and today, these problems become more severe and need urgent efforts to treat.

The master plan of Benghazi covers an area around 31,000 ha, and the gross residential area is about 700 ha. The gross residential density of Benghazi is about 100 ihb./ha (Third generation studies, 2009), and according to the General Authority for Information (2012), the population of Benghazi is 700,000 inhabitants with a growth rate 2.117%. Figure 3a, shows the land-use plan of Benghazi which includes the distribution of the different uses.

Figure 3b shows the classifications of roads in Benghazi according to the second-generation plan, and road network was generally well designed. However, traffic jam and poorly maintained roads are some of the transport problems in this city; also the increase in vehicles ownership has caused a number of problems in the CBD of the city. The road network in the city of Benghazi has been classified into four main categories according to the location of road, traffic volume and the road layout. The first category includes the three radial roads, designed as expressway arterials and prominent in the proposed traffic system of Benghazi. The second category includes the coastal highway and the six ring roads. Most of these roads consist of two lanes without any frontage. These roads are border residential districts and connect the radial roads. The third category includes the collector roads. This category receives the greatest traffic flow in the network as they carry traffic to the city center and the residential areas. Therefore, this category of road is busy with private and light good's vehicles most of the day and with commercial vehicles coming from the harbor in the morning time. Figure 4 illustrates the situation of different roads in Benghazi.

The public road network and bridges have been utilized for a long time since their construction without any proper maintenance, conservation or developmental work being undertaken. This absence of periodic maintenance and the absence of funds have caused real problems including substantial damage to the components of this network and to the road traffic equipment which are necessary for the safety of road-users (Safour, 2011). Today, the roads of Benghazi are in a bad state, and immediate action is needed such establishing a proper strategies and maintenance. The next section will enumerate some basic issues concerning the transport sector in the case study area, Benghazi city.

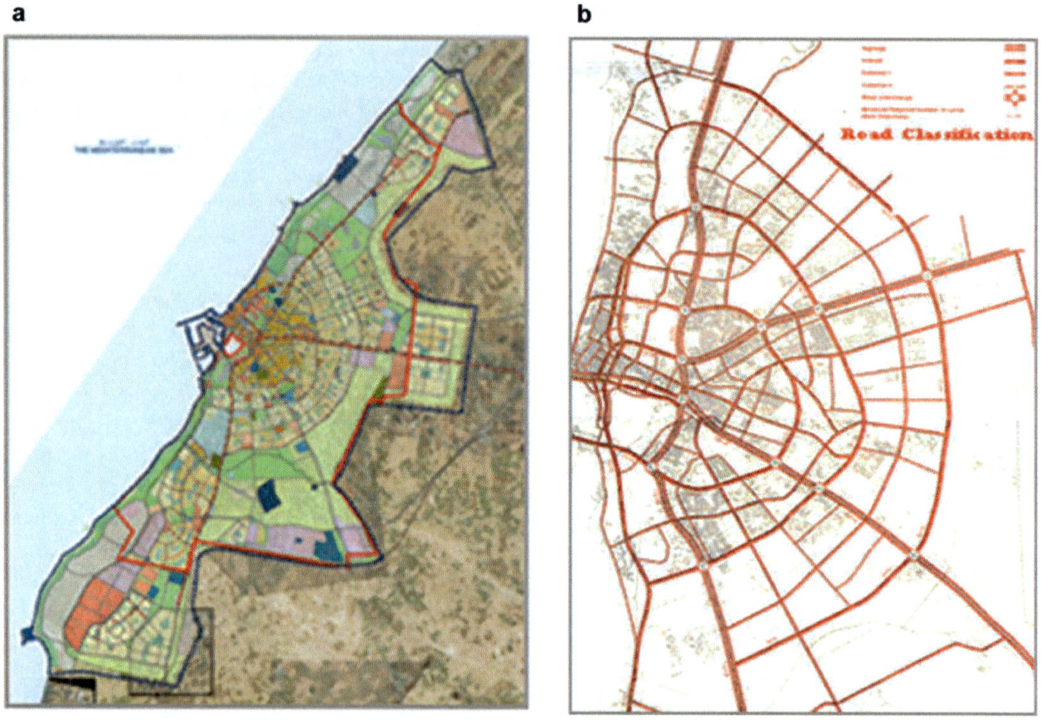

Fig. 3 **a** Land-use plan of Benghazi (TGS, 2009). **b** The classifications of roads in Benghazi (DOX-LIB, 1989)

Fig. 4 Situation of roads in Benghazi

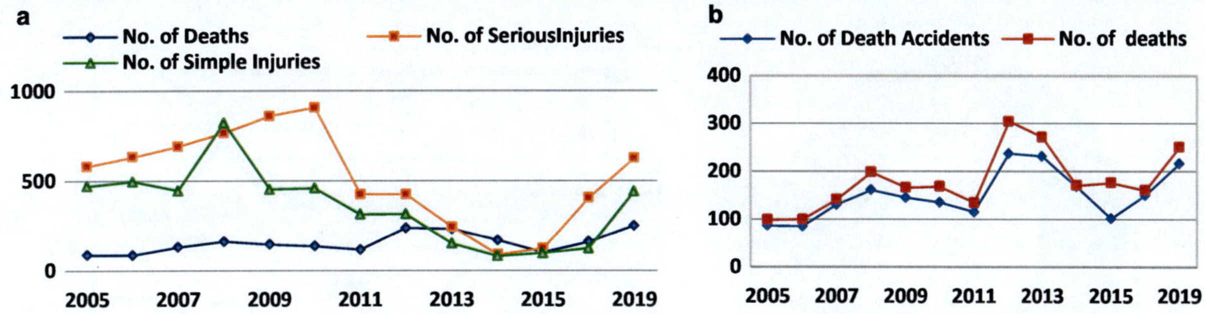

Fig. 5 **a** Number of accidents injuries in Benghazi from 2005 to 2019. **b** The increase in the numbers of deaths accidents in Benghazi from 2005 to 2019

3.1 Why Sustainable Transport Strategies Are Needed in Benghazi Now?

The traditional solution to solving traffic problems such as congestion is constructing new roads. However, the new road construction not only takes up a huge amount of time and money but also causes environmental damage. In Benghazi, the Municipality of Benghazi (MOB) has developed several solutions to reduce traffic problems, such as congestion in the CBD. For instance, changing the level of the roads and increasing the traffic signs. Additionally, in the last decade the MOB has used a new transport policy to reduce the problems of congestion by applying a charging policy for car parking in the CBD area. This policy was aimed at urging people to use other transport modes instead of their cars. However, despite these attempts the problems of congestion are still increasing. The government argues that existing solutions which have been applied are not enough and they are seeking for establishing new strategy that can help to reduce these problems. Another important issue is that Benghazi city has the highest percentage of deaths due to automobile accidents in Libya, and Fig. 5a and b illustrates the number of accidents and injuries in Benghazi from 2005 to 2019 and the traffic accidents and deaths toll in Benghazi during the period from 2005 to 2019, where the highest percentage of accidents and deaths occurred during the year 2012.

3.2 Challenges in the Transport Sector in Benghazi

According to estimations of the International Energy Agency in 2011, the transport sector produced emissions of 6.5 gigatonnes of CO_2, or 23% of world energy-related CO_2 emissions and road transport accounts for about 75% of the CO_2 emissions in the sector. While the increasing of this sector's emissions is strongly due to the increasing of car numbers (IEA, 2011). Figure 6a shows the global CO_2 emissions from fuel combustion in transport sector. In Benghazi, the car-based transport mode is predominated, and most commuters are dependent on private travel mode to move from home to work or other destinations. Furthermore, the traffic congestion causes additional costs for the society in terms of travel delays, longer journey times, road safety and environmental pollution, and Fig. 6b presents the main challenges in the transport sector in Benghazi.

4 Benghazi Proposal Plans and Opportunities for the Future

According to Safour (2011), the Libyan national plan 2005–2030 has a number of objectives such as completing the constructions of the roads network, improving the traffic movement system, implementing light train, metro and railway network, also using number of policies that ensure the efficiency and safety of traffic management. The government is seeking for establishing a new policy that could be effective in reduction of traffic problems and avoiding the amplifying of traffic problems in future.

In Benghazi, the transport decision-makers have also argued that existing solutions that have been applied such as changing the level of roads and increasing the traffic signs are not enough because of the difficulties in changing roads levels or increasing the capacity of road lanes. The new strategy should be linked with the urban planning strategies and transport strategies and plans. To meet these needs, there are some options such as the redistribution of important service locations, improving existing local centers and creating other local centers, such as transit-oriented centers, TOD, instead of centralizing the majority of services in the CBD.

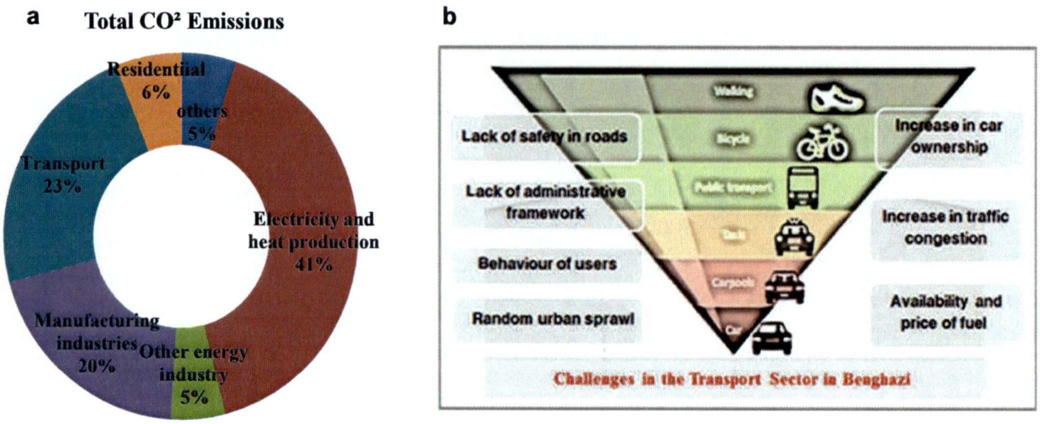

Fig. 6 **a** Global CO_2 emissions from fuel combustion (IEA, 2011). **b** Challenges in the transport sector in Benghazi

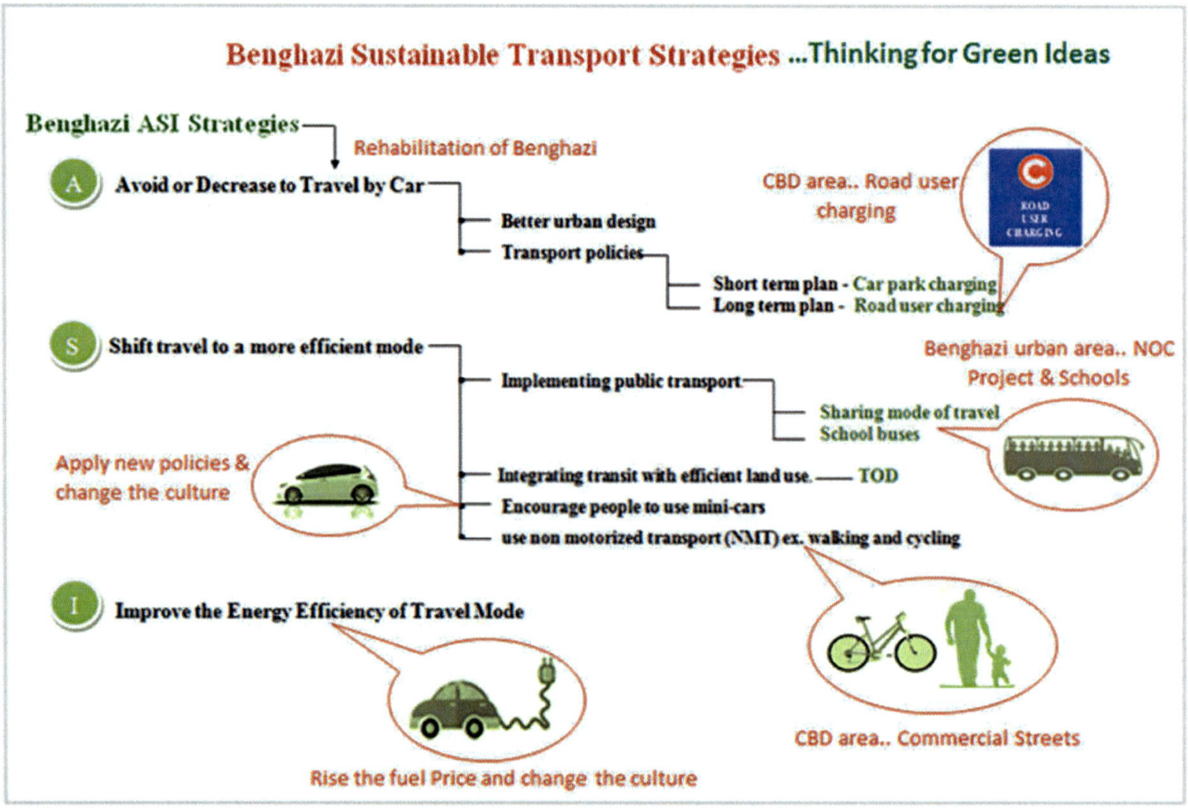

Fig. 7 Suggestion strategies for ASI principles and opportunities for the future of Benghazi

This study suggests number of strategies concerning the sustainable thinking for Benghazi transport system which explains in Fig. 7 by illustrating the thinking for green ideas and the ASI principles and opportunities for the future of Benghazi.

The following part will cover numbers of questions that concern the demand policies and plans that can be applied in Benghazi and the answers that justifies the purposes of these policies.

Fig. 8 Congestion of roads in the CBD area of Benghazi during different peak periods

4.1 Why Road-User Charging Scheme is Needed in Benghazi?

Currently, road-user charging (RUC) policy is considered to be an appropriate policy for managing urban traffic and transport problems. For instance, In Benghazi city, to meet the needs of the transport sector for implementing a good transport strategy that can be used to reduce the congestion in the CBD area and to decrease other existing transport problems, RUC can be implemented as a suitable tool to solve such these problems. Figure 8 shows the congestion in some roads of the CBD in Benghazi during different peak periods.

Safour (2011) mentioned that all over the world, there are a number of road-user charging schemes which have been implemented in order to reduce traffic congestion in the CBD of the city, such as in Singapore, London, Durham and Dubai. Other cities have applied road-user charging schemes to obtain revenue for new infrastructure funds, for instance, in Oslo, Bergen and Trondheim. However, it is essential to state that all of these schemes have achieved a reduction in traffic crossing the charging area in the range of 7–25%.

In addition, a field-based road-user charging experiment in Benghazi was conducted in 2008 with the road users in the central business district (CBD) indicated that RUC policy can have a positive impact in reducing traffic congestion with a reduction in car trips crossing the restricted areas during the peak period by around 35.5%. Because of the reasons stated above, RUC can be applied in, or around, urban areas in Benghazi in order to reduce traffic congestion and other transport problems (Safour & Egbu, 2013).

4.2 Why Public Transport System is Important in Benghazi?

Improve the public transport system by increasing the number of public transport modes and lines and raising the number of buses also encouraging people to use public transport can help in reducing the number of private cars on the roads, organizing the traffic and reducing the negative environmental impacts in Benghazi.

The calculations of the International Road Transport Union (IRU, 2010) concluded that, in *Curitiba*, "*in Brazil, the bus is capable to carry as many passengers as 30 cars, where a car usually carries to work about 1.7–1.8 passengers, while a double (blocked) bus can carry about 200 passengers in rush hours. In terms of the occupied area on the road (dynamic clearance) a blocked bus is equal to 4 cars which carry 7 passengers. A bus, however, can carry 200 passengers. This implies that the priority should be given to public rather than personal passenger transport because, in this case, traffic, accident rate and harmful environmental effect would be reduced and the problems of urban public transport and their solution trends*".

As a result of the above investigations, the authorities of transport in Benghazi should have plans to develop the existing bus network and increase the lines of this network, and

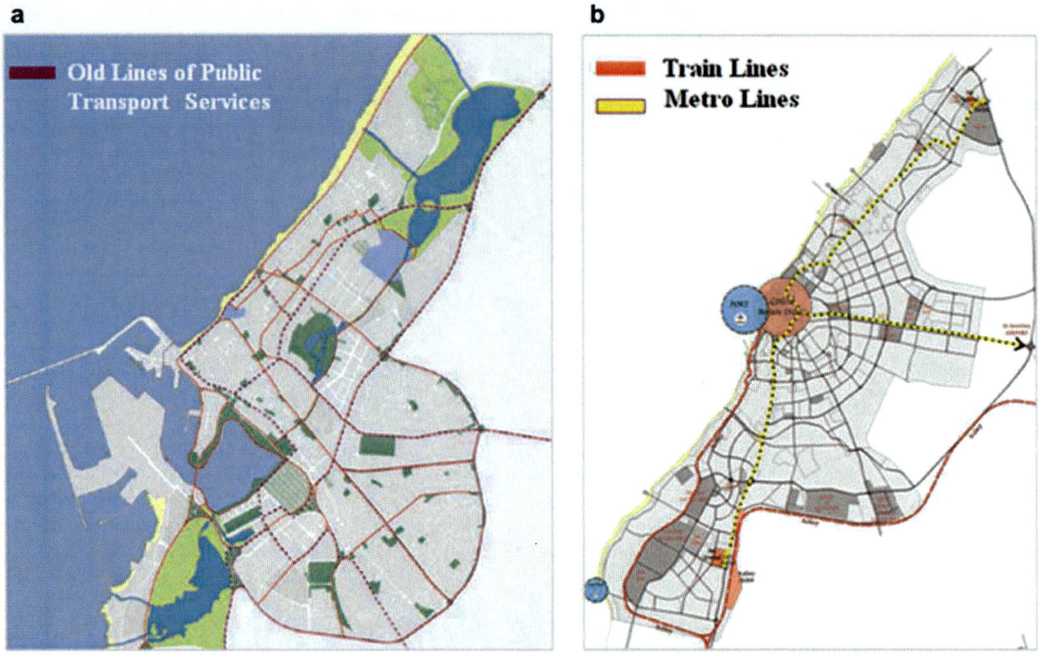

Fig. 9 **a** Old public transport lines (buses) that suggested for tram network (HPCU, 2010). **b** The proposal of light rail and train network in Benghazi (HPCU, 2010)

furthermore, implement a number of public transport projects aiming to improve the life and road conditions as well public transport services. Figure 9a shows the old lines of public transport network in Benghazi, and Fig. 9b offers the proposed light rail and train networks for Benghazi.

4.3 Why Walkable Streets in Benghazi Are Important and How Can We Achieve Them?

To achieve walkable streets in Benghazi, numbers of scenarios have to be implement, in particular the CBD area, which is the main part of Benghazi that includes the historical area, the most service buildings and the main commercial and traditional center and the main problems are congestion, and traffic jam during the morning. Using a policy such as RUC can help the decision-makers in Benghazi to adapt or convert many streets in the CBD to walk streets and establish another street to serve the area by car service lines. For instance, in Benghazi, because of its size (metropolitan city), shape of the city central CBD location (in the middle of the shape), the road network (ring roads), the cordon-based charging system can be considered as the best type for this city.

In a cordon-based charging scheme, drivers who enter the CBD during the restricted time can choose to pay the toll and continue with their current travel pattern, or they can choose another alternative for their travel behavior, such as crossing the cordon before the restricted period; crossing the cordon after the restricted period; changing their destination for certain activities; driving and parking outside the cordon and continuing by another mode; or traveling using public transport. The cordon should delineated around the central area which is around the harbor and has the highest density of residences, shops, offices and most of the public services buildings, and as a result has the busiest roads in the city, especially during peak periods such as in the morning (Safour, 2012).

To sum up, walkable and pedestrian streets for the main and busiest streets, as Gamal St., Omar El Mokhtar St., and Omer Ben Ass St. became crucial demand to ensure livability in the CBD.

4.4 Modal Share and Shuttle Service

For some places or projects which have special conditions, there are various scenarios that can be applied or designed according to the conditions of the place. One of these scenarios is modal share and shuttle service; the modal share and shuttle service has been developed to serve the National Oil Company (2019) business center project in Benghazi near the CBD.

Because of the limitation of the site area and the location of the center, as well the number of the road users and limited space for car parking, also, to keep the site free for

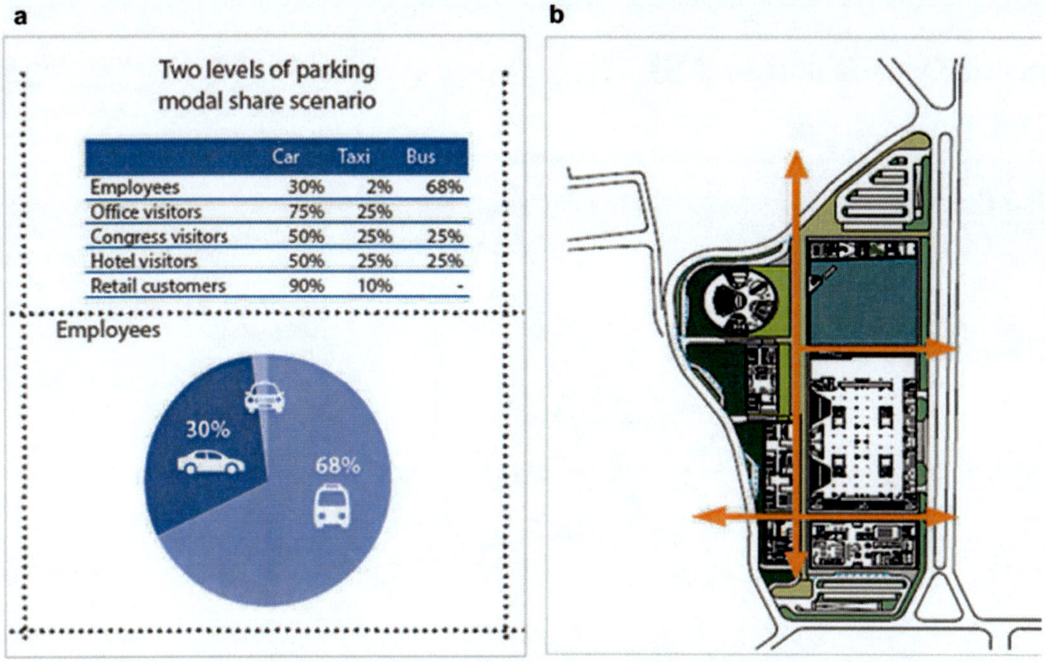

Fig. 10 **a** Percent of car users and shuttles (NOC, 2019). **b** The site of the NOC project (NOC, 2019)

pedestrian and green the company has introduced employees' shuttle buses, where about 68% of employees' trips are carried out by shuttle and parking is provided for 30% of employees, the car parking place has designed to occupy the underground at levels -1 and -2. This scenario is inspired by Istanbul best practices, where employers are expected to either give their employees a free transit subscription, or to offer a private shuttle service that brings about 70% of employees (all general employees) to work every day. Figure 10a shows the percent of the users and the mode of travel for commuter trips, and Fig. 10b presents the site of the NOC project and the main axes of the green pedestrian (NOC, 2019).

4.5 Transit-Oriented Development Approach—TOD

For the cities with high densities, the transit-oriented development (TOD) is considered as a suitable approach to transfer the movement from the traditional automobile or road-based model to create attractive, livable and sustainable urban environments. TOD can be defined as the planning approach that incorporates a high density, mixed use type of development with a walkable radius around any transit node. TOD aims to allocate housing and commercial development close to transit infrastructure that involves various modes of trips to be use instead of automobile trips (Hamand & Petker, 2015). In Benghazi, the new proposal master plan has numbers of sub-centers that aim to reduce the travel to CBD for services trips and to allocate the services buildings close to the residential areas.

Transport authorities should prepare a comprehensive transport plan that involves different transport strategies that integrated with the land-use plan. According to master plan of Benghazi, the sub-centers of the city are the ideal places to implement TOD approach that involves walkable, mixed-use communities around rail stations.

Figure 11 shows different solution that would be applied depends on the existing conditions and proposal plans of Benghazi as well the opportunities for the future of Benghazi. TOD is one of the optimal strategies that needed to develop the transport sector in a city such Benghazi.

5 Conclusion

To conclude, as this paper ascertains the remain situation of transport and mobility in Benghazi, which is the second city in Libya, and hosts several important companies and institutions. The study develops a plan of sustainable transport strategies that can be applied in or around the urban area of Benghazi to reduce transport problems, and this plan could be used as an indicator for achieving healthy and high quality of life in Benghazi community. The plan covers numbers of issues concern the demand policies such as the set of ASI

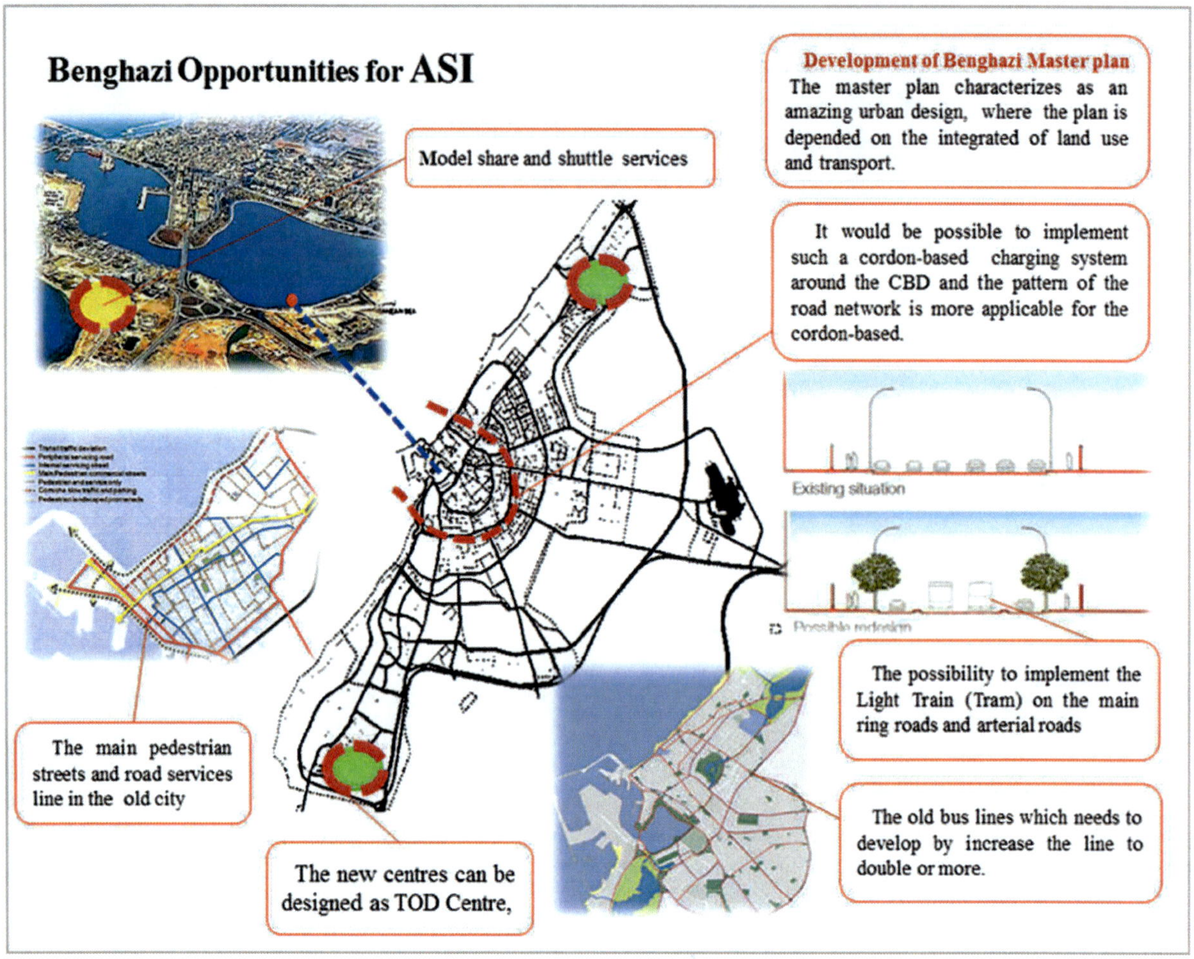

Fig. 11 Opportunities and strategies for the future of Benghazi

strategies. These strategies are called the Avoid, Shift and Improve, where the successful examples of ASI proved that having strategies such as ASI that allow access to the variety of transport modes and options can help road-users cutting down transport costs, improve transport safety, reduce pollution and providing easy and safe access to work, services and opportunities. Furthermore, public transport includes buses; metro, light rail; and other modes, is one essential aspect of the overall system. Another ideal strategy is transit-oriented development (TOD) that aims to allocate housing and commercial development close to transit infrastructure which involves various modes of trips to be used instead of automobile trips. According to the master plan of Benghazi, the sub-centers of the city are the perfect places to implement TOD approach that involves walkable and mixed-use communities around rail stations. Other different solutions that would be applied depends on the existing conditions and proposal plan of Benghazi as well the opportunities for the future of Benghazi (ECOU, 2010).

To implement sustainable transport strategies successfully from the beginning the transport policymakers could use this green plan as indicator to identify the main issues, requirements and needs of the sustainable transport strategies before the process of rehabilitation of Benghazi begins, and this plan will be the first step in the implementation sustainable transport strategies in Libya.

References

Abbas, S., & Imran, Y. (2016, March). Sustainable transportation and urban form. *Iraqi Journal of Architecture Engineering, 1*.

DOX-LIB. (1989). *Benghazi Region, Baladiyat of Binghazi, final report on the master plan. Doxiadis Corporation.* Urban Planning Agency.

ECOU. (2010). *Architectural and Urban Brochure of Benghazi City Centre, development strategies and proposals*. Engineering Consulting Office for Utilities.

G.A.I. (2012). *Statistical book*. General Authority for Information.

G.A.I. (2015). *Statistical book*. General Authority for Information.

G.A.I. (2017). *Statistical book*. General Authority for Information.

HPCU. (2010, July) *Benghazi City Center's architectural and urban charter, Phase 5*. Report 5.4, Higher Popular Committee for Utilities.

IEA. (2011). *CO_2 emissions from Fuel Combustion International Energy Agency*. IEA.

IRU—International Road Transport Union. (2010). https://www.iru.org .

Ishii, N. A. (2013). Global environment facility, investing in the sustainable transport and urban systems. GEF experience. Retrieved September 2019, from www.theGEF.org

McDonald, S., Malys, N., & Maliene, V. (2009). Urban regeneration for sustainable communities: A case study. *Technological and Economic Development of Economy Baltic Journal on Sustainability, 15*(1), 49–59.

Mezghazni, M. O. (2011). Sustainableurban tansport system, UITP Dubai, Dubai, United Arab Emirates.

NOC. (2019). *A shining diamond for Libya project*. National Oil Company Report.

Safour, A. (2011). *Analysis of road user charging impacts on activity travel patterns* (Ph.D. Dissertation). Salford University, Manchester, UK.

Safour, A. (2012, April). *The impacts of transport policies on travel patterns in the cities centres*. Libyan Urban Planning Society.

Safour, A., & Egbu, C. (2013). Road user charging impacts in Libya, perceptions from drivers. In *IPGRC Conference*, April 2013. Salford University, Manchester, UK.

Safour, A., & Elmazeg, E. (2018). Rehabilitation of Benghazi City, ideas to be more sustainable. *Libyan Journal for Engineering Research, 1*(1).

SDG. (n.d.). Sustainable development goals. Retrieved August 2019, from https://www.un.org/sustainabledevelopment/development-agenda/

The World Bank. (2014). *Toward sustainable and energy efficient urban transport*. The International Bank for Reconstruction and Development.

TLDB. (2020). Traffic and Licensing Department of Benghazi.

TOD. (n.d.). Transit orianted development, Chapter 27. Retrieved September 2019, from https://www.researchgate.net/publication/308984324_CHAPTER_27_TRANSIT-ORIENTED_DEVELOPMENT_TOD

UN-Habitat. (2014). A new strategy of sustainable neighborhood planning: Five principles. United Nations Human Settlements Programme.

UPA. (2009). *Third generation studies, Benghazi region*. Baladiyat of Binghazi, final report on the master plan. Urban Planning Agency.

USDOT. (2010). *Livability in transportation guidebook: Planning approaches that promote livability* (FHWA-HEP-10–028).

Valuing Public Perceptions of Biophilia Impact on Human Well-Being: 2 Sustainable Building Case Studies from Greece and India

S. M. Jackson, L. Singh, and T. Doshi

Abstract

This study focusses on valuing the 'green technologies' of designing and building with nature to encourage a wider dimension to the current ratings and evaluations of effectiveness of 'green buildings', by including the perceived impact on human well-being. The authors believe that for buildings to offer a 'sustainable' way of living, they must also include the technologies and intelligence to provide what all of life needs to thrive beyond just surviving. This study aims to give a wider understanding of 'green buildings' beyond reporting on energy, water and waste, to show a more sophisticated, wider evaluation of sustainable buildings by including the value of subjective perception of individuals' experience, and, to contribute to changing existing paradigms about how 'green buildings' are valued. Other studies conclude that leading bodies for 'green building' certification have failed to provide a holistic measure of sustainable buildings. Current environmental measures of 'green buildings' conflict with the values of human health and there are conflicting 'logics' and technologies with little consensus on what defines a sustainable building. The perceived 'value' of the health and well-being benefits of a 'green building' appear to be disregarded as a measure of effectiveness; this chapter challenges that view. Findings from questionnaires and testimonials from the public using two green buildings in different countries suggest that people do believe that they experience physical and emotional health benefits from spending time in green buildings. Specifically, it related to their perception of the reduction of stress and anxiety. This suggests that valuing the 'unmeasurable' perceived benefits of sustainable buildings on health and well-being, equally alongside quantitative audits and environmental measures, could bring combined societal and environmental benefits. More research and evaluation with larger samples, more complex data collection, over longer time periods in different countries is necessary to build on the initial findings of this novel study. Further research could make an important contribution to greater understanding about the positive impacts of biophilia design for healthcare institutions, education buildings, community spaces, workplaces and homes.

Keywords

Biophilia • Bioclimatic • Green building • Sustainability • Health • Nature • Greece • India • Placebo

1 Introduction

This study was instigated after noticing anecdotal feedback from the users of two 'green buildings' in different parts of the world, Greece and India. Both of these buildings were designed using bioclimatic (Swaid et al., 1993) and biophilia (Wilson, 1984) principles and are discussed in Sects. 3.1 and 3.2.

Both buildings were prototypes with no precedence to measure against. This meant neither building was eligible for mainstream Green Building Councils (GBC) (Our Green Building Councils | World Green Building Council, n.d.) ratings and audit mechanisms such as Leadership in Energy and Environmental Design (LEED) or the British Building Research Establishment Environmental Assessment Method (BREEAM). In both cases at separate, unconnected times, the owners and architects have been challenged that the buildings are not really green because they do not have the standard rating. However, in practice, users of both buildings have given feedback about their positive experiences of visiting and working in these places. Most users speak of

S. M. Jackson (✉) · L. Singh
The European Sustainability Academy (ESA), Crete, Greece
e-mail: Sharon.Jackson@EuroSustainability.org

T. Doshi
The Auroma Group, Pondicherry, India

F. Trapani et al. (eds.), *Advanced Studies in Efficient Environmental Design and City Planning*,
Advances in Science, Technology & Innovation, https://doi.org/10.1007/978-3-030-65181-7_16

how they 'felt' in the building with an overriding theme of the perceived positive impact of being connected with nature and the natural materials used in both cases. This disconnection between valuing formal audit standards and user perception is the focus of this study. The authors posit that users' perceptions of green buildings have value that is presently ignored.

To explore this hypothesis, the researchers entered the fields of literature around green building design, construction and effectiveness measurement through the lens of two questions:

LRQ 1. Is the perception of the benefit of green buildings on human health and well-being valued in the context of green building assessment?
LRQ 2. Do the perceived benefits of nature on the human sense of well-being actually improve health and well-being?

A literature review explored studies on the impact of green building on human health and well-being, comfort and enhanced productivity. Internet searches and 'grey' literature have been included as this new area of enquiry is growing quickly. Each study was read and reviewed by at least 2 of the 3 authors.

Focus on green buildings has increased over the past 4 decades with an emergence of voluntary standards and benchmarks to guide design, operation and effectiveness (Van Wyk, 2008; Xie et al., 2017). Global GBCs have ambitious mission statements which include promises to 'transform the building market' and 'build a better world'. However, the generic 'do least harm' approach of their rating methodologies questions whether all their objectives are achievable and also what is ignored in those ratings (Van Wyk, 2008).

The literature soon reveals competing terms and logics of sustainable architecture, sustainable buildings, ecological buildings and green buildings (Guy & Farmer, 2001, p. 141). For the purpose of this study, the terms 'sustainable building' and 'green building' are used interchangeably.

1.1 Green Buildings

The concept of 'green' in the built environment and the elements of 'green buildings' have attracted extensive debate across the globe.

Definitions include green architecture, or green design, as an approach to buildings that minimise the harmful effects of construction projects on human health and the environment (Craven, 2019). A green building is 'an environmentally sustainable building, designed, constructed and operated to minimise the total environmental impacts' (Build Green, n.d.).

The advancement towards green buildings and green retrofit is evolving from sole concern of environmental impacts to recognising the positive and negative health impacts buildings that can have on people (Allen et al., 2015; Croome, 1999).

Research around green buildings and human health is expanding to include the impact of green buildings on comfort, productivity and health in housing and workplaces (Jacobs et al., 2009; Kim et al., 2013; Wolf & Robbins, 2015), in education (Paul & Taylor, 2008) and long-term impacts of well-designed working places on improved health on productivity (Issa et al., 2010; Xie et al., 2017, p. 4).

Studies have explored the broader dimensions of health and well-being associated with green building with a focus on the impacts that the built environment has on occupants' physical and psychological health. Diseases like asthma, obesity, mental illness and cardiovascular disease (Mendell et al., 2011; Bonnefoy et al., 2003; Houtman, 2007; Wee et al., 2013) were found to have close links with the buildings that people live and work in.

Specific studies in the healthcare sector found a 19% decrease in mortality, 70% decrease in blood stream infections and a general increase in quality of patient care and increase in employee satisfaction in 'green hospital' facilities (Jackson et al., 2017).

A 2013 survey across Canada and northern USA reviewed evidence from post-occupancy evaluations of 12 green and 12 conventional office buildings. The study concluded that green offices have better indoor performance compared with conventional ones in terms of well-being, including mood and physiological symptoms (Newsham et al., 2013). The study also suggests that sustainability rating systems require development regarding the effectiveness of auditing of environmental factors related to human health. This echoes other findings (Kok et al., 2012; Miller et al., 2009; Singh et al., 2010).

The health of city dwellers is strongly influenced by physical, social, economic, political and cultural factors (WHO, 1991, p. 11). An increasing number of research studies worldwide suggest that people living in urban environments are generally more overactive and stressed than those in rural areas (Tracada & Caperna, 2012) with evidence of the negative effects of stress on human health. A 2011 study by the University of Oregon supports this notion by revealing that 10% of employee absences could be attributed to buildings that are disconnected from nature. The study showed that employees with natural landscape views as opposed to urban views took an average of 57 h of sick leave per year compared to the 68 h taken by employees with no view (Elzeyadi, 2011; Thatcher & Milner, 2014).

The findings from studies on green buildings and benefits beyond environmental impacts are varied and sometimes conflicting. However, there is a developing new theme with some emerging congruence in findings in respect to the experiences and perceptions of the users, notably the studies which focus on the human need for connection with nature in living and working spaces (Kellert et al., 2008). This is discussed further in Sect. 1.2.

The next section explores the current standards and assessments that measure the value of the effectiveness of green buildings.

1.2 Green Building Standards

It has already been mentioned that an increasing number of studies criticise the effectiveness of green building ratings (Xie et al., 2017, p. 5; Van Wyk, 2008).

Rating systems like LEED and BREEAM each have their own set of factors (indoor air quality, thermal comfort, lighting and noise) which are used to measure the extent of 'greenness' of a building. These rating systems and their effectiveness are a matter of critical discussion with suggestions that standards need revision and repair (Bady, 2008; Schendler & Udall, 2005). Specifically, Del Parcio (2007) criticises LEED for its formulaic approach which leaves no room for the broader architectural implications of a building design.

The shortcomings of the rating systems include the 'process of standardisation' as a 'checklist' approach to environment-friendly features. This does not mean that the building cannot create negative environmental impacts (Guy & Farmer, 2001; Van Wyk, 2008; Xie et al., 2017). Most of the rating methodologies focus on energy saving and resource recycling which can lead to a conflict with improved human comfort and well-being (Pelenur & Cruickshank, 2013).

Some studies about measuring the effectiveness of green buildings hint towards the importance of health and well-being, and vital issues pertaining to human interaction with buildings that have been missed in current green building thinking and practice (Van Wyk, 2008). However, they do not go on to explore 'why'.

Buildings are 'not only a machine for living, it is also a shelter to flee away, place to settle down and first of all, it should be an environment where people can gain physical and mental health, relaxation and freedom' (Xie et al., 2017, p. 92). Most of the terms used in this definition hint towards the unmeasurable, perceived value of human well-being.

1.3 Biophilia—Valuing Nature

The fundamental principle of biophilia is that humans have a biological need for connection with nature and without that connection humans become unwell (Wilson, 1984). In Japan, forest bathing known as Shinrin-yoku (Hansen et al., 2017) was developed in the 1980s by the Japanese Ministry of Agriculture, Forestry, and Fisheries to enhance health and well-being. Researchers at the 'Centre for Environment, Health, and Field Science', Chiba University in Japan, studied 280 subjects across 24 Japanese forests and urban environments in 2010. They found that the forest setting promoted lower concentrations of cortisol as well as lower pulse rate and blood pressure. The result was improved cognition and mood, increased focus and energy, reduced stress, and even increased immune system function.

A research study on urban planning by Carrus et al. (2015) suggests that a shift to urban living environments has not only decreased human exposure to the natural world, and it has also contributed to an increase in the worldwide prevalence of mental disorders, causing changes in psychological functioning.

What has long been intuitively sensed and perceived is now being studied, explained and validated.

The concept of biophilic design in the built environment is gaining momentum in practice which has given rise to more study opportunities. Global trends indicate that biophilic design can have an overwhelmingly positive effect on people, physically, mentally, and especially in health care, to improve patient and employee wellness (Jackson et al., 2017). Therefore, the building sector is keenly embracing biophilia design elements in a bid to create more healthy work and living spaces (Browning et al., 2012; Thatcher & Milner, 2014).

The notion of the built environment impacting on human health and well-being seems well established; however, understanding 'how and why' is still far from being explained (Xie et al., 2017). Moreover, well-being is just one aspect of mental health with other factors still under explored. In the debate of health and well-being and its relationship with the built environment, the concept of biophilia and its impact has been widely discussed. Inherent inclination of humans to connect with nature and its impact on physiological and psychological well-being has been demonstrated in many studies. Contact with nature is proven to reduce stress (Chang & Chen, 2005), improve attention (Kaplan & Kaplan, 1989), help in mental restoration (van den Berg et al., 2007) and increase longevity (Takano et al., 2002). Through deliberate design, this connection can be enhanced by bringing buildings to life for bringing humans back in contact with nature (Kellert, 2005).

Studies exploring how green buildings can promote health and well-being in the healthcare sector have found compelling evidence of the positive benefits of biophilic elements, including increased daylight, running water, plants, trees and non-threatening animals in hospitals, which reduce anxiety and stress levels of patients, staff and visitors (Jackson et al., 2017). Scenes of nature depicted in artwork are shown to reduce post-operative anxiety and discomfort in patients recovering from open heart surgery (Browning et al., 2012; Ulrich & Lunden, 1990) and enhance patient healing by up to 20% (EPA, 2007).

More generally, studies to explore the biophilia theory in working offices conducted by Cardiff University School of Psychology and the University of Exeter (2014) revealed how the smallest 'greening' changes in the workplace such as simply adding plants as interior decoration, can boost productivity which challenges the workplace philosophy that a space with clean desks is more productive (Jackson et al., 2017; Thatcher & Milner, 2014).

The next section explores potential for valuing the 'unmeasurable' of perception on human health.

1.4 Placebo Effect—Valuing Perception

The authors are mindful of the challenges of auditing and 'valuing' subjective issues such as perception.

Enquiries to address the question 'Do the perceived benefits of nature on the human sense of well-being actually improve health and well-being?' led to the field of literature around the 'mind and body healing mechanisms of 'placebo' (Crane, 2016; Harvard Publishing, 2017) and the capacity for people to 'think themselves well'.

The 'placebo effect' is an emerging scientific discipline (Crane, 2016, p. 39) and the concept of a 'feeling' or 'perception' having a real effect on the body is now well established (Spiegel, 2004; Kaptchuck & Miller, 2018; Harvard Publishing, 2017). Our thoughts, feelings and expectations can have a profound impact on our well-being (Vance, 2016).

The concept of self-stimulated healing has existed for thousands of years. More recently, scientific trials have found that under the right circumstances, a placebo can be just as effective as medical treatments (Kaptchuck & Miller, 2018).

Since the seminal studies on the 'placebo effect' by Beecher (1955), the concept of the human brain's ability to heal the human body through 'belief' has confounded healthcare and medical research.

The placebo effect means that a human health improvement occurs without medical intervention or drugs. Placebo test trials usually involve administering a 'sugar pill without telling the patient. The human brain can be a healing tool that convinces the body that a 'fake treatment' is the real thing and this is the underpinning theory of the 'placebo effect'. This effect is widely understood to happen because the person expects the outcome and believes the feeling of well-being (Moerman & Jonas, 2002).

Studies on the 'placebo effect' range from migraine, eating disorders, sports enhancement and improving the symptoms of diseases such as Parkinson's disease. A study by Professor Kaptchuck, Harvard-affiliated Beth Israel Deaconess Medical Center, found that a placebo was 50% as effective as the real drug to reduce pain after a migraine attack. The placebo effect is more than positive thinking and believing a treatment or procedure will work. 'It's about creating a stronger connection between the brain and body and how they work together' (Kaptchuck & Miller, 2018).

The largest sample of 'placebo effect' studies in Britain in 2018, with Manchester University was televised by BBC Horizon (Manchester University, 2018). From a sample of 117 chronic backpain sufferers, 45% of the placebo group said that all pain and symptoms had vanished. More interesting was that after being told that they were in the placebo group, several participants asked to continue taking the 'dummy pills of nothing but ground rice flour' to continue to manage their pain.

So far only a few studies have combined the fields of the 'placebo effect' with the human health and well-being impacts of nature. A study of the impacts of perceived comfort benefit of 'ecolabelling' on actual capabilities gives fresh insight. (Sörqvist et al., 2015). People believed they would experience improved eyesight conditions and better performance from an eco-labelled lamp and that is what happened.

2 Research Aims

The focus of this study is to encourage a wider dimension to the current ratings and evaluations of the effectiveness of 'green buildings' Specifically, the research aims to gain better understanding about 'how' people feel about the impact of green buildings on their health.

This chapter aims to give a wider understanding of green buildings beyond reporting on energy, water and waste, to show a more sophisticated, wider evaluation of sustainable buildings by including the value of subjective perception of individuals' experience, and to contribute to changing existing paradigms about how green buildings are valued.

This disconnection between valuing formal audit standards and user perception is the main focus of this study. The authors posit that user perceptions of green buildings have value that is presently ignored.

A review of literature suggests that 'perceived' benefits of nature on the human sense of well-being can actually improve health and well-being; however, in practice, the perception of the benefit of green buildings on human health and well-being is not valued in the context of green building assessment.

This suggests that the perception of biophilic impact on health and well-being should have a place in the holistic rating of green buildings. This practical study aims to encourage more meaningful and robust evaluation of what the 'perception of biophilic impact on human well-being is in a physiological context'.

To start, we need to first understand more about:

RQ. How do people perceive the impact of biophilia on their health and well-being?

The next section explains the selected research method to address this Research Question.

3 Method

This study is interested in understanding 'how' people feel about the impact of green buildings on their health; therefore, a case study methodology (Yin, 2009) was selected.

The overall approach for this qualitative study is two case studies and a literature review.

The data collection questionnaire was taken from the conclusion of an earlier study of sustainable entrepreneurship in healthcare which recommends further empirical research on how green buildings can promote health and well-being (Jackson et al., 2017). The same questionnaire (Appendix 1) was used at both buildings discussed in Sects. 3.1 and 3.2.

Data was collected from 100 respondents from each building during July to September 2019. Some were completed during the visit and others sent by email after the visits.

A 'visitor satisfaction' questionnaire captured visitors' comments about their perceptions and experiences of the green building in two ways. Firstly, a survey of comfort, using a 1–5 Likert scale and secondly an 'open qualitative question', 'Overall how did you feel during your time at ESA/Sharanam?'.

The comfort 1–5 simple analysis was to assess if any extreme perceptions of comfort might influence the overall feeling of the visitor's experience. For this same reason, the respondents were asked about their journey to the buildings to ensure that adverse travel experiences did not unduly influence their overall perception.

The main research focus was a discourse analysis of the 'open qualitative question'.

Other text was selected for analysis from free flowing, qualitative written comments in the buildings' visitor books' and direct testimonials.

3.1 Case Study: The European Sustainability Academy (ESA) Crete

The European Sustainability Academy (ESA) (European Sustainability Academy, n.d.) was built during 2011, 300 m above sea level, in the hilltop village of Drapanos, Crete and opened in 2012. ESA was created to provide 'Leadership for positive impact', immersive and experiential, action learning (McGill & Brockbank, 2003; Revans, 1980) teaching services. The core focus of ESA is executive education in the field of sustainable business leadership and management.

This experiential learning facility was created with the aim of enhancing learning and sensemaking (Weick, 1995) processes by 'disconnecting' participants from the usual disruptions of their hectic daily lives. Therefore, learners are more receptive to new knowledge, information, ideas and attitudes towards global sustainability challenges (Tracada & Caperna, 2012).

ESA was built during the deep economic crisis in Greece that brought the country to the brink of economic collapse (Kalyvas et al., 2012). During this time, the health of Greek society was impacted by increased chronic stress, depression and suicide rates.

The 400 m^2 building was designed as a novel, bioclimatic, biophilic principles, construction, set in 10,000 m^2 of ancient olive groves, encompassing a freestanding natural rock circle as a main feature.

Although walking distance from the village, ESA is accessed by a 400 m rough track and has the feeling of remoteness and wilderness (Fig. 1).

3.1.1 Design

ESA's bioclimatic architectural design is an example of biophilic design. (Jackson et al., 2017). The construction is built around existing trees and rock formations which directly connect the building with nature and creates an overlap of indoor and outdoor spaces.

The underpinning objective of ESA is to be a low environmental impact building for conducting low environmental impact business training activities.

The young female architects from Chania (piliko.gr) created a deep immersive connection with nature combined with comfort and beauty in design and interiors. The design encourages people to feel the limitations of natural resources and to show them how to adapt and to live with less conspicuous consumption without loss of comfort.

A core aim was to create a sense of 'home from home' for users and encourage people to reconnect with nature in a way that makes the sustainable business training sessions more meaningful and make more sense.

This concept was enhanced by designing a small kitchen and adjoining area with a circular table, chairs and a relaxing sofa at the heart of the building. This is the link between the working areas (lecture room and dining room) with a panorama viewing widow to make a visible connection with the monolith rocks and old olive, oak and carob trees outside.

Visitors are encouraged to notice the smells, wind, earth, trees, insects and to appreciate every small aspect of natural beauty.

Fig. 1 ESA Crete, south facing view featuring the natural rock circle and photovoltaics

Many people have been disconnected and sanitised against nature, especially those from cities, so this can be challenging for them. The sofa area and rock viewing window can be a first safe step towards immersing in nature and appreciating the need to conserve the flora and fauna that make our planet healthy (Sánchez-Bayo & Wyckhuys, 2019).

Night-time planet viewing activities in the dark, non-light-polluted area also connects 'city people' with the night sky that they did not realise existed.

Overall, the building design works in concert with the climate and environmental conditions with a distinct relationship between the building functions and natural elements (Jackson et al., 2017) (Fig. 2).

3.1.2 Materials

ESA was literally raised from the ground using local natural materials (wood, earth, straw-bales, stone) and is powered entirely from the sun via photovoltaic panels. ESA operates with net zero emissions and zero waste to landfill.

The use of eco and natural materials throughout the building means the elimination of indoor pollutant and hazardous materials which are damaging to human health and the natural environment (EPA, 2009; Jackson et al., 2017).

3.1.3 Use

People come from across the world to ESA events ranging from seminars, retreats, awards, galas and TEDx.

Catering for ESA events is locally sourced simple and wholesome food, prepared and served in a contemporary fusion style with local organic wines.

Programme feedback surveys reveal little about the programme content and a great deal about the visitors' feelings about the building and surrounding environment with strong inferences of the impact of the natural setting on their capacity to understand about enacting the concepts of sustainability in real life.

In the early opening stages, ESA attracted students offering to volunteer and just be around the beauty of ESA to help them with their feelings of 'no hope' in the 'dark times' of the Greek economic crisis. These days many of those young people are now running their own sustainable enterprises.

ESA was designed as an international teaching centre, however, beyond the planned low environmental impact and operational efficiency gains, ESA has shown anecdotally to provide a '*deeply fulfilling*' effect on visitors, clients, employees and the local community and '*gratitude*' that was not anticipated (Jackson et al., 2017).

Unexpected responses about how people 'feel' at ESA include the spontaneous message, written on the floor at the entrance to the lecture room by a young Greek man attending a Women's Empowerment event in 2017.

'*From here on you are entering a sacred temple... Love everyone, communicate, listen, thank and hug everyone! #ESAWE4 (Spyros)*'.

Fig. 2 Night-time view of the sofa and kitchen area at the heart of ESA Crete

3.2 Case Study: Sharanam Centre for Rural Development, Pondicherry India

The Sharanam Centre 2000 m^2 building, constructed between February 2007 and May 2014 in coastal Tamil Nadu, South India, belongs to the Rural Development Initiative of The Sarvam Trust, (Sarvam Trust, n.d.) a local NGO based in India, Pondicherry, 150 km south of Chennai.

This area was devastated by the 2004 Indian Ocean earthquake and tsunami and the building was commissioned because the NGO wanted to broaden its outreach in the surrounding district in rural Tamil Nadu comprising of 40 villages entrenched in chronic poverty, violence and alcoholism. The aim of the Sharanam Centre was to act as a catalyst for positive socio-economic change through workshops and training programmes in the fields of health, education, income generation, skill development and community building.

Since it opened more than 15,000 people have visited Sharanam, which is considered to be an exemplar for Eco Design and Social Change, has been recognised by the United Nations Environmental Programme as a model for Sustainable Development in India (State of play for sustainable buildings in India, 2010).

Sharanam is approached via a regenerated eucalyptus grove into an avenue of Palmyra trees. The building entrance is then preceded by a large circular 'sacred grove' with a stepped amphitheatre that can seat 400 people.

The Centre is designed with bioclimatic and biophilic principles, consisting of a 525 m^2 multipurpose hall with reception, administrative offices, meeting rooms, computer rooms, a newspaper office and community radio station, stores, kitchen, dining and a toilet block (Fig. 3).

3.2.1 Bioclimatic Design in Action

The building is in an excessively hot and humid climate with summer temperatures reaching 50 °C and relative humidity of 80%. Design is aimed at creating thermal comfort through bioclimatic principles without using air conditioning. These include north–south orientation, funnelling breeze through angled columns, evaporative cooling through water bodies, cavity walls increasing thermal mass, vertical gardens and rooftop gardens.

To reduce the operational energy during the life cycle of the building and making it energy efficient, the design maximises the use of daylight. Radiant cooling, a low energy active system for thermal comfort, uses cooled water circulating in pipes under the floor to give a comfortable temperature at a fraction of the cost.

To overcome the erratic power supply, the vaulted roof has been acoustically designed to ensure that speakers can be clearly heard by 300 people without using microphones.

3.2.2 Biophilic Design in Action

The primary building material is unfired earth used as rammed earth in the foundations and compressed stabilised earth blocks in walls, columns and roof. This earth was procured from the rainwater reservoir on the same site ensuring zero transportation costs, minimal dependence on expensive and high energy use materials like steel and cement, minimum carbon footprint and low embodied energy.

The earthen vaulted roof was built with 9.5 m wide free spanning arches, without any framework. Through precise

Fig. 3 Sharanam, Pondicherry, serene, biophilia designed, entrance to the main hall

structural calculations, the thickness of the arched roof was brought down from 150 cm to only 9 cm at the keystone, thus, reducing the need for several hundred bags of cement to merely 33 bags.

In terms of ecological landscaping, over 1000 native trees were planted to repair the soil and prevent erosion. This thick tree cover helped in creating a favourable local microclimate through indigenous landscaping. The entire campus has been designed around existing trees and the surrounding flora greatly enhances the indoor air quality.

The building encompasses zero waste principles and is entirely recyclable. To ensure minimal water footprint, rainwater is recycled for secondary domestic purposes such as flushing and refilling pools in the landscaping and wastewater from the kitchen and washrooms is treated and re-used for irrigation (Fig. 4).

3.2.3 Building Use

Sharanam is used daily by a wide array of visitors from uneducated villagers to professionals from urban areas who are interested in a sustainable lifestyle. The activities at Sharanam include workshops and vocational training programmes about: Organic farming and re-forestation, health, hygiene and sanitation, community building and gender equality, teacher training, youth programmes, architectural studies and cultural activities such as theatre, language and creative thinking.

The next section explains the results of this study and perception of the impact of a biophilic designed 'green building' on the health and well-being of people using the facility.

4 Results

The critical question in the 200 surveys was 'Overall how did you feel during your time at ESA/Sharanam?' This was formed as an open question in the hope of unstructured, free flowing, rich open text to give insights into the perception and feelings of the visitors.

4.1 Demographics

A demographics comparison of the 100 respondents from each case study shows a similarity in terms of Profession/Managerial roles (ESA 82%, Sharanam 70%) with a notably larger number of architecture related professionals in the Sharanam data set. The 20–50 core age group of respondents is also similar (ESA 91%, Sharanam 80%). There is a greater difference in the number of female respondents (ESA 78%, Sharanam 44%) largely due to women focused events at ESA during the time of data collection.

4.2 Survey

In order to explore the perceived impact of 'biophilic' characteristics of both buildings on the well-being of visitors, it was first necessary to ascertain a more broad sense of the respondents' perceptions of comfort and physical ambient conditions (Paul & Taylor, 2008) and their experiences directly before arriving at the building.

Fig. 4 Earth brick vaulted roof of Sharanam, Pondicherry

Ten questions about comfort including light, temperature, humidity (Appendix 1) were asked with a 1–5 Likert scale of response. A further question about overall perception of the 'working environment' was asked, also with a 1–5 Likert response. The responses about comfort and working environment were consistent between both buildings with no extremes.

With a view to eliminating responses from people who had been adversely affected by their journey to the buildings in a way which may have impacted on their current state of mind, perception and sensemaking (Weick, 1995), they were asked about their journey. All responses suggest equal levels of satisfaction and no extreme stresses of travel. Therefore, the free flowing text, qualitative responses from all 200 respondents were included for analysis.

4.3 Analysis

Analysis of the data was conducted by two of the authors independently and a similar interpretation of findings was reached. Here follows a combined summary of the analysis and NVIVO scan results for triangulation.

From the emerging factors, dimensions and themes from the 200 respondents, the most dominant narratives are about positive 'state of well-being' with strong themes of 'stress relief' and relaxation. 'Nature', with connection to nature, is the second-largest narrative with the word 'nature' as the third most used word (35 times) equally with 'experience' (35) across all 200 respondents.

The next strongest narrative is 'personal security' and safety (Table 1).

An unanticipated number of responses relating to 'spirituality', 'energy' and 'enlightenment' bring further insight through this study and a new dimension for consideration (Table 1).

More broadly, despite the participants coming from two very different geographical regions with different cultures and demography, the responses from both cases reveal a congruence of 'themes' and 'tone' used by the 200 respondents to explain their personal experiences, perceptions and feelings of visiting the buildings. There are multiple overlaps of emotive words with the most used 'complimentary' word in both cases being 'beautiful' (26 times) and 'gratitude', 'thank' (25 times).

This analysis is discussed further below and in Sects. 4.4 and 5.

To check the researchers' own analysis, the responses were scanned with NVIVO for word frequency, word tree (nodes) and word cloud (Appendix 2). NVIVO word count, without interpretation of the words, corroborated the researchers' findings. The words 'feel' and 'place' were most used (46 and 42 times, respectively) but discounted in the research output as these words were influenced by the question.

The survey revealed many descriptive and technical observations about the architectural designs of the buildings. Although interesting, this is not the focus of this study to explore the value of 'unmeasurable' perceived benefits of sustainable buildings on health and well-being and therefore not included in the data analysis.

Table 1 Themes emerging from ESA and Sharanam visitor responses to the question 'Overall how did you feel during your time at ESA/Sharanam?'

Factors	Dimensions	Themes (words used)	Frequency		
				ESA	
Sharanam	Total				
State of well-being	Stress (relief)	Relaxing	8	7	15
	Relaxation	Peace	3	11	14
		Well-being	5	4	9
		Calm/centred	6	3	9
		Harmony	2	3	5
		Tranquillity	1	4	5
	Healthy	Health	3	2	5
		Nourished	3	2	5
		Balancing	2		2
		Serene	1		1
		Subdued	1		1
Total					**71**
Connection to nature	Natural	Nature	18	17	35
	Eco	Connectivity	2	5	7
	Organic	Earth connect	2	5	7
	Connection	Sky		4	4
		Grounded	3		3
		Insects	1		1
Total					**57**
Personal security	Safety	Home	9	1	10
	Reassurance	Comfortable	1	7	8
		Happiness	3	3	6
		Hope	2	3	5
		Safety	3		3
		Sense of belonging	2		2
		Snugness	1		1
		Simplicity	1		1
		Joy		1	1
Total					**37**
Spirituality	Mindfulness	Soul	4		4
		Enlightened		4	4
	Self-actualisation	Magical	3	1	4
		Finding self		3	3
		Belief/purpose	2		2
		Consciousness		2	2
		Meditative		2	2
		Heaven		1	1
		Sanctuary	1		1
		Introspective		1	1
Total					**24**

4.4 Main Themes Emerging from the Study

4.4.1 State of Well-Being

The actual words 'health' and 'well-being' are not widely used by the respondents (5 and 9 times in total, respectively), instead, they use more descriptive words to describe their feelings of well-being, specifically around 'stress relief' and 'relaxation'. We should keep in mind that both buildings are functional working and educational centres. In the case of ESA, there is a focus on learning output productivity with intense and sometimes challenging interventions. The usual event formats do not include time for individual peace and relaxation. Neither building operates as a 'health spa' or 'vacation retreat'. This makes the visitors' perception of 'anti stress' and 'relaxation' more interesting in this study.

Respondent, ESA Case: '*This is the antithesis of stressful living*' and '*better than a weekend at a health farm!*'.

Respondent, ESA Case: '*Enjoying the sounds of nature whilst working was amazing beyond words*'.

4.4.2 Connection to Nature

Not surprisingly, given the construction principles in both cases, the word 'nature' was used 18 (ESA) and 17 (Sharanam) times by the respondents.

How the visitors describe their 'experiences' in nature and about their perceived interactions with 'nature', landscapes and surroundings is more insightful. These narratives imply a positive sense of well-being impact of biophilia on people. The feeling of 'one with nature', 'integrated with the nature' and 'close to the nature' seems to generate a perception of 'spirituality' and 'peace' in many of the responses. In the case of Sharanam, some respondents specifically perceive the building to be 'sacred'. This is discussed further below and in Sect. 5.

Respondent, Sharanam Case: '*... open to sky and nature, gives soothing feel with presence of beautiful trees around, it's an experience in itself to stay here*'.

Respondent, Sharanam Case: '*you feel like you have been transported to a different place... feels like the descent of the heavens*'.

4.4.3 Personal Security

In this study, the perception of the safety of 'home' and 'belonging' is prevalent in both cases, but more so at ESA. We should keep in mind that ESA is a much smaller building (400 m^2) compared with Sharanam (2000 m^2) and that will most likely have an impact on feelings of 'snugness' and 'security'. As mentioned in the case study 3.1, the design of ESA explicitly included a 'sense of home from home' aim. However, visitors to the larger building use many more words of 'comfort', 7 (Sharanam) compared with 1 (ESA).

Respondents, ESA Case. *'I felt like I was at home for every minute' And 'ESA is the second home I didn't know I had. I will always have the ESA feeling with me'*.

Respondent, Sharanam Case: '*still I can experience the comfort that was created by the use of climactic strategies in the most appropriate way*'.

4.4.4 Spirituality

Themes of 'spiritual well-being' emerged unexpectedly. Perception of 'healing power', 'discovering yourself', are common in both sets of responses. At first glance, specific terms of 'spirituality' appear more abundant in the Indian case to describe a personal and individual sense of well-being, whereas the European respondents refer more to the 'experience' of the place 'sanctuary, magical'.

These results are discussed further in the next section.

5 Discussion

This study proposes that models for auditing the effectiveness of green buildings should include the value of biophilic impact on the well-being of users. Sustainability building assessments have been broadly criticised and in need of updating. Not all green buildings are biophilic and not all green buildings 'do no environmental harm' (Van Wyk, 2008).

The outcomes of this study build on the premise that existing building assessments are inadequate because some of the most effective sustainable buildings, with positive impacts on human well-being, are being ignored by the GBC movement for not being rated in the standard way.

Stress and anxiety are generally understood to be increasing across societies, specifically in urban areas (Tracada & Caperna, 2012) and negatively impacting human well-being. Perceptions of 'stress relief' and 'well-being' enhancement whilst visiting both green buildings are evident in the surveys from this empirical study. This suggests the meaningful 'value of biophilic impact'.

In response to the study questions: Is the perception of the benefit of green buildings on human health and well-being valued in the context of green building assessment? And, Do the perceived benefits of nature on the human sense of well-being actually improve health and well-being? A review of literature suggests that 'perceived' benefits of nature on the human sense of well-being can actually improve health and well-being; however, the perception of the benefit of green buildings on human health and well-being is not valued in the context of green building assessment.

This overarching findings of this study and 200 respondents strongly suggest a positive biophilic impact on the perception of enhanced well-being. These feelings have validity for the individuals and should not be ignored because they cannot be quantified by an external formulaic standard. Therefore, the perceived benefits of green building on health and well-being, though unmeasurable, should have 'value' in the assessments of effectiveness of sustainable buildings.

The findings are compelling; however, the limitations of this research include just two cases, examined over a relatively short time. A further weakness of the survey tool (Appendix 1) was the omission of the question 'do you live in the city or countryside?' It is possible that this will have an impact on the perception of the respondents.

5.1 Biophilia

Throughout the exploration to understand the value of biophilia, the challenge became more evident that a social constructionist paradigm of 'valuing perception' of well-being is rejected by current positivist rating systems. A shift to a more 'relative rather than absolute sense' (Guy & Farmer, 2001, p. 140) may be necessary to address this in the longer term. A massive shift is required in terms of research with many small steps to make that leap towards a 'new world' of understanding (Crane, 2016, p. 48).

The built environment and sustainable architecture should be about attitude and approach rather than an 'absolute' tick-box prescription (Guy & Farmer, 2001, p. 140), and the interpretation of 'green buildings' is wide ranging, encompassing many viewpoints which suggests an 'essentially contested concept' (Cok & Golton, 1995).

It may be that the necessary different approaches to assessment never become combined.

5.2 Placebo

A wide number of studies have shown how people can 'perceive them themselves' well (Manchester University, 2018) and enhanced capabilities can be triggered by belief (Sörqvist et al., 2015). In which, case it is reasonable to suggest that the respondents in the study are experiencing enhanced wellness, calm and reduced stress from the connection with nature that they experience when visiting these green buildings.

There is no doubt that the field of research into the impact of sustainable buildings on human health is broadening with some studies attempting to reinterpret sustainable architecture in a way to reconnect humans with the natural world. Contemporary studies are framed by calls for more sensuous and creative green architecture and the need for 'healing environments' where the occupants can feel relaxed, balanced and one with the world (Guy & Farmer, 2001, p. 140).

The core argument of this paper is that if people believe that ESA and Sharanam make them feel more relaxed and less stressed, applying the evidence of the 'placebo effect', which clearly shows that humans can believe themselves well, suggests that our respondents actually are being healed by these buildings. This is more than just perception and warrants further study to show how this temporal positive impact could be continued and replicated.

One respondent mentions Maslow's Hierarchy of needs (Maslow, 1943). This takes the tone of response to beyond a grounded feeling of physical well-being to a spiritual sense of well-being. This is aligned with some studies calling for buildings to meet combined human physical, biological and spiritual needs, where health, ecology and spirituality are interwoven (Guy & Farmer, 2001, p. 141).

With Maslow's hierarchy of needs theory in mind, the most used words from 200 respondents in Greece and India hint at a chronological story of the users' experience. Starting on arrival and observing the natural landscape, to develop a feeling of connection with nature which brings a sense of being safe and secure. This becomes a sense-making anchor (Weick, 1995) which connects thoughts and feelings to other places where people expect to have similar experiences such as home, sanctuary, sacred place and heaven. Finally, in the 'self-actualisation' stage of Maslow's framework they perceive feelings of healing, rediscovery, empowerment and peace.

The India respondents mostly refer to 'personal spirituality'; however, it is a Greek male visitor who wrote on the floor of ESA: '*From here on you are entering a sacred temple…*'.

Our conclusion summarises the key propositions emerging from this study and recommended next steps of enquiry.

6 Conclusion

This simple study is a starting point for a new exploration into the possibility that people's perception of the health benefits of green buildings might have an actual impact on their sense of health and well-being. The initial findings from this pilot study suggest fertile ground for further exploration with larger samples and more complex data collection over longer time periods in different countries.

The most important proposition emerging from this enquiry is that people across the planet are increasingly suffering from stress and illness, especially in the built environment. The value of biophilia has shown to be an antidote to this in terms of fully biophilia designed green buildings or retrospective low-cost interventions.

The plight of the global natural environment is creating widespread stress and anxiety with feelings of paralysis, anger and depression (Bendell, 2018) regarding climate change concerns, species depletion and ocean pollution (World Wildlife Foundation, 2018).

Sometimes, scientific reports are so careful to be 'scientific' and balanced that the most vital points remain unnoticed and unacted upon. This could be a key reason why despite 20 years of scientific warnings, global warming and oceanic harm has continued largely unabated. The scientific warnings were not understood and felt by humanity at large and therefore did not feature in the sensemaking processes (Weick, 1995) that lead to behaviour changes.

Furthermore, unhealthy, sick and stressed anxious people are unlikely to be able to stand up against the causes of climate change.

From a research perspective, the entrenchment of epistemological stances can block advancement of knowledge and these days of global sustainability crisis requires 'new world' thinking and collaboration with a merging of approaches (Guy & Farmer, 2001, p. 146; Crane, 2016, p. 47).

A sense of 'calm, focussed, connection' is expected and believed in biophilic conditions, and therefore, those feelings are manifested, if only for a temporary period. There is a train of thought that suggests if it works, why the need to measure it? The response to that is in the interest of replicability and to encourage more practical applications of greening of public buildings, schools, hospitals, general places of work and private homes.

Wide numbers of studies have shown how, without the need to build ESA or Sharanam, the same biophilia wellness feelings can be replicated with design choices and retrospective greening such as low-cost solutions including green views, paintings of nature and plants in the workplace. Convincing studies in hospitals have shown these actions to boost positive energy, increase sense of well-being and enhance healing in both physical and mental conditions.

The critical point seems to be that if people 'want' to feel well and 'believe' that nature can heal them, that is, what happens. This study warrants further exploration including longitudinal research across multiple sites to bring this aspect into the assessments of impacts of green buildings.

Furthermore, in terms of empirical measurement of this theory, simple heart rate, pulse rate and breathing rate tests at the start and end of their visit could substantiate the dimension of how green buildings can heal people.

Ongoing research discussions and engagement with WGBC, schools of architecture, construction industry and architect firms are vital to encourage new understanding into practice.

Future survey tools should include clarification of respondents' usual residence in the city or countryside to bring new insight into the different perceptions and response to a biophilic environment.

Finally, the unexpected dimension of 'spiritual health' emerging through this studying is worthy of further exploration to observe if a 'Maslow's hierarchy of needs' type transition to 'self-actualisation' can occur over a period in biophilic buildings.

Disclosure Statement The authors declare S. M. Jackson as the owner of ESA and T. Doshi as an architect of Sharanam.

Appendix 1: Survey Questionnaire

Visitor Satisfaction survey

Event:...Date: ....

Name (optional) ....

Eco-building

We are interested about understanding the impact of the features of green buildings on people's experiences. We would greatly appreciate your feedback about your experience of the ESA/ Sharanam building.

Please rate the ESA/ Sharanam building by ticking the box on the scale of 1-5 basis where 3 is the middle point.

Beautiful	5	4	3	2	1	Ugly
Relaxed	5	4	3	2	1	Tense
Colourful	5	4	3	2	1	Dull
Too bright	5	4	3	2	1	Too dim
Glare	5	4	3	2	1	No glare
Draughty	5	4	3	2	1	Still
Too hot	5	4	3	2	1	Too cold
Too noisy	5	4	3	2	1	Too quiet
Too dry	5	4	3	2	1	Too humid
Too sunny	5	4	3	2	1	Not enough sun
Poor working environment	5	4	3	2	1	xcellent working environment

Overall how did you feel during your time at ESA/ Sharanam?

About you:

How would you describe your work?

1. Administration/ Clerical
2. Professional
3. Managerial
4. Other.......................................

What is your age?

1. 20 or under
2. 21 – 30
3. 31 – 40
4. 41 – 50
5. 51 – 60
6. Over 60

What is your gender? 1. Female 2. Male

Appendix 2: NVIVO Word Cloud from Analysis of Respondents' Text

References

Allen, J. G., MacNaughton, P., Laurent, J. G. C., Flanigan, S. S., Eitland, E. S., & Spengler, J. D. (2015). Green buildings and health. *Current Environment Health Report, 2*, 250–258. https://doi.org/10.1007/s40572-015-0063-y

Arendsen, L., & Henshaw, J. (2018). *HORIZON: The Placebo experiment: Can my brain cure my body*. Research Explorer. The University of Manchester. Retrieved May 1, 2020, from https://www.research.manchester.ac.uk/portal/en/media/horizon-the-placebo-experiment-can-my-brain-cure-my-body(cdbb2e6a-0f9b-4fad-a8fe-c07af2e267a3).html

Bady, S. (2008). *Green building programs more bias than about science, expert argues*. Retrieved September 26, 2019 from https://www.probuilder.com/green-building-programs-more-about-bias-science-expert-argues

Beecher, H. K. (1955). The powerful placebo. *Journal of the American Medical Association, 159*(17), 1602–1606. https://doi.org/10.1001/jama.1955.02960340022006

Bendell, J. (2018). *Deep adaptation: A map for navigating climate tragedy*. IFLAS Occasional Paper 2. Retrieved September 26, 2019 from https://www.iflas.info

Bonnefoy, X., Braubach, M., Moissonnier, B., Monolbaev, K., & Röbbel, N. (2003). Housing and health in Europe: Preliminary results of a Pan-European study. *American Journal of Public Health, 93*(9), 1559–1563.

Browning, W. D., Kallianpurkar, N., Ryan, C. O., Labruto, L., Watson, S., & Knop, T. (2012). *The economics of Biophilia*. Terrapin Bright Green L.LC. Retrieved September 5, 2020 from https://www.terrapinbrightgreen.com/report/economics-of-biophilia/

Build Green. (n.d.). Build Green: Definition. Retrieved May 1, 2020 from https://buildgreen.co.nz/definition.html

Carrus, G., Scopelliti, M., Lafortezza, R., Colangelo, G., Ferrini, F., Salbitano, F., Agrimi, M., Portoghesi, L., Semenzato, P., & Sanesi, G. (2015). Go greener, feel better? The positive effects of biodiversity on the well-being of individuals visiting urban and peri-urban green areas. *Landscape and Urban Planning, 134*, 221–228. https://doi.org/10.1016/j.landurbplan.2014.10.022

Chang, C. Y., & Chen, P. K. (2005). Human response to window views and indoor plants in the workplace. *HortScience, 40*(5), 1354–1359.

Cok, S. J., & Golton, B. L. (1995). Sustainable development concepts and practice in the built environment—A UK perspective. *Fuel and Energy Abstracts, 4*(36), 299.

Crane, G. S. (2016). Harnessing the placebo effect: A new model for mind-body healing mechanisms. *International Journal of Transpersonal Studies, 35*(1), 39–51.

Craven, J. (2019). *When "Green" architecture is more than a color*. https://www.thoughtco.com/what-is-green-architecture-and-green-design-177955

Croome, D. (1999). *Creating the productive workplace*. CRC Press.

Del Percio, S. (2007). What's wrong with LEED? *The Next American City*. Retrieved September 26, 2019 from https://americancity.org/article.php

Elzeyadi, I. (2011). *Daylighting-bias and biophilia: Quantifying the impact of daylighting on occupants health*. US Green Building Council. https://www.usgbc.org/sites/default/files/OR10_Daylighting%20Bias%20and%20Biophilia.pdf

EPA. (2007). *Building healthy hospitals: Case studies overview*. Retrieved May 8, 2015 from https://www.epa.gov/region9/waste/p2/pdf/overview-factsheet-final-071126.pdf

EPA. (2009). *Buildings and their impact on the environment: A statistical summary*. Retrieved May 8, 2015 from https://www.epa.gov/greenbuilding/pubs/gbstats.pdf

European Sustainability Academy. (n.d.). Retrieved May 1, 2020 from: https://eurosustainability.org/

Guy, S., & Farmer, G. (2001). Reinterpreting sustainable architecture: The place of technology. *Journal of Architectural Education, 54*(3), 140–148. https://doi.org/10.1162/10464880152632451

Hansen, M. M., Jones, R., & Tocchini, K. (2017). Shinrin-yoku (forest bathing) and nature therapy: A state-of-the-art review. *International Journal of Environmental Research and Public Health, 14*(8), 851. https://doi.org/10.3390/ijerph14080851

Harvard Publishing. (2017). *The power of the placebo effect*. Harvard Health. Retrieved May 1, 2020, from https://www.health.harvard.edu/mental-health/the-power-of-the-placebo-effect

Houtman, I. (2007). *Economic and scientific policy new forms of physical and psychosocial health risks at work study*. https://doi.org/10.1139/F04-118

Issa, M. H., Rankin, J. H., & Christian, A. J. (2010). Canadian practitioners' perception of research work investigating the cost premiums, long-term costs and health and productivity benefits of green buildings. *Building and Environment, 45*, 1698–1711.

Jackson, S. M., Maleganos, J. K., & Almantariotou, K. (2017). Lessons from sustainable entrepreneurship towards social innovation in healthcare: How green buildings can promote health and wellbeing. In K. Nicolopoulou, M. Karatas-Ozkan, F. Janssen, & J. M. Jermier (Eds.), *Sustainable entrepreneurship and social innovation. Research in sustainability and business* (Chap. 8, pp. 143–169). Routledge.

Jacobs, D. E., Wilson, J., Dixon, S. L., Smith, J., & Evans, A. (2009). The relationship of housing and population health: A 30-year retrospective analysis. *Environmental Health Perspectives, 117*(4), 597–604.

Kalyvas, S. N., Pagoulatos, G., & Tsoukas, H. (2012). *From stagnation to forced adjustment reforms in Greece 1974–2010*. Hurst and Company.

Kaplan, R., & Kaplan, S. (1989). *The experience of nature: A psychological perspective*. Cambridge University Press.

Kaptchuk, T. & Miller, F. (2018). Open label placebo: Can honestly prescribed placebos evoke meaningful therapeutic benefits? *British Medical Journal, 363*. https://doi.org/10.1136/bmj.k3889

Kellert, S. (2005). *Building for life: Designing and understanding the human-nature connection* (p. 24). The University of Chicago Press.

Kellert, S. R., Heerwagen, J., & Mador, M. (2008). *Biophilic design: The theory, science and practice of bringing buildings to life*. Wiley.

Kim, M. J., Oh, M. W., & Kim, J. T. (2013). A method for evaluating the performance of green buildings with a focus on user experience. *Energy and Buildings, 6*, 203–210.

Kok, N., Miller, N., & Morris, P. (2012). The economics of green retrofits. *Journal of Sustainable Real Estate, 4*(1), 4–22.

Maslow, A. H. (1943). A theory of human motivation. *Psychological Review, 50*(4), 370–396.

McGill, I., & Brockbank, A. (2003). *The action learning handbook powerful techniques for education, professional development and training*. Routledge Falmer

Mendell, M. J., Mirer, A. G., Cheung, K., Tong, M., & Douwes, J. (2011). Respiratory and allergic health effects of dampness, mold, and dampness-related agents: A review of the epidemiologic evidence. *Environmental Health Perspectives, 119*(6), 748–756. https://doi.org/10.1289/ehp.1002410

Miller, N., Pogue, D., Gough, Q., & Davis, S. (2009). Green buildings and productivity. *Journal of Sustainable Real Estate, 1*(1), 65–89.

Moerman, D. E., & Jonas, W. B. (2002). Deconstructing the placebo effect and finding the meaning response. *Annals of Internal Medicine, 136*(6), 471–476. https://doi.org/10.7326/0003-4819-136-6-200203190-00011

Newsham, G. R., Birt, B. J., Arsenault, C., Thompson, A. J., Veitch, J. A., Mancini, S., Galasiu, A. D., Gover, B. N., & Macdonald, I. A. (2013). Do 'green buildings' have better indoor environments? New evidence. *Building Research & Information, 41*(4), 415–434.

Our Green Building Councils | World Green Building Council. (n.d.). Retrieved May 1, 2020 from https://www.worldgbc.org/our-green-building-councils

Paul, W. L., & Taylor, P. A. (2008). A comparison of occupant comfort and satisfaction between a green building and a conventional building. *Building and Environment, 43*(11), 1858–1870.

Pelenur, M. J., & Cruickshank, H. J. (2013). Investigating the link between well-being and energy use; An explorative case study between passive and active domestic energy management systems. *Building and Environment, 65*, 26–34. https://doi.org/10.1016/j.buildenv.2013.03.020

Revans, R. W. (1980). *Action learning: New techniques for management*. Blond and Briggs Ltd.

Sánchez-Bayo, F., & Wyckhuys, K. A. (2019). Worldwide decline of the entomofauna: A review of its drivers. *Biological Conservation, 232*, 8–27.

Sarvam Trust. (n.d.). Integral Village Development Programme. Retrieved May 1, 2020 from https://thesarvamtrust.org

Schendler, A. & Udall, R. (2005). *LEED is broken-lets fix it*. Retrieved September 5, 2020 from https://drive.google.com/file/d/0B1pYffuHFkvlY2YwZDdiYjAtNzIwYi00YjNjLTlmMzgtZDYwZTQ2OThlOTAw/view?sort=name&layout=list&num=50

Singh, A., Syal, M., Grady, S. C., & Korkmaz, S. (2010). Effects of green buildings on employee health and productivity. *American Journal of Public Health, 100*(9), 1665–1668.

Sörqvist, P., Haga, A., Holmgren, M., & Hansla, A. (2015). An eco-label effect in the built environment: Performance and comfort effects of labeling a light source environmentally friendly. *Journal of Environmental Psychology, 42*, 123–127.

Spiegel, D. (2004). Placebos in practice: Doctors use them, they work in some conditions, but we don't know how they work. *British Medical Journal, 329*(7472), 927–928. https://doi.org/10.1136/bmj.329.7472.927

State of play for sustainable buildings in India (2010). Retrieved May 1, 2020 from https://www.unenvironment.org/resources/report/state-play-sustainable-buildings-india

Swaid, H., Bar-El, M., & Hoffman, M. E. (1993). A bioclimatic design methodology for urban outdoor spaces. *Theoretical and Applied Climatology, 48*(1), 49–61. https://doi.org/10.1007/BF00864913

Takano, T., Nakamura, K., & Watanabe, M. (2002). Urban residential environments and senior citizens' longevity in megacity areas: the importance of walkable green spaces. *Journal of Epidemiology &*

Community Health, 56(12), 913–918. https://doi.org/10.1136/jech.56.12.913

Thatcher, A., & Milner, K. (2014). Changes in productivity, psychological wellbeing and physical wellbeing from working in a green building. *Work, 49*(3), 381–393. https://doi.org/10.3233/WOR-141876

Tracada, E., & Caperna, A. (2012). *Biourbanism for a healthy city: Biophilia and sustainable urban theories and practices.* Biotechnology/Textile Technology/Fashion Technology of the International Convention on Innovations in Engineering and Technology for Sustainable Development, 3–5 September 2012, in Bannari Amman Institute of Technology, Erode District, Tamil Nadu, India.

Ulrich, R. S., & Lunden, O. (1990, June). Effects of nature and abstract pictures on patients recovering from open heart surgery. In *International Congress of Behavioral Medicine*, Uppsala, Sweden.

University of Exeter. (2014). *Why plants in the office make us more productive*. Retrieved September 30, 2019, from https://www.exeter.ac.uk/news/featurednews/title_409094_en.html

Van den Berg, A. E., Hartig, T., & Staats, H. (2007). Preference for nature in urbanized societies: Stress, restoration, and the pursuit of sustainability. *Journal of Social Issues, 63*(1), 79–96. https://doi.org/10.1111/j.1540-4560.2007.00497.x

Van Wyk, L. (2008). Do green building assessment criteria meet sustainability imperatives: a critical analysis. In *3rd Built Environment Conference (ASOCSA)*, 6–8 July 2008, Cape Town. Retrieved from https://researchspace.csir.co.za/dspace/handle/10204/3070

Vance, E. (2016). *Suggestible you: The curious science of your brain's ability to deceive, transform, and heal*. National Geographic Books.

Wee, L. E., Koh, G. C. H., Yeo, W. X., Chin, R. T., Wong, J., & Seow, B. (2013). Screening for cardiovascular disease risk factors in an urban low-income setting at baseline and post intervention: A prospective intervention study. *European Journal of Preventive Cardiology, 20*(1), 176–188. https://doi.org/10.1177/2047487311433890

Weick, K. E. (1995). *Sensemaking in organizations*. Sage.

WHO, Environmental Health in Urban Development. (1991). Report of Who Expert Committee, World Health Organization.

Wilson, E. O. (1984). *Biophilia*. Harvard University Press.

Wolf, K. L., & Robbins, A. S. (2015). Metro nature, environmental health, and economic value. *Environmental Health Perspectives, 123*(5), 390–398.

World Wildlife Foundation. (2018). *Half of plant and animal species at risk from climate change in world's most important natural places*. Retrieved September 29, 2019 from: https://wwf.panda.org/wwf_news/?324471/Half-of-plant-and-animal-species-at-riskfrom-climate-change-in-worlds-most-important-natural-places

Xie, H., Clements-Croome, D., & Wang, Q. (2017). Move beyond green building: A focus on healthy, comfortable, sustainable and aesthetical architecture. *Intelligent Buildings International, 9*(2), 88–96. https://doi.org/10.1080/17508975.2016.1139536

Yin, R. (2009). *Case study research*. Sage.

Evaluation of Development Strategies and Community Needs in Developing Countries: A Comparative Case Study of Informal Settlements in Asia and Africa

Aya Awad, Debbie Bartlett, and Guido Conaldi

Abstract

From the millennium development goals (MDGs) to the sustainable development goals (SDGs), social inclusion has been highlighted as an aspect to consider while developing urban areas. Progress has been witnessed to various degrees in developing countries. However, low-income communities have generally experienced a set of operational setbacks (Cobbinah et al. in Environ Dev 13:18–32, 2015). For instance, official development proposals have not met community needs, and assistance promised had not been provided (Sachs in The Lancet 379(9832):2206–2211, 2012). Basic needs are fundamental to sustainable development schemes, and Governments should develop strategies to ensure that basic needs are met. As stated in Brundtland report, "*Sustainable development requires meeting the basic needs of all and extending to all the opportunity to satisfy their aspirations for a better life*" (World Commission on Environment and Development in Report of the World Commission on environment and development: our common future, p. 16, 1987). This includes the provision of adequate food, energy, housing, sanitation, employment, water supply and health care. Satisfying these needs at community level will ultimately provide a better life for individuals (Holden et al. in Glob Environ Change 26:130–139, 2014). Informal settlements—or slum areas—are a common occurrence in developing countries. This phenomenon results from the migration of people from rural to urban areas looking for better opportunities to overcome poverty. Informal settlements play a significant role in the housing market as affordable housing schemes are often not attainable within the budget of a large section of the population. Furthermore, affordable accommodation rarely meets the requirements of the community (Hassan in Alexandria Eng J 51(3):229–239, 2012; Naceur in Front Architectural Res 2(4):400–408, 2013). This leads to the potential conflict between informal settlement communities, land government authorities and bordering societies (Lombard and Rakodi in Urban Stud 53(13):2683–2699, 2016). Moore's strategic triangle model of the public policy states that a balanced strategy for an appropriate governance framework should comprise democratic legitimacy and government support, with community endorsement. In addition, adequate operational capacity is required to deliver development schemes that satisfy informal settlement resident's needs (Kavanagh in Gov Finance Rev 4, 2014; Moore in Nonprofit Voluntary Sect Q 29(1_suppl):183–204, 2000). This paper compares informal settlements in Africa and Asia to identify what factors influence community satisfaction and, based on this makes, recommendations to inform future proposals for the redevelopment of informal settlements.

Keywords

Informal settlements • Basic needs • Development strategies • Sustainable development • Developing countries • Sustainable development goals (SDGs)

1 Introduction

Rapid population increase is a significant factor contributing to the informal settlement phenomena in large urban areas. These are also known in the literature as shantytowns, slums, squatter camps, bidonvilles, katchi abadis, ghettos and capamentos (Davis, 2006; Smit et al., 2017) and become the only viable alternative for individuals with limited financial capabilities to meet their housing needs. Urban policies and strategies often exclude informal settlements, although arguably these should be considered alongside formal settlements when formulating long-term city development

A. Awad (✉) · D. Bartlett · G. Conaldi
University of Greenwich, London, UK
e-mail: a.m.awad@Greenwich.ac.uk

F. Trapani et al. (eds.), *Advanced Studies in Efficient Environmental Design and City Planning*, Advances in Science, Technology & Innovation, https://doi.org/10.1007/978-3-030-65181-7_17

plans and generating solutions for people's needs (UN-Habitat, 2008). Abbott (2001) reported that over time, informal settlements develop their own infrastructures, local economy, and local rules; once this has occurred, particular strategies are needed to address the situation.

Following Dovey (2013, p. 86) "*slum is a symptom of poverty; informality is a transgressive practice through which residents manage the conditions of poverty*". As a definition, this underlines the fact that these urban areas are the consequence of communities trying to satisfy their basic needs and overcome poverty (Khalifa, 2015). To highlight the scale of this emergency, the United Nations Human Settlements Programme reported that: "*By 2030, the majority of Africans will be urban residents, and the majority of them are predicted to live in slums and informal settlements unless radical correctives are taken*" (UN-Habitat, 2008, p. ii). The paper will use case studies to compare different informal settlements to identify how the situation can be addressed and contribute to achieving sustainable development goal (SDG)11, "*Make cities and human settlements inclusive, safe, resilient and sustainable*" (UN-Habitat, 2015, p. 6).

2 Informal Settlements

Informal settlements typologies are mostly based on the topography, city transportation infrastructure and morphology. Researchers identify informal settlements through common elements. These include primitive materials for dwellings, which the literature lists alternatingly as institutions' discrimination, the absence of land ownership and lack of basic services. They are recognised as mainly illegal, self-constructed accommodation not complying with planning/building regulations and codes. Such areas usually display various social difficulties, including crime and poverty and are high density with walkable car-free streets. These include squatter camps, transit camps, housing–turned slum, site and services settlements, hybrid multi-structured settlements and townships (Smit et al., 2017; Srinivas, 2015). In some case, fragments of old rural or urban fringe settlements turn into slums over time (UN-Habitat, 2008, 2012).

The residences are incrementally constructed, room by room, or can be multi-functional spaces with poor light, sanitation and ventilation. Usually, one or more of the following conditions are found: inadequate and overcrowded living spaces; lack of clean water and access to sanitation facilities; insecure tenure; insufficient quality and durability of houses. Most are located at the borders of cities, near railways, roads or waterfronts, sometimes built along escarpments or against industrial or institutional walls that provide enclosure (Abdelhalim, 2010; Dovey, 2013; Marutlulle, 2017; UN-Habitat, 2008).

Over the years, informal settlements in developing countries have been considered a problem, as they comprise clusters of the most impoverished and vulnerable people, potentially posing a threat to human development and generating conflict with government authorities and bordering communities. These can include disputes around resources, land and power, goods and materials that help individuals to generate income and maintain their livelihoods under challenging conditions. One of the reasons is informal settlements often forgotten in political plans, without proper infrastructure systems or planning strategies and policies applied by authorities (Marx et al., 2013; PSUP, 2016; UN-Habitat, 2003). Occupants are usually day labourers who deliver facilities for the formal sector or are unemployed or underemployed. In some cases, labourers work in the city for half the year and spend the second half in the countryside. Some may run micro-businesses providing services such as carpenter, markets, small animal production and mechanical workshops (Davis, 2006; Kohli et al., 2012). Sub-groups of residents form and dissolve, changing power dynamics and outsiders categorise residents as not always following the rules and regulations to achieve their aims (Barry, 2000; Kovacic & Giampietro, 2017; Wekesa et al., 2011).

3 Sustainable Development

In the Brundtland report, sustainable development is defined as meeting the basic needs of all and extending to all the opportunity to satisfy their aspirations for a better life (World Commission on Environment & Development, 1987). So, while 'basic needs' may vary geographically, this essentially means food, energy, housing, sanitation, employment, water supply and health care are needed to provide an adequate standard of living for individuals (Blewitt, 2018; Holden et al., 2014). For local and national governments to evidence that they are working towards achieving SDG 11, the situation of those living in informal settlements must be addressed to ensure that their basic needs are met. According to Jabareen (2006) and UN-Habitat (2015), an effective, sustainable and urban system require sufficient diversity, high density, greening, sustainable transportation, mixed land use and passive solar energy, so aiming towards reducing automobile use, pollution, waste have adequate infrastructure and open spaces to create and maintain a livable community orientation. Since the adoption of the UN sustainable development goals, there has been increasing interest in this issues as Goal 11 is to "*Make cities and human settlements inclusive, safe, resilient and sustainable*" (UN-Habitat, 2015, p. 6). This research was undertaken to investigate the current situation in selected informal settlements strategies in order to determine what is needed to meet the basic needs of the residents.

4 Research Method

Moore's strategy triangle model and Alderfer's existence, relatedness and growth (ERG) theory principles were used to understand the phenomena of informal settlements and guide the comparative case study analysis. Specifically, a set of criteria adopted to evaluate the operational capabilities (legal power, stakeholders, institutions strategies/Visions), governance context (political parties, community attitudes, resources, laws/policies) and the resultant public value (accessibility, legitimacy, service quality). A comparative case study was chosen as method because it is suitable to generate theories to enable testing complex research situations. The goal when selecting the cases for the study was to include multiple selection criteria, so maximum variation and convenience sampling were used to include a range of cases in the analysis.

4.1 Characteristics

This research takes a case study-based analytical approach, based on secondary sources, to identify the conditions that characterise informal settlements as the first step in developing an action plan and recommendations for redevelopment and improvements to address residents' basic needs. Also, case studies have been used to profile the causes, the strategic/policy context and the reasons for their persistence. Ten examples were used for this study with the similar socio-economic status: five in Asia (Banganga in Bhopal, Shivaji Nagar and Dharavi in Mumbai, Kathputli in Delhi and Salia in Bhubaneswar), and five in Africa (Kibera and Kukuru in Kenya, Joe Slovo Park in Capetown, Kasoli In Uganda, and Nima in Ghana). The aspects of the cases that were investigated included setup date, location drivers, reasons for the formation, infrastructure (public and private level), households building materials, community characteristic, stakeholder characteristics, interactions, interventions and development schemes or approaches. The data was compiled in table format for each example and then collated into a matrix to enable comparisons (see Table 1).

4.2 Strategies and Policies

4.2.1 The Asian Context

In the fifties and sixties, the Indian government policy was to demolish informal settlements and slums, but this was unsuccessful as people simply returned and rebuilt their homes or shifted to any unoccupied land nearby. During the seventies, these began to be seen as a housing solution with residents issued with photo passes as part of a programme to deliver electricity, sanitation and water facilities. However, these improvements were carried out by government institutions with no community participation or consultation (Burra, 2005). In 1991, a 'Slum Redevelopment' plan introduced to transform Mumbai, under the responsibility of the congress government's Chief Minister, Sharad Pawar. Based on changing the land use zones for economic reasons, this resulted in existing slums either being moved onto open land of no commercial use or further out, to the city edge (Björkman, 2014; Shatkin, 2013).

The Indian government identified two types of informal settlements/slums in 2001, "notified slums", areas in cities or towns notified by the local government or state under the Slum Act and non-notified slums, recognised by the government but not notified under the Act. Both typically have around 300 residents distributed between sixty to seventy homes in building, with poor services and inadequate facilities. The main difference is that notified slums have access to use government facilities with some households having a legal title and so are protected from eviction (Lucci et al., 2018; Nakamura, 2014).

In Bangladesh, the World Bank and other experts have highlighted the major reason for city slums which is the absence of proper planning. The slums here have high populations, so rooms are crowded with inadequate water and sanitation services. Occupants have very low economic status, with about 50% of households below the poverty level and lacking tenure rights so are vulnerable to eviction (Siddique, 2017; UN-Habitat, 2006). Despite this renting out slum units, particularly those near waterways in Dhaka, can be a profitable business. Most of the slum communities are domestic workers, hawkers, rickshaw pullers or work for low pay in factories and offices (The Daily Star, 2009).

Since 1975 government and local authorities have trialed several approaches to address informal settlements in Bangladesh (Mohit, 2012), the Grameen Bank and Bangladesh Rural Advancement Corporation have supported various campaigns to ease slum poverty (UN-Habitat, 2006), and in 2006, the committee for urban local government developed an inclusive development policy to solve everyday issues in Dhaka city. Other approaches have been adopted by NGOs to enhance living standards, for example, the Bangladesh Rural Advancement Committee (BRAC) and Nari Udyog Kendra (NUK) project offer shelter for female clothing workers (Rashid, 2009). Water and sanitation initiatives providing access to clean water for Dhaka slum communities have been developed by Dustha Sastha Kendra (DSK), Assistance for Slum Development (ASD) and Water Aid Bangladesh in partnership with various community-based organisations (CBOs) (Mohit, 2012). More recently a partnership between the United Nations Development Programme (UNDP), UN-Habitat and the Government's urban

Table 1 Informal settlement general characteristics

Factors	Africa					Asia				
	Kibera, Kenya (1)	Mukuru, Kenya (2)	Joe Slovo Park, CapTown (3)	Kasoli, Eastern Uganda (4)	Nima, Ghana (5)	Banganga, Bhopal (6)	Shivaji Nagar, Mumbai (7)	Dharavi, Mumbai (8)	Kathputli slum,Delhi (9)	Salia Sahi (10)
Set up Date:										
>20 years			■							
>30 years		■				■	■			
>40 years										■
>60 years									■	
>70 years	■		■	■						
>100 years								■		
Location Drivers:										
Near City centre	■	■	■		■	■		■	■	■
Prime Location					■				■	
Proximity to immigrants' origin										■
Access to main roads							■	■		
Access to minimal infrastructure							■			
Near transportation services				■			■	■	■	
Deserted Former settlement								■		
Open vacant land	■		■		■	■		■	■	
Near work/ opportunities		■				■				
Formation:										
Cheap place to live near city	■									
Job Opportunities	■	■			■					■
Relocation from another area failed Schemes			■				■			
Evacuation / Incident						■		■	■	
Available renting houses / Living space		■					■			
Occupied by returning veterans				■						
Accommodate different Socio-economic level					■					
Land Owned by Government	■	■	■	■	■	■	■	■	■	■
Land Owned by private sector			■		■					
Infrastructure–Public:										
Roads-unpaved	■	■	■	■	■	■		■	■	■
Roads-paved-not maintained							■			
Water	■	■	■	■	■		■	■	■	■
Sanitation	■	■	■	■		■	■	■	■	■
Waste Deposal -Sewage-Inadequate			■	■		■			■	■
Electricity	■				■		■	■	■	■
Garbage disposal-Collection points		■		■	■			■	■	■

Table 1 continued

Household Level–Private:										
Water (few households-storage tanks)								■	■	
Sanitation										
Waste Deposal - (not maintained) Sewage.										■
Electricity (Illegally-expensive)	■					■	■	■	■	■
Garbage disposal										
Households Building Materials:										
Mud	■				■	■				■
Corrugated iron sheets	■	■	■	■	■	■		■		■
Wood						■		■		■
bricks			■	■			■	■	■	
Cement /Concrete			■	■			■	■	■	
Straws						■				■
Cloth							■			■
Community:										
No Tenure rights/ Illegal Distributed deeds	■	■	■	■	■	■	■	■	■	■
Tenure rights - Old tents		■								
Poor Living condition	■	■	■	■	■	■	■	■	■	■
Labour work		■				■		■	■	■
Self-employed					■	■			■	■
Multiple Ethnics/social levels	■				■	■		■	■	■
Ethnic conflicts.	■									
Live in multipurpose/high dense area	■	■			■			■	■	■
Health problems / diseases/ death	■	■		■		■	■	■		■
Unsafe Crime / Gang Violence	■	■					■	■		■
Losing houses due to mortgage			■							
Resisting Redevelopment/ Relocation schemes			■			■				
Supporting redevelopment schemes				■						
Poor Education (primary, secondary level)			■	■	■					
Eviction threats		■		■			■	■	■	
Food insecurity								■		
Stakeholders Interactions:										
Government initiate strategies	■	■	■	■	■	■	■	■	■	■
Private	■		■	■			■	■	■	
NGOs	■	■	■	■			■	■		
Community	■		■						■	■
Interventions:										
Settlement Residents	■	■	■				■			
Local-Residents		■	■	■			■			

Table 1 continued

Government	■	■	■	■						
NGOs-Local-Grassroots		■	■				■			
NGOs-National	■	■		■						
NGOs-International	■	■		■						
Private Sector							■			
Infrastructure:										
Water	■					■			■	■
Sanitation	■	■				■	■		■	■
Waste Deposal	■								■	■
Electricity	■					■			■	■
Garbage disposal									■	■
Affordable housing	■		■	■	■	■			■	■
Recycling scheme		■				■		■		
Develop community skills/ Economy		■				■	■	■		
Health		■				■	■			
Education.							■			

partnerships in poverty reduction (UPPR) has been offering resources, skills and knowledge to increase slum dwellers incomes to improve the living conditions (UNDP, 2019). The African Context.

Africa has a high and rapidly increasing population projected to rise by sixty per cent by 2050, which will transform the profile of the region (UN-Habitat, 2015). This is reflected in the formal housing market and construction prices (Housing Finance), where a gap between supply and demand has led to a housing crisis, exacerbated by migration. Dispossession of the urban poor and demolition of urban areas without replacing housing is considered a crime towards the informal settlement communities (Dovey, 2015). So, long-term strategies for removal and replacement with formal housing were the dominant strategy in the 1950s and 1960s. Despite the delivery of permanent public housing units, informal settlements have been increasing due to social, economic, aesthetic and environmental reasons (Abbott, 2002; Dovey, 2013).

Unplanned housing in Africa usually takes the form of informal settlements occupied by low-income communities. Generally, the houses do not conform to regulations and standards set by authorities because, for example, of unlicensed land subdivision, illegal urbanisation, unauthorised building and construction of units with incorrect building standards (see Fig. 1). In some cases, families may own the land, and houses may be built of durable materials and have access to electricity and water. In other properties may be transferred legally, but not registered (see Fig. 2). However, even these may violate building regulations and often have inadequate sanitation (Barry, 2000). Informal settlement dwellers often struggle to access education and health systems and addressing these issues effectively has failed, with sections of society not provided with suitable housing (Marutlulle, 2017).

Fig. 1 Building materials in Mansheya Nasir, Egypt

5 Discussion

5.1 General Characteristics

Informal settlement conditions and characteristics might differ based on location, community conditions and governmental positions, but there are common elements (Nolan, 2015). This research took specific features which are location, set update, formation and selection drivers, community, household condition and infrastructure condition, to build a general profile for informal settlements. Infrastructure included sanitation, sewage, water, electricity and waste disposal, as well as the urban fabric and roads.

All the ten case studies showed lack of tenure rights irrespective of whether the households owned or rented their dwelling, as all were occupying government lands. Choice of location was based on job opportunities, vacant lands near the city and transportation services, as well as affordable price or rent. six of the ten case studies indicated that part of the reason the informal settlements took shape was settlement on vacant land; in five cases, this was due to nearby transportation services; in eight cases, because it was near the city centre. Occupants look for a place near to their workplace and city to minimise transportation cost and reach their daily needs within walking distance; most residents do not own a car.

Crime and gang activities are associated with many informal areas due to the lack of police or government presence. Informal settlements may exhibit ethnic problems, as there are people with diverse backgrounds, and the education level is generally a low as the need for money means that people work rather than completing their schooling; in only one case study (7) had education facilities in the area. The communities suffer from health issues and disease due to poor living conditions and lack of nearby health facilities (see Fig. 3). Infrastructure is in poor condition, and facilities are generally public or shared (see Fig. 4). five out of the ten case studies have inadequate waste deposal facilities. The communities need to have access to water either as private water taps or shared standpipes near their households. The data shows that in almost all case studies, water taps require long travelling distance; only in case studies eight and nine do some households have water storage tanks.

Most of the interventions implemented in informal settlements are creation of affordable housing and mostly carried out by the government. In four out of the ten case studies, government agencies delivered the redevelopment schemes. However, non-governmental organisations (NGOs) may play a significant role, although mostly in small-scale projects. For instance, in case studies 1, 2, 3, 4 and 7, development schemes were carried out or funded by NGOs.

Fig. 2 Illegal extensions in Abagaya, Khalifa, Egypt

Fig. 3 Dharavi, Mumbai, India. *Source* McLeod (2016)

5.2 Sustainable Measures in Development Strategies

This research was undertaken as the first step in considering how developing countries can work towards achieving

Fig. 4 Bhopal, India. *Source* Dalbéra (2015)

sustainable development with respect to providing adequate housing for the vulnerable, with least financial resources, currently living in informal settlements or slums. Sustainable urban development and planning are a composite of many factors such as social life, energy, transport, land use, urban design ecology and economics (Dempsey et al., 2009). After that, the social sustainable development can only be achieved with communication and collaboration between civil societies and community participation (Ukko et al., 2018; UN-Habitat, 2015) and within social and cultural frameworks that incorporate a stable economy and social support. According to Turkoglu, (2015, p. 10) "*A well-informed framework equipped with the data including quality of life assessment is essential in enhancing the sustainability process*". In 2015, SDGs were formulated with 169 targets and 304 actions (Blewitt, 2018). These were adopted as the framework for the sustainable development agenda to be achieved by 2030 (United Nations, 2019) and are now being used by governments across the globe to development strategies, including those for slums and informal settlements. Here, we would like to review two development strategies—proposed, respectively, by the Egyptian and Bangladeshi governments—in light of the conditions characterising informal settlements determined by our case study.

In Africa, for example, the Egyptian 2030 vision has been developed to overcome the challenges identified to achieving Egypt's sustainable development strategy, for example, the National Urban Policy (NUP) which will contribute to achieving sustainable development goal (SDG) 11, which deals with sustainable cities and communities (MIIC, 2016). The rationale for the vision is the need to address the persistent spread of urbanisation, and social inequalities and to take a multidimensional approach maximising the benefits of growth, while integrating local views on progress towards implementing the SDGs (Cabinet of Ministers, 2016). Official development assistance (ODA) funded projects set up specifically to contribute to SDG 11, such as the World Bank Inclusive housing finance programs for results, which aims to increase affordable formal housing for low-income households in Egypt and social housing fund (SHF) capability to coordinate policies in the social housing division (MIIC, 2016). However, some countries are struggling in their strategies for instance; in Kenya, the challenge of carrying out sustainable measures is not solely about poverty, but about the lack of practical ideas from organisations and government in terms of recognising the settlements and investing in improving the living situation. Stakeholders involved in the strategies and schemes need to deal with informal settlements, international organisations, NGOs, CBOs, FBOs movements and financial institutions (Panek & Sobotová, 2015; Schouten & Mathenge, 2010). The involvement was in establishing schools, healthcare centres, water kiosks and toilets in different parts of Kibera. However, they were not adequate because of the high number of residents in the villages. Inadequate financing schemes and managerial skills to manage large-scale solid waste are also a problem.

Turning to an example from Asia, the sustainable development goals (SDGs) were integrated into with the Bangladesh government's 7th Five Year Plan (7FYP), covering the period 2016 to 2020. Despite significant progress, a quarter of the Bangladeshi population still lives below the poverty line (Ahmed, 2014; Munir, 2019). The 7FYP tackles

SDGs 1, 2, 3, 4, 11, 13, 14, 16 and 17 through various strategies, policies and initiatives, with an action plan for SDG 12. In 2015, the cabinet division and general economic division-GED developed the National Social Security Strategy (NSSS) to work towards SDG1 (EquityBD, 2018) and the National Health Policy-2011, Population Policy-2012, and Ministry of Social Welfare are working together towards SDG3 (Munir, 2019). There are still significant areas to address before sustainable development can be claimed to be fully achieved, such as reducing inequality in terms of development and income distribution (Rashid, 2009). The income gap is a key contributor to poverty, which leads to slums formation (EquityBD, 2018). Youth unemployment is a particular challenge that Bangladesh is facing with more than 95% working in the informal sector (Munir, 2019; Rashid, 2009).

So, then how are developing countries applying the principles of sustainable development in their plans, policies and strategies to address living conditions in informal settlements and slums? In reality, few governments in developing countries have the resources, power and qualified experts to provide their fast-growing communities with the services, land and facilities to meet the urban sustainable development requirements, for example, schemes that provide adequate schools, transport, proper sanitation and clean water. This would require governments to develop clear development strategies aiming to release the pressure on rapidly expanding cities to enable building smaller settlements integrated with rural lands. This would enable site and service planning in informally developed areas and slums with more appropriate and sustainable housing, providing the basic needs for communities (World Commission on Environment & Development, 1987). Fundamental to this process is the consideration of sustainable development principles. The literature describes these in different ways, but there are seven basic requirements, i.e. long-term perception, distribution of resources, secure ecosystem and resources, polluters pay, settlements the precautionary principle, participation and taking an integrated approach (Hedelin, 2008).

The current framework, the SDGs, provides a different perspective on urban development issues particularly if they are considered as interlinked, rather than the linear way in which they are usually displayed (Rockström & Sukhdev, 2016). To achieve sustainable cities and communities, a set of needs have to be fulfilled and SDG 11: Sustainable cities and communities require delivery of safe, sustainable, resilience and inclusive human settlements and cities (Roorda, 2017). There are ten targets including sustainable and safe transportation systems, integrated urban participatory planning, affordable and safe essential services and housing, disaster resilience, public and green spaces, integrated plans and policies, urban–rural relationships, minimise the environmental impact of the cities, natural and cultural heritage and technical and financial support for resilient and sustainable buildings (The Global Goals, 2019).

However, SDG11 cannot be considered in isolation. To achieve sustainable cities and communities, which includes dealing with informal settlements in developing countries, other SDGs will also be needed to be addressed specifically

- SDG 1 to create an action plan to end poverty.
- SDG 2 to achieve secure food resources, promote sustainable agriculture and improve nutrition.
- SDG 3 to ensure and promote healthy lives and wellbeing.
- SDG 4 inclusive good quality of education.
- SDG 5 to achieve gender equality.
- SDG 6, 7 and 9 are needed to provide clean water and sanitation, access affordable energy sources and build resilience infrastructure.
- SDG 8 to provide opportunities for productive employment for everyone.
- SDG 10 create effective and accountable institution at all levels to reduce inequality.
- SDG 16 to develop partnerships between all involved stakeholders.

This is represented graphically in Fig. 5.

6 Conclusion

Informal settlements result due to the high rate of people immigration from rural areas to the city to overcome poverty. The reasons for people choosing to live in informal areas are usually one or more of the following, looking for a location near work, for job opportunities and to have a better life. Also, it is the result of the unrestrained housing market, pushing a substantial number of people to move to informal housing, as the market is continuously rising with costs which exceed individual income capability (PDP, 2017; Shawkat, 2015). The comparative case studies showed informal settlements typically lack tenure rights (as they are illegally developed on land that residents do not own) and inadequate essential facilities. Although most case studies had access to water and electricity, but this was either public, shared and suffered from regular cuts was accessed/extended illegally by occupants (Huchzermeyer, 2004; Space Syntax, 2010).

In developing countries, governments have shown signs of efforts in setting sustainable strategies for upcoming future development. Research has shown that the sustainable development agenda has been evolving since the concept was first defined in the Brundtland report as "*Sustainable development requires meeting the basic needs of all and*

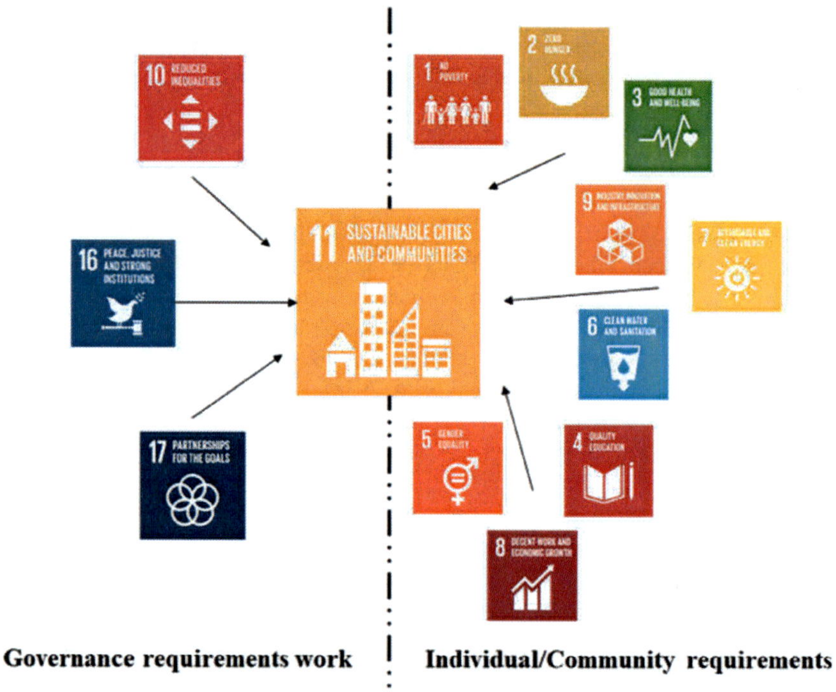

Fig. 5 Sustainable development goals arrangement

extending to all the opportunity to satisfy their aspirations for a better life" (World Commission on Environment & Development, 1987, p. 16). This includes the provision of adequate food, energy, housing, sanitation, employment, water supply and health care. The same basic principles were reframed, in the millennium development goals (MDGs) in the early 2000s (Holden et al., 2014) and, most recently, in the sustainable development goals (SDGs). Progress has been witnessed to varying degrees in developing countries; however, low-income communities experience particular challenges (Cobbinah et al., 2015) in improving life for individuals and in meeting community needs; assistance promised had not been provided.

This is the first stage of a more in-depth research project which is continuing with interviews with both residents of informal settlements and those involved in developing strategies and policies aimed to move towards achieving SDG 11 and in improving the welfare of residents.

References

Abbott, J. (2001). Use of spatial data to support the integration of informal settlements into the formal city. *International Journal of Applied Earth Observation and Geoinformation, 3*(3), 267–277. https://doi.org/10.1016/S0303-2434(01)85033-9

Abbott, J. (2002). An analysis of informal settlement upgrading and critique of existing methodological approaches. *Habitat International, 26*(3), 303–315. https://doi.org/10.1016/S0197-3975(01)00049-2

Abdelhalim. (2010). *Publication: "Participatory upgrading of informal areas: A decision-makers' guide for action"* | Participatory Development Programme in Urban Areas. https://egypt-urban.net/new-publication-participatory-upgrading-of-informal-areas-a-decision-makers-guide-for-action/

Ahmed, I. (2014). Factors in building resilience in urban slums of Dhaka, Bangladesh. *Procedia Economics and Finance, 18*, 745–753. https://doi.org/10.1016/S2212-5671(14)00998-8

Barry, M. M. (2000, September 6). *Informal settlement characteristics in a rural land restitution case: Elandskloof, South Africa* [Text. Article]. https://www.socresonline.org.uk/5/2/barry

Björkman, L. (2014). Becoming a slum: From municipal colony to illegal settlement in liberalization-era Mumbai. *International Journal of Urban and Regional Research, 38*(1), 36–59. https://doi.org/10.1111/1468-2427.12041

Blewitt, J. (2018). *Understanding sustainable development* (3rd ed.). Routledge.

Burra, S. (2005). Towards a pro-poor framework for slum upgrading in Mumbai, India. *Environment and Urbanization, 17*(1), 67–88. https://doi.org/10.1177/095624780501700106

Cabinet of Ministers. (2016). *The cabinet—Egypt's vision 2030*. https://www.cabinet.gov.eg:80/English/GovernmentStrategy/Pages/Egypt%E2%80%99sVision2030.aspx

Cobbinah, P. B., Erdiaw-Kwasie, M. O., & Amoateng, P. (2015). Rethinking sustainable development within the framework of poverty and urbanisation in developing countries. *Environmental Development, 13*, 18–32. https://doi.org/10.1016/j.envdev.2014.11.001

Dalbéra, J.-P. (2015). *Le quartier autour de l'ex-usine Union Carbide à Bhopal (Inde)* [Photo]. https://www.flickr.com/photos/dalbera/16944288732/

Davis, M. (2006). Planet of slums. *New Perspectives Quarterly, 23*(2), 6–11. https://doi.org/10.1111/j.1540-5842.2006.00797.x

Dempsey, N., Bramley, G., Power, S., & Brown, C. (2009). *The social dimension of sustainable development: Defining urban social sustainability—Dempsey-2011—Sustainable development*. Wiley Online Library. https://onlinelibrary.wiley.com/doi/10.1002/sd.417

Dovey, K. (2013). Informalising architecture: The challenge of informal settlements. *Architectural Design, 83*(6), 82–89. https://doi.org/10.1002/ad.1679

Dovey, K. (2015). Sustainable informal settlements? *Procedia—Social and Behavioral Sciences, 179*, 5–13. https://doi.org/10.1016/j.sbspro.2015.02.406

EquityBD. (2018). *Sustainable development goals (SDGs) in Bangladesh: Key challenges and missing links: Focus should be on internal resource mobilization and effective democratic institutions*. Social Watch. https://www.socialwatch.org/node/18086

Hassan, G. F. (2012). Regeneration as an approach for the development of informal settlements in Cairo metropolitan. *Alexandria Engineering Journal, 51*(3), 229–239. https://doi.org/10.1016/j.aej.2012.02.003

Hedelin, B. (2008). *Planning for sustainable use of water* (Doctoral), Karlstad University. https://www.researchgate.net/publication/301199386_Planning_for_Sustainable_Use_of_Water

Holden, E., Linnerud, K., & Banister, D. (2014). Sustainable development: Our common future revisited. *Global Environmental Change, 26*, 130–139. https://doi.org/10.1016/j.gloenvcha.2014.04.006

Huchzermeyer, M. (2004). From "contravention of laws" to "lack of rights": Redefining the problem of informal settlements in South Africa. *Habitat International, 28*(3), 333–347. https://doi.org/10.1016/S0197-3975(03)00058-4

Jabareen, Y. R. (2006). Sustainable urban forms: Their typologies, models, and concepts. *Journal of Planning Education and Research, 26*(1), 38–52. https://doi.org/10.1177/0739456X05285119

Kavanagh, S. (2014). Defining and creating value for the public. *Government Finance Review, 4*.

Khalifa, M. A. (2015). Evolution of informal settlements upgrading strategies in Egypt: From negligence to participatory development. *Ain Shams Engineering Journal, 6*(4), 1151–1159. https://doi.org/10.1016/j.asej.2015.04.008

Kohli, D., Sliuzas, R., Kerle, N., & Stein, A. (2012). An ontology of slums for image-based classification. *Computers, Environment and Urban Systems, 36*(2), 154–163. https://doi.org/10.1016/j.compenvurbsys.2011.11.001

Kovacic, Z., & Giampietro, M. (2017). Between theory and quantification: An integrated analysis of metabolic patterns of informal urban settlements. *Energy Policy, 100*(Supplement C), 377–386. https://doi.org/10.1016/j.enpol.2016.06.047

Lombard, M., & Rakodi, C. (2016). Urban land conflict in the Global South: Towards an analytical framework. *Urban Studies, 53*(13), 2683–2699. https://doi.org/10.1177/0042098016659616

Lucci, P., Bhatkal, T., & Khan, A. (2018). Are we underestimating urban poverty? *World Development, 103*, 297–310. https://doi.org/10.1016/j.worlddev.2017.10.022

Marutlulle, N. K. (2017). Causes of informal settlements in Ekurhuleni Metropolitan municipality: An exploration. *Africa's Public Service Delivery & Performance Review, 5*(1), 11. https://doi.org/10.4102/apsdpr.v5i1.131

Marx, B., Stoker, T., & Suri, T. (2013). The economics of slums in the developing world. *Journal of Economic Perspectives, 27*(4), 187–210. https://doi.org/10.1257/jep.27.4.187

McLeod, S. (2016). *Dharavi slum, Mumbai, India 63/365 #project365 #365 #dharavi #dharavislum #mumbai #india* [Photo]. https://www.flickr.com/photos/mcleod/25487732025/

MIIC. (2016). *Arabic Republic of Egypt: National review sustainable development goals*. UNDP in Egypt. https://www.eg.undp.org/content/egypt/en/home/sdgoverview/post-2015-development-agenda.html

Mohit, M. A. (2012). Bastee settlements of Dhaka City, Bangladesh: A review of policy approaches and challenges ahead. *Procedia—Social and Behavioral Sciences, 36*, 611–622. https://doi.org/10.1016/j.sbspro.2012.03.067

Moore, M. H. (2000). Managing for value: Organizational strategy in for-profit, nonprofit, and governmental organizations. *Nonprofit and Voluntary Sector Quarterly, 29*(1_suppl), 183–204. https://doi.org/10.1177/0899764000291S009

Munir, M. H. (2019). *SDGs and Bangladesh: Progress, challenges and missing links*. Social Watch. https://www.socialwatch.org/node/18286

Naceur, F. (2013). Impact of urban upgrading on perceptions of safety in informal settlements: Case study of Bouakal, Batna. *Frontiers of Architectural Research, 2*(4), 400–408. https://doi.org/10.1016/j.foar.2013.06.004

Nakamura, S. (2014). *Impact of slum formalization on self-help housing construction: A case of slum notification in India—Shohei Nakamura, 2014*. https://journals.sagepub.com/doi/abs/10.1177/0042098013519139

Nolan, L. B. (2015). *Slum definitions in urban India: Implications for the measurement of health inequalities*. https://www.ncbi.nlm.nih.gov/pmc/articles/PMC4746497/

Panek, J., & Sobotová, L. (2015). Community mapping in urban informal settlements: Examples from Nairobi, Kenya. *The Electronic Journal of Information Systems in Developing Countries, 68*, 1–13.

PDP. (2017). *Why informal areas?* Participatory Development Programme in Urban Areas, GIZ Egypt. https://www.egypt-urban.net/1568/, https://www.egypt-urban.net/why-informal-areas/

PSUP. (2016). *Slum almanac 2015–2016*. UN-Habitat. https://unhabitat.org/slum-almanac-2015-2016/

Rashid, S. F. (2009). Strategies to reduce exclusion among populations living in urban slum settlements in Bangladesh. *Journal of Health, Population and Nutrition, 27*(4), 574–589. https://www.ncbi.nlm.nih.gov/pmc/articles/PMC2928105/

Rockström, J., & Sukhdev, P. (2016, June 14). *How food connects all the SDGs*. Stockholm Resilience Centre [Text]. https://www.stockholmresilience.org/research/research-news/2016-06-14-how-food-connects-all-the-sdgs.html

Roorda, N. (2017). *Fundamentals of sustainable development* (2nd ed.). Routledge.

Sachs, J. D. (2012). From millennium development goals to sustainable development goals. *The Lancet, 379*(9832), 2206–2211. https://doi.org/10.1016/S0140-6736(12)60685-0

Schouten, M. A. C., & Mathenge, R. W. (2010). Communal sanitation alternatives for slums: A case study of Kibera, Kenya. *Physics and Chemistry of the Earth, Parts A/b/c, 35*(13), 815–822. https://doi.org/10.1016/j.pce.2010.07.002

Shatkin, G. (2013). *Contesting the Indian city: Global visions and the politics of the local*. Wiley.

Shawkat, Y. (2015). *Government failure to upgrade informal settlements in Egypt: A brief history*. openDemocracy. https://www.opendemocracy.net/north-africa-west-asia/yahia-shawkat/government-failure-upgrade-informal-settlements-egypt-brief-history

Siddique, A. (2017). *Bangladesh's urban underbelly a cause for concern*. Dhaka Tribune. https://www.dhakatribune.com/bangladesh/2017/10/30/bangladeshs-urban-underbelly-cause-concern

Smit, S., Musango, J. K., Kovacic, Z., & Brent, A. C. (2017). Conceptualising slum in an urban African context. *Cities, 62* (Supplement C), 107–119. https://doi.org/10.1016/j.cities.2016.12.018

Space Syntax. (2010). *Slums and informal settlements: An evidence-based approach to sustainable upgrading and development*. Space Syntax. https://spacesyntax.com/media/books-brochures/

Srinivas, H. (2015). *Defining squatter settlements*. GDRC Research Output. https://www.gdrc.org/uem/squatters/define-squatter.html

The Daily Star. (2009, June 20). *Urban slums of Bangladesh*. The Daily Star. https://www.thedailystar.net/news-detail-93293

The Global Goals. (2019). *Goal 11: Sustainable cities and communities*. The Global Goals. https://www.globalgoals.org/11-sustainable-cities-and-communities

Turkoglu, H. (2015). Sustainable development and quality of urban life. *Procedia—Social and Behavioral Sciences*, *202*(December 2014), 10–14. https://doi.org/10.1016/j.sbspro.2015.08.203

Ukko, J., Saunila, M., Rantala, T., & Havukainen, J. (2018). Sustainable development: Implications and definition for open sustainability. *Sustainable Development*. https://doi.org/10.1002/sd.1904

UNDP. (2019). *Bangladesh: Transforming life in the slums*. UNDP. https://www.undp.org/content/undp/en/home/ourwork/ourstories/bangladesh–how-to-transform-life-in-the-slum.html

UN-Habitat. (2003). *The challenge of slums*. Global Report on Human Settlements. https://unhabitat.org/books/the-challenge-of-slums-global-report-on-human-settlements-2003/

UN-Habitat. (2006). Slum trends in Asia. In *Asia-Pacific Ministerial Conference on Housing and Human Settlements*. https://mirror.unhabitat.org/documents/media_centre/APMC/Slum%20trends%20in%20Asia.pdf

UN-Habitat. (2008). *The state of the African cities report 2008*. United Nations Human Settlements Programme (UN-HABITAT). https://unhabitat.org/books/the-state-of-the-african-cities-report-2008/

UN-Habitat. (2012). *Uganda national urban profile | Slum | Uganda*. Scribd. https://www.scribd.com/document/100388437/Uganda-National-Urban-Profile

UN-Habitat. (2015). *Sustainable urban development in Africa*. UN-Habitat. https://unhabitat.org/sustainable-urban-development-in-africa/

United Nations. (2019). *Report of the secretary-general on SDG progress 2019 special edition*. https://sustainabledevelopment.un.org/content/documents/24978Report_of_the_SG_on_SDG_Progress_2019.pdf

Wekesa, B. W., Steyn, G. S., & Otieno, F. A. O. (Fred). (2011). A review of physical and socio-economic characteristics and intervention approaches of informal settlements. *Habitat International*, *35*(2), 238–245. https://doi.org/10.1016/j.habitatint.2010.09.006

World Commission on Environment and Development. (1987). *Report of the World Commission on environment and development: Our common future*. A/42/427 Annex—UN Documents: Gathering a body of global agreements. https://www.un-documents.net/wced-ocf.htm

New, Smart, and Sustainable Cities in Algeria: Realities and Challenges. Case of the City of Sidi Abdellah in Algiers Region

Saouane Med Boudiaf and Zeghichi Sarra

Abstract

Algerian metropolitan cities in particular, the capital of the country Algiers, have a remarkable impact on the back of the country or even the entire territory of Algeria. Indeed, Algiers continues to attract a large flow of investment and people and goods. It knows also programming and implementation of many major development projects that highlight the image of the entire nation. However, in order to reduce the pressure on this city and satisfy its current and future needs, the strategy adopted for the coherent development of its urbanization aims to seek solutions in its immediate and distant environment (in its metropolitan area and its region), within the framework of double scales: its intercommunal (intermunicipal) master plan and its regional membership. As a solution envisaged consists of the creation of new urban centers of support, able to participate in the economic strengthening of the Metropolitan potential of Algiers and to take charge of its social and residential dimensions. The new city is Sidi A. Allah as the first urban relay is part of this strategy of development of the capital Algiers. It is located 30 km from the capital and covers an area of 3000 ha, located in the urban perimeters of the towns Mahelma and Rahmania fixed by their master plans. This city aims, as originally planned through studies, to be a smart city that contains in addition to the function of residence, and at the same time medical, biotechnology, university, research and technological projects (Information and Communication technology and advanced technologies: ICT & AT), conceived in a perspective of sustainable development. However, for a decade, this ambitious project, still under construction, is lagging a delay comparing to what was planned in the context of the studies. And as a result, a review of what has been done to date is needed. This article explains, first, how and to what extent the project of this new city will reduce, in practical terms, the pressure on Algiers and to achieve, at the same time, the perennity and the sustainable development of the urban space of Sidi A. Allah through the critical reading of the superstructure and infrastructure programs conceived for the city. In other words, may Sidi A. Allah be considered as a sustainable city at a time when shortcomings persist, not meeting the international standards set for sustainable cities. In the second step, the present paper highlights the real causes responsible of the delay in the implementation of this project and proposes, at the end, recommendations to remedy the studied situation.

Keywords

Strategies of coherent development of the urbanization of Algiers • Scales of belonging of the territories • Relationship of Algiers • New city of Sidi A. Allah • Sustainable urban project in the new smart city

1 Introduction

Since the Iron Age, man has not stopped making progress until you can skillfully master all the factors of production.

The evolution of the man who went from farming market capitalization, industrialization, and specialization of work has led to the growth of modern cities which, although having had positive repercussions, also caused side effects. Socioeconomic growth exploded to spectacular levels, but alas, also had an unprecedented environmental cost. The population moves to cities hoping to find better chances in terms of education and better-paying jobs.

In some countries, rapid industrialization leads to migrations of people in search of better-paid jobs, from rural areas to urban areas. This trend has started a few years ago and should last at least until 2050. If this situation exacerbates the problems in numerous urban areas, it also offers

S. M. Boudiaf (✉) · Z. Sarra
University of Tébessa, Tebessa, Algeria

F. Trapani et al. (eds.), *Advanced Studies in Efficient Environmental Design and City Planning*, Advances in Science, Technology & Innovation, https://doi.org/10.1007/978-3-030-65181-7_18

urban planners opportunities to create new cities from scratch new neighborhoods.

So far, urban infrastructure and the integration of information and communication technology (ICT) in this infrastructure have evolved piecemeal to meet the needs of "internal growth" when villages transformed into cities and then metropolises, housing an ever-increasing population. Each new building or group of buildings was thus designed and constructed at a time difference.

The strategy to build smart cities and proposed by the specialized group redefines planning and building city infrastructure smart as well as service delivery, the participation of citizens, and the interconnection of systems. The objective is to transform cities into more living environments, sustainable, smart, strong, and resilient, taking into account the disaster resistance and the need to cut gases to protect against crime and to guarantee cybersecurity.

Algeria, and like other countries around the world, has invested in this area, through the creation of smart, sustainable world-class cities that will provide cutting-edge infrastructure and services to businesses and institutions. It is a real benchmark in the field of buildings called "smart and sustainable." The creation of Sidi Abdallah the first smart and sustainable city in Algiers, is part of the national strategy to build an information society and accelerate Algeria's transition to an information and communication technology economy. This is a strategic project of national scope, and it is both a trigger for the ICT sector and a driver of innovation. The city of Sidi Abdallah contains a set of world–class amenities that combines the latest architectural, urban, technical and technological methods in sustainable development, the anticipated performance of these buildings, in terms of energy efficiency (heating, lighting and air conditioning), the preservation of natural resources (water, soil, and vegetation) and the promotion of recycling, make it possible to envisage monetary savings this is what this city should have been thanks to studies by the authorities concerned, but the truth is something else.

2 The Problem of Research

This article explains, first, how and to what extent the project of this new city will reduce, in practical terms, the pressure on Algiers and to achieve, at the same time, the perennity and the sustainable development of the urban space of Sidi A. Allah through the critical reading of the superstructure and infrastructure programs conceived for the city. In other words, may Sidi A. Allah be considered as a sustainable city at a time when shortcomings persist, not meeting the international standards set for sustainable cities.

In the second step, the present paper highlights the real causes responsible of the delay in the implementation of this project and proposes, at the end, recommendations to remedy the studied situation.

3 Research Objectives

- The objective of this research is to assess the success of the city experience on an urban scale, the reliability, the extent of the city's response to the concept of urban sustainability, adaptability, and ease of use in terms of operation or programming.
- Highlight the differences, or the gap between conception and experience, concerning the technology in architecture, as well as proposing solutions to overcome this gap.

4 Research Importance

The research contributes to filling the gaps in the planning and practice of the process which lack the characteristics of sustainability and technology in the field of urban development in Algeria.

5 Methodology

It is the main method that was adopted in the study with the aim of monitoring the dimensions and criteria of the problem or the topic studied in an attempt to understand it, provide an illustrative image through an accurate description of the problem of the studied topic in an attempt to understand it, and provide an illustrative image through an accurate description of the study or topic problem in a quantitative or qualitative form within a specific time period.

One of the advantages of this approach is to provide the researcher with a good amount of basic information and data that paint a general picture of the problem and help other researchers to choose research that they consider worthy of study.

Field Descriptive Method

It is an auxiliary method of the descriptive analytical approach concerned with specific areas and not others in order to cover the shortcomings or provide clarifications about specific cases highlighted by the numbers, so it deals with a phenomenon that is not generalizable to all human societies and all times because the influences that touch the studied phenomenon are subject to diversity and change.

He uses interviews to collect information that the literature or the form could not provide or explain some aspects of the problem.

This approach relies on providing an analytical description of the variables or effects that touch the problem under study.

6 Previous Studies

The development of smart and sustainable cities in Algeria took place at the end of the nineteenth century, and it is the first experiment in Algeria in this field, the city of Sidi Abdullah, and there are no previous studies that classify this city. Is it considered a smart and sustainable city, but many publications consider it a stupid city because it came in many areas, including Le Monde magazine in a press release. Sir Abdellah is a "stupid city" in Algeria.

7 Description of the Study Area

In the present study, we are investigating in the field, using a mainly qualitative method. The particularity of this method lies in the fact that the researcher is involved in a relational interaction with the members of the sample, who constitute for him the source of the truth that he is looking for qualitative study.

Like all scientific research proceeds in four transdisciplinary stages: definition of the problem, methodical choice, collection of the data, and finally analysis of the data and interpretation of the results.

In this title, we first present the characteristics of the environment of the case study than the different stages of an investigation, through the choice of techniques for collecting and using data.

7.1 Presentation of the City of Sidi Abdellah

A few 30 km from Algiers, the New Town of Sidi Abdellah offers the ideal site to allow to engage the new strategy implied by the coherent and controlled urbanization of the Algéroise region (metropolitan area of Algiers). The city is projected on an ideal site for the implementation of the new strategy of spatial planning. Without major agricultural constraints, this site is well located with the transport infrastructure, namely the motorway link (South Ring Road) between Boudouaou and Zéralda and the Zéralda—study of Sidi Abdellah.

The city is projected on an ideal site for the implementation of the new strategy of spatial planning. Without major agricultural constraints, this site is well located with the transport infrastructure, namely the motorway link (South Ring Road) between Boudouaou and Zéralda and the Zéralda—Sidi Abdellah and Birtouta rail link. Two peripheries North and West surround the new city and allow the connection between the Ring 1 (Zéralda—Dar El Beida), the Ring 2 (Zéralda—Boudouaou), and the National Route 67.

The perimeter of the New Town extends over an area of 7000 ha including 3000 ha buildable areas including the communes of Mahelma, Rahmania, and part of the municipalities of Zéralda, Douéra, and Souidania. It will eventually house a population of 350,000 with the creation of 80,000 jobs.

The new city must at the same time participate in national economic development by strengthening the weight of the advanced industrial sectors.

In addition to the integration of existing villages and the establishment of an innovative urban model, the new city contributes mainly to the loosening of the metropolitan area in favor of the periphery and the establishment of a polycentric organization (Fig. 1).

7.2 The Physical Environment (Climate and Precipitation)

The region of Sidi Abdellah is characterized by a sub-humid mediterranean climate. The climatological observation station "Mahelma–Ferme" (code 02-05-11, altitude 145 m, coordinates X: 517.0, $Y = 376.85$, and Z: 150) thus possesses data on precipitation, temperatures, and winds. According to the monthly precipitation regime, all seasons are likely to receive rain; however, the months of October, November, December, and January are the wettest. Winter is the rainiest season with over 40% of total precipitation, and summer is dry with a proportion of 4% of these. The average interannual rainfall is of the order of 60 mm with a total of 80 days of a rainy day (Figs. 2 and 3).

7.3 Seismicity

The seismic map indicates that the whole sector of Algiers is classified as. The study area fits into this context and is, moreover, surrounded by a set of faults. It is located near several epicenters including Nador, to the west, and that of El Affroun to the south, which is highly unstable areas seismically. There are also flexures at the limit of the Sahel, which can influence earthquakes. As for gully dynamics, it is translated in the field by forms that are generated, either by concentrated runoff or by diffuse runoff. This dynamic is expressed on almost the entire hilly area and is favored (Fig. 4).

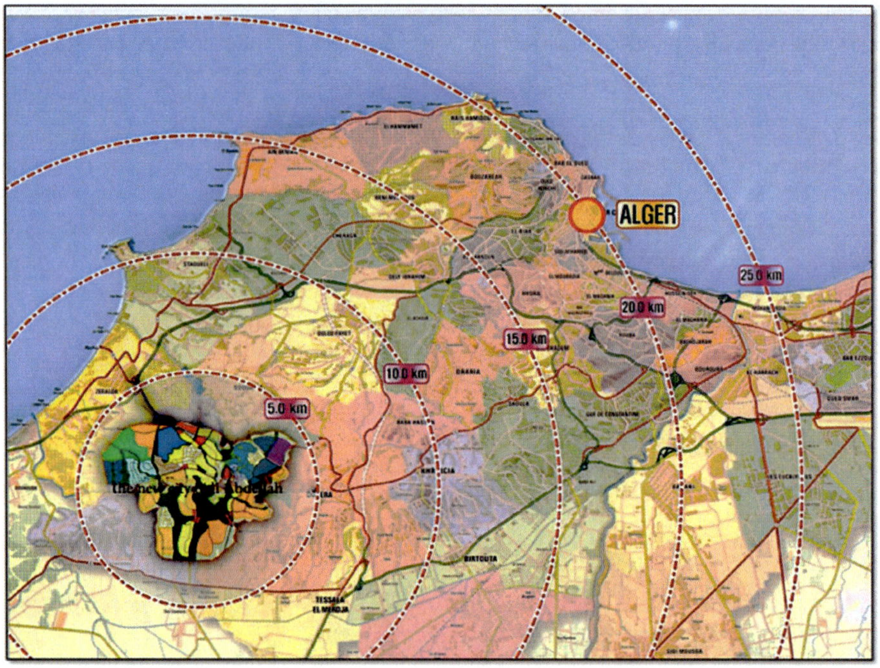

Fig. 1 Plan 01 of situation of the new city of Sidi Abdellah. *Source* Public institutions for the development of the new city of Sidi Abdellah

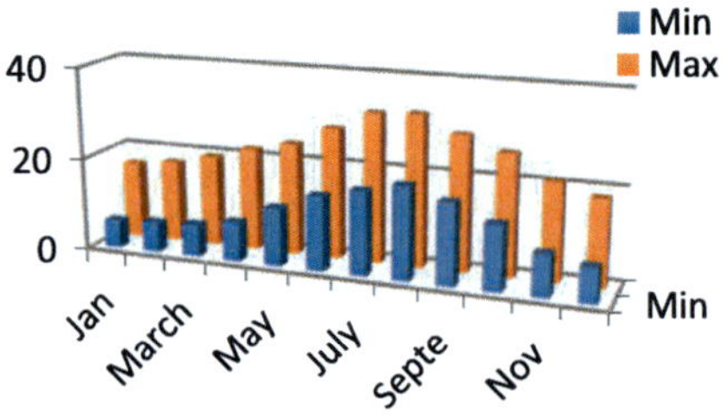

Fig. 2 Average temperatures. *Source* Baraki station hydrographs 2018

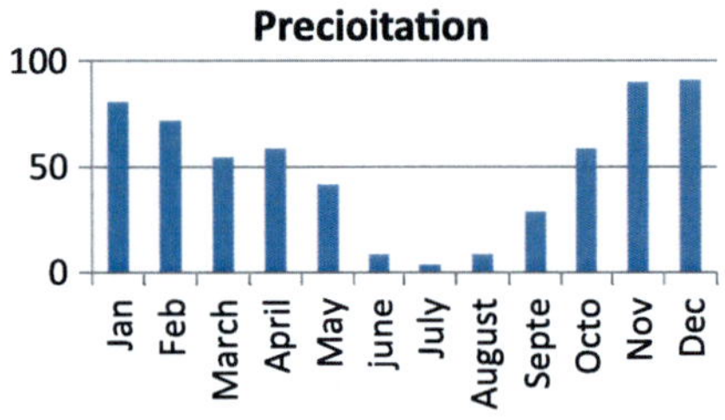

Fig. 3 Annual distribution of precipitation. *Source* Baraki station hydrographs 2018

7.4 The Flora

- The study area is characterized by a hilly landscape with a little forest. The forest area is about 800 ha. At the level of Mahelma commune, we find the Mahelma forest, and the urban and suburban woods. In addition to urban and suburban woods, Rahmania is home to Bousalem Nord and Sud woods.
- The natural landscape of the site is mostly low hills. The most accentuated reliefs are located north of Mahelma and south of Rahmania. The landscape is structured by a ridgeline delimiting a narrower northern part facing the sea and a more rugged southern part facing the Mitidja.

8 The Definition of the Development Plan Related to the Vocation of the New Town

The New Town is conceived around a concept that seeks to harmonize the urban and the landscape: to treat the great natural spaces of the New Town (the landscape) as an inseparable component of the existing urban fabrics and future neighborhoods. The urbanistic party adopted for the new city consists of linking the urban and the landscape. From the beginning, the character of Sidi Abdellah was defined by the search for harmony with the landscapes turned toward the mediterranean coast, centered on the hills and open on the Mitidja and Atlas Blidéen. The concept then results from a work developed on two large entities of the New Town: the northern part, descending toward the sea and the southern part, hanging on the hillsides of the Central Park. The landscape thus asserts itself as a major characteristic of cohesion between neighborhoods and as the central axis of a voluntarist strategy of urban development

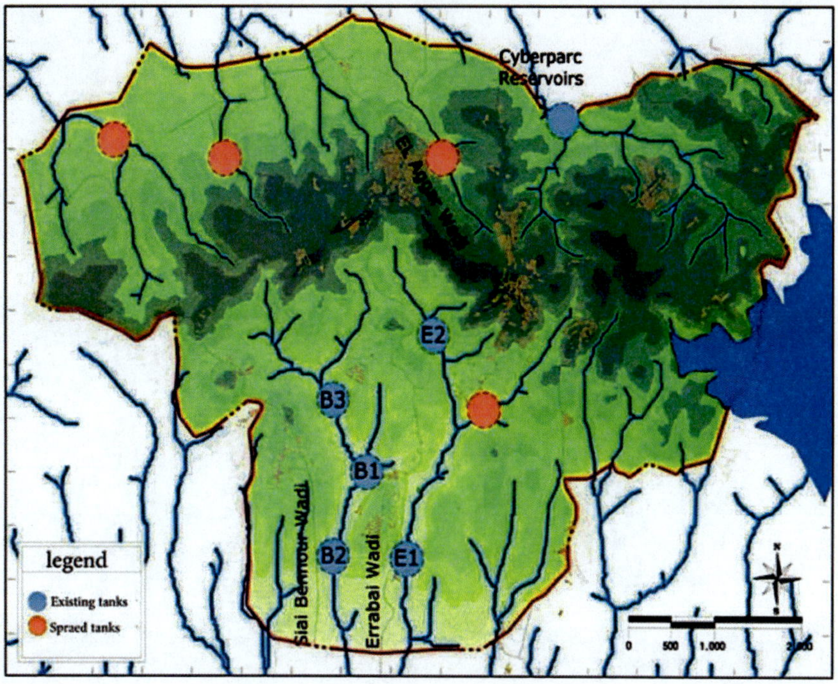

Fig. 4 Map wadis and hollows. *Source* Public institutions for the development of the new city of Sidi Abdellah

content "in the insular form." The concept thus builds on this concept to take into account the principle of urban compactness by building new neighborhoods, poles, business parks, sports or health parks, and clusters energizing the city. The preserved green spaces then constitute limits of urbanization and places of leisure for all. By its very nature, this project is opposed to development in the form of oil and an agglomeration without discontinuity. It requires a special effort from the entire community to preserve the green fabric by integrating it into a network that does not disfigure the landscape but on the contrary strengthens it. Besides, this approach reinforces networking or networking at the city scale. Thus, Sidi Abdellah marries two urban conceptions that hybridize uniquely by the archipelago concept. Urban planning concepts are based on an ecological approach to the mobility of cars and public transport as the central solution to urban mobility. In Sidi Abdellah, the widths of the tracks (35 m for the Avenue Magistrale) attenuate the place of the car and the cultural and social weight that it symbolizes in the Algerian society.

At the same time, public transport is planned from the outset to promote alternative and ecological methods for all social categories. The territory covers the urbanization perimeter: All neighborhoods, functions, and facilities are concerned. For reasons of coherence, unity, and connectivity, the infrastructure and networks of the management plan (primary and secondary roads, and technical networks) are designed and organized on this scale. The New Town is organized in structuring neighborhoods with a wide variety of functions, namely: teaching, research, industry, culture, recreation, health, and housing (Fig. 5).

9 The General Program of the City

The land structure of the site can be summarized as follows (Table 1).

9.1 The Residential Functions

This is to accommodate the planned population (200,000 inhabitants) in the best conditions and based on a balanced distribution at the project level. It is the match between population and jobs that will reduce travel distances and encourage the use of public transport modes. The basic principle is to develop residential neighborhoods (high standing in particular) in spaces offering the most beautiful views and perspectives. High-quality programs are primarily located in sites oriented toward the Mediterranean and Lake Douéra. The quality of the residential environment is linked to the upgrading of infrastructure and equipment of existing villages. Besides, concentrations of public facilities at the hinge between residential neighborhoods strengthen the functional links. The development of residential neighborhoods is planned by minimizing volumes of earthworks to adapt the programs to the natural topography (stepped installations for example). On the one hand, standard

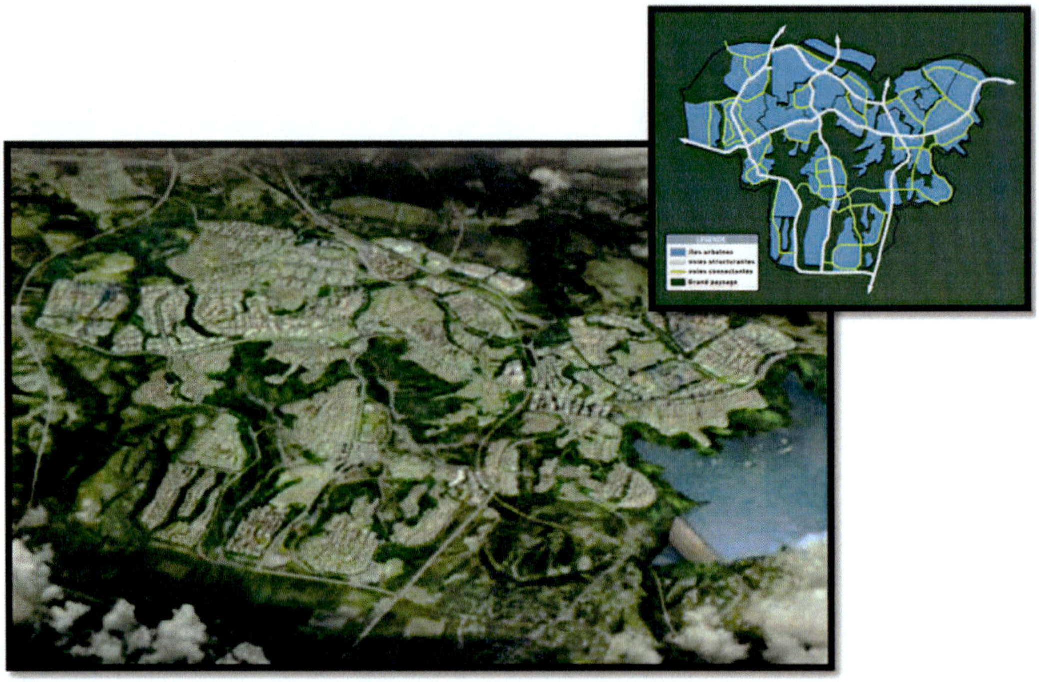

Fig. 5 Plan the development plan of the New Town of Sidi Abdellah. *Source* Public institutions for the development of the new city of Sidi Abdellah

Table 1 Current state of the properties

Ranking	Area (ha)	Proportion (%)	
Public and private land acquired by VNSA	309	10.3	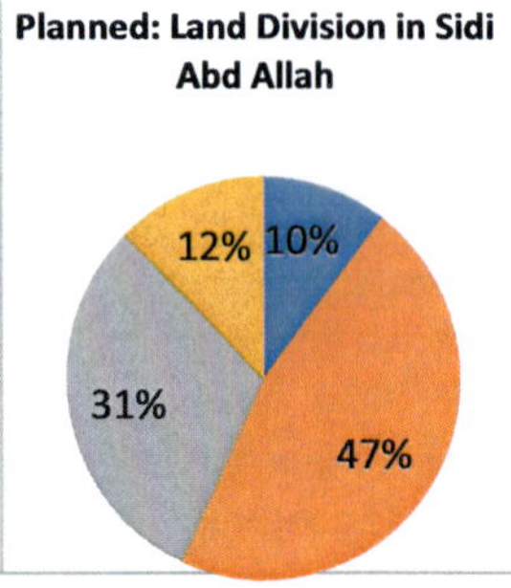
Public lands to acquire	1400	46.7	
Private properties to acquire	919	30.6	
Existing urban units, cyberparc, infrastructures	372	12.4	
Total	3000	100	

Source EPIC VNSA—Juillet 2009

residential spaces are located near valleys and hillsides. Regarding existing villages, residential and social diversity is favored in the context of the complete requalification of fabrics and infrastructures.

Pole Residential

The New Town is divided into 4 structuring neighborhoods organized around the 5 centralities. These include businesses, services, and other tertiary functions. The distribution of neighborhoods also takes into account the natural environment (topography and valleys) and existing villages, thus respecting the structure of the territory (Table 2).

9.2 Functions of the New City

Structural poles of the New Town of Sidi Abdellah are shown in Table 3 (Fig. 6).

Table 2 Public housing programs of the New Town of Sidi Abdellah

Operator	Program consistency	Delivered to date	Remains to deliver
AADL	46,100	10,210	35,890
ENPI	9596	2267	7329
OPGI H.DEY	1000	1000	
Total	56,696	13,477	43,219
Rental and sales program (LV)		46,100 accommodation rent–sale	
Public promotional program (LPP)		6596 public housing promotional	

Source Directorate of Sidi Abdellah, Department of Follow-up Works, April 2018

Table 3 Structural poles of the New Town of Sidi Abdellah

Poles	Number of the project	Area of the pole
Pharmaceutical and biotechnology cluster	Public projects: 05 projets Private projects: 41 projets	83 ha
Health center	Public projects: 01 project Private projects: 07 projects	27.61 ha
Pole of attraction and leisure	Public projects: 2 projets Private projects: 4 projets	852,323 m^2
Heart of city	Public projects: 06 projets Private projects: 01 projet	89.31 ha
Security pole	Public projects: 6 projets	630,000 m^2
Commercial pole	Public projects: 06 projets	210,000 m^2
University center	Public projects: 13 projets	1,190,000 m^2
Technology center	Public projects: 02 projets Private projects: 08 projets	60.2 ha

Source Directorate of Sidi Abdellah, Department of Follow-up Works, April, 2018

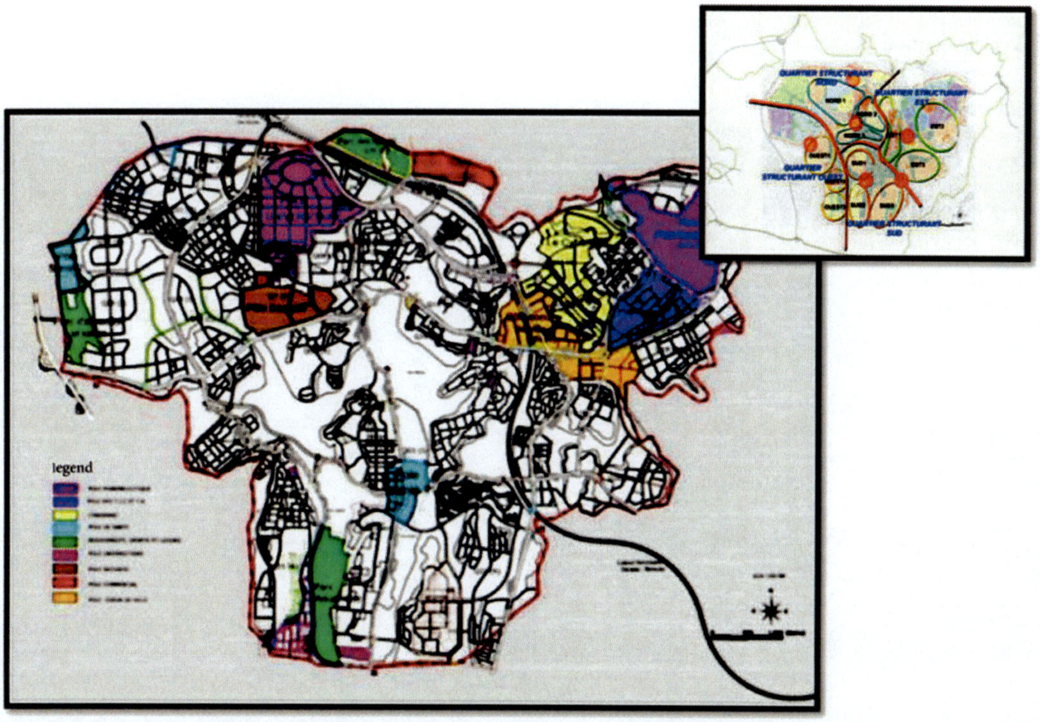

Fig. 6 Structural poles of the New Town of Sidi Abdellah. *Source* Public institutions for the development of the new city of Sidi Abdellah, April 2018

10 The Organization and the Urban Structure

10.1 Accessibilities

The New Town of Sidi Abdellah is part of an articulated network. It is accessible by the ring road 1 "Tipaza–Dar El Beida" currently the main axis of agglomeration in east–west. Rocade 2 Zéralda–Boudouaou will be directly connected to the New Town by the Northern Periphery. The western ring road will connect ring road 1 to national road 67 (Kolea–Birtouta). Besides, facing east–west (Douéra–Zéralda), National Highway 63 is the main structuring axis. It is planned to integrate it into the development plan and to provide it with an efficient public transport system. The Wilaya Path 112 is the other main axis that connects Mitidja to the coast. Its route will also be widened and optimized to improve the quality of north–south urban links. It will also be served by the railway line of a linear 22 km (double electrified track) linking Zéralda, the New Town of Sidi Abdellah to Birtouta: Three stations are programmed inside the urban perimeter.

10.2 Drinking Water Supply

The connection to the drinking water network of the housing and equipment programs of the Ville Nouvelle will be made from the following three (03) sources.

Infrastructures and Utilities

Seawater desalination–

The Daouda plant, which desalinates sea water with a capacity of 120,000 cubic meters, of which 65,000 cubic meters is destinated for the new city of Sidi Abdullah, and the rest will ensure the enhancement of drinking water of the localities with Coastal chains and highlands linking the Boukerdene and Bouroumi dams.

The three springs will supply the 65,000 m^3 capacity of the Sahel complex, which will be spread over the 05 hydraulic complexes located throughout the New Town of Sidi Abdellah.

10.3 The Sanitation

- **Treatment plants**
  - The New Town of Sidi Abdellah is divided into 2 north and south watersheds, for which two purification stations are planned, namely: the north wastewater treatment plant with a capacity of 200,000 population equivalent for an average flow of 30,000 m^3/day.
  - The south wastewater treatment plant with a capacity of 150,000 population equivalents for an average discharge of 25,000 m^3/day.
- **The Networks (Fig. 7)**

  The sanitation network of the New Town of Sidi Abdellah is of the separative type:
  - The rainwater network.
  - The wastewater network.

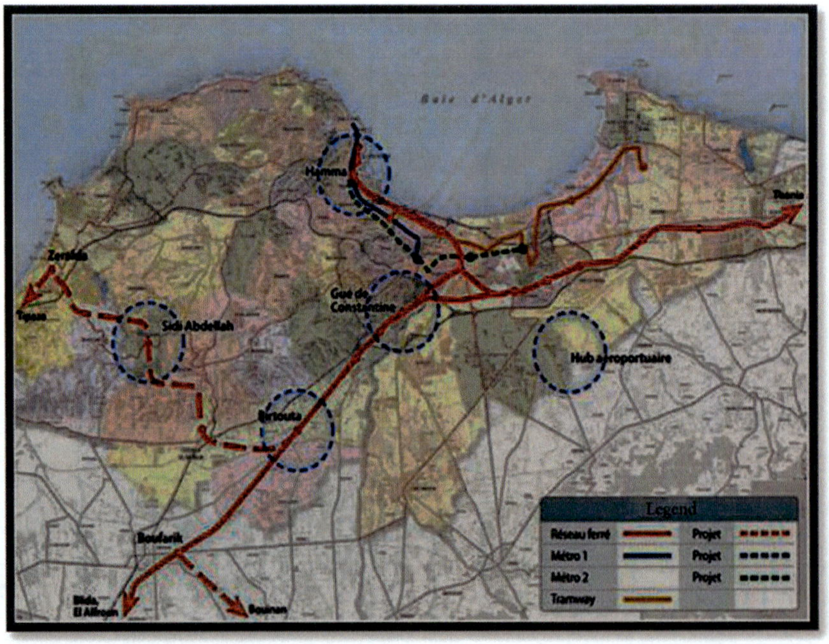

Fig. 7 Map railway network and projects in progress. *Source* Public institutions for the development of the new city of Sidi Abdellah, April 2018

10.4 The Power Supply

• **Electricity**

The new city of Sidi Abdellah will be fed by a powerful network connected to the existing network of Ouled Fayet for the northern part and the existing Boufarik network for the southern part.

• **The gas**

The future gas distribution network divides the site into two (02) supply areas each with one substation: two HP substations in the north to connect to the existing gas pipeline from Souidania. and an HP substation in the south to connect to the existing gas pipeline from Boufarik.

• **Telecommunications**

As a city of excellence in information and communication technologies, Sidi Abdellah must play a leading role in Thiès field. It is planned to create relay stations for mobile communication services, five of which will be installed, as well as ubiquitous communication networks (ICT and TA networks).

Currently, 11 km of optical fibers has been installed (Zéralda–Sidi Bennour).

• **The Treatment of Household Waste**

As part of the household waste treatment, there is a technical landfill on the perimeter of protection with a capacity of 10 million m^3. This equipment, being developed on an area of 90 heetares, is consists of 07 boxes, sorting area and a leachate treatment plant. It should promote an efficient and environmentally friendly processing. Expectations of implementating a program with the completion of the new city.

10.5 The Phasing Plan of the New City (Table 4)

The complete development of the new city in all its components (activities, housing, basic infrastructure, green spaces) is set for the 2030 horizon in accordance with the objectives of the National Plan of Spatial Planning. The period of realization of the investments will start from the execution of the works of viabilization of the new city. Developing a gradual development action plan by following a development plan in four general stages.

11 Analysis of the Results of the Study Area Description

The preliminary investigation took place in two simultaneous stages. A spontaneous step concerned the collection of data on the inventory in terms of spatial configuration (urban file), regulation of comfort factors (interviews with managers and technicians), environmental behavior (observation), and this through guided tours. A documentary and iconographic collection have been created for the city which is the subject of a case study. It includes location maps and technical sheets relating to the city's comfort facilities.

This allowed us to establish the first contact with the population of the city and to question them and questioning the city managers on the indicators of city sustainability and the use of technology, and it is this tool that interests us in this section.

In the second step, we used the inductive method to extract the most important foundations and principles.

Through the theses of the United Nations and organizations that support sustainability, a previous study was published on this subject (a mechanism to achieve urban sustainability through the proposals of the United Nations 2017). These principles have been applied to the city, Sidi Abdellah, in Algiers, to verify the most important sustainability indicators that have been applied. Finally in the city, we looked at some of the technological factors used in the city and assessed whether it is considered a smart city according to the international biography of smart cities.

11.1 The Extent to Which the City of Sidi Abdellah Has Reduced the Pressure and Congestion of the Capital

The city of Sidi Abdullah will contribute to the alleviation of the pressure and congestion, especially after the completion of work in this city by 2030 by housing some 200,000 citizens who were in the capital and provide them with facilities crisis of services and entertainment facilities and provide about 60 thousand homes and provide 85 thousand jobs and decomposition Traffic congestion on the road of the transport line by electric train connecting the capital in the city of Zeralda, passing through the city of Sidi Abdullah in many relays with a capacity of 50 thousand passengers a day, and you can say that this city is a strategy of development strategies Algeria.

11.2 Assess the Extent to Which the Program Is Consistent with the Concept and Principles of Sustainability

The research explains how and to what extent the project of this new city will in practice reduce pressure on Algeria and, at the same time, realize the legacy and sustainable development of urban space in Sidi A. God through the critical reading of superstructure and infrastructure programs that were designed for the city. In other words, Sidi Abdellah can be considered a sustainable city at a time when shortcomings continue, not meeting the specific international standards of

Table 4 Phasing plan of the new city

Stages of urban and urban fabric development (2016–2019)	
Realization of the extension of basic infrastructures relying on investment projects and neighborhoods Public housing as well as primary and secondary viabilities Completion of the technological centers Completion of residential neighborhoods	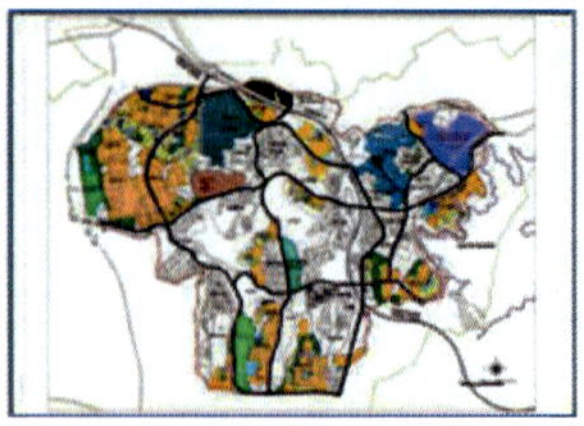
Completion phase (2020–2023)	
Finalization of the development by investments mixed (public and private) to complement housing programs and equipment and the upgrading of existing villages	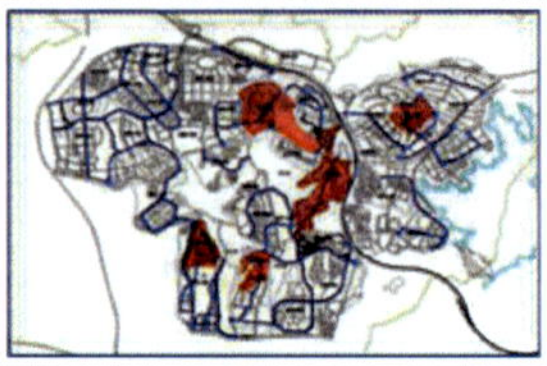
Urban consolidation phase (2024–2025)	
Finalization and reinforcement of the functions developed in the previous phases System of management and balanced development of the city Finalization of the landscape treatment	

Source EPIC VNSA—July 2018

sustainable cities. To achieve this, we have experimented on several samples from the city of Sidi Abdellah. This is an experiment to know the most important principles of sustainability and the answer to the question is Sidi Abdella sustainable city and to know the extent of the city's response, Sidi Abdellah for the concept of the smart city. This experiment was conducted on the Egyptian cities on the mechanism of achieving urban sustainability. In 2017 we returned this experiment to the city of Sidi Abdullah, but the linomplete works in this project we took samples from them. The samples were for the city where they work. They were completed and applied to the following conditions for conducting the experiment and these samples include the following conditions:

- The site should have different population centers.
- The completion of works on this site up to 80%.
- The completion of the implementation of facilities for the site studied by 80%.
- The services are activated for this site.
- The current resident population is about 25% of the population of the city Sidi Abdellah.
- These samples were connected to different networks (electricity communication and sewage water)—review the most important principle of sustainability in these samples (respect for location—planning—building heights—and energy and water consumption and green areas).

- **Respect the site**: This city did not succeed in respecting the site, especially since this project was completed on agricultural lands.
- **Respect urban planning**: Yes, this city respected the urban planning in what contributed to the ease of mobility between the facilities and meet the needs of the citizen.
- **Water and electric power**: The population is fed by underground water, and there is no recycling of this water.

As for the electrical energy in the outdoor areas, there are solar panels above the electric poles, while the residential housing is fed in the traditional way (Fig. 8).

- **Respect the height of buildings**: Yes, the buildings in these samples are respected for what helps in good ventilation and entering the sun (Fig. 9).

- **Building materials**: Most construction materials are local materials such as cement and iron, but these materials are unsustainable and have environmental impacts.
- **Green spaces and facilities of residential neighborhoods**: The site is located on large green areas within the population and also the outer perimeter of these neighborhoods. You find facilities for neighborhoods such as entertainment center, schools, and services (Fig. 10).

Fig. 8 Plan village of Sidi Abdellah. (a) *Source* EPIC VNSA—July 2018. (b) *Source* https://alger-roi.fr and Google Earth 2018

Fig. 9 In these pictures, we observe respect for the heights and the presence of solar panels above the electric poles. *Source* Researcher December 2018

11.3 Application of the United Nations Principles Which Ensure Urban Sustainability in the Studied Sample (Sidi Abdellah, Algeria)

The United States of America adopted the concept of sustainable development and presented it in this set of indicators that reflects the success of countries in achieving sustainable development and first assess the conditions of countries that can be compared. With developed Countries in the fields of sustainable development if they are on the right track towards sustainable development or if they are still hesitant. Sustainable development indicators focus on the main challenges listed in the Agenda 21 recommendations that make up the global environmental framework.

The United Nations has also submitted several reports on human settlements, providing everyone with a basis for achieving sustainable development, from which you will learn the main indicators related to the urban area.

Many organizations have developed systems and programs to create frameworks capable of applying sustainability in building design or town planning. This is what we are going to do by applying the United Nations urban sustainability indicators and by answering the question of the sustainability of the city of Sidi Abdellah in Algeria.

Visits to relevant sites and devices were carried out in the city to study the foundations and principles of urban sustainability in the USA and to ensure the presence or absence of the reality of the sample studied. Some foundations are verified by the observation of the researcher. Some were verified by talking to the agencies involved and interviewing the residents. Others examine graphics such as green spaces and the distribution of good uses, and we explain them in Table 5.

After conducting this experiment, we found in the samples studied and expected in Sidi Abdellah that the sustainability of this project is 44.44%, according to United Nations indicators. The following figure shows the approximate proportions of the urban indicators extracted from Table 5 (Fig. 11).

Fig. 10 In these pictures, we note the existence of leisure facilities, green spaces, and service facilities. *Source* Researcher December 2018

Table 5 Foundations and principles for sustainability of samples

Substandard	Sustainability	Basic standard	Site	
			Exist	There is no
Atmosphere	Indications maintainer	– Reduction of oxide emission		√
		– No use of zone depleting substances		√
		– Separation of the industrial zone from the population	√	
Freshwater	Conserving water	– Harvesting rainwater and recycle		√
		– Cultivation of how water plants		√
		– Provision of sanitation treatment enclosures	√	
		– Provision of adequate water and sanitation services	√	
		– Use already made materials to reduce water use		√
		– Use of sandy water in toilets and irrigation		√
		The existence of a system to detect water leakage faults		√
Land	– Conservations of the environment	– Protection of available material		√
		– To close the gap between the green agenda and the environment		√
	– Respect the site	– Taking into account the possibilities of the site	√	
		– Consider the soil geology and flad surges	√	
Energy consumption patterns	Use and output power	– The use of verb power (ranking material and architectural treatments)		√
		– The existence of power generation plants		√
		– Use of efficient equipment in order to reduce energy consumption		√
	Sustainable management of materials	– Use of local materials manifested in the site		√
		– Independent recycling of materials		√
		– Recycle materials and construction materials	√	
		– The use of building material with high capacity to endurance	√	
	Sustainable waste management	– Provide recycling plants	√	
		– Use waste materials to energy needs		√
		– Minimize construction waste to a minimum		√
	Use of modern technologic methods	– The production of clean energy in modern methods		√
		– Providing telecommunication services and quality distribution	√	
		– Investment and exploitation of environment resources in scientific way		√
		– The use of clean energy in the industry		√
		– Use of intelligent infrastructure		√

(continued)

Table 5 (continued)

Substandard	Sustainability	Basic standard	Site	
			Exist	There is no
Sustainable places	Use sustainable planing	– Availability of utility and infrastructure networks	√	
		– Improve emergency response and reconstruction		√
		– Provide flushing swabs and water surfaces for air conditioning	√	
		– Good distribution of spaces	√	
	Design and operation	– Continue to improve the standard of living	√	
		– Follow long-term strategies in sustainable planning		√
		– Taking into consideration the rates of growth of the demographic growth of population and future supplies		√
	use of efficiency sustainable	– The use of plants above the roofs of buildings		√
		– Reducing the design and operation cost of building energy consumption		√
		– Internal environmental quality of buildings		√
		– Provide light and moisture meters inside the building		√
	Transport/taking into	– Providing efficient and less noisy transport networks using sustainable transport	√	
		– Providing sustainable public transportation	√	
		– Providing parking	√	
		– Providing alternative transportation methods that effectively reduce vehicle usage	√	
		– Strengthening the link between land and transport		√
Poverty	Consideration of the human	– Providing a good standard of living that is sensitive to human poverty oui—achieving justice and social level	√	
		– Providing job opportunities and reducing unemployment	√	
		– Popular participation in planning in order to know the needs		√
		– Raising awareness of sustainability principles		√
		– Provide good health and education services	√	
		– Achieving justice and the social level	√	
Natural hazards	Element security and safety	Enhancing road safety	√	
		– Strengthening road safety	√	
		Secure infrastructure for transportation	√	
Number of sub-bases		Fifty-four—54		
Number of sub-established bases		Twenty-four—24		
Number of sub-bases not realized		Thirty—30		
Percentage of established bases		44.44%		

Source Researcher, December 2018

11.4 The Extent to Which Sidi Abdellah Has Responded to the Concept of Sustainable Development and Its Relation to the Term Smart City

The scheme of Sidi Abdellah is very different from the reality embodied. The officials who adopted the smart city project installed the name wave only because many of the techniques and standards of intelligence are missing here, as the inhabitants of the other social neighborhoods are experiencing the same problems.

There are no advanced security systems. An alarm and no trace of the digital or ecological city are only monolithic buildings that reflect the size of the housing crisis. We tried to know the city's smart standards and dropped them on Sidi Abdellah. We did not find anything linking the name to the reality. Smart city was a term that appeared in the year 2000, which means either "knowledge" city, research center,

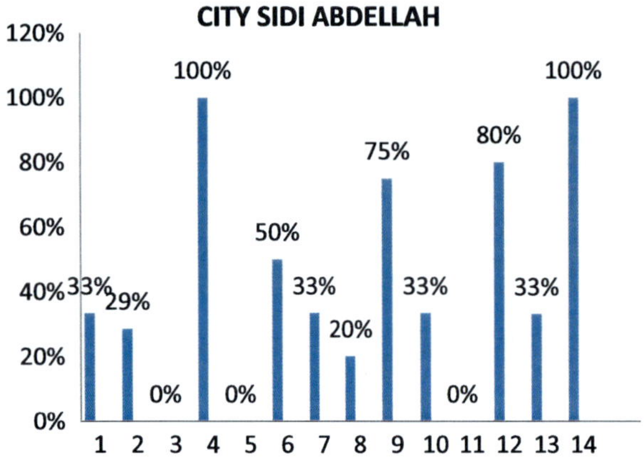

Fig. 11 Percentage of sub-criteria achieved for each major criterion of sustainability for the samples of Sidi Abdellah. *Source* Researcher December 2018

universities or laboratories, planning, and goals. However, the city of Sidi Abdellah was a social housing city aimed at easing the housing crisis only as it seems after a year of inauguration.

The smart city monitors basic infrastructure such as roads, bridges, tunnels, railways, airports, trains, and seaports, as well as communications, water, energy, and major buildings, all to optimize resources and security. Smart city can provide services to citizens to enhance their sense of happiness, health, comfort, and safety, but the first year of the inauguration of Sidi Abdellah did not feel happy residents, but fatigue and exhaustion to obtain the most necessities of life. Sidi Abdella's population-based project does not include sensors to detect and repair the city's water leakage areas, or to monitor air quality to help people with allergies take precautions or use sensors to see that the parking lot is full and guide the driver to other locations.

The stupidity of the city has become apparent, perhaps because of many of the planned works, such as modern residences, a miniature climate, a green belt, 20 km of lake water, the establishment of the fourth biotechnology pole, liver, heart, university mole, and other economic projects under construction. Another decade can be held 2–3. The most important factors and obstacles have hindered the progress of the city of Sidi Abdellah: to change the features of Sidi Abdellah. Perhaps, it will become a smart city with smart people.

The most important factors and obstacles have hindered the progress of the city of Sidi Abdellah.

1. Real Estate (Land). This factor is one of the most important factors that contributed to the obstruction and progress of the city of Sidi Abdellah, especially since this factor is that the majority of land intended for the construction of the project belongs to the property especially for the original citizens who reside in this city and the area of this land is about 40 percent of the total area prepared for the project. This has become a huge problem for the state and the citizen at the same time, and they have been involved in disputes over the years of the last of this project and the solutions that they have taken to solve this problem. It is a change in the city's general program such as the main roads which led to a change in the infrastructure of the city. The competent authorities have taken several measures, including, the development of new laws, including the public interest law for the purchase of lands to end the problems that led to the delay and incompleteness of the project and its extension until 2030.
2. At the end of this study which is part of a theme relating to new smart and sustainable cities in Algeria realities and challenges, we give the main conclusions to which our work has led. This is not a question of giving miracle recipes, but rather of giving rise to some questions and initiating a reflection which today seems more than necessary on the programming, design, and management of a city more conducive to and more sustainble. The contract was terminated with this company for several reasons including land problem and then received the works of the national company Cosidar which also led to the change of the general program of the city which was programmed for the construction of a smart city and contributed to this factor in changing the features of this city from a smart city to a D. This is also a factor in the great loss of the project, especially after the closure of the contracts executed by the Korean company of Mazda to increase the value of the project which is contrary to the principles of smart cities.
3. Political factors. This is the factor that contributed to the transformation of the city of Sidi Abdellah, which is programmed to be the first smart city in Algeria, but the dream turned into a nightmare, especially this factor, which is the change of programmed projects and scientific projects and institutes to residential places, which is known as the project of the President. In Algeria, the country did not find any places for this project. This project increased the city program and became a new city. This is the main reason for the change of the Korean company to the national company. This became a project of the Ministry of Housing and the city after it was not a separate project.
4. For this, our work engages in a problem linked to the evaluation of smart and sustainable cities. How and to what extent the project of this new city will reduce and achieve, at the same time, the sustainability and sustainable development of the urban space of Sidi A Allah through the critical reading of the superstructure and

infrastructure programs designed for the city. In other words, may Sidi Abdullah be seen as a sustainable city at a time when gaps persist, not meeting international standards established for sustainable cities

5. After the study arrived to assess the current state of the city and reach the most important challenges it faced during its completion period and the reasons for what it is now, it has reached the results of a study after an analytical study of the city Sidi Abdullah and the conduct of numerous discussions of the agencies concerned and a questioning of the residents of the city on the sustainability indicators adopted by the United Nations, which supports the United Nations, which supports Urban sustainability and the extent of its implementation in the city Finally, we carried out a general assessment of the use of technology in the city. We have achieved the following results and recommendations:

12 Conclusion

At the end of this study which is part of a theme relating to new smart and sustainable cities in Algeria realities and challenges, we give the main conclusions to which our work has led. This is not a question of giving miracle recipes, but rather of giving rise to some questions and initiating a reflection which today seems more than necessary on the programming, design, and management of a city more conducive to and more sustainable.

For this, our work engages in a problem linked to the evaluation of smart and sustainable cities. How and to what extent the project of this new city will reduce and achieve, at the same time, the sustainability and sustainable development of the urban space of Sidi A. Allah through the critical reading of the superstructure and infrastructure programs designed for the city. In other words, may Sidi A. Allah be seen as a sustainable city at a time when gaps persist, not meeting international standards established for sustainable cities.

After the study arrived to assess the current state of the city and reach the most important challenges it faced during its completion period and the reasons for what it is now, it has reached the results of a study after an analytical study of the city Sidi Abdellah and the conduct of numerous discussions of the agencies concerned and a questioning of the residents of the city on the sustainability indicators adopted by the United Nations, which supports the United Nations, which supports urban sustainability and the extent of its implementation in the city. Finally, we carried out a general assessment of the use of technology in the city. We have achieved the following results and recommendations:

Results

Descriptive study of the region In the first phase of the study, we found that the strategic location of the region which connects the city of Sidi Abdallah to Algeria is one of the most important advantages which reduce the pressure on the city of Algiers, and this was one of the most important goals of this city on the part of the citizen, especially that this city will provide a high level of Maids to the citizen, including health and education. Industry and entertainment ... etc.

Lack of management of the Algerian state and the intervention of the authorities and lack of experience in the studies and the lack of long-term planning led to delay the project.

Non-application of principles Most of the principles of sustainable development, especially the use of renewable energies in this area.

The reason for the delay of this city is due to the intervention of the authority in changing the program of the year for the city.

This study aims to prove that the city is smart, but the contrary says that this city is a new city, this returned to the lack of commitment to studies, and this led to an increase in the value of the project.

Algeria still does not apply the principles of sustainable development. The latter needs more awareness in this area.

From a study, we found that the majority of the lands are agricultural lands and this project drained most of the forest resources to complete this project at the expense of the ocean and also the impact of the project on the neighboring green areas due to the spread pollution caused by pollution.

This study reached several objectives for the achievement of sustainable development, including the waste of energy used in transport, heating, waste minimization, and the use of sustainable building materials.

This study concludes that we must introduce popular participation in the realization of urban planning to know the needs and achieve more sustainability.

In order to measure the extent to which the sustainability principles were applied in this city, we found that 44.44% of the city's sustainability was achieved. This is due to the failure to implement the principles of sustainability and that this city is not sustainable.

Recommendations

Depending on the results obtained from this study, we will present below a set of recommendations intended for those involved in decision making and the inhabitants of the city. These commands will make it possible to upgrade and adequately improve their redesign to meet the needs and aspirations of citizens on one side, and to achieve urban sustainability and the introduction of the concept of intelligence and technology in the city:

- Preparing a strategy for the preservation of air and reducing the depletion of the ozone by reducing the use of air conditioners and heating by developing and respecting the interior design of buildings and taking into consideration all that is specific to the lighting and natural ventilation and completion of rainwater recycling plants.
- Establishing new sources of clean and renewable energy by installing solar panels on the roofs of buildings and benefiting from the dams adjacent to the area for the production of electrical energy.
- Reduce the consumption of buildings by providing a good environment inside the buildings and planting plants on the roofs and installation of moisture measuring devices
- Popular participation in the completion of urban plans and awareness of the importance of applying the principles of sustainability in this city.
- Respect long-term planning strategies, taking into account the demographic growth of the region.
- Strengthen the relationship between land and transport by providing the city with sustainable transport systems such as metro and tramway.
- Improve emergency response and reconstruction.
- Provide the internal districts with monitoring systems to improve the security situation in the city.
- Installation of signs to facilitate mobility within the city.
- Installation of sensors and fault monitoring.
- Providing the city with a high-flow Internet line in the outer spaces.
- Development of protection systems for the city.

References

Aroumougom, J.-C. (2003/02/00083). *New city project Sidi Abdellah, Algeria.* Prior appraisal December 2002.

Chabane, I. J. (EPAU, 2006). Evaluation of the lived quality of airtight environments in a glass curtain wall. *Mémoire de magister.*

Halim, M. A. A. (2017). The mechanism of achieving urban sustainability through the UN's applied to some new Egyptian cities. *Journal of Engineering Sciences.*

Houlin Zhao, S. G. (2016). Building the smart and sustainable cities of tomorrow. *nouvelles de l' ITU News No. 2.*

Hussein, D. (2012). *M Al Dardairy mechanisms of environmental architecture to reach urban sustainability.* Department of Interior Architecture, Faculty of Design, Dammam University, Saudi Arabia

Jean-François Romain, E.-I.-C.-T. (2009, décembre Mercredi 2). *maintenanceandco.* Retrieved 09-15, 2018, from https://www.maintenanceandco.com/files/avis_experts/Batiments_intelligents_comment_allier_confort_des_occupants_et_reduction_des_charges.pdf

KT, S. (n.d.). *Report mission A-C-D of new city Sidi Abdullah.* Ministry of Urban Development and the Environment.

KT, S. A. (n.d.). *Report mission B: The five parts of new city Sidi Abdullah.* Ministry of Urban Development and the Environment.

Marwan, T. (2013). An analytical comparative study of some of the global residential sustainability criteria. *Damascus University of Engineering Sciences, 29*(Second Issue 2013).

Presentation new city of Sidi Abdullah (2018). VNSA.

Presentation report definition of the concept of development linked to the vocation of the new city concerned (2018, April). VNSA.

Report analysis of the project site of the new city (2018). VNSA.

Retrieved 08-12, 2018, from https://drive.google.com/file/d/1tX2_tm_DE_id2yEb-yRyLypiY9cZWCTf/view

Séraphin, X. L. (2009). *Post-occupation study of residential projects for families in dense urban areas.* National Institute for Scientific Research Montrea: Montreal, Library and National Archives of Quebec.

Signoles, R. S. (2017). *journals.openedition.* Retrieved 09-15, 2018, from https://journals.openedition.org/emam/1323

SNAT schema national d'aménagement de region (national pattern of region development) (2018). VNSA.

VNSA. (n.d.). Retrieved 07-01, 2018, from https://www.vnsa.dz/

The Sustainable Strategy to Create Decentralized Economic Hubs for Regional Development

Mohammed Rasmy, Tarek Abdel Latif, and Asmaa Abdelaty Mohamed Ibrahim

Abstract

Objectives—The study aims to facilitate the establishment decentralized economic regional hubs to achieve social equity, enhance economic productivity, and attain the sustainable regional development taking Egypt as a case study. The study is based on the integration between the economic regional hubs and existing small and medium cities in the region to achieve a decentralization strategy for the metropolitan and capital cities. **Methods**—The study uses qualitative methods to analyze the existing centers in Egypt and accordingly proposes a spatial distribution to create decentralized economic hubs. **Results**—The results highlight proposed planning strategies for the decentralized economic hubs that policymakers can apply to achieve socioeconomically sustainable development for the regions. **Conclusion**—The results of this study potentially benefit decision-makers, economists, and planners who aim to implement decentralized economic hubs and achieve sustainable development for a region. This will help increase GDP and attract high-skilled labor to a region and thus would help decentralize political capitals.

Keywords

Decentralized economic hubs • Regional planning development strategies • Urban economic strategies • Decentralization strategy • Integrated regional development

1 Introduction

Nowadays, the centralization of populations in political capitals is the main reason for the unfair distribution of resources and thus leads to less development in the small and medium cities in other regions. This affects the quality of life in big cities such as Paris, London, Tokyo, Beijing, Buenos Aires, and Cairo (World Population Review, 2018). Moreover, approximately 70% of the world's population which are living in urban areas with fewer than 500,000 residents and 30% in rural areas (Angel et al., 2012) with a population of 500,000 or more, which are considered small cities in some regions (Cox, 2014). Countries need to find new economic and environmental resources and establish new centers for sustainable development.

The study proposes a decentralization strategy for regional development in Egypt based on creating economic hubs that will support medium and small cities while providing the missing services for the region, which will be an economic regional capital. This strategy can provide new job opportunities to attract more of the population away from overcrowded populated areas such as political capitals and metropolitan cities.

The study also determines the key enablers for creating decentralized hubs, including financing mechanisms and governance systems by analyzing several successful world capitals and metropolitan cities in the US, China, Europe, and the Middle East.

Figure 1 outlines the research methodology and clarifies the centralization problems caused by immigration from

Article classification—Research paper.

M. Rasmy (✉) · T. A. Latif
Department of Architecture, Faculty of Engineering, Cairo University, Giza, Egypt
e-mail: mohammed.rasmy@tu-dortmund.de

M. Rasmy
TU Dortmund University, Dortmund, Germany

A. A. M. Ibrahim
School of Architecture + Planning MIT, University, Massachusetts, USA

A. A. M. Ibrahim
Effat University, Jeddah, Saudi Arabia

F. Trapani et al. (eds.), *Advanced Studies in Efficient Environmental Design and City Planning*, Advances in Science, Technology & Innovation, https://doi.org/10.1007/978-3-030-65181-7_19

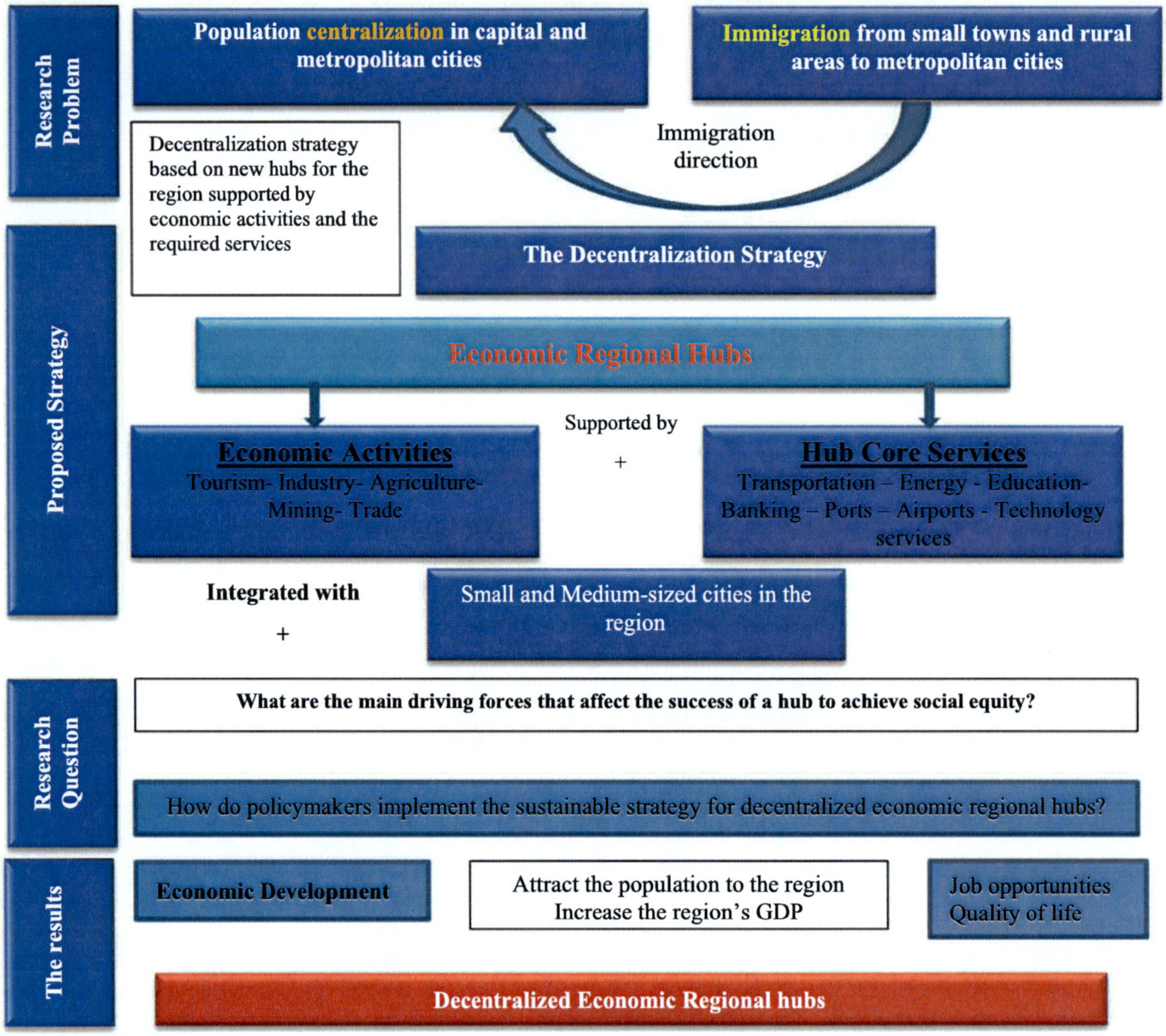

Fig. 1 Methodology. *Source* Authors

rural and small towns to metropolitan and capital cities. Accordingly, the proposed strategy will provide attraction points as part of an economic hub that is integrated with the regional core service and its associated existing economic activities to create decentralized economic regional hubs.

Centralization in Capital Cities

Cities are attractive regions for migrants, considering the greater availability of job opportunities and better facilities. Uncontrolled migration from rural to capital cities and urban areas affects the metropolitan cities detrimentally and is an important reason for overpopulation, loss of agricultural land, relative poverty, and escalating unemployment (Schwan, 2019). The map in Fig. 2 shows that Tokyo, Japan, is considered the largest city in the world, with a population over 37 million inhabitants. Tokyo is considered a multi-function capital as it is both a political and economic capital. The high level of services in Tokyo, such as shopping, local public transportation, and clean streets (World Population Review, 2018), attracts inhabitants. City planning studies should provide a solutions for the problems of pollution and congestion which caused by the agglomeration of manufacturing and factory workers in industrial cities (Liaros, 2019). The world's second most populated capital is New Delhi, with 28 million people (see Fig. 2). The core economic services that attract migrants and visitors are information technology, telecommunications, hotels, banking, media, and tourism. The third-most populated capital is Beijing, with a population of just less than 20 million. Beijing is the political and economic capital of China (see Fig. 2) and one of the most important international financial centers in the world (FMI, 2014). Cairo, the capital of Egypt,

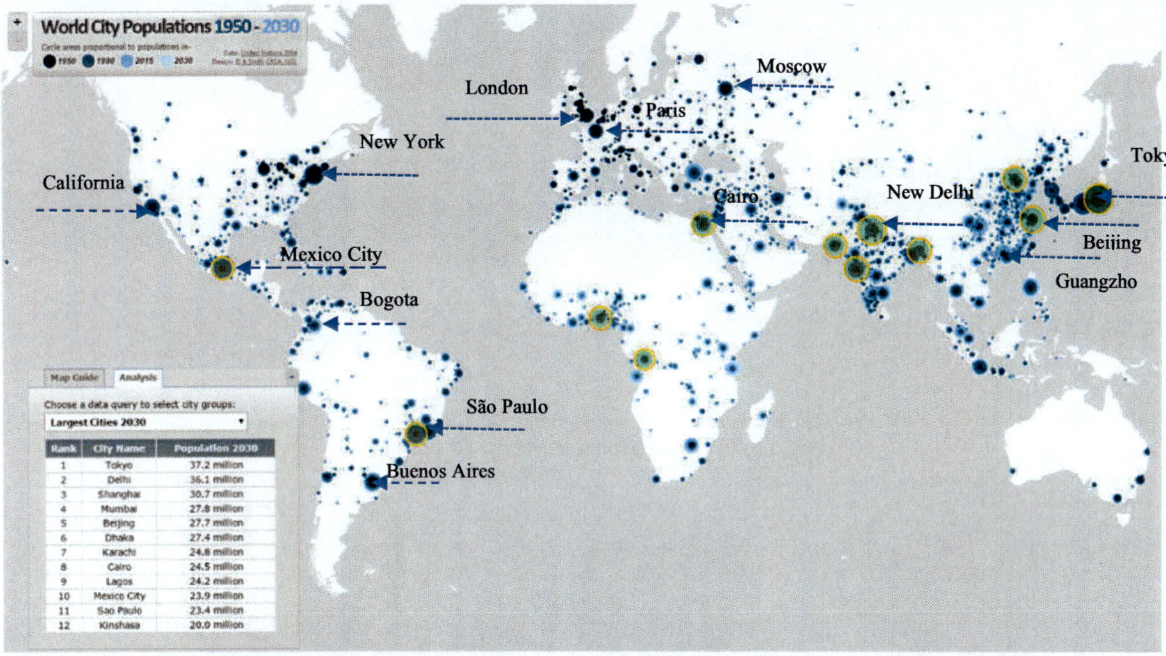

Fig. 2 Centralization of capital and coastal cities. The growing population of the world's cities. The population is centralized in capital cities or coastal cities (Sasol Limited Integrated Report 2018, edited by the authors)

has a population of more than 20 million. It is considered one of the most densely populated capitals, with a population density of 36 persons per square kilometer. Cairo's traffic is overwhelming (Otenyo & Lind, 2006). The planning strategy for the cities is based on the consequences of centralization and agglomeration, ignoring food production. However, the Internet offers possibilities for decentralization. The technology continues to advance which will make many traditional jobs obsolete. Creating resilient places such as technology hub can provide work for people that will satisfy their basic needs so they rely less on jobs that provide an income to satisfy these same basic needs (Liaros, 2019) which will be a very important strategy.

Figure 3 illustrates the existing models. A region has the potential and natural resources for the existing cities and the country's centralized capitals. The capital is the center for all economic activities, which attract inhabitants and investors (Thomas, 1997).

It is important that the decentralized economic hubs can be centered with available resources, including harbors and/or rivers (see Fig. 1), as most of the population is centralized in capital cities. However, such decentralized economic hubs should have economic potential in several areas, such as the tourism, trade, agriculture, and manufacturing industries. The multitude of economic activities will support the hub if one of the activities is dropped (Fig. 4).

The economic hub model is based on economic activities such as agriculture, mining, retail, and manufacturing. The proposed strategy will provide integration for small and medium cities with the new hubs. The main base factors are the political capital, regions, and cities. Over the last century, the city structures shifted from a dominant central business district (CBD) to be one of multiple strategic and regional centers such as hubs which will offer work opportunities and a range of lifestyle.

The main reason for the shift from CBDs to economic hub is a complex mix of social and economic factors driven by competitive advantages, technological advances, and globalization, all largely associated with the knowledge or new economy. Human capital development and innovation are considered an accepted strategy for economic growth. It enhances all inputs, regardless of industry, contributing to national productivity, industrial performance, and competitive advantages (Martinus, 2010). This brought the knowledge component of labor productivity into focus, given its capacity to add value to resources, raise profitability and wages, and reduce costs. The planning for hubs enhances proximity to the knowledge sources also it provides access to social capital networks, which can potentially increase regional labor productivity, supplement national policies, and support greater levels of productivity and national innovation (Martinus, 2010).

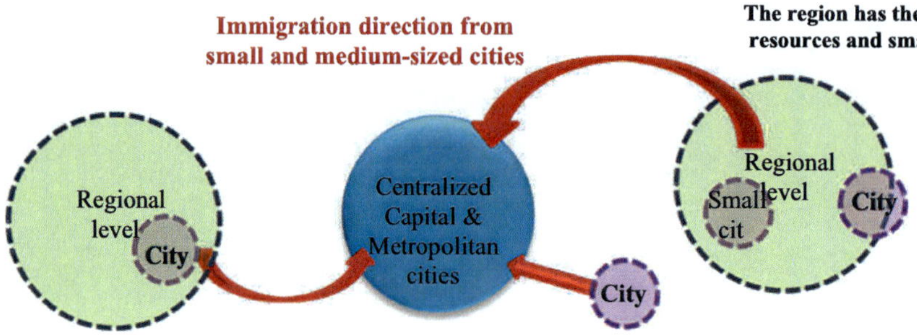

Fig. 3 Existing world development model: immigration from regions to capital cities *Source* Authors

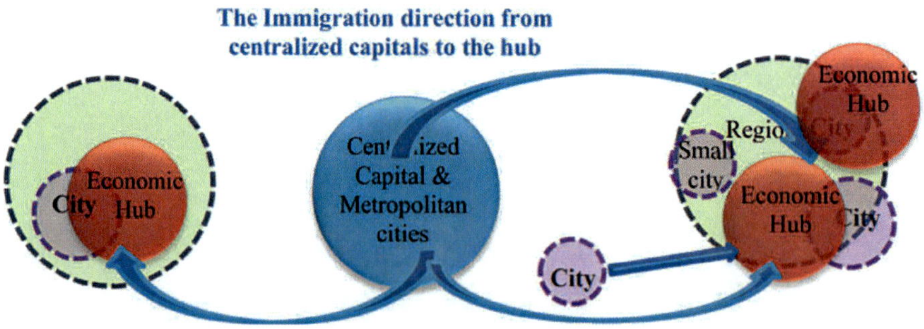

Fig. 4 Proposed model (decentralized economic regional hubs): the economic hubs will attract the population as the attraction points will have integrated economic activities and a high level of services *Source* Authors

2 Decentralization Strategy for the Economic Hub

This paper describes the process whereby decentralized hubs will attract residents from the capital cities. Economic capacity, employment, and the population are concentrated around and in large urban centers. The hub will be centralized and will consist of economic activities. The strategic planning process for the hubs consists of three major components: formulation for the region (including assessing the internal and external environments and setting objectives), evaluating and selecting strategic alternatives for the hub, and implementation and control (Jennings & Disney, 2006).

The main idea is to concentrate on the resources associated with the modern economy such as high-tech industries, ICT-specialized universities, research and development institutes, and flexible IT-skilled labor pools in metropolitan areas and large urban centers which will be the technology hub. The model of the urban labor market aims to explore the market forces behind the equilibrium wages and total employment in a city. It is assumed that the metropolitan area is part of a larger regional economy and is the main reason that households and firms move freely between cities in the region. The demand for labor comes from firms in the city, while supply comes from households living in the city (Glaeser, 2010). The model concentrating on resources will show the effects of changes in demand and supply on the city's equilibrium wages and total employment. The world, in management terms and economically, has become a strong network of prosperous city regions (Glaeser, 2010). The question is that whether such a demographic shift is beneficial from both the economies of scale perspective and in terms of quality of life. This study pursues insights to answer this fundamental question by investigating peripheral locations. The methodology applied (Mills, 1959) allows for an investigation into extreme and contending case scenarios.

2.1 Concept of Decentralized Economic Regional Hubs

Sustainable economic regional hubs serve as new economic centers for each region to provide new economic services for the region (see Fig. 5). Sustainable economic regional hubs can play a pivotal role in the region's cities, as it is a centralization of the economic activities in the region and connects the small and medium cities to attract the population from the political capital and metropolitan cities. The regional capital will provide the missing and needed services using a sustainable developmental strategy, in addition to advanced education, new job opportunities, personal and

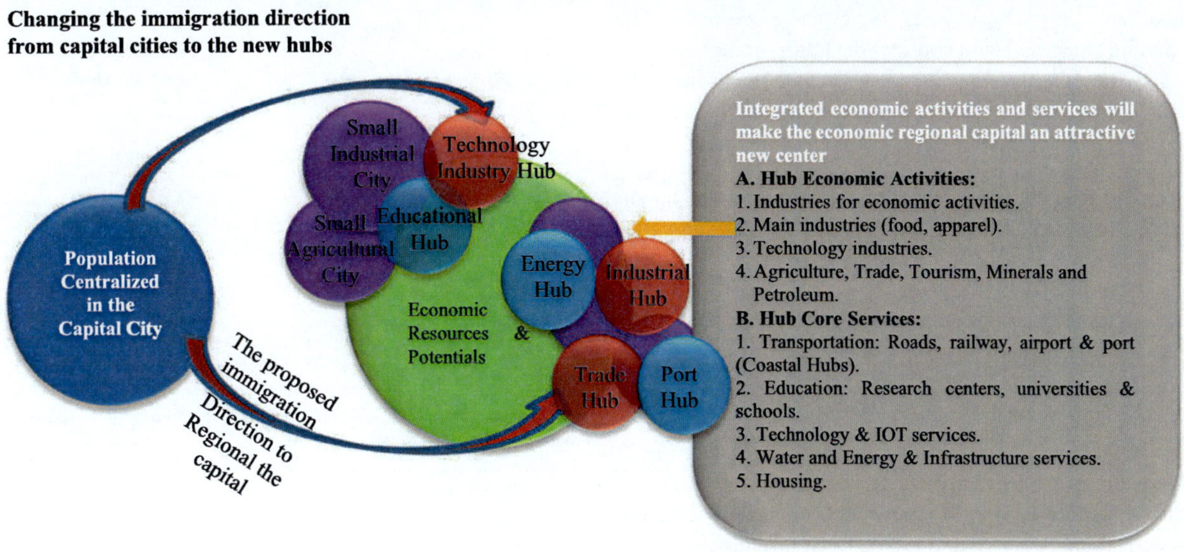

Fig. 5 Decentralized economic regional hubs. *Source* Authors

professional services, entertainment, and cultural participation, along with a host of other amenities.

Decentralized development will be based on the economic hub, as it will be the new attraction CBD for development and will provide new activities and the required core and smart services for the region to attract skilled labor. The region has great economic potential, and a planning analysis can propose the best location for a centralized hub connected to the region's natural resources, which will provide integrated urban development (see Fig. 5).

The integration between the hub and cities' economic activities such as tourism, industries, agriculture, and the region's core services such as transportation, educational activities, research centers and universities, and infrastructure such as water and energy provided in the economic regional capital can allow for the hub to have new smart technology and IOT services. All these economic activities and smart core services will increase the human development index. Accordingly, the new economic hubs will attract highly skilled people.

3 A Hub's Economic Characteristics

Town planning for a decentralization strategy has several models. The proposed decentralized hub model will be a new strategy for future development. The first decentralization strategy planning model was at 1896, which is for the planning of Garden Cities of Tomorrow by Ebenezer Howard's. This model's flow of people into cities that based on the various attractions of city life. The proposal for the development for Garden Cities by Howard was a mix or integration between the best aspects of town and country life, which was acting as an alternative attractor outside existing towns. These would reverse the flow and draw people out (Liaros, 2019). Nowadays, the total population of non-food producers grew dramatically at the devolved urban areas, such as in capital cities. Although Howard sought by redistributing the population in the landscape to relieve congestion, Garden Cities model did not address the relationship between food production and people (Liaros, 2019). Garden city model is reflecting the importance of food production for people. The proposed hubs should consider the location of food production.

In 1930, Le Corbusier suggested tower buildings that disconnect people from the land and natural systems, which also creates a wasteland between the buildings. This model proposed Ville Radius (Radiant City). The huge scale of the populations and buildings housed within them reflects the mechanistic thinking prevalent at the time. This thinking saw humans as separate from nature (Liaros, 2019). The Radiant City which proposed by Le Corbusier would spread across a vast green area, contain prefabricated and identical high-density skyscrapers, and arranged in a Cartesian grid, allowing the city to function as a "living machine" (Liaros, 2019). Based on Le Corbusier, the decentralized hub can be a tower buildings or high-density skyscrapers that include all the required economic activities vertically. In 1932, Frank Lloyd Wright proposed the Broadacre city and basically represents the automobile-dependent suburban sprawl; it was standard practice in the planning of Greenfield subdivisions. The Broadacre city model provides walkable environments, and possibly, it can be with electric carts within the zone which are surrounded by natural systems and large areas of food (Liaros, 2019). This strategy was provided the suburban sprawl by cars at 1932, and accordingly, the planning strategy should consider the transportation methods inside the hub and the connectivity with the surrounded hubs. The empirical studies found associations between the characteristics of planning processes and a range of organizational characteristics and environmental (Jennings & Disney, 2006). Accordingly, these studies propose guidelines for designing the strategic planning process which can be implemented for the new hubs.

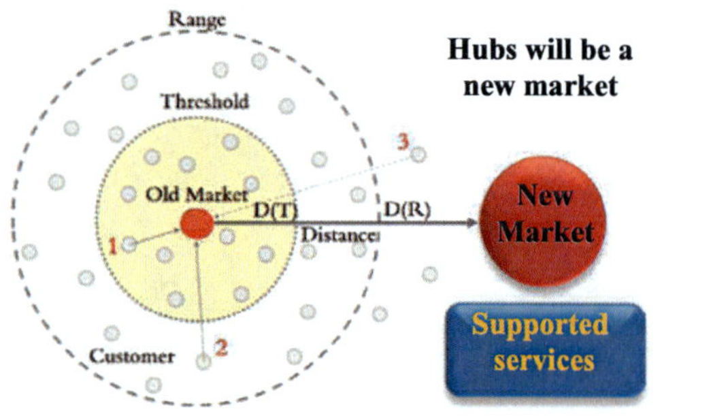

Fig. 6 Central place theory: integration between the old centralized market and the new economic market supporting the missing services will provide regional development. *Source* The author's analysis of the main sketch of the theory

3.1 A Hub as a Centralization of Economic Activities

The central place theory is a geographical theory (see Fig. 7) that explains the location, size, in addition to the number of human settlements in a residential system. Settlements function simply as "central places" that provide services to surrounding areas (Openshaw et al., 2003). As centrality indicates importance, it manifests in the quantity and quality of the different services and functions of the settlement. Accordingly, the important characteristics of a central function are to generate spatial interactions through the movement of labor, ideas, and materials between the hub as a central place and the complimentary region surrounding it (see Fig. 6). August refined the theory of central place by incorporating non-service activities (see Fig. 7). This model postulated one superior center where produce for all goods. The size of these centers is increasing with distance from the central place, and the other small centers tend to be located at the halfway between larger ones (see Fig. 7). Thus, the hub center will have several hexagonal hubs or markets for different products as transport cost is a function of distance; industry X, with lower-cost transportation, which will have a bigger hexagonal hub market area, as the hub will be centralized to connect the required core services such as education (universities, research centers), technology (5G service, IoT), and energy and transportation networks (high-speed rail, ports, airports). The new hub centers have economic activities, and the missing services that will provide new attraction points will be integrated with the existing centers in the city (see Fig. 7).

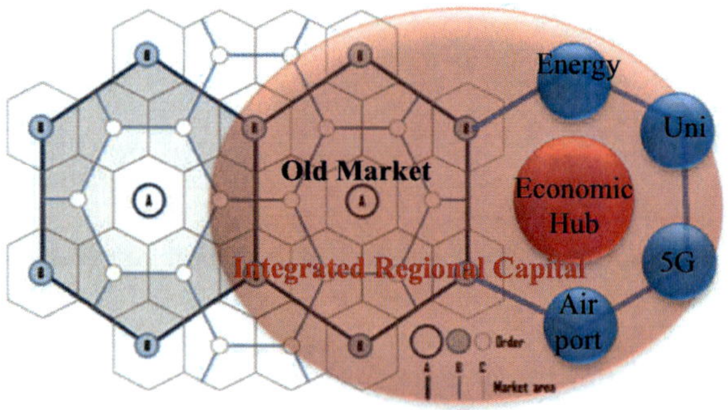

Fig. 7 Modifications of the central place theory to provide an integrated regional capital. *Source* The author's analysis of the main sketch of the theory (*Central Places Theory* (*Market Principle*) | *The Geography of Transport Systems*)

3.2 A Hub Is Preferable at the Peripheries

Core-periphery approaches to the faster growth of boundary cities than that of other cities are part of the broader circular causation models discussed in Higgins and Savoie, which explains why some regions are prosperous while those beyond the center are in decline. Thus, the focus is on the relationship between the core area and the periphery hubs and the regions on the outer areas away from the core. The developed core hubs in the industrialized zone are sustained by exports of raw materials from peripheries and hinterland areas. The nature of the flow from the core to the periphery and vice versa explains the differential development. The term "spread" describes the flows of capital, innovations, and people from the core to the periphery. The backwash effect (flows from the periphery to the core) is highly significant and contributes to the decline of peripheral regions. New economic geography is the core of the core-periphery model. The model clarifies that agglomeration; that is, the clustering of businesses together as a hub yields economic benefits for all businesses in the hub in this region. The focus on these financial interests, this theory examines the relationships between several factors such as transportation costs and economies of scale. Furthermore, the model illustrates which equilibrium can occur either through the concentration or decentralization of communities. The two-region model suggests that if the costs and production in the regions are considered same, then nothing much happens, and on the other hand, if there is a slight difference, the "population will start to concentrate … once started, this process will feed on itself" (Liaros, 2019).

3.3 A Hub Should Be Located Close to Natural Resources

The staple theory originated in Canada, and its main concept is the export of natural resources (see Fig. 8) or staples, which affect both the economy and the social and political systems (Harold Mackintosh, 1st Viscount Mackintosh of Halifax—Wikipedia, 2020). The different staples were fur, fish, timber, grain, oil, agricultural products, and minerals. These staples influenced the rates of settlement, federal-provincial conflicts, and so on. The staple theory demonstrated how an economic hub can exploit a particular staple, which would induce regional growth and become a key factor for regional economic development (see Fig. 9). The intercontinental east–west economic development was based on the trade of staples and cemented Canada's cultural links to Britain and that of Britain to the West Indies according to Harold and Mackintosh (1920). In a domestic economy, such trade relationships exemplify the "hub core-periphery" structure of intraregional relationships. It embeds management practices and transportation functions and supports manufacturing hub growth. This approach sees the concept of the staple as a key factor for economic regional development and society.

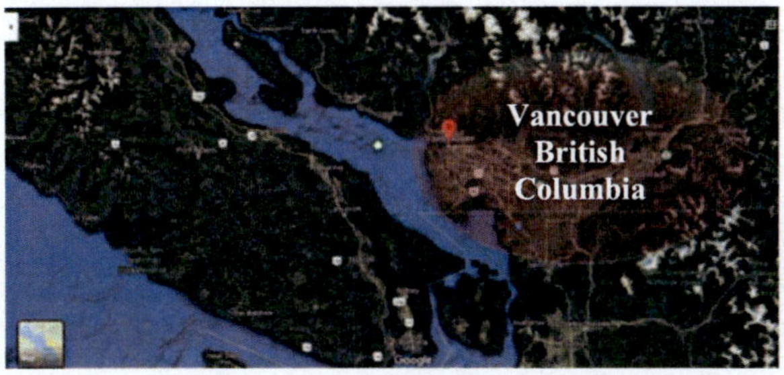

Fig. 8 Location of Vancouver, British Columbia, Canada. *Source* Authors

Fig. 9 Core-periphery model, Vancouver, British Columbia, Canada (*The Ultimate Whistler Tack-on Trip*, 2020)

3.4 Proposed Size of an Economic Hub

Based on the previous analysis of city sizes in urban economics studies and the economic characteristics, the hub is almost will be the same size as a city. For example, New York is the most populated area in the US, with a population of more than 18 million (New York City Metro Area Population 1950–2020 | MacroTrends, 2020), while other small cities have smaller populations, such as Andrews, a city located in Texas. With a 2020 population of 13,466, it is the 197th (New York City Metro Area Population 1950–2020 | MacroTrends, 2020). There are some metropolitan cities, a moderate number of medium-size cities, and many small cities (Glaeser, 2010). The economic forces are the main factor of hubs of various sizes. The agglomeration economies created by the integration between the hub, and the small cities will increase productivity and wages, so workers in larger cities or hubs will earn higher wages (O'Sullivan, 2012). As a hub will grow, the advantages of higher wages are partly offset by several undesirable features of larger cities, including longer commuting times, greater density, and more congestion and pollution. The rise in city sizes (population) affects utility and the standard of living (O'Sullivan, 2012). Based on the city size analysis of metropolitan cities with a minimum of 50,000 people (O'Sullivan, 2012), the minimum proposed economic hub size should start from a population of 50,000 people.

Promoting the sustainable development and improve the quality of life for all people in the world is the World Bank's position mission (Otenyo & Lind, 2006). An alternative framework of planning strategies for development is provided by Lempert and his collaborators that includes using the full potential of human resources in addition influencing the state of mind, or attitudes, in the institutions in which development occurs (Otenyo & Lind, 2006; Glaeser & Gottlieb, 2006; Otenyo & Lind, 2006). Population growth, in addition to the degradation and depletion of the development of new technologies, social restructuring, and natural resources, has been so slow as to be imperceptible during an individual lifespan. However, the global economy has shown exponential growth, transforming human life especially and the character of the planet in the past two centuries (Kelly et al., 2004).

What is the best strategy for the hubs? If the region has 6 million people, is it better to provide 3 hubs with 2 million inhabitants each, or 6 hubs with 1 million inhabitants each? The below diagrams clarify that the optimum size for

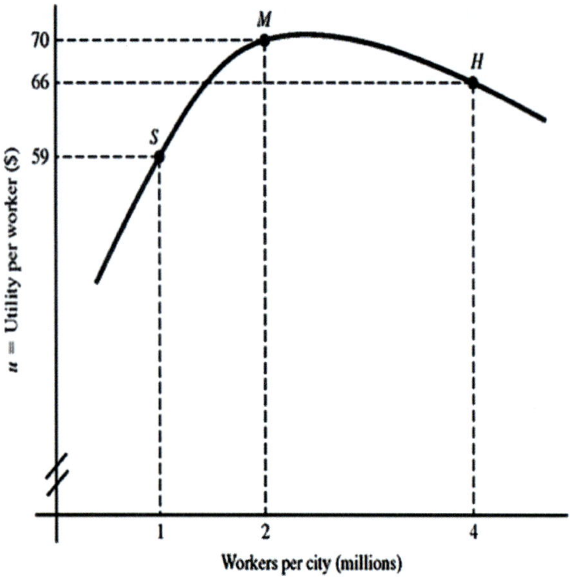

Fig. 10 Utility and hub size (O'Sullivan, 2012)

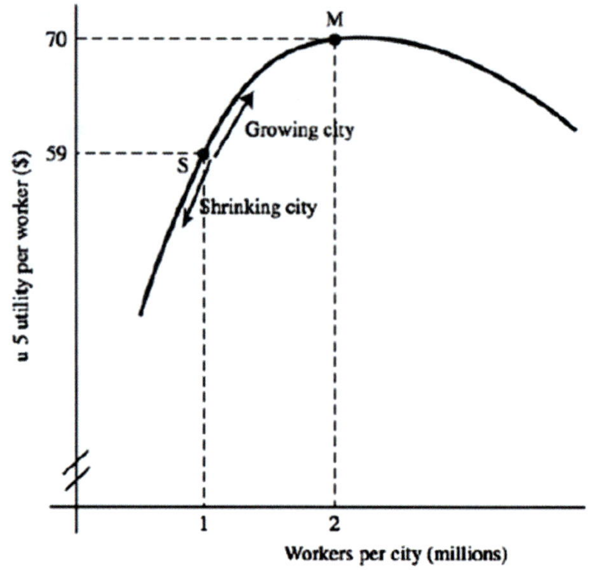

Fig. 11 Cities or hubs are not too small; the utility curve reaches its maximum at 2 million workers in a city (point M) (O'Sullivan, 2012)

workers in a hub is 2 million. An increase in hub or city size increases wages owing to agglomeration economies and increases commuting costs (see Fig. 10). As long as agglomeration economies are stronger, utility increases with hub size. When agglomeration economies are weaker than the diseconomies from commuting, utility decreases as city size increases (O'Sullivan, 2012). The utility curve reaches to the maximum at 2 million workers in a hub or city (point M), so a neighborhood which have 6 million workers, and it will be maximized utility with three cities, each with 2 million workers. The result for six small cities (point S) is unstable; the main reason is the utility curve is positively sloped. The result with two big cities (point L) is stable because the utility curve is negatively sloped (O'Sullivan, 2012) (Figs. 11 and 12).

The utility curve shows the trade-offs from a rise in population. Agglomeration economies will increase productivity and wages at the hub, but diseconomies of scale from increased commuting costs reduce utility. Hubs are unlikely to be too small because such an outcome is not a stable equilibrium. Migration is self-reinforcing because the growing hub becomes more productive while the shrinking hub becomes less productive (O'Sullivan, 2012). Hubs are also overlarge because such an outcome may be a stable equilibrium. In this case, migration is self-correcting because the shrinking hubs become more productive while the growth toward a mega-scale city becomes less productive. Local employment amplifies differences in workforces and populations across cities and hubs. Various activities at the hubs foster experimentation result in innovations in product design and production.

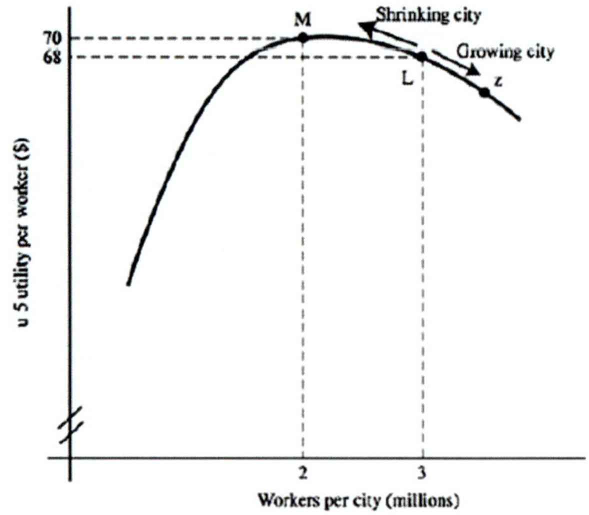

Fig. 12 Cities or hubs may be too large (O'Sullivan, 2012)

3.5 Sustainable Development for Hubs

The main concept of the sustainable development was coined explicitly to suggest that it was possible to achieve industrialization and economic growth without damaging the environment (Fig. 13). It is considered the main target of sustainable development.

The sustainable development of economic hubs is different as successful development should attract inhabitants. That can be achieved by providing economic activities in the region, supported by the region's natural resources, which

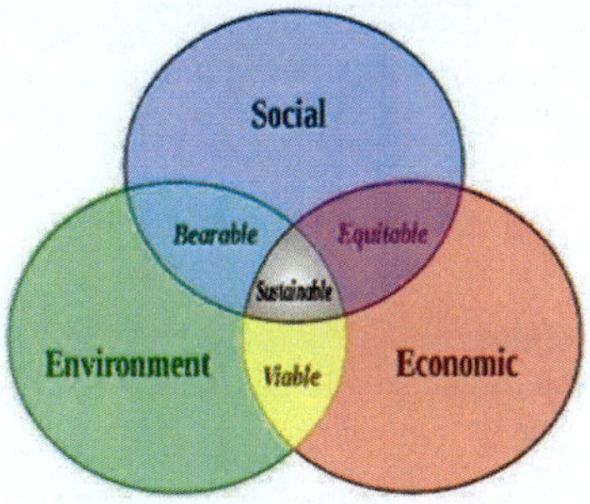

Fig. 13 Venn diagram of sustainable development: the three pillars of sustainability, also known as the triple bottom line (Adams, 2006)

serve the human needs in the region. The new hub should be supported by core services to achieve the kind of sustainable development that will provide for human needs and attract inhabitants. Thus, the region should see population growth (see Fig. 14).

4 Decentralized Hubs: Case Studies

The study discusses the economic impact of a hub and how it attracts a population, and whether decentralized hubs are more efficient for region development. The study selected three international cases for regional development: California, Guangdong, and Hamburg to explore how to implement a hub and how Egypt can apply the results to create a decentralization strategy.

4.1 California, US: Decentralized Technology Hubs

California has several successful integrated hubs, which have the highest GDP in the world, of around 3000 billion $ between 2014 and 2018 (Tables 1 and 2). The authors analyzed California's decentralized hubs (Tables 1 and 2). The most famous hub in California is Silicon Valley, a technology and educational hub. The valley has 30 cities, and moreover, it has five universities, including Stanford University which is considered the backbone University for the valley. Silicon Valley has the headquarters of thousands of technology firms. Silicon Valley is an attraction point for investors, capitalists, skilled labor, and international firms. These attributes affect both the GDP and the size of the region's population. In terms of high-tech production, Silicon Valley is considered the largest economic region in the US. It is located at Santa Clara Valley and expanded to encompass parts of Gilroy and the East Bay. This region is named "silicon" owing to the firms which are manufactured silicon chips and are considered a hub for silicon manufactures. Currently, many high-tech companies are located in the valley. Silicon Valley was created by Hewlett and Packard, which consisted of students at Stanford University and their professor Lewis Terman. The establishment of Stanford University as the educational hub for the region led to the creation of the technology giant. With the help of the Internet, Terman established Stanford Industrial Park and helped with the research for the first few organizations. The firms located in the valley created numerous jobs. In 2006, 225,300 technology-related jobs were created (Blumenthal et al., 2018).

Silicon Valley is a high-tech production hub. In 2012, 46% of venture capital in the US went to the valley (Blumenthal et al., 2018). Consequently, more firms are establishing their branches in the region. The technology hub is driving the economy; the high number of provided jobs exceeded the number of the available homes. In 2017, the San Jose–San Francisco–Oakland had a GDP around $878 billion Fortune 2017 (Dempsey, 2016). This explains its attraction for investors, capitalists, skilled labor, and international firms. These attributes influenced both the GDP and the region's population size (Figs. 15 and 16).

Table 1 clarifies that the increasing GDP is attracting the population and civilian labors. Moreover, the decentralization

Fig. 14 Interaction between a decentralized economic hub and the three pillars to achieve sustainable development. *Source* Authors

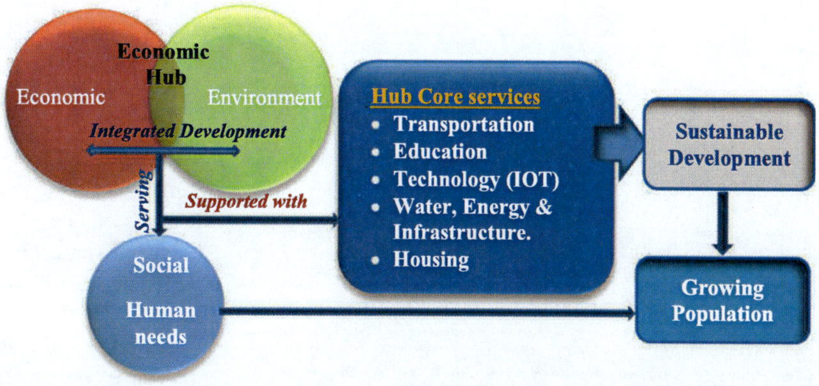

Table 1 Decentralized hubs in the US, California's decentralized hubs have the highest GDP

			Economic			Population	Labor force		
			Value of industry's top-selling products (2017), billion	Nominal GDP (billions USD), 2014–2018	Nominal GDP per capita (thousands USD)	Population (millions)	Civilian labor force	Unemployed 2019(P)	Unemployed percent of labor force 2019(P)
	Regions	Development	VAR01	VAR02	VAR03	VAR04	VAR05	VAR06	VAR07
1	Washington	Capital city	42.16	477	62	7.3	3,512,865	117,089	3.3
2	California	Decentralized hubs	27.00	**3000**	75	40	19,453,554	864,476	4.4
3	Texas	Decentralized hubs	73.00	1755	62	28.3	14,066,138	523,938	3.7
4	Massachusetts	Centralized hub	3.20	516	76	6.8	3,873,413	112,615	2.9
5	Florida	Decentralized hubs	7.58	1043	49	20.6	10,470,763	363,073	3.5
6	New York	Centralized hub	25.00	1675	84	20	9,581,857	398,313	4.2
7	Nevada	Centralized hub	20.00	151	52	2.9	1,552,470	69,224	4.5
8	Michigan	Centralized hub	22.74	505	50	10	48,029	2085	4.3
9	Georgia	Centralized hub	6.69	533	49	10.3	5,130,878	190,687	3.7
SPSS results			**0.68**	**0.828**	**0.224**	**0.893**	**0.852**	**0.83**	**0.205**

The significance of bold at : It is SPSS results and the highest GDP for California as it is 3000.

strategy has the highest GDP of any US region except New York, which is a strong centralized hub in the US economy.

4.2 Guangdong, China: Logistic Hubs

This study examines Guangdong province in China as it has 2 logistics hubs (Guangzhou and Shenzhen), which are also 2 of the 8 global logistics areas selected by the CBRE logistic hub model at 2015. Guangzhou and Shenzhen are key manufacturing centers (*Five China Cities Join Ranks As Global Logistics Hubs | CBRE*, 2015). Moreover, the connectivity between the Guangdong industrial hubs (Guangzhou and Shenzhen) and the strong consumption and trade hubs in Shanghai, Tokyo, Hong Kong, and Singapore is based on a high level of infrastructure, logistics services, transportation networks, trading routes, and free trade zones. It experienced significant growth in the logistics sector, particularly fueled by the "One Belt-One Road" strategic initiative. The implementation of this new initiative will bring a multitude of opportunities for China's logistics sector in land- and sea-based routes (*Five China Cities Join Ranks As Global Logistics Hubs | CBRE*, 2015). Guangdong has long coastlines facing the South China Sea; it is located at Pearl River. It has a connection to Hong Kong and Macau. The connectivity with other countries is provided by the port of Guangzhou, as it is achieved by the emergence of a separatism and degree of self-sufficiency.

The economic foundation of Guangdong was primarily agriculture, which makes Guangdong an agricultural hub. Rice is the main crop in Guangdong, though several other industrial crops were successfully raised, including black pepper, palm oil, sisal, hemp, rubber, and coffee, as well as other crops such as sweet potatoes, peanuts (groundnuts), tea, citrus, litchi, pineapples, and bananas. Since the mid-1980s, the agricultural sector's proportion of the provincial economy has been declining. The pillar industries for the region are metallurgy, plastic products, garments, foodstuffs, textiles, electric power generation, electronics, and household electric appliances (Guangdong Province, 2020).

Guangdong's population growth is nearly double in size since 1980. This increase is largely due to the influx of millions of people who work in the factories which were provided at the coastal cities. The region's population in 2018 was 113,460,000. In the first half of the twentieth century, Guangdong experienced modern growth as Guangzhou developed into a commercial (trade hub), industrial (industrial hub), and transportation center (transportation was a core service for the successful economic activities in the region). Moreover, other mineral deposits were discovered (natural resources) and, however, prompted the development of some heavier industries, such as

Table 2 US decentralized economic hubs and main industries, which attracted firms to the region and increased GDP

	Main regions		Most profitable industry	Region development	Economic hub activity	Key firms and institutions
1	**Washington**	**Washington-Arlington-Alexandria, DC-VA-MD-WV**	**Aerospace**	**Capital city**	**Federal government**	Fannie Mae
2	California	San Jose-Sunnyvale-Santa Clara, CA San Francisco-Oakland-Hayward, CA Los Angeles-Long Beach-An, CA, CA Sacramento-Roseville-Arden-Are, CA San Diego-Carlsbad, CA Riverside-San Bernardino-Ontario, CA	Machinery and mechanical appliances	Decentralized hubs	Technology hub-tourism hub-trade hub silicon valley	Cisco, eBay, and PayPal, Wells Fargo, Gap, and Salesforce, Commercial real-estate giant CBRE Group
3	Texas	Austin-Round Rock, TX Dallas-Fort Worth Arlington, TX San Antonio-New Braunfels, TX Houston-The Woodlands-Sugar L, TX	Mineral products	Decentralized hubs	Educational and energy hubs—universities, creative and individualistic thinkers, firms, energy industry	University of Texas' flagship campus, Southwest Airlines and AT&T. Oil refiner Valero Energy and provider of insurance to military families USAA. Phillips 66 and Halliburton
4	Massachusetts	Boston-Cambridge-Newton, MA-NH	Precision instruments	Centralized hub	Educational hub prestigious universities in the world universities	MIT General Electric and Liberty Mutual
5	Florida	Orlando-Kissimmee-Sanford, FL Tampa-St. Petersburg-Clearwater, FL Jacksonville, FL Miami-Fort Lauderdale-West Palm Beach, FL	Machinery and mechanical appliances	Decentralized hubs	Tourism hub	Health insurer WellCare Health Plans, and Outback Steakhouse operator Bloomin' Brands, railroad operator CSX, homebuilder Lennar
6	New York	New York-Newark-Jersey City, NY-NJ-PA	Precious metals, stones, etc	Centralized hub	Economic hub—trade hub—banking hub—wall street is the financial world—megabanks	JPMorgan Chase, Citigroup, and Goldman Sachs
7	Nevada	Las Vegas-Henderson-Paradise, NV	Accommodation and food services	Centralized hub	Entertainment hub—touristic hub—the largest gambling companies in the world	Las Vegas Sands and MGM Resorts International
8	Michigan	Detroit-Warren-Dearborn, MI	Automotive	Centralized hub	Automotive industry hub	General Motors'
9	Georgia	Atlanta-Sandy Springs-Roswell, GA	Aerospace	Centralized hub	Firms hub	Coca Cola and Delta Air Lines

The significance of bold at : It is for Washington DC as it is the Capital city.

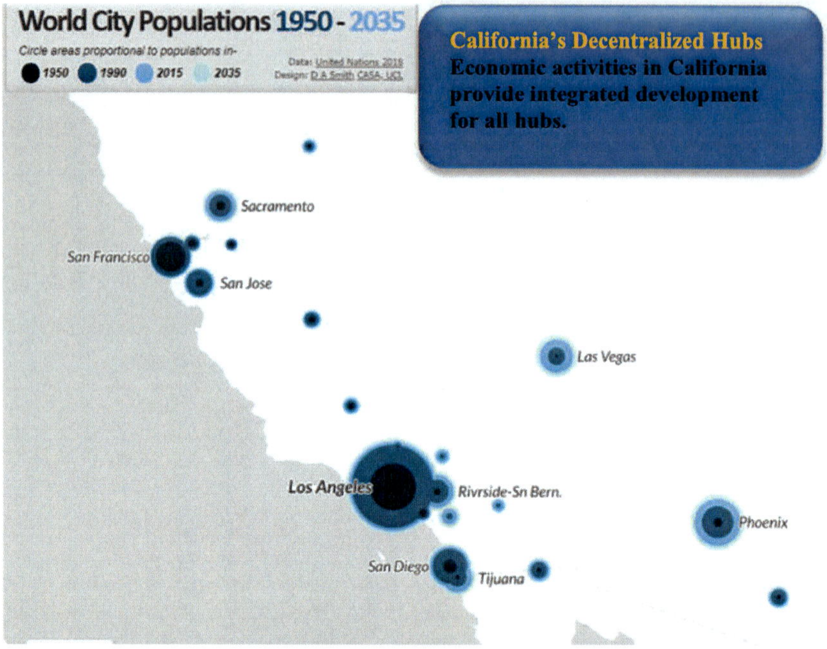

Fig. 15 Populations of California's main decentralized hubs (World City Populations Interactive Map 1950–2035, 2020)

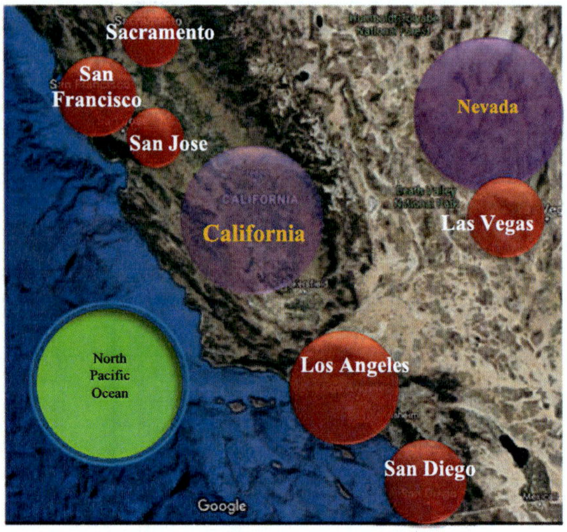

Fig. 16 California's regional economic hub locations

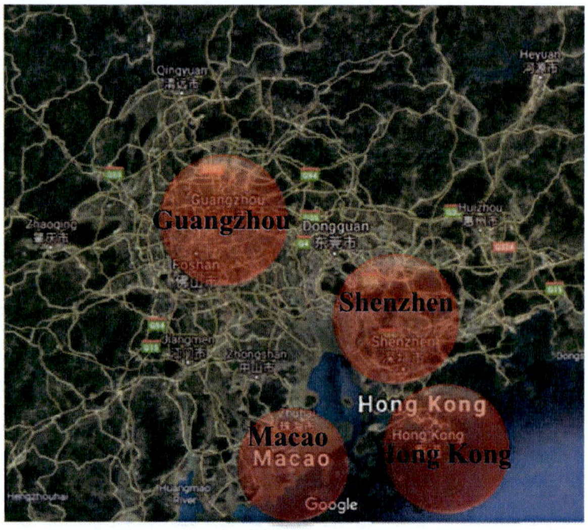

Fig. 17 Guangdong's multi-sector hubs. The economic hubs in Guangdong Province are Guangzhou (industrial and agricultural hub) and Shenzhen (technology hub), which are connected to Hong Kong (trade hub)

petrochemical processing and metal, shipbuilding and ship repairing, and machinery manufacturing (*Guangdong—History | Britannica*, 2020) (Figs. 17 and 18).

The integration in the economic hub provides regional economic capital; that is, Guangdong is supported by the economic hubs of Guangzhou, Shenzhen, and Hong Kong, making it attractive for the population, as with the Shanghai zone. The educational hub is based on dozens of higher learning institutions which located in several cities in the region, notably in Guangzhou, such as the South China University of Technology, Jinan University, Sun Yat-sen University, and the Guangzhou University of Chinese Medicine.

The transportation hub is based on the linked waterways of the Pearl River system. Moreover, several ports and coastal and international shipping routes are linked to more

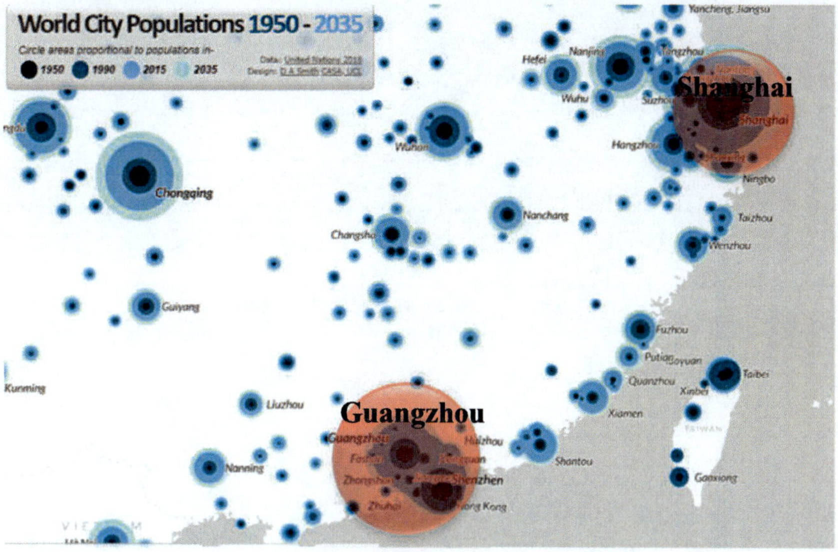

Fig. 18 Guangdong, China: logistics hubs (World City Populations Interactive Map 1950–2035, 2020)

than 100 small and large ports. Leading ports such as Huangpu (Guangzhou's seaport), Shantou, and Zhanjiang are of national significance. Guangdong developed highway networks in China which is considered one of the best in the world; this highway is running primarily along river valleys. Interprovincial links, both for railroads and highways, it is usually run north–south. In the early 1960s, the vital Beijing-Guangzhou railroad was double-tracked (*Guangdong—History | Britannica*, 2020).

4.2.1 Guangzhou Manufacturing Hub

Guangzhou is considered the main manufacturing hub of the Pearl River Delta (Fig. 19) and one of mainland China's leading manufacturing and commercial regions (Yule & Burnell, 2013). Its GDP reached in 2017 to ¥2150 billion (US$318 billion), or ¥150,678 (US$22,317) in per capita terms (Stand Rate-Exhibitor—China Import and Export Fair (Canton Fair), 2017). The city development is becoming greener and is considered one of the most livable cities in China. Guangzhou is an integrated hub of the region's activities and includes the political, economic, educational, cultural, and military hubs of southern China. Guangzhou developed into a multi-functional mega-city after more than 2000 years of changes. It is also a famous tourist hub in South China and an important coastal port city. In its overall strength, after Beijing and Shanghai Guangzhou is ranked third in China (OECD, 2015). Guangzhou port is one of the first ports for import and export trade in China, and Guangzhou's strong commercial hub's influence on the country is embodied by the rapid development of its manufacturing industry. The port of Guangzhou is considered the main seaport of Guangdong province, and currently, it is the largest comprehensive port in South China. As its international maritime trade reaches more than 80 countries over 300 ports in and districts worldwide, the port is economically significant and is a transport center for Guangdong province and the Pearl River Delta region, as well as industries located in neighboring provinces such as Guangxi, Hunan, Hubei, Yunnan, Guangzhou, Jiangxi, and Sichuan (Port of GUANGZHOU-ShippingOnline.Cn, 2020). Guangzhou is known throughout the country for its vigorous free trade hub. In addition, it has always been China's most market-oriented first-tier city, as Guangzhou is considered the lowest cost of doing business. Currently, around 60% of the world's top 500 firms invested in Guangzhou, and some 120 firms established main or regional headquarters in Guangzhou (OECD, 2015).

4.2.2 Shenzhen: The Best Hub in China for IoT Product Manufacturing

Shenzhen is located in south-central Guangdong *sheng* (province), southeastern China. It has a long coast at the South China Sea, it is immediately north of Hong Kong. It marked an economic milestone when it surpassed the 2 trillion yuan GDP mark, and in 2017, the growth was 8.8%. This growth cemented its role as an engine of the Pearl River Delta (*Shenzhen Surpasses*, 2017).

Before 1980, Shenzhen was a small border city; the total inhabitants were 30,000 that served as a customs stop into mainland China from Hong Kong. Since 1980, the city grew at a phenomenal rate, as it attracted a tremendous amount of domestic and foreign capital. Shenzhen became a special economic zone (SEZ) for Chinese economic reform (Ge, 1999). The main reason is Shenzhen was opened to foreign investment, managerial expertise, and technology through the establishment of foreign-owned, joint-venture, and other

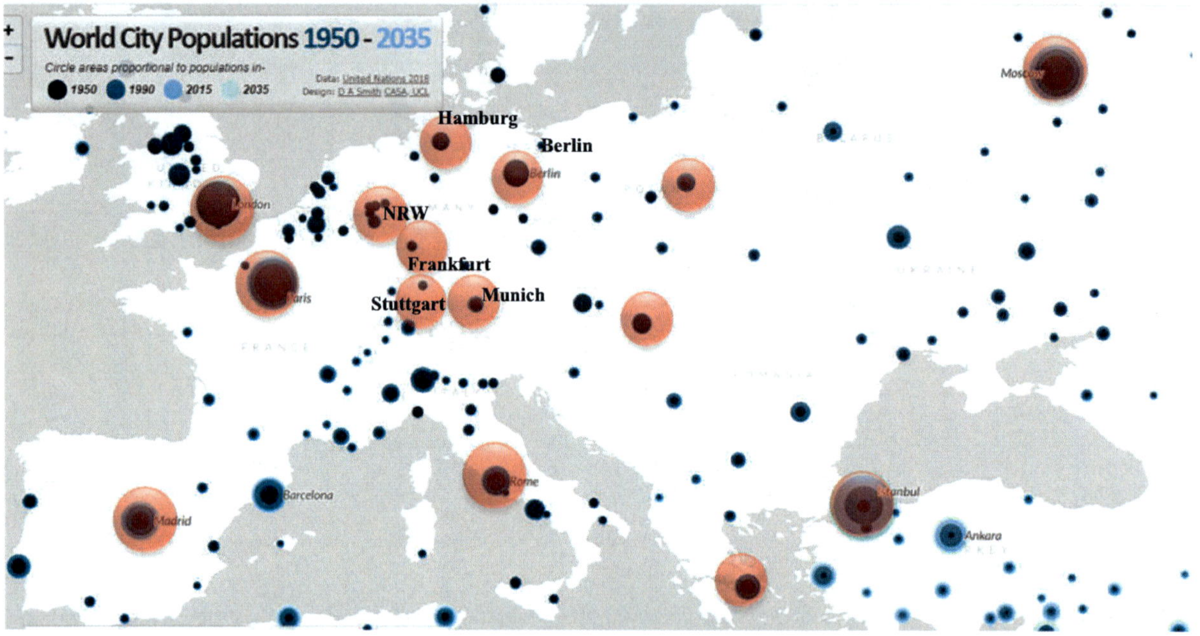

Fig. 19 Europe's main economic hubs. The population map shows that the most attractive zones are those with the highest GDP (refer to Table 3) (*World City Populations Interactive Map 1950–2035*, 2020)

business enterprises without the prior approval of the central government. Moreover, Shenzhen received sizable loans from Chinese banks, which were used to build new schools, roads, housing, and to provide power, water, and communications facilities for its burgeoning population.

Over 40% of the output is from "innovative" businesses such as the telecoms, biotech, and Internet (*Shenzhen Surpasses*, 2017). The rapid growth rate in Shenzhen became known in China as "Shenzhen speed," and the success of the project prompted the Chinese government to convert several other coastal cities as a port hubs into SEZs and that encouraged modernization in different fields (*Shenzhen Surpasses*, 2017).

4.3 Germany: A Decentralized Country in Europe

The level of political and administrative decentralization in European countries offers a successful model for economic development. Strong local governments follow the decentralized development in countries such as Switzerland, Germany, and Spain. During the 1960s, 1970s, and 1980s, developing countries faced highly centralized political and administrative systems. Germany is considered a decentralization country and is supported by several industrial hubs. Germany's process of rapid social transformation and economic, Germany did not suffer from too much centralization (Schade & Ludwig, 2013).

Germany has six main economic hubs among the first 30 hubs in Europe refer to Table 3. The decentralization strategy was very successful for the German economy (see Table 3). The economies of Europe are centralized at capital cities (see Fig. 19).

The analysis below addresses an economy that attracted residents by analyzing Hamburg as an important port hub.

4.3.1 Hamburg Is a Port Hub

Hamburg is located on the Elbe River in northern Germany. It is the largest port in Germany and an industrial and commercial hub in northern Germany. It is connected to the North Sea by the Elbe River and crossed by many canals, while also containing large areas of parkland. The port of Hamburg is a seaport on the Elbe River approximately 110 km from its mouth on the North Sea. Hamburg is the most populous city in Germany after Berlin, as the Hamburg port is one of the largest and busiest ports in Europe. Hamburg has dealings with many nations and has more consulates than any other city in the world except New York City. Shipping and Trade are Hamburg's lifeblood for centuries. In addition, many attractions contributed to a growing tourist industry (*Hamburg | History, Population, Points of Interest, and Facts | Britannica*, 2020), making it a tourism center second only to Berlin.Hamburg has become Germany's major industrial city. All processing and producing industries have representation in Hamburg. The industrial hub handles most of the country's copper, and the Norddeutsche Affinerie, on Veddel, and it is Europe's

Table 3 Europe's main economic hubs and the highest GDP regions

Rank	Metropolitan area	Product (in billions EUR)	Product (in billions $)	Country GDP $	GDP %	Population	Country population	Population %	City main activity
1	Paris, France	681	783.15	2794.70	28.02	12,193,865	67,372,000	18.10	Capital city
2	London, UK	665	764.75	2808.90	27.23	14,187,146	66,040,229	21.48	Capital city
3	Milan, Italy	594	683.1	2086.91	32.73	4,316,398	60,390,560	7.15	Automotive industry and tourism hub
4=	Rhine-Ruhr, Germany	310	356.5	4029.14	8.85	12,190,000	82,887,000	14.71	Technology and tourism hub
4=	Madrid, Spain	310	356.5	1437.05	24.81	6,476,838	46,733,038	13.86	Capital city
6	Rome, Italy	297	341.55	2086.91	16.37	4,353,738	60,390,560	7.21	Capital city
7	Turin, Italy	280	322	2086.91	15.43	2,277,857	60,390,560	3.77	Automotive and aerospace hub
8	Randstad, Netherlands	263	302.45	909.887	33.24	8,219,380	17,305,660	47.50	Capital city
9	Frankfurt, Germany	173	198.95	4029.14	4.94	2,671,358	82,887,000	3.22	European central banking hub (business) and transportation hub
10	Bologna, Italy	170	195.5	2086.91	9.37	944,297	60,390,560	1.56	Industrial hub
11	Florence, Italy	162	186.3	2086.91	8.93	1,007,435	60,390,560	1.67	Industry and tourism hub
12	Munich, Germany	147	169.05	4029.14	4.20	2,879,107	82,887,000	3.47	Industrial hub
13	Berlin, Germany	146	167.9	4029.14	4.17	5,207,915	82,887,000	6.28	Capital city
14	Naples, Italy	140	161	2086.91	7.71	3,107,006	60,390,560	5.14	Port city, tourism, and agricultural industries
15	Barcelona, Spain	137	157.55	1437.05	10.96	5,474,482	46,733,038	11.71	Port city hub
16	Brussels, Belgium	133	152.95	536.055	28.53	2,513,849	11,449,656	21.96	Capital city
17	Hamburg, Germany	132	151.8	4029.14	3.77	3,282,164	82,887,000	3.96	Port city hub
18	Stockholm, Sweden	131	150.65	554.659	27.16	2,325,353	10,215,250	22.76	Capital city
19	Stuttgart, Germany	120	138	4029.14	3.43	2,757,930	82,887,000	3.33	Automotive hub
20	Vienna, Austria	111	127.65	459.401	27.79	2,811,186	8,857,960	31.74	Capital city
21	Copenhagen, Denmark	102	117.3	354.683	33.07	2,014,225	5,806,015	34.69	Capital city
22	Oslo, Norway	96	110.4	441.439	25.01	1,271,127	5,323,933	23.88	Capital city
23	Athens, Greece	95	109.25	218.057	50.10	3,773,559	10,768,193	35.04	Capital city
24	Bolzano, Italy	92	105.8	2086.91	5.07	107,317	60,390,560	0.18	Intensive agriculture boundary hub
25	Dublin, Ireland	84	96.6	366.448	26.36	1,917,677	4,857,000	39.48	Capital city

(continued)

Table 3 (continued)

Rank	Metropolitan area	Product (in billions EUR)	Product (in billions $)	Country GDP $	GDP %	Population	Country population	Population %	City main activity
26	Helsinki, Finland	75	86.25	276.553	31.19	1,638,293	5,522,015	29.67	Capital city
27=	Lyon, France	74	85.1	2794.70	3.05	1,860,112	67,372,000	2.76	High-tech industries
27=	Manchester, UK	74	85.1	2808.90	3.03	3,287,460	66,040,229	4.98	Industrial hub
29	Lille, France	72	82.8	2794.70	2.96	2,612,189	67,372,000	3.88	Food industry and boundary hub

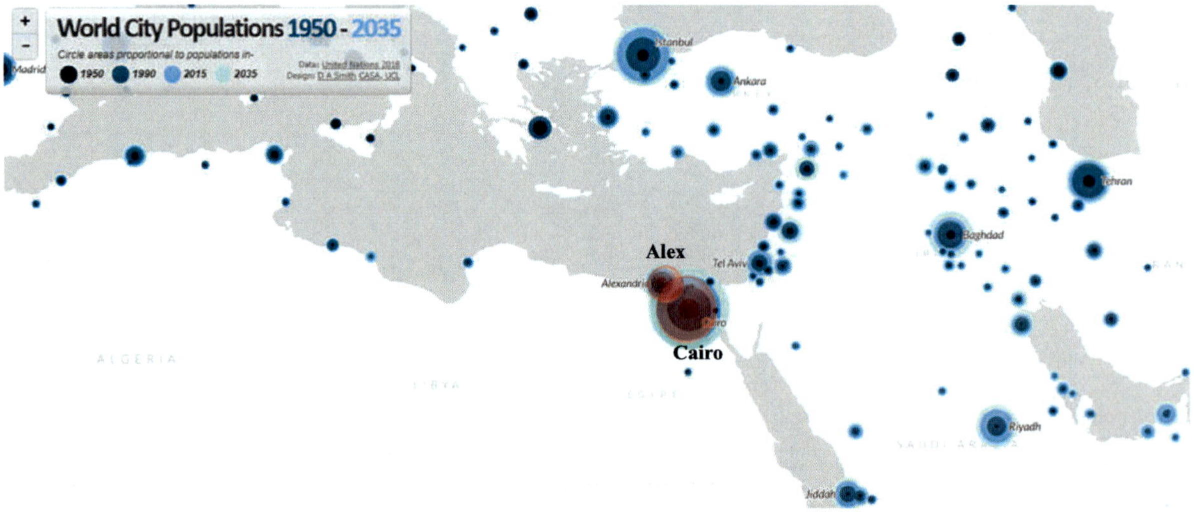

Fig. 20 Centralization of Egypt at Cairo and Alexandria (*World City Populations Interactive Map 1950–2035*, 2020)

second-largest copper works. The shipbuilding industries and chemical steel are the most important industries. Moreover, the nuclear plants at Krümmel and Brunsbüttel (energy hub) provide power at a reasonable cost to the industries bordering the Unterelbe and to parts of Hamburg. In Germany, this urban center handled almost half of West Germany's foreign trade not only in shipping but also through rail and airfreight. The main imports in the area are vegetable oils and fats, coffee, tea, petroleum, tropical fruit, and uncured tobacco. The exports include machinery, electro-technical merchandise, processed crude oil fuel and lubricants, copper, and pharmaceuticals. The greatest economic hub of the European country, since 1960, this urban center gained an excellent position at trade fairs.

The harbor is Hamburg's "gateway to the planet." Nearly 15,000 ships from over 100 countries visit it every year. The city's Übersee-Zentrum is the world's largest roofed warehouse, and therefore, the Waltershof instrumentation terminal is the largest of its kind on the continent.

The harbor and the town square are well-served by the German railway network, and therefore, the hub incorporates a smart system of buses and underground trains, as well as a tunnel (opened in 1977) underneath the Elbe along the Stockholm-Lisbon route to alleviate the long-distance traffic.

5 Proposal of Decentralization Strategies to Create Economic Hubs in Egypt

Egypt is considered the third-most populous on the African continent and the most populated country in the Middle East. Nearly 100% of the country lives in three major regions of the country: Cairo, which will reach 28.5 million by 2035, Alexandria, and the third region is the Nile Delta, which will reach 7.2 million (see Fig. 20). The Egyptian Government tried to encourage migration to the newly irrigated land reclaimed from the west and east deserts, with mixed success. However, the proportion of the population living in

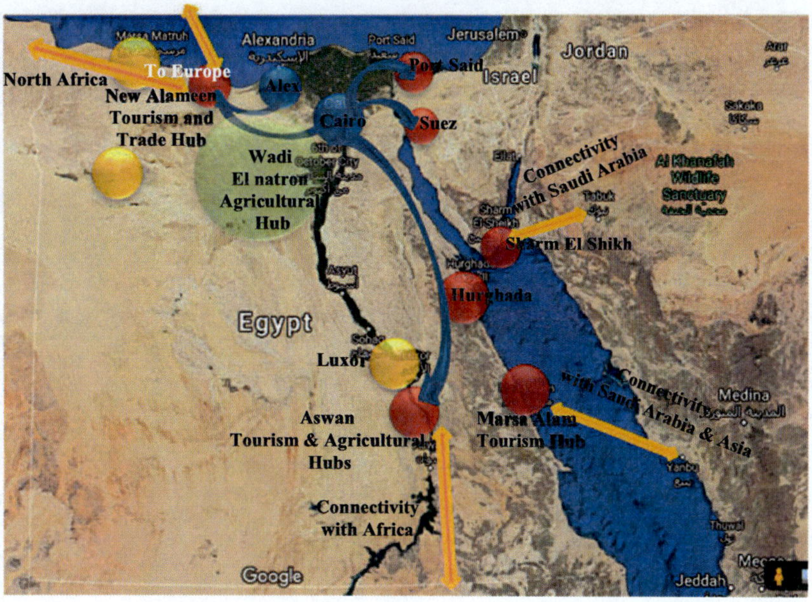

Fig. 21 Decentralization strategy for Cairo and Alexandria: providing boundary hubs at the Red Sea and Mediterranean Sea to provide connectivity with Europe, Asia, North Africa, and South Africa. It will consist of trade, tourism, and agricultural hubs

rural areas continued to decrease as people moved to Cairo as it is the capital of Egypt in search of employment opportunities and a higher standard of living. Intraregional migration in Egypt mostly occurs from south to north; that is, from Upper Egypt to the Lower Egypt Nile Delta and Cairo. Migration from Upper Egypt to Alexandria and Cairo increased considerably during the past decades although Upper Egypt has a lot of potentials, as more people leave rural areas and move to big cities to gain a better quality of life. By both push and pull factors between the regions, these voluntary migrations are caused, which affected Egypt socially, politically, and economically.

This section reflects on the implementation of regional economic hubs in Egypt (see Fig. 21). This research aims to facilitate a decentralization strategy for the population from the capital city of Cairo. The below map show the integrated economic hubs. The hubs were designed to connect the existing cities around the Suez Canal (Port Said, Ismailia, Suez, and Ain Sokhna). Clearly, the Suez Canal is an important link between the eastern and the western political economies.

5.1 Economic Activities in the Suez Canal Zone

The Suez Canal Zone serves as the main trade hub and is centralized between Africa, Asia, and Europe. As a zone, it is surrounded by the Red Sea, the Suez Gulf, the Saint Catherine Mountains, and the agricultural lands of North Sinai. It is also near the Mediterranean Sea, which supports significant international trade in the area. In addition, the Suez Canal Zone is surrounded by natural resources such as metals, oil, and gas, which promotes industrial development. Tourism is a booming industry, as the area has many historical locations. The core services of this hub should include ports, airports, railways, trade, tourism, and various smaller industries.

Technology industries such as silica are already available in the zone. These services, when integrated, will provide Sinai with massive social and economic development opportunities. The hubs should be attached to existing cities to provide the facilities and smart services that the region lacks. The integration between the hubs and old cities will lead to development in Sinai at the eastern side, in Sharm El Sheikh, and the northern side of Europe, whereas the southwestern direction will be linked with Hurghada. The map clearly presents the main concept of economic regional hubs as multi-sector (see Fig. 21). This integration will provide the regional capitals with opportunities to attract Cairo's highly congested population, which is currently more than 20 million inhabitants. Thus, employment opportunities and developing infrastructure will attract the population from the capital city to the hub.

The hubs should be attached to existing cities to provide facilities and smart services lacking in the region. The integration between the hubs and old cities will lead to development in Sinai at the eastern side, in Sharm El Sheikh, and the northern side of Europe, whereas the southwestern direction will be linked with Hurghada. The map clearly presents the main concept of economic regional hubs as multi-sector (see Fig. 21). This integration will provide the regional capitals with opportunities to attract Cairo's highly congested population, which is currently more than 20 million inhabitants. Thus, employment opportunities and a

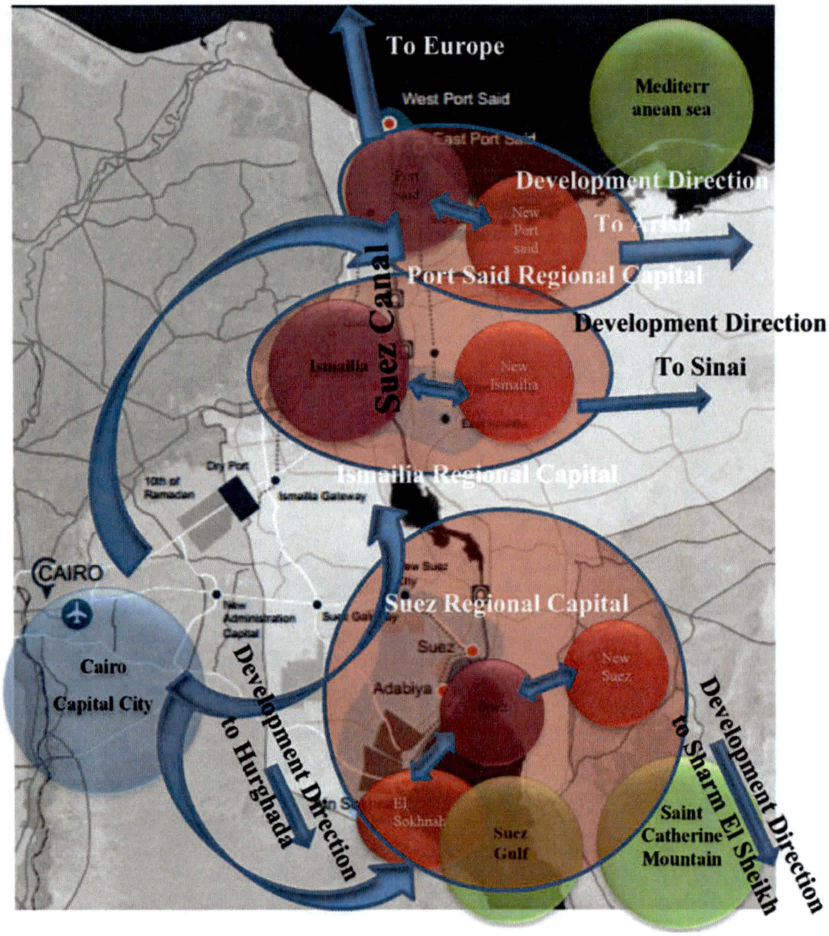

Fig. 22 Regional development strategy for the Suez Canal Zone (multi-sector development). *Source* Authors

developing infrastructure will attract the population from the capital city to the hub (Fig. 22).

6 Conclusion

Decentralized regional economic hubs will provide sustainable development for the region as it will decentralize the population in the capital city and supply the region with services. The hub will be integrated with the existing small and medium cities to create regional economic capitals. The decentralized economic hubs can provide spatial equality, enhance economic productivity, and help attain environmental sustainability. Such hubs will attract new generations, international investments and highly skilled labor to establish resilient and more sustainable communities.

It is very likely that such hubs will evolve to become international capitals for social and economic development. Sustainable economic regional hubs serve as new economic centers providing a range of economic services to large regions throughout the international political economy. The hubs can become pivotal points for the region's cities as it is centralized between multiple sectors of the economy at the regional level and connect small cities with the transportation networks and smart services of larger entities. The regional economic capitals will provide the missing and much-needed services based on a pre-designed strategy for sustainable development.

In addition, these cities will provide the population with access to advanced education, employment opportunities, and personal and professional services. This study considered economic rates, population size, agricultural development, industry services, employment, tourism, economic sustainability, health rates, quality of life measures, healthcare accessibility, quality of education, and availability of natural resources, provision of energy and freshwater, and environmental threats as indicators of GDP growth. The statistic is based on regional development measured in GDP compared with all other regions and countries. The decentralization strategy will be more efficient when it aims to provide economic activities and smart services for all regions. This study discussed the challenges of centralized development in metropolitan and capital cities and accordingly proposed a decentralization strategy for Egypt based on establishing economic hubs to support and integrate with medium and small cities by providing the missing core services for the region.

References

Guangzhou port completes dredging project. (2009). Bairdmaritime.com. https://www.bairdmaritime.com/index.php?option=com_content&view=article&id=1256:guangzhou-port-completes-dredging-project-&catid=68&Itemid=59&el_mcal_month=3&el_mcal_year=2010

New York city metro area population 1950–2020 | MacroTrends. (n.d.). Retrieved August 5, 2020, from https://www.macrotrends.net/cities/23083/new-york-city/population

The ultimate whistler tack-on trip. (2020). Retrieved August 16, 2020, from https://www.ski.com/blog/2-days-in-vancouver-b-c-the-ultimate-whistler-tack-on-trip/

World city populations interactive map 1950–2035. (n.d.). Retrieved August 6, 2020, from https://luminocity3d.org/WorldCity/#5/39.606/-121.992

Central places theory (Market principle) | The geography of transport systems. (n.d.). Retrieved June 5, 2020, from https://transportgeography.org/?page_id=1457

Five China cities join ranks as global logistics hubs. (n.d.). CBRE. Retrieved August 4, 2020, from https://www.cbre.com.cn/en/about/case-studies/five-china-cities-join-ranks-as-global-logistics-hubs-

Adams, W. M. (2006). *The future of sustainability re-thinking environment and development in the twenty-first century*. The World Conservation Union. Scientific Research Publishing. Retrieved August 13, 2020, from https://www.scirp.org/(S(lz5mqp453edsnp55rrgjct55))/reference/ReferencesPapers.aspx?ReferenceID=1637285

Angel, S., Parent, J., Civco, D. L., & Blei, A. M. (2012). *Making room for a planet of cities*. Lincoln Institute of Lang Policy.

Blumenthal, G., Jose, S., Greig, J. M., Aitken, J., Levy, S., Uhl, M., Boland, J., Yost, D., Mccalmont, T. O. M., Mccown, J., & Sager, S. R. (2018). *2018 silicon valley index*. https://jointventure.org/publications/institute-publications/1640-2018-silicon-valley-index

Cox, C. (2014). *An introduction to LTE: LTE, LTE-advanced, SAE, VoLTE and 4G mobile communications* (2nd ed., Vol. 9781118818). https://doi.org/10.1002/9781118818046

Dempsey, C. (2016). *Geography of fortune 1000 companies*. Geo Lounge. https://www.geolounge.com/geography-fortune-1000-companies-2016/

FMI. (2014). *World economic outlook update—An uneven global recovery continues*. https://www.imf.org/en/Publications/WEO/Issues/2016/12/31/An-Uneven-Global-Recovery-Continues

Ge, W. (1999). *Special economic zones and the economic transition in China* (Vol. 5). World Scientific. https://doi.org/10.1142/4041

Glaeser, E. L., & Gottlieb, J. D. (2006). Urban resurgence and the consumer city. *Urban Studies, 43*(8), 1275–1299. https://doi.org/10.1080/00420980600775683

Glaeser, E. (2010). Urban economics. In *The heart of teaching economics: Lessons from leading minds*. https://doi.org/10.12681/rp.18554

Guangdong. (n.d.). *History, Britannica*. Retrieved August 3, 2020, from https://www.britannica.com/place/Guangdong/History

Guangdong Province. (n.d.). Retrieved August 1, 2020, from https://www.china.org.cn/english/features/48080.htm

Hamburg. (n.d.). *History, population, points of interest, & facts, Britannica*. Retrieved June 4, 2020, from https://www.britannica.com/place/Hamburg-Germany

Harold Mackintosh, 1st Viscount Mackintosh of Halifax. (n.d.). *Wikipedia*. Retrieved June 1, 2020, from https://en.wikipedia.org/wiki/Harold_Mackintosh,_1st_Viscount_Mackintosh_of_Halifax

Jennings, D., & Disney, J. J. (2006). Designing the strategic planning process: Does psychological type matter? *Management Decision, 44* (5), 598–614. https://doi.org/10.1108/00251740610668860

Kelly, R., Sirr, L., & Ratcliffe, J. (2004). Future thinking to achieve sustainable development at local level in Ireland. *Foresight, 6*(2), 80–90. https://doi.org/10.1108/14636680410537547

Liaros, S. (2019). Implementing a new human settlement theory: Strategic planning for a network of regenerative villages. *Smart and Sustainable Built Environment*. https://doi.org/10.1108/SASBE-01-2019-0004

Martinus, K. (2010). Planning for production efficiency in knowledge-based development. *Journal of Knowledge Management, 14*(5), 726–743. https://doi.org/10.1108/13673271011074863

Mills, C. W. (1959). The sociological imagination. *Mississippi Valley Historical Review, 46*(2). https://doi.org/10.2307/1891592

OECD. (2015). *National urban policy review: China 2015*. OECD Publishing. https://doi.org/10.1787/9789264230040-en

Openshaw, S., & Veneris, Y. (2003). Numerical experiments with central place theory and spatial interaction modelling. *Environment and Planning A, 35*(8), 1389–1403

Otenyo, E. E., & Lind, N. S. (2006). Part III: managing institutions through planning and decentralization. *Research in Public Policy Analysis and Management, 15*(06), 359–369. https://doi.org/10.1016/S0732-1317(06)15051-4

O'Sullivan, A. (2012). *Urban economics*. McGraw-Hill/Irwin.

Port of Guangzhou. (n.d.). *ShippingOnline.cn*. Retrieved August 4, 2020, from https://www.shippingonline.cn/port/result.asp?id=efi

Schade, J., & Ludwig, R. (2013). *Forest carbon baseline study in Leyte as a federally owned enterprise*. GIZ Supports the German Government in Achieving its Objectives in the Field of International Cooperation for Sustainable Development.

Schwan, G. (2019). *Sustainable development goals*. Gaia. https://doi.org/10.14512/gaia.28.2.1

Shenzhen Surpasses US$338 Billion GDP Mark in 2017, Beats Hong Kong and Singapore's Growth. (n.d.). *South China Morning Post*. Retrieved August 4, 2020, from https://www.scmp.com/news/china/economy/article/2128310/shenzhen-88-cent-hi-tech-growth-roll-hit-y2tr-2017#comments

Stand Rate-Exhibitor. (n.d.). *China import and export fair (Canton Fair)*. Retrieved August 4, 2020, from https://www.cantonfair.org.cn/html/cantonfair/en/exhibitor/2013-12/29146.shtml

Thomas, H. (1997). *The superlative man*. Farrar, Straus, and Giroux.

World Population Review. (2018). *2018 world population by country*. https://worldpopulationreview.com/

Yule, H., & Burnell, A. C. (2013). *Hobson-Jobson: The definitive glossary of British India*. Reprinted by Oxford University Press. https://books.google.com/books?id=F51h6q-bB6sC&printsec=frontcover

Consultative Hubs for Democracy Within Neighborhoods in Muscat. Alternatives for Urban and Social Development Strategies

Rowa Elzain

Abstract

The case of Khoud 6 in Muscat, Oman, manifests the collective power of inhabitants within local neighborhoods. The research presents interventions concerning physical infrastructure and youth capacity development within the neighborhood carried by various men and women committees. This model of decision-making and financing process raise the question to the existent of democratic practices within neighborhoods in a monarchy governing structure. Following a qualitative research approach, the research evaluates the governing structure of the mosque and surrounding facilities in Khoud 6, using local, regional and western principle of participation, social capital and democracy. The research focus on both political and social definitions that are legally binding and socially defined. The result presents a local participatory model within a neighborhood scale alongside culture values that governs urban development. The democratic model practiced in Khoud 6 evaluated alongside Arab philosophers and scholars articulation on democracy provides an environment that is inclusive to the women, men, and children of the neighborhood. Recommendation on policy levels for neighborhoods in the Middle East region in inclusive and democratic decision-making process is outlined. The research concludes with an outlook on democratic processes within neighborhood scale that carries direct impact on the lives of those who live within a defined community along with the hierarchical framework of planning.

Keywords

Participatory neighborhood development • Democratic decision-making • Social capital • Mosque activities

R. Elzain (✉)
MCTspaceLab, Muscat, Oman

1 Introduction

"A great city is more than stone and steel. It is also an act of civic imagination. Fact and Allegory fuse in a civic myth that interprets experience, justifying the city's history and uniqueness. The civic myth is a story of the city's founding, its heroes, and core values. It is important not because it is true, but because believing it gives meaning to civic endeavors. It helps its keepers explain the city to strangers. It reflects the culture of the city—the rules and folkways by which the city operates—of which it is both author and product. It is the city's soul." (Hanson, 2003, p. 15).

Cities and neighborhoods, as outlined by Royce Hanson, is more than a physical structure, it is the social structure that operates and activates these physical boundaries defined by urban planners, architects, and governmental officials. When exploring the role of the citizens within these spaces, question regarding level of their involvements are raised. In various governing structure, numerous citizens' interactions take place. With current trends hitherto sifting toward the use of technology within the city, primary focus still remains on the improvements of quality of peoples live at its center (Conley, 2017). Thus, questioning the role of resident is vital. Are they a source of information and data to feed into a matrix, or an engine cable of developing and steering toward solutions? Recent discourses look into the sharing of this information, focusing on transparency for decision-making by various stakeholders (Hoang et al., 2019). This relies heavily on mutual understanding between actors is a key factor for success (Suurendonk et al., 2010). In the case of Muscat and the lack of facilities within new emerging neighborhoods, some neighborhoods are witnessing the creation of self-organized residents' committees that are taking steps toward the improvements of their livelihoods following a democratic decision-making process that ensures the involvement of various group members, while retaining a mutual understanding and line of communication with legal

F. Trapani et al. (eds.), *Advanced Studies in Efficient Environmental Design and City Planning*,
Advances in Science, Technology & Innovation, https://doi.org/10.1007/978-3-030-65181-7_20

bodies and other stakeholder involved. Hence, this paper aims to understand the formation of such communities.

1.1 Urban Development Within Neighborhoods in Muscat

"Since the early 1970s, Oman's economy has depended mainly on oil revenues." (Al-Farsi, 2013). Through petroleum revenue incentives, the development of Muscat is similar to that of car-based cities as a result of rapid urbanization. However, it is vital to highlight that the petroleum revenue was not the only factor for the horizontal urban expansion of the city, rather, it is simply the catalytic support with rural–urban migration and urban planning policies. In the case of Muscat, the capital of Oman and home to over 30% of the country's population (NCSI, 2019), residential neighborhoods primarily consist of single residential villas that consumes urban lands in the city. According to study carried by Sonja Nebel and Aurel von Richthofen, the individual family house covers over 40% of privately owned land (Nebel & Richthofen, 2016). The root cause for such an urban phenomena dates to the royal decrees number 81/84 and 125/2008 that allocates state-owned lands to Omani citizens for residential purposes as a scheme for housing public welfare. As a result, many are entitled to residential land, mostly 600 m^2 with a negligible fee.

Following the order of the royal decree, the Ministry of Housing is responsible for planning these new residential neighborhoods alongside the distribution of the plots through a random selection system. Master planning, hitherto, is carried following an outline of Weidleplan 1991 Muscat master plan, aiming toward the expansion of the city to the west (in direction of Seeb province) as a relief of the congestion in the eastern side of the city (Nebel & Richthofen, 2016). Nevertheless, since 2017, the Supreme Council of Planning has been working on the "Oman National Spatial Strategy" as a step toward unifying a national and regional development plan. Following distribution of plots to citizens, Muscat Municipality provisional role comes to action. From the provision of basic infrastructure within neighborhoods to laying out regulations and codes for building, this provision of neighborhood infrastructure is usually executed within phases spanning a decade or more. This correlates to the construction of these plots independently by multiple land owners; whenever it is convenient. Such neighborhood takes decades to completely reach its full built-up capacity. The case of Khoud 6 showcases a neighborhood that hitherto reached the built-up capacity of 64% after 11 years from initial construction. Hence, neighborhoods tend to accommodate the missing facilities through communal initiatives in means of temporary and permanent solution till the municipality infrastructural provision reach their neighborhoods. Aiming to understand the role of citizens in urban development within neighborhood, the research explores the relations between democracy, participation, and urban development through the case study of Khoud 6 in Muscat. Thus, an understanding of democracy within regional level and western definition is essential, alongside defining forms participation, and urban development.

1.2 Democracy, Participation, and Democratic Participatory Process

Defining Democracy

The common understanding of the term democracy, perhaps in the field or urban planning, is similar to the definition of UN General Assembly stating "democracy is a universal value based on the freely expressed will of people to determine their political, economic, social and cultural systems and their full participation in all aspects of their lives," (UN General Assembly, 2009). In addition to freedom, rights, and participation, the phrase "democracy" is often referred to as a "democratic process" in decision-making, drawing a relation between democracy and the idea of "majority ruling". Along these terms, John H. Hallowell outlines the idea of the majority ruling by "The principle of the majority rule is founded upon the belief that the widest possible popular discussion and participation in the formulation of policy… to represent not a portion of the society but the whole people." (Hallowell, 1954). However, Robert M. MacIver in his normative empirical definition of democracy, provides a distinction between democracy and majority ruling by "Democracy is not a way of governing, whether by the majority or otherwise, but primarily a way of determining who shall govern and, broadly, to what ends." (MacIver, 1947).

Arab scholar, Mohammed Al-Jabri definition of democracy correlates to prior definition as he states "Democracy is a sound and positive method to regulate relations inside the society in a rational manner, directing the struggle toward the advancement of society as a whole, within the citizens enjoyment of his right" (Al-Jabri, 2009). Nevertheless, Al-Jabri also argues that west definition of democracy referring to "political" and "liberal" freedom as a contradiction to the concept of democracy itself as "… "political freedom and economic freedom", although freedoms of course, are only freedoms for those who have access to them… only the "upper crust" benefits from political and economic freedom." (Al-Jabri, 2009). Vali Nasr outlines the position of "Muslim Democrats" in the Middle East and the Islamic world, as "… their main goal tends to be the more mundane one of crafting viable electoral platforms and stable governing coalitions to serve individual and collective interests… within a democratic arena whose bounds they respect." (Nasr, 2005).

The term democracy in the Arab world is often related to the term "Shura". In an attempt to draw similarities with the two terms, Arab scholars such as Abduh and Jamal Al-Din Al-Afghani state that democracy is merely the western translation to the word "Shura" (Al-Jabri, 2009). Contradictorily, Al-Jabri definition of Shura extracted from verses of Quran, exhibits a clear difference between Shura and democracy as reflected in his statement "Al-Shura as consultation is an exchange of opinion, and seeking the opinion of whoever is capable of giving it… "Taking the opinion" does not commit the "taker", nor are the "givers of the opinion specified… authoritative concept of al-Shura falls under "good morals" and "commendable behavior" but not under "obligations and duties"… Thus, the question of al-Shura in Islam remains as a matter of advice, as a virtue in the ruler and never a question of obligation and duty." (Al-Jabri, 2009).

Participation in Urban Development

In urban development projects, participation, and democratic process are often referred to in the level of interaction, contribution, and involvement with the end users of the proposal, may it be a law or a project. Sherry Arnstein's concept of "Ladder of Citizen Participation" gives an outlook on different powers given to citizens through various interactions. Arnstein ladder "rungs" of participation are categorized into three: nonparticipation, tokenism, and citizen power. Hence, eight levels of citizen participation are contested beginning with "Manipulation" having least power and "Citizen Control" as the uttermost exercise of citizen empowerment. Manipulation refers to as "a vehicle to power holders" where titles of community groups are used as prove for community engagement. On the other side of the ladder, citizen control, suggest a model where citizens are in the managerial position on the execution of proposal. In neighborhoods participatory projects, Arnstein states "A neighborhood corporation with no intermediaries between it and the source of funds is the model most frequently advocated." (Arnstein, 1969).

Through the suggestion of a democratic design methodology and collaborative environment in the planning of neighborhoods, Howell and Simmons delineate crucial issues aiming at a true citizen control, whether it is in consultations or the involvement of the community in general. This is fundamental since "…if you cannot interest people in the substance of what you are trying to do, neither of these is likely to be decisive and any improvement will be short-lived." (Howell & Simmons, 2015). They further elaborate the need of forming a consultative environment where "local people find ways of consulting local people about their views" maybe it be through online discussion or the more effective, exhibitions and seminars as "low-tech" mediums. Strategies that call for citizen engagement; co-production or asset-based community development, are approaches that "… do not frame communities simply as bundles of needs, but rather as assets with problem solving capacity." (Howell & Simmons, 2015). An example is further given through "The Neighborhood Plan" as binding document that of its first kind, to give power to the communities to decide on design and development of proposals within their neighborhoods.

Nevertheless, participation planning in urban development is "no substitute for democratic politics" (Sirianni, 2007). Rather, participatory planning is "… the political, institutional, cultural, and social conditions capable of stimulating ways of territorial governance which, being based on participation, on collaboration between different entities, and on institutional coordination… to achieve results that would not be achievable otherwise" (Ferrao, 2015). Hence, a collaborative and participatory environment, eventually "… would benefit the developer, since it should increase creative ideas, improve ecological and cost efficiencies, and reduce conflict by giving the community a sense of ownership of the planning and design process." (Birkeland, 1999). Similarly, Anthony Downs argues that "The ultimate way to raise neighborhood consciousness would be to transfer major elements of the government authority to the neighborhood level… Neighborhood organizations would undoubtedly be more responsive to small-area concerns than many citywide bureaucracies are now… it would make citizen participation in the neighborhood-level activities far more meaningful." (Downs, 1981, p. 177).

Democratic Participatory Process Within Neighborhoods

Although citizens' participation within urban development, regardless of the urban scale, is suggested as an ideal approach for localizing urban development; however, Richard Box, raises a few concerns that needs to be considered from the perspective of the public authority where he states "The movement back toward local control and citizen self-governance has significant implications for the practice of public administration. If we are metaphorically coasting down the backside of the wave of reform that swept the early twentieth century-moving away from professionalism toward citizen control—then "traditional" public administration, based on administrative power, control, and a sense of positional "legitimacy" is a thing of the past." (Box, 1998, p. 35). Furthermore, he outlines the "Principles of Community Governance" which include various working scales, democratic decision-making process, accountability and rationality, as an ideal proposition for communities as "... the community shaped by these principles can be envisioned by the content of the principles themselves. Such community is a place where action is taken at the "lowest" level consistent with good decision making; free and open discourse and decision making are practiced; there is a direct linkage in accountability from citizens to the creation and implementation of public programs; and public decision making is recognized as an important enterprise that deserves careful deliberation" (Box, 1998, p. 148).

In the discourse of citizen control within smaller scales, the concept of "Social Capital" as outlined by Robert D. Putnam is vital as it plays a crucial role within the inner ties that responds to community interactions. Similar to physical and human capital, referring to material objects and individuals' skills and knowledge respectively, social capital which refers to "...connections among individuals—social networks and the norms of reciprocity and trustworthiness that arise from them." (Putnam, 2000, p. 19). These are essential for urban development as whole, including other factors. As a result, "... some forms of Social Capital are good for democracy and social health; others are destructive.[1] Many cities in all the countries represented in this volume have organized neighborhood-based citizens' groups that meet regularly to accomplish a wide variety of purposes. One careful study of such group in some American cities found that they increased government responsiveness to the organized neighborhoods and enhanced citizen' respect for the government." (Putnam, 2002, p. 9). Social capital plays a fundamental role in the living qualities of urban configuration. Hence, safer neighborhoods tend to exhibit higher levels of social capital as outlined by Jane Jacob book "The death and Life of Great American Cities". She outlines that safer neighborhoods exist in planning that account for higher levels of informal contact among neighbors (Putnam, 2000, p. 308). On the contrary, "... the decline in neighborhood social capital—community monitoring, socializing, mentoring, and organizing—is one important feature of the inner-city crisis, in addition to purely economic factors" (Putnam, 2000, p. 312).

Putnam also delineates concerns which occur in cases where social capital within a larger scale and trust is questioned. Hence, cooperation and discussions between authorities and citizens, "... absence of credible sanctions against defection: How can each be confident that the other will keep his word in the face of temptation to shirk? More complex contexts, like modern government (modern markets), bring the added complication of monitoring: How can one agent know whether another did in fact make a "good faith effort" to keep his word, in the face of multiple uncertainties and countervailing pressures? Both accurate information and reliable enforcement are essential to the successful cooperation" (Putnam et al., 1993, p. 165). However, as suggested by Putnam in viable solutions for such situation within a scale of neighborhood communities, "... overcoming dilemmas of collective action and the self-defeating opportunism that they spawn depends on the broader social context within which any particular game is played. Voluntary cooperation is easier in the community that has inherited a substantial stock of social capital, in the form of norms of reciprocity and networks of civic engagement" (Putnam et al. 1993, p. 167).

[1]High levels of social capital could result on the misuse of trust leading to manipulation within a social network.

1.3 Research Methodology

Understanding of the relation between community participation and democratic process raise a question on their influence in urban development. Within the various urban scale; cities, towns, neighborhoods, and quarters, the neighborhood offers an experimental field, where the effects of policies and laws are visible and tangible. Thus, the research takes the neighborhood scale in Muscat as a case study for the examination on how neighborhoods organize themselves in solving urban issues that relate to the urbanization. The research looks at how the neighborhood governs themselves to solve urbanization issues. Hence, these three questions were asked:

- What are the current urban issues within neighborhoods in Muscat?
- How do citizens interact with the legal framework structure responsible for neighborhood development?
- What are the local initiatives carried by citizens for the development of their neighborhoods?

The methodological approach taken for this paper is of a descriptive one. Focusing on the observation of the neighborhood resident and their public spaces development within the past three year as a documentation and basis for further research. The observation was carried throughout focus group discussion within various members of the neighborhood focusing on public space development projects and workshop carried with collaboration with MCTspaceLab. The aim of MCTspaceLab collaboration with Khoud 6 was the upgrade of public spaces within the neighborhood. Using a framework that focuses on awareness for the first phase, MCTspaceLab conducted wood construction, planting, and technology workshops with adolescent of the neighborhoods. However, the focus group discussion that took place between these workshops opened up a platform for new ideas of improving the neighborhood current facilities and infrastructure. The neighborhood residents managed to independently take they suggest idea forward which showcased a high level on internal communal organization. Hence, observations outline the patterns of project development by the neighborhood residents from ideas, thoughts, and discussions to tangible outcomes. Further research through semi-structured interviews and small focus group discussion was carried as a source for primary data collection to understand the decision-making process and collective efforts required for such projects within the neighborhood. Interviewees of representatives from Ministry of Housing, Muscat Municipality and Shura Council provided in-depth understanding of neighborhoods interaction with the legal bodies that shape and regulate their neighborhoods. Secondary data collection was mainly gathered from review of legal documents, media coverage, and laws that were essential for the outline of practiced governing structures within Muscat.

2 Muscat Governing Structure

Understanding governing structure of Oman correlates to understanding the practiced religion. Following the "Ibadism" as a sect in Islam, Oman's religion is asserted by the Basic Statute of the State (BSS). As Sulaiman Al-Farsi states "The 1996 BSS stipulates that Shari'a is the basis of the legislation, and it preserves the freedom to practice religious rights in accordance with tradition provided that such freedom does not breach public order." (Al-Farsi, 2013). The Ibadism sect of Islam in Oman plays a political role as "Ibadis, by contrast, take Shura as a political agenda in which the ruler (Imam) is elected, thereby denying legitimacy to the idea of hereditary rule." (Al-Farsi, 2013). As a result, the Omani political system; dating from the eighth century, is considered the "oldest participatory systems in the Arab world" and is a "good case study for exploring the extent to which the notion of Shura offers an alternative to Western models of democracy" (Al-Farsi, 2013). In Muscat, the participation of citizens takes place in the Shura Council and the Municipal Council in Muscat Governorate. Figure 1 shows the actors involved in neighborhood development in Muscat and their direct/in-direct relation in terms of consultation, regulation, and monitoring. The remaining two entities that deal with neighborhood and citizens are the Shura Council and Muscat Municipal Council. The Shura council is responsible of monitoring the ministries that relate to direct citizen services. While the local municipal council, Muscat, reviews recommendations through consultation process within the Wilayat (province).

The Shura Council, one of Oman parliamentary houses, consists of 85 members, elected every four years, within the 61 Wilayat in Oman since 1991. The Shura Council scope of work was limited to conducting studies on development and providing recommendations on policy proposal to ministries that relates to citizen development. Since 2011, a royal decree (No. 99/2011) assigned the council with more legislative and auditing powers. Current scope of work includes, but not limited, to suggestion and ratification of laws, in addition, monitoring and evaluations of ministries. Using a set of eight parliamentary oversights, the Shura Council conduct their tasks within eight parliamentary specialized committees. The structure of the Shura Council consists of a president, two vice-presidents, six members forming the "Council Office", and finally, the remaining members distributed in the eight parliamentary specialized committees. Members of the Shura Council, having the highest power on urban and housing issue, have offices in neighborhoods within their representing Wilayat, dedicated to interactions with citizens in regards to discussions, complains and requirements. Wilayat Seeb, considered the highest Omani-populated province in Muscat governorate, has currently two offices for the Wilayat Seeb Shura Council member. Their role is to bring feedback from citizens to the council in addition to working directly with the responsible authorities within the Wilayat, such as, Muscat Municipality Seeb Administration and Wali Seeb.

The Municipal Council in Muscat Governorate consists of appointed representatives from seven ministries, elected members from the Wilayats in Muscat governorate, and two individuals from the "people of opinion". The first form municipal council establishment dates to 1939 with limited authoritative role; however, in 2011, an official Royal Decree number 116/2011 along with the Diwan Resolution No. 115/2012 set the beginning of Muscat Municipal Council (2019). Similar to Shura Council, the Municipal Council running term is four calendar years with the option of renewal as per the provisions of the law. The tasks assigned to the council reaffirm the provision of views and recommendations regarding the development of regulations and municipal services within the governorate. These tasks include, but not limited to, recommendations on public health, suggesting names for districts, locations, and roads, and communication with local communities along with private sector on suggestions that aim on developing the governorate. In Muscat, the council holds internal meetings at their hall at the Municipality, and in addition, members are available for discussion with citizens at the local municipality administration in each Wilayat.

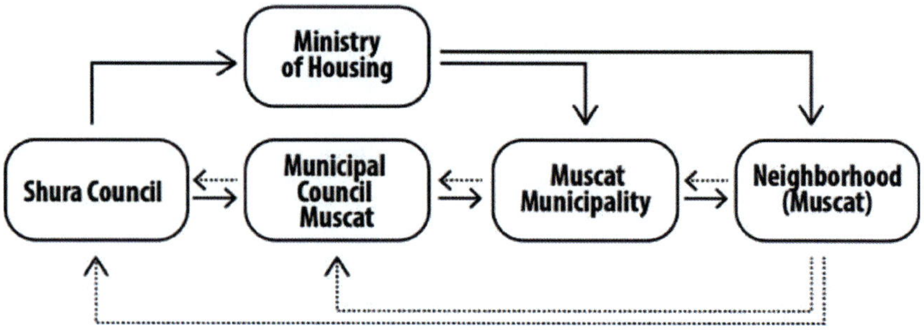

Fig. 1 Relation of actors in neighborhood development in Muscat

The regulating body for services within neighborhood is *Muscat Municipality*. Their primary responsibilities within the residential neighborhoods are provision of street networks, street lighting, sewage systems, public cleaning services and waste management, building regulations,[2] and other hygienic services. Due to the rapid horizontal expansion of the city, new neighborhoods pose high pressure on Muscat Municipality on the provision of municipal services such as street networks, street lighting, and parks. With the limited financial capacity of the municipality, provision of these services is prioritized as street network, followed by street lighting, and eventually the development of public space if the neighborhood fulfills certain criteria; a matured neighborhood with high density. The limitation on public space development is associated with two primary issues. First, the development of public space, especially parks, required high levels of maintenance and consumes great amount of water resource.[3] Secondly, development within this realm is considered an investment that does not generate revenue. Hence, projects with high initial investments and low revenues fall low on the priority list. As a result, recent discussions on municipal levels are raised on the extent of providing policies that enable communities within neighborhoods to take ownership in the development of public spaces. Thus, high return value for the initial investment by the community is established as they are the end users. Furthermore, the municipality is working on partnership projects with the private sector in the provisional and financial support of municipal services. Current examples are showcased in the proposed Hilton Garden by ASAAS and the recent partnership with Bank Muscat for Al-Khoud Park.

A legal system of community management association is the *Property Management and Community of Owners (PMCO)* which provides a local civil association as part of the Royal Decree number 48/84 regulated by Ministry of Housing. The PMCO is a legally binding community association that practices its power within its internal members. The PMCO was a response to prior recurring issues within multi-ownership plots. Hence, the PMCO aims on preserving communal assets within the plot. Through the association, each owner reserves his real estate rights within a legally binding document to ensure a safe real estate that is well maintained by the members of the association. In such multi-ownership plots, the "Real Estate Trust" is responsible for informing the owners on the association formation policy. The owners of plot are then required to set the "primary laws" on fees, property management committee members and assign an association president as a representative, and a vice president responsible for carrying administrative and regulation work with official bodies. The Property Management Committee, which falls under the association, is responsible for either partnering with a third party for property regulation, or designing the agenda required for carrying the task. Organizational structures such as the community of owners association set a legal precedent for neighborhood as means to establishing legal associations that aims on development of communal spaces.

As Al-Farsi mentioned, such councils for consultation and management are fundamental in the Omani context; hence, "The process of consultative participation is practiced at different levels, as Sheikhs (tribal leaders) in their local Sablah (public hall) discuss tribal problems and find solutions for them" (Al-Farsi, 2013). In urban development relating to housing, the roles of Shura Council, Muscat Municipal Council working along with Muscat Municipality contributes to a certain level of land democratization within neighborhoods. The example of The Property Management and Community Owners exhibits the potential of associations' forming that hold legal power within themselves, to improve their immediate surrounding environment in the level of shared communal assets. However, one question remains, how can communities within neighborhoods, use such structures and institution for the development of their neighborhood?

[2]Muscat Building Regulation as Local Order No. 23/92, that delineates the buildings offsets within plots boundary walls along within building facades both of which contribute to the streetscape of neighborhoods.
[3]Oman geographic location along with its current issues on the hydrological cycle has resulted in high dependency on desalinated seawater counting for over 80% of water sources in the country (Diam, 2018).

3 Case Study—Khoud 6 (Hayi Al Maarifa)

Taking the case of Khoud 6, also known as Hayi Al Maarifa, an analysis of how a community organize themselves to carry official actions for the development of their neighborhood is outlined. The research data was gathered through interviews and focus group discussion with various members of the community in aim of understanding the neighborhood development process from all aspects. Through civil committees established by the neighborhood residents, the neighborhood managed to carry various projects using a collaborative process. Located in Wilayat Seeb, "Khoud 6" neighborhood is part of greater Al-Khoud area which covers over 42 km^2 of the Wilayat in Muscat Governorate. The naming of the neighborhood refers to its phase, being the 6th phase of Al-Khoud expansion (Nebel & Richthofen, 2016). Geographically, the neighborhood borders Wadi Al-Khoud from the northern side, a seasonal rainwater drainage to Al-Khoud dam. The neighborhood size is around 3.5 km^2, with over 2600 plots and home for over 10,000 inhabitants; referring to 60% of plots being built with a household median of 6 individuals per plot (Elzain et al., 2017). The neighborhoods urban structure is "mainly of 2–3 story detached residential villas and 6 stories residential-commercial building. These residential units are mainly homes of young families, university students and skilled Omani and non-Omani bachelors" (Elzain & Al Othmani, 2019). Similar to other new neighborhoods within Muscat, residential/commercial plots in Khoud 6 were distributed through the land entitlement scheme in 2008, shortly after, the provision of primary road network by Muscat Municipality began. Construction of residential/commercial villas within the neighborhood starts soon after with 120 built-up buildings in 2009 to 1670 buildings by mid-2019.

3.1 Khoud 6 Neighborhood as "Hayi Al Maarifa"

The current known name of the neighborhoods is "Hayi Al Maarifa" which translates to "Knowledge Neighborhood". The process of the neighborhood naming took place during one of the Mosque Construction Supervision committee meetings in early 2014 where a suggested name from one of the members was approved by the committee members. Hence, since 2014 the name of the neighborhood was known within the neighborhood. Looking at recent norms of neighborhood naming in Muscat, the local municipalities would either give neighborhoods their name, or in some cases, the neighborhood inhabitants can select a name for their neighborhood; therefore, the community chose to register the name that they believe is a better representation of their neighborhood. The name of the neighborhood was officially registered in the local municipality in 2017 by the Activities Committee. The selected name of the neighborhood, Hayi Al Maarifa, was considered ideal as it clearly communicates the neighborhood pursuit of science, knowledge, and culture with special focus on youth. In addition, it also aims on communicating to others that this community has its clear mission and an object in pursuing knowledge. Although the name of the neighborhood is currently registered and official, yet, during some of the interviews with resident of the neighborhood, the name "Khoud 6" is still used to communicate to other residents of Muscat on the location of the neighborhood as it still not known within the larger city.

3.2 Hayi Al Maarifa Neighborhood Committees

The neighborhood committee, unlike the PMCO, is not officially registered, hence they do not carry legal power that can be enforced within its member. The committee, which consists of volunteers, was initiated by the community as a response to the need for establishing a structure that is responsible for the supervision of the mosque development. Thus, over the past 10 years, as Fig. 2 shows, hitherto 10 committees are established in aim to develop the neighborhood and improve its livelihood and social life. These committees are run by both men and women of the neighborhood, where seven are organized by men and three are organized by women. The communication within these committees is carried through open discussion within the mosque, Quran School, public hall, but more frequently, through social media channels. The neighborhood currently has over 1500 followers on their social media account; in Twitter, YouTube, and Instagram. A review of their photographic posts on the social media channels of the neighborhood exhibits high level of the photographic posts (29% of total posts) relating to "Islamic educations" activities carried in the neighborhood. The second common theme discussed are skill development activities (22% of total posts) followed by "recreational activities" covering 17% of the photographic posts. The remaining posts range from competitions and awards, lectures by Islamic scholars, social work, and community discussions. Communication with legal authorities by these committees are often carried by the men coordination committee; whether it is to Ministry of housing, Muscat Municipality—Seeb, Wali Seeb, and Municipal Council members—Seeb.

3.3 Hayi Al Maarifa Neighborhood: Men Committees

The primary committees currently operated by men within Hayi Al Maarifa are the Activity Committee, Endowment

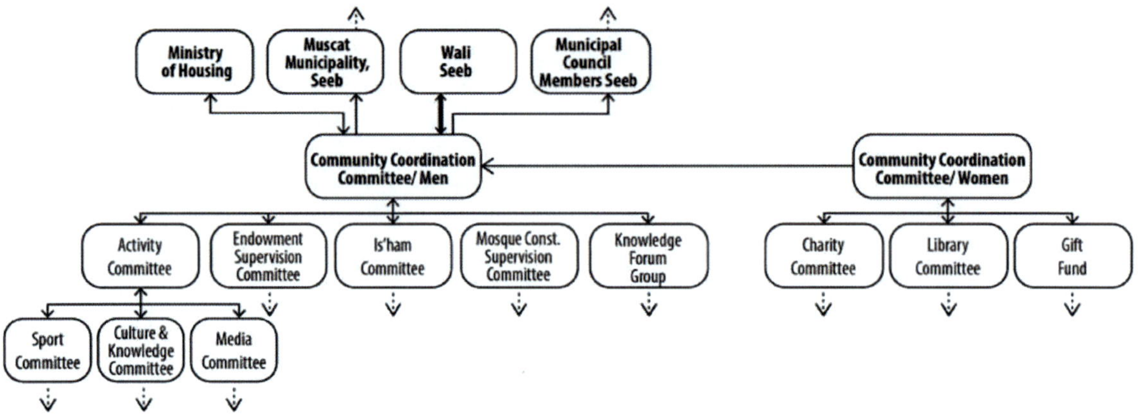

Fig. 2 Hayi Al Maarifa neighborhood committee structure

Committee, Is'ham Committee, Mosque Construction Supervision Committee, and "Knowledge Forum Group", with seven, five, fifteen, three, hundred, and twenty members, respectively. These committees are created by a member of the neighborhood that carries a "representative role." The community representative also takes forward and follows up the community suggestions with local municipality, ministry of housing, and other expert when required. The committees are established in accordance to the requirements by the neighborhood residents. The committees operate in an interconnected manner where they rely on each other for the execution of various tasks in accordance with their field. Three of the primary achievements completed by these committees are: Hayi Al Maarifa Mosque construction, Hayi Al Maarifa Mosque Administration Structure, Quran School Administration and Wadi plots allocation.

Mosque Design and Construction

In Islamic and Arabic regions, mosques[4] within neighborhoods play a fundamental role and are often the spatial and social node of a community. Similar is the case in Muscat, where two ministries are responsible for the provision of land and supports with provisional cost of the physical structure, Ministry of Housing and Ministry of Religion Affairs and Endowment, respectively (Supreme Council of Town Planning, 2010). When the first resident that came to Khoud 6 was faced with a neighborhood that lacked such a facility, the residents organized themselves to follow up with this issue. As a given preset, most Omani residents have a specific vision of the neighborhood mosque and its surrounding and requirements. In the case of Khoud 6, the primary requirement that was agreed upon by the resident for the mosque was a spacious mosque with high capacity. The current building consists of two floors with an addition of a public hall suitable for various communal activities.

The initial planning of the neighborhood mosque started in 2009. A committee for the mosque construction was organized within the neighborhood to tackle the development of the mosque. The assigned representative of the neighborhood followed up with Ministry of Religion Affairs and Endowment for the approval of the idea then with Ministry of Housing for the allocation of the land. The committee was responsible for the design and the assigning of the contracts and consultants to follow with the execution of the mosque. As the neighborhood residents had a clear perception of spatial requirements for a mosque, the committee sent a request to the Ministry of Housing to increase the parking lot capacity through the inclusion of some of the unbuilt residential plots to the mosque area. Following a negotiation with the assigned owners of the surrounding unbuilt plots, neighborhood residents, and Ministry of Housing, the owners of the plots were given a replacement of their plot within the neighborhood.

During the construction period of the mosque, the neighborhood residents used temporary alternatives. In 2011, a portable cabin that was donated by a resident of the neighborhood, was used as a prayer hall till the number

[4]In this paper, Mosque is used to refer to Ja'ma. According to the Urban Planning Regulation book, a Masjid is a place for worship for all daily prayers and should serve a maximum of 2500 individuals and with a service radius of 500–700 m. The Ja'ma is, similar to Masjid, a worship place with the exception of carrying Jum'a (Friday) prayers and Eid prayer. In addition, the Ja'ma should have additional amenities within its boundary such as women prayer area, Quran school, Imam accommodation, and commercial shop. Hence, the capacity area for the Ja'ma is 10,000–20,000 individuals with a service radius of 1–2 km.

neighborhood residents increased, and a large tent was used to host prayers for the neighborhood in 2012. By 2012, the construction of the mosques was complete. Funding of the mosque throughout the construction process was mostly covered by the neighborhood residents counting for over 50% of funds and materials donated by the residents. The remaining funds were collected from Ministry of Religion Affairs and Endowment, donations from private companies, and donations from others who wished to support the mosque construction. The Endowment Supervision Committee was assigned for the collection of funds using Waqf[5] and various tools to support the gathering of the funds from written proposals to companies and signage for other interested individuals. Residents were encouraged to support with the funding through the purchase of "stocks"; each stock cost 200 Omani Rials. The final constructed mosque; called Ja'mi Hai Al Maarifa' is currently the center of activities, gathering, initiatives and discussions within the neighborhood. The surrounding landscape of the mosque consists of 20 palm trees donated by the residents. The seasonal harvest of dates within the residents of neighborhood.

Hayi Al Maarifa Mosque Administration

Along with the Mosque Administration, the Activity Committee consisting of seven members, plays a role in the organization of activities and communication with the residents in upcoming plans of the mosque. The Knowledge Forum Group, consisting of 120 members, is held on Monday evening within the mosque open discussion with all members of the community. The primary responsibility of the Mosque Administration is the organizational and managerial aspects of the mosque from opening hours to Friday's speeches along with organized lectures and classes in the mosque. These tasks are discussed on weekly or monthly basis through internal meetings. Through the Communication Committee, the Mosque Administration informs residents of the neighborhood on proposed schedules of classes, activities, and key discussions for the upcoming Monday Knowledge Forum Group. Communication with the residents is conducted via the mosque mobile phone through messages on a mobile messenger, WhatsApp, directly sent to individual contacts of registered neighborhood members.

The selection of members in the Activity Committee and Mosque Administration took place at one of the Monday Knowledge Forum Group meeting at the mosque. Through mutual agreement with the attendees of the meeting, the selected committee members are assigned along with general guidelines of activities they should carry. In addition to the assigning of the committee members, the Monday Knowledge Forum Group meetings provides a platform for discussion on educational activities (neighborhood clean-ups, Iftar Sa'im), urban issues, new projects, and suggestion. The process on which project is selected is through voting that take place after each suggested proposition has been discussed adequately during the session. Attendees of these meetings are welcome from all various age groups with the inclusion of youth and adolescent. Younger age groups are encouraged to share their perspective in addition to sharing poems or speeches they which to perform as it is seen as a way to support the youth in taking actions and responsibilities in activities that aim on developing their neighborhood and surrounding.

The Knowledge Forum group beneficial outcomes for the administration and committees are visible through the list of various action taken following discussions that take place during the sessions. These discussions are not limited to activities, rather, social issues such as concerns facing current youth, e.g., required skill development workshop to support the human resources of the neighborhood. The forum is also the first networking event for new residents in the neighborhood where they could understand the essence of how the community organize themselves in addition to sharing their view point to further the neighborhoods' visions. The Knowledge Forum group meetings are also used to invite expert guest speakers to give a talk on current issues taking place in the neighborhood. An expert in public sewage system was invited during the sewage system installation period in the neighborhood. In addition, members from Municipal Council or Shura Council, and Civil Servants attend the forum to share information regarding themes that relate to their field to open the discussion with the community. Finally, during these sessions, nominations and assigning of individuals responsible in carrying task on upcoming events such as trips for men and the youth, walking activity with hundred participants to walk four kilometers as a way to encourage neighborhood residents to walk.

Quran School Administration

During the construction of the mosque, the neighborhood sought to expand the built-up facilities through dedicating a building just for Quran teaching and extracurricular activities. The community purchased a plot with close proximity to the mosque and began with the construction and funding similar to the mosque under the supervision of the Endowment Committee. Although the school started in 2012 with just 70 participants, it was officially opened in 2016. The

[5]Waqf is a "… religious endowment, recognized by Shariah as religious, pious or charitable donation." Also, "make endowment of houses and keep its gains in the way of God. By this, we can understand that waqf or endowment is to preserve or upkeep the original wealth, whether land or house, etc., and give its profit to the needy." (Yalawae & Tahir, 2008).

school consists of a twin villa of which one villa is used for male students and the other for female students. The school has over twenty classes; ten classes for men and ten classes for women, with a shared library and multi-purpose hall with twenty paid school teachers (ten male teachers and ten female teachers), most of whom are residents of the neighborhood. Currently, the school has over 500 students from age five onwards.

The primary activity carried in the Quran School is the teaching of Quran for all ages and gender; however, secondary activities such as Arabic calligraphy, English classes, support after school teaching, and martial arts are also included in the curriculum of the school. The Quran School also provides a summer school from 9 am to 2 pm, during which the participants attend classes in Fa'qh, Seera, hadith in addition to Arabic, English, personal and skill development classes, and educational competitions. The summer school also provides workshop in collaboration with other organizations relating to crafting, carpentry, farming, first aid, and robotics. Some lectures and activities are carried in the public hall such as workshops, lectures, competitions, ceremonies, charity markets, and other activities.

The success of the Quran school is visible through the number of events and activities the school carried out independently. The classes in the school provide opportunities for different groups to learn with an allocated teacher for their age and capabilities. Women activities increased significantly as the school provided the required space for them. The current school administration has plans to expand with the cultural and educational center and a Fa'qh school. The school administration plans to increase workshops that focus on personal and skill development. Similar to the case of the Mosque Administration, communication with parents and residents within the neighborhood is carried through social media channels.

Wadi Plots Allocation

One of the small projects the community committees managed to achieve is the stopping of land distribution along the Wadi. In 2014, the residents of the neighborhood noticed construction activities within plots bordering the Wadi, thus, within a process that took between six to twelve months, the community managed to organize themselves to stop the land allocation. The community's aim from carrying this task was to provide access to natural topography for recreational purposes where all neighborhood members could benefit of. The community committee in collaboration with neighborhood residents prepared a letter signed by neighborhood residents directed to the Ministry of Housing requesting to stop land distribution along the Wadi. Although more than 75% of plots bordering the Wadi were already distributed, the remaining 25% of area bordering the neighborhood is currently unbuilt. Looking at the current population density of neighborhood and lack of public facilities, the unbuilt area bordering the Wadi is currently available for the public to develop as a neighborhood park.

3.4 Hayi Al Maarifa Neighborhood: Women Committees

Similar to men's committees in Hayi Al Maarifa, the women committees divided into three committees, the Charity Committee, Library Committee, and Gift Fund Committee, consisting of four, three, and three members, respectively. These committees work independently of the men's committee in terms of organization and structure. Communication within committees is carried through home gathering or group discussion on mobile messengers. The committees are also in charge of the female Quran School. In addition, these committees work in an interconnected manner similar to the men's committees in regards to project execution and activities. Unlike the men's committees, the women committee was established recently as they were provided with the required space to conduct various activities with neighborhood residents. One of the outstanding achievements of these committees is the "Money Clay Pot" that is under the process of documentation by the Library Committee.

Charity Committee

The Charity Committee is responsible for various activities within the neighborhood and it consists of four members that overlook the female Quran School administration, funds collection, and management, in addition to organization of recreational trips and skill development programs. Funds for committee operations are collected through various activities and donation. The primary source of funds is collected through the school fees and annual donations from members within the neighborhood. The female Quran School, similar to the male Quran School, has ten teachers who provide classes to various age groups starting from five year old onward. It also provides classes in various topics throughout the year and activities similar to the male Quran School with the addition of Arabic classes to non-Arabic speakers. During the summer break, the female Quran School provides a "Summer school" similar to the male school. In addition to Quran classes, the summer school provides classes and activities on personal skills development and sessions of "self-appreciation" for young girls.

In addition to managing the Quran School, the Charity Committee works on supporting four families within the neighborhood with forty Omani Rials allowance on monthly bases. Furthermore, employment opportunities in the female Quran School are prioritized to households within the neighborhood that require further financial support. Another activity carried by the committee is the organization of recreational trips for young female children and adolescents throughout the year. This year, the young children visited a "Kids Mall" while the female adolescent went on a "Farm day trip".

Library Committee

As most committees in the neighborhood, the Library committee consists of three members working voluntarily. The primary responsibilities of the Library Committee are developing and managing the library program in addition to collaborating with other local libraries for sponsorship opportunities. Furthermore, the committee is currently working on a storybook that documents the annual activity of breaking of the "Money clay pots" carried a few weeks before the month of Ramadan in Hayi Al Maarifa. The planned timeframe for publishing the book is mid-end of the year 2020. Although this activity is currently carried in various neighborhoods in Muscat; however, it was initially organized and carried annually by Hayi Al Maarifa Charity Committee. The breaking of the Money clay pot activity targets the youth of the neighborhood (between age four to thirteen) as a stimulating way to encourage the youth to donate small portions of their allowance to the goodwill and charitable projects. The collected funds from the event are occasionally given to families in need.

Gift Fund Committee

In means of welcoming and engaging with new residents to the community, the gift committee, managed by three committee members, is responsible for the collection of funds from approximately 60 residents (households) in triennial intervals. The funds are then given according to the occasion. In the case of a mother giving birth to a new child, sixty Omani Rials are given to the mother on behalf of the community. Similarly, in the case of a new family moving into the neighborhood, a hundred Omani Rial is given, respectively.

4 Discussion

The case of Khoud 6 neighborhood (Hayi Al Maarifa) presents an opportunity to analyze and extract the criteria for understanding how residents of new neighborhoods could manage and fund projects that service their community without following traditional planning structures. Faced with urban sprawl issues, under a local governing structure that emphasize on consultation and discussion, Hayi Al Maarifa community were able to take control for shaping their surrounding in a structured manner that relies of social network rather than governing regulations and laws. Politically, the neighborhood practices numerous tools for the discussion, decision-making, and nomination that reflect a democratic-consultative structure to ensure the engagement and involvement of the residents and the views in the development of their neighborhood from a physical and social aspect. Socially, the neighborhood relies heavily on social networks and trust for the financial support and maintenance of interventions within the neighborhood. Hence, in order to understand the democratic-consultative model carried in Hayi Al Maarifa, further elaboration on the decision-making process (including committees) and their keys to success shall be outlined.

4.1 Consultative Hubs for Democracy Within Neighborhoods in Muscat

As Al-Jabri mentions, democracy in its uttermost goal is a way in which a society, in a ration manner, agrees on certain values and criteria for the development of the society as whole. The case of Hayi Al Maarifa, a neighborhood resulted from urban sprawl, exhibits how families with different backgrounds could come together to enhance their livelihoods within their close surrounding.

The Democratic-Consultative Decision-Making Process

Why? Where? How? Through these three questions one could outline the decision-making process carried by Hayi Al Maarifa. "Why?" refers to neighborhoods goal, the reason they invest their time in their neighborhood and ideas they envision for their neighborhood. "Where?" reflects the local platform for discussion, the place where they meet to discuss the future of their neighborhoods, issues and concerns. Finally, "How?" is the means of executing their visions and ideas, the physical and financial resources required to complete these tasks.

Why?… invest time?

Unlike most new neighborhoods in Muscat, Hayi Al Maarifa was one of the first to identify themselves with a name that does not reflect a tribe or a family name, rather a vision for a neighborhood. As a relatively young neighborhood consisting of young families, the future of the youth is a crucial topic that reflects the vision of the neighborhood. Through various means, the neighborhood invests highly on the development of their kids in terms of knowledge, skills, and personal development. This shared vision represents the common values that the neighborhood follows and commonly agrees upon. A common value that was suggested, selected, and approved by the community.

Where?... When?

The Knowledge Forum Group meetings that are conducted on weekly basis represent the communication platform for discussion and selections of interventions and projects. These meetings are essential for the creation of new committees, nominating members, and finally voting and selection. The involvement of all age groups in these open discussions ensures high level of transparency within parents and the youth within the neighborhood. The involvement of these actors further support the longevity of the projects as all members are aware of the task and the end goal aimed to be achieved. The consultations with expert and officials within the forum are a mean for ensuring the communication of accurate information and opening up the discussion on potential collaborations.

How?... Whom?

Common issues within community initiatives or civil governance are the financing of projects and the management of funds. In the case of Hayi Al Maarifa, the neighborhood overcame these issues through assigning a committee that focuses on gathering funds from private companies and individual donations. With a financial capital on hand, the progression of projects is uninterrupted as the decision-makers (the neighborhood) are also the financer and the monitor. To avoid any form of corruption in the fund collection, the neighborhood, similar to the PMCO, uses a bank account that is exclusively used for the neighborhood projects.

Hayi Al Maarifa Success Factor

As Putnam outlines, social capital and level of trust are highly relevant in situations where no legal authoritarian structure is practiced. The case of Hayi Al Maarifa exhibits how a neighborhood that came together for the initial construction of a mosque could evolve to a neighborhood that strives toward knowledge and education. Although the PMCO provides an example of how a community can form a legal structure for the enhancement of their common goods, the case of a larger neighborhood with over 10,000 inhabitant relies on other factors such as social capital that ensure quicker actions carried in an honest manner. In addition, through the internal management of funds by the neighborhood committees, projects came to completion within short periods of time, hence, providing the neighborhood residents with a physical output of their donations and investments. As a result, the quick and visible output was key for the embedding the trust within neighborhood residents and encouraging them to continue with the financial support.

5 Recommendations

The case of Hayi Al Maarifa demonstrates the potentials of civil societies to take the lead in the urban development of their surround context. Parallel with the Omani governing structure, some of the neighboring countries with urban sprawl concerns could benefit from the "Democratic-consultative decision-making process" practiced by Hayi Al Maarifa residents. Adaption of this approach should consider the regular discussion platform, the setting of common values, and finally, the localization of funds managements and monitoring to ensure a corruption-free and transparent collaborative environment. Such a model is therefore beneficial for that follows a similar cultural and legal structure of Muscat, referring to the practiced religion and consultative legal structure. However, further recommendation on policies is also required to ensure the sustainability of these community committee. A vital recommendation on this end is the sharing of information and communication on urban development proposal by the local municipality that carries direct impact on the neighborhood residents, starting from the planning phase.

Limitations

Decisive factors to considering in the proposed decision-making process in Hayi Al Maarifa are the social background of the residents, type of residents, and the local regulation that fostered such activities. The majority of the neighborhood resident is Omani nationals who own the house they live in. Although they might originate from different regions of the Sultanate, yet, they share a common language, religion, and culture. In the case of a multi-national neighborhood, residents might have to compromise further to reach a common goal and set common values. Similarly, as the neighborhood is young (eleven years old), the majority of the families that moved in are young families with children. As a result, most residents share common issues relating to adolescents and children. If the situation called for multi-generation residents within the neighborhood, activities that include all various age groups would need to be considered. Finally, local regulation in the Sultanate calls for a consultative atmosphere, hence, discussion would by the norm. In the case of strict monarchy system, such approach would not be possible.

6 Conclusion

The research outlines the definitions of democracy and participation within decision-making processes alongside neighborhood urban development. Following an understanding of local governing structures, the case study of

Khoud 6 in Muscat exhibits a democratic-consultative governing structure where neighborhood residents are key actors for neighborhood development. The research provides an outlook on democratic processes within neighborhood scale that carries direct impact on the lives of those who live and within a defined community along with the hierarchical framework of planning. Hence, one wonders whether democratic decision-making process is a constant norm that carries its prerequisites, or whether it is a dynamic concept reshaped and reconfigured in accordance to social, cultural, religious, and technological shift and interpretations.

References

Al-Farsi, S. (2013). *Democracy and youth in the Middle East*. I.B. Tauris & Co.

Al-Jabri, M. (2009). *Democracy, human rights and law in Islamic thought*. I.B.Tauris & Co. Ltd.

Arnstein, S. R. (1969). A ladder of citizen participation. *Journal of the American Planning Association, 35*(4), 216–224.

ASAAS. (2016, March 14). *ASAAS unveils plans for Hilton Garden Inn Muscat & Urban Park in Al Khuwair*. ASAAS. https://asaas.om/en/asaas-unveils-plans-for-hilton-garden-inn-muscat-urban-park-in-al-khuwair/

Birkeland, J. (1999). Community participation in urban project assessment (an ecofeminist analysis). In B. Martin (Ed.), *Technology and public participation* (pp. 113–142). University of Wollongong.

Box, R. C. (1998). *Citizen governance: Leading American communities into the 21st century*. Thousand Oaks, CA: Sage.

Conley, C. (2017). *Making smart decisions about smart cities*. https://www.aclunc.org/sites/default/files/20171115-Making_Smart_Decisions_About_Smart_Cities.pdf

Diam. (2018). *Annual report 2018* (p. 17) [Online]. Diam: Public Authority for Water. Available at: https://www.diam.om/getattachment/b9538383-b5d1-4011-83f1-9d46e1aabeeb/Annual-Report-2018

Downs, A. (1981). *Neighborhoods and urban development*. Brookings Institution Press.

Elzain, R., & Al Othmani, R. (2019). Approach towards localization of public spaces development in Muscat. *Resourceedings, 2*, 14. https://doi.org/10.21625/resourceedings.v2i1.449

Elzain, R., Salheen, M., & Jessen, J. (2017). *Rethinking unbuilt spaces within neighborhoods in Muscat* (Master thesis), University of Stuttgart, Germany and Ain Shams University, Egypt. https://iusd.asu.edu.eg/wp-content/uploads/2018/04/IUSD-V-Rowa-Elzain_-Reduced.pdf

Ferrao, J. (2015). The critical neighbourhood initiative program as policy project: A personal account. In L. Vasconcelos (Ed.), *Sustainability in the 21st century—The power of dialogue* (pp. 235–247). MARGov Project—Collaborative Governance of Marine Protected Areas.

Hallowell, J. (1954). *The moral foundation of democracy*. University of Chicago Press.

Hanson, R. (2003). *Civic culture and urban change*. Wayne State University Press.

Hoang, G., Dupont, L., & Camargo, M. (2019). Application of decision-making methods in smart city projects: A systematic literature review. *Smart Cities, 2*(3), 433–452. https://doi.org/10.3390/smartcities2030027

Howell, J. & Simmons, R. (2015). *Designing democracy* (pp. 71–88). Design Commission at Policy Connect.

MacIver, R. M. (1947). *The web of government* (et seq, 410). https://archive.org/details/in.ernet.dli.2015.274488/page/n209

Muscat Municipal Council. (2019). *Municipal Council*. https://www.mmc.gov.om/Pages.aspx?PAID=2&MID=8&PGID=3

Muscat Municipality. (2019). *Muscat municipality brief history*. https://www.mm.gov.om/Page.aspx?PAID=1#Details&MID=6&PGID=112

Nasr, V. (2005). The rise of "Muslim democracy" . *Journal of Democracy, 16*(2), 13–27.

NCSI. (2019). *Monthly statistical bulletin: June 30* (pp. 2, 32). NCSI.

Nebel, S., & Richthofen, A. (2016). *Urban Oman*. LIT Verlag GmbH & Co.

Putnam, R. (2000). *Bowling alone*. Touchstone.

Putnam, R. (2002). *Democracies in flux*. Oxford University Press.

Putnam, R., Leonardi, R., & Nanetti, R. (1993). *Making democracy work: Civic traditions in modern Italy*. Princeton University Press.

Sirianni, C. (2007). Neighborhood planning as a collaborative democratic design. *Journal of the American Planning Association, 73*(4), 373–387.

Supreme Council of Town Planning. (2010). *Urban planning regulations* (p. 119). Ministry of Housing.

Suurendonk, M., Otter, A. A., Barrett, P., Amaratunga, D., Haigh, R. P., Keraminiyage, K., & Pathirage, C. (2010). Stimulating value creation in the initial phase of urban developments. In *Conference; Building a Better World; 2010-05-10; 2010-05-13* (pp. 1–12). International Council for Research and Innovation in Building and Construction (CIB).

Times of Oman. (2019, June 17). *A cooler, greener capital city*. Times of Oman. https://timesofoman.com/article/1469396

United Nation. (2009). *General assembly: Third committee* (A/64/439/Add.2 (Part II) A/RES/64/155).

Yalawae, A. & Tahir, I. (2008, September 10–13). *The role of Islamic institution in achieving equality and human development: Waqf or endowment* [Paper presentation]. In 5th Annual Conference of the HDCA, New Dehli, India.

Hortas Comunitárias in the Favelas of Rio de Janeiro: How New Ideas Are Adopted and Spread Across Communities

Mariano J. Magalhães

Abstract

This paper applies the Diffusion of Innovation Theory popularized by Everett Rogers to examine the adoption and dissemination of community gardens in favelas in Rio de Janeiro. The study involved interviewing adopters of *hortas comunitárias* in ten communities. The findings suggest that the five traits that Rogers puts forward to explain diffusion of innovation fit reasonably well the adoption and dissemination of *hortas*. The results also suggest that, with minimal support from the municipal government, experiments in Urban Agriculture in general can become more widespread, which is especially important as food insecurity increases in the wake of the coronavirus pandemic. Leveraging the social networks and developing robust linkages with existing NGOs that support urban agroecology are key to the dissemination of this socioenvironmental innovation.

Keywords

Agroecology • Sustainability • Community gardens • Diffusion of innovation • Brazil

1 Introduction

The city of Rio de Janeiro is home to over 1000 favelas housing over 1.2 million residents—about 1 in 4 residents of Rio live in favelas. The international press tends to portray favelas as the dark side of Rio, teeming with violence and controlled by drug gangs, sporting substandard housing, and widespread squalor. The media in Brazil also paints a very grim picture of these communities. The reality, however, is much more nuanced. While some of these communities do indeed fit this description, in the vast majority of favelas residents live in sturdily built houses with running water and electricity, surrounded by a robust community, with a plethora of businesses, churches, and community centers.

One common thread running throughout almost all Rio favelas, however, is the absence of the state. What this means is that these communities cannot rely on the government to serve their basic needs. Instead, these communities organize, identify their most pressing challenges, and use local resources and their creativity to find solutions. Since most of these communities face similar problems, successful solutions in one community often find their way into others. This paper examines how communities adopt innovative ideas to confront these challenges and the process by which these ideas spread across communities. For the purpose of this research, I chose *hortas comunitárias* (community gardens) as the innovation to examine.[1]

The research project consisted of visiting leaders of favelas, during August of 2019, to investigate a process known as diffusion of innovation. The Diffusion of Innovation Theory was first discussed in 1903 by French sociologist Gabriel Tarde (Toews, 2003) followed by Ryan and Gross (1943) who introduced the adopter categories that were later used in the current theory popularized by Everett Rogers. Katz (1957) is also credited for first introducing the notion of opinion leaders, opinion followers, and examining how the media interacts to influence these two groups. The Diffusion of Innovation Theory is often regarded as a valuable change model for guiding technological innovation where the innovation is modified to meet the needs across all levels of adopters. It also stresses the importance of communication and peer networking within the adoption process.

[1]The original plan was to examine a combination of *hortas comunitárias* and recycling programs but I had much more success gaining access to communities and residents who adopted the former innovation.

M. J. Magalhães (✉)
Political Science, Augustana College, Rock Island, IL, USA
e-mail: marianomagalhaes@augustana.edu

F. Trapani et al. (eds.), *Advanced Studies in Efficient Environmental Design and City Planning*, Advances in Science, Technology & Innovation, https://doi.org/10.1007/978-3-030-65181-7_21

Diffusion of innovation occurs when "an innovation is communicated through certain channels over time among the members of a social system" (Rogers, 2003, p. 5). Diffusion of innovation can be found at many levels. At the national level, in the 1990s, democracy spread across Africa (Diamond, 1999) and, over the course of 2011, popular uprisings against authoritarian regimes advanced throughout the Middle East and North Africa (Lesch & Haas, 2012). In the USA, individual states often serve as social laboratories with states copying their neighbor's policy ideas in areas such as smoking regulations and organic food certification. Weyland (2007) wrote about how similar pension reform traveled in Latin America from one country to another following democratization in the 1980s and 1990s. Sugiyama (2016) described how education and health reform spread in Brazil from one municipality to another in the 1990s.

Most of the research in this area, especially by social scientists, involves government entities. Scholars, for example, often examine how ideas spread from one government to another. For scholars of diffusion in the USA, this means examining how a policy innovation adopted by one state gets emulated by other states (e.g., see Mossberger, 1999). Scholars also investigate how ideas get picked up (or not) by local communities or spread across communities (Wellin, 1955).

The conceptual framework chosen to investigate the adoption and diffusion of *hortas comunitárias* is offered by Rogers (1962, 2003) in his book *Diffusion of Innovations*. Rogers highlights five characteristics of innovations that explain why ideas are adopted and the different rates at which new ideas spread. The first trait is relative advantage, which is the degree to which an innovation is perceived as better than the idea it supersedes. Second is compatibility. This trait refers to the degree to which an innovation is perceived as being consistent with the existing values, past experiences, and needs of potential adopters. Third is complexity or the degree to which an innovation is perceived as difficult to understand and use. The fifth trait is trialability or the degree to which an innovation may be experimented with on a limited basis. And the final trait is observability, which is the degree to which results of an innovation are visible to others.

Although Rogers' model is classic and widely established, it has several limitations regarding its predictive power related to the dissemination of an innovation (Bass, 1969). Bass, therefore, proposed a model to forecast the sales of new products–one type of dissemination of an innovation. Under the mathematical formula used by the Bass model, the number of adopters of a new product during a time period is almost identical to the number of sales throughout most of the diffusion process. The Bass model has been revised and implemented in forecasting innovation diffusion in multiple fields (Mahajan, 1990). While the Bass model has potential to predict the distribution of the adoption curve, especially as it relates to the sales of new products, Rogers' model serves as a more comprehensive framework for understanding the diffusion process of an innovation and the underlying factors driving the diffusion.

2 Research Design

The investigation conducted in Rio involved primarily interviews with individuals in favela communities across the city. These individuals were either community leaders or residents who had already established a community garden or were in the process of adopting this innovation. Interviewees in these communities ranged from older residents to very young innovators, men and women, as well as people with a range of knowledge of how to build and manage a *horta*. In addition to these interviews, I also spoke with members of NGOs who are pro-urban agriculture, scholars who have studied urban gardens and members of the *Conselho de Segurança Alimentar* (Food Security Council), an entity at the municipal level of government that brings together members of civil society and city officials to discuss policy related to food security.

For the project, I interviewed fifteen individuals. The interviews with community residents/leaders were semi-structured, with open-ended questions centered around the background of the respondent, the history of his/her community, a description of the *horta* project, why the person chose to adopt a *horta* (and the evolution of the *horta* over time), the engagement of the community in the project, the challenges in adopting and sustaining a garden, and any outside assistance provided (especially by NGOs and the government). The questions put to non-community members were also semi-structured and overlapped with those above, with an emphasis on the challenges and opportunities associated with the adoption of *hortas comunitárias*.

3 Results

The conceptual framework provided by Rogers to examine the adoption and diffusion of innovation clearly helped understand the growth and spread of *hortas* in Rio de Janeiro in low-income communities. The five traits of innovations mostly fit with what was uncovered from interviews. This section examines each of the five traits that Rogers argues influence the adoption of an innovation.

Relative advantage is the degree to which an innovation is perceived as better than the idea it supersedes. In the context of *hortas*, based on the interviews, I took this to mean mainly how the space occupied by the community garden was being used prior to planting the garden. And here the respondents were pretty clear that the use of the space to plant either medicinal herbs, edible plants or trees (or some combination of all three, depending on the amount of space), was substantially more beneficial to the community than how the space was being "used" before. Community members talked about the following "uses" of the spaces:

- In most cases, residents of the community would use the space as a trash dump. Out of the 10 people interviewed, nine explicitly mentioned that their desire to start a garden was borne from the need to make people stop depositing their trash in that particular space.
- Half of the respondents talked about how the space was used as a *boca de fumo*, a location where drugs were consumed and bought/sold. In other words, there was a desire to eliminate a space that encouraged the sale and use of drugs, especially to young children and teenagers, as a way to combat the presence of drugs in the community.
- In roughly half of the interviews, people's desire to start a garden came from the fact that the space was a *terreno baldio*, in other words, not being used for anything at all. Thus, compared to a vacant lot, having a space that produces food or medicine was much better. In a couple of cases, respondents talked about using the space for a garden to discourage new dwellings being built, emphasizing that the community could not accommodate more homes.[2]

The idea of using the space for a *horta* was clearly better than the way it was being used prior to planting the garden.

The second trait that influences the adoption of an innovation is **compatibility**. This trait refers to the degree to which an innovation is perceived as being consistent with the existing values, past experiences, and needs of potential adopters. The values of the adopters I took to mean two things. First, the values of the person I was actually interviewing since she/he was the actual individual choosing to start a community garden. Second, since the garden was to serve the community, the nature of the community itself was key, which was much more difficult to ascertain and the answers were often rather fuzzy.

Almost everyone interviewed clearly valued nature, planting, and the benefits of cheaper and healthier food options. Much of this value attached to gardens came from the fact that many of these individuals were from the interior (from the *roça*, as one elderly gentleman put it) where they planted and cultivated any number of vegetables, herbs and trees. Some had parents who lived in a rural area before moving to Rio and instilled in their child from an early age the importance of nature. It was actually quite common to hear in their words real passion for what they were doing because of this background.

In addition to the value they attached to nature and planting, what was also striking was the value that all the respondents placed on being members of their community. In other words, these folks could have easily made gardens in their own homes, where they would personally benefit from their efforts. But they chose to start a community garden. These adopters strongly identified with the communities they lived in. It was striking, though not surprising, that well over half of the respondents were activists in their communities before beginning their efforts to establish a community garden. One respondent, a middle-aged woman from Salgueiro, used to teach capoeira, a Brazilian martial art created by African slaves, to youth in the community as an alternative to kids getting involved in the drug gangs. Another respondent, a 23-year-old woman from Morro da Providência, had plans to create a recycling program for her community prior to settling on a *horta*. A third respondent, a middle-aged man from Asa Branca, had been the president of the community Resident's Association for many years before embarking on a community garden project. These resident activists were critical in the decision to pursue community gardens.

Even more than values, past experiences influenced the respondents' commitment to planting a community garden. Undoubtedly, past experience greatly influences what a person values, so it would be difficult to disentangle the impact of each on a person's decision to adopt a *horta*. Thus, these two factors were/are closely intertwined in the adoption of this innovation. Expanding the concept of past experiences to the community, the answers were very similar in the sense that these adopters, especially older residents, also grew up in a rural area or had parents who did and who had passed on their knowledge and passion for planting to their children so it was part and parcel of the community itself. In a few communities, especially smaller communities, respondents also gave the sense that the community had strong collective solidarity, where there would be significant support by most members for a project that would bring benefits to the larger group.

The third component of compatibility refers to how the innovation fits the needs of the adopter. There is an obvious need in poor communities for cheaper and healthier food choices. Although most of the community gardens were quite small, almost all of the adopters talked about how any resident could pick whatever they wanted from the garden. It was also common for the adopters to insist that people ask first, so that picking herbs and vegetables could be done

[2]The numbers add up to more than the actual number of interviews because respondents often cited two or three uses of the land prior to planting the *horta*.

using the right techniques and in a manner that would be sustainable for the garden. Another aspect of need being met was that almost everyone would offer seeds and seedlings to anyone who wanted to plant them at home. They explicitly mentioned how the recipients could use the medicinal herbs to save on the cost of medication and/or plant *plantas comestiveis*, which would be cheaper than buying the same items at a grocery store.

Adopters also were asked about community and household gardens as a means to generate income for the community or a household, by collecting the items produced and selling them at a market. Most of the respondents said it was not something they necessarily encouraged or thought the households that wanted seeds to plant themselves would necessarily do either. The simple adoption of a garden was a cost-saving measure in and of itself. Indeed, one respondent, a highly energetic and passionate young man, argued that saving money by planting was by itself a way of generating income because the money that is saved can then be used to pay for other household expenses.

The third trait is **complexity** or the degree to which an innovation is perceived as difficult to understand and use. This trait overlaps, in a way, with compatibility because the vast majority of the adopters, with the exception of two, had experience with and knowledge of planting going into the project. In other words, this combination of experience and knowledge meant that, for people in most communities, creating a *horta* was easy.

In two cases, however, in the communities of Asa Branca and Salgueiro, the respondents offered interesting responses to this question. In Asa Branca, the adopter had no prior knowledge of gardening, except for participating in *mutirões* in other communities to help start a garden. This person clearly stated that he did not know much at all about gardening, planting, or cultivating, and openly said "how hard could it be" and "if it doesn't work, we'll plant some more." In Salgueiro, the respondent also openly admitted to a lack of knowledge of how to start and maintain a garden. This person confessed to killing some plants of her own because she overwatered them. She had no idea that there were plants that did not need constant watering. In her case, however, she took a couple of courses to learn more about what, where, and how to garden.

Ultimately, though, there was a strong sense that, as one respondent put it, "gardening isn't that hard." And this perception of the ease of planting plays a key role in the potential adoption of gardens, community, or household– even if the reality is more difficult. People perceive having gardens as an easy task. It is this perception of gardening that partly explains why Rio's community gardens are adopted and how this innovation can be disseminated fairly rapidly across communities.

The fourth feature is **trialability** or the degree to which an innovation may be experimented with on a limited basis. Gardens of any sort appear to fit this description. This seems especially true in these locations, because the space used for community gardens is typically fairly small. In other words, there is not a huge investment in terms of space so people are not dedicating large swaths of space to try this out. Smaller spaces mean it is easier (and cheaper) to experiment and see whether the garden will produce results. The largest space for a garden that I saw was about half the size of a basketball court (in Asa Branca) and the smallest was about 15 ft^2 (in Salgueiro). Household gardens require an even smaller area, as small as the front yard or backyard or even on the roof, as was the case with the *Teto Verde* (Green Roof) project in Manguinohs.

What also makes these *hortas* relatively easy to experiment with is the cost of starting and maintaining a *horta*. When asked about the cost associated with gardens, almost every respondent said the biggest cost was their time. In other words, the main expense is the time necessary to get the *horta* up and running and then maintaining it, weeding and watering especially. Of course, there were other expenses, including seeds, seedlings, fertilizer, good soil, and gardening implements. But these adopters, while pointing out these costs, were quick to stress that many of these items were either donated (by residents and NGOs, especially) or came to them in the form of *mutirões*.

A *mutirão*, which roughly translates as communal effort in English, is a gathering of people to accomplish a task for a community. This gathering includes residents of the community itself and often people and organizations from outside the community to lend a hand on a project or series of projects in a community. In this collective effort, people work toward a common goal, and manual labor is provided for free. *Mutirões* are widely used in Rio's favelas to describe collaborative efforts to revitalize community spaces.

There was not a single adopter who did not have at least one *mutirão* related to some aspect of the *horta comunitária* project–whether to prepare the space for planting, planting seeds and seedlings, and expanding existing gardens. In these *mutirões*, people volunteer their expertise, donate seeds and fertilizer, and offer up equipment to help with the gardening project. Due to the small size of spaces and the low startup/maintenance cost of the *hortas* (aside from the time commitment), this type of project fits the bill as an innovation that can be tried on a limited basis to see if it yields positive outcomes.

The final characteristic that Rogers contends influences the adoption of innovations is **observability**, which is the degree to which results of an innovation are visible to others. Again, this was a fairly evident attribute of community gardens. Every adopter emphasized how the gardens were seen by residents every day—on the way to work, taking kids to school, and returning from the market. Most of these gardens were in public spaces visible to anyone who walks by. A number of adopters mentioned how people walking by would stop and ask about the garden, find out what they needed to do to start their own home garden, ask for some seeds, and encourage the adopter to "keep up the good work."

In sum, my research suggests that *hortas comunitarias* in Rio strongly match the factors that Rogers sees influencing the decision of people adopting an innovative idea.

Another crucial aspect of Rogers' diffusion theory concerns the communication channels that connect individuals or groups with knowledge of the innovation with the individuals or groups lacking that knowledge. Rogers contends that the main channels are the mass media and interpersonal connections (Rogers, 2003, p. 17). Using this conceptual framework as a lens, respondents were asked how they learned about *hortas comunitárias*. The result was a bit disappointing but not surprising. The majority of adopters had some knowledge of gardens from personal experience or by professional background (one interview was with a gentleman in the community of Taboinha who was a gardener by profession). Or they would say they did not know where they got the idea from. In light of these vague responses, respondents were asked how they disseminated news about their projects. This question yielded a variety of promising answers, reflecting a myriad of channels of communication. The only discernible pattern was that most projects were disseminated via the Web, either as a Web site or a Facebook group or both. Some projects used other social media platforms, such as Snapchat, Twitter, and Instagram, to share news of the *hortas* (and other projects). The use of technology was critical to disseminate the idea of *hortas* throughout these communities. It is unclear without additional research if these tools were successfully spreading the idea of *hortas* or if the social media posts were preaching to the choir, that the users/viewers of these platforms already know about *hortas*.

One communication channel that stands out is from Morro da Providência, where the adopter, a 23-year-old resident of the community, created an *empresa social* (social business) that sells its services to firms and private schools to increase knowledge of urban gardens and the tools needed to successfully adopt and maintain them. This individual primarily used the *empresa* as a strategy to gain financial resources to run and expand the *hortas* and the *agrofloresta* that she helped establish in her community. Albeit not deliberately, this became a channel by which this innovative idea could be diffused to other communities or households.

Another common response was the connection to formal social networks, or *redes* in Portuguese. Again, although it was unclear just how influential they were/are in the adoption of *hortas comunitárias*, these networks were key players in promoting urban agriculture in these communities. Some notable examples are Rede Carioca de Agricultura Urbana (CAU) and the Sustainable Favela Network (SFN). Rede Carioca is a "social movement that brings together people and organizations to defend agroecology in cities" (https://www.facebook.com/pg/redecau/about/?ref=page_internal).

The Sustainable Favela Network was "designed to build and reinforce solidarity networks, bring visibility and develop joint actions to support the expansion of community-based initiatives that strengthen environmental sustainability and social resilience in favelas across the Rio de Janeiro Metropolitan Region" (https://catcomm.org/sustainable-favela-network/). This network has played a critical role in spreading awareness and know-how regarding community gardens and urban agriculture more broadly.[3] The development of the network began in 2012 and in 2017 over one hundred community initiatives had been mapped across the region, including thirty nine community gardens (https://catcomm.org/sfn-launch/). Since the beginning of 2018, the network has held several exchanges. Each exchange is hosted by a community and features a number of activities, workshops, and courses centered on one or two themes. The 2018 exchanges consisted only of members of the network. For the 2019 on-site exchanges, anyone from outside the network was invited to come as well. Several dozen people have participated in these day-long events.

For instance, at the fourth network exchange of 2019, held in the community of Vila Kennedy, over 40 people participated in the day-long event. One of the activities during the visit was a workshop conducted by Environmental Agents[4] from the community on best practices in and the importance of gardening, using the garden at the Kennedy Brothers Community Center.

Under the guidance of the Catalytic Communities, an NGO headquartered in Rio, the network also recently created working groups connected to the socioenvironmental challenges most commonly faced by favela communities, including a Gardens and Reforestation Working Group. One of the objectives of the working groups is to create joint projects that can be funded to reach across communities and thus expand on what any individual project can do.

It was this working group that proposed holding one of the on-site exchanges at the Quilombo do Camorim, which had been devastated by torrential rains during the first half of 2019. As a result of the rainfall, the community garden had

[3]It is important to note that SFN addresses much more than just the challenge of environmental sustainability in these communities.

[4]The city of Rio de Janeiro offers training to city residents on environmental education through its Center for Environmental Education.

been completely destroyed. The working group members decided that combining an on-site network exchange, with workshops, with a *mutirão* to help rebuild the space of the community garden would represent an excellent opportunity. As a participant in the *mutirão*, I saw firsthand the rebuilding and replanting of the garden, with resources provided by a number of participants. Two of the key participants were members of Planta na Rua, an urban gardening collective that works citywide, who brought equipment, seeds, and saplings and directed the replanting of the community garden. Over the course of the day they taught a group of children how to build a garden, what fertilizer to use, and what should be planted and where. This hands-on lesson for children, ranging in ages from as young as six to as old as mid-teens, helped select the seeds and also planted them in the garden.

If we expand gardens beyond the community level to the individual household level, then clearly the most important and effective channel of communication is interpersonal. In other words, the idea of a *horta* is communicated from those with experience with/knowledge of them to those without by word of mouth. Indeed, Rogers contends that person-to-person contact is the most common form of communicating innovative ideas. All of the adopters were able to cite numerous examples of residents, friends, and family members who learned about their *hortas* personally, from seeing the *horta* and inquiring about how they could start their own household garden.

One final aspect of Rogers' diffusion of innovation theory is the existence of "innovation champions." According to Rogers, people who fit this description increase the chance of successfully spreading an innovation. Everyone interviewed would undoubtedly be considered an innovation champion for *hortas*. The initial research design did not call for an independent verification of the extent to which these individuals championed the cause of *hortas comunitárias*, or *hortas* in any form. However, all of the adopters spoke passionately about their commitment to these projects, of the many benefits derived from gardens (of any shape and with whatever is planted) to people and households. Nonetheless, one interviewee stands out for his devotion to this idea. The horta adopter from the Pavuna community showed an enthusiasm and fervor for the projects he was engaged in that, to be fair, went beyond just gardens, that was unmatched in other people. This individual not only developed a community garden in a converted warehouse but also worked on projects throughout the city to create green spaces in public areas.

It is not surprising that these adopters would be considered champions of this innovative idea. Building a community garden requires passion, because there are significant challenges due to the intense effort that is required to coax initial assistance from others (in the community and beyond), and acquire even minimal amounts of financial resources for upkeep. As many told me, it was an ongoing challenge to find the time to maintain the gardens–often by themselves.

4 Limitations

One of the main aspects limiting my findings is the sample size. I was only able to interview a tiny fraction of *comunidades* in Rio, which, as noted in the introduction, numbers over one thousand. Many of these communities are large enough to house a significant number of *hortas*. For example, the community of Rocinha, the largest in Rio, has over 100,000 residents. In other words, to make sweeping generalizations based on a dozen or so examples would be foolish. What this study can do is provide tentative findings that can be tested by expanding the sample size considerably.

Another potential problem involves the project's conceptual framework. Rogers uses a framework for understanding the diffusion of innovation that deals with technological innovation. *Hortas comunitárias* may not be a good match, because community gardens are not a technological innovation, and thus the dynamics of innovation diffusion and the independent variables with the greatest impact might be different than the ones Rogers presents in his work. Furthermore, Rogers' model assumes an environment with abundant resources, as was pointed out by Dr. Theresa Williamson, the founder and director of Catalytic Communities, an NGO in Rio that has served the interests of favela communities for almost twenty years. In the case of favela communities, there is a significant dearth of resources typically associated with the adoption of innovative ideas, the main resource of course being money. Indeed, one of the main features of favela communities is their ability to find solutions to the many challenges they face using minimal resources or using whatever resources they possess creatively to minimize these challenges.

5 Conclusion

This project investigated why communities or individuals adopt the idea of a community (or household) garden using a conceptual model offered by Everett Rogers. The project also sought to uncover the methods by which the idea of a *horta* spread within and across communities. While the findings are tentative and need additional testing, the results suggest that with some revisions the conceptual model can be applied to this phenomenon.

The advantages of adopting *hortas* are numerous and the resources necessary to successfully carry out these projects, especially on the micro-level of households, are relatively easy to obtain. Because of these dual features, fostering the growth of *hortas* in these communities could be a primary objective of the municipal government in Rio. With few resources, the government could easily promote the spread of *hortas* throughout communities. With one exception, the people interviewed for this research were adamant that the government did nothing to help. In fact, in one case, the municipal government in Rio de Janeiro did more damage than good by insisting on building and maintaining the *horta* in a specific way when residents in the community, with some level of knowledge, knew it would not work.

The benefits of community gardens go beyond those mentioned above. In the conclusion to her study of urban agriculture in Rio de Janeiro, Rekow (2015) notes:

> There is a wealth of evidence that suggests that UA [Urban Agriculture] is an effective means not only for helping to alleviate poverty, but for bringing social stability and therapeutic relief to the residents of favela communities. Other benefits of UA is that it can be delivered at a relatively low cost, while at the same time mitigating environmental risk and ecological scarcity, all without expanding a development footprint. Gardens also can provide safer, cleaner, and more productive social space that can diminish feelings of lack, anxiety and stress. (p. 6)

In sum, the promotion of community gardens would greatly benefit these communities. This is particularly true now with the coronavirus pandemic sweeping through these areas, leading to more food insecurity. Since March of this year, millions of Brazilians have seen their incomes reduced or disappear completely due to the pandemic. Shortages of food are becoming more widespread in favela communities (Goulart et al., 2020).

However, given the current political climate in Rio de Janeiro, with the mayor a vocal opponent of favela communities, it would be unreasonable to expect him to become an ally of favela residents. Instead, additional studies can seek to uncover more precisely what communication channels are most effective in disseminating the idea of *hortas* and leveraging these channels for greater success. In addition, it is critical that communities work together in conjunction with the many social networks that exist in support of these projects. Anecdotally, often when asked about which social networks the respondents were familiar with or worked with, the answers were often very different. Two exceptions were Rede Carioca and the Sustainable Favela Network, which most knew about. Not infrequently, I would mention a network I had learned about from a previous respondent, and the person would say they had not heard about it.

The field of social innovation offers a useful framework for addressing this challenge and a potential template for institutionalizing innovative ideas. Social innovation is a growing field among policy-makers, especially in industrial democracies. Social innovation refers to "solutions [that] come from ordinary people in their own localities responding creatively and innovatively to the pressing challenges they and their communities are experiencing" (TEPSIE, 2014, p. 8). The "social" in social innovation refers to values, needs, well-being, and community impact. Many initiatives are responses to the day-to-day social problems of people and groups and provide a means of accessing rights, goods, and services. Community gardens are an excellent example of a social innovation.

Typically, the sustainability of social innovation depends heavily on the state (Sammut, 2020; Sinclair, 2018). In Brazil, and elsewhere in the developing world, however, poor communities that develop innovative solutions to unmet social needs cannot rely on partnerships with the state for a variety of reasons, not the least of which is the tendency of the state to systematically ignore the needs of these communities. In large part because of the absence of the state in these communities, Brazil, according to one study, is considered a "cradle" of social innovation (Cipolla and Afonso). This study notes that initiatives are largely based on interpersonal face-to-face relationships and encounters–the kind that occur in small groups, on a local scale. This is certainly true in the case of community gardens.

Although the role of the government in sustaining social innovations in Brazil is minimal, other intermediary organizations and networks are viewed by scholars as keys to spreading and scaling these solutions to other communities (TEPSIE, 2014). Special attention is devoted to partnerships with higher education institutions. An example of this is the case of the *Teto Verde* (Green Roof) project in the community of Manguinhos, which developed a close relationship with the faculty in the Environmental Studies program at a university in the neighboring city of Niteroi. Another example of an intermediary institution that seeks to promote and disseminate innovative ideas at the grassroots level in Rio de Janeiro is the Oswaldo Cruz Foundation. Fiocruz has operated since the early 1900s and part of its mission is to "disseminate and share knowledge and technologies aimed ... [at] the promotion of health and quality of life of the population" (https://portal.fiocruz.br/en/institutional-profile).

However, a review of numerous social innovations across Europe suggests that these partnerships are insufficient for widespread dissemination of social innovation. The authors contend that social innovations usually need to be mainstreamed through the market or with support from the state. As discussed above, social innovators in Rio cannot rely on the state for any kind of meaningful support. On other hand, there is significant evidence that social innovators can exploit market opportunities to set up and grow a social enterprise that will help diffuse the innovative idea to other communities (TEPSIE, 2014). Social entrepreneurship has grown exponentially among the Brazilian lower class since the early 1990s. The case of *Horta Inteligente* in Morro da Providência is a perfect example of social entrepreneurship in action. As mentioned above, the adopter of the garden created a social enterprise that develops workshops on gardening for firms and private school in Rio. The proceeds from these workshops are funneled back into community garden projects in Morro.

In the end, what is critical is citizen engagement in finding solutions to local problems. A deep level of citizen engagement, which these adopters of community gardens exemplify, offers the baseline for participation that enables a better understanding of community needs so that appropriate solutions can be developed. The diversity of these adopters, and how they differed in establishing their gardens, demonstrates the power of all kinds of people, from all sorts of backgrounds and experiences, to devise innovative solutions to the challenges that their communities face.

References

Bass, F. (1969). A new product growth model for consumer durables. *Management Science, 15*, 215–227. https://doi.org/10.1287/mnsc.15.5.215

Cipolla, C. & Afonso, R. (n.d.). *Social innovations in Brazil: How do social innovations flourish?* https://www.socialinnovationatlas.net/fileadmin/PDF/einzeln/02_SI-in-World-Regions/02_13_SI-in-Brazil_Cipolla-Afonso.pdf

Diamond, L. (1999). *The spirit of democracy: The struggle to build free societies throughout the world*. St. Martin's Griffin.

Gourlart, G., et al. (2020, April 2). Em meio à pandemia, cariocas lutam para pôr comida na mesa, mas cresce número de casas com geladeiras vazias. *Extra Digital*. https://extra.globo.com/noticias/rio/em-meio-pandemia-cariocas-lutam-para-por-comida-na-mesa-mas-cresce-numero-de-casas-com-geladeiras-vazias-24345400.html

Katz, E. (1957). The two-step flow of communication: An up-to date report on a hypothesis. *The Public Opinion Quarterly, 21*(1), 61–78. https://doi.org/10.1086/266687

Lesch, D. & Haas, M. (2012). *The Arab spring: Change and resistance in the Middle East*. Westview Press.

Mahajan, V., et al. (1990). New product diffusion models in marketing: A review and directions for research. *Journal of Marketing, 54*, 1–26. https://doi.org/10.1177/002224299005400101

Mossberger, K. (1999). State-federal diffusion and policy learning: From enterprise zones to empowerment zones. *The Journal of Federalism, 29*(3), 31–50. https://doi.org/10.1093/oxfordjournals.pubjof.a030036

Rekow, L. (2015). *Fighting insecurity: Experiments in urban agriculture in the favelas of Rio de Janeiro*. Field Actions Science Reports [Online]. https://journals.openedition.org/factsreports/4009

Rogers, E. (1962). *Diffusion of innovations*. (1st ed.). Free Press.

Rogers, E. (2003). *Diffusion of innovations* (5th ed.). Free Press.

Ryan, B., & Gross, N. (1943). The diffusion of hybrid seed corn in two Iowa communities. *Rural Sociology, 8*, 15–28.

Sammut, N., et al. (2020). The status of sustainable social innovation in Malta. *Sustainability, 12*(10), 1–10. https://doi.org/10.3390/su12104238

Sinclair, S., et al. (2018). Social innovation, social enterprise, and local public services: Undertaking transformation? *Social Policy & Administration, 52*(7), 1317–1331. https://doi.org/10.1111/spol.12389

Sugiyama, N. (2016) *Diffusion of good government: Social sector reforms in Brazil*. University of Notre Dame Press.

TEPSIE. (2014). *Social innovation theory and research: A summary of the findings from TEPSIE*. A deliverable of the project: The theoretical, empirical and policy foundations for building social innovation in Europe (TEPSIE), European Commission—7th Framework Programme, European Commission, DG Research.

Toews, D. (2003). The new tarde: Sociology after the end of the social. *Theory, Culture & Society, 20*(5), 81–98. https://doi.org/10.1177/02632764030205004

Wellin, E. (1955). Water boiling in a Peruvian town. In B. D. Paul (Ed.), *Health, culture and community*. (pp. 71–103). Russell Sage.

Weyland, K. (2007). *Bounded rationality and policy diffusion: Social sector reform in Latin America*. Princeton University Press.

An Overview of 100 Resilient Cities Network—The Case of Amman

Sameh Shamout and Paola Boarin

Abstract

Amman was chosen to join the 100 Resilient Cities Network (100RC) in December 2014 to design and implement an urban resilience strategy that was announced three years later to assist strengthen the city and improve people's quality of life. This paper discusses what urban resilience means, introduces the 100 Resilient Cities Network and its contribution to the case of Amman to build its first resilient policy and provides an overview of the development and implementation process. The strategy is then analysed using the word frequency method. A web-based questionnaire is also used to get feedback from local experts on the Strategy and some of the difficulties surrounding it. The survey is offered to approximately 200 local professionals, policymakers and academics with a background on the different urban problems in the context of Amman. The results of the questionnaire highlight several key points that may be used to further enhance the emerging "Resilient Amman" initiative. First, in the case of Amman, boosting urban resilience is critical due to the city's rapid population growth, which intensifies other stresses and shocks. Second, further effort is needed to improve stakeholder engagement in the project by enhancing stakeholder selection criteria in terms of background and specialisation variety. Third, because many local practitioners are unaware of the Amman Resilience Strategy, additional work is needed in terms of participatory procedures and results distribution among stakeholders and the general public in order to increase the project's influence on the city's future.

S. Shamout (✉) · P. Boarin
Faculty of Creative Arts and Industries (CAI),
School of Architecture and Planning, Future Cities Research Hub,
The University of Auckland, Auckland, New Zealand
e-mail: s.shamout@auckland.ac.nz

Keywords

Urban resilience • 100 resilient cities • Amman • Resilience challenges

1 Introduction

Similar to other Jordanian cities, Amman has been facing significant problems and urban challenges that hinder its sustainable development and growth. Those urban-related issues are compounded by the instability of the region that has led to waves of influxes of refugees, with many cities having to host a large number of people in a very short time (Greater Amman Municipality, 2017; Ministry of Planning and International Cooperation, 2017). Accordingly, the population of Jordan has increased by about 87% in the last ten years, reaching 9.5 million in 2015 (Greater Amman Municipality, 2017). Amman has attracted domestic migration and received many different waves of refugees and asylum seekers from the surrounding unstable conflict areas including Palestine, Lebanon, Iraq, Libya, Syria, Yemen and other areas in the region (Potter et al., 2009). The population of Amman has doubled from two million in the year 2004 to four million in the year 2015 and is expected to reach over six million by the year 2025 (Greater Amman Municipality, 2017; Department of Statistics, 2018). This unexpected massive population increase has added more strains on the city services and the resources of the country and has caused an unplanned and fast urban expansion. Palestinian refugee camps that were established around Amman, for example, have become informal urban areas with a poor quality of life (Alnsour & Meaton, 2014). Polit (2019) discusses resilience from an urban point of view and confirms that when city systems become bigger, their urban resilience decreases, and then, the city becomes more vulnerable to external shocks and internal pressures. However, these permanent stresses and shocks could give cities the opportunities to learn how to

F. Trapani et al. (eds.), *Advanced Studies in Efficient Environmental Design and City Planning*,
Advances in Science, Technology & Innovation, https://doi.org/10.1007/978-3-030-65181-7_22

confront those challenges, adapt to them and grow, and ultimately, it develops resilience towards those challenges. In the case of Amman, studying its urban resilience appears to be important in order to assist the city survive and retain its functions, make the best use of its limited resources, and continue to provide crucial services to both its residents and hosted refugees. The urban resilience concept was first introduced in Jordan in the local urban context of Amman, when the city was selected to join the 100 Resilient Cities Network (100RC) in December 2014 to develop and implement its first urban resilience strategy. Three years later, the Greater Amman Municipality released Amman's first urban resilience strategy, being the first-ever strategy of its kind for a Jordanian city (Greater Amman Municipality, 2017).

This paper discusses the experience of Amman in participating in the 100 Resilient Network (100 RC) by discussing and reflecting on some of the most relevant aspects of the Amman Resilience Strategy.

2 The Main Objectives of This Study

This research aims to introduce the Amman Resilience Strategy and evaluate its key phases of development, commenting on stakeholders' engagement in the plan's creation and local professionals' perceptions of the strategy in order to identify weaknesses and opportunities to improve urban resilience. It also seeks to determine how important urban resilience is in the case of Amman.

3 The Significance of This Study

Amman is considered a role model for planners and architects in other Jordanian cities. Ensuring its successful engagement in what can be called an "urban resilience movement" could be strategic for boosting urban resilience on a country scale, especially in light of the development and implementation of the Amman Resilience Strategy, the first local urban resilience strategy in the country. Given its position as the country's capital, as largest city in terms of population (Department of Statistics, 2018), as largest refugee host (UNHCR, 2019a) and as an important regional hub, it is imperative to reflect on the Amman Resilience Strategy and draw the attention of local and international researchers on the importance of this strategy for a potential urban resilience movement in Jordan. Unfortunately, since its first release in 2017, the strategy has not been discussed and analysed in scholarly literature or professional fora, limiting the impact of this vital tool on both the city and its inhabitants. This paper aims at filling this gap, thus contributing to urban resilience research.

4 Method

This study uses a literature review and a web-based questionnaire as primary research methods, further supported by the word-frequency method.

Following a brief summary of the 100 Resilient Cities Network, the literature review briefly reviews the meanings of resilience with a specific focus on urban resilience (100RC). It then discusses Amman's journey in building the Amman Resilience Strategy, including the major stages and views from key stakeholders. Lastly, it reflects on emerging Amman's main weaknesses.

In the web-based questionnaire about the Amman Resilience Strategy, three main topics are investigated: (1) whether urban resilience should be an essential issue for the case of Amman or not; (2) people's perceptions of the Amman Resilience Strategy; and (3) stakeholders' involvement (or lack thereof) in the development of the strategy. The Greater Amman Municipality (a governmental organisation) and the Jordanian Engineers Association (JEA) (a non-governmental organisation) contributed to the research by inviting their affiliates and members to undertake the questionnaire. The third-party organisations were informed that participants had to be familiar with Amman's built environment and its related issues in order to take part in the web-based questionnaire, thus ensuring the relevance and accuracy of data. The survey was offered to approximately 200 local professionals, policymakers and academics.

The word frequency is used as a secondary research method. Counting the frequency of words within a text and visualising trends may help to discover relationships between particular concepts and enable new readings previously inhibited (Clement, 2008). Highlighting the frequency of words in the policy helps to identify the terms receiving the attention from decision-makers, and therefore, understanding whether the use of those terms is reflected on the real projects being developed for the city or not. The study analyses the 2017 published document of the Amman Resilience Strategy using the software NVivo 12 to show the most frequent words used in the text and visualises them in a word cloud.

5 What Is Urban Resilience?

Resilience is an expansive concept that has several definitions and applications across many different academic fields. It has roots in math and physics (Norris et al., 2008) and has been used in ecology (Holling, 1973), psychology (Coutu, 2002), biology (Kitano, 2004), organisations (Gilbert et al., 2012), engineering systems (Robbins et al., 2012) and societies (Longstaff et al., 2010). Although there are many studies on resilience, there is no single universally recognised definition for the concept yet (Sharifi & Yamagata, 2014). Overall, the idea might refer to complex adaptive systems with self-organisation.

Defining urban resilience also changes depending on the context. Many scholars define it with particular attention to disaster resilience by addressing the unexpected challenges urban areas might face, such as natural disasters (Cutter et al., 2008; Mileti, 1999; National Infrastructure Advisory Council (NIAC), 2012; National Research Council, 2012; Stevenson et al., 2015; Yamagata & Maruyama, 2016). Others, in their definition of the concept, include, beside disasters, other ongoing stresses that may be disturbing people's normal life in cities (Resilient Design Institute (RDI), n.d.; Urban Resilience Hub, n.d.; Da Silva & Morera, 2014; Hughes & Healy, 2014). The concept should go further than disaster resilience, moving towards the urban areas' capacities to adapting with or withstanding any chronic problems or changes that affect their functions to meeting people's basic needs. Including both ongoing stresses and potential shocks (Fig. 1) when defining urban resilience seems more appropriate, especially for contexts where the built environment is exposed to more stresses on a daily basis than unexpected adverse events or shocks. Examining these two main dimensions requires the identification of the main urban-related challenges and stressors that affect our cities to find out how resilience could be boosted in the built environment.

Incorporating resilience in the urban context may result in city systems that, according to the Urban Resilience Hub launched by the UN-Habitat (Urban Resilience Hub, n.d.), can sustain continuity through all stresses and shocks while systems adapt positively, and transform towards the achievement of sustainability. Also, the Rockefeller Foundation describes the resilient city's systems as systems that are survival, adaptable and able to grow regardless of any stresses and shocks they are exposed to (Da Silva & Morera, 2014).

It is argued that that the concept of resilience to the period of the 2000s and 2010s is what the concept of sustainability was to the period of 1980s and 1990s (Foster, 2010; McPhearson, 2014; Saunders & Becker, 2015; Stumpp, 2013). Both resilience and sustainability concepts have been disparately used in the last recent years, with some scholars considering them as similar concepts while others claiming they are not related (Fig. 1).

6 100 Resilient Cities Network (100RC)

The Rockefeller Foundation established the initiative of '100 Resilient Cities', aiming to help cities all over the world enhance their resilience to social, economic and physical challenges and problems in the twenty-first century (Spaans & Waterhout, 2017). The 100RC programme defines urban resilience as 'the capacity of individuals, communities, institutions, businesses and systems within a city to survive, adapt and grow no matter what kinds of chronic stresses and acute shocks they experience' (100 Resilient Cities, n.d.).

Amman joined the 100RC in December 2014 to develop and implement its first urban resilience strategy. Three years later, the Greater Amman Municipality released Amman's first urban resilience strategy, being the first-ever strategy of its kind for a Jordanian city (Greater Amman Municipality, 2017).

7 Why Boosting Urban Resilience Is Essential for Amman?

As a response to many global problems, pressures and challenges that have emerged, such as climate change-related events, conflicts and shortage of resources, there is a need to enhance the resilience of our cities. In the context of Jordan, there are many reasons for the concept of resilience to be prioritised in policymaking, urban development and city planning (Shamout et al., 2020). Results of the web-based questionnaire show that about half of the 150 participants, who responded to the question asking about the importance of boosting urban resilience in the context of Jordan, see resilience as a highly crucial urban issue for Jordan, especially in the case of Amman (Fig. 2).

Many challenges were discussed as reasons why boosting urban resilience should be considered essential in the context of Jordan.

Jordan has experienced rapid population growth caused by several waves including domestic migration (Potter et al., 2009) and the influx of refugees and asylum seekers (Ministry of Planning and International Cooperation, 2019; Seeberg, 2020). Regarding the refugee crisis, one questionnaire participant stated that the country has been suffering from its surrounding unstable neighbouring region that is still adding burdens at many levels. In the urban context, for example, prices of real estate have dramatically increased, resulting in rising rents and demands for water, electricity and other basic needs. The reason is that most refugees do not live in refugee camps, as some might think. Most of them are settling in the northern part of Jordan in urban, peri-urban and rural areas, and 30% of them are settling in Amman (UNHCR, 2019b). Accordingly, Amman became the second-largest host of refugees per capita worldwide (Greater Amman Municipality, 2017). The crisis is still adding pressure, overstretching the city's infrastructure, including traffic, public transportation and essential services such as water and energy. According to some participants, one compelling reason behind this pressure on infrastructure, public services, and the increasing demand for water and energy is the increased demand for low-cost housing.

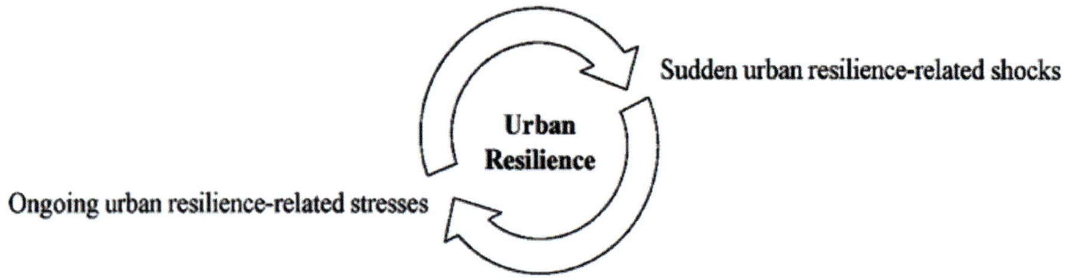

Fig. 1 Two main dimensions required to understand urban resilience

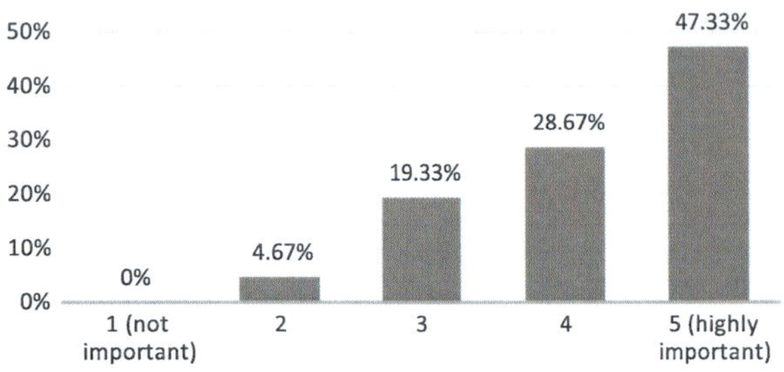

Fig. 2 Results from the questionnaire regarding the importance of urban resilience in the context of Jordan

They added that in some cases, there are more than one refugee family stay in one apartment to reduce rent on each family, and therefore, pressuring the city services. Due to all these problems, one participant claimed that the government had recently been focusing on finding strategies to provide basic needs, rather than developing a long-term strategy for the state. Some participants emphasised that this rapid population growth, as well as the significant economic burden and all the other burdens, should call for a resilience movement at the country's scale. The increased crime rate was also mentioned as a reason for the call of resilience. According to participants, the education sector was also affected as the number of students in public schools increased, affecting the quality of education. Kumaraswamy and Singh (2017) argue that Jordan faces a population pressure that has a strong correlation with the refugee crisis. Refugees did not only intensify current challenges but also created new problems for the country.

Other reasons were mentioned by participants when enquired about why boosting resilience should be essential for the case of Jordan, as follows:

- The economic crisis: the economy of the country was affected many times in the last decades by the instability of its surrounding region, e.g. the 1990 Gulf War (Swaidan & Nica, 2002);
- Climate change: climate change-related events such as flash floods have been disturbing people's lives and the built environment in Jordan, including Greater Amman. Amman was frequently hit by heavy rainfall that led to flash floods because many areas in the city were not prepared enough for such kind of heavy rain. Greater Amman Municipality (2017) stated that some unusual flash floods occurred in Amman, causing damage to vehicles, buildings and infrastructure and loss of lives, such as the flash floods in November 2015;
- Lack of energy resources: meeting the increasing energy demand, especially electricity is a reason why resilience should be essential for Amman as this sector is adversely affecting the country's economy, environment and social life (Jaber et al., 2015). Shamout et al. (2019) emphasised the need for developing measures for boosting resilience in the existing built environment to address the everyday stresses and potential shocks which the country faces or could face, first highlighting the problem of the energy resources shortages.

There are many reasons why urban resilience should be a priority for national and local policymaking and urban development plans and strategies. The participants to the web-based questionnaire mentioned a few of those challenges and problems only, but this would be an area for further investigation.

Although most participants in the web-based questionnaire agreed on the importance of boosting urban resilience in Amman, the results of another question in the survey showed that over half of them had never heard about Amman Resilience Strategy, and 20% were not sure if they heard about it or not (Fig. 3). Taking into account that participants work in the built environment industry, the result emphasises the distance between the Amman Resilience Strategy and local professionals, researchers and academics, highlighting that not enough communication and outreach was done by the Greater Amman Municipality, thus not engaging with local stakeholders (Fig. 3).

8 Amman's Journey in Developing Its First Urban Resilience Strategy

The Amman Resilience Strategy developed goals, pillars and actions aiming to establish a vision for Amman to become resilient. It was mainly divided into the following three stages (Fig. 5):

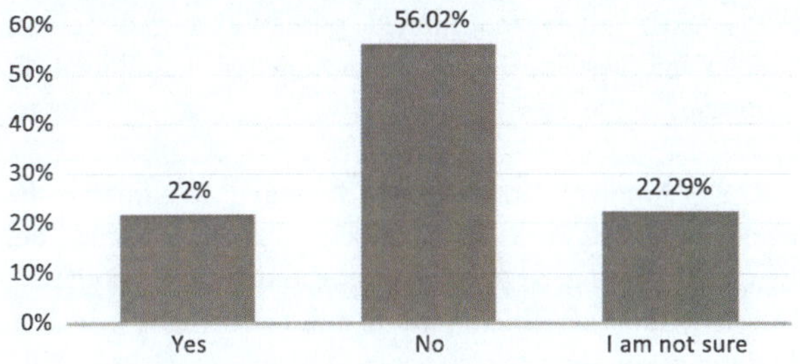

Fig. 3 Results from the questionnaire regarding the awareness of the Amman Resilience Strategy or not

- Phase I: Preliminary resilience assessment (PRA);
- Phase II: Amman resilience strategy;
- Phase III: Implementation.

8.1 Phase I—Preliminary Resilience Assessment (PRA)

Phase I aimed to understand the resilience-related issues in Amman. It identified areas that needed to be addressed for building the city's resilience. The result of this phase was the preliminary resilience assessment of Amman, and the identification of critical areas addressed later in phase 2.

Held by the Greater Amman Municipality in 2016, the Agenda-Setting Workshop and some other focus group workshops brought stakeholders together, presented them with the resilience concept, introduced them to the 100 Resilient Cities initiative and, eventually, identified Amman's critical resilience challenges. The five discovery areas identified in the PRA in August 2016 are as follows: efficient institutional financial planning; comprehensive urban development; effective legislation; resourceful environment; and robust community. In addition, a cross-cutting discovery area was identified, i.e. improving Amman's ability to cope with the sudden increase in population. This further stresses the impact that population growth has on the resilience of Amman.

8.2 Phase II—Amman Resilience Strategy

In its resilience strategy, the Greater Amman Municipality adopted the city resilience framework to help identify the city's resilience weaknesses, strengths and opportunities. Figure 4 shows the city resilience framework, which is divided into four main dimensions and 12 drivers (Da Silva & Morera, 2014). This framework is the 100RC's illustration of what makes a city resilient.

The Amman Resilience Strategy identified opportunities and initiatives that will help improve the city's resilience. It prioritised them using the 'Resilience Lens' tool and produced a shortlist of opportunities that have implementation potential and resilience value (Greater Amman Municipality, 2017). It also included recommended actions and projects.

8.3 Phase III—Implementation

Phase III was about 'operationalising resilience', i.e. making the proposed plans happen (Greater Amman Municipality, 2017). It aimed to put the Amman Resilience Strategy into practice by implementing its proposed actions. After the release of the strategy, the Greater Amman Municipality planned to start preparing a detailed implementation proposal in collaboration with their partners and relevant stakeholders (Greater Amman Municipality, 2017) (Fig. 5).

9 Amman's Significant Weaknesses

The Amman Resilience Strategy identified significant weaknesses in the performance of the Governorate of Amman on particular issues, dividing them into two main groups, stresses and shocks (Table 1). During the first phase of the strategy, Amman's Resilience Team in the Greater Amman Municipality used qualitative approaches to identify

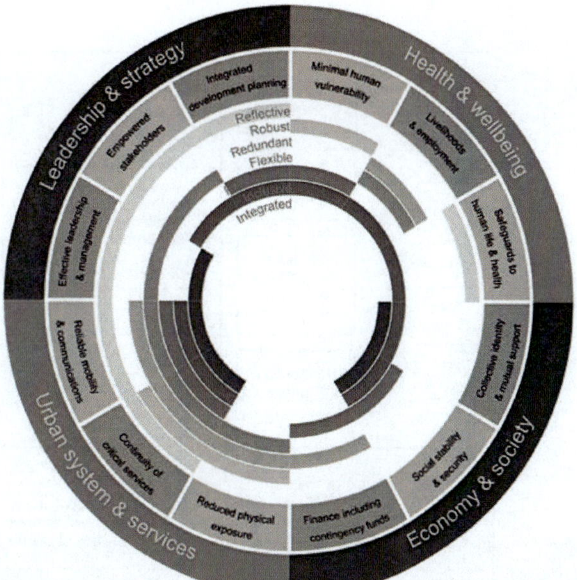

Fig. 4 City resilience framework. *Source* Da Silva and Morera (2014)

Amman's principal stresses and shocks. The team conducted an Agenda-Setting Workshop that was attended by around 100 people, five focus group workshops and fourteen individual interviews. After reviewing the activities that determined the city's main weaknesses, a number of points can be highlighted:

- The Greater Amman Municipality organised workshops to obtain data from selected individuals. Amman's Resilience Team invited approximately 100 related stakeholders from different sectors, including some governmental bodies, the private sector, academic institutions and NGOs. The participants were engaged in some exercises and activities where their perceptions on top stresses and shocks facing the city of Amman were summarised. However, some questions may arise about the stakeholders' selection criteria in terms of diversity of background and specialisations;
- Economic crisis, terrorism, infrastructure failure, drought, heatwaves and flash flood were identified as infrequent top shocks;
- High energy costs, water scarcity, lack of natural resources, limited access to financial assistance, refugees' influx, changing demographics, urbanisation, transportation and other challenges were listed as top ongoing stresses;
- Based on both the participating stakeholders and working groups, fewer shocks (6) were listed compared to stresses (12). Understanding whether we should prioritise to work on stresses over shocks will help provide the city resilience project in Amman with a more practical approach, especially given the limited possibilities and current economic, political and environmental conditions in Jordan.

The Amman Resilience Strategy identified the resilience-related stresses and shocks affecting Amman, but these challenges need to be examined further and prioritised. The city's level of preparedness to these challenges as well as their impact on the city also needs to be measured, especially when considering urban resilience. It is essential to consider those points to establish a more robust understanding of Amman's main urban resilience-related challenges to ensure the efficient development of the strategy's actions and implementation to maximise the resilience-related benefits to the city of Amman and its inhabitants (Table 1).

10 Amman Resilience Office and the Development of the City Plan

One of the most significant achievements of Amman's experience with 100 Resilient Cities Network is the establishment of the first office for resilience in the country. The Greater Amman Municipality decided to appoint four main people to be the core team of this office: Chief Resilience Officer, Project Manager, Planner, Administrative Assistant and Public Relations. Their tasks are a combination of data analysis, research, documentation, correspondences, developing the resilience strategy, logistics, decision making and solving debates. They are also responsible for liaising

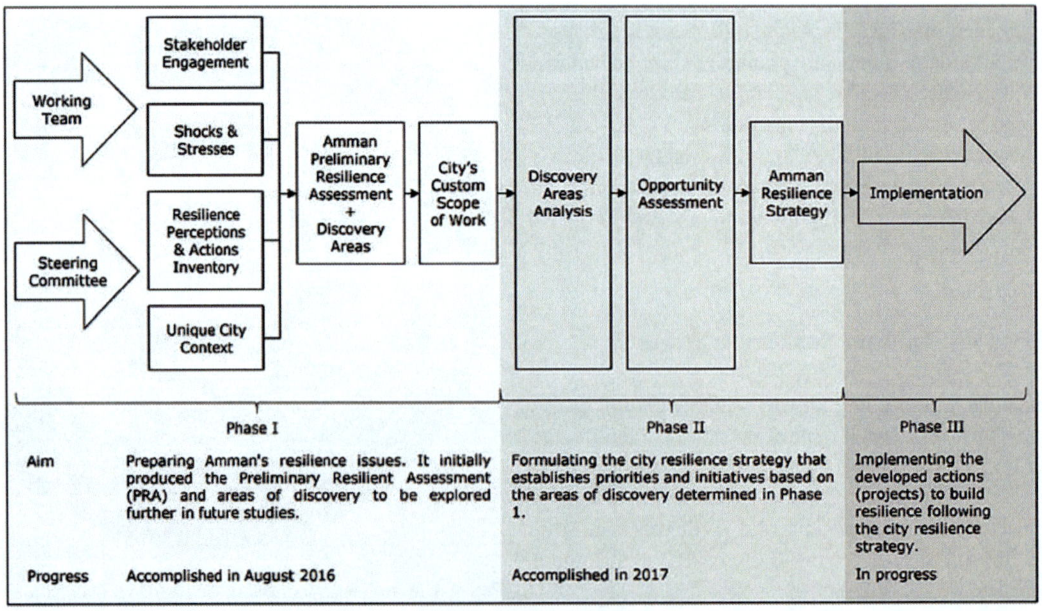

Fig. 5 Summary of the main phases of Amman Resilience Strategy. *Source* Elaboration of the author based on Amman Resilience Strategy

Table 1 Stresses and shocks affecting Amman (Greater Amman Municipality, 2017)

No	Acute shocks	Chronic stresses
1	Economic crisis	High energy costs
2	Terrorism	Water shortage
3	Major infrastructure failure	Lack of natural resources
4	Drought	Limited access to financial assistance
5	Heatwaves	A major influx of refugees and/or asylum seekers
6	Flash/surface flood	Changing demographics
7		Low-quality services especially in health and education
8		Urbanisation
9		Unemployment
10		Lack of affordable housing
11		Lack of diverse livelihood opportunities
12		Traffic congestion/lack of public transport system

Note The list is taken from Amman Resilience Strategy and challenges are not presented in order of priority

between the Amman Resilience Office and 100RC, Stakeholder Working Groups, the Steering Committee and city decision-makers. It could be argued that only four core team members are not enough to fulfil their required tasks, taking into account the significance of such a project, especially for the case of Amman in the current time. Therefore, having a larger team that includes more stakeholder from different backgrounds would be an efficient solution as it will save time and recruitment costs. In addition to the potential value that the stakeholders can add to the project, enhancing their engagement in the project will also strengthen community resilience so that the stakeholders will become more aware of the actual situation of the city's resilience state, weaknesses and potentials. A number of participants in the web-based questionnaire of this study agreed that by increasing stakeholders' engagement in the 'Resilient Amman' project, the engaged stakeholders would be more comprehensively exposed to the city's challenges, and therefore, they could help to raise people's awareness of the Amman's current resilience state.

11 Stakeholder Engagement

In the preparation process of the Amman Resilience Strategy, a stakeholder was defined as anyone who has a significant influence on developing the Amman Resilience Strategy and/or implementing it. The related stakeholders in the 'Resilient Amman' project were governmental bodies, city officials, local communities, NGO's and academic institutions. After reviewing the stakeholders' engagement plan, a few points can be highlighted:

- It seems that the selection process of stakeholders did not follow specific criteria. Selecting potential stakeholders is very important as they may have negative or positive impacts on the resilience strategy. In other words, instead of contributing to the development of the 'Resilient Amman' project, some stakeholders might become an obstacle to define plans, projects and activities related to the resilience strategy. They could provide misleading recommendations that do not come from a professional background. Therefore, there should be clear criteria that consider many factors, such as the stakeholders' academic background, knowledge and local and international experience. It is essential for stakeholders to be selected based on specific criteria in both the policymaking and the implementation of the Amman Resilience Strategy. The selected stakeholders were allocated during Phase I to join each or both groups, steering and working groups, according to their field of experience to make the best use of stakeholders' input in the strategy. Among the four main discussion groups held in Phase I, the 15 stakeholders involved in the second group that worked on improving infrastructure, accessibility and linkages of Amman can be grouped as the following. First, two stakeholders were from professional associations: Jordanian Contractors Association and Jordanian Engineers Association. There are around 13 other civil society organisations acting as professional associations (Comprehensive Guide to Civil Society Organisations in Jordan, n.d.), so engaging with two of these associations only seems reductive. Second, three out of the 15 stakeholders were governmental: Land Transport Regulatory Commission, Housing and Urban Development Corporation and Ministry of Public Works and Housing. Three others were governmentally related providing essential services to Amman's citizens: the Greater Amman Municipality, Miyahuna, Jordan Electric Power Company and Airport International Group (AIG). Four stakeholders were included from the private sector working in the urban and building industry: A/E Business Council, Yaghmour Architects, Turath, and Project Jordan. Based on the companies' profiles, the reasons why they had been chosen is unclear. Only Jordan University was engaged in the project as a stakeholder, although there are more than ten other universities in Amman. Engaging a more significant number of academics would have been valuable to the development of this project. Similarly, only one local initiative was involved, i.e. the Ma'an Nasel. Therefore, the establishment of a more diverse and larger stakeholders' team with professionals from industry, government, private, educational sectors would have been significant, especially as the core team is relatively small compared to what needs to be done in this project. The engagement of individuals was also not enough. Many researchers and practitioners could contribute to the city's resilience vision with their experiences and perceptions, especially the ones who already live in the city and experience its resilience-related challenges. There were indeed few individual participants included in the first preparation stages of the strategy. There are no tools to engage future individuals or to facilitate the stakeholders' participation in the city's plans and projects in general. It was mentioned that a communication plan was developed for engaging the public in the Resilience Team's activities,[1] including social media outreach and city-wide events. However, very few social events were held, and they were not worth mentioning in the strategy.

- Besides the need for clearer selection criteria, the stakeholders' contribution depends on how their engagement level in the projects is managed and whether it is well organised or not. Therefore, it is essential for their engagement to be efficiently planned. The stakeholders' involvement is managed by the Chief Resilience Officer and the office core team. It seems that the stakeholders were mainly engaged in the early development stages of the strategy to get their feedback, input and support on the resilience-related issues facing Amman. Some participants to the web-based questionnaire in this study agreed on this point and claimed that stakeholders should also be engaged in the initiatives/actions that will be adopted to address the challenges, including the implementation stages and all the possible difficulties that might face the municipality.
- Engaging younger citizens was a further step towards enhancing social resilience, and a separate workshop was held by the municipality to capture such insights. Involving younger generations is very important to build future resilience, as they are the people most exposed to the city's challenges on a daily basis. After a group discussion with some students and young people, three main issues were highlighted for Amman:
  - Urban planning is not flexible enough to handle rapid changes;
  - Lack of public transportation;
  - The municipal regulations and laws are not sufficiently enforced. Shamout et al. (2019) emphasised the issue of the insufficient enforcement of the buildings regulations and municipal laws, highlighting that this is a significant obstacle and challenge at the urban scale.

Based on the discussion above, we suggest that consultation with stakeholders should still be possible as that would be beneficial for the development of the Amman Resilience strategy as well as its implementation stages. It is essential to understand who are the ones who would be more advantageous to the 'Resilient Amman'. To answer this question, participants to the web-based questionnaire in this study were asked to rank the stakeholders in order of priority, from the most beneficial to the least influential (Table 2). Around 30% of the 151 participants who responded to this question gave the municipalities the first rank. Participants were given a chance to add more stakeholders under the 'others' category. Most suggestions included local citizens (Table 2).

12 Analysing Amman Resilience Strategy—Word Frequency Method

The text of the first resilience strategy for Amman was analysed using the software NVivo 12 to show the most frequently used words that were then visualised in a word cloud (Fig. 6). Resilience was not only introduced in the Amman Resilience Strategy as the core concept in terms of context, but the term was also very often used in the strategy, making it the third most frequent word in the document (357 count—1.86% weighted percentage) after the terms 'Amman' (452 count—2.35% weighted percentage) and 'city' (393 count—2.05% weighted percentage). The word 'sustainability' is repeated 39 times only with a weighted percentage of 0.2%, and this might align with the findings of some researchers about resilience replacing sustainability in urban policymaking (Foster, 2010; McPhearson, 2014; Saunders & Becker, 2015; Stumpp, 2013). From this analysis, it seems that resilience is becoming an essential aspect to policymakers and is starting to emerge in conversations on the development of urban-related policies and plans. It is significant to start seeing such frequent use of and reference to resilience in a strategic urban policy such as the Amman Resilience Strategy, as that might become a potential

[1]Data was obtained from the Greater Amman Municipality's Resilience Office.

Table 2 Stakeholders' ranking based on participants' answers

Stakeholders	Ranking						
	1	2	3	4	5	6	7
Municipalities (%)	30.4	27.1	11.2	9.9	15.8	5.3	0.0
Governmental bodies (%)	28.4	19.8	13.9	15.2	8.6	13.2	0.6
Professional associations (%)	17.2	20.5	30.4	17.8	8.6	5.3	0.0
Academic institutions (%)	9.9	20.5	18.5	28.4	16.5	5.3	0.6
NGOs (%)	5.3	4.6	9.2	12.5	32.4	35.1	0.6
Local communities (%)	6.6	5.9	15.8	15.8	17.8	35.1	2.6
Others (%)	1.9	1.3	0.6	0.0	0.0	0.6	95.3

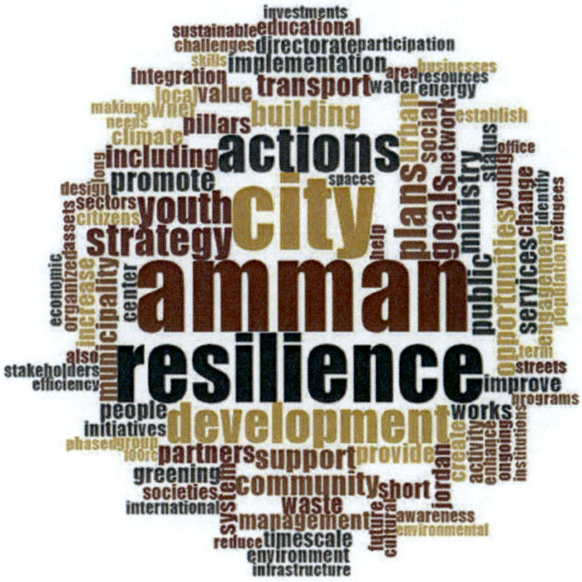

Fig. 6 Word cloud showing the most frequently used words in Amman Resilience Strategy generated by NVivo 12

positive driver for urban policymaking in the wider Jordan country given the importance of the case of Amman. However, many questions can be raised on the excessive use of the resilience concept in a local urban resilience strategy, especially when there is no similar policy at the national level. Furthermore, the next and more important step is to investigate whether there is an actual connection between what is included and aimed for in policy and actual plans, in order for 'resilience' not be used as a buzzword only.

13 Conclusions

This study investigated the experience of Amman in participating in the 100 Resilient Network (100 RC) by discussing and reflecting on some of the most relevant aspects of the Amman Resilience Strategy. It has highlighted different issues in regard to the importance of boosting urban resilience in the case of Amman, the stakeholders' participation in the development of the 'Resilient Amman' project and the local practitioners' awareness of Amman Resilience Strategy. Highlights from the study can be summarised as follows:

- The population pressure in Jordan should call for an 'urban resilience movement' in the country, especially in Amman, after reviewing some of the urban-related challenges that emerged or compounded from the rapid population growth. According to Kumaraswamy and Singh (2017), 'the insufficient response of a country to meet the basic needs of its growing population is commonly recognised as population pressure and is normally due to the exponential growth of population without a corresponding increase in its resources'.
- Although the Amman Resilience Strategy has managed to identify the principal stresses and shocks affecting the city, these challenges need to be examined and prioritised. The city's level of preparedness to these challenges as well as their impact on the city also needs to be measured. These points can be considered as potential research gaps that need to be further investigated.
- It cannot be denied that Amman Resilience Strategy as the first local Strategy of its kind in Jordan is a forward step towards a more resilient Jordan. However, it is essential to investigate how the resilience concept is used in other local and national policies and urban strategies to find ways to enhance urban resilience in Jordan in terms of policy and municipal regulation.
- According to the Greater Amman Municipality, the Amman Resilience Strategy is a live document, and it is subject to change. This can be considered a call for researchers and professionals to take the initiative to develop the strategy further as the core team might consider their recommendation over in the development of 'Amman Resilient' project in the coming years.
- The Agenda-Setting Workshop and the other workshops and activities held by the Greater Amman Municipality to engage stakeholders discussed the weaknesses and

strengths of Amman and other aspects of the strategy. It seems that the stakeholders were involved in the early development stages of the strategy only. A reconsideration of the engagement of potential stakeholders may have a positive impact on the future of the strategy, especially at the implementation stages. More activities and workshops may be organised to ensure successful engagement of a wider array of stakeholders.

- Although stakeholders' engagement exists in the 'Resilient Amman' project to some extent, their engagement in terms of diversity of background and specialisations and their participation in the preparation and planning of the policy and the ongoing implementation stage do not seem to be enough, and it needs to be further enhanced. It is essential to have a variety of stakeholders to be involved in large strategic projects such as the Resilience Amman Strategy from the design and preparation stages to the implementation and operation stages. In general, local stakeholders can provide a better understanding to the current state of the city, primarily when a broad range of opinions and perspectives is engaged in the process of transforming Amman to a resilient city. 'Most of those stakeholders are looking forward to a better future to their city, its citizens and the hosted refugees', said one participant to the web-based questionnaire of this study. The related stakeholders can contribute to building a resilience strategy that is for Amman as a whole and implement projects with more specific and practical functions. Transparency may also be ensured through their engagement in the development of 'Resilient Amman' projects and thereby bridging the gap between the government and the community and build trust. Their involvement can also raise people's awareness and build capacity in the local community on the issues of urban resilience and, therefore, contributing to a more resilient future through activating the community and promoting its cohesion.
- Identifying who could be involved as a stakeholder in the resilience-related projects in Amman is also crucial for the success of both the planning and implementation of Amman Resilience Strategy. More academic institutions and local initiatives should be engaged in future.

Acknowledgements The University of Auckland Human Participants Ethics Committee (UAHPEC) approved the web-based questionnaire on 20 March 2019, for three years (Reference Number 022561). Authors would like to thank all participants who took part in this study and gave their time and knowledge to complete the web-based questionnaire. Authors would also like to thank the Greater Amman Municipality's Resilience Office for providing the required data and the third-party organisations, the Greater Amman Municipality and the Jordanian Engineers Association (JEA), who helped to distribute the web-based questionnaire.

References

Alnsour, J., & Meaton, J. (2014). Housing conditions in Palestinian refugee camps, Jordan. *Cities, 36*, 65–73. https://doi.org/10.1016/j.cities.2013.10.002

Clement, T. E. (2008). 'A thing not beginning and not ending': using digital tools to distant-read Gertrude Stein's the making of Americans. *Literary and Linguistic Computing, 23*(3), 361–381. https://doi.org/10.1093/llc/fqn020

Comprehensive guide to civil society organizations in Jordan Professional Associations. (n.d.). https://www.civilsociety-jo.net/en/organizations/5/professional-associations. Accessed June 13, 2019.

Coutu, D. L. (2002). How resilience works. *Harvard Business Review, 80*(5), 46–56.

Cutter, S. L., Barnes, L., Berry, M., Burton, C., Evans, E., Tate, E., & Webb, J. (2008). Community and regional resilience: Perspectives from hazards, disasters, and emergency management. *Geography, 1*(7), 2301–2306.

Da Silva, J., & Morera, B. (2014). *City resilience framework*.

Department of Statistics. (2018). Jordan in figures.

Foster. (2010). *Regional resilience: How do we know it when we see it?* University Buffalo Regional Institute. https://www.gwu.edu. Accessed November 28, 2017.

Gilbert, C., Eyring, M., & Foster, R. N. (2012). Two routes to resilience: Rebuild your core while you reinvent your business model (spotlight on how to manage disruption) (cover story). *Harvard Business Review, 90*.

Greater Amman Municipality. (2017). Amman Resilience Strategy (pp. 1–148).

Holling, C. S. (1973). Resilience and stability of ecological systems. *Annual Review of Ecology and Systematics, 4*(1), 1–23

Hughes, J. F., & Healy, K. (2014). *Measuring the resilience of transport infrastructure* (No. 546).

Jaber, J. O., Elkarmi, F., Alasis, E., & Kostas, A. (2015). Employment of renewable energy in Jordan: Current status, SWOT and problem analysis. *Renewable and Sustainable Energy Reviews, 49*, 490–499. https://doi.org/10.1016/j.rser.2015.04.050

Kitano, H. (2004). Biological robustness. *Nature Reviews Genetics, 5*(11), 826–837.

Kumaraswamy, P. R., & Singh, M. (2017). Population pressure in Jordan and the role of Syrian refugees. *Migration and Development, 6*(3), 412–427. https://doi.org/10.1080/21632324.2016.1141500

Longstaff, P. H., Armstrong, N. J., Perrin, K., Parker, W. M., & Hidek, M. A. (2010). Building resilient communities: A preliminary framework for assessment. *Homeland Security Affairs, 6*(3), 1–23.

McPhearson, T. (2014). The rise of resilience: Linking resilience and sustainability in city planning. *The Nature of Cities, 10*.

Mileti, D. (1999). *Disasters by design: A reassessment of natural hazards in the United States*. Joseph Henry Press.

Ministry of Planning and International Cooperation. (2017). Jordan's way to sustainable development first national voluntary review on the implementation of the 2030 agenda, New York.

Ministry of Planning and International Cooperation. (2019). Jordan response plan for the Syrian crisis 2019. Hashemite Kingdom of Jordan.

National Infrastructure Advisory Council (NIAC). (2012). Regional Resilience Working Group.

National Research Council. (2012). Community resilience: A national imperative.

Norris, F. H., Stevens, S. P., Pfefferbaum, B., Wyche, K. F., & Pfefferbaum, R. L. (2008). Community resilience as a metaphor, theory, set of capacities, and strategy for disaster readiness. *American Journal of Community Psychology, 41*(1–2), 127–150. https://doi.org/10.1007/s10464-007-9156-6

Polit, D. J. (2019). Resilient thinking: An important tool to contribute to sustainable development. In *Advanced studies in energy efficiency and built environment for developing countries* (pp. 73–79). Springer.

Potter, R. B., Darmame, K., Barham, N., & Nortcliff, S. (2009). "Ever-growing Amman", Jordan: Urban expansion, social polarisation and contemporary urban planning issues. *Habitat International, 33*(1), 81–92. https://doi.org/10.1016/J.HABITATINT.2008.05.005

Resilient cities what is urban resilience? (n.d.). https://100resilientcities.org/resources/

Resilient Design Institute (RDI). (n.d.). *What is resilience*? https://www.resilientdesign.org/what-is-resilience/

Robbins, J., Krishnan, K., Allspaw, J., & Limoncelli, T. A. (2012). Resilience engineering: Learning to embrace failure. *Queue, 10*(9), 20–28. https://doi.org/10.1145/2367376.2371297

Saunders, W. S. A., & Becker, J. S. (2015). A discussion of resilience and sustainability: Land use planning recovery from the Canterbury earthquake sequence, New Zealand. *International Journal of Disaster Risk Reduction, 14*, 73–81. https://doi.org/10.1016/J.IJDRR.2015.01.013

Seeberg, P. (2020). Syrian refugees in Jordan and their integration in the labour market: Jordanian migration diplomacy and EU incentives. *Mediterranean Politics*, 1–20. https://doi.org/10.1080/13629395.2020.1758452

Shamout, S., Boarin, P., & Melis, A. (2019). Energy retrofit of existing building stock in Amman: State of the art, obstacles and opportunities. In *Advanced studies in energy efficiency and built environment for developing countries* (pp. 133–145). Springer.

Shamout, S., Boarin, P., & Wilkinson, S. (2020). From a sustainability movement to the need for an urban resilience movement: A policy analysis for more resilient and sustainable cities in Jordan. *Sustainable Production and Consumption*. https://doi.org/10.1016/j.spc.2020.08.015

Sharifi, A., & Yamagata, Y. (2014). Resilient urban planning: Major principles and criteria. *Energy Procedia, 61*, 1491–1495. https://doi.org/10.1016/j.egypro.2014.12.154

Spaans, M., & Waterhout, B. (2017). Building up resilience in cities worldwide—Rotterdam as participant in the 100 resilient cities programme. *Cities, 61*, 109–116. https://doi.org/10.1016/J.CITIES.2016.05.011

Stevenson, J., Vargo, J., Ivory, V., et al. (2015). National science challenges: Resilience to nature's challenges short-term project working paper, delivery 1. In *Resilience benchmarking & monitoring review*

Stumpp, E. M. (2013). New in town? On resilience and "Resilient Cities." *Cities, 32*, 164–166. https://doi.org/10.1016/j.cities.2013.01.003

Swaidan, Z., & Nica, M. (2002). The 1991 Gulf war and Jordan's economy. *Middle East, 6*(2), 70–77.

UNHCR. (2019a). External statistical report on UNHCR registered refugees.

UNHCR. (2019b). Total persons of concern. In Situation Syria regional refugee response.

Urban resilience hub what is urban resilience? (n.d.). https://urbanresiliencehub.org/what-is-urban-resilience/. Accessed January 4, 2019.

Yamagata, Y., & Maruyama, H. (2016). *Urban resilience: A transformative approach*. (p. 2016). Springer.

Preservation of a Porous City; Strategies for Tirana Urban Fabric

Irina Branko

Abstract

Every day and more, the sustainable development of the metropolitan area is becoming a keyword in Albanian reality as well. The country, and in a particular way Tirana the capital, is facing many challenges as per adapting its old and new developments toward more sustainable and resilient urbanism. Declared the capital city only in 1920, mostly has been low-rise and porous development, with a strong physical, and also visual communication to the surrounding natural environment. It continued to be so even during the socialist regime. The city has undergone some drastic changes these last 30 years, mainly resulting in footprint enlargement and an intense and chaotic process of urban filling. In 2017, the General Local Plan was finally provided for Tirana, having as its main strategies: (1) The definition of the city boundaries, by implementing a Ring Peripheral Park; (2) The densification of the urban fabric, thus having a leading focus on the buildings. Looking at the latest developments, also the presence of a broad pressure for new buildings, the preservation of porosity, and urban green space becomes a priority. Through the applicative methods, by analyzing the most representative, existing, and planned urban fabrics, this paper aims first to evaluate, for later emphasize the importance of the urban green space. Preservation of existing spatial networks and their connections to the natural environment could contribute to a desirable level of sustainable development not only in terms of the qualitative physical environment but also in terms of a qualitative life as well. As a result, possible strategies and recommendations are proposed and focused on in the future implementations of the existing regulatory plan.

Keywords

Sustainability • Landscape • Green infrastructure • Urban fabric • Porous

1 Introduction

All over the world, the tendency of demographic movement toward the urban areas is a continuous process. More than half of humanity already lives in urban areas, and the trend is growing. By 2050, an estimated 70% of the world population would live in megacities, exceeding a population of 10 million people (Renner, 2018). The concentration of the demography in the cities is the cause of many problems related to environment and ecology, generation of pollutions such as gas emission, disconnection of the population from the natural environment. The general result is a decrease in quality of life.

The tendency of inner migration toward the cities is a noticed phenomenon in Albanian reality as well. This movement, results to be more intense toward central Albania, was two of the most important cities which are, *Tirana* the capital, and the city of *Durrës*, the biggest port city in the country. Although there is an increased population in such locations, in terms of numbers and dimensions, it cannot still be comparable to those of megacities of Europe. However, quite many problems do exist. The abovementioned problems in the biggest cities are starting to be present and are accompanied with another one—the increase in the number of shrinking cities.

The distribution of population we face today is the product of the global tendencies in detachment of rural life, but also of the political changes Albania has undergone during the early '90. The return, from a socialist regime to a democratic system, impacted the socio-economic life of the population, especially in the province, giving a powerful impact on territorial development through these last

I. Branko (✉)
Department of Urbanism, Faculty of Architecture and Urbanism, Polytechnic University of Tirana, Tirana, Albania
e-mail: branko_irina@yahoo.com

F. Trapani et al. (eds.), *Advanced Studies in Efficient Environmental Design and City Planning*, Advances in Science, Technology & Innovation, https://doi.org/10.1007/978-3-030-65181-7_23

30 years. Thus, this particularity has given the country a specific territorial transformation, bringing problems of various kinds.

This study has its focus on the urban area of Tirana and its foreseen development by the Regulatory Plan, or as it is known as the General Local Plan—Tirana 2030 (Later on, the paper will use the abbreviation GLP TR030). There have been many studies on Tirana, on its creation, history by identifying the different layers, their morphological characteristics, the city footprint changes due to urban sprawl, thus studies with a primary focus on the built environment. In this paper, the open space—in particular, the green space within the urban area will be the subject of analyses and study as one of the main elements defining the porosity, environmental quality, and therefore the quality of life for its inhabitants.

2 Porosity Meaning and Importance of Study

Porosity per se seems to have quite a complex definition during its analysis as it is related to many aspects of a city. From a physical point of view, we could define porosity as a relation between buildings and the void. Not necessarily a perfect ratio between these two main elements would provide the quality of the city and the qualitative life in the city. For achieving the quality, there are other aspects to be considered in the definition of porosity, besides this relationship, which is better represented by urban morphology. Directly connected to the porosity of a build environment, there are the configuration of spaces and voids, their possible connections, the altimetry of buildings, the materiality, also the human perception of the space, and also many other aspects. There have been many studies completed on porosity and the use of this "metaphor" in urban development. To understand porosity, it is essential to consider the different aspects of porosity and the theoretical debate over the concept. In the past, the metaphor of porosity was not used frequently in urban studies, and on those occasions when it was used, it had no fixed meaning. In other words, the metaphor has been associated with a range of phenomena (Pessoa et al., 2015). The same authors continue:

> In 1924, German philosopher Walter Benjamin describes porosity as a metaphor for spatial experience of the city of Naples; Paola Vigano describes porosity as a concept involving several spheres such as ecological, mobility/social, epistemological, and mathematical; Luc Adolphe has created a porosity coefficient using the volumes of open spaces and build spaces in a city as variables, which plays a role in understanding the environmental performance of an urban area; Berndardo Secchi relates metaphor of porosity with the shrinking cities and the disappearing of industrial city in Europe, while Stavros Stavrides refers to the fuzzy boundaries between public and private spaces. As the authors emphasize, they are all different concepts in defining the porosity in the city, but in all cases, the authors include both physical and social dimension.

As is mentioned above, this study will focus on one specific aspect considered important in defining the porosity in our context—The urban green space. It has already proven that a good relationship between the built environment and qualitative green space assures both (1) absorption of problematics including the environmental ones the city produces and (2) assuring quality of life, the social life included. This study does not focus on the analyses of existing fabrics, but on the new approach foreseen and indicators provided by the GLP TR30, recently approved (On 14.04.2017). Thus, the study tries to make an evaluation of future scenarios and confront the results from a theoretical perspective.

During the twenty-first century, theories have emerged such as "Landscape Urbanism' which argues on the importance of landscape on urban development. As per these theories, the landscape design should be guiding the city design. GLP TR030 seems to have brought such a theoretical approach. It is needed to understand how this theory is applied, to quantitative and qualitative characteristics, thus be able to evaluate the proposed porosity scale it assures in the city for the next remaining years.

Observing the Albanian context, where there is high pressure of construction, it is necessary to underline the importance of the green spaces in the city. Such presence, if organized as an infrastructural network, definitely will have a positive impact on environmental sustainability, thus assuring of air ventilation, temperature control through solar radiation, control of relative humidity, stormwater management, and quality of air. Other benefits of a green network that we could list are as follows: (1) Ecological preservation and continuity of the wildlife within the city limits; (2) Improvement of social cohesion; (3) Inhabitants psychological wellbeing; (4) Health improvement, through the motivation of its residents, increasing their physical activities; (5) Educational aspects by increasing the interaction between the environment and the community (Stiles, n.d.).

3 Background

The change of the political system, from a socialist regime to a free-market economy, during the late 80s in Albania, was accompanied by changes. The most distinctive ones were the massive migration of the population and changes in territorial development. All these movements were directed mainly toward the neighboring countries (Greece and Italy), but many others were from rural areas to urban areas. Of course, offering more services and opportunities in the job market,

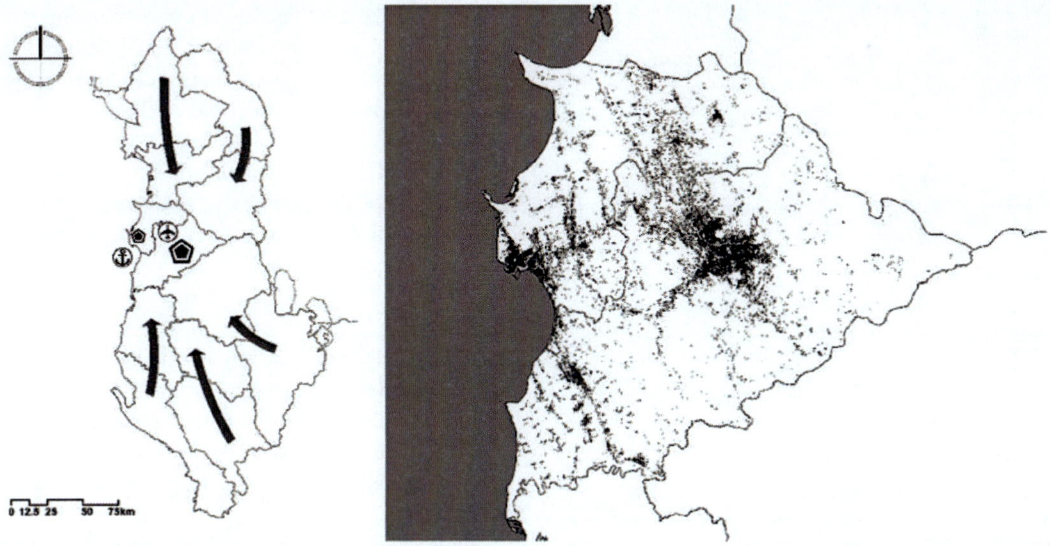

Fig. 1 a Schematic illustration of demographic movements. *Source* NTPA (2018); Elaborated: Author. **b** Urban sprawl representation in *Tirana—Durrës* Area. *Source* NTPA (2018); Elaborated: Author

the vastest cities, *Durrës and Tirana*, became the movement's target (Fig. 1a, b). This last one declared that the capital of the country only in 1920 has changed the population drastically from 200,000 inhabitants to approximately 800,000 inhabitants nowadays. As per the territorial breakdown has grown by nearly 50%.

After the fall of the socialist regime, unfortunately, the existing urban morphologies of the cities started to transform, by formal and informal developments. During the first years of democracy, the informality began to be a widespread phenomenon both by occupying the empty spaces within the urban footprint, but mostly, the new natural and semi-natural areas. The high rate of migration was accompanied by a high demand for housing, which resulted in the construction industry's development. This sector carried out its activity mainly within the urban footprint. Regulatory plans and building legislations existed at that time (Aliaj et al., 2003), but the new state structures were not ready for these drastic changes and high pressure of construction. Moreover, that private property during the socialist regime had been returned to the state and at these conditions, and the lack of appropriate property legislation was not able to control the urban sprawl and urban filling process as well. Inside the urban area, in a short time, public spaces were filled with constructions, and peripheric agricultural and natural land began to get covered in the city peri-urban area. The urban area started to get transformed and showed two main results: (1) A new fabric, which sometimes could be described as a set of patchwork patterns, and (2) Urban sprawl phenomena. For almost 25–30 years, the central and local governments have failed in managing this development due to many reasons, the failure in migration control, and failure in managing the pressure of the construction industry, sometimes lack laws, but mostly lack in their appropriate implementation.

In 2014, due to a territorial reform developed by the Albanian Government, there was an in-deep analysis of the territory conducted with the main focus to control both the build and non-build environment. The reform centralized the administration in 61 municipalities from 373 previews ones (National Territory Planning Agency, 2016), drafted national strategies, and started the completion of General Local Plans for all municipalities. It is since April 2017 that the municipality of Tirana has its GLP named GLP TR30, which according to the current legislation, includes the urban and rural territory of 14 administrative units (Fig. 2), and approximately an area of 111,100 Ha (Fig. 3). The main primary focus of this new territorial reform was the urban sprawl control, preservation of the agricultural and natural land, subsequently and also development of strategies for the future.

From all the analyses conducted, it results that the Municipality of Tirana is very rich in its natural resources. In this municipality, there are two important protected areas, part of the European Emerald Network (1) National Natural *Park of Dajti Mountain,* with a surface of 29,384 ha, and (2) *Mountain with Pits—Bizë—Martanesh* with an area of 25,266.42 ha. The region is rich in water resources and hydrography. The city itself has as boundaries flowing waters of *the Erzen River* in the south, *the Tirana River* in the north, and is divided in two by the *Lana River* (Municipality of Tirana, 2017a, b). It is noticeable that the pressure of the construction industry during this period,

Fig. 2 Municipality of Tirana and its administrative units map. *Source* NTPA (2018); Elaborated: Author

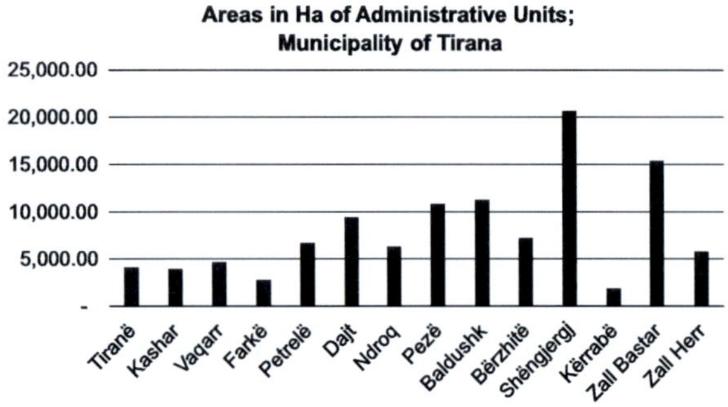

Fig. 3 Surfaces of administrative units in numbers. Elaborated: Author

including formal and informal developments, has not ever considered the preservation and/or good management of these natural resources (Fig. 4). However, by trying to understand the actual city footprint, we conclude that topography has played an important role in the development.

In this general framework, it was clear that the preservation of the ecosystems should become a priority in the sustainable territorial development taken into consideration also by GLP TR030. This plan, approved in 2017 and conceived by studio Stefano Boeri Architects, has presented three scenarios of Tirana development: (1) development of a linear city, (2) the concentric growth, and (3) densification within the same footprint. The last one is also the final scenario chosen for Tirana development until 2030 (Municipality of Tirana, 2017a, b). This scenario is accompanied by a denser urban area that would accommodate the growing population of the capital.

According to this concept, a further footprint enlargement of the urban area would be prevented by the implementation of the ring peripheral park, also known as Metrobosco. The ring peripheral park would further contribute to two large developments, the connection of the surrounding ecosystems such as natural parks, agricultural land, riverbanks, and new oases proposed. The concept brings also the development of ecological corridors as illustrated in (Fig. 5a–c).

The strategies of the General Local Plan state the importance of preservation of what is natural and further development of the natural and semi-natural environments. The method of achievement is by the implementation of 13 strategic projects among which particularly emphasize (Municipality of Tirana, 2017a, b):

(1) Ring peripheral park implementation;
(2) The implementation of natural oasis;
(3) The implementation of third and fourth green rings;
(4) The implementation of world park;
(5) The implementation of three green and blue rings.

While all the strategic projects are still in the drafting process, evidence shows that the construction industry has intensified its pressure. From 2016 to the second quarter of 2019, there is an increase in the number of construction permits, constructed areas, and value of investments (INSTAT, 2019). As shown in the graphs below (Fig. 6a–c), for these last three years, the total number of constructions permits for the district of Tirana has been 894; with a total of 2,368,742 m^2 with an investment value 90,615,930 Mio All (Albanian Lek) (INSTAT, 2019).

The construction permits illustrated include three categories of construction: (1) Engineering works; (2) Residential buildings; (3) Non-residential buildings and are spread in all Administrative Units of Tirana Municipality. Part of the municipality is also Tirana Administrative Unit representing the area where there is the greatest urbanization. As it can be noticed through the abovementioned figures, it is an irrefutable fact the growing trend of constructions in this municipality. Most of the construction permits do correspond to categories (2) and (3). It turns out clear from GLP TR30 the concept of densification, what is not very clear is the phases to go through for receiving the right result for the territory and its inhabitants. Therefore, this ambiguity needs to be further studied to prevent further mistakes in the territorial development.

4 Method and Study Area

The analysis was conducted by following the steps below.

4.1 Collection and Analysis

The collection of data from the GLP TR030 was a significant step having a direct impact on the planned green space. The

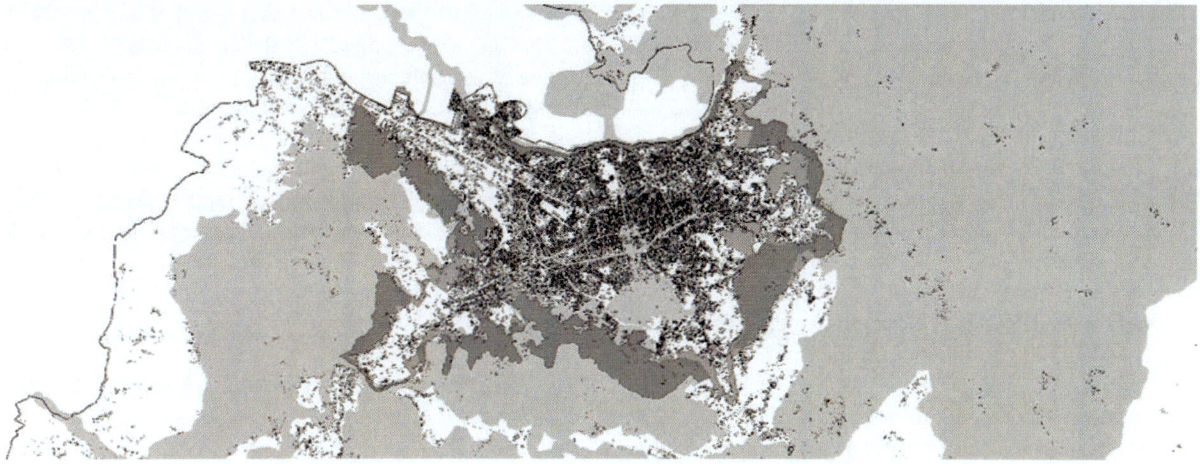

Fig. 4 Urbanization of Tirana, its Green Crown (The Green Crown of Tirana is the area of a national importance in the territorial planning, defined and protected as per the Albanian laws (CM, The definition of case and area of national importance in the territorial development, identified as the Green Crown of Tirana, 2014).), River Banks, and the proposed Perimetral Park by TR030. *Source* Municipality of Tirana (2017a, b), NTPA (2018); Elaborated: Author

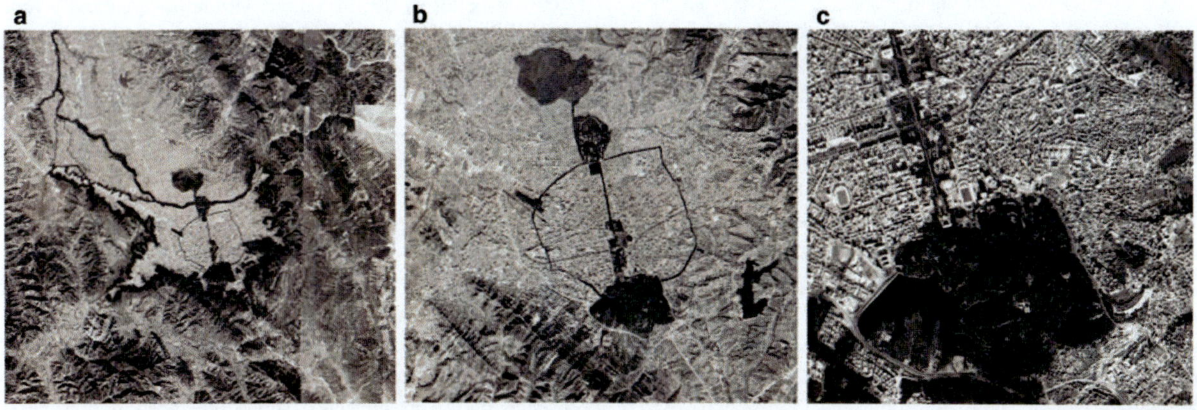

Fig. 5 **a** Perimetral park. **b** Green fourth ring. **c** Green boulevard and park; Stefano Boeri concept for GLP TR030. *Source* Boeri (2017), Calzavara (2016)

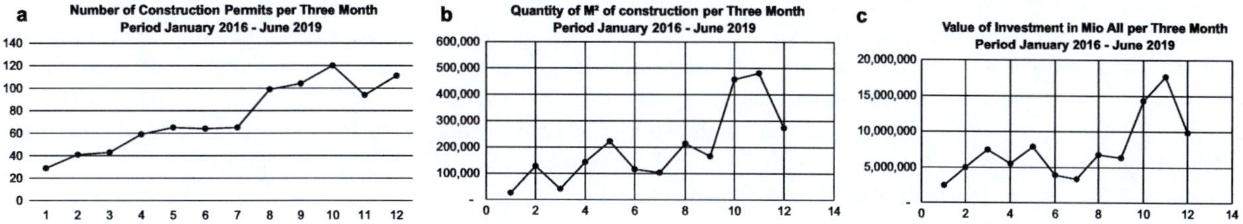

Fig. 6 **a** Number of construction permits; **b** quantity of construction in M^2; **c** value of investments in Mio All. *Source* INSTAT (2019); Elaborated: Author

data collection includes: (1) The strategies of TR030; (2) The regulations and indicators defining the land use proposed; (3) Land use map; (4) Data on existing green areas; (5) Situation on existing green spaces; (6) First strategic draft of ring peripheral park (Municipality of Tirana, 2017a, b). For a better understanding of the territory, it is crucial to understand its future extension and the new administrative boundaries of the urban area. It is also important having a clear understanding of the population foreseen, as the city is conceived basically on its needs.

Afterward, the philosophy of the proposal needs to be analyzed, how it is related to its applicative possibilities in general and as per the green spaces in particular.

4.2 Prepare Working Materials

In the proposal, the ring peripheral park defines the outer boundary of the urban area. As the study has its main focus inside the city limits, an important process was the result achieved by overlapping (1) Orthophoto representing actual situation (ASIG State Authority for the Information, 2019), proposed land use map (Municipality of Tirana, 2017a, b), and the ring peripheral park (Municipality of Tirana, 2017a, b). This represented a great interest, as per the understanding of the process of transformation of the existing territory.

4.3 Defining Areas of Interest—Case Studies

The next step was defining the areas of interest which by choice are adjacent to natural or semi-natural land use areas. These areas representing our case studies, contained high potential referring to the green space development, especially in the future, as per the landscape urbanism theories application. Although areas with a strong connection with the ring peripheral park, they do represent different fabrics of the city. The first case study includes the main historical axes with a north–south direction and its intersection with urban blocks along Lana River. It represents not only the historical axes but also the administrative and socio-economic life of the capital as well as the future development of the city to its northern part. In terms of porosity, it represents one the more porous areas in the capital, and many studies have foreseen these axes as connective green lines between existing Artificial Lake Park (*Parku i Liqenit Artificial*) south of the city and the future *Paskuqan* Park (*Parku i Paskuqanit*) north of the city. The Lana River divides the city in two and represents one of the existing ecological corridors in the city (Fig. 2).

The second case study represents new fabrics developed on the west side of the urban development, mainly representing by informal settlements and formal neighborhoods. What is more important to be emphasized for this area is the containing large amount of empty spaces. It represents the transformation between the agricultural, semi-natural to an urban fabric.

The third case study represents another particular fabric of the city, the ex-industrial areas, non-functional, representing once the city edge and now an abandoned area surrounded by neighborhoods, furthermore at a close distance with the proposed ring peripheral park and the natural oases. This last one is planned to be implemented near the *Farka Reservuar* on the southeastern part of the city.

It is important to emphasize that a grid 800 m × 800 m has been applied serving as a guideline on the map. This was achieved by offsetting the two main city axes. The dimension of our case studies was multiple of this unit. This was considered an important tool, as the analysis needed to take into consideration not only the quantity of green space but also its frequency. The grid also defines the territories within the radius of 400 m—an indicator assumed to have been considered during the drafting of GLP TR30, as per its documentation (Municipality of Tirana, 2017a, b).

4.4 Simplified Application of the Planed Green Space

For all the three case studies, the green area predicted by the plan has been schematically applied. Besides the indexes and quantities given, the graphic representation was taken into consideration, as a simple method of understanding the foreseen green space in each structural unit.[1] While the GLP TR030 is being in use in Tirana Municipality, mostly represented by interventions and constructions in particular sites, it should be stated that no standards have been defined for each of the five territorial systems.[2] These last ones are defined by Albanian territory development law.

4.5 Comparison of the Results with Other Standards

The results received were compared to an actual standard in the UK (Stiles, n.d.). As per these standards ANGST (Natural England's Accessible Natural Greenspace Standart), the presence of the green space should be present in several aspects and at different dimensions. Thus, they should vary 300 m to 10 km in terms of distance, and 2 ha to 500 ha in terms of quantity (Stiles, n.d.). For a better graphic understanding of this fact, another schematic map has been elaborated, containing distances in km from the city center. This system applied was referring to the center of Tirana and was evaluated suitable, as it should considered that the city has a concentric development from its origin. Its main arteries are an intersection of radiuses and rings.

---

[1]Structural Unit is the smallest area in which the administrative territory of a local unit is subdivided for development purposes (CM, Law Number 107; For the planification and territorial development, 2014).

[2]As per the Albanian legislation, the local government units, for planification purposes are divided in five Territorial Systems as the following: (1) Urban System, (2) Natural System, (3) Agricultural System, (4) Water System and (5) Infrastructural System (CM, Decision of Council of Ministers, Number 671, 2015).

5 Results

From the mapping process, it is clear that the urban sprawl had overcome the previous urban boundary. Tirana and the adjacent Administrative Units are conceived in such a way, they do function almost as one single area. The GLP TR030, more specific The Land Use Map, does not include the ring peripheral park or any other planned park. An overlapping process was necessary to understand the new city limits and also its relation with the park. A total of 202 structural subunits from other Administrative Units resulted being developed under the ring peripheral park footprint, and the city limit had gone beyond its previous border by including other structural subunits from adjacent administrative units (Fig. 13). This was a significant part of the process for understanding the population of the urban area foreseen by the GLP TR030 and make the necessary calculations afterward. By analyzing all GLP documents (Municipality of Tirana, 2017a, b), we arrived at the following conclusions and numbers:

a. The total population foreseen for the municipality of Tirana following the trend is 1,590,696 inhabitants. The new GLP TR30 has planned a city able to welcome this maximum population (Fig. 7).
b. The total green area forced by the GLP regulations is 3,052,731 m^2 or 305 Ha (Fig. 7).
c. Other data, as per the calculation of the population within the city boundary and green space foreseen within the same area, are as shown in the following Fig. 8. As it can be seen, the main conclusion of this analysis is the standard of 1.91 m^2 green space calculated per inhabitant in GLP TR30. From (Fig. 7), it could be extracted that the same standard applies to the whole municipality, including rural and urban areas.

As per the ring peripheral park per se, from the actual documentations and referring to the land use under its footprint, it is noticed that not all it is planned as an open space; moreover, the proposed land use continues the infill process. From this analysis, there is a total of 324.38 ha mainly of residential use, and also a total of 139.70 ha of other land uses, infrastructural are included (Figs. 9 and 10a–d).

The study of the ring peripheral park is important to be studied and defined properly as it represents a continuous open space.

Maximum Population Municipality	Maximum Green Area Municipality (M²)
1,590,696	3,052,731

Fig. 7 Table of population and green area foreseen in the GLP TR030. *Source* Municipality of Tirana (2017a, b); Elaborated: Author

As per the area within the city limits, it is also analyzed in terms of indexes given by the GLP TR030, and as described above, the analysis was applied on the areas chosen to be studied. In this paper, we would further use the terms case study 1; case study 2; and case study 3, and the results will be shown for each of them.

(Figure 12) shows all the summarized indexes given by GLP TR030, some of which, such as construction areas, green area per inhabitant are the summary of all indexes, meanwhile, indexes such as intensity of the construction or utilization coefficient are average values. In terms of values, this analysis gives clearly the general characteristics of each case study, the densification scale of each area, defines clearly the residential areas versus other functional areas, and it gives information on quantities of additional constructions by comparison of planned refereeing to what is existing, and furthermore, future density and quantity of green spaces foreseen.

Another important fact that we need to bring out is the existing green space quantity, as per the data given in the GLP TR030 documents (Municipality of Tirana, 2017a, b). The total green existing space counts around 346,184 m^2, and it is as per green area categories which are described in (Fig. 11). This information was digitalized from a not editable copy taken at the GLP TR030 documentation and is helpful in understanding the quantitative and qualitative aspect of existing green space (Fig. 12).

As mentioned before, the three case studies are multiple of 800 m × 800 m quad; therefore, by analyzing the values, we are able at the same time value the quantity of green open space—therefore, the standard used and compare it with international models. In (Fig. 14), it is illustrated: (1) The study areas; (2) The ring peripheric park, and (3) Case studies 1, 2, and 3 (Fig. 13).

While the numbers give quantitative understanding, the quality of them is another aspect. As described, a schematic graphic design would help to interpret an approximate possible result. For this reason, a simplified map was processed (Fig. 14), which helps in the understanding of the foreseen green spaces and their distribution for each structural subunit. As per the size of these spaces, we can mention the largest value being approximately 10,000 m^2 (100 m 100 m), while the smallest is 230 m^2 (15 m × 15 m), thus demonstrating quite low values.

6 Discussions

The results obtained by this study are necessary to build a holistic understanding of the planned and non-planned developments in the city related to its porosity preservation. As has been earlier defined, this study refers to porosity used as a metaphor for the green space within the city limits.

	Administrative Unit	No. of Structural Units Under Perimetral Park	Area of Structural Units (Ha)	Maximum Population Capacity for UUPP	Maximum Population Capacity rest of Tirana	Total Green / Unit (M²)	Total Green / Unit for the rest of Tirana (M²)
1	Tiranë	43	1,045.20	80,275	1,160,928	113,188	2,236,668
2	Farkë	42	1,310.79	21,017	-	52,511	-
3	Kashar	58	739.02	7,448	-	18,622	-
4	Dajt	32	1,112.75	20,695	-	40,642	-
5	Vaqarr	27	2,076.86	12,147	-	30,366	-
	TOTAL 1	202	6,284.62	141,582	1,160,928	255,329	2,236,668
				TOTAL 2	1,302,510	TOTAL	2,491,997
	MAIN CONCLUSION					STANDARD M²/IN	1.91

Fig. 8 Table of population and green area foreseen in the GLP TR030 within the city limits. *Source* Municipality of Tirana (2017a, b); Elaborated: Author

Calculated Area of Perimetral Park - Metrobosco (Ha)			
1	System	Water	217.26
2	System	Natural	1,018.64
3	System	Urban	324.38
4	System	Agricultural	863.80
5	System	Infrastructure; Others	139.70
		TOTAL	2,563.78

Fig. 9 Table of land use system under the perimetral park footprint. *Source* Municipality of Tirana (2017a, b); Elaborated: Author

From the analysis, it is clear that the ring peripheric park contains buildings in parts of it. Due to difficulties in being able to provide the open, non-build space in the existing fabric, the presence of some constructions could be comprehended. However, the additional construction area within the perimetral park footprint is difficult to be understood and be interpreted. Usually, these kinds of interventions are similar to a green belt implementation—which could be defined as an open space where construction is restricted.

From a land use point of view, and due to the different indexes, it is difficult to understand whether the perimetral park would function as a green park, it will serve as an open restricted open space or will be a combination of these two elements. In all cases, a further detailed study is necessary to get completed as per the integration of the functions proposed by the land use map and the ring peripheric park.

On the other hand, in terms of porosity, this peripheric open space—represented mostly by natural and semi-natural areas should be able to penetrate the urban fabric for contributing in green infrastructure implementations. Somehow, it is evaluated that "footprints" of such solutions should have been readable in the GLP TR30 land use planes, where the porosity scale gradually should have been transformed from peri-urban to urban areas. In the graphic representation of the green areas for each structural subunit, the opposite is illustrated. The presence of the green is higher where the city is denser. This fact correlates mostly with just a mathematical relationship between green space and population and sounds correct at a first sight. In the conditions of an existing urban form, such achievement should carefully get analyzed, as there might be limited possibilities of application. Further morphological studies could concentrate on this particularity.

In the indexes given, there are significant areas of the city, mainly central areas, which have no specifications at all as per the development indexes. With all the constructions being processed in these central areas, it is necessary to emphasize the importance of detailed studies related to porosity conservation to reduce the level of abusiveness, especially in the context of total lack of standards.

The approach of densification within the city boundaries takes clearly in consideration the prevention of urban sprawl, but it needs to be studied furthermore in terms of quantity and quality of the urban space within the boundary limits, therefore within the ring peripheric park footprint.

From the analysis of the indexes given by GLP TR030, it is clear that the plan allows a growing intensity index varying: (1) from 1.55 to 2.62 in CS 1; (2) from 0.96 to 2.07 in CS 2; and from (3) from 0.7 to 0.73 in CS 3. These numbers give an additional 8,098,753 m^2 just for the areas subject of study, while the green space is calculated at approximately 1.91 m^2/inhabitant, showing an insufficient quantity in terms of numbers and quality of life. As mentioned above, the green spaces surfaces planned result to be between 10,000 and 230 m^2. Larger areas within the city fabric seem to be absent.

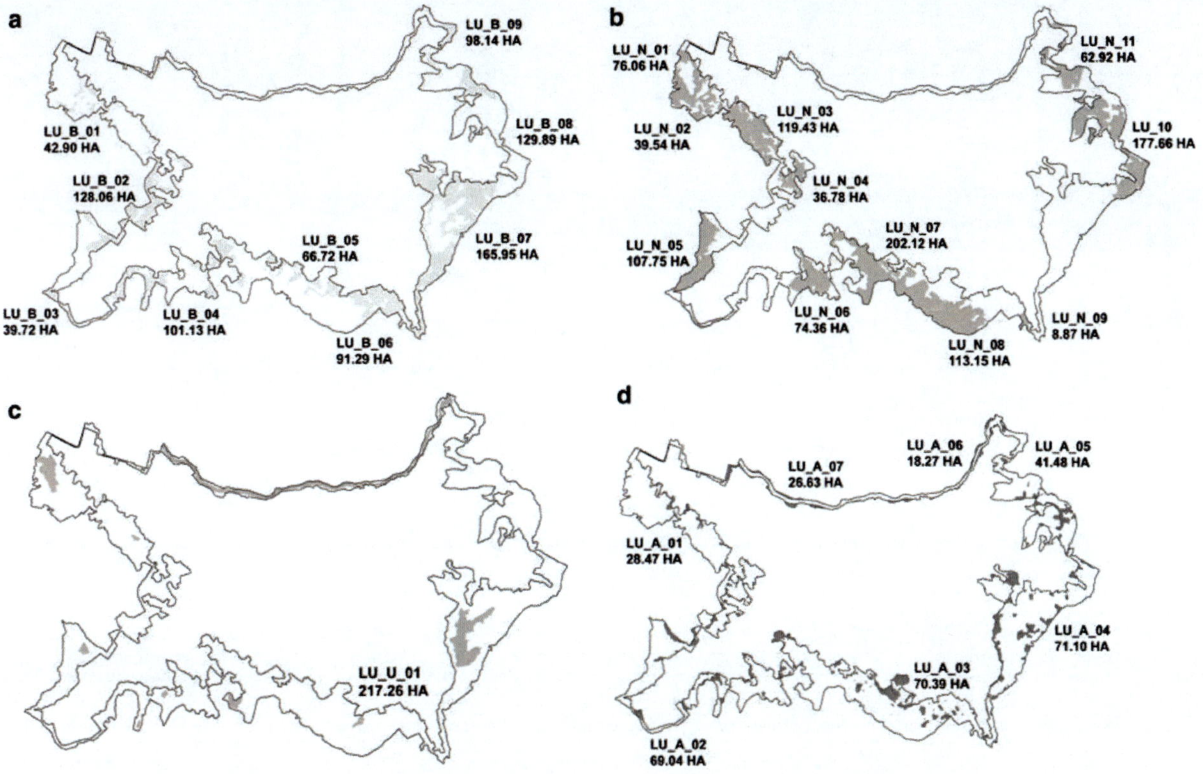

Fig. 10 **a** Agricultural land use under the PP footprint; **b** natural land use under the PP footprint. *Source* Municipality of Tirana (2017a, b); Elaborated: Author; **c** water land use under the PP footprint; **d** residential land use under the PP footprint. *Source* Municipality of Tirana (2017a, b); Elaborated: Author

No. Sector	Position into the City	Total Area (M²)	Occupied Area (M²)	Street Area (ML)	Factic Area (M²)	Grass Area (M²)	Flower Area (M²)	Street Plaza (M²)	Green Borders (M²)	Water Basins (M²)	Area of Graves & Scating Areas (M²)	Decorative Trees (pieces)	Decorative Shrubs (pieces)	Waste bins (pieces)	Benches (pieces)	Decorative Vases (pieces)
Sector 1	**Center**	**70,238**	3,139	-	67,099	54,383	1,489	10,361	476	369	-	1,300	1,843	105	246	30
Sector 2	**Lana River - Dēshmorēt e Kombit BLV - Hospital**	**55,838**	-	-	55,838	50,608	-	2,374	-	246	2,610	1,235	974	45	182	-
Sector 3	**Rina Park - The Piramid - Palace of Congress**	**65,489**	3,060	2,230	62,429	41,543	1,513	16,992	314	1,930	-	641	1,953	81	115	122
Sector 4	**The Presidence - Stadium Selman Stërmasi- Q.Stafa - S.Frashëri**	**34,691**	-	7,433	34,691	20,795	1,330	10,456	154	1,094	-	2,078	1,099	55	189	-
Sector 5	**Kavaja Street - The combinat - New Ring - Lapraka**	**54,323**	-	11,813	54,323	29,020	200	24,427	58	590	-	5,375	8,401	25	214	-
Sector 6	**Lana River - Dēshmorēt e Kombit BLV - Arrey building**	**59,571**	-	-	59,571	59,571	-	-	-	-	-	1,162	1,268	18	105	-
Sector 7	**Green of Streets**	**6,034**	-	27,000	6,034	3,118	52	293	2,502	-	-	9,111	277	2	92	7
		346,184	**6,199**	**48,476**	**339,985**	**259,038**	**4,584**	**64,903**	**3,504**	**4,229**	**2,610**	**20,724**	**14,661**	**263**	**1,014**	**159**

Municipality of Tirana, APR, 2016

Fig. 11 Summarized table of existing green spaces. *Source* Municipality of Tirana (2017a, b); Elaborated: Author

	Structural Unit	Area	Existing Building Footprint	Existing Building Total Area of Construction	Existing Intensity	Construction Area Primary Category	Construction Area Secondary Category	Construction Area Tertiary Category	Construction Area Quaternary Category	Total area for Living	Construction Intensity	Land Use Coeficient of Utilization for Buildings	Number of Floors	Structures Maximum Height	Maximum Population Capacity	Green Area / Inhabitant (M²)
CASE STUDY 1	TR su & FA su	1,015.8	1,700,885	6,884,188	1.55	8,044,469	1,260,961	661,974	124,605	7,658,218	2.62	42.82	7.30	24.21	205,542	421,961
	Total Existing Construction Area			6,884,188		Total Proposed Construction Area			10,092,009	3,207,821						
		Existing Intensity			1.55			Proposed Intensity			2.62					
												Standart of M² Green Area / Inhabitant in CASE STUDY 2				2.05
CASE STUDY 2	TR su & KA su	361.7	623,086	2,423,824	0.96	3,055,215	414,143	125,574	11,408	5,067,198	2.07	40.33	6.11	20.47	127,465	162,575
	Total Existing Construction Area			2,423,824		Total Proposed Construction Area			3,606,340	1,182,516						
		Existing Intensity			0.96			Proposed Intensity			2.07					
										627,133	1.65159	Standart of M² Green Area / Inhabitant in CASE STUDY 2				1.28
CASE STUDY 3	TR su & FA su	411.5	198,071	398,040	0.07	3,235,237	714,841	119,016	37,363	1,161,853	0.73	29.25	2.65	9.55	26,191	42,597
	Total Existing Construction Area			398,040		Total Proposed Construction Area			4,106,457	3,708,417						
		Existing Intensity			0.07			Proposed Intensity			0.73					
										8,098,753		Standart of M² Green Area / Inhabitant in CASE STUDY 3				1.63

Fig. 12 Summarized table of important indexes in the existing and foreseen fabric. *Source* Municipality of Tirana (2017a, b); Elaborated: Author

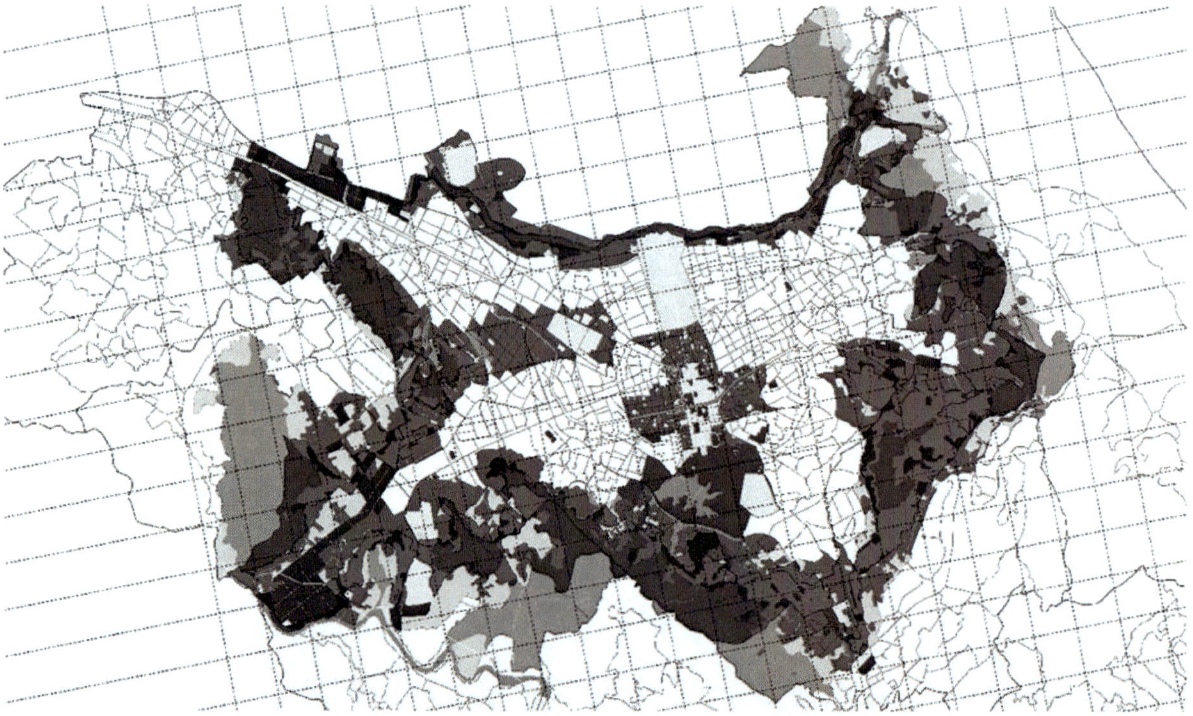

Fig. 13 Land use area of the case studies; ring peripheric park; CS 1; CS 2; CS 3. *Source* Municipality of Tirana (2017a, b); Elaborated: Author

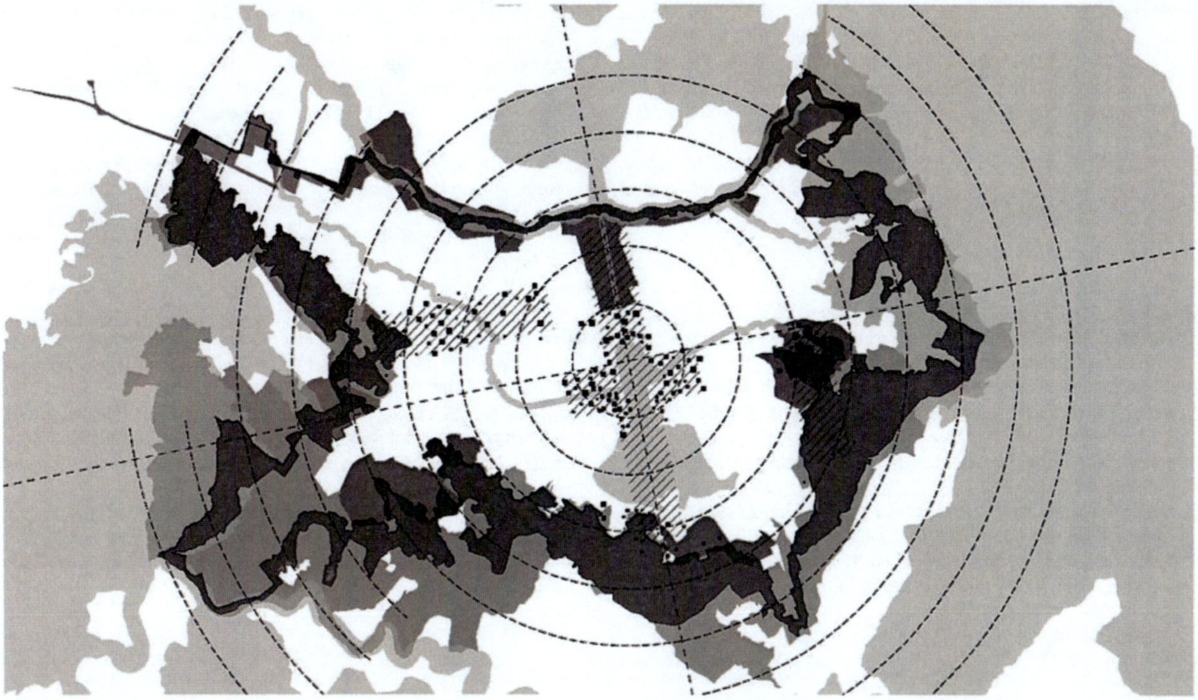

Fig. 14 Schematic representation of green spaces foreseen by GLP TR30 for the CS 1; CS 2; CS3. *Source* Municipality of Tirana (2017a, b); Elaborated: Author

7 Conclusions

As a conclusion, it emerges the need for detailed strategies related to green space. In the absence of urban standards, it is necessary to define each type of green space and give quantitative and qualitative specifications. Meanwhile, international standards would be a helpful reference for achieving such a goal. These would help in creating a clearer view of what are the necessary steps of phasing the development processes within the city limits.

Afterward, the implementation of strategies on developing green networks and green infrastructure into the city is quite a task the involved actors should be paying attention to, as it is the new way for the twenty-first-century city development. Through the institutional strategies, a new model should get presented, a model which quite close to sustainability and livability concepts.

Acknowledgements This research is made with the assistance of digital materials and documents made available by Albanian Institutions, Natural Territory Planning Agency (NTPA), and Municipality of Tirana, and therefore, the author reserves a special thanks to both institutions. Above all, the author would like to thank the Polytechnic University of Tirana, Faculty of Architecture and Urbanism for supporting, and always encouraging the research work of its members.

References

Aliaj, B., Lulo, K., & Myftiu, G. (2003). *Tirana, the challenge of urban development.* Gorenjski Tisk.

ASIG State Authority for the Information. (2019). *Services.* Retrieved from www.geoportal.asig.gov.al: https://geoportal.asig.gov.al/map/?auto=true

Bélaner, P. (2017). *Landscape as infrastructure.* Routledge.

Boeri, S. (2017). *Boeri; Stefano Boeri Architetti.* Retrieved from Stefano Boeri Architetti: https://www.stefanoboeriarchitetti.net/en/project/tirana-030-2/

Calzavara, M. (2016, December). *Tirana Green City.* Retrieved from Abitare: https://www.abitare.it/it/habitat/urban-design/2016/12/18/il-progetto-di-stefano-boeri-per-tirana/

CM. (2014). *Law Number 107; For the planification and territorial development.* Center for Official Publishings; Albanian Government.

CM. (2014). *The definition of case and area of national importance in the territorial development, identified as the Green Crown of Tirana.* Center for Official Publishings; Albanian Government.

CM. (2015). *Decision of council of ministers, Number 671.* Center for Official Publishings, Albanian Government.

Corner, J., & Hirsch, A. B. (2014). *The landscape imagination; collected essays of James Corner 1990–2010.* Princeton Architectural Press.

Dahl, P. -J. (2016). Exloring design potentials in porous urban space: Split vision urbanism HK through Montage. In *Proceedings Beyound ism: The Landscape of Landscape Urbanism.* Alnarp, Sweeden.

De Oliveira, F. L. (2017). *Green wedge urbanism; History, theory and contemporary practice*. Bloomsbury Academic.

Gavrilidis, A. A., Ciocanea, C. M., Nita, M. R., Onose, D. A., & Nastase, I. I. (2016). *Urban landscape quality index—Planning tool for evaluating urban landscapes and improving the quality of life*. Elsevier.

Gavrilidis, A. A., Nita, M. R., Onose, D. A., Badiu, D. L., & Iulia, N. I. (2017). *Methodological framework for urban sprawl control through sustainable planning of urban green infrastructure*. Elsevier.

INSTAT. (2019). *CENSUS 2011*. Retrieved from www.instat.gov.al: https://www.instat.gov.al/al/search/?query=CENSUS+2011

Marson, A. (2015). *Ridesigning the territories of diffused urbanisation [original title: Riprogettare i Territori dell'Urbanizzazione Diffusa]*. Quodlibet Studio.

Municipality of Tirana. (2017a). *General local plan of tirana municipality, development plan, 2017 [original title: Plani i Përgjithshëm vendor i Bashkisë Tiranë, Plani i Zhvillimit]*. Retrieved from www.tirana.al: https://tirana.al/faqe/plani-i-pergjithshem-vendor

Municipality of Tirana. (2017b). *General local plan of tirana municipality, the territorial strategy [original title: Plani i Përgjithshëm vendor i Bashkisë Tiranë, Strategjia Territoriale]*. Retrieved from www.tirana.al: https://tirana.al/faqe/plani-i-pergjithshem-vendor

National Territory Planning Agency. (2016). *Albania 2030, The National General Plan, The first national document on territorial Planning 2015–2030*. Tirana.

Ng, E., Wong, H. -H., & Han, M. (2006). Permeability, porosity and better ventilated design for high density cities. In *PLEA2006—The 23rd Conference on Passive and Low Energy Architecture*. Geneva.

NTPA. (2018, May). *Albania; the basic maps of the territory*. Retrieved from www.planifikimi.gov.al: https://planifikimi.gov.al/index.php?id=njoftime0

Pessoa, I. M., Tasan-Kok, T., & Altes, W. K. (2015). *Brasilian urban porosity: Treat or threat?* ICE Publishing.

Renner, R. (2018). *Urban being; anatomy and identity of the city*. Niggli.

Stiles, R. (n.d.). *A guideline for making space; joint strategy, activity 3.3*. Retrieved from Central Europe and Regional Environmental Center Website: https://www.urbanspace.rec.org/

Tadi, M., Zadeh, H. M., Biraghi, C., & Brioschi, L. (2017). Urban porosity. A morphological key category for the optimization of te CAS's environmental and energy performance. *Journal of Engineering Technology*.

Waldheim, C. (2006). *The landscape urbanism reader*. Princeton Architectural Press.

Waldheim, C. (2016). *Landscape as urbanism*. Princeton University Press.

Urban Landscapes and Green Infrastructure

Green Spaces and Environmental Justice: Measuring the Accessibility and Fair Distribution of Public Green Spaces in the Town of Al-Mughayyer

Lara Daradkeh, Siba AlGharaibih, Rania Shawaqfeh, and Anne Gharaibeh

Abstract

Access to green spaces in urban environments promotes social equity and improves the quality of life. In this context, the public good emphasizes the importance of accessibility and the possibility of walking to a public green space for a variety of leisure options. Many researchers focus only on quantitative methods to measure environmental justice. But for more accurate results, a multimethod approach must be used that also integrates qualitative data to get a fuller assessment. No previous research exists on this topic in the Jordanian context. This paper examines environmental justice by measuring the accessibility and fair distribution of public green spaces (active and passive) in Al-Mughayyer town to determine the area served by the existing green open spaces and examine injustice aspects and where they exist. The research questions addressed equitable distribution and accessibility. Using GIS network analysis, the research focused on qualitative data from Al-Mughayyer residents using interviews. The interviews focused on residents' demands for green open spaces, accessibility, availability, and arising problems. The results detected the unfairness of urban public green space distribution in terms of active recreation, while passive recreation provided by natural green spaces is accessible and distributed equally to all. This necessitated carrying out a suitability analysis to find the best places for new green public spaces. The suitable locations for future green open spaces might support and benefit government officials, developers, and urban planners in determining the requirements for green open spaces, public demands, and accessibility issues.

Keywords

Environmental justice • Public green spaces • Network analysis • Suitability analysis • Expert choice

1 Introduction

A well-established urban green space strengthens the essential connections between social communities and the environment. Countries are authorizing policies to reduce inequalities in urban communities between the districts and neighborhoods by ensuring environmental justice (Cetin, 2015). These inequalities have occurred due to very low green spaces per capita, lack of established urban green spaces, and uncontrolled urban growth (Fainstein, 2014).

According to the Jordanian department of statistics (2018), Jordan's population is estimated at 10.36 million people (DOS, 2018). Such a massive rise in population results in a higher need for open urban spaces as there is a direct link between creating more urban green space and overall public health (Rigolon et al., 2018; Wolch et al., 2014). In Jordan, where the per capita of green spaces is very low compared to international standards (Aljafari, 2014), according to the Jordanian buildings, cities, and villages planning regulation, No. 136, (2016), the per capita of green spaces in Jordan is 2 m^2, while the World Health Organization (WHO, 2016) regulates the per capita of green spaces as 9 m^2.

There are two types of entertainment that urban green spaces can provide: passive and active (Cetin, 2015; Dahmann et al., 2010). In this study, only public green spaces are considered because they can provide both active and passive recreation opportunities while private green spaces only provide passive recreation. Also, private urban green spaces are not affordable for all (Makhzoumi, 2007). Rural areas usually have more access to green spaces because of the high proportions of domestic gardens (Al-Kofahi et al.,

L. Daradkeh (✉) · S. AlGharaibih · R. Shawaqfeh · A. Gharaibeh
Department of City Planning and Design, College of Architecture and Design, Jordan University of Science and Technology, Irbid, Jordan

F. Trapani et al. (eds.), *Advanced Studies in Efficient Environmental Design and City Planning*, Advances in Science, Technology & Innovation, https://doi.org/10.1007/978-3-030-65181-7_24

2019). Al-Mughayyer as a rural town has a high proportion of domestic gardens reaching 83% of housing, more than 70% of these gardens are productive utilitarian gardens planted (Al-Kofahi et al., 2019). This indicates that Al-Mughayyer residents use their home gardens usually for agricultural purposes, and if used for entertainment, it would only be passive.

This research seeks to examine the public green spaces in Al-Mughayyer and identify where injustices exist by observing whether patterns of inequality of green space distribution exist and offer potential solutions to overcome this, by answering the following overarching questions: In Al-Mughayyer, what is the distribution of public green spaces (passive\active) and is their accessibility considered fair for its population? Do existing green spaces meet their requirements? To understand these guiding questions the following must be understood:

- Are public green spaces distributed equitably within city limits? How can Al-Mughayyer overcome unequal accessibility?
- How accessible are public green spaces for all residents in Al-Mughayyer?
- Are active green spaces accessible to males, females, elderly, and children? If not, who are the less beneficiary population and why? And which communities need more access to public green spaces?
- What are the types of green space activities residents need?
- Does Al-Mughayyer need more public green spaces, if so where will they be best placed?

These questions will be estimated by studying each of these variables: distribution of public green spaces both active and passive, access to public green spaces, main users accessing the green spaces, and the ratio of residents served by these areas. Access to public green spaces, in this research, refers to the distance traveled from a residential area to public green spaces, which are publicly owned and accessible places that everyone can use daily (e.g., urban parks, public gardens, and natural green space).

Many researchers calculated green space distribution according to land use and population data using a GIS-based land suitability analysis and network analysis (Kabisch et al., 2016; Mahmoud, 2011; Peschardt et al., 2012). But the weakness of this approach is that it focuses only on the quantitative distribution of green spaces in a relationship to the associated population (Kabisch et al., 2016). For more accurate results, a multimethod approach must be used that also integrated qualitative data to get a fuller assessment (Balram et al., 2009).

2 Literature Review

2.1 Green Spaces

Green spaces can be defined as any piece of land that is retained in a non-built state and intended for public or private use (Wolch et al., 2014). These spaces can be public (reserves, public parks, green trails, and sports fields) or private (domestic gardens and private green roofs) (Al-Kofahi et al., 2019; Lee et al. 2015; Wolch et al., 2014). Green spaces can provide active or passive recreations (Cetin, 2015). The difference between the two is that active spaces encourage users to move and expend energy by becoming physically active, while the passive spaces provide more of a relaxation area and the users do not have to engage in physical activities. It is important to achieve a balance between active and passive spaces for a healthy lifestyle (Backman & Crompton, 1990).

Public green spaces have many benefits such as providing social cohesion and solidarity that fuels social development, which supports economic goals and activities (Cetin, 2015; Lee et al., 2015). Public green space provides areas for physical activities to take place, which enhances overall public health both mentally and physically by decreasing obesity, reducing stress levels, and reducing the risks of many chronic illnesses (Cetin, 2015; Kabisch et al., 2016; Peschardt et al., 2012; Rigolon et al., 2018; Wolch et al., 2014). While most research focused on the health benefits of parks and green spaces, there could be health risks as well, including exposure to air pollutants near these parks and safety attention in parks located near areas with heavy traffic (Wolch et al., 2014).

2.2 Environmental Justice

Environmental justice is the recognition of the differences between people in the distribution of costs and benefits (Rigolon et al., 2018). Some scholars indicate that environmental injustice is the non-use of parks and gardens, despite their existence. This can be attributed to two reasons: (1) metaphorical boundaries include the social and cultural background; (2) physical boundaries, including the geographical distance between the user and the green space (Lindsey et al., 2001; Wolch et al., 2014).

Many researchers consider race, sex, age, and class as variables that limit the use of public parks (Lindsey et al., 2001; Rigolon et al., 2018; Wolch et al., 2014). However, Lindsey et al. (2001) classified spaces into "open mind spaces" and "single mind spaces," while the first is accepting the differences between people and serves the equity, the second concerns a specific type of activity for a certain group of people and thus serves equality.

2.3 Environmental Justice Applications in the Literature

Urban planning researchers have used various methods for measuring environmental justice by studying the accessibility and distribution of green spaces (Moseley et al., 2013; Wolch et al., 2014). Green space planning is usually based on quantitative standards that require the availability of a specific area of green space per population (Moseley et al., 2013). Despite rising literature, there is no agreement among scholars about the methods that measure green spaces accessibility and distribution. The majority of studies have used geographic information systems software (GIS) (Euclidean and Network analysis) (Cetin, 2015; Lindsey et al., 2001; Moseley et al., 2013; Wolch et al., 2014). The accessibility of green spaces can be measured by collecting spatial data on city communities that are served within a ten-minute walk of a park as the basis for the analysis (Meerow & Newell, 2017).

One more method that has been used is the space syntax method, which is used to analyze the characteristics of space, composition, visibility, accessibility, and their functional links to the physical, social, and psychological environments (Ascensão et al., 2019). Another method called the least cost method provides a better understanding of the process of green space provision and a larger separation of user analysis (Moseley et al., 2013). Other researchers developed a more integrative method to green space planning and management, by considering the size and quality of green space (Moseley et al., 2013), and most of the researchers only take into account what should be the minimum size the green space and did not address what are the motives for using or not using green spaces. As for the quality of green spaces, it is measured according to the features of the green spaces and in what way this affects user activity (Rigolon et al., 2018).

2.4 Green Spaces in Jordan

In Jordan, streets are considered as public spaces for socializing for many male teenagers, older adults sit on the doorsteps along the sidewalk (Abu-Ghazzeh, 1998). There is a difference when comparing Jordan to western countries in terms of determinants that prevent access to public green spaces. Where many studies in the USA, China, Denmark, and Germany confirm that the economic situation and racism limit access to green spaces (Rigolon et al., 2018; Wolch et al., 2014). In Jordan, sex and age are the determinants of access to green spaces (Abu-Ghazzeh, 1998). Measuring environmental justice in Jordan faces a problem due to the lack of quality data. According to AlKofahi et al. (2018), data about green spaces in Irbid are not available, and the quality of data is very important to evaluate environmental justice.

3 Study Area

Al-Mughayyer village is in the northeastern part of the Irbid Governorate in Jordan, located at 32.6 °N and 35.9 °E, between two canyons. The population is estimated at 17.455 in 2017. It covers an area of 16.6 km^2. One-third of the area is planned, and the rest consists of agricultural land and canyons with a slope of 450 m. The town has 15 water springs, 7 of these springs used to pump water. The area has one recreational park, two sports stadiums, and natural green spaces distributed throughout the entire region.

4 Methodological Design Flow

Figure 1 presents the methodological design flow of the research. The study was carried out in three phases: literature review, data acquisition, and database development, along with data analysis.

4.1 Methodological Approach

A combination of quantitative and qualitative approaches was applied, including participatory mapping, interviews, and GIS spatial analysis. This category refers to the application of qualitative methods along with GIS analysis, leading to the application of mixed methods for merging different sources of data (Balram et al., 2009).

4.1.1 Interviews

Interviews were used to capture data from green spaces users about their perception and opinions of green spaces and the way they use public green spaces. Interviews provided data about the respondent's stated preferences in relation to the green space. It consisted of 21 open-ended questions and Appendix 1 identifies the interview questions.

Researchers stated the questions, not objectives. Interview questions were open-ended to allow the participants to respond freely without any external effect. All participants had to provide verbal consent before they were interviewed, they were also reminded of their rights to stop the interview at any moment if they felt uncomfortable. The data analysis used a continually reflective data collection and analytic

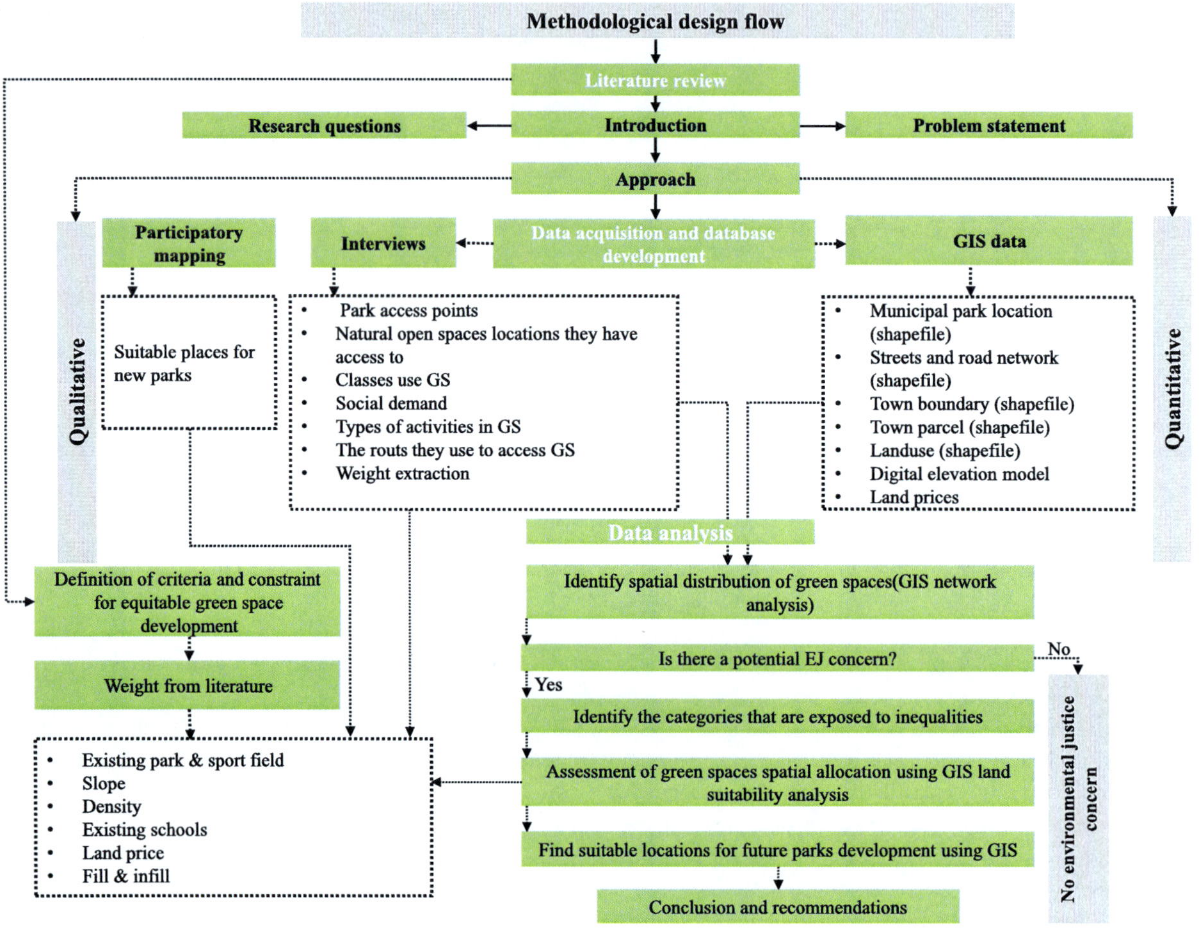

Fig. 1 Methodological design flow

process, where potential explanations for the patterns observed were to rely on literature to support findings for more reasonable and practical explanations.

Sampling

The sample size was based on the saturation system, which intended to include all age groups, genders, and covers different areas of the town (Mason, 2010). According to Rohrbach (2016), the sample size must be around 18–42 (elicit landscape value, development preferences, and special places). The sample size in this research was 42 persons, and four interviews were excluded, so the final participant number is 38. To achieve the highest number of participants completing the interview, the following distribution approaches were employed: (1) On-site participation (completion of the interview with people who are visiting the public green space). (2) The door-to-door approach by surveying people at their homes. (3) Municipality members' participation. (4) People in street participation.

4.1.2 Participatory Mapping

This approach was used to obtain easily understandable and acceptable results for the town residents. Some interview respondents were asked to complete a simple mapping exercise to capture data on their suggestions for new parks. The map used was a simple A2 map where the main landmarks of the town were identified along with notifying them of where they are standing at the time on the same map. Out of the total number of 42 interview participants, 38 accepted the mapping exercise. Four were excluded from the sample group because they could not read the map, and these maps were geo-referenced, digitized, and imported into GIS to be combined with suitability analysis.

4.1.3 GIS-Based Spatial Analysis

The network analysis approach is more accurate and suitable for environmental justice analysis, such as urban green space access. This tool defines the served area, taking into account the street's network and evaluates accessibility. The service area can be used to identify how many people can access the park.

Suitability analysis was used to find the best new location for active green spaces. A GIS database for suitability analysis was developed by using criteria and sub-criteria-based opinions taken from researchers and interviews participants through Expert Choice 11.5 software. Each layer was weighted once from our point of view and once from the interview participants' point of view, the final weight was the average of both the same as Cinderby (1999) study.

4.2 Data Collection

The primary GIS datasets for this study are listed in Table 1. Data for this study were mostly obtained through four sources: (1) Greater Irbid Municipality shapefiles, (2) Al-Mughayyer Municipality AutoCAD, (3) Researchers maps digitizing, and (4) Interviews and participatory maps.

5 Results

5.1 Public Green Spaces Service Areas (Network Analysis)

The walkable standard distance to any park using network analysis should preferably not exceed a 10 min' walk, which is about the equivalent to 250 m (Layton, 2017), While the walkable distance to natural green spaces according to Accessible Natural Greenspace Standard ANGSt (2010) should not exceed 500 m. Green spaces accessibility was analyzed under four metrics of travel distance as indicated in Table 2, where Fig. 2 depicts the served and unserved area and the dark color indicates high accessibility. 94.5% of the total municipal area was out of the park service range, and 86% of the total municipal area was out of the sports fields service range.

Network analysis was used to calculate the number of people who live within the service range crossing with the density layer to calculate the per capita. 660 houses are served by the park (park area equal 1.34 dunam) located in the north of Al-Mughayyer consisting of a small playground with insufficient equipment for playing and sitting areas. According to the Jordanian Department of Statistics, the average family size is (4.8) in 2018. By multiplying the number of houses with the average family size, it turns out that the park serves 3168 people. The park per capita equals 0.42 m^2/person. The same calculation was used to measure sports fields per capita. The sports fields serve 4828.8 people, and the per capita equals 1.7 m^2/person. Natural green spaces almost serve all the planned areas as shown in Fig. 2c.

Table 1 Data used in this study

Data	Purpose	Source
Road network	Network analysis	Digitizing
Al-Mughayyer boundary	Network analysis	Greater Irbid municipality
Park and sports fields location and boundary	Network And suitability analysis	Digitizing
Houses shapefile	Density map and per capita	AutoCAD
Natural green spaces locations	Network analysis	DEM (hydrological toolbox)
Slope	Suitability analysis	DEM (USGS)
School locations shapefiles	Suitability analysis	Digitizing
Land price	Suitability analysis	Department of Lands and survey / digitizing
Fill and infill	Suitability analysis	Digitizing
Residents preferable locations for new parks	Suitability analysis	Participatory mapping
Weight	Suitability analysis	Interviews

Table 2 Network analysis results

Distance (m)	Park served area (m^2)	Tow sports fields served area (m^2)	Distance (m)	Natural green spaces served area (m^2)
750–1000	401,720	1,129,242	1500–2000	469,524
500–750	238,788	611,524	1000–1500	1,281,775
250–500	209,580	488,209	500–1000	2,315,243
0–250	89,432	165.565	0–500	3,591,830
Served area	939,523	2,394,541	Total	7,658,372
Total area	1340	7500/1000	Total area	–
Served (%)	5.5%	14%	Served (%)	40%
Per capita	0.42 m^2/person	1.7 m^2/person	Per capita	–

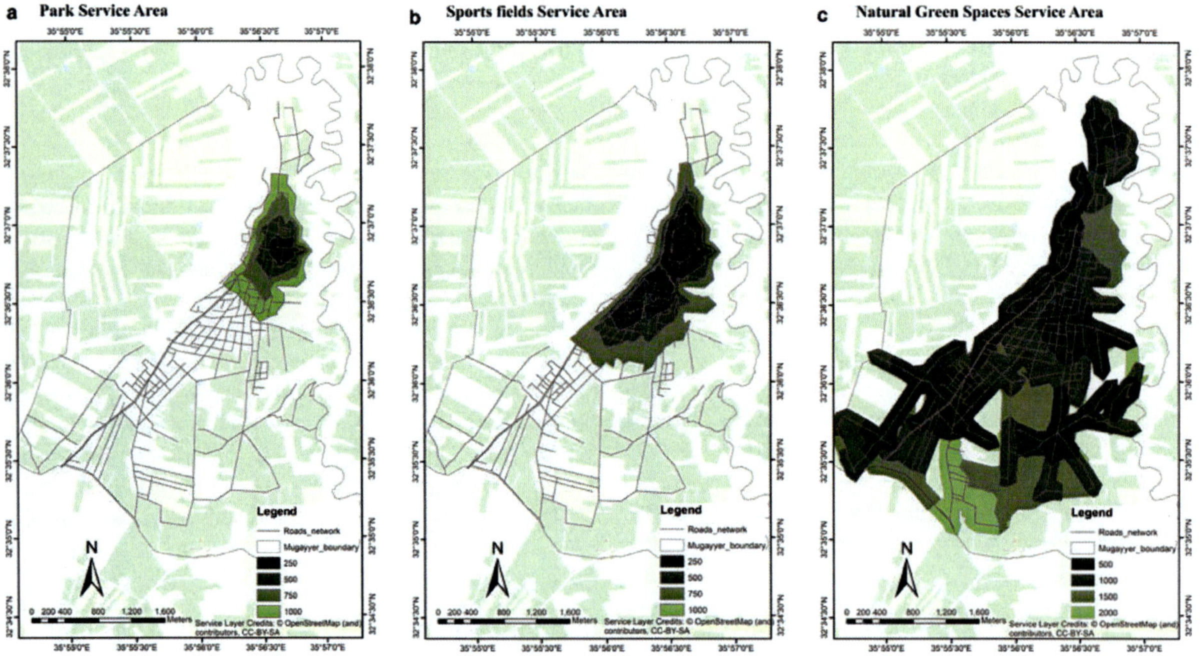

Fig. 2 **a** Park service area. **b**. Sports fields service area. **c**. Natural green spaces service area

5.1.1 Network Analysis Summary

The natural green spaces service range covers 40% of the total municipal area and covers all the planned areas, accessible and distributed equally to all. The park service range covers 5.5% of the total municipal area, not accessible, and it is not distributed equally to all. Furthermore, the park per capita is 0.42 m^2/person. Finally, the sports field service range covers 14% of the total municipal area, not accessible and it is not distributed equally to all residents. The sports field per capita equals 1.7 m^2/person.

5.2 Interviews

In-depth interviews with Al-Mughayyer residents indicated that the public green spaces have other problems besides not serving all residents and low per capita. When people were asked about what they think green spaces meant? They answered:

- Aya said, *"A place where all my children can play freely and safely" she has three children and their ages 3 years, 7 and 12 years.*

- Ahmad 19-year-old said, *" Any place he can do physical activities in".*
- Layan 16-year-old said, *" The place where there is access to trees, grass, and natural views."*

The conclusion of these answers is: first, some people refrain from using the park due to the cultural issues, especially that it is located in a neighborhood with a relatively bad reputation, and the residents of the neighborhood are using the park in a wrong way. Second, they like to visit parks in large groups, and the size of the existing park is not suitable for groups. Third, children prefer going to Petra Street and other parks in Irbid city where there are more choices of play equipment that meet different ages and needs.

As for sports fields, they only serve males. All females who participated in the interview agree that they want to use sports fields, but they cannot do so because there is no place for them and cannot share existing sports fields with males because of the existing culture. The males also complained of the dusty playgrounds that sheep graze in it, leaving behind dung and dirt.

Finally, only young males can access natural green spaces on foot. Furthermore, there is no public transportation serving these areas at all. Most of the activities in these areas are sitting, relaxing, and BBQ. However, the majority of people prefer visiting local natural green spaces than visiting other places, and they visit it weekly in good weather.

5.2.1 Interviews Summary

From the interview, we can conclude that there is no fair distribution of public (active) green spaces, not accessible for all, and do not meet all residents' requirements, also the less beneficiary population are females, elderly, and children.

5.3 Future Possible Green Space Locations Results (Municipality New Park Location Proposal)

The municipality divided the lands on the south side, according to the request of the residents, after the division, an irregular 0.5 dunams land was left (surplus) and the municipality proposed it as a public park. The municipality faces funding problems to provide a new park and most of the lands are privately owned and the municipality has only some lands with steep slopes at the canyons. The municipality employee said: *"public green spaces and recreational spaces have little or no attention at all as schools or health centers."*

5.4 Suitability Analysis Results

Several areas of the town are located 1 km from the existing park along with the street network. Determining where the best location for new parks within the areas identified as being underserved by current parks and sports fields using suitability analysis. First of all, it was important to define the optimal area for new parks. According to the Jordanian regulation of cities and villages No. 136 of 2016 and its amendments, the per capita of green spaces equal 2 m^2/person. The optimal size was calculated as in the formula below.

$$\begin{aligned}\text{Optimal new parks size} &= \text{number of unserved populations} \\ &\quad * \text{per capita} \\ &= (\text{number of houses} * \text{Jordanian family size} * 2) \\ &= 912 * 4.8 * 2 = 8.8 \text{ dunams}\end{aligned}$$

Before constructing a model, the distances from the new park and certain features were identified. In most cases, the new locations should be close to certain features, such as schools, but in other cases, it must be far enough away from some features, such as steep slopes and the existing park.

For more accurate results each layer was classified from 0 to 9, using GIS reclassify application and converted into raster form, the cell size is 26 according to used DEM cell size, these classes were changed according to layer properties, for example, density, slope, and land price layer were classified into equal intervals while for fill and infill layer it must be 0 for fill land and 9 for infill land. Appendix 2 shows the classification for each layer.

Using Expert Choice software, each layer was weighted once from our point of view and once from the interview participants' point of view. The final weight was the average of both as given in Table 3. Each variable weight was

Table 3 Weight criteria for variables

Variables	Experts	Participants	Average
Existing park and sport field	33.9	14	24
Density	29.3	36.5	32
Fill and infill	17.5	14.5	16
Existing schools	12.1	21.3	17
Slope	3.9	11.9	8
Land price	3.4	1.7	3

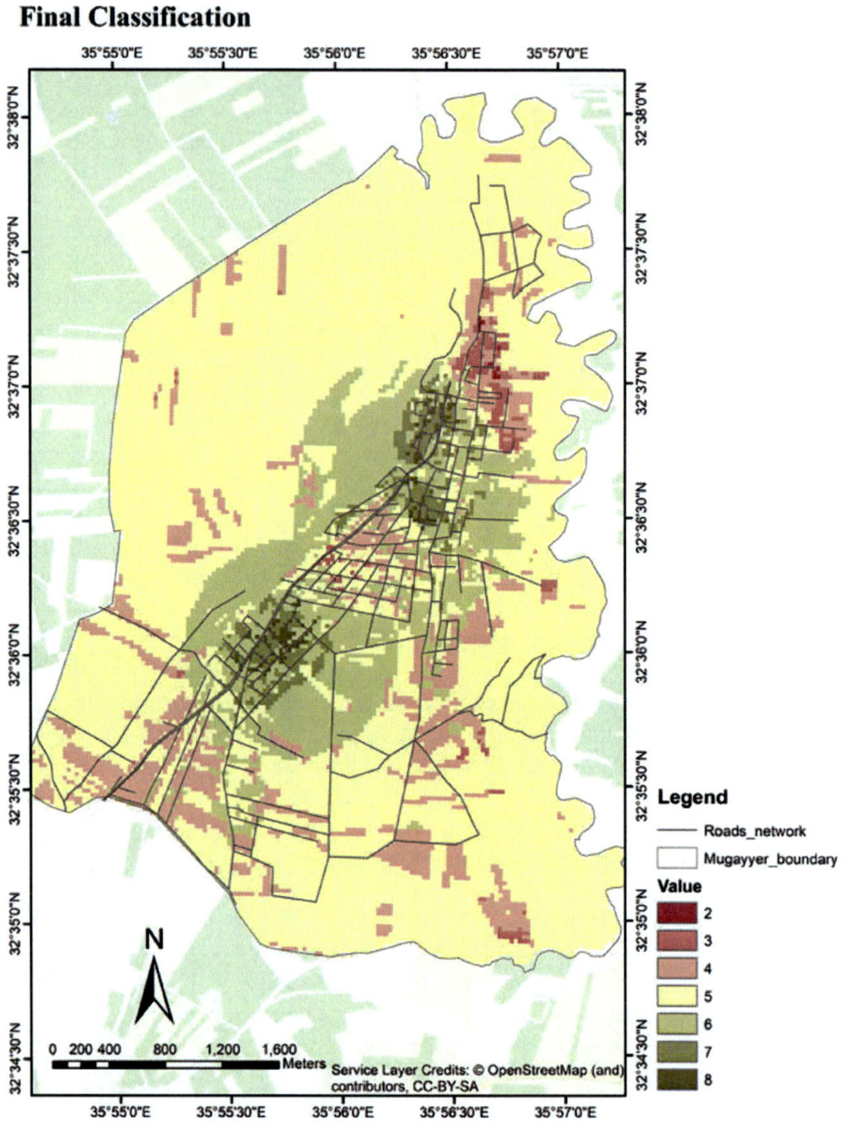

Fig. 3 Final classification after overlay process

inserted in the GIS overly weight tool to evaluate the most suitable land for public parks based on inserted weights. Figure 3 shows the final classification after applying the overly weight tool.

5.4.1 Suitable Locations Selection Criteria

More than one parcel got ranked 8 as shown in Fig. 3. But to be more specified, three criteria were chosen to select the most suitable parcels based on first the highest weight, second street network access, and finally the closeness to the locations determined by participatory mapping. Figure 4 shows the best five locations for new parks, and Table 4 identifies each location's characteristics.

5.4.2 Suitability Analysis Results Summary

More than one parcel ranked 8 and met the criteria used. Five sites have been identified to be suitable for new parks. Each parcel was checked to be infill land and the total area of all spots together is 35.9 dunam.

Fig. 4 Best five locations for new parks

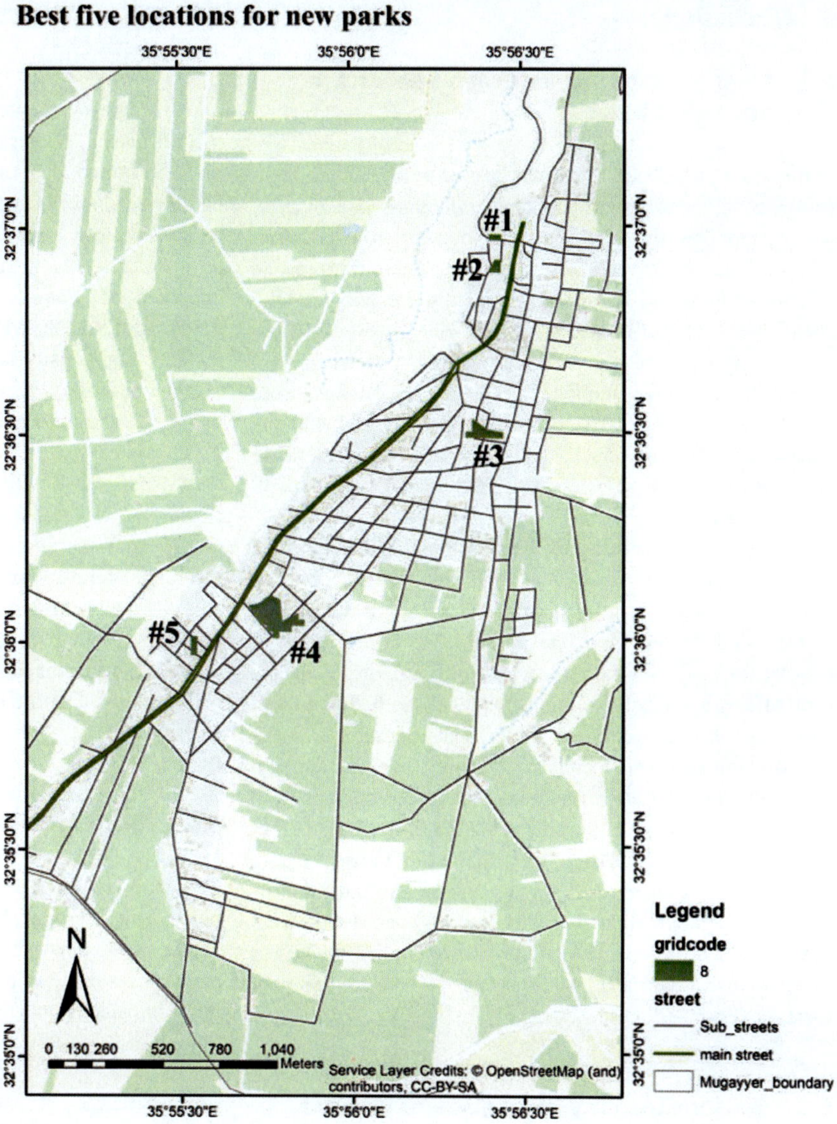

Table 4 Suitable locations characteristics

Classification	Location				
	#1 1.6 dunam	#2 2.4 dunam	#3 7.7 dunam	#4 21.8 dunam	#5 2.4 dunam
Park	9	7	7	9	9
Slope	8	8	8–9	8–9	9
Density	7	8–9	8	7	7
Fill–infill	9	9	9	9	9
Price	9	9	5	6	7
Schools	5	7	9	9	7

6 Discussion

6.1 Main Reasons Active Green Spaces Are not Being Used

In this study, the results show that people who do not benefit from active green spaces are mostly females and elderly, which supports the study carried out by Abu-Ghazzeh (1998). The results are not surprising, according to the existing culture and traditions. Especially for females who spend more time at home. While the elderly showed their satisfaction with the current green spaces, but when they were asked about the frequency of visiting the answers were once or twice a year. These answers are illogical, but the reason for these answers is that the elderly do not tend to discredit the place they belong to, and they feel satisfied with fewer things.

Furthermore, according to the classification of the green space by Lindsey et al. (2001) (single mind and open mind spaces), the active green spaces in the town are considered single mind spaces, these spaces serve specific groups of people, for example, the park serves only children, and the sports fields serve only young males, with the fact that these spaces are not accessed for all residents according to the network analysis. This means these spaces face physical and cultural boundaries and these boundaries were mentioned in two studies (Lindsey et al., 2001; Wolch et al., 2014).

According to the children, they do not like to use the park because there is not enough playing equipment, but from the reviewed literature, there are other reasons even if they were not mentioned like the attractiveness, size, quality, the access to grass and trees and safety issues related to the park location (Peschardt et al., 2012).

6.2 Reasons for Why People Prefer Natural Green Spaces

Natural green spaces serve 45% of the total area (according to network analysis), it serves all the planned areas, which means all residents have access to these spaces. Natural green spaces in contrast to active green spaces are open mind spaces as mentioned in Lindsey et al. (2001). Providing passive recreation spaces erase the differences between people because everyone can relax and enjoy nature. On the other hand, these spaces are attractive with good views and can serve a large number of people, especially that most of the interview participants said that they prefer to go to public areas within groups.

6.3 Do the New Spots Achieve Environmental Justice?

In Al-Mughayyer, two barriers prevent achieving environmental justice, physical and cultural barriers. To overcome physical boundaries, three variables were discussed size, fair distribution, and access. The optimal size needed for new parks is 8.8 dunam and the total area for all spots together is 35.9 dunam, which means there is enough space to serve unserved populations from the existing park and sports fields. Also, to achieve the fair distribution, the existing park distance and density layers were highly weighted, to ensure that the new spots serve non-beneficiaries from the existing green spaces while reaching more people at the same time. To ensure easy access, each spot was checked to be in slope class from 8 to 9 with a good street network access.

For the cultural boundaries, comes the designer role to provide safe and proper spaces designed for females, males, disabled, children, and the elderly to be able to use and enjoy. Finally, to avoid bias and achieve social justice, the residents participate in the weighting process and their opinion was taken into consideration fifty-fifty by researchers to capture data on their suggestions for new park locations.

6.4 Policy Implications

The municipality's intention to increase the provision of public green spaces is a step in the right direction; their choice of the new location was very similar to the suitability analysis results, but they did not take into account that the new land size is only 1/2 dunam while the needed size is 8.8 dunam to serve all residents.

6.5 Discussion of Methods

This research faces many limitations. At first, more time for interviews was needed, and not all data needed are available, such as street network data, it was digitized and drawn, also there is no information about the existence of sidewalks which means that these streets may not be suitable for walking in the first place.

Furthermore, the data for green spaces are not available as mentioned in (Al-Kofahi et al., 2019) and the quality of the available data for suitability analysis was very poor. The land price map is not 100% accurate, because the land department's Web site gives the land price according to the number of the basin regardless of the land number itself. Also, the density map considers all types of buildings (homes, mosques, health centers, and schools) because it was digitized through Google Earth.

7 Conclusion and Recommendations

The benefits of having access to green spaces, parks, community gardens, and nature are becoming more and more obvious and supported through studies in several fields. As the world becomes increasingly urbanized, providing adequate access to natural green spaces becomes more difficult and more important. The literature shows that having access to green spaces and "natural" areas can be beneficial at many levels. Benefits provided by green spaces contribute to the overall resilience of the urban system and the human security of the individuals living within it. Because of these benefits, some researchers have become interested in the ways green spaces are distributed in cities and who have more access to their benefits. This has become a small but growing section of the literature relating to environmental justice that focuses on inequalities in access to environmental benefits. Many studies have found that green spaces and other environmental benefits are distributed unequally in cities and some groups are being farther from green spaces and having access to less natural space, tree cover, and biodiversity. Al-Mughayyer has a widespread natural green and provided an interesting location to study because of the existence of natural green space (provide passive recreation) and the absence of adequate urban green spaces (provide active and passive recreation). Therefore, this study was able to examine inequalities currently present in the distribution and access to green spaces throughout the town.

The research attempted to examine the mixed-methods approach to support a multidimensional analysis of access to urban green spaces in terms of distance as well as the perceptions and behavior of various social groups. A review of the literature showed that accessibility had investigated based on distance or travel time, but these approaches focused only on the quantitative distribution of green space in relation to the associated population and for more accurate results a multimethod approach must be used that also integrated qualitative data would help to get a fuller assessment because access is a broad and multidimensional concept that cannot be thoroughly investigated using only GIS approach since this provides a narrow definition of access. This says nothing related to access impressions and behaviors. The participatory mapping and interviews were eye-openers to the fact that not all citizens have equal access to green spaces, and through using the participatory GIS (land suitability analysis along with participatory mapping), the study came up with five sites which considered suitable for new parks to be implemented by the municipality in future work. This study can open the way for many urban studies related to towns where natural green spaces are available but are under-serviced in terms of public green spaces and study the main reasons behind this and find appropriate ways to overcome these problems.

After thoroughly analyzing the accessibility and distribution of public green spaces in Al-Mughayyer town, several recommendations can be directed to the decision-makers in the municipality of the town. As a start, the municipality must work to classify the land and provide pieces of land to create new green public spaces with the attention that these founded public spaces need to be in decent sizes and provide activities for everyone and working on the best spatial distribution for these parks. They also need to include the public in making these decisions for highlighting what they need and what can be done to increase social equity, then providing these areas with proper services to encourage people to use them, which can also create new job opportunities, as well as proper transportation that is highly recommended for the existing public spaces along with the new ones. Many factors can be taken into account in suggesting these new parks such as mosques locations, not to forget that the new garden design should consider the cultural aspects in the region to suit all groups, especially women and the elderly.

This research focuses only on public green spaces and does not include private ones. Future research can analyze both public and private green spaces to examine whether environmental inequity exists, for a more detailed study of accessibility, social, environmental, and pedestrian measures, data on the region or neighborhood scales, such as crime and safety, well-being, and personal space should be added.

Appendix 1 (Interview Questions)

Dimensions	Interview questions
User's opinion	• What would you think green space is? • What spaces are considered active green space in Al-Mughayyer? • Do you face any problems when using public green spaces? • What changes would you suggest for your town's public green spaces? • How satisfied are you with the green spaces in your town? • Is aging or disabled population in your family participating in visiting green spaces and if so, do these groups have accessibility for green spaces? • If there is an intention to build a new park, arrange the following criteria as the most important for you: slope/density/the existing park/school /land price/fill-infill areas?
Visit details	• How often do you use your local public green spaces? Or do you prefer to go outside of the town? • What motivates you to visit a green space? • Does the proximity of the green space affect the frequency of your visit? • Who do you usually attend these spaces with?
Access details	• How long does your usual journey to the green space take? • Is local access to green spaces important to you? • How do you usually travel to public green spaces? • How long would you like for your journey to take to your preferred green space? • Would you prefer to go to one particular space only because it is closer to you?
Activities details	• What kind of activities are you able to perform in the existing green spaces? • What activities do you think you would like to be able to do? • What other uses of the space you would like to make? • Which place provides more social experience for you? • What are the activities you do when visiting public green spaces?

Appendix 2 (Classification Maps)

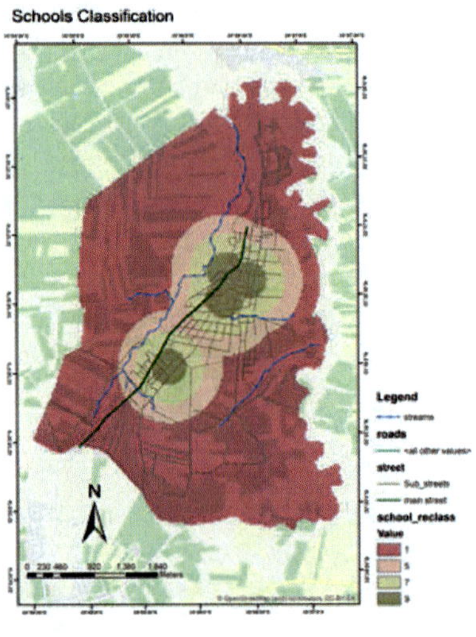

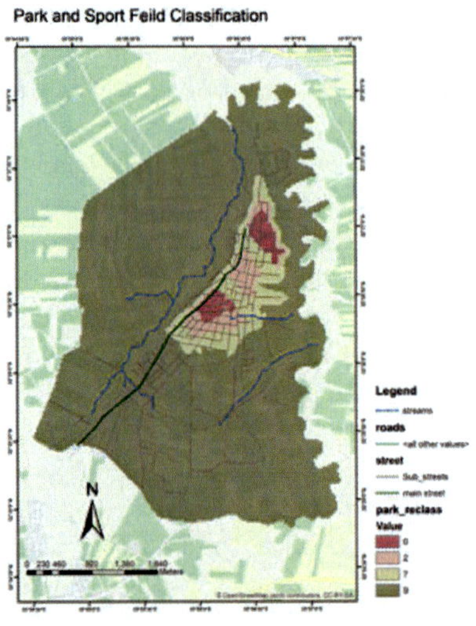

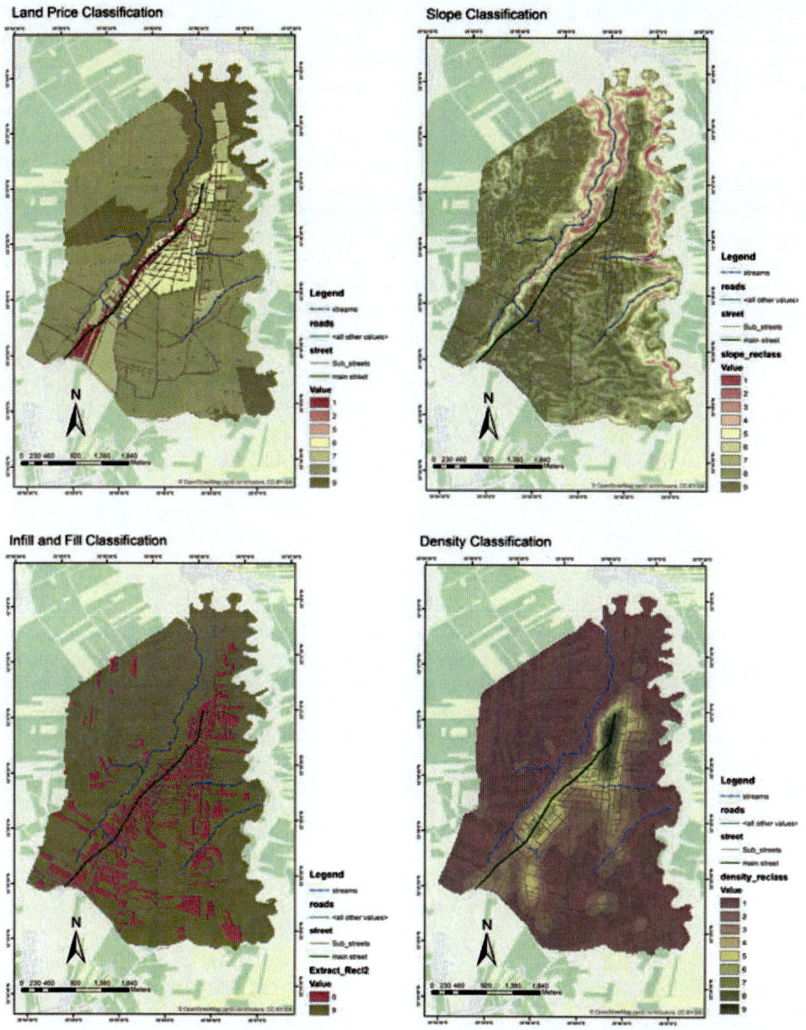

References

Abu-Ghazzeh, T. M. (1998). Reclaiming public space: the ecology of neighborhood open spaces in the town of Abu-Nuseir, Jordan. *Environment and Behavior, 36*(3), 799–831. https://doi.org/10.1016/S0169-2046(96)00343-X

Aljafari, M. (2014). *Emerging public spaces in the City of Amman, Jordan: An analysis of everyday life practices.*

Al-Kofahi, S. D., Gharaibeh, A. A., Bsoul, E. Y., Othman, Y. A., & Hilaire, R. S. (2019). Investigating domestic gardens' densities, spatial distribution and types among city districts. *Urban Ecosystems, 36*(1), 1–15. https://doi.org/10.1007/s11252-019-0833-7

AlKofahi, S., Jamhawi, M., & Hajahjah, Z. (2018). Investigating the current status of geospatial data and urban growth indicators in Jordan and Irbid municipality: Implications for urban and environmental planning. *Environment, Development and Sustainability, 20*(3), 1067–1083. https://doi.org/10.1007/s10668-017-9923-y

ANGSt. (2010). *Nature nearby: Accessible natural greenspace guidance*. Natural England: Peterborough. Retrieved from https://www.naturalengland.org.uk/regions/east_of_england/ourwork/gi/accessiblenaturalgreenspacestandardangst.aspx

Ascensão, A., Costa, L., Fernandes, C., Morais, F., & Ruivo, C. (2019). 3D space syntax analysis: Attributes to be applied in landscape architecture projects. *Urban Science, 3*(1), 20. https://doi.org/10.3390/urbansci3010020

Backman, S. J., & Crompton, J. L. (1990). Differentiating between active and passive discontinuers of two leisure activities. *Journal of Leisure Research, 22*(3), 197–212. https://doi.org/10.1080/00222216.1990.11969825

Balram, S., Dragicevic, S., & Feick, R. (2009). Collaborative GIS for spatial decision support and visualization. *Journal of Environmental Management, 90*(6), 1963–2117. https://doi.org/10.1016/j.jenvman.2008.07.014

C.F.R. (2016). *Buildings, cities and villages planning regulation*, 136 C.F.R.

Cetin, M. (2015). Using GIS analysis to assess urban green space in terms of accessibility: Case study in Kutahya. *International Journal of Sustainable Development and World Ecology, 22*, 6. https://doi.org/10.1080/13504509.2015.1061066

Cinderby, S. (1999). Participatory geographic information systems (GIS): The future of environmental GIS. *International Journal of Environment, 11*(3), 304–315.

Dahmann, N., Wolch, J., Joassart, M., Pascale, R., & Kim, J. M. (2010). The active city? Disparities in provision of urban public recreation resources. *Health and place, 16*(3), 431–445. https://doi.org/10.1016/j.healthplace.2009.11.005

DOS. (2018). Jordanian Department of statistics 30/4/2019. Retrieved from https://dosweb.dos.gov.jo

Fainstein, S. S. (2014). The just city. *International Journal of Urban Sciences, 18*(1), 1–18. https://doi.org/10.1080/12265934.2013.834643

Kabisch, N., Strohbach, M., Haase, D., & Kronenberge, J. (2016). Urban green space availability in European cities. *Ecological Indicators, 70*, 11. https://doi.org/10.1016/j.ecolind.2016.02.029

Layton, R. (2017). Walkability standards: Test of common assumptions *design concepts*.

Lee, A., Jordan, H., & Horsley, J. (2015). *Value of urban green spaces in promoting healthy living and wellbeing: Prospects for planning* (Vol. 8).

Lindsey, G., Maraj, M., & Kuan, SonCheong. (2001). Access, equity, and urban greenways: An exploratory investigation. *The Professional Geographer, 53*(3), 332–346. https://doi.org/10.1111/0033-0124.00288

Mahmoud, A. H. A. (2011). Development of sustainable urban green areas in Egyptian new cities: The case of El-Sadat City. *Landscape and Urban Planning, 101*(2), 157–170. https://doi.org/10.1016/j.landurbplan.2011.02.008

Makhzoumi, J., & Zako, R. (2007). *The Beirut Dozen: Traditional domestic garden as spatial and cultural mediator*.

Mason, M. (2010). Sample size and saturation in PhD studies using qualitative interviews. Paper presented at the Forum qualitative Sozialforschung/Forum: qualitative social research.

Meerow, S., & Newell, J. P. (2017). Spatial planning for multifunctional green infrastructure: Growing resilience in Detroit. *Landscape and Urban Planning, 159*, 14. https://doi.org/10.1016/j.landurbplan.2016.10.005

Moseley, D., Marzano, M., Chetcutia, J., & Watts, K. (2013). Green networks for people: Application of a functional approach to support the planning and management of greenspace. *Landscape and Urban Planning, 116*, 1–12. https://doi.org/10.1016/j.landurbplan.2013.04.004

Peschardt, K. K., Schipperijn, J., & Stigsdotter, U. K. (2012). Use of small public urban green spaces (SPUGS). *Urban Forestry and Urban Greening, 11*(3), 235–244. https://doi.org/10.1016/j.ufug.2012.04.002

Rigolon, A., Browning, M., & Jennings, V. (2018). Inequities in the quality of urban park systems: An environmental justice. *Landscape and Urban Planning, 178*, 14. https://doi.org/10.1016/j.landurbplan.2018.05.026

Rohrbach, B. (2016). *Assessing participatory mapping and multi-criteria evaluation for land-use change analysis*. The University of Zurich.

WHO. (2016). *Urban green space and health: Intervention impacts and effectiveness*. Retrieved from World Health Organization: https://www.euro.who.int/pubrequest

Wolch, J. R., Byrne, J., & Newellc, J. P. (2014). Urban green space, public health, and environmental justice: The challenge of making cities 'just green enough.' *Landscape and Urban Planning, 125*, 234–244. https://doi.org/10.1016/j.landurbplan.2014.01.017

Urban Greenery as a Resource for Urban Environment

Riccardo Pollo, Matteo Giovanardi, and Andreina Mariani

Abstract

In many suburban districts, urban greenery is an important component in terms of quantity and quality in the urban landscape. The benefits on the environment on the quality of human life and the economic impact on the surrounding built heritage make the natural environment a collective asset for the entire ecosystem. However, the lack of human and economic resources on which today's public administrations can rely makes the management of such spaces problematic, often transforming them into poorly cured and unused places. In this regard, maintenance activities, such as mowing and pruning, as well as a series of continuous operations produce a significant amount of material to be disposed, constituting a monetary and environmental cost for the community. In order to compare different scenarios of reuse of green waste from an economic, environmental, and social point of view, the case study of the "Le Vallette" district in Turin was analyzed. On the basis of the concepts on which the material flow analysis (MFA) is based, the flows that can be produced each year from the maintenance activities of the green areas were evaluated. The qualitative and quantitative analysis carried out opened a series of reflections on the current green waste treatment chain. In this way, the concept of urban green is reinterpreted from a landscape element to an economic and environmental resource.

Keywords

Green urban management • Urban metabolism • Material flow analysis

R. Pollo · M. Giovanardi (✉) · A. Mariani
Department of Architecture and Design (DAD), Polytechnic of Turin, Turin, Italy
e-mail: matteo.giovanardi@polito.it

1 Introduction

Urban greenery, especially in suburban districts, represents a significant share of the territory with obvious ecosystemic benefits. However, the green system requires a significant organizational and economic effort for management and maintenance, which must be planned over the years. At a national level, official reports (ISPRA, 2016) show that the use of plans and planning tools such as the "Piano del Verde" is only present in one out of ten chief towns. The management of urban greenery, where not carefully planned, is carried out by guaranteeing the minimum activities for the safety of the sites. In this regard, Law 10/13 "Regulations for the development of urban green areas" does not require public administrations to use such instruments.

Therefore, the issue of the enhancement and planning of the management of public green heritage is necessary in order to allow a more sustainable use of the resources available to us. In this context, the management and maintenance of urban greenery is combined with the themes of the contemporary debate on the sustainability of cities, such as urban agriculture, the promotion of the circular economy and the smart city.

The proposals for urban habitat and self-sufficient city of Vicente Guallart, a city made of interconnected units at different scales, local and global, represent the ideological synthesis of a possible approach to a circular economy on an urban scale (Lo Sasso, 2011). A city that is careful to produce without waste and to re-insert as much matter and energy as possible into the production cycles necessary for its life (Guallart, 2014).

The objective of the study is to compare alternative scenarios for the reuse of urban green waste. Different approaches and technologies have been evaluated in order to enhance the green waste in terms of material and energy flows generated by the management and maintenance process (EU, 2010). Normative and logistical aspects enter

F. Trapani et al. (eds.), *Advanced Studies in Efficient Environmental Design and City Planning*, Advances in Science, Technology & Innovation, https://doi.org/10.1007/978-3-030-65181-7_25

forcefully into the topics, introducing the theme of the scale and length of the supply chain.

Urban green maintenance activities, both horizontal and vertical, such as mowing lawns and pruning trees, produce a considerable amount of organic material that can be recovered and reintroduced into production cycles such as soil improver (Tong et al., 2018), biogas or fuel.

At a national level, data on the treatment of municipal green waste is often difficult to assess. A local assessment of urban green wastes production is even harder. It is possible to identify two different approaches to estimate the production of green waste from maintenance activities. The first, based on the frequency of maintenance activities and the specific growth of different tree species, associates a parametric quantitative value based on the extent of the land area. These parameters are used on a daily basis to draw up contracts for maintenance activities between the public and private sectors.

The second approach, on the other hand, counts the production of greenery that reaches the end of the supply chain. In this way the quantified material is the one that is really treated every year by the specialized plants. This method, although useful to compare the real impact on the likes to other forms of waste, does not allow to evaluate the potential quantity that can be produced by an optimization of maintenance activities. Starting from this last approach we can define some aspects. According to the estimates of the Italian Composting Consortium (CIC), in Italy, out of 4.7 million tons of total organic waste sent for aerobic digestion, 1.5 million tons comes from mowing and pruning and represent about 40% of Italian separate waste collection (CIC, 2017).

In addition to the analysis of the amount of green waste, the issue requires reflection on how to treat it. As underlined by several authors (Drechsel et al., 2015; Viljoen & Bohn, 2014), schemes for the collection and treatment of urban organic waste, green maintenance residues and household organic waste have until recently been based exclusively on large plants and services. Only recently, following European and national legislation have forms of recovery with small plants been encouraged. In particular, in Italy, with the legislation on community composting, MATTM Decree no. 266 of December 29, 2016 and Dgls 152/06, the aim was to encourage small-scale activities managed for the benefit of individual members who carry out agricultural activities, including urban ones. This scenario could also be considered interesting for the effects on the involvement of local communities in the cycles of the circular economy and in the management and enjoyment of greenery.

2 Urban Metabolism and Material Flow Analysis

Cities operate according to metabolic processes that absorb, transform, and release forms of energy and matter (Pincetl et al., 2012; Zhang et al., 2013). To evaluate any process, it is necessary to identify the inputs and outputs associated with them (Wolman, 1965; Liang et al., 2012). The term "urban metabolism" represents the quantitative accounts of the overall inputs and outputs of energy, materials, and substances into and out of cities (Shahrokni et al., 2015). Such concept has been widely adopted in the field of urban studies with the goal of environmental mitigation. Moreover, it has a deep interdisciplinary nature and encompasses a wide range of methods and disciplines such as the materials flows analysis (MFA), life-cycle analysis, energy analysis, and ecological footprint.

With the aim of encouraging circular processes in the use and management of green spaces, the quantification of material flows inevitably becomes a primary aspect to be investigated. These material and energy flows, from public or private bodies, represent a significant part of the urban metabolism. In this regard, it is possible to analyze individual processes using the principles of MFA. Such a method is necessary to handle data from observation, accounting to statistical analysis, simulation, and decision support (Dijst et al., 2018). The MFA approach, developed in the industrial sector and then applied to different sectors, allows to evaluate the quantity and direction of material flows within the boundaries of a given system (Baccini & Baader, 1996; Allesch & Brunner, 2015).

The evaluation of the different material and energy resources that interact with each other is necessary to better understand the environmental and economic impacts of long and complex processes. Today, municipal green waste from the maintenance of green areas is reused, according to the EU rules, to produce biogas or soil improver compost, allocating to municipal solid waste incineration (MSWI) only the not recyclable fraction of solid municipal waste (SMW). In both processes, although different in terms of methods, the organic material from greenery is mixed with other organic wastes from households and commercial activities to allow the optimal transformation (Lebersorger, 2011). The processing capacity of the material makes vegetable waste a by-product that can be easily reintroduced into the closed process of a circular economy. However, the virtuous closure of the cycle still finds some regulatory and economic obstacles. The survey carried out among the main companies of the sector in the city has shown that the circular process is hampered by economic issues. For example, the amount of greenery treated each year is much smaller than the actual production capacity. Many greenery maintenance companies' reasons prefer to adopt mowing methods that do not involve waste collection such as mulching to avoid disposal costs. At the same time, the use and sale of natural compost does not find the same appreciation on the market, again for economic reasons, as synthetic fertilizers.

The UM investigations are often carried out at a regional or urban scale not allowing the evaluation of specific policies and project at a neighborhood scale. Nevertheless the shift at a more granular spatial scale of the UM analysis would allow to support the decision making in the urban planning and design at the neighborhood scale. Such a shift has been underlined by literature as a promising development of UM research and urban planning practices in the field (Kennet, 2013).

In order to compare and evaluate in a comprehensive way alternatives policy strategies for urban green management and design, we consider necessary, firstly to quantify the associated flows of matter. Such evaluation takes into account the production of waste from the green areas and the relationship with inhabitants and with locally based as well as regional models of reuse and treatment, composting, biogas production, etc. In this regard, we have chosen a case study that could support, compared to the current management model and due to the high presence of green spaces, different planning scenarios. The analysis has been made on a neighborhood of nearly 8.000 people and can be applied to any other urban district. The quite reliable and updated input data used in the analysis is mainly from municipal statistics referred to the neighborhood. Where not explicitly stated, quantities have been estimated on the basis of data available in the literature, compared with municipal, regional, and national reports on organic waste treatment in Italy. The greenery surfaces and the number and species of trees have been calculated by a GIS model.

3 Case Study: Neighborhood "Le Vallette," Turin

The case study has been selected following two distinct criteria. The first refers to the urban nature of the neighborhood, which can be identified as a distinct part of the city of Turin both for cultural identity and architectural morphology. The second has been its size and the large amount of green spaces. The neighborhood and its greenery is, indeed, large enough to generate a material flow to allow the adoption of specific strategies at neighborhood scale and the comparison of different scenarios. Moreover, the provision of such a large green areas would allow the hypothesis to devote a part of it to urban agriculture. In such a way, we can imagine a metabolic flow from green areas management wastes to local agriculture as well as a significant flow contributing to the urban raw material provision to the biogas production by the regional industrial system facilities.

The urban agriculture and community composting is rooted on the experiences carried out in the past decades in many countries, wealthy as well in the south of the world (Oberlin & Szántó, 2011). Many successful examples of community composting are described in literature, in urban areas as well as in the countryside (Slater et al., 2010).

As we note above, research on urban metabolism at neighborhood scale has been suggested and developed in recent years (Kellett et al., 2013; Shahrokni et al., 2015). Many authors identify the neighborhood scale as the basic level to estimate the urban metabolic flows with the scope to identify the flows and compare the environmental mitigation strategies.

The study area is the district of Turin called Lucento-Vallette, built as an iconic complex of Public Residential Building of the sixties. The 800,000 m^2 district on the northwest outskirts of Turin has been designed between the 1950s and 1960s looking at the Scandinavian examples of self-sufficient urban areas, equipped with all facilities and surrounded by greenery. The project was contributed by the major exponents of the Turin architectural scene of the time: Gino Levi Montalcini, Ottorino Aloisio, Roberto Gabetti and Aimaro Isola and many others. This part of the city, identified with the name "Le Vallette," is still today a peripheral area characterized not only by the quality of the architecture, but also by the presence of significant green areas, even if not fully recognized by the real estate market as an environmental and economic value (Fig. 1).

The neighborhood is inhabited by about 8300 people with a large share of the elderly, slightly higher than the average number of Turin, and with a fair presence of population of recent immigration. The per capita green area is about 42 m^2, twice as the average for the city. The green character of the district can become a precious element and a resource for its enhancement.

4 Development of Analysis

The analysis has been carried out starting from the identification of green areas capable of "producing" green waste. Consistent with the MFA methodology, the first phases of the research were used to define the system's inputs. Data on the consistency of existing green areas was obtained from the geographic information system (GIS) of the municipality of Turin. The GIS methods have been underlined by literature as an important tool for the UM and MFA, above all in the field of energy supply and demand models, as well as

Fig. 1 Path, green area (**a**) and urban garden (**b**) in Le Vallette

transport of goods and people models (Dijst et al., 2018). The cartographic data that has been made available by the municipality for a wide range of purposes allows to make, in fact, a series of qualitative and quantitative analyses on urban green areas (www.geoportale.it). By using the GIS software, it has been possible to estimate the extension for each type of green area identified by a quite an accurate database. The area analyzed counts about 350,000 m^2, divided into different categories: generic green areas, green areas with trees, sports green areas, green areas for traffic distribution, and green areas that cannot be used (Fig. 2).

Fig. 2 Characterization of green spaces by GIS mapping (ArcGis by ESRI)

Urban green areas have been classified with the scope of associating the maintenance activity and green wastes produced every year to each class. The open data from the municipality of Turin allows to evaluate not only the horizontal green areas, but also the trees present. In the area analyzed, there are about 800 tall and medium-stem trees, divided by type as shown in Table 1. Also, with regard to vertical green areas, standard maintenance activities for the calculation of waste quantities was considered on an annual basis. The data reported in Table 1 with reference to the type of maintenance activities and annual frequency has been estimated on the basis of specialized literature (Baldan, 2015; Blengini & Fanti, 2009; Saer et al., 2013) and interviews with operators involved in the maintenance of public green areas and the reuse of waste for energy purposes within the study (ACEA).

Each maintenance activity is associated with a quantity of waste material produced as shown in Table 2.

As it is well known, these quantities are related to the vegetative cycle of the plants and are therefore variable according to the climate, the specific site, and the weather conditions of the year. However, it was considered acceptable, in the first instance, to use average data provided by a recent study by the Laboratory of Environmental Systems Analysis & Management (LASA) of the University of Padua (Baldan, 2015).

Once the input data was defined, the research shifted to green waste treatment methods. As already mentioned, there are two main types of treatment of green waste: the "anaerobic" one, aimed mainly at the production of biogas and the "aerobic" one useful to produce soil improver compost. Starting from Scenario A, which represents the business as usual scenario in the context analyzed, the consequences generated by favoring one treatment over the

Table 1 Type and frequency of green maintenance activities

Class	Quantity	Rate (%)	Maintenance activity	Frequency
Green areas	219.500 m^2	62	Mowing with harvesting	6/year
Tree-lined green areas	91.071 m^2	26	Mowing with harvesting	6/year
Green areas not usable	7.244 m^2	2	Mowing without harvesting (mulching)	3/year
Green sports areas	17.757 m^2	5	Mowing with harvesting	12/year
Green traffic areas	16.034 m^2	5	Mowing without harvesting (mulching)	8/year
Total green areas	351.606 m^2	–		
Deciduous trees	742	92	Foliage harvesting - pruning	3/year–0.2/year
Evergreen trees	65	8	Pruning	0.125/year
Total trees	807	–		

Table 2 Quantity of waste generated by maintenance activities

Green waste produced	Quantity
Mowing with harvesting	2 kg/m^2 year
Mowing without harvesting (mulching)	–
Foliage harvesting	50 kg/tree year
Light pruning	30 kg/tree year
Light pruning	150 kg/tree year
Heavy pruning	450 kg/tree year

other were assessed in terms of energy and compost produced. The energy production has been evaluated in terms of thermal energy and electricity by biogas through current technologies. The three scenarios are thus defined:

Scenario A. Which corresponds to the current management model, estimated on the basis of the percentages of waste treatments by the published regional government reports, where a portion of the vegetables collected is sent to composting, a part conveyed to the process of anaerobic digestion and a residual percentage incinerated (Città Metropolitana di Torino, 2017; ISPRA, 2019);

Scenario B. An "Anaerobic" model in which it is proposed to increase the share of green waste collected and sent for anaerobic digestion treatment for the production of biogas and the composting of residual sludge;

Scenario C. A scenario of collection and treatment on a local basis with the start of recovery and community composting processes with reuse in urban agriculture at a local scale.

For the setting of the three scenarios, the quantities of waste collection from mowing and pruning have been calculated on the basis of the maintenance programs by the municipality of Turin and by agronomic literature (Baldan, 2015) The split of the portion sent to aerobic as well as anaerobic treatment has been done according to the figures by the aggregated regional government data published (Città Metropolitana di Torino, 2017) (Table 3).

As shown in Fig. 3, the scenario A is characterized by a slightly long and complex chain, which involves part of the waste treatment in sites located outside the metropolitan city and, in some cases, outside the region. The Scenario B has a more rational industrial character of the treatments, while the Scenario C adopts a community-based approach to the urban agriculture (De Zeeuw, 2015; Mougeot, 2000), involving in an active way the inhabitants (Viljoen et al., 2014, Caputo, 2012). It should be noted that the waste, sludge, from biogas production (Scenario B) is in part reused for compost soil improver production.

In the Scenario C, we adopt the hypothesis of implementing a local waste collection from the maintenance of the green area in order to implement an aerobic treatment plant located at the edge of the area devoted to the reuse of biowaste, green waste plus household organic waste (the whole biowaste). The compost produced would be used to feed the community's horticultural crops.

On the basis of the average figures by the literature, relating to the production of both compost and energy that can be activated by the treatment of organic waste, the estimate products for each scenario have been calculated. In particular, the reference has been made to: the figures from the study of the University of Padua (Baldan, 2015), the data

Table 3 Waste treatment options considered in the analysis

Waste treatment	u.m	Scenario A	Scenario B	Scenario C
Landfill	t (%)	14 (2%)	–	–
Anaerobic digestion	t (%)	347 (50%)	694 (100%)	–
Aerobic digestion	t (%)	333 (48%)	–	694 (100%)

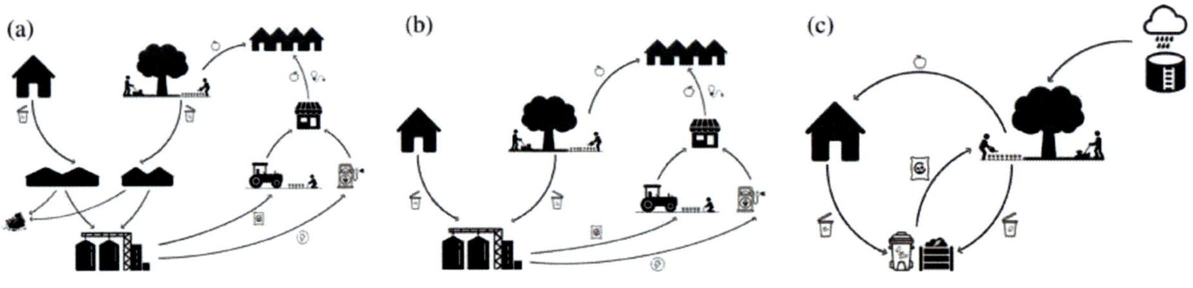

Fig. 3 Flowchart of the three scenarios analyzed: Scenario A (**a**), Scenario B (**b**), Scenario C (**c**)

provided by Italian biogas producer associations (Francescato, 2007) and the data provided by the municipal waste collection company of Turin (AMIAT) for the year 2015 regarding the biodegradable solid urban waste. For the production of compost, reference was made to the values presented by the study Sustainable Compost Application in Agriculture (ECN, 2008). To calculate the horticulture garden surface feeder by the compost, we adopted the index by ECN (2008).

The output of the three alternative scenarios has been estimated in terms of material obtained, biogas and compostable soil improver, energy from biogas, thermal and electrical kWh, and arable land supported by the use of compost.

5 Results and Perspectives

The evaluations carried out on the three scenarios have shown the achievable potential of each chain.

Within the common goal of limiting direct landfill material as much as possible, the analysis shows in principle the results of two different approaches. The analysis carried out allows a quantitative comparison only between the Scenarios A and B being not possible to directly measure against the Scenario C. Such a comparison would be too complex and not meaningful due to the wide range of the variables in different context. The quantitative analysis carried through the method described in the paragraph 4 allows the following considerations, although is mandatory to remind the need of reliable figures on average production of biogas as well as compost to allow such material flow analysis.

The Scenario A, “business as usual”, shows quite a high performance of the waste management system driven by the fulfillment of the EU Directives goals relating the urban waste sorting threshold and the minimization of landfilled or incinerated fraction. Nevertheless, the chain is quite long and complex and it is difficult to trace the wastes from the production to the treatment plant. The duty for the traceability is in charge to the firm managing the maintenance. The chain is fully top-down and there is no involvement of the inhabitants in the wastes production by the reuse of harvesting, mowing and pruning activities. In the hypothesis to consider in our accounting also the household biowaste, the citizen has only charges without any direct benefit, real, or perceived (Manfredi et al., 2011; Lang et al., 2006).

The output of the Scenario B is more relevant for the energy production and is based on the development of the biogas by waste industry. This sector has had in the past years a massive development in Italy and a technological shift. Nevertheless it represents a market driven solution pushed by tax tariffs and by rules. The production of energy is **quantity** relevant, and the technology is more and more efficient but not competitive with other forms of energy generation.

The Scenario C is an alternative bottom-up approach linking the waste by the greenery maintenance to the production of compost feeding the urban gardening. Above all if combined with the collection of household biowaste this process would allow a large provision of amendment and fertilizer. Due to the extension of green areas and the number of residents, such production would exceed the demand to satisfy the local needs. Nevertheless, such an approach, although the technological advances in community composting plants and quality of the compost, requires additional resources to be implemented and an awareness and participation of the community itself.

In both the improvement scenarios, B and C, we can measure the benefits in terms of reducing the use of fossil fuels, for energy production replacing natural gas with biogas or for the avoided production of synthetic fertilizers replaced by compost (Bordoni, 2009; Francescati et al., 2007; Piccini, 2007). The energy production by anaerobic digestion plants is significant but certainly not comparable to that obtained with traditional or renewable energy production plants. At the same time, the wide spreading of compost from organic waste is still not comparable to synthetic fertilizer, although the first one is at the center of continuous developments and improvements in the field of agriculture (Cofie et al. 2006) (Table 4).

Moreover, the balance between the three scenarios' evaluations from an economic and environmental point of view requires the comparison of the two methods with respect to the technologies and types of installations used, and it is possible to make a number of considerations about the length of the supply chain and its effects on the social issues. As far as the social impacts on the study area are concerned, the results of the evaluations carried out considerably differentiate the two improvement models (B and C). In the case of the anaerobic digestion chain, the current collection and sorting mechanisms do not change from the Scenario A, except in the potential shortening of the chain, while composting at local level, Scenario C, shows a strong potential for the production of soil improvers and fertilizer that can be reused on site by leveraging the availability of public and private green spaces.

The evaluation of the cultivable space potentially supported by the production of soil improver, therefore, seems to indicate a scenario of local development with the creation of opportunities for work and participatory economy (Lahec, 2019).

Even if a minority in urban context, community composting practices are wide spreading. On the other hand, the industrial sector of the production of installations for the decentralized production of quality compost is growing in quality and quantity. The link with the development of urban agriculture initiatives can lead to social benefits through the mechanisms of participation both in the collection of waste, and in the improvement of its quality, and in the development of cultivation for self-consumption.

In the case of the Le Vallette district, the potential for the production of soil improvers and fertilizer would exceed the needs of the area. This circumstance may suggest the adoption of mixed solutions between the improvement scenarios. Moreover, the strong dependence of both product quality and emissions on the technological performance of installations, both large scale and local, requires considerable attention to the engineering and management aspects of the digestion plants, both anaerobic, for the production of biogas, and aerobic, to obtain compost.

6 Conclusion

In regard to the methodological issues we can observe that a more granular evaluation of the material flows by technologies and tools like the GIS, big data sources or sensors appears to be useful to foster environmentally conscious choices. A neighborhood scale like in the case study can be adopted as a field of investigation and evaluation of alternative management scenarios. Making explicit the outputs of the scenarios, quantitative as well as qualitative, can help in the decision making in urban planning. The literature in urban metabolic studies underlines (Broto, 2012; Lehec, 2019; Bahers & Giacché, 2019; Passaro et al., 2015) the nonlinear character of the processes described and the complexity of urban systems requiring mixed strategies and responses to the inhabitants needs. Moreover, the availability from agricultural and industrial disciplines of data and information is needed to define indicators enforcing such analysis. This issue underlines the interdisciplinary nature of such investigations of the urban system.

Table 4 Outputs from green waste treatment

Generation from green waste		Scenario A	Scenario B	Scenario C
Total harvesting material		694 t	694 t	694 t
Landfill	–	14 t	–	–
Compost	0.30 t/t	100 t	70 t	208 t
Biogas (pruning/mowing)[a]	7.5/50 Nm^3/t	17.350 Nm^3	34.700 Nm^3	–
Electrical energy[b]	10 kWh/Nm^3	52.050 kWh	104.000 kWh	–
Thermal energy[b]	10 kWh/Nm^3	52.050 kWh	104.000 kWh	–
Cultivable area[c]	0.67 m^2/kg year	66.816 m^2	46.400 m^2	139.200 m^2

[a] CH4 part
[b] Production by a co-generation process. Data from ACEA, Pinerolo. Efficiency by CH4 production 0.3
[c] Cultivable area means the extension of agricultural greenery that can be fertilized and soiled by the production of compost. From the literature data hypothesized 1.5 kg/m^2

The data obtained shows that the share of waste material from mowing and pruning, especially in the case study presented, may represent a significant share of urban metabolism. From a circular economy point of view, now at the center of European research actions and policies, the analysis of urban organic waste treatment chains stimulates environmental, social, and economic considerations. The current measures, aimed at achieving the objectives of separate waste collection, do not yet make it possible to accurately measure the origin and quality of the raw material. The evaluations carried out highlight the potential of both improvement scenarios for the energy and agricultural valorization of waste, whose supply chains have seen a significant development of application models, legislation, and technology in recent years. In particular, scenarios mainly aimed at decentralized compost production allow to support large urban cultivable portions.

With regard to the analysis of environmental impacts, it is necessary to refer to a detailed analysis able to quantify, with respect to the specific plant adopted, the means used for transport and the characteristics of the materials delivered. In this way a life-cycle assessment (LCA) analysis (Lebersorger et al., 2011) would be necessary to assess the relative emissions and the quality of the material produced.

Moreover, the studies conducted (Blengini & Fanti, 2009) have shown a strong dependence, in terms of harmful emissions (GWP), on the efficiency of the equipment and the effectiveness of filtration devices. With the exception of incineration, which is not only environmentally critical but also unsuitable for this type of waste, the emission-related effects of both aerobic and anaerobic digestion are balanced by the production of energy and material that can be reused in crops.

The industrial biogas chain may require higher economic costs related to collection, storage, treatment, and transport than the short composting chain on a local scale. In addition, the use of large areas, as in the case study, horticultural crops, would reduce the burden of green maintenance, now borne by local government and condominiums.

A further element for the valorization of the green resource is the application of traceability technologies also in the part of the organic waste chain upstream of anaerobic treatment and composting, contributing to an evaluation of urban metabolism flows and the benefits achievable in terms of energy, materials, and food production. In other words, through the detection of green maintenance activities it is possible to both make evident the economic importance of waste and optimize its management. In this sense, and with the support of advanced technologies, even in scenarios of integration between industrial supply chains and local production, it is possible to think of a new form of self-sufficiency of some parts of the city.

The refunctionalization of urban greenery, from a simple and indistinct green area to an area destined for other uses, such as agriculture, seems feasible, especially in reality as in the case study. On a social and urbanistic level, these scenarios suggest new urban "landscapes" characterized by productive functions integrated with recreational and environmental mitigation functions in which the destination to forms of cultivation, supported by closed waste recovery cycles, can flourish and give a new meaning to areas currently underutilized and devoid of character.

As clearly stated by the literature in the field, the adoption of a model treatment of the greenery wastes, also in association with household food waste, is strongly dependent from the local conditions, in the supply side, the waste, as well as the demand side, the material to be converted in energy, through the biogas production installations, or in nutrients for crops, through the composting practices (Broto, 2012). Future developments in the research about UM at the neighborhood scale will require an interdisciplinary cooperation between architects and environmental engineers, botanist, and agronomist providing a sound basis for standards and planning at a territorial scale in a circular economy perspective.

Acknowledgements We thanks the Cooperatives P. G. Frassati and Cooperative Agriforest S.C. for the information and technical support provided during the research. Thanks to Davide Mainero, engineer at ACEA Pinerolo, for the information about the biogas production.

References

Allesch, A., & Brunner, P. H. (2015). Material flow analysis as a decision support tool for waste management: A literature review. *Journal of Industrial Ecology, 19*(5), 753–764

Baccini, P., & Bader, H. P. (1996). *Regionaler stoffhaushalt: erfassung, bewertung und steuerung*. (pp. 1–420). Spektrum Akademischer Verlag.

Bahers, J. B., & Giacchè, G. (2019). Towards a metabolic rift analysis: The case of urban agriculture and organic waste management in Rennes (France). *Geoforum, 98*, 97–107

Baldan, D. (2015). *Linee Guida per un'economia circolare nella manutenzione del verde.* LASA—Laboratorio di Analisi dei Sistemi Ambientali.

Blengini, G. A., & Fanti, M. (2009). *Life cycle assessment di scenari alternativi per il trattamento del FORSU.* Lecture at *XI Conferenza internazionale del compostaggio*, Rimini. https://www.compost.it/attachments/390_XI%20Conferenza%20Compostaggio%2009_M%20Fanti_G%20A%20Blengini.pdf

Bordoni, A., Romagnoli, E., Foppa Pedretti E., Toscano G., Rossini G., Cozzolino E. (2009). *La filiera del biogas. Aspetti salienti dello stato dell'arte e prospettive*, Regione Marche, consulted on 02.03.2019 at https://www.laboratoriobiomasse.it/media/docs/downloads/103-1.pdf

Broto, V. C., Allen, A., & Rapoport, E. (2012). Interdisciplinary perspectives on urban metabolism. *Journal of Industrial Ecology, 16*(6), 851–861

Caputo, S. (2012). The purpose of urban food production in developed countries. In A. Viljoen & J. Wiskerke (Eds.), *Sustainable food planning: Evolving theory and practice.* (pp. 259–270). Wageningen Academic Publishers.

Città Metropolitana di Torino. (2017). *Rapporto sullo stato del sistema di gestione dei rifiuti* (pp. 160). https://www.cittametropolitana.torino.it/cms/ambiente/rifiuti/osservatorio-rifiuti/rapporto-sistema-gestione-rifiuti/rapporto-sistema-gestione-rifiuti

Cofie, O., Bradford, A., & Drechsel, P. (2006). Recycling of urban organic waste for urban agriculture. In *Cities farming for the future: Urban agriculture for green and productive cities* (pp. 210–229).

Cox, J., Giorgi, S., Sharp, V., Strange, K., Wilson, D. C., & Blakey, N. (2010). Household waste prevention—a review of evidence. *Waste Management and Research, 28*(3), 193–219

Dijst, M., Worrel, E., Bocker, L., Brunner, P., Davoudi, S., Geertman, S., Harmsen, R., Helbich, M., Holtsag, A.A.M., Lenz, B., Lyons, G., Mokhtarian, P.L., Newman, P., Perrels, A., Ribeiro, A.P., Carreòn, J.R., Thomson, G., Urge-Vorsatz, D., & Zeyringer, M. (2018). *Exploring urban metabolism—Towards an interdisciplinary perspective. Resources, Conservation and Recycling* (Vol. 132, pp. 190–203).

Drechsel, P., Keraita, B., Cofie, O. O., & Nikiema, J. (2015). Productive and safe use of urban organic wastes and wastewater in urban food production systems in low-income countries. *Cities and Agriculture–Developing Resilient Urban Food Systems*, 162–191.

ECN—European Compost Network. (2008). Sustainable compost application in agricolture. *ECN-INFO 02/2010*, ITZ. https://www.compost.it/attachments/467_ECN_Info%20paper_02_2010_Sustainable_Use_of_Compost_in_Agriculture_LTZ-Project.pdf

EU. (2010). Comunicazione della commissione al consiglio e al parlamento europeo relativa alle prossime misure in materia di gestione dei rifiuti organici nell'Unione europea. *SEC, 2010*, 577

Francescati, W., & Antonini, E. (2007). *Energia elettrica e calore dal biogas*, AIEL. Associazione Italiana Energie Agroforestali, consulted on 12.04.2019 at https://www.ducabruzzi.gov.it/materialididattici/Tiso/energie%20rinnovabili/AIEL_BIOGAS.pdf

Guallart, V. (2014). *The self sufficient city*. Actar Publishers.

ISPRA. (2019). *Rapporto Rifiuti Urbani - Edizione 2018, Rapporti 297/2018*. ISPRA—Istituto Superiore per la Protezione e la Ricerca Ambientale. https://www.isprambiente.gov.it/it/archivio/eventi/2019/12/rapporto-rifiuti-urbani-edizione-2019

ISPRA. (2016). *Comitato per lo sviluppo del verde pubblico*. https://www.isprambiente.gov.it/files/comitato-verde-pubblico/Relazione2016_rev15.pdf

Kellett, R., Christen, A., Coops, N. C., van der Laan, M., Crawford, B., Tooke, T. R., & Olchovski, I. (2013). A systems approach to carbon cycling and emissions modeling at an urban neighborhood scale. *Landscape and Urban Planning, 110*, 48–58

Lang, D. J., Binder, C. R., Stauffacher, M., Ziegler, C., Schleiss, K., & Scholz, R. W. (2006). Material and money flows as a means for industry analysis of recycling schemes: A case study of regional bio-waste management. *Resources, Conservation and Recycling, 49*(2), 159–190

Lebersorger, S., & Beigl, P. (2011). Municipal solid waste generation in municipalities: Quantifying impacts of household structure, commercial waste and domestic fuel. *Waste Management, 31*(9–10), 1907–1915

Legge 10/2013. (2013). "*Norme per lo sviluppo degli spazi verdi urbani*". Linee guida per la gestione del verde urbano e prime indicazioni per una pianificazione sostenibile, Ministero dell'Ambiente e della Tutela del Territorio e del Mare, Comitato per lo sviluppo del verde pubblico. https://www.conaf.it/sites/default/files/Linee%20guida%20pubblicate%202017.pdf

Lehec, É. (2019). Vers un service composite de gestion du métabolisme urbain. Ce que compostage industriel et compostage en pied d'immeuble ont en partage. *Flux, 2*, 95–111

Liang, S., & Zhang, T. (2012). Comparing urban solid waste recycling from the viewpoint of urban metabolism based on physical input–output model: a case of Suzhou in China. *Waste Management, 32*(1), 220–225

Lo Sasso, M. (2011). Il progetto come prodotto di ricerca scientifica. *Techne, 2*, 78–85.

Manfredi, S., & Pant, R., (2011). *Supporting environmentally sound decisions for bio-waste management a practical guide to life cycle thinking (LCT) and life cycle assessment (LCA), joint research centre (JRC),* Institute for Environment and Sustainability (IES), Sustainability Assessment Unit., consultato il 05.04.2019 al https://eplca.jrc.ec.europa.eu/uploads/waste-Guidance-on-LCT-LCA-applied-to-BIO-WASTE-Management-Final-ONLINE.pdf, viewed 090419

Mougeot, L. J. (2000). Urban agriculture: Definition, presence, potentials and risks. *Growing Cities, Growing Food: Urban Agriculture on the Policy Agenda, 1*, 99

Oberlin, A. S., & Szántó, G. L. (2011). Community level composting in a developing country: case study of KIWODET, Tanzania. *Waste Management and Research, 29*(10), 1071–1077

Passaro, A., & Francese, D. (2015). Rigenerazione urbana e Bioregionalismo. *TECHNE: Journal of Technology for Architecture & Environment, 10*, 249–257.

Piccinini, S. (2007, May). Le tecnologie di produzione del biogas. In *Proceedings of the Seminar on IL Biogas: Modello di Calcolo a Supporto Della Fattibilità Tecnico-Economica (CRPA), Reggio Emilia, Italy* (Vol. 30).

Pincetl, S., Bunje, P., & Holmes, T. (2012). An expanded urban metabolism method: Toward a systems approach for assessing urban energy processes and causes. *Landscape and Urban Planning, 107*(3), 193–202

Saer, A., Lansing, S., Davitt, N. H., & Graves, R. E. (2013). Life cycle assessment of a food waste composting system: Environmental impact hotspots. *Journal of Cleaner Production, 52*, 234–244

Shahrokni, H., Lazarevic, D., & Brandt, N. (2015). Smart urban metabolism: Towards a real-time understanding of the energy and material flows of a city and its citizens. *Journal of Urban Technology, 22*(1), 65–86

Slater, R., Frederickson, J., Yoxon, M. (2010). Unlocking the potential of community composting: Full project report. *Report to the Department for Environment, Food and Rural Affairs (Defra)*, London.

Tong, J., Sun, X., Li, S., Qu, B., & Wan, L. (2018). Reutilization of green waste as compost for soil improvement in the afforested land of the Beijing plain. *Sustainability, 10*(7), 2376

Viljoen, A., & Bohn, K. (2014). *Second nature and urban agriculture: Designing productive cities*. Oxford Routledge

Viljoen, A., Schlesinger, J., Bohn, K., Drescher, A. (2014). Agriculture in urban design and spatial planning. In: *Cities and agriculture: Developing resilient urban food systems* (pp. 89–107).

Wolman, A. (1965). The metabolism of cities. *Scientific American, 213* (3), 179–190

De Zeeuw, H., & Drechsel, P. (2015). *Cities and Agriculture—Developing resilient urban food systems*. (p. 2015). Routledge.

Zhang, Y., Liu, H., & Chen, B. (2013). Comprehensive evaluation of the structural characteristics of an urban metabolic system: Model development and a case study of Beijing. *Ecological Modelling, 252*, 106–113

The Necessary Integration of Green Infrastructures in the Public Transport Infrastructure Design

Ferdinando Trapani, Roberta Carrara, Guglielmo Di Chiara, Lorenza Maria Ferrara, Massimiliano Giudice, and Gianluigi Pirrera

Abstract

The spatial planning of the new infrastructures for the urban mobility can provide the opportunity for the development of the urban green areas toward the condition of a *healthy city*. Elements such as architectural, heritage, and landscape centralities can lead to build a converging point between the two infrastructural systems of *green* and *mobility*. The development of the new urban railway lines in the city of Palermo (Italy) offers the chance to apply an integrated approach to the field. The local potentials of the biodiversity on one side and the fragmented landscape and environment systems on the other side build the framework for the development of a case study oriented to the enhancement of the people's health and of the environment's ecological functions. With the aim to produce a green–blue mobility infrastructure, the study takes as founding elements the many landscape-environmental characters of the territory like historic parks and gardens (Florence Chart), EC SPA, SCI and Natural urban reserves, and the existing water networks. In particular, the comparison between the green–blue linear networks of the territory before and after the modern urbanization has allowed to develop the structural map of "reconnections" as the base for the new infrastructure. Furthermore, taking into consideration the Italian regulation, the "monumental" trees have been adopted as urban pillars of the case study project for their environmental, cultural, and landscape value. The results of the study proved that the decision-making process for the "fast" mobility can positively interact with the "slow" mobility systems through the landscape system itself. In fact, the landscape, characterized by an intrinsic tendency to dynamics, can be interpreted as a "landscape in motion" connecting the existing urban tissue with cultivated, natural, and historic green areas. In conclusion, key aspects such as mobility, nature, culture, and social dimension have become milestones for the integrated spatial planning process.

Keywords

Green infrastructures • Healthy city • Smart city • Participatory planning • Urban landscape • Social green planning

1 Introduction and Methodology

Within the studies on the complexity of spacial structural phenomena oriented to implementing tactic and/or strategic policies and practices (i.e., mobility, green–blue infrastructures, health, smart grids, resilience, etc.), the first question that has to be stated is the type of the research approach that shall be adopted. The answer refers to the trans-disciplinarity as a reference model (Lang et al., 2012) (Fig. 1) converging on its broader meaning that takes into account the different approaches to the issue of complexity in the research fields. "Trans-disciplinarity is a reflexive, integrative method driven by scientific principles aiming at the solution or transition of societal problems and concurrently of related scientific problems by differentiating and integrating knowledge from various scientific and societal bodies of knowledge"(Lang et al., 2012, 26–27).

In the schematization of the different trans-disciplinary approaches to the complexity of the research, it emerges that there are at least three phases:

F. Trapani
Urban Planning, Università Degli Studi Di Palermo, Palermo, Italy

R. Carrara · L. M. Ferrara · M. Giudice
Palermo, Italy

G. Di Chiara (✉)
Politecnico Di Milano, Milan, Italy
e-mail: guglielmo.dichiara@gmail.com

G. Pirrera
AIPIN (Italian Association for Water and Soil Bio-Engineering), Palermo, Italy

F. Trapani et al. (eds.), *Advanced Studies in Efficient Environmental Design and City Planning*, Advances in Science, Technology & Innovation, https://doi.org/10.1007/978-3-030-65181-7_26

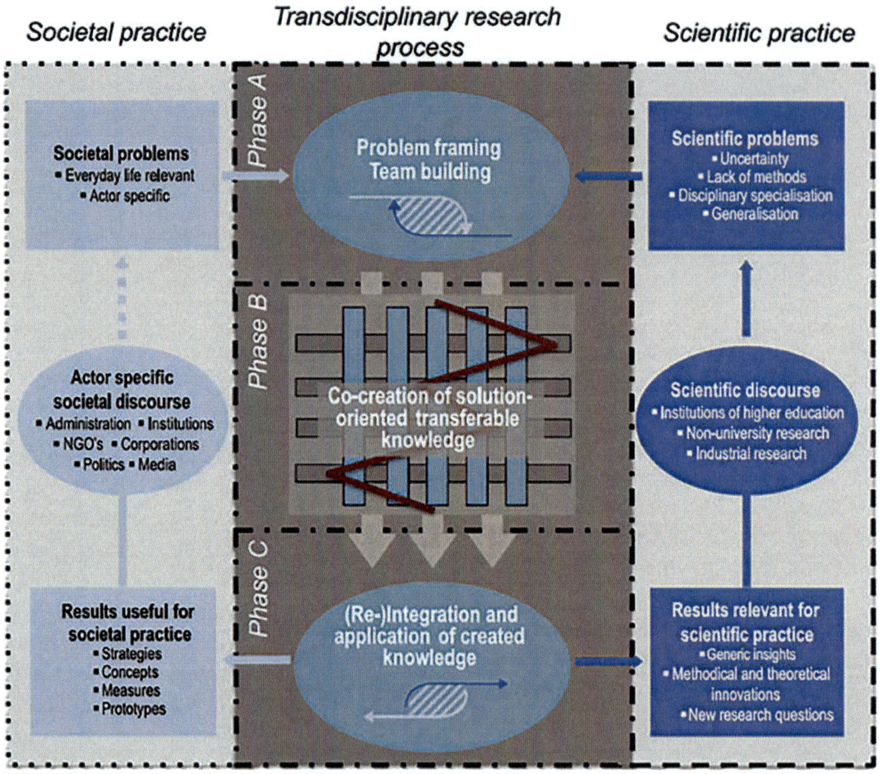

Fig. 1 Conceptual model of an ideal–typical trans-disciplinary research process (Lang et al., 2012, 28)

A. Problem Framing—Team Building;
B. Co-creation of solution-oriented and transferable knowledge through collaborative research;
C. (Re-) integrating and applying the co-created knowledge.

These three phases of the scientific process that use the research model with a trans-disciplinary approach constitute the operational background in which the dimensions of research converge at all levels and types of expertise (even non-academic) and of the social dimension from policy decision-makers to users.

The "trans-disciplinary approach" model can be applicable also in the case of the analysis of the policies on the sustainable mobility (Bergmann & Jahn, 2008) which is confirmed to be considered a way to capture the intertwining of technology, expertise, and social sectors supporting the policy decision-makers.

According to the results of the studies on the research methodology,[1] it emerges that, within the analysis of the decisional processes of government of the cities and territories that undertake sustainability objectives in mobility, the trans-disciplinary approach must deal on one side with the technical, technological, and analysis aspects of environmental impacts and on the other side with of the entire social and economic process.

The sectoral studies that research on the issues of the sustainable mobility and its energy systems with the aim to define dedicated policies do not give definitive conclusions. In fact, on one hand, it is certain that we must abandon the traction systems based on petrol and diesel. On the other hand, the importance of increasing the use of the electric traction systems is commonly accepted despite the unsolved problem of the batteries reuse. In this perspective, the only solution appears to be related to the hydrogen systems (EEA, 2016a, b).

In the same way, the research concerning the construction processes of the urban and regional green infrastructures (Bergmann & Jahn, 2008; Lennon, 2014) is currently considering the interdisciplinary approach to the urban and regional complexity at the center of the policy decision-making process. In this sense, in the planning processes, we need to find the support of the scientific research on the environmental dimension and/or the technical experience, considering at the same time the social and cultural dimension for the governance of the cities and the regions in a sustainable way.

[1]Comparing a large literature in the mobility and trans-disciplinary approaches (Banister & Marshall, 2007; Boscacci et al., 2001; Geels & Smit, 2000; Jenks & Dempsey, 2005; Morandi et al., 2013; Poli, 2011; Staricco, 2013).

According to Pareglio (2008, 5), the transport problems—considered together with the issue of permeability of soils and energy—due to the strong concentration of the infrastructural offer in the rubberized sector are caused by the "unstructured" urban systems of the Italian cities. This implies the need to possess: "a clear long-term vision, a strong and continuous political action, an effective coordination between instruments and between institutions, a real willingness to change rooted habits and a significant endowment of resources."

Another sector of analysis is represented by the studies on the *healthy city* that are the principal research bases for the value of every urban and design planning proposal (Ashton, 1992; de Leeuw & Simos, 2017) for every kind of city (Davies, 2015). Furthermore, public health is one of the UN Agenda 2030 goals (UN, 2015) and in particular among the design ratings of linear infrastructures (ISI, 2015).

The paper proposes a method of analysis of the urban sustainable mobility problems that can be developed as an approach based on the data that frame the theme of daily habits filtered through the qualities and structural characteristics of public space understood as a space of the city related/immersed in the space of human life (Gehl & Svarre, 2013) focusing on achieving the objectives set by the European Commission for the construction of the green infrastructures.

2 Literature on Green Infrastructures and Mobility

Taking into consideration the work of Rousse and Bunster-Ossa (2013), *infrastructure* can be defined as a "structured connection" or "network" necessary for a society, or to an economy to function, and it can be distinguished into *hard* and *soft* infrastructure. While hard (or gray) infrastructures refer to transportation and physical networks, soft (or green) infrastructures "encompass the institutional systems," that aim at meeting the needs of the community. Therefore, green infrastructures (GI) help to improve the quality of life of the community, encouraging growth for health and community cohesion, among others (Rouse and Bunster-Ossa, 2013).

On the other hand, within a social and spatial context of constant development of the urban territories—through gray infrastructures—ecological systems have been divided bringing to a weakening of the ecological system and producing *land patterns* based on the spatial configuration of the *land mosaic* (Forman, 2014). On this perspective, green infrastructures offer a relevant opportunity to reconnect the downgraded ecosystems improving social cohesion. Indeed, the definition of GI provided by the European Commission states a stronger connection between ecosystem services (ES)[2] and social benefits: "a strategically planned network of natural and semi-natural areas with other environmental features designed and managed to deliver a wide range of ecosystem services. It incorporates green spaces (or blue if aquatic ecosystems are concerned) and other physical features in terrestrial (including coastal) and marine areas. On land, GI are present in rural and urban settings. […] The aim of GI is to promote ecosystem health and resilience, contribute to biodiversity conservation and enhance ES.

The given definition undoubtedly refers to the strong and necessary relation between GI and ES. Indeed, the GI are conceived as planned networks which affect both rural and urban contexts, connecting the ecosystems. From now on, it is therefore necessary to refer to the GI as a tool to obtain ecosystem services itself.

Referring to the social aspects, two major focuses may be brought in developing green infrastructural systems: public health and social participation. These two elements play a key-role in enhancing the positive impact of GI on the environment. Besides, public health can bring to a better quality of life. In this sense, the "dodecalogue of the bio-engineering for the cities" (Scalora & Pirrera, 2016) can help in supporting the relation between GI and social life by integrating in the GI planning the strategy stated by the twelve points for the biocity (Scalora & Pirrera, 2016). On the other hand, social participation can significantly contribute to the effectiveness of the GI strategies with some points of uncertainty. In fact, social participation may bring to a better quality the urban planning and the urban enhancement projects, considered in single case studies or in broader projects. The successful outcome in the process of the participatory design depends on some critical aspects: spontaneity in the participation, sharing of the process and sense of belonging (Scalora & Pirrera, 2016).

Facing climate change, GI can contribute to fight its negative impact on the urban environment when connected to their ability to improve the conditions of the urban climate and to significantly influence the urban hydrology. Indeed, green–blue infrastructures help in adapting urbanized territories to climate, moderating the expected increases in extreme precipitations or temperatures. By reducing the soil imperviousness, GI can substantially bring advantages in the storm-water management, lowering the sewer overflow, helping water infiltration and limiting water surface run-off, reducing the urban heat island effect on urban temperatures, etc. (Foster et al., 2011).

Considered that, the present condition of the urban rivers in Italy is highly critical as public policies (and private

[2]Ecosystem services are services provided by nature, such as water regulation, that benefit the environment and humans" (European Commission, 2013).

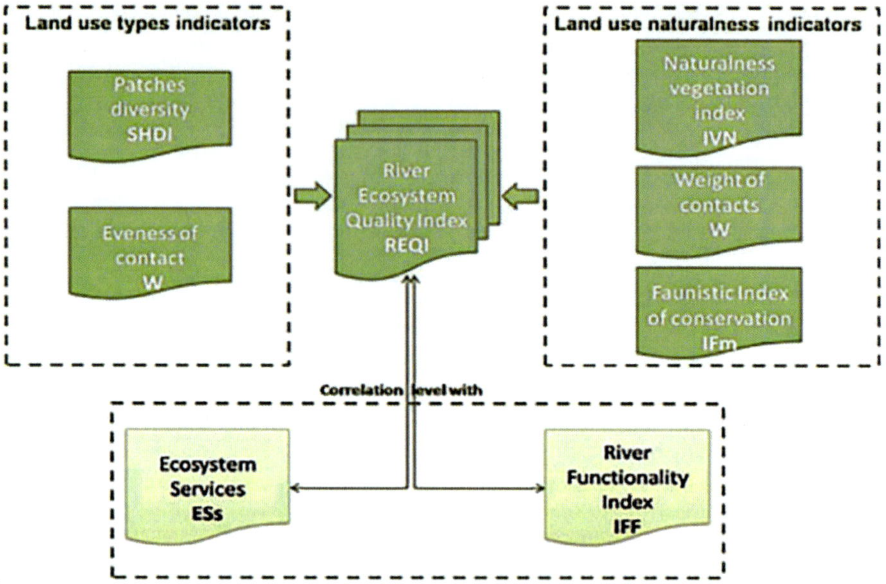

Fig. 2 Connection between water quality and ecosystemic services (Santolini et al., 2015)

initiatives)—through gray engineering projects—have been downgrading riverbeds and their levees, exposing to risk public health and security. In this direction, the 2000/60 CE Directive for Waters has enhanced the principle of hydrology remediation through the benefits of bioengineering actions. A method of assessing the connection between an index of water quality river (REQI), two land-use indicators (naturalness and types) and ESs has been developed by Santolini et al. (2015) (Fig. 2).

Pollution, weeds, hydraulic regulation, waterproofing thus determine the need to support the necessary ecosystemic functions that must guide the planning process to counter the net loss of the ecosystemic services in order to cope with the objectives of the ecosystemic quality and the natural capital of the river.

In conclusion, the single approach to the infrastructure design based on hard-gray engineering solutions can no longer exist by itself as "this approach alone can no longer provide the climate resilience" (Browder et al., 2019) necessary to cope with the new challenges imposed by the climate change.

3 Methodology

The methodology developed to elaborate the research involves tools which are able to refer to the different scales of complexity of the design process. The analysis of the spacial system has resulted to be fundamental in outlining the main features of the context, giving relevant hints for a strong design of the green–blue infrastructures. On the other side, the focus of the research can be oriented toward the historical and cultural dimension: elements like well-known ancient villas, popular parks, and famous trees are some of the tools taken into the study to generate social value.

At the same time, the territory has been analyzed at the different scales: the major environmental systems, the urban green spaces, the water network, and the monumental trees have been the founding elements to develop the case study proposal.

The framing of the current law system and guidelines is another analysis tool that has contributed significantly to the development of the research, as soon as the chance of the integrated infrastructures can bring effectiveness to the principles contained in the current environment–social regulations.

The interest for the international regulations and the interest for the existing green–blue infrastructures have lead to consider *water* and *green quality* as founding elements for the *public health (*which are three of the Agenda 2030 goals (UN, 2015) (Fig. 3). In fact, the improvement of the quality of water and of the greenery with benefit for the environment and the animals [in line with the concept of "One Health" as defined by the One Health Initiative Task Force (OHITF)] can give strength to the "the collaborative efforts of multiple disciplines working locally, nationally, and globally, to attain optimal health for people, animals and our environment."

The multidisciplinary approach has been a guiding tool that leaded throughout the study. In particular, for transport infrastructures, multidisciplinarity in the design and planning phase is expected to be applied through the Envision Rating (ISI, 2015) (Fig. 4) in which five categories of values must be taken into consideration, including public health (i.e.,

Fig. 3 Sustainable development goals, SDGs (UN, 2015)

better quality of life). As a result, a structural map has been developed as a base for the proposals set out in the master plan.

4 The Case Study: Palermo, Italy

4.1 The Outer Belt

Referring to the case study, the city of Palermo, in terms of landscape and naturalistic quality of the city, two fundamental reference systems are evident: the sea and a belt of mountains. The mountains, since ancient times, enclosed a plain that earned the name of "Conca d'Oro" (golden valley), a naturally fertile territory which saw a significant development through the wide agricultural system introduced by the Arabs. During the Norman domination, Palermo acquired the reputation as a "city full of gardens surrounded by a larger garden" (De Seta & Di Mauro, 1980) and as an "area of an ancient and almost mythical predominance of the trees" (Bevilacqua, 1988). Today, we can still identify this belt, naturally generated by the orographic conformation of the territory. From a naturalistic point of view, Palermo contains several valuable areas that are part of the Natura 2000 network, as well as geosites and important bird areas (IBAs). This first screening draws up a charter of the Environmental Restrictions (Fig. 5), where the so-called nodes of the Sicilian Ecological Network are indicated. They are present in the area of Palermo with Special Areas of Conservation (SAC), Sites of Community Interest (SCI), and Special Protection Areas (SPA), as well as the three "suburban" natural reserves of Palermo (Isola delle Femmine, Capo Gallo and Mount Pellegrino). The RES represents in fact a first attempt for the enhancement of the connections of the areas of greater naturalistic value, which however has not found in the years an adequate development to face the problems imposed by the climate change.

4.2 Historic Green in the City: Gardens and Monumental Trees

Once the external belt has been outlined, the analysis focused on the historical and cultural dimension of the existing urban green areas, useful to define the inner belt (Fig. 6). These are divided into two fundamental elements: historic gardens and monumental trees.[3] The importance of the historic gardens is rooted in the Charter of Florence, drafted by the International Committee of Historic Gardens ICOMOS-IFLA in 1981.

[3]According to the Italian regulation, a monumental tree (as a single or as a group of trees) can be so defined when it is:

(a) a tall trunk tree isolated or as part of a natural or artificial forest formation wherever located; a typical secular tree which can be considered as rare example of majesty and longevity by age or size; a tree of particular naturalistic value due to botanical rarity and peculiarity of the species; a tree that has relevance from the historical, cultural, documentary, or local traditions point of view or which bear a precise reference to events or memories.

(b) a row of trees of particular landscape, monumental, historical, and cultural value, including those inserted in urban centers;

(c) a tall tree inserted in particular architectural complexes of historical and cultural importance, such as villas, monasteries, churches, botanical gardens, and private historical residences".

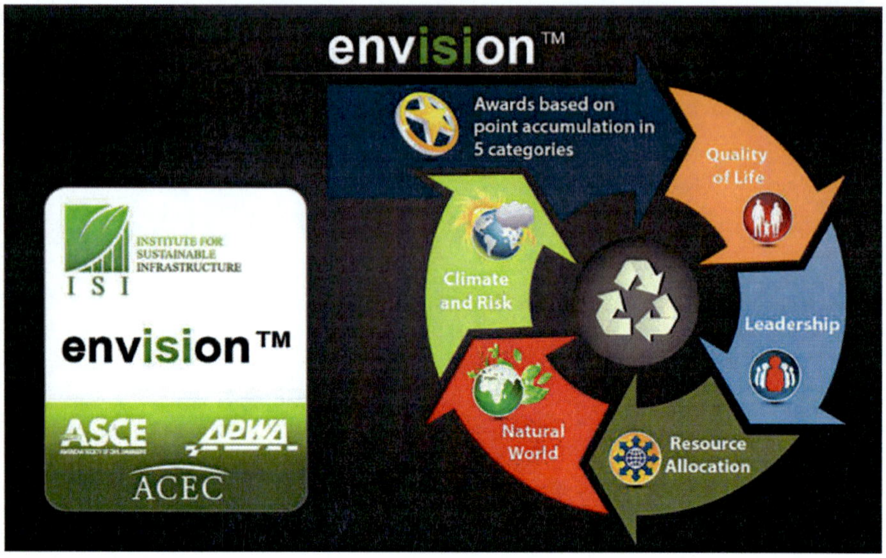

Fig. 4 Envision Rating (ISI, 2015)

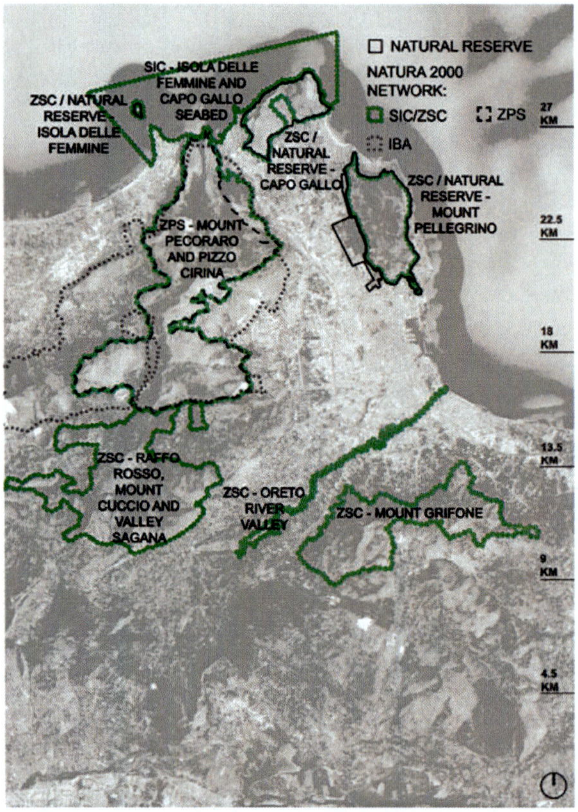

Fig. 5 Charter of the environmental restrictions

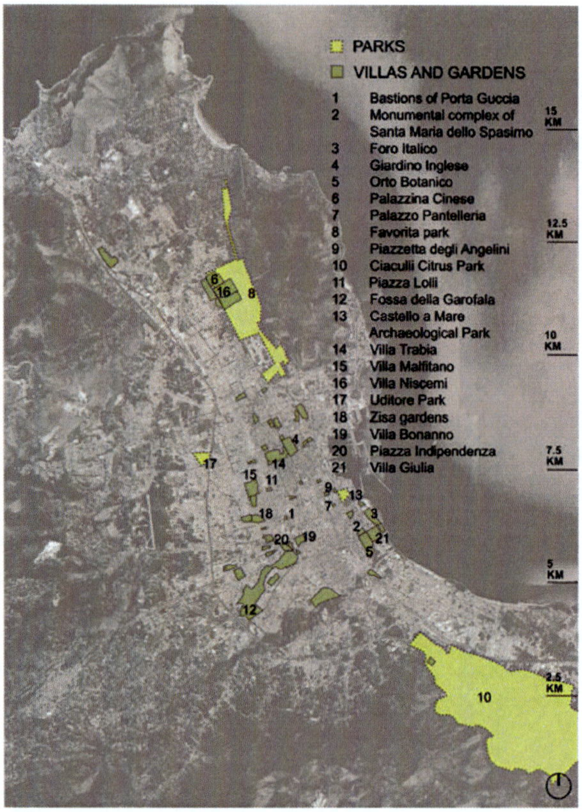

Fig. 6 Map of the historic green

The historic garden as a material artifact is a work of art, and as such, it is a cultural asset, an architectural and environmental resource, and heritage of the entire community that benefits from it. In the context of gardens and historic parks, private villas are also taken into consideration, as they delineate human cultural resilience.

This system generates an inner ring that follows the spatial axis of *Monte Grifone / Monte Pellegrino*, marked by the succession of the villas of *Piana dei Colli*, the urban parks of Villa Trabia and Giardino Inglese, reaching the agricultural area of *Parco di Ciaculli* further east, and then heading toward the great vegetation patch of the Botanical Garden and *Villa Giulia*. From there, it settles on the coast by the vast green area of the urban park of the Foro Italico and, passing over the archaeological park of the *Castello al Mare*, follows the rows of the port area and the large patches of vegetation intercepted toward *Monte Pellegrino*, thus generating a closed ring.

The monumental trees are the second element of the historic green system (Fig. 7), the true patriarchs and witnesses of the history of the city, as some of them exist before its foundation. Some are true and proper punctual points, which by their own delineate a precise historical and cultural meaning, while others acquire importance as being part of an identity system, for the ecological value they possess, for their size or for their botanical rarity.

In Palermo, there are 38 recognized monumental trees (DM n.757 of 04/19/2019), including "the tree of the judge Falcone" (*Ficus macrophylla* Desf. Ex Pers.), a symbol of one of the strongholds in the battle against the Mafia, and therefore important for the community. Among others, the "Cypress of San Benedetto" (*Cupressus sempervirens* L.), in the highest part of the Convent of Santa Maria di Gesù on Mount Grifone, presents a high religious value. In fact, according to the tradition, this tree was originated from the miraculous rooting of the "stick" which San Benedetto il Moro, the saint protector of Palermo together with Santa Rosalia, used to carry. Others are the millennial olive tree (*Olea europaea* L.) of the Favorita park, nicknamed "The Patriarch," or the *Ficus macrophylla* of Piazza Marina, recognized as the largest tree of his kind in Europe.

As a conclusion, the green and blue belts and corridors have been based on the system of historic green, with monumental trees as structural pillars, to reshape cultural resilience for both humans and vegetation.

4.3 The Historical Water Network

The series of hydraulic interventions that have taken place over time have canceled the existing *water network* (Reticolo Idrico Minore, RIM). The first step was therefore to study the disappeared water network through a historical analysis, to identify the current wetlands (those that today we would call biolakes), and then the current superficial ones, through the flows generated by the level curves. The next step has been to design the future one, enhancing the identified wetlands that could be used to purify water in areas required for temporary flooding.

The hydrography of the "Conca d'Oro" in Palermo has been modified over the course of history due to the

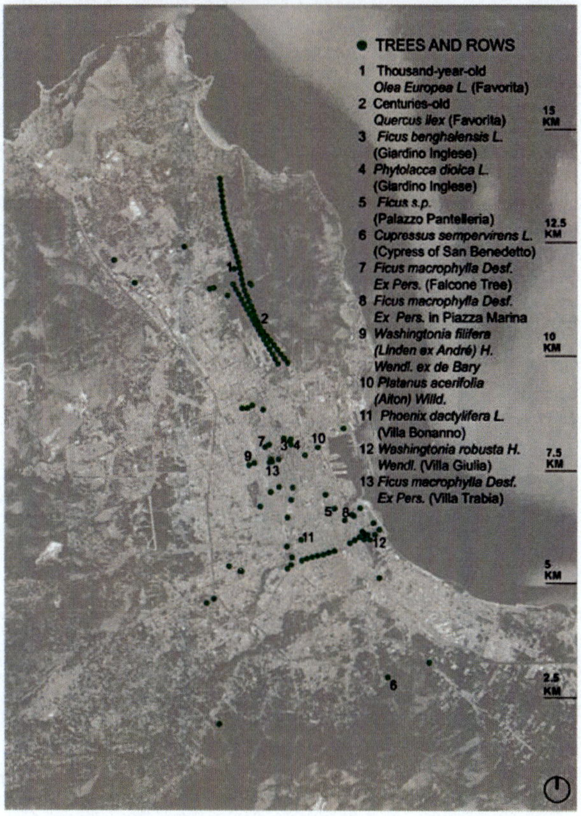

Fig. 7 Map of the monumental trees

remediation dictated by urbanization. With regard to the blue corridors, therefore, it was opportune to understand the history and the transformations, retracing the historical water network and the level curves dating back to the Arab era (Fig. 8). According to Todaro (2006), the first urban settlement was built between two rivers that today are no longer visible, the Kemonia and the Papireto, and it was given by Greeks the name of *Panormos*, "all port," while the river Oreto (the only river that still flows on the surface until today) was situated outside the urban core.

There were also many seasonal torrents, which contributed to create swamps or other water-rich areas, such as the northern territory between San Lorenzo and Mondello, one of the areas of greater concentration of the *Qanats*, hydraulic engineering works carried out during the Arab era in the subsoil of the city, to bring water to the surface by intercepting the natural groundwater (Todaro, 1988). Moreover, there were and still are today gorges called "Valloni," a sort of bad weather streams that channel the alluvial slopes of the mountains.

Over the centuries, there have been several witnessed floods that led to a series of interventions on the urban water network, with the aim of draining and making urban areas healthy. One of these involved the deviation of the urban

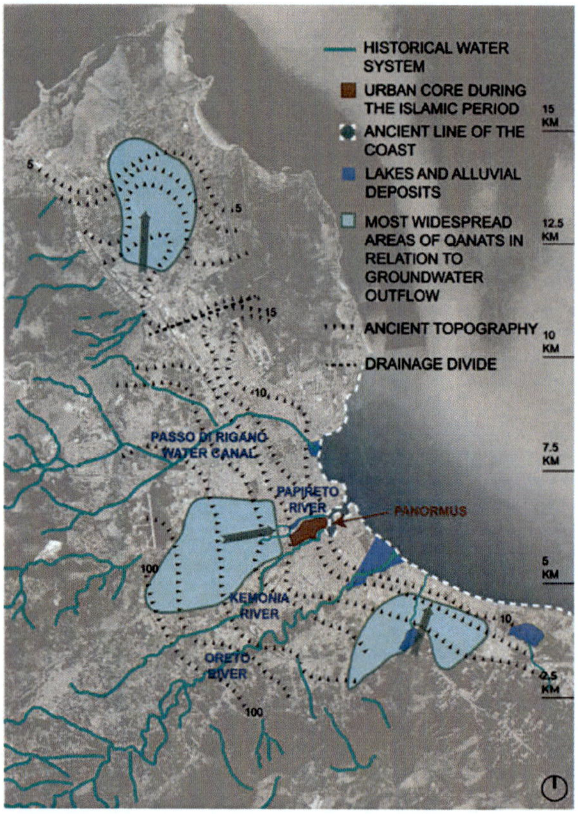

Fig. 8 Reconstruction of the historical water network

Fig. 9 Reconstruction of the flooding in 1931

river Kemonia, which was largely diverted toward the Oreto, and in the burial of the Papireto, which was channeled underground and led up to the Cala (Fig. 9). Between 21 and February 23, 1931, a storm broke out with an extent that no longer occurred, at least until today, caused by 50 incessant hours of rain, in addition to a strong north wind, which caused the destruction of fields and buildings. The main rivers flowed again inside the original bed: the Oreto river overflowed, once again passing under the historic Admiral's Bridge (Ponte dell'Ammiraglio) and producing a flooding of the sea; the underground torrent Cannizzaro—which was the river Kemonia after being buried—began to flow again between the urban streets, until it reached the area of Papireto, equally flooded due to the morphological depression of the bed of the homonymous river (AII, 1987).

The interventions on the water network, dated 1987 (Fig. 10), defined a new main position of the blue corridors within the urban area, determining the design of new channels, whose areas that can still be flooded.

In conclusion, the result of the analysis of the water networks has been useful to plan a renaturalization of the river channels.

4.4 The Water Network Today

Figure 11 outlines the current water network within the basin boundary, overlayed on the risks outlined by the *Piano di Assetto idrogeologico* (PAI) which, despite the floods of 1931, still includes urban flooding areas, such as those of the two buried rivers Kemonia and Papireto, which in the blue infrastructure leave traces of paleo-river bed in the current Cassarà Park / *Fossa della Garofala* and *Danisinni*.

These axes, in the context of the interventions for the enhancement of the Natural Capital, would not only be ideal, but real, since even the *Piano Generale Rischio Alluvioni Regionale* (General Plan for Flood Risk Management) imposes new strategies for non-structural interventions, such as urban de-paving.

4.5 The Structural Map

The overlap of the elements of the "green system" and those of the "blue system" generated a structural map (Fig. 12) in which we deduced all the elements to consider for the

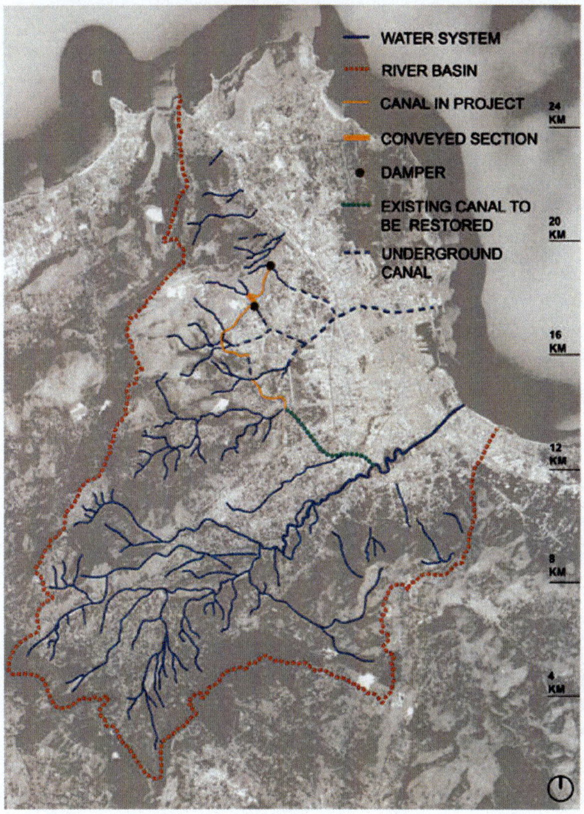

Fig. 10 The water network in 1987

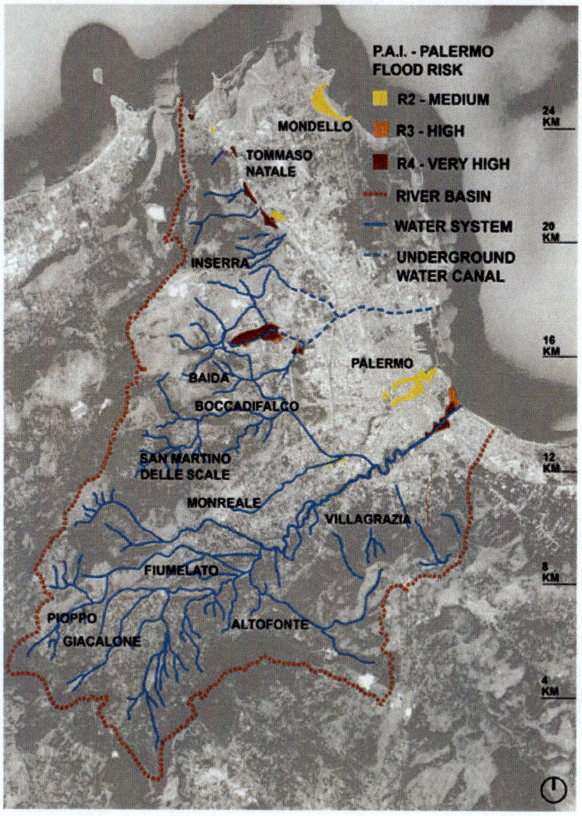

Fig. 11 PAI and water network today

reconnection of the ecological systems, on a macrosystemic and local scale, with the infrastructure system.

4.6 Green Policies in Palermo: Success and Failures

Urban environmental policies are always determined by each specific experience in the development of the planning process. This takes on special meanings in the city of Palermo, whose territory is almost completely man-made.

A huge plain open to the sea and surrounded by a mountainous arch that over the centuries developed in the so-called Conca D'Oro—a green macrosystem healthily influencing the climate and contributing to the city's landscape, cultural, urban, and economic splendor. Its waterways, reaching the sea from the nearby mountains, dictated the founding principles of Palermo, that until the sixteenth century shows a development determined by the main roads, running almost parallel to the rivers Papireto and Kemonia, in a west–east direction, while the secondary, transversal roads, served as a link between the first and the watercourses.

Starting from the second half of the sixteenth century, a period began for the city of Palermo in which relations with the hinterland would increase more and more. Thus, an expansion of the city took off in the north–south direction, which over the years will extend well beyond the city walls, culminating in the opening of Via Maqueda.

Nonetheless, the demographic growth of the city up to the late eighteenth century did not go hand in hand with an adequate geographical expansion—so much so that, starting from the second half of the nineteenth century, it was deemed urgent an urban and environmental intervention—which indeed started with the Giarrusso Plan (1885 Piano Regolatore Generale) and, therefore with the opening of a second road axis along the north–south direction: Via Roma. These interventions—also dictated by the new demand for "civic beauty" and, at the same time, the need of wealthy classes to show off—gave the chance to improve the hygienic-sanitary conditions of the city center by reducing the density of the built surface and, somehow, starting the policies of "gutting," then advocated by many.

During the first half of the twentieth century, the will to confront the urban-environmental and hygienic-sanitary issues gripping the city finally led to the competition

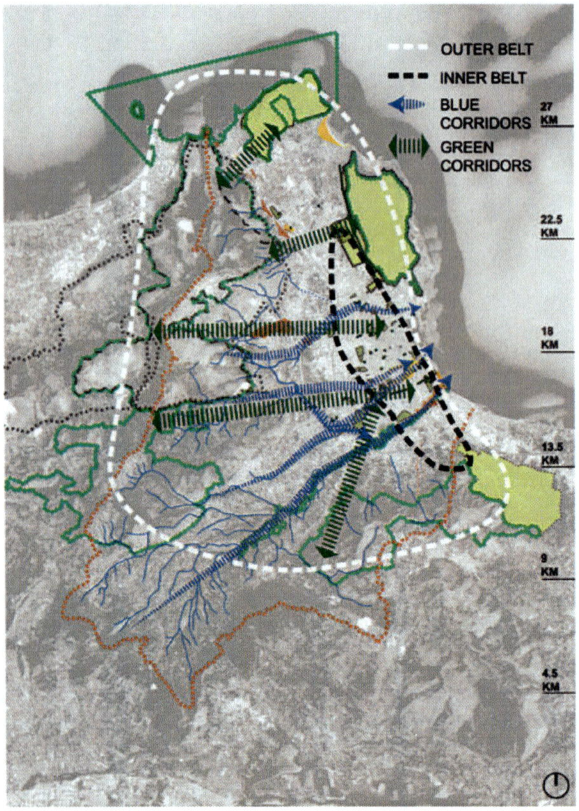

Fig. 12 Structural map

announcement for the 1939 Piano Regolatore Generale (PRG), with two main issues to be addressed: first, the demographic "thinning" of the ancient city and the imperative to adapt to the new urban needs, that meant—besides other interventions—the opening of further roads in the north–south direction; second, the location for the new districts for the inhabitants "cleared" from the old city center, as well as for those needed to satisfy the demographic increase unjustifiably assumed by local politicians.

Among the joint winners of the competition for the new plan, the group headed by Professor Edoardo Caracciolo did look ahead of the times and, among other things, seemed to call for special attention to the environmental aspects and to the question of *who* would be ultimately allowed to put his … "*hands on the city.*"

In the Caracciolo's group project, the question of the old city center is framed within a deep-seated respect for pre-existing buildings, providing for an urban and environmental enhancement through a reasonable, measured "thinning"—thus avoiding further disruptions of the kind started with the Giarrusso Plan and its variants—and the opening of two new streets in a north–south direction through which, differently from what had happened with Via Roma and Via Maqueda, it was intended to reinforce and respect the pre-established topographical trend. Here, Caracciolo's predictions show his profound knowledge of the city. The two new roads would have had to satisfy the ever-increasing needs of a city that had by now opted to extend in the south-north direction—they had become necessary and could not be postponed, given the future envisaged for the sixteenth-century Via Maqueda axis: minimal interventions, "very delicate touch-ups" and its pedestrianization. Caracciolo's vision for Via Maqueda seems now vindicated, despite the vehicular flows are enormously increased and the two new "streets" have not been realized planned east of Via Roma and west of Via Maqueda, the latter now to be considered completely anachronistic.

The attitude of the group led by Edoardo Caracciolo toward the second major issue to be addressed through the 1939 PRG (General Master Plan)—namely the social question and the areas of expansion for the new residences—favored development toward the south and southeast outskirts of the city, even moving beyond the Oreto river for working-class dwellings, while to the north, the residences for the wealthy were expected to run parallel to Via Libertà, where the aristocracy and the upper middle-class had already settled with several prestigious and refined villas surrounded by gardens and extremely well-kept parks.

The Commission of experts drafting the Notice of Competition (which composition inappropriately coincided with the adjudication panel) foresaw an unrealistic demographic increase, to which even the winners had to somehow follow up. The new areas of expansion for residential construction were expected next to the then-urban area, paying great attention to historical gardens and parks and almost eliminating the forecast of residual, easily attackable green areas. Thus, even the new arterial roads of penetration to the city became green corridors of connection between the gardens, parks, and the green macrosystem of the Favorita and Monte Pellegrino, from the Piana dei Colli and from the mountainous arc that, interrupted by the two green systems of Tommaso Natale and the Valle dell'Oreto, from Monte Gallo arrives at Monte Grifone and beyond with Favara the citrus groves of Ciaculli.

The Second World War, primarily, and the poorly managed autonomy of the Sicilian Region, during the post-war years, did the rest. The war itself forced the abandonment of the PRG in favor of a Reconstruction Plan (1946) which—although Caracciolo did collaborate to it—had to give way to the obvious post-war priorities and, again, to as many as five variants to the plan that finally distorted it; the second determined an unpredictable migratory phenomenon toward the capital (between 1951 and 1955 Palermo rose from 488,000 to 555,000 inhabitants) which aggravated housing needs—still today a social emergency and a "source" for the underworld and the (bad) politics—and forces new rules.

In the 1950s, once again Edoardo Caracciolo, with few others, did invoke a new plan and new rules to limit the many urban irregularities, perpetrated not only by private individuals. These were the years in which, rather than starting a process of "thinning out" the old center, the voids were used for the construction of new public structures, and, again, social housing was built in areas distant from the urban center. The latter—maybe as social housing was then thought as in itself "above suspicion"—ended up determining an inordinate increase in the land rents of the territory between them.

Now the premises for the total failure of urban and environmental policies were actually on the table—on the one hand the gradual abandonment of the old city center by its residents; on the other the fragmentation of the aforementioned macrosystem of greenery, that Caracciolo had tried to preserve in all possible ways. For the city of Palermo, the preconditions for a structural crisis, both ethical and moral, political and socio-economic, were defined during the 1970s and 1980s decades, the very years that in the narratives of certain social groups supporting those developments still feature as the Golden Years of growth and well-being for the city. They actually brought about a massive identity crisis which, even today, few are prepared to challenge. A whole political–social system, often colluding with unhealthy powers, no longer aimed to plan for the future, but rather to "spend and realize" (and more often than not, to waste) in the shortest possible time, regardless of the real needs and the very "measures" of the city.

In the second half of the 1950s, pressures for a new city-plan found their way to the Prefectural Commissioner Giuseppe Salerno, so that in 1956 work began on the new PRG through the appointment of a Draft Committee—in which sat once again Edoardo Caracciolo—chaired by the director of the city's Technical Office V. Nicoletti, with the collaboration of professionals and connoisseurs of the city, plus other specialist consultants.

The new PRG led a troubled life, due to a diabolical bureaucratic process which increased is vulnerability, registering about 3000 appeals (alas, not only by private individuals and real estate companies or by groups directly linked to the shady politics of Lima and Ciancimino), mostly upturned, which gave rise to hundreds of *variazioni* and to the withdrawal of the entire Draft Committee! The PRG (or general master plan), emptied in content, was approved in 1962 once completely distorted. Many of the prescriptions and constraints aimed at the preservation of the green macrosystem of the "Conca D'Oro" and of its mending with the gardens and the historical parks incorporated by the urbanization were removed or sidelined. The attack on Palermo's "green system"—praised throughout the centuries by illustrious travelers, and a natural source of well-being for the city—was justified by the loss of its economic functions after all, only 10% of the population now lived on agriculture. From this point of view, the roads of penetration, which in the 1939 and 1946 plans should have served the function of mending the greenery, became mere road axes of urban connection, often over-sized, leading to a dangerous fragmentation of the green, now weakened and made even more vulnerable.

A key factor in this regulatory plan process—not so much providing a rationale for its misadventures, as rather obscuring and mystifying its actual trajectory—was the economic "*Boom*" of 1950s–1960s Italy.

These are the years of the second post-war period, and the country, on the one hand, is affected by an exciting infrastructural growth and by the great achievements of Italian civil engineering—to which key manufacturing sectors such as automotive industry contribute, to some extent indeed governing the whole process—culminating in the construction of the "Autostrada del Sole"; on the other, it is profoundly affected by the new cultural phenomena of Rai (i.e., State Television) and film industry. To these two novelties, uniting and "remaking" Italy, Sicily seemed to look passively; while Palermo—even worse—blatantly misread them.

The private car, elevated to status-symbol throughout Italy, for Sicilians becomes at the same time the only means to go to work, in a region with an unbridgeable public transport and infrastructural gap. The city of Palermo will to some extent be able to do even worse than that and, in place of offering adequate services of local public transport (LPT) to accompany the growth of the city, seemed to actually foster a whole process of municipal disengagement as to regards LPT (See Lega Ambiente's publication *Pendolaria*).

The period between twentieth and twenty-first century is for Italy one of overall crisis as regards infrastructure and urban planning-Palermo then starts the studies for the General Variation to the 1962 PRG while, with regard to environmental and LPT issues, politicians quarrel (or pretend to!) and actively prevent those citizens willing to participate in the debate to meaningfully do as much.

Over the following years, a decent metropolitan network —one capable of adequately serving the development direction of the city and of lightening surface flows—should have been put in place, and was not. These are the years in which mobility policies that conflict with specialist planning are adopted, to such an extent that the sixteenth-century axis of Via Maqueda, for which Caracciolo already foresaw pedestrianization in 1939 (but not before, as already mentioned, the opening of two new parallel roads, one to the west and one to the east), today it is pedestrianized, thus contradicting the relevant General Plan (PGTU). Today sector plans seem to be "tailor-made" to justify aprioristic choices (e.g., the tramlines and the PUMS). Policies seem strictly dictated by cash requirements, rather than by real environmental options (see LTZ, Blue Zones …); years in which the demographic increase forecast is highly unclear, so that published studies are often conflicting with the official data proposed by local institutions.

Such a moral climate, which has lasted for over 60 years, would be capable to suffocate any breath of hope; and yet a new desire for redemption seems to be now in place—a sort of new "falling in love" with one's own city.

5 Proposals

The development of the urban mobility infrastructures offers the opportunity for a better implementation of the planning strategies for the urban green. The policies for the improvement of the mobility system can reorganize the cities integrating new green–blue networks bringing resilience through environmental mitigation and compensation, in connection with the different cultural layers of a city. Indeed, a discussion on three different levels is necessary: architecture, city monuments, and landscape centralities; greater green systems and green corridors; greater urban attractors at the urban and regional scale.

In the case of the Italian law, municipalities themselves have to implement urban policies toward sustainable solutions for the urban green at the different scales through design strategies such as green belts and the enhancement of the monumental trees according to the Italian definition (art. 7 law 10/2013). With regard to the green strategy within a mobility infrastructure plan, designers and policymakers have to take into account the existing green, both public and private. Therefore, green–blue belts can be developed starting from the existing green areas—public parks, private historic gardens, urban ecological patches (Florence Chart 1981), SPA and SCI (European Commission, Habitats Directive 92/43/EEC), protected urban green areas and monumental trees.

The tools and the means for reaching strong and well-developed green–blue infrastructures integrated in the mobility system are several, and only an integrated approach to the issue can bring to successful solutions that enhance the natural capital of the city, helping social health and restoring the urban environmental systems. In this sense, the monumental trees play a key-role in strengthening the green networks within the current fragile urban environment. They may be ideally considered as cornerstones in the green infrastructures, due to their multiple-level values. In fact, a monumental tree present values not only on their physical–biological aspect (longevity, height, *habitus*, rarity of the botanical specie, etc.), but also on the socio-cultural side relating to the context where it grows. In Italy, the indexing of the monumental trees is regulated by the Law n. 10/2013, while the seven categories useful to define a *monumental* tree have been stated in the ministerial order 23/10/2014.

Another strong focus in the combination of the mobility system and the green–blue infrastructures is the general idea to consider landscape as the dynamic feature of the urban environment. The "landscape in motion" reconnects and enhances the existing green spaces. In relation to the mobility system, the landscape assumes linear features (green ways) made possible by the systematic reconnection of the urban green elements such as urban agricultural spaces, historic private and public gardens and the monumental trees. The linear green infrastructures can be in this sense considered as a proactive entity, like a "gardener in motion" which selects and eradicates the existing invasive species found along his path (*Ailantus altissima, Pennisetum setaceum, Robynia pseudoacacia,* etc.) replacing them with the indigenous ones and with those species which have bioremediation abilities (as an advantage to rainwater–soil and air pollution). Soils can particularly benefit from the plants—and water biorestoration within the promotion of a virtuous system which is based on the circular economy process of recycling vegetation wastes coming from urban and agriculture pruning, beached *posidonia* seaweed, digging soil, etc. They can be transformed in techno-soils conceived and used as proto-soils for the urban green (Pirrera & Ferrara, 2018).

The planning process of the ecological networks—conceived in this way—has to be sided by the social participation activities *ex-ante* (*débat public*) and during the design phase (social green planning).

Considering this framework, the project of green–blue infrastructures for the urban area of Palermo offered the chance to test and develop the different guidelines that literature and policies have been tracing, creating a new experimental case study. On the occasion of the new tram lines planning, green–blue infrastructures can take over a relevant role in enhancing the existing urban green spaces at the metropolitan and district scale, bringing benefits at the same time to the public health and to the quality of life. In particular, the strategic guidelines for the integrated green and mobility plan for Palermo include: structural green design elements such as a main green axis (green sword), an inner and an outer green belt and a secondary system of blue corridors. The "green sword" is the founding element of the green plan: it is based on the main urban axis of Via della Libertà (boulevard coming from the nineteenth-century expansion), which crosses two of the most consolidated public green spaces (Giardino Inglese, Villa Trabia), running north–south and reaching the private green spaces of the baroque villas of the Piana dei Colli, to the north of the city. The name "green sword" comes from the transversal crossing with the actual major roads, stating a symbolic element of resilience and green relief of the "concrete city." Along his path, the green cross coincides with the planned mobility system of the new tramlines. This green regenerative axis runs north until reaching two relevant protected natural areas, Monte Gallo and Monte Pellegrino.

On the second level, two green belts are enclosing the urban area to the west, crossing the green sword in different points. The marine coast encloses the city to the east. The inner belt, which corresponds to the urban underground railway line Francia–Brancaccio, joins some of the structural elements of the strategic master plan: the monumental trees (seen as stars, or strong environmental value points) and the blue-green rays (or corridors). The inner belt crosses and connects together several other areas of environmental value, such as the historic park of Maredolce, the Oreto river park, historic gardens and the royal park of Favorita, among others. On the other hand, the outer green belt relates to the metropolitan scale connecting north–south the SCI areas (Monte Grifone and Monte Pellegrino, among others) and running along the main by-pass road of the city (v. Regione Siciliana). It also crosses the most important green corridor of the river Oreto which flows west–east, as well as several covered torrents running in the underground of Palermo. At the same time, the green–blue infrastructural system integrated in the planning of the new tram lines intersect a "green-way" infrastructure, connecting the city of Palermo with Monreale, to the west.

The choice of the plant species adopted have to take into account the different elements related to the context, such as culture (Unesco world Heritage Arab-Northman tour), environmental values (native species, as predicted by the L.10/2013, relating to the Rete Natura 2000 of Palermo, giving a homage to the idea of biodiversity which was born on the Monte Pellegrino in Palermo thanks to the Academic G. E. Hutschinson in 1959), urban agriculture species (olive trees, orange trees, etc.), the phyto-remediation ability of the plants (necessary to clean storm water, soil, and air), and the ability to give new habitat to the animal species.

In conclusion, the development of the green–blue infrastructures integrated to the mobility system can have a significant improvement when developed starting from the existing green space of the urban territory. The different types of approach to urban green (historic, ecological, and bioengineering) will help in strengthening the ability of the interconnected vegetation to positively impact on the urban environment and on the community (Fig. 13).

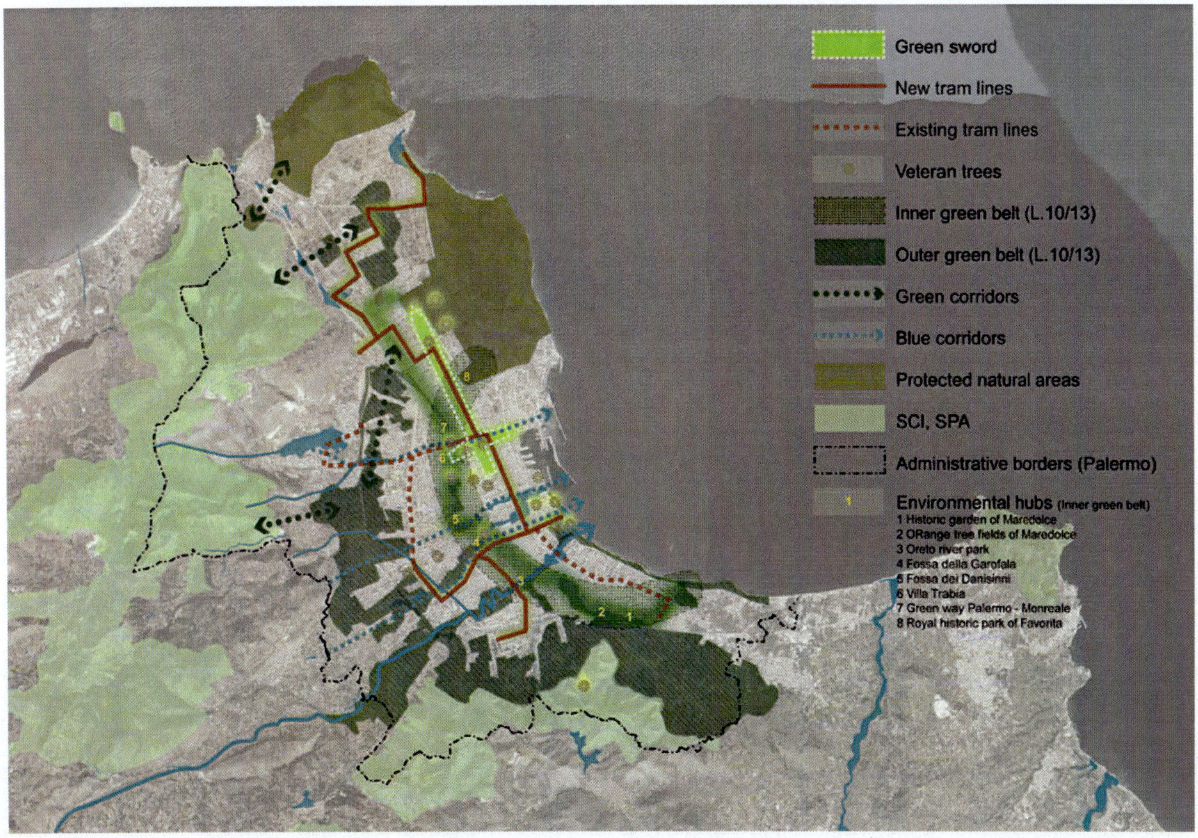

Fig. 13 Master plan for the green–blue infrastructures integrated in the new tram lines plan of Palermo, Italy. Study edited by: G. Pirrera, G. Trapani, M. Giudice, L. M. Ferrara, G. Di Chiara

6 Conclusions

Usually the most advanced European cities involving and testing sustainable urban mobility policies analyze:

1. Economic benefits;
2. Use of public transport;
3. Present and future use of vehicles (private/public);
4. The social, cultural, and environmental attractors of the city;
5. The characteristics and needs of citizens using the urban areas;
6. Ways of moving people;
7. Public activities taking place in the city;
8. Demands of urban users.

According to some operators in the international sector (EEA, 2016c), by putting these factors at the center of the analysis, it is possible to identify the root causes of certain models of social behavior and to use this knowledge for the development of the city.

Based on the EC Green Paper on Sustainable Mobility (COM, 2007), Pareglio (2008, 9) recalled the contribution of Federmobilità (2008) on some measures for the Italian situation: renewing and developing collective transport (especially on iron), promoting slow (non-motorized) mobility and alternatives to non-individual use of the car (especially car ownership), control access and adopt dissuasive tools for motorized mobility (such as road pricing and differentiated parking rates), resorting to new forms of urban logistics (such as transit points combined with last mile logistics), developing and spreading less polluting and energy-intensive engines, using information technology for mobility management, reinforcing information and education to urban mobility and finally to coordinate and integrate interventions at different territorial scales, between the different planning and programming tools action, and within the same mobility policy.

Through these interventions the idea to address urban taxation policies in urban plans becomes more possible, although it is necessary that these interventions run in synergy with the regional planning of infrastructures for the mobility both of people and goods. Furthermore, in view of the great costs of planning and realization, and considering the necessary time for the analysis of the demand and the social sharing of the technically identified alternative scenarios, it is important to mention the contribution of environmental analysts for the definition of the alterations suffered from natural and anthropic ecosystems due to the use of energy deriving from combustion and all the other impacts on the quality of welfare, well-being, and living in cities and territories.

The application of the methodology to the case study has proved that considering the different aspects of analysis may be fundamental to develop a result oriented to social and environmental effectiveness.

On the other side, the research shows that green–blue infrastructures have provided a feasible answer to the failure of past urban planning policies by integrating environmental values in the urban mobility planning process. Besides, a system of values oriented to the social affection for the open spaces (through the monumental trees and the urban parks) has been created, providing health and ecosystem services for inhabitants and wildlife. GI, applied to mobility infrastructure, offered the chance to renaturate, with selected species, the damaged landscape system rebuilding a network of open spaces of values.

The health of the planet and of the human beings that inhabit it, in accordance with recent UN approaches, should justify both the costs for trans-disciplinary research and those to face the real participation of all the actors and beneficiaries of sustainable mobility policies in order to block the current tendency to climate change and recover the lost ground putting resilience strategies in an environmental, social, and economic/financial sense at the center of the ordinary urban and territorial development policies.

References

Ashton, J. (1992). *Healthy cities*. Open University Press.

Associazione Idrotecnica Italiana. (1987, February 23). *Palermo tra due Alluvioni [Conference session]*. Associazione Idrotecnica Italiana, Sezione per la Sicilia Occidentale, Palermo, Italy. (mimeo). https://books.google.de/books?id=RDVQzQEACAAJ

Banister, D., & Marshall, S. (Eds.). (2007). *Land use and transport planning. European perspectives on integrated policies*. Elsevier.

Bergmann, M., & Jahn, T. (2008). CITY: Mobil': A model for integration in sustainability research. In G. Hirsch Hadorn, H. Hoffmann-Riem, S. Biber-Klemm, W. Grossenbacher-Mansuy, D. Joye, C. Pohl, U. Wiesmann, & E. Zemp (Eds.), *Handbook of transdisciplinary research*. (pp. 89–102). Springer.

Bevilacqua, P. (1988). Il paesaggio degli alberi nel mezzogiorno d'Italia e in Sicilia (tra XVIII e XX secolo). *Annali Cervi, X*. 259–306.

Boscacci, F., Camagni, R., Corda, G. P., & Moretti, A. (2001). *Mobilità, modelli insediativi ed efficienza territoriale*. FrancoAngeli.

Browder, G., Ozment, S., Bescos, I. R., & Gartner, T. (2019). *Integrating green. Creating next generation infrastructure*. World Bank Group. https://hdl.handle.net/10986/31430

European Commission. (2013). *Green infrastructure (GI)—Enhancing Europe's natural capital—Communication from the commission to the European parliament*. https://ec.europa.eu/environment/nature/ecosystems/index_en.htm

Davies, W. K. D. (2015). *Theme cities: Solutions for urban problems*. Springer.

De Leeuw, E., & Simos, J. (2017). *Healthy cities: The theory, policy, and practice of value-based urban planning*. Springer.

De Seta, C., & Di Mauro, L. (1980). *Le città nella storia d'Italia. Palermo*. Laterza.

European Environmental Agency. (2016a). *Report No 15/2016, annual european union greenhouse gas inventory 1990–2014 and inventory report 2016*. https://www.eea.europa.eu/publications/european-union-greenhouse-gasinventory-2016

European Environmental Agency. (2016b). *Explaining road transport emissions—A non-technical guide*. https://www.eea.europa.eu/publications/explaining-road-transport-emissions

European Environmental Agency. (2016c). *Towards clean and smart mobility Transport and environment in Europe*. European Environmental Agency Signals. https://www.eea.europa.eu/highlights/towards-clean-and-smart-mobility

Federmobilità. (2008, February 28). *Quali prospettive per la mobilità urbana*. (mimeo). www.federmobilita.it

Forman, R. T. T. (2014). *Urban ecology: Science of cities*. Cambridge University Press.

Foster, J., Lowe, A., & Winkelman, S. (2011). *The value of green infrastructure for urban climate adaptation*. The Center for Clean Air Policy.

Geels, F. W., & Smit, W. A. (2000). Failed technology futures: Pitfalls and lessons from a historical survey. *Futures, 32*(9), 867–885

Gehl, J., & Svarre, B. (2013). *How to study public space*. Island Press.

Institute for Sustainable Infrastructure. (2015). *Envision rating system for sustainable infrastructure*. https://sustainableinfrastructure.org/

Jenks, M., & Dempsey, N. (Eds.). (2005). *Future forms and design for sustainable cities*. Architectural Press.

Lang, D. J., Wiek, A., Bergmann, M., Stauffacher, M., Martens, P., Moll, P., Swilling, M., & Thomas, C. J. (2012). Transdisciplinary research in sustainability science: Practice, principles, and challenges. *Sustainable Science, 7*(Supplement 1), 25–43. https://doi.org/10.1007/s11625-011-0149-x

Lennon, M. (2014). Green infrastructure and planning policy: A critical assessment. Local Environment. *The International Journal of Justice and Sustainability, 20*(8), 957–980.

Morandi, C., Rolando, A., & Di Vita, S. (2013). ICT: interfacce tra persone e luoghi. *TeMA Journal of Land Use, Mobility and Environment, 6*(1), 19–33

Commission of European Communities. (2007). *Green Paper. Towards a new culture for urban mobility, no. 551*. https://eur-lex.europa.eu/LexUriServ/LexUriServ.do?uri=COM:2007:0551:FIN:EN:PDF

Pareglio, S. (2008, April 17–29). *Ambiente ed energia: entrando nel merito delle scelte del nuovo piano [Paper presentation]*. INU IstitutoNazionale di Urbanistica—XXVI National Congress.

Pirrera, G., & Ferrara, L. M. (2018, November 21–23). New scenarios from ecological restoration work to contain the risk of desertification in Custonaci [Paper presentation]. In *10th AEIP-APENA-EFIB-ECOMED conference: Soil and Water Bioengineering in the Mediterranean ecoregion, Madrid, Spain. ECOMEDBIO EFIB Award-European Soil and Water Bioengineering for "Better communication in the congress"*.

Poli, C. (2011). *Mobility and environment. Humanists versus engineers in urban policy and professional education*. Springer.

Rouse, D. C., & Bunster-Ossa, I. F. (2013). *Green infrastructure: A landscape approach*. American Planning Association Planning Advisory Service (No. 571).

Santolini, R., Morri, E., Pasini, G., Giovagnoli, G., Morolli, C., & Salmoiraghi, G. (2015). Assessing the quality of riparian areas: the case of river ecosystem quality index applied to the Marecchia river (Italy). *International Journal of River Basin Management, 13*(1), 1–16. https://doi.org/10.1080/15715124.2014.945091

Scalora, G., & Pirrera, G. (2016). *Infrastrutture verdi e partecipazione sociale. Un modello bioispirato di rigenerazione urbana*. Libellula.

Staricco, L. (2013). Smart mobility: Opportunità e condizioni. *Tema, 3*, 341–354.

Todaro, P. (1988). *Il sottosuolo di Palermo*. Flaccovio Editore.

Todaro, P. (2006). Giardini islamici - Sistemi di captazione e gestione dell'acqua nella Piana di Palermo nel Medioevo. *Paysage 16*.

United Nations. (2015). *Transforming our world: The 2030 Agenda for Sustainable Development*. https://sustainabledevelopment.un.org/post2015/transformingourworld/publication

Understanding the Social Appropriation of Public Green Spaces in a Disadvantaged Neighbourhood. A Conceptual Model Related to Urban Sustainability

Jairo Agustín Reyes-Plata and Marí del Carmen Villanueva-Vilchis

Abstract

Social appropriation can contribute to keeping public green spaces as a benefit for social welfare and sustainability in cities. However, it is not clear which social and physical factors are related to appropriation and what it implies for sustainability. Based on a literature review, a conceptual model on social appropriation was elaborated and statistically validated (multivariate analysis of hierarchical linear regression) for a disadvantaged neighbourhood ($N = 257$), located in one of the most populated cities in Central Mexico. The study identified three variables significantly related ($R^2 = 0.852$, $p < 0.001$) to social appropriation: (i) age, (ii) security perception and (iii) physical state of vegetation and pavement. The study concluded that social appropriation of green public spaces is part of a complex phenomenon, in which tangible and intangible variables related to physical attributes and population influence the provision of environmental and social benefits. From a sustainability perspective, the benefit of maintaining green spaces relies on social appropriation and constitutes a condition to reach urban sustainability.

Keywords

Public green spaces • Social appropriation • Urban sustainability • Social welfare

J. A. Reyes-Plata (✉) · M. del CarmenVillanueva-Vilchis
Escuela Nacional de Estudios Superiores, Unidad León,
Universidad Nacional Autónoma de México, León, Mexico
e-mail: jreyes@enes.unam.mx

1 Introduction

The city's green spaces provide environmental, economic, social and cultural benefits that can have repercussions on the social well-being and quality of life of the population. It is strategically important for generating sustainable urban environments, especially if we consider that 66% of the world's population will live in cities by 2050 (Naciones Unidas, 2019). In Mexico, 8 out of 10 inhabitants already lived in urban areas in 2010 (INEGI, 2010).

Cities are certainly centres of opportunities for work, education and a wide range of services, all important elements for habitability and social well-being. However, they are also a space of threats and risk due to air pollution, water pollution, excessive noise, traffic or physical inactivity. All of the negative externalities not only generate disturbances on the ecological balance of cities but also cause high levels of stress that disrupt the physical, psychological and emotional health of the inhabitants. It also has an impact on their work, social and family life all of which have a considerable effect on their quality of life (WHO, 2012). Maintaining optimal environmental conditions for social welfare and the enjoyment of life are above all insoluble objectives of sustainable development.

The United Nations General Assembly launched 17 goals to achieve sustainable development by 2015, with a target date of 2030. In the Agenda, Goal 11, entitled "Sustainable Cities and Communities", stated that cities and human settlements should be inclusive, safe and resilient, particularly for women and children, older people and people with disabilities (Naciones Unidas, 2019). In this sense, the declaration of Quito (2016) on the New Urban Agenda points out the demand to promote the creation of inclusive, safe, peaceful and plural cities where the specific needs of those social sectors in situations of vulnerability are recognized. To achieve this purpose, it is considered essential to foster a sense of belonging to the city, among other factors, through the appropriation of well-connected and distributed public

F. Trapani et al. (eds.), *Advanced Studies in Efficient Environmental Design and City Planning*,
Advances in Science, Technology & Innovation, https://doi.org/10.1007/978-3-030-65181-7_27

spaces in the city that are safe, inclusive, accessible, green and of good quality. These spaces should also contribute to the free manifestation of cultural expressions as well as to the improvement of social interaction and intergenerational coexistence in two senses. First, in the way people interrelate today, among the members of their own communities, and second, towards the future, through the living conditions inherited for future generations.

In particular, the public green space has qualities that can generate multiple benefits for the inhabitants, many of which derive directly from their use. However, the appropriation of public green spaces does not occur automatically, nor is the maintenance of its benefits guaranteed in the long term, nor its equitable distribution to all sectors of society, which is why it is important to identify the factors that favour the appropriation of green spaces with a view to maintaining the benefits they generate. This text aims to present a conceptual model on the social appropriation of public green spaces based on the analysis of a case study in the city of León, Guanajuato, Mexico, and thus reflect on the implications of the social appropriation of green spaces for the urban sustainability.

2 Urban Sustainability

An approach based on sustainability implies recognizing the contributions of maintaining a social system linked to the green spaces of the city considering the individual and collective effects that they would have on social welfare and starting from the fact that social appropriation of the green space is an unavoidable condition (Kirbert et al., 2010).

In this work, we start from the premise that sustainability addresses the interaction between natural and social systems, as well as the mechanisms that affect the satisfaction of the needs of present and future generations, the substantial reduction of poverty and the conservation of life support systems on Earth (Kates, 2011). The maintenance of key characteristics and components for long-term functioning, in accordance to the principles of social and environmental justice and equity, demand the search for mechanisms to achieve an environment with adequate living conditions for all sectors of the population.

From the point of view of urban sustainability, the city is conceived as a complex system, in which a multitude of interrelated and interdependent objects converge temporally and spatially, integrating harmoniously the environmental, economic, social and cultural aspects of urban development. Shmelev (2012) proposes that the development of a sustainable city requires a multidimensional approach in which waste management, the sustainable use of energy, ecodesign and landscape architecture are required, as well as the preservation of cultural and natural heritage, the psychology of interaction with the environment, green areas and biodiversity, quality of life and health. All of the above, in order to have a holistic approach that makes the city habitable, referring to habitability as the quality that keeps a place to satisfy the needs and aspirations of the person (Lezama & Domínguez, 2006; Castro, 1999 cited by Moreno, 2008).

Thus, the adequacy of the environment has two purposes: improvement in the quality of life and social welfare. Regarding the first, Pérez Maldonado (1999, cited by Moreno, 2008) argues that the quality of urban life involves the biological and psychosocial comfort of those who live in the city, sensations that are determined by the degree of satisfaction with the use of services and the perception of the urban space as healthy, safe and visually pleasing. As for social welfare, it encompasses collective needs, problems and aspirations, which have their origin in the basic needs of the individual.

From a psychological point of view, Blanco et al. (2005) mention that well-being comes from the valuation we make of circumstances and the way we interact within society in a desire to relate to others, to have enough for survival, to integrate with society and to live in harmony with nature, which would imply identifying objective and therefore measurable variables in association with others of a subjective nature, which are subject to scales of valuation based on the perception of the individual.

Well-being, as a process of individual and collective action, can be measured in principle by social integration (Mazaira et al., 2011). That is, the feeling of belonging that an individual has towards a group and that is evidenced by the ties that he tends to have with family, friends or neighbours. If the individual feels that he belongs to a group, the bond he builds with others will enjoy reciprocal trust and positive attitude, leading to social acceptance. The benefit is first individual, but it goes out into the collective sphere. In sustainability, well-being is an end to which one aspires, but which acquires meaning when it transcends the individual sphere. The effects of well-being are not measured only by what it means for the individual but for society as a whole.

The public space is, in that sense, the natural place where social interactions occur, reason for which its appropriation constitutes a key element of the social dimension of urban sustainability. It deals with the way in which a city is lived in and experienced and the consequences that it has in the construction of the social capital of a community. It can be understood as the sum of the resources, real or virtual, that an individual or group accumulates in time by virtue of the possession of a durable network or more or less institutionalized relations of mutual knowledge or recognition (Bourdiue, 1999, quoted by Vargas & Merino, 2013).

Since it is linked to a more equitable, egalitarian and democratic access to the natural or socially generated wealth as well as to the satisfaction of the needs both material and immaterial that the inhabitants of a city have. It is for this reason that in this work we analyse the factors that promote the social appropriation of the green space and analyse its link with urban sustainability.

2.1 Social Appropriation of Space

We approach the study of the social appropriation of public green space from the theoretical perspective of environmental and social psychology, according to which the social appropriation of space is a dialectical process through which people and spaces are linked (Pol, 2002; Vidal & Pol, 2005). This connection is reflected in the relationships and ties that are established individually and collectively through the experiences and daily activities that favour recreation and health care, which in green spaces can be done leading to an appreciation of them for their benefits to human welfare, one of the unavoidable ends of sustainability.

The social appropriation of space, therefore, is understood as the "feeling of owning and managing a space" (Pol, 2002: 124) by the habitual use that individuals make of the place or by their symbolic identification with it. This process contemplates two complementary dimensions: action-transformation and symbolic identification.

Pol (1996) points out that action-transformation is fundamental for the appropriation of space to occur, since through action on space, people, groups or collectivities incorporate the environment in their cognitive and affective processes in an active and updated manner giving space a meaning. Action allows people to interact and act in space and in doing so they can transform it through a series of practices (symbolic interaction) with which imprints are left on it or signs endowed with meaning. This symbolic interaction allows people to recognize themselves in the environment and attribute to themselves the characteristics of space as defining their own identity, leading to symbolic identification with space, as shown in Fig. 1.

According to Vidal et al. (2004), actions can be of three types: (1) daily actions (those that people perform daily in the place), (2) actions of belonging to the place (to be informed of what is happening, knowledge of activities that are organized there) and (3) future actions (interest of people to know the projects for the improvement of the place, to be involved in the identification and solution of problems). In this way, appropriate space favours the continuity and stability of the person as well as the construction of the identity and cohesion of the group. It generates an attachment to the place and with it the production of social capital, the result of a network of social relations that can guarantee a fair distribution of the benefits provided by the green spaces of the city to all the population (Bourdiue, 1999, quoted by Vargas & Merino, 2013).

When we speak of public green spaces, the most important dimension of the appropriation process is symbolic identification with the place since the action-transformation possibilities of individuals are limited in terms of the level of intervention and necessary resources that would be required. In this sense, it is important to understand how public green space can become for people a place of their own that they endow with meaning.

The meaning of space derives mainly from the experience that people have with it, including the emotional aspect. The emotional experience of green space also depends on the perception that people have of their physical and functional characteristics which become factors that can favour their appropriation as will be seen below.

2.2 Factors Favouring the Appropriation of Public Green Space

Some theoretical approaches place special emphasis on the evaluation of tangible aspects related to the perception of spatial components starting from the premise that physical characteristics (such as appearance, lighting, cleanliness, state of the pavement and state of trees and plants) constitute the primary context that structures the place and that serves as a stage for the realization of activities or experiences that

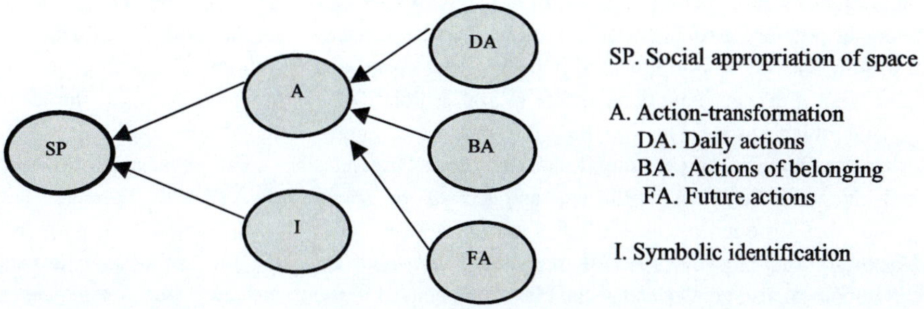

Fig. 1 Theoretical model of social appropriation, proposed by Vidal et al. (2004)

allow or not the social appropriation of the green space. Different studies conclude that there is a positive relationship between physical quality and satisfaction with the public space of parks according to people's assessment of their cleanliness, accessibility, aesthetics or comfort (Baum et al., 2010; Hur & Morrow-Jones, 2008; Kamphuis et al., 2010; Lovejoy et al., 2010). Other studies support the idea that there is a causal relationship between the physical characteristics of green spaces (accessibility, safety, attractiveness, quality and type of facilities) and the use people make of them (Vargas & Merino, 2013). It is worth mentioning the tendency for architectural and urban design to play a relevant role in preventing or exacerbating conditions of insecurity or perception of risk in space which has led to greater attention being paid to the state or quality of the spatial elements of open space such as vegetation, lighting or the characteristics of buildings. Likewise, the functionality of the space is linked to the social needs and practices that are developed in them, which come from the perception regarding the possibilities of what can be done with the space and/or in it and that derive in the realization of certain types of activities.

There are other theoretical approaches that explain the phenomenon of spatial appropriation as a result of a much more complex relationship of multifactorial nature highlighting the importance of the social environment in the process of the appropriation of space. For example, some studies indicate that residential satisfaction and the propensity to move out of a neighbourhood is influenced by the presence of antisocial behaviour and potentially criminal or marginal activities, circumstances that not only lead to a perception of risk and insecurity in space but that would be indicative of the degradation of the social fabric (Cheshire et al., 2013, cited by Martínez & Ibarra, 2017). High-risk behaviour can have important consequences at both individual and group-level reducing opportunities for exchange and strengthening social ties which would be reflected in high levels of community disorganization and poor social cohesion (Vargas & Merino, 2013).

Dempsey (2009) mentions that social cohesion is associated among other aspects with social interaction, the formation of bonds between community members, a sense of community, participation in organized activities and a sense of belonging. With respect to the latter, interactions between individuals or groups and place lead to the construction of a personal memory associated with it, as well as to its identification and the development of a feeling that space can have what it takes to have a better life (Vidal & Pol, 2005).

Undoubtedly, the forms in which this process of appropriation occurs depend on cultural models, roles or lifestyles and characteristics such as the age and sex of the social groups that surround or inhabit the places. In a case study, MacIntyre and Ellaway (1998) demonstrated that there are differences in age, gender and social class with respect to the activities and uses they make of green areas, which undoubtedly affects their social appropriation.

Based on these theoretical antecedents, a conceptual model was constructed on the social appropriation of public green space, which is shown in Fig. 2, and which allowed the identification of the variables involved to explain the predictors of the social appropriation of public green space as well as the consequences associated with urban sustainability.

It is important to highlight that this research empirically validated only the part of the model referring to the physical and social factors that favour social appropriation: physical aspect of the green space, social cohesion, perception of security and risk as well as sociodemographic characteristics. However, we consider it important to provide a broader view of the complexity of studying this phenomenon in order to understand its contribution to urban sustainability from its social dimension.

3 Methodology

3.1 Area and Unit of Study

The study area is located in the neighbourhood of "Parques La Noria" on the south-western edge of the urban area of León, the fourth most populated municipality in Mexico with 1,578,626 inhabitants and a figure equivalent to 26% of the total population of the State of Guanajuato (INEGI, 2015). León is located in one of the most dynamic economic zones in the centre of the country and has a population density estimated at 1,334 inhabitants per square kilometre, with an urban population of 93.11% (INEGI, 2015) of which approximately 42% lived in poverty or social backwardness in 2010 (INEGI, 2015).

The neighbourhood "Parque La Noria" is an economically depressed area with social problems of violence. Its population is characterized by a high degree of urban marginalization, a classification used by the National Council of Population (CONAPO by its acronym in Spanish: Consejo Nacional de Población, 2010) to express the deficiencies suffered by the group in terms of income, health, education and housing. The neighbourhood is made up of six main blocks, subdivided into six or eight sets of single-family dwellings of one or two levels in height. In addition, it has donation areas designated as public spaces for recreation in the central and southern parts of the neighbourhood.

The physical unit of study is constituted by a residential cluster called "parque" (English: Park). The dwellings in each "parque" are organized around a central garden of 1000 m^2 on average, as can be seen in Fig. 3. Because of its surface area, the central garden is considered a

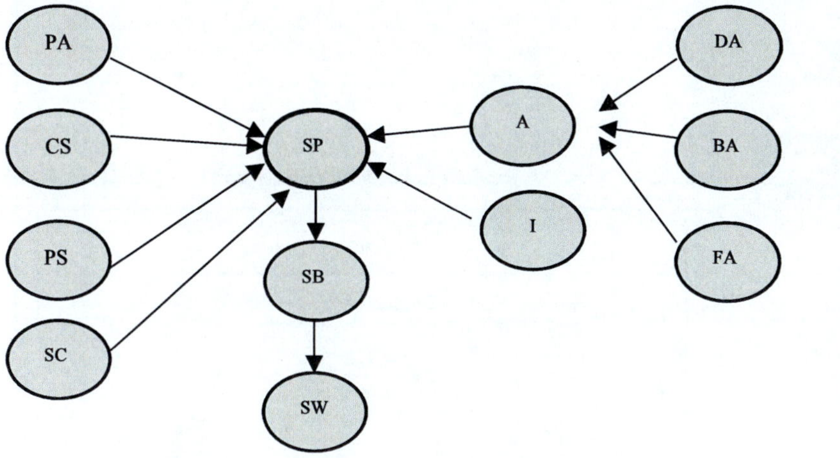

Fig. 2 Theoretical model of social appropriation of public green space and urban sustainability. Own elaboration based on MacIntyre and Ellaway (1998), Vidal et al. (2004), Dempsey (2009), Vargas and Merino (2013)

neighbourhood park, according to the Municipal Program of Green Spaces of León (Ayuntamiento, 2016). Each housing block has an access, closed or open, which allows direct entry from a primary road to the interior through a large esplanade, used mainly for parking. There are between 70 and 80 dwellings per "park".

For the purposes of this study, a multistage sample was carried out, taking as an eligible population the 32 housing blocks that make up Parques la Noria, which are made up of 2496 dwellings. Thirteen blocks were chosen at random from which 300 dwellings were taken; the last unit of study was constituted by a representative of each dwelling that met the inclusion criteria corresponding to being over 16 years of age and being a permanent resident.

3.2 Gathering Information

A structured questionnaire was elaborated with 27 questions on social and physical factors that can influence the appropriation of green spaces. As shown in Table 1, the variables included in the instrument were: physical aspect of the green space (PA), social cohesion (CS), perception of security and risk (PS), as well as sociodemographic characteristics (SC), taking as reference the reagents included in instruments made by Vargas and Merino (2013) as well as Vidal et al. (2004) on residential satisfaction and social appropriation of space, respectively.

The measurement scale for each reagent was 10 points, being 1 the lowest value and 10 the highest, therefore, for the PA dimension, the scale goes from 5 to 50, while for both the CS dimension and the PS dimension, it goes from 9 to 90. The verification of the internal consistency of the instrument used was carried out through a pilot test on a subsample equivalent to 150 people in "Parques La Noria". The Cronbach alpha values showed a good internal consistency of the scale in general (0.878) as well as of each of its dimensions, the lowest being the one related to the perception of insecurity (0.520) and the highest the one of social cohesion (0.879).

3.3 Statistical Analysis

We obtained descriptive measures of sociodemographic variables, proportions for qualitative, and mean variables and standard deviation for quantitative variables. A bivariate student t analysis was carried out for independent samples in order to verify significant differences between the mean of the scale of appropriation by variables of the dichotomized sociodemographic variables. In addition, a multivariate analysis of hierarchical linear regression was carried out to determine the association between physical and social variables with the social appropriation of green areas.

4 Results

4.1 Sociodemographic Aspects

A total of 257 people were interviewed, 67.7% of whom were women. Most of the people said to have more than 10 years residing in the place. We conducted the survey from May 6, 2018 to May 13, 2018.

More than half of those interviewed had basic schooling (primary and secondary), and less than 10% had a bachelor's

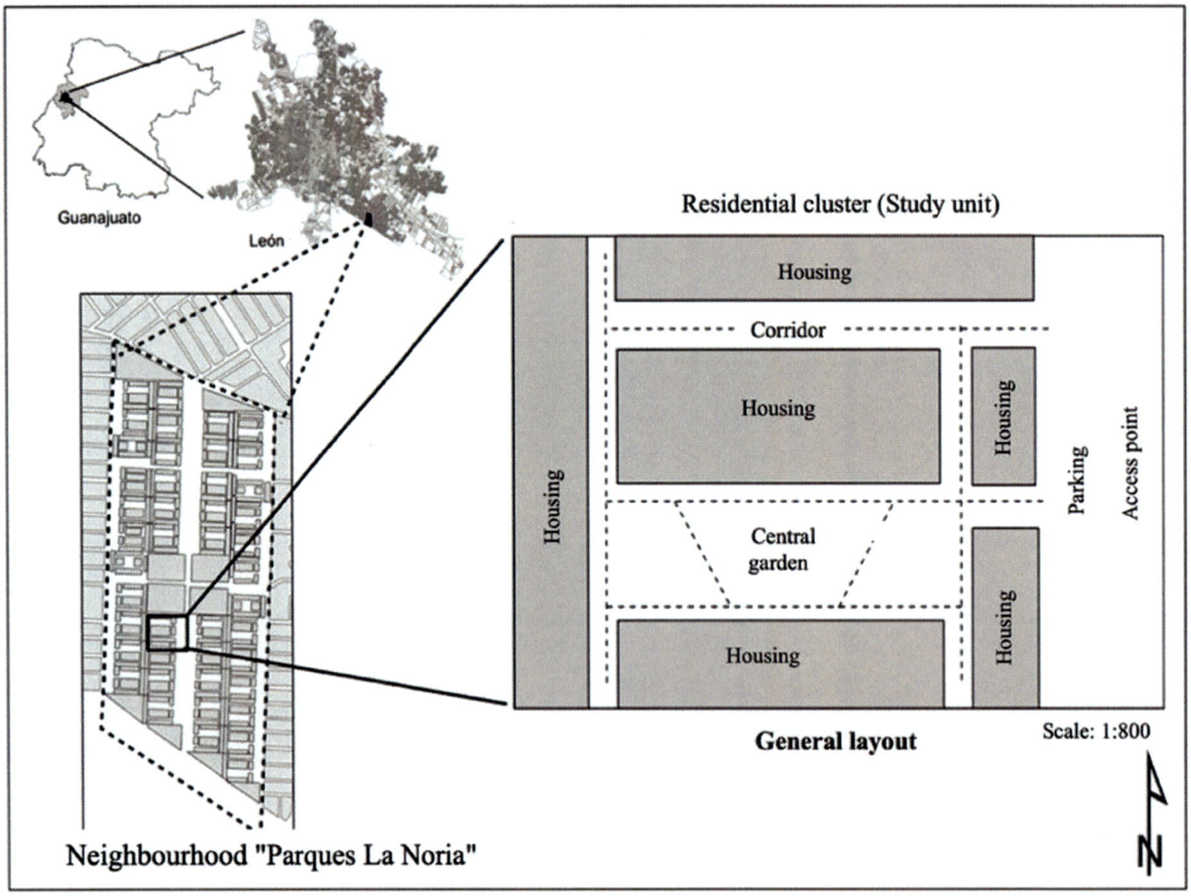

Fig. 3 Type of housing block in neighbourhood "Parques La Noria", León, Guanajuato, Mexico (Own elaboration)

degree or more. Finally, almost 45.0% of people are over 40 years of age (Table 2).

Subsequently, we searched for differences in the appropriation of space according to sociodemographic variables (Table 3). The student *t* test for independent samples revealed that there are only differences with respect to schooling ($t = 2.38$, $p = 0.018$). In this sense, the basic education group showed a higher score in appropriation (Mean = 12.51 SD = 5.4) than the "higher and higher secondary education" group (mean = 10.81 SD = 4.9).

4.2 The Model for the Evaluation of the Appropriation of Public Green Spaces

In order to identify those variables that are significantly related to the social appropriation of green space, a multiple regression model was carried out using the hierarchical method from which three models were obtained as a result. In the first, the sociodemographic variables were included as independent variables, in the second, the variables related to the perception of the physical aspect of the central garden, and in the third, the ones referring to the perception of insecurity and risk. The dependent variable in the three analyses was the social appropriation of the green space. Social cohesion variables were eliminated since none were statistically associated with the dependent variable. The results are summarized in Table 4.

The results of the first model showed a significant relationship between the appropriation of green space and some of the sociodemographic variables, the physical aspect of green space and the perception of insecurity as shown in Table 4. Age was the only variable that maintained consistency in models two ($p < 0.002$) and three ($p < 0.036$). Overall, this model explains 78.0% of the variation in appropriation resulting statistically significant ($p < 0.001$).

In the second model, the physical aspects of the public green space are included as predictors; the results show evidence that a negative perception of cleanliness, pavement condition and vegetation become important in the appropriation of the site. This result is consistent in models two and three, both in the sign and in the statistical significance

Table 1 General characteristics of the questionnaire on the appropriation of green spaces (central gardens)

Dimension	Questions	Cronbach alpha rate	Cronbach alpha rate by each dimension
Physical characteristics of public space (5 questions)	In general, how you would rate the appearance of the public spaces in this park?	0.685	0.753
	In general, how you would rate the state of the lighting (light poles and lamps) that are in the public spaces in this park?	0.796	
	In general, how you would rate the cleanliness of the public spaces in this park?	0.677	
	In general, how you would rate the pavement status of public spaces in this park?	0.689	
	In general, how you would rate the conditions of trees and other plants in the public spaces in this park?	0.689	
Social cohesion (9 questions)	The people in the park where I live are willing to help other neighbours	0.860	0.879
	The people in the park where I live get along with each other	0.860	
	The neighbours of my park are reliable	0.853	
	My neighbours share the same moral values as me	0.860	
	I like to participate in neighbourhood meetings with my neighbours	0.881	
	I would like to ask my neighbours for help or advice	0.868	
	In my park the neighbours are organized to carry out activities that benefit everyone	0.873	
	In general, the relationship among the neighbours in this park is good	0.854	
	I would ask for a monetary loan from my neighbours	0.880	
Perception of insecurity (9 questions)	There are a lot of criminals or groups of criminals in my park	0.404	0.520
	There are a lot of junkies in my park	0.376	
	People are afraid of being mugged in my park	0.431	
	There are a lot of thieves on the streets and in the houses in my neighbourhood	0.435	
	I feel safe in my park overnight	0.607	
	Police surveillance in my park is enough	0.549	
	I feel safe in my park during the day	0.587	
	Violence against women occurs frequently in my park	0.481	
	There is drug sale on the streets in my park	0.453	
Appropriation of space (4 questions)	I feel the garden as something of my own	0.817	0.839
	I feel like the garden is part of my daily life	0.767	
	I feel attached to the garden	0.745	
	I like living near the garden	0.846	

of the parameter. It is important to notice that model two explains 84.5% of the variation in the social appropriation of green spaces being statistically significant ($p < 0.001$).

Regarding the aspects of insecurity and low risk corresponding to the group of variables included as part of model three, this work found that the perception of insecurity during the day significantly impacts the appropriation of public green space particularly the central garden. This model also explains 85.2% of the variation in appropriation and is statistically significant ($p < 0.001$).

Figure 4 graphs the conceptual model of the factors that determine the appropriation of public green spaces in the study area, "Parque La Noria". This model integrates both tangible and intangible variables of the public green space, but also relative to the characteristics of the population.

Table 2 Distribution of sociodemographic variables of the study population

Variable	*N*	%
Gender		
Female	174	67.7
Male	83	32.3
Residence time		
≤ 1 year	23	8.9
2–5 years old	53	20.6
6–10 years old	46	17.9
11–20 years old	116	45.2
More than 20 years	19	7.4
Schooling		
Elementary school	67	26.1
Secondary school	107	41.6
High school	65	25.3
Bachelor's degree	14	5.5
Graduate	4	1.5
Age		
18 years old	33	12.9
19–29 years old	61	23.8
30–39 years old	46	17.9
40–49 years old	73	28.4
50–59 years old	30	11.6
≥ 60 years old	14	5.4

Source Direct

Table 3 Difference in means of the scale of appropriation by sociodemographic variables

Variable	*N*	Mean	SD	*t*	Sig. (95% confidence)
Gender					
Female	163 (63.4%)	12.12	8.6	0.101	0.920
Male	83 (32.2%)	12.04	5.3		
No response	11 (4.2%)				
Years of residence					
Less than 10 years old	122 (47.5%)	12.23	5.2	0.298	0.766
10 years or more	135 (52.5%)	12.03	5.6		
No response	0				
Schooling					
Basic education	169 (65.7%)	12.51	5.4	2.38	0.018
Higher and higher middle education	74 (28.8%)	10.81	4.9		
No response	14 (5.5%)				
Age					
Less than 40 years old	140 (54.4%)	12.26	5.3	0.906	0.366
40 years or older	103 (40.1)	11.62	5.5		
No response	14 (5.5%)				

SD standard deviation
Source Direct

Table 4 Results of the hierarchical multiple linear regression for the conceptual model

Variables	Model 1 summary		Model 2 summary		Model 3 summary	
	R^2 (adj) = 0.780 $p < 0.001$		R^2 (adj) = 0.845 $p < 0.001$		R^2 (adj) = 0.852 $p < 0.001$	
Sociodemographic characteristics						
	[*p*]	CI 95%	[*p*]	CI 95%	[*p*]	CI 95%
Years of residence	0.025	0.237, 3.525	0.454	–0.889, 1.983	0.468	–0.890, 1.930
Schooling	0.000	1.355, 4.121	0.565	–0.889, 1.626	0.648	–1.665, 1.038
Age	0.000	2.042, 4.945	0.002	0.788, 3.306	0.036	0.091, 2.697
Physical characteristics of public green space						
Lighting state			0.623	–0.172, 0.287	0.937	–0.240, 0.222
Cleaning			0.009	0.098, 0.676	0.050	0.000, 0.588
Pavement state			0.012	0.093, 0.741	0.050	0.001, 0.642
Vegetation state			0.002	0.193, 0.815	0.004	0.152, 0.765
Security and low risk						
Police surveillance					0.282	–0.106, 0.360
Daytime security					0.007	0.084, 0.523
Presence of criminals or criminals					0.656	–0.149, 0.235
Violence against women					0.085	–0.028, 0.422

Note Only the significant variables for each regression are presented
Source Direct

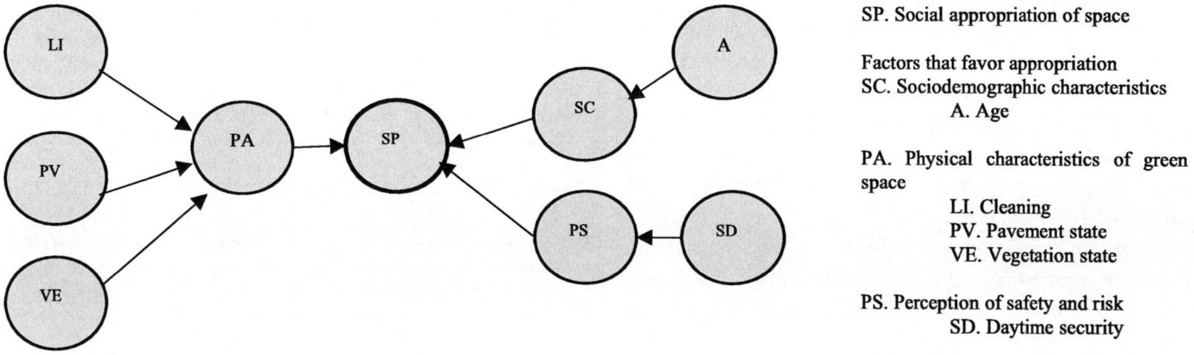

Fig. 4 Conceptual model for the social appropriation of green space

As it can be deduced from the analysis presented above, the perception of the place seems to be associated with the age of the people, who with the passing of the years could become more sensitive to the changes that the appearance of the physical space undergoes, particularly in relation to the cleanliness and the state of the pavement and vegetation. Of these three, vegetation is expected to be an essential element in the conceptualization of a green area and for the provision of the environmental services that these spaces provide, a fundamental aspect in determining their contribution to urban sustainability. In addition, these spaces make places attractive and contribute to the creation of an atmosphere that can favour their use or, on the contrary, discourage it. However, it is still necessary to specify how these aspects may be specifically influencing the appropriation of green spaces.

The perception of green spaces involves a much more complex process, as it can be determined by the environment in which these places are immersed, not only in physical terms but also in social terms. The human context determines the character and dynamics of use, determining the meaning of the place, thus facilitating a process of social appropriation that derives in the enjoyment by all the population of the benefits provided by green spaces which is a fundamental aspect to achieve the social sustainability of cities.

5 Conclusions

The recent integration of the concept of sustainability into the scientific field requires clarification of its theoretical and methodological implications. The results of the empirical test applied provide evidence to understand the factors involved in the appropriation of public green spaces, particularly in disadvantaged conditions of poverty and violence. This study has found that the appropriation of green space relies on a broader complex and multidimensional process which relates to sociodemographic, physical and safety perception domains into a nested system of tangible and intangible attributes and qualities of public green spaces. Specific attributes embrace the state of places (e.g. pavement, vegetation and cleaning), people's age and temporal changes. Interestingly, the maintenance of these attributes places a challenge to increase the opportunities of accessing benefits that public green spaces provide.

Furthermore, access to public green space contributes to social appropriation and consequently it can lead to develop a sense of belonging and attachment to place, favouring respectful behaviours with the environments and care of the environment. It has implications on social and environmental domains through the provision of cultural and ecological services which are determining aspects to reach the city's sustainability. Undoubtedly, the ways in which this process of appropriation takes place depend on the cultural models, roles or lifestyles of the social groups that surround or inhabit the places.

Some studies have found a positive relationship between the appropriation of space and the development of responsible ecological behaviours (Corral-Verdugo & de Queiroz Pinheiro, 2004). However, this relationship has not yet been clearly identified for this study. According to Calderón-Carrillo and Bustos Aguayo (2007), responsible ecological behaviour originates as a way of maintaining the place and its elements, with which one feels an identification, activating defence or protection mechanisms while trying to avoid destroying and even drastically modifying the elements of the place. It means that preserving places and their elements is to preserve individual and collective identity. Other authors have questioned that ecological behaviour is only reduced to the care of the physical environment, proposing instead the concept of sustainable behaviour to refer to the set of effective, deliberate and anticipated actions that result in the preservation of natural resources, including the integrity of animal and plant species, as well as in the improvement of the individual and social welfare of current and future human generations. Responding skillfully to demands for the care of the physical and social environment, starting from setting a specific purpose or intention, are aspects that would characterize a sustainable behaviour. The individual can anticipate the effects of their present actions to the future, conscious of the fact that it is in the future where his present actions will take effect. Behind every action lies a feeling of solidarity that manifests itself in altruistic tendencies and actions deployed in response to concern for others, so the action has to be directed to ensure the welfare of all people, both current and future generations. Thus, appropriation of a green space is only the beginning of a process that can lead to a behaviour of care of the environment and with it the sustainability of the city.

Acknowledgements This work was supported by UNAM PAPIIT, IA402418.A special acknowledgement to Alan Sánchez Vázquez on his work of the translation of this article.A special acknowledgement to Leonardo Arturo Rivas Rivas for style corrections.

References

Ayuntamiento de León, H. (2016, April 5). *Programa Municipal de Espacios Verdes Urbanos del Municipio de León, P.O. Guanjauto, 5 de abril de 2016*. https://periodico.guanajuato.gob.mx/downloadfile?dir=files_migrados&file=PO_55_2da_Parte_20160405_1638_3.pdf

Baum, S., Arthurson, K., & Rickson, K. (2010). Happy people in mixed-up places: The association between the degree and type of local socioeconomic mix and expressions of neighbourhood satisfaction. *Urban Studies, 47*(3), 467–485. https://doi.org/10.1177/0042098009351941

Blanco, A., & Díaz, D. (2005). El bienestar social: su concepto y medición. *Psicotherma, 17*(4), 582–589

Calderón-Carrillo, M., & Bustos Aguayo, J. M. (2007). Apropiación y conducta proambiental en un poblado periurbano de la ciudad de México, *Revista Electrónica Internacional de la Unión Latinoamericana de Entidades de Psicología*, Número 10 Julio 2007.

Consejo Nacional de Población (CONAPO). (2010). *Índice de marginación urbana 2010*. https://www.conapo.gob.mx/es/CONAPO/Capitulo_1_Marginacion_Urbana_2010

Corral-Verdugo, V., & de Queiroz Pinheiro, J. (2004). Aproximaciones al estudio de la conducta sustentable. *Medio Ambiente y Comportamiento Humano, 5*(1y2), 1–26.

Dempsey, N. (2009). Are good-quality environments socially cohesive? Measuring quality and cohesion in urban neighbourhoods. *Town Planning Review, 80*(3), 315–345. https://doi.org/10.3828/tpr.80.3.5

Hur, M., & Morrow-Jones, H. (2008). Factors that influence resident's satisfaction with neighborhoods. *Environment and Behavior, 40*(5), 619–635. https://doi.org/10.1177/0013916507307483

Kamphuis, C., Mackenbach, J., Giskes, K., Huisman, M., Brug, J., & van Lenthe, F. (2010). Why do poor people perceive poor neighbourhoods? The role of objective neighbourhood features and psychosocial factors. *Health and Place, 16*(4), 744–754. https://doi.org/10.1016/j.healthplace.2010.03.006

Kates, R. (2011). What kind of a science is sustainability science? *Proceedings of the National Academy of Sciences of the United*

States of America, 108(49), 19449–19450. https://doi.org/10.1073/pnas.1116097108

Kirbert, Ch., Monroe, M., Perterson, A., Plate, R., & Thiele, L. (2010) *Working toward Sustainability. Ethical decision making in a technological world.* Wiley Inc.

Lezama, J. L., & Domínguez, J. (2006). Medio Ambiente y sustentabilidad urbana. *Papeles de Población, 12*(49), julio-septiembre, 153–176.

Lovejoy, K., Handy, S., & Mokhtarian, P. (2010). Neighborhood satisfaction in suburban versus traditional environments: An evaluation of contributing characteristics in eight California neighborhoods. *Landscape and Urban Planning, 97*(1), 37–48. https://doi.org/10.1016/j.landurbplan.2010.04.010

Macintyre, S., Ellaway, A., Der, G., Ford, G., & Hunt, K. (1998). Do housing tenure and car access predict health because they are simply markers of income or self-esteem? A Scottish study. *Journal of Epidemiology and Community Health, 52*(10), 657–664. https://doi.org/10.1136/jech.52.10.657

Martínez, A., & Ibarra, J. (2017). Los determinantes de la satisfacción residencial en México. *Estudios Demográficos y Urbanos, 32*(95), 283–313.

Mazaira, Z., Becerra, F., & Alonso, I. (2011). Propuesta de un procedimiento metodológico para medir el bienestar social de la familia en Cuba. *Ciencia y Sociedad,* Vol. XXXVI, número 2, abril-junio, 212–238.

Moreno, S. (2008). La habitabilidad urbana como condición de calidad de vida. *Palapa, III* (II), julio-diciembre, 47–54.

Instituto Nacional de Geografía y Estadística (INEGI). (2015). *Encuesta Intercensal 2015. Criterios de agrupación para generar los tabulados básicos : estatal/municipal.* https://internet.contenidos.inegi.org.mx/contenidos/productos//prod_serv/contenidos/espanol/bvinegi/productos/nueva_estruc/702825086657.pdf

Pol, E. (1996). La apropiación del espacio. In Iñiguez, L. In Pol, E. (Eds.), *Cognición, representación y apropiación del espacio. Monografies Psico-Socio Ambientals nº 9.* (pp. 45–62) Publicacions Universitat de Barcelona.

Pol, E. (2002). El modelo dual de la apropiación del espacio, In R. García-Mira, J. M. Sabucedo, & J. Romay (Eds.), *Psicología y medio ambiente. Aspectos psicosociales, educativos y metodológicos.* (pp. 123–132). Asociación Galega de Estudios e Investigación Psicosocial-Publiedisa.

Instituto Nacional de Geografía y Estadística (INEGI). (2010). *Población Rural y Urbana.* https://cuentame.inegi.org.mx/poblacion/rur_urb.aspx?tema=P

Shmelev, S. (2012). *Ecological economics. Sustainability in practice.* Springer.

Naciones Unidas. (2019, September 10). *Objetivo 11: Lograr que las ciudades y los asentamientos humanos sean inclusivos, seguros, resilientes y sostenibles.* https://www.un.org/sustainabledevelopment/es/cities/#5daa7124851e142e8

Vargas, D., & Marino, M. (2013). Public spaces in Mexico as social cohesion promoters: a structural modeling perspective. *Well-Being and Social Policy, 9*(1), 157–177

Vidal, M. T., & Pol, E. (2005). La apropiación del espacio: una propuesta teórica para comprender la vinculación entre las personas y los lugares. *Anuario De Psicología, 36*(3), 281–297

Vidal, T., Pol, E., Guàrdia, J., & Peró, M. (2004). Un modelo de apropiación del espacio mediante ecuaciones estructurales. *Medio Ambiente y Comportamiento Humano, 5*(1y2), 27–52.

World Health Organization (WHO). (2012). Health Indicators of sustainable cities in the Context of the Rio+20 UN. In *Conference on Sustainable Development, Initial findings from a WHO Expert Consultation*, May 17–18, 2012.

Multifunctional Assessment of Green Infrastructure for Sustainable City Planning

M. Pietsch, M. Makala, R.-U. Syrbe, and J. Louda

Abstract

Green infrastructure provides numerous ecosystem services that protect the health of residents and enhance their quality of life in cities. Some sets of ecosystem services, called bundles, always appear together in time and space. Sometimes the maximization of one ecosystem service or bundle results in a decline of other services. Therefore, it is necessary to understand relationships among multiple services and the underlying mechanisms. Synergies are possible if multiple ecosystem services are enhanced, but trade-offs appear if services decrease due to the increase of others. In several projects, a framework for a multifunctional approach was developed and tested for the selected ecosystem services of recreation, climate regulation and habitat provision. Visions, objectives and individual measures can be derived or impacts on individual or numerous ecosystem services can be assessed. The paper presents one example to implement the framework in future sustainable city planning.

Keywords

Ecosystem services • Assessment • Green infrastructure • Sustainability • Urban planning

M. Pietsch (✉) · M. Makala
Anhalt University of Applied Sciences, Bernburg, Germany
e-mail: matthias.pietsch@hs-anhalt.de

R.-U. Syrbe
Leibniz Institute of Ecological Urban and Regional Development, Dresden, Germany

J. Louda
Jan-Evangelista-Purkyně-University, Ustí nab Labem, Czechia

1 Introduction

Urban ecosystems provide numerous ecosystem services (ES) classified as regulating services (e.g. climate regulation), cultural services (e.g. recreation) or provisioning services (e.g. food), protecting the health of urban residents and enhancing their quality of life (Heiland et al., 2015; Kowarik et al., 2017). For the purpose of this article, we understand ecosystem services as "the direct and indirect contributions of ecosystems to human well-being" (TEEB, 2010). The ecosystem services should be taken into account in landscape management (Braat & De Groot, 2012; De Vreese, 2016; Farley, 2012; Albert et al., 2012) to secure the quality of life, especially in cities. In the context of urban environment and planning processes, many cities are applying the concept of green infrastructure. According to European Commission (2013), green infrastructure (GI) is defined as "a strategically planned network of natural and semi-natural areas with other environmental features designed and managed to deliver a wide range of ecosystem services. It incorporates green spaces (or blue if aquatic ecosystems are concerned) and other physical features in terrestrial (including coastal) and marine areas". With a proper strategy in planning and management of green infrastructure, environmental pressures can be reduced by matching ecosystem service supply with its demand (Syrbe & Grunewald, 2017; Syrbe & Walz, 2012). Green infrastructure provides multiple benefits (so-called multifunctionality) on the same spatial area (European Environment Agency, 2017). These benefits may be: air quality regulation, erosion protection, water flow regulation, pollination, water purification, carbon sequestration and storage, recreational experiences, etc. Moreover urban nature helps to strengthen social cohesion (Jennings & Bamkole, 2019; Pauleit et al. 2017) and to reduce stress. Parks, rivers and other green space types are freely accessible and can be used by anyone regardless of social status or cultural background (Kowarik et al., 2017). On the other hand, the way of use and maintenance of urban green

F. Trapani et al. (eds.), *Advanced Studies in Efficient Environmental Design and City Planning*, Advances in Science, Technology & Innovation, https://doi.org/10.1007/978-3-030-65181-7_28

infrastructure has a vital impact on the provided ecosystem services and their values. Assessment of the existing services is necessary as valuable information for decision-makers and planners (Bastian et al., 2013). Multifunctional assessment is especially needed to analyse the impact if use or maintenance should be changed in the future. In this paper, a multifunctional approach is described based on several examples from the Czech Republic. The framework was developed and tested by German-Czech student projects of Anhalt University of Applied Sciences, Leibniz Institute of Ecological Urban and Regional Development and the Jan-Evangelista-Purkyně-University in Ustí nad Labem. Research areas were mapped and the existing ES have been assessed (e.g. climate regulation, recreation, habitat provision). On this basis, also future scenarios were developed to estimate the impact on the concerning ES. The method is based on existing frameworks (Burkhard et al., 2012) and helps to improve the appreciation of landscape structures and provided ecosystem services by decision-makers, politicians or planners (Pietsch et al., 2019). Integrating the multifunctional assessment in decision support systems may enhance sustainable city planning in the future (De Groot et al., 2010).

2 Green Infrastructure and Ecosystem Services

2.1 Urban Nature Versus Green Infrastructure

Urban nature refers to "the totality of natural elements occurring on public and private land in the city" (Kowarik et al., 2017). It includes remnants of pristine ecosystems and rural cultural landscapes as well as designed urban green spaces or novel urban ecosystems (see Fig. 1). The term green infrastructure has been adopted by different disciplines that leads to slightly different concepts. It is possible to group them under two concepts:

- Green infrastructure at urban scale or
- Green infrastructure at landscape scale (regional, national and transnational) (European Environment Agency, 2011).

For landscape architects, "GI is an approach to land use, underpinned by the concept of ecosystem services. Green assets such as parks, coastlines or embankments have generally been thought of in terms of their single functions—the approach that recognizes their vast range of functions and their interconnectivity is called green infrastructure" (Landscape Institute, 2009).

In this paper, the term green infrastructure refers to the definition of urban nature and their elements (see Fig. 2). It can perform a wide range of services for health and well-being. These services include:

- provisioning (e.g. food, wood)
- regulating (e.g. air quality, climate control, erosion control)
- cultural (non-material benefits for people like aesthetic qualities and recreational experiences) (TEEB 2010).

Green infrastructure assets (e.g. parks) include natural or designed elements which provide social, economic and environmental benefits. A useful approach to map and assess different types of green infrastructure assets is to classify them according to the spatial scale and the role they can play to provide the mentioned ecosystem services.

2.2 Multi-functionality

Most landscapes and structures, especially in cities, produce multiple ecosystem services: e.g. climate regulation, habitat provision or recreational experiences. Some sets of ES called bundles (Kowarik et al., 2017; Turner & Gardner, 2015) appear together in time and space. Sometimes, the maximization of one ecosystem service or bundle results in declines of other services. Therefore, it is necessary to understand relationships among multiple services and the underlying mechanisms (Bennet et al., 2009; Heiland et al., 2015; Turner & Gardner, 2015). Synergies are possible if multiple ecosystem services are enhanced, but trade-offs may appear if services decreases due to the increase of others (Rodriguez et al., 2016). In city planning, individual solutions like climate change adaption are often needed. Planting trees or alleys for CO_2 sequestration and a reduction of heat stress are common measures. Such a monofunctional approach helps to find specific solutions and to identify service beneficiaries (Kowarik et al., 2017). But these components are part of a full range of ecosystem service bundles. This would cause adverse impacts if one ecosystem service weakened another. For instance, the reduction of mowing frequency (extensification) for lawns in park areas will, on the one hand, increase the number of plant species and thus habitat quality. On the other hand, usage options like playing football or having a barbecue are no longer possible in the same way. Figure 3 shows an example of monofunctional and multifunctional approaches to regard the effects of city trees. The individual ecosystem services/disservices, benefits/damages and values (positive/negative) to certain segments of society are shown in the lines (monofunctional approach). Synergies of different ecosystem services and trade-offs between positive and negative effects are presented between in Fig. 3 (Kowarik et al., 2017).

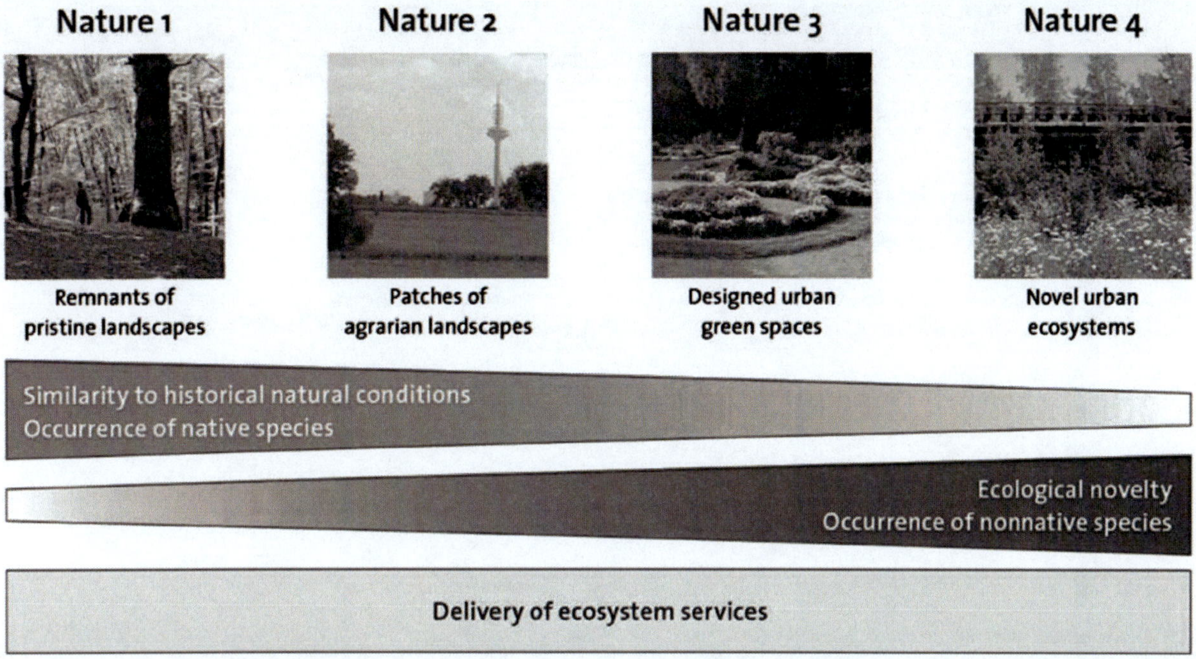

Fig. 1 Urban nature as generic term for different types of nature (Kowarik et al., 2017)

Fig. 2 Examples of urban green infrastructure (left: square with old trees and lawn; right: alley)

"The whole is more than the sum of its parts" (Aristotle, cf. Kowarik et al., 2016). In the context of sustainable urban planning, it is necessary to balance the need for ecosystem services that green infrastructure can provide against the demand. Especially in the planning and maintenance of public green spaces and green infrastructure elements, this approach can make a significant contribution. The spatial relationship between user needs and offers as well as seasonal changes have to be taken into account, as these have an impact on the achievable services as well as the needs (Bastian et al., 2012).

3 Methodology

The methodology for multifunctional assessment of green infrastructure used in this paper has six main steps:

- Selection of suitable ecosystem services (1)
- Collecting information in the field (2)
- Monofunctional assessment of each ecosystem service (3)
- Multifunctional assessment of selected ecosystem services (4)

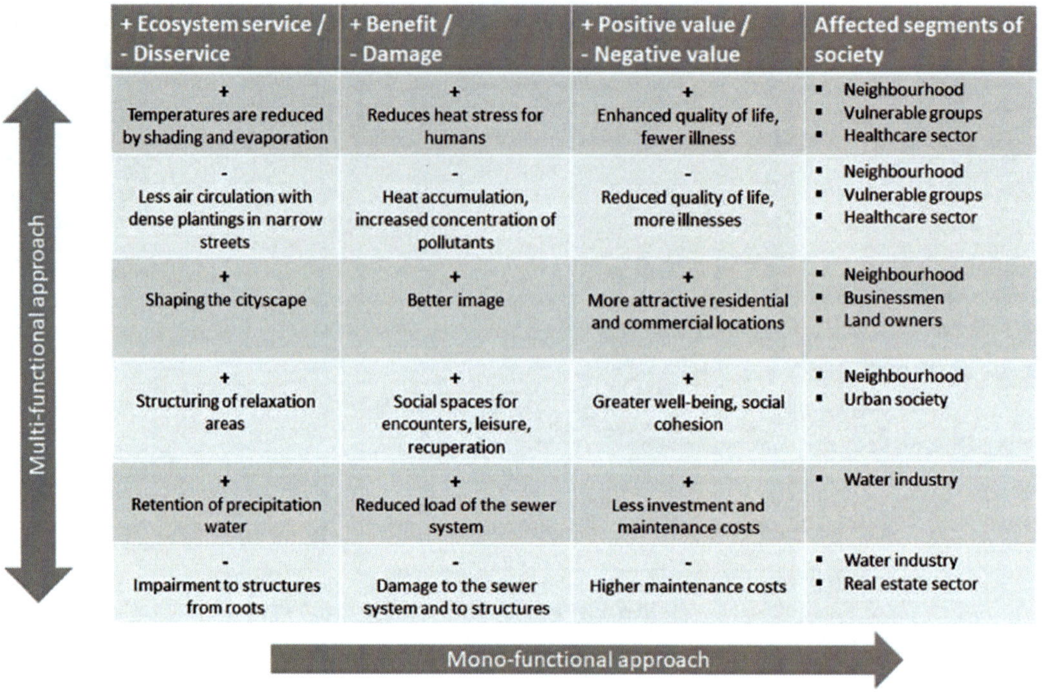

Fig. 3 Monofunctional and multifunctional approach to urban ecosystem services using the example of city trees. Adapted from Kowarik et al. (2017)

- Development of future scenarios (5)
- Impact evaluation of future scenarios (6).

Děčín is a medium-sized city (ca 50,000 inhabitants) surrounded by protected landscape areas, however in the central part of city, there is a lack of public green spaces for short-time recreation, which could serve in the same time as habitats for wildlife species. Because of ongoing climate change, the city is also suffering by extremely high temperatures during last summers. Based on these reasons, the ecosystem services recreation, habitat provision and climate regulation have been selected due to their importance in the study areas (1). Using a survey, urban green assets (e.g. the park area), elements (e.g. tourist infrastructure) and additional required information (e.g. usage possibilities, tree cover) were collected during fieldwork for the selected study areas (2). In the result, individual areas were delimited for the entire investigation area. The selected ecosystem services were evaluated for each asset using different criteria and an evaluation matrix (Burkhard et al., 2012) (monofunctional assessment) (3). For each individual ecosystem service, the values have been summed up and reclassified into three classes from 1 = low to 3 = high. Then, aggregated values are generated combining and summing up the single values for the multifunctional assessment (4). Lastly, the results are aggregated to the three final classes 1 = low to 3 = high for each asset. Future scenarios have been drafted by redesigning assets, adding elements or functions to the mapped structures by the participants of the project (5). Finally, the impact of the scenarios has been assessed using the same methods (6). The results are presented in tables and thematic maps.

3.1 Survey

A survey was developed based on the findings of the projects MaGICLandscapes (Interreg Central Europe Project MaGICLandscapes, 2019) and BIDELIN. Quantitative and qualitative information about actual land use based on Corine Land Cover codes, green and blue structures and covering, recreational infrastructure (e.g. playground, cycle path), possible uses (e.g. jogging, football), habitat structures (e.g. nesting aids), impairments (e.g. noise), native and invasive species, degree of sealing and design elements were collected and marked in the survey. A separate data entry form was drawn up for each asset.

3.2 Methods for the Assessment of Recreation, Habitat Provision and Climate Regulation

For each ecosystem service, individual criteria have been defined that allow for a quantitative and qualitative

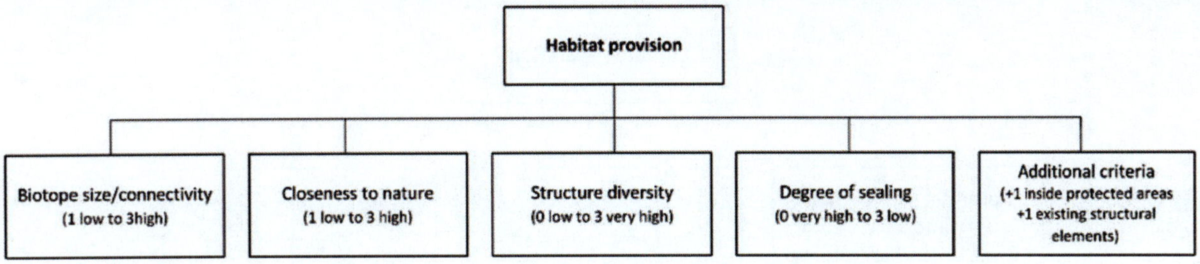

Fig. 4 Criteria for the assessment of habitat provision for mapped assets. Adapted from Bartůněk et al. (2018)

assessment, where necessary and meaningful, additional criteria have been developed that take into account the current characteristics of each individual area. Each method is briefly described below.

Habitat provision

The habitat provision was evaluated based on the criteria degree of sealing, closeness to nature, biotope size and connectivity, structure diversity, existence of protected areas and additional structural elements (e.g. nesting aids, tree holes) (Bastian & Schreiber 1999) (see Fig. 4).

Biotope size and conncetivity was evaluated from 1 low (e.g. small isolated patches) to 3 high (e.g. well connected park areas). Closeness to nature was assessed on the basis of current use as well as naturalness. Areas that are strongly influenced by humans (e.g. road spaces) receive a low value of 1 and near-natural areas a high value of 3. The structural diversity is assessed in the categories low (e.g. intensive grassland) to high (e.g. diverse woody structures). The degree of sealing can be classified from low to high based on the percentage. If areas within protected areas or additional structural elements were present, an appreciation by maximum one value level was made within the criteria of closeness to nature or structural diversity.

Climate regulation

Climate regulation was assessed using the criteria potential overheating intensity and cold-air productivity using a well-known method in landscape planning by Mosimann et al. (1999). To implement the spatial relationship of assets, the method of climate air flow paths for the bioclimatic regulatory function was adapted (von Haaren et al., 2004) (see Fig. 5).

Areas with a potential risk of overheating were rated from − 3 very high (e.g. sealed areas) to 0 non-existent (e.g. water bodies). Values from 0 non-existent (e.g. sealed areas) to + 3 very high (e.g. green spaces) were assigned for cold-air production. Urban climate air flow paths were considered depending on their functionality and width from + 3 (100 m) to + 1 (20 m). Barriers were included in the evaluation with − 1 (e.g. dams).

Recreation

The assessment was made by the criteria diversity of recreational infrastructure, accessibility to the public, share of visible green infrastructure, quality of existing recreational infrastructure (e.g. high quality of recreation infrastructure, well-kept plantations) and negative impacts (e.g. noise, garbage, pollution, dysfunctional recreation infrastructure) (see Fig. 6).

Accessibility was evaluated from fully accessible 3 to totally blocked 0 (restricted). The overall share of green is a combination of degree of sealing and green-covering. Unsealed areas are assigned the highest value level of 3, while fully sealed areas without trees are assigned level 0. Depending on the area under investigation, evaluation classes are formed on the basis of the number of available equipment (e.g. benches, playing facilities). In this way, intensively used areas can be distinguished from extensively used areas. Quality of paths, damage to existing equipment, maintenance condition of planting areas, garbage, noise or odour pollution are included in the evaluation either positively or negatively.

4 Case Study Děčín (Czech Republic)

The described framework was developed and tested in several cities. In this paper, the example of Děčín as a city in the Czech Republic near the border to Germany is used. The example was chosen because there are different types of green infrastructure in the test areas. The development of future scenarios was very simple and easy to understand due to the initial situation. Two research areas in the city centre were selected for the tests (see Fig. 7): (i) The castle surroundings with an historic park area and a part of the Elbe River (Zamek Děčín) (about 34 ha) and (ii) a transition area between the city and its adjacent landscape (Nove Mesto Děčín) (about 27 ha).

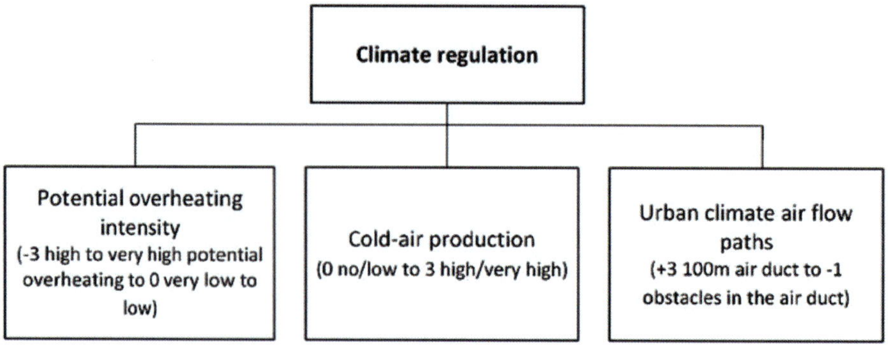

Fig. 5 Criteria for the assessment of climate regulation for mapped assets. Adapted from Bartůněk et al. (2018)

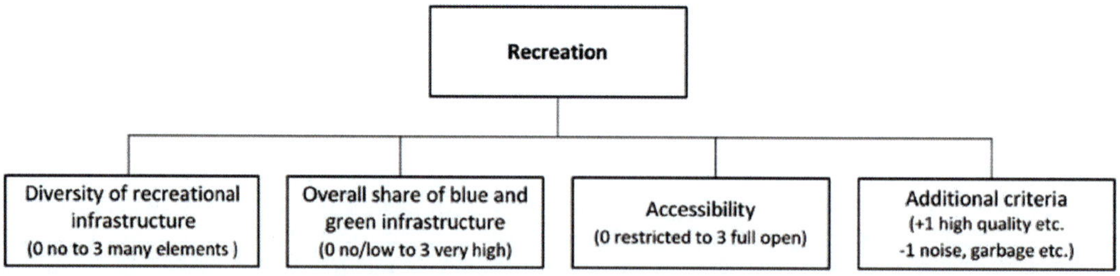

Fig. 6 Criteria for the assessment of recreation for mapped assets. Adapted from Bartůněk et al. (2018)

Fig. 7 Research areas in the city centre of Děčín (Czech Republic) (left: overview city of Děčín; right: two research areas Zamek Děčín (left) and Nove Mesto Děčín (right))

5 Results

5.1 Field Work

Mapping assets influence the evaluation process, planning and management strategies. Spatial relationships and time aspects should be considered while mapping and collecting specific elements (Bastian et al., 2012). In the example, the assets are divided by recognizable structures and functions in the field (e.g. entire park area). The field mapping results are different than the outcomes of remote sensing techniques. This is helpful to develop new planning or management strategies (e.g. adapted maintenance) (see Fig. 8).

5.2 Monofunctional Assessment

In the first step, each individual ecosystem service was evaluated on the basis of the established methods. This was done at different levels of scale. Some criteria were assessed at the local level (e.g. structural diversity), others had to be assessed at the level of the whole urban area (e.g. air duct).

Fig. 8 Mapped assets for the research area Nove Mesto Děčín

Time aspects (e.g. seasons) were not taken into account. Figures 9 and 10 show the example of habitat provision for each area with the results of the individual evaluation for each criterion, possible additional criteria, the total value added and the agreed value level. Seven of eighteen assets got high, eleven medium and two low values. On this basis, high-quality areas and areas with deficits can be identified (see Figs. 9 and 10).

5.3 Multifunctional Assessment

The multifunctional assessment sums up the aggregated values of each assessed ES. These values visualize the "top" assets that must be protected and the "worst" assets with potential for further development. In the example, the park areas have the highest values while settlement areas (e.g. hospital, villas with private gardens) get the lowest values (see Figs. 11 and 12). Based on the evaluation, potentials can be identified and developed in combination with the monofunctional approach.

5.4 Impact Evaluation of Planning and Design Strategies

After the evaluation of the individual areas recorded in the terrain, future scenarios with individual measures were developed for the respective areas by the project participants. An increase in the supply of living space through the creation of new structures or qualitative improvements to existing areas, for example through unsealing or adapted maintenance, can be represented and evaluated in a spatially concrete way. Likewise, improvements in the equipment or quality of tourist infrastructures. By using the tables, each measure or idea can be examined in terms of its impact on each individual ecosystem service as well as on all ecosystem services studied. In general, this procedure can be carried out at different levels of scale:

- defining visions (e.g. eliminate all existing negative impacts) including monofunctional or multifunctional assessment,
- developing strategies for groups of assets (e.g. protection of assets with manifold ecosystem services, enhancements for all assets with an aggregated value of a minimum of 2),
- designing individual assets and analysing the impacts on ES using the multifunctional assessment,
- finding new maintenance strategies for individual or a set cof assets to enhance the quality of live and biodiversity,

Asset	Description	Degree of sealing	Closeness to nature	Biotope size/ connectivity	Structure diversity	Additional criterion a)	Additional criterion b)	Summed value	Aggregated value
1	park with creek and old trees	3	2	3	2	0,5	1	11,5	3
2	park with a pond	3	2	3	2	0,5	1	11,5	3
3	park with monument and benches	3	2	3	2	0,5	1	11,5	3
4	little wild of a park	3	2	3	3	0,5	0	11,5	3
5	single standing multi living houses	2	1	2	1	0	1	7	2
6	broad leave forest-park	3	3	3	3	1	1	14	3
7	villas with private garden	2	1	2	1	0	1	7	2
8	parking lots with lawn	1	1	2	1	0	0	5	2
9	hospital	0	1	1	0	0	0	2	1
10	urban wasteland	3	2	2	2	0	1	10	3
11	green urban way with natural playground	3	2	2	2	0	0	9	2
12	private gardening	2	1	2	1	0	0	6	2
13	residential estate	1	1	1	1	0,5	1	5,5	2
14	green urban area with long term succession	3	3	2	3	0	1	12	3
15	mostly old allotment	2	2	1	2	0	0	7	2
16	single houses and gardens	1	1	1	1	0	0	4	1
17	single family houses and villas	2	1	1	1	0	1	6	2
18	allotment and fenced shrubs	3	2	1	2	0	0	8	2

a) existence of structural elements
b) inside protected areas

Aggregation

Summed value	aggregated value
0 - 4,5	1 - low
>4,5 - 9	2 - medium
>9 - 14	3 - high

Fig. 9 Monofunctional assessment of the ES habitat provision for each asset in Nove Mesto Děčín. Adapted from Bartůněk et al. (2018)

- implementing the framework in decision-making processes (e.g. public, stakeholder, politicians) to assess different design or management strategies at the local level or
- prioritization of measures or adapted management.

Figure 13 shows a combination of planning strategies and individual measures (see explanation below the table). For example, the creation of new green structures as well as nesting aids and structural elements are planned for area no. 8. The impact on each individual ecosystem service can be determined and presented (see impact column). This can lead to changes within an ecosystem service or to a change in the overall assessment (see column changes in aggregated value). In this example, areas 15 and 18 show improvements in one ecosystem service without showing an increase in the aggregated value. In contrast, the improvements in areas 4, 10, 11 and 14 also lead to an increase in the total value. At the same time, negative effects within the respective ES can be analysed and presented. It is conceivable, for example that the establishment of new recreational infrastructures or an intensification of the maintenance of green spaces will lead to an increase in the respective ES. At the same time, negative effects on habitat quality can be associated with this, as valuable structural elements may be lost or the adverse effects of noise may increase. On this basis, optimized individual measures as well as packages of measures can be defined in a spatially concrete manner. Figure 14 shows the results for the ecosystem service recreation.

6 Discussion

Green infrastructure provides a wide range of ecosystem services and it supports human well-being and biodiversity in cities (Heiland et al., 2015, 2016; Kowarik et al. 2017). In sustainable planning, social, economic and ecological aspects must be taken into consideration. Mapping and assessment of ecosystems services at different levels and a framework to implement the results in planning, management and decision-making are necessary (Kowarik et al. 2017; Rößler 2015; Renk 2019).

Using the example presented, an approach for the multifunctional analysis of ecosystem services of selected urban areas was shown using the city of Děčín. For this purpose, three ecosystem services were selected, since they are representative in the study areas. For the demarcation of homogeneous subareas, field surveys were carried out and necessary information was collected by means of a data entry form. In the course of the mapping work, it became

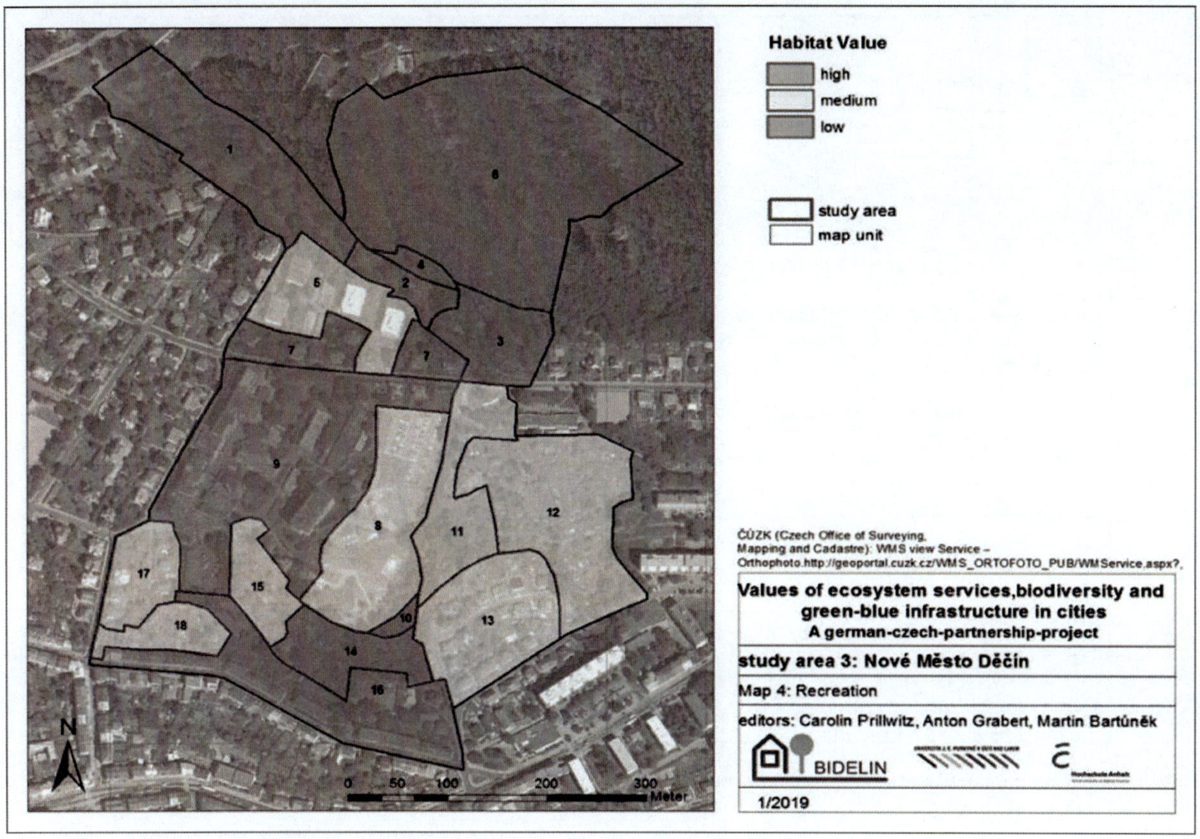

Fig. 10 Visualization of monofunctional assessment of the ES habitat provision for each asset in Nove Mesto Děčín (Bartůněk et al., 2018) (high = Nr. 1, 2, 3, 4, 6, 10,14; medium = Nr. 5, 8, 11, 12, 13, 15, 17, 18; low = Nr. 7, 9, 16)

clear that especially the recording of the current use shows considerable deficiencies. While existing recreational infrastructure, its quality and activities at the time of the survey can be recorded, a selective survey is still necessary. Thus, it is not possible to determine the full range of possible uses or to determine the complete range of uses, which leads to considerable deficits in the evaluation. The situation is similar with the recording of the degree of sealing or the green portion. Especially for larger areas an estimation in the terrain is difficult.

On the other hand, an assessment of climate regulation was carried out. As the areas delimited in the terrain have a high degree of detail, an urbanistic view is more appropriate for the methodology used. Against this background, it is more reasonable to deviate from previously delimited subareas in the evaluation of the ecosystem services mentioned above.

Another challenge is the definition of the evaluation framework, especially with regard to the equipment. A purely quantitative recording and evaluation does not seem to make sense here. For this reason, supplementary qualitative parameters such as path quality were used. However, how the equipment is to be evaluated in principle must be determined location-specifically depending on the type of area (e.g. park) and the study area.

The use of existing data sources, such as land use plans, protected area data, open street map, noise protection plans and hiking maps can significantly improve the assessment. In this context, the creation of a working map with all available data with a subsequent field check is useful.

The derivation of action strategies and measures was carried out by the project staff. The cooperation with stakeholders, citizens or decision-makers would be necessary in the future. In particular, the review of the procedure and the evaluation framework could be improved.

Generally speaking however, the approach presented can be adapted and extended very flexibly. The integration of further ecosystem services, the use of other methods or the extension of the existing ones as well as an application at different scale levels is possible. An implementation in existing GI systems or online tools is feasible and could thus be used for decision support in urban and open space planning.

Asset	Description	Habitat provision	Recreation	Climate regulation	Summed value	Aggregated value
1	park with creek and old trees	3	3	3	9	3
2	park with a pond	3	3	2	8	3
3	park with monument and benches	3	3	2	8	3
4	little wild of a park	3	2	2	7	2
5	single standing multi living houses	2	2	1	5	2
6	broad leave forest-park	3	3	3	9	3
7	villas with private garden	2	1	1	4	1
8	parking lots with lawn	2	2	2	6	2
9	hospital	1	2	1	4	1
10	urban wasteland	3	2	2	7	2
11	green urban way with natural playground	2	2	2	6	2
12	private gardening	2	1	2	5	2
13	residential estate	2	1	1	4	1
14	green urban area with long term succession	3	1	2	6	2
15	mostly old allotment	2	1	2	5	2
16	single houses and gardens	1	1	1	3	1
17	single family houses and villas	2	1	1	4	1
18	allotment and fenced shrubs	2	1	2	5	2

Aggregation

Summed value	aggregated value
0 - 3	1 - low
4 - 6	2 - medium
7 - 9	3 - high

Fig. 11 Multifunctional assessment of all ES in Nove Mesto Děčín. Adapted from Bartůněk et al. (2018)

Fig. 12 Visualization of multifunctional assessment of all ES in Nove Mesto Děčín (Bartůněk et al., 2018) (high = Nr. 1, 2, 3, 6; medium = Nr. 4, 5, 8, 10, 11, 12, 14, 15, 18; low = Nr. 7, 9, 13, 16, 17)

Asset	Habitat provision		Recreation		Climate regulation		Changes in aggregated value
	Strategies / measures	Impact	Strategies/measures	Impact	Strategies / measures	Impact	
1	a)	no	a)	no	a)	no	→
2	a)	no	a), b)	no	j)	no	→
3	a)	no	a)	no	j)	no	→
4	a)	no	c), d)	2 → 3	j)	no	↑
5	j)	no	e)	no	a)	no	→
6	a)	no	a), c)	no	j)	no	→
7	j)	no	j)	no	j)	no	→
8	e), f)	no	j)	no	j)	no	→
9	j)	no	j)	no	j)	no	→
10	j)	no	c), d)	no	h)	2 → 3	↑
11	f)	2 → 3	c), e)	2 → 3	j)	no	↑
12	j)	no	j)	no	j)	no	→
13	j)	no	j)	no	a), h)	no	→
14	a)	no	b), c), d), g)	1 → 3	h)	no	↑
15	f), i)	2 → 3	b), c), d), g)	1 → 2	a)	no	→
16	f)	no	j)	no	j)	no	→
17	j)	no	j)	no	j)	no	→
18	f), i)	2 → 3	b), c), d)	1 → 2	j)	no	→

Strategies / measures

a) protection of existing structures and functions
b) eliminate negative impacts
c) improve way quality
d) add recreation infrastructure
e) add green infrastructure elements
f) add nesting aids and addtional landscape elements
g) improve accessibility
h) enhance climate air-ducts
i) improve connectivity of biotopes
j) no enhancement

Symbol	Change
↓	decrease
→	no
↑	increase

Fig. 13 Development of planning strategies and individual measures with integrated impact assessment (Bartůněk et al., 2018)

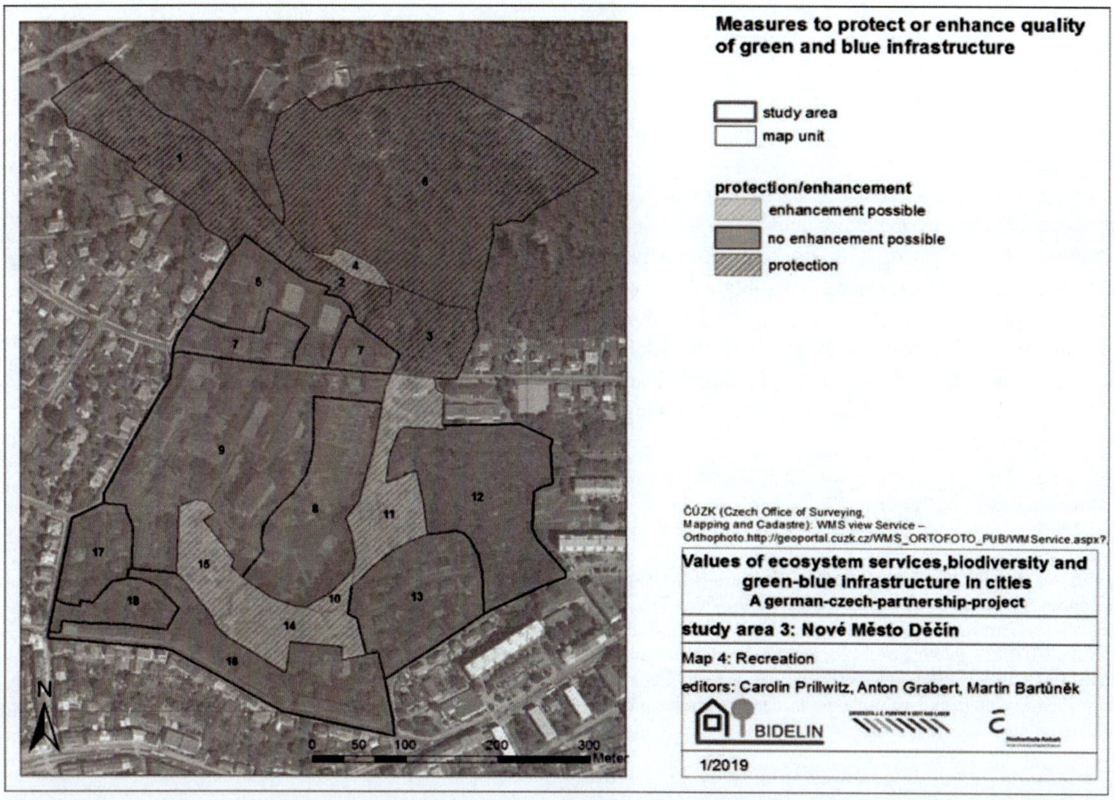

Fig. 14 Visualization of planning strategies for Nove Mesto Děčín (recreation) (Bartůněk et al., 2018) (enhancement possible = Nr. 4, 10, 11, 14, 15; no enhancement possible = Nr. 5, 7, 8, 9, 12, 13, 16, 17, 18; protection = Nr. 1, 2, 3, 6)

7 Conclusions

The presented concept can be used to make it clear to stakeholders within which socio-economic and biophysical framework they can operate. An economic valuation which has been sometimes used and favoured for ecosystem service assessment is on the one hand easier to understand for decision-makers but, on the other hand, often critically discussed especially concerning ethical aspects. The presented framework uses existing well-known and accepted methods. It is used in landscape planning in Germany for many years. In comparison to the landscape planning procedure, the synergistic evaluation of different ecosystem services for mapped assets provides a transparent tool for participation and decision-making processes. The impact evaluation of future alternatives and the assessment of existing ecosystem services and disservices provide additional advantages and motivation for such transdisciplinary work. An integration or adaption of additional assessment methods in the framework is possible allowing to enlarge the range of ecosystem services under consideration. The framework can be used at different scale levels and with different data sources.

Planners and decision-makers can use a comparison of each ecosystem service, the aggregated value between the existing situation and future scenarios to get necessary information for their work. Based on the results, scenarios for improvements in specific assets can be developed and the impact on one ecosystem service (monofunctional approach) and on several ecosystem services (multifunctional approach) become recognizable. Benefits and damages as well as positive and negative values can be visualized in a very transparent way.

The framework helps predict new strategies and implements the evaluation of synergies and impacts. It can be implemented into online tools or used for interactive discussion or participation processes. This might be a way to enhance transparency for urban planning in the future. Of course, a regular application should include additional methods, especially for the assessment of cultural ecosystem services in order to effectively support future planning processes (e.g. aesthetics).

Acknowledgements This research was realized within the project BIDELIN—The value of ecosystem services, biodiversity and blue-green infrastructures in cities, exemplified by Dresden, Liberec and Děčín (reg. no. 102282320), financed by the European Union from the European Regional Development Fund within the Programme Interreg V-A—Germany/Saxony-Czech Republic. We thank the students of the BIDELIN international course on ecosystem services who conducted the map survey, did the basic calculations and prepared the maps for their inventive and valuable work, namely Martin Bartůněk, Anton Grabert and Carolin Prillwitz.

References

Albert, C., Von Haaren, C., & Galler, C. (2012) Ökosystemdienstleistungen. Alter Wein in neuen Schläuchen oder ein Impuls für die Landschaftsplanung? [Old wine in new bottles or a stimulus for landscape planning?]. *Nat.schutz Landsch.plan, 44*, 142–148.

Bartůněk, M., Grabert, A., & Prillwitz, C. (2018). *Values of ecosystem services and blue-green infrastructure in cities—German-Czech-partnership project*. Anhalt University of Applied Sciences.

Bastian, O., Haase, D., Grunewald, K. (2012). Ecosystem properties, potentials and services–The EPPS conceptual framework and an urban application example. Ecological Indicators 21:7–16

Bastian, O., Syrbe, R.-U., Rosenberg, M., Rahe, D., & Grunewald, K. (2013). The five pillar EPPS framework for quantifying, mapping and managing ecosystem services. *Ecosystem Services* 15–24.

Batian, O., & Schreiber, K.-F. (1999). *Analyse und ökologische Bewertung der Landschaft*. (2nd ed.). Spektrum Akademischer Verlag.

Bennett, E. M., Peterson, G. D., & Gordon, L. J. (2009). Understanding relationships among multiple ecosystem services. *Ecology Letters, 12*, 1394–1404

Braat, L. C., & De Groot, R. (2012). The ecosystem services agenda: Bridging the worlds of natural science and economics, conservation and development, and public and private policy. *Ecosystem Services, 1*(1), 4–15

Burkhard, B., Kroll, F., Nedkov, S., & Müller, F. (2012). Mapping supply, demand, and budget of ecosystem services. *Ecological Indicators, 21*, 17–29

De Groot, R. S., Alkemade, R., Braat, L., Hein, L., & Willemen, L. (2010). Challenges in integrating the concept of ecosystem services and values in landscape planning, management and decision making. *Ecological Complexity, 7*, 260–272

De Vreese, R., Leys, M., Fontaine, C. M., & Dendoncker, N. (2016). Social mapping of perceived ecosystem services supply—The role ofsocial landscape metrics and social hotspots for integrated ecosystemservices assessment, landscape planning and management. *Ecological Indicators, 66*, 517–533

European Commission. (2013). Communication from the Commission to the European Parliament, the Council, the European Economic and Social Committee and the Committee of the Regions. Green Infrastructure (GI) — Enhancing Europe's Natural Capital. (https://eur-lex.europa.eu/legal-content/EN/TXT/HTML/?uri=CELEX:52013DC0249from=EN)

European Environment Agency. (2011). Green infrastructure and territorial cohesion. The concept of green infrastructure and its integration into policies using monitoring systems, Copenhagen (https://www.eea.europa.eu/publications/green-infrastructure-and-territorial-cohesion).

European Environment Agency. (2017). What is green infrastructure (https://www.eea.europa.eu/themes/sustainability-transitions/urban-environment/urban-green-infrastructure/what-is-green-infrastructure)

Farley, J. (2012). Ecosystem services: The economics debate. Ecosystem Services 2012, 1(1), 40–49

Heiland, S., Bredow, L., Hokema, D., Nowak, D., Rittel, K., Wanka-Pail, E., & Wilke, T. (2015). Gesundheitsförderung durch städtische Grünräume – Aufgaben für Naturschutz, Landschafts- und Freiraumplanung. *Natur Und Landschaft, 1*, 2–7

Heiland, S., Kahl, R., Sander, H., & Schliep, R. (2016). Ökosystemleistungen in der kommunalen Landschaftsplanung. *Möglichkeiten*

Der Integration. Naturschutz Und Landschaftsplanung, 48(10), 313–320

Interreg Central Europe Project MaGICLandscapes (Ed.). (2019). Green infrastructure handbook—Conceptual and theoretical background, terms and definitions. Project Output O.T1.1, Dresden. With contributions from: John, H., Marrs, C., Neubert, M., Alberico, S., Bovo, G., Ciadamidaro, S., Danzinger, F., Erlebach, M., Freudl, D., Grasso, S., Hahn, A., Jała, Z., Lasala, I., Minciardi, M., Rossi, G. L., Skokanová, H., Slach, T., Uhlemann, K., Vayr, P., Wojnarowicz, D., Wrbka, T. Published online https://www.interreg-central.eu/Content.Node/MaGICLandscapes.html

Jennings V., & Bamkole O. (2019). The relationship between social cohesion and urban green space: An avenue for health promotion. *International Journal of Environmental Research and Public Health, 16*(3), 452.

Kowarik, I., Bartz, R., Brenck, M., & Hansjürgens, B. (2017). *Natural capital Germany—Ecosystem services in the city—Protecting health and enhancing quality of life*. Leipzig.

Landscape Institute. (2009). Green infrastructure: connected and multifunctional landscapes — position document (www.landscapewpstorage01.blob.core.windows.net/www-landscapeinstitute-org/2016/03/GreenInfrastructurepositionstatement13May09.pdf)

Mosimann, T., Frey, T., & Trute, P. (1999). Schutzgut Klima/Luft in der Landschaftsplanung. *Informationsdienst Naturschutz Niedersachsen, 19*(4), 201–216

Pauleit, S., Hansen, R., Rall, E. L., Zölch, T., Andersson, E., Luz, A. C., Szaraz, L., Tosics, I., & Vierikko, K. (2017). Urban landscapes and green infrastructure. In *Oxford research encyclopaedia of environmental science* (pp. 1–53).

Pietsch, M., Uhrig, N., Sayadyan, H., & Kim, D. Y. (2019). Analyzing ecosystem services in Armenia using ArcGIS online: A case study of the Gerhard monastery and the kayak province. *Journal of Digital Landscape Architecture, 4–2019*, 363–371

Renk, J. (2019). Grüne Infrastruktur - Umsetzungsstrategien für die kommunale Planungspraxis. Naturschutz und Landschaftsplanung, Band 51, Heft 2, 62–65

Rößler, S. (2015). Klimagerechte Stadtentwicklung durch grüne Infrastruktur. Raumforschung und Raumordnung, Band 73, Heft 2, 2015, 123–132.

Rodriguez, J. P., Beard, T. J., Bennett, E. M., Cumming, G. S., Cork, S. J., Agard, J., Dodbson, A. P., & Peterson, G. D. (2016). Trade-offs across space, time and ecosystem services. *Ecological Society, 11*, 28

Syrbe, R.-U., & Grunewald, K. (2017). Ecosystem service supply and demand—The challenge to balance spatial mismatches. *International Journal of Biodiversity Science, Ecosystem Services and Management, 13*(2), 148–161.

Syrbe, R.-U., & Walz, U. (2012). Spatial indicators for the assessment of ecosystem services: Providing benefiting and connecting areas and landscape metrics. *Ecological Indicators, 21*, 80–88

TEEB. (2010). The economics of ecosystems and biodiversity ecological and economic foundations. In P. Kumar (Ed.), *Earthscan*. London and Washington.

Turner, M. G., & Gardner, R. H. (2015). *Landscape ecology in theory and practice*. (2nd ed.). Springer.

Von Haaren, C. (2004). *Landschaftsplanung*. Verlag Eugen Ulmer.

Biophilic Institutions and Governance: Biophilic Urbanism Initiatives (BUIs) Fostering Green Urban Features in Emerging and Developing Cities

Virginia Carter and Cristián Henríquez

Abstract

One of the goals promoted by the '2030 Agenda for Sustainable Development' is to make cities more inclusive, safe, resilient, and sustainable, responding to the challenges that cities face today, including the increase of urban temperature because of climate change effects, and the deficit of urban green spaces. Within this context, green urbanism aims to create more sustainable cities in terms of greenness, compactness, energy efficiency, zero emissions, and waste production, having as a result sustainable places, communities, and lifestyles. The approach arises to address some of the challenges mentioned above, through seven features expressed in emerging archetypal cities, where the 'biophilic city' is one of them. Biophilic cities are the materialization of the biophilic urbanism approach, which pursues the strategic integration of nature in cities through planning and design to achieve more livable cities in terms of the well-being and health of their inhabitants, as well as the creation of resilient urban environments. Firstly, this research compiled initiatives performed under the biophilic urbanism, defined here as biophilic urbanism initiatives (BUIs), where the main differences in its implementation can be traced back to income differences between countries. Secondly, it presents opportunities, as well as challenges for greening emerging and developing cities, based on one of the five categories of indicators associated with biophilic cities, 'biophilic institutions and governance.' The indicators comprising this group aim to describe the extent to which local governments are involved in encouraging green features and biophilic design. Finally, this research draws contributions already implemented by a group of Latin American cities located in Argentina, Chile, Colombia, and Peru. One of our main findings is that these initiatives have not yet been widely implemented in emerging and developing countries, compared with advanced economies, such as the USA, Germany, and Singapore, among others.

Keywords

Biophilic urbanism initiatives (BUIs) • Biophilic institutions and governance • Biophilic cities • Green urban features • Biophilic urbanism • Emerging and developing economies

V. Carter (✉)
Department of Geography, Ghent University, Ghent, Belgium
e-mail: mariavirginia.cartergamberini@ugent.be

C. Henríquez
Geography Institute, Pontificia Universidad Católica de Chile, Santiago, Chile

C. Henríquez
CEDEUS and Centro de Cambio Global UC, Santiago, Chile

1 Introduction

Urban areas, especially in emerging and developing countries, face several challenges today, i.e., climate change effects derived from local temperature increases and changes in rainfall patterns (Masson-Delmotte et al., 2018); projections of urbanization during the next decades (United Nations [UN], 2015) from 50% in 2018 to 60% by the year 2030 (UN, 2018a); lack of urban green spaces (Rigolon et al., 2018); and the vulnerability in the presence and quality of ecosystem services (Chen et al., 2015).

Accordingly, 193 countries adopted for the year 2030 the Agenda for Sustainable Development, which includes the Goal 11 (SDG11) of 'Sustainable Cities and Communities.' This goal aims to make cities inclusive, safe, resilient, and sustainable, encouraging governments to protect and safeguard the world's natural heritage (target 11.4), and to provide universal access to safe, inclusive, and accessible, green and public spaces (target 11.7) (UN, 2015). Consequently, the inclusion of nature as a centerpiece in the planning process of cities is needed (Beatley, 2011).

F. Trapani et al. (eds.), *Advanced Studies in Efficient Environmental Design and City Planning*, Advances in Science, Technology & Innovation, https://doi.org/10.1007/978-3-030-65181-7_29

Latin America (LA), as part of the emerging and developing regions, is no stranger to this situation. Here, climate change is a critical environmental challenge with physical and economic consequences (Organisation for Economic Co-operation and Development [OECD], 2019). LA is positioned as the most urbanized region in the developing world, with 80% of its population living in cities and with projections to reach 90% by 2050 (Pauchard & Barbosa, 2013). Here, there remain significant challenges related to the performance of SDG11 in terms of green and public space provision. The WHO suggested a standard of a minimum of 9 m^2 green areas per capita and an ideal value of 50 m^2 green areas per capita (Badiu et al., 2016), and Latin American cities are far from those numbers. Some examples are Buenos Aires in Argentina, with 1.90 m^2 per capita (Morar et al., 2014), or Santiago, in Chile, with 3.46 m^2 per capita (Organisation for Economic Co-operation and Development (OECD), 2013). For all these reasons, urban transformations for achieving sustainable cities (in terms of city greenness and people well-being), and the subsequent participation of governmental stakeholders, are essential for promoting adequate green spaces, and other public infrastructure for cycling and walking (Sachs et al., 2019). Achieving progress in terms of sustainability planning is strongly related to the prevailing governance system (Wheeler, 2013), and local governments, as performers of the planning and design of cities, are key actors in the context of a sustainable urban transformation (Mc Cormick et al., 2013). Hence, local governments will be fundamental in the process of influencing new structures toward ambitious environmental objectives.

Based on the biophilia concept, the biophilic urbanism approach is associated with the introduction of nature into cities, through planning and design (Beatley, 2017) aiming for more sustainable and resilient urban environments to improve the quality of life of their inhabitants (Cook, 2016; Jennings et al., 2019; Littke, 2016; Russo & Cirella, 2017; World Health Organization [WHO], 2016). Biophilic urbanism becomes materialized through biophilic cities, which are cities full of nature and natural systems, visible and accessible to everyone (Beatley, 2011).

This article presents the geographies of biophilic urbanism initiatives (BUIs) implemented at national and local levels worldwide, defined here as those initiatives accomplished in the context of city planning actions, which emphasizes in several, overlapping ways the contact of urban dwellers with nature. This research focuses on initiatives published in scientific literature associated with biophilia and biophilic urbanism fields, linked to specific countries and cities. Our analysis was based on the creation of a self-proposed classification of BUIs, grounded on the primary purposes, identified after analyzing the complete ensemble of initiatives collected. Secondly, this article shows projects promoted by local governments, supporting and encouraging the implementation of biophilic urbanism for greening cities under the indicators grouped in the category named 'institutions and biophilic governance.' This paper presents initiatives already performed by advanced economies, and emerging and developing governments worldwide, as well as those published by LA countries on their National Systems of Spatial and Environmental Information Web sites. Specifically, the research shows initiatives performed in Argentina, Colombia, Chile, and Peru, as emerging and developing study cases.

1.1 Green Urbanism as a Framework of Biophilic Urbanism

Green urbanism, as an interdisciplinary approach, promotes the greening of cities to create more sustainable urban environments in terms of their compactness, energy efficiency, zero emissions, and waste production, while promoting healthy communities and lifestyles (Beatley & Newman, 2013; Lehmann, 2011a, b; Newman & Matan, 2013). Based on the interaction among three pillars: (1) energy and materials, (2) water and biodiversity, and (3) urban planning and transport, fifteen principles are emerged for integrating green urbanism into urban development. Some of the principles related to biophilic urbanism goals are (a) landscape, gardens, and biodiversity; (b) green buildings and districts, using passive design principles; (c) livability, healthy communities, and mixed-use programs; d) local food and short supply chains; (e) urban governance, leadership, and best practice; (f) education, research, and knowledge; and (g) strategies for cities in developing countries.

To achieve more environmentally friendly cities, continuous interaction among these principles is essential (Lehmann, 2011b). Biophilic cities capture the previous principles to shape and prepare cities toward high density when biophilic features are introduced by reducing the urban heat island effect, increasing the carbon sequestration, and finally promoting healthy communities (Newman et al., 2017). Building upon the mentioned principles, seven archetypal cities have been put forward within the broader approach of green urbanism. These are the renewable energy city, the bioregional carbon neutral city, the distributed city, the eco-efficient city, the place-based city, the sustainable transport city, and the biophilic city (Newman & Matan, 2013).

1.2 Biophilia and Biophilic Urbanism

Biophilia was initially studied in the field of psychology and preventive medicine and was primarily associated with improvements experienced by patients exposed to contact with nature (Grinde & Patil, 2009; Kaplan, 1995; Ulrich, 1984). Wilson (1984) popularized the concept from a conservationist and evolutionary perspective, defining it as the innate need of human beings to be in permanent contact with nature (Beatley, 2009; Jones, 2013; Kahn, 1997). Kellert (1993) did a seminal work related to the connection between the innate need of contact with nature and its incorporation into the built environment; but it was Beatley (2009, 2011, 2012) who linked biophilia with urban spaces, coining the concept of biophilic urbanism and the biophilic city.

Biophilic urbanism, as a holistic approach, suggests a broad relationship between nature and cities, namely (1) the commitment of inhabitants and governments with the promotion and insertion of nature, (2) wildlife conservation, (3) motivation to spend time outdoors, (4) restoration and renovation of green spaces, urban natural reserves, and green infrastructure, and (5) mental and physical health (Beatley, 2011; Reeve et al., 2015; Xue et al., 2019). A biophilic city is a city with an abundance of nature, lived from outdoors, which aims to create connections among nature and its inhabitants, and that frequently looks for the integration and recovering of all the natural features within the city (Beatley, 2011, 2017).

Concepts such as 'green infrastructure (GI),' 'ecosystem services (ES),' 'urban green infrastructure (UGI),' and 'nature-based solutions (NBS)' arise as complementary approaches, promoted by the meaning of the biophilic urbanism, which also helps to better understand the relevance of the biophilic urbanism approach, and its materialization through biophilic cities. Green infrastructure is defined by the European Commission (2014, p. 7) as 'a strategically planned network of high quality natural and seminatural areas with other environmental features, which is designed and managed to deliver a wide range of ecosystem services and protect biodiversity in both rural and urban setting.' Ecosystem services are the benefits and functions provided by the ecosystems for the people (Cortinovis & Geneletti, 2018; European Commission, 2015; Sadler et al., 2018; Woodruff & BenDor, 2016). UGI, defined as the conceptualization of the connected green space in urban environments (Davies & Lafortezza, 2017), is an undeniable central provider of ES (Artmann et al., 2017; Badiu et al., 2016). Finally, NBS are actions, inspired or supported by nature, that emerge as the operationalization of the ecosystem services. These actions seek to insert the approach of ES in spatial planning concerns (Scott & Lennon, 2016); face challenges suffered by societies, e.g., climate change, food security, or natural disasters, and protect human well-being (Cohen-Shacham et al., 2016; Fink & Kaltenegger, 2016; Nesshöver et al., 2017; A. Russo et al., 2017).

1.3 The Biophilic City

Biophilic cities are characterized by the commitment of their local governments and their inhabitants to place nature in the first place in its design, planning, and management (Beatley, 2011). Biophilic cities seek to generate a connection between inhabitants and nature in different scales, 'from the intimate immersion in the natural environment to the visual and acoustic connection' (Huang, 2017, p. 28). To be considered a biophilic city, cities need infrastructure, such as green parks, roofs, and walls; green alleys; ecological behaviors; and inhabitants and governments interested in increasing their knowledge about nature. They also require the commitment of the authorities through budgets, legislation, and education programs, as well as river restoration initiatives, biophilic neighborhoods, and community garden programs.

Last but not least, they need institutions that promote the enhancement of nature, such as NGOs or museums (Beatley, 2009; Beatley & Newman, 2013). Beatley (2011) proposed a list of twenty indicators that characterize a biophilic city. These indicators are assembled into four groups, namely: (a) biophilic conditions and infrastructure, (b) biophilic activities, (c) biophilic attitudes and knowledge, and (d) biophilic institutions and governance. Table 1 depicts the indicators included in the category 'institutions and biophilic governance.'

Table 1 Indicators for a biophilic city included in the category institutions and biophilic governance (based on Beatley, 2011)

Indicator ID number	Indicator name
1	Adoption of a local biodiversity action plan or strategy
2	Priority given to environmental education
3	Percent of local budget devoted to nature conservation, recreation, education, and related activities
4	Adoption of green building and planning codes, grant programs, density bonuses, green space initiatives, and dark-sky lighting standards
5	Number of city-supported biophilic pilot projects and initiatives
6	Extent of local biophilic support organizations, such as an active natural history museum or botanical garden

The way governments structure their funding, budget, laws, and codes, as either barriers or incentives is essential to pursue more resilient and sustainable cities (Beatley, 2011; Evans et al., 2005) and to achieve biophilic governance.

2 Methodology

In order to generate and analyze a comprehensive dataset of worldwide implemented BUI and projects explicitly linked to biophilic governance, we followed a five-step procedure. First, we searched for BUIs in the scientific literature comprising major international journals and books. The selectivity criteria to be part of this dataset were for an initiative to be linked to a specific city. Second, we analyzed the initiatives and proposed a classification of BUIs. We created this classification based on the primary purpose of each BUI (see Table 2). Third, every project in the dataset was associated with the country based on the city information, and fourth, the countries were classified by income using the International Monetary Fund (IMF) classification. Finally, initiatives related to governments encouraging biophilic

Table 2 Categories of biophilic urbanism initiatives (BUIs) defined

BUIs categories	BUIs code	Types of BUIs identified in each category
Biodiversity enhancement	BE	Aquariums, Biodiversity Atlas, Biodiversity Guides, Biodiversity Maps, Biophilic Design Manuals, Biophilic Cities Network, Biophilic Research, Botanical Gardens, Desert Parks, Environmental Projects, Festivals, Museums, Guidelines, Groups for Biodiversity Conservation, Inventories, NGOs, Play Spaces, Protection of Mountain or Topographic Views, Research Centres, Science Programs, Studies, Zoo
Biophilic schools	BS	Environmental Education Project, Green Schools, Forest Schools, School Outdoor Classes, Walking School Bus Program
Buildings with green design elements	BG	Eco-villages, Forest Cities, Grass Roofs, Green Atria, Green Bridges, Green Courtyards, Green Walls (living walls), Roof Gardens and Parks, Vegetated Rooftops
Community activities	CA	Allotment Gardens, Associations, City Farms, Clubs, Community Forests, Community Gardens, Community Orchards, Community Roots, Cooperative Blocks, Corps, Ecological Communities or Neighborhoods, Farms, Foundations, Permaculture Projects, Restoration Groups, Sidewalk Gardens, Street Gardens, Urban Farms
Former infrastructure supporting nature insertion in the city	FI	Parks in Formers Airports, Elevated Freight Lines, Railway Lines or Yards
Government plans, policies, programs, strategies, or visions	GP	Agencies, Biotope Area Factor, City Visions, Compact Growth Districts, Eco-District, Goals, Initiatives, Plans, Policies, Programs, Strategies, Targets
Government regulations	GR	Acts, Codes, Laws, Mandates, Ordinances, Requirements, Resolutions, Schemes, Standards
Government support	GS	Bonuses, City Tax Incentives, Funding, Incentives Programs, Subsidies
Green belts, greenways, or networks	GN	Green Belts, Coastal Trails, Greenways or Networks, River Ring Networks, Urban Trail Networks
Green infrastructure	GI	Ecological Reserves, Green Alleys, Parks, Pocket Parks, Squares, Street Trees
Pedestrian preference zone	PZ	Car-Restricted Zones, Compact Pedestrian Zones, Pedestrian Districts, Pedestrian Squares
Restoration	RT	Forestation, Gardening Projects, Greening Projects, Planting Projects, River or Streams Restorations, Riverfront Parks, Waterfront Promenades
Wildlife enhancement	WE	Animal Conservation Projects, Bird Atlas, Fauna Bridges or Highways, Fauna-Friendly Structures, Guidelines, Lagoons, Marine Reserves, Nature Centers, Wild Street Projects, Wildlife Refuges

governance were gathered from the BUI dataset and compared with others not included in scientific publications.

2.1 Compilation of BUIs

Between October 2018 and April 2019, a BUI dataset was created using the repository of scientific research Scopus. This research considered information linked exclusively to biophilic projects implemented at the city level, not including initiatives implemented at higher spatial levels (national, regional, or provincial). A broader search was conducted by querying the following three combinations of concepts: (a) 'biophilic cities,' 'biophilic city,' and 'biophilia' (resulting in 203 findings); (b) 'biophilic cities' and 'biophilic city' (resulting in 592 findings); and finally, (c) 'biophilic urbanism' (resulting in 216 findings). After merging and removing the duplicate documents, the query yielded a total of 629 documents, of which their abstract or introduction was read. Only 198 documents were linked with the biophilic approach, and after a full-text screening, 127 documents contained information about biophilic projects linked to a city.

These initiatives, accomplished in the context of city planning, in some cases driven also by the community, were grouped in categories grounded on its primary purpose, emphasizing in several and overlapping ways the contact of urban dwellers with nature. The initiatives range from green infrastructure (i.e., parks, squares, roofs green, urban gardens, green belts), governmental plans, policies, and subsidies for supporting green features, to a broad set of community-type activities such as urban farms, biophilic schools, among others. Hence, initiatives performed under the biophilic urbanism were defined here as biophilic urbanism initiatives (BUIs).

2.2 Classification of BUI Gathered

As a second step, we defined 13 categories of BUIs. First, we classified projects according to their commonalities. Then, we named the categories based on their primary objective, and finally, each BUI gathered was associated with one category. Table 2 shows the type of initiatives identified for each category.

To compare this information, cities already implementing biophilic urbanism projects were associated with countries, and then, the countries were classified by income. Based on the GNI per capita (for 2018), the World Bank Group (WBG) (2020) classifies countries of the world according to their income (annual gross national income USD $ per capita), defining the following four categories: (a) low ($995 or less), (b) lower-middle (between $996 and $3,895), (c) upper-middle (from $3,896 to $12,055), and (d) high ($12,056 or more) (UN, 2018b). Based on this classification, the IMF (2019) groups countries as 'Advanced Economies' (AE) (high-income countries) and 'Emerging and Developing Economies' (EDE) (upper-middle countries, low-middle countries, and low-income countries).

2.3 Institutions and Biophilic Governance Initiatives

Based on the biophilic indicators depicted in Table 1, and using the dataset of BUIs gathered, we searched for homologous projects related to the involvement of local governments in encouraging green features and biophilic design within cities that had already been performed worldwide. We specifically searched for projects under the following types: biodiversity action plan; biodiversity program; biodiversity strategy; biophilic pilot initiatives; biophilic pilot projects; biophilic schools; density bonuses; environmental education; environmental programs; grant program bonuses; green buildings; green schools; green space initiatives; lighting standards; nature conservation; and nature restoration planning codes and planning mandates. The data associated with biophilic governance was finally found in 22 documents from the 127 described in Sect. 2.1. In order to show projects not included in scientific publications and linked explicitly to mandates or codes for greening cities, gray literature was revised. Information from official governmental Latin American Web sites was gathered, and study cases from Argentina, Chile, Colombia, and Peru were found.

3 Results and Discussion

3.1 General Overview

According to the information in our dataset, a total of 1,252 biophilic urbanism initiatives (BUIs) implemented worldwide were found. These initiatives were distributed among 52 countries and 280 cities (see Appendix). From the total of countries involved in BUI, 32 correspond to the group of AEs (high-income countries), and 20 falls into EDE classification (upper-middle-income, low-middle-income, and low-income countries).

Regarding the total of BUI gathered, AE implemented 1,159 initiatives (92%), and merely 93 BUIs (8%) were performed by the group of EDE. Figure 1 displays the total number of BUIs implemented by category and country classification, and Fig. 2 shows the spatial distribution of the BUIs gathered.

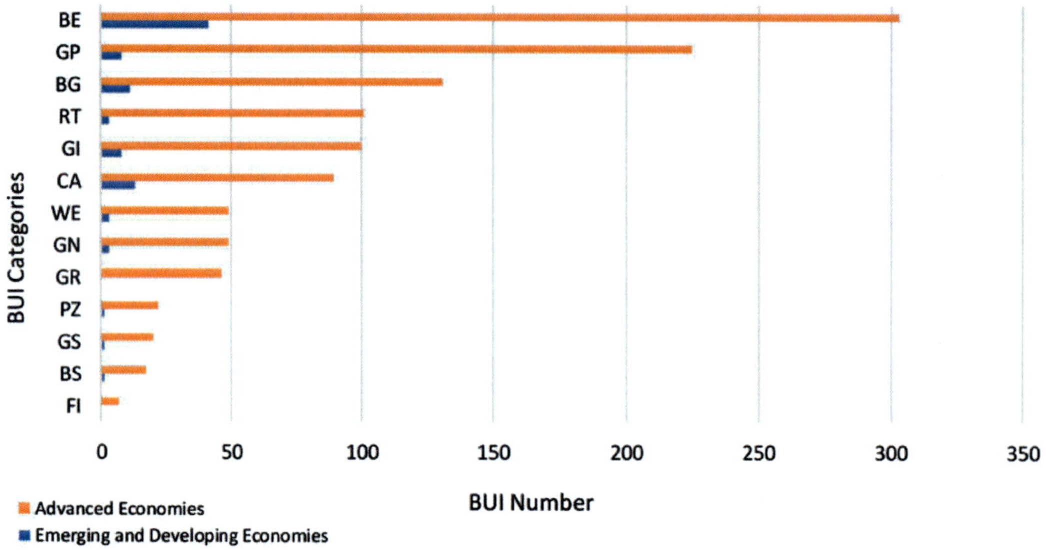

Fig. 1 Number of biophilic urbanism initiatives (BUIs) by category and country classification

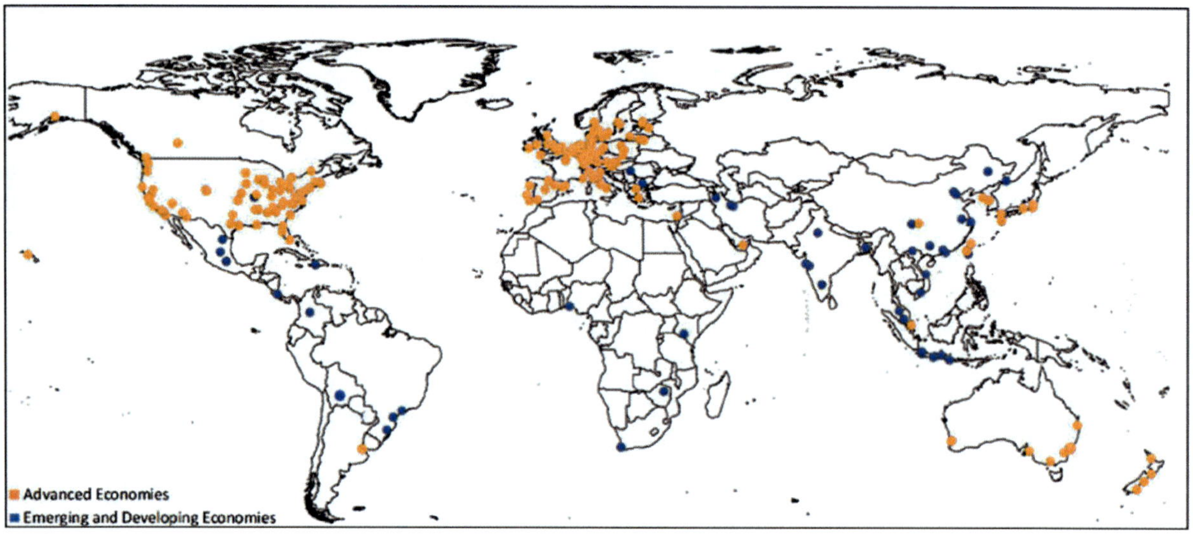

Fig. 2 Geographies of BUIs

Countries implementing the largest number of initiatives are coincidentally countries with a high GNI within each group, except India (see Table 3).

3.2 Biophilic Urbanism Projects Under the Biophilic Governance Approach

The data gathered, linked to biophilic governance initiatives, ended up in 152 initiatives distributed among 27 countries and 72 cities, where seven countries and eight cities were part of the EDE group. These countries were Brazil, Haiti, India, Indonesia, Malaysia, Mexico, and South Africa. Only three countries are part of LA, being those Brazil, Haiti, and Mexico. At the city level, four cities are part of LAC (Curitiba, Mexico City, Port-au-Prince, and Rio de Janeiro). Table 4 shows the total number of initiatives performed by EDE countries, gathered by each indicator in the category 'institutions and biophilic governance.'

Specifically related to the 'biophilic governance' BUIs, only ten initiatives (7% from the total of 152) belong to EDE countries, and LA countries performed five. No initiatives

Table 3 Countries implementing the largest number of BUIs

Country name	Number of initiatives found	GNI USD$ 2018
Advanced economies		
USA	461	62,850
Australia	84	53,190
The Netherlands	73	51,280
UK	68	41,330
Germany	65	47,450
Singapore	61	58,770
Emerging and developing economies		
Brazil	17	9,140
China	16	9,470
India	11	2,020
Mexico	10	9,180

Table 4 Number of 'biophilic governance initiatives' found by indicator

Indicator ID number	Indicator name	Number of initiatives found
1	Adoption of green building and planning codes, grant programs, density bonuses, green space initiatives, and dark-sky lighting standards	90
2	Adoption of a local biodiversity action plan or strategy	9
3	Extent of local biophilic support organizations, such as an active natural history museum or botanical garden	9
4	Number of city-supported biophilic pilot projects and initiatives	3
5	Percent of local budget devoted to nature conservation, recreation, education, and related activities	23
6	Priority given to environmental education	18

Table 5 Number of 'biophilic governance initiatives' by indicator

Project name	Country	City	Indicator ID number associated
Green plan	Mexico	Mexico City	1
City tax incentives for green rooftops			
Green corridors plan	Brazil	Rio de Janeiro	1
Rio botanical garden			3
Botanical gardens	India	Bangalore	3
Vaga Vivas project	Brazil	Curitiba	4
Restoration biodiversity initiatives—Martissant Park	Haiti	Port-au-Prince	5

included in the indicators 'adoption of a local biodiversity action plan or strategy' and 'priority given to environmental education' were found in LA. Table 5 depicts the specific name of the initiatives found in LA.

Related to the projects performed by AE, the countries showing the highest number of initiatives gathered under the category 'institutions and biophilic governance' were the USA (n = 56), Germany (n = 19), and Singapore (n = 10). From the total spectrum, Table 6 shows the specific type of initiatives identified for each indicator, by city.

After analyzing the number of initiatives performed by the EDE countries compared with the AE group in terms of the urban population (potentially benefited by biophilic urbanism), the distribution of initiatives observed is hugely uneven. The EDE countries, even having the 77% of the world urban population (WBG, 2020b), concentrates

Table 6 'Biophilic governance initiatives' found by indicator

Indicator name		Type of initiatives	Cities implementing initiatives **(in EDE cities)**
Adoption of green building and planning codes, grant programs, density bonuses, green space initiatives, and dark-sky lighting standards	Green space initiatives	**Green Corridors Plan**—Ecological Network Plan	Brisbane, Edmonton, Helsinki, London, **Rio de Janeiro**, Wellington
		Urban Forest Plan—Urban Forest Program—Urban Forest Strategy	Berlin, Brisbane, Houston, Los Angeles, Melbourne, New York, Phoenix, Sacramento, Vaxjo, Vienna, Zürich
		Tree Policy—Tree Planting Program—Tree and Shade Master Plan—Tree Monitoring Program—Tree Planting Initiative—Heritage Trees Scheme	Berlin, Brisbane, Erlangen, Freiburg, London, Milwaukee, Phoenix, Singapore
		Municipal Funding for Green Design	Chicago, Faenza, Portland, Vienna
		Green Alley Program—Green Roof Program—Green Roof Strategy—Green Walls Policy	Berlin, Chicago, Frankfurt, Kassel, Linz, Munich, Portland, Toronto
	Green building and planning codes	**Green Plan**—Green Policy—Green Strategy and Pocket Parks Program	Amsterdam, Berlin, Birmingham, Chicago, Copenhagen, Freiburg, Houston, Linz, **Mexico City**, Oslo, Philadelphia, Portland, San Francisco, Singapore, Tokyo, Toronto, Vancouver
		Green Roof by Law—Green Building Policy—Municipal Code Promoting Green Features—Standard for Green Roofs	Cologne, Portland, Port Coquitlam, Seattle, Toronto
		Zoning Codes for Urban Agriculture	Chicago, San Francisco
	Density bonuses	Bonuses for Greening—Green Density Bonuses—**City Tax Incentives for Green Rooftops**	Arlington County, Chicago, **Mexico City**, New York, Portland, Seattle, Vienna
	Grant programs	Incentive Scheme—Eco-Roof Incentive Program—Funding for the installation of green roofs—Municipal Funding—Skyrise Greenery	London, Saarbrücken, Singapore, Toronto
Adoption of a local biodiversity action plan or strategy		**Biodiversity Action Plan**—Biodiversity Strategy—Urban Biodiversity Strategy—Urban Forest Biodiversity Program	Adelaide, Berlin, **Cape Town**, Chicago, Dublin, Helsinki, London, Paris, Wellington
Extent of local biophilic support organizations, such as an active natural history museum or botanical garden		**Botanical Garden**	**Bangalore**, Cleveland, Key Biscayne, Malmö, Richmond, **Rio de Janeiro**, Singapore, Vitoria-Gasteiz
Number of city-supported biophilic pilot projects and initiatives		Pilot Green Walls—**Green Public Spaces**—Temporary Biodiversity Gardens	**Curitiba**, Fremantle, Paris
Percent of local budget devoted to nature conservation, recreation, education, and related activities		**River Restoration**—Streams Restoration—River Reconnection	Albufeira, Denver, Dortmund, Freiburg, Heidelberg, **Johor Bahru**, Leicester, London, Milwaukee, New York, Newark, Nice, Norfolk, North Charleston, **Port-au-Prince**, Portland, Richmond, San Luis Obispo, Seattle, Singapore, Vitoria-Gasteiz, Washington DC
Priority given to environmental education		Biophilic Schools—Environmental Education Project—Green Schools	Albertslund, Auckland, Badung, Charlottesville, Copenhagen, Denver, Freiburg, Helsinki, Kolding, Milwaukee, New York, Norfolk, Perth, San Francisco, Singapore, Washington DC, Zürich

Table 7 Examples of 'biophilic governance initiatives' promoted by governments in Latin America

Country	Spatial scale scope	Agreement-Decree-Law-Ordinance No. and Name	Promoter institution	Type	Brief description
Argentina	Buenos Aires (City)	Law No. 4.428 (2012). Green Roofs and Terraces. Legislatura de la Ciudad Autónoma de Buenos Aires. (2012)	Buenos Aires City Government	Not compulsory	Owners of buildings that implement and maintain green roofs have a reduction in the amount of lighting, sweeping, and cleaning
	Cordoba (City)	Ordinance No. 12.548 (2016). Incorporate into the building code the regulatory framework for the incorporation of green roofs and walls within the city (2016). Concejo Deliberante de la Ciudad de Córdoba. (2016)	Cordoba City Government	Compulsory	Forces to apply green roofs to new and already built residential and industrial buildings with terraces of more than 400 m^2. Also for large industrial and institutional buildings of more than 600 m^2 covered
Chile	Country	Decree No. 58, (2019). Modifies the General Ordinance on Urban Planning and Construction, to promote the construction of green roofs. Ministerio de Vivienda y Urbanismo (MINVU). (2019)	Ministry of Housing and Urbanism	Compulsory	The national urban planning and construction law was modified to allow a more extensive area available for terraces, the so-called green roofs and other ornamental elements, as well as solar panels. Previously, 20% of the surface was allowed for that kind of infrastructure. Now, 75% is allowed
Colombia	Bogota (Bogota District)	Agreement No. 418 (2009). The implementation of sustainable architectural technologies is promoted, such as green roofs or terraces. Consejo De Bogotá D.C. (2009)	Bogota City Government	Compulsory	Progressive and adequate implementation of roofs, green terraces, among other technologies, in new or existing public real estate projects, as a measure of adaptation and mitigation to climate change
Peru	Ate District (Lima Province)	Ordinance No. 342-MDA. Green Roof Program (2014). Municipalidad Distrital de Ate. (2014)	Municipality of Ate	Not compulsory	Created to encourage neighbors to install and progressively covered with vegetation roof surfaces in houses and buildings, among others
	Barranco District	Ordinance No. 427-MDB. Roof and Green Wall Program in Barranco District (2015). Municipalidad Distrital de Barranco. (2015)	Municipality of Barranco	Not compulsory	Intended to sensitize, involve, and encourage neighbors, owners of land premises in the district, to install progressively, vegetation on the surfaces of roofs and terraces of their homes and buildings, allowing gardens and plant cultivation

merely the 6% of the biophilic initiatives of governance, and 8% of the total BUIs gathered by this research.

While there is a consensus about higher-income countries likely to achieve better environmental quality, environmental indicators (Dutt, 2009; Mukherjee & Chakraborty, 2013), and better governance (Dutt, 2009; Lee et al., 2005), the previous analysis highlights the need to foster more active governance in EDE countries and cities.

3.3 Latin America Study Cases

As was mentioned in Sect. 3.2, compulsory initiatives promoted by governments for encouraging biophilic urbanism are not frequently implemented (being 11% of the total of BUIs gathered). Given this, they become relevant, especially in EDE countries which present a lack of urban green spaces. Table 7 shows examples of biophilic initiatives promoted by governments already implemented in LA.

Chile and Argentina are ranked at the top of the richest countries of the region, with 14,670 USD$ and 12,390 USD $, per capita, respectively (WBG, 2020a). Nevertheless, even among this group of countries, there are no records of municipal funding for green design or bonuses for greening bonuses. To conclude, few other initiatives were found at the city level in Chile, but mostly related to ordinances at the communal spatial scale, for regulating the design, maintenance, and renovation of green areas in communes, or prohibit the cutting of trees.

4 Conclusions

This chapter aimed to show the state of the art of biophilic urbanism implementation, linked explicitly to the governance for greening cities. Even though biophilic urbanism implementation is broadly related to high-income countries, strategic initiatives such as the promulgation of mandates and ordinances for encouraging and promoting the greening of cities and territories are imperative in EDE countries. This may be possible, either compulsorily by law, or through free processes linked with subsidies delivered by local or national governments.

We recommended the evaluation of some initial actions for increasing the biophilic urbanism performance in emerging and developing countries and cities: first, at the national spatial scale, through a compulsory implementation of initiatives such as green roofs in new private buildings, as well as public buildings such as governmental installations, hospitals, and schools, among others, and second, from the national spatial scale, aimed the provision of direct subsidies to cities and communes, suffering a scarcity of green infrastructure, urban green spaces, and low budgets.

To conclude, this research identified some avenues for future research, linked with the identification of the current governance required for implementing biophilic urbanism in emerging and developing economies, either inside the urban planning and design processes, as well as the governance for promoting biophilic urbanism initiatives related to urban health such as urban agriculture, ecosystem services, and educational policies linked to biophilic school programs.

Acknowledgements This paper was supported by the FONDECYT Regular Project No. 1180268, ANID/Chile.

Appendix

Cities implementing biophilic urbanism initiatives (BUIs) by group

BUI groups	Cities implementing BUI	Total of BUI
Biodiversity enhancement	Adelaide, Amherst, Amsterdam, Athens, Atlanta, Auckland, Austin, Baltimore, Bangalore, Bedford, Beijing, Bekasi, Berlin, Birmingham, Bogota, Boston, Brisbane, Buenos Aires, Cambridge, Cape Town, Changwon, Charlottesville, Chicago, Chóngqìng, Cleveland, Columbus, Copenhagen, Cornwall, Curitiba, Curridabat, Davis, De Ronde Venen, Denver, Dhaka, Dunedin, Durham, Edinburgh, Edmonton, Fremantle, Fukuoka, Gainesville, Garden Grove, Gothenburg, Greensboro, Guangzhou, Haifa, Hamilton, Harbin, Harmony, Hashtgerd, Hong Kong, Honolulu, Houston, Ikeja, Iowa, Kagoshima, Kalamaria, Karlsruhe, Kaunas, Key Biscayne, Klaipėda, Kolding, Kristianstad, Kuala Lumpur, Leipzig, Lisbon, Livorno, Łodz, London, Long Beach, Los Angeles, Luton, Maastricht, Malmö, Mashhad, Melbourne, Menglun, Milton Keynes, Milwaukee, Monterey, Muncie, Nanchung, Napoli, New York, Norfolk, North Charleston, Nottingham, Oberlin, Oregon City, Oslo, Oxford, Palmares do Sul, Paris, Perth, Pezinok, Philadelphia, Phoenix, Pittsburgh, Plovdiv, Plymouth, Portland, Potosi, Pune, Randwick, Reston, Richmond, Rio de Janeiro, Rome, Rotterdam, Saint Louis, San Diego, San Francisco, San Luis Potosí, San Nicolás de los Garza, Santa Cruz, Sao Paulo, Scheemda, Scottsdale, Seattle, Shanghai, Sheffield, Singapore, Stockholm, Surabaya, Surakarta, Sydney, Tabriz, Tainan, Taipei, Tampa, Tehran, The Hague, Tianjin, Timisoara, Tochigi, Tokyo, Toronto, Toruń, Uppsala, Utrecht, Vancouver, Vienna, Vilnius, Vitoria-Gasteiz, Vladivostok, Vösendorf, Washington DC, Wellington, Winschoten, Wollongong, Wulai, Yogyakarta, Yokohama, Zürich	344
Biophilic schools	Albertslund, Auckland, Badung, Charlottesville, Copenhagen, Denver, Freiburg, Helsinki, Kolding, Milwaukee, New York, Norfolk, Perth, San Francisco, Singapore, Washington DC, Zürich	18
Buildings with green design elements	Abu Dhabi, Adelaide, Alphen, Amsterdam, Antwerp, Austin, Berlin, Birmingham, Blumau, Boston, Bremen, Brighton, Brisbane, Charlottesville, Chicago, Delft, Dordrecht, Drachten, Dublin, Dunkerque, Enschede, Frankfurt, Frederikshavn, Fukuoka, Graz, Haarlem, Harare, Ho Chi Minh, Honolulu, Katowice, Leiden, Leusden, Lisbon, Liuzhou, London, Madrid, Malmö, Melbourne, Mexico City, Milan, Milwaukee, Minneapolis, Mississauga, Montreal, Munich, Nanjing, Neemrana, New York, Norwich, Oostmalle, Oslo, Paris, Perth, Pittsburgh, Portland, Saarbrücken, San Bruno, San Francisco, Seattle, Singapore, Stockholm, Stuttgart, Sydney, Tarragona, The Hague, Toronto, Utrecht, Vienna, Washington DC, Wollongong, Zevenaar, Zürich	142

(continued)

BUI groups	Cities implementing BUI	Total of BUI
Community activities	Adelaide, Amsterdam, Auckland, Baltimore, Berlin, Birmingham, Bogor, Boston, Boulder, Bristol, Canberra, Charlottesville, Chicago, Copenhagen, Curitiba, Davis, Dunedin, Freiburg, Gothenburg, Hamilton, Harmony, Helsinki, Houston, Kansas City, Leiden, Lincolnshire, London, Madison, Melbourne, Milwaukee, Minneapolis, Montreal, Mumbai, Nairobi, Neemrana, New York, Philadelphia, Phoenix, Portland, Pune, Richmond, Rotterdam, Sacramento, San Francisco, Santa Monica, Seattle, Singapore, Skövde, Sukabumi, Sydney, The Hague, Tucson, Utrecht, Vancouver, Vitoria-Gasteiz, Wellington	102
Former infrastructure supporting nature insertion within the city	Austin, Berlin, New York, Paris, Vitoria-Gasteiz	7
Government plans, policies, programs, strategies, or visions	Aarhus, Abu Dhabi, Adelaide, Amsterdam, Anchorage, Arcata, Arnhem, Auckland, Austin, Barcelona, Berlin, Birmingham, Bologna, Boston, Boulder, Brisbane, Buenos Aires, Cape Town, Charleston, Chattanooga, Chengdu, Chicago, Christchurch, Copenhagen, Danang, Delft, Denver, Dongtan, Dublin, Edmonton, Erlangen, Frankfurt, Freiburg, Fremantle, Hamilton, Hannover, Heidelberg, Helsinki, Houston, Kaohsiung, Kassel, Laguna Beach, Leicester, Leiden, Linz, London, Long Beach, Los Angeles, Malmö, Melbourne, Mexico City, Milwaukee, Montreal, Munich, Münster, Nagoya, Nairobi, New York, Oregon City, Oslo, Paris, Penrith, Perth, Philadelphia, Portland, Richmond, Rio de Janeiro, Rotterdam, Sacramento, Saint Louis, San Francisco, Sao Paulo, Scottsdale, Seattle, Seoul, Singapore, Stockholm, Sydney, Tokyo, Toronto, Utrecht, Vancouver, Vaxjo, Vienna, Vitoria-Gasteiz, Washington DC, Wellington, Zürich	233
Government regulations	Auckland, Austin, Baltimore, Basel, Berlin, Chicago, Copenhagen, Davis, Dunkerque, Freiburg, Graz, Greensboro, London, Malmö, Port Coquitlam, Portland, San Francisco, Seattle, Singapore, Stockholm, Stuttgart, Toronto, Washington DC	46
Government support	Arlington, Chicago, Cologne, Faenza, London, Mexico City, New York, Philadelphia, Portland, Saarbrücken, Seattle, Singapore, Vienna	21
Green belts, greenways, or networks	Almere, Amersfoort, Anchorage, Arnhem, Atlanta, Birmingham, Bologna, Boulder, Brisbane, Cape Town, Chengdu, Christchurch, Curitiba, Davis, De Bilt, Hannover, Helsinki, Leusden, London, Milan, Milwaukee, Newark, Oslo, Rijnwoude, Saint Louis, San Francisco, Soest, Stockholm, Sydney, Utrecht, Vancouver, Vitoria-Gasteiz, Waddinxveen, Wellington, Zeist, Zoetermeer, Zürich	52
Green infrastructure	Amsterdam, Anchorage, Arnhem, Barcelona, Berlin, Birmingham, Brisbane, Buenos Aires, Canberra, Chicago, Christchurch, Curitiba, Duisburg-Nord, Fremantle, Fukuoka, George Town, Hamburg, Hamilton, Key Biscayne, Leiden, Lingewaard, London, Merseyside, Milwaukee, Minneapolis, Montreal, Nairobi, New York, Odense, Overbetuwe, Paris, Philadelphia, Portland, Rio de Janeiro, Rome, Rotterdam, Saarbrücken, San Diego, San Francisco, Seattle, Seville, Singapore, Songdo, Stockholm, Sydney, Tianjin, Toronto, Vancouver, Venice, Vienna, Wellington, Zaragoza	108
Pedestrian preference zone	Barcelona, Bologna, Boulder, Brisbane, Copenhagen, Dublin, Freiburg, Groningen, Heidelberg, Houten, Leicester, Leiden, Minneapolis, Münster, New York, Paris, Portland, Pune, Sacramento, Stockholm, Utrecht, Vienna	23
Restoration	Adelaide, Albufeira, Asheville, Auckland, Berkeley, Boston, Brisbane, Cape Town, Charlottesville, Chicago, Christchurch, Cleveland, Denver, Detroit, Dortmund, Freiburg, Fremantle, Graz, Groningen, Guimarães, Heidelberg, Houston, Johor Bahru, Leicester, London, Los Angeles, Madrid, Malmö, Melbourne, Milwaukee, Montcada i Reixac, New York, Newark, Nice, Norfolk, North Charleston, Oslo, Perth, Philadelphia, Pittsburgh, Port-au-Prince, Portland, Richmond, Rotterdam, San Francisco, San Jose, San Luis Obispo, Seattle, Seoul, Singapore, Slochteren, Stockholm, Tallinn, Tartu, Todmorden, Tokyo, Toronto, Vancouver, Vitoria-Gasteiz, Washington DC, Wellington, Zürich	104
Wildlife enhancement	Anchorage, Asheville, Austin, Bangalore, Birmingham, Brisbane, Bristol, Buenos Aires, Chengdu, Chicago, Edmonton, Fremantle, Hong Kong, Houston, London, New Orleans, New York, Oslo, Philadelphia, Portland, Richmond, Saint Louis, San Diego, San Francisco, Seattle, Singapore, Toronto, Vancouver, Vitoria-Gasteiz, Wellington	52

References

Artmann, M., Kohler, M., Meinel, G., Gan, J., & Ioja, C. (2017). How smart growth and green infrastructure can mutually support each other—A conceptual framework for compact and green cities. *Ecological Indicators*. https://doi.org/10.1016/j.jclepro.2019.01.185

Badiu, D. L., Iojă, C. I., Pătroescu, M., Breuste, J., Artmann, M., Niţă, M. R., Grădinaru, S. R., Hossu, C. A., & Onose, D. A. (2016). Is urban green space per capita a valuable target to achieve cities' sustainability goals? Romania as a case study. *Ecological Indicators, 70*, 53–66. https://doi.org/10.1016/j.ecolind.2016.05.044

Beatley, T. (2011). *Biophilic cities*. Island Press. https://doi.org/10.5822/978-1-59726-986-5

Beatley, T., & Newman, P. (2013). Biophilic cities are sustainable, resilient cities. *Sustainability, 5*(8), 3328–3345. https://doi.org/10.3390/su5083328

Beatley, T. (2009). Biophilic urbanism: Inviting nature back to our communities and into our lives. *William and Mary Environmental Law and Policy Review, 34*, 209–238. https://scholarship.law.wm.edu/wmelpr/vol34/iss1/6

Beatley, T. (2012). Green cities of Europe: Global lessons on green urbanism. In *Green cities of Europe: Global lessons on green urbanism*. https://doi.org/10.5822/978-1-61091-175-7

Beatley, T. (2017). *Handbook of biophilic city planning and design*. Island Press. https://books.google.be/books?id=4V_ECwAAQBAJ

Chen, C., Noble, I., Hellmann, J., Coffee, J., Murillo, M., & Chawla, N. (2015). *University of Notre Dame Global Adaptation Index*. https://gain.nd.edu/assets/254377/nd_gain_technical_document_2015.pdf

Cohen-Shacham, E., Walters, G., Maginnis, S., & Janzen, C. (2016). *Nature-based Solutions to address global societal challenges*. https://doi.org/10.2305/IUCN.CH.2016.13.en

Concejo Deliberante de la Ciudad de Córdoba. (2016). *Ordenanza N° 12.548. Incorporar al Código de edificación el marco regulatorio para la incorporación de cubiertas y muros verdes dentro de la Ciudad de Córdoba*.

Consejo De Bogotá, D. C. (2009). *Acuerdo 418. Por el cual se promueve la implementación de tecnologías arquitectónicas sustentables, como techos o terrazas verdes, entre otras en el D. C. y se dictan otras disposiciones*. https://www.alcaldiabogota.gov.co/sisjur/normas/Norma1.jsp?i=38262

Cook, E. (2016). *Biophilic urbanism: Making cities sustainable through ecological design*. https://doi.org/10.15242/iicbe.dir1216404

Cortinovis, C., & Geneletti, D. (2018). Ecosystem services in urban plans: What is there, and what is still needed for better decisions. *Land Use Policy, 70*, 298–312. https://doi.org/10.1016/j.landusepol.2017.10.017

Davies, C., & Lafortezza, R. (2017). Urban green infrastructure in Europe: Is greenspace planning and policy compliant? *Land Use Policy, 69*, 93–101. https://doi.org/10.1016/j.landusepol.2017.08.018

Dutt, K. (2009). Governance, institutions and the environment-income relationship: a cross-country study. Environment, Development and Sustainability,11(4), 705–723. https://doi.org/10.1007/s10668-007-9138-8

European Commission. (2015). *Ecosystem services and biodiversity*. https://ec.europa.eu/environment/integration/research/newsalert/pdf/ecosystem_services_biodiversity_IR11_en.pdf

European Commission. (2014). *Building a green infrastructure for Europe*. https://doi.org/10.2779/54125

Evans, B., Joas, M., Sundback, S., & Theobald, K. (2005). Governing sustainable cities. In *Governing sustainable cities*.

Fink, H. S., & Kaltenegger, I. (2016). Integration of mother nature into smart buildings. In *Integration of nature and technology for smart cities* (pp. 225–261). Springer International Publishing. https://doi.org/10.1007/978-3-319-25715-0_13

Grinde, B., & Patil, G. (2009). Biophilia: Does visual contact with nature impact on health and well-being? *International Journal of Environmental Research and Public Health, 6*(9), 2332–2343. https://doi.org/10.3390/ijerph6092332

Huang, G. (2017). Indexing the human-nature relationship in cities. *Upland Journal of Urban Planning, Landscape and Environmental Design, 2*(2), 25–35. https://doi.org/10.6092/2531-9906/5255

International Monetary Fund (IMF). (2019). *World economic outlook: Growth slowdown, precarious recovery*. https://www.imf.org/en/Publications/WEO/Issues/2019/03/28/world-economic-outlook-april-2019

Jennings, V., Browning, M. H. E. M., & Rigolon, A. (2019). Urban green space at the nexus of environmental justice and health equity. In *Urban green spaces: public health and sustainability in the United States* (pp. 47–69). Springer International Publishing. https://doi.org/10.1007/978-3-030-10469-6_4

Jones, D. R. (2013). 'The Biophilic University': A de-familiarizing organizational metaphor for ecological sustainability? *Journal of Cleaner Production, 48*, 148–165. https://doi.org/10.1016/j.jclepro.2013.02.019

Kahn, P. H. (1997). Developmental psychology and the biophilia hypothesis: Children's affiliation with nature. *Developmental Review, 17*(1), 1–61. https://doi.org/10.1006/drev.1996.0430

Kaplan, S. (1995). The restorative benefits of nature: Toward an integrative framework. *Journal of Environmental Psychology, 15*(3), 169–182. https://doi.org/10.1016/0272-4944(95)90001-2

Kellert, S. R. (1993). The biological basis for human values of nature. In E. O. Wilson & S. R. Kellert (Eds.), *The Biophilia hypothesis* (pp. 42–69). Island Press.

Lee, H.-H., Chung, R., & Mo, C. (2005). On the relationship between economic growth and environmental sustainability. In *5th ministerial conference on environment and development in Asia and the Pacific*.

Legislatura de la Ciudad Autónoma de Buenos Aires. (2012). *Ley 4428 - Techos o Terrazas Verdes - Implementación*. https://www2.cedom.gob.ar/es/legislacion/normas/leyes/ley4428.html

Lehmann, S. (2011). Transforming the city for sustainability: The principles of green urbanism. *Journal of Green Building, 6*, 104–113. https://doi.org/10.3992/jgb.6.1.104

Lehmann, S. (2011b). What is green urbanism? Holistic principles to transform cities for sustainability. In *Climate change—Research and technology for adaptation and mitigation*. https://doi.org/10.5772/23957

Littke, H. (2016). Becoming biophilic: Challenges and opportunities for biophilic urbanism in urban planning policy. *Smart and Sustainable Built Environment, 5*(1), 15–24. https://doi.org/10.1108/SASBE-10-2015-0036

Masson-Delmotte, V., Zhai, P., Pörtner, H. O., Roberts, D., Skea, J., Shukla, P. R., Pirani, A., Moufouma-Okia, W., Péan, C., Pidcock, R., Connors, S., Matthews, J. B., Chen, Y., Zhou, X., Gomis, M. I., Lonnoy, E., Maycock, T., Tignor, M., & Waterfield, T. (2018). Summary for policymakers. In *Global warming of 1.5°C. An IPCC special report on the impacts of global warming of 1.5°C above pre-industrial levels and related global greenhouse gas emission pathways, in the context of strengthening the global response to. In Ipcc - Sr15*. https://report.ipcc.ch/sr15/pdf/sr15_spm_final.pdf%0A; www.ipcc.ch/report/sr15/

Mc Cormick, K., Anderberg, S., Coenen, L., & Neij, L. (2013). Advancing sustainable urban transformation. *Journal of Cleaner Production, 50*, 1–11. https://doi.org/10.1016/j.jclepro.2013.01.003

Ministerio de Vivienda y Urbanismo (MINVU). (2019). *Decreto 58. Modifica Decreto Supremo N° 47, de Vivienda y Urbanismo, de 1992, Ordenanza General de Urbanismo y Construcciones, para fomentar la construcción de techos verdes*. https://www.leychile.cl/Navegar?idNorma=1129304&buscar=Modifica+Decreto+Supremo+No+47%2C+de+Vivienda+y+Urbanismo%2C+de+1992%2C+Ordenanza+General+de+Urbanismo+y+Construcciones%2C+para+fomentar+la+construcción+de+techos+verdes

Morar, T., Radoslav, R., Spiridon, L., & Păcurar, L. (2014). Assessing pedestrian accessibility to green space using gis. *Transylvanian Review of Administrative Sciences, 10*, 116–139. https://rtsa.ro/tras/index.php/tras/article/view/94/90

Mukherjee, S., & Chakraborty, D. (2013). Is environmental sustainability influenced by socioeconomic and sociopolitical factors? Cross-country empirical evidence. *Sustainable Development, 21*(6), 353–371. https://doi.org/10.1002/sd.502

Municipalidad Distrital de Ate. (2014). *Ordenanza Municipal N° 342-MDA. Crean el Programa Techo Verde en el distrito de Ate*. https://sinia.minam.gob.pe/normas/crean-programa-techo-verde-distrito-ate

Municipalidad Distrital de Barranco. (2015). *Ordenanza Municipal N° 427-MDB. Ordenanza que crea el Programa Techo y Muro Verde en el Distrito de Barranco*. https://sinia.minam.gob.pe/normas/ordenanza-que-crea-programa-techo-muro-verde-distrito-barranco

Nesshöver, C., Assmuth, T., Irvine, K. N., Rusch, G. M., Waylen, K. A., Delbaere, B., Haase, D., Jones-Walters, L., Keune, H., Kovacs, E., Krauze, K., Külvik, M., Rey, F., van Dijk, J., Vistad, O. I., Wilkinson, M. E., & Wittmer, H. (2017). The science, policy and practice of nature-based solutions: An interdisciplinary perspective. *Science of The Total Environment, 579*, 1215–1227. https://doi.org/10.1016/j.scitotenv.2016.11.106

Newman, P., Beatley, T., & Boyer, H. (2017). Resilient cities overcoming fossil fuel dependence. *Island Press/center for Resource Economics*. https://doi.org/10.5822/978-1-61091-686-8

Newman, P., & Matan, A. (2013). Green URBANISM in Asia. *World Scientific*. https://doi.org/10.1142/8596

Organisation for Economic Co-operation and Development (OECD). (2013). *Urban policy reviews, the Chilean urban system and its challenges*. OECD Publishing. https://doi.org/10.1787/9789264191808-en

Organisation for Economic Co-operation and Development (OECD). (2019). *Latin American economic outlook 2019*. OECD Publishing. https://doi.org/10.1787/g2g9ff18-en

Pauchard, A., & Barbosa, O. (2013). Regional assessment of latin America: Rapid urban development and social economic inequity threaten biodiversity hotspots. In T. Elmqvist, M. Fragkias, J. Goodness, B. Güneralp, P. J. Marcotullio, R. I. McDonald, S. Parnell, M. Schewenius, M. Sendstad, K. C. Seto, & C. Wilkinson (Eds.), *Urbanization, biodiversity and ecosystem services: Challenges and opportunities: A global assessment* (pp. 589–608). Springer. https://doi.org/10.1007/978-94-007-7088-1_28

Reeve, A., Desha, C., Hargreaves, D., & Hargroves, K. J. (2015). Biophilic urbanism: Contributions to holistic urban greening for urban renewal. *Smart and Sustainable Built Environment, 4*(2), 215–233. https://doi.org/10.1108/SASBE-11-2014-0057

Rigolon, A., Browning, M., Lee, K., & Shin, S. (2018). Access to urban green space in cities of the global South: A systematic literature review. *Urban Science, 2*(3), 67. https://doi.org/10.3390/urbansci2030067

Russo, A., & Cirella, G. T. (2017). Biophilic cities: Planning for sustainable and smart urban environments. In *Smart cities movement in BRICS* (Vol. 2018, Issue March, pp. 153–159).

Russo, A., Ignatieva, M., Cirella, G., Belelli Marchesini, L., Krestov, P., Korzhov, E., Kalita, V., & Pavlovsky, V. (2017). Biophilia: Nature-based solutions for sustainable cities. In *Conference: Three pillars of landscape architecture: Design, planning and management. New visions.*

Sachs, J. D., Schmidt-Traub, G., Mazzucato, M., Messner, D., Nakicenovic, N., & Rockström, J. (2019). Six transformations to achieve the sustainable development goals. *Nature Sustainability, 2* (9), 805–814. https://doi.org/10.1038/s41893-019-0352-9

Sadler, J., Grayson, N., Hale, J., Locret-Collet, M., Hunt, D., Bouch, C., & Rogers, C. (2018). *The little book of ecosystem services in the city.*

Scott, M., & Lennon, M. (2016). Nature-based solutions for the contemporary city. *Planning Theory and Practice, 17*(2), 267–270. https://doi.org/10.1080/14649357.2016.1158907

Ulrich, R. S. (1984). View through a window may influence recovery from surgery. *Science, 224*(4647), 420–421. https://doi.org/10.1126/science.6143402

United Nations (UN). (2018a). *The world's cities in 2018—Data booklet*. https://www.un.org/en/events/citiesday/assets/pdf/the_worlds_cities_in_2018_data_booklet.pdf

United Nations (UN). (2018b). *World economic situation and prospects 2018*.

United Nations [UN]. (2015). *General assembly resolution A/RES/70/1. Transforming our world, the 2030 agenda for sustainable development*. https://sustainabledevelopment.un.org/post2015/transformingourworld

Wheeler, S. (2013). Planning for sustainability: Creating livable Equitable and ecological communities, second edition. *Planning for Sustainability Creating Livable Equitable and Ecological Communities, Second Edition*, 1–401. https://doi.org/10.4324/9780203134559

Wilson, E. O. (1984). *Biophilia*. Harvard University Press.

Woodruff, S., & BenDor, T. (2016). Ecosystem services in urban planning: Comparative paradigms and guidelines for high quality plans. *Landscape and Urban Planning, 152*, 90–100. https://doi.org/10.1016/j.landurbplan.2016.04.003

World Bank Group [WBG]. (2020b). *Urban population 2018*. https://data.worldbank.org/indicator/SP.URB.TOTL.IN.ZS?most_recent_value_desc=true.

World Bank Group [WBG]. (2020a). *GNI per capita, Atlas method (current US$) 2018*. https://data.worldbank.org/indicator/NY.GNP.PCAP.CD

World Health Organization (WHO). (2016). *Urban green spaces and health*. https://www.euro.who.int/en/health-topics/environment-and-health/urban-health/publications/2016/urban-green-spaces-and-health-a-review-of-evidence-2016

Xue, F., Lau, S., Gou, Z., Song, Y., & Jiang, B. (2019). Incorporating biophilia into green building rating tools for promoting health and wellbeing. *Environmental Impact Assessment Review, 76*, 98–112. https://doi.org/10.1016/j.eiar.2019.02.004

Neighborhood Urban Green Spaces for Senior Citizens and Children, Case Study: Delhi

Anita Mandal

Abstract

Due to rapid urbanization trend, the distance between city inhabitants and nature is increasing day by day and it is necessary to bridge this gap by promoting urban green spaces in the city. Green spaces are important for sustained well-being of people from different age groups, ensuring diverse benefits including improved health, social bonding, recreation, and sheer experience of nature while being in the city. During the past few decades, the Master Plan of Delhi has made efforts in improving and increasing the provision planned green spaces. Delhi is the greenest city in India with 19% green cover having different types of green space at different levels such as regional, city, district, community, neighborhood, and housing play area. Now, the research question is that, are the neighborhood parks accessible to senior citizens and children, if they are accessible then, are they usable to senior citizens and children? How much is the usage? This paper describes how the parks are designed? And what are its components? What issues are coming in using the parks? The aim of the study is to analyze the utilization pattern of neighborhood urban green spaces by our target age groups and develop a plan for better utilization in Delhi. Only the neighborhood level parks (daily use parks) are taken into account for the study. There are three parameters used for the calculation of urban green spaces usage: (i) number of users, (ii) length of stay (time spent), and (iii) frequency of visits. There are three parameters used for calculating the usability of the selected parks: (i) ease of accessibility, (ii) provision of facilities, (iii) maintenance. To assess the usage of neighborhood green spaces in Delhi, the primary survey of senior citizens and children was conducted. Analysis of the collected data shows that the sector-6, Dwarka neighborhood park has a higher number of users than the sector-8, Dwarka neighborhood park due to availability of facilities, distance traveled by the users, and the population living near to the park. Issues related to maintenance and availability of facilities in both the parks are also observed.

Keywords

Urban • Green spaces • Neighborhood • Senior citizens • Children • Delhi

1 Introduction

Due to rapid urbanization trend, the distance between city inhabitants and nature is increasing day by day. In order to reduce this distance, it is necessary to promote green spaces in the city. In the past few decades, many environmental activists and society have placed a much greater emphasis on urban green spaces in order to improve the quality of life in urban areas. Green spaces have been incorporated in urban planning of Delhi since its earliest planning. These green spaces integrate economic, environmental, and social aspects together. Green spaces are important to human health, balance out built structure, social gathering, and recreation and interaction with nature. People of the different group have their specific needs, which in turn are interrelated. Among these, children are in their growth phase and considered to be the weakest age group. Senior citizens have limited capabilities and are more susceptible to disease, syndromes, and sickness in comparison to younger adults. They also face other social problems such as retirement, loneliness, and agism.

A. Mandal (✉)
Master in Environmental Planning, School of Planning and Architecture, New Delhi, India

A. Mandal
Bachelor of Planning, School of Planning and Architecture, New Delhi, India

F. Trapani et al. (eds.), *Advanced Studies in Efficient Environmental Design and City Planning*, Advances in Science, Technology & Innovation, https://doi.org/10.1007/978-3-030-65181-7_30

To boost the greening efforts, 18 greening agencies have been discovered which majorly includes government agencies and municipal bodies. Some of them are Municipal Corporation of Delhi (MCD), Delhi Development Department (DDA), Public Works Department (PWD), Central Public Works Department (CPWD), Delhi Parks and Garden Society (DPGS), North Delhi Municipal Corporation (NDMC), Education Department, Transport Department, Flood Control Department, Development Department, Delhi Metro Rail Corporation (DMRC), etc. These agencies were assigned with certain plantation objectives annually which were done by them using their own resources. Most of the agencies have their own nurseries and infrastructure. All these greening activities are coordinated by the Forest Department. Ceremonial Plantations were carried out at regular intervals for giving proper awareness among the citizens (Sinha, 2013).

Recreation is the activity or activities engaged in during leisure time. In a deeper psychological sense, recreation refers to the human emotional and inspirational experiences arising out of the recreation act. Father of the playground movement, Joseph Lee, defined "recreation for adults as the renewal of life and for children as the gaining of life" (Jewell, 1997). Another author Rodel E. Sinamban defined "Recreation is the expenditure of time with an intent to gain some refreshment. It is a break from the monotony and a diversion from the daily routine. It is a positive change from the stereotypical lifestyle and involves active participation in some entertaining activity" (Rodel, 2010). Recreation activities render an essential approach to refill people's energy and prepare them for the next task. It also restores people's senses and makes them feel refreshed and pleasant (Wang, 2012).

Census of India has divided the total population into three different age groups, i.e., 0–14 years, 15–59 years, 60 years, and above. The "senior citizens" population, i.e., people of age 60 years and above and "children" population, i.e., people of age 0–14 are important groups among the city population whose requirements are different. The percentage of the elderly population (persons aged + 60) is expected to increase from 5.9% in 2001 to 10.7% in 2021. At the same time, population of children (aged 0–14 years) would decrease from 29.5% in 2001 to 21.75% in 2021. The percentage of the elderly population (persons aged + 60) was increased from 4.8% in 1991 to 5.9 in 2001. On the other side, the population of children (aged 0–14 years) decreased from 34.8% in 1991 to 29.5% in 2001 (Table 1).

"The age-group wise distribution of population should form the basis for providing special facilities to the elderly such as old age homes, low floor buses, special seats in buses, and special seats in public toilets and ramps in public buildings." Master Plan of Delhi-2021 (Delhi Development Authority, 2010a, b).

This research study examines the policies and provisions of green spaces for senior citizens and children in the current planning practices, analyzes the usage and usability of neighborhood green spaces by targeted age group in case study green spaces. This study discusses guidelines and recommendations for both existing and future neighborhood green spaces taking into consideration the requirements of senior citizens and children.

2 Need of the Study

As discussed in the previous section, the senior citizens and children segment of Delhi's population is increasing. The increasing population requires emphasis and a different approach to design for better mobility of the selected target age group. Children who are generally more active being in their growth phase and are considered to be the weakest age group and elderly, who often have limited capabilities and are more susceptible to disease, syndromes, and sickness in comparison to younger adults. They also face other social problems such as retirement, loneliness, and agism, and for whom the provision of green space is inadequate. Future green spaces should be innovative and targeted in their design.

According to MPD-2021, Delhi has not yet seen a shift in emphasis from quantity of green space to quality, a shift may be required for future plans. New green spaces will need to stress on use and quality rather than quantity, that will require flexibility in design.

Table 1 Elderly and children population in Delhi (1991–2021)

Year	Population in lakhs	
	Elderly (60+ year)	Children (0–14 years)
1991	4.5 (4.8%)	32.8 (34.8%)
2001	8.1 (5.9%)	40.6 (29.5%)
2011	14.2 (7.8%)	45.0 (24.7%)
2021	24.6 (10.7%)	49.9 (21.7%)

Figures (in bracket) indicate percentages to the total population
Source Census of India and projections by DDA Sub-Group (Delhi Development Authority, 2010a, b)

3 Literature Study

This section of the paper is divided into five subsections. First part is about the basic concept of green spaces and their types, functions, benefits, usability, and supporting elements. The second section talks about the provision of green spaces in MPD-2021: Delhi. The third and fourth part is describing the requirements of senior citizens and children. Last and fifth part explains the neighborhood concept.

3.1 Green Spaces

3.1.1 Concept of Green Spaces

United State Environmental Protection Agency defines urban green space as any open piece of land that is not developed (has no buildings or other built structures) and can be accessed by the public. Green space (land that is partly or completely covered with grass, trees, shrubs, or other vegetation) includes parks, community gardens, and cemeteries.

University College London (UCL) Institute of Health Equity "They include areas of greenery such as local parks, public gardens and playing fields, but also 'spaces' such as streets where there are trees planted, and cycleways. 'Accessible' green space is considered to be that which is located close to residents' homes, easy to walk to, physically accessible, safe to use, and provides well-maintained facilities" (UCL Institute of Health equity, 2014).

The Delhi Master Plan states, "green spaces are generally considered as the developed parks and green belts."'

3.1.2 Types of Green Spaces

Types of urban green spaces as per provision at hierarchy level are as follows:

- Local level—Totlots, housing area, neighborhood parks
- City/zonal level—Community, district, city parks.
- Regional level—Regional parks.

3.1.3 Functions of Green Spaces (Vijai Shanker Singh, 2010)

The functions of the urban green space can broadly group into three major categories:

- Utilitarian: The utilitarian functions refer to the more tangible aspects like optical and thermal comforts inside the green spaces, providing access to these spaces.
- Social: These functions usually involve the direct and active participation of the human being in these spaces. These may include social interaction and recreation, etc. It is the green spaces where the infants and children come into contact with society and its culture.
- Perceptual: There are two important aspects to the perceptual functions. First, the green spaces may be used as one of the evaluative criteria by residents or visitors. The second aspect relates to the even more intangible aspects of creating a distinct identity of a given residential area.

3.1.4 Usability of Urban Green Spaces

Usage of the urban green spaces is given by number of users, length of stay (time spent), and frequency of visits and its usability is dependent primarily on three factors as follows:

- Ease of accessibility
- Provision of facilities
- Maintenance.

3.1.5 Benefits of Urban Green Spaces (Center for Urban Green Spaces, 2012; The Benefits of Green Space, 2015)

The green spaces in urban areas play an important role in the urban living of a large population. The benefits and uses broadly classified as environmental, economic and social and trifurcated as follows:

(a) Social and psychological benefits: As per WHO, health is not only the absence of illness but means mental, physical, and social well-being and the green spaces provide ample opportunity for social, physical, and psychological benefits with recreation and health.
   - Recreation and well-being: Resource for relaxation and social interaction, interaction with nature and environment, lower crime and enhance self-esteem, provide safe play space for children, provides a perfect outing place for elderly people and family as a whole, contribute to physical, mental, and social development (Brodhead, 2009).
   - Human health: Reduce stress, decrease in respiratory illness, increase in work productivity (Barton, 2009).

(b) Environmental benefits: The environmental benefits of the green spaces can be summarized as ecological, pollution control, and biodiversity and nature conservation (Brodhead, 2009):
   - Ecological benefits: Regulation of urban climate, reduce temperature, reducing urban heat island effect, groundwater recharge during monsoon, increase precipitation/rainfall (Consullting, 2013).
   - Pollution control: Reducing pollutants and smoke particles, reducing noise levels (Consullting, 2013).
   - Biodiversity and nature conservation: Protection of species, conservation of plants, soil, and water quality (Consullting, 2013).

(c) Economic and aesthetic benefits: The economic and aesthetic benefits of the green spaces can be summarized as energy-saving and property value (Brodhead, 2009):

- Energy saving: Reducing overall energy consumption for heating and cooling reduces the cost of cooling of the building (Haq, 2011).
- Property value rise: Increase in property values and returns, increase scenic value, employment, and tourism boost (Haq, 2011).

3.1.6 Supporting Elements of a Park (Christiansen, n.d.; Riverside General Plan 2025; Amended, 2012)

The supporting elements of park change with its hierarchy. But in the study, the focus is on neighborhood level parks. The elements of neighborhood parks are as follows:

(a) Pedestrian circulation: It is important to plan for specific pedestrian needs when developing a system of walkways in a park. Pedestrian routes may be utilitarian linkages between activity areas and other supporting facilities. Nature walk and jogging paths are examples of pedestrian ways that are directly activity-related. The proper walking surface is important to pedestrians' comfort and satisfaction.

(b) Lighting: The provision of an adequate level and quality of illumination in a park is a basic support consideration. Lighting can be provided for five functions such as (i) activity lighting: necessary for providing illumination to do a specific task or performance like washing hands and playing a ball game, (ii) safety lighting: provides necessary illumination during normal situations to protect human well-being and reduce the chances of accidents or harm to people, it is for walks, steps, ramps, and road crossing, (iii) security lighting: provides a degree of protection to property, equipment, and goods, (iv) emergency lighting: is needed when normal lighting provisions fail and people need temporary supplementary illumination to evacuate essential tasks, and (v) aesthetic lighting: is used to enhance the beauty and attractiveness of a visual element. It may be used to illuminate fountains, statues, trees and shrubs (Dillon Consultant Limited, 2014).

(c) Water system: The provision of a sufficient and safe supply of water in a park or recreational area is essential for public consumption and use, fire protection, removal of sewage and park maintenance and operation. Park visitors need water for drinking, washing, and flushing. In addition, water is used for park irrigation, sprinkling, and cleaning. Water fountain increases the aesthetics of the park.

(d) Waste water and solid waste: A proper control of human and domestic wastes (leaves of trees) in park areas is an essential factor. A well-planned and operated park system is necessary to protect the health and convenience of the users, protect vegetation from harmful pollution. A park sewerage plan should include proper and efficient collection, storage, treatment, and disposal of the solid.

(e) Plantation and play area: Trees and other shrubs add to the aesthetics of the park. Also, give shades to persons using parks. They are mostly used by senior citizens. Swings, sliders, sand-pit, and mini-basketball increase the usability of the park.

(f) Washroom or toilets: They are also important amenities of a park for the cleanliness and health reason (Dillon Consultant Limited, 2014).

3.2 Provision for Green Spaces in MPD

In MPD-2001, neighborhood parks were proposed to be planned with 7.5 ha for 15,000 population, whereas in MPD-2021, it is proposed to be planned with 3.5 ha for 10,000 population. Comparison of three Delhi Master Plans (Table 2).

The green area and green area per person are decreasing from 1962 to 2001, but Master Plan of Delhi-2021 made an effort in increasing these green spaces. But it is not talking about the special provision in the green spaces for senior citizens. Master Plan has also included neighborhood-level parks from 1962, which can be used by the senior citizens and children.

3.3 Children

(a) **Definition of a child** (Lemish, 2011)

A child is biologically defined as a human being between the stages of birth to puberty and legally referred as a minor who has not attained the age of maturity. The age of majority accepted and specified by each country differs from fourteen to eighteen. The UN Convention on the rights to children defines a child as a majority is attained earlier. The Indian law, as specified in the National Action Plan for children, also defines a child as defined by the UN Convention on the rights of the child. The child undergoes different stages of development as they grow and based on the stages of development the growth of the child has been classified into different phases, like infancy, early childhood, middle childhood, and adolescents.

Table 2 Comparison of three Delhi Master Plans (1962, 2001, 2021)

Category	MPD-1962	MPD-2001	MPD-2021	UNIT
Catering population	53,00,000	82,00,000	230,00,000	Nos.
Total urban area	44,777	44,777	148,300	Ha
Green area	9101	8722	22,245–29,660	Ha
Green area %	20.3	19.5	15–20	(approx)
Urban green per person	17.2	9.7	9.7–12.9	m^2

Source MPD-1962, 2001, 2021 (Delhi Development Authority, 1962, 1981, 2010a, b)

Note MPD-2001 Additional 40 Lakh Population to be accommodated in urban extension with open space at 9.7 m^2 per person. MPD-2021 has not given any thought to the recreational facilities for senior citizens and children who need the spaces more. The standards are only quantity oriented and not quality (Table 3)

Table 3 Increase in hierarchy of urban green spaces in MPDs (1962, 2001, 2021)

Level	Hierarchy	MPD-1962	MPD-2001	MPD-2021
Shown in land-use plan (Master Plan)	Regional park	Yes	Yes	Yes
	City park	No	Yes	Yes
	District park	Yes	Yes	Yes
	Community park	No	No	Yes
Part of the residential land use	Neighborhood park	Yes	Yes	Yes
	Housing park	Yes	Yes	Yes

Source MPDs-1962, 2001, 2021 (Delhi Development Authority, 1962, 1981, 2010a, b)

(b) **Development stages of the child** (Child Development Institute, 1999; SKIP, n.d.)

The child in his various growth stages undergoes different types of development like physical, emotional, cognitive, and psychological. This section tries to understand the different types of development in children. Biological and psychological changes that take place in human beings between birth and end of adolescence pertains to child development. As the individual grows by age, he or she progresses from dependency to increasing autonomy. Based on the growth stages, the development periods are classified as: (1) infant and toddler (ages 0–2 year), (2) early childhood (ages 3–6 years), (3) middle-age childhood (ages 7–12 years), and (4) adolescence (ages 13–18 years).

(c) **Requirements of the children** (Child Development Institute, 1999)

The needs of the children vary with the various growth stages undergoes. The needs are as follows:

- Infant and toddler (Ages 0–2 year): At this stage, the child depends mostly on their parents for their emotional and physical needs. Child's attachment to caregivers and other significant people in their life is important to their healthy emotional and psychological development. They have a determined understanding of their world. They need predictable care giving and routines. They also require nurturing and emotional connections.
- Early childhood (ages 3–6 years): At this stage, child remains mostly with their mother and elders within the home environments. Their movements are generally in and around the house. They prefer to ride their tricycles near their houses. The child wrapped up in itself and is interested in individuals play. The activity does not change between a boy and a girl in this age group. The mobility is very low and passive in nature.
- Middle-age childhood (Ages 7–12 years): At this age, the child starts expanding their social network outside their family. Becoming more independent, their focus is more on their friends and the world outside their family. However, they still need emotional support and guidance of their family. They remain near their home environment and take part in mini cricket, riding bicycles, ball games, running skipping and see-saw.
- Adolescence (ages 13–18 years): They are becoming more independent. They involve themselves in selected activities such as flying kites and playing cricket, football, volleyball, basketball, badminton, table tennis, and tennis. They may also be concerned about having not much time for friends or activities.

3.4 Senior Citizen

(a) **Definition of a Senior citizens**
National Human Rights Commission defines "Old-Age" as associated with declining faculties, both mental and physical, and reduced social commitments (including sports participation). In India, the people who attained the age of sixty years and above are defined as elderly for availing old-age benefits (Commision, 2011).

In Germany, Prince Bismarck, the chancellor of the German Empire, announced 65 as the age for citizens' participation in the pension plan. On the other hand, many Japanese believe that the elderly should be characterized by loss of functional independence and defined persons over 70 or 75 years of age as senior citizens (Hajime Orimo, 2006).

In the USA also, people who are more than sixty years of age are usually counted as senior citizens. They often have limited capabilities and are more susceptible to disease, syndromes, and sickness in comparison to younger adults. They also face other social problems such as retirement, loneliness, and agism (Eryn Pleson, 2014).

(b) **Requirements of the senior citizens**
In this age group, the males and females mostly involve themselves in passive recreation like walking, looking after kids, relaxing, gossiping, and so on. (Kiran, 2014) Some of them are jogging, running, exercise, etc. Some of them not able to walk, talk, and hear properly (Takano, 2002).

3.5 Neighborhood

Often the term "neighborhood" is used to describe the subdivisions of urban settlements. It is a vicinity in which people reside and can access their basic and daily facility within walkable distance. Facilities such as school, commercial area, recreational areas, and streets connecting them.

(a) **Clarence Arthur Perry, New York Planner**

Clarence Arthur Perry was the first planner given a definition of the neighborhood unit' in 1929. Perry described six basic features of good neighborhood design which were organized around several institutional, social, and physical design ideals (Perry, 1929). These are:

- Major arterials and through traffic routes should provide boundaries of the neighborhood and should not pass through residential neighborhoods.
- Use of cul-de-sacs, curved layout and light-duty surfacing for designing and constructing interior street to encourage a quiet, safe, and low-volume traffic movement and preservation of the residential atmosphere.
- Neighborhood population should support its elementary school.
- The centrally located elementary school should be the focal point of the neighborhood, along with other institutions that have service areas coincident with the neighborhood boundaries.
- The neighborhood radius should be maximum of one quarter-mile, thus precluding a walk of more than that distance for any elementary school child and
- Shopping districts should be sited at the edge of neighborhoods preferably at major street intersections.

In the below concept (Fig. 1) Clarence Perry had taken 400 m or one quarter-mile radius of walking distance for an elementary school child. But it is concluded from previous section that there are factors affecting the provisions of green space and walking roads are not straight (i.e., walking roads are distance not displacement), so the 200 m radius of walking distance is taken up for this study.

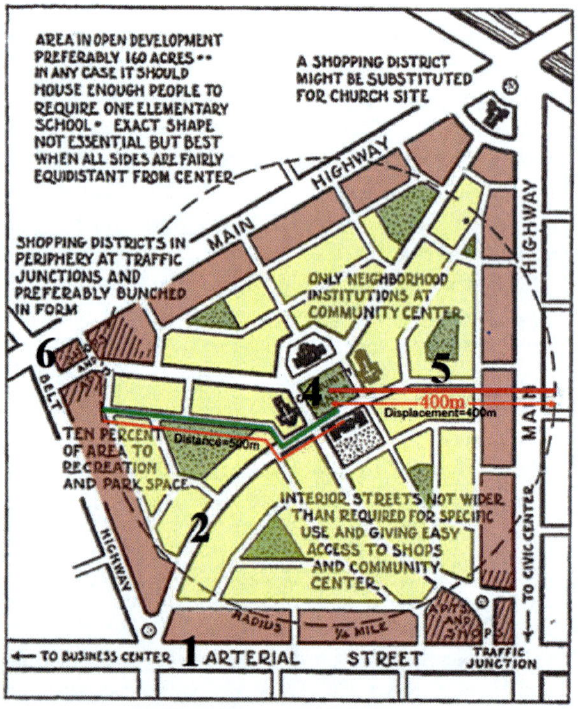

NEIGHBORHOOD UNIT PRINCIPLES
Reproduced from New York Regional Plan, volume 7

Fig. 1 Neighborhood concept by Clarence Perry (1929). *Source* http://www.itpi.org.in/files/jul10_11.pdf

4 Methodology and Data Collection

This section of paper is talking about selection of area with some criteria for detailed study of neighborhood-level parks. The type of data to be collected and format of the survey also discussed. First of all two areas are selected as per the following criteria.

4.1 Criteria for Selection of Case Study Areas

The study areas have been selected for the detailed analysis on the basis of following criteria:

- Type of development: For this study, two types of development at neighborhood level have been taken. These are areas having (1) plotted housing: Dwarka sector-8 (2) group housing: Dwarka sector-6.
- Age of development: The neighborhood should be at least 2 years older, so that individuals experience the problems and share them.
- Type of structure: The neighborhood which have different type of structure, i.e., high rise and low rise are selected for the detailed study. (1) High rise: Dwarka sector-6 and (2) low rise: Dwarka sector-8.

4.2 Methods Used to Collect Data

The study is based on a quantitative as well as a qualitative approach. Secondary data such as maps, plans, and population of the study area is collected from different organizations' Web site and office. Primary data through primary survey is collected for the particular research study. There were three types of primary survey: (1) observation survey, (2) park user survey, and (3) residents survey. There are three parameters used for the calculation of urban green spaces usage: (i) number of users, (ii) length of stay (time spent), and (iii) frequency of visits. There are three parameters used for calculating the usability of the selected parks: (i) ease of accessibility, (ii) provision of facilities, (iii) maintenance. Data is collected for weekday as well as weekend for comparison, data of three consecutive Fridays and three Sundays in the month of February 2015. Finally, a comparative analysis has been done for both the sectors.

To determine the utilization of green spaces in Delhi, the main emphasis is given to the primary surveys. Secondary data was collected to understand the existing green spaces system in Delhi.

- Secondary data: Secondary data was collected to understand the existing hierarchy of green spaces in Delhi, their standards and factors determining these standards, and also information about green spaces areas for hierarchy, and information about administration of the green spaces was collected through secondary sources.
- Primary data: As the study concentrates more on the primary sources of data, so following types of surveys were conducted to gather information regarding design characteristics and usage pattern of neighborhood green spaces.
- Observation survey: To study the design characteristics of various green spaces by senior citizens and children.
- User survey: To study the usage pattern of various green spaces by senior citizens and children.
- Residents survey: to study the household characteristics. This survey was conducted in different study area. The selection of households was done by the random sample survey. The data collected through this survey was regarding the household characteristics, their leisure time activities, their habits of green space utilization.

Dwarka sector-6 and sector-8 have been selected for detailed study which have group and plotted housing, respectively (Fig. 2).

4.3 The Sample Size

In neighborhood concept, Clarence Perry had taken 400 m or one quarter-mile radius of walking distance for elementary school child. But it is concluded from previous section that there are factor affecting the provisions of green space and walking roads are not straight (i.e., walking roads are distance not displacement), so the 200 m radius of walking distance is taken up for this study. Population data within 200 m radius of park is collected and then 2% sample size is taken up for the residents survey.

4.3.1 Dwarka Sector-6, Neighborhood Park

- Population near to the park: 3000.
- Residents' survey sample size: The survey of the residents living within 200 m radius of walking distance around the park has been done for detailed study. The sample size is 2%.
- Users' survey sample size: The survey of users coming within particular park. The sample size is approx. 50%.

4.3.2 Dwarka Sector-8, Neighborhood Park

- Population near to the park: 1500.
- Residents' survey sample size: The survey of the residents living within 200 m radius of walking distance around the park has been done for detailed study. The sample size is 2%.

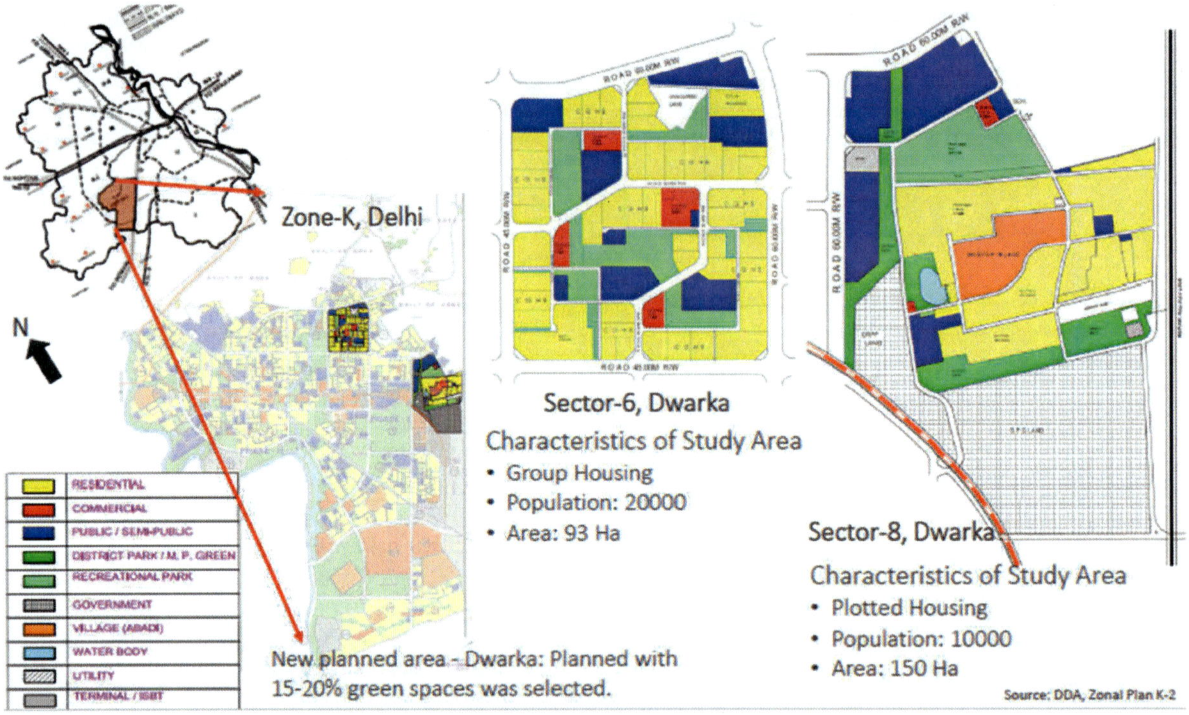

Fig. 2 Data collection and selected case study. *Source* DDA, Zonal Plan K-2 (Delhi Development Authority, 2010a, b)

- Users' survey sample size: The survey of users coming within particular park. The sample size is approx. 50%.

5 Data Analysis

There were three types of primary survey: (1) observation survey, (2) park user survey, and (3) residents survey. There are three parameters used for the calculation of urban green spaces usage: (i) number of users, (ii) length of stay (time spent), and (iii) frequency of visits. There are three parameters used for calculating the usability of the selected parks: (i) ease of accessibility, (ii) provision of facilities, and (iii) maintenance. Data is analyzed by giving weightage to each of the parameters according to it accessibility, availability, and condition.

5.1 Observation Survey

In observation survey ease of accessibility, presence and lack of facilities, maintenance of facilities, and land-use around the parks are observed. These facilities are plants/trees, shed, dust bin, jogging tracks, toilets, water, etc. A comparative analysis has been done for both the parks.

5.1.1 Comparison of Ease of Accessibility of the Parks

These are facilities available and provided around the parks (Table 4).

The ease of accessibility in sector-6 and sector-8 park is 58% and 16%, respectively.

5.1.2 Comparison of Presence of Facilities in the Parks

These are facilities available and provided within the parks (Table 5).

The availability of facility in sector-6 and sector-8 park is 58% and 16%, respectively. The comparison of facilities provided shows that sector-6 has major issues of lightening, water only and but, it is better than sector-8 park. Although sector-8 has no other facility, it has the facility of water supply. One of the major issues in sector-8 park is of safety, reason being the drunkards visit every day.

5.1.3 Comparison of Maintenance of Facilities in the Parks

These are facilities available and provided within the parks (Table 6).

The maintenance of facilities in sector-6 and sector-8 park is 58% and 16%, respectively. The weightage percentage indicates that neither of parks maintained fully, there is a

Table 4 Comparative table of ease of accessibility of the parks

	S. no.	Facilities	Neighborhood parks		Maximum weightage
			Sector-6	Sector-8	
Ease of accessibility	1	Ramp	No (0)	No (0)	2
	2	Gates	Yes (2)	Yes (2)	2
	3	Footpath availability	Yes (2)	No (0)	2
	4	Footpath walkability-surface	Yes (2)	No (0)	2
	5	Footpath cleanliness	No (0)	No (0)	2
	6	Footpath height	Not adequate (1)	No (0)	2
	7	Total weightage	7 (58%)	2 (16%)	12

Source Primary Survey, 2015
Note Number in bracket shows the weightage given according to the ease and availability of facility around the parks. 2—good, 1—satisfactory, 0—bad)

Table 5 Comparative table of presence of facilities in parks

	S. no.	Facilities	Neighborhood parks		Maximum weightage
			Sector-6	Sector-8	
Status of facilities	1	Plants/trees	Not adequate (1)	Not adequate (1)	2
	2	Grass	Yes (2)	No (0)	2
	3	Shed	No (0)	No (0)	2
	4	Water	No (0)	Yes (2)	2
	5	Dustbins	Yes (2)	No (0)	2
	6	Toilets	Yes (2)	Yes (2)	2
	7	Lighting	No (0)	No (0)	2
	8	Benches	Not adequate (1)	No (0)	2
	9	Jogging tracks	Yes (2)	No (0)	2
	10	Swings/slider	No (0)	No (0)	2
	11	Total	10 (50%)	5 (25%)	20

Source Primary Survey, 2015

Table 6 Comparative table of maintenance of facilities in parks Sector-6 and 8, 2015

	S. no.	Facilities	Neighborhood parks		Maximum weightage
			Sector-6	Sector-8	
Maintenance of facilities	1	Plants/trees	Less maintained (1)	Less maintained (1)	2
	2	Vegetation	Less maintained (1)	No (0)	2
	3	Shed	No (0)	No (0)	2
	4	Water	No (0)	No (0)	2
	5	Dustbins	Yes (2)	No (0)	2
	6	Toilets	No (0)	Yes (2)	2
	7	Lighting	No (0)	No (0)	2
	8	Benches	Less maintained (1)	No (0)	2
	9	Jogging tracks	Less maintained (1)	No (0)	2
	10	Swings/sliders	No (0)	No (0)	2
	11	Overall cleanliness	Less maintained (1)	Less maintained (1)	2
	12	Total weightage	7 (31%)	4 (18%)	22

Source Primary Survey, 2015

lack of maintenance in both the parks. The comparison of maintenance of facilities shows that, sector-6 is again better than sector-8 in terms of facility maintained. The reason is that sector-8 is not provided with those facilities.

5.1.4 Total Weightage of All the Above Three Parameters

These are facilities available and provided within the parks (Table 7).

5.1.5 Land use Around Both the Parks

Sector-6 park (Fig. 3) is surrounded by residential, commercial and semipublic land use. Whereas sector-8 park (Fig. 4) is surrounded by residential and another proposed playground.

5.2 Park User Survey—A Comparative Analysis

The usage analysis consists of data collected on number of users, age group of users, time spent in the park and frequency of using the park.

5.2.1 Total Number of People Using Park (All Age Group)

The number of users in the given green spaces varies. Well-connected sector-6 park is used more as compared to sector-8 park. In both the parks, user number is higher at the evening time and low in the afternoon time. Although it is weekend the maximum number of users (all age group) is 93 and 51 only (Fig. 5).

5.2.2 Age Group of People Using Park (All Age Group)

The age group of users using the open area varies. The sector-6 park is mostly used by senior citizens (Fig. 6).

5.2.3 Time Spent at the Park (Minutes) (Only Senior Citizen and Children)

The users spent more than 1 h in both the parks (Fig. 7).

5.2.4 Frequency of Using the Park (Only Senior Citizen and Children)

Mostly users come daily in both the parks (Fig. 8).

Table 7 Comparative table of all three parameters

	S. no.	Facilities	Neighborhood parks		Maximum weightage
			Sector-6	Sector-8	
Usability	1	Ease of accessibility ramp	7	2	12
	2	Status of facilities	10	5	20
	3	Maintenance of facilities	7	4	22
	4	Total weightage	24 (44%)	11 (20%)	54

Sector-6 park has 44% usability, while on the contrary sector-8 park has 20% usability

Fig. 3 Land use around the park in sector-6. *Source* Secondary and Primary Survey, 2015

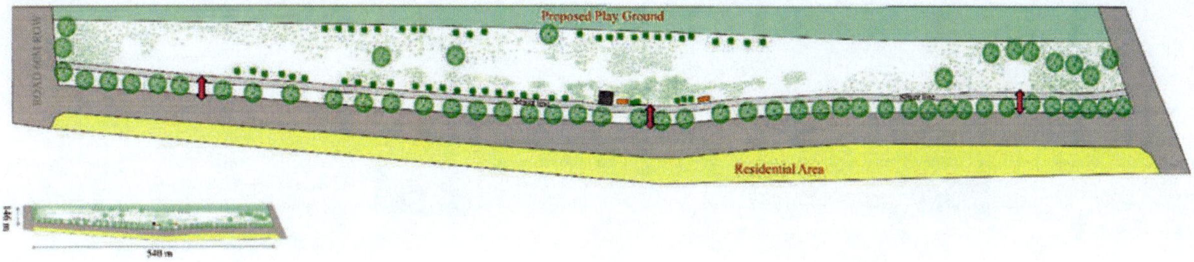

Fig. 4 Land use around the park in sector-8. *Source* Secondary and Primary Survey, 2015

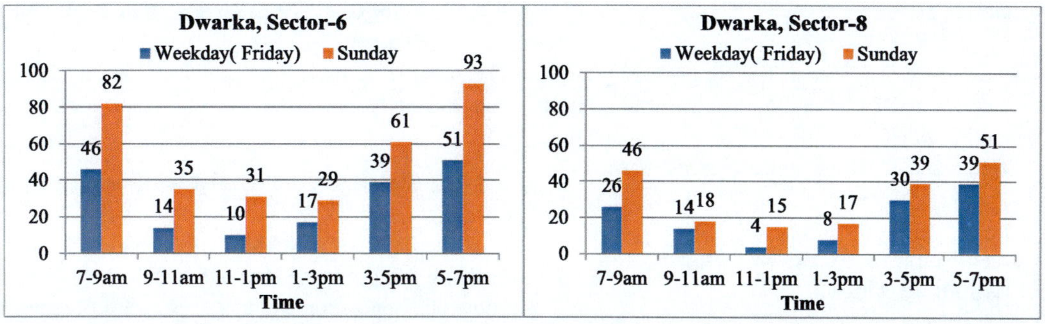

Fig. 5 Total number of people using park, 2015. *Source* Primary Survey, 2015

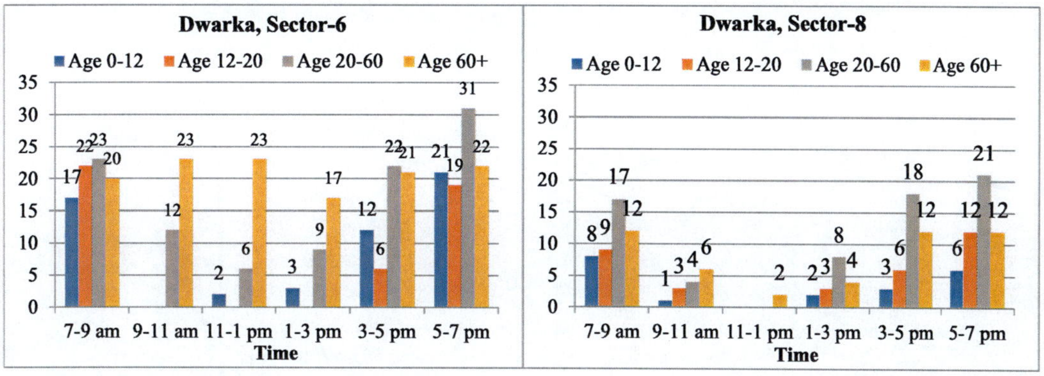

Fig. 6 Age group of people using park, 2015. *Source* Primary Survey, 2015

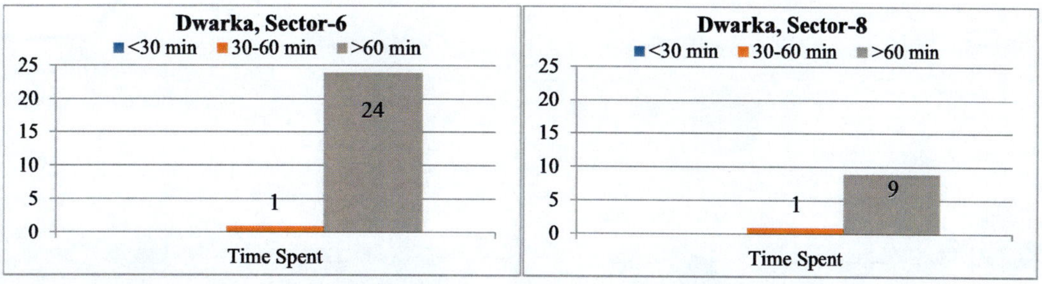

Fig. 7 Time spent at the park. *Source* Primary Survey, 2015

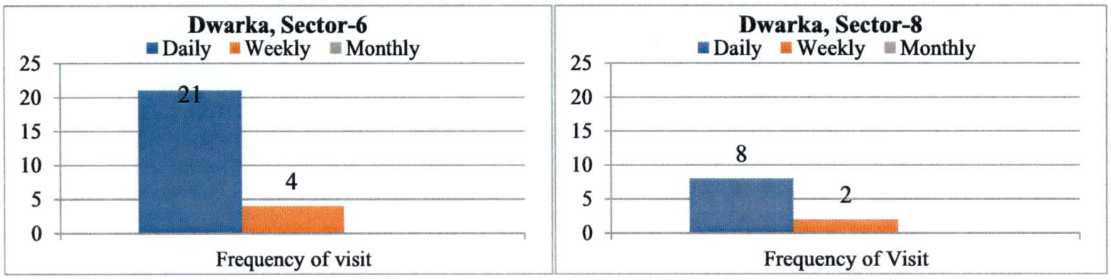

Fig. 8 Frequency of using the park. *Source* Primary Survey, 2015

5.3 Residents Survey

From residents survey, it is found that out of 3000 population 20% are children and 12% are senior citizens in sector-6 case. On the other side in sector-8, out of 1500 population, 21% are children and 12% are senior citizens as shown in Fig. 9.

5.3.1 General Characteristics and Population of the Study Areas

From the user survey, it is noticed that out of 3000 population only 3% area park users in sector-6 case. While in case of sector-8 out of 1500 population only 4% area park users and rest all area non-users.

5.3.2 Activities Performed at the Park (All Age Group)

The number of users for "walking" activity is highest for both the study areas (Fig. 10).

5.4 A Summary of Analysis of Study Areas

Characteristics of the study area, neighborhood park and land use around it are detailed in below Table 8. Now we know that population within 200 m radius of sector-6 park is 3000 and the population within 200 m radius of sector-8 park is 1500. Secondly population of children and senior citizens is also collected within 200 m radius of both parks which is also detailed in Table 8.

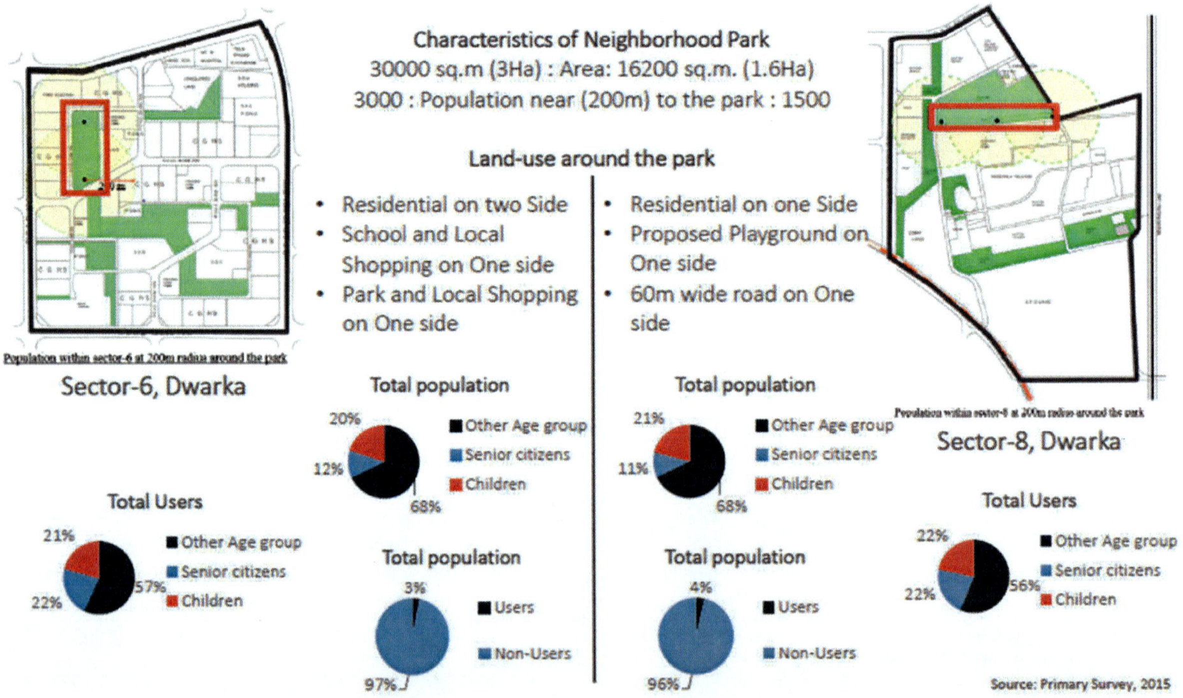

Fig. 9 Comparison of total population, total user, and division of users. *Source* Primary Survey, 2015

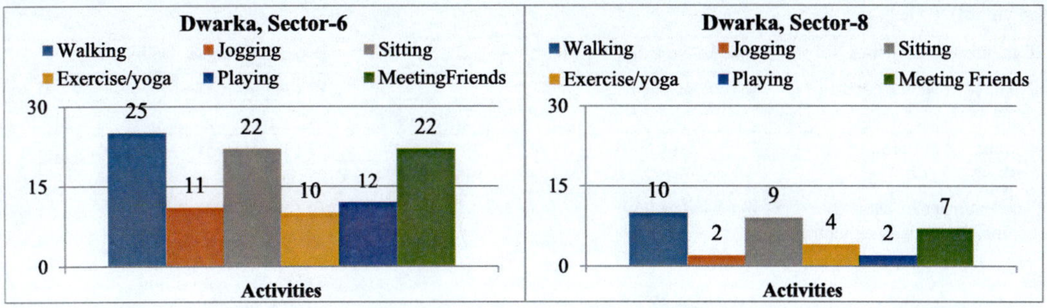

Fig. 10 Activities performed at the park, 2015. *Source* Primary Survey, 2015

5.4.1 General Characteristics, Usage, and Usability of Both the Study Areas

The maximum number of people using the park at a time (in two hours) is approx. 100 (3.3% of 3000) in sector-6 and 55 (3.6% of 1500) in sector-8. Usage of the parks in both sectors is almost 4%, which is very low.

5.4.2 Comparison of Aspects and Its Issues in Neighborhood Parks

People are using these parks for different activities, even facility is not provided. The usage of parks depends on the facility provided, safety, development pattern, and land use around it (Table 9).

6 Results and Finding

6.1 Site-Specific Issues and Problems

Out of total users, almost 20–25% park users are senior citizens and children Fig. 9. Out of the total population around the park 20–25% are children and 10–15% are senior

Table 8 General characteristics of both the study areas, 2015

S. no.	General characteristics	Dwarka Sector-6, Delhi	Dwarka Sector-8, Delhi
1	*Characteristics of study area*		
1.1	Type of development	Plotted housing	Group housing
1.2	Location	Sector-6, Dwarka (Zone: K-2)	Sector-8, Dwarka (Zone: K-2)
1.3	Area (ha)	93	150
1.4	Population	20,000	10,000
2	*Characteristics of neighborhood parks*		
2.1	Area	30,000 m^2 (3 ha)	16,200 m^2 (1.6 ha)
2.2	Dimension	100 m × 300 m	540 m × 30 m
3	*Land use around the park*		
		• Residential—two sides • School and local shopping—one side • Park and local shopping—one side	• Residential—one side • Proposed playground—one side • 60 m wide road—one side
4	*Population*		
4.1	Population near to the park	3000	1500
4.2	Senior citizen population	0.12 × 3000 = 360 (it is concluded from primary survey that 12% population is of senior citizens)	0.11 × 1500 = 165 (it is concluded from primary survey that 11% population is of senior citizens)
4.3	Children population (0–12 years)	0.20 × 3000 = 600 (it is concluded from primary survey that 20% population is of children)	0.21 × 1500 = 315 (it is concluded from primary survey that 21% population is of children)
5	*Usage of parks*		
5.1	Total maximum number of users (all age group)	100 (3.3% of 3000)	55 (3.6% of 1500)

(continued)

Table 8 (continued)

S. no.	General characteristics	Dwarka Sector-6, Delhi	Dwarka Sector-8, Delhi
5.2	Number of senior citizens using park	22 (6% of 360)	12 (7% of 165)
5.3	Number of children using park (0–12 years)	21 (3.5% of 600)	12 (4% of 315)
5.4	Total number of users (only senior citizens and children)	43 (4.4% of 960)	24 (5% of 480)
6	*Usability of parks*		
6.1	Ease of accessibility ramp	7 (58% of 12)	2 (16% of 12)
6.2	Status of facilities	10 (50% of 20)	5 (25% of 20)
6.3	Maintenance of facilities	7 (31% of 22)	4 (18% of 22)
6.4	Total weightage	24 (44% of 54)	11 (20% of 54)

Source Secondary and Primary Survey, 2015

Table 9 Comparative table for aspects in neighborhood parks, 2015

S. no.	Indicators	Neighborhood parks		Remarks
		Sector-6	Sector-8	
1	Type of development	Group housing	Plotted housing	Usage varies with type of development
2	Land use of the area	Residential	Residential	
3	Area of the park	3 ha	1.6 ha	Size is varying
4	Accessibility	Three-side road, three gates Footpaths are not walkable and clean, height is more	Three-side road, two gates Footpaths are not walkable and clean, height is more	Not maintained
5	Facilities provided	Plantation	Less plantation	Both the parks are some facilities
		Dustbins	Water	
		Toilets	Toilets	
		Benches		
		Jogging tracks		
6	Activities performed	Recreation	Recreation	Parks are used different activities, even facility is not provided
		Playing	Playing	
		Sitting	Sitting	
		Walking	Walking	
		Jogging	Jogging	
		Meeting friends	Meeting friends	
		Yoga/gathering		
7	Maintenance	Partially maintained	Not maintained	Lack of maintenance
8	Nuisance	Lack of water, lightening, swings	Lack of lightening, swings	
9	User opinion	Need facilities such as water, benches, shed, swings for children and improvement in maintenance	Need facilities like safety, plantation, shed, benches, and improvement in maintenance	Maintenance is the first requirement

Source Secondary and Primary survey, 2015

Table 10 Site-specific issues and problems

S. no.	Neighborhood parks	Issues and problems
1	Sector-6	• Lack of ease of accessibility and connectivity • Lack of basic facilities such as drinking water, adequate benches, and lighting • Salty water is available for the plants and trees, which is not good for their survival • Lack of maintenance of facilities • Lack of other facilities such as sitting/walking and swings/sliders
2	Sector-8	• Lack of ease of accessibility and connectivity • Lack of basic facilities such as benches, dustbins, lighting, and grasses • Lack of maintenance of facilities • Lack of other facilities such as sitting/walking and swings/sliders • Safety is major problem, reason being drunkard come for playing cards

Source Primary Survey, 2015

citizens Fig. 9. Usage of the park is approx. 4%. Usability of park is 44% in sector-6 and 20% in sector-8 (Table 8).

7 Discussion

General recommendation for neighborhood green spaces are listed below:

- Provision of at least one entrance gate.
- Provision of ramps connecting gates and access road for a disabled user.
- Provision of basic facilities such as toilet, drinking water, and other facilities such as lighting, sheds, benches, and jogging/walking tracks.
- Provision of CCTV cameras for safety and security.
- Provision of allocated space for providing special facilities to senior citizens and children in the park.
- Providing special programs at regular intervals.
- Provision of 3–6 labors for maintenance and cleanliness of the park.

Table no. 9.4 of Master Plan of Delhi-2021 is specifying the activities permitted in the Green Belt, Regional Park, City Park, District Park and community. It is recommended that Master Plan should also specify the activities and facilities of the neighborhood parks which is used on daily basis by the senior citizens and children. Following are the guidelines which Master Plan of Delhi should include (Table 11).

More parks can be studied and their issues and potential can be identified and allocated space senior citizens and children can be defined as per the requirement of particular Park.

8 Conclusions

The studied parks are lacking in many facilities and their maintenance; however, they have the potential of allocated space for senior citizens and children activities and the following facilities should be provided in both the parks.

- Provision of ramps connecting gates and access road for a disabled user.
- Provision of basic facilities such as toilet and drinking water and other facilities such as lighting, sheds, benches, and jogging/walking tracks.

Table 11 General recommendations

S. no.	Proposed guideline	For children	For senior citizens
1	Allocated space in the park	As per requirement	As per requirement
2	Facilities/activities such as	Playing, swings/sliders, mini-hand ball, basketball, riding cycle, and exercise.	Sitting, yoga, daily discussion, exercising, religious and cultural activities.
3	Special programs such as at regular intervals	Charitable programs, sports activities and health check-up camps, volunteering of senior citizens for maintenance of plantation in the park	

Source Concluded from Primary Survey, 2015

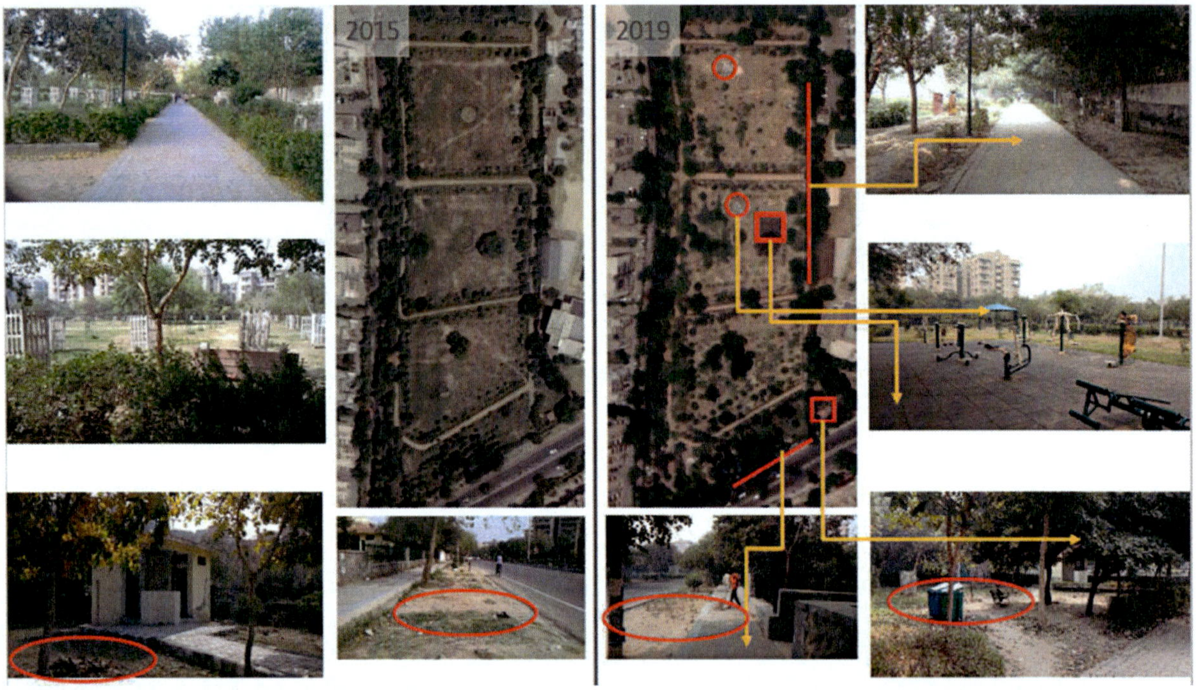

Fig. 11 Sector-6, before and now. *Source* Primary Survey, 2015 and 2019

Fig. 12 Sector-8, before and now. *Source* Primary Survey, 2015 and 2019

- Provision of CCTV cameras for safety and security.
- Provision of allocated space for providing special facilities to senior citizens and children in the park.
- Providing special programs at regular intervals.
- 15–20% park space should be given for senior citizens activities such as daily discussion, exercising, religious, and cultural activities. 20–25% park space should be given for children (0–12 years) playing facilities such as swings/sliders, mini-hand ball, and basketball.

9 Before and Now

In 2017, Master Plan of Delhi has brought a amendment, "Multi-Gym would be permissible in parks having an area of one hectare" (MPD-2021, pg. 9–9). A observation survey is again done in October 2019 to see the impact of amendment (Figs. 11 and 12).

After the amendment of the MPD-2021, more studies can be done because it has brought positive changes within and surroundings of the parks in both cases. The gym/exercise, special area for yoga, sheds, drinking water, toilets, benches, jogging tracks, and swings/sliders facilities are now added/available at both the parks due to which more children started coming in the parks.

References

Barton, S. (2009, January). Human benefits of green spaces. In *Sustainable landscapes series*. Retrieved January 22, 2015.

Brodhead, F. (2009). *Green space development: A literature review of research on the benefits of urban green space and what green space can become*. West Broadway Development Corporation. Retrieved December 24, 2014.

Center for Urban Green Spaces. (2012, March). *Congress on urban green spaces*. Department of Environment. Aravali Foundation for Education. Retrieved December 24, 2014.

Child Development Institute. (1999). Retrieved February 4, 2015, from http://childdevelopmentinfo.com/agesstages/

Christiansen, M. L. (n.d.). *Park planning handbook, fundamental of physical planning for parks and recreational areas*. Wiley.

Commision, N. H. (2011). *Know your rights: Elderly people*. National Human Rights Commission, Faridkot House. Retrieved January 22, 2015.

Consullting, B. (2013). *Green spaces: The benefits for London*. City of London Economic Development. Retrieved December 25, 2014, from www.cityoflondon.gov.uk/economicresearch

Delhi Development Authority. (1962). *Master Plan for Delhi- 1962*. Delhi Development Authority, New Delhi.

Delhi Development Authority. (1981). *Master Plan for Delhi-2001*. Delhi Development Authority, New Delhi.

Delhi Development Authority. (2010a). *Zonal development plan: Zone K-II*. Delhi.

Delhi Development Authority. (2010b). *Master plan of Delhi-2021*. Delhi Development Authority.

Dillon Consultant Limited. (2014). *Green space master plan*. Camrose.

Eryn Pleson, L. M. (2014, February). Understanding older adults' usage of community green spaces in Taipei, Taiwan. *International Journal of Environmental Research and Public Health*.

Hajime Orimo, H. I. (2006, February). *Reviewing the definition of "elderly"*.

Haq, S. M. (2011). Urban green spaces and an integrative approach to sustainable environment. *Journal of Environmental Protection*, 601–608. Retrieved December 25, 2015.

Jewell, D. L. (1997). *Reflections on leisure, play, and recreation*. Southern Illinois University.

Kiran, B. S. (2014, April). Recreational activities for senior citizens. *IOSR Journal of Humanities and Social Science, 19*(4), 24–30. Retrieved January 22, 2015.

Lemish, B. K. (2011). *Communicating with children: Principles and practices to nurture, inspire, excite, educate and heal*. UNICEF. Retrieved February 4, 2015.

Perry, C. A. (1929). Neighborhoor and community planning. *Regional Survey of New York and its Environs, VII*.

Riverside general plan 2025; Amended. (2012, November). *Riverside general plan 2025*. Retrieved February 2015 6, 2015.

Rodel, S. E. (2010). Recreational activities. Angeles City.

Sinha, R. S. (2013). *Urban forestry: Urbanisation and greening of Indian cities—Efforts for Green Delhi*. Retrieved December 24, 2014.

SKIP. (n.d.). *Child development and behavoiur: A resource for supporters of parents (Module 2)*. Aotearoa/New Zealand. Retrieved February 5, 2015.

Takano, K. N. (2002). *Urban residential environments and senior citizens' longevity in megacity areas: the importance of walkable green spaces*. J Epidemiol Community Health. Division of Public Health, Graduate School of Tokyo Medical and Dental University, Yushima. Retrieved January 18, 2015.

The Benefits of Green Space. (2015). *The benefits of green space*. Design Agency Creative. Retrieved January 22, 2015, from http://www.operationgroundwork.org.uk/takeaction/

UCL Institute of Health Equity. (2014, September). *Local action on health inequalities: Improving Access to green spaces*. Public Health England Publications. Retrieved February 8, 2015.

Vijai Shanker Singh, D. N. (2010). *Urban forests and open green spaces: Lessons for Jaipur, Rajasthan, India*. Rajasthan State Pollution Control Board. Retrieved December 24, 2014.

Wang, C.-L. L.-Y. (2012, December). The factors affecting life satisfaction: Recreation benefits and quality of life perspectives. *Sports and Exercise Research, 14*(4), 407–418.

The Use of Landscape as a Method of Integrating Informal Settlements in Tirana

Irina Branko

Abstract

The development of informal buildings has been a widespread phenomenon in the Albania during the last 30 years. Tirana, the capital of the country, has been the most affected one. This new type of development is spread all over the perimeter of the capital filling different territories by changing their land use from natural, agricultural, industrial, recreational, etc., to residential. It does have a powerful impact on the city due to its spontaneity and also carries on many problems related mostly to the lack of infrastructure, services, public spaces, and therefore lack of social cohesion and quality of life in these areas. There have been some developments during the last years, a Perimetral Park is proposed to be implemented defining the city limits which happens to be between nature and informal settlements. Considering this and the potentials of the natural and semi-natural environment have, this paper aims to develop further a sustainable approach that of regeneration and integration of the informal settlements, through the use of landscape. After the study and evaluation of all the databases, embracing landscape theories, a set of indicators and steps to follow for possible intervention will get developed. A theoretical methodology generated from this research could contribute to further studies on the issue for having an elevated attention from local authorities on the edge of the city by integrating the informality to the consolidated urban fabric, thus contributing toward sustainable development of Tirana.

Keywords

Informal settlements • Land use • Perimetral Park • Landscape • Sustainability

I. Branko (✉)
Department of Urbanism, Faculty of Architecture and Urbanism, Polytechnic University of Tirana, Tirana, Albania

1 Introduction

Informality is a widespread phenomenon all over the world, especially in developing countries. It is a process that involves many sectors, but that of territorial development as well, and it is mostly known as informal settlement phenomena. According to UN-Habitat, nowadays, 1.6 billion people live in inadequate housing globally, of which 1 (one) billion in slums and informal settlements. It means that about one in four people in cities lives in conditions that harm their health, safety, prosperity, and opportunities. Lack of access to basics services is a common constraint in informal settlements and slums (UN-HABITAT; For a Better Urban Future, 2018).

Defined a developing country, Albania as well has produced its informal settlement models in the territory, especially during the change of the political systems and all the transitory periods corresponding to the last 25–30 years. This particular phenomenon is evident in all the significant cities and more dominant in Tirana, the capital, as it represents the socio-economic pole in the country with many dynamic developments. Characterized by extended informal interventions also a high dynamic in the transformation of its territory, informal areas of Tirana are our object of study. The city of Tirana, nowadays, is still characterized by continuous urban growth. The developments of these last three decades, especially the informal areas, carry many problems that are related to the ecological, social, and economic aspects.

On a global scale, since 2015, the United Nations have outlined the sustainable development goals (SDG), which represent 17 urgent calls for action that need to be taken by 2030. These include health and education improvements, inequalities reduction, economic growth encouragement, and all while tackling climate change and working to preserve our ecosystem (UN, 2015). At the national and local levels, strategies as per the sustainable territorial development have also been proposed. However, there is no clear strategy for

F. Trapani et al. (eds.), *Advanced Studies in Efficient Environmental Design and City Planning*, Advances in Science, Technology & Innovation, https://doi.org/10.1007/978-3-030-65181-7_31

the informal areas, which represents more than 40% of the city footprint. In such a background, the future of the informal settlements in Tirana needs to be firstly understood and developed after. As will be later illustrated, this is a necessary intervention. Some of the objectives of this paper are to evaluate the informal areas and understand what planned. The main goal of this study is the introduction of a theoretical model of regeneration and integration of the informal settlements by the use of landscape. The explanation of this new model of intervention would lay on three main dimensions that of ecological, social, and economic aspects as representants of the "magic triangle," the sustainability model often established in practice (Bott et al., 2017).

2 Background and Case Study

2.1 History and Facts

During the socialist regime in Albania,[1] 70% of its citizens were living in rural areas all over the country. People were working and building their economy on agriculture and livestock activities, as per the natural character of the territory. The system failure in the late '80s and the economy deficits created turned to a total lack of opportunities, especially in the rural areas. The whole picture created was the cause of the elevated migration process of the population from these areas toward other countries, but also toward the important cities which offered more economic opportunities. This process was characterized mainly by movements from remote rural areas from the east side toward the western part, especially the central-western area, which is better identified by the *Tiranë—Durrës* linear development (Fig. 1).

The direct impact of this intense demographic movement was dualistic territorial development. On one side, it caused the shrinking effect of more peripheral cities and abandonment of rural areas. On the other hand, it caused rapid urbanization in most of the remarkable areas of the Albanian territory.

There was too much demographic pressure during the period in which this development got introduced. Lack of appropriate laws on land property, lack of experience of managing the territory, weakness of existing institutions, and sometimes lack of institutions were the main reasons for the territorial informality to develop causing also, the "urban sprawl" effect. The pattern of the chaotic development of Tirana during the early '90s is the best indication of a difficult political, economic, and social transition from a centralized economy to a free-market society (Aliaj et al., 2003).

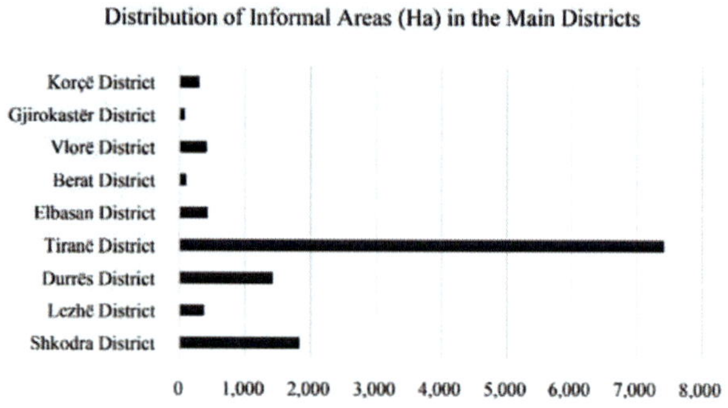

Fig. 1 Geographical distribution of informal areas. *Source* State Cadastre Agency (2018); Elaborated: Author

The cities of *Tirana, Durrës*, and the area in between have been developed and changed throughout the years. By development, phenomena such as change of the population, change of build, and natural environment have also occurred. Figure 2a–d illustrates the footprints of the urban areas through four periods starting from the Italian occupation to nowadays. The footprint enlargement and the transformation to build the environment of what previously was a natural and semi-natural environment is evident. Most of them have come as a result of informal development. According to the data collected from Institute of Statistics (INSTAT, 2019), during these last 30 years, just the city of Tirana itself results twice as big in terms of territory and four times as large in population.

Nowadays, the trend of migration and population movement toward the capital is continuing, but the informality development has almost stopped as a territory development process. However, there is a large number of areas covered with informal buildings, as previously mentioned, carry many problems that are related to the social, ecological, and economic aspects.

2.2 Definition of Informal Settlements

According to UN-Habitat, the definition of informal settlements is:

> Residential areas where (1) Inhabitants have no security of tenure vis-à-vis the land or dwellings they inhabit, with modalities ranging from squatting to informal rental housing, (2) The neighborhoods usually lack, or are cut off from, basic services and city infrastructure and (3) the housing may not comply with current planning and building regulations, situated in geographically and environmentally hazardous areas and may lack a municipal permit.
>
> Also, informal settlements can be a form of real estate speculation for all income levels of urban residents, affluent,

[1]The Socialist Regime in Albania was a mono-party economic and political system that ruled from 1946 to 1990. A system characterized by common ownership and every sector of the economy relying on government. During this period, everything was lead by the Party of Labor of Albania and the supreme leader, dictator E. Hoxha.

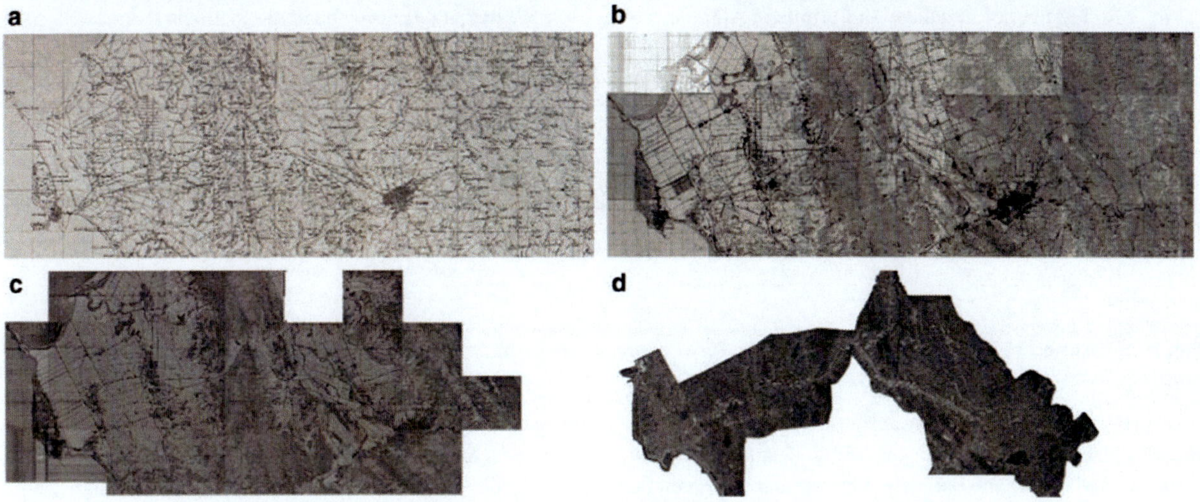

Fig. 2 **a** Map of Tirana-Durrës region during the Italian occupation period; **b** Map of Tirana-Durrës region during socialist period in the late '50s; **c** Map of Tirana-Durrës region during socialist period in the late '80s; **d** Map of Tirana-Durrës region in 2018. *Source* ASIG State Authority for the Information (2019)

and poor. Among them, slums are the poorest and most dilapidated form of informal settlements.

Thus, informality should not be understood as an income-based denomination that stigmatizes the poor. Rather, informal settlement estimates should be based on technical compliance relevant to all income levels related to the above criteria.

For example, a valid municipal permit could be a reliable indication of formality (UN-HABITAT; For a Better Urban Future, 2018).

In the Albanian context, informal settlements partially correspond to this definition. They do not have the characteristics of a slum, but those of a building with permanent materials, and we could easily define them as individual self-organized developments. In general, in the areas, there is not too much flexibility in terms of investments made and material used since most of them came as reinforced concrete structures with brick walls. The buildings vary from 1 floor to 3–4 floors and often are investments supported by the immigration income.

2.3 Today Legal Situation

It is important to emphasize that the informality areas in the Albanian context happened due to lack of planning instruments and often the occupation of private or public land. These areas have evolved as urban forms over the years. The first tendency was that of filling the territory.

After, the settlements have undergone continuous changes and transformations, which most of the time corresponded with having additional floors. Nowadays, it seems like they have reached a stable form.

It is since 2006 that many laws and legal instruments passed, allow the legalization of the informal buildings by their owners. Sometimes the developers owned the land, many times they had occupied it. In these last cases, the legal landowners are being compensated first by economic value and after would be expropriated. The legalization process is still ongoing.

This process somehow confirmed the acceptance of the informality existence and has legally made them a legal part of the city. In Fig. 3a, b) progresses of this process is shown. Here could be noticed the completed number of legalized constructions for each district for the 2006–2018 period of time.

It is important to emphasize that the term informal settlements used through this study, do not illustrate their legal status nowadays, rather than their origin. It helps to identify them from the rest of the city, due to their previous legal status and particularities.

3 Importance of Study

The informal settlements due occupy a high percentage of the city footprint, and as previously mentioned, they are an adjacent part of the city by now, yet not integrated. There are many studies on informal settlements, mostly describing the specificity of development in Albania. In real terms, almost nothing has been done for regenerating and integrating. The General Local Plan, Tirana 2030 (later in this paper will be called GLP TR30) approved lately also, does not introduce a clear or detailed strategy on these particular areas (Municipality of Tirana, 2017a, 2017b, 2017c).

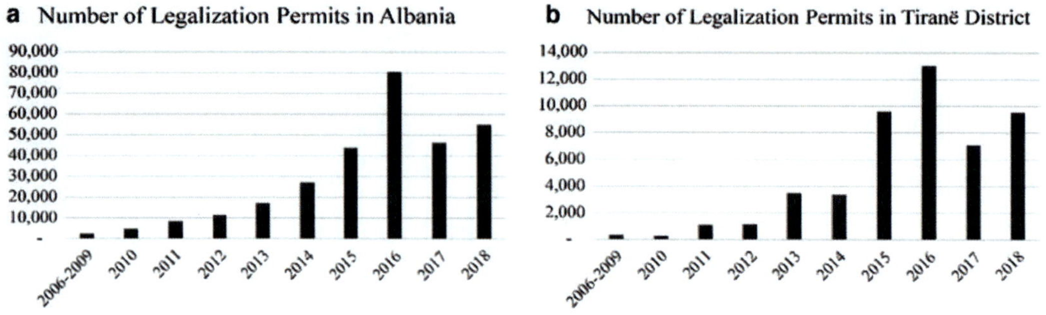

Fig. 3 a, b Number of legalization permits in Albania; Tirana. *Source* Information (State Cadastre Agency, 2018); Elaborated: Author

For the first time, the future-city development are introduced planning strategies regarding banning of the urban-sprawl effect. The Perimeter Park similar to a green belt would be implemented around the build environment footprint. It will not serve just as a natural and recreational area to the city, but is also supposed to achieve the desired banning effect (Municipality of Tirana, 2017a, 2017b, 2017c).

It is very obvious that the informal settlements areas have been developed concentrically along the perimeter of the planned city. As shown schematically in the figure below (Fig. 4), they are positioned in between the Perimetral Park and the planned city. Considering the Perimetral Park perimeter around 95,000 m of length, this strategy constitutes a future asset to the areas in terms of offering a qualitative space. Due to their morphology of urban form, the informal areas do construct a barrier between the city and nature. By having such position in the territory definitely, centripetal forces would not get generated between different areas.

To summarize, the importance of the study does not stop just to the regeneration and integration of informal areas, but goes far beyond. It will have additional importance in trying to create a better connection even between the planned city and the countryside, and vice-versa.

4 Introducing the Theoretical Approach—Understanding Landscape Urbanism

In the opening years of the twenty-first century, that seemingly old-fashioned term landscape has curiously come back into vogue. This reappearance of landscape in the larger cultural imagination is due, in part to the remarkable rise of environmentalism and a global ecological awareness, to the growth of tourism and the associated needs of regions to retain a sense of unique identity and to the impacts upon rural areas by massive urban growth (Corner, 2006).

In the holistic framework of environment conservation and ecology infrastructure development, urban theories have arisen, stating the potential of landscape in rethinking the urban development of the cities (Waldheim, 2016). This is partially due to also the failure of the modern design which based its theoretical framework on the best possible spatial synthesis between function and form (Ricci, 2018). Many authors such as Edward Soja and later by Bernardo Secchi describe the nowadays city being in crisis. It is described as a crisis laid on three dimensions: environmental, social relations, particularly in terms of differences between "rich" and "poor," and accessibility everywhere by single people or group of people, caused by high evidence of spatial injustice (Secchi, 2011).

In such framework "Landscape Urbanism" has been introduced. Charles Waldheim coined the term "Landscape Urbanism" to describe the practices of many designers for whom landscape had replaced architectural form as the primary medium of city making …. Waldheim saw Landscape Urbanism, like landscape architecture, as an interstitial design discipline, operating in the spaces between buildings, infrastructural systems, and natural ecologies. In these contexts, landscape urbanism became a useful lens through which to view those "unseen," the residual *terrain vogues* ... (Shane, 2006). This approach attempts to create an environment that is not so much an object that has been "designed," as it is an ecology of various systems and elements that set in a motion a diverse network of interaction. Landscape urbanism is here both instigator and accelerator, working across vast surfaces of potential (Corner, 2006).

The approach situates the emergence of landscape as a medium of urbanism in a variety of sites. (1) Most often the sites associated with rethinking the urban through landscape are found at the limits to a more strictly architectonic order for the shape of the city. (2) Landscape has been found relevant for sites in which a strictly architectural order of the city has been rendered obsolete or inadequate through social, technological, or environmental change. (3) Landscape has also been found useful as a way of thinking through urban form in the wake of macroeconomic transformations. This includes so-called shrinking cities as well as the countless

Fig. 4 Map of green belt proposed by GLP TR30 and informal areas 2007–2008; 2015. *Source* Municipality of Tirana (2017a, 2017b, 2017c), NTPA (2018); Elaborated: Author

individual sites of brownfield abandonment left in the wake of economic transformations. (4) Thus, landscape as a medium of urbanism has often been invoked to absorb and in some ways mitigate various impacts associated with social, environmental, and economic crises. (5) It has equally been found relevant for thinking through sites at the intersection of large, complex ecological, and infrastructural systems. (6) Most recently, landscape has been found relevant to questions of green infrastructure in the informal city and in response to questions of risk and resilience, adaptation and change.

The cumulative effect of these sites and subjects has been to foreground the potential for landscape as a medium and model for the city as a collective spatial project. In its most ambitious formulation, this suggests the potentials for the landscape architect as urbanist of our age. In this role, the landscape architect assumes responsibility for the shape of the city, its built form, and not simply ecological and infrastructural exceptions to its architectonic structure. Rather, landscape thinking enables a more synthetic understanding of the shape of the city, understood in relation to its performance in social, ecological, and economic terms (Waldheim, 2016).

Such urban theory emphasizes the potential of landscape in urban development especially that organized in the macro-scale or as an infrastructure in the city. With the characteristics, we could read from the informal settlements area, also findings made in the strategies foreseen in the GLP TR30, this approach is hypothesized that could be the winning one for their regeneration and integration to the rest of city.

5 Method of Study

For achieving the objectives of the study, the work is organized considering the parallelism and evaluation process of the three following main phases.

5.1 Understanding the Landscape Potential in the Planning Process

The city development is at an intersection nowadays, in terms of its future evolution and especially in terms of quality it offers. In the Albanian context, so far, quality is not a measurable indicator. Many theoretical approaches are being introduced to professionals lately. Therefore, it is important to evaluate their adaptability in local situations and contexts.

This phase of the research is necessary for the understanding of significant landscape aspects and make practical use of them. For this reason, through the theoretical approach, simple indicators related to planning have been generated. These indicators will be after used for structuring a matrix (Fig. 5; Structure). It is important to emphasize that the indicators of the matrix are being organized according to the three main pillars of sustainability, ecological, social, and economic aspects.

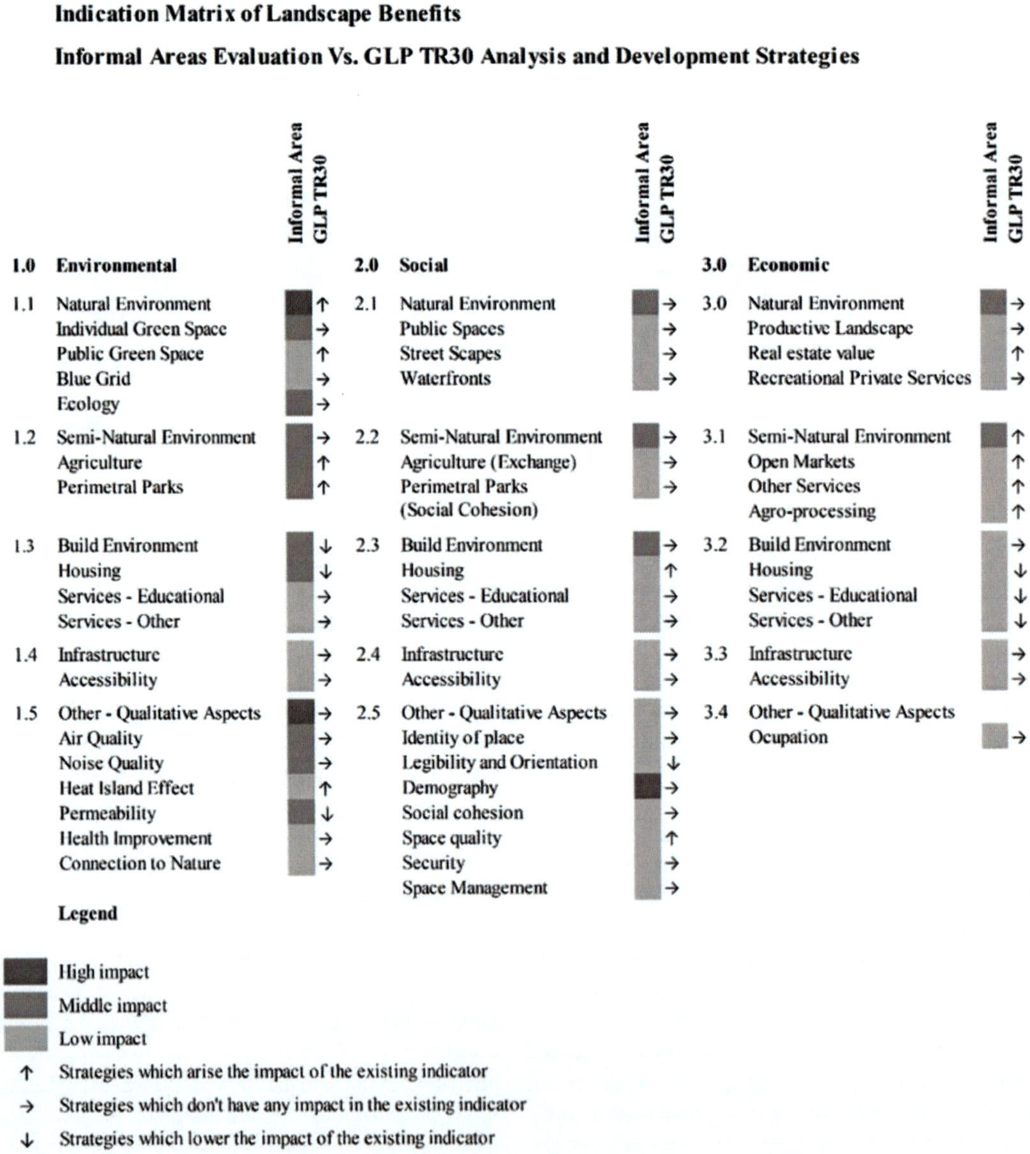

Fig. 5 Indicators matrix; evaluation of landscape potentials. *Source* Municipality of Tirana (2017a, 2017b, 2017c); Elaborated: Author

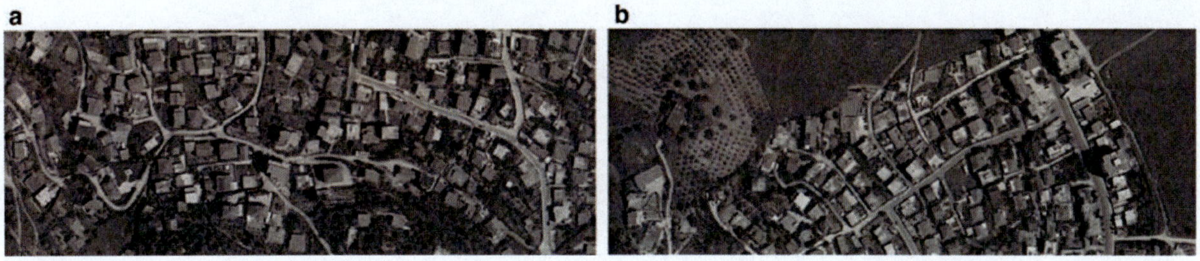

Fig. 6 a, b Orthophoto; informal areas in Tirana; single units characteristics. *Source* ASIG State Authority for the Information (2019)

5.2 Identification of Potential and Problems in the Areas

This phase includes close observation of the informal settlements and the areas they occupy in the city. During this phase of the study, the existing analysis made, existing land use maps, aerial views belonging to different periods, orthophotos, furthermore sites visits, and photos are analyzed. The understanding of issues related to the landscape, such as the space, space network, space functionalities, space qualities, lifestyle, ecology, economic activities, services, and every other aspect included in the build-in matrix, has importance for creating a physiognomy of each singular area. Parts of the study are also talkings and discussions with the local families as the inhabitant's origin was the fundamental purpose of such meetings. The evaluation of each indicator was later compiled, by determining the impact had in the informal areas (Fig. 5; Evaluation of Existing Informal Settlements).

5.3 Undertanding Strategies in the GLP TR30, Foreseen for the Future-City Development

This phase includes the collection and reading of all data, existing and planned maps, reports, evidence, analyzation, and strategies foreseen at the national and local level for a better understanding of the impact the GLP TR30 strategies have on the areas of interest. The evaluation of the impact strategies, which have on the predefined indicators, is being made. In the end, the completed matrix will give a better understanding of how all GLP TR30 strategies are related to the landscape urbanism approach, what provided for the city in general, and for the informal settlements in particular. Finally, the indicators provided by the landscape theories in the first step, evaluation of the existing also of the impact of future strategies considering ecological, social, and economic dimensions, are cross-checked in one single matrix named indicators matrix (Fig. 5; Evaluation of GLP TR30 Strategies Impact).

6 Results

The results are better expressed in the indicators matrix; however, for a better understanding, the following need also be interpreted.

6.1 Findings on Informal Settlements Areas

Single Unit—The main particularities of a single unit are (1) Self-development constructions placed in previews natural or agricultural territory; (2) The building usually situated in the middle of a land parcel, therefore assuring an individual space; (3) Intimate private space due to the presence of high and non-transparent barriers; (4) Private yard with the main use of private parking and gardening. Olive-grooves, vineyards, fruit trees, vegetables, and flowers are part of every single individual yard (Figs. 6a, b and 7a–d).

Groups—The positioning of the individual units tends to create rows that differ from area to area. The documents of the GLP TR30 has analysis and classification of their typo-morphology as per their relation to the territory. As per this classification, there are four main typologies (a) One which follows the infrastructure, (b) Surface cover, (c) Rural, and (d) Mountainous. Other main particularities of these groups illustrated in Fig. 8a–d. Their description could be summarized as follows: (1) Large extension in the territory; (2) Mostly a cul-de-sac street pattern; (3) Street space gained as a result of single unit high, non-transparent walls; (4) Street space with only distributive function; (5) Lack of any other type of public space; (6) Lack of services; (7) Demography, from the GLP TR30 Analysis,[2] it results that the people in the informal areas create groups in the territory as per their origin (came from the same province).

[2]GLP *TR30 "The Technical Evaluation Report on the Analysis and Evaluation of the Territory, Metabolic Analysis of the Flows and in-depth Analysis of the Territory—this fact also confirmed by the discussions with inhabitants.*

Fig. 7 **a–d** Streetscape photos; informal areas in Tirana. *Source* GOOGLE (2019)

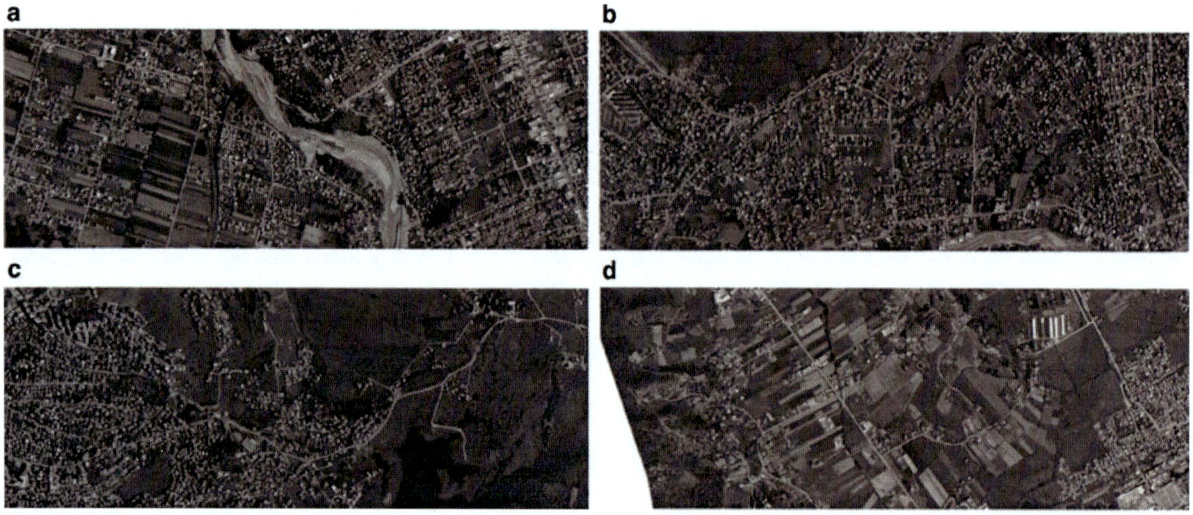

Fig. 8 **a–d** Orthophoto; informal areas in Tirana; groups characteristics. *Source* ASIG State Authority for the Information (2019)

Macro-scale—It could be read from the maps (1) Mostly the filling of the territory is present in flat terrain or soft slopes, this corresponding to previews agricultural or natural land (Fig. 9a, b); (2) The naturally hilly terrain has contributed to the footprint the informal areas. Due to the morphology of the territory, the informal areas are defragmented somehow naturally by green, natural wedges.

6.2 Findings on the Main Strategies Foreseen

Strategies at National Level—In terms of future developments foreseen, there is a strong focus on (1) The agricultural and livestock economic model, as it seems like one of the most significant assets of the country and (2) Preservation of natural and semi-natural environment. The national strategy focuses on the use of resources all over the country and inhibition of internal migration for being able to have a scattered population and have a sustainable use of the national resources.

Strategies at Local Level—The attention toward the natural and semi-natural areas are a key-topics for foreseen GLP TR30 strategies as well. It is best illustrated through two decisions: (1) The declaration of the natural area surrounding Tirana, as an area of national interest, but also with (2) The implementation of the Perimetral Park as earlier explained, comes as a new approach in terms of the control of the territory and resources (Fig. 10).

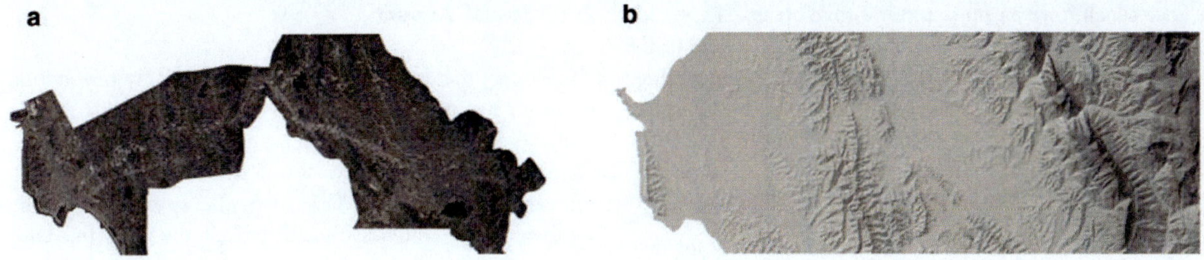

Fig. 9 **a**, **b** Orthophoto of Tirana-Durrës Area; digital model of the territory. Source ASIG State Authority for the Information (2019)

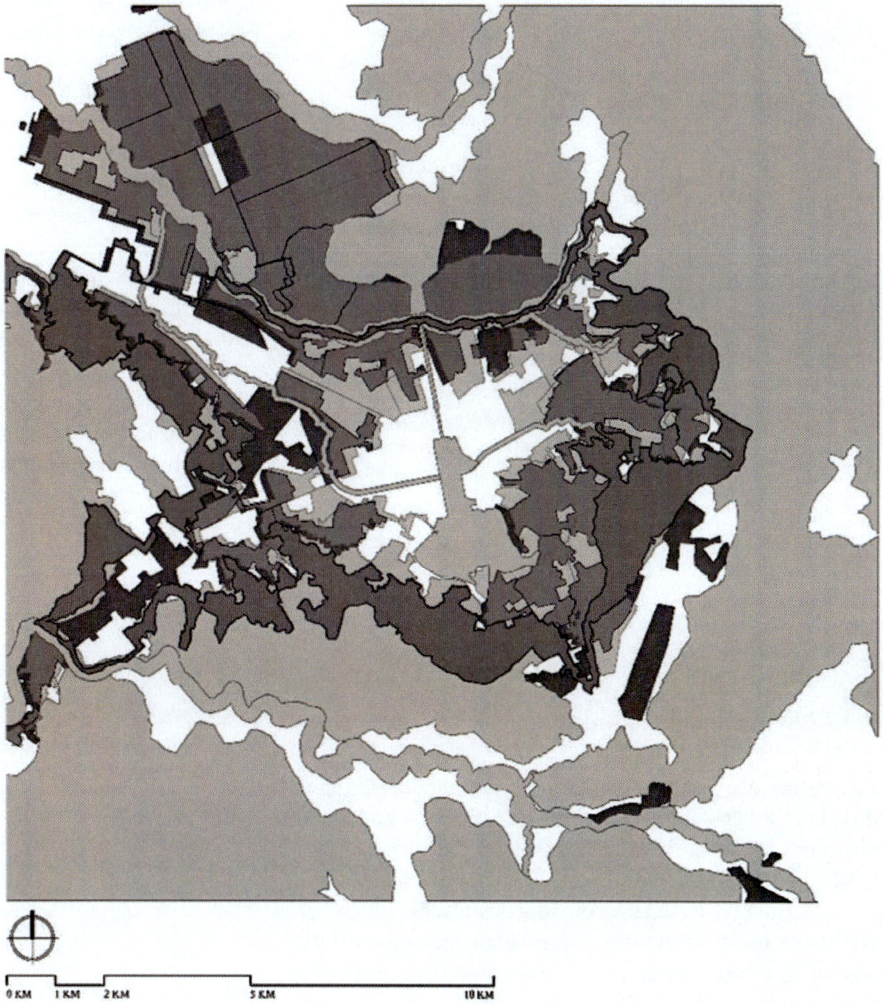

Fig. 10 Map of "Green Crown" and informal areas 2007–2008; 2015. *Source* Municipality of Tirana (2017a, 2017b, 2017c), NTPA (2018); Elaborated: Author

This park, with a total area of 2563.78 Ha, would serve as natural and semi-natural space for finally defining the boundaries for the city development (Municipality of Tirana, 2017a, 2017b, 2017c). According to GLP TR30, the further development of the city would be carried out by the densification of the existing fabric. Although still at the strategy level, it comes as a temptation toward banning the "urban sprawl" effect. There are no specific findings regarding the transitory areas. It is referred here to those between the Perimetral Park and the formal city, therefore, area of our interest.

Other strategies—GLP TR30 proposes many contemporary and strategies of interest for development. We are listing just a few, which have an impact at informal areas (1) Creation of a polycentric city—the creation of five epicenters in the city peripheries; (2) Sustainable development based on three primary pillars ecology, social, and economic development; (3) City accessible for all (Municipality of Tirana, 2017a, 2017b, 2017c).

Densification—GLP TR30 has as main strategies the bounding of "urban sprawl" and the densification of the city. Parts of informal areas due undergo such process, therefore, to that of filling the territory.

7 Discussions

7.1 Today Evaluation

It is for 30 years now that the informal settlements are present in Tirana. While the process of legalization its ongoing, some infrastructural investments have been completed, which remain at the level of sewerage systems and the asphalting of the roads, somehow the quality of accessibility has increased, but not in a meaningful way the quality of life in these areas.

Occupying more than 40% of the territory, this part of the city should not be out of our attention. We as planners need… to give a meaning, narrative, and uses—even temporary—to spaces that already have got some given forms, and to turn them into places where living is both pleasant and affordable. This phase requires designers to adopt new points of view on the future, but also and above all present, and a new idea of projecting a physical space (Ricci, 2018).

7.2 Environmental Aspect

As schematically shown in the indicators matrix, the area does have potential in terms of the natural and semi-natural environment. These areas previously agricultural and natural are fragmented now, and most of them not efficiently used, often creating voids. In terms of productive landscapes, there is the presence of private gardens, an indicator of the inhabitant life-styles. At a local level, strategies on having additional but fragmented structured green spaces due exist. However, what it does not come clear, is how this future natural environment will co-exist and co-operate with the different parts of the city in terms of continuity, creation of an ecologic matrix, and offer fluidity. The problematic gets more complicated with the continuous densification foreseen, therefore, the urban filling process which threatens the most vulnerable parts of the territory and causes the consumption of the space in between buildings.

7.3 Social Aspect

It is evident that in these areas, there is life only beyond the walls, in the private space. The rest has only distributive character and does not serve as a gathering and shared space; consequently, there is low social cohesion. This effect is exacerbated also by the lack of public spaces and public services. As a self-constructed part of the city, the built environment contains mainly private houses. Through these individual constructions, the previous landscape is fragmented in such a way that the territory has lost their identities of place[3] and also their genius-loci,[4] the spirit of the place… The character of a place is to a high extend determined by how this standing and rising is concretized. This also holds for entire settlements, such as towns. When a town pleases us because of its distinct character, it is usually because a majority of its buildings are related to the earth and sky in the same way; they seem to express a common form of life, a common way of being on the earth. Thus, they constitute a genius-loci which allows for human identification (Norberg-Schulz, 1980).

We can denote that landscapes respond to the needs of identification and recognition, spatial anchoring, rooting, and collective values. The idea of landscape responds to the ultimate reasons of inhabiting a place and represents "an essential aspect of life situation of the population of cultures and which represents a fundamental component of Europe's cultural and natural heritage." For offering the quality of life in the city, the identity of the place should be replaced.

7.4 Demography

The informal areas have a diverse population, and as mentioned before, it can create a sense of neighborhood due to large groups of people originating from the same place. In all the analysis of the built environment, there is a strong sense of preserving existing natural elements and also basing the life on products of the territory.

[3]Place is evidently an integral part of existence. With the word "place" we mean a totality made up of concrete things having material substance, shape, texture, and color. Together these things determine an "environmental character" which is the essence of a place. A place is therefore a qualitative, "total" phenomenon, which we cannot reduce to any of its properties, such as spatial relationship, without losing its concrete nature out of sight (Norberg-Schulz, 1980).

[4]Genius-loci is a Roman concept. According to ancient Roman belief, every "independent" being has its genius, its guardian spirit. This spirit gives life to people and places, accompanies them from birth to death, and determines their character or essence (Norberg-Schulz, 1980).

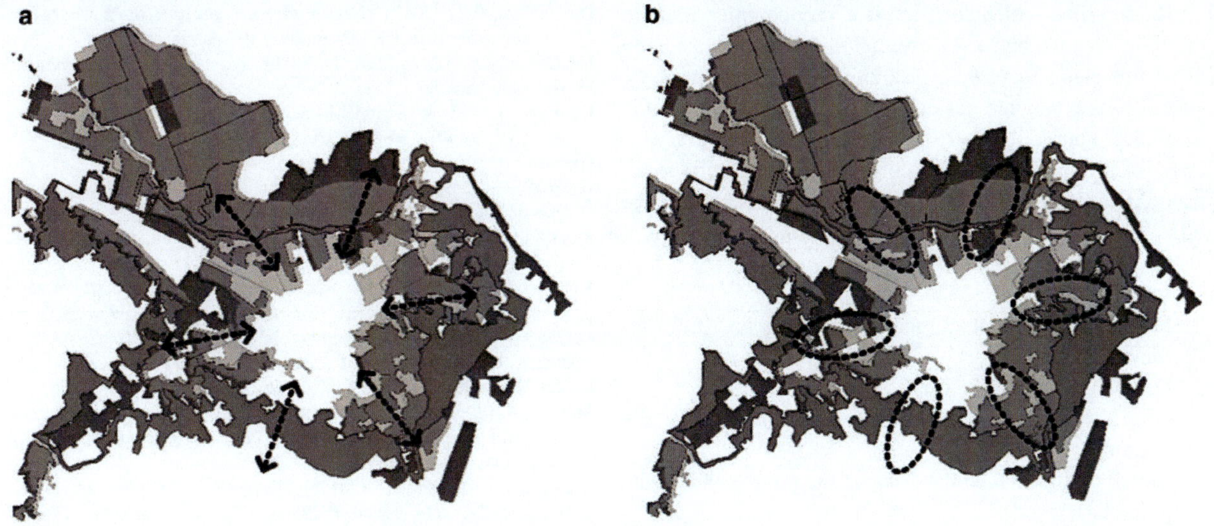

Fig. 11 **a**, **b** Maps of spatial connections and the distribution of nodes. *Source* Municipality of Tirana (2017a, 2017b, 2017c); Elaborated: Author

7.5 Risks or Benefits

The implementation of the planned Perimetral Park might be accompanied with an increase of the real estate value, therefore, with additional pressure from the construction industry. This could be also supported by the GLP TR30 in several areas. The risk of loss of natural territories versus build environment is high although new developments, if done correctly might bring integration and prosperity in the area.

7.6 Economic Aspect

There is not too much database on how the population of informal settlements areas due organize their living beside the fact that gardening covers a small part of the economy. These areas have extended at a distance that they sometimes have embraced the pre-existing rural areas; therefore, a part of the population is historically connected with the agriculture and agro-processing. A detailed demographic database would give a better understanding of the people, territory, its products, and how all are related to each other.

However, the GLP TR30 strategies, which support agricultural, livestock economic model, and brand creation from local products, do create a legal framework that strongly connects the area with the territory productions, contributing this way not only in the economy, but also in the sense of belongingness.

As the indicators matrix and other findings show, landscape urbanism is capable of offering ecological, social, and economic qualities through its instruments such as the implementation of (1) Productive Landscapes; (2) Natural Landscapes; (3) Public Spaces; (4) Recreational Spaces; (5) Ecological Corridors; (6) Cultural Landscape. Through the use of landscape, it is possible to create such an infrastructure that relates to the city and the countryside but also … collaboration between organizations of various sectors and citizens (Public, Private, People, and Partnerships) (Fig. 11a) (Garbarini, 2018). Within the landscape, we can propose and articulation of nodes that can become catalysts of regenerative processes both in the places in question and surrounding areas. Exploiting the large scale of the landscape, we can work with different surfaces, devices, and policies, creating a sort of re-alignment, in which landscape replaces urban planning and architecture (Corner, 2006) (Fig. 11b).

8 Conclusions

The main objective of this paper is the contribution in developing a theoretical model of intervention by the use of landscape as a flexible medium for the regeneration and integration of the informal settlement areas to the city. This study comes as a contribution to distract the attention of conventional planning and emphasizes the new approach of landscape urbanism to a local level.

To achieve the desired result, it was necessary to analyze at different scales the potentials of all forms of the territory, potential strategies planned from authorities, and evaluate them through the lens of landscape. In the presence of a large number of potentials, as resulted, it could be concluded that landscape urbanism can activate all the existing ecological, social, and economic resources, thus complying with the sustainable development goals.

The informal settlements area is a very complex issue in the local context, and therefore, this research is considered just a first step toward the possible interventions. For being able to have tangible results and implementation of the theoretical approach, other steps to follow are necessary.

Step 1—Creation of planning concepts by integrating rural, natural, informal, and formal city being coherent with the GLP TR30 strategies. The following further analysis is a significant part of this phase:

1. Identification of cultural geographic areas based on territorial identity;
2. Detailed demographic data;
3. Origin, lifestyle, family structure, and position in the territory;
4. Creation of a detailed database on the private and public properties;
5. Understanding the urban morphological mapping of spatial typologies.

Step 2—Communication with the population of these areas for presenting ideas and development models. Such a process is necessary for creating real development models. Throughout this phase, necessary explanations and presentations must be completed, so that bottom-up approaches could generate.

Step 3—Detailed planning process and pilot project implementation.

Acknowledgements This research was realized with the assistance of digital materials and documents made available by Albanian Institutions, NTPA (Natural Territory Planning Agency), and Municipality of Tirana. A special thank you, the author would like to give to both of these institutions for being cooperative, assisting with a database, and trusting in the research work of the universities for further development in the city planning. Above all the author would like to thank the Polytechnic University of Tirana for supporting and always encouraging the research work.

References

Aliaj, B., Lulo, K., & Myftiu, G. (2003). *Tirana, the challenge of Urban Development*. Gorenjski Tisk.

ASIG State Authority for the Information. (2019). *Services*. www.geoportal.asig.gov.al: https://geoportal.asig.gov.al/map/?auto=true

Bélaner, P. (2017). *Landscape as infrastructure*. Routledge.

Bott, H., Grassl, G. C., & Anders, S. (2017). *Sustainable urban planning*. DETAIL.

Corner, J. (2006). Terra Fluxus. In C. Waildheim (Ed.), *The landscape urbanism reader* (pp. 021–033). Princeton Architectural Press.

Corner, J., & Hirsch, A. B. (2014). *The landscape imagination; collected essays of James Corner 1990–2010*. Princeton Architectural Press.

Council of Europe. (2006). *Landscape and sustainable development; challenges of the european landscape convention*. Council of Europe Publishing.

De Oliveira, F. L. (2017). *Green wedge urbanism: History, theory and contemporary practice*. Bloomsbury Academic.

De Oliveira, F. L., & Mell, I. (2019). *Planning cities with nature; theories*. Springer.

Deming, E. M., & Swaffield, S. (2011). *Landscape architecture research; inquiry, strategy, design*. Wiley.

Garbarini, G. (2018). *Smart landScape*. LISt Lab.

GOOGLE. (2019). *Google maps*. www.google.com: https://www.google.com/maps/

INSTAT. (2019). *CENSUS 2011*. www.instat.gov.al: https://www.instat.gov.al/al/search/?query=CENSUS+2011

Konijnendijk, C. C., Nilsson, K., & Randrup, T. B. (2005). *Urban forests and trees*. Springer.

Lefebvre, H. (1991). *The production of space*. Blacwell.

Lynch, K. (1981). *A theory of good city form*. MIT Press.

Lynch, K. (1990). *The Image of the city*. MIT Press.

Marson, A. (2015). *Ridesigning the territories of diffused urbanisation [original title: Riprogettare i Territori dell'Urbanizzazione Diffusa]*. Quodlibet Studio.

Municipality of Tirana. (2017a). *General local plan of Tirana Municipality, Development Plan, 2017 [original title: Plani i Përgjithshëm vendor i Bashkisë Tiranë, Plani i Zhvillimit]*. www.tirana.al: https://tirana.al/faqe/plani-i-pergjithshem-vendor

Municipality of Tirana. (2017b). *General local plan of Tirana Municipality, The Territorial Strategy [original title: Plani i Përgjithshëm vendor i Bashkisë Tiranë, Strategjia Territoriale]*. www.tirana.al: https://tirana.al/faqe/plani-i-pergjithshem-vendor

Municipality of Tirana. (2017c). *Planning and territorial development*. www.tirana.al: https://www.tirana.al/sherbimi/planifikim-territori#

National Territory Planning Agency. (2016). *Albania 2030, The National General Plan, The first national document on territorial Planning 2015–2030*. Tirana.

Norberg, C. (1980). *Genius Loci: Towards a phenomenology of architecture*. Rizzoli.

NTPA. (2018, May). *Albania: The basic maps of the territory*. www.planifikimi.gov.al: https://planifikimi.gov.al/index.php?id=njoftime0

Pandakovic, D., & Dal Sasso, A. (2016). *To know how to see the landscape; [original: Sper vedere il Paesaggio]*. Cittá Studi Edizioni.

Ricci, M. (2018). Landscape 4.0; forward. In G. Garbarini (Ed.), *Smart landscape* (pp. 4–7). LISt Lab.

Secchi, B. (2011). The new urban question: Environment, transportation and social inequalities. In *CRIOS* (pp. 89–99).

Shane, G. (2006). The emergence of landscape urbanism. In C. Waldheim (Ed.), *The landscape urbanism reader* (pp. 055–067). Princeton Architectural Press.

State Cadastre Agency. (2018). www.ashk.gov.al: https://www.ashk.gov.al/

Strappa, G. (2018). *Observations on urban growth*. Franco Angeli.

Thompson, C. W., Aspinall, P., & Bell, S. (2010). *Innovative approaches to researching landscape and health*. Routledge.

UN. (2015). *Sustainable development goals*. www.un.org: https://www.un.org/sustainabledevelopment/

UN-HABITAT; For a Better Urban Future. (2018). *Adequate housing and slum upgrading*. UN-Habitat.

Viljoen, A., & Bohn, K. (2014). *Second nature urban agriculture; designing productive cities*. Routledge.

Waldheim, C. (2006). *The landscape urbanism reader*. Princeton Architectural Press.

Waldheim, C. (2016). *Landscape as urbanism*. Princeton University Press.

The Climate and the Environment

Supporting Climate Adaptation Measures in Small- to Medium-Sized Austrian Cities Using Climate Modelling

Sandro M. Oswald, Brigitta Hollosi, Maja Žuvela-Aloise, Linda See, Stefan Guggenberger, Wolfgang Hafner, Gundula Prokop, Alexander Storch, and Wolfgang Schieder

Abstract

This study outlines the results of current climate conditions for the city of Mödling, Klagenfurt and Salzburg located in Austria. Furthermore, it will analyse specific measures regarding climate adaptation, which are potentially realizable. To characterize the current climate conditions with the average number of summer days per year (SD y^{-1}), we use the microscale urban climate model MUKLIMO_3, in combination with a trilinear interpolation method. For the reference model simulation, we use atmospheric observations from 1981 to 2010 provided by one weather station located in a rural site near each city. We combined different land use data sets with various spatial resolutions to evaluate precise properties for the land use classification. Furthermore, we compared changes in SD y^{-1} to the reference simulation to evaluate the effectiveness of the here proposed adaptation strategies. Specifically, we consider two types of adaptation measures: (i) an increase in the shortwave reflectivity of impervious and sealed areas and (ii) an increased number of low-vegetated areas (i.e. lawns on streets and at roof level) and high-vegetated areas (i.e. bushes and trees). The results of the individual simulations indicate a substantially decreased SD y^{-1} by up to − 7.3 (− 10.9%) for the case of highly reflective roofs and up to − 7.7 (− 22.7%) for the case of an increased number of trees. A combination of both pathways leads to a decrease in SD y^{-1} by up to − 19.4 (− 29.3%), indicating a clear potential to mitigate the heat burden in each city.

Keywords

Urban heat island • Climate adaptation measures • Microscale climate modelling • White city • Green city

S. M. Oswald (✉) · B. Hollosi · M. Žuvela-Aloise
Zentralanstalt für Meteorologie und Geodynamik (ZAMG), Vienna, Austria
e-mail: sandro.oswald@zamg.ac.at

L. See
International Institute for Applied Systems Analysis (IIASA), Laxenburg, Austria

S. Guggenberger · W. Hafner
International Project Management Agency Klagenfurt on Lake Wörthersee GmbH (IPAK), Klagenfurt, Austria

G. Prokop · A. Storch · W. Schieder
Environment Agency (UBA), Vienna, Austria

1 Introduction

Since 2007, more than the half of the world's population resides in urban environments, and by 2050, the urban population will increase to 68% (UN, 2018). In the last decade, the increased urbanization has affected the local urban heat load due to the higher number of sealed areas. Hence, the local surface energy balance is disrupted and urban areas yield higher ambient air temperature than surrounding rural areas. This effect is commonly known as the so-called urban heat island (UHI) effect (Landsberg, 1981; Oke, 1982). Larger European cities and other megacities have been analysed in a comprehensive way regarding UHIs (e.g. Dessai (2003) and D'Ippoliti et al. (2010)), but smaller cities are also affected. Therefore, climate adaptation measures mitigating urban heat loads for their city's inhabitants are being considered by urban planners and local governments. The United Nations predicts a substantial rise in urbanization (UN, 2018) and continued global warming, although the magnitude depends on the scenario (Masson-Delmotte et al., 2018). Hence, negative UHI effects and higher air temperature in dense urban environments will probably intensify and impact the quality of life of city inhabitants through increases in heat stress and thermal discomfort (UN, 2019).

Several studies, which have used urban climate models, simulated only parts of a city or short periods of time such as one to three months (Müller et al., 2013). Due to the high computational power, other studies have focussed their

F. Trapani et al. (eds.), *Advanced Studies in Efficient Environmental Design and City Planning*, Advances in Science, Technology & Innovation, https://doi.org/10.1007/978-3-030-65181-7_32

simulations only for ten or 20 years to analyse heat wave days and/or UHI patterns for entire cities such as Brussels (Lauwaet et al., 2015, 2016). However, to identify specific effects from the UHI, climate change and large meteorological variations need to be considered over a long time period, e.g. at least 30 years. Thus to combat negative UHI effects, successful strategies are needed to stabilize or even mitigate increases in ambient air temperature T_a in urban environments and to reduce future heat loads in cities (Taha et al., 1988; McPherson, 1994; Hassid et al., 2000; Orehounig et al., 2014). In the last decades, various studies have shown climate adaptation measures, e.g. highly reflective surfaces (i.e. coloured or painted streets and/or roofs with bright colours) to decrease surface temperatures (Bretz et al., 1998; Taha 1997; Santamouris et al., 2011), or more vegetated areas to amplify the evapotranspiration (Alexandri & Jones, 2008; Gill et al., 2007; Qiu et al., 2013) to counteract future heat burdens.

Some studies have shown that increasing the albedo of sealed areas can lower T_a (see, e.g., Synnefa et al. (2008)). In Santamouris (2014) and Straka and Sodoudi (2019), cool roofs, with an increase in the albedo of 10%, show the potential to reduce the urban canyon T_a by 0.2 °C on average. Another study by Jandaghian and Akbari (2018) showed that enhanced sun reflectivity at all sealed areas (varying from 25 to 45% for street, roof and wall, respectively) can lead to decreases in T_a by more than 2 °C in urban areas. Air quality in terms of nitrogen dioxide or ozone can be improved additionally.

Considering instead more vegetation in cities, a recent study by Zhang et al. (2019) showed that the influence on T_a in an urban canyon through green roofs exists depending on the green coverage on the roof and the building height. In particular, T_a decreased by up to 0.44 °C at a 50% coverage and a building height of 10 m. Moreover, Lindberg et al. (2016) and Ali and Patnaik (2019) showed that by increasing tree coverage, the ambient air temperature and the mean radiant temperature can be effectively lowered at the local scale. However, in contrast, Vaz Monteiro et al. (2016) showed that the presence of very small green spaces (i.e. areas less than 0.5 ha) did not result in substantial reductions in T_a. However, various adaptation measures as mentioned before can be combined to increase the adaptability against (extreme) heat waves in urban environments, as, e.g., shown in Laukkonen et al. (2009), Voskamp and de Ven (2015) and Žuvela-Aloise et al. (2017).

Hence, the adaptation measures discussed here were chosen to represent realistic adjustments to the current situation in each of the three cities and include higher albedo values for (a) roofs, (b) walls and (c) streets; increases in (d) ornamental lawns, (e) vegetated roofs and (f) the number of trees within urban environments; and (g) a decrease in impervious and sealed areas.

Today, heat load changes in urban areas over longer time periods can be simulated with various microscale climate models. As mentioned before, modelling approaches must be practically simplified due to computational requirements for a 30-year time period. Some approaches use combined dynamic and statistical methods such as different weather pattern classifications (Hoffmann & Schlünzen, 2013), while others use more simplified physics that takes seasonal UHI trends into account (e.g. UrbClim) (Ridder et al., 2015). Here, the simulations have been performed with the microscale urban climate model MUKLIMO_3 (Sievers, 1995), developed by the German Meteorological Service (DWD) to determine the impact of the adaptation measures in isolation and in combination. Furthermore, the so-called cuboid method (Früh et al., 2011) is used to calculate the average number of summer days per year SD y^{-1} which characterize heat load changes on a longer temporal timescale. The essential land use and land cover properties to simulate these changes in a proper way are taken from different national and international land monitoring services with additional data sets from each city government.

In this study, we evaluate different adaptation measures for their effectiveness in the three cities of Mödling (PS[1] 21,000), Klagenfurt (PS 100,993) and Salzburg (PS 152,000) located in Austria. It focusses on the main aspect by evaluating the impact of seven different climate adaptation measures and their combination regarding the difference in SD y^{-1} compared to the 1981–2010 reference simulation. This climate index was chosen due to the knowledge of the Expert Team on Climate Change Detection and Indices (ETCCDI) (Karl et al., 1999).

2 Climate Model and Material

2.1 Urban Climate Model

In this study, we use the urban climate model MUKLIMO_3 with the version from 10 October 2014 (Sievers, 1995, 2012, 2016). This model simulates various meteorological parameters including humidity, wind flow and ambient air temperature in cities and also the surrounding rural areas on a 3-D grid.

The calculation of SD y^{-1} is derived with the MUKLIMO_3 model using a combination of daily maximum temperature fields. Microscale urban climate simulations undertaken at a high spatial resolution are computationally expensive, especially if the variability of meteorological parameters over longer time scales (i.e. 30 years or more) needs to be considered. Therefore, the cuboid method was

[1]Population size.

developed by Früh et al. (2011), which is a dynamic, statistical downscaling technique that uses a trilinear interpolation scheme. The method combines long-term regional climate simulations or measurements with short-term microscale simulations (e.g. one idealized diurnal cycle). Near-surface values of T_a, relative humidity RH and wind speed V are assumed to be the three main parameters needed for capturing the weather conditions when an urban heat load occurs in a city. Assuming a linear relationship between T_a, RH and V, a trilinear interpolation can be used to calculate any combination of each meteorological parameter within the cuboid as a weighted average of the values simulated for the surrounding cuboid corner points. The interpolated fields of the air temperature extremes (i.e. the minimum and maximum T_a) are, therefore, calculated by the cuboid method for each day over a 30-year period. These data are further used for statistical analysis, such as in the calculation of SD y^{-1} (see also Žuvela-Aloise et al. (2014)). For this purpose, eight different representative weather conditions representing the requirement, where urban heat loads occur, are calculated by idealized daily MUKLIMO_3 simulations. Vertical profiles of T_a, RH, V and short- and long-wave radiation on a one-dimensional grid have been used as meteorological inputs. To consider local meteorological conditions, the dominant wind directions are needed as inputs for the cuboid method to consider the wind direction dependence on the air temperature fields (see Sect. 2.2).

2.2 Regional and Meteorological Conditions

As Austria lies at the eastern end of the Alps, the three cities have their own regional and climatic characteristics. Salzburg lies at 430 m above sea level on the northern end of Austria near to Germany. Conditions specific to Salzburg include the mountain–valley breeze from the south during the daytime, leading to an airflow with low air temperature from the mountains towards the city. The monitoring station of Salzburg is located near the airport (47° 48′N, 13° 00′E) with prevalent wind directions of 340° (NNW) and 150° (SSE).

Klagenfurt (447 m above sea level) lies on the southern end of the Alps and also has specific meteorological conditions due to Lake Wörthersee at the western end of the city. During the daytime, there is an onshore breeze from the lake to the city centre causing an important flow of fresh air. The weather station is also located at the airport (46° 39′N, 14° 20′E) with dominant wind directions of 335° (NNW) and 125° (ESE).

The lowest city with 218 m above sea level is Mödling. The city is near the capital Vienna, and the largest climatological influence is from the pannonic climate, which is characterized by flat areas and a continental climate. The reference weather station is located in the south of the model domain with measured wind directions of 295° (WNW) and 120° (ESE) at 48° 02′N, 16° 17′E (in small village called Gumpoldskirchen).

Each of these weather stations has continuous meteorological observations of mean daily 2 m T_a, 2 m RH and 10 m V over a 30-year period. From this, climate indices were derived using a trilinear interpolation method between the 16 simulations (i.e. for each wind direction); further information can be found in Früh et al. (2011), Geletic et al. (2019) and Žuvela-Aloise et al. (2014).

2.3 Land Use and Land Cover

Detailed land cover and land use information are essential for precise climate simulations. Here, we use a few data sets from the Copernicus Land Monitoring Service portal, which was initiated by the European Environment Agency (EEA). Specifically, we used the land use information in the form of an urbanized classification, so-called Urban Atlas (UA), and other high-resolution layers available from Copernicus such as the tree density and the fraction of sealed areas at a resolution of 20 m (European Union, 2018). The UA is available for several cities with more than 50,000 inhabitants. In this study, the information from 2012 was used, as these layers are updated every six to eight years. In addition to these data sets, detailed area zoning plans, including a registry of public trees (including crown diameter and stem height), were made available by each city government.

The land use classification of the UA was combined with the tree cover density layer from Copernicus and the information received from the city governments including the surrounding districts to describe the urban properties for each land use class. As a second land cover information, we used the Land Information System Austria (LISA) covering larger cities in Austria with a resolution lower than 1 m (Geoville GmbH, 2016). This data set provides land cover data generated from satellite images containing more than ten different classes, e.g. buildings, sealed areas, bare soil and broadleaved and conifer trees. Figure 1a, b and c shows the slightly modified land use classification for each city through the aforementioned precise land cover information. These land use data sets and the related parameter such as average building fraction and tree height were used as inputs to MUKLIMO_3.

3 Results

3.1 Reference Simulations

As a reference simulation and for comparison with the measurements, we use the calculated climate index summer

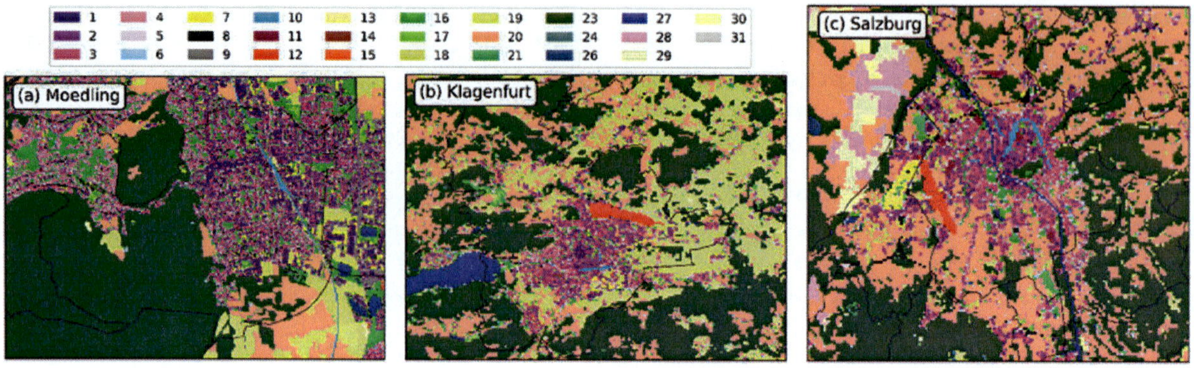

Fig. 1 Land use classification of **a** Mödling (20 m spatial resolution), **b** Klagenfurt and **c** Salzburg (100 m spatial resolution, respectively) based on land use and land cover data of Urban Atlas and Land Information System Austria. The black lines mark the district boundaries. Land use classification: 1—Continuous urban fabric; 2—discontinuous dense urban fabric; 3—discont. medium-density urban fabric; 4—discont. low-density urban fabric; 5—discont. very low-density urban fabric; 6—isolated structures; 7—military units; 8—transit roads; 9—other roads; 10—railways; 11—business areas; 12—airport; 13—mineral extraction; 14—construction sites; 15—land without current use; 16—green urban areas; 17—sports and leisure facilities; 18—annual crops; 20—pastures; 21—golf course; 23—forests; 24—herbaceous vegetation; 26—wetlands; 27—water. Note that classes 28 until 31 are used for grid cells within Germany (panel (**c**) upper left corner) which are not modified regarding adaption measures

days per year SD y^{-1} for 1981–2010. Therefore, routine measurements from the rural weather stations of each city are used to provide the maximum air temperature (T_{max}), which is used to determine the number of days where $T_{max} \geq 25$ °C (SD).

Figure 2a–c shows the model results and their spatial distribution of SD y^{-1} for our three study domains for 1981–2010, respectively. In Table 1, values for SD y^{-1} are shown, given through the measurements in rural areas near the city compared to the reference simulation. For the number of summer days in Mödling, simulations show a difference of + 7% compared to observations while the relative differences (Bias) of Klagenfurt and Salzburg provide good results with −1% and −2%, respectively.

3.2 Climate Adaptation Measures

As introduced in Sect. 1, 'cooling' effects inside urban areas are necessary and showed (various) potentials to mitigate negative UHI patterns. Therefore, we compare not only individual adaptation measures but also a combination of measures including highly reflective surfaces (referred hereinafter as the 'White City') and highly evaporating

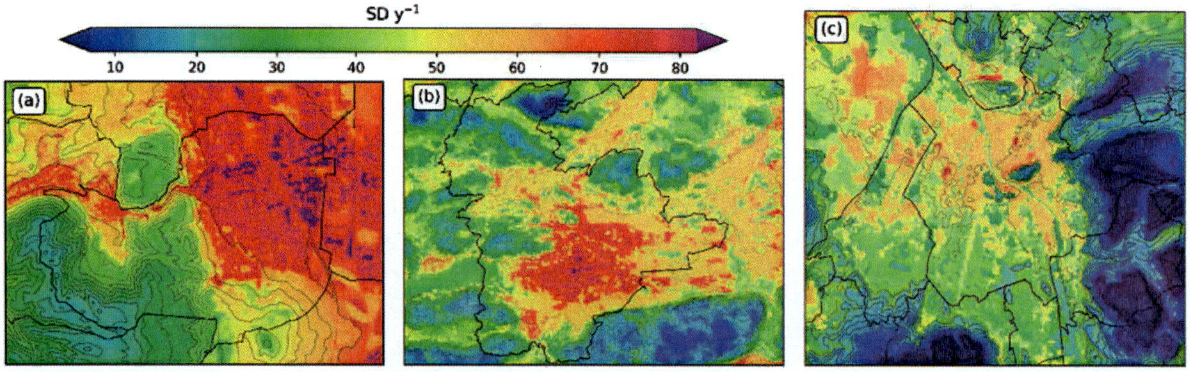

Fig. 2 Reference simulations of the average number of summer days per year (SD y^{-1}) for **a** Mödling, **b** Klagenfurt and **c** Salzburg for the time period 1981–2010

Table 1 Results of the reference simulations of MUKLIMO_3 compared to measurements of the weather station Gumpoldskirchen near Mödling, Klagenfurt Airport for Klagenfurt and Salzburg Airport for Salzburg regarding the average number of summer days per year for the time period 1981–2010

	Mödling	Klagenfurt	Salzburg
Observations	62.1	62.8	54.8
Simulations	66.8	62.4	53.7
Bias (%)	+ 7	− 1%	− 2%

The relative difference (Bias) is given in the third line

surfaces/areas (referred hereinafter as the 'Green City') with the reference case as the simulation of the climate for 1981–2010.

3.3 'White City'

First, we consider a possible scenario with changes in urban albedo, i.e. increased values of albedo for impervious areas such as streets/pavements, walls and roofs. To implement these possible measures in the model simulations, we increased the short-wave reflectivity of walls α_w (originally 0.3) and roofs α_r (originally 0.2) to 0.5, and the value for streets α_{st} (originally 0.2) was increased to 0.4 within LU classes from continuous urban fabric (1) to industrial areas (7), 11 and 17 (see Fig. 1) in order to limit the increase in short-wave reflections and glare. Table 2 shows a clear decrease in SD y^{-1}, where α_r yielded the lowest values as the local minimum (i.e. the maximum cooling effect). Klagenfurt reaches values of up to − 8.4%, whereas Salzburg is little bit lower with − 10.9%. In the case of increasing α_{st} (line 3 in Table 2), the minimum value has the same result as α_r of − 4.4% for Mödling, whereas Klagenfurt and Salzburg have slightly decreased cooling effects (1–2%) compared to the brightly coloured roof. In the case of enhanced sunlight reflectivity on walls, the values of α_w reached the weakest effect of the first part of the adaptation measure pathways (around 50–75% of α_r and α_{st}).

Furthermore, a combination of α_r, α_w and α_{st} ('White City') has a cumulative impact on the urban heat load and, hence, on negative UHI effects. The minimum values of each city show a strong decrease across the considered climate indices with − 9.7% in Mödling, − 15.3% in Klagenfurt and − 20.2% in Salzburg for ΔSD y^{-1} (see Table 2). The complete model results and the related potential cooling effect of these combined 'White City' adaptation measures are given in Fig. 3a, d, g for each city, respectively. The number of summer days was reduced the most under this assumption in dense urban environments.

3.4 'Green City'

Other frequently suggested climate adaptation measures include adding green roofs or increasing the amount of low vegetation. To include such measures in our analysis, we have considered four different green adaptations that can potentially mitigate UHI effects in each of the three cities.

Table 2 Absolute minimum and relative (%) difference values of summer days per year (ΔSD y^{-1}) produced by the model simulations including change in roof albedo α_r, wall albedo α_w, street albedo α_{st}, fraction of impervious areas σ_i, number of trees σ_t, fraction of low-vegetated areas σ_m; the 'White City' (lines 1–3 combined), the 'Green City' (lines 4–7 combined); and a simulation with all measures combined) relative to the 1981–2010 reference simulation

Adaptation measure	Mödling	Klagenfurt	Salzburg
Increase α_r (0.2 → 0.5)	-3.7 (-4.4%)	-6.4 (-8.4%)	-7.3 (-10.9%)
Increase α_w (0.2 → 0.5)	-2.8 (-3.3%)	-3.3 (-4.3%)	-4.7 (-7.1%)
Increase α_{st} (0.2 → 0.4)	-3.6 (-4.4%)	-5.6 (-7.3%)	-6.6 (-7.8%)
Decrease σ_i (-30%)	-1.8 (-2.2%)	-3.0 (-4.2%)	-3.4 (-5.2%)
Set green roofs (50%)	-1.8 (-2.6%)	-4.7 (-6.5%)	-6.2 (-9.0%)
Increase σ_t (+50%)	-6.8 (-11.1%)	-3.5 (-8.2%)	-7.7 (-22.7%)
Increase σ_m (+15%)	-1.9 (-2.7%)	-2.7 (-4.9%)	-2.9 (-4.6%)
'White City'	-7.8 (-9.7%)	-12.3 (-15.3%)	-14.0 (-20.2%)
'Green City'	-6.9 (-11.2%)	-9.7 (-14.3%)	-9.0 (-18.8%)
Combination	-9.9 (-13.2%)	-16.2 (-20.8%)	-19.4 (-29.3%)

The increase or decrease in percentages of σ_i, σ_t and σ_m refers to relative changes in the relevant land use classes

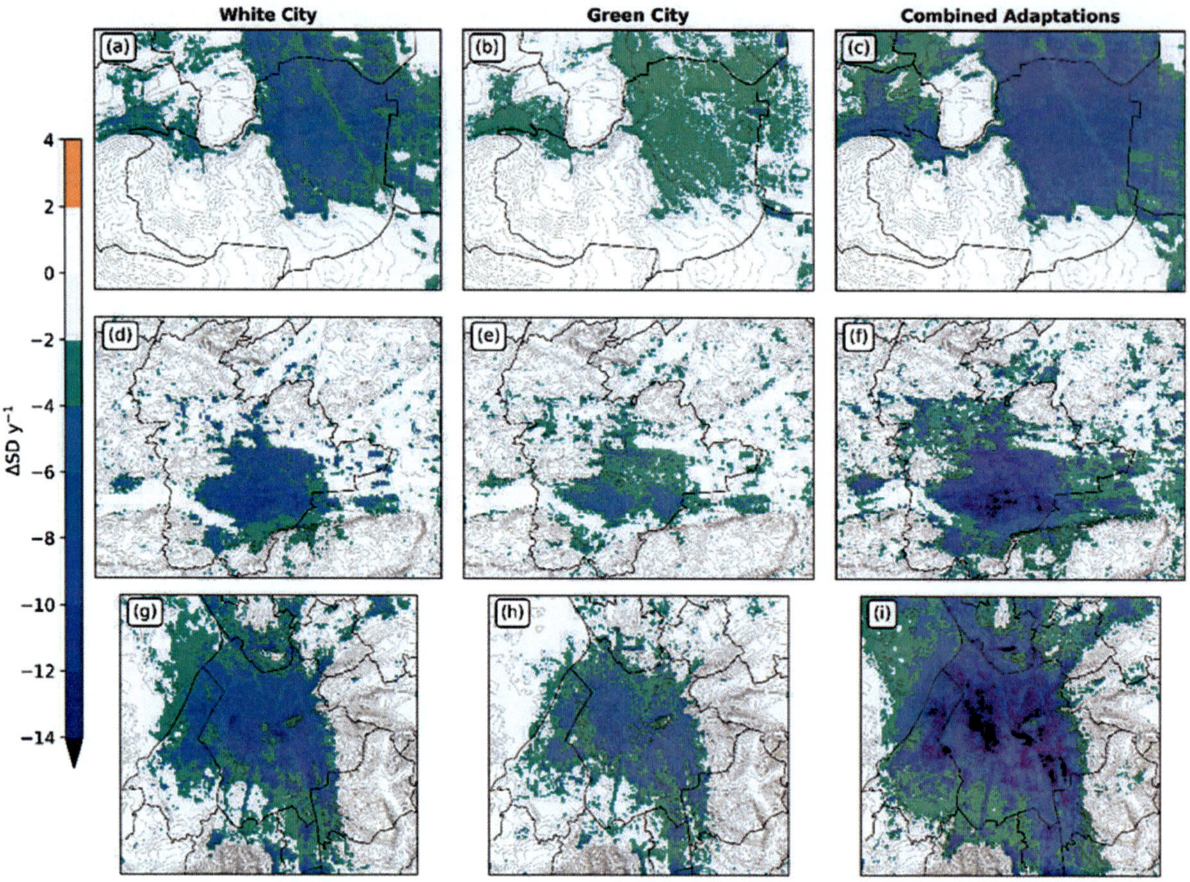

Fig. 3 Model results for the city of Mödling (panels **a–c**), Klagenfurt (panels **d–f**) and Salzburg (panels **g–i**). Each column shows the combined measures of adaptation corresponding to the 'White City' (changing the albedo values of the roofs, street and walls), the 'Green City' (changing the number of trees, vegetated areas and green roofs) and an overall 'Combined Adaptation' ('White City' and 'Green City' together). Results are presented as the difference in the average number of summer days per year (ΔSD y^{-1}) relative to the 1981–2010 reference simulation (see Fig. 2)

First, we consider a reduction in sealed surfaces σ_i, especially in public areas managed by the city government. Under such a scenario, MUKLIMO_3 would increase areas with the proportion of low-vegetated areas σ_m while keeping the remaining unvegetated and pervious surfaces in the same way as σ_i decreased. Therefore, we reduced impervious areas by − 30% of the given value and increased σ_m at the same rate as σ_i as a compensation in the LU classes 1 to 6 but not in LU classes 7 and 11 (since military and businesses cannot implement such changes because they require impervious surfaces) to contain the percentage of unvegetated, pervious areas. The second measure implements green roofs (small shrubs with a substrate depth of 25 cm) on buildings of LU classes with medium dense urban fabric (3) to very low dense urban fabric (5), 7 and 11 (as the addition of green roofs to buildings that are older (i.e. LU classes 1 and 2) would result in tremendous costs). For the third measure, we consider an increase in the number of trees by 50% in each city, which means that in those LU classes which have a $\sigma_t > 0$, the tree coverage is increased by 50%. Furthermore, we increase σ_m, i.e. unvegetated, pervious areas vanish (i.e. meadows rather than stone paving), while the amount of sealed surfaces is the same. Lines 4–7 in Table 2 indicate that such measures would yield a slight to an extensive decrease in SD y^{-1} of − 2.2% for decreasing σ_i to − 11.1% for increasing σ_t in Mödling, − 4.2% for decreasing σ_i to − 8.2% for increasing σ_t in Klagenfurt and − 4.6% for increasing σ_m to − 22.7% for increasing σ_t in Salzburg as the local minimum.

Finally, the 'Green City' was simulated, which combines the adaptation measures related to evaporative cooling, and is shown in Fig. 3b, e, h. The results show that, over the entire extent of each city, there is a cumulative reduction in SD y^{-1}. However, compared to the 'White City', these combined measures appear to be less effective. However, the local minimum values indicate a reduction in the magnitude of summer days per year with − 11.2% in Mödling (which is a higher reduction compared to the 'White City' with − 9.7%), − 14.3% in Klagenfurt and − 18.8% in Salzburg (see Table 2).

3.5 All Adaptation Measures in Combination

As a pooling climate adaptation measure, we combined all previous adaptation steps to see how large the potential effect would be in mitigating the UHI. When white and green roofs are combined in a scenario, half of the roof areas are assigned a value 0.5 for the albedo while the remaining half will have green roofs as outlined in Sect. 3.4. Figure 3c, f, i shows the largest reduction in SD y^{-1} with a local minimum of − 9.9 in Mödling, − 16.2 in Klagenfurt and − 19.4 in Salzburg; i.e. the number of summer days could be reduced with such combined measures by up to − 29.3% in the western part of Salzburg.

These results from the model show that different scenarios, and hence different climate adaptation measures, produce a larger mitigation potential for reducing the UHI than others. This potential for each of the considered urban environments ranged from − 2.2 to − 22.7% in terms of reducing heat stress, but a reduction as large as − 29.3% can be produced when several measures are combined.

4 Conclusions

In this paper, we investigated the effect of different climate adaptation measures on decreasing the effects of the urban heat island (UHI) in Austria for three cities: Mödling, Salzburg and Klagenfurt. In combination with the cuboid method, the MUKLIMO_3 urban climate model was used to estimate changes in climatological indices in the long term, such as the average number of summer days per year (SD y^{-1}), which can be used to characterize heat exposure.

The model used land use (LU) information to capture differences in urban and rural areas, which included data from the Urban Atlas (UA) and the Land Information System Austria (LISA).

We evaluate (and compare) the effectiveness of different climate adaptation measures with a simulation for the 1981–2010 time period, which is a reference. Adaptation is comprised of two main strategies. The first is to increase the value of the albedo (or reflectivity) of sealed surfaces. In model terms, this is an increase in the albedo of the walls from 0.3 to 0.5, an increase of roof reflectivity from 0.2 to 0.5 and an increase of 0.2 to 0.4 for the albedo of streets and pavements. The second involves changes to surfaces with high evaporation. In model terms, this involves increasing the number of trees by 50%, reducing by 30% the sealed surfaces, adding green roofs through a 50% increase and transforming some of the unvegetated or pervious surfaces to low-vegetated areas such as grassland. The measures' properties are taken from previous studies (Gross, 2012) involving energy balance modelling and evapotranspiration modelling, and our results are in good agreement with previous studies (Ali & Patnaik, 2019; Lindberg et al., 2016; Santamouris, 2014; Straka & Sodoudi, 2019; Vaz Monteiro et al., 2016).

However, uncertainties and limitations exist within MUKLIMO_3 considering the fact that trees and buildings cannot be modelled in one grid cell. Therefore, the percentage of trees near buildings is compensated for by low vegetation (and their properties) due to the calculated fraction of any kind of vegetation within each land use class. Hence, the evaporation in such grid cells can be underestimated. Another limitation is the lack of street orientations as they have a strong influence on the heat load during the daytime (i.e. solar radiation impinge on the street) and cold-air corridors from rural areas towards the city. In MUKLIMO_3, just the percentage of buildings and streets and the resulting building shadow are considered. The last aspect that needs to be considered is the low number of weather stations within each model domain, which are used to compare the simulation results. Hence, the spatial distribution of SD and urban heat loads cannot be verified sufficiently.

However, the strongest decrease in summer days per year was found when increasing the albedo or reflectivity of roofs. Here, reductions of up to − 4.4% in Mödling, up to − 8.4% in Klagenfurt and up to − 10.9% in Salzburg have been found. When considering vegetation as a measure of adaptation, the largest difference was found by increasing the number of trees by up to − 11.1%, up to − 8.2% and up to − 22.7% for SD y^{-1} in the three cities, respectively. We also considered the pooling of all seven aforementioned climate adaptation measures; such a combined scenario produced a substantial reduction up to − 20.8% in Klagenfurt and − 29.3% in Salzburg as a local minimum.

The results show that ambient air temperature, and hence heat stress, can be reduced by implementing a range of different measures for climate adaptation, in particular when a combination of measures is used together. This can lead to an effective strategy to reduce the urban heat burden and improve the thermal comfort of urban areas.

Conflict of Interest The article 'Using urban climate modelling and improved land use classifications to support climate change adaptation in urban environments: a case study for the city of Klagenfurt, Austria' contains some of the same results as this study for Klagenfurt and is already published in the Urban Climate Journal. However, this paper allows three cities to be compared and hence is different than the paper published previously.

References

Alexandri, E., & Jones, P. (2008). Temperature decreases in an urban canyon due to green walls and green roofs in diverse climates. *Building and Environment, 43*, 480–493. https://doi.org/10.1016/j.buildenv.2006.10.055

Ali, S., & Patnaik, S. (2019). Assessment of the impact of urban tree canopy on microclimate in Bhopal: A devised low-cost traverse methodology. *Urban Climate, 27*, 430–445. https://doi.org/10.1016/j.uclim.2019.01.004

Bretz, S., Akbari, H., & Rosenfeld, A. (1998). Practical issues for using solar-reflective materials to mitigate urban heat islands. *Atmospheric Environment, 32*, 95–101. https://doi.org/10.1016/s1352-2310(97)00182-9

Dessai, S. (2003). Heat stress and mortality in Lisbon Part II. An assessment of the potential impacts of climate change. *International Journal of Biometeorology, 48*, 37–44. https://doi.org/10.1007/s00484-003-0180-4

D'Ippoliti, D., Michelozzi, P., Marino, C., de'Donato, F., Menne, B., Katsouyanni, K., et al. (2010). The impact of heat waves on mortality in 9 European cities: results from the EuroHEAT project. *Environmental Health, 9*. https://doi.org/10.1186/1476-069x-9-37

European Union. (2018). *Copernicus land monitoring service. European Environment Agency (EEA)*.

Früh, B., Becker, P., Deutschländer, T., Hessel, J.-D., Kossmann, M., Mieskes, I., et al. (2011). Estimation of climate-change impacts on the urban heat load using an urban climate model and regional climate projections. *Journal of Applied Meteorology and Climatology, 50*, 167–184. https://doi.org/10.1175/2010jamc2377.1

Geletic, J., Lehnert, M., Dobrovolny, P., & Zuvela-Aloise, M. (2019). Spatial modelling of summer climate indices based on local climate zones: expected changes in the future climate of Brno, Czech Republic. *Climatic Change*, 1–16. https://doi.org/10.1007/s10584-018-2353-5

Geoville GmbH. (2016). *Land Information System Austria (LISA)*.

Gill, S. E., Handley, J. F., Ennos, A. R., & Pauleit, S. (2007). Adapting cities for climate change: The role of the green infrastructure. *Built Environment, 33*, 115–133. https://doi.org/10.2148/benv.33.1.115

Gross, G. (2012). Numerical simulation of greening effects for idealised roofs with regional climate forcing. *Meteorologische Zeitschrift, 21*, 173–181. https://doi.org/10.1127/0941-2948/2012/0291

Hassid, S., Santamouris, M., Papanikolaou, N., Linardi, A., Klitsikas, N., Georgakis, C., & Assimakopoulos, D. N. (2000). The effect of the Athens heat island on air conditioning load. *Energy and Buildings, 32*, 131–141. https://doi.org/10.1016/s0378-7788(99)00045-6

Hoffmann, P., & Schlünzen, K. H. (2013). *Weather pattern classification to represent the urban heat island in present and future climate. American Meteorological Society: Journal of Applied Meteorology and Climatology*. https://doi.org/10.1175/jamc-d-12-065.1

Jandaghian, Z., & Akbari, H. (2018). *The Effect of increasing surface Albedo on urban climate and air quality: A detailed study for Sacramento, Houston and Chicago*. Climate 6 (2), 19, MDPI. https://doi.org/10.3390/cli6020019

Karl, T., Nicholls, N., & Ghazi, A. (1999). CLIVAR/GCOS/WMO Workshop on Indices and indicators for climate extremes workshop summary. In *Weather and climate extremes* (S. 3–7). Springer Netherlands. https://doi.org/10.1007/978-94-015-9265-9_2

Landsberg, H. E. (1981). *The urban climate*. Elsevier Science.

Laukkonen, J., Blanco, P. K., Lenhart, J., Keiner, M., Cavric, B., & Kinuthia-Njenga, C. (2009). Combining climate change adaptation and mitigation measures at the local level. *Habitat International, 33*, 287–292. https://doi.org/10.1016/j.habitatint.2008.10.003

Lauwaet, D., Hooyberghs, H., Maiheu, B., Lefebvre, W., Driesen, G., Looy, S. V., & Ridder, K. D. (2015). Detailed urban heat island projections for cities worldwide: Dynamical downscaling {CMIP}5 global climate models. *Climate MDPI*. https://doi.org/10.3390/cli3020391

Lauwaet, D., Ridder, K.D., Saeed, S., Brisson, E., Chatterjee, F., Lipzig, N.V., et al. (2016). *Assessing the current and future urban heat island of Brussels*. Elsevier. https://doi.org/10.1016/j.uclim.2015.11.008

Lindberg, F., Thorsson, S., Rayner, D., & Lau, K. (2016). The impact of urban planning strategies on heat stress in a climate-change perspective. *Sustainable Cities and Society, 25*, 1–12. https://doi.org/10.1016/j.scs.2016.04.004

Masson-Delmotte, V., Zhai, P., Pörtner, H.-O., Roberts, D., Skea, J., Shukla, P., et al. (2018). *Global warming of 1.5 °C*. Intergovernmental Panel on Climate Change. https://www.ipcc.ch/site/assets/uploads/sites/2/2019/06/SR15_Full_Report_Low_Res.pdf

McPherson, E. G. (1994). Cooling urban heat islands with sustainable landscapes. In R. H. Platt, R. A. Rowntree, P. C. Muick (Eds.), *The ecological city: Preserving and restoring urban biodiversity* (pp. 151–171). University of Massachusetts Press.

Müller, N., Kuttler, W., & Barlag, A.-B. (2013). *Counteracting urban climate change: Adaptation measures and their effect on thermal comfort*. Springer Science and Business Media: Theoretical and Applied Climatology 115 (1-2), 243-257. https://doi.org/10.1007/s00704-013-0890-4

Oke, T. R. (1982). The energetic basis of the urban heat island. *Quarterly Journal of the Royal Meteorological Society, 108*, 1–24. https://doi.org/10.1002/qj.49710845502

Orehounig, K., Mahdavi, A., Doppelbauer, E.-M., Loibl, W., & Tötzer, T. (2014). Projections of design implications on energy performance of future cities: A case study from Vienna. *Sustainable Cities and Society, 12*, 92–101. https://doi.org/10.1016/j.scs.2014.03.001

Qiu, G.-Y., Hong-yong, L. I., Zhang, Q.-T., Wan, C. H., Liang, X.-J., & Li, X.-Z. (2013). Effects of evapotranspiration on mitigation of urban temperature by vegetation and urban agriculture. *Journal of Integrative Agriculture, 12*, 1307–1315. https://doi.org/10.1016/s2095-3119(13)60543-2

Ridder, K. D., Lauwaet, D., & Maiheu, B. (2015). UrbClim A fast urban boundary layer climate model. *Urban Climate, Elsevier*. https://doi.org/10.1016/j.uclim.2015.01.001

Santamouris, M. (2014). Cooling the cities—A review of reflective and green roof mitigation technologies to fight heat island and improve comfort in urban environments. *Solar Energy, 103*, 682–703. https://doi.org/10.1016/j.solener.2012.07.003

Santamouris, M., Synnefa, A., & Karlessi, T. (2011). Using advanced cool materials in the urban built environment to mitigate heat islands and improve thermal comfort conditions. *Solar Energy, 85*, 3085–3102. https://doi.org/10.1016/j.solener.2010.12.023

Sievers, U. (1995). Generalization of the streamfunction–vorticity method to three dimensions. *Meteorologische Zeitschrift, 3*, 3–15

Sievers, U. (2012). Das kleinskalige Stroemungsmodell MUKLIMO_3 Teil 1: Theoretische Grundlagen, PC-Basisversion und Validierung. *Berichte des Deutschen Wetterdienstes, 240*.

Sievers, U. (2016). Das kleinskalige Stroemungsmodell MUKLIMO_3 Teil 2: Thermodynamische Erweiterungen. *Berichte des Deutschen Wetterdienstes, 248*.

Straka, M., & Sodoudi, S. (2019). Evaluating climate change adaptation strategies and scenarios of enhanced vertical and horizontal compactness at urban scale (a case study for Berlin). *Landscape and Urban Planning, 183*, 68–78. https://doi.org/10.1016/j.landurbplan.2018.11.006

Synnefa, A., Dandou, A., Santamouris, M., Tombrou, M., & Soulakellis, N. (2008). *On the use of cool materials as a heat island mitigation strategy. American Meteorological Society:*

Journal of Applied Meteorology and Climatology. https://doi.org/10.1175/2008jamc1830.1

Taha, H. (1997). Urban climates and heat islands: albedo, evapotranspiration, and anthropogenic heat. *Energy and Buildings, 25*, 99–103. https://doi.org/10.1016/s0378-7788(96)00999-1

Taha, H., Akbari, H., Rosenfeld, A., & Huang, J. (1988). Residential cooling loads and the urban heat island—the effects of albedo. *Building and Environment, 23*, 271–283. https://doi.org/10.1016/0360-1323(88)90033-9

UN. (2018). World Urbanization Prospects. *United Nations*.

UN. (2019). Sustainable development goals: Goal 11. *United Nations*.

Vaz Monteiro, M., Doick, K., Handley, P., & Peace, A. (2016). The impact of greenspace size on the extent of local nocturnal air temperature cooling in London. *Urban Forestry & Urban Greening, 16*, 160–169. https://doi.org/10.1016/j.ufug.2016.02.008

Voskamp, I. M., & Ven, F. H. (2015). Planning support system for climate adaptation: Composing effective sets of blue-green measures to reduce urban vulnerability to extreme weather events. *Building and Environment, 83*, 159–167. https://doi.org/10.1016/j.buildenv.2014.07.018

Zhang, G., He, B.-J., Zhu, Z., & Dewancker, B. J. (2019). Impact of Morphological characteristics of green roofs on pedestrian cooling in subtropical climates. *International Journal of Environmental Research and Public Health, MDPI*. https://doi.org/10.3390/ijerph16020179

Žuvela-Aloise, M., Andre, K., Schwaiger, H., Bird, D. N., & Gallaun, H. (2017). Modelling reduction of urban heat load in Vienna by modifying surface properties of roofs. *Theoretical and Applied Climatology, 131*, 1005–1018. https://doi.org/10.1007/s00704-016-2024-2

Žuvela-Aloise, M., Koch, R., Neureiter, A., Böhm, R., & Buchholz, S. (2014). Reconstructing urban climate of Vienna based on historical maps dating to the early instrumental period. *Urban Climate, 10*, 490–508. https://doi.org/10.1016/j.uclim.2014.04.002

Local Climate Action Planning: An Evaluation of Michigan Climate Action Plans Through Text-Based Analysis

Ayse Ozcan and Eric J. Strauss

Abstract

This study analyzes how well local governments understand the terminology of climate change and prepare for climate change mitigation and adaptation by reviewing Climate Action Plans (CAPs) from five cities (Ann Arbor, East Lansing, Detroit, Traverse City, Ypsilanti). The goal is to better understand the roles of CAPs in climate change adaptation and engaging in sustainable development. The content analysis is based on eight key categories which include greenhouse gas (GHG) emissions, sustainability, renewable energy, responsibility (education and awareness), land use and transportation policies, international policies, and development. With text-based analysis, it is possible to determine the most frequent terms or components cited in the listed CAPs. The eight categories are basically identified by the following two topics: (1) carbon management and (2) urban infrastructure. Thus, this paper examines and evaluates policy perception of Michigan CAPs based on objectives of climate action planning.

Keywords

Climate change • Local climate action planning • Climate urbanism • Climate action plans (CAPs)

1 Introduction

Climate change is one of the biggest challenges of the twenty-first century and has shown itself to be one of the core obstacles to sustainable development (IPCC 2020a; OECD, 2010). When the United Nations Framework Convention on Climate Change (UNFCCC) entered into force in 1994, it became the fundamental guideline aiding countries' policies on climate change. After the UNFCCC, the Kyoto Protocol and the Paris Climate Agreement have become the other important documents leading global climate policies. Intergovernmental Panel on Climate Change (IPCC) reports also provide crucial inputs to international climate change negotiations and climate action planning. The meetings of the Conference of the Parties (COP), the supreme decision-making body of the UNFCCC, have served as milestone events in bringing to light the seriousness of climate change. The results of these meetings have led to the signing of the Paris Agreement. The last meeting of COP 25 was held in Madrid in December 2019, pointing out the importance of capacity building under the Paris Agreement and the ability of participating countries to take effective climate action (UNFCCC COP25, 2019). Developing national and local climate plans are two commitments of these agreements. Each climate plan is specific to the participating countries' goals for reducing GHGs, by considering their domestic circumstances and capabilities (UNFCCC, 2020).

'Local Climate Action Planning (LCAP)' is one of the new effective strategies used by local governments to create a framework for climate-sensitive urban planning. Climate-sensitive urban planning primarily draws attention to the measuring and monitoring of carbon and other greenhouse gases (GHGs) in cities (Bulkeley & Newell, 2015; Rice, 2010). The idea of climate planning and its relationship to the emergence of a new urban policy is also integrated with the 2030 Agenda for Sustainable Development Goals (SDGs) (UN, 2016, 2019). Climate action is the thirteenth goal of the UN 2030 Agenda, which aims to provide a global response and cooperation to climate change. Climate Action Plans (CAPs) are comprehensive road maps showing the specific activities that a city or agency can adopt to reduce GHGs emissions.

A. Ozcan (✉) · E. J. Strauss
SPDC Urban and Regional Planning, Micigan State University, East Lansing, MI 48823, USA
e-mail: ayozcan@msu.edu

E. J. Strauss
e-mail: strausse@msu.edu

F. Trapani et al. (eds.), *Advanced Studies in Efficient Environmental Design and City Planning*, Advances in Science, Technology & Innovation, https://doi.org/10.1007/978-3-030-65181-7_33

CAPs build on the information gathered by greenhouse gas inventories and focus on activities that can achieve emission reductions in the most cost-effective manner (ILG, 2018). CAPs also include broad policy goals, as well as specific programmatic and policy recommendations to be used as GHGs reduction measures (Pattison & Kawall, 2018).

Most cities in the world have been developing their own CAPs and strategies over the past 20 years. Some of them have become members of national and transnational city networks, ranging from the US Mayors Climate Protection Agreement[1] and the Swedish Klimatkommunerna (Climate Municipalities) network, to the C40 Cities, an international network of the world's largest cities committed to combating climate change (World Bank, 2010).

Historically, cities in the USA have embraced the idea that they can meet the climate action challenge, having produced plans for the last 100 years. State-level climate policies have been created in response to the lack of national policy. By the end of 2009, there were over 200 US municipalities that had or were in the process of adopting or undergoing climate-related plans, including CAPs and sustainability plans (ICLEI, 2009). Since then, more than 600 US cities have developed their own CAPs (C2ES, 2020; Millard-Ball, 2012).

The development of CAPs is becoming an increasingly important local policy. This process is more than a set of actions managed by the municipality. It is a comprehensive political structure involving professionals, urban planners, practitioners, politicians, non-governmental organizations and citizens. Because CAPs also call for the adopting measures and analyzing impacts that are often beyond the control of municipalities. Therefore, it is still controversial whether or not the plans produce effective results. The fundamental question is, 'Do communities have the capacity, tools and goals in place to take the necessary actions to address climate change impacts?'.

A CAP consists of different levels of political and institutional components. Some contain policies designed only to lead certain individuals, while others have a much clearer set of political goals and objectives (Bassett & Shandas, 2010). While national standards are useful for developing a CAP, using local standards that emphasize community hazard reduction and adaptation to climate change can be considered a more successful approach (Frazier, 2013). The existence of local political leadership and public participation is integral in determining the most effective CAP for a municipality.

[1]The US Mayors Climate Protection Agreement was founded to reduce US GHG emissions through local government leadership and action. Within this framework over 1000 signatory cities have committed to large-scale reductions in emissions for their respective communities using a range of climate policies and plans (US Mayors Conference of Mayors, 2018).

Understanding how CAPs target emissions and influence specific emission reduction is of theoretical and policy interest. Evaluating CAPs can help explain these policy differences in management of GHGs between cities. It can improve our understanding how local governments are determining their own environmental and energy-related problems. This study is based on the roles and functions of CAPs and their influence on local climate action planning and regulatory decisions.

1.1 Aim of the Study and Its Objectives

This is a descriptive study that aims to examine Michigan's CAPs for the five listed cities (Ann Arbor, East Lansing, Detroit, Traverse City, Ypsilanti). Through text-based (or text mining) analysis, we determined the different factors on how CAPs are weighted and what purpose they serve in achieving sustainable development goals. Text mining analysis is based on the terms (or key concepts) selected within the logic of climate urbanism and local climate action planning.

Local CAPs are evaluated on two main topics: (1) carbon management and (2) urban infrastructure. The key policy terms of the analysis were determined on these two topics, and the term frequency was created accordingly. We reviewed selected Michigan CAPs by using policy indicators such as GHGs emissions, renewable energy, sustainability, sustainable development, green buildings, land use and transportation policies, education and society, and international policies. These indicators have also been used by the UN Agenda 2030 in order to evaluate the policy perception of local CAPs. The term frequency, key policy terms density, and policy definitions based on UN Sustainable Development Goals for 2030 could be a theoretical road map to measure the 'policy perception of the selected CAPs.' Henceforth, it is possible to determine the policy trends for Michigan's climate action planning and identify its deficiencies for further research.

In this context, the aim of the research is to examine how well local climate policies analyze the impacts of climate change in their CAPs. Finally, we wanted to see how successful a CAP has been and how the objectives of a CAP responded to identified problems.

The population for this research is from all of the Michigan cities that have adopted a local CAP or sustainability plan. First, we searched websites on climate policy documents that have joined the US Conference of Mayors Climate Protection Agreement. Next, the required information for CAPs was found in C2ES (Center for Climate and Energy Solutions) and related Michigan official websites. After systematic searching, all Michigan CAPs were collected, and selected ones were examined. The characteristics of Michigan's CAPs are dependent variables of the research. UN 2030 Agenda Goals are the independent variables which lead to the core policies of CAPs. U.S. Census database (Census 2000) and the American Community Survey were used for the descriptive statistics of the research.

1.2 The Literature Review

In the literature, research specifically addressing local climate action planning has not yet answered clearly as to whether CAPs have a causal impact on levels of GHGs emissions. Studies discussing the relationship between the functions of CAPs and the policies implemented are not sufficient, and the steps of climate policy are not limited to developing a stand-alone CAP. Developing a CAP is the first form of the policy perception in how to take responsibility for climate change impacts in cities. We think that there is still a gap in the literature about why local CAPs are needed and what kind of information they should provide to local governments.

Furthermore, researchers have discussed less on identifying any causal impacts of planning, and instead explored the types of cities that pursue CAPs (Hanak et al., 2008; Zahran et al. Zahran, Brody, et al., 2008; Zahran et al. Zahran, Grover, et al., 2008); some studies have evaluated the quality and contents of adopted CAPs (Wheeler, 2008; Bassett & Shandas, 2010; Boswell et al. 2012; Tavel, 2012). Additionally, some have preferred climate planning as a tool to help explain theoretical debates in disciplines such as geography, economy, law, sociology, political science (Betsill & Bulkeley, 2004).

The important works and reports of international organizations such as the UN, IPCC, OECD, and IEA have made great contributions to the literature within the context of green studies and climate change. One of those important outputs is the UNFCCC adopted document, which earlier stated is the foundational guide for countries' policies on climate change and an important tool for international negotiations.

After the UNFCCC, the Kyoto Protocol, adopted in 1997, and the Paris Climate Agreement, adopted by 196 countries in December 2015, are the other important documents leading international climate change policies. The Agreement requests each country to outline and communicate their post-2020 climate actions (UNFCCC-Paris Agreement, 2015; UNFCCC, 2020).

IPCC[2] reports also provide crucial inputs to international climate change negotiations and climate action planning. The IPCC prepares comprehensive Assessment Reports on climate change, its impacts and future risks and policy options for adaptation and mitigation (IPCC, 2020a).

OECD's green growth reports are other reference sources which could be a guide for climate action planning. From the OECD perspective, green growth indicators are markers or milestones toward greening growth and reducing GHG emissions. The first set of OECD urban green growth indicators is the *environmental and resource efficiency*, which represents the output volume per natural resource unit. Determining the indicators was influenced by the idea of capturing the basic characteristics of a low-carbon and resource-saving economy (OECD, 2011, 2015).

2 Local Climate Action Planning

Urban planners today are faced with two opposite approaches: (1) globalizing environmental policies and (2) decentralization of policy development responsibilities (Corburn, 2009). In this context, climate change policies must be structured on a mutually supportive relationship between global approaches and local policies.

Although the effects of climate tendencies are experienced 'locally', scientists and academics focus on global climate impacts and their economic costs at the macro-level (Musco, 2010: 67–68). The burdens of climate change are not equally distributed in urban areas (EEA, 2008), and there are large differences from city to city. For example, poor people living in unfavorable areas will be exposed to a higher threat due to climate change (e.g., closer to rivers with a larger danger of flooding), and consequently, will not have the resources to adapt to these negative impacts (Musco, 2010: 71; UN, 2011). Even though climate adaptation and mitigation strategies are very different from region to region, or from city to city, local governments mostly have control on sources of pollution in their communities and the capacity to cut greenhouse gas emissions by 30–50%. Examples of such pollution reduction can come from better management of buildings, energy, transportation, waste and water management. Municipal policies have been directly managing and influencing all these sectors. In this regard, climate change strategies and CAPs can be applicable solutions to assess all local policies in place to combat climate change (mainly in terms of CO_2 emissions equivalent). There is no need to define a global climate-planning mechanism to be used in all local authorities; however, this mechanism is a crucial issue that outlines a common frame in which all public and private actors play a role in responding to this challenge.

In the USA, several states and cities over the last 15–20 years have begun to take responsibility on climate change impacts (Wheeler, 2008, pp. 481–482, C2ES, 2020).

[2]The objective of IPCC, created by the World Meteorological Organization (WMO) and the United Nations Environment Program (UNEP) in 1988, is to provide scientific information to governments at all levels that they can use to develop climate policies and programs. The IPCC currently has 195 members (IPCC, 2020b).

More than 600 cities started to develop a climate initiative or policy in the 2000s. The 2018 survey by the Alliance for a Sustainable Future, which is a joint effort of the US Conference of Mayors (USCM)[3] and the Center for Climate and Energy Solutions (C2ES),[4] finds that mayors are a key force behind US action to reduce GHGs emissions responsible for the growing effects of climate change. The survey of 158 cities nationwide investigated the actions that these cities are taking or considering to limiting emissions. According to the survey, 60% of cities have started or significantly developed a climate initiative or policy within the past 5 years. (C2ES, 2020). The key issues of local climate planning especially arise during the design of buildings, the rebuilding of urban infrastructures, and the management of renewable energy resources (Bulkeley & Moser, 2007).

Our research focus is to examine the characteristics of Michigan CAPs and evaluate the climate policy perception of CAPs. With this study, it may be possible to determine which cities are taking a more responsive path to the policies that local climate action plans envision for climate policy. In addition, it may also be identified local factors and sectors that CAPs focus on.

Within the scope of the study, a different research based on stakeholder analysis has been planned to discuss the roles of decision-makers who are responsible for policy development in local climate action planning. This research is aimed to measure the efficiency of local decision-making for Michigan climate action planning. It is still ongoing, but will be finished by the end of this year. We have conducted semi-structured interviews with city staff and stakeholders for each city, but we couldn't have done interviews because of a global coronavirus pandemic that started in March.

3 Michigan Climate Action Planning

3.1 General Overview

As of July 2019, 74.6% of the Michigan population lives in urban areas, while 25.4% live in rural areas. Of the roughly 10 million people living in Michigan, 6.5 million live in urban areas. Cities in the study are among the top 10 cities in terms of population size: Detroit (713,777), Ann Arbor (113,934), East Lansing (48,579), Ypsilanti (19,345) and Traverse City (14,674) (US Census Bureau, 2012; US Census Bureau, 2019).

Michigan's distribution of energy-related CO_2 emissions by sectors is shown in Table 1. Electric power (55.7 MMT CO2) and transportation (49.5 MMT CO2) are the first sectors in terms of total CO2 emissions at the state level. These are followed by industrial (18.0) and commercial (10.6) sectors. The residential and transportation sectors total emissions are 12.4% and 32.4%, while the Industrial and commercial sectors are 11.8% and 6.9%, respectively.

In Michigan, climate planning has a history of about 12–13 years and is quite new. In 2007, thirty mayors from Michigan cities signed the US Conference of Mayors Climate Protection Agreement (MCAN, 2019). In March 2009, the Michigan Climate Action Council released the Michigan CAP, emphasizing how the state can take responsibility in the reduction of carbon emissions, while recognizing the risks that climate change will pose to the state (Georgetown Climate Center, 2020). The Compact of Mayors was launched in 2014, and the mayors of four cities (Ann Arbor, Grand Rapids, Muskegon and South Haven) committed to a set of meaningful targets for carbon emissions reductions and transparent reporting (MCAN, 2019). A variety of state agencies have also been undertaking efforts to assess climate vulnerabilities and to identify adaptation policies. For example, Michigan Climate and Health Adaptation Program is working with communities, organizations, and other local health institutions to help enhance climate resilience in the public health system. After the onset of the Coronavirus Pandemic (COVID-19), it will be much more important to develop a public healthcare system responding clearly to climate impacts.

There are currently eleven cities and counties in Michigan with CAPs, and two more cities have plans in progress. At least nine communities have set clean energy or greenhouse gas reduction goals, and two have set 100% renewable energy goals (MCAN, 2019). The Ann Arbor CAP suggests mid-term and long-term targets for community wide GHGs emissions reductions. The mid-term target is to reduce CO_2 emissions 25% by 2025, while the long-term target is to reduce CO_2 emissions 90% by 2050. The city has already accomplished 20% renewable energy for municipal operations (City of Ann Arbor CAP, 2012).

When examining CAPs, we can see each one has specific core policies. Michigan CAPs and their core sectors or action categories are as follows:

- *Infrastructure*: Possible damage to public facilities that may affect economic and social development of cities.
- *Energy*: Supply and demand will be affected by the new climate scenario, particularly because of the increase of urban energy need during winter and summer.

[3]The United States Conference of Mayors is the official non-partisan organization of cities with populations of 30,000 or more. Each city is represented in the Conference by its chief elected official, the mayor (USCM, 2020).

[4]The Center for Climate and Energy Solutions (C2ES) is an independent, nonprofit organization working to forge practical solutions to climate change. Its mission is to advance strong policy and action to reduce greenhouse gas emissions, promote clean energy, and strengthen resilience to climate impacts (C2ES, 2020).

Table 1 State energy-related carbon dioxide emissions by sector (2017)

	Commercial	Electric Power	Residential	Industrial	Transportation	Total
Million metric tons of carbon dioxide (MMT CO_2)						
Michigan	10.6	55.7	18.9	18.0	49.5	152.7
Shares						
	6.9%	36.5%	12.4%	11.8%	32.4%	
Metric tons of energy-related CO_2 emissions per person (2017)				15.3		

Source EPA (2017), US Census Bureau (2010)

- *Building Efficiency*: Most of GHGs emissions are indirect CO2 emissions caused from electricity use and infrastructure-related output in buildings.
- *Water*: The quality and amount of water resources is a basic issue within the context of a changing climate for all cities in Michigan.
- *Agriculture and Land Use*: The impact on crops, livestock management, and the location of production will be relevant.
- *Human health and well-being*: Michigan as a state is experiencing more extreme climate events, and weather-related diseases could increase.

All of these policies are among the core policies of the UN 2030 Agenda, which is a universal call to action to reach 17 sustainable development goals, especially for ending poverty and hunger. The third goal of the Agenda is directly related to human health policies, ensuring healthy lives and promoting well-being for people all over the world. After the COVID-19 pandemic, public healthcare strategies based on climate change impacts must have more focus on communicable diseases. The COVID-19 pandemic has brought public health into focus, and we must have a better understanding of the relationship between climate change and public health.

In Michigan, the primary healthcare system must be strengthened, as the percentage of the population that does not have health insurance is 6.1% (American Community Survey 2014–2018). Although Michigan Climate and Health Adaptation Program is focusing on the public health consequences of climate change, it still needs to develop strategies identifying and targeting vulnerable populations who are most sensitive to diseases.

3.2 Content Analysis of the Michigan Climate Action Plans (CAPs)

Michigan as a state has been working on climate action planning since 2007. The Michigan Climate Action Council, which was established on November 14, 2007, prepared and published the state CAP in March 2009. The plan analyzes the impacts of climate change in terms of the following issues: (1) energy supply, (2) market-based policies, (3) residential, commercial and industrial, (4) transportation and land use, (5) agriculture, forestry, and waste management, (6) cross-cutting issues (Michigan State CAP 2009, pp. 3–34). The CAPs examined in the study are presented in Table 2.

The nature of Michigan CAPs varies from city to city. Some CAPs are brief and some others are remarkably detailed and comprehensive, even if many of the proposed initiatives and policies have not been implemented. For example, some CAPs are hundreds of pages and recommend specific policies, and may include, but are not limited to renewable energy, infrastructure, and green buildings.

In this study, CAPs' objectives are evaluated in the framework of two issues addressed within the context of climate planning: (1) carbon management and (2) urban infrastructure. We have done our research in the following five cities: Detroit, Ann Arbor, East Lansing, Traverse City, and Ypsilanti (see Table 2).

3.2.1 The Key Policy Terms of the Content Analysis

The selected five CAPs were examined by reading through the basic concepts, policies, and strategies according to the two main issues we mentioned above. In this sense, the

Table 2 Selected five Michigan CAPs

Year of the Plan	City	Name of the Plan
July 2012	Ann Arbor	City of Ann Arbor Climate Action Plan
October 2017	Detroit	Detroit Climate Action Plan
April 2012	East Lansing	East Lansing Climate Sustainability Plan
February 2011	Traverse City	City of Traverse City Climate Action Plan
July 2012	Ypsilanti	City of Ypsilanti Climate Action Plan

Table 3 Key policy terms of content analysis

1	Greenhouse Gas Emissions/GHGs **Greenhouse; Greenhouse Gas; GHGs**
2	**Sustainability** Sustainability; Sustainable
3	**Renewables /Energy efficiency** Renewable (including the terms of solar and wind); Energy; Efficiency
4	**Responsibility: Education, Community and Awareness** Education; Community; Awareness
5	**Land Use and Transportation** Land and Land Use; Transportation
6	**International Documents/Agreements** United Nations; Kyoto; Paris; IPCC; ICLEI
7	**Buildings/Environmental Design** Buildings; Environmental Design-LEED (including the terms of green buildings, green roofs, infrastructure, environmental)
8	**Development** Development; Sustainable Development

content of the CAPs was subjected to a text-based analysis. Through this process of text-based analysis, it is possible to identify the most important issues cited in the CAPs.

Table 3 shows the key terms that characterize the policies set forth in the CAPs. Since discussing all the objectives of each plan is beyond the scope of the study, the key terms/concepts explaining the core policies or strategies within the context of climate-sensitive planning are determined as the key components.

The content analysis includes eight policy areas. To measure the content of CAPs, we used the term frequency technique for each plan, and we attempted to reveal a conceptual density map with the results obtained from the key terms and their closest sub-terms. For example, the frequency of the term 'greenhouse gas emissions' was determined by searching three different sub-terms such as greenhouse, greenhouse gas, and GHGs. The simplest form of term frequency technique we use here is to count the number of uses of each key term in each plan.

3.2.2 Text-Based Analysis of the Selected Five CAPs

Text-Based Analysis

Text-based or text mining analysis is the process of extracting interesting and significant patterns to examine information from textual data sources (Talib et al., 2016: 414). The process includes the following steps: (1) documents collection, (2) retrieve and preprocess documents, (3) information extraction (analyzing text) within two sub-steps; clustering and summarization, (4) management information system, (5) knowledgeable or applicable conclusions within the relevant topic (Zhong et al., 2012).

There are different text mining techniques that can be applied for examining the text patterns and their mining process (Gupta & Lehal, 2009). The techniques that are mainly addressed and used in the literature are: (1) document classification, (2) information retrieval (keyword search/querying and indexing), (3) document clustering (steps/policies/phrases clustering), (4) natural language processing (spelling correction, grammatical parsing, and word sense disambiguation), (5) link analysis (relationship extraction), and (6) web mining (site analysis) (Talib et al 2016, p. 415; He, 2013; Sagayam et al. 2012; Dang & Ahmad, 2015; Weiss et al. 2010; Agrawal & Batra, 2013; Fan et al. 2006).

In this study, we used the information retrieval technique, and the techniques of document clustering and information extraction to analyze the selected CAPs. We used term frequency within the information retrieval. The term frequency shows how often a term, word, or concept occurs in a document. It also indicates the significance of a determined term (or concept) within the overall document. The term frequency value provides the calculation of keyword density (of key policy terms density).

For calculating key policy terms density, the following formula is used:

$$\text{Keyword density} = \text{term frequency}/\text{total word number} \times 100$$

The example below would result in the following weights for the key policy term 'GHGs':

$$\text{Ann Arbor}: \ 318/1830 \times 100 = 0.17\%$$
$$\text{Detroit}: \ 125/663 \times 100 = 0.19\%$$

This calculation shows a higher relevance of the Detroit CAP since the key terms have a larger relative frequency than Ann Arbor.

The Term Frequency and Michigan CAPs

In the Michigan CAPs, the strategies have been sorted into climate action areas: (1) city's baseline emissions, existing conditions and targets for emission reductions, (2) buildings and infrastructure, (3) land use and transportation, (4) responsibility for cultural change within the communities. Additionally, the strategies have been given 'tags' that indicate the nature of each policy.

Here, we focus on the 8 (eight) areas corresponding to the actions addressed in the state-level policies. Each area is evaluated by examining the CAPs in searching the frequency of key terms.

Through the text-based analysis, it is possible to identify the most frequent terms or the key policy components cited in the CAPs. According to the classification of this study, the most frequent terms are *renewables and energy efficiency*, *greenhouse gas emissions (GHGs)*, followed by buildings and environmental design, responsibility (including the terms of education, community, awareness), sustainability, development, international documents (agreements). Each CAP illustrates the need for more education when it comes to climate change (see Fig. 1).

While the conceptual density map of Michigan CAPs based on carbon management and urban infrastructure was created, similar terms (or concepts) were evaluated together and not assessed separately. In this context, a calculation was made on the 8 categories and its sub-terms shown in Table 3. However, the Fig. 1 only provides an assessment of the 8 key terms or action categories, and the frequency of the sub-terms is not reflected in the Fig. 1 separately.

As shown in Table 3, the key term of the first category is linked to the following three sub-terms: *Greenhouse, greenhouse gas,* and *GHGs*. For the first category, Fig. 1 shows that all five CAPs have a comprehensive approach to reduce CO_2 emissions within the context of renewable energy policy when comparing other categories. The Ann Arbor CAP is the plan that has the highest term frequency value of this category. The number of uses of this term in the plan is determined to be 318. The second is Traverse City CAP (210), and the third is Ypsilanti CAP (195). These plans are followed by Detroit (125) and East Lansing (25).

The second key term is *sustainability* which includes the sub-terms *sustainability* and *sustainable.* It presents a few correlations relating to development, climate change, renewable energy, green infrastructure, research and technology. Ann Arbor keeps its first position for this category as the plan having the highest frequency value which is 107 uses. The Ann Arbor CAP is followed by Ypsilanti (56), Detroit (45), East Lansing (37), Traverse City (21).

The third key term is *renewables and energy efficiency*. It can be defined as the interrelation of renewables, energy, and efficiency. As shown in Fig. 1 below, Ann Arbor also ranks first in this category with a total of 728 uses. This term is used 645 times in the Traverse City CAP. These plans are followed by East Lansing (199), Ypsilanti (195), Detroit (137).

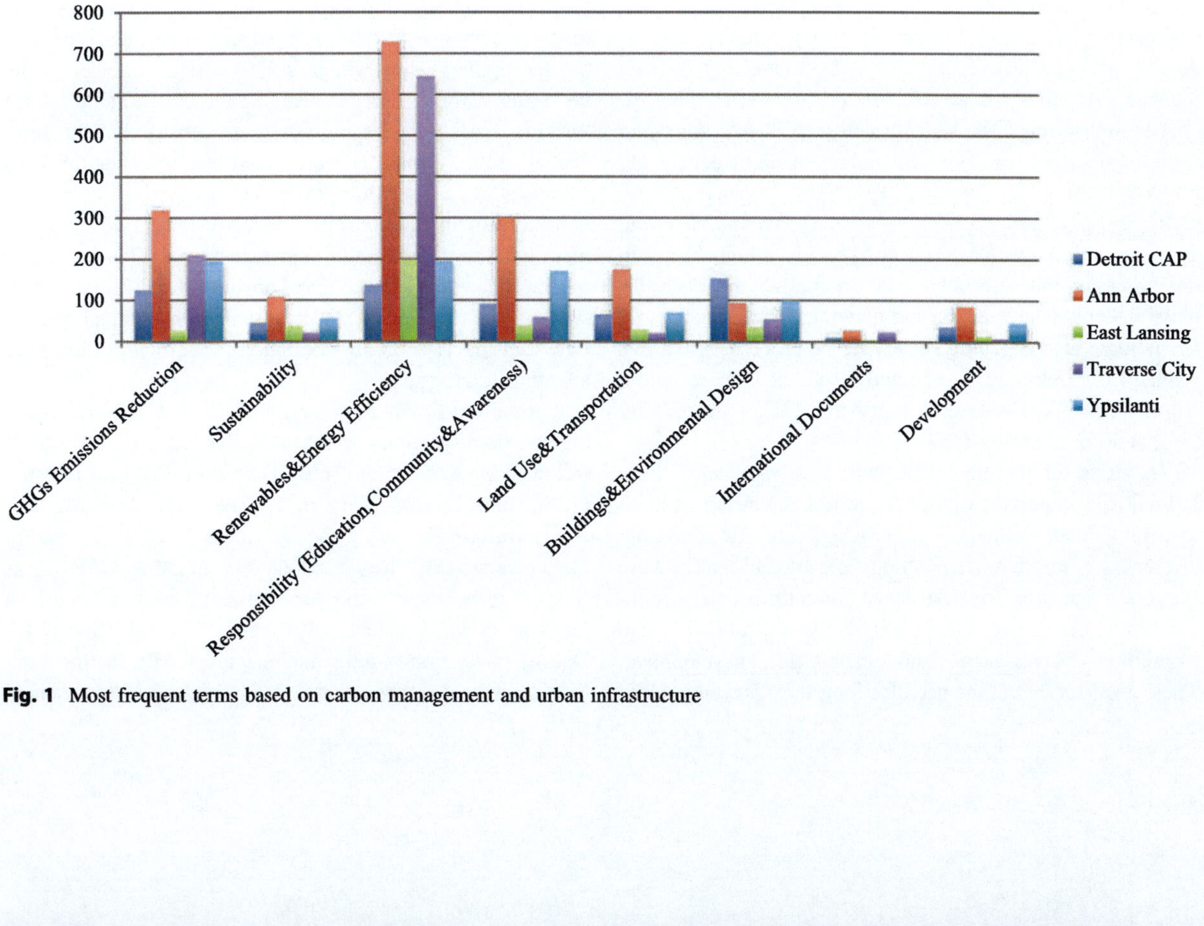

Fig. 1 Most frequent terms based on carbon management and urban infrastructure

The key term for the fourth category is responsibility. This term connects the following three sub-terms: Education, Community and Awareness. It is also noticeable that the Ann Arbor CAP still maintains its first place for the term responsibility, with the highest term frequency value which is determined to be 300 uses. The second is Ypsilanti CAP (170), the third is Detroit CAP (91), and these plans are followed by Traverse City (59) and East Lansing (38), respectively.

The key term for the fifth category is land use and transportation, which links the two sub-terms 'transportation, land and land use.' These terms present several correlations relating to sustainable development, climate change, renewable energy, productivity, awareness, research, and technology. However, the analysis of all the correlated terms is not given here because it extends beyond the scope of the study. The term frequency value in this category is slightly lower than the other categories. The CAPs are listed as follows: Ann Arbor (174), Ypsilanti (70), Detroit (66), East Lansing (29) and Traverse City (21). Ann Arbor takes first in this category.

International documents or agreements are the key terms of the sixth category. In this category, the analysis is based on four sub-terms: United Nations, Kyoto, Paris, IPCC, and ICLEI. These terms have been chosen by considering the leading actors in the international system regarding climate change. The terms are not commonly used in each plan, so the key terms and sub-terms of this category are the lowest term frequency value. For example, in the Ypsilanti CAP, the terms have not been determined. This may be a result of approaches based mainly on local policies and objectives. The CAPs are listed for the term frequency value as follows: Ann Arbor (26), Traverse City (23), Detroit (10), East Lansing (5), and Ypsilanti (0).

The key term of the seventh category is buildings and environmental design. The components of this category are important terms that combine green infrastructure, renewable energy, green roofs, green design, sustainability and efficiency. Although CAPs do not provide a high figure in this category, all selected CAPs emphasize the need for massive knowledge creation and dissemination, research and technology, also engaging society in the process. As shown in Fig. 33.1, Detroit ranks first with a total of 153 uses and it is followed by Ypsilanti (99), Ann Arbor (93), Traverse City (55), and East Lansing (35).

The term of the final category is development, highlighting the importance of development regarding sustainability, climate change, land use, capacity building, awareness, research and technology and education. The term reinforces the urge for a stronger commitment to sustainability and climate change. It is linked to the following two sub-terms: Development and sustainable development. These terms point out the need for long-term precautions. In this context, the number of uses of this category in CAPs is as follows: Ann Arbor (84), Ypsilanti (45), Detroit (36), East Lansing (13), and Traverse City (8).

The most frequent terms in all five CAPs are renewable energy and greenhouse gas emissions (GHGs). Thus, we see that the selected CAPs have a strong understanding of how crucial the reduction of carbon emissions is, and about the direct correlation between GHGs and the challenges we face for a sustainable future.

If we quantify the *conceptual density map of CAPs* in percentage terms, it could help to better understand the perception of climate action policy of selected CAPs. The weight of determined policy terms is represented by a different percentage in each CAP. Table 4 explains that the core policy of three CAPs (Detroit, Traverse City, Ypsilanti) is to reduce GHGs emissions (also see Table 4). The most important policy is the renewables and energy efficiency for Ann Arbor CAP that has 40% of total uses of key terms. East Lansing is the only Plan that is named as a Sustainable Plan which combines sustainability and climate change policy. So its focus is also more renewables than GHGs emissions in conceptual sense. It has 52% of total uses of all key terms.

When we compare each plan according to the key terms density (=term frequency/total word number × 100), it is possible to see clearly their policy preferences. Ann Arbor CAP, for example, is the plan that has the highest number of uses (1,830 uses) in which the terms including 8 categories. The first core policy of the plan is renewable energy and energy efficiency.

In all plans, there are two core policies that come to the forefront: to reduction GHGs emissions and to increase renewable energy. For the term 'GHGs emissions', Ypsilanti has the highest value which is 23% of total number of the key terms. Other CAPs can also be followed in Table 4. East Lansing CAP has the highest term density for the term 'sustainability' which is represented by 10%. For the term '*renewables*', we see that Traverse City CAP has the highest value (62%) in the conceptual density map. Energy is the most important policy the plan focuses on. This is closely related to the scope of the East Lansing CAP (52%) and Ann Arbor CAP (40%) and the details of each political sphere. East Lansing CAP is also designed in a structure that pays attention to energy.

For the fourth key term '*responsibility*', Ypsilanti has the highest density among cities with the value of 20%. It is followed by Ann Arbor (16%), Detroit (14%), East Lansing (10%), and Traverse City (6%). The term '*land use and transportation*' has very similar values in all CAPs except for Traverse City. Key term density in other CAPs is as follows, respectively: Ann Arbor (10%), Detroit (10%), East Lansing (8%), Ypsilanti (8%), and Traverse City (2%). Detroit ranks first among the selected CAPs for the term '*buildings and environmental design*' which is represented

Table 4 Percentage statement of policy terms

Key policy terms	Ann Arbor		Detroit		East Lansing		Traverse City		Ypsilanti	
	Weight ratio percentage %									
	%	Uses	%	Uses	%	Uses	%	Uses	%	Uses
GHGs	17	318	19	125	7	25	20	210	23	195
Sustainability	6	107	7	45	10	37	2	21	7	56
Renewables	40	728	21	137	52	199	62	645	23	195
Responsibility	16	300	14	91	10	38	6	59	20	170
Land use and transportation	10	174	10	66	8	29	2	21	8	70
International documents and agreements	1.4	26	1,5	10	1.3	5	2	23	Not determined	
Buildings and environmental design	5	93	23	153	9	35	5	55	12	99
Development	5	84	5	36	3.4	13	0.8	8	5.4	45
Total Number of Key Terms	1,830		663		381		1,042		830	

by 23% in the Plan. The other CAPs have similar values, and those can be followed in Table 4.

The category of *international documents* is the least used term in CAPs. Although adaptation to climate change is one of the main objectives of sustainable development, the term '*development*' has also been the least cited terms in the CAPs. Ann Arbor and East Lansing have included climate change policies within more comprehensive sustainability and renewables policies, though often with a loss of focus on climate change per section. Despite all this, it should be noted that all Michigan CAPs for the three categories including 'buildings and environmental design', 'development' and 'responsibility' have some deficiencies for improvement.

3.2.3 Policy Design of CAPs

Michigan CAPs have focused on GHGs mitigation rather than climate change adaptation. It should be noted that climate regulatory instruments from the participating municipalities must be considered when evaluating key terms of CAPs. For example, Ann Arbor has a longer historical experience in energy planning.[5] The key term 'renewables and energy efficiency' describes the core policy of the Ann Arbor plan. It is the most detailed and longest CAP, with a total of 188 pages. It is followed by: Traverse (83 pages), Detroit (80 pages), Ypsilanti (76 pages), East Lansing (22 pages).

The core policies and climate action categories vary from one CAP to the next. Table 5 shows each city's core policies and how they relate to the UN Sustainable Development Goals (SDGs) for 2030.

As shown in Table 5, Michigan selected CAPs have a strong political perception of the UN SDGs Agenda for 2030. However, this perception includes categories of action that are prioritized for some goals. For example, a total of 25 subcategory actions were set in Ann Arbor's priority policy titled 'energy and buildings'. The political priorities of the CAPs are directly related to the content of the SDGs. Ypsilanti CAP offers a climate policy regarding 9 goals of the UN SDGs Agenda, although it is not a comprehensive plan.

Goal 13 of the Agenda which is Climate Action is the primary policy of all CAPs. Excluding Traverse City CAP, all selected CAPs identify action categories that are directly line with the following SDGs: Goal 6. Clean Water and Sanitation, Goal 7. Affordable and Clean Energy, Goal 9. Industry, Innovation and Infrastructure, Goal 11. Sustainable Cities and Communities, Goal 12. Responsible Consumption and Production. Traverse City CAP has not set an action category for Goal 6. However, Goal 3. Good Health and Well-being is a policy in which other municipalities, except East Lansing, set action categories. The Goal 16. Peace, Justice and Strong Institutions is another policy brought by Detroit, Ypsilanti, and East Lansing to the local climate policy agenda. Goal 4. Quality Education, on the other hand, is only a policy that Ypsilanti has put on its agenda.

The synergies between SDGs and the CAPs climate action policies illustrate a major theme, urban climate action policies influence each other in complex ways, so effective CAPs cannot develop on piecemeal solutions. Policies based on SDGs that support GHGs reduction, renewable energy, mass transit, develop sustainable cities and communities, develop green infrastructure, encourage mixed-use development, and promote higher density will support each other's effectiveness.

[5]Ann Arbor has been experiencing to develop new energy policies since 1981. Municipality established the Energy Advisory Board and first Energy Plan completed in 1981. In 2005, Green Energy Challenge was issued by Mayor and the following year (2006) City Council passed Energy Challenge Goals (City of Ann Arbor CAP 2012).

Table 5 Core policies and action categories of Michigan's selected CAPs

Climate action plan	The core policies and categories	The UN sustainable development goals 2030 (Relevant Goal/Goals)
Ann Arbor CAP (1881 pages)	• Energy and Buildings • Land Use and Access • Resource Management • Community and Health	**Goal 13. Climate Action** Goal 3. Good Health and Well-being Goal 6. Clean Water and Sanitation Goal 7. Affordable and Clean Energy Goal 9. Industry, Innovation and Infrastructure Goal 11. Sustainable Cities and Communities Goal 12. Responsible Consumption and Production
Traverse City CAP (831 pages)	• Reduce Unnecessary Electrical Loads (Office Operations) • Purchasing Energy Efficient Equipment (Office Operations) • Resource Conservation Policies or Initiatives (Office Operations) • Optimize Interior Lighting (Facilities) • Optimize HVAC Performance (Facilities) • Optimize Exterior Lighting (Facilities) • Water System Upgrades and Conservation Measures (Facilities) • Utilize Renewable Energy (Facilities) • Promote Smart Transportation Options (Employee Commute and Fleet)	**Goal 13. Climate Action** Goal 3. Good Health and Well-being Goal 7. Affordable and Clean Energy Goal 9. Industry, Innovation and Infrastructure Goal 11. Sustainable Cities and Communities Goal 12. Responsible Consumption and Production
Detroit CAP (801 pages)	• Solid Waste and Waste Management • Public Health • Businesses and Institutions • Parks, Public Spaces and Water Infrastructure • Homes and Neighborhoods	**Goal 13. Climate Action** Goal 3. Good Health and Well-being Goal 6. Clean Water and Sanitation Goal 7. Affordable and Clean Energy Goal 9. Industry, Innovation and Infrastructure Goal 11. Sustainable Cities and Communities Goal 12. Responsible Consumption and Production Goal 16. Peace, Justice and Strong Institutions
Ypsilanti CAP (761 pages)	• Buildings • Infrastructure • Land Use and Transportation • Culture Change (including community sustainability library, residential energy education, sustainability education in Schools, and so on…	**Goal 13. Climate Action** Goal 3. Good Health and Well-being Goal 4. Quality Education Goal 6. Clean Water and Sanitation Goal 7. Affordable and Clean Energy Goal 9. Industry, Innovation and Infrastructure Goal 11. Sustainable Cities and Communities Goal 12. Responsible Consumption and Production Goal 16. Peace, Justice and Strong Institutions

(continued)

Table 5 (continued)

Climate action plan	The core policies and categories	The UN sustainable development goals 2030 (Relevant Goal/Goals)
East Lansing (221 pages)	• Waste Management • Energy Efficiency • Renewable Energy • Transportation • Water Management	**Goal 13. Climate Action** Goal 6. Clean Water and Sanitation Goal 7. Affordable and Clean Energy Goal 9. Industry, Innovation and Infrastructure Goal 11. Sustainable Cities and Communities Goal 12. Responsible Consumption and Production Goal 16. Peace, Justice and Strong Institutions

4 Conclusion

Establishing an effective and applicable CAP is the initial step in responding to climate change for developing countries and developed countries, alike. The CAP is the most effective instrument integrating relevant local adaptation and mitigation strategies into global climate policies. In this framework, local governments serve as the key actors in urban planning, and are central to climate change policies and efforts. They control the majority of buildings, construction, transportation improvements, and land-use decisions in the USA and all over the world. In short, CAPs are comprised of a series of tools aimed at transforming urban planning and design policies.

All CAPs' responses are the results of different motivations—all targeting the reduction of GHGs, some analyzing their responsibilities to protect the environment, others only focused on reducing greenhouse gas emissions (GHGs) to mitigate their communities' impact on climate change. Some pay great attention to improve urban infrastructure based on renewable energy and urban air quality. Others give more importance to the sustainable local development.

Understanding how CAPs target emissions or influence specific emission reduction projects is of interest in theory and policy. CAPs explain a local policy orientation in a climatic sense. A carefully crafted policy orientation can show that cities and local governments are capable and credible institutions in addressing climate change. In other words, a CAP is a local political document that lays out economic, social, environmental, and technological goals in adapting to climate change impacts. Therefore, it is valuable to understand the perception of policy by a text-based analysis of CAPs.

Michigan has been working on climate action planning since 2007. The state CAP was prepared and published in 2009 by Michigan Climate Action Council. In this study, we evaluated five Michigan CAPs as an example of local climate action planning through text mining analysis. CAPs' objectives are evaluated in the framework of two main issues addressed local climate planning and UN 2030 Agenda: (1) carbon management and (2) urban infrastructure. We have done our research in the following five cities: Detroit, Ann Arbor, East Lansing, Traverse City and Ypsilanti. In addition, the core policies and climate action categories of CAPs are examined in the context of UN SDGs for 2030. Although the CAPs have a strong political perception of the UN SDGs for 2030, action categories of the plans are prioritized for some goals. However, the political priorities of the CAPs are directly related to the content of the SDGs.

In this study, text mining analysis explains the key policy terms or concepts relating to goals of CAPs and climate change planning such as greenhouse gas reduction, renewable energy and energy efficiency, green buildings and design, sustainability, land use and transportation, awareness and knowledge. To measure the content of CAPs we used the term frequency technique for each plan, and we attempted to reveal a conceptual density map with the results obtained from the key terms and their closest sub-terms. According to the classification of this study, the most frequent terms and core policies are *renewables and energy efficiency*, *greenhouse gas emissions (GHGs)*, followed by buildings and environmental design, responsibility (including the terms of education, community, awareness), sustainability, development, international documents (or agreements).

In conclusion, examining the CAPs' key terms and policies has provided an overall picture about the content of Michigan local climate action planning. The existence of local CAPs shows that elected officials and planners in Michigan are taking climate change seriously.

References

2018American Community Survey 5-Year Estimates(December 2019). Retrieved May 1, 2020 fromhttps:, , www.census.gov, programs-surveys, acs, technical-documentation, table-and-geography-changes, , 5-year.html., .

2014–2018 American Community Survey 5-Year Estimates (December 2019). Retrieved May 1, 2020 from https://www.census.gov/programs-surveys/acs/technical-documentation/table-and-geography-changes/2018/5-year.html

Agrawal, R., & Batra, M. (2013). A detailed study on text mining techniques. *International Journal of Soft Computing and Engineering (IJSCE).*, *2*(6), 2231–2307

Bassett, E., & Shandas, V. (2010). Innovation and climate action planning: perspectives from municipal plans. *Journal of the American Planning Association, 76*(4), 435–450. https://doi.org/10.1080/01944363.2010.509703

Betsill, M., & Bulkeley, H. (2004). Transnational networks and global environmental governance: The cities for climate protection program. *International Studies Quarterly, 48*(2), 471–493. https://doi.org/10.1111/j.0020-8833.2004.00310.x

Boswell, M. R., Greve, A. I., & Seale, T. L. (2012). *Local climate action planning*. Island Press.

Bulkeley, H. & Moser, S. C. (2007). Responding to climate change: Governance and social action beyond Kyoto. *Global Environmental Politics, 7*(2), 1–10 (2007)

Bulkeley, H., & Newell, P. (2015). *Governing climate change*. Routledge.

Bulkeley, H., & Betsill, M. (2005). Rethinking sustainable cities: Multi-level governance and the urban politics of climate change. *Environmental Politics, 14*(1), 42–63

C2ES (2020). The Center for Climate and Energy Solutions. Mayors leading the way on climate how cities large and small are taking action. Retrieved March 19, 2019 from https://www.c2es.org/site/assets/uploads/2018/09/mayors-leading-way-on-climate-2018.pdf

Calthorpe, P. (2011). *Urbanism in the age of climate change*. Island Press.

City of Ann Arbor Climate Action Plan. (2012). Retrieved March 19, 2019 from https://www.a2gov.org/departments/systems-planning/planning-areas/energy/Documents/CityofAnnArborClimateActionPlan_low%20res_12_17_12.pdf

City of East Lansing Sustainability Plan. (2012, April). Retrieved March 19, 2019 from https://www.cityofeastlansing.com/DocumentCenter/View/635/City-of-East-Lansing-Climate-Sustainability-Plan-PDF?bidId=

City of Traverse City Climate Action Plan. (2011, February). By Solomon Townsend, Barton Kirk and Michael Powers. SEEDS, Inc.

City of Ypsilanti Climate Action Plan. (2012, July). Retrieved March 19, 2019 from https://cityofypsilanti.com/DocumentCenter/View/309/Climate-Action-Plan-PDF

Corburn, J. (2009). Cities, climate change and urban heat island mitigation: Localizing global environment. *Urban Studies, 46*(2), 413–427

Dang, D.S., & Ahmad, P. H. (2015). A review of text mining techniques associated with various application areas. *International Journal of Science and Research (IJSR), 4*(2), 2461–2466.

Detroit Climate Action Plan. (2017). Retrieved March 19, 2019 from https://toolkit.climate.gov/reports/detroit-climate-action-plan

EEA (European Environmental Agency). (2008). *Impact of Europe's Changing Climate-2008 indicator-based assessment.* Retrieved March 22, 2019 from https://ec.europa.eu/jrc/sites/jrcsh/files/jrc_reference_report_2008_09_climate_change.pdf. Joint EEA-JRC-WHO Report No. JRC47756.

EPA. (2018). State climate and energy program: State examples climate change action plans. Retrieved April 22, 2020 from https://19january2017snapshot.epa.gov/statelocalclimate/state-climate-and-energy-program_.html

EPA (Environmental Protection Agency). (2017). *Energy-related CO_2 emissions data tables*. Retrieved April 24, 2019 from https://www.eia.gov/environment/emissions/state/

Fan, W., Wallace, L., Rich, S., & Zhang, Z. (2006). Tapping the power of text mining. *Communications of the ACM, 49*(9), 76–82. https://doi.org/10.1145/1151030.1151032

Frazier, T., et al. (2013). Spatial and temporal quantification of resilience at the community scale. *Applied Geography, 42*, 95–107. https://doi.org/10.1016/j.apgeog.2013.05.004

Georgetown Climate Center. (2020). *Preparing for climate change in Michigan*. Retrieved March 22, 2020 from https://www.georgetownclimate.org/adaptation/state-information/michigan/overview.html

Gupta, V., & Lehal, G. S. (2009). A survey of text mining techniques and applications. *Journal of Emerging Technologies in web Intelligence, 1*(1), 60–76. https://doi.org/10.4304/jetwi.1.1.60-76

Hanak, E., Bedsworth, L., et al. (2008). *Climate policy at the local level: A survey of California's cities and counties*. San Francisco, Public Policy Institute of California. Retrieved March 22, 2020 from https://www.ppic.org/content/pubs/report/R_1108EHR.pdf

He, W. (2013). Examining students' online interaction in a live video streaming environment using data mining and text mining. *Computers in Human Behavior, 29*(1), 90–102. https://doi.org/10.1016/j.chb.2012.07.020

ICLEI. (2009). *U.S. local sustainability plans and climate action plans*. Retrieved April 22, 2019 from https://cityofithaca.org/DocumentCenter/View/1643/US-Local-Sustainability-and-Climate-Action-Plans

ILG (Institute for Local Governments). (2018). *Climate action plans*. Retrieved April 22, 2019 from https://www.ca-ilg.org/climate-action-plans

IPCC. (2020b). *About IPCC*. Retrieved April 2, 2020 fromhttps://www.ipcc.ch/about/

IPCC (Intergovernmental Panel on Climate Change). (2007). AR4 *Climate change 2007 synthesis report: Summary for policy makers*. Core Writing Team, Pachauri, R. K. & Reisinger, A. (Eds.). IPCC. Retrieved April 22, 2019 from https://www.ipcc.ch/site/assets/uploads/2018/02/ar4_syr_spm.pdf

IPCC Special Report (2020a). *Climate change and land: Summary for policy makers*. Retrieved April 2, 2020 from https://www.ipcc.ch/site/assets/uploads/sites/4/2020/02/SPM_Updated-Jan20.pdf

Long, J., & Rice, J. L. (2018). From sustainable urbanism to climate urbanism. *Urban Studies, 56*(5), 992–1008. https://doi.org/10.1177/0042098018770846

MCAN (Michigan Climate Action Network). (2019). *Michigan Cities taking action on climate change*. Retrieved April 14, 2019 from https://www.miclimateaction.org/michigan_cities_taking_action_on_climate

Meerow, S., & Stults, M. (2016). Comparing conceptualizations of urban climate resilience in theory and practice. *Sustainability, 8*(7), 701–717. https://doi.org/10.3390/su8070701

Michigan CAP. (2009, March). *Michigan climate action plan*. Michigan Department of Environmental Quality, Retrieved April 14, 2019 from https://www.michigan.gov/documents/deq/deq-miclimateactionplan-part1_276563_7.pdf

Millard-Ball, A. (2012). Do city climate plans reduce emissions? *Journal of Urban Economics, 71*(3), 289–311. https://doi.org/10.1016/j.jue.2011.12.004

Musco, F. (2010). Policy design for sustainable integrated planning: From local agenda 21 to climate protection. In M. van Staden &

F. Musco (Eds.), *Local governments and climate change* (pp. 59–76). Springer Science+Business Media

OECD (2010). *Cities and climate change* (pp. 17–20). OECD Publishing. Retrieved April 14, 2019 from https://www.oecd-ilibrary.org/governance/cities-and-climate-change_9789264091375-en

OECD. (2011). Urban environmental indicators for green cities: A tentative indicator set. Working Party on Environmental Indicators, ENV/EPOC/WPEI(2011) 6. OECD.

OECD. (2015). *Towards green growth? Tracking progress*. Retrieved April 14, 2019 from https://www.oecd.org/env/towards-green-growth-9789264234437-en.htm

Pattison, A., & Kawall, J. (2018). Equitable local climate action planning: Sustainable & affordable housing. *Ethics, Policy & Environment, 21*(1), 17–20. https://doi.org/10.1080/21550085.2018.1447900

Rice, J. L. (2010). Climate, carbon, and territory: Greenhouse gas mitigation in Seattle, Washington. *Annals of the Association of American Geographers, 100*(4), 929–937. https://doi.org/10.1080/00045608.2010.502434

Sagayam, R., Srinivasan, S., & Roshni, S. (2012). A Survey of text mining: Retrieval, extraction and indexing techniques. *International Journal of Computational Engineering Research, 2*(5), 1443–1446. ISSN 2250-3005(online).

Talib, R., Hanif,M. K., Ayesha, S., & Fatima, F. (2016). Text mining: Techniques, applications and issues. *International Journal of Advanced Computer Science and Applications, 7*(11), 414–418. https://doi.org/10.14569/IJACSA.2016.071153

Tavel, M. (2012). Climate-responsive urbanism: Historic Precedents and Progressive. In H. J Neis, G. Brown, J. M. Gurr, & J. A. Schmidt (Eds.), *Generative process, patterns, and the urban challenge*. Portland: PUARL Press.

UN. (2011). *Cities and climate change: Global report on human settlements* (p. 19). United Nations Human Settlements Program. https://unhabitat.org/wp-content/uploads/2012/06/GRHS2011-0.pdf . Accessed March 19, 2019.

UN. (2016). *Transforming our world: The 2030 agenda for sustainable development*. https://sustainableurbanization.un.org/post2015/transformingourworld/publication

UN SDGs Report (Sustainable Development Goals Report). (2019). https://unstats.un.org/sdgs/report/2019/The-Sustainable-Development-Goals-Report-2019.pdf

UNCCC (United Nations Framework Convention on Climate Change). (1992). https://unfccc.int/files/essential_background/background_publications_htmlpdf/application/pdf/conveng.pdf

UNFCC Paris Agreement. (2015). https://unfccc.int/files/essential_background/convention/application/pdf/english_paris_agreement.pdf

UNFCCC. (2020). Conference of the parties. https://unfccc.int/process/bodies/supreme-bodies/conference-of-the-parties-cop

UNFCCC COP 25. (2019, December). Capacity-building under the Convention. https://unfccc.int/sites/default/files/resource/sbi2019_L29E.pdf. Accessed April 24, 2020.

US Census Bureau. (2010). *Urban area list by state 2010*. Retrieved January 1, 2020 from https://www.census.gov/programs-surveys/geography/guidance/geo-areas/urban-rural/2010-urban-rural.html

US Census Bureau. (2019). *State population totals and components of Change: 2010–2019*. Retrieved April 12, 2020 from https://www.census.gov/data/tables/time-series/demo/popest/2010s-state-total.html

USA Census Bureau. (2012, July). *Population and housing unit counts: 2010-Michigan-2010*. Retrieved April 12, 2020 from https://www.census.gov/library/publications/2012/dec/cph-2.html

USCM (The U.S. Conference of Mayors). (2018). *Climate protection agreement*. Retrieved May 12, 2020 from https://www.usmayors.org/mayors-climate-protection-center/

USCM (The U.S. Conference of Mayors). (2020). *About the conference*. Retrieved May 1, 2020 from https://www.usmayors.org/

Weiss, S. M., Indurkhya, N., Zhang, T., Damerau, F. (2010). *Text mining: Predictive methods for analyzing unstructured information*. Springer Science and Business Media. https://doi.org/10.1007/978-0-387-34555-0

Wheeler, S. M. (2008). State and municipal climate change plans: The first generation. *Journal of the American Planning Association, 74* (4), 481–496. https://doi.org/10.1080/01944360802377973

World Bank. (2010). *A city-wide approach to carbon finance*. Retrieved January 1, 2020 from https://documents.worldbank.org/curated/en/389951468340161434/A-city-wide-approach-to-carbon-finance

Zahran, S., Brody,S. D., Vedlitz, A.et al. (2008a). Vulnerability and capacity: explaining local commitment to climate change policy in the United States. *Environment and Planning C, 26*(3), 544–562. https://doi.org/10.1068%2Fc2g

Zahran, S., Grover, H., Brody, S. D., &Vedlitz, A. (2008b). Risk, stress, and capacity: Explaining metropolitan commitment to climate protection. *Urban Affairs Review, 43*, 475–496. https://doi.org/10.1177%2F1078087407304688

Zhong, N., Li, Y., & Wu, S. T. (2012). Effective pattern discovery for text mining. *IEEE Transactions on Knowledge and Data Engineering, 24*(1), 30–44. https://doi.org/10.1109/TKDE.2010.211

Mobile Environmental Measurement Laboratory Using Electric Vehicles for Smart Cities

H. Gracia-León, E. Galvis Restrepo, A. Uribe-Jongbloed, and L. Rodríguez Urrego

Abstract

Nowadays, the capital cities and the main cities of the world have serious problems in transport and mobility, and this has a high impact on air quality due to the effects of traffic, such as traffic jams, traffic lights, and a large number of vehicles that work with an internal combustion engine. According to this, the alliance between the Ean University and the Mitsubishi Motors Company arose in 2018 to formulate a project that would impact electric mobility and increase the environmental benefits of it based on smart cities principles. This development, which was inaugurated in May 2019, shows the design of an environmental measurement system based on IoT, LTE, and data analytics that allows data to be taken with vehicles in motion and in real time for the generation of heat maps of the roads of the city. The design, installation, and dashboards for monitoring heat maps are registered in this paper. As well as, the expected impacts on environmental, social, and public policy issues applied to Bogotá city.

Keywords

Mobile laboratory • Smart cities • Electric vehicles • Environmental measurements

H. Gracia-León · E. G. Restrepo · A. Uribe-Jongbloed · L. R. Urrego (✉)
Indevos Research Group, Ean University, Bogotá D.C., Colombia
e-mail: lrodriguezu@universidadean.edu.co

H. Gracia-León
e-mail: hgracia@universidadean.edu.co

E. G. Restrepo
e-mail: egalvis@universidadean.edu.co

A. Uribe-Jongbloed
e-mail: auribejo@universidadean.edu.co

1 Introduction

Climate change is one of the greatest challenges that humans have ever faced. Currently, the emergence of new technologies avoids the disproportionate use of fossil fuels which is an undeniable reality. So, the electric mobility is one of the most important environmental-friendly technologies. To encourage the use of electric mobility, it is necessary to break paradigms, and at the same time demystify potentials problems of electric vehicles (EVs) in the cities. Actually, VE are one of the most promising solutions in terms of traffic, medium pollution environmental and sustainable mobility in order to avoid the worst of climate change as they will bear the brunt of its effects.

Currently, addressing the issue of environmental sustainability is a recurring action taken by cities, researchers, NGOs, multinational, the business community, and governments. These actions are not only temporary trend but also the development and execution of long-term sustainable projects. This trend of global consensus is consistent with the UN Sustainability Development Goals (United Nations).

In Colombia, there are several challenges on the issue of sustainable development. According to "Millennium Development Goals Report 2015" (United Nations, 2015) carried out the United Nations Development Program (UNDP), Colombia meets certain established objectives but encourages the national government to implement a regional policy that addresses the high disparity between urban and rural sectors to reduce the ratio of people without access to sustainable projects. Consequently, to meet objectives addressed by the UN for the country (Sustainable Development Goals, Colombia, SDGC) (ODS Colombia, 2016), and moreover recognizing the positive impact generated by the implementation of sustainable systems, it is also required that the academic community take a step forward with the promotion of projects that encourages the design, development, and execution of sustainable systems.

F. Trapani et al. (eds.), *Advanced Studies in Efficient Environmental Design and City Planning*, Advances in Science, Technology & Innovation, https://doi.org/10.1007/978-3-030-65181-7_34

Although there are innovations in Colombia regarding the implementation of sustainable systems, the analysis for its design is not very ambitious and can be evidenced in the approach of the objective related to sustainable development (SDG) (ODS Colombia, 2016). It is here that the synergy between measuring devices and the use of renewable energies for their operation become vital in the search to find relevant tools for analysis and development of sustainable systems. These topics are related to the concept of smart cities, which are worldwide urban developments that have been defined by the Inter-American Development Bank (IDB) as "those that place people at the center of development, incorporate Information and Communication Technologies in urban management and use these elements as tools to stimulate the statement of an efficient government that includes collaborative planning processes and citizen participation"(Bouskela et al., 2016).

Nowadays, the city of Bogotá has an environmental monitoring network composed of 15 static stations with DASIBI and OPSIS measurement systems, which make permanent reports of air quality in the city, as a result of these reports, it has been identified a problem regarding the amount of PM 10 particulate material directly related to the wind speed of the city. (Gaitán et al., 2007).

On the other hand, the electric vehicle has become an option for reducing greenhouse gas emissions, mainly due to the arrival of lithium ion batteries and their development (Hauch et al.). A growing number of studies and projects related to EVs have emerged in the last 20 years. There is even the expectation that EVs reach the equivalent of 10% of the world market until 2020 (Rosolem et al., 2012) managing to capture the eyes of the most important companies in the world in these technologies seeking to reach all markets and countries by this time. An electric motor is a class of devices that converts electrical energy into mechanical energy through electromagnetic interactions. In the motor, the mechanical torque is developed by the interaction of conductors carrying current in a direction at right angles to a magnetic fiel (Chan, 1993). The development and evolution of the VEs depend on the technological evolution of the batteries (World Nuclear Association, n.d.). The efforts are currently aimed at improving efficiency, consumption, autonomy, and charging speed, which is why in a context of applied research pioneer companies in the market strive to achieve the best batteries with greater autonomy, useful life, greater efficiency, and lower market cost.

The present work proposes a versatile electric mobility system, which involves strategies and activities that seek to positively impact the use of EVs as well as in-situ monitoring of some environmental variables of cities. The project has been developed by the Ean University in Bogotá, Colombia within the framework of an institutional project called Smart Block Ean (Rojas et al., 2018). The project from its structure has been raised from several approaches, namely academic, social, economic, and environmental. In the project, the novel integration of the usability of electric cars is carried out, with the adoption of applied research strategies toward measuring environmental parameters, as well as time measurements and road traffic. This integration also considers strategies for the use of EVs and the social appropriation of these technologies by the academic community of the university.

The main contribution of this work relies on a synergistic system in which two VEs (Mitsubishi-I MIEV) are used as an extension of a mobile monitoring station. With these cars, it is intended to carry out applied research based on an environmental mapping of the streets in Bogotá city. The measurements are taken in real time, while these vehicles are used by the community, the university has made an alliance with a car sharing company to allow the use of the vehicles by the students and the institution workers. In addition to this, research tasks are proposed regarding the autonomy, use, and efficiency of the vehicle in the city, to demystify a bit about the use of electric cars today and thus, also promote its use.

Finally, in the proposed project, the establishment of mobile stations, taking advantage of the EVs to receive pollution at the height of people and on the city roads, having direct contact with the sources of emission by transport and the actual exposure of the population. The work is organized as follows: Sect. 2 describes the methods used for the construction of the prototype, the components, and installation inside the vehicles and the advantage of using this type of laboratory to support the construction of smart cities. In Sect. 3, some of the results obtained with the development of the project are presented; and Sects. 4 and 5 discuss the results and presents the conclusions.

2 Methods

Currently there is a need to be able to carry out a timely analysis of the concentration of air pollutants and microclimate variables. In the case of study of Bogotá city, it should be considered that it has an approximate urban area of 307 km^2, which includes residential, commercial, and industrial areas. Particularly, the Capital District has the Air Quality Monitoring Network of Bogotá. The monitoring network consists of 13 fixed stations and a mobile one to collect information (Gaitán et al., 2007). In this district network, a mathematical interpolation model is used to estimate the concentration of air pollutants at intermediate points across the city, which lends itself to some inaccuracies given the size of the city.

On the other hand, it is noteworthy a study carried out by Google related to determine air quality in cities by using a mapping process on the streets of four cities around the world (Alexeeff et al., 2018). The study was launched in 2018 and does not include any city in South America. Finally, the work of the Ean University in Bogotá is highlighted with the project called " Live Ci Project," which is a research developed in a parallel way with similar techniques to those implemented by Google. The project seeks the measurement of greenhouse gases in Bogotá, with the hypothesis of the use of EVs and how a combustion vehicle directly interferes with the taking of environmental measures.

Consequently, in the initiative to implement EVs to setting up a mobile environmental measurement laboratory using EVs for Bogotá, a system has been designed which incorporates an electric vehicle, sensors for measuring the air pollutants concentration and environmental variables, as well as an electronic system for global positioning system, record of mentioned variables and a human–machine interface (see Fig. 1).The procedure for the global positioning (GPS) sensor is stated and tests were carried out in different locations dynamically, and the GPS has been used with a tolerance of 4 mts2 and at the end when a point shares the location the point shows the average of the location. Once sensors are characterized and calibrated, the design and implementation of the electronic device is achieved.

The application of methods for measuring toxic concentrations in the environment (Isakov et al., 2007) is a great help within the project, in addition to benchmarks in which the applications of similar work (Wang et al., 2009) make the proposed design of the project feasible and also considering the mathematical models (Chen et al., 2019; Muntean, 1999) and application of commercial products (Amanatidis et al., 2013). In this work, four phases are considered in the design and implementation processes.

In this work, four phases are considered in the design and implementation processes. As a first phase, the assembly and programming of the devices in the university laboratory were developed, there, with the appropriate equipment, performance tests were performed, and reading/recording of the variables. In a second phase, the connection to the communications network was tested for sending data to a server located using cloud computing technology. In this phase, the test was initially done with a connection to the local wireless network and then to make a mobile wireless test, taking the device out of the laboratory. Once a satisfactory operation of the device was obtained, the VE is adjusted and installed. For installation purposes in the vehicle, different aspects such as connection, aesthetics, ergonomics, and usability are considered. Once the adjustments are finished, in a third phase, field tests are carried out using the mobile wireless connection at different points previously determined across the city.

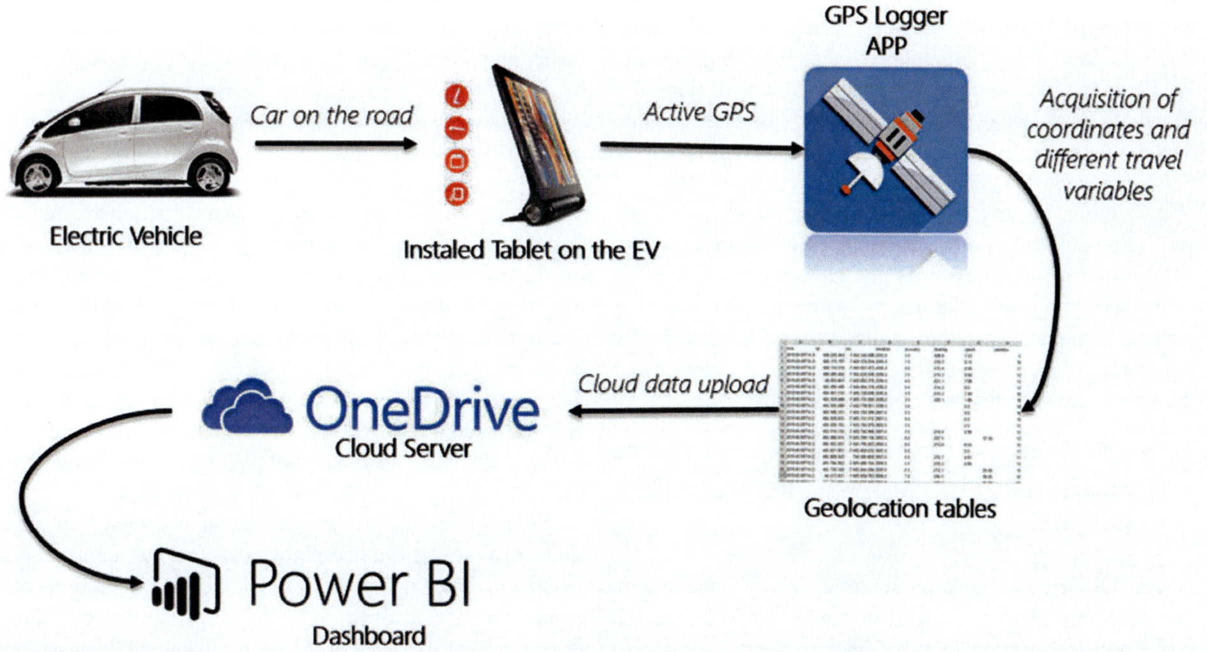

Fig. 1 Information flux

Table 1 Variable, measurement, and sensor

Variable	Measurement	Sensor
Noise	Intensity of noise in the streets. (dB)	Microphone MEMS (ADM401)
CO2	Carbon emissions in the streets (ppm)	CCS811/BME280
Temperature	Air temperature on city roads, considering traffic on the road and location (°C)	CCS811/BME280
Luminosity	Lighting level within the city roads day and night (Luxes)	VEML7700 o TSL2561
Humidity	humidity outside EV (%)	CCS811/BME280
Particulate matter	Particulate matter in the environment (TVO)	CCS811/BME280

2.1 Mobile Data Acquisition Hardware

In this project, the mobile acquisition device (DaqM) is the core of the system, and this device is responsible for reading signal distribution within the mobile laboratory. The device was developed using commercially available microcontroller and open-source microprocessor-based development cards. Some commercially available industrial quality sensors have been considered in the selection. According to the designed system, the measured variables are carbon dioxide, temperature, luminosity, noise, relative humidity, and particulate matter. Variables and its units measured in the mobile environmental laboratory using EVs are shown in Table 1.

For the operation of the monitoring system, it was necessary to develop an algorithm in the C++ programming language to characterize the acquired sensors, with this, monitoring tests were performed on variables locally. In the configuration of the wireless machine-to-machine communication system between devices, several communication channels have been programmed, namely connection via Wi-Fi with cloud computing server and connections via Bluetooth with some sensors to make easier installation inside the vehicle.

The designed system also considers basic tests of charging times, battery capacity, price per charge, autonomy within the city, behavior of the vehicle on usual routes, and finally tests to define the cost per km traveled by the VE were performed. In addition, a software platform was developed to data acquisition, visualization (dashboard) and monitoring of environmental variables, as well as routes and time spent from destination to destination.

2.2 Measurement of Environmental Variables and GPS

Having an electric vehicle on the streets gives to the system an excellent outlook to activate its circulation as a research tool. In this project, there is a device to control the global position of the vehicle; in addition, CO_2 levels are measured by the places where the vehicle passes. In the system, there is a datalogger of variables such as noise levels, brightness levels, temperature, the amount of energy used per car, and carbon dioxide levels. For instance, to analyze the level of CO_2 in a street with traffic jam, temperature in different areas of land use, noise levels across main streets of the city, hours in which more CO_2 is generated, and levels of increased noise pollution, among others.

With the information of the measured variables, the monitoring system performs environmental analysis of the city's roads. For the information processing, big data techniques are used, which allow to locate the measurements within a cross-map where they are correlated, and thus identify problems and early warnings caused by mobility to environment, likewise, environmental and health awareness. One of the main objectives is making the electrical vehicle a mobile laboratory that continuously develops tests within the city streets. So, the mobile environmental measurement laboratory automatically gets outcomes such as heat maps and particulate matter graphics information on test streets (see Fig. 2).

3 Results

The project has already yielded some preliminary results through which trends have been identified within the routes of the mobile environmental laboratory. Some of these trends occur mainly in vehicular congestion, traffic lights, and the busiest areas. Although the results are preliminary, as can be seen in the heat maps of Fig. 2, two specific zones with high averages at the decibel level were identified, one zone is due to a construction, and the second zone to a traffic lights. However, there are not yet significant results for decision making since there is not enough data.

Although the electric vehicle serves the university community to move from one place to another, we consider it a laboratory, since it is in the ability to perform measurements, corrections, and analysis on the fly, allowing the acquisition in real time of variables of the environment, while providing a service to the community. The application of technological

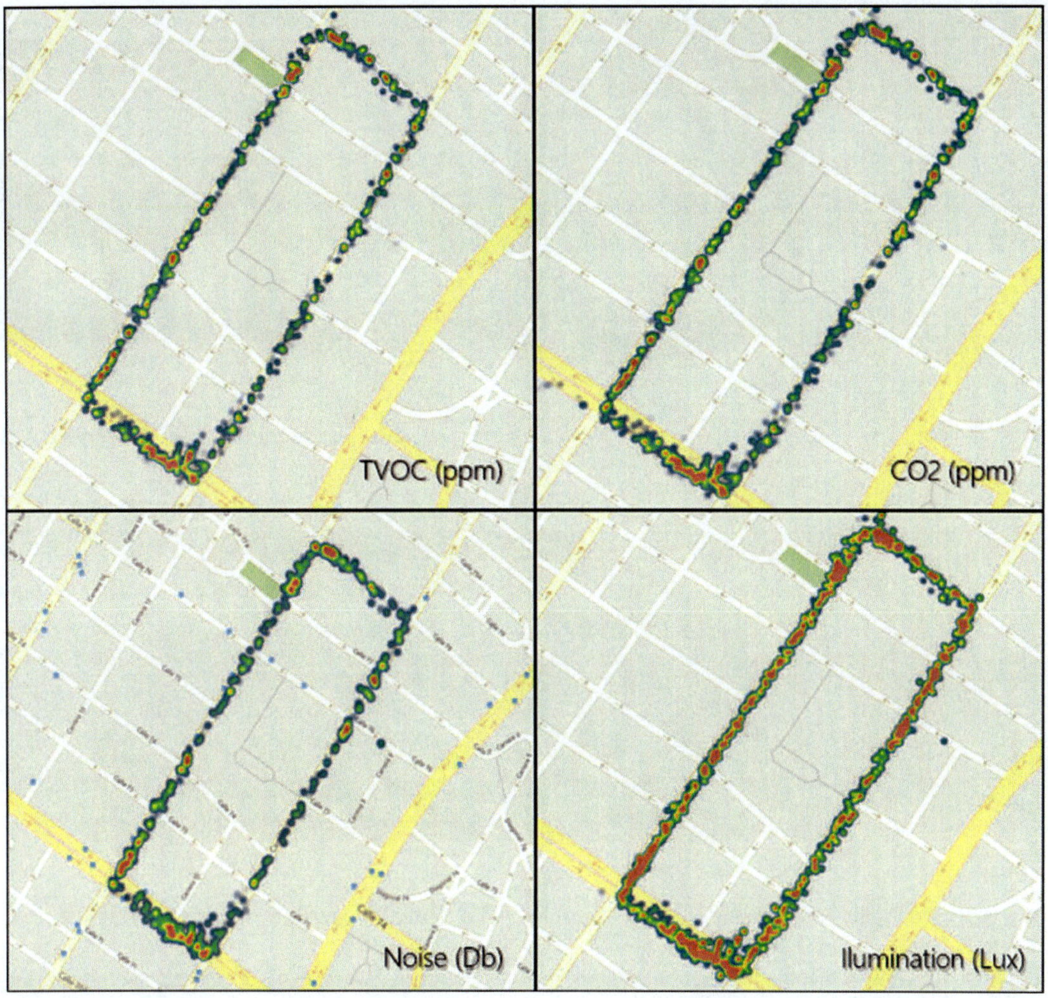

Fig. 2 Test results heat map

trends such as the Internet of things and big data allow great versatility and robustness in this type of solutions, generating spaces to reimagine all these processes, improving their efficiency and facilitating their application.

In Fig. 2, you can identify consistent results that show relationships between the generation of particulate matter and CO_2. Noise problems can be found graphically within the traffic lights that the route has. It is worth noting that in lighting issues the tests were performed during the day, so the graph represents high range values because there was always daylight. These results are preliminary and untreated but present a giant possibility when developing a real-time monitoring system for the city, with a greater number of mobile laboratories, data, and routes.

Regarding the results of the first tests, it is important to narrow down the dimension of the test, because the first measurements were made in a 2.5 km stretch in the Bogotá city surrounding four headquarters of the Ean University. Although the results are preliminary, the impact of this laboratory in the future will be reflected in issues such as the paving of new urban roads, street lighting according to levels of lighting and types of luminaires, urban arborization and the types thereof, the location of vehicular electric recharging points, the behavior of the batteries in the EVs and the environmental characteristics of the roads, their certifications, and location of bicycle routes.

3.1 Housing and Installation

The presented housing was designed to enable the installation of the sensors on the EV. It has five holes to anchor the housing to the vehicle, spaces to allow the flow of air inside the device, and a transparent covering to allow the passage of light. Figure 3 shows the CAD design of the housing. We are currently optimizing the housing in

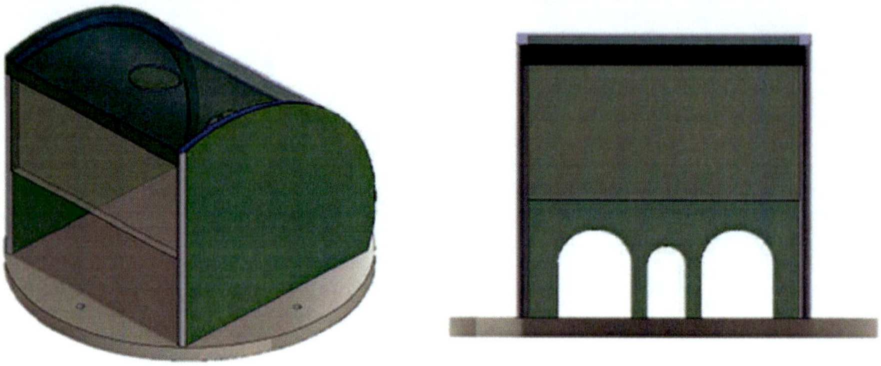

Fig. 3 Sensor housing

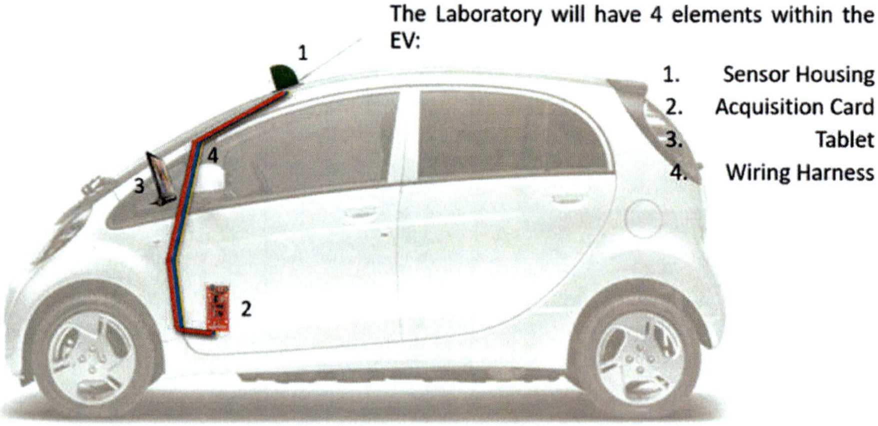

Fig. 4 Mobile laboratory installation

installation, operation, air aerodynamics, and sensor measurements.

In addition, it was defined that the installation and wiring of the sensors within the EV would be carried out with three components: display tablet, control card, and the housing for the capture of measurements and sensorization (Fig. 4).

4 Discussion

With the work carried out and the results obtained it is important, even more in Latin American countries, the implementation of sensitization and appropriation of technologies for its scalability and massification, with this it is taken into account that although the project, based on a scientific experiment must be supported by a series of trainings and disclosures to the population that refer to this knowledge within society. In the case of the aforementioned project, the team proposed two initial strategies for this work.

4.1 Social Knowledge Dissemination

Although the work carried out and the project are a clear example that should be implemented, massification and scalability to Latin American cities, it is important to carry out appropriation and awareness work toward the population that allows adopting this knowledge, technologies, and applications within society. In the case of the project, Ean University proposed two initial strategies for this social work.

4.2 Student Urban Mobility Service

The use of EVs by university students includes a structured reservation system that facilitates the management of vehicles resulting from an agreement with a private Carsharing venture called "Emobi" which rents only electric vehicles (Emobi, n.d.)

Fig. 5 Real ambient mobile laboratory

5 Conclusions

5.1 Mobile Environmental Laboratory as Part of a Smart City

The construction of this environmental laboratory, product of this project with capabilities to perform environmental measurements, calibration corrections, and analysis on the fly, allows the acquisition in real time of environmental variables of the environment. The use of EVs, taking into account its advantage of zero emission generation is the ideal element for analysis within cities, since there are no emissions or noise, nor excessive vibration, there is no interference in the measurements. Within the execution of the project presented in this article, capabilities were found to generate inputs in the identification of problems in lighting and noise issues, enabling its intervention in alliance with entities such as the National Public Lighting Association and the District Secretary of Environment. This is why the scalability of these types of solutions is vital for the transformation of cities into smart cities, to be evaluated, intervened, and revolutionized according to the population.

In the future, it is planned to implement this technique within freight transport and review the feasibility of a conversion analysis to hybrid or electric technology to suppress the amount of emissions generated by the sector, in addition to this, this laboratory will be implemented within a company of car sharing with the purpose of agglutinating as much data as possible and generating more extensive heat maps of the city.

Finally, it would be very interesting to make a scalability of this laboratory to as many cities as possible in order to impact the quantification of air quality within these laboratories and generate policies and projects that direct cities to true smart cities (Fig. 5).

Acknowledgements This project has been funded by the Ean University and supported throughout its process by Motorysa Colombia (Motorysa Corporativo, n.d.), entities committed to fostering an ecosystem of clean technologies within the cities of the world.

References

Alexeeff, S. E., Roy, A., Shan, J., Liu, X., Messier, K., Apte, J. S., et al. (2018). High-resolution mapping of traffic related air pollution with Google street view cars and incidence of cardiovascular events within neighborhoods in Oakland, CA. *Environmental Health, 17* (1), 38. https://doi.org/10.1186/s12940-018-0382-1

Amanatidis, S., Ntziachristos, L., Samaras, Z., Janka, K., & Tikkanen, J. (2013). *Applicability of the Pegasor particle sensor to measure particle number, mass and PM emissions* (No. 2013-24-0167). SAE Technical Paper.

Bouskela, M., Casseb, M., Bassi, S., De Luca, C., & Facchina, M. (2016). *La ruta hacia las smart cities: Migrando de una gestión tradicional a la ciudad inteligente* (Vol. 454). Inter-American Development Bank.

Castro, B. H. R. D., & Ferreira, T. T. (2010). Veículos elétricos: aspectos básicos, perspectivas e oportunidades (pp. 267–310). *BNDES Setorial, n. 32, set. 2010.*

Chan, C. C. (1993). An overview of electric vehicle technology. *Proceedings of the IEEE, 81*(9), 1202–1213. https://doi.org/10.1109/5.237530

Chen, C., Yao, A., Yao, C., Wang, B., Lu, H., Feng, J., & Feng, L. (2019). Study of the characteristics of PM and the correlation of soot and smoke opacity on the diesel methanol dual fuel engine. *Applied Thermal Engineering, 148*, 391–403. https://doi.org/10.1016/j.applthermaleng.2018.11.062

Emobi Emobi Carsharing. (n.d.). Alquiler de carros eléctricos por Minuto o por Día - emobi

Gaitán, M., Cancino, J., & Behrentz, E. (2007). Análisis del estado de la calidad del aire en Bogotá. *Revista De Ingeniería, 26*, 81–92

Isakov, V., Touma, J. S., & Khlystov, A. (2007). A method of assessing air toxics concentrations in urban areas using mobile platform measurements. *Journal of the Air & Waste Management Association, 57*(11), 1286–1295. https://doi.org/10.3155/1047-3289.57.11.1286

Motorysa Corporativo. (n.d.). https://www.motorysa.com/

Muntean, G. G. (1999). A theoretical model for the correlation of smoke number to dry particulate concentration in diesel exhaust. *SAE Transactions*, 316–322.

ODS Colombia. (2016). *Objetivos de Desarrollo Sostenible*. Colombia

Rojas, A. E., Mejía-Moncayo, C., & Rodríguez-Urrego, L. (2018, May). Smart block EAN: Ten scalable initiatives for a smart city. In *International Conference on Computational Science and Its Applications* (pp. 87–99). Springer

Rosolem, M. F. N. C., Beck, R. F., Santos, G. R., & Arioli, V. T. (2012). Bateria de lítio-íon: conceitos básicos e potencialidades. *Cad. CPqD Tecnologia, Campinas*, *8*(2), 59–72.

United Nations. (2015). Millennium development goals 2015

Wang, M., Zhu, T., Zheng, J., Zhang, R. Y., Zhang, S. Q., Xie, X. X., et al. (2009). Use of a mobile laboratory to evaluate changes in on-road air pollutants during the Beijing 2008 Summer Olympics. *1 foldr Import 2019-10-08 Batch 8*.

World Nuclear Association. (n.d.). Electricidad y Coches

Biophilic Water Criteria: Exploring a Technique to Develop an Environmentally Sustainable Biophilic Design Framework

Niranjika Wijesooriya, Arianna Brambilla, and Lina Markauskaite

Abstract

The built environment is obligated to respond to the global environmental movement, considering its significant environmental impacts. Environmentally Sustainable Design (ESD), one of the key strategies in this response, is based on frameworks that encourage designers to provide evidence of performances against predetermined targets. ESD frameworks are often used in sustainable design studios as educational tools, highly influenced by Green Building Rating Tools (GBRT) used in the industry. These tools primarily promote technologically advanced design solutions with limited opportunities for enhancing human–nature connectedness. In contrast, biophilic design has been identified as the missing link to reconnect humans to nature, as it promotes the use of natural processes. If these natural processes can ensure the building performances required by ESD, the design could harness the benefits of human–nature connectedness and enhanced building performance. Thus, a systematic technique to incorporate biophilic approach into the ESD framework is required, that is replicable and easily used in architectural design education. This paper outlines a systematic technique to develop an ESD framework for water efficiency that incorporates biophilic design and sustainability performance, applicable within architectural design education. The methodology relied on a two-step approach: (1) the development of the biophilic water criteria framework and (2) the evaluation of the criteria against four different and widely used GBRTs. This study revealed that the technique could assist in developing a framework to guide towards a biophilic and sustainable design. Furthermore, the evaluation showed that the proposed criteria were compatible with the GBRT. The proposed technique has a great potential to be adopted as a guide in architectural design education to bridge ESD and biophilia.

Keywords

Environmentally sustainable design • Human–nature connectedness • Biophilic design • Biophilia • Water efficiency • Green building rating tools • Architecrual design education

1 Introduction

The rising environmental concern requires adopting more sustainable approaches in every sector of human activities. The building sector, recognised to be one of the largest consumers of resources, responded through the development of Environmentally Sustainable Designs (ESD). The ESD approach differs from that of a conventional design by requiring to fulfil a set of predetermined targets and providing evidence. This approach requires a design framework or a success matrix that outlines the different criteria that a building must meet to be classified as sustainable. Architectural design education was also compelled to adopt this new approach through the introduction of sustainable architecture studios, which could incorporate ESD target frameworks. Almonte (2009) identified a crucial need for a new pedagogical methodology that facilitates the 'transfer of knowledge between sustainability-related building sciences and creative design in the architectural curriculum'. This transfer requires an understanding of how ESD is practised

N. Wijesooriya (✉) · A. Brambilla
School of Architecture, Design and Planning, The University of Sydney, Darlington, Sydney, Australia
e-mail: niranjika.wijesooriya@sydney.edu.au

A. Brambilla
e-mail: arianna.brambilla@sydney.edu.au

L. Markauskaite
School of Education and Social Work, The University of Sydney, Camperdown, Sydney, Australia
e-mail: lina.marauskaite@sydney.edu.au

F. Trapani et al. (eds.), *Advanced Studies in Efficient Environmental Design and City Planning*, Advances in Science, Technology & Innovation, https://doi.org/10.1007/978-3-030-65181-7_35

in the profession and how this transfer could be facilitated to achieve the anticipated targets.

Within the profession, Green Building Rating Tools (GBRT) dominate the ESD, providing the basis for the target framework and certifying a building's success in terms of sustainability (Xue et al., 2019; Gou & Xie, 2017; Illankoon et al., 2017; Cole, 2006). With the increasing demand for more sustainable buildings, GBRTs also evolved in various shapes and forms, revolutionising the building industry (Yudelson, 2008) and integrating a life cycle approach to sustainability. With GBRT's inception, hundreds of GBRTs were developed and adopted by different countries and organisations (Li et al., 2017; Illankoon et al., 2017; Sev, 2011; Cole, 2006), such as UK Building Research Establishment Environmental Assessment Method (BREEAM), Singapore Green Mark, US Leadership in Energy and Environmental Design (LEED) and China green building label (GBL) (Gou & Xie, 2017). The fundamental goal of these systems is to minimise the building's environmental impacts.

Buildings designed in response to GBRTs criteria are highly efficient and characterised by optimised building performances, yet the impact on human performance is being challenged (Kayihan, 2018). Indeed, the architecture of these buildings reflect the highly sophisticated technological features embedded in the ESD framework, with very limited opportunities to enhance human–nature connectedness. Thus, these buildings generally do not foster biophilia, which is the inherent connection between nature and humans that plays a pivotal role in improving physical and psychological well-being (Wijesooriya & Brambilla, 2021; Wilson, 1984), among many other benefits. The ability of existing GBRTs to support a holistic building outcome by merely providing evidence-based performance certifications is being questioned (Gou & Xie, 2017; Xue et al., 2019). Yeang and Spector (2011) argued that these GBRTs are usually based on checklists, focusing on building performance and overlooking the perspective of health and the environment.

Recently, we have witnessed a paradigm shift in research towards the human factors within sustainable design with a focus on biophilic design (Xue et al., 2019; Kayihan et al., 2018; Gou et al., 2014; Lam, 2011), moving towards a human-centred approach. Concurrently, GBRTs are also trying to incorporate biophilic design principles through their credit award criteria (BCA, 2012; BRE, 2013; USGBC, 2013) for enhancing indoor–outdoor relationships. However, Kayihan et al. (2018) highlighted that sustainability in traditional thinking is still more concerned about achieving targets, neglecting the human–nature connectedness in the building.

Biophilic design was identified as a strategy to enhance human–nature connectedness in the building sector (Kayihan, 2018; Abdelaal & Soebarto, 2019). Kellert (2018) presented a list of 70 attributes within six principles, developing a comprehensive framework to guide architectural design towards biophilic outcomes. A later development of this biophilic framework introduced a shorter version with three experiences, namely *direct nature experience*, *indirect nature experience* and *experience of space and place* and 25 attributes (Kellert & Calabrese, 2015). Browning et al. (2014) developed another comprehensive framework for biophilic design based on 14 patterns across three main categories: *nature in the space*, *natural analogues* and *nature of the space*. Xue et al. (2019) recently proposed a biophilic framework that also considers the existing GBRTs with six criteria: *biophilic infrastructure*, *sensorial design*, *biophilic setting and performance*, *transportation connectivity*, *work–live–play integration* and *green space place making*. However, these frameworks were focused on health and well-being, and human performance (Kayihan, 2018), while the building performance, usually promoted by GBRTs, is not explicitly apparent. It seems that both frameworks demonstrated their shortfalls in bridging human–nature connectedness and building performance, overlooking the potential benefits of the biophilic design principles. The impacts of these frameworks are more visible in architectural design education, where they have been sparsely used as education tools. A systemic technique, that is easily replicable is crucial to facilitate students to develop frameworks, that can bridge ESD and biophilic design.

Therefore, this study proposes a framework based on natural elements that can be visually experienced in buildings, while serving a second purpose of also being a performance criterion suitable for inclusion in more traditional ESD approaches. *Water*, *Sun*, *Air* and *Earth* was suggested in environmental philosophy as a basis to understand nature, a perception change that can also assist in overcoming environmental crisis (De Silva, 2016; Callicott & McRae, 2014; Callicott & Ames, 1989). Therefore, they were identified as primary natural elements to be incorporated in this study. We have inlcuded *Flora* as a fifth element that resonate with the use of vegetation in and around the building. This definition of nature as elements is commonly found in both Eastern philosophies (Hardy, 1853; Kalupahana, 1976; Upham, 1829) and Western philosophies. In this study, we used the following classification:

- *Flora*—plants, trees and all forms of vegetation within the building,
- *Water*—presence of water in the building for aesthetic and utilitarian purposes,
- *Sun*—visual attributes of daylight and perceived heat in the building,
- *Air*—elements related to the ventilation and air quality performance creating a sense of space and light, and
- *Earth*—materiality of the building that contributes to the biophilic experience.

This paper explores the possibility of developing a framework for *Water* as a natural element that combines the visual experience with the building performance. GBRTs usually include a parameter for water efficiency, and we, therefore, analysed these criteria in relation to the nature element, *Water*. Four key GBRTs were considered: LEED, BREEAM, WBS (well building standard) and LBC (living building challenge). Through a comprehensive analysis performed on the selected GBRTs, a list of Water Performance Criteria (WPC) was identified. Accordingly, Biophilic Water Criteria (BWC) was proposed, analysed through the sensory design strategies and evaluated against the selected GBRTs for potential associations and applicability through design elements.

The ultimate goal was to provide evidence that the natural elements constituting the basis of biophilia can be integrated into GBRTs to enhance both the building performance and the human–nature connectedness of the design outcome with the use of a systematic technique.

2 Methodology

The methodology included two phases: (1) Development of the BWC framework and (2) Evaluation of the proposed BWC against currently used GBRTs to ascertain their compatibility. The first is based on a bottom-up identification process, while the second relied on a comparative analysis.

2.1 Development of Biophilia Water Criteria Framework

Biophilic water criteria development required a systematic methodology that could be easily replicable to apply to different contexts as a suitable method to develop general biophilic design frameworks. The methodology comprising four steps is shown in Fig. 1. Step (1) identifies the comprehensive list of WPC, Step (2) identifies sensory design strategies for biophilic design, Step (3) identifies nature-based processes that can associate visual strategy and performance and Step (4) develops the BWC.

Step (1) Identification of the WPC performed through a criterion analysis following the methodology adopted by Al-Quawasmi et al. (2019) in a similar study on 11 GBRTs covering water efficiency criteria.
Step (2) Identification of visual design strategies for biophilic design through a literature search on biophilic frameworks and their applications.
Step (3) Identification of nature-based processes that can be used to achieve the building performance indicated in the WPC and that can be associated with sensory design strategies.
Step (4) Development of the BWC by associating the sensory design strategies and nature-based processes that support water performance.

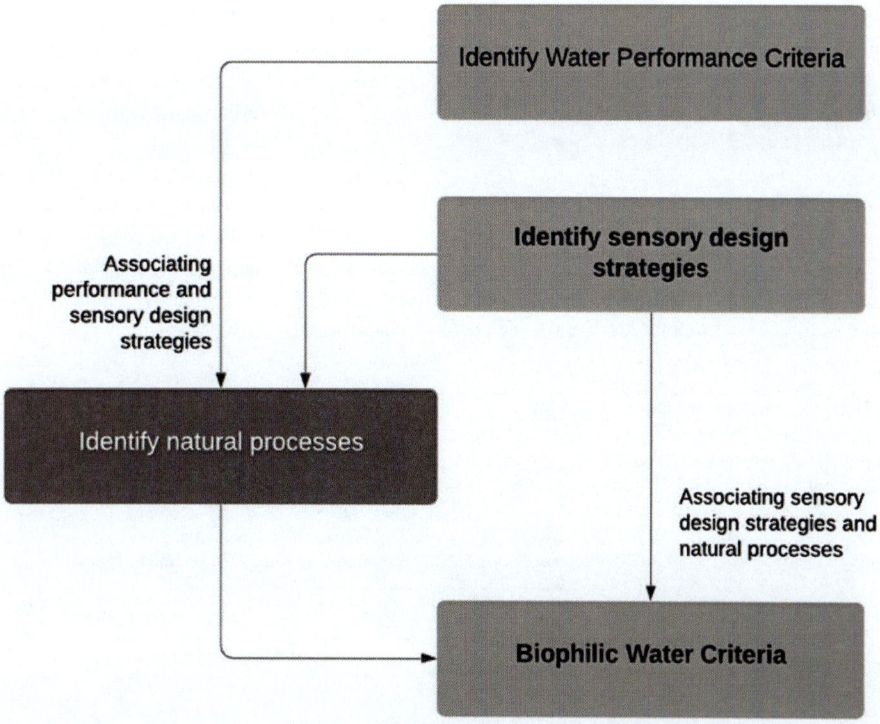

Fig. 1 Biophilic water criteria development process

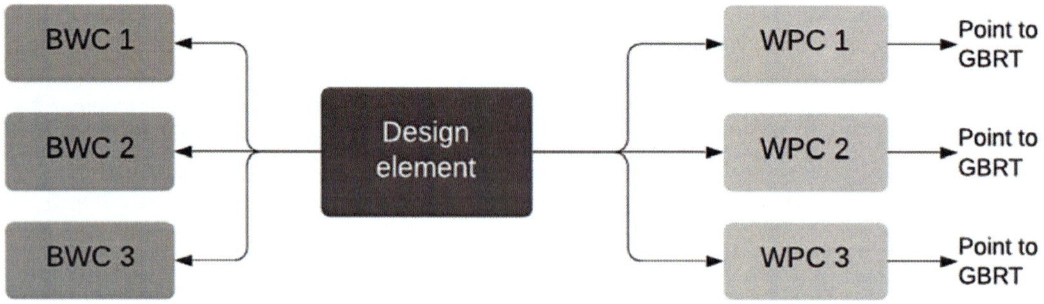

Fig. 2 Association mapping system used for evaluation linking BWC to WPC through design elements

2.2 Biophilic Water Criteria Evaluation against traditional GBRTs Criteria

The identified BWC was analysed and evaluated against the traditional GBRTs *Water* criteria through the use of an association mapping technique. This method allowed us to unveil the existing relationships and links between the design elements adopted, the BWC and the WPC (Fig. 2).

The mapping consisted of a systematic association of common elements that can fulfil the criteria within both the BWC and WPC. Whenever a association was found, a credit was marked in the GBRT that provided the criteria for the WPC. For example, a constructed wetland can be linked to both BWC and WPC; therefore, a credit was associated with the GBRT *Water* criteria *maintaining water quality*, which is the performance it supports.

Finally, a radar diagram was used to compare the full credit value within the GBRT and the credits that can be achieved through the BWC. This comparison provideed an evaluation of the applicability of BWC to contribute to increased building performances.

2.3 Selection of the GBRTs

There are numerous GBRTs currently developed and in use around the world. Consistent with the exploratory scope of the study, four frameworks were selected to cover a broad spectrum of varied criteria within the existing literature (Xue et al., 2019, Illankoon et al., 2017; Chen et al., 2015): LEED, BREEAM, Living Building Challenge (LBC) and Wells Building Standard (WBS).

LEED and BREEAM are two GBRTs with broad adoption, not only in their geographic area of origin (the USA and UK, respectively) but also internationally (Shan & Hwang, 2018; Xue et al., 2019). LEED was extensively investigated in the literature, as a pioneer in popularising the green building ratings. LEED V4 is the current version in use, introduced in 2013, that attempts to address the shortfalls of earlier versions by incorporating criteria for healthy buildings and people (USGBC, 2013; Xue et al., 2019).

BREEAM is also a popular and well-studied GBRT. BREEAM NC 2016, the latest version, was used for the study that also focused on a holistic approach to green buildings (BRE, 2013).

WBS is the leading GBRT with an emphasis on human health and well-being, identified by Morton (2015) as the first tool that incorporates those aspects. The tool is available online, and WBS version 2 was accessed on 3 August 2019 for the study.

LBC (ILFI, 2016) was selected based on its different approach, focused on a more holistic perspective that includes the occupancy stage. The credit system was defined as petals with 20 diverse criteria. LBC version 3.1 was used for the study. Table 1 summarises the details of the selected GBRTs.

Table 1 Details of selected green building rating tools adapted by Xue et al. (2019)

GBRT	Guide used	Buildings certified	Countries in use	Year
LEED BD + C V4	LEED V4 Reference Guide for Building Design and Construction	More than 94,000	165 countries and territories	2013
BREEAM International NC 2016	BREEAM International New Construction 2018 Technical Manual	566,727	79 countries and territories	2018
WBS V2.1	Well Building Standard V2 Polit	1166	43 countries	2018
LBC V3.1	Living Building Challenge 3.1	386	23 countries	

3 Results of Phase 1: Biophilic Water Criteria Framework Development

As outlined in Fig. 1, the BWC development adopted a four-step method discussed further in the sections below.

3.1 Water in GBRTs: Identifying a Comprehensive List of Water Performance Criteria

Water was consistently found within every GBRT as a primary sustainability criterion. The scarcity of this precious resource has emphasised water conservation in buildings (Al-Qawasmi et al., 2019), and the importance of water-saving strategies is underlined by the fact that two of the 17 United Nations sustainable development goals focused on water use (UN, 2019). With its broad presence in certification systems, water-related criteria have been widely investigated (Usman & Abdullah, 2018; Wong, Cheung & Mui, 2013). The literature review revealed that the presence of water efficiency measures across the GBRTs created a positive impact on successful water conservation strategies within buildings (Yasutoshi et al., 2013).

Al-Qawasmi et al. (2019) recently analysed eleven key GBRTs to understand the links and overlaps between the different frameworks, generating an extensive list termed *Comprehensive list of water criteria.* The study revealed that there are 81 criteria with 65 different indicators used to improve water efficiency in buildings. Table 2 reports these criteria and their relative weight on the GBRTs, expressed as a percentage of the total available credits.

Greenmark Singapore had the highest weight, while DGNB of Germany had the lowest. On average, 12.85% of criteria in these GBRTs were allocated for water-related performance measures in buildings. The variation of the weight can be due to contextual factors such as local water scarcity, climate and the specific objective of GBRT.

The *Comprehensive list of water criteria* (Al-Qawasmi et al., 2019) include *rainwater harvesting and management, surface water pollution, water recycle and reuse, water monitoring and control* and *outdoor water conservation* as environmental criteria. *Indoor water conservation, operations and management, water quality and health* and *reduce water consumption* were categorised as social criteria. However, their analysis omitted WBS and LBC, overlooking two major GBRTs, that may have missed some criteira.

In this current study, we used a similar approach to identify the *Water* performance list from the four GBRTs selected, reported in Table 3. The analysis included the presence of the *Water* criteria, as well as their potential impacts on the other performance criteria. For example, *heat island reduction* was selected from LEED since a water-covered area can impact this performance criterion.

The WPC include 26 performance criteria under *managing ecological impact, water conservation, water monitoring and control, air quality management, health and well-being promotion* and *water quality management.* This list provided an overview of the different performance criteria associated with the natural element of *Water.*

3.2 Identifying the Sensory Design Strategies

Water had an essential role in biophilic design, providing both a direct and indirect nature experience (Kellert & Calabrese, 2015). The biophilic approach was focused on providing a sensory experience dominated by visual perception (Wilson, 1984), as it establishes the first connection between humans and nature, which is then supported and further developed by the other sensory perceptions, in particular the auditory one. Existing biophilic design

Table 2 Comparison of water-related criteria in 11 GBRTs

GBRT	Country	Total criteria	Water criteria	% Water criteria	Indicators
Gstar	Australia	34	4	11.76	3
DGNB	Germany	141	2	1.42	2
BEAM	Hong Kong	67	11	16.42	10
CASBEE	Japan	101	8	7.92	7
GSEED	Korea	54	4	7.41	4
GSAS	Qatar	58	7	12.07	6
GMark	Singapore	22	7	31.82	4
Pearl	UAE	86	12	13.95	8
BREEAM	UK	52	8	15.38	5
LEED	USA	55	8	14.55	6
SBTool	International	115	10	8.70	10
		Total	81		65

Table 3 Comprehensive list of water performance criteria (WPC)

Managing Ecological Impact	Air Quality Management
Restore sites	Heat island reduction
Managing impacts on ecology	Thermal comfort
Ecological change and enhancement	Moisture Management
Long-term ecological management and maintenance	
Urban agriculture	**Health and Well-being Promotion**
Habitat exchange	Drinking water promotion
Water Conservation	Hand washing
Outdoor water use reduction	Safe and healthy surrounding
Indoor water reduction	
Cooling tower and process water use	**Water Quality Management**
Water-efficient equipment	Fundamental water quality
Water leak detection	Water contaminants
Rainwater management	Legionella control
Net positive water	Enhanced water quality
Water Monitoring and Control	Water quality consistency
Water metering	
Water monitoring	

frameworks contain a significant number of visual criteria (Browning et al., 2014; Kellert, 2011; Kellert & Calabrese, 2015). They generally fall into *direct use of water* (e.g. water-covered areas), *use of water features* and *indirect use of water* as by providing *views of water*, *use of water in images*, *artwork* and *installations*. The use of water in non-visual sensory experience, such as sound, is also suggested by Kellert (2011) for the biophilic attributes. Strang (2005) argued that 'there is probably more poetry, more literature and more art describing the form of water than any other aspect of the environment' indicating that indirect nature experiences with water can have an equal impact on human perception as direct experiences, supporting the use of indirect sensory elements as a sensory design strategy.

Therefore, for the development of the BWC, we used the following sensory design strategies:

- Water-covered areas
- Water features
- Water views
- Indirect sensory element.

3.3 Identifying Natural Processes

The WPC criteria were investigated to find evidence of natural processes that could be used to achieve the related performance. This step also allowed us to identify the nature-based technologies that can be adopted to develop the BWC.

Ostroumov (2017) conducted an extensive analysis of different natural processes used to purify water. Many of these techniques were allowed to make water elements visible. These 'microorganisms, phytoplankton, higher plants, invertebrates and fishes are involved in self-purification of aquatic ecosystems and the formation of water quality' (Ostroumov, 2017). Inclusion of plants can easily be used for water purification (Liang et al., 2003; Liu et al., 2019; Tang et al., 2019) either on a large scale or in compacted spaces (Vrhovšek et al., 1996).

Water used as a strategy for heat reduction was consistently investigated (de Gracia & Cabeza, 2015), especially in relation to its phase change properties. Water walls and roofs were used in the design for this purpose or to serve as a passive cooling strategy, contributing at the same time to the building's visual appearance. Direct use of water elements also contribute to heat reduction by evaporation as well as psychological thermal comfort.

Therefore, to develop the BWC, we used the following natural processes:

- Wetland water purification
- Water for heat reduction.

3.4 Proposed Biophilic Water Criteria (BWC)

To develop the BWC, we associated the sensory design strategies and natural processes, as shown in Fig. 3.

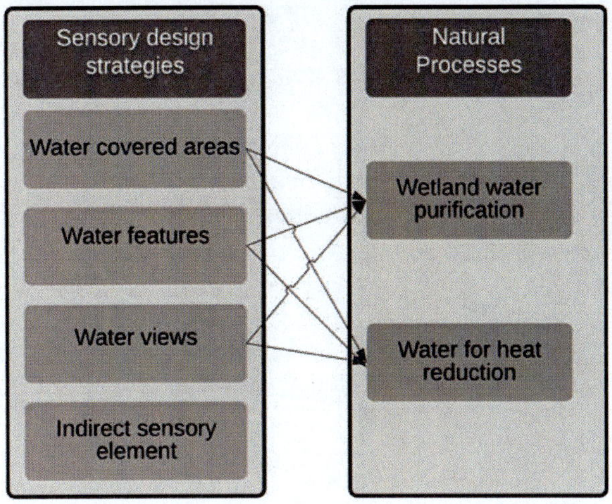

Fig. 3 Associating sensory design strategies with natural processes

As illustrated in Fig. 3, we established all the potential associations and then proposed criteria that could match those associations. Some criteria were proposed to match several associations such as the '*visible water elements for heat reduction*' that associate water elements with both *natural water purification* and *water for heat reduction*. Complete BWC items are given in Table 4.

4 Phase 2: Evaluation

The criteria identified in the previous step were linked to specific design elements. These design elements are associated with the four GBRTs criteria to evaluate the extent to which BWC can lead to successfully meeting building performance criteria. These steps are discussed in detail below.

4.1 Associating Biophilic Criteria with Performance Criteria Through Design Elements

A literature review was used to identify the design elements usually adopted to integrate *Water* as a visual strategy in the final architectural project. Specifically focusing on supporting the biophilic experience, Ryan (2016) suggested some key design elements that can be incorporated into buildings. They are rain chains, birdbaths, stepwells, modular waterfalls, fountains, reflection ponds, imagery and elements framing views of an existing water body.

Bringing water into buildings was recently rediscovered as a result of technological advancement, from the simplest water fountains to sophisticated lifelike waterfalls or digitally controlled interactive water features in urban public spaces. Changi Airport in Singapore is a clear example: the recently opened waterfall became an attractive feature to encourage tourism. Apart from these direct elements, architectural features mimicking water patterns, colours, textures and images also have been identified as contributing to the biophilic experience (Kellert and Calebrese, 2015).

Based on the natural processes discussed earlier, constructed wetland was a design element used for plant-based water purification (Vrhovšek et al., 1996). Similarly, heat reduction was achieved through water walls and water roofs. LEED (USGBC, 2013) provides guidance on potential strategies that also include many ways to improve water efficiency. Notably, the use of 'permeable paving', consisting of porous above-ground materials can become a natural element that assists in improving the performance of *Water*. The design strategies also included green infrastructure for water management that could be combined with the water covering as well. The use of plants and water was further explored recently in urban agriculture through hydroponics; Rodríguez-Delfín et al. (2017) highlight the potential social and economic benefits that an interesting water experience could provide. This idea could be further expanded by introducing water and plants for improved building performance as well.

In conclusion, we used the following design elements in the association mapping:

- Indoor direct water elements (modular waterfall, fountain, reflection pond)
- Outdoor direct water elements (rain chain, birdbath, stepwell, modular waterfall, fountain, reflection pond, wetland)
- Water imagery, artworks, installations

Table 4 Biophilic water criteria (BWC)

Biophilic water criteria			
W1	Enhanced water-covered area	W5	Visible features used for water quality management
W2	Raising awareness through water features	W6	Low maintenance natural landscape water features
W3	Visible water elements for heat reduction	W7	Indirect sensory experience through water
W4	Views of water within the building		

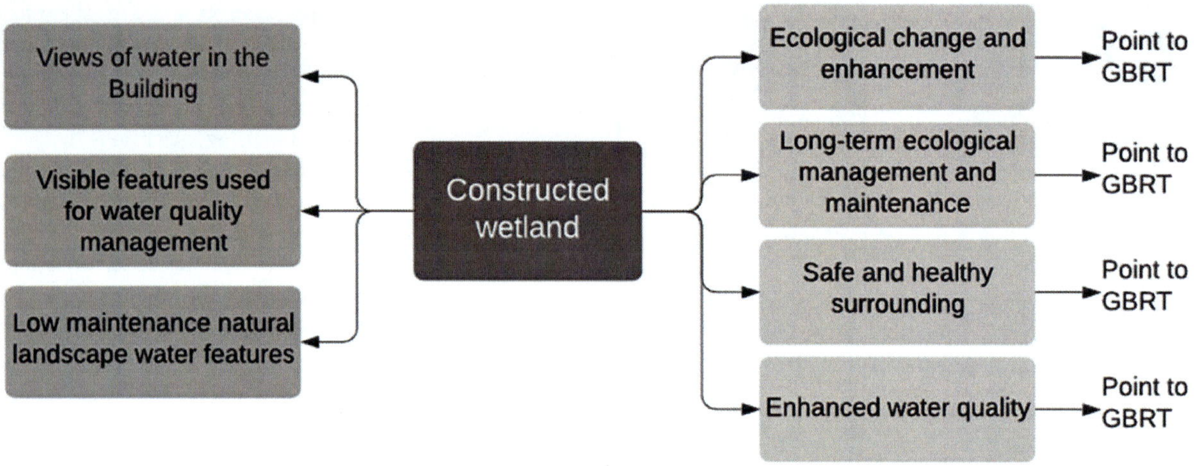

Fig. 4 Typical BWC-WPC association map for design elements

- Water patterns, colours and textures in architectural features
- Constructed wetlands
- Water walls and roofs
- Permeable paving
- Hydroponics.

These design elements were used as the link between WPC and BWC in the association map shown in Fig. 4.

Figure 4 shows the associations between WPC and BWC based on these design elements. This association mapping exercise provided a deep understanding of different design elements in terms of their performance in a building. For example, the use of a constructed wetland could provide an association for *enhanced water area* and *water quality management*. Furthermore, this tool can become useful in education to show how design elements could perform in terms of both biophilic experience and typical building performance. By further expanding the design elements, this could also expedite the design process by giving different design and strategy options to students as a starting point to develop their own design.

4.2 Evaluation Against Green Building Rating Tools

The four selected GBRT's *Water* criteria were analysed with regard to the potential credits that could be achieved by linking both biophilic and performance criteria in a unique design element. The BWC were benchmarked against the GBRTs using the association mapping tool.

LEED results (Fig. 5) show that the *restore sites* and *rainwater management* criteria provided the greatest opportunity to obtain certification points while *cooling tower and process water use* indicated no potential credits, mainly due to the high-tech nature of these criteria, far from the biophilic dimension.

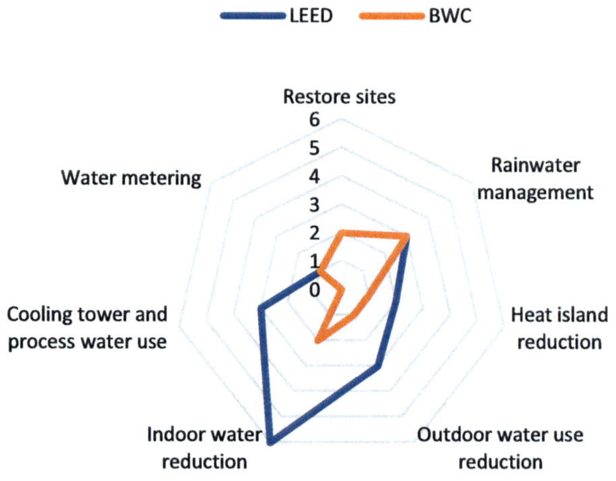

Fig. 5 Comparison of credit scoring for LEED criteria using BWC

The performance criteria in BREEAM that relate to water are quite broadly found across *heat management* and *water and landscape ecology*, and the comparison is given in Fig. 6. *Landscaping and ecological management* showed high potential for the proposed biophilic criteria, allowing for maximum points. *Water leak detection* seemed to be an area that had no potential to be scored through the newly proposed criteria. This area may be an interesting one to explore with regard to whether an exposed water distribution system can be used that makes leaks visible and easy to detect. *Safe and healthy surroundings* can also be promoted using the BWC with water quality-enhanced water features.

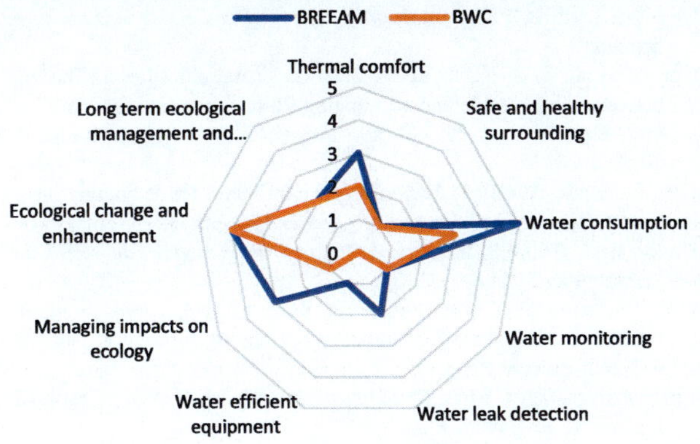

Fig. 6 Comparison of credit scoring for BREEAM criteria using BWC

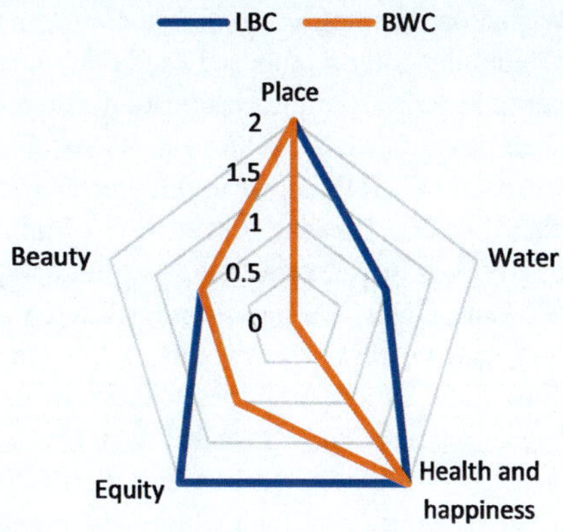

Fig. 8 Comparison of credit scoring for LBC criteria using BWC

Within the well building standard, points for water were awarded under five criteria, as shown in Fig. 7. However, there were three more prerequisites, *fundamental water quality*, *water contaminants* and *legionella controls*, that provided a strong basis for health and sanitation aspects of water use in the building.

Drinking water promotion, *hand washing*, *enhanced water quality* and *water quality consistency* represented potential opportunities to include visible elements while fulfilling the key building performance of the GBRT; thus, credit points were given to these criteria. *Moisture management* was not highly dependent on the visual elements and may not contribute substantially to the experiential quality. Therefore, *moisture control* was only given one point out of three with less potential achievements. The WBS had either one or fewer points in each category and highly relied on the evidence provided. Even though the potential was identified, actually achieving these criteria and providing strong evidence were challenges within a biophilic approach.

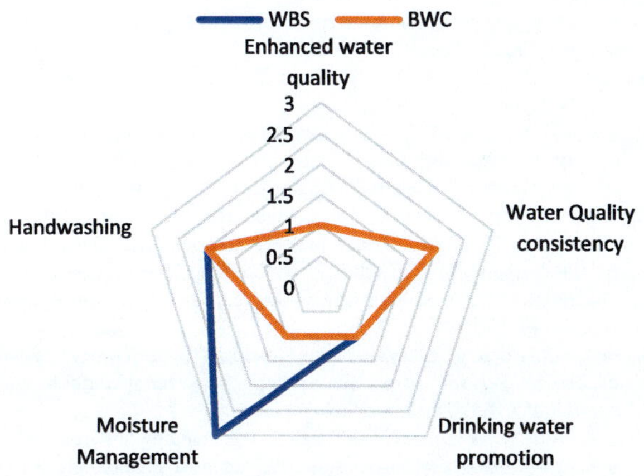

Fig. 7 Comparison of credit scoring for WBS criteria using BWC

The LBC has a unique format for a GBRT, providing points as petals. With its holistic approach, the potential *Water* elements identified by the association map can be found across different sectors. However, to make the comparison aligned with the other three selected GBRTs, the key criteria were selected, and water-related petals were given credit under them. This allocation provided a basis to understand the achievement of water-based biophilic performance across the total system. As shown in Fig. 8, *place, beauty, health and happiness* provided opportunities for biophilic water elements, while *Water* criteria did not show potential. Notably, *Water* criteria were based on a phenomenon of *net positive water* that has no association with the biophilic experience, indicating that, although there is a separate petal for biophilic environments, the inclusion of biophilic visual elements across the other petals were not promoted.

5 Conclusions

In this study, we used a systematic technique and developed a BWC, and we benchmarked it against four widely adopted GBRTs.

Using natural processes as the linking point between sensory design strategies and performance, played a major role in the technique, guiding a biophilic design without compromising the performance of the building. The comparison between the BWC and the four GBRTs revealed that each of the natural processes may interact differently with the building performance, in which the natural element and the performance may be in two different dimensions. For example, a water wall can provide a design element for *Water* contributing towards the visual experience where the heat reduction may not contribute towards the water performance, but on thermal performance, which would be included within *Sun* element as we classified for the study.

The association mapping using design elements revealed that many biophilic criteria proposed can be associated with a performance criteria from the certification schemes, highlighting their importance in achieving sustainable design strategies. All four GBRTs showcased a greater potential to achieve their water efficiency targets by adopting these proposed BWC. It was interesting to note that potential credit achievement was not significantly different between the four different GBRTs. One would expect the LBC to provide more biophilic opportunities with its focus on biophilic design, yet, the study revealed that the two other performance-based GBRTs, LEED and BREEAM, possessed an equal opportunity for inclusion of biophilic elements. The adopted technique was successful by deriving a framework for *Water* in a biophilic approach providing a method to develop similar criteria for other biophilic elements such as *Air* or *Earth*. With the growing concerns around achieving sustainability, a replicable systemtic technique to develop a ESD framework that encourages human–nature connectedness, while assuring building performances, may contribute immensely towards facilitating students in architectural design education.

References

Abdelaal, M. S., & Soebarto, V. (2019). Biophilia and Salutogenesis as restorative design approaches in healthcare architecture. *Architectural Science Review, 62*(3), 195–205

Al-Qawasmi, J., Asif, M., El Fattah, A. A., & Babsail, M. O. (2019). Water efficiency and management in sustainable building rating systems: Examining variation in criteria usage. *Sustainability, 11* (8), 2416

Altomonte, S. (2009). Environmental education for sustainable architecture. *Review of European Studies, 1*(2), 12–21

BCA. (2012). *BCA Green Mark Certification Standard for New Buildings. Building and Construction Authority, Singapore*. https://www.bca.gov.sg/EnvSusLegislation/others/GM_Certification_Std2012.pdf

BRE. Building Research Establishment. (2013). *BREEAM international new construction technical manual 2013*. https://www.breeam.com/discover/technical-standards/newconstruction/

Browning, W. D., Ryan, C. O., & Clancy, J. O. (2014). *14 patterns of biophilic design*. Terrapin Bright Green, LLC.

Callicott, J. B., & Ames, R. T. (Eds.). (1989). *Nature in Asian traditions of thought: Essays in environmental philosophy*. SUNY press.

Callicott, J. B., & McRae, J. (Eds.). (2014). *Environmental philosophy in Asian traditions of thought*. SUNY Press.

Chen, X., Yang, H., & Lu, L. (2015). A comprehensive review on passive design approaches in green building rating tools. *Renewable and Sustainable Energy Reviews, 50*, 1425–1436. https://doi.org/10.1016/j.rser.2015.06.003

Cole, R. J. (2006). Shared markets: Coexisting building environmental assessment methods. *Building Research & Information, 34*(4), 357–371

de Gracia, A., & Cabeza, L. F. (2015). Phase change materials and thermal energy storage for buildings. *Energy and Buildings, 103*, 414–419

De Silva, P. (2016). *Environmental philosophy and ethics in Buddhism*. Springer.

Gou, Z., Lau, S. S. Y., & Ye, H. (2014). Visual alliesthesia: The gap between comfortable and stimulating illuminance settings. *Building and Environment, 82*, 42–49. https://doi.org/10.1016/j.buildenv.2014.08.001

Gou, Z., & Xie, X. (2017). Evolving green building: Triple bottom line or regenerative design? *Journal of Cleaner Production, 153*, 600–607

Hardy, R. S. (1853). *A manual of Budhism, in its modern development*. Partridge and Oakey.

ILFI International Living Future Institute. (2016). *Living building challenge V3.1. Vol. 3.1*. https://www2.living-future.org/LBC4.0?RD_Scheduler=LBC4

Kalupahana, D. J. (1976). *Buddhist philosophy: A historical analysis*. University of Hawaii Press.

Kayıhan, K. S. (2018). Examination of biophilia phenomenon in the context of sustainable architecture. In: S. Fırat, J. Kinuthia, A. Abu-Tair (Eds.), *Proceedings of 3rd International Sustainable Buildings Symposium (ISBS 2017)*. ISBS 2017. *Lecture Notes in Civil Engineering* (Vol. 6). Springer

Kayihan, S. K., Güney, O. S., & Ünal, F. C. (2018). Biophilia as the main design question in architectural design studio teaching. *Megaron, 13*(1), 1–12

Kellert, S. R. (2011). Dimensions, elements and attributes of biophilic design. In: S. R. Kellert, J. Heerwagen, M. Mador (Eds.), *Biophilic design: The theory, science, and practice of bringing buildings to life* (pp.3–19).Wiley.

Kellert, S. R. (2018). *Nature by design: The practice of biophilic design*. Yale University Press.

Kellert, S., & Calabrese, E. (2015). *The practice of biophilic design*. Terrapin Bright LLC.

cLam, K. P. (2011). The human dimension in product and process modelling for green building design. In K. Yeang & A. Spector (Eds.), *Green design: From theory to practice* (pp. 79–88). Black Dog.

Li, Y., Chen, X., Wang, X., Xu, Y., & Chen, P. H. (2017). A review of studies on green building assessment methods by comparative analysis. *Energy and Buildings, 146*, 152–159

Liang, W., Wu, Z. B., Cheng, S. P., Zhou, Q. H., & Hu, H. Y. (2003). Roles of substrate microorganisms and urease activities in wastewater purification in a constructed wetland system. *Ecological Engineering, 21*(2–3), 191–195

Liu, G., He, T., Liu, Y., Chen, Z., Li, L., Huang, Q., & Liu, J. (2019). Study on the purification effect of aeration-enhanced horizontal subsurface-flow constructed wetland on polluted urban river water. *Environmental Science and Pollution Research, 26*(13), 12867–12880

llankoon, I. C. S., Tam, V. W., Le, K. N., & Shen, L. (2017). Key credit criteria among international green building rating tools. *Journal of Cleaner Production, 164*, 209–220.

Morton, J. (2015). Support occupant health with the WELL building standard. *Buildings, 109*(9), 19–26

Ostroumov, S. A. (2017). Water quality and conditioning in natural ecosystems: Biomachinery theory of self-purification of water. *Russian Journal of General Chemistry, 87*(13), 3199–3204

Rodríguez-Delfín, A., Gruda, N., Eigenbrod, C., Orsini, F., & Gianquinto, G. (2017). Soil based and simplified hydroponics rooftop gardens. In: F. Orsini, M. Dubbeling, H. de Zeeuw, & G. Gianquinto (Eds.), *Rooftop urban agriculture. Urban agriculture* (pp. 61–81). Springer.

Ryan, C. (2016, June 13).Water for wellbeing: Perspectives on design challenges. *Terrapin Bright Green*. https://www.terrapinbrightgreen.com/blog/2016/06/water-wellbeing/

Sev, A. (2011). A comparative analysis of building environmental assessment tools and suggestions for regional adaptations. *Civil Engineering and Environmental Systems, 28*(3), 231–245

Shan, M., & Hwang, B. G. (2018). Green building rating systems: Global reviews of practices and research efforts. *Sustainable Cities and Society, 39*, 172–180

Strang, V. (2005). Common senses: Water, sensory experience and the generation of meaning. *Journal of Material Culture, 10*(1), 92–120

Tang, V. T., Rene, E. R., & Li, Q. (2019). Application and performance of eco-bag revetment for water purification in lake environments: A case study from rural China. *Journal of Water Supply: Research and Technology-Aqua, 68*(2), 149–156

UN–United Nation. (2019). Sustainable development goals. United Nations. Available online: https://sustainabledevelopment.un.org/?menu=1300. Accessed on 7 June 2019

Upham, E. (1829). *The history and Doctrine of Budhism: Popularly illustrated: with notices of the Kappooism, Or Demon Worship, and of the Bali, Or Planetary Incantations, of Ceylon*. R. Ackermann.

USGBC. (2013.) LEED V4 Reference guide for building design and construction. U.S. Green Building Council, Washington, DC.

Usman, A. M., & Abdullah, K. (2018). Consolidation of water management and efficiency parameters for development of green building rating system. *Journal of Built Environment, Technology and Engineering, 4*, 284–294

Vrhovšek, D., Kukanja, V., & Bulc, T. (1996). Constructed wetland for industrial waste water treatment. *Water Research, 30*(10), 2287–2292

Wilson, E. O. (1984). *Biophilia*. Harvard University Press.

Wong, L. T., Cheung, C. T., & Mui, K. W. (2013). Energy efficiency benchmarks of example roof-tank water supply system for high-rise low-cost housings in Hong Kong. In *Proceedings of the 39th International Symposium CIB W062 on Water Supply and Drainage for Buildings* (pp. 349–359), Nagano, Japan.

Xue, F., Lau, S. S., Gou, Z., Song, Y., & Jiang, B. (2019). Incorporating biophilia into green building rating tools for promoting health and wellbeing. *Environmental Impact Assessment Review, 76*, 98–112

Yasutoshi, S., Satoshi, D., & Kanako, T. (2013). CO_2 Emission factor for rainwater and reclaimed water used in buildings in Japan. *Water, 5*, 394–404.

Yeang, K., & Spector, A. (2011). *Green design: From theory to practice*. Black Dog.

Yudelson, J. (2008). *The green building revolution*. Island Press.

Wijesooriya, N., & Brambilla, A. (2021). Bridging biophilic design and environmentally sustainable design: A critical review. *Journal of Cleaner Production, 283*, 124591. https://doi.org/10.1016/j.jclepro.2020.124591.

Design Indicators Based on Nature and Social Interactions to Enhance Wellness for Patients in Healthcare Facilities

Tarek Abou Ouf, Abeer Makram, and Shahira Assem Abdel Razek

Abstract

Healthcare facilities are complex and complicated structures that affect their inhabitants heavily. The effects may become both physical and psychological and address patients, users, and visitors. However, recent studies have shown that "coming back to nature" affects hospital stays significantly. Patients that are given access to nature and continuous connection to family, friends, and specialized social workers may eventually recover faster and become dismissed earlier. This research is based on an analytical overview of previous literature in relevance to the promotion of health and healing in healthcare facilities which have higher access to daylight and natural views. It accentuates previous research and emphasizes the importance of these factors as well as exterior connectivity for patients and their essentiality in accelerating the healing process. The research also presents the design indicators and criteria that future hospitals should follow to enhance the productivity of these institutions and promote wellness for patients. The results of this research came out as a group of indicators that address the effect of design factors on overall building and interior design of healthcare facilities.

Keywords

Hospital stay • Recovery • Stress factors • Design of healthcare facilities • Design indicators

1 Introduction

Patients in healthcare facilities often feel anxious and unsafe, leading to a delay in their overall healing and well-being. Previous research has confirmed that exposure, for long periods of time, to stress, can considerably reduce immune response in humans.

That is why hospitals should set a primary focus on the patient and develop standards and criteria to achieve an optimum healing environment, while taking into consideration, emotional, psychological, and physiological aspects in order to decrease the level of anxiety and stress as well as promote better healing, shorten the length of stay and support emotions through focusing on bringing nature, landscapes, and scenery to the patient for its adverse and positive effect on the nervous system; worn out and negative feelings decrease and feelings of enjoyment and pleasure increase. Studies have proven that watching pleasurable natural scenery affects physiological aspects like blood pressure and heart function (Ulrich, 1991). Recently, (Kuo, 2015) proposed a "central pathway" between nature and health which may inadvertently lead to enhanced immune function.

At the beginning of the early 1980s, medical research concluded that it was of utmost importance, to focus on patients' needs, not only from the medical service perspective but also from the physical, sensory, and psychological comfort. One of these points was set on improving way-finding systems in healthcare facilities and increasing the clarity of meanings communicated by space design (Totaforti, 2018).

With the beginning of the twenty-first century, the design of healthcare facilities took on a turn to include human aspects, in an attempt to support the productivity of these institutions. The length and capacity of stay of patients became an issue that was addressed to promote better health and living and enhance the quality of life of individuals undergoing hospitalization. With this idea in mind, healthcare facilities started taking on a more hospitable role and

T. A. Ouf
Department of Islamic Architecture, Umm El Qura University, Mecca, Kingdom of Saudi Arabia

A. Makram · S. A. Abdel Razek (✉)
Department of Architecture, Faculty of Engineering, Delta University for Science and Technology, International Coastal Road, Gamasa City, Mansoura, Dakhliya, Egypt

F. Trapani et al. (eds.), *Advanced Studies in Efficient Environmental Design and City Planning*, Advances in Science, Technology & Innovation, https://doi.org/10.1007/978-3-030-65181-7_36

the emergence of hotel-like hospitals started to arise. This concept helped designers model a comfortable hospital environment that focused on patients' needs and overall well-being.

In the next section, a literature review of the theories formulated on human benefit from nature and social interactions will be addressed, a description of two case studies of children hospitals will then be analyzed, and the methodology undertaken for this research after that will be underlined. The section following the methodology will focus on the results and discussion reached and will end with concluding the design indicators compiled for the design of healthcare facilities in relation to social and nature interactions.

2 Literature Review

The association between health and physical surroundings has long been established, and the theories of "Kaplan and Kaplan's; Attention Restoration Theory (ART)" (Kaplan, 1995) and "Ulrich's, psycho-evolutionary model" (Ulrich et al., 1991) are only two of the theories that deal with the influences affecting recovery from stress and attention fatigue from exposure to natural elements. Both theories are built around the effects that nature has on the restoration of human emotions' attention. Roger Ulrich, in his well-known study, published in 1983 (Ulrich, 1983) discussed the influence of natural views from even a hospital window after surgery or urban nature (the setting for that study was a park with trees) and concluded that it was beneficial in aiding and influencing the boost of physical recovery following adverse health conditions or surgery. The study also focused on the social well-being of the patient, feelings of safety and security and a sense of adjustment as well as their state of mind and mental fatigue (Netherlands, 2004).

Visual exposure to actual or simulated natural views results in efficient recovery from stress, emotionally and psychologically, within a period of three to five minutes at most, or as quickly as 20 seconds, in some bodily systems, and physically, over subjection to longer periods of natural views (Ulrich et al., 2006) According to a study done in 1984, it was found that just looking at greenery and nature reduces hospitalization time by 8%. In times of fatigue and stress, nature acts as a mediator, which can provide a sense of familiarity and connection with the everyday challenges faced, and this is also particularly important in healthcare settings, where patients and staff are continuously confronted with life-changing and often challenging events. That is why it is important to subject patients to natural elements and views, including interior gardens, water elements, and urban scenery and landscapes.

Studies in the 1990s concluded that people, in general, when asked where they preferred going when upset, tired, or stressed, found that a majority chose to go to a place of nature, whether natural or designed (Cooper Marcus & Francis, 1998).

In 1997, McRae conducted several consumer surveys on former hospital patients, and these surveys chose a sample that differed by age, location of hospital, and medical problems of the patients, but still found that the one thing common between the majority surveyed was their preference to access to nature in the physical environment of the healthcare facility, gardens, views of nature, balconies, indoor plants, and landscape scenery and pictures (MacRae, 1997).

This remark was not only true of patients, but of staff as well, nurses and members of staff claimed that they preferred being surrounded by natural open settings (Huisman, 2012), whether it was through bringing nature into the building, or through designing hospitals around open outdoor space or combining both aspects together through integrated hospital design that promotes physical activity and social interaction (Ulrich, 1984).

Nowadays, the evidence is even more clear as to the effect of nature on people's well-being, one study held in 2018 showed that spending just 120 min a week in natural settings may help boost immunity and relieve stress (White, 2019), while another one, published this year, 2020, discussed the benefits of visiting nature more than once a week helped promote health and well-being (Martin, 2020).

2.1 Designing for Patient Wellness: The Concept of Health Care Facilities Design and Development According to Patient Wellbeing

The term "Biophilia," coined by Edward O. Wilson in 1984, was defined as the need of incorporating natural elements to human needs, and it emphasized that humans need connections to the natural environment in order to survive and excel. This term, in turn, led to the coining of a new term that focuses on aspects of the natural world and its contribution to human health and productivity: biophilic design. Biophilic design is based on the idea that humans need constant connections to nature, even in modern settings, to improve their mental and physical well-being, through everyday common techniques such as opening up views to natural landscapes or placing planters at workstations. This conclusion led, in turn, to experimenting with biophilic design in healthcare settings to evaluate its impact on the hospital environment, as well as, its users. The principles of biophilic design include incorporating natural materials, natural light, views of nature and plants, and vegetation into the overall design setting.

Integrating biophilic design and access to nature in the design of healthcare facilities aims at improving patient health and satisfaction, while combining sustainability and human health goals (Silvis, 2013).

Given all the above, the current research sheds light on two of the principles of biophilic design; (a) natural light and its effect on patients and hospitalization rate and (b) views of nature, plants, and vegetation in the hospital setting (whether through using them inside the hospital itself and incorporated in the design setting of the healthcare facility, or by designing gardens embedded in the healthcare environment), and this research also examines the effect of social interactions on patient's well-being and state.

2.2 Natural Daylighting and Healing Process

2.2.1 Natural Lighting

In an attempt to better explain the importance of daylight and views of nature to healthcare personnel, architects and design experts at Green Build 2012 explained that subjection to natural daylight has been recorded to improve alertness, focus, and function, as well as alleviate depression. Natural daylight, which helps promote production of vitamin D (one of the main vitamins responsible for heart health, bone growth, and other biological effects), thus has a direct and positive impact on health. Natural daylight also helps regulate and modulate patients' heart rhythm, while maximizing user comfort, as well as providing an ambient and more relaxing indoor environment, with a better performance and higher productivity yield.

The presence of access to natural lighting in healthcare facilities is also advantageous to patients who have been admitted for long hospital stays; through offering them the ability to see the time of the day, through window watching as well as observing the weather and natural landscape along with natural light. This connection helps them feel that they are not isolated. A study, recently done, at the University of Texas in the United States of America, portrayed that comforting colors and natural views helped in promoting the healing process of patients. Daylighting not only reduced the length of stay and admission, but also helped in alleviating the feelings of isolation felt on hospital premises (Iyendo & Alibaba, 2014) This can be achieved through efficient placement of windows in the initial design, to incorporate windows with a view of the outside, in every space in the building, where possible. This design allows natural daylighting, through the design of windows, to become the main source of lighting for most of the spaces inside hospital buildings and create an interconnected link between the inside and the outside as well. This will ultimately lead to considering optimizing window placement to efficiently utilize exposure to daylight in patient rooms while designing hospital layouts. Design settings may also include outdoor decks, courtyards, and a rooftop track, which can be seen and observed from the inside and help in engagement with nature in the wellness setting.

The orientation of the building acts as a major factor in the aid of healing of healthcare facility patients. Optimum orientation utilizes airflow and sunlight into the building, helping the designer make better decisions as to the width and dimensions of windows and relevant openings (Aripin, 2006, 2007). Optimum orientation also helps admit natural light without glare and distortion through placing the building on the east–west axis and having the large elevation facing the north (Iyendo & Alibaba, 2014). It is also worth mentioning that most components of healthcare facilities need natural light that may be provided through skylights, large atria, or clerestory windows.

One of the main physical attributes, of hospital rooms design and layout, to patients, both physiologically and psychologically, is the presence of exterior viewing windows; however, the design and placement of efficient and effective windows depend on elements other than the view itself but rather on orientation, environmental considerations, and specific building considerations. The importance of windows to patients lies in the fact that most hospitalized patients have limited movement and thus depend on windows to provide them with daylight and other natural aspects. Internally, lighting should be focused and directed toward the patients' bed, in a manner as to not provide discomfort to the patient. Table 3 discusses the size of windows in accordance with the UK criteria for hospitals (Department of Health, 2013).

2.2.2 Design Example Adopted to Optimize Access to Daylight and Views of Nature

Hospitals in most European countries are required to have narrower building configurations and operable windows, and this enables them to have widespread access to daylight as well as utilize natural ventilation, when exterior weather conditions permit, and interior needs can be met without mechanical conditioning. Hospitals in these countries are typically laid out with narrow wings and/or contain courtyards and atria to provide occupants with

- Almost universal access to daylight,
- Views to nature and
- Natural ventilation.

Typical hospital typologies such as the street typology, terrace typology, perforated block, or the campus plan promoting these connections in patient care, work spaces, and public areas are based on narrow building footprints or articulated and penetrated plan forms. In some countries in Europe, for instance in Germany, the connection to the outside and providing a view of nature, in primary staff working areas, is mandatory by law due to its necessity for a safe and healthy environment (Behringer, 2011).

Research also concluded that patients who stayed in lucidly lit rooms were discharged earlier in comparison with those who stayed in dimly lit or duller rooms. It was found that patients hospitalized for severe depression and whose rooms were poorly lit or overlooking spaces in shadow had an average increase in the length of their stays by around 3.67 days more than those who were admitted to sunny rooms or with a better outside view. This study also found that patients exposed to more intense sunlight suffered from less perceived stress, complained of less pain, and were given 22% less painkiller medications during the course of an hour (resulting in an overall cost reduction of 20% for pain medications) (Ulrich et al., 2004). It also found that natural light helped improve sleep, decrease agitation, and aid in the treatment of hyperbilirubinemia in infants (Ulrich et al., 2004).

2.3 Gardens in Healthcare Environment

As mentioned before, evidence-based studies have found that gardens and places of nature are often getaways where people go when they are feeling upset, stressed, or tired. Gardens, in general, provide a relief to most pressing matters and bestow a calming and relaxing effect on visitors. Psychological research shows that green as a color has a calming, soothing, and healing effect as it symbolizes nature.

Gardens, in healthcare facilities, have been proven to, not only, reduce stress and improve patient outcomes, but also supply healthful and regenerative calming natural views, through:

- Fostering access to social support (enabling meeting with family and friends in other contexts other than the patients' room or ward)
- Enabling patients with a sense of control regarding stressful settings as well as providing positive chances of escape (Cooper Marcus & Francis, 1998; Ulrich, 2001).

After assessment of post-occupancy data evaluations of hospital gardens in the state of California, USA, researchers found that patients were not the only ones who used these gardens to promote healing; nurses, healthcare workers as well as other staff members also used these gardens as a much needed escape to recuperate from everyday stress (Cooper Marcus & Barnes, 1999). Results from other post departure research found that admitted patients and family members, who used hospital gardens, reported reduced stress levels and positive mood changes (Whitehouse, 2001), leading to heightened satisfaction with overall quality of care from both patients and family members (Ulrich et al., 2004). "Healing gardens," as they were termed, offered a way for patients and healthcare workers and personnel to step back, relax, recover, and revive. Patients and staff, alike, reported that using these hospital gardens helped them to better focus and calm down, while families of patients claimed that the time spent in these gardens allowed them to better cope and help loved ones.

In a healthcare therapeutic garden, based on several studies, acknowledgeable positive results of hospital recovery and discharge are related to the patient's leisure time with outdoor open nature, as well as, the patient's presence in green open spaces and landscapes. These gardens may either be shared or private; however, the private ones are considered to be the best type of meditation gardens, leading to positive healing impact, due to the fact that the user has his own space and utilizes his own time to access the most healing benefits of the garden (Momtaz, 2017).

Strategies for the design of "healing and therapeutic gardens" vary from simple garden layouts to gardens that feature a home to diverse systems and habitats. This diminishes the "sterile" sensation of a hospital, which may in itself act as a stressor to patients. This "sterility" can be mitigated, creating something more pleasing to the senses through seeing lush green surroundings, listening to the sound of flowing water, and rustling leaves or even through the natural smell of plants (Pouya & Demirel, 2015).

As a whole, the design of healthcare gardens prevails with green materials; hardscaping is minimized to only one-third of the space occupied and plant materials dominate the garden. With the spread of greenery, the minimization of hardscapes, and the softening of landscapes, patients may be encouraged to exercise; walking has been found to lower severity of depression. The different opportunities present in the garden, aiding in elevated mobility and accentuated exercise, may support different choices and facilitate useful distraction through direct connections with nature (views, scenery, shade, or sunlight) or indirect connections (rustle of leaves, soft breeze, sound of water or birds) (Momtaz, 2017).

Thus, "therapeutic gardens":

- Aid activity, mobility, and movement
- Aid in the decrease of stress
- Aid in the building of convivial and emotional bonds
- Accentuate sensory benefits from natural elements (sun, wind, rain, breeze, smell of soil, trees, shrubs, and flowers, the sound of birdsong, the rustle of leaves, trees and water and the sight of scenery, butterflies, moonlight, and more (Pouya & Demirel, 2015).

Healing gardens have many benefits and when they cannot be incorporated into a design, plants, water, and other natural elements with restorative qualities can take their place. Hospitals are often disorienting and confining. Patients are often under a constant reminder of their illness with various protocols and equipment. Plants (especially entire gardens) reduce the feeling of isolation. The gardens encourage patients to feel part of a larger group and offer an opportunity to focus on living plants that represent hope, growth, and renewal (Beggs, 2015).

With the above design decisions in mind, it is recommended that the leading landscape to the healthcare facility itself and the entrance be equipped with plants and water elements that act as a mini-oasis to help calm the various nervous thoughts and feelings that arise before entering the facility and allow for a more relaxed waiting period before being admitted to the physician or hospitalized (Beggs, 2015), and this recommended design criteria result from the evidence that nature extends beyond visually aesthetic pleasure, but also encompass a range of psychological well-being benefits as well (Ulrich, 1984). Some of these features may be

- A garden that is soothing to patients, family, and staff and that is easily accessible
- Aquaria in waiting areas (often distinguished with high levels of stress and anxiety)
- Atria with green elements and details
- Fountains or the sound of waterfalls to intensify the soft sound of water, and
- Calming art that is accessibly mounted for access to bedridden patients (Ulrich, 2003).

Well-designed hospital gardens emphasize the importance and essentiality of social interaction and its benefits which through providing pleasant views, which aims at decreasing stress and improving therapeutic results. Designing a hospital garden should focus on location, accessibility, patients' needs, and priorities.

2.4 Integrate Nature in Interior Design

The above design parts are all subject to availability of an element which may contribute to the healing process, whether through availability of a space allocated for the design of a garden inside the facility, through which an integrated design can be implemented. However, present functioning healthcare facilities may not have the leisure of a space that may be converted into a garden. That is when the use of indirect scenery may be of use. This can be achieved through imagery of nature. The same aspect of gardens arises in this indirect layout, just as gardens aimed at engaging the senses, so does indirect subjection to nature aim. Color, smell, touch, and sound of nature as perceived may be incorporated into the facility's setting. Horticultural therapy emphasizes the positive impact that close contact and care for plants have on health and healing outcomes. The addition of fragrant plants, variations in color and texture, and opportunities to become closer to elements of nature enhance patients' sense of calm, act as a positive distraction to his current state, and bring back experience-based memories of nature (Ulrich et al., 2004).

These elements may be, for example,

- Green walls, green roofs, plants in rooms (Totaforti, 2018)
- Images of nature, videos, pictures, or slides
- Virtual connection with nature, represented through use of modern technology, whether augmented or virtual reality
- Accentuating the senses, through the sound of leaves or that of birds or the smell of flowers in the air.

2.5 Social Support and Positive Social Interaction

2.5.1 Social Contacts

It is without doubt that social relationships and interactions are essential for health and prosperity, and that the absence of strong connections is directly related to development and progression of several diseases as well as adverse health-threatening conducts such as smoking, drinking, substance abuse, and even gang involvement (Beggs, 2015).

Approximately, 49% of all family-related visits to patients in healthcare facilities took place in the patients' room, and family members and loved ones spend most of the hospitalization time at the patient's bedside, up to several hours a day. "Close" family members' visits had a significant effect on patients' well-being and their mental status. Although it cannot be stated that all patients improved after visits from family members, some experienced a decline in their mental status, it can be stated that those who improved had access to public areas that contained simple green common spaces with trees or grass (Netherlands, 2004) and had access to appropriate facilities which encouraged social interactions and a chance for social support, this included, provision of appropriate and comfortable seating in patient rooms, overnight accommodations, for close family members, or next of kin, inpatient rooms and even a place to accommodate small groups that wish to visit (Huisman et al., 2012). Design interventions also include comfortable, pleasant waiting areas; convenient access to food, drink, telephones, and restrooms; and accessible areas for sitting and socialization.

2.5.2 Social Support

The importance of social support lies in the caring and tangible assistance and emotional endorsement received by the person (White et al., 2019). It has been concluded, from an occupational research, that those who receive higher social support often feel less stress and better health. On the other hand, introverts and those who are socially isolated experience higher rates of illness and reduced recovery. This is evident in cardiac patients whom received a higher level of social support and was noticed to recover quicker from heart attacks and better survival rates. It has also been witnessed that patients recovering from myocardial infarction had better recovery outcomes when receiving social support. The same was concluded for patients recovering from metastatic cancer. Evidence-based benefits of social support across health-relevant contexts suggest that focusing on a design that promotes social support will in turn reduce stress and improve outcomes. (Ulrich et al., 1991) This can arise from the intersection between public spaces and a homely environment with both the comfort and amiability provided by each space (Mogensen, 2011).

2.5.3 Social Interaction

Social interaction is essential to health and well-being, and community interaction zones where individuals can bond and get inspiration and support from others can, in turn, achieve a social bond between those who share similar views. This is an environment where people gather with a mission to alleviate the overall wellness of the community through positive social interaction. Support, whether parental, familial, or social, plays an important role in helping patients recover from illness. Evidence-based research shows that increasing of social interaction levels can be achieved by providing lounges and waiting rooms with comfortable movable furniture arranged in small flexible groups (Hilal, 2006).

Variety of Spaces

A variety of spaces while designing healthcare facilities may better promote the healing process, where the patient is given different choices, ranging from different outdoor spaces in which to explore or indoor spaces with natural views and works of art. This, in turn, results in an increased sense of control, which leads to lower stress levels. Areas of solitary occupancy may allow the patient to "get away in a controlled area" from the sterilized environments of the hospital. Adjunct areas for family members or support staff to congregate may also help provide social and emotional support to the patient (Pouya & Demirel, 2015).

It is also worth establishing that single-bed inpatient rooms also make it easier for visits from family and friends and helps promote intimate and serious conversations without disturbing others. With this in mind, the social support needed by the families of hospitalized patients is often forgotten, the feelings of anxiety, worry, and stress that arises from not knowing what is wrong with a loved one is sometimes overlooked, and the stress from the patient from suffering from serious and life-threatening illnesses is transferred to families who often also require social support in order to adapt. This may be derived from other family members and intimate friends but may also be highly advantageous from social interaction with families of patients with similar experiences (Ulrich, 2003).

2.5.4 Orientation

Research shows that the design of seating arrangements and patterns played a powerful hold over the amount of social interaction between patients in a patient dayroom area setting. Chairs, which had been positioned shoulder-to-shoulder along the walls, strongly suppressed social interaction, while those that were arranged around small tables in the middle of the room increased interaction, especially among socially inclined patients (Huisman et al., 2012).

The above can also be applied to the design of "healing gardens" that encourages conversations and social networks between groups of patients and visitors (Cooper Marcus & Barnes, 1999). These design features may also be made use of in waiting areas and dayrooms; this can be achieved through creating transparency in public spaces by designing floor to ceiling windows or by engaging patients and visitors in artwork and graphics.

2.6 Examples of Hospitals Accommodating Nature and Social Interaction

The following two examples are of hospitals that adopted design criteria to promote healing through providing daylight, natural scenery, and social interaction and support. The first one "The Dell Children's Hospital" is located in Austin, Texas, USA, and is the first hospital to achieve the Platinum LEED. The design of the hospital embraces the concepts of sustainability and utilizes these concepts in creating a healing environment through design features like natural lighting, evident in over 80% of the hospital's interiors, while natural ventilation and monitoring systems are embedded in the design to ensure high indoor air quality. The hospital is also designed with natural wood materials, bricks, and stone, giving it a natural look. There is also an onsite garden, with most of the hospital having its view on it. The interiors also feature over 700 original pieces of art (Table 1).

The other hospital is in Melbourne, Australia, "The Royal Children's Hospital." The hospital building's formal

Table 1 Dell Children's Medical Center

<table>
<tr><th colspan="3">1. Example (1) DELL CHILDREN'S HOSPITAL</th></tr>
<tr><td colspan="3">"Dell Children's hospital" utilizes natural and aesthetic factors in order to promote healing both psychologically and physiologically to the children they care for, as well as, to the families through providing an environment to help people get well.

The hospital has an articulated building footprint and incorporates several courtyards to increase the perimeter wall. As a result, 36 percent of the building space is located within the daylight zone. (Aripin, 2007)

Artwork and sculptures, are carefully placed throughout the hospital, along the walls and reception areas. Design specialists have described the healing garden at the Dell Children's grounds as "breathtaking".

The design of the hospital aims to optimize access to daylight, views to nature and social interaction (Central, 2019) as shown in Figure 36.1.</td></tr>
<tr><td>1</td><td>A hub and spoke design incorporates seven interior courtyards that bring natural light throughout the building</td><td>Dell Children's Medical Center of Central Texas
(https://www.architecturalrecord.com/articles/12178-dell-childrens-medical-center-of-central-texas, 2019), (https://segd.org/dell-childrens-medical-center, 2019).</td></tr>
<tr><td>2</td><td>The courtyards represent the ecosystems found in the hospital's</td><td rowspan="8">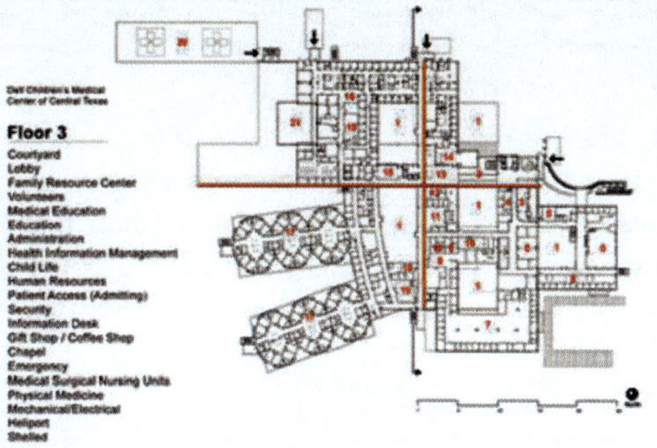

Figure 36.1: Plan of Dell's Children Hospital showing number of courtyards
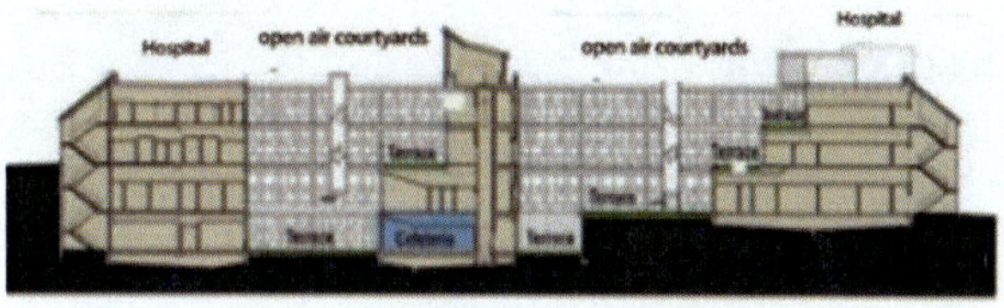

Figure 36.2: Section of Dell's Children Hospital
Figure 3: Total area: 180,962 SF, Daylit area: 65,374 = 36% (13)
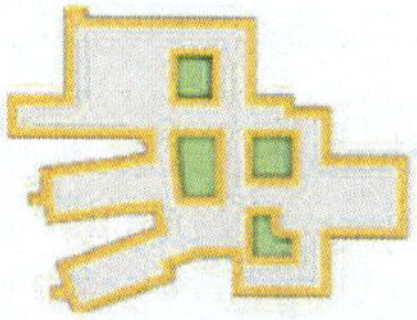</td></tr>
<tr><td>3</td><td>A multi-level "Healing Garden" with an embedded labyrinth, a reflecting pond and bridge.</td></tr>
<tr><td>4</td><td>"Sensory Garden" that uses plants and other elements with unique textures, colors, smells and tastes to stimulate the senses to promote rapid recovery.</td></tr>
<tr><td>5</td><td>"Butterfly Gardens" planted with vibrant florals that attract butterflies to better awaken the senses with the sound of fluttering and vivid colors.</td></tr>
<tr><td>6</td><td>Global and Local art, with pieces chosen as much for the collection's clinical healing power as they are for their beauty.</td></tr>
<tr><td>7</td><td>A natural palette of indigenous plants, in line with the overarching theme of our "green" building to complement the landscaping of the gardens</td></tr>
<tr><td>8</td><td>Family Lounges are located directly outside each patient unit on 2nd, 3rd & 4th floors. They're available 24 hours a day.</td></tr>
<tr><td>9</td><td>The Family Resource Center is located near the main entrance on the 3rd floor. The Center offers resources ranging from computers to books and DVDs, and a medical librarian is available during daytime hours.</td></tr>
</table>

10	Circular eight-room pods have a nursing station just outside the doors of every patient room.
11	Family Sleeping Accommodations are part of every child's room.
12	Playrooms for kids and siblings 10 and under are located on the 2nd, 3rd and 4th floors.
Dell Children's Medical Center	

Figures 36.4 & 36.5: Interior Courtyard providing play area for children and views for rooms facing the courtyard (Chaudhury, et al., 2003).

Figures 36.6 & 36.7: Healing Garden

arrangement, along with the internal and external spatial distribution, was assembled to promote a regenerative and restorative healing environment for children and their families. The hospital is built around a large park and is coined to be a "park in a hospital, and a hospital in a park" (Bensalem, 2002) (Table 2).

3 Methodology

3.1 Data Collection

This research is based on an analytical overview of several theories that people in general are affected directly by their interaction with both nature and social interacommunications, which has led to a focus that patients in healthcare facilities would heal better when subjected to nature, daylight, social support, and interactions.

3.2 Methodology Used

The methods used are reviews of eminent published research, evident in the literature review, and analysis of pre-established and designed healthcare facilities that follow the above literature. This is carried out through deductive and analytical methodology. It also aims to emphasize the benefits of this material to analyze the indicators needed for the design of these facilities to empower better functioning and aid in rapid healing and recovery of patients through decreasing stress and anxiety and increasing relaxation and positive social interaction.

4 Results and Discussion

4.1 Design Indicators Based on Nature and Social Interaction

After a thorough look at the above research, a set of design indicators may be discussed. Table 3 shows the indicators concluded that the indicators were grouped to address the effect of design factors on overall building and interior design of healthcare facilities.

5 Conclusion

This research contributes to the promotion of the aspects needed to improve patients' overall well-being, based on results from previous psychological studies, through developing design indicators and solutions to be able to incorporate them in the process of architectural design through two important elements, namely the connection to nature and social interaction. The research is targeted to help introduce indicators that may be embedded in the code of hospital design to promote healing, well-being, and faster discharge rates.

The research concludes that efficient design of the building and its structure, to aid in interaction with nature,

Table 2 The Royal Children's Hospital in Melbourne

2. Example (2) THE ROYAL CHILDREN'S HOSPITAL IN MELBOURNE

The inpatient building of the royal children's hospital in melbourne is designed in a star shape, connecting the rooms to the park. Around 20% of the rooms look out onto the courtyards, while remaining 80 percent have park views (Bensalem, 2002)

The star shape has these wedge-shaped fingers resulting in a more spacious corridor, ideally a place for parents and kids to come out and socialize. An open space lounge is placed at the end of one of the wedges for families to spend time and has access to an open view of the park.

The use of narrow footprints for the clinical buildings of the hospital, sheds ample light and makes natural sunlight accessible to all areas of the hospital.

The exterior glass of the hospital is specially designed to allow the activities on the ground floors to be witnessed from the patient's bed. 85% of all patient in rooms are of single occupancy, designed to be calming and comforting. Procedures involving the medical practice is typically conducted far away from the in patient room, whenever possible, so that the room remains a place of rest and family time. Optimized design to access to daylight, views to nature and social interaction: Figures 36.9-36.13.

1	The hospital is surrounded by the parkland of Royal Park, with views of trees and lots of natural light.
2	The hospital is family-centered with improved facilities and places for accommodation and rest for parents
3	It includes a fun, interactive, child-friendly zone, both indoors and outdoors.
4	The place making and wayfinding techniques used is ideally situated to make it easier for caregivers, and visitors to find their way around the premises without the stress of losing their way
5	The building is of northern orientation to be able to make optimum use of natural light, while more than seventy five percent of all the inpatient rooms have a view of the park located on the grounds.
6	Open access to the garden courtyards as well as to Royal Park.
7	Maximizes access to daylight, gardens and views throughout.
8	Single rooms with en-suite bathrooms, allowing children and families extra space and privacy.
9	Children's playgrounds have been included, some set in central courtyards that flood rooms with daylight.
10	The hospital's central hub is called Main Street, with direct links to all areas of the hospital, it encompasses a two-floor aquarium, and large-scaled works of art and sculptures.
11	A central 'street' allows intuitive way-finding and creates a social heart for the hospital.
12	The hospital's central hub is called Main Street

Royal Children's Hospital, Melbourne

The Royal Children's Hospital in Melbourne Australia

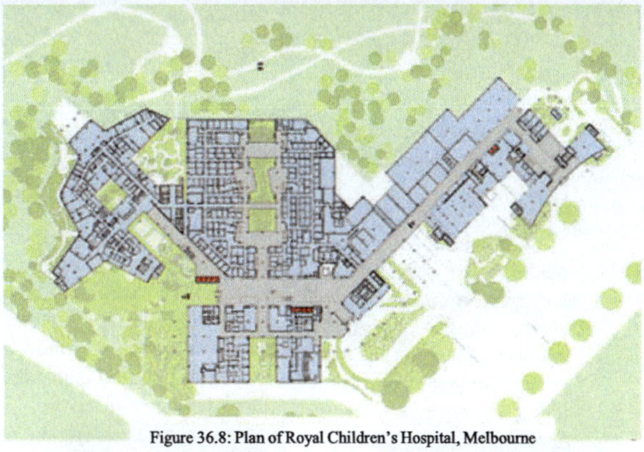

Figure 36.8: Plan of Royal Children's Hospital, Melbourne

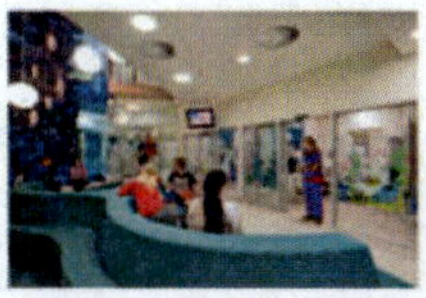

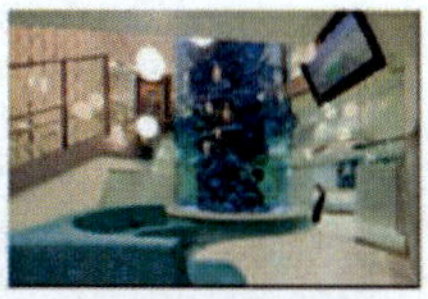

Figures 36.9 & 36.10: The outpatient courtyard featuring the aquariam and other activities

Figures 36.11& 36.12: The multi –faith courtyard

Figure 36.13:Patient Bedroom and family zone

Table 3 Wellness design elements based on nature and social interactions

Wellness Design Elements Based on Nature and Social Interactions			
Design Indicators Scale	***Design Factor/issues***	***Design Indicators***	
Building Design	***-Visual access to nature:*** Viewing nature in indoor and outdoor (external view) settings like a meditation garden, greenery, flowers, or water.	***SITE LOCATION:*** Visual access to nature &direct views of nature	Location with External views: green space and good natural views. Optimize the surrounding scenery.
	- Being able to get fresh air/ go outside. **- Optimize access to sunlight** - Ensure spaces for patients are filled with daylight through orientation and shape of a building which dictates the potential availability for daylighting. (Behringer, 2011)	***FORMING MASS:*** Narrow building footprints or articulated and penetrated plan forms (Behringer, 2011) courtyards and atria, configurations and operable windows that enable them to have widespread access to daylight and utilize natural ventilation.	Master plan: Narrow plan design: maximizing patient's exposure to fresh air and daylight (Behringer, 2011)  Figure 36.14: Typical hospital typologies used in Europe: street typology, perforated building block and campus plan.
	- The orientation decisions made require unique skin and apertures design for eastern, southern, northern and western exposures such as exterior shading systems including vertical or horizontal blinds, sliding shutters and low-emission glazing.	***ORIENTATION PLAN:*** The design of a daylight hospital (Behringer, 2011) must address local climate conditions. The orientation and shape of the inpatient units are designed to maximize rooms facing daylight with no direct sun. The majority of the reviewed hospital orient their inpatient units along the east/west axis such as the Dell Children's Hospital in Texas and the KIZ Children's Clinic in Austria.	 Figure 36.15: Dell Children's Hospital, Texas Latitude 30°18'8.93"N Longitude 97°42'24.29"W(17) 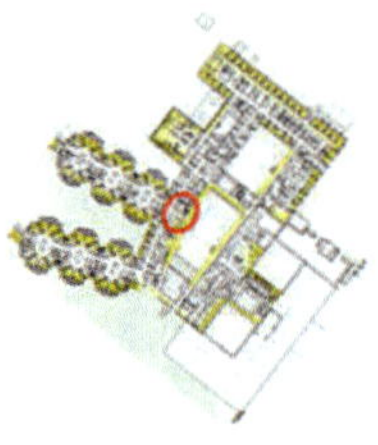 Figure 36.16: Orientation of inpatient rooms
	- Openings allow the exchange of light and air and provide visual connections to the outside.	***Building Height:***	Horizontal solutions offer better connection with the surrounding environment.
	-Location of hospital beds near to windows, ensures provision of maximum access to daylight without the feeling of discomfort, while supplying an outdoor view for patients lying on the bed. (Joarder, 2011)	***The building skin:*** ***Window Design:*** The building skin creates a filter between the interior and exterior. It also must control sunlight penetration to minimize unwanted heat gain, lighting and glare. (Joarder, 2011)	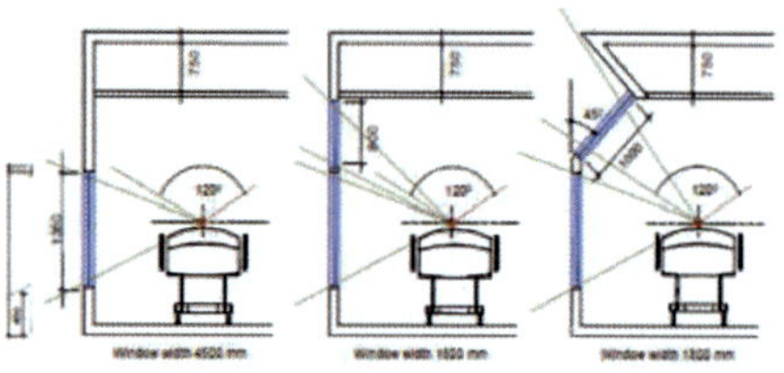 Figure 36.17: "Sky window configurations performed better than traditional typical standard hospital window configurations" (Joarder, 2011)
		Urban Areas, building cluster healing spaces: Garden, balcony, Open window, Courtyard or atrium containing interior landscape and fountains.	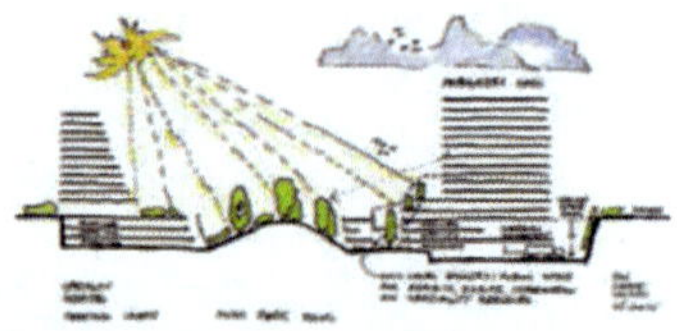 Figure 36.18: Courtyards and, Gardens were arranged to catch the sun's rays.

Interior design	***Social support &interaction*** - Physical access to Nature to help recovery through accessible gardens. - Contact with other people: friendly, personal spaces and invite to walk or other forms of physical activity.	***- Access to Nature :*** accessible gardens with sitting areas - pleasant waiting areas - building cluster social interaction spaces: the layout of the nursing unit: Nurses and patients are grouped closer together. (Central nurses' station).	 Figures 36.19 & 36.20: Outdoor seating & walking areas 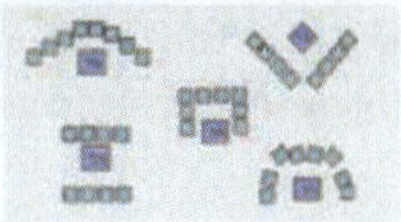 Figure 36.21: Nursing unit layout [(29)]
	Visual access to nature: -Provide the illusion of nature and help patients, family and friends relax in a stressful situation. -Mimic natural surfaces and textures.	***Accommodation in patient spaces or waiting and treatment areas:*** Green walls, plants in rooms. **The artificial connection with nature.** Biophilic design : - Views of nature and plants and vegetation into the overall design setting - Virtual connection with nature(VR) and virtual skylight - The nature feature pictures.	 Figures 36.22 – 36.25: The Northfield Hospital Radiology Suite in Minnesota features a virtual skylight (https://medium.com/the-healthy-city/mother-nature-takes-a-peek-into-medical-facilities-f62d4b76c6c, 2018) 
	Social support &interaction - social interaction in smaller rooms -Staying in touch with friends/relatives/ work. -Access to therapeutic help, being able to talk freely to someone (Social worker/volunteer, etc.) - Visual relationship between room and staff -Continuation of daily activities (Home, work, friends and family and celebrate birthdays).	***social interaction spaces:*** -Single bedrooms: -Family zone in patient room /Family lounges/ dayrooms / waiting rooms with comfortable movable furniture arranged. -Nurse station closed patient rooms: patient and team interactions and satisfaction. -Counseling room, Quiet room. - Phone/ email/TV/radio. - The different design solutions and decisions offers better views to the outside by the patients, it also aids nurses in being able to see the patient and better connect with them.	Figures 36. 26 – 36.28: Miami Valley Hospital conventional room (http://www.nbbj.com/work/miami-valley-hospital-heart-and-orthopedic-center/, 2019) Excellent visibility and optimal connection to daylight. 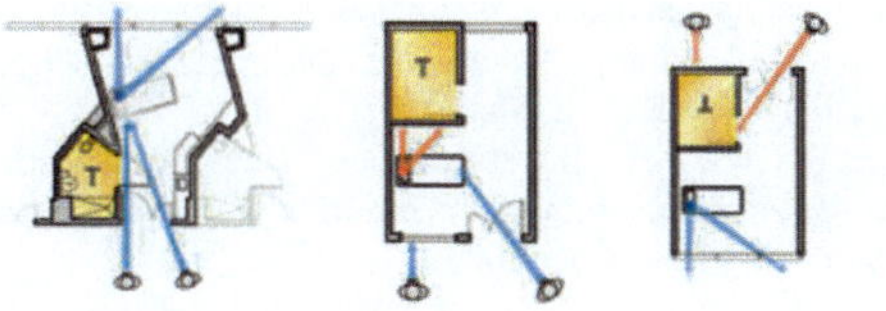 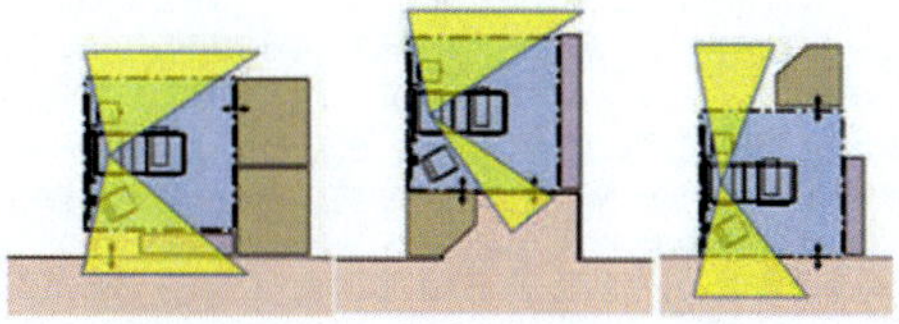 Figure 36.29: Layout in terms of viewing and daylighting potentiality. (Joarder, 2011)

should be carried out in the mass forming stage through optimum orientation and placement of efficient and effective windows. The urban design and coordination of elements should permeate the blocks and interact with them, as well as provide spaces to achieve daily social interaction and support. During the design process, a transparent setting should be in place, to enable monitoring of patients, whether in their rooms or in intensive care units and central care units. A special focus should be given to the design of individual inpatient rooms with an appropriate accommodating surface that allows for interviews with family while providing a suitable natural or stimulated environment. The necessary technology to enable social communication with family and friends during a patients hospital stay should be provided and taken into consideration during the initial design to make for the appropriate wiring and cables, while at the same time, designing physical social spaces in general for social interaction with families, friends or other patients, and medical staff. It is also concluded that in areas where natural views are not possible, designing artificial connection with nature in places of accommodation and patient spaces or waiting and treatment areas should be present. Nursing units and doctors' rooms are advised to be integrated within inpatient areas to achieve the desired social interaction, whether with the patients or with their families.

References

Aripin, S. (2006). Healing architecture: A study on the physical aspects of healing environment in hospital design. In *40th Annual Conference of the Architectural Science Association (ANZAScA)* (p. 344), Adelaide, Australia.

Aripin, S. (2007). Healing architecture': Daylight in hospital design. In *Conference on Sustainable Building South East Asia, Malaysia* (p. 177), 7 November 2007.

Architectural Record. (2019, February 2). Retrieved from architecturalrecord.com: https://www.architecturalrecord.com/articles/12178-dell-childrens-medical-center-of-central-texas

Beggs, J. L. (2015). *Healing through Architecture, Master of Architecture.* University of Waterloo.

Behringer, E. (2011). The daylight imperative. *Master of Architecture, the Graduate School of Clemson University, All Theses. 1120.* Tiger Prints. https://tigerprints.clemson.edu/all_theses/1120

Bensalem, S. (2002). Sustainable Healthcare Architecture : Designing a Healing Environment. Center for Sustainable Development.

Central, H. (2019, June 3). *Healthy parks healthy people, nature heals—Melbourne's new children's hospital.* Retrieved from https://www.hphpcentral.com/article/royal-childrens-hospital.: https://www.hphpcentral.com/article/royal-childrens-hospital.

Chaudhury, H., & Mahmood, A., et al. (2003). *The use of single patient room VS, Multiple occupancy rooms in acute care environments.* Simon Fraser University.

Cooper Marcus, C., & Francis, C. (Eds.). (1998). *People places: Design guidelines for urban open space.* (2nd ed.). Wiley.

Cooper Marcus, C., & Barnes, M. (Eds.). (1999). *Healing gardens: Therapeutic benefits and design recommendations.* Wiley.

Department of Health. (2013). *In-patient care, health building note 04–01: Adult in-patient facilities.* Crown.

Dell's Children Medical Center. (2019, February 2). Retrieved from https://segd.org/dell-childrens-medical-center

Hilal, Ö. (2006). Healing design: A holistic approach to social interaction in pediatric intensive care units in the United States and Turkey. In *1st International CIB Endorsed METU Postgraduate Conference: Built Environment & Information Technologies* (pp. 243–260), Ankara, Turkey.

Huisman, E. R. C. M., et al. (2012). Healing environment: A review of the impact of physical environmental factors on users. *Building and Environment, 58*, 70–80

Iyendo, T. O, & Alibaba, H. Z. (2014). Enhancing the hospital healing environment through art and day-lighting for user's therapeutic process. *International Journal of Arts and Commerce, 3*(9).

Joarder, M. (2011). Incorporation of therapeutic effect of daylight in the architectural design of in-patient rooms to reduce patient length of stay (LoS) in hospitals, A Doctoral Thesis, pp. 75, 83, 213. Bangladesh University for Engineering and Technology, Dhaka, Bangladesh

Kaplan, S. (1995). The restorative benefits of nature: Toward an integrative framework. *Journal of Environmental Psychology*, 169–182.

Kuo, M. (2015). How might contact with nature promote human health? Promising mechanisms and a possible central pathway. *Frontiers in Psychology.*

MacRae, S. (1997). Consumer Perceptions of the Healthcare Environment: An Investigation to Determine What Matters. *Cambridge, MA: The Center for Health Design, Martinez, CA, and The Picker Institute.*

Martin, L., et al. (2020). Nature contact, nature connectedness and associations with health, wellbeing and pro-environmental behaviours. *Journal of Environmental Psychology, 68*, 101389

Mogensen, J. (2011). *Health caring architecture—Spaces for social support, Master Thesis.* Aalborg University, Aalborg, Denmark.

Momtaz, R. I. (2017). Healing gardens—A review of design guidelines. *International Journal of Current Engineering and Technology, 7* (5).

Netherlands., H. C. (2004). Nature and health: The influence of nature on social, psychological and physical well-being. *The Hague: Health Council of the Netherlands/Dutch Advisory Council for Research on Spatial Planning.*

Pouya, S., & Demirel, Ö. (2015). What is a healing garden? Akdeniz University, Turkey. *Akdeniz Üniversitesi Ziraat Fakültesi Dergisi*, 5–10.

Retrieved from https://medium.com/the-healthy-city/mother-nature-takes-a-peek-into-medical-facilities-f62d4b76c6c (2018).

Retrieved from https://www.nbbj.com/work/miami-valley-hospital-heart-and-orthopedic-center/ (2019).

Silvis, J. (2013, July). Designing access to nature in healthcare. *Healthcare Design Magazine.*

Totaforti, S. (2018). Applying the benefits of biophilic theory to hospital design. *Springer open Journal.*

Ulrich, R. S., et al. (1991). Stress recovery during exposure to natural and urban environments. *Journal of Environmental Psychology*, 201–230.

Ulrich, R. S., et al. (2004). *The role of the physical environment in the hospital of the 21st century: A once-in-a-lifetime opportunity.* Center for Health Design.

Ulrich, R. S., et al. (2006). The environments impact on stress. In S. O. Maarberry (Ed.), ACHE Management Series (pp. 37–61).

Ulrich, R. (1983). Aesthetic and affective response in natural environments. *Behaviour and the natural environments*, 85–125.

Ulrich, R. (1984). View through a window may influence recovery from surgery. *Sciences* 224, 420–421.

Ulrich, R. (2001). Effects of healthcare environmental design on medical outcomes. *World Congress on Design and Health: International Academy for Design and Health*, pp. 49–59, Stockholm, Sweden

Ulrich, R. (2003). Evidence Based Environmental Design for Improving Medical Outcomes, Healing by Design Building for Healthcare in 21st Century. Retrieved from http://www.muhchealing.mcgill.ca/english/Speakers/ulrich_p.html

White, M. P., et al. (2019). Spending at least 120 minutes a week in nature is associated with good health and wellbeing. *Scientific Reports, 9*, 7730

Whitehouse, S. e. (2001). Evaluating a Children's Hospital Garden Environment: Utilization and Consumer Satisfaction. *Journal of Environmental Psychology*, Vol. 21, 301–314.

Effects of Climate Change on Vegetation in the Province of Macerata (Central Italy)

Matteo Gentilucci, Maurizio Barbieri, Marco Materazzi, and Gilberto Pambianchi

Abstract

Climate change will have a major influence on vegetation, including crops. Thus, it is essential for a conscious spatial planning to analyse the environmental changes taking place. This research is focused on a small part of the west coast in Central Italy, the province of Macerata. The aim of this research is to highlight climate change and to relate it to the environmental changes taking place. The data of 30 rain gauges and 40 weather stations for temperature from 1961 to 2014 were analysed in order to assess climate change. The data were divided into two standard periods, 1961–1990 and 1991–2014, and through GIS software were calculated and spatialized some climatic and bioclimatic indices, in addition to the study of the hydrological water balance for the province of Macerata. Then, it was assessed the land use as well as the Landsat TM imagery, in the last 30 years. Through the GIS software, the differences were quantitatively evaluated, both for climatic indices and for vegetation through math between rasters. However, a substantial relationship has only been identified for normalized difference vegetation index (NDVI) in August for the years 1991, 2001, and 2011. These results were in agreement, and in fact, a decrease of precipitation in some months and an increase of maximum temperature lead to a decrease in vegetation. In conclusion, climate change can have a major impact on the natural environment and it is essential to know it, in order not to passively undergo the changes induced by the lack of human care for its habitat.

Keywords

Climate change • GIS • Vegetation • Satellite imagery

1 Introduction

Climate change is increasingly at the centre of the global environmental debate due to its significant impact in many fields. However, climate change is usually analysed from a global perspective, although local impacts may be different. This study assesses climate change in the province of Macerata (Fig. 1), a small area of Central Italy that overlooks the east coast, bounded by the Adriatic Sea (an arm of the Mediterranean Sea) on one side.

Climate change was investigated over the last 55 years from 1961 to 2016 for the purpose of assessing it through some indices, while the interaction with the surrounding environment evaluated by the available satellite maps was from 1991 to 2011. In fact, it could be very interesting to evaluate the changes in vegetation in relation to the climate. There is a very extensive literature on global (Alexander et al. 2006) and local climate change (González-Rouco et al. 2000). The results of research carried out on climate in Europe (Klein Tank & Können, 2003; Schär et al. 2004) and Italy (Brunetti et al. 2002; Ventura et al. 2002) are in agreement, with an increase of temperatures, dictated mainly by greenhouse gas emissions (Pognant et al., 2018), and a decrease of precipitation. For the particular area in question, there are few climate studies, including precipitation change (Gentilucci et al., 2018a; Gentilucci, Materazzi, et al., 2019) and analysis of extreme indices (Soldini & Darvini, 2017). In order to identify climatic differences better in a standardized way, a number of climatic and bioclimatic indices have been investigated: the aridity index of FAO-UNEP (Middleton & Thomas, 1997), a standard formula created by the United Nations Environmental Programme (UNEP) in order to evaluate the moisture availability of a land excluding the soil capability to absorb

M. Gentilucci (✉) · M. Materazzi · G. Pambianchi
School of Sciences and Technologies, University of Camerino, 62032 Camerino, Italy
e-mail: matteo.gentilucci@unicam.it

M. Barbieri
Department of Earth Sciences, Sapienza University of Rome, 00185 Rome, Italy

F. Trapani et al. (eds.), *Advanced Studies in Efficient Environmental Design and City Planning*, Advances in Science, Technology & Innovation, https://doi.org/10.1007/978-3-030-65181-7_37

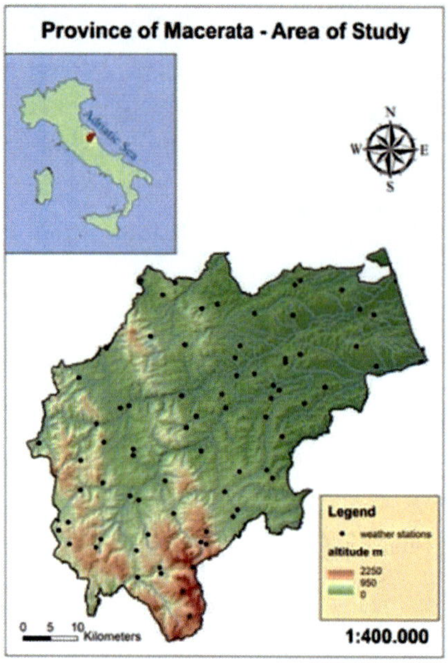

Fig. 1 Area of study (without weather stations outside study area), province of Macerata, Central Italy (Gentilucci et al., 2018a)

and hold water; the ombrothermic index Ios2 (Rivas-Martínez et al., 2011) allows identification of whether a climate is Temperate or Mediterranean; the Huglin heliothermal index (Huglin, 1978) measures the amount of heat for vineyards necessary in order to assess its adaptation to the territory investigated. From the point of view of the theme of desertification, there are numerous studies in the Mediterranean area that have highlighted this problem (Feoli et al., 2003; Piccione et al., 2009) also in relation to the effect of the water balance on crops (Gentilucci, Barbieri, et al., 2019), in addition to possible solutions to mitigate this risk (Basso et al., 2010; Piccione et al., 2013). In the same way in Italy, there are some studies dealing with bioclimatic indices (Canu et al., 2015; Pesaresi et al., 2014)). However, there is a lack of studies that can compare the bioclimatic changes which occurred between different standard periods (1961–1990 and 1991–2014). In this context, in order to make the research even more complete, was evaluated the general response of vegetation to climate change. Land use map and satellite maps in false colours were used to assess the correlation of vegetation or agricultural areas variations with climate change. There are some interesting studies that correlate vegetation and climate through satellite maps. Some are focused on forest response at high latitude (Masek, 2001), and others estimate the change in land use (Feranec et al., 2010; Shalaby & Tateishi, 2007; Ulbricht & Heckendorff, 1998) by evaluating normalized difference vegetation index (NDVI) (Hill et al., 2008; Pettorelli et al., 2005). These studies usually focus on areas of the world where the impacts are strongest or would be expected to be primarily the semi-arid regions. However, assessing the problem of variations in vegetation in relation to climate can also be of fundamental importance also in temperate zones. Practical actions must be prepared well in advance in order to have the desired effect. Therefore, there are three aims of this research: assessment of climate change even through bioclimatic indices between two different standard periods (1961–1990 and 1991–2014), study of the hydrogeological water balance, and analysis of satellite images and land use map to investigate whether climate change has had an effect on vegetation.

2 Materials and Methods

Climate data from 1961 to 2014 were collected from the Geophysical Observatory of Macerata, the Department of Civil Protection Marche Region and the Hydrological Yearbooks. The climatic data collected were monthly and daily for the maximum, average, and minimum temperature and monthly for total rainfall. The accuracy of the temperature sensor is ± 0.2 °C and ± 0.1 mm for rain gauges in accordance with WMO prescriptions (Llansó, 2003). All these data have been subjected to an accurate method of quality control process, divided into validation (Gentilucci et al., 2018) and homogenization. The data were then interpolated through GIS software with the geostatistical technique of ordinary cokriging, with altitude as the independent variable (Gentilucci et al., 2018). This procedure has led to maps of average annual precipitation and temperature for each standard period investigated (1961–1990 and 1991–2014). The climatic data were then included in a series of equations which can synthesize certain climatic characteristics not directly deduced from the analysis of temperatures or precipitation. Thus, the following climatic and bioclimatic indices allow further evaluations of the data collected:

- Aridity index of FAO-UNEP: it is a standard formula created by the United Nations Environmental Programme (UNEP) in order to evaluate the moisture availability of a land excluding the soil capability to absorb and hold water. This numerical indicator that evaluates the availability of water in a given location is calculated with the following formula:

$$Ia = \frac{P}{ET_0} \tag{1}$$

P = mean annual rainfall (mm).

ET_0 = reference evapotranspiration calculated with the Hargreaves–Samani method (Hargreaves & Samani, 1982)

$$ET_0 = \left(C_i(T_{\text{mean}} + 17,78)\sqrt{(T_{\max} - T_{\min})Ra}\right) * N \quad (2)$$

ET_0 = daily reference evapotranspiration; T_{max} = monthly maximum temperature °C; T_{mean}. = monthly average temperature °C; T_{min} = monthly minimum temperature °C; N = number of days of each month considered; Ra = extraterrestrial solar radiation in evaporation equivalent of water (mm/day^{-1}); C_i = 0.0023 is the empirical coefficient originally proposed by (Hargreaves & Samani, 1982).

In order to rank the result of the aridity index maps, the classification of (Middleton & Thomas, 1997) has been adopted (Table 1).

This index has been calculated to evaluate the risk of territorial desertification for the standard periods: 1961–1990 and 1991–2014.

- The ombrothermic index (Ios2) of Rivas Martinez: belongs to the first stage of the bioclimatic classification proposed by Rivas Martinez. It is obtained by calculating the ratio between the mean rainfall of the two hottest months, selected in the three months of the summer season, and the mean temperature of the same two months.

$$\text{Ios2} = P(\text{month1} + \text{month2})/T(\text{month1} + \text{month2}) \quad (3)$$

$P(\text{month1} + \text{month2})$ = sum of the rainfall (mm) of the two hottest months; $T(\text{month1} + \text{month2})$ = sum of the temperature (°C) of the two hottest months.

If the rainfall in the bimester does not exceed twice the value of temperatures in the same interval time, the climate can be considered Mediterranean; otherwise, it is Temperate. In this context, a bioclimatic index is also reported to assess the adaptation of the vine to the territory (heliothermal index). In fact the area of the province of Macerata is the location of numerous quality wine, and this index shows the possible implications of climate change on the environment and local economy.

- The heliothermal index (IH) is a viticulture climate index developed by (Huglin, 1978); the aim of this index is to simplify heat sum for the grapevine, in order to identify the best areas for its growth. IH should be calculated from 1 April to 30 September (i.e. for what is typically considered the period of growth for the grapes). It needs daily temperatures, both maximum and minimum; furthermore, there is also a day length coefficient to be considered according to latitude. Huglin has experimentally observed that this index allows a better correlation between optimal temperature sum and sugar content of the grapevine compared with the Amerine–Winkler index. This index is calculated as follows:

$$IH = \sum_{01/04}^{30/09} \frac{(T_{ave} - 10) + (T_{\max} - 10)}{2} * K \quad (4)$$

T_{ave} = average daily temperature °C; T_{max}. = average maximum daily temperature °C; K = day length coefficient that varies from 1.02 to 1.05 between 40° and 50° of latitude.

An assessment of the hydrological balance is essential in a context of an analysis of climate change, since it has major influence on the adaptability of vegetation. Then, to get a complete overview on the tendency to dryness of the territory of the province of Macerata, the annual water balance has been analysed. The potential evapotranspiration of the area was calculated using the Hargreaves–Samani method and was deducted from the precipitation in order to identify the potential annual water balance.

$$\text{WB} = P - \text{ET}_0 \quad (5)$$

Subsequently, two land use maps of CORINE Land Cover (CLC) project were compared, one for 1990 and one for 2012. The third level of the code CLC was analysed, and the comparison assessed the variations of the areas in square kilometres. In this way, it has been possible to evaluate changing in urbanization, agriculture, forestation, construction and management of water bodies (Feranec et al., 2010).

Finally, satellite images from the Landsat TM 4–5 satellite were collected by the USGS Website (https://earthexplorer.usgs.gov/) and imported into the GIS software. Landsat images were collected at regular ten years intervals, from 1991 to 2011. The Landsat images have a resolution of 30 m, and the period chosen is August, because it represents

Table 1 Middleton and Thomas classification based upon the aridity index (Middleton & Thomas, 1997)

Aridity index	Type of climate
AI < 0.05	Hyperarid
0.05 ≤ AI < 0.2	Arid
0.19 ≤ AI < 0.49	Semi-arid
0.5 ≤ AI < 0.65	Dry sub-humid
AI ≥ 0.65	Humid

the most lush vegetation period and has the greatest mean NDVI values (Alatorre & Beguerìa, 2010). In this study, the standard "false colour" composite was used, and thus, vegetation appears red. Three bands were isolated (4, 3, 2) that could highlight the differences in vegetation between the various periods. NDVI was calculated by GIS software, to compare the various periods (1991; 2001; 2011), in order to assess the presence of live vegetation in the area:

$$\mathrm{NDVI} = \left(\frac{(\mathrm{IR} - \mathrm{R})}{(\mathrm{IR} + \mathrm{R})}\right) \quad (6)$$

IR = pixel values from the infrared band.
R = pixel values from the red band.

Through the mathematics between the rasters, the different maps have been compared, both spatially and quantitatively. At the same time, changes in maximum temperature and precipitation were assessed to investigate the correlation between vegetation and climate variations. Data from all the weather stations available in this area for the three years (1991, 2001, and 2011) have been collected.

3 Results and Discussion

The province of Macerata has a wide variety of environments, which affect temperatures and rainfall. Therefore, it is essential to interpolate the data to have them spatialized throughout the provincial territory. The interpolation investigated in a previous research (Gentilucci et al., 2018) shows for temperatures a significant average increase in the last period, equal to 1.2 °C (Fig. 2c). Spatially, there is a significant increase of temperature in the last period for the territories just behind the coast (East) and for the Sibillini mountains (South West) (Fig. 2a, b).

For precipitation, there is a slight decrease on average (Gentilucci et al., 2018) in the annual amount, from 1961–1990 to 1991–2014 (Fig. 3c). From a spatial point of view, the situation is different from temperatures, in fact, there is a little increase of precipitation for the coast and surrounding territories, while the decrease is focused on the hilly and mountainous band (West part of the Province).

The Macerata Province is located on the edge of the Mediterranean climate, and in this territory, the analysis of climate and bioclimatic indices can highlight hidden changes which are very important in a long-term perspective. These indices define some important limits for quantifying the amount of climate change. In this case 3, climate and bioclimatic indices were calculated, chosen for their importance in the evaluation of climate and effects on crops: aridity index, ombrothermic index, and heliothermal index. Then, the first characteristic of the climate analysed was the aridity index (AI), which indicates the degree of dryness of a location. The aridity index shows great differences from 1961–1990 to 1991–2014. There is a strong decrease of the AI in the last period, even if the territory of the province of Macerata can be defined in both periods as "humid" because on average it exceeds the value of 0.65 (Fig. 4c). The map (Fig. 4a, b) shows a decrease of AI above all for the period

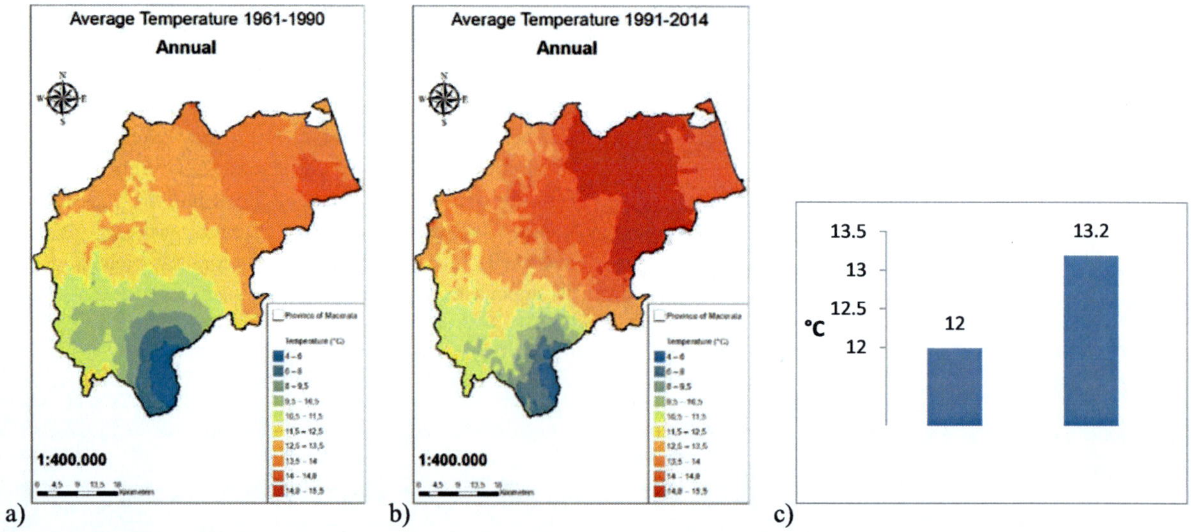

Fig. 2 Temperature for periods: **a** 1961–1990; **b** 1991–2014 (Gentilucci et al., 2018); **c** Average changes of temperatures between standard periods (Gentilucci et al., 2018)

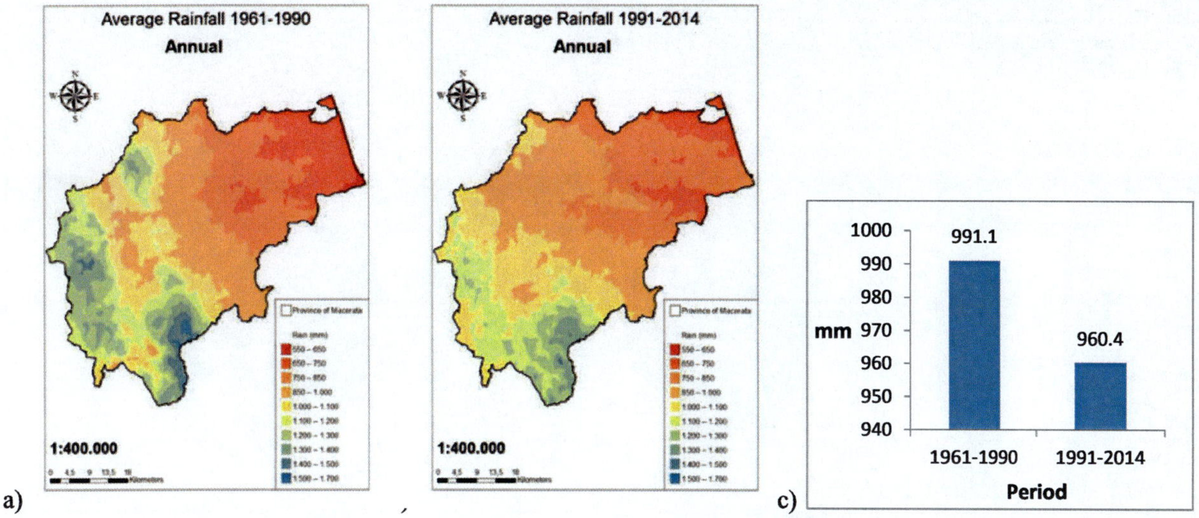

Fig. 3 Precipitation for periods: **a** 1961–1990; **b** 1991–2014 (Gentilucci et al., 2018); **c** Average changes of precipitation between standard periods (Gentilucci et al., 2018)

1991–2014, in which some coastal zones can be defined dry sub-humid following the (Middleton & Thomas, 1997) classification.

In order to achieve a better climate and bioclimatic classification, it is necessary to assess whether a type of climate, in this area, is Mediterranean or Temperate. In fact, Temperate and Mediterranean climates differ, but even more sensitive is their effect on vegetation, with a prevalence of xerophytic and mesophytic species (Pirintsos et al., 2011) in the Mediterranean climate. In this case, the ombrothermic index (Ios2) of Rivas Martinez allows the distinction to be made. The ombrothermic index (Fig. 5a, b) shows that in the period 1961–1990 the Mediterranean climate does not exist, while in 1991–2014, it occupies about one third of the territory to the east, i.e. coast and surrounding areas. Thus, now part of the province has a Mediterranean climate, while from the hills towards the interior, the climate becomes Temperate. This index is a clear example of climate change in the last 30 years.

Climate change inevitably has a great impact on crops, so it is interesting to evaluate how a typical crop in this area

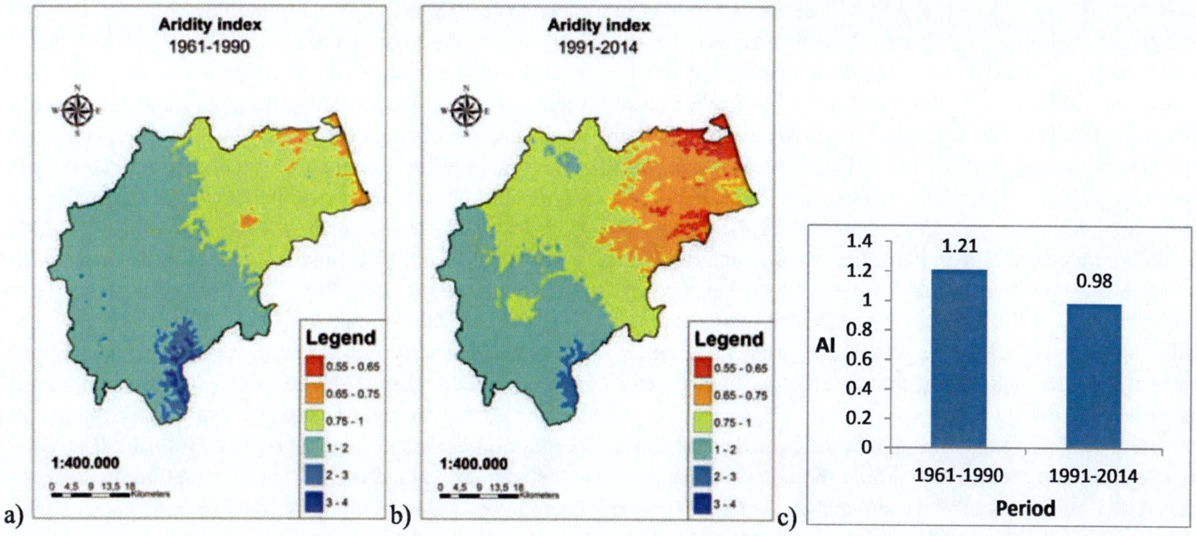

Fig. 4 Aridity index FAO-UNEP for the two periods: **a** 1961–1990; **b** 1991–2014; **c** Average changes of aridity index (AI) between standard periods

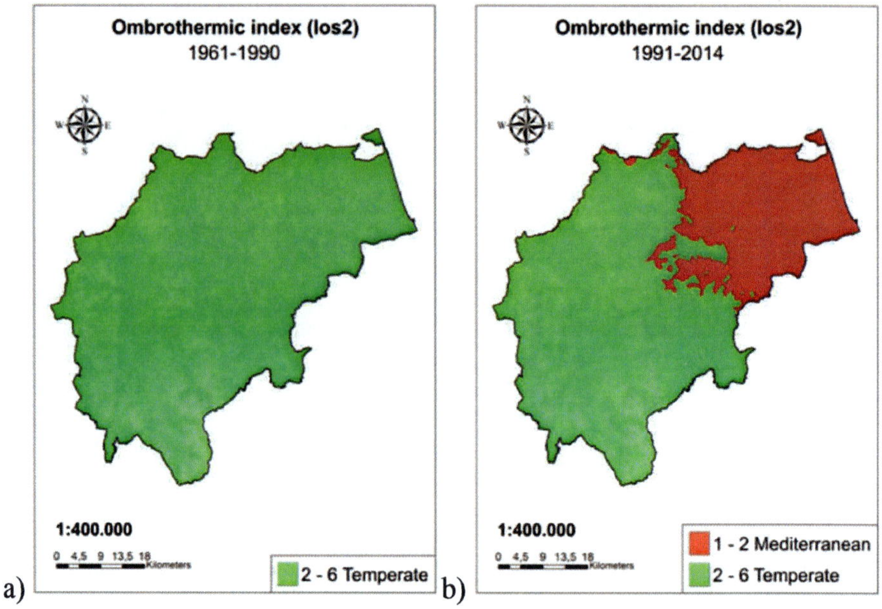

Fig. 5 Map of ombrothermic index (Ios2) for all standard periods investigated: **a** 1961–1990; **b** 1991–2014

responds to climate change. Wine production is an excellence and one of the most profitable activities in the territory of the province of Macerata. In this study was assessed the influence of climate change on vine growth, through the heliothermal index of Huglin. Huglin supports that the limit for the grapevine is in the range from 1400 and 2400 HI. This range is displayed as a dotted marker in the maps (Fig. 6a, b); however, the hilly belt behind the coast is excluded from the optimum range in the period 1991–2014, while up to now, it has been the most productive part in terms of quality wine.

Another parameter that may adversely affect plant production is the water stress, and thus, it becomes essential to assess the hydrological balance. In order to calculate the annual water balance for the two periods investigated (1961–1990 and 1991–2014), was performed a mathematical operation between rasters (subtraction between precipitation and potential evapotranspiration). The results of the analysis were evident, with negative values that have expanded from a small area, up to more than half of the provincial territory. In fact, in 1961–1990, the average annual water balance was positive, while in 1991–2014 has become negative (Fig. 7a, b). Obviously, this is a potential situation different from the real one, in which plant protection systems from water stress (closure of the pores of the stomata, etc.) must be considered.

Interpolations were performed for the climatic parameters temperature and precipitation, while the other indices were calculated on the basis of the interpolation of temperature and precipitation. The goodness of the interpolation has been evaluated by cross validation on the basis of four statistical indices: Root mean square error (RMSE), average standard error (ASE), root mean square error standardized (RMSSE), and mean standardized error (MSE) (Table 2).

All these preliminary investigations have allowed us to have a clear overview of the context. The further progress in this study is represented by the analysis of the actual repercussions on the territory, which is independent of the paper result. Land use change was primarily assessed, through a comparison of the CORINE land cover (CLC) of 1990 with the CLC of 2012. The result, not clearly identifiable on the map, has been summarized on a histogram, in order to assess the differences of land use over a period of more than 20 years. Analysing the third level of CORINE land cover, the biggest difference in square kilometres of surface can be identified in seven different codes: 211 (non-irrigated arable land), 231 (pastures), 241 (annual crops associated with permanent crops), 242 (complex cultivation patterns), 243 (land principally occupied by agriculture, with significant areas of natural vegetation), 321 (natural grasslands), and 324 (transitional woodland shrub). In 2012, the area with the codes 211 and 321 grew, while in 1990, the areas with codes 231, 241, 242, 243, 324 were larger (Fig. 8). In addition, the first level of the CORINE land cover has also been studied to analyse changes in land use in general. This further survey showed an increase in artificial surfaces (21.2 km^2) and forest (56.6 km^2), while there was a decrease in agricultural areas (80 km^2) from 1990 to 2012.

In addition to land use, it is very interesting to assess the overall vegetation present, with observations at 10-year intervals, to evaluate a possible correlation with climate change. Satellite images were analysed by calculating the

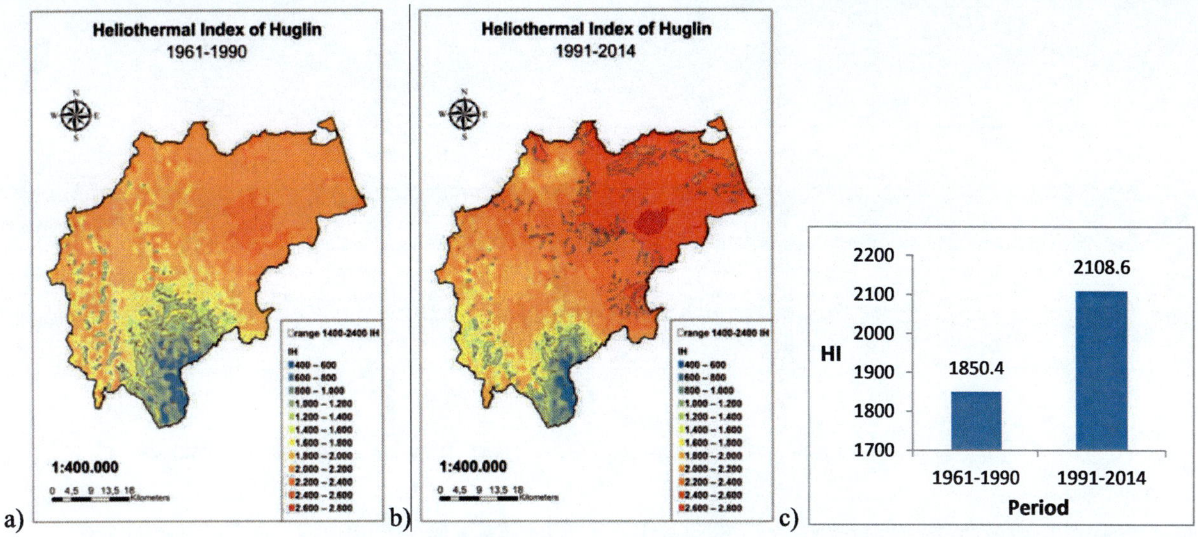

Fig. 6 Helithermal index of Huglin for the periods: **a** 1961–1990 and **b** 1991–2014; **c** Average changes of heliothermal index (IH) between standard periods

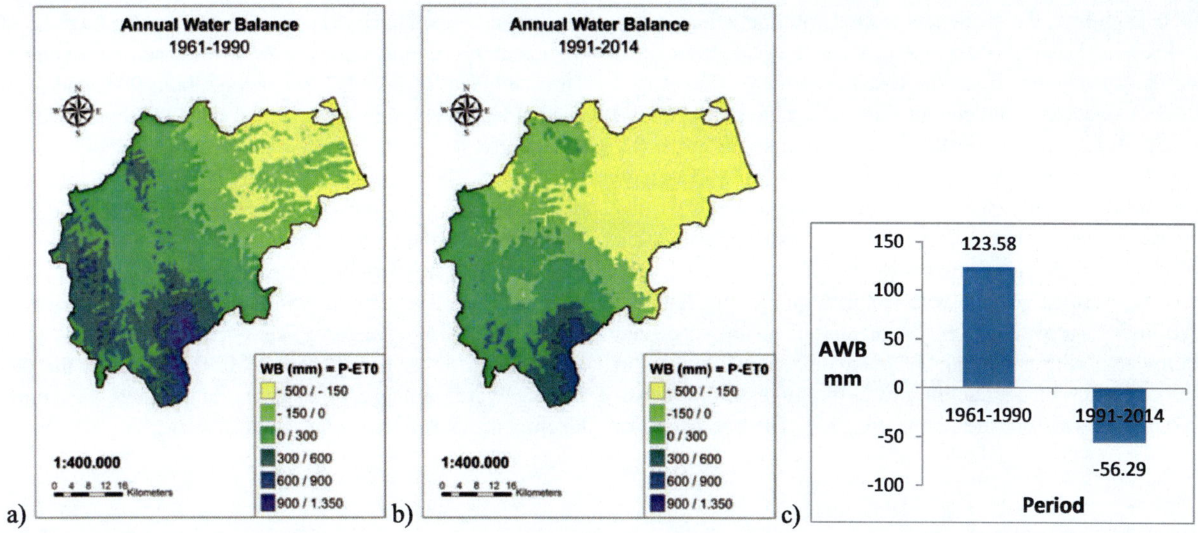

Fig. 7 Annual water balance for the periods: **a** 1961–1990; **b** 1991–2014; **c** Average changes of annual water balance (AWB) between standard periods

Table 2 Cross validation results for the interpolations

Parameters	Period	RMSE	ASE	RMSSE	MSE
Temperature	1961–1990	0.97	0.91	0.99	0.02
Temperature	1991–2014	1.46	1.29	1.07	-0.16
Precipitation	1961–1990	132.36	132.77	0.96	0.11
Precipitation	1991–2014	102.00	112.32	1.03	-0.10

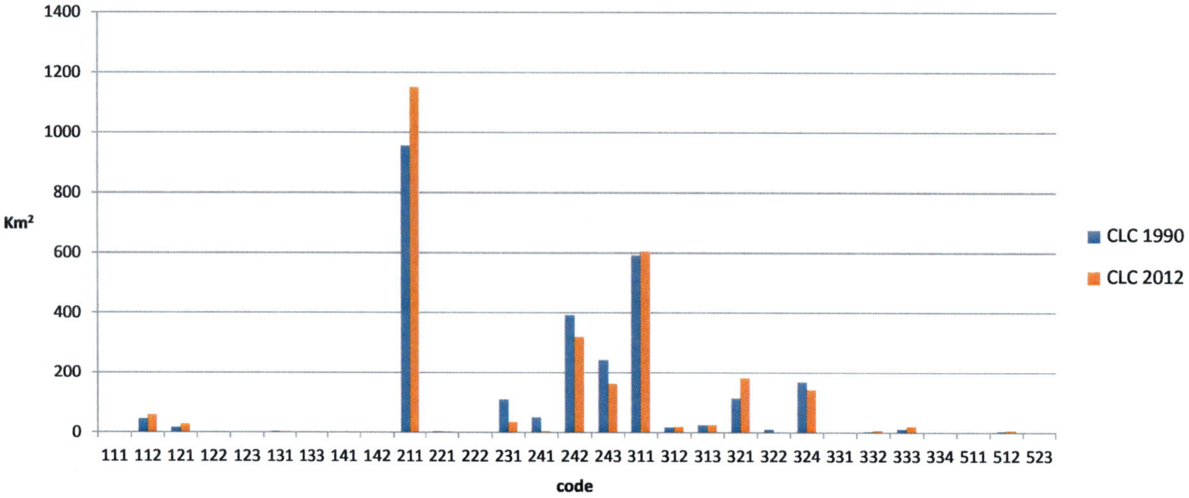

Fig. 8 Histogram comparing the areas between the CORINE land cover 1990 map with the CORINE land cover 2012 map

NDVI, evaluating the difference band TM 4 near infrared (IR) and TM 3 visible red (R). Additionally, for each image (1991, 2001, and 2011), a careful analysis of the areas was made, and first, the graph was created with the values of the index NDVI on the x-axis and the areas in square kilometres on the y-axis (Fig. 9), on the basis of the maps (Fig. 10).

Subsequently, the quantity of area above the NDVI value of 0.2 was analysed (Fig. 11), because the value 0.2 can be considered as the boundary between shrubs and grasslands and barren areas of rock, sand, water, or snow (Gascon et al., 2016). Also, in this case, a spatial evaluation of the situation was necessary by means of maps (Fig. 12).

Afterwards, a relation with temperature and precipitation was investigated; however, for precipitation, only a certain correlation with the months of May, June, and July prior to the collection of the satellite image (in August) was plausible, as shown from the graph (Fig. 13). The coefficient of determination has a good value, and the trend line fit well the points.

As regards the temperature, a certain correlation with the maximum temperature was observed. In fact, the increase of the maximum temperature causes a decrease of vegetation. The coefficient of correlation is good, about 90%, although this evidence should be investigated with additional satellite data (Fig. 14).

In these last maps (Fig. 15.) through a subtraction between rasters has been highlighted on paper, the differences between the periods (August 1991, August 2001, and August 2011) in terms of presence of live vegetation. The lowest value of vegetation was recorded in 2001, above all in the lower hilly area surrounding the sea and in the foothills. On the other hand, between 1991 and 2011, the differences are not very evident, and they are more heterogeneous from a spatial point of view.

Fig. 9 NDVI index in x-axis and km^2 for each value of NDVI index in y-axis for the province of Macerata, in the years 1991, 2001, and 2011

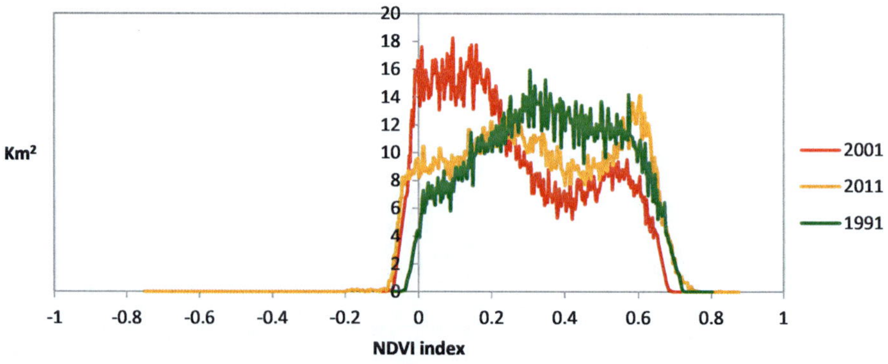

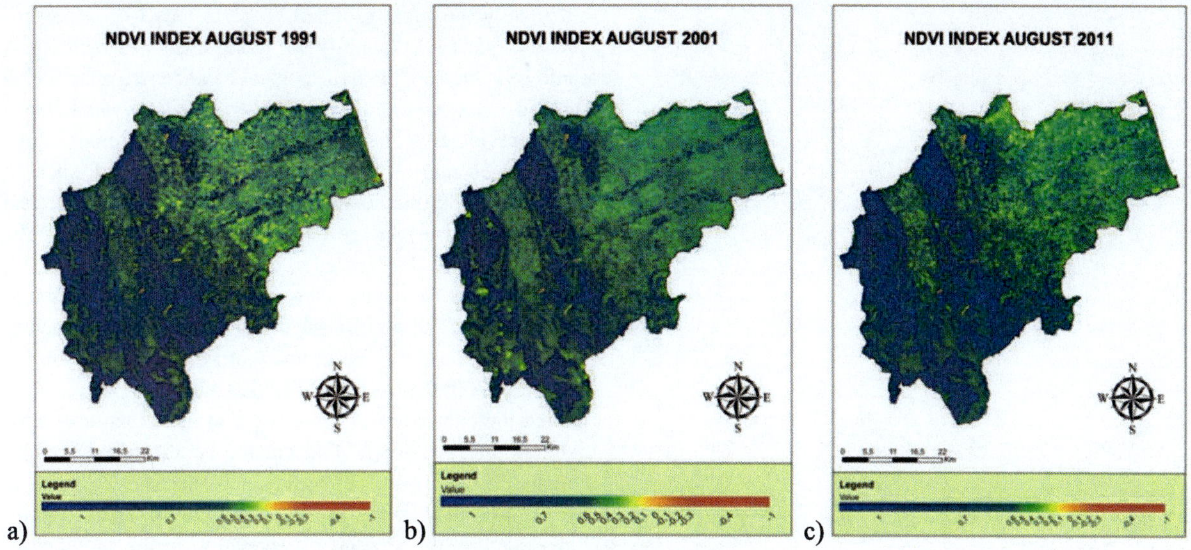

Fig. 10 NDVI index for the province of Macerata, in the years: **a** 1991; **b** 2001; **c** 2011

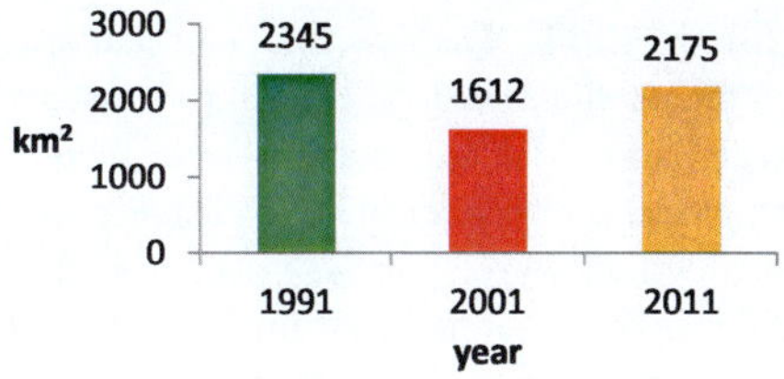

Fig. 11 Area in km^2 of NDVI index above the value 0.2 for the province of Macerata, in the years 1991, 2001, and 2011

4 Conclusions

The function of climatic and bioclimatic indices is to classify a given phenomenon or its variation by a value that cannot be expressed by direct measurement of climatic variables. The indices analysed provide information on soil moisture, the type of climate, and the potential adaptability to the territory of a given plant species. All these values are of great importance for vegetation and deserve to be analysed

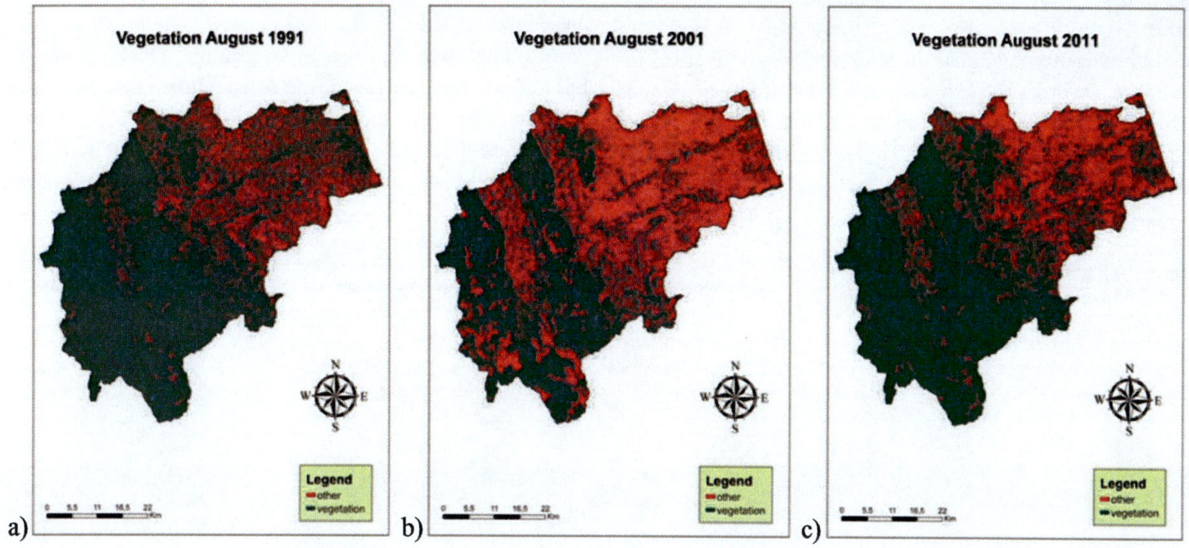

Fig. 12 NDVI index above the value 0.2 (vegetation) for the province of Macerata, in the years: **a** 1991; **b** 2001; **c** 2011

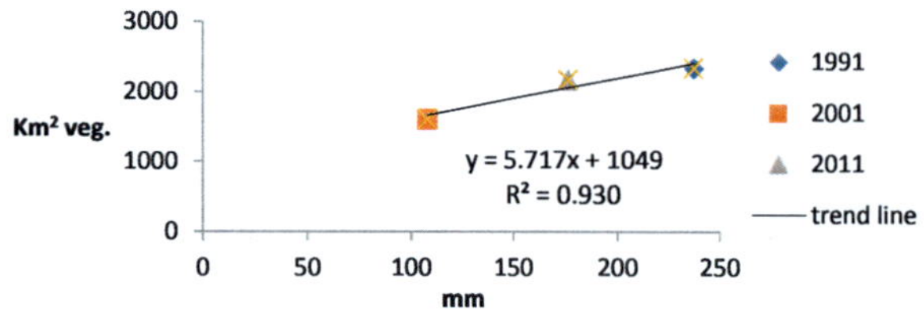

Fig. 13 Ratio of May, June, and July rainfall to the area with vegetation in the province of Macerata, in the years 1991, 2001, and 2011

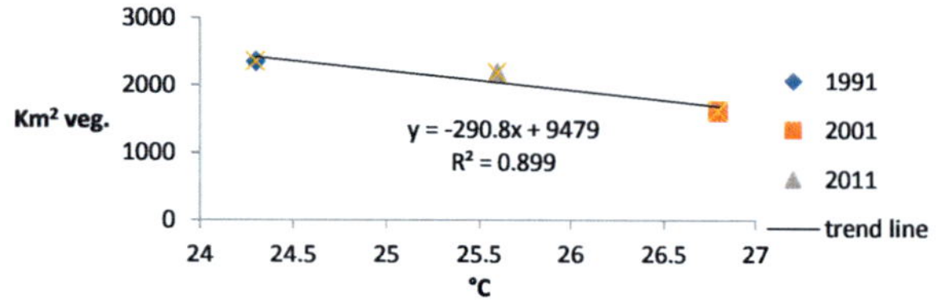

Fig. 14 Ratio of May, June, and July maximum temperature to the area with vegetation in the province of Macerata, in the years 1991, 2001, and 2011

in order to have a complete characterization of the investigated territory. In particular, this study had the purpose of analysing climate change through the comparison of two different standard periods (1961–1990 and 1991–2014). The differences that emerged from this study are evidence of a sudden change in the climate. In fact, the last period is characterized by a lower AI (0.98) compared with the previous period 1961–1990 (AI = 1.21), and even if the general classification of the territory remains "humid", there are some parts of territories near the coast that become "dry sub-humid" at present. The ombrothermic index also highlights the appearance in the last period of Mediterranean-type climate in hilly and coastal areas. Furthermore, in order to assess the possible effects of climate change on a valuable crop like that of the vine, the heliothermal index was calculated. This index showed excessive warming in the areas behind the coast in 1991–2014, historically the most profitable area in terms of wine production. Finally, to quantify the water stress to which plants could be subjected, the hydrological balance was analysed. It showed areas subject to a strongly negative water balance and a provincial average that went from +124 mm in 1961–1990 to −56 mm in 1991–2014. This unequivocal potential scenario was further explored, in order to assess the real effects of climate change on land use and vegetation. The data from the CORINE land cover project were collected, and the results were surprising; in fact, in 2012 compared to 1990, the areas of non-irrigated arable land grew indicating an increase in aridity. Furthermore, it was very interesting to evaluate the first level of CORINE land cover which highlighted the expansion of forest areas at the expense of agricultural ones; although this cannot be directly linked to climate change, it can be considered a success, compared to the deforestation trend in Europe (Feranec et al., 2010). Subsequently, the amount of vegetation in the province of Macerata was assessed using satellite images. The calculation of NDVI, with observations ten years apart, showed a sharp decline in vegetation from 1991 to 2001, with a significant rise in 2011. The possible correlations between vegetation fluctuations and the climatic variables investigated (precipitation and temperature) were then evaluated. Relationships are not very sensitive to changes on a ten-year scale, but a certain correlation was observed, between the precipitation in May, June, and July and the vegetation existing in August. The precipitation has also been evaluated for the single months and for the couples; however, it is precisely by joining the months of May, June, and July that the maximum relation with the presence of vegetation is obtained. Maximum temperature has the same correlation in the same months, but with a countertrend compared with precipitation. Thus, an increase in the maximum temperature corresponds to a decrease in vegetation. Therefore, in the light of this first analysis, it can be concluded that vegetation appears to be much more sensitive to seasonal changes than to medium-term ones (10–20 years). These trends, if confirmed, could lead to confirmations of studies on the subject

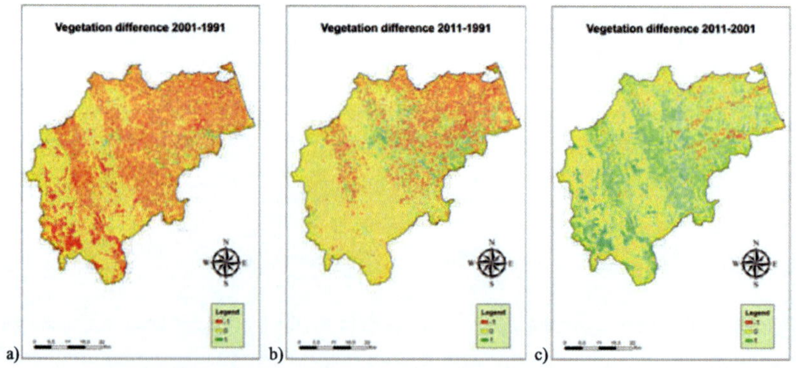

Fig. 15 Map of difference in vegetation between, **a** 2001–1991, **b** 2011–1991, and **c** 2011–2001

carried out in Europe (Bounoua et al., 2000). This study certainly has practical implications that are consequential to monitoring, as the problems encountered with regard to both bioclimatic indices and vegetation should be evaluated. This should of course be followed by a practical strategy for mitigating the risks of desertification or water supply or loss of biodiversity, possibly through the creation of reservoirs and the choice of sustainable farming methods. In addition, the most suitable crops for a given territory could be assessed, trying to maintain the native biodiversity of the study area. However, this first report on the evolution of vegetation in the province of Macerata should be further expanded, with satellite images relating to additional years and different periods of the year, in order to detect the exact relationship between climatic differences and amount of vegetation, or whether there has been in comparison with the past a selection of a different type of vegetation.

References

Alatorre, L. C., & Beguería, S. (2010). Analysis of spatial and temporal evolution of the NDVI on vegetated and degraded areas in the central Spanish Pyrenees. na.

Alexander, L. V., Zhang, X., Peterson, T. C., Caesar, J., Gleason, B., Tank, A. K., Haylock, M., Collins, D., Trewin, B., Rahimzadeh, F., Tagipour, A., RupaKamar, K., Revadekar, J., Griffiths, G., Vincent, L., Stephensons, D. B., Burn, J., Aguilar, E., Brunet, M., … Vazquez-Aguirre, J. L. (2006). Global observed changes in daily climate extremes of temperature and precipitation. *Journal of Geophysical Research: Atmospheres, 111*(D5), 1–22. https://doi.org/10.1029/2005JD006290

Basso, B., De Simone, L., Ferrara, A., Cammarano, D., Cafiero, G., Yeh, M. L., & Chou, T. Y. (2010). Analysis of contributing factors to desertification and mitigation measures in Basilicata region. *Italian Journal of Agronomy*, 33–44. https://doi.org/10.4081/ija.2010.s3.33

Bounoua, L., Collatz, G. J., Los, S. O., Sellers, P. J., Dazlich, D. A., Tucker, C. J., & Randall, D. A. (2000). Sensitivity of climate to changes in NDVI. *Journal of Climate, 13*(13), 2277–2292. https://doi.org/10.1175/1520-0442(2000)013%3c2277:SOCTCI%3e2.0.CO;2

Brunetti, M., Maugeri, M., & Nanni, T. (2002). Atmospheric circulation and precipitation in Italy for the last 50 years. *International Journal of Climatology: A Journal of the Royal Meteorological Society, 22*(12), 1455–1471. https://doi.org/10.1002/joc.805

Canu, S., Rosati, L., Fiori, M., Motroni, A., Filigheddu, R., & Farris, E. (2015). Bioclimate map of Sardinia (Italy). *Journal of Maps, 11*(5), 711–718. https://doi.org/10.1080/17445647.2014.988187

Feoli, E., Giacomich, P., Mignozzi, K., Oztürk, M., & Scimone, M. (2003). Monitoring desertification risk with an index integrating climatic and remotely-sensed data: An example from the coastal area of Turkey. *Management of Environmental Quality*. https://doi.org/10.1108/14777830310460351

Feranec, J., Jaffrain, G., Soukup, T., & Hazeu, G. (2010). Determining changes and flows in European landscapes 1990–2000 using CORINE land cover data. *Applied Geography, 30*(1), 19–35. https://doi.org/10.1016/j.apgeog.2009.07.003

Gascon, M., Cirach, M., Martínez, D., Dadvand, P., Valentín, A., Plasència, A., & Nieuwenhuijsen, M. J. (2016). Normalized difference vegetation index (NDVI) as a marker of surrounding greenness in epidemiological studies: The case of Barcelona city. *Urban Forestry & Urban Greening, 19*, 88–94. https://doi.org/10.1016/j.ufug.2016.07.001

Gentilucci, M., Barbieri, M., & Burt, P. (2019b). Climate and territorial suitability for the vineyards developed using GIS techniques. In *Exploring the Nexus of Geoecology, Geography, Geoarcheology and Geotourism: Advances and Applications for Sustainable Development in Environmental Sciences and Agroforestry Research*. CAJG 2018. Advances in Science, Technology & Innovation (IEREK Interdisciplinary Series for Sustainable Development) (pp. 11–13). Springer.https://doi.org/10.1007/978-3-030-01683-8_3

Gentilucci, M., Barbieri, M., & Burt, P. (2018). Climatic variations in Macerata province (Central Italy). *Water, 10*(8), 1104. https://doi.org/10.3390/w10081104

Gentilucci, M., Barbieri, M., Burt, P., & D'Aprile, F. (2018). Preliminary data validation and reconstruction of temperature and precipitation in Central Italy. *Geosciences, 8*(6), 202. https://doi.org/10.3390/geosciences8060202

Gentilucci, M., Bisci, C., Burt, P., Fazzini, M., & Vaccaro, C. (2018a). Interpolation of Rainfall Through Polynomial Regression in the Marche Region (Central Italy). In A. Mansourian, P. Pilesjö, L. Harrie, R. van Lammeren (Eds.), *Geospatial Technologies for All AGILE 2018*. Lecture Notes in Geoinformation and Cartography (pp. 55–73). Springer. https://doi.org/10.1007/978-3-319-78208-9_3

Gentilucci, M., Materazzi, M., Pambianchi, G., Burt, P., & Guerriero, G. (2019). Assessment of variations in the temperature-rainfall trend in the province of Macerata (Central Italy), comparing the last three climatological standard normals (1961–1990; 1971–2000; 1981–2010) for biosustainability studies. *Environmental Processes, 6*(2), 391–412. https://doi.org/10.1007/s40710-019-00369-8

González-Rouco, J. F., Heyen, H., Zorita, E., & Valero, F. (2000). Agreement between observed rainfall trends and climate change simulations in the southwest of Europe. *Journal of Climate, 13*(17), 3057–3065. https://doi.org/10.1175/1520-0442(2000)013%3c3057:ABORTA%3e2.0.CO;2

Hargreaves, G. H., & Samani, Z. A. (1982). Estimating potential evapotranspiration. *Journal of the Irrigation and Drainage Division, 108*, 223–230

Hill, J., Stellmes, M., Udelhoven, T., Röder, A., & Sommer, S. (2008). Mediterranean desertification and land degradation: mapping related land use change syndromes based on satellite observations. *Global and Planetary Change, 64*(3–4), 146–157. https://doi.org/10.1016/j.gloplacha.2008.10.005

Huglin, P. (1978). Nouveau mode d'évaluation des possibilités héliothermiques d'un milieu viticole. In *Symposium International sur l'Écologie de la Vigne*, l, Constança, Roumanie, 1978. Ministère de l'Agriculture et de l'Industrie Alimentaire, 89–98.

Klein Tank, A. M. G., & Können, G. P. (2003). Trends in indices of daily temperature and precipitation extremes in Europe, 1946–99. *Journal of Climate, 16*(22), 3665–3680. https://doi.org/10.1175/1520-0442(2003)016%3c3665:TIIODT%3e2.0.CO;2

Llansó, P. (Ed.). (2003). *Guidelines on climate observation: Networks and systems*. World Meteorological Organization.

Masek, J. G. (2001). Stability of boreal forest stands during recent climate change: Evidence from Landsat satellite imagery. *Journal of Biogeography, 28*(8), 967–976. https://doi.org/10.1046/j.1365-2699.2001.00612.x

Middleton, N., & Thomas, D. (1997). *World Atlas of desertification*. (2nd ed.). Arnold, Hodder Headline.

Pesaresi, S., Galdenzi, D., Biondi, E., & Casavecchia, S. (2014). Bioclimate of Italy: Application of the worldwide bioclimatic classification system. *Journal of Maps, 10*(4), 538–553. https://doi.org/10.1080/17445647.2014.891472

Pettorelli, N., Vik, J. O., Mysterud, A., Gaillard, J. M., Tucker, C. J., & Stenseth, N. C. (2005). Using the satellite-derived NDVI to assess ecological responses to environmental change. *Trends in Ecology and Evolution, 20*(9), 503–510. https://doi.org/10.1016/j.tree.2005.05.011

Piccione, V., Castro, R., Duro, A., Piccione, A., Rapicavoli, R. V., & Veneziano, V. (2013). Ruolo del Parco Regionale Naturale dell'Etna (Sicilia) nella mitigazione del rischio desertificazione. *Boll. Accad. Gioenia Sc. Nat, 46*(376), 62–91

Piccione, V., Veneziano, V., Campisi, S., & Malacrinò V. (2009). Rischio Desertificazione Regione Sicilia (Protocollo Medalus). Mappe di sensibilità e incidenza territoriale a scala comunale del processo in divenire. *Quaderni di Botanica Ambientale e Applicata, 20*(1), 3–250.

Pirintsos, S. A., Paoli, L., Loppi, S., & Kotzabasis, K. (2011). Photosynthetic performance of lichen transplants as early indicator of climatic stress along an altitudinal gradient in the arid Mediterranean area. *Climate Change, 107*(3–4), 305–328. https://doi.org/10.1007/s10584-010-9989-0

Pognant, F., Bo, M., Nguyen, C. V., Salizzoni, P., & Clerico, M. (2018). Design, modelling and assessment of emission scenarios resulting from a network of wood biomass boilers. *Environmental Modeling & Assessment, 23*(2), 157–164. https://doi.org/10.1007/s10666-017-9563-5

Rivas-Martínez, S., Rivas-Sáenz, S., & Penas, A. S. (2011). Worldwide bioclimatic classification system. *Global Geobotany, 1*, 1–634. https://doi.org/10.5616/gg110001

Schär, C., Vidale, P. L., Lüthi, D., Frei, C., Häberli, C., Liniger, M. A., & Appenzeller, C. (2004). The role of increasing temperature variability in European summer heatwaves. *Nature, 427*(6972), 332. https://doi.org/10.1038/nature02300

Shalaby, A., & Tateishi, R. (2007). Remote sensing and GIS for mapping and monitoring land cover and land-use changes in the Northwestern coastal zone of Egypt. *Applied Geography, 27*(1), 28–41. https://doi.org/10.1016/j.apgeog.2006.09.004

Soldini, L., & Darvini, G. (2017). Extreme rainfall statistics in the Marche region, Italy. *Hydrology Research, 48*(3), 686–700. https://doi.org/10.2166/nh.2017.091

Ulbricht, K. A., & Heckendorff, W. D. (1998). Satellite images for recognition of landscape and landuse changes. *ISPRS Journal Photogrammetry, 53*(4), 235–243. https://doi.org/10.1016/S0924-2716(98)00006-9

Ventura, F., Pisa, P. R., & Ardizzoni, E. (2002). Temperature and precipitation trends in Bologna (Italy) from 1952 to 1999. *Atmospheric Research, 61*(3), 203–214. https://doi.org/10.1016/S0169-8095(01)00135-1

Reviving Cities' Urban Fabric

Urban Fabric Patterns on Urban Wetland

Carolina Rojas, Sebastian Díaz, and Juan Munizaga

Abstract

Latin American cities have experienced rapid urbanization and with it the social and environmental challenges caused by urban sprawl. Over the last decades, this increase has led to strong wetland area loss and fragmentation. It has negatively impacted its ecological value and biodiversity, and ignored their ecosystems services. This study analyzes wetland urbanization patterns in the Metropolitan area of Concepción (South of Chile). It proposes a characterization of the urban fabrics using the urbanization variables of density of population and dwellings, distances to roads and city center, and % covered by roads and green areas surrounding the wetland, in order to evaluate which urban variable explains the increase in urbanization, reducing recreation services for the well-being of residents from the different urban fabrics. In Chilean cities, urban wetlands are typically built on with dwellings that are not friendly with the natural environment. The urbanization intensity is modeled with a linear regression ordinary least squares (OLS), and predictions or a model of the dependent variable in terms of its relationships to a set of explanatory urban variables is generated. The built-up area (BA) is the dependent variable, and the variables of densities, distances, green spaces, and roads are independent. A discussion about the effect of spatial urbanization on urban wetlands is achieved, and the urbanization intensity is determined. The study adds to the discussion about spatial patterns and urban processes in urban landscapes; the results are inputs for natural planning and designing for people who intend to use urban wetlands in cities as nature-based solutions.

Keywords

Urban fabric • Urban wetland • Urbanization • Regression model • Sustainable cities

1 Introduction

Currently, cities are increasingly urbanized, encroaching on several global biodiversity hotspots in the surrounding metropolitan areas. More than 80% of the population lives in cities, and by 2050, that figure is expected to reach 90% (UN, 2014). It is generally agreed that the global trends of urbanization are: (i) urban areas expanding faster than the urban population, leading to an increase in land build cities; (ii) urban areas modifying their local and regional climate; (iii) built-up area expansion drawing heavily on natural resources; (iv) expansion occurring in areas adjacent to biodiversity hotspots, and faster in low-elevation, biodiversity-rich coastal zones; and (v) future expansion occurring in areas of limited economic development and institutional capacity for the protection of biodiversity (Elmqvist et al., 2013). Consequently, natural landscapes have been completely transformed by urbanization. In fact, the recent IPBES (2018) report mentioned the increase of urban growth and population, new kilometers of paved roads, among others as key factors of special concern in developing countries. Indeed, social and environmental problems are caused by the process of urbanization, as conurbations will likely continue to expand and invade adjacent biodiversity hotspots, especially in the case of cities

C. Rojas (✉)
Centro de Desarrollo, Instituto de Estudios Urbanos y Territoriales Pontificia Universidad Católica de Chile, Urbano Sustentable CEDEUS, Santiago, Chile
e-mail: carolina.rojas@uc.cl

S. Díaz
Centro de Desarrollo, Urbano Sustentable CEDEUS, Santiago, Chile
e-mail: sebastiandiaz@udec.cl

J. Munizaga
Laboratorio Reducción de Riesgos de Desastres Universidad de Concepción, Concepción, Chile
e-mail: juanmunizaga@udec.cl

F. Trapani et al. (eds.), *Advanced Studies in Efficient Environmental Design and City Planning*,
Advances in Science, Technology & Innovation, https://doi.org/10.1007/978-3-030-65181-7_38

with more than 200,000 inhabitants (Science, 2016). Therefore, there is an urgent need to understand how biodiversity and ecosystem services to human societies interact in urban cities of developing countries.

Wetlands are one of the habitats most threatened by the unsustainable growth of Metropolitan areas. Latin America leads in wetland loss at an international level, with 59% (Darrah, 2019). Wetlands provide a space for recreation, promoting mental, and physical health (de Groot et al., 2012), making it a possible contribution to livable cities. Wetland disturbance by the adjacent urban fabric is poorly understood in the literature, but recently we know that the urban fabric has a strong influence on urban metabolism (Thompson & Newman, 2018).

Therefore, we aim to address the following question:

(i) What are the urban fabrics surrounding urban wetlands like? Is there any spatial relationship between population density or other urban variables of the urban fabric with proximity to wetlands? The objective of this study is to characterize the urban fabric and spatial patterns from variables surrounding the *Los Batros* wetland located in *San Pedro de la Paz*, a mid-sized Chilean city.

We were using a statistical spatial model or linear regression ordinary least squares (OLS). OLS was applied to detect the most intense urbanized area, as well as generate predictions. The model uses a dependent variable (built-up areas surrounding the wetland) in terms of its relationships to a set of explanatory or independents urban variables (density of population and dwellings, distances to the city center and roads, percentage of covered by roads, and green areas).

1.1 Case Study

Los Batros wetland is in San Pedro de la Paz, which is an urban subcenter (1 million people), located approximately 5 km from the city of Concepción in the Metropolitan Area of Concepción MAC (Fig. 1). San Pedro is a suburban city, which up to 2015 presented an urban growth rate of 0.6 km^2 per year (Martínez et al., 2016). The population has increased by more than 130,000 (INE, latest census 2017). Los Batros wetland is one of the natural ecosystems in MAC most affected by urbanization. It is composed of the water body named *Estero Los Batros*, flooding and low areas of grasslands, and through water flow to the Laguna Chica and Laguna Grande lagoons in San Pedro. Vegetation biodiversity is characterized by sections of permanent flooding, which can surpass 1.5 m in-depth, and broad homogenous zones covered in emergent aquatic macrophytes, over 2 m in height. Among these, the native plant Schoenoplectus californicus (totora or batro) stands out.

Currently, Los Batros has an approximate area of 133 ha (2014 Rapid Eye boundaries). First analysis shows the comparison between the wetland's current area and its known original perimeter, detected in 1976 by Ilabaca (1994) to be a total of 506 ha. In 40 years, Los Batros has lost 73% of its surface (373 ha) from different drivers. Regarding the increase of the occupied built-up area, 68 hectares were recently urbanized (2000s), equivalent to 14% of Los Batros original surface (see Fig. 1). Therefore, a relevant surface of the urban expansion has taken place over the wetland, which has been reduced by artificial filling for new built-up areas that have specifically been filled in since the 1990s for residential condominiums, mainly for middle class residents.

2 Methods

2.1 Wetland Perimeter

The *Los Batros* wetland was detected with a rapid eye high-resolution satellite image remote sensing method with a support vector machine SVM technique in "radial basin function" using images from March 2014. Rapid eye has five spectral bands (blue, green, red, red edge, and near-infrared/NIR) and 5 m resolution, and the most important being red edge and infrared. The preparation data includes radiometric and atmospheric corrections, normalizing their values between 0 and 1. To delimit Los Batros perimeter, the hydrophilic vegetation was used as criteria, including hydric soils permanently saturated or temporarily flooded, saturated cover with hydrophilic vegetation with water mirror, and land with insufficient drainage permanently or temporarily flooded or saturated cover with dispersing hydrophilic vegetation. The land use cover categories were also detected, identifying built-up areas, undercover soil, water bodies, native forest, grasslands, plantations, mixed forest, beaches, and sands. The wetland surface was verified in fieldwork, with 150 points inside the wetland and 500 points on the rest of categories.

2.2 Urban Variables

This study will develop a characterization of the urban fabrics from urbanization variables surrounding the urban wetland perimeter within a buffer of 1500 m, in order to determine how the urban model was promoted and integrated with the natural ecosystem. They are measured in 250 m cells.

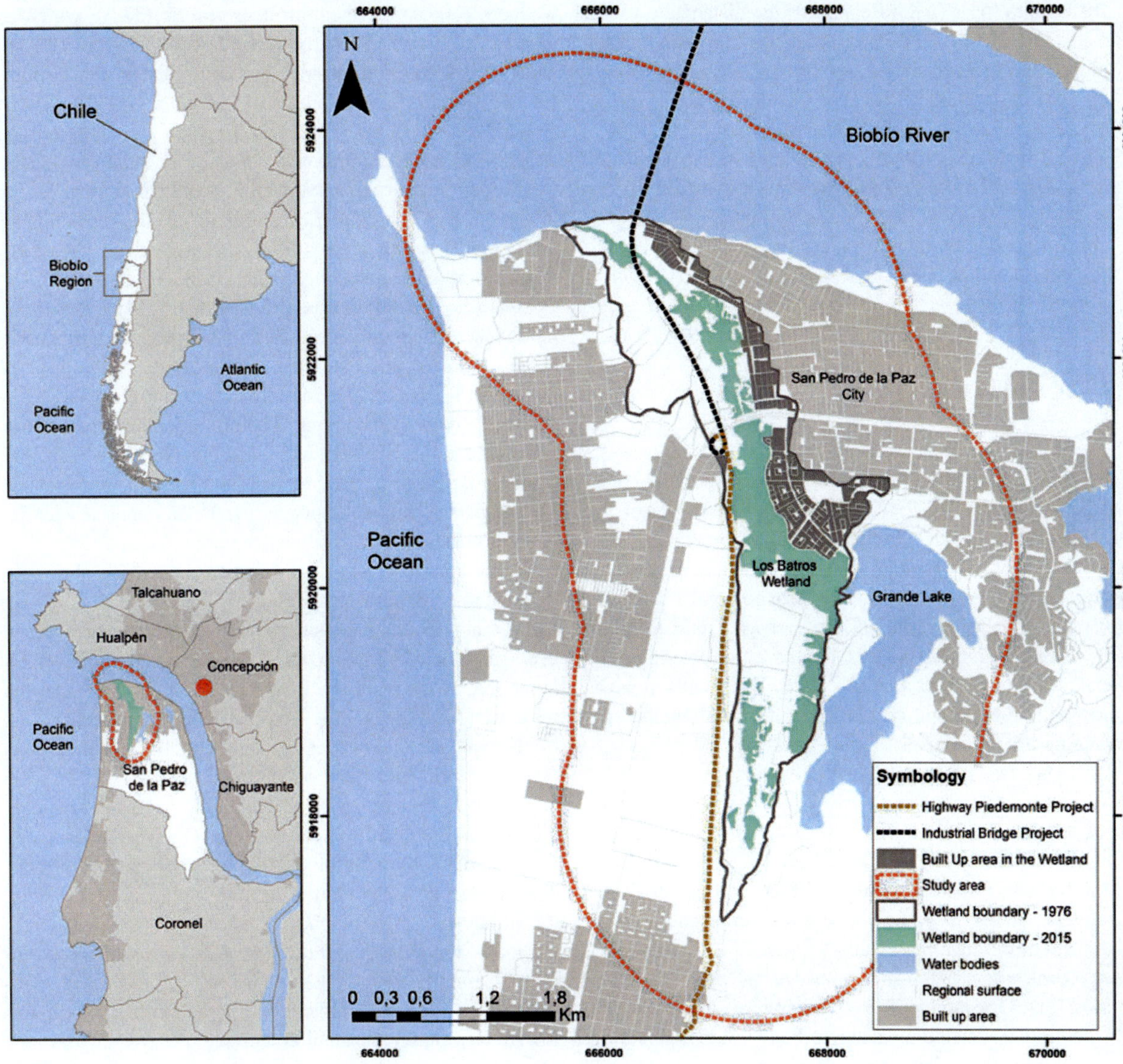

Fig. 1 Localization of Los Batros urban wetland

The urbanization variables, generally used in the prediction model of urbanization from geography and urban studies are:

- Built-up area (BA): Corresponds to a percentage of area covered by houses and buildings.
- Urban Fabric (UF): Urban forms and infrastructure classifications: (1) Discontinuous low density residential fabric, (2) continuous medium dense urban residential fabric, (3) dispersed discontinuous residential fabric, and (4) discontinuous residential fabric in blocks. To improve the delimitation by photointerpretation, we use a digital globe satellite image from February 2016 (0.3 m), available at the ArcGIS 10.2 online server.
- Population density (PD): Normally used in statistical models of urban growth, it is calculated by an estimation of density of population from the results of the 2017 census divided by surface census blocks.
- Dwelling density (DD): Also calculated by houses based on the results of the census 2017.
- Distance to city center (central Business District (CBD)) (DC): Represents the Euclidean distances between census blocks to the business center of San Pedro de la Paz.

- Distance to roads (DR): Represents the distance to roads, calculated by the Euclidean distance between census blocks and main roads.
- Roads (R): Represent the percentage of transport infrastructure covered by roads and streets.
- Green urban areas (GA): Derived from urban land uses map, GA corresponds to a percentage of open spaces classified in formal (recognition of planning and government) and informal green areas covered by any kind of vegetation and/or trees.

2.3 OLS Modeling

The urban variables are used to carry out statistical analysis and model spatial relationship with linear regression ordinary least squares (OLS). OLS was applied to detect the most intense or urbanized area, as well as generate predictions or model a dependent variable in terms of its relationships to a set of explanatory urban variables. In the statistical model, the variables were mapped on a grid cell analysis 250 × 250 m and 6.25 ha in size, to determine the intensification of the urbanized area. The built-up areas surrounding the wetland are the dependent variable and the remaining urban variables are independent.

3 Results

In the wetland's influence area, there are a total of 72,903 inhabitants living in 25,851 dwellings, representing more than 55% of the total population (of San Pedro de la Paz?). The recent urbanization is composed of different urban fabrics: (1) Discontinuous low density residential fabric, (2) continuous medium dense urban residential fabric, (3) dispersed discontinuous residential fabric, and (4) discontinuous residential fabric in blocks. The total residential fabric represents 25% of the studied area, with low density residential fabric being the most representative (47%), characterized by recent construction (between the 1990s and 2000s), scattered in distribution, associated with enclosed gated communities with large detached homes just over the wetland. Next is the medium dense fabric (37%), which is represented by older, mostly public housing (1970s, 1980s), generally, attached multi-unit housing. Next is the discontinuous dispersed urban fabric (10%) located across from the wetland on the coast (Boca Sur), where we find public housing, as well as improvised housing on land previously used for agriculture. Finally, we find the discontinuous fabric in blocks (6%), or the most compact of all, made up of consolidated urban areas (1960s), characterized by apartment blocks and attached housing (Table 1 and Fig. 2).

According to Thompson and Newman (2018), the urban fabrics have inherently more sustainable properties that need to be optimized and extended to other parts of the city. In fact walking, transit, and automobile urban fabrics also have different urban metabolisms. Therefore, in the urban fabric surrounding the wetland, we can detect a more walkable medium dense urban residential fabric and a discontinues and dispersed fabric associated with cars and residential fabrics in blocks more inclined to public transport.

In the spatial relationship between the urban fabric and the patterns of urban variables by cells (see Fig. 3), we found that population density (inhabitants per hectare) has an average value of 32 inhab./ha. It reaches a maximum of 498 inhab./ha in the medium dense residential fabric and the discontinuous blocks. The lowest densities are located west of the wetland, mainly associated with the discontinuous disperse fabric and forest plantations.

Dwelling density has an average of 9 houses per hectare, with a maximum of 146 houses per ha. On average, each cell is covered 8% by buildings, with the maximum being 74%. The greatest concentration of housing units is found among the public housing complexes and the new private neighborhoods. On average, roads cover 6%, with a maximum of 44%. Naturally, the highest values of percentage of roadways is concentrated around Route 160 (main road), followed by denser residential complexes. Meanwhile, the coverage of green urban areas is quite low, averaging 22%, and presenting a maximum of 100% in the area surrounded by a forest plantation. Most of the land does not have green spaces, especially the area west of the wetland, due to the prevalence of residential urban fabric characterized by large expanses of public housing and closed housing complexes that are lacking in green spaces. The oldest housing

Table 1 Urban fabric

Urban fabric	Area (Ha)	% of total
Discontinuous low density residential fabric	282.4	47.4
Medium dense urban residential fabric	219.1	36.7
Dispersed discontinuous residential fabric	56.6	9.5
Discontinuous residential fabric in blocks	38.2	6.4
Total surface	596.2	100

Source Authors' Elaboration

Fig. 2 Urban fabrics

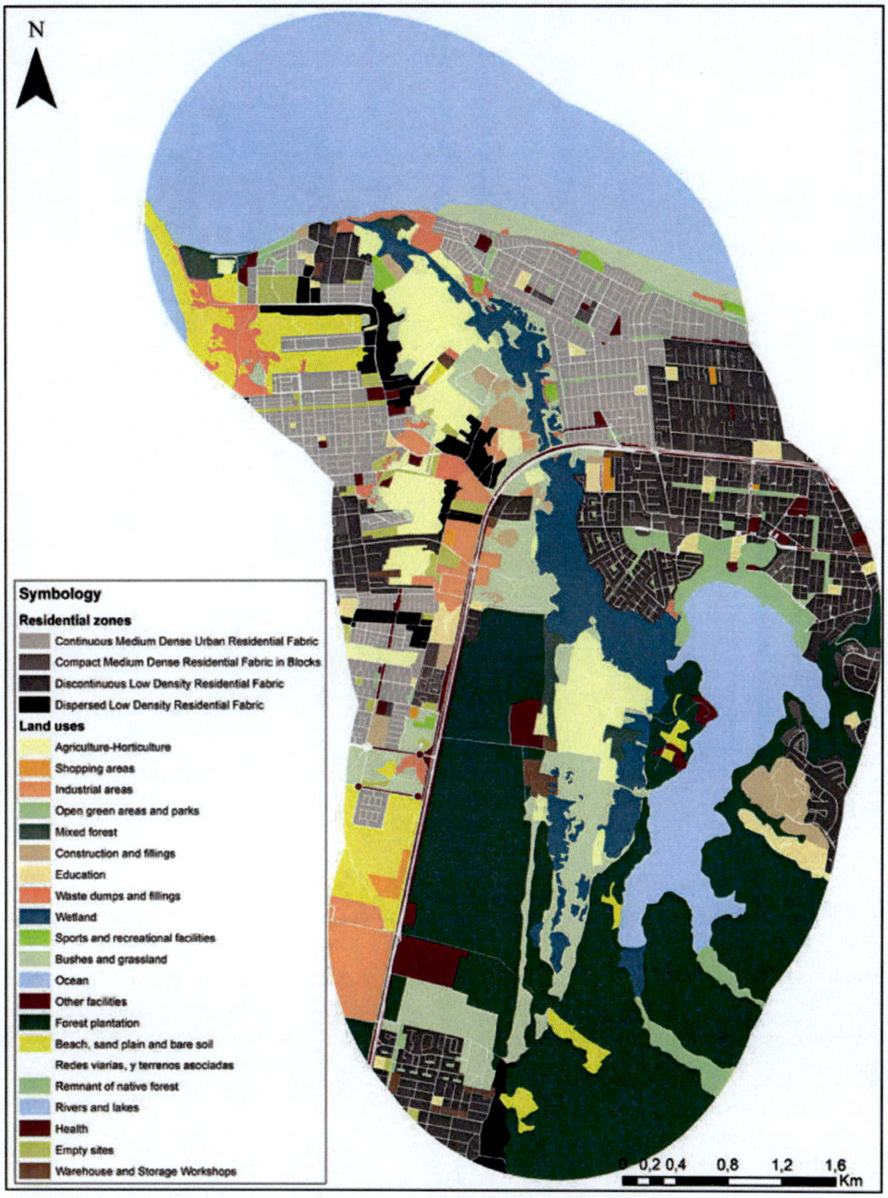

complexes, surrounding the Laguna Grande, present the highest percentages of green spaces. Here, we find a fragment of Los Batros Wetland used as a part of the *Laguna Grande* Park.

The behavior of distances variable shows that the average distance to the center (DC Distance to CBD) is 2890 m, with the outlying areas presenting a maximum of 5420 m. The distance to highways is an average 1115 m, and it is directly related to the Route 160 network. This distance increases as houses move away from Route 160, reaching maximum values of 3093 m (see Fig. 3).

Using Pearson's correlations to define the non-correlational independent urban variables, the analysis shows high degrees of relationship between population density and dwelling density (0.97), so we decided to eliminate the latter. GA does not present a high correlation, while the percentage of roads (R) positively correlates with BA (0.72). DR shows a normal relationship, as does DC. Built-up area is the least correlated with urban fabrics (0.79) for that reason it is the most appropriate for OLS analysis, in accordance with other studies (Basnou et al., 2015; Rojas et al., 2015).

The OLS model uses the BA as the dependent variable, and the remaining non-correlation variables as the independent variables: PD, DC, DR, R, and GA. The model presents an adjusted R^2 of 0.54, which is consistent when

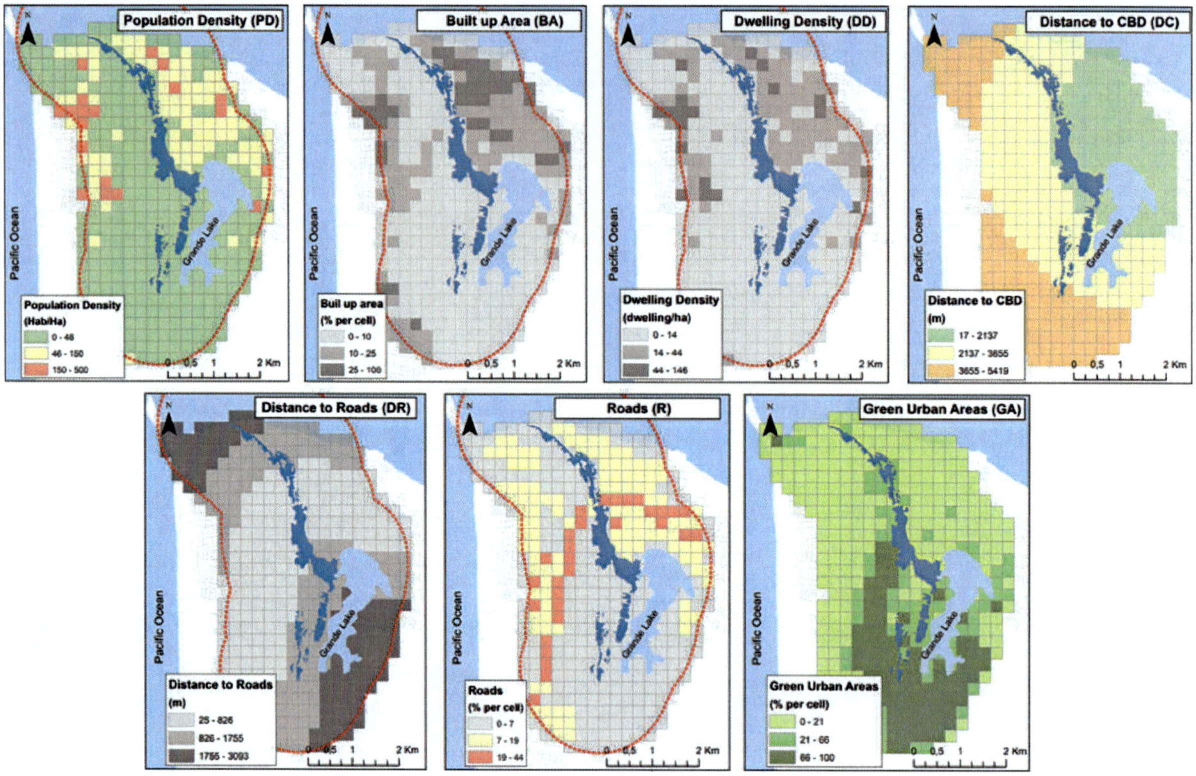

Fig. 3 Urban variables

explaining 50% of the urbanization around the wetland, with a 95% confidence. The values of the VIF coefficient are less than 7.5 and F of 106.46 with p 0.00000, showing a high significance level for the model, and Jarque–Bera statistic shows a significant p-value (0.000000*), meaning a non-normal residual distribution (Table 2).

The variables GA, DR, and DC present a negative relationship with urbanization. Therefore, the built-up volume decreases slightly if density and GA decrease, and the distance or proximity to roads and center increases. However, there are positive relationships with PD and R, consistent with the previous model for explaining urban sprawl at the

Table 2 OLS regression diagnostics

	Values
Number of observation	433
Akaike's information criterion (AICc):	2948.028774
Multiple R^2	0.55
Adjusted R^2	0.549681
Joint F-statistic	106.464122
Prob(>F), (5,427) degrees of freedom:	0.000000*
Joint Wald statistic	484.419583
Prob(>chi-squared), (5) degrees of freedom	0.000000
Koenker (BP) Statistic	61.293840
Prob(>chi-squared), (5) degrees of freedom:	0.000000*
Jarque–Bera statistic:	393.387210
Prob(>chi-squared), (2) degrees of freedom:	0.000000*

Source Authors' Elaboration

Table 3 OLS model summary

Variable	Coefficient	StdError	*t*-Statistic	Probability	Robust_SE	Robust_t	Robust_Pr	VIF
Intercept	8.388953	1.19779	7.003692	0.000000*	1.344257	6.240588	0.000000*	–
PD	0.092714	0.007355	12.605977	0.000000*	0.013669	6.782887	0.000000*	1.27308
GA	− 0.063488	0.01057	− 6.006549	0.000000*	0.010034	− 6.327447	0.000000*	1.266806
R	0.278614	0.055515	5.018675	0.000001*	0.083868	3.322064	0.000984*	1.54404
DR	− 0.001484	0.000513	− 2.892949	0.004017*	0.000508	− 2.921934	0.003670*	1.312454
DC	− 0.00043	0.000289	− 1.485794	0.138085	0.000328	− 1.311708	0.190331	1.129557

Source Authors' Elaboration

metropolitan level (Rojas et al., 2014). Therefore, the built-up area is denser when there are more residential dwellings and the presence of the transportation network by private transport. According to Rbst-Pr, the most significant variables are PD, GA, DR, and R. PD was expected to be consistent due to results of the previous automat cellular

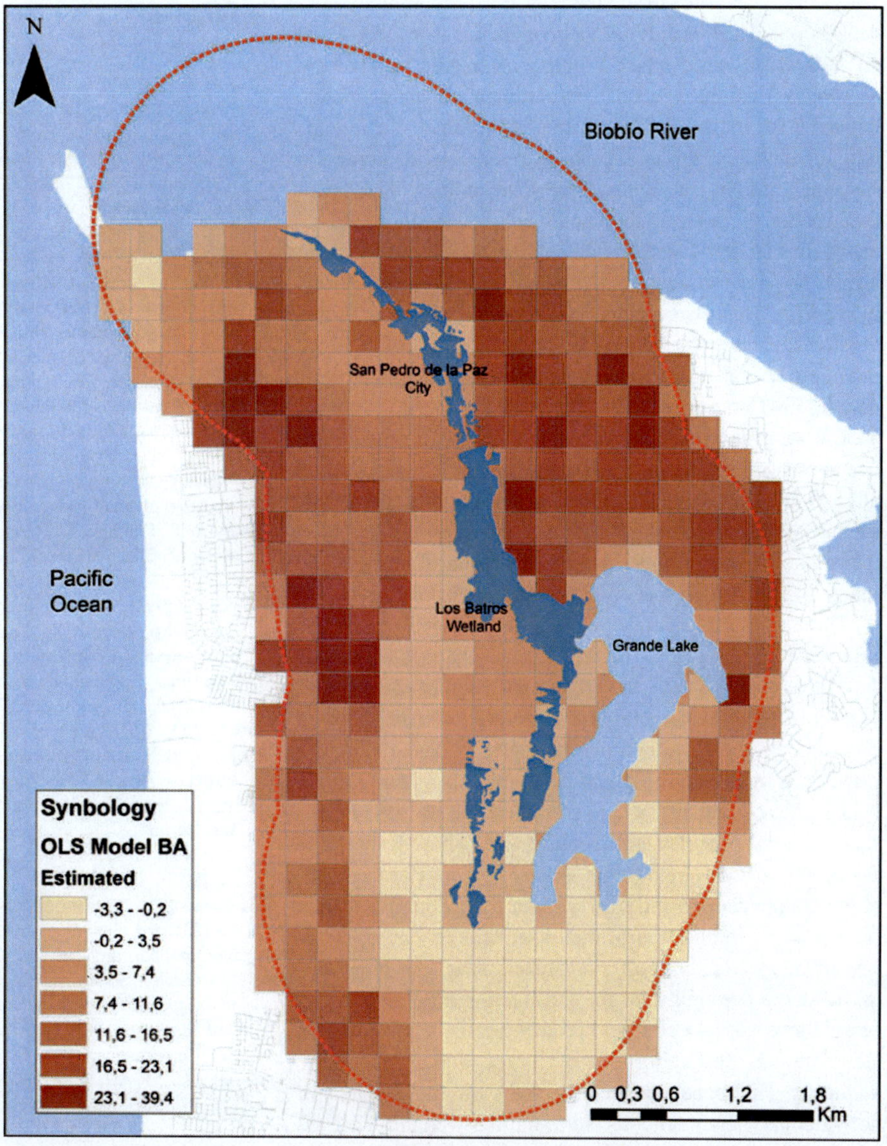

Fig. 4 OLS results

model done to predict the urbanization (Maldonado et al., 2012) (Table 3).

Figure 4, shows the distribution of the OLS estimated coefficients by cells in brown pantone, where the beige cells are under the predicted values and therefore represent less urbanized areas. The graduation of brown cells represents average values, and is therefore close to the estimated value, corresponding to urbanization on the border of the wetland interacting with agricultural and forest plant zones. Finally, the darker brown cells represent values over the models' prediction; these urbanized areas coincide with the denser urban fabrics.

4 Conclusions

The wetland area of Los Batros is in a neighborhood that is densely populated, mainly with discontinuous low density and medium density residential dwellings, which were not planned to integrate with the wetland. Considering that the urban fabric is one of the most important factors of sustainable cities, the new urbanization surrounding the wetland should promote a sustainable model that is less dependent on cars. Traditional urbanization of old neighborhoods includes medium density and walkable areas together with natural spaces (wetlands and lakes); however, discontinuous and dispersed urban fabric is increasing. This study aims to provide a conceptual and methodological contribution to understanding the spatial urbanization patterns over and around natural spaces. For decades in Concepción, the urban wetland has lost surface area, having its surface and adjacent land transformed to be occupied by different urban residential fabrics. They have been affected by the pressure of urban planning, transport infrastructure projects, and real estate value for housing; the urban wetland of Concepción is seen as wastelands ready to be drained and filled completely (Pauchard et al., 2006).

These results confirm that for at least the last 20 years, the urban wetland has been intervened, and changes in its surface have been made by the urban expansion of the low and medium density residential use. A medium density, garden city was the most appropriate urban fabric for the area, but now, it is necessary to build an opportunity for a more sustainable community, aware of natural spaces. This means that as urban fabrics get denser, they should be more accessible and integrated with natural spaces. For example, if the people can see and walk around the wetland, it would be a positive implication of the ecosystem service of recreation that a wetland can offer urbanized areas. The integrated wetland is a possibility for the surrounding neighborhood to learn about and take action to protect the space, avoiding its deterioration. Restoration of Los Batros wetland is urgent, requiring landscape design especially on the borders of low-income neighborhoods surrounding wetland. This is essential for the promotion of the recreation ecosystem service and the preservation of the space, allowing residents to be in contact with nature. It would also reduce the effect of urban sprawl and urban flooding. Integrating the urban wetland into a benefit for the urban population in its surroundings would also improve environmental justice. The transformation focused of urban fabric is also an opportunity to create a regenerative city (Thompson & Newman, 2018). Our model is available as input for the design of a nature-based solution for the city.

Acknowledgements ANID Fondecyt 1190251 "Efectos de urbanización y accesibilidad en los Humedales Urbanos del área metropolitana de Concepción (URBANCOST II)" (2019-2022)".FONDAP CONICYT 15110020.

References

Basnou, C., Iguzquiza, J., & Pino, J. (2015). Landscape and urban planning examining the role of landscape structure and dynamics in alien plant invasion from urban Mediterranean coastal habitats. *Landscape and Urban Planning, 136*, 156–164. https://doi.org/10.1016/j.landurbplan.2014.12.001

Darrah, S., Shennan–Farpón, Y., Loh, J., Davidson, N., Finlayson, C., Royal, C., Gardner, M., & Walpole, M. (2019). Improvements to the wetland extent trends (WET) index as a tool for monitoring natural and human-made wetlands. *Ecological Indicators*, *99*, 294–298. https://doi.org/10.1016/j.ecolind.2018.12.032

De Groot, R., Fisher, B., Christie, M., Aronson, J., Braat, L., Gowdy, J., Haines-Young, R., Maltby, E., Neuville, A., Polasky, S., Portela, R., & Ring, I. (2012) Integrating the ecological and economic dimensions in biodiversity and ecosystem service valuation. In P. Kumar (Ed.), *The economics of ecosystems and biodiversity* (p. 9–40). Ecological and Economic Foundations. Taylor and Francis. https://www.teebweb.org/our-publications/teeb-study-reports/ecological-and-economic-foundations/

Elmqvist, T., Fragkias, M., Goodness, J., Guneralp, B., Marcotullio, P., McDonald, R., Parnell, S., Schewenius, M., Sendstad, M., Seto, K., & Wilkinson, C. (2013). *Urbanization*. Springer.

Ilabaca P. (1994). Geomorfología y Riesgos Naturales en la LLanura de San Pedro. *Revista de Chile Terra Australis*, *39*, 7–17. https://bibliotecadigital.ciren.cl/handle/123456789/22760

INE Instituto Nacional de Estadisticas. (2017). *Census data base*. https://redatam-ine.ine.cl/redbin/RpWebEngine.exe/Portal?BASE=CENSO_2017&lang=esp

IPBES. (2018). The IPBES regional assessment report on biodiversity and ecosystem services for the Americas. In J. Rice, C. S. Seixas, M. E. Zaccagnini, M. Bedoya-Gaitán, & N. Valderrama (Eds), *Secretariat of the intergovernmental science-policy platform on biodiversity and ecosystem services*. https://doi.org/10.5281/zenodo.3236252

Maldonado, D., Gajardo, A., Rojas, C., Valdebenito, P., & Cantergiani, C. (2012). Cellular automata model for the urban growth of the metropolitan area of concepción (CAMAC). In N. N. Pinto, J. Dourado, & A. Natálio (Eds.), *Proceedings of CAMUSS, the International Symposium on cellular automata modeling for urban and spatial systems* (pp. 309–312). Oporto, 8–10/nov/2012, Coimbra. Department of Civil Engineering of the University of Coimbra.

Martínez, C., Rojas, C., Rojas, O., Quezada, J., Lopez, P., & Ruiz, V. (2016). Crecimiento urbano sobre geoformas costeras de la llanura de San Pedro, area metropolitana de Concepcion. In R. Hidalgo, D. Santana, V. Alvarado, F. Arenas, A. Salazar, C. Valdebenito, & L. Álvarez (Eds.), *En las costas del neolineralismo* (pp. 287–312). Geolibros Pontificia Universidad Católica.

Mcdonnell, M. J., & Macgregor-fors, I. (2016). The ecological future of cities, *Science*; *352*(6288).

Pauchard, A., Aguayo, M., Peña, E., & Urrutia, R. (2006). Multiple effects of urbanization on the biodiversity of developing countries: The case of a fast—Growing metropolitan area (Concepción, Chile). *Biological Conservation, 127*, 272–281. https://doi.org/10.1016/j.biocon.2005.05.015

Rojas, C., Plata, W., Valdebenito, P., Muñiz, I., & De la Fuente, H. (2014). La Dinámica de expansión urbana del Área Metropolitana de Concepción. In J. Williams, R. Hidalgo, P. Brand, & L. Pérez (Eds), *Metropolizaciones Colombia–Chile: Experiencias de Bogotá, Medellín, Santiago y Concepción* (pp. 39–56). Línea Editorial 1 Investigaciones–Universidad Nacional de Colombia.

Rojas, C., Sepúlveda-Zúñiga, E., Barbosa, O., Rojas, O., & Martínez, C. (2015). Patrones de urbanización en la biodiversidad de humedales urbanos en Concepción metropolitano. *Revista de Geografía Norte Grande* 2015, *61*, 181–204. https://scielo.conicyt.cl/scielo.php?script=sci_arttext&pid=S0718-34022015000200010&lng=es&nrm=iso. https://doi.org/10.4067/S0718-34022015000200010. ISSN 0718-3402

Science (2016). Rise of the city. *Science, 352*(6288), 906–908. https://doi.org/10.1126/science.352.6288.906

Thompson, G., & Newman, P. (2018). Urban fabrics and urban metabolism—From sustainable to regerative cities. *Resources, Conservation and Recycling, 132*, 218–229. https://doi.org/10.1016/j.resconrec.2017.01.010

United Nations. (2014). *World urbanization prospects: The 2014 revision, Highlights (ST/ESA/SER.A/352)*. New York, United States. https://doi.org/10.4054/DemRes.2005.12.9

Shoreline Alteration Rate and Urbanization in the Burullus Region

Amany Ragheb, Shahira Assem Abdel Razek, and Yasmin Moanis

Abstract

Egyptian cities have faced several challenges since the start of this new millennium; rapid population growth, urban decay at the expense of available shelter, and services, degradation of infrastructure and land, downgrade of the environment, social structure, and the economy in general, are all factors that have led to unsustainable living. The Burullus region, located in the Northern part of the Egypt delta, is an active commercial and administrative hub. It is one of the five protectorate lakes of Egypt which accommodates both a recreational summer resort and a poorer all-year residence. The lake has faced, across time, several changes in respect to the surrounding cities, which in turn impacted dramatically the shore. The changes that happened to the land use and land footprint have changed the shoreline and will, eventually, lead to many environmental and urban problems as well. Many actions should be taken when thinking about future extensions of cities that have natural edges like Lake Burullus and huge agricultural lands, and the aforementioned challenges will face existing communities, as well as, extend with respect to the surrounding areas leading consequently to the change of the identity of the area at large. This paper focuses on the importance of urban development and urban solutions of the cities surrounding Lake Burullus as well as the shoreline protection and implementation policies to optimize deterioration of shoreline, taking into consideration the importance of rural regeneration and its impact on this area. It also highlights a list of actions for successful lake upgrading steps based on the experiences reviewed and concludes that the success of the development of shoreline depends on several factors and that metropolitan planning should cover problems across urban and peri-urban areas and address multijurisdictional issues.

Keywords

Egyptian cities • Burullus region • Implantation policies • Urban development • Informal settlements • Waterfront • Shoreline-Egypt

1 Introduction

Over the last years, the term "sustainable development" has become a common phrase. Developers, mainly urban planners, have, gradually, started showing interest in incorporating their work with the globally uttered concern for sustainability, to develop the idea of growth and development. Urban planning and design strategies that support sustainability can help in achieving the optimum design for developing principles, strategies, policies, programs, and projects (Table 1).

Urban planning encompasses not only areas which are mainly land but also water bodies that are attached or located within an urban realm. Many coastal lakes around the world suffer from different ecological and urbanization problems, the lakes of Egypt being no exception. Egypt consists of many unique lakes that are of great importance, some of which have been declared a natural protectorate by ministerial decrees, and only recently have authorities started to develop the coastal shoreline of some of those lakes to maintain and improve the lakes' conditions.

The importance of the northern delta lakes arises from the fact that they are the most productive lakes in Egypt, supplying Egypt with 37% of its overall fishing production. They also have international importance, as it hosts sites of wintering waterbird, providing valuable habitat for them and maintaining ecological equilibrium.

A. Ragheb (✉) · S. A. A. Razek · Y. Moanis
Faculty of Engineering, Department of Architectural Engineering, Delta University for Science &Technology, Gamasa City, Mansoura, Dakhliya, Egypt

F. Trapani et al. (eds.), *Advanced Studies in Efficient Environmental Design and City Planning*, Advances in Science, Technology & Innovation, https://doi.org/10.1007/978-3-030-65181-7_39

Table 1 Vast researches conducted in the Burullus region

Topics covered by researchers
Lake biodiversity
Lake Burullus wildlife
Exploitation of the ecosystems services
Nitrogen and phosphorus species of Lake Burullus water
Impact of environmental conditions on the biodiversity of Mediterranean Sea lagoon
Using satellite remote sensing in monitoring lakes
Climate changes and the lake
lake carbon level

Source Researchers

Lake Burullus is one of the northern lakes of Egypt but however, despite being declared a protectorate by Prime Ministerial Decree 1444/1998, has lost an estimated 37% of its open-water area and 85% of its marsh area in the past 40 years, largely as a result of ongoing drainage and reclamation of the lake's eastern, western, and southern margins. It is considered to be the least polluted, but it is still suffering from pollution and is subject to increasing quantities of agricultural drainage water which contributes significantly to eutrophication and pollution (Hoda Baraka, 2012, Egypt independent). The present study will carry out the following:

- Evaluate the current status of the land use and land cover of surrounding Burullus Lake
- Summarize the main problems facing the development and management of this urban district by the act of urbanization
- Introduce a conceptual strategy to help in protecting Burullus urban districts.

The absence of an integrated urban vision which takes into consideration lake management and maintenance as well as the surrounding land can lead to a catastrophic problem concerning the urban fabric that surrounds the lakes. That is why deep design decisions regarding the protection of the lake from the urban designer team should take place to overcome the act of increasing population and urbanization in order to protect those lakes.

2 Research Methodology

The research aims to evaluate the current state of the Lake Burullus region from the point of view of sustainable urban developers and the current state related to urbanization and population and how to overcome the future deterioration of the lake conditions and the deterioration of the patch of the agricultural land throughout:

- **Inductive approach**
  Through a brief theoretical study for the idea of lake development and management and the study of the environmental states of the lake as well.
- **Deductive and analytical approach**
  Through a case study of the Burullus Lake as a whole and the northern part of it (Baltim district), Baltim district is considered to be the biggest urban spot surrounding the northern part of the Lake Burullus, focusing on the land use and land cover, and emphasizing the ways that change and deteriorate the lake.

3 Nile Delta Shoreline Challenges

The Nile Delta shoreline covers over 150 Km from east Nile branch to west of Damietta Nile branch, and this huge shoreline has suffered for a long time from erosion due to water flow and sedimentation discharges that were carried out by the Nile River's two branches to the sea; however, it has now become reduced due to climate change and the construction of dams.

Those lakes have an ecological importance due to the variety and bio-diverse species inhabiting the lakes and their hydrologic attributes. And the change which has occurred in the shoreline of the Nile Delta has diverted the problem to the adjacent urban areas. (Table 2).

4 Data Resources

As this study is carried out by urban planners, the source of data is based on real on-site visits to indicate the real state of the lake as well as land use and land cover for Baltim district. Data gathered by the site visits and the satellite images and maps were analyzed and managed by the researchers to help in illustrating the problem, questions, and objectives of the paper topic (Table 3).

Table 2 Problems and challenges facing delta lakes

Problems and challenges facing delta lakes	
1	Encroachment of the lakes over the years due to urbanization
2	Further reduction of the shoreline of the lakes to protect the new outline and create a promenade along part of the lake
3	Delta lakes suffer from many environmental issues like local inhabitant loss, and water level diversity with 70% of their original area
4	Delta lakes suffer from pollution as they collect great amounts of industrial and civic and agricultural wastewater without any management or treatment (El-Hamid et al., 2017)
5	The appearance of the lake surface plants and corals which cover a big patch of the lakes
6	The illegal fishing acts on the lakes and the over-fishing cause the decrease in the quality of the water life of the lake
7	The accumulation and sedimentation on the Bougaz of the lakes is considered one of the big threats that face those lakes
8	The absence of environmental awareness for the fishermen about their fishing environment how to take care of it and also the absence of the regulating rules by the governorates adjacent to those lakes

Source Researchers

Table 3 Main details about Burullus lake

Location	Kafr-El Sheik Governorate
Arabic name	بحيرة البرلس
English name	BurulluslLake
Average depth	75–100 cm
Surface area	460 km^2
Max. width	6–21 km (3.7–13.0 mi)
Max. length	54 km (34 mi)
North border	Mediterranean sea
South border	Agricultural land

Source Researchers

5 Case Study Application: Lake Burullus

The Egyptian coastal lakes have witnessed a massive change in their water bodies due to severe urbanization activities, Burullus Lake is one of the most vulnerable lakes (due to the shallowness of its water body, which in turn means that any changes in the quality of the lake are easily spotted and identified), Lake Burullus is also the second-largest lake on the Mediterranean Sea, the lake is connected to the Mediterranean Sea via Boughaz El Burullus at the north-eastern part of the Lake, and many researches about the lake have been conducted focusing on the environmental, biological, and fishing environment and its natural resources (Abdel-Fattah et al., 2018).

This study explores the impact of the change in land-use and land-cover variations and its impact on the agricultural land and built environment that surrounds the Lake Burullus which leads by default to some urban problems.The Burullus Lake is one of the sites of the Ramsar Convention on Wetlands of International Importance in 1971 and was declared a national protected area.

The case study of the northern Burullus Lake was chosen according to the many potentials of the lake that could help improve the environment on both the environmental and the built environment scales.

The dramatic change of the land use and the land cover of Lake Burullus should be considered by many studies as it is an essential factor in sustainable development and an important factor to understand the relation between the lake and the surrounding built environment to achieve environmental sustainability. It is also essential to understand the phenomena of erosion (natural and man-made) of land cover of the lake especially the East to the West: Baltim, El-Hamoul, El-Riad, Sidi Salem, and Metobes land (Fig. 1).

A great concern by the changes in the size and quality of the wetlands took place after the massive change in the urbanization and agriculture land management. Wetlands, as an environmental tool, play a great role in the environmental balance as it works as an atmospheric greenhouse, emitting

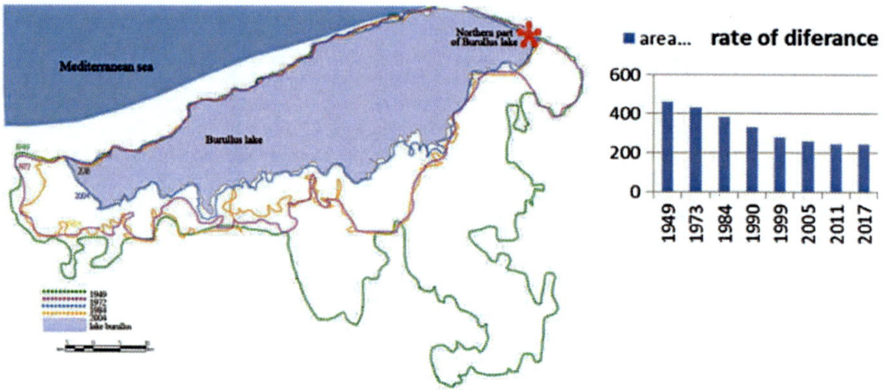

Fig. 1 Transformation of Burullus shoreline through time 1949/2017. *Source* Researchers

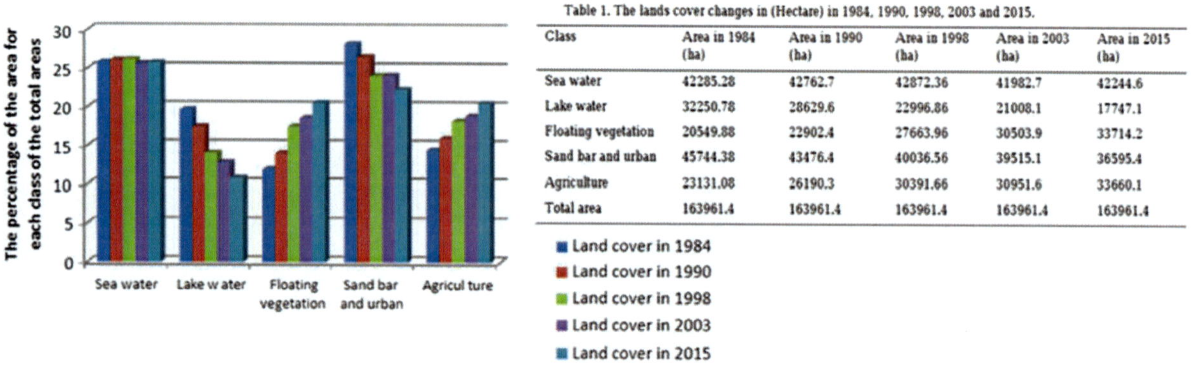

Table 1. The lands cover changes in (Hectare) in 1984, 1990, 1998, 2003 and 2015.

Class	Area in 1984 (ha)	Area in 1990 (ha)	Area in 1998 (ha)	Area in 2003 (ha)	Area in 2015 (ha)
Sea water	42285.28	42762.7	42872.36	41982.7	42244.6
Lake water	32250.78	28629.6	22996.86	21008.1	17747.1
Floating vegetation	20549.88	22902.4	27663.96	30503.9	33714.2
Sand bar and urban	45744.38	43476.4	40036.56	39515.1	36595.4
Agriculture	23131.08	26190.3	30391.66	30951.6	33660.1
Total area	163961.4	163961.4	163961.4	163961.4	163961.4

Fig. 2 Changes on the land cover of Lake Burullus. *Source* Hickmat Hossen and Abdelazim Negm/procedia

gasses such as methane and rich sources of carbon and nitrogen as well. The mapping and monitoring of those lakes should take place to calculate the change and to study how to stop it and how to manage the urban planning problem as well. (Fig. 2).

6 Lake Burullus Surrounding Districts

The changes of the shoreline of Lake Burullus, especially the northern part of it, were part of the improvement plan to enhance the quality of the shorelines of the detached cities (Baltim, El-Hamoul, El-Riad, Sidi Salem, and Metobes,). After the construction of the international coastal road that links both sides of northern parts of Lake Burullus, the economic opportunities of the city were heightened, and as a result, the produced urban fabric and the patch of the agricultural land changed widely.

The Lake Burullus basin mainly consists of three parts, namely eastern, middle, and western. The new land use and land cover that took place and the reusing of the agricultural land affected the urban life of the cities surrounding the Lake Burullus (El-Hamid et al., 2017) (Figs. 3, 4, 5, and 6).

7 Baltim

Baltim lies on the north of Kafr-El-Sheikh governorate opposite to the Mediterranean Sea, Burullus Lake, and el Hamual town. It lies over one of the sand dunes opposite to Lake Burullus, and Baltim is an active administrative and commercial node (unhabitat-2010), that is why the deterioration of Lake Burullus and the deterioration of Baltim shoreline have reached Baltim and made it suffer from the many problems related to the environment, urban, social, and economic level which, in turn, caused more pressures on the local authorities. On the other hand, the rapid urbanization caused agricultural land retreat and turned many lands into informal settlements (Fig. 7)

Baltim, as a city, was chosen as an illustration of how those factors affect the quality of urban development, and it also shows, through analysis, the real effect of urbanization on the smaller scale. The diagrams below show the

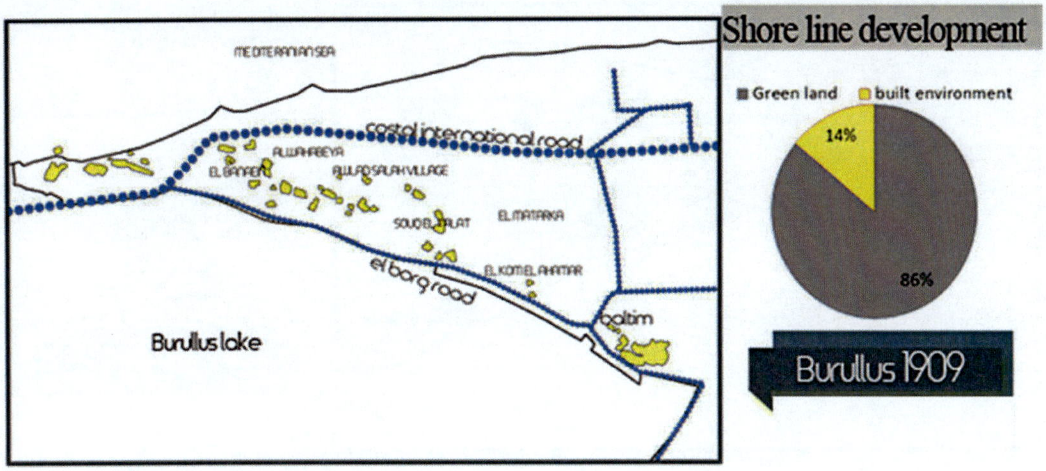

Fig. 3 Shoreline of Burulus at 1909. *Source* Researchers

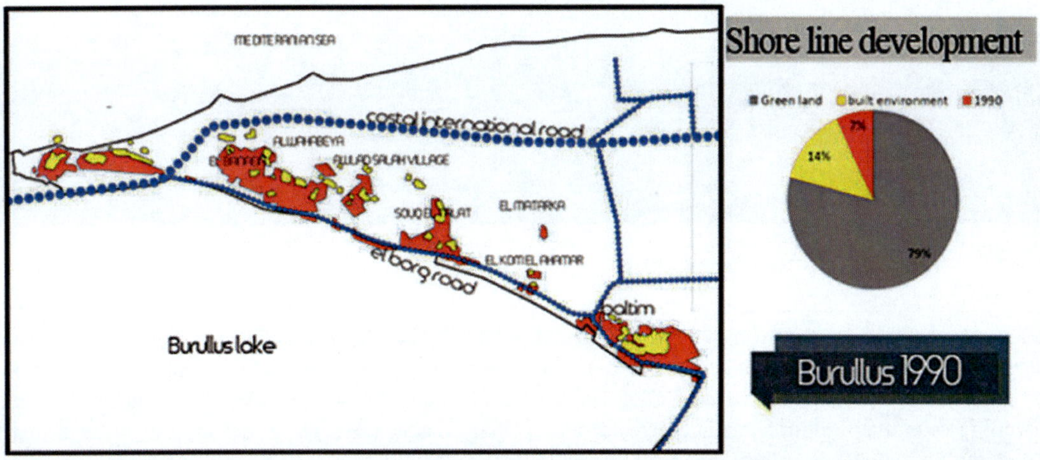

Fig. 4 Shoreline of Burulus at 1990. *Source* Researchers

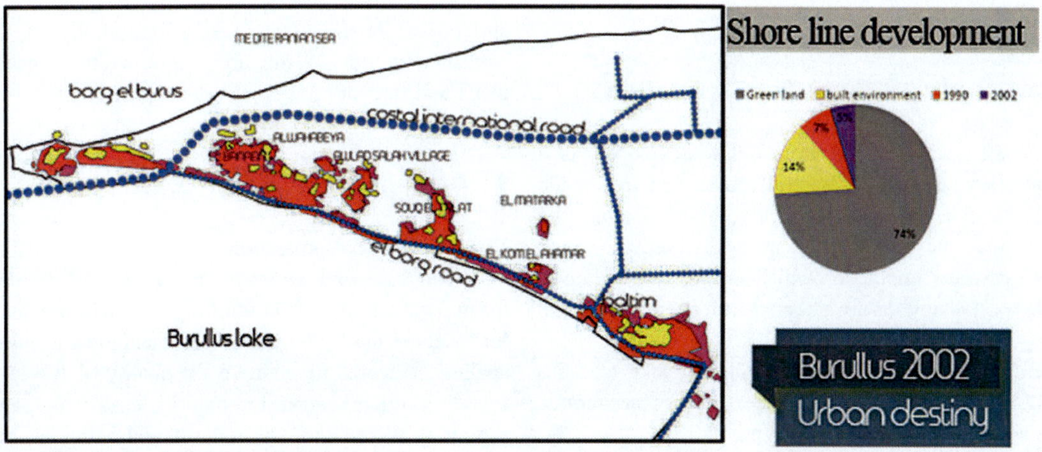

Fig. 5 Shoreline of Burulus at 2002. *Source* Researchers

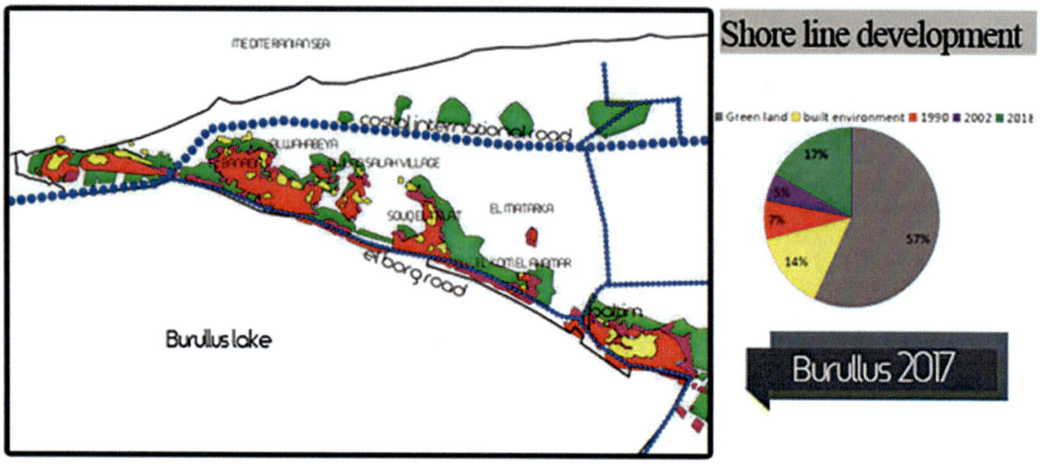

Fig. 6 Shoreline of Burulus at 2017. *Source* Researchers

Fig. 7 Part of Baltim shoreline 2017. *Source* Researchers

deterioration and the changes that occurred on the part of Lake Burullus that faces Baltim region and can also give a hint on how far the deterioration has reached other smaller cities as well. Despite that, Baltim has a huge area of agricultural land, most of which have inhabitants which work in fishing, and the economy of that area is based mainly on the marine life of Baltim and el Burullus Lake (Figs. 8, 9, 10, 11, and 12).

8 Analysis of Coastlines Transformation

Over the past years, coastal erosion has developed as a global problem due to both natural causes and man-made ones. Coastal erosion has also been recognized along the delta coastline, and around lakes as well, coastal change influences seawater imposition and flooding, and without a defensive coastal buffer, seawater could affect environmental life (Figs. 13 and 14).

Therefore, sustainable urban planning for lake land-use and land-cover changes is crucial for the socioeconomic development of coastal cities in Egypt.

A study for the shoreline of the northern part of Lake Burullus was extracted from satellite images to show thedifferences and the effect of shoreline changes into surrounding urban districts changes monitored from 1909/2016 to examine the changes and to answer the question of why that erosion takes place and how to manage that change? (Fig. 15).

And duo to the deterioration and changes that affect the shoreline, the green footprint of the agricultural land and the urban development of the area, this changes decreases the agricultural land and increasing informality of the built environment due to the absence of well-designed urban plans and the appearance of urban sprawls.

9 Planning Concept of Coastal Line

History of coastal protection

Urban lakes and wetlands in general work as a reservoir for rainwater to help in controlling the environmental supply for local use and agricultural needs, and even if it is small or shallow, it works to improve the quality of life in the surrounding urban areas, through increasing the amenities, providing recreational and educational activities, and even contributing to mitigate the urban climatic change. (Naselli-Flores, 2008) (Figs. 16 and 17).

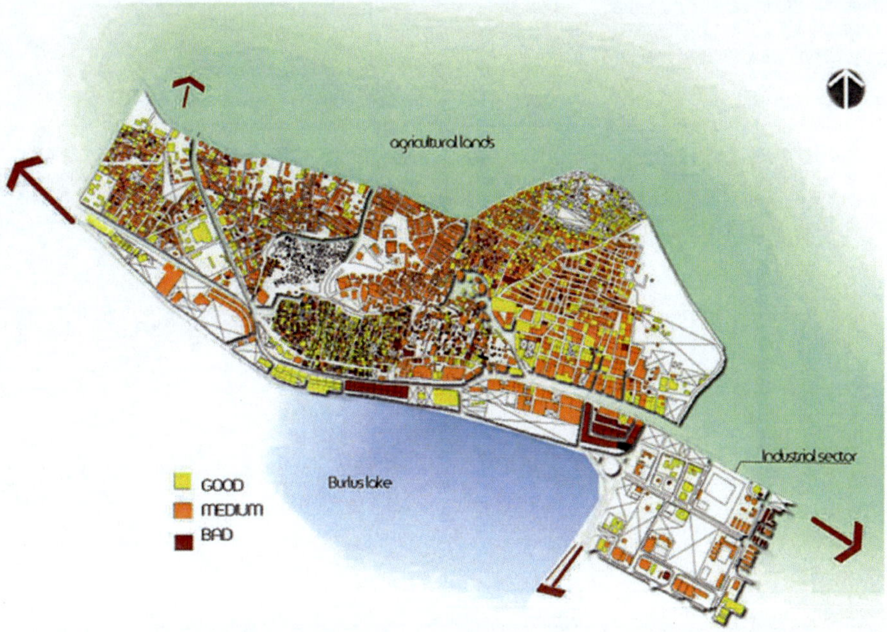

Fig. 8 Study case on the urban fabric (building conditions) of Baltim city 2017. *Source* Researchers

Fig. 9 Study case on the urban fabric (land-use variations) of Baltim city 2017. *Source* Researchers

Lake protection regulations (LPR) deal mostly with urban lakes, to guide the developers on how and where they should start. Historically, there has been limited leadership on how to propose an effective and efficient LPR which can protect lake resources, preserve the quality of recreational experience, and accommodate property rights of landowners. Usually, lake protection regulators focus on the moderately narrow line surrounding the lake where the new designed plan could take place (Krysel et al., 2003), assuming that lakes are so toughly influenced by overflow from their watersheds, they frequently need to be managed from a watershed viewpoint (Prok & Donald, 2004) (Table 4).

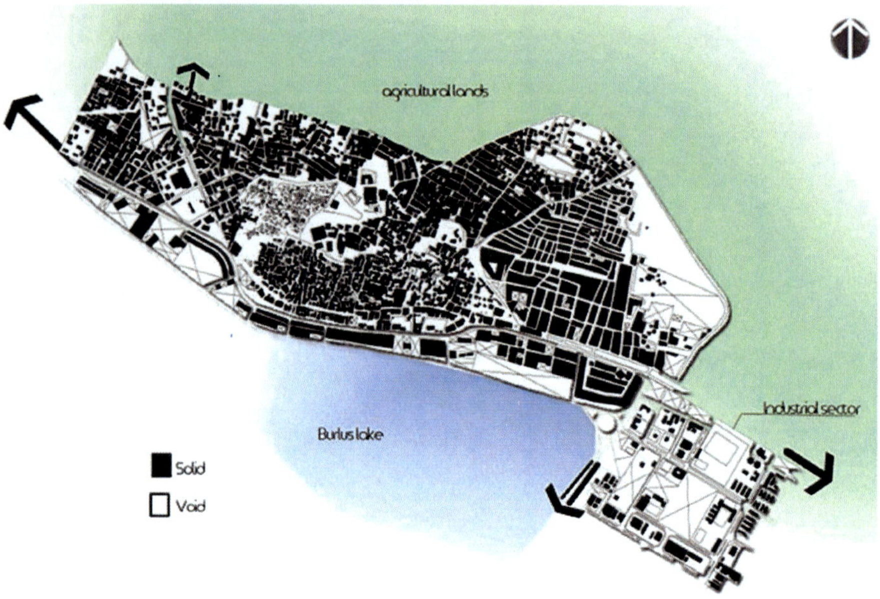

Fig. 10 Study case on the urban fabric (solid and void) of Baltim city 2017. *Source* Researchers

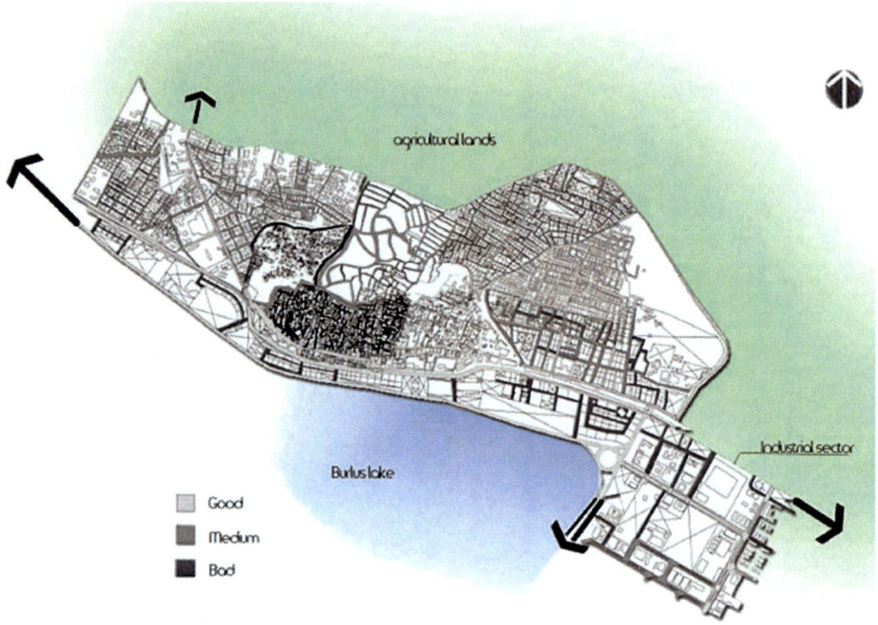

Fig. 11 Study case on the urban fabric (road conditions) of Baltim city 2017. *Source* Researchers

The delta coastal region, especially around the Burullus Lake, is gifted with a lot of resources. However, the environmental problems in the Burullus coastal region are becoming serious issues of concern, which call for immediate attention. The Burullus coast is characterized by a high-lying topography in most areas. The country's coastal boundary with the sea is classified into three broad regions, mainly morphological, vegetation, and beach type differences. Heraty (1993).

Unplanned and uncontrolled development of the coast has real potential to damage social, economic, and environmental interests of residents of the area. Coastal zones are

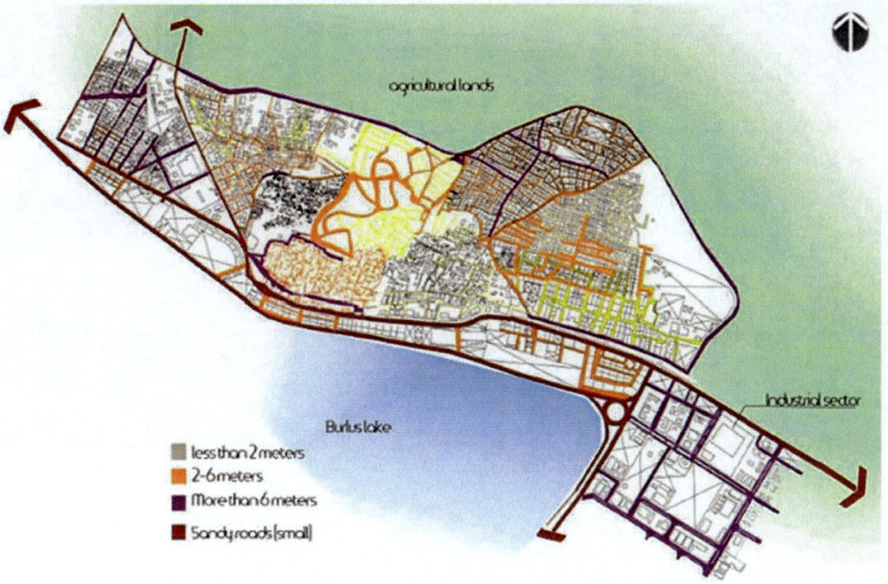

Fig. 12 Study case on the urban fabric (road width) of Baltim city 2017. *Source* Researchers

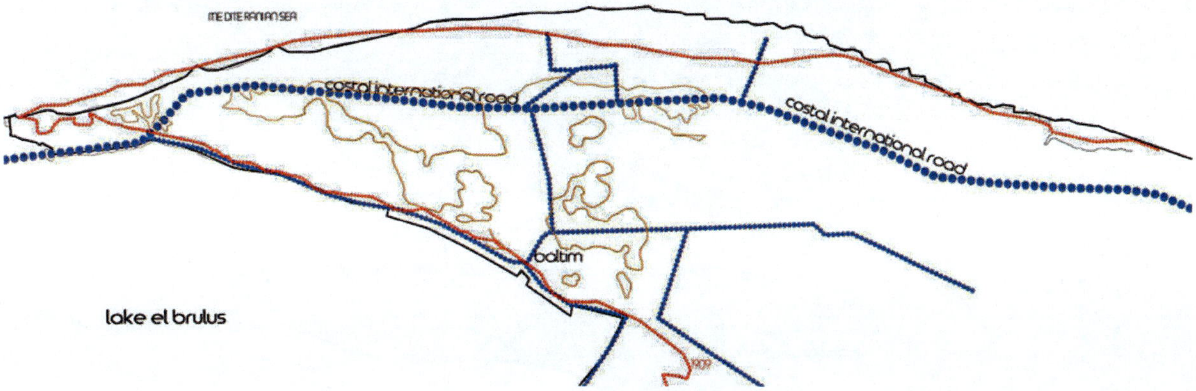

Fig. 13 Transformations of the coastline for the northern part of the lake from 1909/2017 (satellite map modified by researcher). *Source* Researchers

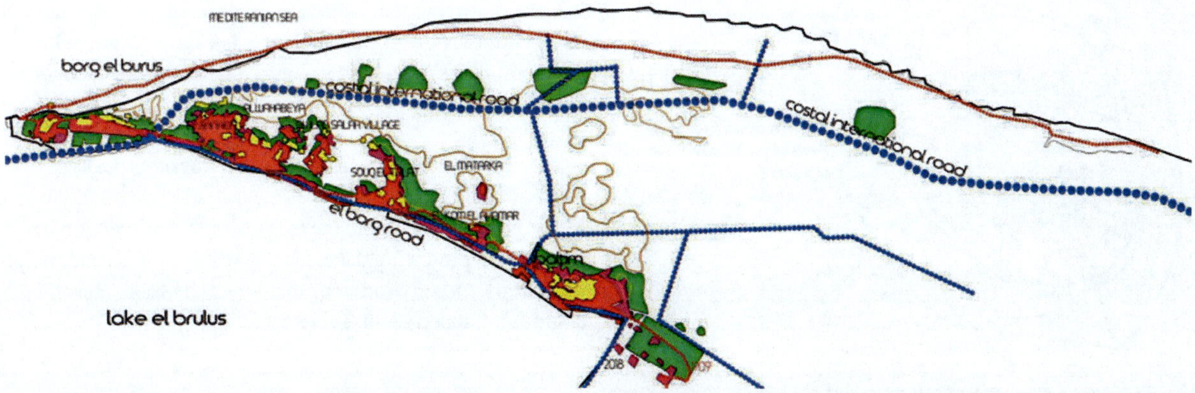

Fig. 14 Transformations of the coastline and its effect on the urban fabric of attached cities for the northern part of the lake from 1909/2017 (satellite map modified by researcher). *Source* Researchers

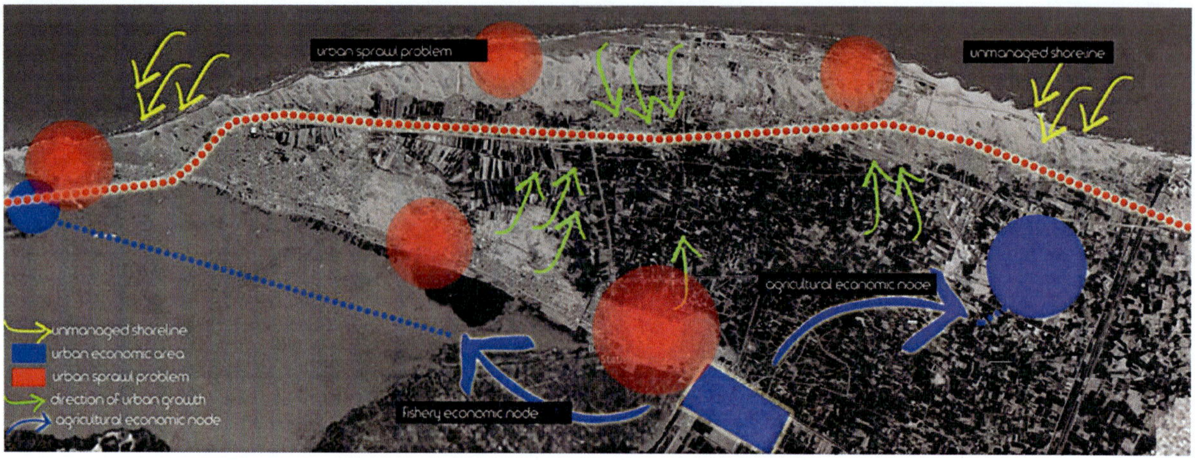

Fig. 15 Problems facing the lake. *Source* Researchers

Fig. 16 Typical development pattern around el Burullus Lake. *Source* Researchers

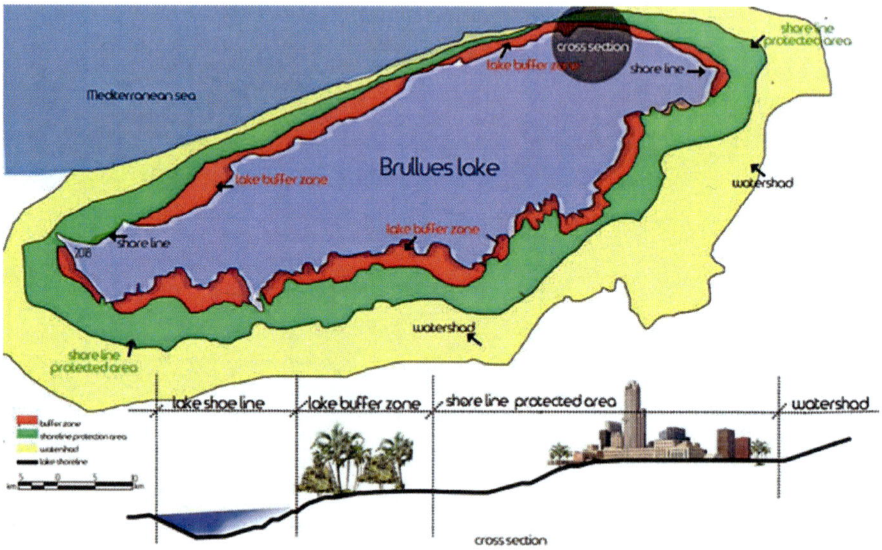

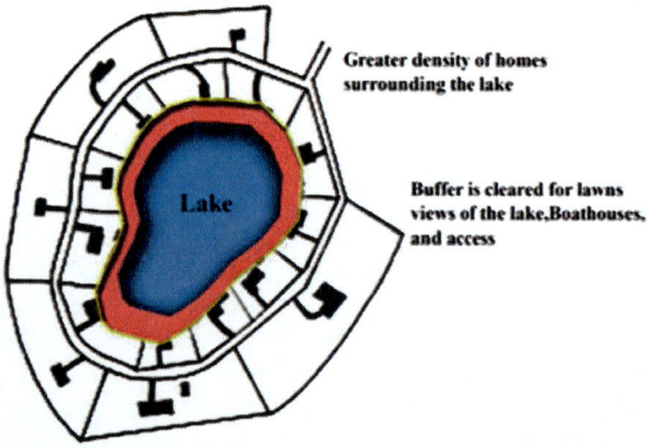

Fig. 17 Typical development pattern around el Burullus Lake. *Source* Ragheb (2018)

uniquely sensitive and vulnerable areas, and ensuring a balance between the natural environment and the human interventions is a great challenge because of the inherent "vulnerabilities" associated with coastal areas (Engel & Pederson, 1998).

– **Techniques proposed to protect Burullus Lake**

Many ways and strategic techniques of lake protection could help while protecting lakes, but putting into consideration the factors and environment that affect the Lake Burullus, it could be planned as in Table 5.

Table 4 Four main zones of lake

The four zones of lake protection			
The shoreline	Shoreline buffer	Shoreland protection	Watershed
Is the point where the high watermark touches the land and is considered to be the most valuable piece of land and has special treatment from the real estate developers as it has both good access to the lake water surface and clear view of the lake (Boak & Turner, 2005)	Is the area surrounding the lake with a common range of width between 15 and 45 m, and when well designed and maintained, it acts as a shield to protect the wildlife and reduce erosion. Shoreline buffer zones work as an aesthetic, recreational, and educational hub for lakefront development and surrounding communities as well (Engel & Pederson, 1998)	Is the space that comes after the shoreline buffer and ranges from 60 to 305 m, and it plays a great role in regulating the nature of development on the adjacent lots of urban lakes. And it mostly embraces residential development (Wenger, 1999)	Is the fourth protection zone that covers the watershed as a whole, as the buffer zone of lakes may not always be enough to protect the lake, it carries ample urban development to create communities on those large zones (Heraty, 1993)

Source Researchers

Table 5 Proposed principles for coastal development for Lake Burullus

Principles for coastal development	
1	Protect and mangle the natural systems present at Lake Burullus.
2	Identify the threat that faces the lake and try to reduce the weaknesses.
3	Apply comprehensive assessments to the region and site lower risk by exceeding standards for siting and construction..
4	Implement positive practices from dynamic coastal conditions.
5	Use more market-based decisions regarding agriculture, fishing tourism, and urbanism for fruitful development.
6	Mention the social and economic equity worries and the public's eight of access and use with private property rights.
7	Protect the sources of water to the lake by deep study of the environmental impacts.
9	Regulate the commercial fishes of this lake, and manage the process of fishery by studying the biology, dynamics, and life cycle of it. And that will help in improving the economic life of the surrounding district that mainly works on fishing.
10	Water quality should be improved to maintain the case of Lake Burullus and to reduce environmental effects
11	Prohibit the destroying of the agricultural land on the northern part of the lake to avoid the urban sprawls formed by the act of urbanization
12	Clear the Boughazes of the lake regularly to avoid the change of the rate of water flow

Source Ragheb (2018)

10 Proposed Shoreline Development Plan

Developing and maintaining the shoreline of any community is essential to any country development plan, as the alteration that occurred along those shorelines happened due to the environment may also affect, in the long run, the land cover and land use of the adjacent area. What maybe called a "living" or "vibrant shoreline" may actually stimulate natural coastal process, improving the attitude of growth around those lakes and shorelines to maintain the case of the surrounding areas (Fig. 18).

A shoreline development plan offers a large scale of assessments of the risk accompanying with the shoreline evolution and presents a strategic framework to point at those risks and how to treat them in a sustainable manner which will also help in improving the urban fabric adjacent to that area as well. Our study introduces a proposed framework and strategy to develop Lake Burullus for future sustainable treatment.

Table 6 Plan proposed for the lake development and management

Plan proposed for the lake development and management	
Long term	Short term
Improve the main roads, and introduce new parallel roads to the main international coastal road, to reach to the depth of lands on both sides	Improve the urban fabric of the adjacent communities each one according to the type and the problem statement
Improve the quality of social and economic life for the residents, and increase the awareness of the importance of the agricultural and the wetland as well	Improve the industrial sector to enhance the economic life of the area for both fishing and agriculture
Treat el Bougaz zone, and remove the accumulations made by mud to secure the Lake Burullus	Improve the quality of Lake Burullus water by stopping the invasion and decreasing the alterations made by man-made
Improve the area of manufacturing the boats, and improve the area for the fisherman to work and live	

Source Greater Cairo urban development strategy, 2012, part 1 future vision and strategic directions

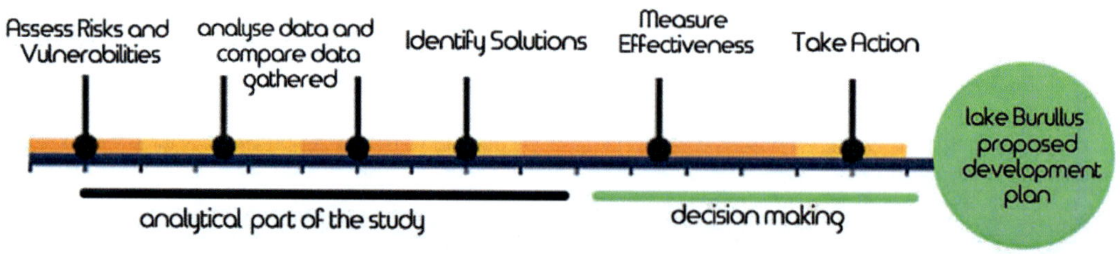

Fig. 18 Steps the study takes to reach for the proposed solution. *Source* Researchers

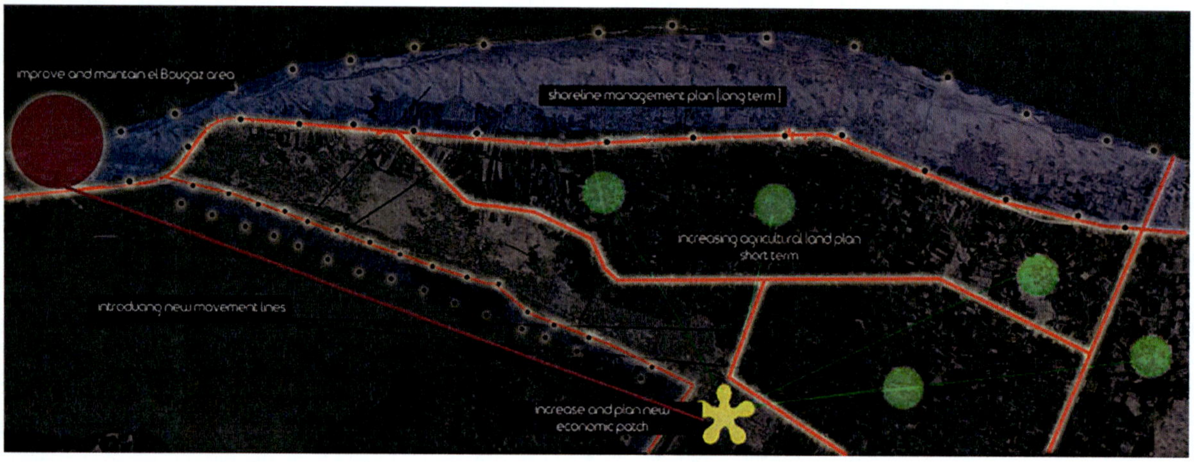

Fig. 19 Diagram for the proposed solution. *Source* Researchers

The target area for the study was chosen to be the northern part of the lake Burullus, due to its importance and the consideration of having the most condensed urban fabric and the most vibrant part of the lake. The northern part has a main road passing and linking the northern part to the other land "international coastal road" and also plays a role in improving the area by linking those parts all together, one of the main problems of the area is the deterioration of agricultural land due to urbanization, and one significant issue is the absence of certain policies and regulations to control those agricultural lands in the country's development plan (Fig. 19).

Throughout out the study, it was concluded that the effective planning process should involve the following stages:

- Engage property owner,
- Set project goals,

- Determine a timeline,
- Determine permitting requirements,
- Determine site conditions and develop project design,
- Develop a monitoring plan, and
- Plan for project construction (Table 6).

11 Research Findings

Shoreline are considered to be important ecological areas which play a role in balancing the ecosystem due to their vulnerability regards the easy changes that occur due to environmental actions. Successful management or development of any shoreline requires some long-term and short-term plans and adaptive management, and monitoring the project overtime will help in stating if the development is either progressing or recessing (it may also conclude if the diversity of the environment and man-made changes have influenced it or not).

The study area could be transformed to help and to improve the social, economic, urban, and environmental life of the surrounding areas, and the results from this study can provide information for development planning to well organiz
e for future development.

Recommendations of Burullus Lake scale: Preserve the shoreline environment, through stabilizing the environmental–ecological balance, which should be well managed, in order, to protect the lake from future deterioration throughout. This may be achieved through creating buffer zones, water tests, and increasing awareness of how to protect the lake.

Baltim district scale: Holistic and integrated improvement and management plans should be taken into consideration by the decision-makers while focusing on the laws and legislations present to help in deterring the deterioration of the agricultural land patch and how to regulate the economic actions within the range of Baltim district with no harm to the lake itself.

Environmental scale: The rapid erosion of the coast of Burullus is caused by several processes that act together; these are natural processes that occur frequently (cyclones, waves, and tides), anthropogenic activities such as human settlement and aquaculture that remove mangroves and other coastal vegetation should stop, and sustainable processes should take their place.

Economic scale: Improve and design the industrial sector that faces the Lake Burullus from the northern part to help in improving the economic life, while introducing more roads and factories to support both fishers' life and agricultural products.

Agricultural scale: More care should be taken when treating the agricultural land, beginning with legislations, from the community or by the users themselves, through providing the way of using those lands for more income to the inhabitants and also by utilizing public awareness campaigns.

Fishing scale: Create a healthy environment for the fishermen that facilitates fishing and helps in increasing that source of economic income, while protecting the unique species of fish that started to disappear with all the environmental and man-made changes.

References

Abdel-Fattah, S., Negm, A. M., & Bek, M. A. (2018). Summary, conclusions, and recommendations for Egyptian coastal lakes and wetlands: Climate change and biodiversity. In *Egyptian coastal lakes and wetlands: Part II* (pp. 261–270). Springer. https://doi.org/10.1007/698_2018_333.

Boak, E., & Turner, I. (2005). Shoreline definition and detection: *A Review. Journal of Coastal Research. 21*, 688–703. https://doi.org/10.2112/03-0071.1.

El-Hamid, H. T. A., Hegazy, T. A., Ibrahim, M. S., & Khalid, M. (2017). Assessment of water quality of the Northern Delta Lakes, Egypt. *Journal of Environmental Sciences, 46*(1), 21–34

Engel, S., & Pederson Jr, J. L. (1998). The construction, aesthetics, and effects of lakeshore development: A literature review.

Heraty, M. (1993). Riparian buffer programs: A guide to developing and implementing a riparian buffer program as an urban stormwater best management practice. *Wetlands and Watersheds.*

Krysel, C., Boyer, E. M., Parson, C., & Welle, P. (2003). Lakeshore property values and water quality: Evidence from property sales in the Mississippi Headwaters Region. Submitted to the Legislative Commission on Minnesota Resources by the Mississippi Headwaters Board and Bemidji State University.

Naselli-Flores, L. U. I. G. I. (2008). Urban lakes: ecosystems at risk, worthy of the best care. In *Proceedings of Taal2007: the 12th world lake conference* (Vol. 1333, p. 1337).

Prok, J., & Donald, P. (2004). *Guide to lake protection and management*. Published by the Freshwater Society in cooperation with the Minnesota Pollution Control Agency.

Ragheb, A. A. (2018). Sustainable Urban Planning for Protecting Salt Lakes; case Study Bardawil Lake, North Sinai, Egypt. *International journal of Scientific & Engineering Research, 9.*

Wenger, S. (1999). *A review of the scientific literature on Riparian buffer width, extent and vegetation.* University of Georgia.

Investigating the Children's Satisfaction in Cairo Toward Achieving Child-Friendly City

Haidy Ehab, Abeer Elshater, and Ahmed Toimah

Abstract

Children are the eventual future of the city. Children who grow up feeling truly at home here as secure, energetic, engaged citizens will be grown-ups who add to making an incredible city for the age to come. Children are additionally among our most powerless natives, and their hardships or prosperity is the best pointers of what our city resembles as a home for us all. The scientific research documented a prove that children are the most critical elements in their surroundings. Children in Egypt live in a polluted, unsafe, and overcrowded urban environment that provides very few opportunities for learning and recreation which consequently result in them being unable to gain skills for urban competence. The literature describes Egypt in a way that does not encourage children to explore their surroundings or learn the needed social, physical, and physiological skills required to be good citizens. This research discusses the children's satisfaction toward the built urban environment in Egypt. The aim is to investigate the possibilities of the application of the approach of the child-friendly city (CFC) in Egypt. The results show that the outdoor environment is appropriately sufficient for children's needs. The contribution presents an evaluation of the concept of CFC for the urban communities and how the investigated points can integrate children as a participatory part in the urban planning and design process. The research follows the inductive path with open-ended questions asking children and their parents about the place in Heliopolis, Cairo. Interviewees were questioned about their satisfaction and their experience in the public realm. The findings extracted a solution to address children's inconvenience.

Keywords

Child-friendly city • Human's needs • Urban environment • Physical activities • Children's participation

1 Introduction

Children are the eventual future of the city. Children who grow up feeling truly at home here as secure, energetic, engaged citizens will be grown-ups who add to making an incredible city for the age to come. Children are additionally among our most powerless natives, and their hardships or prosperity is the best pointers of what our city resembles as a home for us all. Urban communities and cites can be either positive or negative places for children, depending on their social status within society. Cities, towns, and neighborhoods should be the place where children can socialize, observe how communities work, and play a role in shaping the culture of their communities. There also should be places where children feel sheltered and supported. On the other hand, cities should provide children's needs and commitments as set out in the Convention on the Rights of Children. Therefore, there should be sustainable systems and local capacity. This will only be possible by creating environments that are based on cooperation and partnerships at a variety of different levels across national and local borders, across social domains and groups (Malone, 2006).

Around the world, numerous things have affected children's well-being and prosperity such as industrialization, increasing population, ecological degradation, and war (Gleeson & Sipe, 2006). UNEECIF believes, "*all children have a right to survive, thrive and fulfill their potential to the benefit of a better world*" (UNICEF, 2017). Within this context, this paper will discuss why the built urban environment in Egypt is not appropriately sufficient for children's needs as does not take into consideration the aspect of

H. Ehab (✉) · A. Elshater · A. Toimah
Department of Urban Planning, Ain Shams University, Cairo, Egypt

F. Trapani et al. (eds.), *Advanced Studies in Efficient Environmental Design and City Planning*, Advances in Science, Technology & Innovation, https://doi.org/10.1007/978-3-030-65181-7_40

involving children as an important age-group in the process of city planning and design.

This paper investigates the possibilities application and approaches of CFC in Egypt for a better society and community. The concept CFC here considers children as citizens of today, not merely the adults of tomorrow. This paper also explores the status of why Egypt does not meet the essentials criteria to be considered a CFC until now. The concluded remarks reach what are the difficulties faced by urban communities to apply this approach to achieving CFC. The results also will show that the urban communities work on solving these problems to achieve their goal for a CFC. The research is based on the UN estimations that by the year 2025, most of the world population will live in cities, of which 60% are children (Christensen & O'Brien, 2003).

The life and experience, for millions of children, are partly shaped by their environments. Around 25% of the population of western countries is younger than 18, and it is clear that the inclinations, needs, and encounters of these children are not quite the same as those of adults (Frank, 2006). These inclinations and needs are based on the children's recognition of their environment. Accordingly, understanding how children perceive, see, and relate to their social and physical surroundings is fundamental to provide their needs in the design and planning procedure (Al Arasi, 2013). Globally, according to World Bank statistics, the population of children by age of 0–14 is 26%, as in some countries it increases to 48%, which indicates the importance of children in cities (World Bank, 2018) (Fig. 1).

In their regular day-to-day existence, children to a great extent remain inside and identify with three settings—their homes, schools, and recreational organizations. These conditions have been made by grown-ups and assigned by them as "places for children" (Rasmussen, 2004). Scientific research proved that children's most important elements of their surrounding urban communities are their homes, their relatives' homes, commercial areas, and playgrounds as well as independent mobility. These elements can mostly be viewed as urban planners and urban designer's jobs to provide. Children constitute a large part of the urban population. Besides, they are a significant piece of urban areas: They live and develop in these communities. Yet, they are rarely considered in the design and planning procedures, and they are mostly rejected from the basic leadership that affects their neighborhoods, environments, and the places they use and visit daily basis. Decision-makers consider children as clients of places such as open parks and schools, yet adults are the ones who control and shape these places (Torres, 2009).

Since children are simply finding out about their environment and the world, their condition of the world they live in will significantly impact almost all aspects of their lives. This puts a lot of pressure on planners, who have to accommodate children's needs in urban design to ensure that these places will be child-friendly. Four significant issues are basic to the creation and upkeep of a child-friendly community: integration, greenspace, safety, and access. The advantages of a child-friendly community range from improving the nature of social collaborations, the advancement of more advantageous ways of life, and to long-term sustainability of natural spaces (McAllister, 2008).

Children are at the center of development agendas. Accordingly, initiatives such as child-friendly cities without slums, civic stewardship can improve children's lives (Gleeson & Sipe, 2006; UNICEF, 2004).

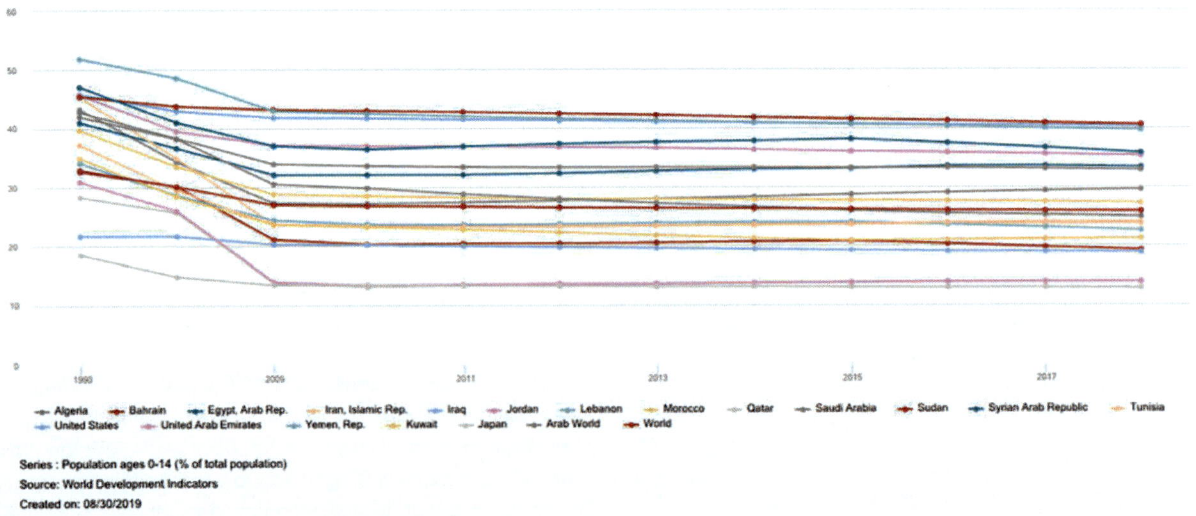

Fig. 1 Population age 0–14. *Source* World Bank (2018)

Policies of The UN Convention on the Rights of the Children focus on improving the future of children, yet they do not consider their present lives, and issues that arise as children's rights to life should incorporate their rights to live decent lives in both the present and the future (van Gils, 2012).

2 Literature Review

In this part of the paper, we will address the term child-friendly city, which is formed and occurs according to some elements that are said to be the nine building blocks for the child-friendly city development and also address the aims of this approach and how to focus on children to make the city belong to them, and they belong to it in order to reach the significance of children's participation in the decision-making of the city urban design and planning process where they become commensurate with their needs and ambitions and meet all the perceptions that revolve in their minds.

2.1 The Approach of Child-Friendly City

The idea of child-friendly city is applied equally to the governance of all communities which incorporate children (UNICEF, 2004). There are some recognitions of several important trends emerged in the child-friendly city initiatives; the fast diversion and urbanization of worldwide social orders; the developing duties of municipal and community for their populaces with regard to decentralization; and therefore, the expanding significance of urban areas and towns inside national political and financial frameworks which will be a procedure for proceeding the resident's highest quality of life all of them. (UNICEF, 2004). CFC is focused on the fullest implementation of the Convention on the Rights of the Child (UNICEF, 2004) (Fig. 2).

The child-friendly city has been identified as UNICEF said "…. a city, or any local system of governance, committed to fulfilling children's rights. It is a city where the voices, needs, priorities, and rights of children are an integral part of public policies, programs, and decisions. It is, as a result, a city that is fit for all" (UNICEF, 2017). The building

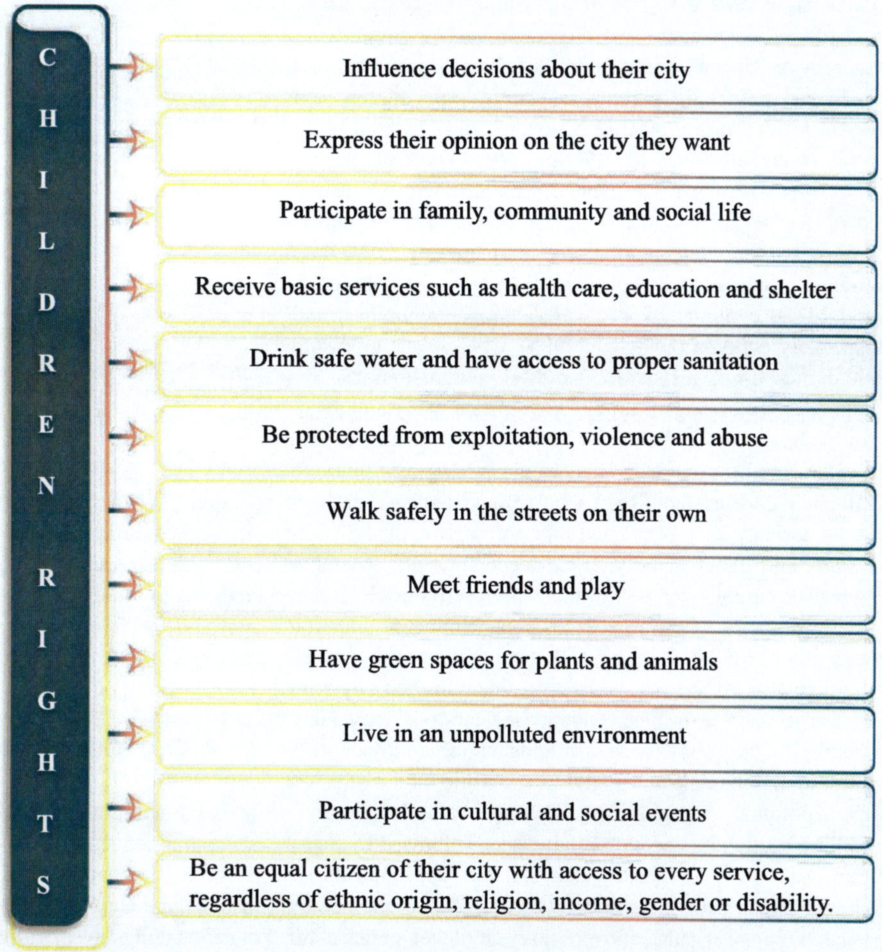

Fig. 2 Diagram of the children's rights. *Source* UNICEF (2004)

of a child-friendly city is the way toward actualizing the Convention on the Rights of the Child under leading of local government. The CFC's aim is to improve children's lives now by recognizing and understanding their privileges—and subsequently change for the better networks today and what will happen in the future.

Building child-friendly cities is a practical procedure that must connect effectively with children and their genuine lives. The building process can develop from or pull together other child-friendly initiatives: child-friendly hospitals and schools; environmental projects to guarantee children safe water and hygiene. Children themselves, or child-led organizations, other non-governmental organizations, or human rights institutions—a children's ombudsman—can initiate a campaign (UNICEF, 2004).

3 Child-Friendly City in a Child-Friendly World

The child-friendly city is an arrangement of good nearby administration that focuses on the full implementation of the Convention on the Rights of the Child. Enormous urban communities, medium-sized cities like smaller networks—even in provincial settings—are all called to guarantee that their governance is given the priority to children and involved in decision-making processes, and let them be in basic leadership forms. (UNICEF, 2004). The CFC framework gives an expansive methodology that will be custom fitted as per neighborhood needs, goals, and practices. Adjusting the framework is a participatory procedure including every single concerned partner—neighborhood specialists, common society, specialists, communities, and especially children (UNICEF, 2004). CFC means to ensure children's privileges to fundamental administrations, for example, well-being, training, cover, safe water, conventional sanitation, and security from brutality, misuse, and abuse. Similarly, it tries to engage young residents to influence choices about their city, express their assumption about the city they need, and take an interest in the community and public activity. It elevates children's privileges to walk securely in the boulevards without anyone else, meet companions and play, live in an uncontaminated situation with green public spaces, take an interest in social, get-togethers and be an equivalent resident of their city with access to each assistance, without separation of any kind. During the previous decade, a various number of urban areas in cities and municipalities all over the world have settled on the political decision to become "child-friendly". Child-friendly cities programs have been embraced by many European cities to sensitize mayors and city councils on child rights, ensure that children are included in the nearby local political agenda, and promote city-level policies for children. When studying the approach of the child-friendly citym, we found that to build a medium that adapts to the child and its requirements to be a friendly city there are nine elements to build and develop the city to be a child-friendly city, and the following drawing shows these nine elements and the concept of each of them (Fig. 3).

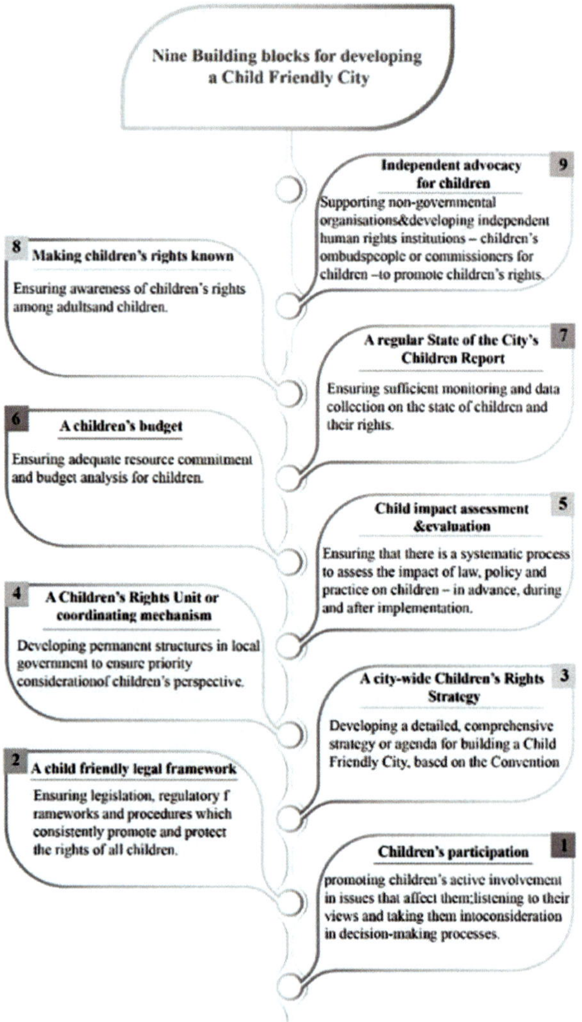

Fig. 3 Building blocks for developing a child-friendly city. *Source* UNICEF (2004)

4 A City with Children in Mind

To begin with, in light of the fact that it is a legal obligation: under the convention, states are committed under universal law to ensure that the children ultimately benefit from a basic idea in all activities related to children. Perceiving and understanding every other appropriate for children are

Fig. 4 Children in a city agenda, *Source* The authors based on UNICEF (2004)

additionally legal commitments, adopted when the state sanctioned the convention (UNICEF, 2004) (Fig. 4).

5 Children as a Part of the Community

Children's lives are influenced by the surrounding environments. Children's well-being is connected to the nature of water, air, and nourishment. Likewise, the physical environments of school and home impact a children's wellbeing through exposures to lead, merge, clamor, or encompassing light. (Committee on Environmental Health, 2009).

The environment significantly impacts a child's social, physical, and mental health. Yet, little efforts on creating child-friendly communities have been made, mostly children are neglected in land use planning in two different ways:

Firstly, they are rarely considered when it comes to design (Matthews & Limb, 1999). Moreover, they are offered a few opportunities in terms of participating in decision-making (Bartlett, 2005). There is a lack of planning with and for children.

Children's involvement in participatory processes has been appeared to positively affect their self-improvement and to profit other children just like grown-ups. In any case, individuals new to children's cooperation may stress that it is something hard to do or that it will give kids rights over those of their elders who ought to be perceived as the decision-makers in society (Jamieson et al., 2011). The participation of children alludes to their involvement in discussions that educate choices about their lives and their society. It goes past children being present, available, to asking those in control to create opportunities for children to have influence (Jamieson et al., 2011). Children's involvement will never turn into a reality without holding governments completely to guarantee that children are empowered to claim their privileges to be heard and be taken seriously in all decisions that affect them (Percy-Smith & Thomas, 2010).

Children too are the focus of the urban designer interest (Elshater, 2018). It is assessed by the UN estimations that by the year 2025, most of the world population will live in cities, of which 60% are children. For many children, the forms of their regular everyday life and experience are (to some extent) constituted by city environment surroundings. As the views of urban designers differed about what the city is and what are the components of the city and what are the elements through which the city is formed and integrated. Thus, the societies will be friendly to the child and the environment, and if the society becomes friendly to the children, child-friendly city will be formed. The timeline indicates the literature of children from the views of urban designers and planners, so we can conclude from it the importance of children in the city (Fig. 5).

6 Children in Cairo: The Case Study

Children in Egypt live in a polluted, unsafe, and overcrowded urban environment that provides very few opportunities for learning and recreation which consequently result in their being unable to gain the necessary spatial, physical, social, and analytical skills required for urban competence. The built environment in Egypt does not encourage children to explore their environment or learn the needed social, physical, and physiological skills required to be good citizens for their country's future and hence the world's future. The design of the city in Egypt adversely affects the children's lives and their behaviors. In this paper, we will take Heliopolis as a case study area which consists of four zones (*Shayakhat* in Arabic) with children population for ages from 0 to 14 in the year 2018 which reaches 136,620 persons (Central Agency for Public Mobilization & Statistics, 2019) and in the following chart shows the division of the number of children in each Shayakhat as shown in (Figs. 6 and 7).

6.1 Data Collection and Findings

The research outline consists of three phases to reach the final results (Fig. 8). This study investigates quantitative information collected from a survey powered by SurveyMonkey.[1] This survey investigates the collected data

[1]https://www.surveymonkey.com/r/QTZ35DN.

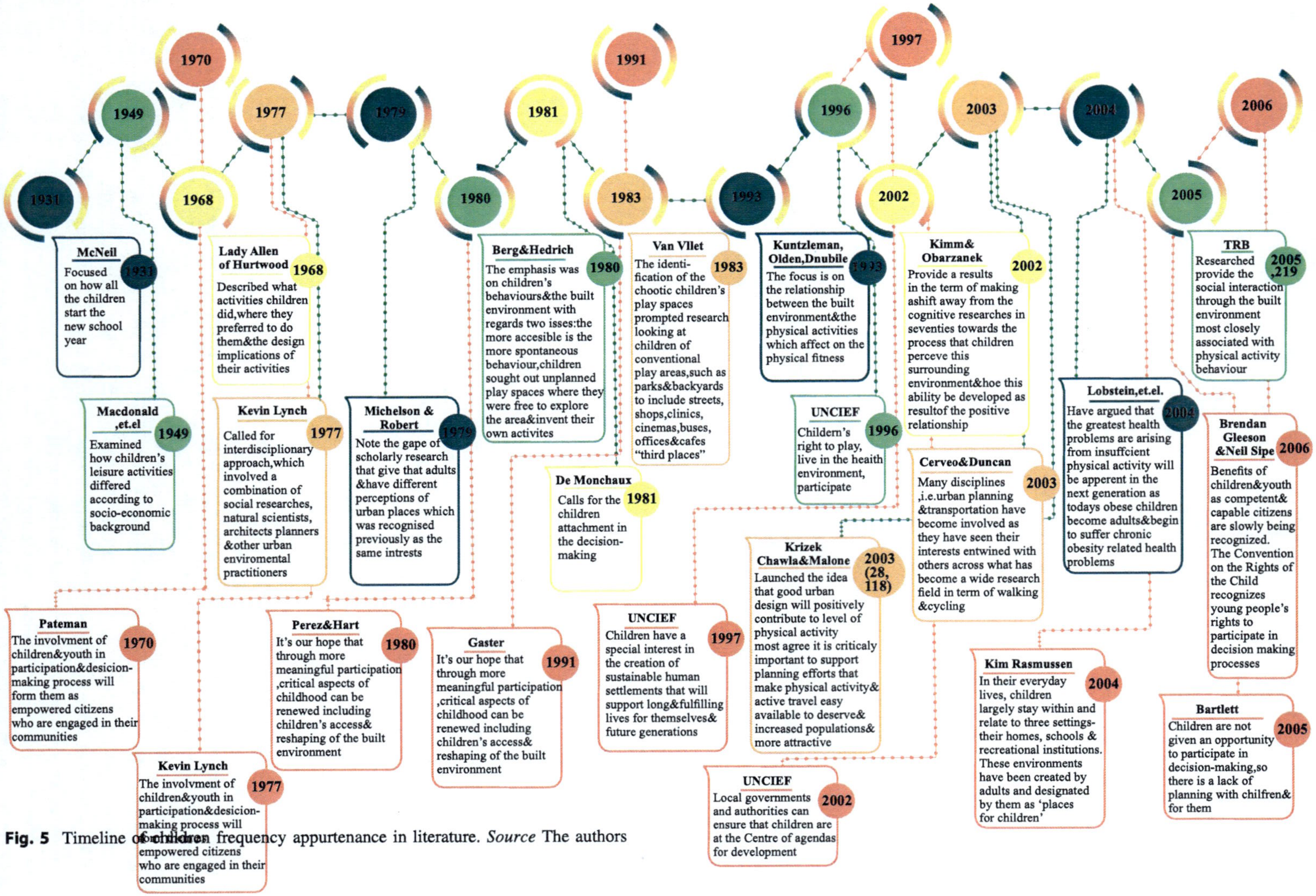

Fig. 5 Timeline of children frequency appurtenance in literature. *Source* The authors

Fig. 6 Location of the case study area and Shayakhat classification. *Source* Google map with modification by the authors

Fig. 7 Heliopolis population statistical analysis. *Source* CAPMAS (Central Agency for Public Mobilization & Statistics, 2019)

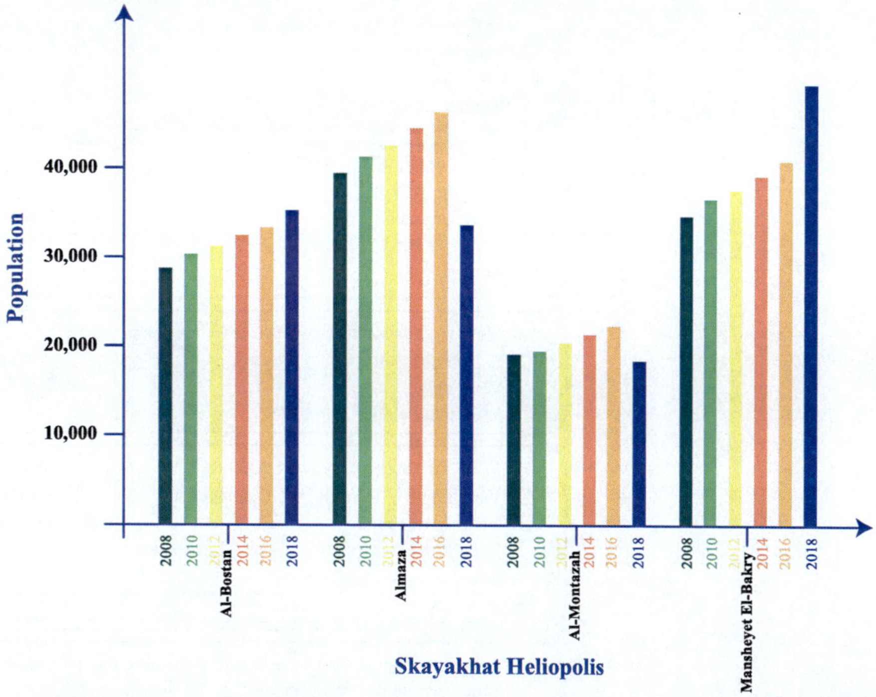

Shayakhat Heliopolis	2008	2010	2012	2014	2016	2018
Al-Bostan	28821	30126	31346	32610	33784	35264
Almaza	39315	41096	42760	44480	46087	33778
Al-Montazah	19030	19895	20697	21537	22309	18257
Mansheyet El-Bakry	34799	36373	37846	39372	40788	49321

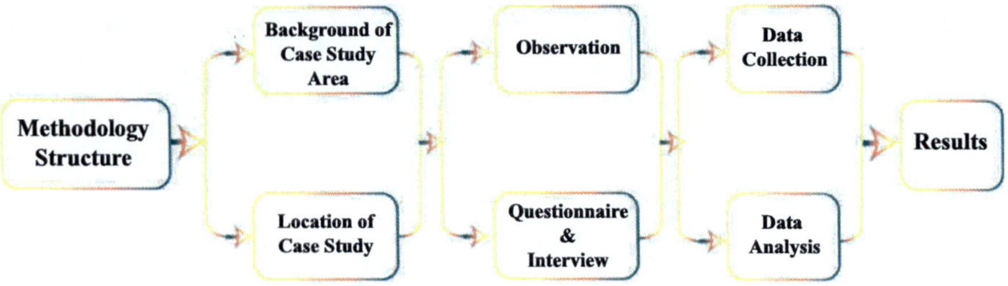

Fig. 8 Research outline. *Source* The authors

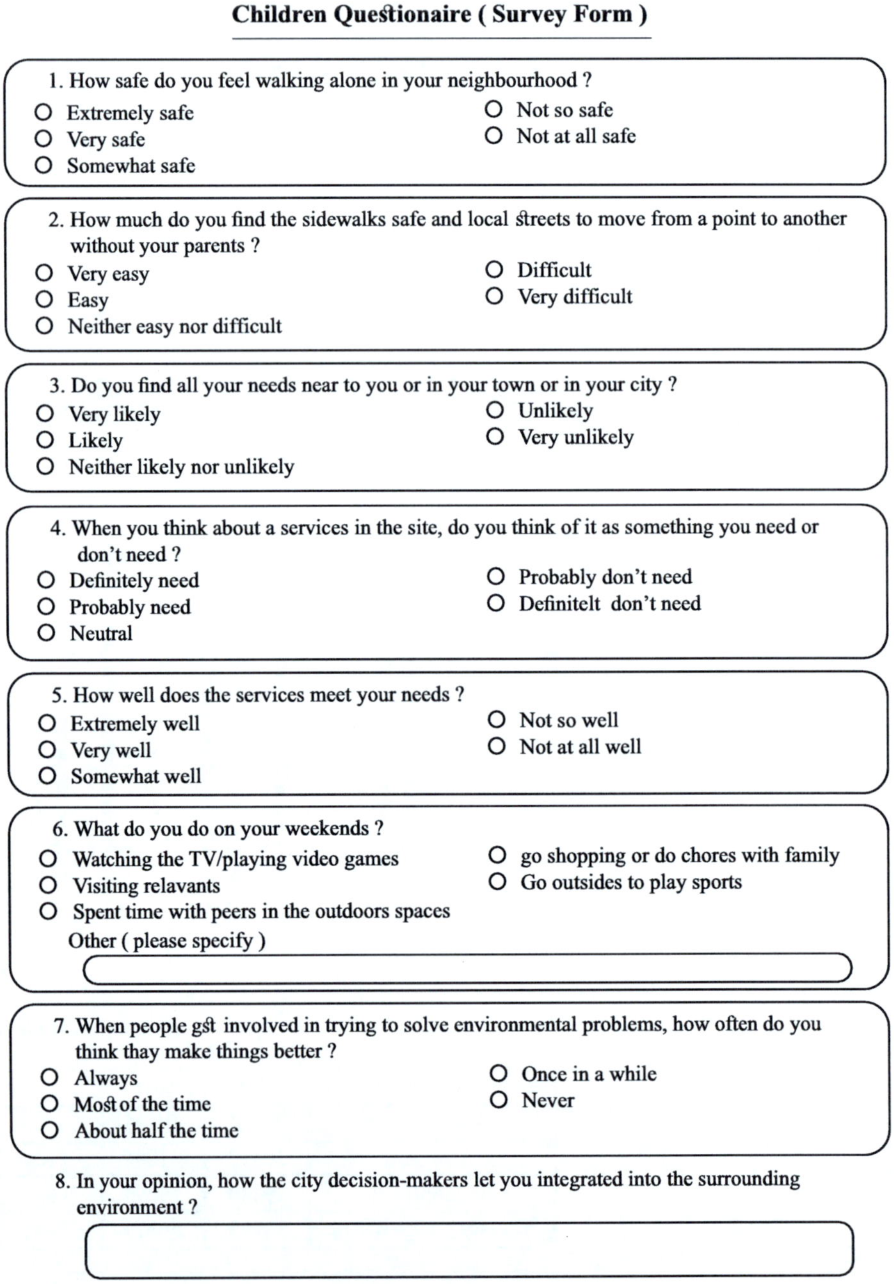

Children Queštionaire (Survey Form)

1. How safe do you feel walking alone in your neighbourhood ?
   - O Extremely safe
   - O Very safe
   - O Somewhat safe
   - O Not so safe
   - O Not at all safe

2. How much do you find the sidewalks safe and local štreets to move from a point to another without your parents ?
   - O Very easy
   - O Easy
   - O Neither easy nor difficult
   - O Difficult
   - O Very difficult

3. Do you find all your needs near to you or in your town or in your city ?
   - O Very likely
   - O Likely
   - O Neither likely nor unlikely
   - O Unlikely
   - O Very unlikely

4. When you think about a services in the site, do you think of it as something you need or don't need ?
   - O Definitely need
   - O Probably need
   - O Neutral
   - O Probably don't need
   - O Definitelt don't need

5. How well does the services meet your needs ?
   - O Extremely well
   - O Very well
   - O Somewhat well
   - O Not so well
   - O Not at all well

6. What do you do on your weekends ?
   - O Watching the TV/playing video games
   - O Visiting relavants
   - O Spent time with peers in the outdoors spaces
   - O go shopping or do chores with family
   - O Go outsides to play sports

   Other (please specify)

7. When people gšt involved in trying to solve environmental problems, how often do you think thay make things better ?
   - O Always
   - O Mošt of the time
   - O About half the time
   - O Once in a while
   - O Never

8. In your opinion, how the city decision-makers let you integrated into the surrounding environment ?

Fig. 9 Children questionnaire (survey form). *Source* The authors from children by asking them about their life in public places, their activities, and needs. The survey also collected data from interviewing with children's parents to determine how their children interact with their surroundings. The materials of data collection operate these interviews with children's parents asking what they need in the study area and how satisfied they are. A question in both the survey and questionnaire collected data about what their requirement is to be happy and to feel friendly with the outdoor lived environment.

As part of this scientific research to arrive at the best analysis of the survey, we calculated the sample size to be analyzed for the best results according to the 2018 population. We have a sample size of the population that reaches 136,620 persons. The confidence level was estimated to be 90% with a margin of error of 15%. The sample size was calculated to receive a response from 31 persons. In interviews conducted with children aged from 10 to 13 and from 14 to 18 years old, it was scheduled to receive results from a randomly selected sample to investigate their ability to respond and to get better results.

Then, after we choose the sample size of children, we put the questionnaire of children as a survey powered with SurveyMonkey as shown in (Fig. 9).

The results gained from the survey with children show that the answers to the questionnaire vary from one to another due to differences in age group and lifestyle. One of the main questions that we have was trying to come up with a proper analysis that was how safe do they feel while navigating their neighborhood alone. The responses we received from the sample show that 45.16% of children feel somewhat safe and 32.26% of children feel not so safe. In contrast, 12.90% of children feel very safe and 6.45% of children feel extremely safe. Few numbers of children (3.23%) feel not at all safe as shown in (Fig. 10).

The results stipulate about children's safety when navigating local streets on their own, and we found that their answers state that 48.39% of children find it neither easy nor difficult, 22.58% of children find it easy, 12.90%of children find it difficult, 9.68% of children find it difficult and finally 6.45% of children find it very easy as shown in (Fig. 11).

The survey extends to investigate how well does the services meet their needs, and we found that their answers

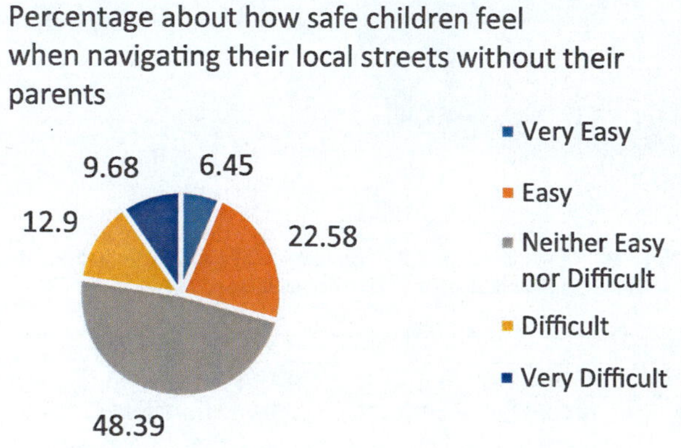

Fig. 11 Diagram of the percentage of how much children find sidewalks safe and local streets to move from a point to another without their parents. *Source* The authors

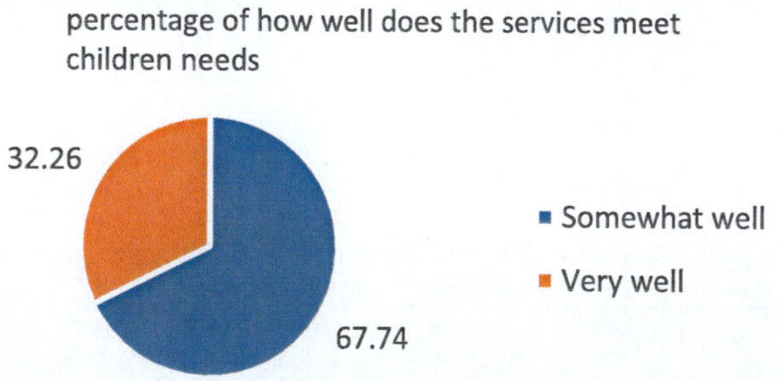

Fig. 12 Diagram of the percentage of how well does the services meet children needs. *Source* Author

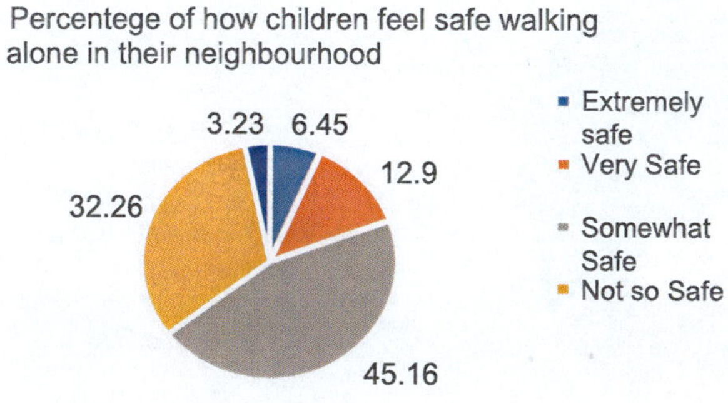

Fig. 10 Percentage of how children feel safe walking alone in neighborhood. *Source* The authors

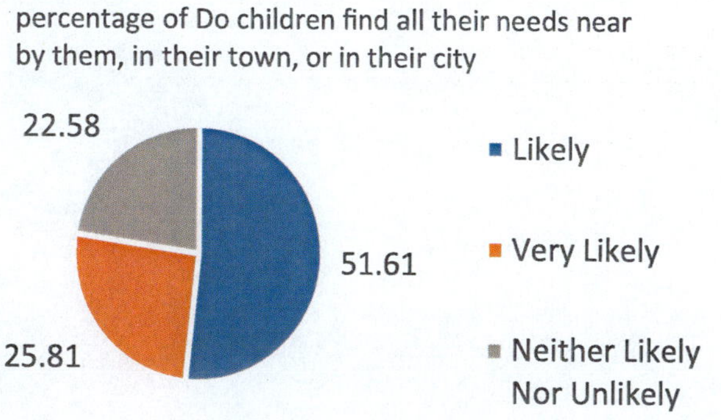

Fig. 13 Diagram of the percentage of doing children find all their needs near to them or in their town or their city. *Source* The authors

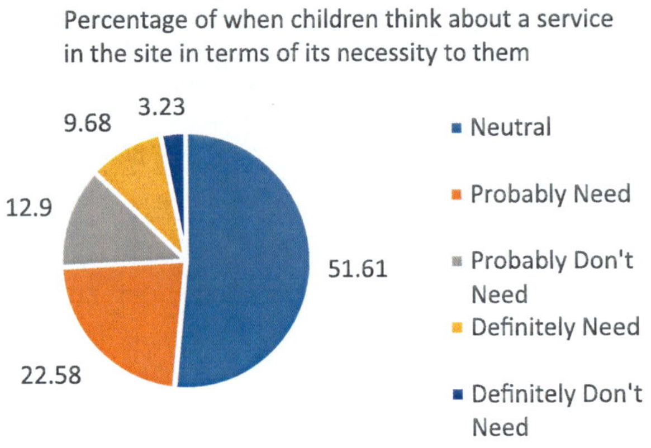

Fig. 14 Diagram to measure children satisfaction level about the provided services in the site. *Source* The authors

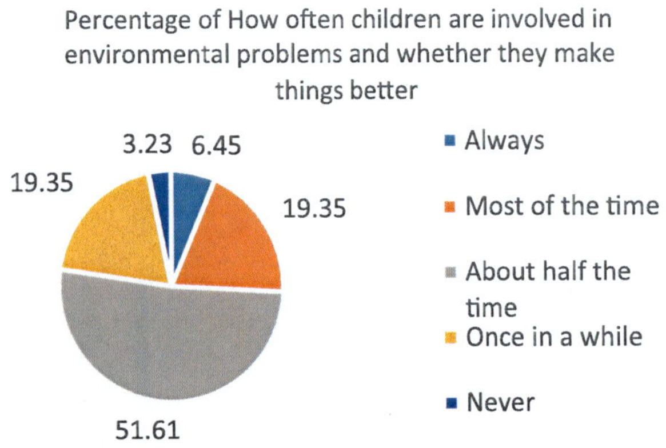

Fig. 15 Diagram to show how children feel about their involvement in solving environmental problems. *Source* The authors

state that 67.74% of children find that services meet their needs somewhat well and 32.26% of children find that services meet their needs very well as shown in (Fig. 12).

After going deeper in the survey to reach satisfactory results by asking children about do they find all their needs near to them or in their town or their city, and we found that 51.61% of children likely find their needs and 25.81% of children find their needs very likely and 22.58% of children find their needs neither likely nor unlikely as shown in (Fig. 13).

Also, the survey examines when children think about services in the site in terms of its necessity to them and the results show that about 51.61% of children find it neutral and 22.58% of children find it probably needs and 12.90% of children find it probably does not need and 9.68% of children find it needs and finally 3.23% of children find it does not need as shown in (Fig. 14).

This survey leads to exploring when people get involved in trying to solve environmental problems, how often do children think they make things better. The results show that around 51.61% of children are involved about half of the time in environmental problems, 19.35% of children are involved most of the time, 19.35% of them get involved once in a while, 6.45% of children are always involved and finally, 3.23% of children never try to get involved to solve get at all, as shown in (Fig. 15).

In light of the survey, we also asked a very important question for children about what the children do on their weekends and we had several answers as shown in (Fig. 16).

Finally, we ask children one of the most important questions which were about their opinion about how well they are integrated into their surrounding environment, and we got several answers such as city decision-makers to post

Fig. 16 Diagram of the percentage of what children do on their weekends. *Source* Author

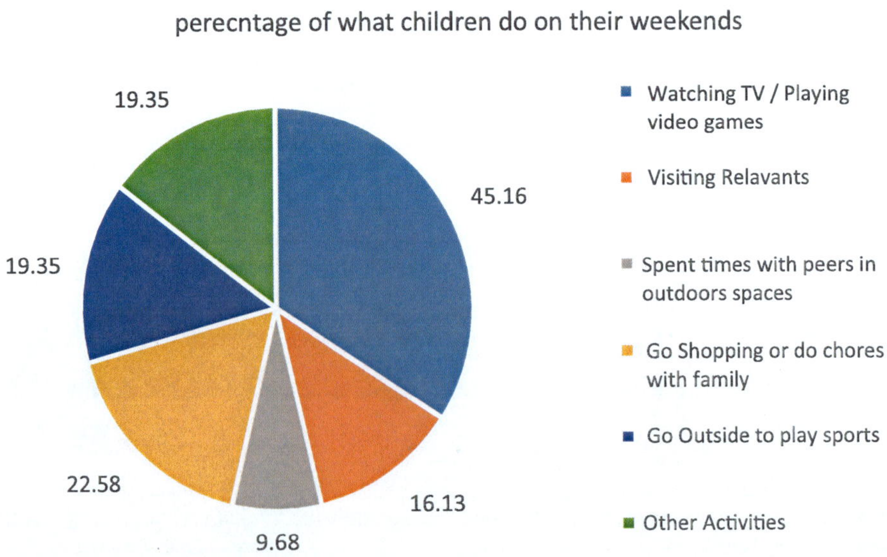

Fig. 17 Parent's interview form. *Source* The authors

Interview Form

Parents Name :.................................. **Date of Interview :................................**

No.of Children they have :............... **Age of each child :.................................**

Type of Interview :............................

Questionnaire	Yes	No	Neither nor
Do you feel repose if you let your child get out of the house alone ? **Comment :..** **...**			
Do you think that your child find he/she needs near to them ? **Comment :..** **...**			
How well does the services surrounding you meet your child needs? **Comment :..** **...**			
Do you find the sidewalks and local streets safe for your child to move from a point to another alone? **Comment :..** **...**			
Do you know what your child needs to find to be happy ? **Comment :..** **...**			
Do you encourage your child to practice activities and hobbies that develope their skills and increase the propotion of their intelligence ? **Comment :..** **...**			

In your opinion , how the city decision-makers let your child integrated into the surrounding environment?

Comment :...

..

..

an online survey or post votes, making the place more organized and clean and safer to roam in, they saw that city decision-makers have not helped them yet, such as not organizing garbage dumps, not assigning pavements to them, not protecting them from cars until their parents check in on them, they still need more public services and attention to this age group, because there is more security around them, they need some improvement in taking more action and more other things.

After that, we make an interview with children's parents asking them some questions about their children as mentioned in the interview form (Fig. 17).

Through these interviews with the parents of the children, we found that they have some opinions about what is related to their children, where they can do activities are not available in their surroundings, for example, the presence of places with safe green areas where they can play with them and their friends and the existence of bicycle paths where they can practice cycling and they also see that the sidewalks and local streets are not safe enough for them to let their children go outside alone.

7 Conclusion

Through this paper, we have found that children are considered to be one of the most important elements in society, but they are not taken into account when making decisions in the city. Also after researching and studying the nine elements of the child-friendly city, we found that here in Egypt there is no child-friendly city or even a child-friendly community and to be sure we have done a study on the area

(Heliopolis District) to measure initially the satisfaction of children from the area where they live and the environment they are dealing with.

The paper concludes by arguing that the concept of a child-friendly city can be achieved through the realization of the rights of the child. On this basis, we figured out that the importance of children's satisfaction as an influential element in society and through which the nine elements through the child-friendly city is built.

Then we did in the field study phase of Heliopolis as a study area and through the observations seen by the researcher, we find that the area is densely populated and the streets are over-crowded, also we asked some questions to the children who live in Heliopolis based on the choice of the appropriate sample size which is 31 children have also identified the age group of 10–13 and 14–18 years old for the ability of this age to respond and give results, and we found that the opinions of children vary according to the age group and the way of thinking of each child and his/her inclinations and other elements that make up the child's personality.

We found that children have requirements and activities they want to do in the city where they live, but the city has to have some factors to be friendly to them, for example, the streets must be safe for them so that they can do what they want without fear, as well as some activities that are not available, in the area so as not to make him think about going somewhere else just because it has this activity that the child is looking for as well as needs.

Also, after interviewing the parents of the children, some questions were asked in the context of the public debate to find out how worried parents are of leaving their children in the streets alone because of their anxiety about them, and also that the children always want someone to listen to their opinions and this is also what we are talking about until we find a way to get to connect children and decision-makers to create a child-friendly environment in the City of Cairo.

To conclude, children are the future of the nation and the city. Also, we must study everything that children need and what they want and seek to provide all of that. On the other hand, we have to make a research about how to make the Egyptian government put in their consideration involving children in making decisions and involve them in to promote the city to become a child-friendly city.

References

Al Arasi, H. (2013). A study on children's perception on their local living environment. Faculty of Geo-Information Science and Earth Observation of the University of the Twente.

Bartlett, S. (2005). Integrating children's rights into municipal action: A review of progress and lessons learned. *Children, Youth, and Environments, 15*(2), 18–40. Retrieved from https://www.colorado.edu/journals/cye/

Central Agency for Public Mobilization and Statistics. (2019). Population estimates: Centers & shayakhat. Cairo, Egypt. Retrieved from https://www.capmas.gov.eg/Pages/StaticPages.aspx?page_id=7188

Christensen, P., & O'Brien, M. (2003). Children in the city; Introducing new perspectives. *Children in the City; Home, neighborhood, and community*. (pp. 1–12). RoutledgeFalmer.

Elshater, A. (2018). What can the urban designer do for children? Normative principles of child–friendly communities for Responsive third places. *Journal of Urban Design, 23*(3), 432–455

Frank, K. I. (2006, May). The potential of youth participation in planning. *Journal of Planning Literature, 20*(4), 351–371. https://doi.org/10.1177/0885412205286016

Gleeson, B., & Sipe, N. (2006). *Creating child friendly: Reinstating kids in the city*. Routledge.

Jamieson, L., Bray, R., Viviers, A., Lake, L., Pendlebury, S., & Smith, C. (2011). *South African child gauge 2010–2011*. Children's Institute, University of Cape Town. Retrieved from www.ci.org.za

Malone, K. (2006). *Building a child-friendly community*. Children's Research Workshops Pilot Study Report.

Matthews, H., & Limb, M. (1999, March 1). Defining an agenda for the geography of children: review and prospect. *Progress in Human Geography, 23*(1), 61–90. https://doi.org/10.1191/030913299670961492

McAllister, C. (2008). Child-friendly cities and land use planning: implications for children's health. *Environments Journal, 35*(3).

Percy-Smith, B., & Thomas, N. (2010). *A handbook of children and young people's participation*. Routledge.

Rasmussen, K. (2004). *Places for children-children's places*. Sage.

Committee on Environmental Health. (2009). The built environment: designing communities to promote physical activity in children. *Pediatrics: Offical Journal of the American Academy of Pediatrics, 123*, 1591–1598. https://doi.org/10.1542/peds.2009-0750

Torres, J. (2009). *Children & cities: Planning to grow together*. The Vanier Institute of the Family.

UNICEF. (2004). *Building child-friendly cities; A framework for action*. UNICEF Innocenti Research Centre.

UNICEF. (2017). *UNICEF*. Retrieved from UNICEF for every child: https://www.unicef.org

van Gils, J. (2012). Introducing the child-friendly city network: One of the most effective programs in the world today to transform our cities into places of well-being. In *Quality of Childhood Group at the European Parliament* (pp. 172–189). MEP Karin Kadenbach.

World Bank. (2018). "The World Bank" Population ages 0–14 (% of Total). The World Bank. Retrieved 2017, from https://data.worldbank.org/indicator/SP.POP.0014.TO.ZS?end=2018&start=2015&view=map&year=2018

CSR and Postal Service Sustainable Development and Its Impact on Urban Environment at the Example of Courier Service Operator Solutions in Europe

Marcin Janowski

Abstract

This paper aims at analyzing actions taken by courier service operators following the concept of CSR. The author compares activities related to the reduction in air pollution and road traffic, mostly in highly urbanized areas. This is a particularly important topic taking into account the increase in the number of courier shipments related, e.g., to the development of e-commerce market. The analysis regarding the CSR concept implementation methods among courier service operators rendering services on the European market will help indicate changes in the area of postal service sustainable development. The research methods applied in this study include *inter alia* the critical analysis of literature as well as evaluation and comparison of secondary research results from various sources. The character of research work is illustrative.

Keywords

Corporate social responsibility • Sustainable development • Courier services • Friendly city logistics • Urban environment

1 Introduction

The main of the work is to draw attention to the risks associated with the operation of transport companies and ways of reducing them by means of the concept of corporate of social responsibility. The paper contains a comparative analysis of selected activities of courier service providers in terms of CSR. The research methods used in the paper are primary and secondary analyses of available data on air pollution and generated noise in the human environment.

The concept of corporate social responsibility (CSR) has become an important element in the sustainable development strategy of the majority of leading enterprises on the market. It reflects concern not only for the enterprise business performance, but also for its business environment understood as the natural environments, cities, and local community. Significant impact on urban environment is exerted by companies making use of road or air transport, such as courier service operators.

The courier service market in Europe has been increasing year by year. The inhabitants in the Old Continent more frequently purchase goods online, which results from the development of virtual infrastructure and easy access thereto. Recently, e-commerce purchase has dominated the amount of all courier shipments. In order to meet the new challenges, courier service operators look for new solutions that satisfy the demand for courier services, while taking care of the environment, mainly in the city where the negative impact of courier vehicles will be minimized. The number of courier vehicles delivering shipments has been increasing, but there are more vehicles which generate less pollution and noise and take up less space in the urbanized area. Replacing traditional trucks with electric cars, bicycles, or drones has become a fact. In order to reach the transport goals determined by the European Union, we need time; however, enterprises such as courier operators take the initiative on their own and seek to achieve these goals. Such approach is justifiable not only socially but also economically since good models and innovative solutions exert positive impact on the client environment, creating a strong and valuable brand.

The courier service enterprises achieve their pro-social and pro-ecological roles goals following the strategy of corporate social responsibility. This is the strategy involving voluntary activities for groups of stakeholders (employees, local communities) and following the sustainable

M. Janowski (✉)
Department of Economics, Faculty of Management and Economics of Services, University of Szczecin, Szczecin, Poland
e-mail: marcin.janowski@wzieu.pl

F. Trapani et al. (eds.), *Advanced Studies in Efficient Environmental Design and City Planning*, Advances in Science, Technology & Innovation, https://doi.org/10.1007/978-3-030-65181-7_41

development of economy which involves taking into account the effects of its activity for the environment and reaction thereto.

2 Impact of Transport Activity on Urban Environment

The activities of courier service operators are based, to a large extent, on road transport, and its impact is mostly noticeable for the residents of urbanized areas. This type of transport is responsible for ca. 25% of greenhouse gas emissions in the European Union. Road transport is also the main source of potential noise and pollution such as nitrogen dioxide NO_2 or particulate matter PM which exert extremely negative impact on the health and environment of the inhabitants in Europe. Another aspect refers to greenhouse gas emissions, including in particular carbon dioxide CO_2. It is important due to the fact that CO_2 is the main negative result of using fossil fuels (including crude) to power the transport fleet of companies, including these related to courier service market. Moreover, according to experts of the EU Eurostat, CO_2 is responsible for 80% of all greenhouse gas emissions in the European Union (Forbes, 2018). Although the average CO_2 emission in the EU Member States dropped by 2.5% compared to 2017, this level is still unsatisfactory. Studies indicate that in the case of eight EU Member States, this level has increased. The emission test results in particular countries are presented in Table 1.

Table 1 Percentage changes and share in CO_2 emissions in the EU Member States in 2018

Percentage changes in CO_2 emissions compared to previous year	Country	Share in CO_2 emissions in 2018 (%)
9.0↓	Portugal	1.4
8.1 ↓	Bulgaria	1.3
6.8↓	Ireland	1.1
5.4↓	Germany	22.5
4.5↓	Netherlands	4.7
4.3↓	Croatia	0.5
3.6↓	Greece	2.1
3.5↓	Italy	10
3.5↓	France	10
3.2↓	Spain	7.7
2.9↓	Sweden	1.0
1.1↓	Austria	1.7
0.8↓	Cyprus	0.2
0.8↓	Hungary	1.4
0.4↓	Slovenia	0.4
0.3↓	Belgium	2.4
0.3↓	Romania	2.2
0.3↓	UK	11.4
0.2↓	Denmark	1.0
0.1↓	Czechia	3.0
0.6↑	*Lithuania*	*0.4*
1.9↑	*Finland*	*1.3*
2.4↑	*Slovakia*	*0.9*
3.5↑	*Poland*	*10.3*
3.7↑	*Luxembourg*	*0.3*
4.5↑	*Estonia*	*0.5*
6.7↑	*Malta*	*0.0*
8.5↑	*Latvia*	*0.2*

Source Own elaboration based on Eurostat (2019)

The above-mentioned urban environment pollution with nitrogen dioxide and particular matter also indicates that road transport contributes, to a large extent, to the pollution and is responsible for over 60% of pollution in the case of NO_2 and 45% in the case of PM (EEA, 2018). Under the European standards as much as 30% of the inhabitants in Europe are exposed to living in urbanized areas where the level of air pollution is higher than the acceptable level. Taking into account, more stringent guidelines provided by World Health Organization (WHO), almost all inhabitants in Europe, namely as much as 98% (Soholt, 2016), are exposed to health-hazardous pollutant emissions exceeding the acceptable level. It is worth indicating that the noise generated by road transport is affecting ¼ of the Europeans. Due to the noise of over 55 decibels, 20 million European citizens complain about its nuisance in their places of living, 8 million suffer from sleep disturbance, and as a result, 43,000 per annum are hospitalized. Meanwhile, the demand for transport in the EU has been increasing. The European Commission estimates that freight transport which refers *inter alia* to courier shipment transport will increase in 2050, by 80% compared to the level from 2013 (Bruyninckx, 2016). Such situation arises from, e.g., the increased consumerism of the EU citizens, easier access to wide, global markets and the related development of e-commerce perceived recently as the driving force for the courier service market.

In order to reduce the negative impact of transport development—including road transport, the European Union seeks to reduce, e.g., the greenhouse gas emissions, assuming 80–95% reduction by 2050. Taking into account road transport which refers to the majority of courier services, the EU assumed a number of indirect goals including:

- reduction of crude oil consumption in transport by 70% until 2050 compared to 2008,

- reduction of average level of CO_2 emissions from new vans up to 157 g/km by 2020,
- reduction of average level of CO_2 emissions from new passenger cars up to 95 g/km by 2020,
- increase in the share of energy from renewable sources in transport to min. 10% in each EU Member States (EEA, 2019).

As a result of the effects of transport adverse impact on the lives of citizens in the European cities and changes proposed by the EU, the courier service operators seek new methods to reduce the difficulties related to running business activity based mainly on road transport. One of such methods may refer to the concept of corporate social responsibility which is more and more frequently implemented in the activities of leading courier service operators.

3 Corporate Social Responsibility—Concept Definition and Characteristics

One of the methods to reduce the adverse effects of transport enterprises business activity is to become involved in the strategy of corporate social responsibility (CSR). There are numerous definitions of the CSR strategy; some indicate the importance of activates provided for groups engaged in the activity of economic organizations. Under this criterion, CSR is a dynamic concept of interdisciplinary organization of social and economic, as well as ethical and cultural life, and reasonable approach to appropriate actions and optimized economic thinking. The idea aims to protect against economic relativism … and first of all, preserve human dignity, moral and ethical principles, respect for the law and free competition, and all forms of social integration (Wolska, 2018). CSR can also be defined as enterprise approach to follow the moral principles and to agree to be accountable to the society and employees. Enterprises are particularly accountable to the following stakeholders:

- internal groups (owners, co-owners, management board, employees),
- external groups (business partners, clients),
- public authority, local government and pressure groups (e.g., the media, offices, creditors),
- local communities,
- non-profit organizations as well as social, ecological, and consumer organizations (Kryk, 2019).

An important aspect of the CSR strategy refers to the principle of its voluntariness, and consequently, the corporate social responsibility is a concept thanks to which enterprises, while developing their strategy, voluntarily include the social interest and environmental protection as well as relations with various stakeholder groups. The responsibility does not only involve meeting the formal and legal requirements, but also voluntarily increased investments in human resources, environmental protection, and relations with business environment. CSR is a process in which enterprises manage their relations with various stakeholders who may have impact on the success of their business activity; therefore, the process shall be treated as investment and not cost (Woźniak–Miszewska, 2011). CSR is a management concept thanks to which enterprises, by performing their tasks, are engaged in six related areas: economic, ethic, legal, sociological, philanthropic, and environmental. The engagement in corporate social responsibility involves not only providing goods of appropriate high quality which guarantees max. usefulness for the stakeholders, but also achieving social goals, which will contribute to the increase in the value of organization, by improving the competitive advantage on the market and more effective implementation of long-term corporate goals. The basic areas of CSR influence are presented in Fig. 1.

As any concept, also the concept of corporate social responsibility has its supporters and opponents. The CSR opportunists frequently quote financial aspects, deny the idea of entrepreneurship, and emphasize the conflict of interest. The consolidated list of for and against arguments regarding the application of corporate social responsibility in economic organizations is presented in Table 2.

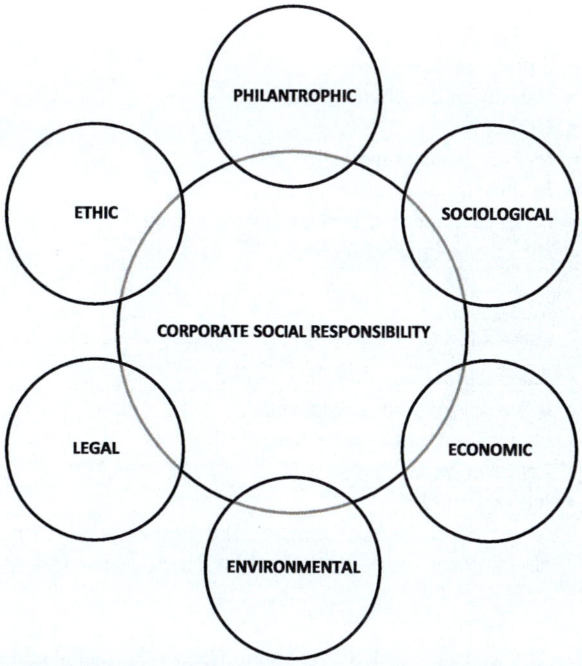

Fig. 1 Areas of CSR influence. *Source* Own elaboration

Table 2 For and against arguments regarding the application of corporate social responsibility

Positive arguments	Negative arguments
Internal stakeholders; i.e., employees should be treated appropriately since each organization is responsible for them	Organization has no legal obligations as for "selfless" support apart from statutory obligations
Since business entities are partners in the social and economic environment together with the government and society, they should participate in meeting not only the economic needs, but also social and environmental ones	Conflict of interest resulting from the achievement of social and economic goals
Actions for the benefit of others build positive relations and image of enterprise, which translates into the organization vested interest	Engagement in actions not related to the economic activity decrease the effectiveness of enterprise operations

Source Own elaboration

The above-mentioned arguments are general. More detailed characteristics of positive aspects resulting from their application were presented by A. Sokół, indicating benefits divided into particular groups:

Enterprises:

- better integration of employees around corporate objectives,
- conflict-free management,
- engagement of employees in the management process,
- higher level of employees' responsibility,
- better sales,
- higher profit,
- improved image of organization,
- improved competitive standing on the market;

Employees:

- higher remuneration,
- stability of employment,
- flexicurity,
- less accidents at work,
- lower employment turnover,
- better timeliness of remuneration payment,
- social protection of families;

Society:

- increased consumption in economy,
- increased level of investments,
- higher level of tax collection,
- improved productivity,
- increase in GDP,
- lower spending on handling social issues: unemployment, environment protection, health-related costs (Sokół, 2011).

The concept of CSR is more frequently included in business activity run by courier service operators since it perfectly matches the nature of such service. By building positive relations with employees and the community as well as improving the condition of natural environment, the courier service operators are building trust to their brands.

4 Methods of CSR Implementation into Business Activities of Selected Courier Service Operators in Europe

The specificity of courier service transport provides numerous possibilities to reduce environmental pollution by replacing the transport fleet with more modern vehicles, of lower exhaust emission indexes, or by replacing combustion vehicles with new ecological electric, hydrogen transport, as per eco-driving principles. The largest operators introduce changes in the shipment delivery on a large scale, change the traditional air and road transport into electric drone transport, as in the case of DHL, UPS or Deutsche Post. Smaller, local couriers also try to include the concept of CSR in their business activity by delivering shipments very often using low-emission scooters or zero-emission bicycles. By analyzing the information provided on websites of the leading operators, we can see that each of them emphasizes the implementation of initiatives related to corporate social responsibility. For example, DHL operator is engaged in numerous activities related to CSR. On the website, we can find information at "CSR-based activities at DHL Parcel." The information provides that for DHL social responsibility is an important element in the corporate strategy. It is even underlined by the company motto "*Living Responsibility.*" The CSR-based activities refer to four basic areas:

- environmental protection, implemented under "*GoGreen*" program,
- support during natural disasters, implemented under "*GoHelp*" program,
- support for education, "*GoTeach*" program,

- support for employee voluntary activity "*Global Volunteer Day, Living Responsibility Fund.*"

Moreover, DHL operator developed and provided a report related to the reduction of exhaust emissions: "*Mission 2050: Zero Emissions,*" and report regarding strategy and actions based on CSR: "*Corporate Responsibility Report 2017.*"

The tasks of CSR in *GoGreen* program involve reducing CO_2 emissions coming from DHL business activity. The operator agreed to reduce CO_2 emissions in 2025 by 50% compared to 2007. Moreover, by 2025 zero-emission delivery and collection of shipments from clients will be increased to 70% and 80% of employees will have knowledge on green energy confirmed by certificates (DHL, 2019).

The *GoHelp* program involves engaging DHL operator in UN humanitarian actions by providing the organization with logistic solutions and engaging employees in voluntary activities. DHL provides airplane fleet for the transportation of water and food in places affected by disasters (hurricanes, earthquakes).

The engagement in *GoTeach* program involves drawing the society attention to increasing opportunities for the young to find jobs, especially the young from troubled communities. The operator cooperates with 48 organizations forming the whole system "Teach for All"; in 2017; support was provided for over 900 000 people around the world.

Actions within *Global Volunteer Day, Living Responsibility Fund* involve engaging DHL employees in voluntary actions related to social work and above-mentioned programs, mainly *GoHelp* and *GoTeach.*

Information on CSR on the operator's website indicates that the operator is aware of the benefits resulting from this concept and fully derives from them. In report *Corporate Responsibility Report 2017*, DHL emphasizes that in 2017 the operator's brand was included in the 100 most valuable brands, ranking 70th (Deutsche Post, 2017).

Another global operator, FedEx is also taking actions related to corporate social responsibility. On the website, we can find reference to a department called social responsibility. Programs taken by FedEx within CSR include *FedEx Cares, EarthSmart, Charitable Giving Guidelines*. FedEx also develops and provides report describing achievements in the area of exhaust emissions reduction and charitable activities: *2018 Global Citizenship Report*. The main goals of CSR strategy by FedEx include:

- reduction by 2020 of emissions intensity by 30% compared to 2005; at present, it was possible to reach 22%,
- by 2030, obtain 30% of fuel for airplanes from alternative sources; the first delivery of such fuel is scheduled for 2019,
- increase in fuel economy of FedEx vehicles by 50% compared to 2005; improvement of 35% has already been observed compared to 2055 (FedEx, 2019).

Moreover, the operator is also engaged in pro-ecological actions and charitable activities.

The activities by FedEx also indicate significant engagement in the concept of corporate social responsibility; however, they are more focused on actions related to sustainable development in the context of transportation.

DPD operator has also included the concept of CSR as part of the company strategy. On the website, you can find information on actions taken in this respect. In 2016, DPD introduced *Driving Change* program aimed at supporting local communities, promoting entrepreneurship and acting for the sustainable development regarding environmental protection and convenient urban logistics. DPD developed a network of drop-off/pick-up points called DPD Pickup, which helps to reduce troubles related to courier transport for the residents of large agglomerations. The operator agreed to reduce CO_2 emissions by 2010 by 10% per each parcel. DPD is also a partner to offset project, Green Binders dealing with processing waste into material for road surface production (DPD, 2017).

UPS is also included in a group of couriers making use of the strategy of corporate social responsibility. On the website, we can find reference to the description of activities within sustainable development and social engagement. The defined actions include information regarding current solutions applied by UPS in this area. The operator's actions within CSR include:

- "*Urban Logistics*" solutions,
- solutions for sustainable transport,
- activities of "*UPS Foundation.*"

In the case of actions within *Urban Logistics*, UPS introduced a number of transport solutions featuring zero-emission transport, such as unmanned delivery of shipments (using robots), eBikes, i.e., replacing traditional vehicles with electric bicycles, developing bicycle logistics system. UPS has assumed that by 2020, 25% of fuels will come from alternative sources, by 2025, CO_2 emissions will be reduced by 12%, 25% of electric energy will come from renewable sources by 2025. Within the activities of *UPS Foundation*, in 2017, UPS employees worked as volunteers 2.9 h for social purposes, and 57.4 million dollars were paid in as donation for the people in need. The same year, members of foundation planted 2.7 million trees (UPS, 2019). In conclusion, UPS places strong emphasis on corporate social responsibility in their activities. The operator clearly focuses on innovation-based activities within the area of sustainable development.

The local operators do not have comprehensive solutions within the area of CSR. It is fully justifiable since the volume of operator and its operational scope determine the implementation or non-implementation of the corporate social responsibility strategy. However, single elements which can be included in the activities of CSR usually occur among these operators; they often refer to logistics solutions, in particular transport-related solutions.

5 Conclusion

The impact of transport services presented in this study, in particular, courier services exerted on urban environment indicate that the inhabitants of the European Union are facing significant challenge regarding the reduction of greenhouse gas emissions, pollution, and noise. Meeting the increasing demand for courier services with the reduction of negative effects of transport-related activity is a difficult and complicated task. However, the issues related to corporate social responsibility and methods to implement the concept in the activities of leading courier service operators in Europe prove that some enterprises voluntarily seek to perform tasks which fall within the concept of sustainable development. The concept of CSR is clearly visible among the so-called integrators of courier service market, operating on international scale; smaller local enterprises also seek to perform socially responsible activities. From the economic perspective, the implementation of CSR strategy may bring measurable benefits for enterprises applying the strategy. Such activities have positive influence on corporate image, are the tools of promotion, and help to become distinguished from the competition. However, we can get the impression that the awareness of CSR concept is too little, especially among the demand side of the market. For many clients, the concept of CSR is unknown or unimportant upon selecting services of a particular operator. The primary studies conducted by the author among the clients of courier service companies showed that the strategy of CSR was perceived as important by 24% of the surveyed (Janowski, 2019). Therefore, in order to help enterprises gain more benefits regarding the development of strong brand, the concept awareness in the society must be increased. Their clients should be aware that enterprises provide significant resources for activities which do not bring them direct economic benefits but bring significant benefits to the inhabitants of urbanized areas. The activities to promote the concept of corporate social responsibility contribute to effective implementation of the aforementioned sustainable development goals, the European Union scheduled to be implemented by 2050.

Acknowledgements The project is financed within the framework of the program of the Minister of Science and Higher Education under the name “Regional Excellence Initiative” in the years 2019–2022; project number 001/RID/2018/19; the amount of financing PLN 10,684,000.00.

Reference

Deutsche Post AG. (2017). *Corporate Responsibility Report 2017* (p. 11). https://www.dpdhl.com/content/dam/dpdhl/en/media-center/responsibility/dpdhl-corporate-responsibility-report-2017.pdf. 5 Jan 2019

DHL (2019). *Our Mission 2050: Zero Emissions*. https://www.dpdhl.com/content/dam/dpdhl/en/media-center/responsibility/dpdhl-flyer-gogreen-zero-emissions.pdf. 12 July 2019.

DPD. (2017). https://www.dpd.com.pl/klienci-biznesowi/O-DPD/Centrum-Prasowe/Informacje-prasowe/01-2017-CSR-w-DPD-Polska-podsumowanie-roku. 5 Jan 2019

EEA. (2018). *Air quality in Europe—Report 2018*, No. 12/2018. European Environment Agency, Copenhagen, p. 17. https://doi.org/10.2800/777411.

EEA (2019), *European Union emission inventory report 1990–2017*, EEA Report No. 08/2019, Copenhagen, pp. 19–20. https://doi.org/10.2800/78220.

Eurostat. (2019). *In 2018, CO_2 emissions in the EU decreased compared with 2017*, Eurostat, 81/2019. https://ec.europa.eu/eurostat/documents/2995521/9779945/8-08052019-AP-EN.pdf/9594d125-9163-446c-b650-b2b00c531d2b. 11 Sept 2019.

FedEx. (2019). https://www.fedex.com/pl/about/sustainability/environment.html. 5 Jan 2019.

Forbes. (2018). *Emisja CO2 w Europie spada, choć nie w Polsce. Najnowsze dane*, Forbes.pl, https://www.forbes.pl/gospodarka/emisja-co2-w-ue-w-2018-r/f925k0n. 27 Aug 2019.

Janowski, M. (2019). *Marka jako czynnik kształtowania przewagi konkurencyjnej operatorów usług kurierskich w Polsce* [Brand as an instrument for developing the competitive advantage of courier service operators in Poland] (p. 175). Faculty of Management and Economics of Services, University of Szczecin.

Kryk B. (2004). *Koncepcja społecznej odpowiedzialności przedsiębiorstw za środowisko przyrodnicze*. https://www.mikroekonomia.net/system/publication_files/1053/original/22.pdf?1315235905. 3 Jan 2019

Signals, E. (2016). Towards clean and smart mobility: Transport and environment in Europe. *European Environment Agency*, 5.

Soholt, H. (2016). *People first, for greener and livable cities* (p. 35), Towards clean and smart mobility. Transport and environment in Europe, Signals, European Environment Agency, Copenhagen.

Sokół, A. (2011). Społeczna Odpowiedzialność Biznesu (CSR), w lokalnej świadomości mieszkańców województwa zachodniopomorskiego na przykładzie studentów ekonomii. Wybrane aspekty społecznej odpowiedzialności biznesu inwestorów zagranicznych, Społeczna odpowiedzialność biznesu jako wyraz reorientacji działania współczesnych firm, Stowarzyszenie Kreatywni dla Szczecina, Szczecin, p. 61.

UPS. (2019). *Road to sustainable urban logistics*. https://sustainability.ups.com/media/UPS_The_Road_to_Sustainable_Urban_Logistics.pdf/. 30 Aug 2019.

Wolska, G. (2018). *Kontrowersje wokół idei społecznej odpowiedzialności biznesu i jej myśli przewodniej „po kryzysie" w Polsce*, Studia Ekonomiczne. Zeszyty Naukowe Uniwersytetu Ekonomicznego w Katowicach, red. U. Zagóra – Jonszta, D. Bochańczyk – Kupka, Wydawnictwo Uniwersytetu Ekonomicznego w Katowicach, Katowice, p. 106.

Woźniak–Miszewska, M. (2011). *Wybrane aspekty społecznej odpowiedzialności biznesu inwestorów zagranicznych*, Społeczna odpowiedzialność biznesu jako wyraz reorientacji działania współczesnych firm, Stowarzyszenie Kreatywni dla Szczecina, Szczecin, p. 39.

Public Spaces and Urban Transition in Tirana City

Andi Shameti

Abstract

Due to relations differences in the features and concepts that characterize open spaces and urban development, it is difficult to define the urban transformation of Tirana city. The central area provides a different function types and way of use and usually a higher number of physical changes, at least in developing countries and transition economies. These rapid processes have overlooked the need for data of actual public spaces information related to transition period and need to define standards for data comparability. The main objective of this paper is to analyse the relationship and classification of the public space transformation, according to their measurable ability and to highlight the effect and impact that these characteristics, as the vegetation has in the sustainability for a better quality of life, based on a resilient approach for data framework as well, and to compare the changes for positively adapting and transforming towards sustainability. Moreover, spatial data for land use changes and functionality are of some considerable importance for the authorities while planning and managing services. Data variables are also presented through diversity using spatial analysis and including the based typology recently. The results show the high differences between the classifications and changes during the transition period and are allocated in the most main areas, while residential zones appear as those negative values on the surface. These demonstrate the lack of criteria use to planning and a volatility development. Today's trend seems to embrace a theory that was created about the complexity of the city as an organism that can be adapted to achieve sustainability, but this remains to be evaluated and monitored.

A. Shameti (✉)
Faculty of Architecture and Urbanism, Polytechnic University of Tirana, Tirana, Albania

Keywords

Public space • Classification • Urban • Transformation • Diversity • GIS

1 Introduction

In today's world, most of the percentage of the world population lives in urban areas. According to the World Bank which underlines that over 80% of world GDP is able to be generated in developed urban centres and that urbanization is seen as an element that, if it is administered correctly, can bring sustainable growth. These can be done through innovation, finding new ideas and increasing productivity. Nowadays, in most of developed cities, socio-economic and environmental changes are seen as incentives for ideas to regenerate and to increase the standard of living conditions and to mitigate the economic effects. Different expert and actors of the planification process are moving rapidly to make plans for the growth and needs of basic services that expanding populations requires. They also encourage investments that create a healthy development, social and friendly with the environment, and more attention has been paid to research into the assessment of open spaces, green areas and their components as well as their impact on the environment. (The World Bank, n.d.)

2 Literature Review

The use and maintenance of a city's public environments often need to be managed to balance collective and individual interests, discussing accessibility in the sense of being "open" (Lynch, 1972). The significance and use of public space may be different over time and the way in which they are act. Ignoring the fundamental urban design aspect such as connections accessibility and mixed uses can result in the

F. Trapani et al. (eds.), *Advanced Studies in Efficient Environmental Design and City Planning*, Advances in Science, Technology & Innovation, https://doi.org/10.1007/978-3-030-65181-7_42

creation of less sustainable and less socially equitable (The Urban Task Force, 1999). The city plans often have little impact on the creation or management of social spaces, and on the other side, the government can play the role of being the main coordinator of public and private actions (Devereux & Littlefield, 2017). Consequently, the use of space can happen to be developed by the way citizens organize their private needs. Public spaces are essential part of urban landscapes with their own capacities of configuration and functionality. However of natural form or not, they provide in different ways the quality of life (Burke and Ewan, 1998). Chiesura (2004) suggests indicators for developing urban cities and their sustainability, which is used by urban planners and planners on the part of urban planners and planners who should consider the availability of public space and their greenery in their research and plans by providing psychological and recreational services (Antrop, 2000), which should direct the city to greater sustainability. (Thompson, 2002) described open spaces as cultural diversity manifest, which interact with natural elements and save the history and memory of the places memories. They fulfil the social needs and they also provide their support at different levels for ecosystems in urban centres (Soulé, 2007). A Better understanding of the needs and the impact of urban development and their important characteristics to the life quality would be needed, such as economic and work activities, people daily habits and leisure activities and the natural. (Van Herzele & Wiedemann, 2003) developed and used indicators to further meet the sustainable green spaces in urban centres by using the GIS model. To monitor the use of green space, a preferred methodology is used, which divides the information into five important levels based on the use of indicators, i.e. "citizen-based"; "functional levels"; "preconditions for use"; "variety of qualities"; and "multiple use". Also, the recognition of the different typologies of public spaces, the assessment of influences and changes in the system are of importance in the relationship created by the dynamics of urban development. They can help guide and encourage more sustainable solutions and strategies for a quality and healthy life, a friendly and environmentally friendly society with positive impacts on urban development (United Nations Sustainable Development, 2015).

3 Tirana Public Space

3.1 Research Problematics

Tirana which is the capital and the main city of Albania in different aspects like that of the population, economic and as owner of the physical territory, see Fig. 1. Often the so-called metropolis is experiencing an intensive expansion and transformation of urban spaces, due to immigration from the other area of the country and from rural to urban areas. Thus changing the socio-economic structure of the country. About one-third of the country population carries out their daily activities in Tirana. The rapid urbanization has had a significant effect on the spatial configuration of the city that is reflected in most of the territory as a chaotic pattern, created by the various layers of development periods and without a clarity of development over the last two decades. The phenomenon of this uncontrolled development is with an influence in the transformation of urban open spaces and the inadequacy of sustainable planning signals that this trend can continue in the future. With an inheritance of urban open space before 1990s and the different reality nowadays. The city is going through the stress of population growth and the living pressures, where the central area are the primarily related to the changes from factors and the various important elements, see Fig. 2.

The importance of the process today is based on an urban development that has a rapid growth and has enabled an amorphous environment with dynamic development that includes a variety of strategies and actors. Often with the multiple functional orientation and in a good part of them, the instability of physically and functionally of the public space directly affects the quality of life. The main objectives contain actions that improve the standard of living and its quality. Carmona and Tiesdell (2007) theory is to preserve and improve, making changes more compatible and by unifying new developments.

In Albania, neither analysis research nor assessment of the effects which have had an impact on the evaluation of public space been made. The lack of which in the case of a big city like Tirana has clearly defined the formation of some urban areas with little or no green areas, such as a result of uncontrolled development. Unfortunately, the current situation shows that the ability to assess different parts of the city by measuring the impact of changes occurring in public spaces from the element of urban development and their influence is not fully recognized by decision-makers and urban planners.

3.2 Objectives of the Research

This paper aims to analyse the effect of urban development and morphological transformations, as well as the relationship with the public space typologies. Specifically the objective is to recognize the areas and define the types of public spaces of Tirana city also to reveal all the aspects which are important in the dynamics process of the urban transformation, leading us to understand and evaluate the relation between public space characteristics and the transition of the urban form, according to their measurable ability and based on a resilient approach for data classification as

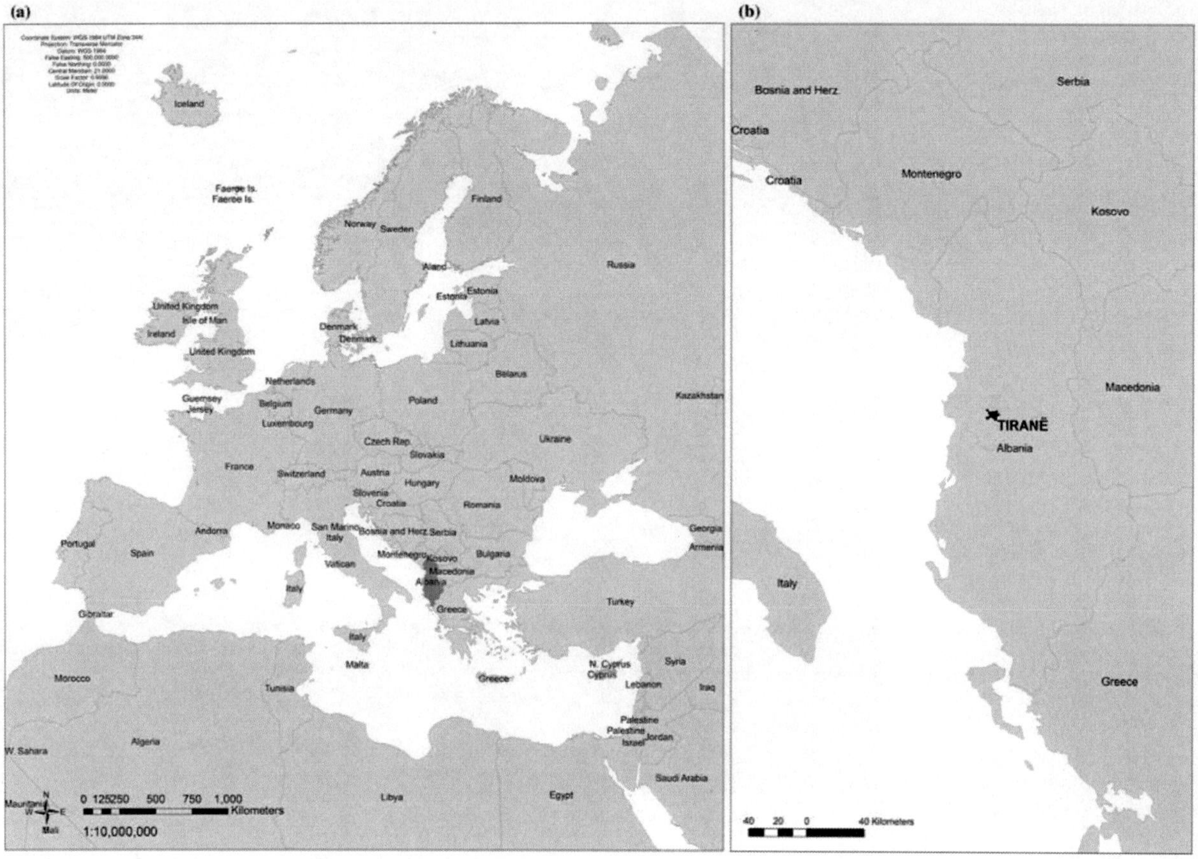

Fig. 1 **a** Map of Albania, **b** Map of Tirana. *Source* Municipality of Tirana—URBAPLAN

well. The importance of the process is to signify a better and more careful actions in decision-making, by comparing the changes for positively adapting and to transforming towards sustainability.

3.3 Research Hypothesis and Development Scenario

The public spaces in the city centre have a formation that allows many different uses which have a strong influenced by socio-economic relations. Open spaces by their characteristics, function and quality, often have a positive–negative impact to the development of urban environment and their dynamic orientation. These factors reflect their impact on the resident's life and activate them in decision-making for the transformation and development of their environments. This is why these types of space can have a direct impact to the sustainability of the city's development.

Our research derives from a hypothesis that the transformation of public spaces and their characteristics have an important role that reflects the way of urban development and in their relationship. We want to highlight the effect and impact that these characteristics, as the vegetation has in the sustainability and orientation of development for a better quality of life. Therefore, their transformations and orientations are viewed as an element that can be examined and evaluated through indicators. Building a clear framework to monitor the trend and analyse the values can determine the impact scale and dynamic scenarios that have positive and sustainable effects on the development of these areas. We also expect that this research may produce some results that allow us to compare knowledge and to have the opportunity to find out how the relationship between public spaces and the transformation and to contribute in sustainable development.

4 The Importance of Today

4.1 Context Analyse

The city of Tirana is a good case study, to examine different samples of open and public spaces according to the types of

Fig. 2 Map of Tirana's metropolitan area. *Source* Municipality of Tirana—URBAPLAN

typologies and to analyse the transformation of the surface coverage over the years. The study is limited to the urban centre area, which have a closed approached with the functionality alternation, physical and typological frequent changings during the development (Fig. 2).

The history of the city of Tirana goes back about 200 years, but the study looks at it after the 1970s plans. During the history, the centre of the city has suffered important transformations through different periods, such as Eastern, Western and Eastern, leaving its influence also in the way of urban space development. Replacements of the main city structure from Ottoman and Italian times during the years of the totalitarian regime are to establish a declaration of the actual status. Almost the largest population growth reached during the period 1990–2000 round 200% (Institute of statistics (INSTAT), 2018), as a result from the transition of communist regime to liberalization of the country's market and economy, as well as opening up with all other foreign cultures. The city had a great expansion, where the old and central part underwent a densification and modification of the previous structures. This made recruiting private property, which had changed its status during the regime from private to public. Also, this process had its impact on the privatization of public space and in other cases as a tendency to possess a property.

The nucleus of the city that is approaching its 100th anniversary is also the urban area that has been transformed more and more. The development after the 1990 to the present is characterized by different elements, but in the most important one we emphasize the economic one that has the greatest role in the urban form. Mainly the typology of the area consists mostly of the institutions and housing, but the balance is changing rapidly, from small or large activities that are undergoing rapid development.

This is easily shown to be due to rapid growth, by observation of what has not been eliminated or who has left traces in the urban space. How did the city, chew this change? How has the public space to the city changed from a significant presence in the past? Some of these spaces relate to the largest and most frequent transformation. Small but numerous changes such as expansions to the ground floors of buildings, shop fronts, in-depth schemes, which in general have an effect that can dramatically transform a place in a few hours, into a better space or otherwise.

The local government institutions seek to promote the relationship of a high-density city with the quality of life of its inhabitants of Tirana, with new plans for advertising and promotion of new parks, green corridors and outdoor recreational areas, (National Territorial Planning Agency, (NTPA), 2015) see Fig. 3. However, their efforts may face

Fig. 3 Green Crown Strategy and New vision of Tirana 030 Plan

some difficulty from the excessive amount of impermeable surfaces from a large number of lack management, their non-compliance with the needs and the failure to realize the previous projects. There are many areas "like those that are developing from separate properties that require good connections and no change or misuse" all these have the potential to be developed in a sustainable manner that can meet the future development needs. The change and elimination that has occurred in some public spaces are negative examples of the urban development. As these areas have reduced the areas and benefits and have greatly increased the needs for such spaces. The transformations are daily and fast, and the overall development is growing. This means that in a short time decision is made and intervened in a large number, which affects and directly affects free spaces. As during the formation and transformation of the open urban space, the actors that are most involved in this process are politicians, a series of technical and managerial professionals, the municipality, peoples who live in the area or manly exercise their activity. By improving the way of their development and the function, it becomes possible to fill public spaces and the urban environment with added features and values, to make them better, vivid, cheerful and pleasant, functional, or unpleasant or unsafe. Also, the development of a good and effective product depends on the collaboration of many components and actors together, where weaknesses in the process are during decision-making or how the development is done and how often an area changes.

5 Methodology—Classification and Analysing the Transformation

5.1 Instrument for Data Collection

This research is related to public spaces in order to identify the important characteristics/elements that affect their transformation during the urban development process and its sustainability, by exploring the impact and the influences during the transition, also the important terminology of public space, according to history, economy, ecology and social aspect. From one point of view, the examination and expression of the use of virtual–spatial relationships occurs through the creation of the mapping process as suggested by. The interconnection of different techniques through observation and the creation of schemes in the form of maps are consolidated methods for the study of environmental relationships and behaviours (Ittelson et al., 1970). The spatial analysis of the function, performed through GIS, shows us a spatial relationship and the formation of spaces and the way they are used, which indicates the dimension of the space, the geometry of the space, the average per unit and identity they occupy. Types of public spaces were able to be traced through field investigations and other supplementary materials, examining the material and creating a geographic information coverage. Also, the level of change was investigated using GIS technology. The spaces taken in the study are selected through a selection, concerning the typology and their mode of operation, their temporal changes, those of the morphological and urban characteristics. All information and data will be grouped by periods, the main important are before 1990, between 1990 and 2000, and subsequent ones until 2018.

The analysis of different examples is conducted for recognizing the impact of the intervention in different ways to the open urban space of physical condition and social–economical context. This will determine the transformation rate by typology, local and theoretical characteristics distingue in categories function and catchment hierarchy. The transformations that have taken place in the country and from other examples abroad will be part of our study, which will give a comprehensive picture of the change process so that we can conclude the theoretical definitions for decision-making in similar situations. Different materials of aerial photographs, cartographic maps from 1980, 1993 to those of orthoimages in 2007, 2015 and 2018 see Fig. 4. They were of 1:25,000 scale, up to 20 to 8 cm resolution orthoimage. Some inspire data helped in analysing city plans, photographs and historical documents. Fault for (rms) was from 0.5 to 0.3 for 1980 and 2007, for each of the periods. Also, they are passed to the system of zone 34°N in UTM. Analysis was done in ArcGIS 10.2 (ArcInfo), Spatial Analyst, (ESRI).

5.2 Data Collection

Nineteen different urban open space categories were defined: agriculture, archaeology historic memorial, found neighbourhood space, gardens, greens and playground, linear greenways, markets, military, open land, park, pedestrian walkway, playing fields, pocket of space, public, school campus, sport, squares plazas, street parking transportation and vacant. Each typology has been geographically indetermined by the polygons that have been created by the digitalization from the orthoimages for each study period and then the attribute tables are enriched with the relative functions. Type and area colonists are created for each polygon for each year. Different basic categories of open spaces type are reflected according to the study of (Esbah & Deniz, 2007). The greatest help in understanding the functioning of these spaces is given to us by the attributes of the form of ownership and by the permitted accessibility, without excluding their management. The areas that occupy the public space are identified and defined, whatever their type, or is affected by public disputes, sports, education, drainage spaces, those of squares, buildings, degraded land, etc. By calculating the ratio between the total surfaces and those of the different periods and their previous functions on the last ones, in order to obtain the average of the differences between them, we can get a rating for the entire period under review. The same procedure will be followed to obtain the percentages of variability of the polygons between them in the total time.

5.3 Data Analyses

In the content of the context investigated, it includes the physical characteristics of the "public space" and the transitions from one state to another over time. In the following description of the context, they are divided into two main points, in the physical expression of urban morphology, by accessing the forms and the dimension of spaces, function and structure buildings, which are clearly evidenced in the city centre. The research on urban form has several variables, one of the most important beings that of urban morphology, a component of the formation and transformation of public space as well. Therefore, they will be divided into categories by characteristics based on different levels of study of the context.

On the scale of the city, mixed development periods are recorded as traditional, Italian, socialist, contemporary. Another scale is a defined area like that of the city centre, where the changes over the years are clearly seen, as well as the identification of the public space categories. After, there is the possibility to define the general characteristics in both main groups and also evaluates the sustainability of their development. The development is influenced by social and economic issues, potential market area; economical and social facilities; and population distribution. Create and shape a framework to allow the analysis of open urban spaces, finalizing a sort of ranking based on the impact they have on the territory and the way in which these vital forms act to influence the public space of the city (Table 1).

6 Results

The loss of physical surface of public space is a consequence of economic development and liberalization which had an immediate effect on new construction, the transition period from 1990 to 2007 and the 2007 to 2018, see Fig. 5, The vegetation in public spaces by categories, see Fig. 6, 7a, b, 8a, b, 9a, b.

Through the results obtained, we can make the acquired knowledge comparable and consequently understand how

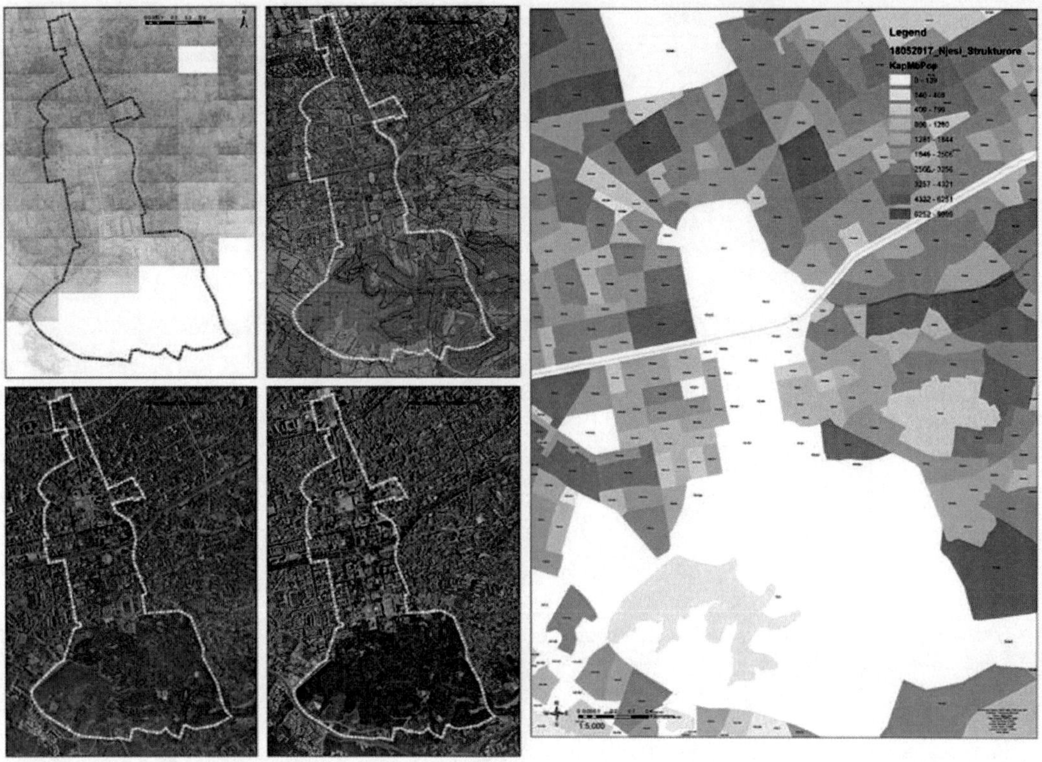

Fig. 4 Tirana central area:1980–2018 and population density

Table 1 Data analyses

Categories' function	Typology						
Agriculture, Archaeology historic memorial, Found neighbourhood Space, Gardens, Greens and playground, Linear greenways, Markets, Military, Open land, Park, Pedestrian walkway, Playing Fields, Pocket of space, Public, School campus, Sport, Squares plazas, Street parking transportation, Vacant	Civic space; Conspicuous spaces; External private space; Interchange space; Internal private space; Internal private space; Internalized "public" space; Movement space; Natural/semi-natural; urban space; Private open space; Public "private" space; Public open space; Retail space; Service space; Third place spaces; Undefined space; User selecting spaces; Visible private space;	**The space–time analysis**, based on GIS, for assessing the impact of the transformation of public spaces	**Comparative elements: Green**, density and socio-economic, demographic distribution	Physical change	Negative	Density and intensity Distribution	Changing rate
					Positive	Green % Positive / Negative	
				Functional change	Number of changes for same function but renewing	Density and intensity	
					Number of changes for each function	Spatial distribution	

Fig. 5 Loss of physical surface on public space

we can contribute to improving the relationship between the characteristics of urban development, from a physical form to the political, socio-economic and the ecological aspect of urban space. Understanding what the elements of a successful public space are will enable us to configure better creations and avoiding those with problems (Shaftoe, 2008). Public spaces are significative component in the cities' sustainability. If we are to reach the goal of minimizing negative influences, it is necessary to evaluate the find of types and the differences that public spaces have between them and their involvement to the overall system. A way for monitoring will be possible through the comparison of indicators and which can also be used for the evaluation of sustainable over the time. (Maiello et al., 2011) can give and help us to improving the sustainability of public space types, therefore such approaches should be considered as they contribute to a better ecological relationship in the intensive urban development. See Fig. 10a, b,

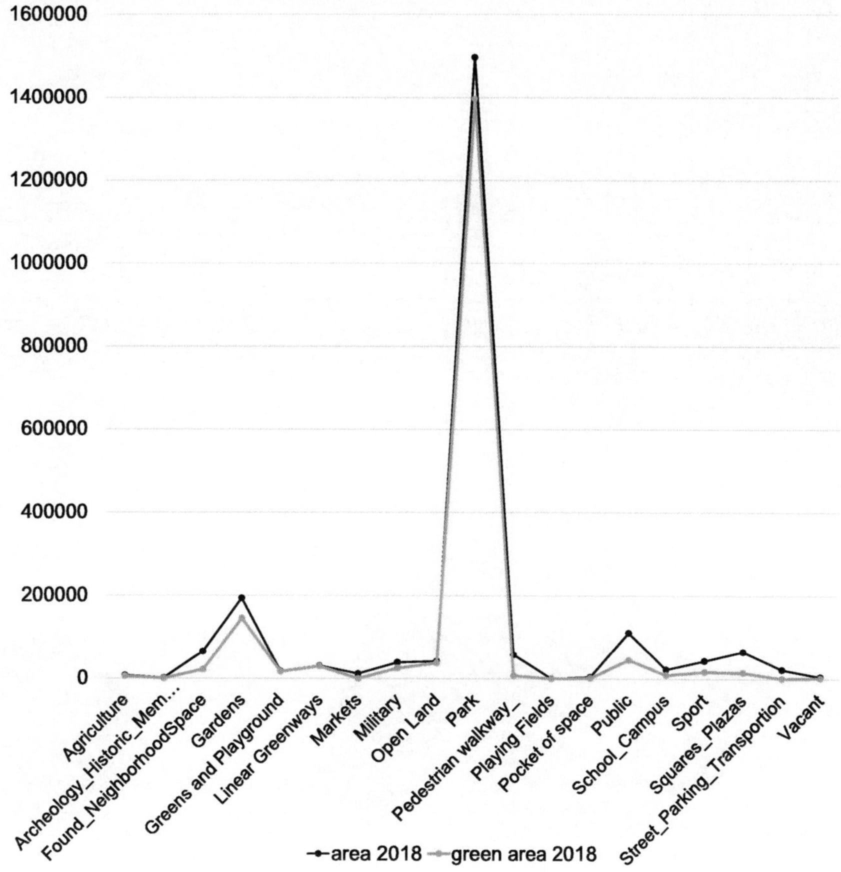

Fig. 6 Green area by categories in 2018

7 Discussion

If we would describe the conclusions in general, from the use of public spaces in Tirana, they would be useful for planners and designers in Albania or someone else, where users feel comfortable or evolve methods that enable people to create relationships between spaces without the concern of engaging in its use. Public spaces clearly reflect what is foreseen by the local and national plans of the country, influenced by economic and social needs and often this leaves without discussion on the topic of green spaces and ecological structure. Once the city's residents feel comfortable with the old public spaces due to the long history of living in those areas and because they have been part of their construction, the nostalgia created on these spaces and what they represent is an important element to consider, as well as their relationship with vegetation. The new plans of the city encourage densification and try to use it as a way of sources to trigger economy, and as a consequence this has led to the change of physical face of Tirana. From a global point of view, this has an influence on the green area and the urban landscape.

8 Conclusion

Today these public spaces have gradually modified and became part of retail or by commercial buildings influences that have introduce on them their characteristics and pretended the public domain and his development, by ensuring negative consequences in the whole public. The other aspect is the balanced type of use that makes the public space so functional, interesting, lively. By using a classification, we

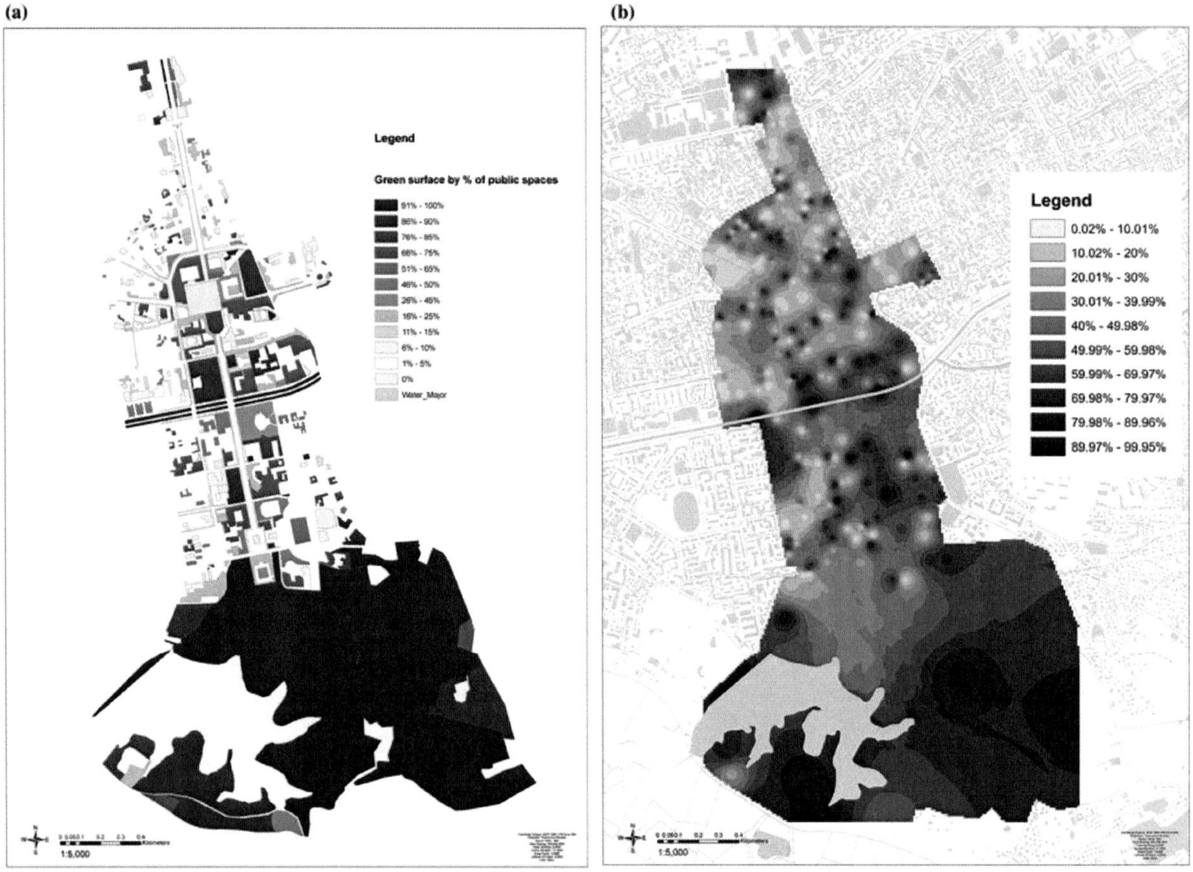

Fig. 7 **a** Green area as a percentage in 2018 and **b** weight distribution of public space by green % and size in 2018

will be able to analyse the prevailing situation and identify the changes that offer benefits and cause negative effects with the physical and social environment in our public spaces. This type of analysis will lead us to providing knowledge of the correlation between the factors of the urban form development as, socio-economic dimension and the quality through the vegetation of public space in Tirana. It was very important to have multiple analysis techniques and to take each one into consideration, to be successful in evaluating the changes and the ability of the public space. As shown in this study, another very powerful and essential tool spatial analyses is the use of GIS technology. We could have avoided number changes and the way of transformation, or finding new vision to view the relationship, together with the help of data knowledge for the final decisions. Based on the results, we realized the importance and geographic advantages that GIS and gravitational model give us. From this analysis, now we possess the knowledge of what and how is the uses of the public spaces in Tirana city, to target for new plans and how they can be sustainable. With this significant geographic and competitive advantage, we will have a decreased chance for future inefficient changes and will see how the urban transformation have determined the space viability by various parameters. A better understanding can lead us to the needs and future potential-sensitive areas. The selected areas are uncovered by green surfaces and the decrease physical and quality of these spaces. The results show the high differences of changes between the characteristics and allocation during the transition period, while residential zones appear as those negative values on the surface, determining the spatial distribution from the rating and demonstrating the relationship between sustainability

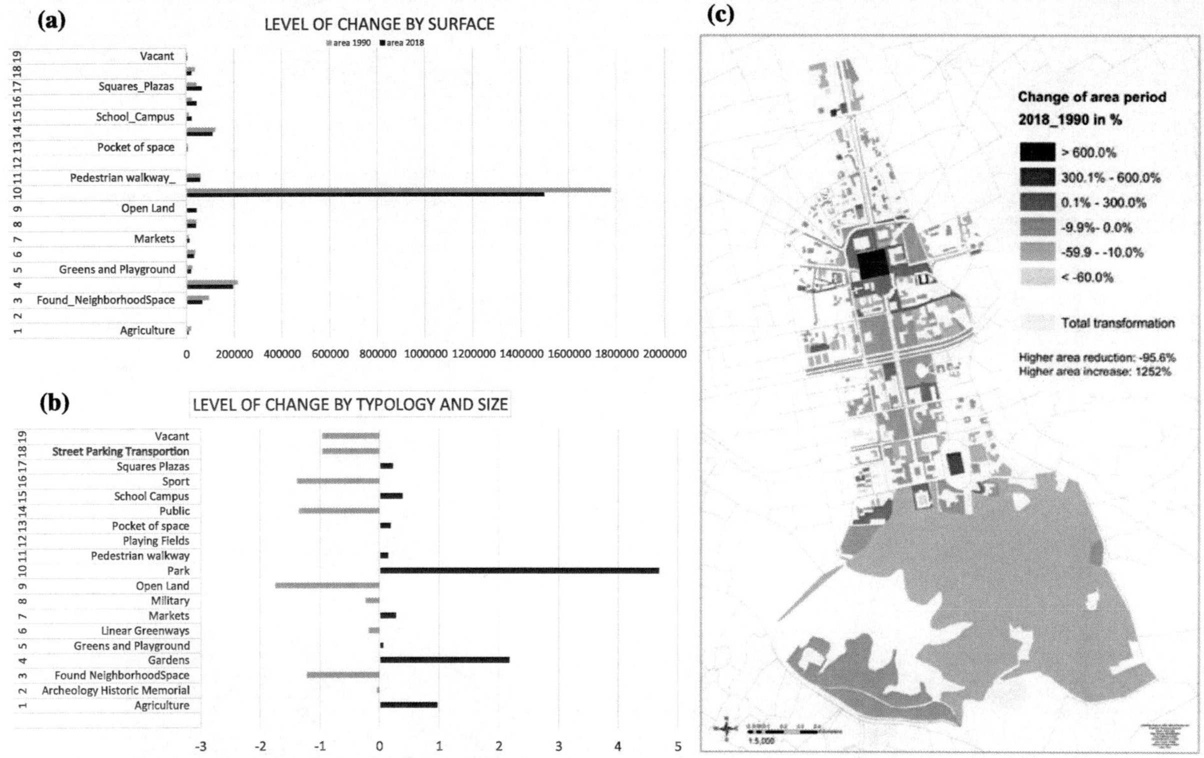

Fig. 8 a, b Level of change by type and dimension, c ranking of the transformation by type and dimension

Fig. 9 a, b, Level of change by green surface, c ranking of the transformation by green surfaces

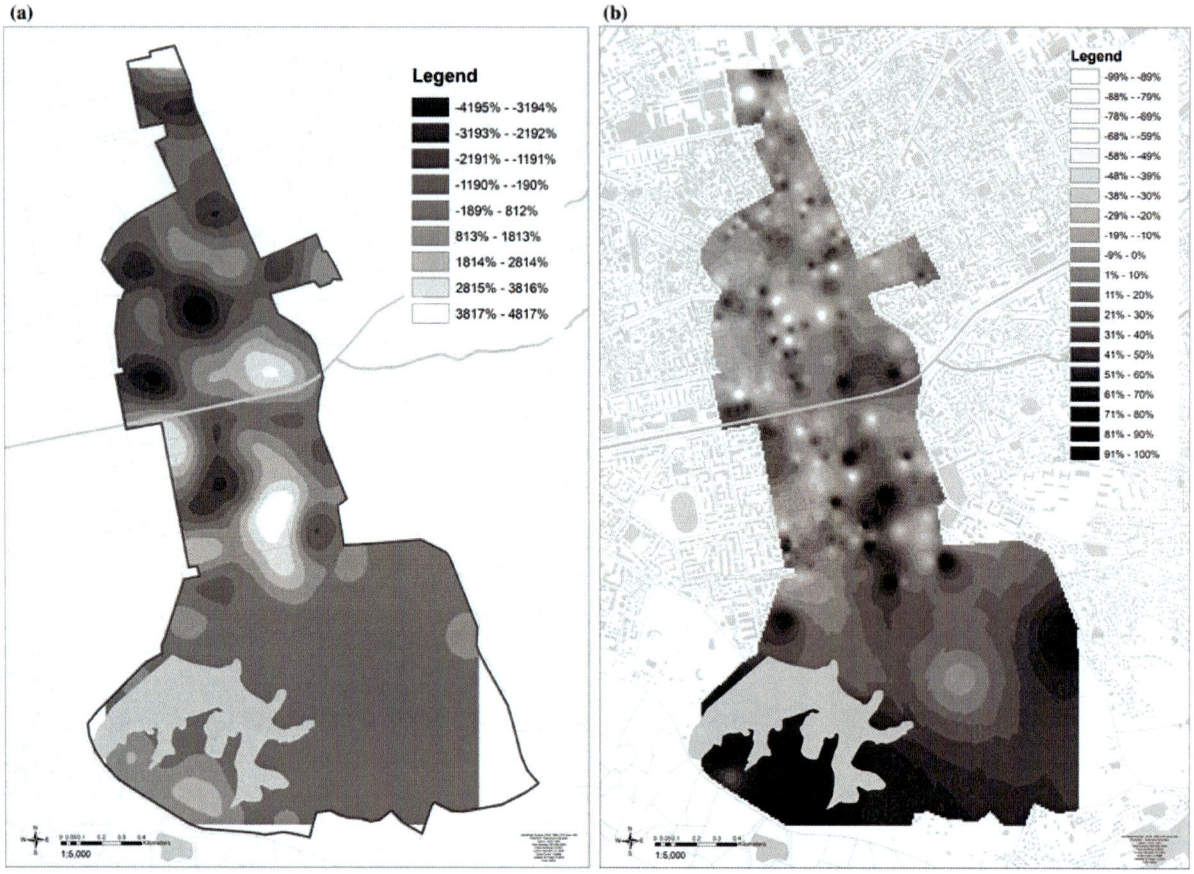

Fig. 10 **a** Ranking of the hotspots by spatial density and **b** the weight of the comparison by green surface and the size

and quality principles. Today's trend seems to embrace a theory that was created about the complexity of the city as an organism that can be adapted to achieve sustainability, but this remains to be evaluated and monitored. Further we will make it possible through comparing the indicators and additional parameters that have an impact on sustainable development. To increase knowledge, it is also recommended to set standards for data comparability.

Acknowledgements I would like to thank the INDASEUROPE, Ltd. for their assistance and ongoing support for this project and for providing customer demographic information.

References

Antrop, M. (2000). Changing patterns in the urbanized countryside of Western Europe. *Landscape Ecology, 15*, 257–270

Carmona, M., & Tiesdell, S. (2007). *Urban design reader: Architecture. Urban & land use planning*. Architectural Press.

Chiesura, A. (2004). The role of urban parks for the sustainable city. *Landscape and Urban Planning, 68*, 129–138

Devereux, M., & Littlefield, D. (2017). *A literature review on the privatisation of public space: Project Report.* UWE. https://eprints.uwe.ac.uk/31529

Esbah, H., & Deniz, B. (2007). Effects of land use development on urban open spaces. *Journal of Applied Sciences, 7*, 1138–1144

ESRI. (2013). *Environmental science research institute* (The ArcGIS Book ed., Vol. ArcGIS 10.1 [computer software]). (ESRI, Ed.) Redlands, California.

ESRI. (2017). *Understanding spatial analysis. Guided lessons based on real-world problems.* (Vols. Book, The ArcGIS). ESRI.

Institute of statistics (INSTAT). (2018). *Albania in figures.* INSTAT.

Ittelson, W. H., Rivlin, L. G., & Prohansky, H. M. (1970). *The use of behavioural maps in environmental psychology.* Holt, Rinehart & Winston.

Burke, J., & Ewan, J. (1998, February 17). Sonoran preserve master plan. An open space plan for the Phoenix Sonoran Desert. The Herberger Center for Design Excellence.

Lynch, K. (1972). *City sense and city design.* MIT Press.

Maiello, A., Battaglia, M., Daddi, T., & Frey, M. (2011). Urban sustainability and knowledge: Theoretical Heterogeneity and need of a transdisciplinary framework. *A Tale of Four Towns. Futures, 43*, 1164–1174

National Territorial Planning Agency, (NTPA). (2015). *Territorial planning and development in Albania: A guideline.* NTPA.

Shaftoe, H. (2008). Convivial urban spaces: Creating effective public places. *Earthscan.*

Soulé, M. E. (2007). Land use planning and wildlife maintenance: Guidelines for conserving wildlife in an urban landscape. *Journal of the American Planning Association, 57*, 313–323

Force, T. U. T. (1999). *Towards an urban renaissance*. Taylor and Francis.

The world bank. (n.d.). https://www.worldbank.org/en/topic/urbandevelopment/overview.

Thompson, C. (2002). Urban open space in the 21st century. *Landscape and Urban Planning, 60*, 59–72

United Nations Sustainable Development. (2015). *Agenda 21*. Available online: https://sustainabledevelopment.un.org/content/documents/Agenda21.pdf

Van Herzele, A., & Wiedemann, T. (2003). A monitoring tool for the provision of accessible and attractive urban green spaces. *Landscape and Urban Planning, 63*, 109–126

Tirana in the Crossroads: Urban Sprawl or Natural Resources

Armand Vokshi

Abstract

This research tends to analyze the city of Tirana, the capital of Albania from 1921. This city is the largest urban area in Albania, and especially since 1991, after the communist dictatorship, the city has seen strong demographic growth. Currently, uninterruptedly, the city is undergoing rapid urban development, which is radically transforming its identity landscape into the historic central, but especially suburban part of the city. The case study is an increase in the suburbs of the city in relation to the green and fragile perimeter of the city. The main issues addressed are two key moments that intertwine with each other as the phenomena of informal construction of the nineties as well as the birth of urban areas for "rich people." Both of these types of interventions have caused extensive damage to green areas around the city. The natural landscape around the city with high environmental values is in danger of transformation. The methodology followed focuses on researching the interventions that are reflected in Tirana's local strategic plan TR2030. It was approved two years ago in 2017 and was designed by a group of architects led by Stefano Boeri. The TR2030 is an attempt by the local government to establish a new relationship with the green environment around the city. The main proposed element is the creation of an "orbital forest" as a belt around the city. It must create the final border of the urban city of Tirana. The institutional mechanisms used are not effective so far. The present conclusions try to show the problems encountered during the implementation of the orbital forest in relation to urban developments in Tirana in the new suburban areas, as well as their real relationship regarding the approved proposal for the "orbital forest." The main problem that the general local plan is currently facing in its implementation, is the lack of a legal, economic and social mechanism for the compensation of these territories, to make this type of intervention feasible in practice. All the positive points discussed above are utopian in reality, in the event that immediate measures are not taken to delve into the logic of the transfer of rights over property and the law of compensation

Keywords

Tirana urban sprawl • Natural resources • Orbital Forest • Urban city limit

1 Urban Development of Tirana

Tirana lies in a very interesting territory in terms of geographical, environmental and panoramic conditions. Currently, uninterruptedly, the city is undergoing rapid urban development, which is radically transforming its identity landscape into the historic central, but especially suburban part of the city. Apparently, the attraction to build benefited from the immobilizer point of view, at the moment, is much stronger than the orthodox theories of modern city planning. Local decision-making institutions are overwhelmed with different demands, to build everywhere and with all kinds of urban typologies. The relationship that Tirana has with the green environment in the city, but also outside of its perimeter, is from a long period in an unjustifiable relation with the part built in total (Fig. 1).

In fact, at the time the capital was chosen, in 1921, Tirana was a simple and quiet urban center with around 15,000–20,000 inhabitants and well integrated with the natural and agrarian environment. This area of central Albania had begun to recover during the Ottoman period in the late 1600s, while it also influenced the oriental urban configuration of the city until it became the capital. Gherardo Bosio, the Italian architect who came up with the decisive plan of

A. Vokshi (✉)
Department of Urbanism, Faculty of Architecture and Urbanism, Politecnical University of Tirana, Albania

F. Trapani et al. (eds.), *Advanced Studies in Efficient Environmental Design and City Planning*, Advances in Science, Technology & Innovation, https://doi.org/10.1007/978-3-030-65181-7_43

Fig. 1 Political map of the Balkan Peninsula

city configuration in the forties, would call the existing Ottoman consolidated part of the historic city until 1939 a "Garden City" (*Città giardino*). This also influenced his project for urban plan design, where different parts of the new areas merged with the natural and agrarian environment, in a fluid and well-connected way. They were naturally associated with existing parts always according to the concept of "the garden city." (Figs. 2 and 3).

Bosio defined the modern city according to three main rules that are worth mentioning:

- Establishment of a circular circulation system of modern city development
- Establishment of a strong boulevard in the form of a boulevard of key state administrative functions
- Strong monocentric center densification
- Integral connection with existing urban areas, as well as with the natural territory around the city.

The urban developments that took place during the communist dictatorship in 1945–90 somehow respected the first two points, but radically changed the third and fourth points conceived by Bosio during the fascist occupation.

First, during this period, the idea of a dense city in the center, built along the main axis that extended the

Fig. 2 Tirana, Scheme of the regulatory plan in 1939–42. Archive of Bosio, Pelago–Firenze

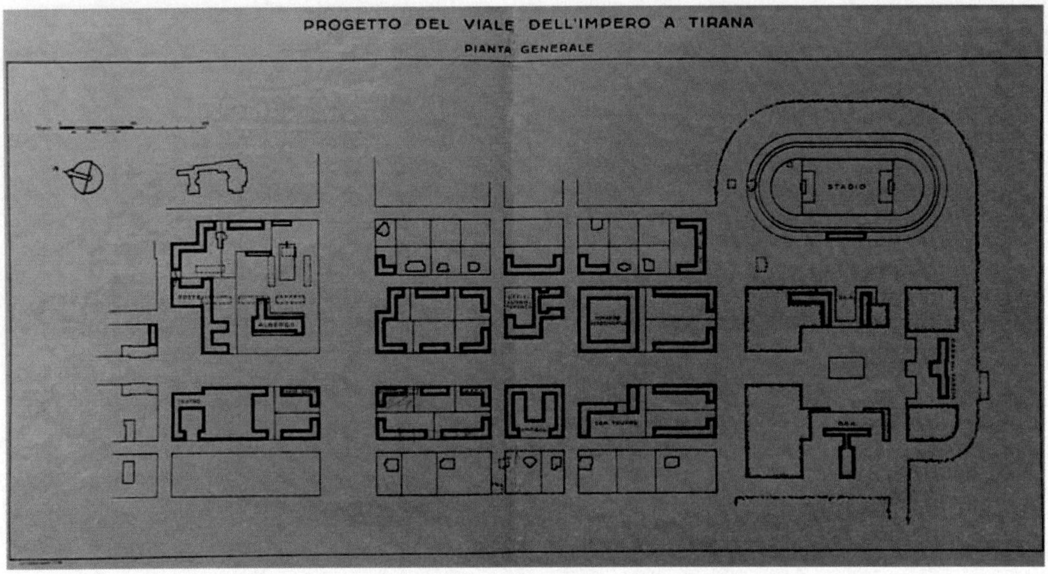

Fig. 3 Tirana, Boulevard project (Viale dell'Impero), in 1939–42. Archive of Bosio, Pelago–Firenze

administrative city, replaced the large geometric parks and arranged in the center of the city. Large parts of the buildings designated by the previous plan turned into green and recreational spaces, following the patterns of real socialist cities in Eastern Europe of the Communist Axis. As the city opened in the center, the opposite happened to the suburbs. The dilapidated villa area which formerly fumigated the natural part with the built one already had a new configuration with new design parameters. The residential quarters had the same typology and architectural configuration imposed by the regime. They consisted of 4–5-story residential buildings, according to the schemes defined from the urban and architectural point of view and their position relative to the center or finish line of the urban areas, and this had no significance for the urban plan. The partitioning between the built and natural or rural territory changed radically, severely (Fig. 4).

It is worth mentioning the arrangement of rivers and the realization of about 100 artificial lakes in the agricultural part of Tirana, some of which were also directly involved in the urban territory of the city, as strong elements of landscape and recreation.

1.1 Tirana—City Development After 90s

The main robes of mass destruction of relations with the green mass and of the natural resources around the city began after the fall of the communist dictatorship in the 1990s. From an extreme measure of control of the territory under the auspices of the state during the dictatorship, it shifts to an irresponsibility and a critical uncontrollable situation of interference in the natural and cultivated suburbs of the city. We can say that this period may be shared with two realities:

up to 2010 informal constructions

after 2010 with the spreading of villas systems for rich people.

The first reality of our analysis concerns the loss of control over the territory and urban planning directly after the fall of communism. So we said above that through close surveillance over the territory during the dictatorship, the country went into total anarchy, where families in difficult conditions from remote rural areas occupied different parts of the planted or natural land and they built their dwellings. Their modest is without a building permit and without a master plan for their accommodation. Their size is quite large and occupies a surface roughly the same as that of the city configured until the 1990s. Most of them focused on the northern part of Tirana (Fig. 5).

The second reality of analysis has to do with that of urban ghettoization that began to crystallize after 2010. Ghettoization is already happening in two directions in the far suburbs:

firstly with illegal constructions that were carried out informally in 1990–2010, where the poor people coming from rural areas were concentrated. These normally occupied the green territories and turned them into urban areas.

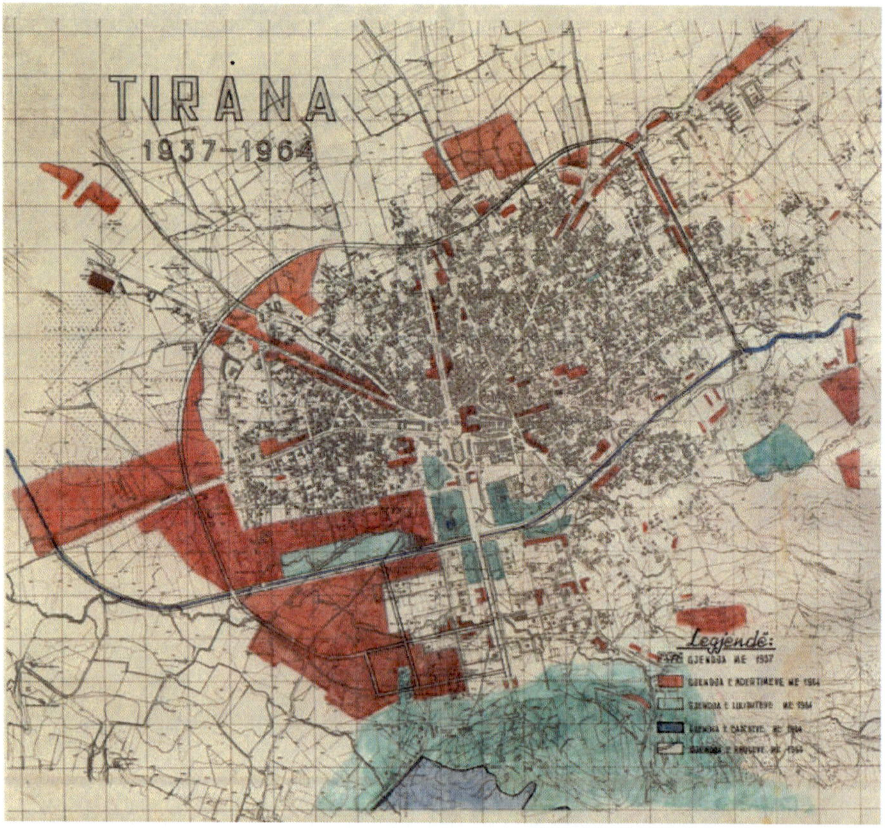

Fig. 4 Tirana, Scheme of the regulatory plan in 1964. Archive of AQTN, Tirana

Fig. 5 Tirana, informal area

Their features are as follows:

- significant lack of minimum urban conditions such as roads, sewers
- disrespect for minimum urban conditions: intensity, distance, etc.
- Poor quality of construction
- lack of public services.
- the second isolated urban mini-centers for the wealthy, which began to be in fashion after 2010. They have started to be designed especially in the natural or cultivated green areas on the south side of Tirana, in the area of Farka. The occupation of green territories serves to accommodate, with dedicated studies and building permits, the so-called elite quartiers and is back in trend now where "migration" of the city toward the green areas outside the city perimeter (Fig. 6).

Although the interventions are different, the way they are treated is the same:

- Residential kuartiere, surrounded by guards and impenetrable
- located in panoramic areas and well-studied by urban conditions
- high quality of construction but not of the aesthetic side. Usually, they have 2/3 typology of villa facilities, which are repeated in series without a personalized identity
- lack of public services. (prevalent private ones offered by the investor himself) (Fig. 7)

Both of these types of interventions have in fact brought about the same problem: that of occupying the natural and cultivated part in a very bad relation with the territory and the city. This has made Tirana a very large and unjustified stretch of construction. This has not been effective in maximizing the use of built-up and built-up territory within the city.

Both of these developments have turned the sad social realities of their closure within themselves with issues that will come to the fore in the near future.

In 2014, seeing the phenomenon of rapidly catching green areas beyond the urban perimeter of the city, the Albanian government made a swift decision thatforests, parks and remaining green areas around the city be included as protected areas in which no construction is allowed. This would in some way hinder the advancement of dispersed and uncontrolled construction (Fig. 8).

1.2 General Local Plan of Tirana 2030, Led by Stefano Boeri

In the local master plan, which was adopted in 2016, there were some points that were directly related to the section beyond the city perimeter. The construction of suburban parks was intended to reshape the identity of the territory, preserving high value ecological areas, reevaluating cultural areas and promoting tourism, recreation and the local economy (Fig. 9).

Parks as planned were to establish in accordance with the following rules:

Fig. 6 Tirana, "Rolling Hills" area

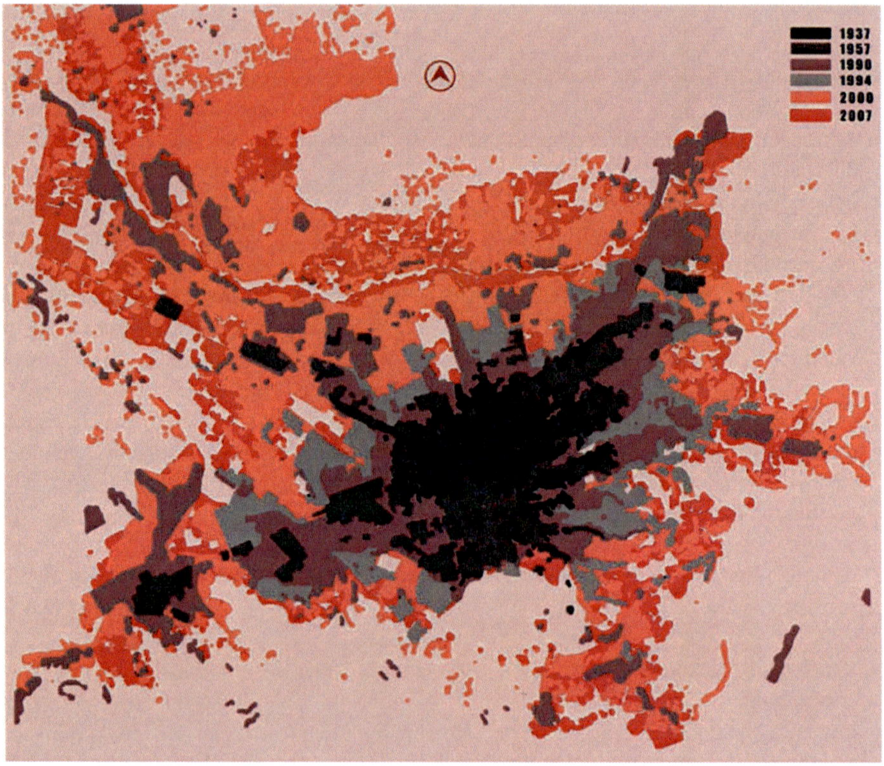

Fig. 7 Demographic development of Tirana elaborated by A.Vokshi

(a) Identification of the territorial area that will constitute the definition of the suburban park, where specific rules should apply regarding: construction, agriculture, ecological conservation, accessibility and recreational, cultural and tourist uses.
(b) In parks, the local economy, tourism and ecology must be reassessed, protected, interconnected and promoted in order to reinforce the park's identity.
(c) Accessibility to the park and the internal interconnection network should be addressed through consolidation, rehabilitation and maintenance of existing roads or alleys, coupled with bicycle lanes and speed control (limited speeds, etc.).
(d) The park should be provided with lightweight infrastructure (such as benches, desks, chats, resting places, recreation) and signage (information, maps) that guarantee easy and uninterrupted use of the territory. Lightweight infrastructure should be located at strategic points near reservoirs, panoramic points (hilltops, plateaus) or near crossroads or intersections.
e) Accessibility to the park should be guaranteed through park entrances: providing parking spaces, public transport interconnections, near villages or urban areas to ensure easy access to services.
(f) Park-dwelling structures are recommended to be built in accordance with the tradition–rural (rural) typology of village buildings in order to reinforce the identity of the territory and preserve historic villages (Figs. 10 and 11).

In accordance with this article and the strategy of this local general plan, upon its approval, the planning authorities shall identify the areas and units in which the development of suburban parks will be carried out through public investment or through economic strategies agreed upon public–private partnership (Fig. 12).

The natural territories considered to be treated as suburban parks of the city of Tirana by this plan and this regulation are as follows:

(a) Tirana artificial lake, Great Park
(b) The hills of the olive trees of Sauk
(c) Kashar Lake and perimeter park
(d) Farka Lake and perimeter park
(e) Metrobosco or the urban and suburban restrictive perimeter park.

1.3 The New Orbital Forest

The New Orbital Forest (Boeri, 2015) project is the idea to develop around Tirana a great forest that combines existing

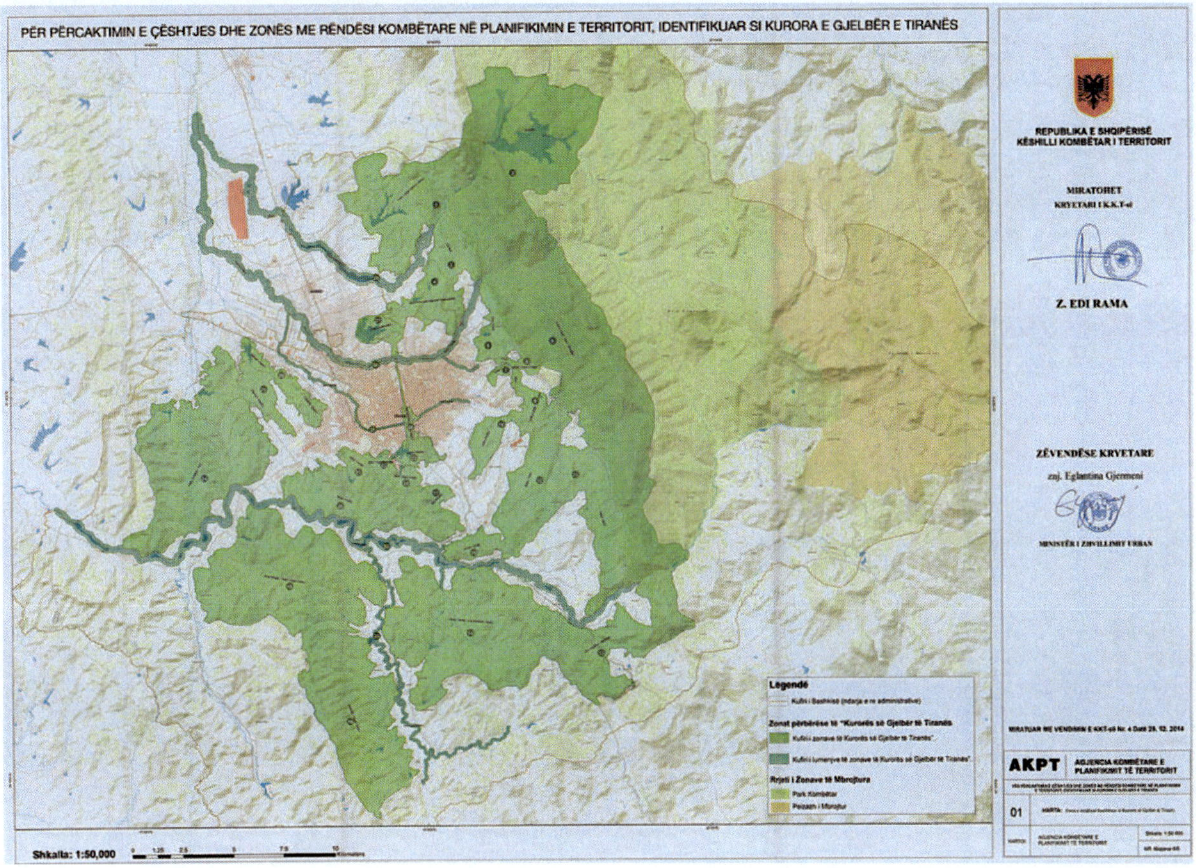

Fig. 8 Tirana, scheme of the green protected zone in 2014, AKPT

Fig. 9 Tirana, Scheme of the general local plan of Tirana 2030

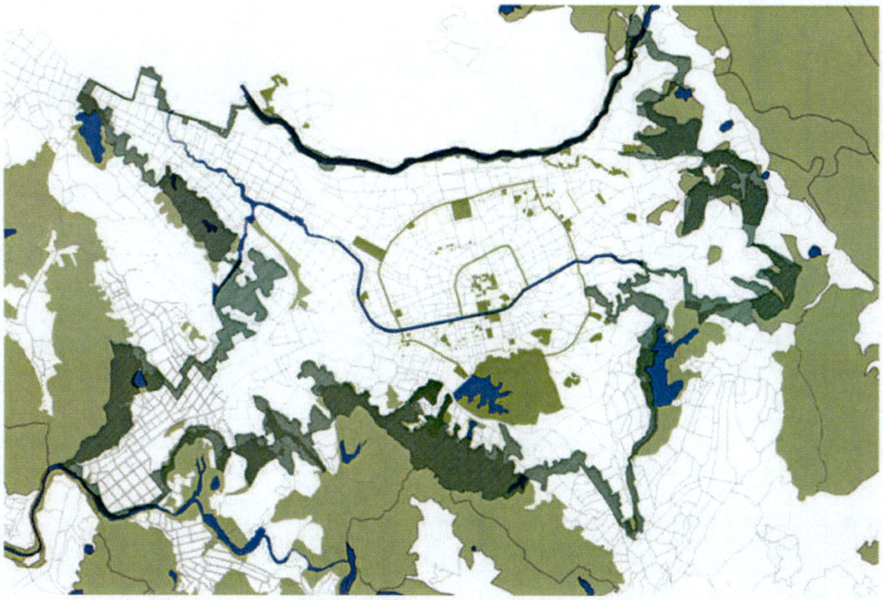

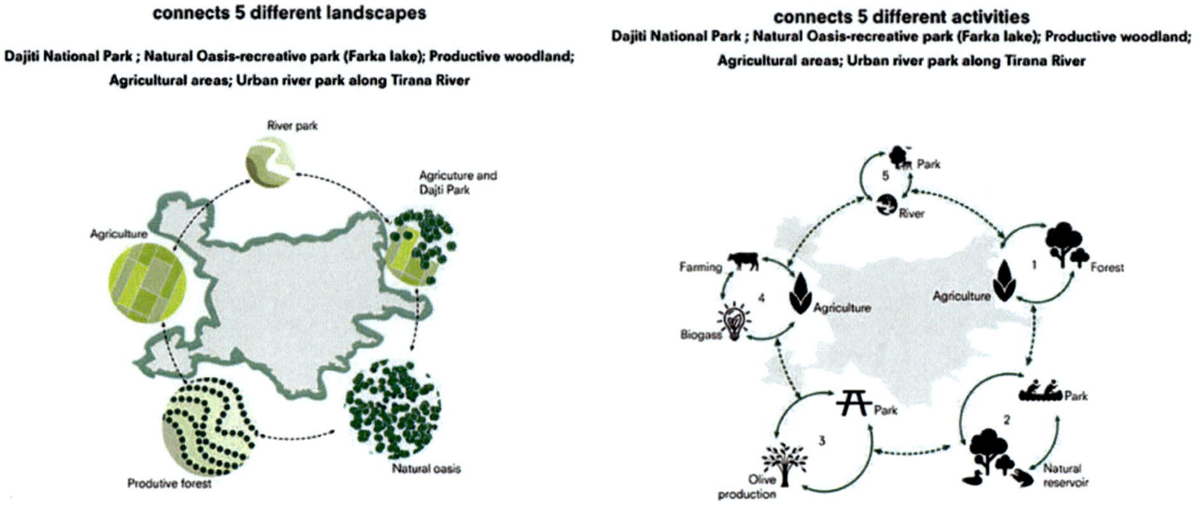

Fig. 10 Tirana, Scheme of the general local plan of Tirana 2030

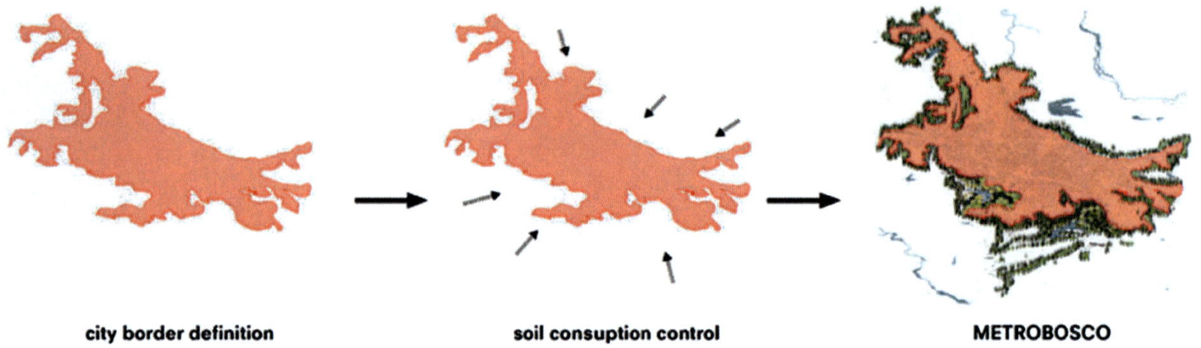

Fig. 11 Tirana, Scheme of the general local plan of Tirana 2030

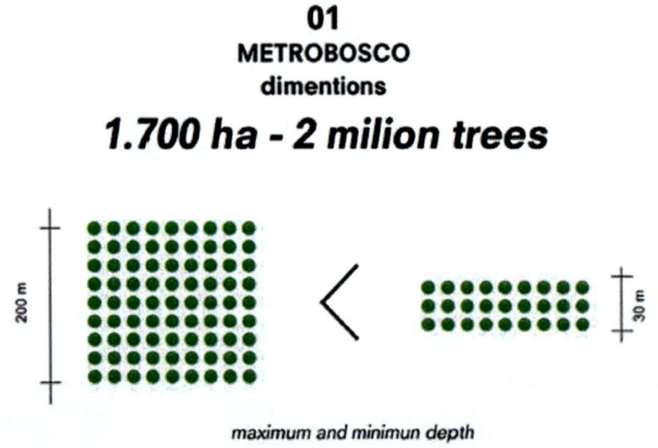

Fig. 12 Tirana, Scheme of the general local plan of Tirana 2030

peri-urban parks, agricultural areas, productive forests, riparian zones around the lakes and that defines at the same time a new shady border, a vegetable boundary and a natural limit to urban growth. It is a green belt, extended around Tirana's fifth ring, a network of green ways and also a green system that connects the agricultural areas along Durana axis to Dajiti National Park. The New Orbital Forest will lead the planting of a green belt of 2 million trees and will connect 14.671 ha of parks, agricultural fields, productive forest and other form of vegetation around the city on both public and private properties (Fig. 13).

Tirana's Orbital Forest connects five different landscapes:

Dajti National Park.

Dajti National Park, already considered as a protected area, is mainly made up of wooden land with some areas near the water reservoirs that are intended for small-scale familiar agriculture. It constitutes the perfect habitat for activities connected with sustainable tourism such as hiking and trekking (Fig. 14).

Natural Oasis—recreative park (Farka Lake).

The area that surrounds Farka Lake is the point wherein the green belt penetrates more profoundly in the city thus

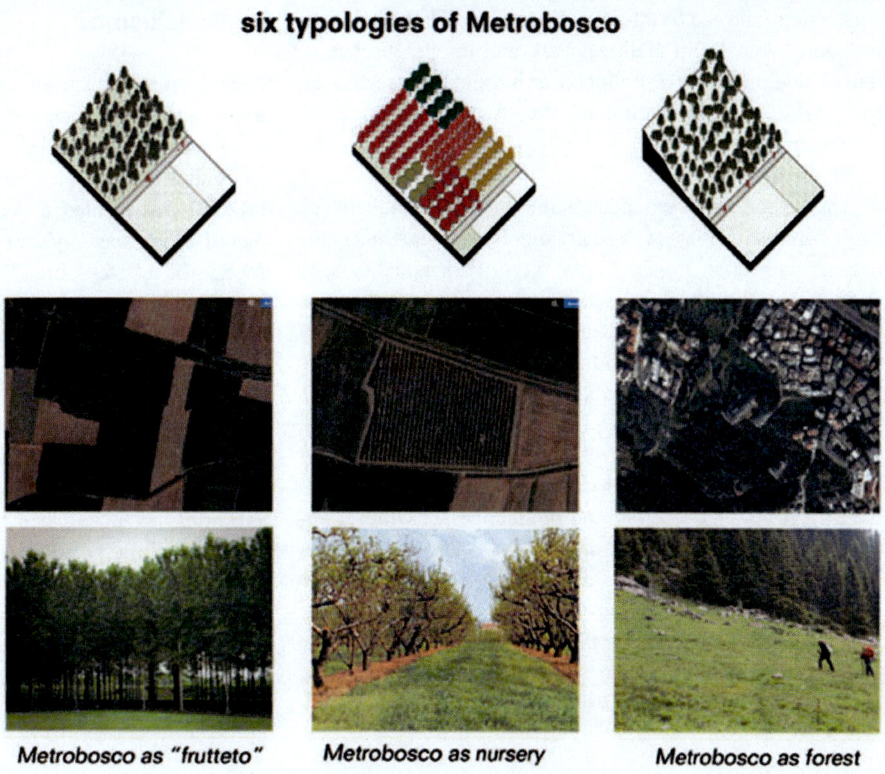

Fig. 13 Tirana, Scheme of the general local plan of Tirana 2030

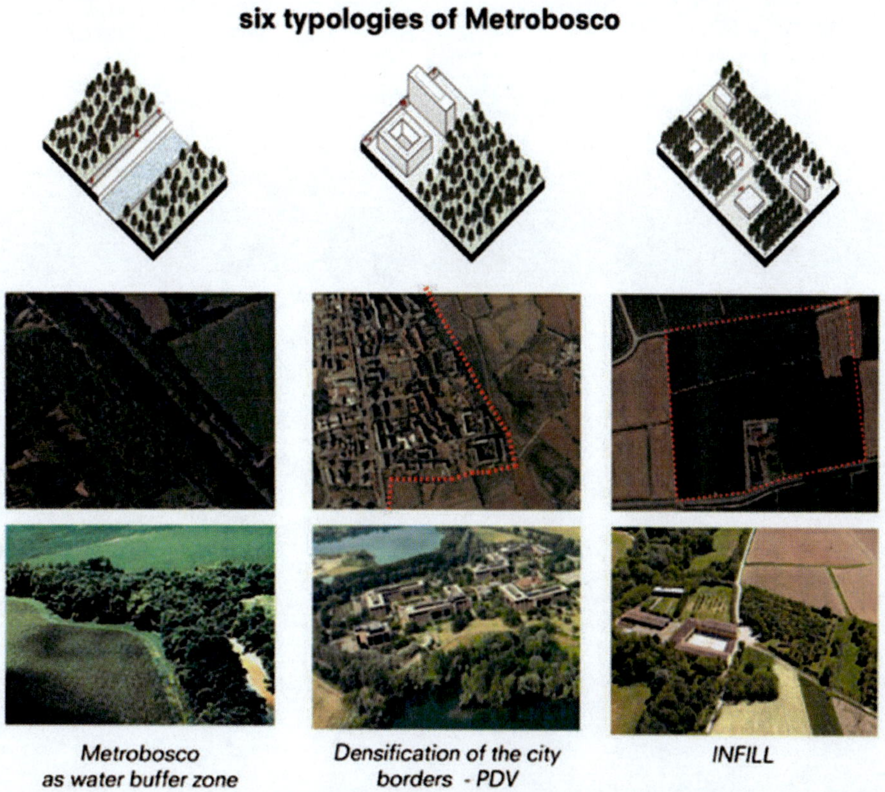

Fig. 14 Tirana, Scheme of the general local plan of Tirana 2030

connecting with the fourth ring. For its natural character, its proximity with Dajiti National Park and for its absence of remarkable problems of pollution or hotspot, this sector can be considered as a natural reserve, a protected area for savage species.

Productive woodland.

This sector, that is extended in the southern area, between the ex-municipalities of Vaqarr and Farka constitutes the natural connection for the Erzeni River Park and the Lake Park, that is, the city center. It is characterized by hilly areas that in the past used to be intended for agricultural activity and it is strongly influenced by the presence of Sharra's landfill that requires important environmental mitigation actions.

Agricultural areas.

This area, delimited by the Tirana-Durres axis—SH2/SH56, is part of a big and productive agricultural territory that also involves the municipality of Kavajes. Other particular elements that it is advisable to take into consideration are the ongoing project for the Farka Lake and the water treatment plant. As for the last one as well as for Lana River, it is necessary to envisage naturalization and environmental mitigation actions (Fig. 15).

2 Conclusions

Tirana, after the strong urban developments in its urban texture, needs to change its intervention strategy, especially in the suburbs. As we said, it shows the strongest problems accumulated in the past years. The city's overall local plan, TR2030, has created a good basis for defining a line that should block the city's exponential growth and the introduction of a "green ring." This ring should serve as a new step of environmental change in the suburbs of the city to reevaluate not only informal areas but limiting the development of urban sprawl and construction without any criteria.

The main problem that the general local plan is currently facing in its implementation and is the lack of a legal, economic and social mechanism for the compensation of these territories, to make this type of intervention feasible in practice. All the positive points discussed above are utopian in reality, in the event that immediate measures are not taken to delve into the logic of the transfer of rights over property and the law of compensation.

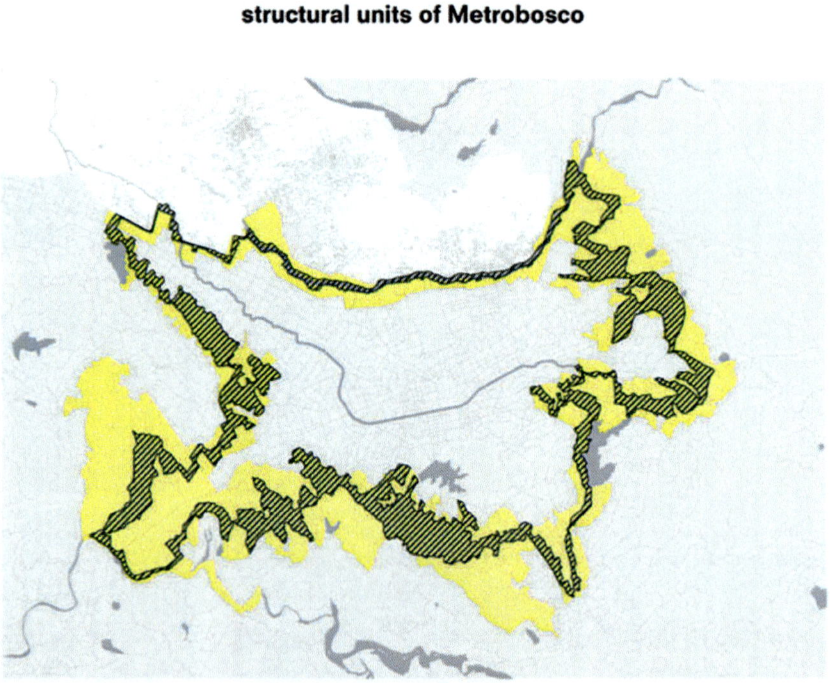

Fig. 15 Tirana, Scheme of the general local plan of Tirana 2030

References

Arkitektura dhe Diktatura. (n.d.). Tirana: Marin Barleti.

Art & Trashegimi. (2012). Bashkia Tirane.

Bace, A. E. (2016). *Historia e Arkitektures Shqiptare*. Kristalina-kh.

Biaggini, A. (2005). *Storia dell'Albania contemporanea*. Bompiani.

Boeri, S. (2015). *General LocalPlan of Tirana 2030*. Tirana.

Carlo, C. (1996). *Gherardo Bosio, architettofiorentino*. Angelo Pontecorboli Editore.

Ciucci, G. (2002). *Gli architetti e il fascismo*. Einaudi.

di San, J., & Savino, F. (1965). *La politica dell'Italia in Albania nelle testimonianze del Luogotenente del Re*. Cappelli Editore.

Fregolent, L., & Tonin, S. (2015). *Growing compact*. Francoangeli.

Frida Pashako, M. P. (2016). L'interpretazione dello spazio urbano e architettonico dell'asse strutturante di Tirana. Mantova: Universitas Studiorum S.r.l. - Casa Editrice. Kolevica, P. (1997).

Giacomelli, M. (2012). *Architetti e ingegneri italianiin Albania*. Edifir.

Giusti, M. A. (2006). *Albania, Architettura e Città*. Maschietto Editore.

Ilardi, M. (1999). *Negli spazi vuoti della metropoli*. Bollati Boringhieri.

Kolevica, P. (2018). Zeri im. Tirana: Kuteli '04.

Purini, F. (2005). *La città uguale*. Il Poligrafo.

Sotir Dhamo, G. T. (2016). *Tirana qyteti i munguar*. Polis Press.

Vokshi, A. (2015). *Tracce dell'architettura italianain Albania*. DNA.

Vokshi, A. (2020). *Tirana arkitektura 1920–2020*. GentGrafik

Dialogues of Sustainable Urbanization: Housing, Transport, and Well-Being

Comparison of Ground-Borne Vibration Induced by Railway Within Peak and Non-peak Hours

Mohd Khairul Afzan Mohd Lazi, Muhammad Akram Adnan, Norliana Sulaiman, and Mohamad Darwish Shahrir

Abstract

Over the years, the Malaysian government is committed to improve public transportation system by expanding its railway transportation system especially throughout the areas within Kuala Lumpur and Klang Valley. This is to reduce traffic congestion and at the same time provide passengers with a more comfortable and time-saving means of transportation. Although this system is relatively more eco-friendly for the public mass movement and saves energy consumption, however, the ground vibration induced by the trains greatly contributes towards human annoyance. The ground-borne vibration produced by the train operation has significant effects especially towards the people who live in the affected residential areas. This research compared the ground-borne vibration induced by the railway traffic between peak hours and non-peak hours in four case studies using seismograph to obtain the ground vibration data and laser meter detector to obtain the speed of train. The relationship between the speed of train and the distance from source showed an inverse relationship, as the distance from source increased, the ground vibration will decrease. As an attempt to estimate the ground-borne vibration induced by the railway, a multiple linear regression model had been developed. The result of this research reveals that the magnitude of vibration during non-peak hour was higher than the magnitude of vibration during peak hour of train operation. This could serve as a starting basis in predicting the ground-borne vibration from the railway and thus contributes towards a sustainable development in the dense, populated metropolitan like Klang Valley, Malaysia.

Keywords

Ground-borne vibration • Speed of train • Distance to source • Human annoyance • Modelling

1 Introduction

In Malaysia, there has been a rapid expansion of railway transportation. This is one of the government's initiatives to improve passengers' means of travelling. In the beginning of this century, the introduction of high-speed trains is the most advance development in transportation industry (Galvin & Dominguez, 2007). The technological advancement of trains is due to the increase of demands to reduce traffic congestion especially in the area with high population density. However, the introduction of high-speed train technologies also contributes to some environmental effects towards humans and may even be disruptive towards sensitive equipments in its surrounding.

According to Sulaiman, ground-borne vibration (GBV) is normally used to defined human made vibrations of the ground, such as air blasting, quarry bombing, vibrations caused by mining works and construction. Ground-borne vibration does not necessarily have to pose any problems to the ground itself. Many natural phenomena that occur can be generally associated with ground vibration, for example, earthquake, landslide and water waves in the oceans. Ground-borne vibration is normally generated by man-made activities that build up energy and will excite the ground underneath the soil until bedrock to form waves of vibration.

The problem from the ground-borne vibration by railway operations incurs major concerns due to short- and long-term effect from the vibration towards human perception. Short-term vibration can be described as a source of nuisance and discomfort for humans. It is well known that noise, or unwanted sound, can interfere with speech

M. K. A. Mohd Lazi (✉) · M. A. Adnan · N. Sulaiman · M. D. Shahrir
School of Civil Engineering, College of Engineering, Universiti Teknologi MARA (UiTM), Shah Alam, Selangor, Malaysia

F. Trapani et al. (eds.), *Advanced Studies in Efficient Environmental Design and City Planning*, Advances in Science, Technology & Innovation, https://doi.org/10.1007/978-3-030-65181-7_44

communication, affect concentration and cause sleep disturbance. The presence of rattles and vibrations can temper with human's comfort and may have adverse effect in terms of human's quality of life general. It also can cause increase in human stress level as noise and vibration are considered to be nuisance (Lombaert et al., 2015; Wanming et al., 2010).

One of the effects towards humans is sleep disturbances where it can affect heart rate of a person and may cause heart-related diseases that may lead to fatality (Croy et al., 2013). The effect of vibration to humans living in the affected area is a major concern in this study. It may affect the quality of life of a person and impose psychological problems such as annoyance, sleep deprivation, stress and increase the risk of fatigue. In a worst-case scenario, it may lead to permanent injuries after a long exposure to the vibration such as back, nerve and joint injury.

Ground-borne vibration by railway may not cause an immediate harm towards the surrounding buildings. However, cyclic vibrations may lead to failure in building structures. When the repetition of vibration exceeds a certain threshold of a building, it will cause the structure to experience a degree of tremors that in turn and will affect the interior building equipment. Furthermore, these vibrations also may interrupt the operation of sensitive equipments such as in hospital if it is located nearby a railway track and may even cause discomfort to the patients inside the hospital.

There were several studies and models have been developed to predict ground-borne vibrations induced by railway traffic worldwide. Among the factors considered in the models were x, y and z components of the velocity vector (DIN 4150-3, 1999), train-type specific vibration level, track quality, building amplification (Madshus et al., 1996), sleeper vertical vibration, energy sum of the transfer functions (Jones & Block, 1996), distance (Jiang & Zhang, 2004; Suhairy, 2000), wheel force applied (Christoffer et al., 2006), measured velocity, referred velocity (Hanson et al., 2006), vibration dose value and velocity (Paneiro et al., 2015).

Up to date, there is no studies found to compare the ground-borne vibrations impact from peak and non-peak hour of railway traffic operation. This is one of the motivation of this research studies. This study attempts to develop models for estimating ground-borne vibration using linear regression. The objectives of this study are.

a. To analyse the relationship of ground vibration induced by railway due to the speed of trains and distance from track using seismograph and laser gun.
b. To compare the occurrence of ground vibration of trains between peak and non-peak hours where it will affect human response living in the surrounding.
c. To develop a regression model to estimate the ground-borne vibration induced by railway.

2 Methodology

This study was conducted along Padang Jawa, Shah Alam to Bukit Badak, Klang KTM line. The rail track is a double-railway track involving two (2) lanes; one heading to Pelabuhan Klang and another to Kuala Lumpur. In this study, it involved four (4) different sites to distinguish the vibration magnitude within these sites along the rail track. The locations were very strategic because along the track, there were numerous residential areas that had been affected by the occurrence of ground-borne vibration. The equipments used to collect data are Mini-Seis seismograph and laser gun. Figure 1 shows the flow chart of the process during study.

Only passenger train or commuter train was considered to achieve the objectives of this study. Data collection process started from 6.30 am until 7.30 pm with appropriate time

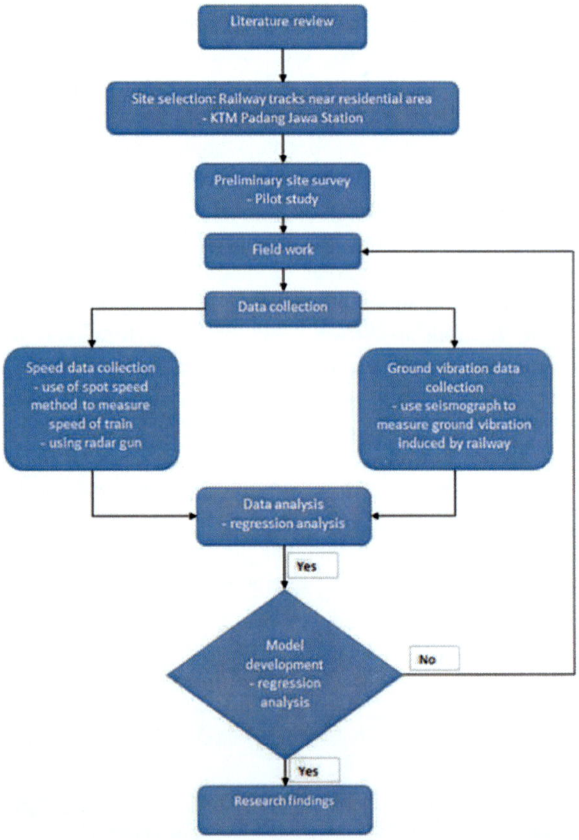

Fig. 1 Flow chart of study

Table 1 Schedule of data collection during peak hour and non-peak hour

Time	Location of seismograph and distance taken	Condition
6.30 am – 8.30 am	Location 1 = D1 & D2	Peak hour
9.00 am – 11.00 am	Location 1 = D1 & D2	Non-peak hour
3.00 pm – 5.00 pm	Location 2 = D3 & D4	Non-peak hour
5.30 pm – 7.30 pm	Location 2 = D3 & D4	Peak hour

margin to differentiate between peak and non-peak hour data. Data collection process was also conducted during clear weather condition without rain. Table 1 shows the schedule of data collection during peak and non-peak hour.

Figure 2 shows the typical schematic diagram of how the equipments had been set up on site during data collection. The seismograph was placed at two different points according to the schedule from the railway track. The distance between the sensor to the tracks had been recorded. Other data parameters such as time of train passing by the study point, speed of train, coordinate of seismograph had also been recorded. Location 1 indicates that the seismograph was placed closer to the rail track, while location 2 was more further from the rail track.

2.1 Result and Discussion

From Table 2, the mean of peak particle velocity (PPV) generated from train was 1.5110 mm/s. By comparing this value to Malaysian Department of Environment standard, "The Planning Guidelines for Vibration Limits and Control", it has violated the maximum permissible value that human can bear within residential area. The value lies between Curve 8 and Curve 16. This value is in the recommended range for industrial areas during day time and is not suitable for residential areas. The generation of ground vibration had caused a major disturbance towards human perception and comfort.

The distance between sensor to rail track was different for every site. There were two points with different distance for every site where the source will be located near to rail track and nearby residential area. The mean distance between sensor to rail track for this study was 10.114 m. The distance needs to be distinguished into types which are: before mean distance value which is near to railway track ($d < 10.114$ m) and after mean distance value which is further away from the railway track ($d > 10.114$ m) because it influenced the magnitude of PPV produced by train during its operating time. The magnitude of ground-borne vibration generated by train will be different even though the train speed is the same, for the distance between source and track. The purpose of distinguishing the distance into two different types is to eliminate bias for each vibration levels using only train

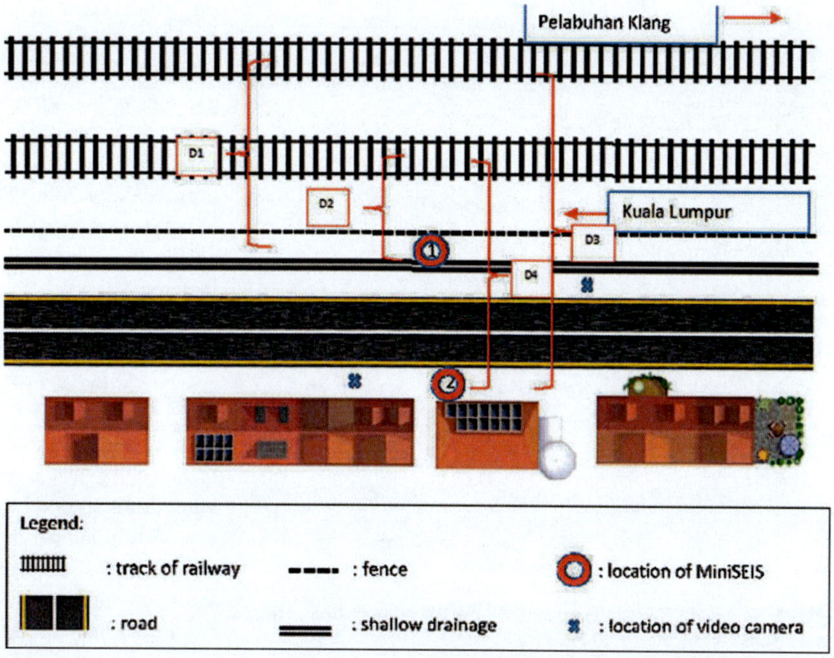

Fig. 2 Typical schematic diagram showing equipment set-up on site

Table 2 Descriptive statistic of all four sites

Variable	N	Mean	St. Dev	Minimum	Maximum
PPV (mm/s)	173	1.5110	1.1428	0.2223	7.2390
PPV Hz (Hz)	173	28.08	14.63	9.80	85.30
Speed (km/h)	173	45.07	16.93	16.00	87.00
Distance from sensor (m)	173	10.114	4.007	3.350	16.430

passages occurring on near or further away from track. This will prevent the skewing of data when plotting graphs due to the distance between tracks to source.

2.1.1 Relationship Between Speed of Train and Peak Particle Velocity (PPV)

Figures 3 and 4 show a positive relationship graph, whereby as the train speed increased, magnitude of PPV also increased. Based on the graphs above, it was found that when all the train speed at all four sites and PPV were combined and plotted together, there was a large scatter of data. It should be noted that a proportion of this may have been caused by differences in train loading. Train load heavily depends on the number of passengers boarding the trains. Furthermore, the data consists of data during peak and non-peak hours. The most significant difference occurred during peak hour of train operation where the train usage is at its highest demand. According to Ministry of Transport Malaysia, the statistics of the ridership of KTM Komuter in Malaysia was approximately 24 million passenger per annum, which translated to nearly 250,000 passengers daily.

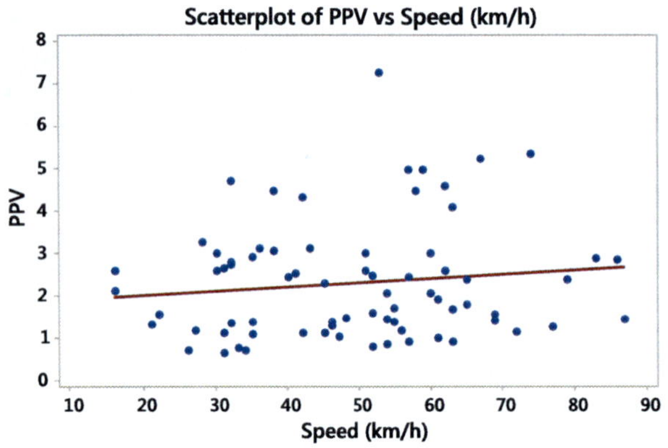

Fig. 3 Graph of PPV (mm/s) versus train speed (km/h) for nearby track ($d < 10.114$ m)

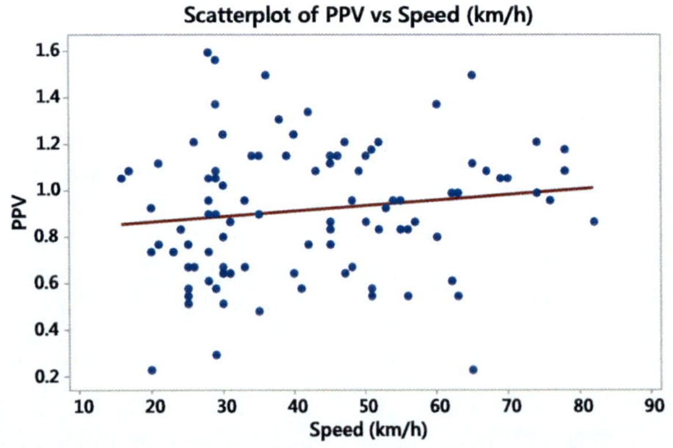

Fig. 4 Graph of PPV (mm/s) versus train speed (km/h) for further from track ($d > 10.114$ m)

During peak hour of train operation whether during morning or evening, the issues and problems related to overcrowding and congestion are likely to happen every day. During this time, the load of train carriage will be higher than during non-peak hour. This increasing in load of train due to numerous number of passengers had influenced the magnitude of PPV induced by railway. Despite this, a tentative best-fit line between train speed and PPV is shown in black line for both graphs. Although the lines showed a positive trend between ground-borne vibration induced by railway and speed of train, the scatter was too large to conclude a definitive correlation due to various factors such as train load and other factors that are also not considered in this study such as the geometrical characteristic of soil, property of soil and many more.

Study by Galvín and Domínguez (2009) stated that parameters that govern the level of ground-borne vibration induced by railway resulting from moving train is the correlation between train velocity and soil wave propagation velocity with PPV. In conclusion, the relationship between peak particle velocity (PPV) and speed of train is, as the train speed increases, the magnitude of PPV also increases.

2.1.2 Relationship Between Distance from Mini SEIS to Track and Peak Particle Velocity (PPV)

Figures 5 and 6 show an inverse relationship graph between distances against peak particle velocity (PPV). As the distance from source increased, PPV level will decrease. This was clearly shown by one study (Connolly et al. 2014), in comparison of different type of trains towards ground-borne vibration induced by high-speed train, when the distance from track increased, the PPV level reduced quickly.

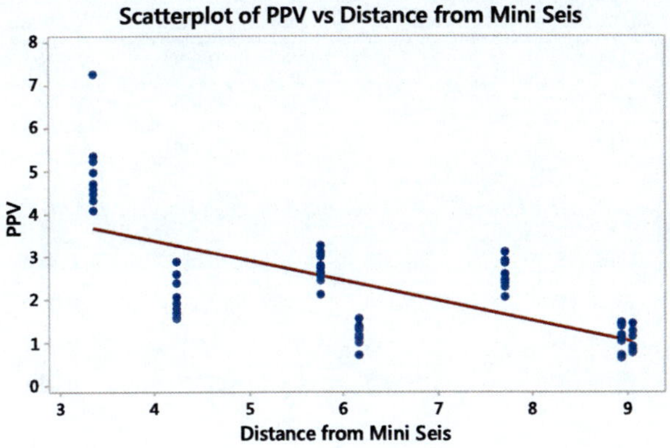

Fig. 5 Graph of PPV (mm/s) versus distance rail track from Mini-Seis (m) for near track

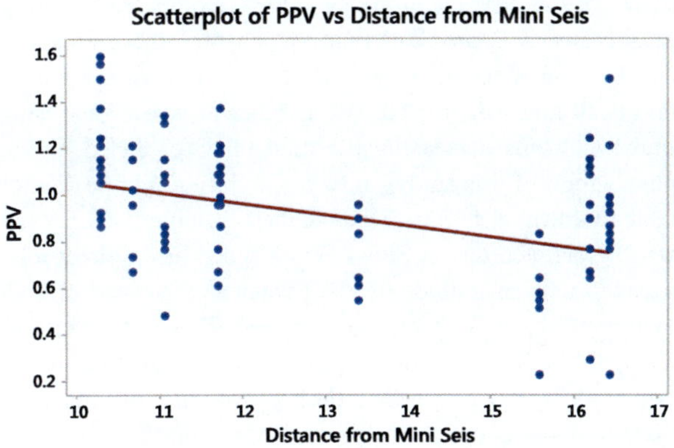

Fig. 6 Graph of PPV (mm/s) versus distance rail track from Mini-Seis (m) for far track

This decreasing in magnitude of PPV as the distance increased is caused by various factors. One of them is due to geometrical attenuation. According to Persson, geometrical attenuation is a vibration decay caused when waves transmitted by a source, the vibration tends to decay as the distance from source increases. The transmission of waves is affected by attenuation along transmission path beneath the ground. Other than geometric attenuation spreading, attenuation can also be through energy loss within material or reflection and refraction at boundaries. The attenuation of ground-borne vibration brought by railway generally depends on discontinuities such as joints, cracks and boundaries between different soil layers and also the topography of soil (Eitzenberger, 2008).

Next, the other factors that would influence in the reduction of PPV level as the distance increases is damping. Damping occurs when energy is dissipated due to friction between individual soil particle as the transmission of vibration progresses within the soil. According to Degrande and Schillemans (2001), the falling off of magnitudes of PPV with distance due to radioactivity and material damping in the soil can be obviously seen in high-speed trains recorded at various distances from track.

Apart from that, the influence of different wave forms during transmission of ground-borne vibration within ground also could affect in the decreasing of vibration amplitude as distance from source increases. This attenuation is frequency dependent and is more significant for higher frequencies of PPV during wave transmission.

2.1.3 Regression Model to Predict Ground-Borne Vibration Induced by Railway

Linear regression models have been used broadly by researchers for predictions of ground-borne vibration phenomenon, such as by (Wanming et al. 2010). In this study, linear regression model was used to forecast ground-borne vibration made by railway. In general, this model can be used to estimate magnitude of PPV whether during peak and non-peak hour of train operation. One sample of linear regression models for forecasting ground-borne vibration made by railway had been established and as exposed in the Eq. (1):

$$\mathrm{PPV} = 3.089 + 0.00951V - 0.1984D \qquad (1)$$

where, D = Distance from track (m).

V = Speed of train (km/h).

From the regression analysis, independent variable train speed, V and distance from track, D were significant for forecasting ground-borne vibration induced by railway, where the p-value is less than 0.05 in Table 3.

Table 4 shows the analysis of variance (ANOVA). The p-value for prediction model was less than 0.05, so the regression model is significant in forecasting ground-borne vibration.

2.1.4 Comparison of PPV Within Peak and Non-peak Hours of Train Operation

Significant differences can be seen between magnitude of PPV generated during peak and non-peak hours. The PPV level during peak hour of train operation is still high because it is influenced by the increasing of train load due number of passengers boarding the trains during peak hour. To investigate further on the significant difference of the PPV level between peak hour and non-peak hour whether PPV level during peak hour is higher than non-peak hour, a paired two-sample for means t-test was conducted. The hypotheses are as below:

Null hypothesis, $H_0: \mu_1 - \mu_2 = 0$.

Table 3 Regression analysis for linear model predicting ground-borne vibration induced by railway

Term	Coefficient	Standard error coefficient	*T*-value	*P*-value	*R*-square
Constant	3.089	0.241	12.79	0.000	52.28%
Speed, *V* (km/h)	0.00951	0.00359	2.65	0.009	
Distance, *D* (m)	−0.1984	0.0152	−13.07	0.000	

Table 4 ANOVA for linear regression model for prediction of ground-borne vibration induced by railway

Source	DF	Adj SS	Adj MS	*F*-value	*P*-value
Regression	2	117.447	58.724	93.16	0.000
Speed, *V* (km/h)	1	4.414	4.414	7.00	0.009
Distance, *D* (m)	1	107.663	107.663	170.79	0.000

The PPV level during peak hour of train operation is not higher than the PPV level during non-peak hour.

Alternative hypothesis, $H_1: \mu_1 - \mu_2 = 0$.

The PPV level during peak hour of train operation is higher than the PPV level during non-peak hour.

μ_1: mean of PPV peak hour (mm/s).

μ_2: mean of PPV non-peak hour (mm/s).

Difference: μ_1 (PPV peak hour) − μ_2 (PPV non-peak hour).

Estimate for difference: −0.176.

95% lower bound for difference: −0.458.

T-Test of difference = 0 (vs ≠): *T*-value = −1.04 *P*-value = 0.849 DF = 171.

Table 5 shows the *T*-test results from Minitab output for all four sites during peak and non-peak hours. The condition to reject H_0 is when *p*-value is less than α-value. α-value is set to be 0.05. For the difference in PPV level between peak and non-peak hours, the interpretation of result is as below:

p-value = 0.849 > α-value = 0.05.

Since the condition is not satisfied, it can be concluded that H_0 cannot be rejected and is accepted. Thus, alternative hypothesis, H_1 is not accepted. There was not enough evidence to conclude that at 5% significant level whether the PPV level during peak hour of train operation is higher than the PPV level during non-peak hour.

In this study, the PPV produced during non-peak hour is higher than during peak hour of train operation. Although during peak hour, the train load will increase due to the increase in passengers, and the speed of train is still lower than during non-peak hours. The additional weight of train will influence in higher generation of ground-borne vibration. As the train load increases, the magnitude of PPV also increases. This is because during peak hours, number of passengers will increase and the train needs to comply to a certain speed limit due to safety reasons.

Apart from that, in this study, during non-peak hours, the train tends to travel with higher velocity compared to during peak hours. As stated earlier in this chapter, the train speed also have an effect on the generation of PPV, although during non-peak hours there were less number of passengers using trains and thus decreasing the train loading. Based on the observation of this study, most trains during non-peak hour tend to travel faster with the highest velocity of 86 km/h which was recorded at Site 4. Train speed has a direct relationship with magnitude of PPV, whereas the speed of train increased, the PPV level also increased. Therefore, it can be decided that the generation of ground-borne vibration during non-peak hour is higher than during peak hour of train daily operation and is due to the speed of the trains.

3 Conclusion

According to all data collected at site, residents that live along Padang Jawa to Bukit Badak Line had been significantly affected by the disturbance due to ground-borne vibration brought by railway operations. The average PPV generated by the train operations in this study had violated the recommended limits for residential area based on "The Planning Guidelines for Vibration Limits and Control" by Malaysian Department of Environment. These vibrations had caused in human disturbance and deemed undesirable among people who live nearby rail tracks.

Ground-borne vibrations can be perceived as dependant on various factors such as type of locomotive, quality of track, speed of train, distance from source, type and build-up

Table 5 T-test result for all sites for comparing PPV level within peak and non-peak hours

Sample	N	Mean	St. Dev	SE mean
PPV peak hour (mm/s)	80	1.416	0.994	0.11
PPV non-peak hour (mm/s)	94	1.59	1.25	0.13

in the ground. Two parameters that influenced the generation of ground-borne vibration induced by railway which are speed of train and distance from source had been studied. The relationship between these parameters with peak particle velocity also can be concluded. Train speed has a direct relationship with the PPV, whereas when the speed of train increases, the magnitude of PPV also increases. Apart from that, distance from source has an inverse relationship with PPV. The further the distance from source, the lower the magnitude of PPV.

In comparison of the PPV level between peak hours and non-peak hours in this study, the magnitude of PPV during non-peak hour is higher than during peak hour of train operation. In this study, during peak hour, although the load of the commuter increases due to higher number of passengers using the train, it did not cause the PPV values to be higher. Train speed during non-peak hour is very high compared to during peak hour, and during this time, the train tends to move faster with high velocity, thus increasing magnitude of PPV. As a conclusion, in this study, the trains during non-peak hour generated higher PPV compared to during peak hour.

The result of this research could serve as a starting basis in predicting the ground-borne vibration from the commuter train and thus contributes towards a sustainable development in the dense, populated metropolitan like Klang Valley, Malaysia.

Acknowledgements The authors would like to thank the authorities from the Faculty of Civil Engineering, Universiti Teknologi MARA (UiTM), for their constant support and encouragement. Appreciation goes to the Research Management Institute (RMI, UiTM) and Ministry of Higher Education Malaysia (MOHE) for the financial supports FRGS Grant scheme: RMI File No: 600-IRMI/FRGS 5/3 (008/2019), Sponsorship File No : FRGS/1/2018/SKK06/UITM/02/5.

References

Christoffer, W., Bahrekazemi, M., & Bodare, A. (2006). Validation of an empirical model for prediction of train-induced ground vibrations. *Soil Dynamics and Earthquake Engineering, 26*.

Connolly, D. P., Kouroussis, G., Woodward, P. K., Alves Costa, P., Verlinden, O., & Forde, M. C. (2014). Field testing and analysis of high speed rail vibrations. *Soil Dynamics and Earthquake Engineering, 67*, 102–118. https://doi.org/10.1016/j.soildyn.2014.08.013

Croy, I., Smith, M. G., & Waye, K. P. (2013). Effects of train noise and vibration on human heart rate during sleep: An experimental study. *British Medical Journal Open, 3*(5), 1–9. https://doi.org/10.1136/bmjopen-2013-002655

Degrande, G., & Schillemans, L. (2001). Free field vibrations during the passage of a Thalys high-speed train at variable speed. *Journal of Sound and Vibration, 247*(1), 131–144. https://doi.org/10.1006/jsvi.2001.3718

Deutsches Institut für Normung. (1999). *DIN 4150-3: Structural vibrations—Part 3: Effects of vibration on structures.*

Eitzenberger, A. (2008). *Train-induced vibrations in tunnels—A review*. Department of Civil, Mining and Environmental Engineering, 100.

Galvín, P., & Domínguez, J. (2007). Analysis of ground motion due to moving surface loads induced by high-speed trains. *Engineering Analysis with Boundary Elements*. https://doi.org/10.1016/j.enganabound.2007.03.003

Galvín, P., & Domínguez, J. (2009). Experimental and numerical analyses of vibrations induced by high-speed trains on the Córdoba-Málaga line. *Soil Dynamics and Earthquake Engineering, 29*(4), 641–657

Hanson, C. E, Tower, D. A., & Mesiter, L. D. (2006). Transit noise and vibration impact assessment. *Federal Transit Administration*, FTA-VA-90-1003-06.

Jiang, T., & Zhang, X. (2004). In situ experiments and numerical assessment of environmental vibration induced by urban viaduct rail transit in Shanghai. In: *13th World Conference on Earthquake Engineering*, Canada.

Jones, J. C., & Block, J. R. (1996). Prediction of ground vibration from freight trains. *Journal of Sound and Vibration.*

Lombaert, G., Degrande, G., François, S., & Thompson, D. J. (2015). Ground-Borne Vibration due to railway traffic: A review of excitation mechanisms, prediction methods and mitigation measures. In: J. Nielsen et al. (Eds.), *Noise and vibration mitigation for rail transportation systems. Notes on numerical fluid mechanics and multidisciplinary design* (Vol. 126). Springer, Berlin, Heidelberg

Madshus, C., Bessason, B., & Harvik, L. (1996). Prediction model for low frequency vibration from high speed railways on sot ground. *Journal of Sound and Vibration.*

Paneiro, G., Durao, F. O., Costa e Silva, P., & Neves, F. (2015). Prediction of ground vibration amplitudes due to urban railway traffic using quantitative and qualitative field data. *Transportation Research Part D.*

Suhairy, S. A. (2000). *Prediction of ground vibration from railways*. SP Swedish National Testing and Research Institute.

Wanming, Z., Zhenxing, H., & Xiaolin, S. (2010). Prediction of high-speed train induced ground vibration based on train-track-ground system model. *Earthquake Engineering and Engineering Vibration.* https://doi.org/10.1007/s11803-010-0036-y

Empirical Model of Ground-Borne Vibration Imposed by Road Traffic on Soft Ground

Norliana Sulaiman and Muhammad Akram Adnan

Abstract

A novelty prediction of ground-borne vibration had been performed based on road traffic at soft ground area within Selangor, Malaysia. Malaysia has developed a guideline to describe the ground vibration threshold. However, no specific technique has been introduced on how to measure the ground vibration level. Moreover, important parameters such as the soil dynamics factor were not considered in the existing ground-borne vibration model for road traffic. The purpose of this research paper was to develop an empirical model that can predict the ground-borne vibration from nearby arterial single carriageway. The model incorporates soil dynamics, pavement characteristics, and traffic parameters as independent variables. Additionally, this study was undertaken to complement the established Malaysian guidelines using a model developed from empirical multiple regression analysis. Existing models were tested with local field data to evaluate the reliability. The results indicated inaccurate prediction when it used with local field data. The sensitivity analysis has shown that the vehicle flow rate is highly sensitive when predicting soft ground vibration. With the finding, highway planner and geotechnical engineer can use the developed model in order to assess the impact of the ground-borne vibration in an effective and systematic manner. The findings of this study can serve as a starting point towards developing national guidelines, for a more sustainable road development and environmental safety.

Keywords

Ground-borne vibration • Single carriageway • Soft ground • Multiple linear regression models • Traffic engineering parameter

1 Introduction

Nowadays, many roadside areas have grown into commercial strip developments, which normally occupy in limited and constrain land nearby to the road with the area behind them exist residential, institutional and industrial. Ground-borne vibrations (GBV) due to the expansion of roads and highways that travelling through the dense populated area are a main environmental concern in metropolitan areas. The ground vibration pollution can annoy people, interfere sensitive operation and equipment and damaging historical building structure (Hunaidi et al., 2000; Henwood & Haramy, 2002; Phyl et al., 2004; Xu & Hong, 2008; Kliukas et al., 2008; Mhanna et al., 2012; Tuan Chik et al., 2013; Del Grosso & Basso, 2014; Coquel & Fillol, 2017) especially in soft ground area. Accurate prediction of ground vibration in this soft ground area before and after development can lead to choose the right mitigation and countermeasure approaches.

Many factors contribute to the reduction or increment of amplitude and frequency of traffic vibration including the dynamic properties of ground, type of vehicle, road pavement condition, traffic volume, traffic speed and distance from road (Agostinacchio et al., 2014; Henwood & Haramy, 2002; Hunaidi, 2000; Hunaidi & Tremblay, 1997; Ju, 2009; Lak et al., 2011). Several empirical models have been developed to interrelate these factors in order to predict the GBV caused by road traffic (Klæboed et al., 2003; Long, 1993; Turunen-Rise et al., 2003; Watts, 1990). An empirical study showed by Watts (1990) predicted annoyance and maximum vertical peak particle velocity (PPV) at a building

N. Sulaiman · M. A. Adnan (✉)
School of Civil Engineering, College of Engineering, Universiti Teknologi MARA (UiTM), Shah Alam, Selangor, Malaysia

F. Trapani et al. (eds.), *Advanced Studies in Efficient Environmental Design and City Planning*, Advances in Science, Technology & Innovation, https://doi.org/10.1007/978-3-030-65181-7_45

foundation by regression analysis. The model is only suitable in determining whether there is any ground vibrations exist from road surface defect. Factors considered in the empirical predictive model are surface defect height or depth, heavy good vehicle speed, ground scaling factor, wheel path effect and surface defect distance from building foundation. Then the prediction of maximum vertical velocity at a building foundation was calculated by combining those factors. Long (1993) developed an empirical equation and evaluated the impact of vibration from a proposed highway to a residential area. The equation takes into account only the distance from steady traffic with no large truck as a contribution factor for particle velocity amplitude. Turunen-Rise et al. (2003) selected statistical 95 percentile value of maximum velocity or acceleration as an estimator of the vibration strength for a given traffic environment. The vibration measurements are determined for a single event vehicle pass-by and statistically as a mean of many event pass-by on the dwelling floor where the highest of weighted velocity or acceleration occurs. The empirical method introduced provides a relationship between the strength of people reaction and vibration exposure.

In the light of above discussion, none of previous studies have been found developed empirical model based on all parameters in one regression model. None of them include shear wave velocity or shear modulus as an important predictor in the prediction model at soft ground area. The goal of this paper is to formulate the GBV due to repetitive movement of land transport such as movement of vehicle that imposed to the community and surrounding structure nearby the road transport facilities. The specific objective of this study is to develop empirical prediction model of GBV by using multiple linear regression analysis at soft ground area. The empirically developed model will combine the soil dynamic parameters, traffic parameters and pavement parameters.

2 Research Methodology

2.1 Test Sites

The location of three study areas in Selangor, Malaysia, is shown in Fig. 1. Two test areas of soft ground consisted of soft marine clay (Meru-s4 and Sabak Bernam-s1 site), and another test area of soft ground was silty clay (Sepang-s6 site).

2.2 Shear Wave Velocity

The in-place shear wave velocity (Vs) profile of soil site was measured by using spectral analysis surface waves (SASW) method which was applied to the soil without requiring a borehole (Wair et al., 2012). SASW testing consists of determining the surface wave at the site and interpreting the dispersion curve to acquire the corresponding shear wave velocity profile (Ali Zomorodian & Hunaidi, 2006; Ismail, 2012). In the field testing, hammer and weight were used to impact the soil while determining the passing Rayleigh surface waves with pair of 1 Hz geophone sensors that were linked to a dynamic signal analyzer. This data was managed in the program WinSASW to develop site-specific dispersion curves, which were then reversed to estimate shear wave velocity profiles (Fig. 2).

2.3 Ground Vibration Measurement

After the wave being generated by the road pavement surface, the ground-borne vibration spreads through the surrounding ground of the road shoulder. The ground vibration measured by using seismograph as in Fig. 3. The component consists of geophone transducer, microphone and Mini-Seis monitor as a data logger. The geophone transducer and microphone were placed at the shoulder of the road with different distance from edge of road line marking. The seismic reading radial, transverse and vertical was recorded by the Mini-Seis monitor in velocity (mm/s) unit. Peak particle velocity (PPV) recorded was the highest reading among radial, transverse and vertical. The distance from road main stream to the geophone transducer was also measured and recorded. Vibration data is extracted from the data analysis software provided. In this study for model development process, vibration data is analysed as 95th percentile within 5-min interval. The 95th percentile vibration is where the 95% of the recorded vehicle vibration are propagating at or below this value surrounding shoulder of roadway.

2.4 Vehicle Speed and Vehicle Class Measurement

Vehicle speed is collected by using Automatic Traffic Counter (ATC) at arterial single carriageway. ATC consists of two road tubes as shown in Fig. 4, and the two road tubes are installed across the road. Vehicles that passed through both of the road tube give pulsed and the speed recorded by the ATC. Proper traffic management is needed to make sure the road tube is placed correctly and safely on the road, during the installation. Speed of vehicle is then analysed as mean speed within 5-min interval. The speed data is synchronized with ground vibration data for model development process. For synchronization, the vehicle speed and vibration at the road shoulder also recorded by using video camera.

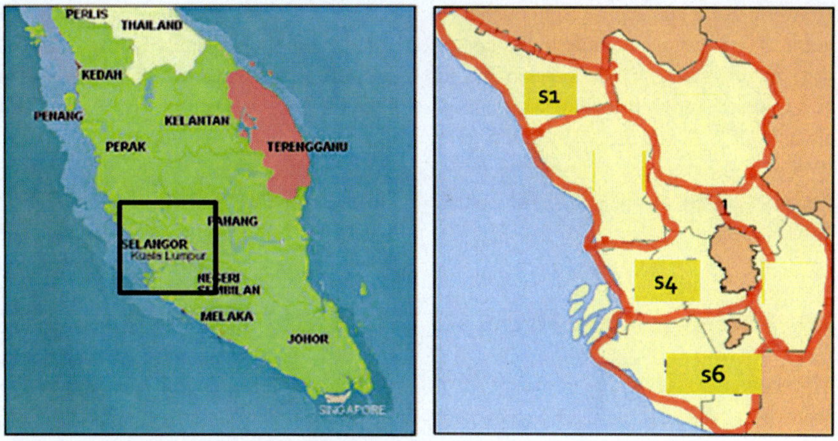

Fig. 1 Location of three study areas in Selangor, Malaysia

Fig. 2 SASW method performed at site study area (Sulaiman, 2016)

Fig. 3 Component of seismograph with automatic traffic counter (ATC) install on road

Fig. 4 Road tube installed across the road

The vehicle classification extracted from TRAXpro software was based on Federal Highway Administration (FHWA) vehicle classification scheme. The data extracted was used to calculate the traffic flow rate. Generally, traffic flow rates are stated in units of vehicles per hour. It represents vehicle flows that exist in a period of time less than one hour. In order to acquire equivalence in measurement for volume in a various vehicle traffic conditions, the technique used in this study was to convert the mix types of vehicle into one standard and constant unit in passenger car equivalent (PCE). It was done by multiplying the total vehicle type with the passenger car equivalents value acquired from Malaysian Highway Capacity Manual (2011) for two-lane highways.

2.5 Pavement Measurement

The pavement width was measured from centre of the roadway to the edge of road pavement layer by using roller meter and pavement thickness measured after coring of the pavement layer. Distance from vibration source is measured from centreline of roadway to the seismograph sensor (Henwood & Haramy, 2002; Kunimatsu et al., 2014).

3 Model Parameters

Figure 5a, b and c shows the shear wave velocity profile of soil at the study area. Two sites of soft ground areas consist of soft marine clay (Meru and Sabak Bernam), and another one site is silty clay (Sepang) area. Low shear wave velocity profile is investigated and measured at soft ground area. If the shear wave velocity of material is less than 183 m/s, then sites are classified as soft ground (Madshus et al., 1996). All sites in this study are classified as soft ground area. Slightly higher shear wave velocity was measured at depth less than 2 m. This is maybe due to the compacted sub-grade at the shoulder of the road. The shear wave velocity then increase gradually with depth. The result acquired by using SASW are reliable if less than 7 m (Lak et al., 2011). Larger depth considered as not reliable due to inadequate energy of excitation source.

Figure 6 shows a plot of 95th percentile vertical vibration against distance. The trend shows slightly higher vibration produced when it is far from vibration source. At soft ground area, the frequency spectrum principally is lower. Normally, ground vibration at low frequency decreases small with respect to distance and magnifies most due to building resonances (Madshus et al., 1996).

Figure 7 shows a plot of vertical vibration in 95th percentile against mean speed of vehicle. The scatter plot shows increasing trend of vertical vibration when the mean speed increases. At soft ground area, the Rayleigh wave velocity is low. If the vehicle speed is higher than Rayleigh wave velocity, the ground vibration produced will be higher.

To validate the vehicle speed and ground vibration data, laser gun and video camera were used to record and capture the synchronization event at site location. Figure 8 shows different vehicle such as lorry, car and motorcycle produced different ground vibration level on road shoulder at one of the soft ground study area.

Figure 9 shows a plot of 95th percentile vertical vibration against vehicle flow rate. The trend shows increment of vertical vibration when the vehicle flow rate increases.

Table 1 shows the pavement thickness (PT) and pavement width (PW) for site study. The pavement width was measured from centre line of the road pavement to the edge of pavement. Pavement thickness is measured from the coring of the pavement layer.

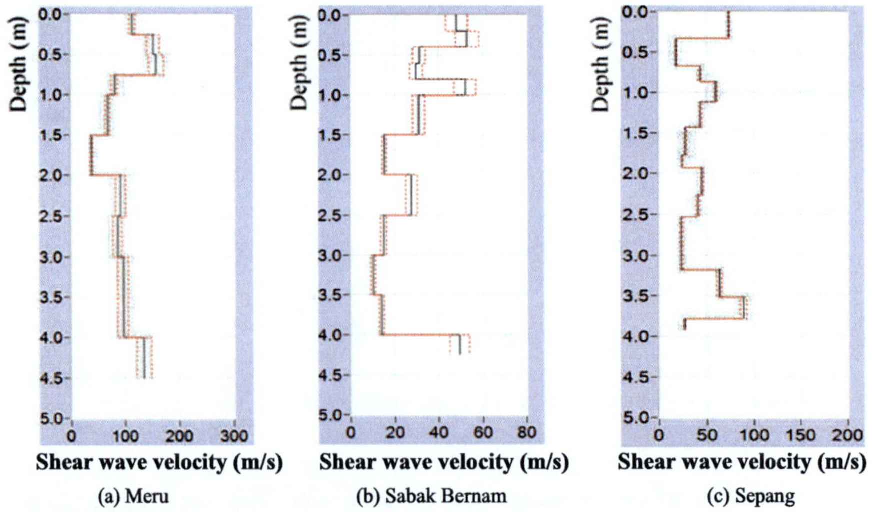

Fig. 5 Shear wave velocity profile of site study

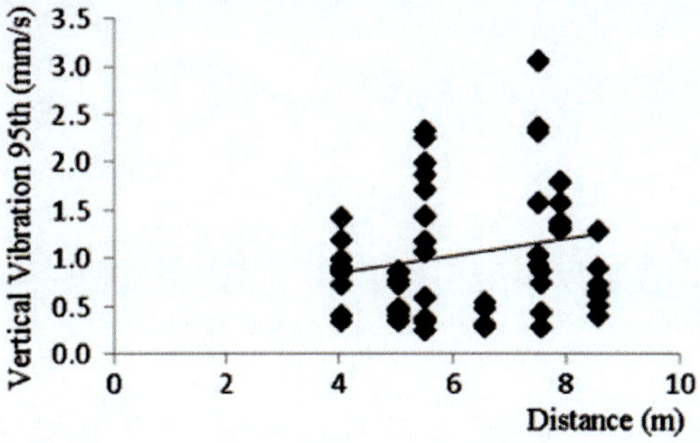

Fig. 6 95th percentile vertical vibration with distance

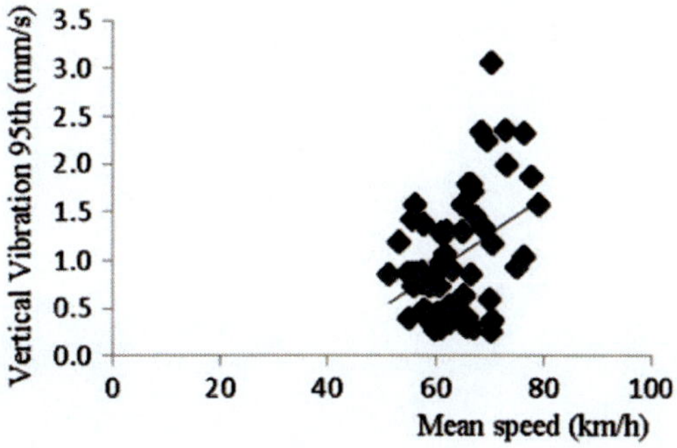

Fig. 7 95th percentile vertical vibration with mean speed

4 Model Development

After all the parameter data gathered from all sites, they were arranged as a row data in advanced statistical software Minitab. The data was sorted and screened to identify outliers that might disturb its distribution. This outliers maybe happen during the data collection or data entry process. The outliers were then removed to avoid distortion in model development process. Table 2 shows the mean, median, standard deviation, minimum and maximum value, skewness and kurtosis for model parameter after removal of outliers. Generally, the value of skewness and kurtosis for most of variable was near to zero, which means the data is normally distributed.

Afterwards, the correlation analysis was employed to measure the degree of the linear relationship between two variables. Table 3 shows the correlation matrix. It showed that shear wave velocity ($1/V_s$), pavement width and thickness (PW/PT) and vehicle flow rate (FL) were significantly correlated with vibration in high r-value and a p-value < 0.05 compared to the other variable, mean speed (S) and distance (D_s).

The vibration model includes variables such as the shear wave velocity (V_s), vehicle speed (S), pavement width and thickness (PW/PT), vehicle flow rate (FL) and distance (D_s) as the predictor. The hypothesis for predicting the models can be listed as follows:

H_0 the predictor cannot be used for predicting in the model.

H_1 the predictor can be used for predicting in the model.

From Table 4, V_s, S, PW/PT, FL and D_s were considerably independent variables for forecasting vibration with the p-value less than 0.05; thus, the H_1 was accepted and H_0 was rejected. Therefore, these predictors can be incorporated in the model for approximating the model.

The analysis of variance (ANOVA) is shown in Table 5, and hypothesis of it can be identified as:

H_0 the model cannot be used for prediction VIB_{soft}.

H_1 the model can be used for prediction VIB_{soft}.

The ANOVA of the VIB_{soft} model output is shown in Table 5 with F-statistic 79.15 and p-value less than 0.05. This offers good evidence to reject null hypothesis, and the regression model gives valuable information for predicting VIB_{soft}.

From Table 5, the final equation for predicting the VIB_{soft} was developed as shown in Eq. (1):

$$\begin{aligned}\text{VIB}_{\text{soft}} &= 2.379 + 5.994/V_s - 0.025S \\ &\quad - 1.046\ \text{PW/PT} + 0.001\ \text{FL} + 0.042D_s\end{aligned} \quad (1)$$

where VIB_{soft} is the 95th percentile vertical vibration (mm/s); V_s is the soil shear wave velocity; S is the mean speed of vehicle (km/h); PW/PT is the pavement width (m) and pavement thickness (cm); vehicle flow rate (PCE/h); D_s is the distance from seismograph sensor to the vibration source (m).

From regression analysis, the co-efficiency level in the predictor S and PW/PT has negative sign, suggesting that an increase in a vehicle speed and a ratio of pavement width and pavement thickness will lead to decrease in the soft ground vibration, VIB_{soft}. The predictor Fl and D_s depict positive sign, thus implies the increase in vehicle flow rate and distance will increase the soft ground vibration. The more vehicles on road actually will affect the vehicle speed; however, this will increase the ground vibration level. Increasing vehicle flow rate on road will cause the vehicle speed reduce, and this will increase the ground vibration. The earlier result of correlation value shows that the vehicle flow rate gave the highest correlation value with the vibration and very significant in explaining the vibration level near the roadway. While predictor V_s also has positive sign but inversely relationship, meaning that increasing soil shear wave velocity will decrease the ground vibration. As been known in literature, area with low shear wave velocity will tend to have higher ground vibration.

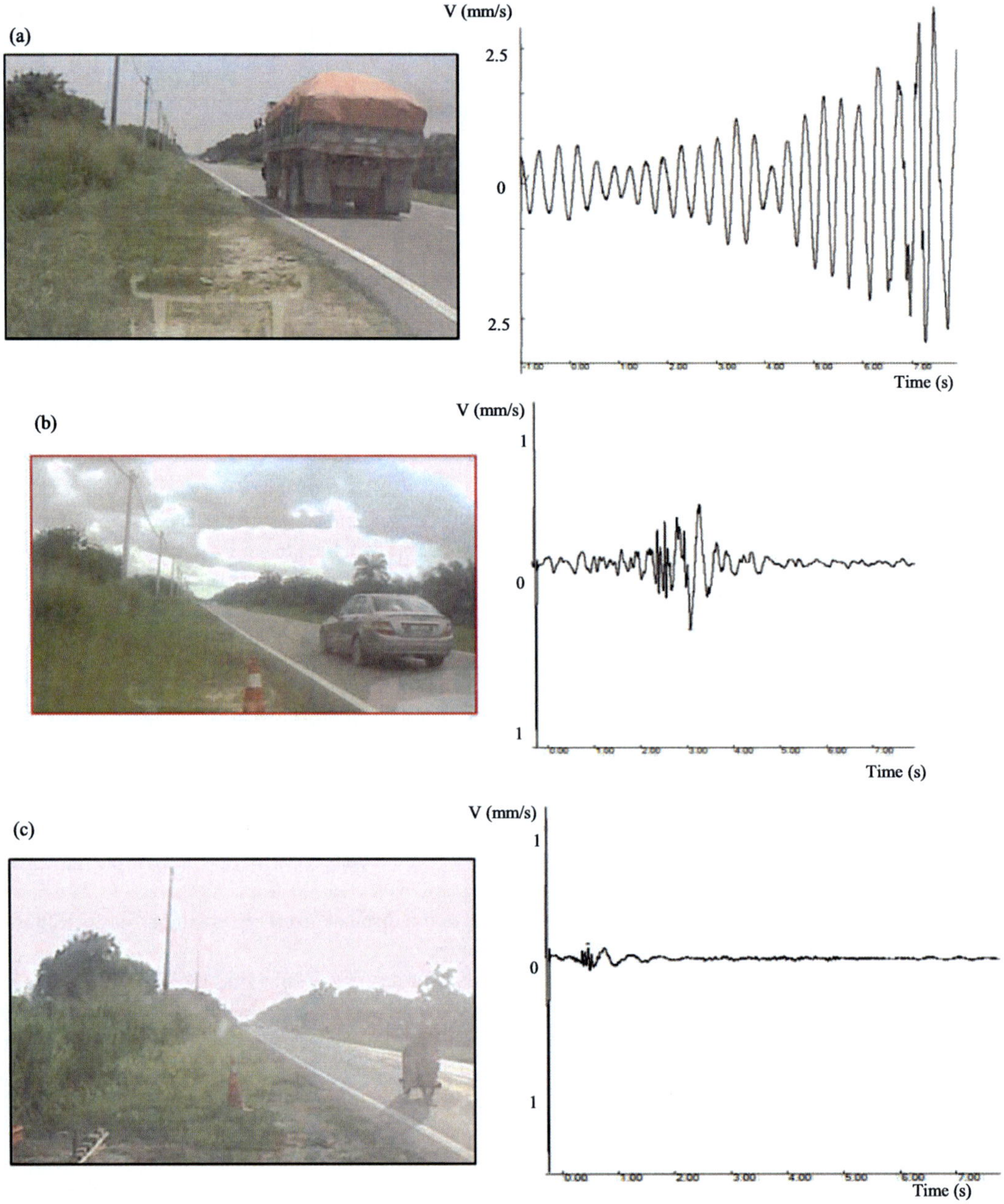

Fig. 8 Vehicle with vibration level and speed: lorry with speed 52 km/h and vibration 2.5083 mm/s (**a**), car with speed 67 km/h and vibration 0.2858 mm/s (**b**) and motorcycle with speed 59 km/h and vibration 0.1588 mm/s (**c**)

4.1 Regression Model Assumptions Justification

The next step was to examine the residual plot whether the model was adequate, and the assumption of regression has been met. The residual plots are the difference between the fitted response and the observed response value. Figure 10 shows the residuals against fitted values plot for the VIB_{soft} models. The figure shows all point in the residual plot distributed randomly at about the horizontal line with zero residual. Since no indication of missing term, outliers or inconstant variance occurs, the assumption of regression models is accepted.

To determine whether a data set is well-modelled for a normal distribution, a normality tests are used. Both graphical and numerical (statistical) methods can be used for evaluating normality (Ab Mughni et al., 2016; Park, 2002).

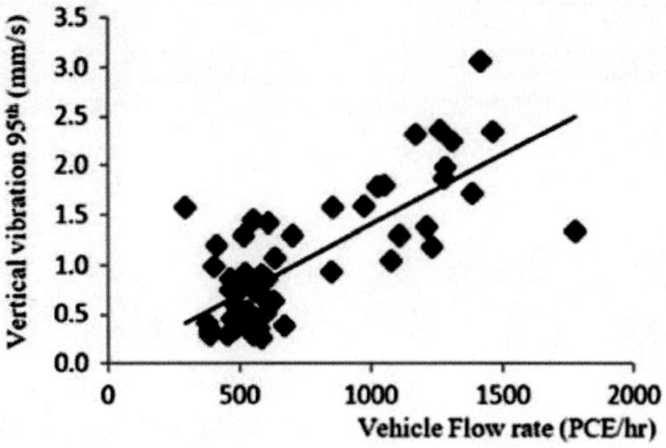

Fig. 9 95th percentile vertical vibration with vehicle flowrate

Table 1 Pavement thickness and pavement width of study sites

Site	Thickness (cm)	Width (m)
Meru	5.5	6.3
Sabak Bernam	8.7	6.5
Sepang	5	5.5

Table 2 Descriptive statistic for soft ground vibration model (VIB_{soft})

Variables	Mean	Median	St. Dev	Min	Max	Skewness	Kurtosis
Shear wave velocity ($1/V_s$) (m/s)	0.0262	0.0220	0.0213	0.0070	0.0970	1.87	2.71
Mean speed (*S*) (km/h)	64.3	63.5	6.6	51.5	79.1	0.34	−0.66
PW/PT	0.951	1.11	0.175	0.673	1.110	−0.20	−1.96
Flow rate (FL) (PCE/hr)	724	570.1	317.4	374.4	1461	0.98	−0.052
Distance (D_s) (m)	6.3	5.5	1.5	4.05	8.55	0.05	−1.37

Table 3 Correlation matrix among the variables

	r-value	*p*-value
$1/V_s$	0.508	0.000
S	0.341	0.000
PW/PT	−0.604	0.000
FL	0.851	0.000
D_s	0.140	0.038

Table 4 Regression analysis for soft ground vibration (VIB_{soft}) model

Predictor	Coefficient	SE Coeff	*T*-value	*p*-value	*R*-square
Constant	2.379	0.5648	4.21	0.000	62%
$1/V_s$	5.994	1.4390	1.44	0.000	
S	−0.025	0.0063	−3.90	0.000	
PW/PT	−1.046	0.2472	−4.23	0.000	
FL	0.001	0.0001	10.50	0.000	
D_s	0.042	0.0193	2.17	0.031	

Table 5 ANOVA for model VIB_{soft}

	DF	SS	MS	*F*	*P*
Regression	5	69.509	13.902	79.15	0.00
Error	243	42.680	0.176	–	–
Total	248	112.190	–	–	–

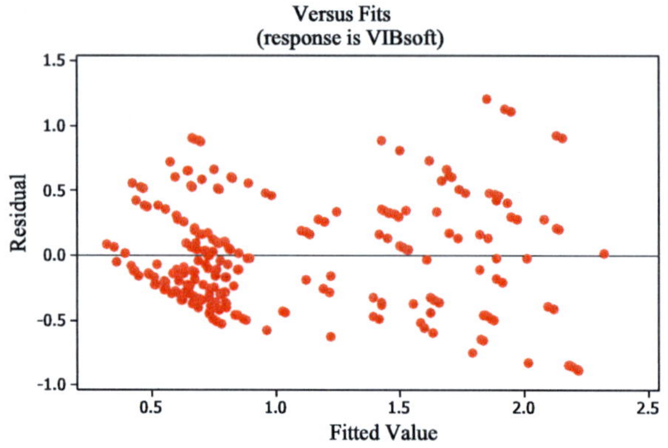

Fig. 10 Graph of residual versus fitted value for VIB_{soft}

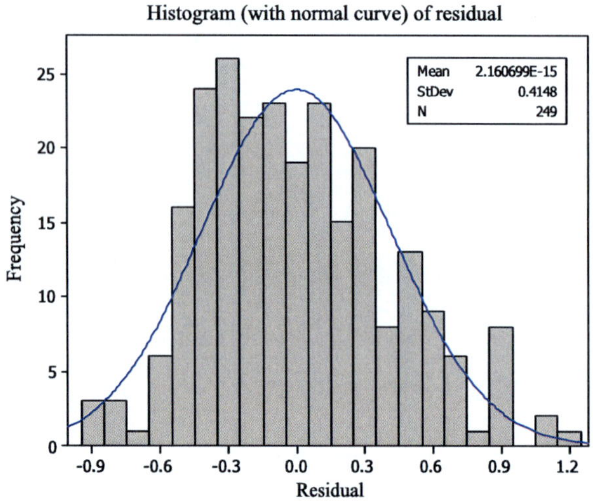

Fig. 11 Histogram of residual with normal curve for VIBsoft

For this test, by using graphical method, histogram is used for evaluating normality. On the other hand, numerical method which consists of skewness and kurtosis is also used for finding normality. Figure 11 shows the normal curve of VIB_{soft} residual. From the residual histogram, it shows that the data was normally distributed and can be accepted.

Table 6 shows the skewness and kurtosis value of the VIB_{soft} residual. For a normal distribution in skewness, the value is 0. According to George and Mallery (2010), value for asymmetry and kurtosis is between − 1 and + 1. They are considered tolerable in proving a normal distribution. The table displays the skewness and kurtosis value for residual VIB_{soft} which is close to 0, which means that the residual data can be accepted for model development.

4.2 Regression Model Validations

In this study, the independent data set was used to check the validation of the regression model. Independent data that not used in the development stage was used for validation (Adnan, 2007). The purpose of consuming independent data set is to recognize how good the model might achieve on the new data set. If the outcomes are reliable, then it can be decided that the regression models reliable, dependable and are appropriate under wider conditions (Abbas et al., 2011; Zainuddin et al., 2013). Figure 12 shows the connection between the measured VIB_{soft} and predicted VIB_{soft}. In the graph, a 45° line is drawn, where the points lie near to the line and the scattered points were revealed to be randomly distributed along the 45°; hence, the model can be accepted (Myatt, 2007).

Referring to Fig. 12, it can be seen that the newly developed VIB_{soft} models in this research is more accurate. Nevertheless, for further examination the validity of the developed model is tested in more precise way, more measures may be used, i.e. the root mean squared error (RMSE), mean absolute error (MAE), and mean absolute percentage error (MAPE) analysis conducted (Adnan, 2007).

Table 7 summarizes the evaluations of RMSE, MAE and MAPE of the VIB_{soft} predicted model. It was showed in Table 7 that the RMSE deviation from the empirical value of VIB_{soft} was 0.3151 mm/s. The MAE deviation from the empirical value of VIB_{soft} was 0.25 mm/s. The MAPE for VIB_{soft} from the empirical value was 30%. Thus, it can be decided that the vibration model can be tolerable to forecast the vibration near the roadway shoulder due to a small

Table 6 Skewness and kurtosis value for residual VIB_{soft}

Variable	Skewness	Kurtosis
Residual VIB_{soft}	0.45	−0.11

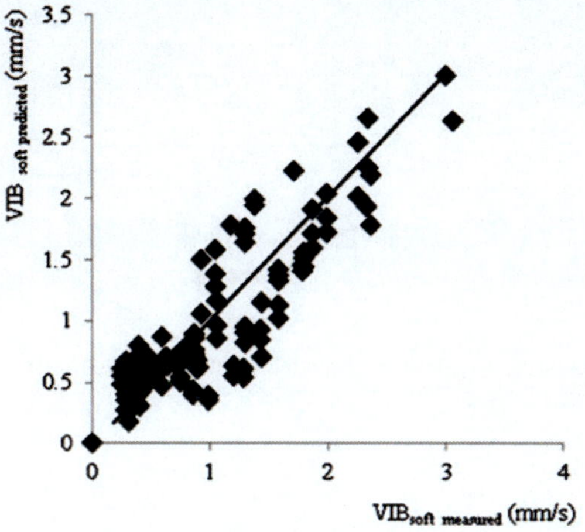

Fig. 12 Predicted VIB_{soft} vs measured VIB_{soft} from independent database

Table 7 Validation analysis result for VIB_{soft}

Model	RMSE (mm/s)	MAE (mm/s)	MAPE (%)
VIB_{soft}	0.3151	0.2500	30.0

divergence value of the RMSE, MAE and MAPE (Adnan, 2007).

Then, the validation database was tested by using paired t-test to compare the VIB_{soft} from empirical result and VIB_{soft} from predicted model. A hypothesis test is conducted as follows:

H_0 = the mean difference of both model equals zero.

H_1 = the mean difference of both model does not equals zero.

From Table 8, the p-value for model was larger than 0.05; hence, the null hypothesis (H_0) was not rejected at the 5% level of significance. It showed that the predicted models did not diverge much from the empirical values.

Data collected from the field was tested with the Watts (1990) and Long (1993) model to evaluate the reliability of the models. The methods used are through comparison of the discrepancy of the Watts and Long models with empirical result. To evaluate the accuracy of the models, mean absolute percentage error (MAPE), mean absolute error (MAE) and mean squared error (MSE) measures were used. While error is defined as difference between the empirical value and model proposed by the Watts and Long methods. Figure 13 shows the relationship among the empirical vibration (VIB_{soft}) and predicted VIB_{WATTS}, proposed by Watts, whereas Fig. 14 shows the relationship among the empirical vibration (VIB_{soft}) versus Long model (VIB_{LONG}).

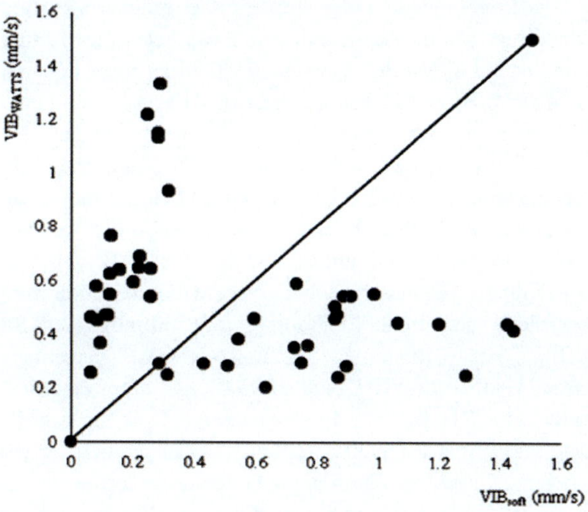

Fig. 13 Predicted VIB_{WATTS} versus empirical vibration, VIB_{soft}

Table 8 Statistical assessment of model

Test	Vibration
t-statistic	−0.130
p-value	0.898

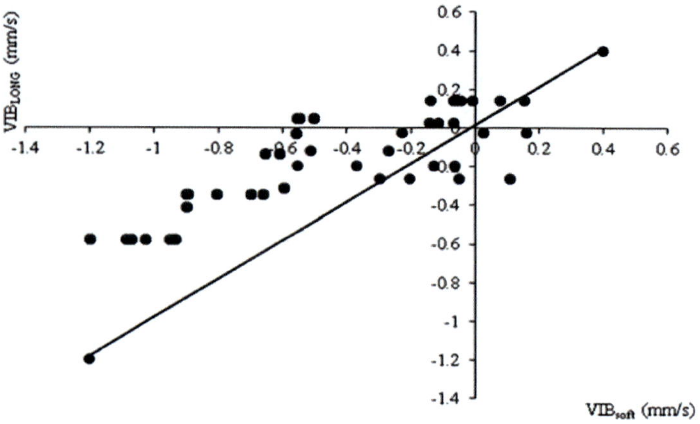

Fig. 14 Predicted VIB$_{LONG}$ versus empirical vibration, VIB$_{soft}$

Table 9 Discrepancy of the Watts and Long models with empirical data

Model	RMSE	MAE	MAPE
VIB$_{WATTS}$	0.557	0.500	67.4
VIB$_{LONG}$	0.441	0.408	103.4

It can be seen that the Watts model inaccurate estimates the ground vibration when it is used on local condition. While Long model underestimates the ground vibration when it is used on local condition. Nevertheless, for further examination to test the Watts model, the root mean squared error (RMSE), mean absolute error (MAE) and mean absolute percentage error (MAPE) analysis were conducted.

Table 9 summarizes a discrepancy of the Watts and Long models with empirical data. It is showed that the root mean squared error (RMSE) deviation from the empirical value of VIB$_{WATTS}$ is 0.557 mm/s, while for VIB$_{LONG}$ is 0.441 mm/s. The mean absolute error (MAE) deviation from empirical value of the VIB$_{WATTS}$ is 0.500 mm/s and for VIB$_{LONG}$ is 0.408 mm/s. The mean absolute percentage error (MAPE) for VIB$_{WATTS}$ and VIB$_{LONG}$ from empirical value is 67.4 and 103.4, respectively. Thus, it can be determined that the Watts and Long model predicted gives too far estimates on vibration prediction using local data. As a result, Watts and Long model is not able to produce accurate estimation when using in local ground and traffic condition.

4.3 Sensitivity Analysis

Generally, sensitivity analysis is a process of evaluation on the dependent variable behaviour with respect to changes in independent variable or parameters values. The sensitivity analysis involved in this study will use values within a range of empirical database for each input parameter and thus plots the relationship as the parameter varies. Sensitivity analyses are an advantageous in defining the future direction of data collection activities and to help identify which data is comparatively sensitive that needs further field characterization for more reliable results.

The levels of sensitivity range threshold for soft ground vibration were based on the range of minimum and maximum of empirical data which are then segregated into three ranges. Low range threshold is less than 0.96 mm/s. Medium threshold is between 0.96 and 1.66 mm/s, and high range threshold is more than 1.66 mm/s. Low sensitivity if the ground vibrations stay within the range. Medium sensitivity if ground vibrations jump to one range. High sensitivity if vibrations jump to more than one range. The three cases in this sensitivity analysis study were based on the results of the minimum, mean and maximum values of the descriptive statistic of the related parameter from the database that used in the development of the models. The three cases are as follows: Case 1, minimum parameter value; Case 2, mean parameter value; and Case 3, maximum parameter value.

When considering at the sensitivity result of soft ground vibration model in Fig. 15, the most sensitive parameter found was the vehicle flow rate (FL). On the other hand, all the remaining parameter (V_s, S, PW/PT and D_s) is considered as medium sensitivity. Therefore, vehicle flow rate (FL) needs closure attention when dealing with data collection or analysis of soft ground vibration in the future.

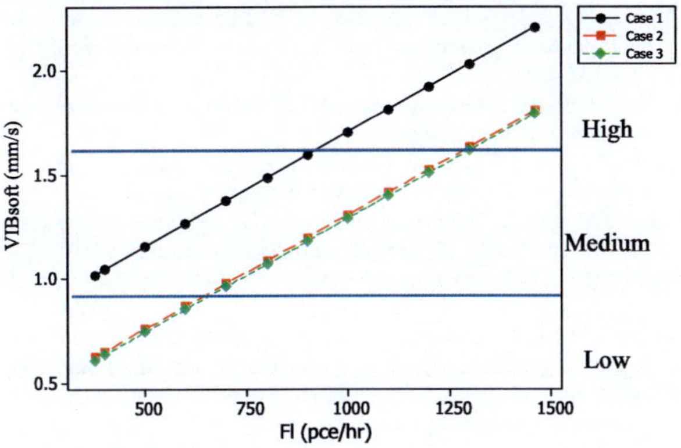

Fig. 15 Sensitivity output for VIB_{soft} at various vehicle flow rates

5 Summary and conclusions

By using multiple linear regressions, a novel methodology on predicting the soft ground vibration on single carriageway has been developed successfully. Based on the result of the empirical research on the ground vibration, statistically significant relationship can be established. The developed model which considers soil dynamic, traffic and pavement parameter in this study discovered good model adequacy.

According to the suggested regression model and analysis, it can be inferred that the soil shear wave velocity plays a role in the wave propagation surrounding the roadway. Ground areas with very low soil shear wave velocity needs to be studied and gained a full understanding of how and why the ground vibration increase before any new road constructions begin. Existing road constructed on very low soil shear wave velocity needs to be evaluated again on the needs of vibration countermeasure works if the ground vibration levels exceed the standard level. These to ensure reduction of the ground vibration impact surrounding the roadway.

About the traffic parameter, increment of vehicle speed will reduce the vibration level due to fast impact of vehicle on the ground and lower flow rate of vehicle using the road. Increment of vehicle flow rate will increase the ground vibration magnitude; this scenario reduced the vehicle speed. Based on sensitivity analysis, vehicle flow rate is found as highly sensitivity in predicting soft ground vibration. Hence, this parameter needs closure attention when dealing with data collection or analysis of soft ground vibration in the future.

For the pavement width and thickness parameter, increment of these factors can lead to increase the ground vibration absorption and then reduce the vibration impact. The pavement thickness and width also should be designed and constructed according to the standard for different types of road.

The recommendation for upcoming work is to increase the site data for other area in Malaysia in order to gain a model with a broader scope and ranges. The model developed in this study was based on empirical field data. Therefore, further exploration and investigation for numerical study is recommended. The model in this study was developed based on single carriageway. In order to apply the models for dual carriageway, further investigation is needed to justify the use of the developed models.

The endurance of this work may led to a very fascinating knowledge on ground vibration surrounding road area and transportation engineering. The outcome of this study is also essential and hopes the equation can be applied to improve the existing standards and guidelines on road vibration in Malaysia.

Acknowledgement The authors would like to acknowledge the authorities from Civil Engineering Faculty, Universiti Teknologi MARA (UiTM), for their inspiration and continuous support. Gratefulness goes to the Research Management Institute (RMI, UiTM) and Ministry of Higher Education Malaysia (MOHE) for the monetary supports FRGS Grant scheme: RMI File No: 600-IRMI/FRGS 5/3 (008/2019), Sponsorship File No: FRGS/1/2018/SKK06/UITM/02/5.

References

Ab Mughni, B., Adnan, M. A., Zamalik, S. Z., Jamali, F., Mohammad, M., Karim, Z. A., & Sulaiman, N. (2016). Evaluation of operating speed at multilane highway along Jalan Meru: Case study of reliability of posted speed limit. In *InCIEC 2015* (pp. 995–1006). Springer.

Abbas, S. K., Adnan, M. A., & Endut, I. R. (2011). Exploration of 85th percentile operating speed model on horizontal curve: A case study for two lane rural highway. In: *6th International Symposium on Highway Capacity and Quality of Service Stockholm*, Sweeden.

Adnan, M. A. (2007). *Development of entrance ramp merging density model based on an urban expressway traffic condition.* Unpublished Ph.D.'s thesis. School of Civil Engineering, University Sains Malaysia, Malaysia.

Agostinacchio, M., Ciampa, G., & Olita, S. (2014). The vibration induced by surface irregularities in road pavements—A Mathlab® approach. *European Transportation Research Review, 6*, 267–275

Ali Zomorodian, S. M., & Hunaidi, O. (2006). Inversion of SASW dispersion curves based on maximum flexibility coefficients in the wave number domain. *Soil Dynamics and Earthquake Engineering, 26*, 735–752

Coquela, G., & Fillola, C. (2017). Analysis of ground-borne noise and vibration levels generated by buses. *Procedia Engineering, 199*, 2699–2704

Del Grosso, A. E., & Basso, P. (2014). Monitoring of vibrations for the protection of architectural heritage. In: *7th European workshop on structural health monitoring*, July 2014. Nantes, France.

George, D., & Mallery, P. (2010). *SPSS for windows step by step: A simple Guide and reference (17.0 ed.).* Pearson.

Henwood, J. T., & Haramy, K. Y. (2002). Vibrations induced by construction traffic: A historic case study. Retrieved on 29 July 2015 https://www.dot.ca.gov/hq/esc/geotech/geo_support/geophysics_geology/pdf/geophysics_2002/043henwood_construction_vibrations.pdf

Hunaidi, O. (2000). *Traffic vibrations in buildings. Construction technology update No. 39*. Institutes for Research in Construction, National Research Council of Canada.

Hunaidi, O., & Tremblay, M. (1997). Traffic-Induced Building Vibration in Montreal. *Canadian Journal of Civil Engineering* 24 (5) (1997) 736–753.

Hunaidi, O., Guan, W., & Nicks, J. (2000). Building vibrations and dynamic pavement loads induced by transit buses. *Soil Dynamics and Earthquake Engineering, 19*, 435–453

Ismail, N. N. (2012). Refinement of beam forming technique for practical and reliable stiffness profiling. Master Thesis. Chung–Ang University, Korea.

Ju, S. H. (2009). Finite element investigation of traffic induced vibrations. *Journal of Sound and Vibration, 321*, 837–853

Klæboed, R., Turunen-Rise, I. H., Harvik, L., & Madshus, C. (2003). Vibration in dwellings from road and rail traffic-part II: A new Norwegian measurement standard and classification system. *Applied Acoustics, 64*, 89–109

Kliukas, R., Jaras, A., & Kacianauskas, R. (2008). Investigation of traffic-induced vibration in Vilnius Arch-Cathedral Belfy. *Transport, 23*(4), 323–329

Kunimatsu, S., Kitamura, Y., Yokota, A., Uchida, H., Shimura, M., Sano, Y., Osafune, T., Iwabuki, H., Ishida, R., & Hirao, Y. (2014). Research activities on INCE/J RTV (Road Traffic Vibration)—Model—Part: 2 prediction of ground-borne vibration induced by traffic from cutting and banking structure roads. *Inter-Noise 2014, Melborne Australia*.

Lak, M. A., Degrande, G., & Lombaert, G. (2011). The effect of road uneveness on the dynamic vehicle response and ground-borne vibration due to road traffic. *Soil Dynamics and Earthquake Engineering, 31*, 1357–1377

Long, L. T. (1993). Measurements of seismic road vibrations. In *Proceeding of the Third International Conference on Case Histories in Geotechnical Engineering*, Paper # 4. 10 (pp. 677–680).

Madshus, C., Bessason, C., & Harvik, L. (1996). Prediction model for low frequency vibration from high speed railways on soft ground. *Journal of Sound and Vibration, 193*(1), 195–203

Mhanna, M., Sadek, M., & Shahrour, I. (2012). Numerical modeling of traffic-induced ground vibration. *Computers and Geotechnic, 39*, 116–123

Malaysian Highway Capacity Manual. (2011). *Highway planning unit*. Ministry of Works Malaysia.

Myatt, G. J. (2007). *Making sense of data. A practical guide to exploratory data analysis and data mining*. Wiley.

Park, H. M. (2002). *Univariate analysis and normality test using SAS, STATA, and SPSS-Univariate analysis and normality test*. The Trustees of Indiana University. https://www.indiana.edu/~statmath .

Sulaiman, N. (2016). Critical review of ground-borne vibration and impact assessment: Principle, measurement and modelling. *Pertanika Journal of Scholarly Research Reviews (PJSRR), 2*(1), 22–23 (Universiti Putra Malaysia Press).

Tuan Chik, T. N., Asiew, R. A., Wan Ibrahim, M. H., & Yusoff, N. A. (2013). Dynamic performance on multi storey due to ground borne vibrations input from passing vehicles. *International Journal of Integrated Engineering, 5*(2), 51–58

Turunen-Rise, I. H., Brekke, A., Harvik, L., Madshus, C., & Klæboed, R. (2003). Vibration in dwellings from road and rail traffic-part i: a new norwegian measurement standard and classification system. *Applied Acoustics, 64*, 71–87

Wair, B. R., De Jong, J. T., & Shantz, T. (2012). *Guidelines for estimation of shear wave velocity profile*. PEER Report 2012/08. Pacific Earthquake Engineering Research Center, Headquarters at the University of California, California.

Watts, G. R. (1990). *Traffic induced vibrations in buildings*. Research report 246. Transport and Road Research Laboratory.

Xu, Y. L., & Hong, X. J. (2008). Stochastic modeling of traffic-induced building vibration. *Journal of Sound and Vibration, 313*(1–2), 149–170

Zainuddin, N. I., Adnan, M. A., & Md Diah, J. (2013). Optimization of speed hump geometric design: Case study on residential streets in Malaysia. *Journal of Transportation Engineering, 140*(3), 05013002

Construction of Housing for Low-Cost in the Town of Al-Mansoura Northern Badia of Jordan

Majd Al-Homoud and Salem Al-Oun

Abstract

Jordan faced a shortage in low-income housing while facing excess in other types of housing. Issues that caused supply imbalance are construction, land, and financing. The government was not able to tackle such issues. Partnership with the private sector may encourage to bridge the supply for the growing low-income housing sector. The present study purpose is to examine the potential provision of adequate low-cost housing for a group of Al-Noaim Bedouins who are in the town of Al-Mansoura. This sociocultural group still dwells *Bayt Esh-Sha'r* housing type to contest their livelihood. *Bayt Esh-Sha'r* is the traditional basic Bedu home. Challenges of the study area include cultural understanding, living needs satisfaction, and essential services provision, in addition to establishment of livelihood sources. The study was based on interviews with of 27 households in addition to ten local contractors. The outcomes of the field work suggested to build the project in two stages over 15 years with a total of 100 units supplied for 27 extended families. The following strategies were used to reduce the project cost: (a) landform and selection, (b) developing typologies for the residential units, and (c) developing wall sections and construction while maintaining the symbolic social meanings and psychological wellness.

Keywords

Low income • Construction • Sustainability • Green • Housing • Jordan • Bedouins • Nomadic • Culture

M. Al-Homoud (✉)
School of Architecture, German Jordanian University, Amman, Jordan
e-mail: majd.alhomoud@gju.edu.jo

S. Al-Oun
Business Management Department, Al Al-Bayt University, Al-Mafraq, Jordan

1 Introduction

The objective of the present study is to examine the possibility of providing proper housing for a group of low-income families of Al-Noaim Bedouin tribe who still inhabit the traditional basic and original Bedu home (*Bayt Esh-Sha'r*) at Al-Mansoura town. Such provision should include affordable prices, customs, and traditions issues, in addition to provision of physical need. Project challenges consist of satisfying cultural references, living needs, basic services, and income sources. Most of the population of the town receives aid from the Ministry of Social Development. The tribe of Al-Noaim used to travel to Irbid city and western regions during spring seasons for grazing land. However, their kinship ties to Ruweished region made them migrate and settle in Al-Mansoura town after consequent droughts. Ownership as it is an important aspect of the tradition; it serves as a measure of social prestige.

The importance of the study comes from the need for a deep understanding of Bedouin cultural group. Bedouins make 10% of Jordan's population and reside in the Badia, which composes 81.3% of Jordan (Bickerton et al., 2020; Prince Ghazi bin Muhammad, 1999). The study area is in Northern Badia, in Mafraq governorate. Jordan is divided to four regions: the highlands, Jordan Valley, Aqaba Gulf, and arid lands of Badia (Bickerton et al., 2020; Al-Oun, 1997; Kanaǹ and Estatiyyah, 1992). Badia covers 72,600 km^2, which is administratively divided to three Badia areas: northern, central, and southern Badia (Al-Oun, 1997).

Provision of shelter is vital to low-income households. Housing has higher priority than education or healthcare in developing countries (Ferguson, 2008). Finance for housing received little attention, but governments find ways to provide directly or indirectly. Lack of finance jeopardizes low-income households regarding satisfying their needs. Meanwhile, social responsibility for low-income housing development makes an alternative approach. Such alternative approaches focus on the social and cultural need of users

F. Trapani et al. (eds.), *Advanced Studies in Efficient Environmental Design and City Planning*, Advances in Science, Technology & Innovation, https://doi.org/10.1007/978-3-030-65181-7_46

rather than the physical needs. This can happen only if developers stress the process more than the product. Further, developers need to integrate users' preferences and needs throughout the production process to achieve congruence with the market. Housing market is interdependent with finance, land, real estate, construction, industry, and management (Balchin & Rhoden, 1998; Harvey, 2000).

For many decades, housing product was a priority for the government in Jordan. The product was mainly multi-family apartment projects, which was an outcome of the population demand. Absence of synchronization between the public and private sectors flunks to bridge the gap between quantity and quality of housing production. As an outcome, products suffered from cost cut that produced low-quality housing: small area and low-quality construction material. Different studies suggested attempts to encourage private sector's investment and participation in the development of low-income housing: to overcome obstacles by developing procedures, to reduce the cost of housing production, to increase profitability of investors, and to provide residential unit that satisfy the needs of low-income users (Al-Hasanat, 2003; Damra, 2003; Akkawi, 1995; MoP, 1987).

Alternative strategies for the development of low-income housing include: (1) accessible site and lands that require low-cost for site construction; (3) construction of the largest number of housing units in order to cut in the cost of services, infrastructure, wall construction, and land; (3) replacement of common construction material with local material; (4) configuration of simple shapes in order to reduce height and floor area; (5) reshaping plan and space to reduce wasted space such as the reception and dining areas despite their cultural symbolism; (6) employment of real estate concepts including management and marketing; (7) co-operation among key players through public/private partnership; and (8) facilitation of banks procedures including: payback period, interest rates, guarantees, amount and type of loans (Al-Homoud et al., 2009a).

The demand for apartment prototype dramatically increased after the 1990 Gulf War, as it satisfied the demand of sudden influxes to the major cities of Jordan (World Bank, 2018; Al-Homoud et al., 2009a). In addition, the arrival of Iraqis in 2003 after the US attack on Iraq caused rapid and widespread price pressures on construction, land, and housing costs (Al-Homoud et al., 2009a; Khammash & Alkhas, 2009). Further, Jordan hosted by 2013 about 1.4 million registered Syrian refugees, 85% of whom lived in different cities (Al-Homoud & Ghanem, 2019; MoP, 2015; UNHCR, 2014). Such sudden demand negatively affected the quality and quantity of housing production (MoP, 2015; UNHCR, 2014; Al-Homoud et al., 2009a, 2009b).

As an outcome, the rate of population growth in Jordan offered demand for residential production, while the supply did not match the demand (World Bank, 2018; Al-Homoud et al., 2009a; Khammash & Alkhas, 2009). Developers focused on extravagance developments; they targeted upper income levels. They over-saturated the market with up-scale housing production (World Bank, 2018; Al-Homoud et al., 2009a; Global Property Guide, 2008). This made a mismatch between demands and supply. A surplus in the supply was an outcome of a continuous rise in property prices that made them no longer affordable to middle income (World Bank, 2018; Al-Homoud et al., 2009a; Khammash & Alkhas, 2009). In 2001, Jordan experienced lack in low-income housing, especially for those who enter the category of 10%–55% of the national income curve (Al-Homoud et al., 2009a; Khammash & Alkhas, 2009). Around 20% of the housing stock is unoccupied. Factors that made the housing sector ineffective included: (1) financing, (2) land possession, and (3) construction (World Bank, 2018; Al-Homoud et al., 2009a). Cost of land makes about 30–60% of the unit cost, construction about 20–40%, while onsite infrastructure, profit margin, and taxes and fees make about 10% of the total cost (World Bank, 2018). Housing and Urban Development Corporation predicted the construction of 45,112 units annually, while the housing unit supply in 2007 was 34,668 units (Al-Homoud et al., 2009a; Khammash & Alkhas, 2009).

The government was not being able to materialize reforms for land and construction markets, which influenced the affordable housing provision. Further, Jordan Government programs do not support development of affordable housing or subsidies to low-income housing (World Bank, 2018). However, in the last decade, developers began to focus on middle and low-income market. This was supported by the collaboration of the leading banks who offered mortgage with low interest rates and long payment periods (World Bank, 2018; Al-Homoud et al., 2009a; Khammash & Alkhas, 2009; Colliers International Research, 2008; Global Property Guide, 2008).

Issues for low-income housing include price, size, location, and tenure (World Bank, 2018). Under pressures of household incomes and housing financing, licensed average unit size in the capital city of Amman declined from 208 m^2 in 2000 to 169 m^2 in 2006 (Al-Homoud et al., 2009a; Khammash & Alkhas, 2009). Only, 30% of households can afford houses above 100m^2 with 30% cut of their monthly income (World Bank, 2018). Prices of properties are more than affordability capacity of the average Jordanian. In addition, mortgage market and banks terms on housing loans did not support the poorer population. Generally, financial institutions offered mortgage loans with interest rates between 8.5 and 10% over 20-year period, which made it impossible for the majority who are low income to benefit from loans (Al-Homoud et al., 2009a). The government in collaboration with the private sector initiated several schemes for the low-income population, developed subsidies, and offered credit facilities. Although rental is more affordable than ownership, only 30–40% of households afford rent below the market average, which jeopardizes quality and location (World Bank, 2018).

2 Literature Background

Housing is not just the occupancy in a dwelling, it is a combination of physical, financial, legal, and social issues. It is more than a basic shelter; it reflects way of life, beliefs, and ideals (Rapoport, 1969, 1991). Housing design gives form of idiom to the ideal reflection of the environment and the needed qualities (Rapoport, 1977, 1985). Designer's center of attention is on product rather than on process that consider users (Marcus & Sarkissian, 1986; Sweis et al., 2013). Developers are worried about cost and architects are disturbed by image with little coordination with end users. Both developers and users perceive housing as an investment, a full profit for developers and a lifetime for users. This makes housing a consumable object not a product, with decreased quality (Rian & Sassone, 2012; Wentling, 1990).

Analysis of housing marketing opportunities include: (1) identifying segments; (2) pricing, distribution, and market size, while maintaining quality; (3) targeting demands and customers' needs; (4) realizing social changes in concepts and in services; and (5) support latest demands (Kotler & Armstrong, 1999; Miles et al., 2001; Schmitz & Deborah, 2001). Usually developers focus on costs or aesthetics; they focus on users' attractions to gain profit. Buildings are rarely evaluated based on end users' preferences and needs because the immediate clients are usually the public and private agents (Al-Oun et al., 2010; Al-Homoud et al., 2009a). Al-Oun et al. (2010) recommended that developers should invest in selecting the location for their developments, design units with flexibility, and provide more typologies.

In Jordan, there are lots of sociocultural values that lead to the increase in the housing unit cost (USAID, 2005; Al-Hasanat, 2003; Al-Omari, 2003; Akkawi, 1995; Barakat & Tuffaha, 1995; MoP, 1987; Sims, 1990; Basboos, 1995): (1) Wasted guest area: users assign 25–33% of the area for the reception room. (2) Large number of rooms: three bedrooms or more. (3) Replicate spaces for gender domains: privacy demands separation in the living, dining, guest spaces, entrances, and bathrooms. (4) Ownership: detached single-family home that satisfies sociospatial needs is a priority. (5) A housing for lifetime: spending lifetime savings on housing purchase regardless of the real household needs. (6) Negative attitudes toward borrowing from financial institutions: people refuse to deal with bank loans for religious reasons; rather, they borrow from their relatives, which does not match the needed cost.

A study by Al-Oun et al. (2010) suggested criteria to increase users' choices in the market based on interviewed tenants and local developers as follows: (1) Housing unit cost: depends on size and distant from the ground, unit cost decreases as you go up. (2) Site location: considering buildings on two streets as an advantage that provides setbacks from adjacent buildings that offer safety. (3) Building entrance: creating sense of privacy and territoriality, while considering how upper floors are affected by articulation of such entrances. (4) Interior space layout: being aware of change in family cycle and not providing only for large size families to reduce cost of change on the interior to cater users' needs. (5) Housing unit type: being flexible in the available stock and not to be only concerned with losing consumers who search for lifetime housing. (6) Visual image: investing in individualistic character of buildings to enhance personalization for users. (7) Open spaces: realizing setbacks as a source of noise for upper floor users. (8) Sense of community: emphasizing tenants' management and maintenance by establishing tenants committee.

3 Cultural Framework and Sociocultural Structure of Bedouins

Culture controls way of living which forms from shared concepts such as habits and communication and transfer from one generation to another through language or other tools. It is composed of: (1) norms that administer acceptable and unacceptable behavior. (2) Legends that explain fundamental shared values. (3) Icons and indices that have shared meaning. (4) Rituals that designate standards and codes. (5) Property which provides meaning to the scope of culture (Culture & Lectures, 2000; News in Review, 1997; Choudhury, 1993). Meanwhile, subculture is composed by defined groups within the framework of the larger culture that meet the values and beliefs of the members. Subcultural values vary within the framework of general culture (Al-Beejawi, 1997; Choudhury, 1993; Kellett, 1999; Sociology at Hewett, 2000).

Bedu society in Jordan is a cultural group that controls most life matters by the law of customs, which is a strict code. Customs reflect the value of tribal obedience and honor by defending it against all other groups through shrewdness and revenge (Bickerton et al., 2020; Weir, 1990). Bedu in Jordan responded to urbanization and their lifestyle changed from traditional to modern one, which encouraged them to settle and live in permanent structure with more physical and spatial comfort compared to traditional *Bayt Esh-Sha'r* (Abed, 2001; Al-Homoud et al., 2008; Al-Mousa, 1996). Although Bedu shifted to a settled life, they did not alter their lifestyle; rather, they still have a combination of traditional and modern lifestyle (Al-Mousa,

1996; Al-Oun, 1997). Bedu way of life in Jordan depends on four basic elements (Abu-Jaber et al., 1987; Bickerton et al., 2020; Jabbor, 1988): (1) the Badia that is the desert environment; (2) the camel that is used for transportation; (3) the Bedu individual who is the son of the Badia; and (4) *Bayt Esh-Sha'r* where the Bedu live.

Bedu society is based on tribal institute that is dominated by collectivism and vicious loyalty to the tribe (Al-Oun, 1997; Al-Ozaizat, 1980; Bickerton et al., 2020). Their constitution is based on *asheirah*, which is subdivided into clan (*hamoulah*) and main families (*fakhdh*) (Al-Oun, 1997). The *asheirah* of the northeast Badia include Bani Khaled, As-Sirhan, Al- 'Isa, Ahl Al-Jabal, and As-Sardiyyah (Al-Mousa, 1996; Al-Oun, 1997). Bedu society is patriarchal; members of the tribe claim that they descended from a familiar male ancestor. The chief is the sheikh who descends from a noble family and is characterized by generosity and bravery. The sheikh has a significant power restricted by customs, primacy, and the advice of the elderly council (*majlis*). He rules by virtue of his character by conciliation, compromise, and negotiation (*shoura*), rather than by influence (Jabbor, 1988).

Bedu lifestyle has shifted to a settled one for the following reasons: governmental strategies toward settlement; economic growth; enhanced transportation and media; interaction with urbanist; less reliance on livestock and farming education and work; and residence in permanent homes (Brandenburg, 1998; Al-Fawal, 1983; Ismael, 1983). These factors reduced the average number of nomadic Bedu as a proportion of the total Arab population from 10% in 1960 to 1% in the late twentieth century (Abu-Jaber et al., 1987; Bickerton et al., 2020). Though other sources of income have a growing importance, traditional Bedu economy is still largely dependent on livestock and on its products (Al-Mousa, 1996; Al-Oun, 1997; Bickerton et al., 2020). Bedu lifestyle is eminent by loyalty, customs, collectivism, and identity.

4 Spatial Structure of Bedouins

Collective and individual home identity is reflected spatially and sociospatially (Duncan, 1985). *Bayt Esh-Sha'r* is the basic and original Bedu home because it suits their traditional life that is based on constant moving. *Bayt Esh-Sha'r* was considered a basic necessity for Bedu identity, such as hospitality, psychological comfort, openness, mobility, durability, occasions, flexibility, lightness, simplicity and beauty, authenticity, continuity, social communication, heritage, identity, assurance, external work, and climatic suitability. It is also considered an effective shelter that provides privacy, and it is handy for the construction and deconstruction that is required by continuous traveling (Abed, 2001; Al-Homoud et al., 2003, 2008, 2010; Kellett, 1999). Livestock ownership is the traditional household occupation that involved traveling and searching for grazing land and water resources for animals. This requires a home that is mobile and light with easy and quick assembly capacity, such as *Bayt Esh-Sha'r* (Weir, 1990). Occupation usually has a direct effect on income; Bedouin traditional occupation, livestock ownership, offers a seasonal income. Traditional *Bayt Esh-Sha'r* suits fluctuations in income, as it is made from hair taken from the livestock by family females (Jabbor, 1988; Weir, 1990). Income fluctuation does not allow Bedu to save money for home construction, since the first expenditure priorities are food, drink, hospitality, and clothing. Rich Bedu has large *Bayt Esh-Sha'r* used to host guests as an emphasis of their traditional hospitality (Abed, 2001; Al-Homoud et al., 2003, 2008, 2010; Kellett, 1999).

Bedu lifestyles are geared toward using the internal spaces, which is affected by traditional male and female household occupations. Householding is considered the traditional female occupation. A housewife usually spends most of her time inside to perform the daily work such as baking, cooking, spinning, and weaving. Diversity of a housewife's work made it difficult to assign one space for each activity; overlapping of such activities became essential. Mobility dictates the overlapped use of one space for more than one activity (Abed, 2001; Al-Homoud et al., 2008). Bedu women can participate in the family discussion; if the guest is a stranger, women will participate from behind the scenes; otherwise, they can join the males of the family in the discussion. Women, within the Bedu code of honor, are allowed to receive male guests in the *Madheif*, if the male householder is not available (Abed, 2001; Al-Homoud et al., 2003, 2008, 2010; Al-Mousa, 1996).

Privacy is mainly regulated through the Bedu values. Privacy is achieved through hierarchical zones in *Bayt Esh-Sha'r* (Abed, 2001; Al-Homoud et al., 2008). Bedu uses outdoor haram *Bayt Esh-Sha'r* for privacy, territory, children playing, and for animals (Al-Homoud et al., 2003, 2008, 2010; Al-Ozaizat, 1980). Bedu uses distances that separate *Bayt Esh-Sha'r* from all sides and from the front side as physical expression for privacy. Another reflection of privacy is on the location of the entrance, which is situated at the corner of *Bayt Esh-Sha'r* (Al-Mousa, 1996). *Bayt Esh-Sha'r* is divided to two major interior spaces: *Madheif* and *Mharam*, simply furnished with carpets and pillows. Its external spaces reflect territory of *Bayt Esh-Sha'r* (*Haram Bayt Esh-Sha'r*) where the Bedu keep their animals. The major receiving space *Madheif* is always open as a symbol for hospitality (Abed, 2001; Al-Homoud et al., 2008). *Bayt Esh-Sha'r* has different sizes based on the number of the central column (*Wassit*): *Kharboosh*, *Gatbah*, *Shawa*, *Mthoulath*, *Mrouba'*, *Mkhoumas*, *Msoudas*, *Msouba'*, and *Mthouman*. It is made out of different material: goat and

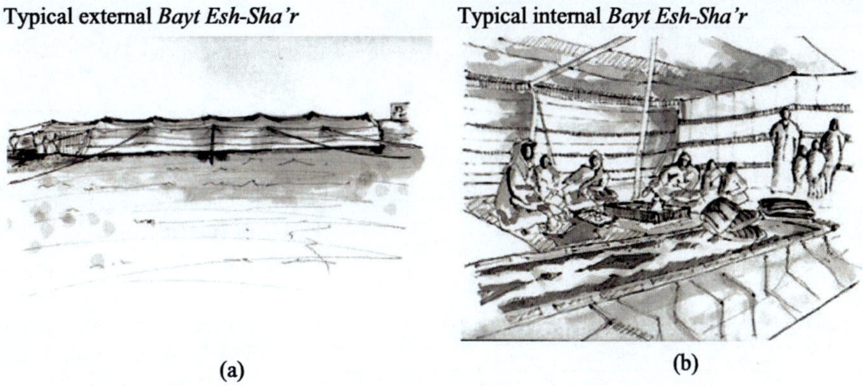

Fig. 1 Traditional *Bayt Esh-Sha'r*—**a** exterior (left) and **b** interior (right). *Source* Authors (2007)

camel hair, cotton, and *Kheish*, see Fig. 1 (Abed, 2001; Al-Homoud et al., 2003, 2010; Al-Mousa, 1996).

Although new prototypes reflected new Bedu demands, *Bayt Esh-Sha'r* remained part of the home design. Traditionally, *Bayt Esh-Sha'r* was functional, while in modern typologies, stayed symbolic and used in ceremonies. In transitional typologies, *Bayt Esh-Sha'r* need and use depended on the number of rooms and family size (Al-Homoud et al., 2003, 2008, 2010).

5 Research Methods and Field Interviews Outcomes

The study was based on face-to-face field interviews. To elicit data about the social responsibility of the architecture and low-cost housing typology and construction for the Bedouins in Jordan field, interviews were conducted with of 27 families. For this part of the study, the unit of analysis was the head of the household from Al-Mansoura town. In addition, to elicit data about alternative design for wall sections, material, and prices face-to-face field, interviews were conducted with ten local contractors.

Field work showed that ownership of property was a very important aspect of the tradition. House serves as a gauge of social status and prestige. About 33% of the population had repeated marriage with average family size of nine members, and the age groups of unmarried children were up to 25 years. This means that the average age of marriage is not more than twenty-five years. This suggests that the cycle of the family composition does not exceed 25 years. Only 19% of men have the desire and ability to work, and 63% women have the desire and ability to work.

Most of the sample suggested the need for flat open land when selecting a site for the project. About 55% of the sample indicated the need for a school, 82% indicated a need for a mosque, 29% indicated the need for health center 29%, and 60% indicated a need for a post office. About 67% of the sample suggested keeping a place for children play within the house area and not in public shared space. Also, 77% have stressed the need to establish boundaries between neighboring houses. About, 60% confirmed the necessity to maintain moderate social relations with inhabitants of the town.

As for the housing units, 82% indicated a need for a yard or garden, 85% focused on need for a fence, 82% focused on the need for back yard for livestock breeding, and 93% of the sample suggested housing unit layout toward the east. *Bayt Esh-Sha'r* was dominant house typology with *Mthoulath* and *Qatbah* sizes, and some live in *Kharboush*.

Most of the sample suggested the use of cement as a construction material, but about 7% suggested the use of white stone. The field survey concluded the availability of three housing typologies in the area: traditional *Bayt Esh-Sha'r* house (traditional type), see Fig. 2; mud house (transitional type), see Fig. 3, and modern house type, Figs. 4 and 5.

6 Research Setting—Al-Mansoura Town

Jordan area is 88,780 km^2 with a population of 10.18 million and a natural growth rate of 1.37%—year 2019. Urban population makes about 91.5% of the total population. Amman, the capital city, population is 1.28 million, which makes 12.5% of the total population of Jordan. Further, 25.96% of the total population of Jordan dwells in three major cities: Amman, Irbid, Zarqa, and Russeifa (about 2,644,239) (Worldometer, 2020). The research setting is in the emerging town of Al-Mansoura is in Mafraq Governorate which has a population of 593,900 (DoS, 2020a). The town is located on the major highway between Mafraq and Irbid cities, 12 km Northwest of Mafraq city, see Fig. 6. It has a population of 5675 and 1068 households (DoS, 2020b). It has services such as water, electricity, health center and a primary school, post office, and two mosques.

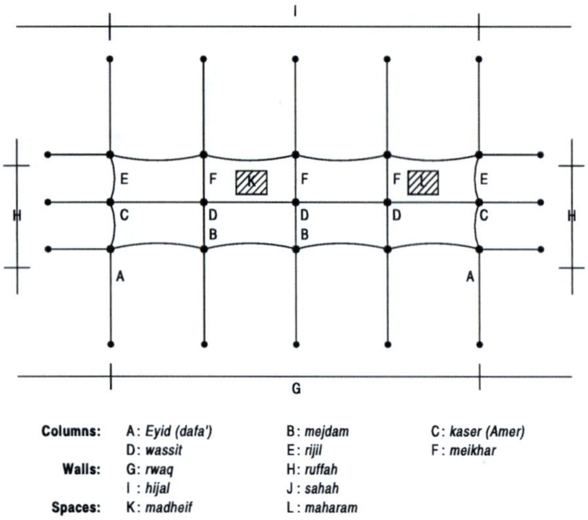

Fig. 2 *Bayt Esh-Sha'r* layout (traditional house type). *Source* Authors

Fig. 3 Mud house (transitional type). *Source* Authors

This emerging town is in Mafraq Governorate which has a population of 593,900 (DoS, 2020a, 2020b).

7 Projections and Outcomes

Results from filed work indicated the need for the construction of the project in two phases based on a 15-year family lifecycle that was explored throughout the interview. Projections suggested 100 residential units distributed on 27 families. Field work showed that ownership of property was a very important aspect of the tradition.

The proposed project cost reduction was based on field work and previous research. The following procedures were utilized as a design method for the development of the project and its components (Al-Homoud et al., 2009a; Al-Oun et al., 2005):

1. First—Site planning and design: Selecting the most appropriate location in terms of total land cost that included the consideration of the following criteria: the cost of infrastructure, as it get affected by land form and topography; land price; and accessibility to public transportation, see Fig. 7.
2. Second—Housing unit design: Identifying housing unit typologies with different floor areas.
3. Third—Wall section design: Providing the best cross section of the construction wall by investigating multiple wall sections and construction, while keeping in mind the social value and psychological symbolism on such sociocultural group. Interviewed subject indicated the need for material that sustain their self-pride and dignity, as well as social acceptance. Most of them requested the

Fig. 4 Modern house plan. *Source* Authors

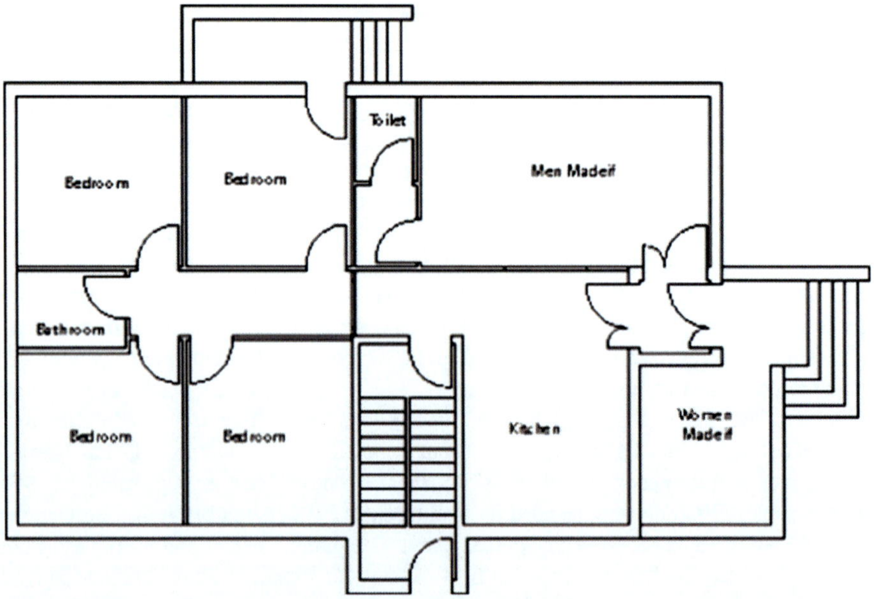

Fig. 5 Modern house—**a** exterior (left) and **b** interior (right). *Source* Authors

Fig. 6 Map of Al-Mansoura town. *Source* Google Maps (2020)

use of limestone for the main facades as it reflect social status in Jordan. The challenge to us was to use such image with low cost. Through field work and interviews with local contractors, we found that we can use a stone veneer which is usually disposed and not used by builders and contractors.

8 Planning and Design Strategies

Challenges for housing such cultural group include economic, policies, and cultural determinants. Therefore, it was important to balance cost with other determinants. Design strategies stemmed from deep understanding of the lifestyle and meanings associated with the Bedouin culture for the surveyed families from Al-Noaim tribe in Al-Mansoura town. It reflected internal and external space layouts and the bonding to the family and the clan:

(1) **Site planning and design**: planning criteria considered the following aspects: (a) Site selection: Location was selected where the tribe currently lives, assessment of land price in the surrounding area, social contact with the rest of the clan through a major spine, and land slope to reduce the cost of site work. (b) Spatial distribution: adapted to the collective social structure of families and their future expansion, provision of adjacent pastoral areas, and water harvesting based on topography mapping to increase impact on vegetation and livestock, see Fig. 8.

Site design: site design criteria considered the following aspects: (a) Defining an area of 100 dunums that house 100 units with the future expansion of 33 families based on lifecycle, cultural domains, and identity. (b) Clustering: grouping of father and sons while maintaining a common area to house *Bayt Esh-Sha'r* and additional housing units when needed based on lifecycle. It considered future expansion of new units, spatial and domain hierarchy, and group and family privacy. It allocated areas for livestock and local farming activities, see Fig. 9.

(2) **Housing unit design**: unit design criteria considered the following visions: (a) evolution of the site,

Fig. 7 Topography map for the selected site at Al-Mansoura town. *Source* Authors

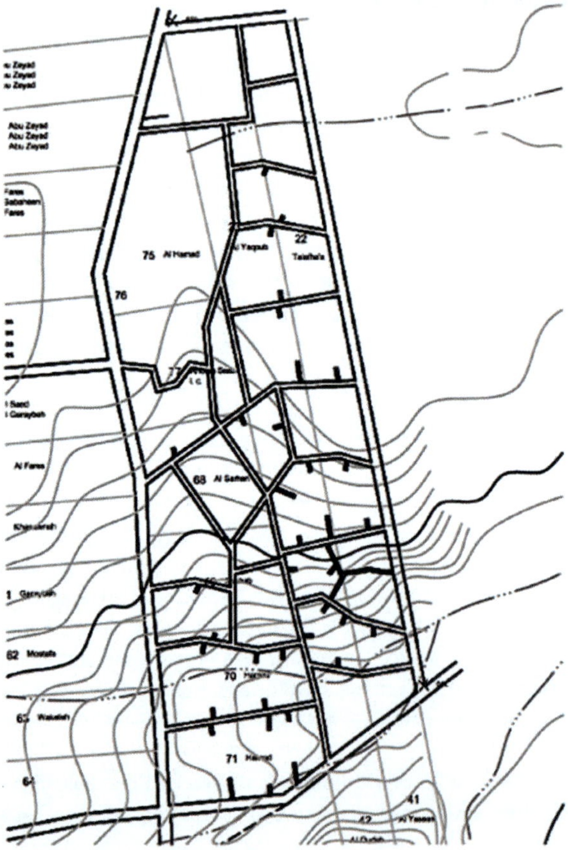

Fig. 8 Proposed site planning entities—land ownership and topography. *Source* Authors

(b) alternatives for the future, (c) household space, (d) socioeconomic characteristics, and (e) adjacent spaces. In addition, the design details considered the following issues: (1) Allotment of urban farming: Provision of open land for eatable vegetation in the yard to increase attachment and production. (2) Provision of two *Madheif* in response to cultural gender separation and seasonal migration within the house. (3) Provision of PV and Solar collectors on the roofs. (4) Provision of thermal walls in the sunny directions to gain energy during winter days, see Fig. 10.

Housing unit design responded to cultural rituals requirements such as hospitality and privacy. Such rituals were reflected on the internal and external spatial layouts of the family and the clan that was based on understanding the traditional *Bayt Esh-Sha'r* layout. Externally: orientation was to the east for symbolic and environmental issues. Internally: Spatial and privacy hierarchies of the traditional layout were reflected in the design to define visual, auditory, and physical interactions. The interior was divided into two major spaces based on privacy needs: reception (*Madheif*) and family private domain (*Mharam*). Solutions considered the number of space divisions (*Rebaat*) of *Bayt Esh-Sha'r*, which correlates to the number of central columns (*Wasit*). Such space division commensurate their cultural values. Housing unit typologies were developed based on the

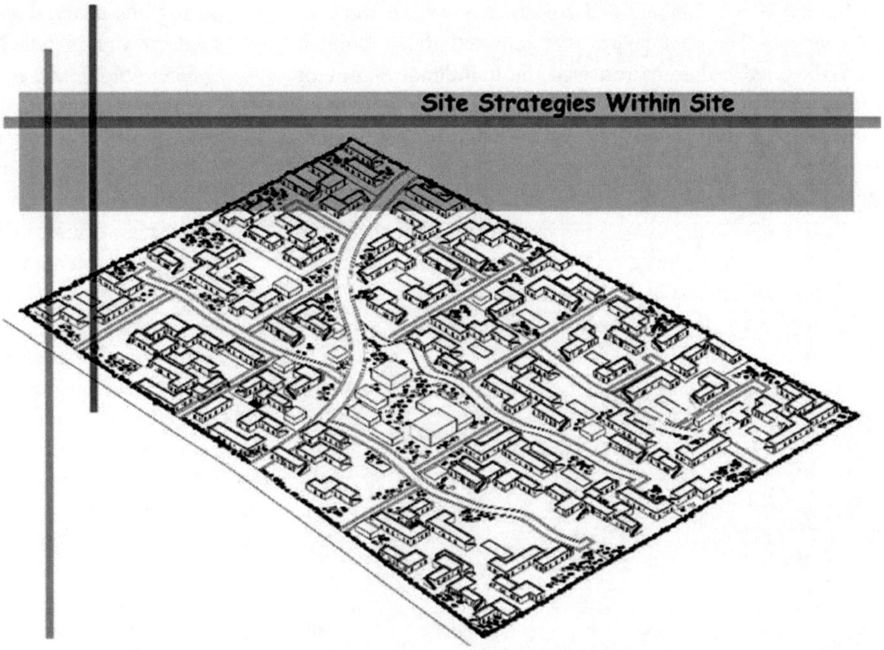

Fig. 9 Proposed site design. *Source* Authors

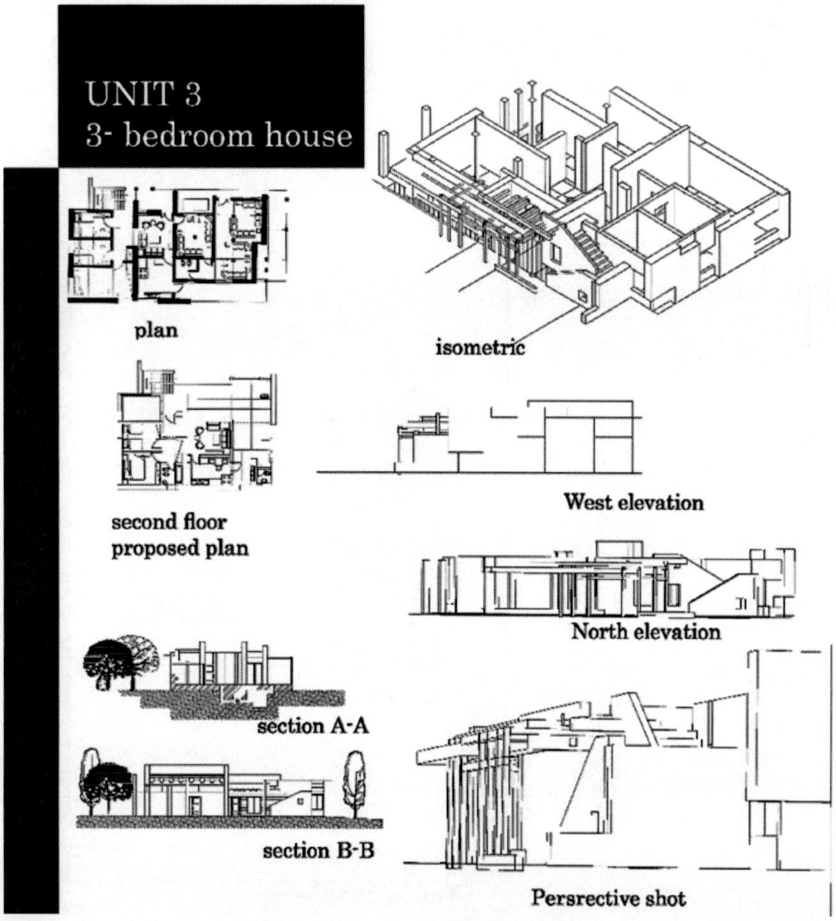

Fig. 10 Proposed housing units design. *Source* Authors

traditional *Wasit* numbers that construct *Bayt Esh-Sha'r* as follows: two-bedroom apartments reflected the traditional *Qatbah*; three bedrooms reflected the traditional *Shawa* or *Mthoulath*; and four bedrooms reflected the traditional *Mroubaa*. These typologies tried to respond to the different family size. Further, the design considered the proportions of the traditional typologies.

(3) **Wall section design:** Construction materials were selected with so much care about its social impact. Most people in Jordan consider white limestone as a prestigious material for exterior walls, which is mainly used for high income houses. Field work indicated that concrete block and reinforced concrete seem to be the common construction materials in most villages for both modern and transitional homes, although old homes are built of mud or rough basalt stones, see Fig. 11. Several modern homes are built out of white limestone. Houses were surrounded with low rough basalt stones fences that define the boundary of each house lot.

Fig. 11 Present and prevailing wall construction in the Badia area. *Source* Authors

The proposed wall section design suggested the use of adobe, stone, and concrete layers in the wall construction. In addition, alternative wall sections designs were investigated based on the field work and interviews with local contractors. Alternatives considered construction material and sections based on social value, cost, and thermal capacity, and sustainability. The proposed best solution for the wall section with a total width of 45 cm composed of white limestone sheer (4 cm) (exterior face), two adobe layers (15 cm each), vacuum, (5 cm), cement (2 cm) (interior face). The cost of which (per one-meter wall length) is 14JD, see Fig. 11. This cost reduction was about 30% of the original wall section construction cost for the same wall width, which is 45-50JD. The original wall section is composed of 25 cm of white limestone at the exterior, two layers of concrete brick, insulation material and cement face lifting at the interior, see Fig. 12.

Fig. 12 Proposed wall section (left) and original wall section (right). *Source* Authors

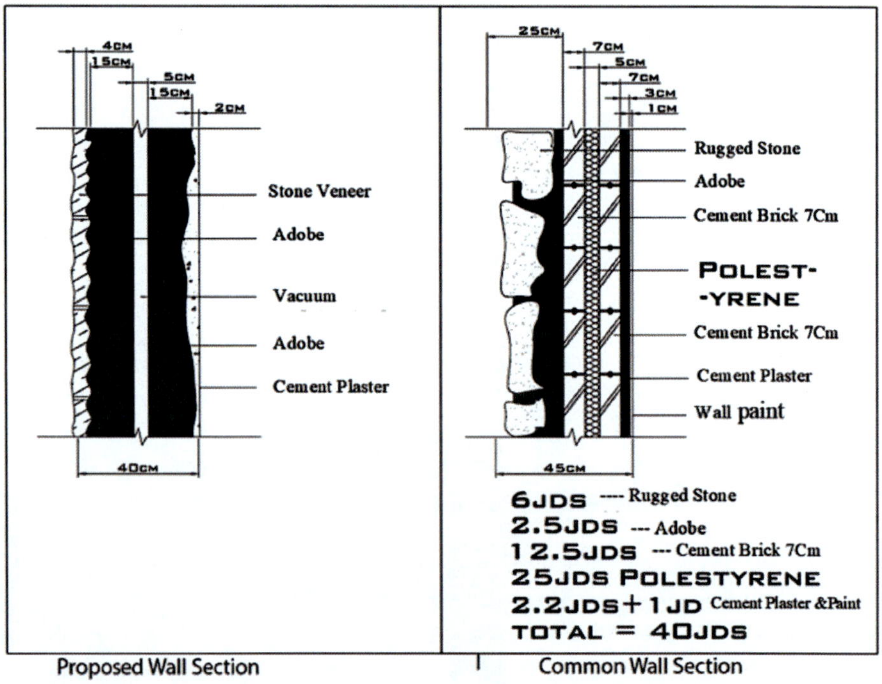

9 Conclusion

The proposed reduction of cost for such projects for low-income sociocultural groups include: (1) Site selection and sustainable planning—cost cut through efficient site selection that requires less site construction, infrastructure, land shape, land fill, and land prices, in addition to site location that provides ease of access being accessible by public transportation and being linked to the rest of the nearby population. (2) Identifying of housing unit patterns with observed cultural values. (3) Identification of multiple typologies for wall sections based on local and sustainable constructions with the preservation of social implications. The outcome of the proposed housing project was based on a holistic research approach that pinpointed critical issues of concern and depicted them as integral layers in the design, see Fig. 13. The study suggested that alternative solutions require socially responsible architecture and construction based on understanding the specific ethnic group we are working with. The study proposes that low-cost housing typologies can be informed and infused by the cultural needs of the community at hand. Future studies suggest the need to

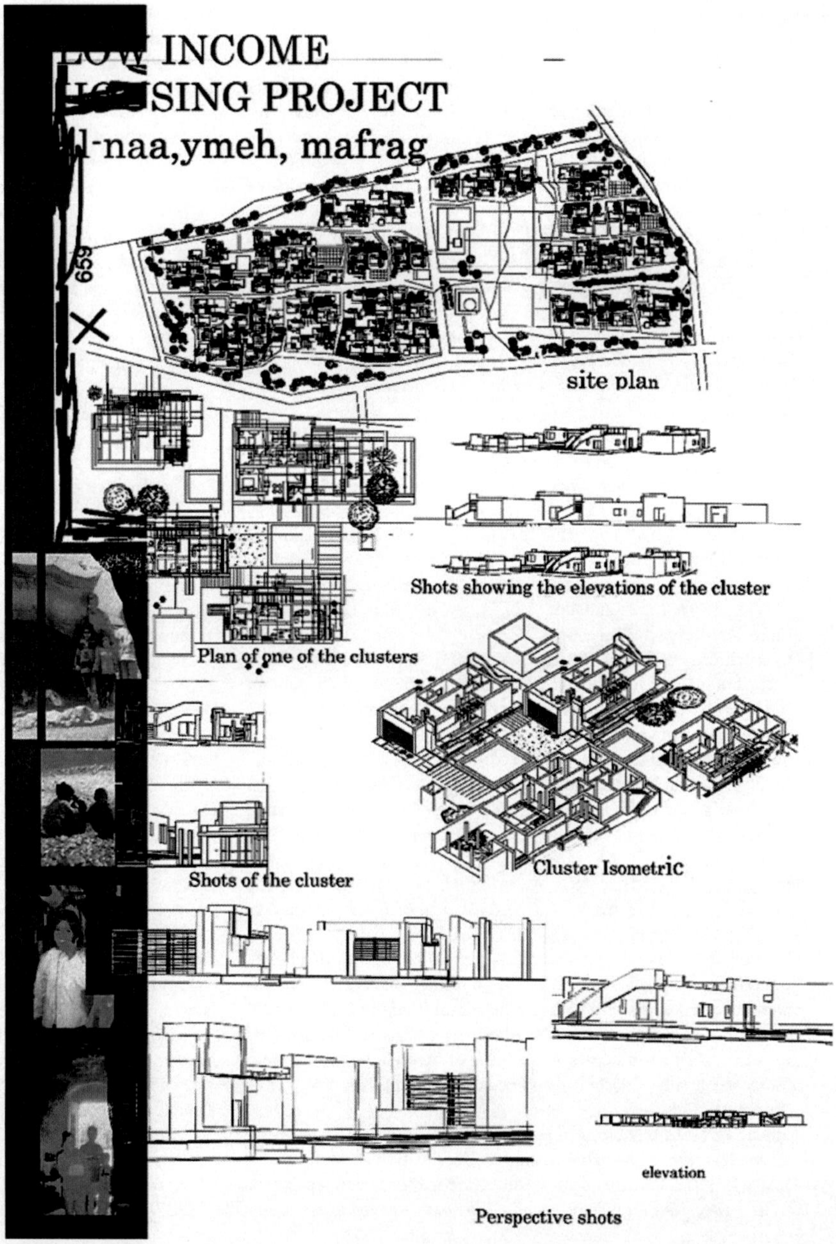

Fig. 13 Proposed housing project layout and design. *Source* Authors

study other cultural groups around Jordan and investigate other low-cost typologies. Future studies should cover the different geographies and climate regions around Jordan. Such studies can inform governmental agencies, planners, and developers of the right approach toward the design and construction of low-cost housing for a very specific group of people.

Acknowledgements Thanks goes to the team of my students at third year design studio at Jordan University of Science and Technology for their help in the field work and design development.

References

Abed, A. (2001). *Change in traditional Bedu home layout as a function of lifestyle: From bayt ash-sha'r to villa*. Thesis, Jordan University of Science and Technology.

Abu-Jaber, K., Gharaibeh, F. A., & Hill, A. (1987). *The Badia of Jordan: The process of change*. Publication of the University of Jordan.

Akkawi, H. (1995). The national survey of housing 1994. In *Proceedings of Symposium on the Assessment of the Housing Sector: Housing and Urban Development Corporation*, Amman, Jordan (pp. 7–25).

Al-Beejawi, B. (1997). *Affect of similarity in cultural status on general characteristics of public buildings*. Thesis, Jordan University.

Al-Fawal, M. (1983). *Social structure in Bedu societies*. Dar Al-Fekr Al-Arabi.

Al-Hasanat, S. (2003). Housing information system. In *Proceedings of Sixth Annual Symposium to Assess Housing Sector, Public Institution for Housing and Urban Development,* Amman, Jordan (pp. 107–113).

Al-Homoud, M., & Ghanem, H. (2019, December). Regeneration of Amman Center—Social Acceptance of Syrian Migrants in Downtown Amman. In *Proceeding of Second International Conference on Green Urbanism* (2nd ed., Vol. 1, pp. 68–81). University of Rome Tor Vergata, Rome, Italy.

Al-Homoud, M., Al-Oun, S., & Abed, A. (2003). Reflective lifestyle on Badu homes in Jordan. In *Proceedings of the 34th Annual Conference of the Environmental Design Research Association* (pp. 41–55), Minneapolis, Minnesota, U.S.A.

Al-Homoud, M., Al-Oun S., & Abed, A. (2008). Socio-cultural, physical, and architectural attributes of Bayt Ash-sha'r in transition in Jordan's Badia. In Dutton, R. W., & Shahbaz, M. H. (Eds.), *Jordan's arid Badia: Deepening our understanding* (pp. 265–294). Smith-Gordon.

Al-Homoud, M., Al-Oun, S., & Al-Hindawi, M. (2009). The low-income housing market in Jordan. *International Journal of Housing Markets and Analysis, 2*(3), 233–252

Al-Homoud, M., Al-Oun, S., Smadi, A., & Al-Hindawi, M. (2009). Exploring sales advertising in the housing market in Jordan. *International Journal of Housing Markets and Analysis, 2*(1), 39–56

Al-Homoud, M., Al-Oun, S., & Abed, A. (2010). Socio-spatial measures of Bedouin homes as an effect of lifestyle in Northeastern Badia of Jordan—Constancy and change. *Al-Manarah, 16*(3), 27–71 (AABU Publication).

Al-Mousa, H. (1996). *Desert habitat in the north Badia: A study in the urban structure*. M.Sc. Thesis, Jordan University.

Al-Omari, Z. (2003). Reality of housing sector as the private sectors see it. In *Proceedings of Sixth Annual Seminar on Housing Sector Assessment, Housing and Urban Development Corporation*, Amman, Jordan (pp. 131–141).

Al-Oun, S. (1997). *Livestock marketing in the Jordanian Badia*. Masters Thesis, University of Durham.

Al-Oun, S., & Al-Homoud, M. (2005). *Housing the disadvantaged: Low-income housing initiatives for the Bedouin cultural demands at Al-Mansourah Village*. Housing and Urban Design Corporation.

Al-Oun, S., & Al-Homoud, M. (2005). *Housing the Disadvantaged: Low-Income Housing Initiatives for the Bedouin Cultural Demands at Al-MansourahVillage*. Habitat of Low- income Workshop, Housing and Urban Design Corporation

Al-Oun, S., Al-Homoud, M., & Abdel-Jawad, J. (2010, July). Fitness between house choice versus market production in Jordan. In *Proceedings of ENHR 2010 Conference—Urban Dynamics and Housing Change: Crossing into the 2nd Decade of the 3rd Millennium*, Istanbul, Turkey.

Al-Ozaizat, Y. (1980). *Arabs heritage*. Jordanian Armed Forces

Balchin, P., & Rhoden, M. (1998). *Housing: The essential foundation*. Routledge.

Barakat, N., & Tuffaha, M. (1995). *Private sector's role in housing limited and low income households* (pp. 3–12). Housing Sector Assessment Seminar, Housing and Urban Development Corporation.

Basboos, A. (1995). Housing financing, housing sector assessment seminar. In *Proceedings of Housing and Urban Development Corporation*, Amman, Jordan (pp. 2–19).

Bickerton, I. J., Abu Jaber, K. S., & Irvine, E. V. (2020, April 20). Jordan. https://www.britannica.com

Brandenburg, C., et al. (1998). Development anthropology: A primary survey. In R. Dutton (Ed.), *Arid land resources and their management: Jordan desert margin*. (pp. 247–265). Kegan Paul International.

Choudhury, I. (1993). *Effect of cultural differences on residential satisfaction with referring reference to outside space*. University of Wisconsin-Milwaukee.

Colliers International Research. (2008). *Jordan—Real estate overview, fourth quarter*.

Culture and Lectures. (2000). *Culture and popular culture*. https://courses.bus.ualberta.ca/consumer-behavior/lectures/culturesld001.html

Damra, M. (2003). The role of statistical data on the formation of housing policies. In *Sixth Annual Symposium to Assess Housing Sector, Public Institution for Housing and Urban Development*, Amman, Jordan (pp 78–83).

DoS (Department of Statistics). (2020a, April 15). *Population statistics for 2018*. https://jorinfo.dos.gov.jo

DoS (Department of Statistics). (2020b, April 15). *Population statistics for 2015*. https://jorinfo.dos.gov.jo

Duncan, J. (1985). The house as symbol of social structure: Notes on the language of objects among collective groups. In I. Altman & C. Werner (Eds.), *Home environments*. (pp. 133–151). Prenum Press.

Ferguson, B. (2008). A value chain framework for affordable housing in emerging countries, *Global Urban Development Magazine, 4* (2).

Global Property Guide. (2008). *Annual report*. Amman, Jordan.

Harvey, J. (2000). *Real estate investment*. Liverpool University Press.

Ismael, F. (1983). *Change and development in desert societies* (2nd ed.). Dar Al-Ma'refah Al-Jame'yah

Jabbor, J. (1988). *Bedouins and the Badia: Images about Nomadic life in the Syrian Badia*. Dar Al-Elem Lelmalayeen

Kanan, R., & Estatiyyah, M. (1992). *The guidance of climatic design*. Ministry of Energy.

Kellett, P. (1999). Cultural values and housing behavior in spontaneous Jerusalem. *Journal of Architecture and Planning Research, 16*(3), 205–224

Khammash, T., & Alkhas, Z. (2009). *Real estate sector report* (June). ABC Investments, Research Department.

Kotler, P., & Armstrong, G. (1999). *Principles of marketing*. Prentice-Hall of India.

Marcus, C., & Sarkissian, W. (1986). *Housing as if people mattered: Site design guidelines for medium-density family housing*. University of California Press.

Miles, M., Berens, G., & Weiss, M. (2001). *Real Estate Development: Principles and process*. Urban Land Institute.

Ministry of Planning (MoP). (1987). *Technical report of the housing strategy* (pp. 12, 112)

Ministry of Planning (MoP). (2015). *Jordan response plan*. Ministry of Planning and International Cooperation.

Prince Ghazi bin Muhammad. (1999). *Those people of Jordan: at the beginning of the twenty-first century*. The Royal Court.

Rapoport, A. (1969). *House form and culture*. Prentice Hall.

Rapoport, A. (1977). *Human aspects of urban form: Towards a man-environment approach to urban form and design*. Pergamon Press.

Rapoport, A. (1985). Thinking about home environment. In I. Altman & C. Werner (Eds.), *Home Environments*. (pp. 255–285). Plenum.

Rapoport, A. (1991). Housing: Symbol, structure, site. In: L. Taylor (Ed.), *Housing and culture* (pp. 14–15). Rizzoli

Rian, I. M., & Sassone, M. (2012, December). Flexible housing, a healthy housing: A brief discussion about the merits of flexibility in designing healthy accommodation. In *Proceeding of 2nd International Conference on Inhabiting the Future* (p. 1088).

Schmitz, A., & Deborah, B. (2001). *Real estate market analysis*. Urban Land Institute Washington.

Sims, D. (1990). Owner-builder housing in Jordan. *Habitat International, 14*(1), 123–135

Sociology at Hewett. (2000). *Overview of subculture theories*. https://www.hewet.norfolk.sch.uk/curric/soc/crimesubculture.htm

Sweis, G., Sweis, R. J., Imam, R., & Kassab, Gh. (2013). Customer satisfaction in apartment buildings: The case of Jordan. *Life Science Journal, 10*(12), 237–245

UNHCR. (2014). *Syrian refugees living outside camps in Jordan*. UNHCR Jordan Operation.

USAID. (2005). *Jordan—case study: Housing bonds spur investors*. USAID.

Weir, S. (1990). *The Bedouin*. British Museum Publications.

Wentling, J. W. (1990). *Housing by lifestyle: The component method of residential design*. Mcgraw-Hill Publishing Company.

News in Review. (1997). What's in a Culture? http://cbc.ca/insidecbc/newsinterview/nov97/scotland/w.1/html.

World Bank. (2018). *Jordan housing sector review. Economic and sector work*. World Bank Urban, Rural and Social Development (GSURR), Middle East and North Africa (MNA). World Bank Group.

Worldometer. (2020, April 15). https://www.worldometers.info

The Utilization of Human-Oriented Transportation Systems as Regeneration Tool for 'Healthy Neighbourhood'

Oleksandr Galychyn and Azraa Eltayeb

Abstract

Recent studies have defined the 'healthy neighbourhood' as the social or socio-economic unit within a healthy district or spatial unit within a human-oriented transportation system. However, those views have failed to notice both standard elements of organic and city transportation policy and other dimensions of organic transportation, excluding spatial one. In this study, the social-ecological system (SES) and human-oriented transportation system (HOTS) frameworks will be compared to each other in terms of structure, application and dynamics to draw a conclusion about the suitability of HOTS in framework to describe the complex socio-technological system such as 'healthy neighbourhood'. Additionally, the structure and the multidisciplinary process that occur within the 'healthy neighbourhood' will be analysed in terms of HOTS framework. Finally, the indicators of pattern and size of 'healthy neighbourhood' in terms of HOTS framework will be suggested. Thus, a healthy interdisciplinary neighbourhood will be captured. This research will be the first attempt to shift from the traditional 'unit' perspective to the network models capable of unfolding the internal socio-economic and technical processes, uncovering the internal organization and functions of healthy neighbourhoods.

Keywords

Sustainable neighbourhood • Transit-oriented development • Human-oriented transportation • Social and transportation networks

O. Galychyn (✉) · A. Eltayeb
Department of Urban and Regional Planning, Mimar Sinan Fine Arts University, Istanbul, Turkey

1 Introduction

Although the definition of neighbourhood, or local community, was formulated as early as 1916 by Robert Park (Anderson, Bulatao, & Cohen, 2004), the concept of the healthy district as a 'healthy' unit of the local community was not introduced until 1972 by Suttles (Anderson et al., 2004). However, this research was limited to the urban sociology referring to this unit as a spatially concentrated group of people in comparison to Ratzel's more comprehensive unit of analysis, namely social system on both local and global (city and country) levels (nation, or folks) (Smith, 1980). In the urban sociology context, the traditional neighbourhood was defined (Anderson et al., 2004) (Galychyn & Ustundag, 2017). Also, social scientists generally rely on 'virtual' boundaries defined by administrative agencies[1] that more suited for governance rather than put efforts into understanding its multi-dimensional structure, which in turn renders an operational definition of the neighbourhood (Anderson et al., 2004; Galychyn & Ustundag, 2017). Physical boundaries do not exist even in health research ranging by the level of abstraction (Anderson et al., 2004). Research in this camp of study is aimed to analyse the impact of individual/group social and physical features and their spatial distribution on disparities in health among the ethnic group rather than to render the more robust design of neighbourhood itself (Anderson et al., 2004; Galychyn & Ustundag, 2017). Those studies suggest that the concept of a healthy neighbourhood differs in regards to the social composition of a neighbourhood without having a ready

[1]Administrative agencies are a governmental regulatory body that controls and supervises a particular activity or area of public interest and administrators and enforces a specific body of law related to that activity or interest. Administrative agencies refer to commissions, corporations, boards, departments such as the Central Intelligence Agency, the national labour relations board, and the Securities and Exchange Commission.

F. Trapani et al. (eds.), *Advanced Studies in Efficient Environmental Design and City Planning*, Advances in Science, Technology & Innovation, https://doi.org/10.1007/978-3-030-65181-7_47

operational definition at hand (Anderson et al., 2004). Developed by architect Leonid Vesnin in 1922, micro-district type of neighbourhood (Zosimov, 1976) has a strict physical boundary defined by streets, that function also as a mean of physical isolation, calculated by the dimensions of sustainability and local context (Anderson et al., 2004; Galychyn & Ustundag, 2017). However, in reality, micro-districts are developments of mixed land uses around the Palace of Culture with priority to the health residents (Galychyn & Ustundag, 2017). In other words, it is a sustainable spatial unit within an urban system (Anderson et al., 2004; Galychyn & Ustundag, 2017). However, being a public land make it dependent on a supply of public utilities for supply and sustainable planning, namely physical regulations according to the sustainable indicators of micro-district (State Building Norms, 1992; Zosimov, 1976). The local environmental policies and land-use regulation, as well as the metabolism of the neighbourhood (have not been developed across the space of Former Soviet Union), have not been considered within this type of planning (Galychyn & Ustundag, 2017; Zosimov, 1976). This type can be described as spatial unit characterized by a mix of land uses similar as TOD[2] type referred as a compact island (Glazyeva, Sukhacheva, & Firsova, 2014) due to the pattern (geometry) and physically isolated social networks, or causes of social exclusion (social dimension of sustainability) (Eremin, Ilyin, & Lomova, 2014; Galychyn & Ustundag, 2017; Varlamov, n.d.).

Consequently, this type of neighbourhood has lost the original meaning of neighbourhood concept leaving out of focus social transformations (Galychyn & Ustundag, 2017); therefore, it is far from being a healthy development in terms of social and physical infrastructure and human-oriented design (Galychyn & Ustundag, 2017). In other words, until the recent time, neighbourhoods have been defined in a spatial sense as a dispersed pattern of service zones (network's structure) for local facilities or spatially fixed units. Until recently, the studies in their approach to urban issues have given much importance to the spatial aspects of neighbourhoods (Anderson et al., 2004; Galychyn & Ustundag, 2017; Gashenko, 2015). However, the newly introduced theory called Green Urbanism defined healthy neighbourhoods as the socio-economical unit, namely identifying the healthy neighbourhood as a mix of housing types, prices and ownership forms that have to avoid gentrification and provide affordable housing for all and also ensure 'ageing in place' (Galychyn & Ustundag, 2017; Lehmann, 2010). To put it, he inserted those catchment areas of local facilities into an economic dimension of sustainability, namely concentration and dispersion of all-income households and public spaces within service zones of local facilities.

Despite having incorporated the 'reuse and recycling' of local materials and water purification strategies physical and spatial regulation has been left out of focus (Galychyn & Ustundag, 2017) not to mention that the essence of neighbourhood concept formulated in 1916 has already reached its demise (Anderson et al., 2004; Galychyn & Ustundag, 2017). The last theory Organic Urbanism has solved this problem by combing Lehmann's definition of the healthy neighbourhood with the micro-district concept, waste-to-energy systems and Cerdà's housing design and pacification measures along the street (Galychyn & Ustundag, 2017). However, this definition has failed to grasp a multi-dimensional nature of the concept healthy neighbourhood (Galychyn & Ustundag, 2017) and the dynamic nature of structure and design under both standard elements of the organic city and transportation policy (Galychyn, Ahmeti, & Ustundag, 2019; Ustundag, 2012) and other dimensions of organic transportation (Galychyn et al., 2019) not to mention that spatial fixation of the neighbourhood is subject to change under the influence of those dimensions (Galychyn & Ustundag, 2017; Galychyn et al., 2019). In addition, a function of the neighbourhood into a broader environment and in the relationship with other structural elements of human-oriented transportation system like TOD, the surface area of the organic hub, has not been considered (Galychyn & Ustundag, 2017; Galychyn et al., 2019).

In the first section, the methodology used in this study will be briefly explained. In the second section, we will introduce the human-oriented transportation system (HOTS) as a theoretical framework to study 'healthy neighbourhood'. In the third section, a concept of the healthy neighbourhood will be developed, and the indicators that capture the organization and functions of 'healthy neighbourhood' in terms of HOTS framework will be determined. In the results and conclusion sections, we will discuss the benefits of this concept for society and about the use of network models as a common model suitable for all 'healthy neighbourhood'.

2 Methodology

In this study, an analytical comparison of the human-oriented transportation system (HOTS) and organic transportation framework are used to study healthy neighbourhoods. Those HOTS and OT frameworks were taken from the paper on organic transportation networks (Galychyn et al., 2019) and the organic transportation approach (Galychyn & Ustundag, 2017; Ustundag, 2012), respectively.

[2]Transit-oriented development (TOD), this term refers to compact, walkable, pedestrian-oriented, mixed-use developments centred around efficient public transport such as train system.

We use both frameworks to analytical compare the recent 'healthy neighbourhood' concept proposed by Galychyn. The indicators were formulated from analysis of the organization and functions of the neighbourhood using the frameworks mentioned above. Dimensions are referred to as the top-tier category with the second-tier components characterizing such indicators as security systems, physical isolation, and open green space per person. Sequentially, features at the lower third-tier belong to the second-tier components.

To be precise, those empirical indicators located at the third-tier in this ontological HOTS framework are units (products and services) that are contained within physical or other dimensions of HOTS that may be consumed or used in different forms. For example, the sun, water, agricultural areas, and so forth belong to the local resources' category. The HOTS framework is used as a theoretical framework to formulate a new definition of 'healthy neighbourhood'. In other words, a general model suitable for all 'healthy neighbourhoods' is proposed.

3 Healthy Neighbourhood Towards Human-Oriented Design

Organic transportation can be archived by incorporation in the planning process of the transport infrastructure project elements of an organic city and transportation policy: physical, spatial and governmental regulation, public participation, management and governance, education and training, timeframe and socio-cultural dimensions (Galychyn et al., 2019; Ustundag, 2012). Those elements can be incorporated into the scale of both districts and cities (Galychyn et al., 2019). In the city scale, the implementation of those elements is not only limited to the transportation master planning process that was partly archived in Yalova (Turkey) due to failure to integrate management and governance component to the city design (Galychyn & Ustundag, 2017; Ustundag, 2012) but also the successful Green Infrastructure Projects such as 10-Year Project for Green Tokyo due to the effective utilization of aspects of management and governance, socio-cultural dimensions, governmental regulations and involvement of educational organizations (Tokyo Metropolitan Government, 2007). In other words, all strategies predetermined in the project were implemented by formulating the multidisciplinary process along with flexible regulations to coordinate processes. Also, one research has been done to compare different types of TOD by physical regulations of TOD without considering spatial regulations and socio-cultural dimensions to compare the human-oriented (social) design of each TOD (Galychyn & Ustundag, 2017).

In another study, management and governance, education and training of TOD were added to develop the conceptual model of this structural element of an organic city and access its function within a human-oriented transportation system (Galychyn et al., 2019).

However, it failed to understand the role of public participation and spatial regulations in shaping TOD (Galychyn et al., 2019), and complex processes that occur between the elements during its timeframe (lifecycle) (Galychyn et al., 2019; Metzger, 2000). In other words, the system without all possible environmental impacts and benefits of all process that has been accessed (Metzger, 2000) does not seem to be generalized conceptual model towards sustainable design and planning of node (structural element) of the human-oriented transportation system (Farreny et al., 2011; Galychyn et al., 2019). Initially, this framework was designed to determine policy actions needed for a shift from motorized transportation network towards human-oriented transportation system (HOTS), for integration at the later stage into standard transportation master planning process (Ustundag, 2012). However, it is suitable for other structural units of modern megapolises such as TOD according to the recent study (Galychyn et al., 2019). HOTS framework in similarly to social-ecological system framework (SES) composed of top-tier categories (broad, abstract, qualitative subsystems) and narrow second-tier subdivisions (attributes) within elements of top-tier category, or characteristics (subcategories) that can be defined operationally for subsequent measurement (McGinnis & Ostrom, 2014). The third-tier category (empirical variables) that are essential factors that are shaping conditions under which second-tier categories function within upper-tier categories (elements) (McGinnis & Ostrom, 2014). Unlike SES, this category has not been developed for HOTS due to the same elements belonging to the different clusters, dynamic relationships among those elements that indicate (Galychyn et al., 2019; McGinnis & Ostrom, 2014). However, HOTS incorporate both metabolic (positive) and catabolic (negative) factors to prevent/cause system crashes and shocks (such as market failure, sprawl, environmental degradation and depletion of resources) (Galychyn et al., 2019) while SES incorporates only positive first-tier and second-tier variables (McGinnis & Ostrom, 2014). Besides, the SES framework, or governance model, applicable only to social and socio-ecological systems (McGinnis & Ostrom, 2014). On the other hand, HOTS is a social-ecological-technical system characterized by the natural dynamic of ecosystems such as river basin and dynamic process of complex technological systems such as transportation/neighbourhood infrastructure (Galychyn & Ustundag, 2017; Galychyn et al., 2019; McGinnis & Ostrom, 2014). Transportation networks, neighbourhoods, TODs are social-technical systems despite being viewed as only technical (till 1991) (Calthorpe et al., 2008; Galychyn et al., 2019) or socio-ecological systems[3] to re-conceptualize them as sustainable systems (Lehmann, 2010; McGinnis & Ostrom, 2014). HOTs is a conceptual framework for capturing a mechanism of transportation network growth by observation of concurrent changes in the typology of motorized transportation network towards integrated public-soft transportation network and towards human-oriented transportation system at later stage across the dimensions of sustainability (social, environmental, economic) and organic transportation (global participation model, government regulations management and governance, cultural, spatial and built environment, physical and underground dimensions) under internal and external geographical constraints inside the ecological aspect of sustainability (Galychyn et al., 2019).

At higher hierarchical level global public participation model, government regulations management and governance, cultural dimensions of organic transportation and economic and social dimension of sustainability converge via social networks to form a societal system. In contrast, the spatial and built environment and physical dimensions under internal and external geographical constraints converge through flow energy and materials to form an urban system (Galychyn et al., 2019). Those systems by overlapping cultural and global participation dimensions shape the human-oriented transportation system (HOTS) (Galychyn et al., 2019). On the other hand, SES shapes policy actions shaped by governance and resource systems and their units (resources units and actors) (McGinnis & Ostrom, 2014). The framework covers both local and global levels in the spatial sense, where 'local' is assigned to the nodes: TODs, healthy neighbourhoods and organic hubs, while 'global' is assigned to HOTs as a whole (Galychyn et al., 2019). Unlike SES, two specific decision models with the organic unit or the central governance unit located in the heart of the safety-driven decision system (instead of comfort-driven) are closely intertwined with public participation on local and global levels (Galychyn & Ustundag, 2017; Galychyn et al., 2019).

The decisions are made within a spatial context (socio-cultural and physical context combined). In other words, it corresponds to the two of four top-tier categories of SES referred to as a governance system with actors operating within its boundaries (management and governance, public participation) (Galychyn et al., 2019; McGinnis & Ostrom, 2014). Resource systems (multiple resource units) encompass design (a combination of physical and spatial regulations) within the ecological dimension of sustainability (Galychyn et al., 2019; McGinnis & Ostrom, 2014). Thus, HOTS framework can be applied to more complex and broader systems with a technical component like organic transportation networks including their nodes: neighbourhoods and organic hubs (i.e. a combination of two neighbours: TOD and metropolicentre[4] nodes). Exogenous influences (McGinnis & Ostrom, 2014) from ecosystems outside of HOTS as well as dynamics of elements within overlapping dimensions and social and urban subsystems of HOTS (Galychyn et al., 2019). As a result, the HOTS concept builds on the foundation of the HOTS framework, Green Urbanism concept, complexity research, transdisciplinarity and urban metabolism (Galychyn et al., 2019). The healthy neighbourhood is a secondary node of HOTS that exists in local level in a spatial sense (Galychyn & Ustundag, 2017; Galychyn et al., 2019). The study on organic transportation networks (Galychyn et al., 2019) has emphasised that healthy neighbourhood should be a part of TOD, the surface area of organic hub, to archive maximum spatial integration (distance, public mode choice and compactness) with integrated public-soft transportation network, management and governance (local decision model), education and training (cooperation of actors, training of staff in the related businesses, fundraising) (Galychyn et al., 2019).

Upward regeneration of neighbourhood means that sustainable neighbourhood planning (Calthorpe et al., 2008; Farreny et al., 2011) triggers the regeneration of human-oriented transportation system as a whole. The neighbourhood should also be understood similar to the transportation system as an element subjected to the adaptation to three phenomena: sprawl, bicycle, pedestrian-oriented design (Galychyn et al., 2019). Thus, the concept of the neighbourhood will be modified and enriched in terms of OT and HOTS framework considering its function within an organic hub (Galychyn et al., 2019).

4 System Analysis (Neighbourhood Analysis) Within OT and HOTS Frameworks

The last updated definition of the healthy neighbourhood was interpreted as a micro-district characterized by mix of housing types to shape the octagonal blocks chamfered in the corners by prices that shared among residents where

[3]The SES framework is a foundation for integrating set of important ecological and social variables that are needed for the analysis of an SESs.

[4]Metropolicentre reffers to a multi-dimensional and mixex-use project that consist of four levels. Underground level for pedestrian crossing which connected to the lobby of subway station and spatially connected to the surface level (TOD). First level includes pedestrian crossing, services for dwellers and private companies. Second level is a tunnel for exiting users and (Bus Rapid Transit) BRT. Third level incorporated underground parking and utilities.

local materials are recycled, energy captured by organic waste streams and clear water utilized surrounded by pacified road (priority to bicycle traffic) (Galychyn & Ustundag, 2017). This definition of the neighbourhood must be analysed under OT and HOTS frameworks as a structural component of the organic hub within HOTS and, simultaneously, the secondary node within HOTS.

Firstly, the neighbourhood phenomenon will be inserted into the OT framework instead of 'viable city' (Galychyn et al., 2019; Ustundag, 2012). According to the definition, design (a combination of physical and spatial regulations) can be easily identified (Galychyn & Ustundag, 2017; Galychyn et al., 2019). Other standard elements of organic city and transportation policy cannot be traced from this definition. At this point, physical and spatial regulations (Galychyn & Ustundag, 2017; Galychyn et al., 2019; Ustundag, 2012) should be explained. The physical regulations (physical arrangement of space) at local level encompass the resident-oriented facilities (Bagliani & Dansero, 2011), pedestrian-oriented design (Galychyn & Ustundag, 2017; Galychyn et al., 2019), security systems (Bagliani & Dansero, 2011), means of physical isolation (physical organization of boundary) (Bagliani & Dansero, 2011; Galychyn & Ustundag, 2017), and traffic-calming measures (Galychyn & Ustundag, 2017; Galychyn et al., 2019). Spatial regulations emphasize occupied space, density, diversity, design parameters and compactness indicators that are influenced by social dimension (encompass governmental regulations, management and governance, public participation) (Bagliani & Dansero, 2011; Galychyn et al., 2019). Governmental regulations indicate local environmental policies (for example, government resource policies and tax incentives to encourage public transit ridership) (Galychyn et al., 2019; Galychyn, 2017b). Local actors of TOD have represented the public participation element (Galychyn & Ustundag, 2017; Galychyn et al., 2019).

Also, management and governance incorporate the ownership structure and related disparities in maintenance (Galychyn et al., 2019). Consequently, education and training indicate training of staff in the related businesses, cooperation between residents and businesses, and fundraising scheme for infrastructure (Galychyn et al., 2019; Tokyo Metropolitan Government, 2007). Finally, the cultural dimension indicates the racial perception of inhabitants of the analysed system (Anderson et al., 2004; Galychyn et al., 2019) as well as the perception of shared ownerships forms within the system in question (Galychyn et al., 2019; Metzger, 2000).

The organic transportation approach is re-conceptualizes healthy neighbourhood as one of the structural elements of HOTS (Galychyn et al., 2019). The simplified diagram of OT framework does not indicate the relationship between elements as well as hierarchy and nature of those relationships (Galychyn et al., 2019; Ustundag, 2012). However, to understand multidisciplinary processes and how to connect them to the neighbourhood level and existing definition, the relationships should be defined (Fig. 1). The compositional relationship can be observed between spatial and physical regulations (Galychyn & Ustundag, 2017; Galychyn et al., 2019; Ustundag, 2012). In general, spatial regulations are formed by adding physical regulations to the social dimension. For example, social transformations are archived by the social design (Galychyn et al., 2019), namely pattern, or geometry of housing modified in accordance with as well as social (affordability, equality and attitude to the security), economical (income level, network effect, path dependence and jurisdictional organization) (Galychyn et al., 2019; Hansen, 1959) and physical aspects (physical isolation, security system and function and type of residential facilities (Bagliani & Dansero, 2011; Galychyn et al., 2019). These social and physical transformations are initiated by sets actors: local administration, scientists, planners and residents, the organic unit which produces safety-driven decision towards the social design through the Kaleidoscopic method (four dimensions of diamond with the safety parameter placed in the centre) (Galychyn & Ustundag, 2017; Galychyn et al., 2019). The hierarchical relationship between spatial regulations and physical regulations indicates that higher hierarchy of spatial dimension exhibits features of degree (hierarchy of nodes) and eigenvector (agglomeration) (Galychyn et al., 2019; McGinnis & Ostrom, 2014). In other words, those features indicate that this framework underlines the network structure. Therefore, the neighbourhood structure will be re-conceptualized as the network consisting from links (bicycle/pedestrian paths) and nodes (origin-destinations) (Galychyn et al., 2019;

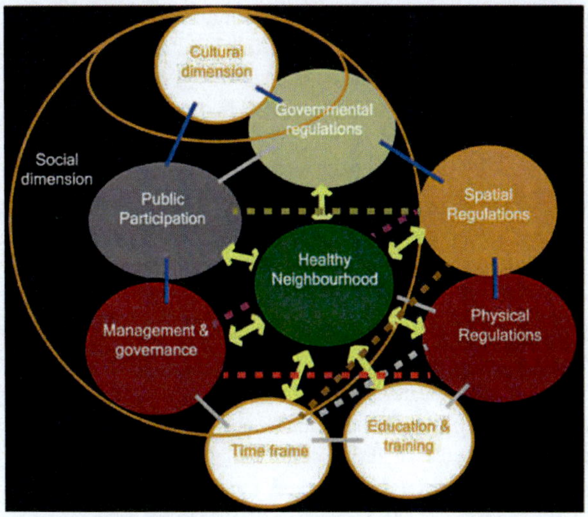

Fig. 1 Organic framework (OT) of a healthy neighbourhood. (*Source* Developed by Galychyn and colleagues, 2019)

Galychyn, 2017a) since the neighbourhood is a complex system with a hub (palace of culture, local administration or 'citizen's space' Galychyn & Ustundag, 2017; Metzger, 2000; Zosimov, 1976) and, hub-and-spoke model (John-Paul, Claude, & Brian, 2017).

The neighbourhood structure will be discussed in details when the HOTS framework will be applied to define the healthy neighbourhood. Governmental regulations influence spatial inequalities in resources for urban dwellers (Paterson et al., 2003; Victoria Transport Policy Institute, 2015) and linear metabolisms model with loss of land, methane and compost (Davis, Polit, & Lamour, 2006). Similarly, this relationship indicates a higher degree of governmental regulations category. The timeframe corresponds to the stages of the life cycle of the product (neighbourhood): growth, stability, decline stagnation and renewal (Galychyn et al., 2019; Metzger, 2000; Ustundag, 2012). During the lifecycle changes in elements of the organic city and transportation, policy occurs due actions of the safety-driven decision system's actors (public participation) (Galychyn & Ustundag, 2017; Galychyn et al., 2019) which extract spatial units (colonization of spatial unit area along with depletion of resources) from the system (transportation network, neighbourhood or city) (Galychyn et al., 2019; McGinnis & Ostrom, 2014; Smith, 1980) during which spatial parameters of the system are subject to changes (occupied space, density, diversity, design parameters and compactness indicators) (Galychyn et al., 2019) produces physical units in the physical space of the system (Galychyn et al., 2019; McGinnis & Ostrom, 2014), accompanied by changes restricted by governmental regulations and authority of each actor over-extraction/production/maintenance of units within system accompanied by a difference in the quality of extraction/production/maintenance between of different actors within the same system (Galychyn et al., 2019; McGinnis & Ostrom, 2014). Spatial, physical units (pedestrian-oriented design) due to their complexity are shared in their extraction/production/maintenance between actors (Galychyn & Ustundag, 2017; Galychyn et al., 2019; McGinnis & Ostrom, 2014) restricted by cultural barriers (racial and shared property perceptions) (Galychyn et al., 2019; Metzger, 2000) and training of actors in regards to the cooperation form towards effective delivery of those sophisticated units (training, integrated projects, etc.) (Davis et al., 2006; Galychyn et al., 2019). However, in the OT framework, unlike in HOTS, changes in the relationship between elements do not occur (Galychyn & Ustundag, 2017; Galychyn et al., 2019; Ustundag, 2012). This confirms that education and training influence physical regulations through cooperation forms between actors to maintain/upgrade physical unit within the system (Davis et al., 2006; Galychyn et al., 2019; McGinnis & Ostrom, 2014; Ustundag, 2012). The safety-driven decision, on the other hand, determines jurisdiction organization and maintenance quality (management and governance) following the socio-cultural dimensions (Galychyn & Ustundag, 2017; Galychyn et al., 2019; McGinnis & Ostrom, 2014). Governmental regulations influence public opinion through popular culture images (culture dimensions) towards comfort-driven decision system (public participation) (Galychyn & Ustundag, 2017; Galychyn et al., 2019), ethnic disparities in health (Anderson et al., 2004; Galychyn et al., 2019) and perceptions of shared spaces and the mix of ownership forms (cultural dimension) (Galychyn et al., 2019; Metzger, 2000). Since the two frameworks are interrelated with HOTS been more complex and dynamic (Galychyn & Ustundag, 2017; Galychyn et al., 2019; Ustundag, 2012), the dimensions of organic transportations and the dynamic of relationships between elements will be explained. HOTS, unlike OT, incorporate both dimensions of sustainability and organic transportation (Galychyn & Ustundag, 2017; Galychyn et al., 2019; Ustundag, 2012).

The social dimension of sustainability in OT following the spatiotemporal sphere of influence of the social networks (Bagliani & Dansero, 2011; Browning & Soller, 2014; Galychyn & Ustundag, 2017; Galychyn et al., 2019) which correspond to the hierarchical (blue line) and ordinary relationships (grey line) between dimensions. On the other hand, the social dimension of HOTS can be represented as the social network across education and training, physical, management and governance, spatial and underground dimensions which after public participation intervention, namely local/global public participation model, generate organic hub, within the human-oriented transportation system (Galychyn et al., 2019). Connecting concepts following their definitions and instances from the transportation networks literature determine the exact coverage and influence of the social sphere of sustainability (Galychyn et al., 2019; McGinnis & Ostrom, 2014). Dynamics of this framework represented not only by changes in the physical state of the product (Galychyn & Ustundag, 2017; Galychyn et al., 2019; Metzger, 2000) by also the change in its metabolism (represented by catabolic and metabolic processes between attributes) (Galychyn et al., 2019). The dimension of governmental regulation has the same meaning as in OT framework except for the compositional relationship between this dimension and ecological and economical dimensions of sustainability and spatial dimension of organic transportation (environmental policies, network effect, path dependence encompass all three dimensions while cost-saving encompasses only cost savings and user prices only governmental and economic one (Galychyn et al., 2019).

An underground dimension has been added to capture advantages of metropolicentre to hide services facilities (to be reserve space of the surface area to the resident's needs)

(Galychyn et al., 2019; Poddubny, 2008) and to revitalize the economy of TOD (increase number of working residents in the service sector and income level within TOD) (Galychyn et al., 2019) as well as increase quality of extraction/production/maintenance of spatial, physical and spatial-physical units within TOD (Galychyn et al., 2019; McGinnis & Ostrom, 2014) along with a reduction of unit numbers and distances between them to reduce outputs (solid waste and wastewater, energy for heat island effect) (Davis et al., 2006; Galychyn, 2013; Galychyn et al., 2019; McGinnis & Ostrom, 2014) and create a whole community (mini city) (Farreny et al., 2011; Lehmann, 2010). Here, new dimension built environment has been added as the surface area of TOD after space has been secured (Galychyn et al., 2019) where the decentralized pattern in the provision of services is arranged following the complexity of units (Galychyn & Ustundag, 2017; Galychyn et al., 2019; McGinnis & Ostrom, 2014) and social transformation during the lifecycle of TOD (Galychyn & Ustundag, 2017; Galychyn et al., 2019) and reduced the number of actors through spatial co-location of users and producers (residents and businesses) to facilitate innovation (Galychyn et al., 2019), and users and services (user of biogas and biogas plants) (Davis et al., 2006), combining those culturally defined units (Galychyn et al., 2019; Smith, 1980) with units characterized by experience/engagement in the relevant activity (Galychyn et al., 2019; McGinnis & Ostrom, 2014). Definition of cultural dimension does not change except its restricted by media influence on the public opinion that organic unit receives as information to generate socially optimum decision towards safety (Galychyn et al., 2019).

Public participation dimension adapted to both local and global level in a spatial sense but without underlying a transcalar sphere of influence of each actor of the public participation (Bagliani & Dansero, 2011; Galychyn et al., 2019) as well as a hierarchy of links between actors of the local participation model (Galychyn et al., 2019). Therefore, the OT framework is more suited for drawing governance models compared to HOTS. Environmental dimension defined as organic transportation metabolism (energy and materials exchange) between HOTS and ecosystems (for example; hinterlands) (Bagliani & Dansero, 2011; Galychyn et al., 2019); therefore, the most complex dimension of organic transportation limited only to the ecological footprint (Bagliani & Dansero, 2011; Farreny et al., 2011). The economic dimension of sustainability is defined by the triangle of top-tier attributes: network effect (network externalities)-user prices-cost savings restricted by historical path dependence of network that results in the profit of network industry and the increase of private R&D potential of a system (Galychyn et al., 2019). Spatial dimension serves to maintain preferential attachment, or self-organization, among the interconnected elements: occupied space, density (hierarchy of nodes), diversity (mix of land uses), design and compactness indicators (distribution of income and taxes and land lost cost subject to the zoning) under geographical constraints in the biophysical subdimension of the ecological dimension of sustainability (Bagliani & Dansero, 2011; Galychyn et al., 2019; Hansen, 1959). The intersection between social and cultural and ecological dimension represents biophysical subdimension (Bagliani & Dansero, 2011; Galychyn et al., 2019); therefore, the hierarchical relationship between dimensions as well as connections between attributes is clearly defined compared to the OT framework.

In addition, compactness indicators have been added to capture the hierarchical relationship between spatial and ecological dimensions (Galychyn et al., 2019; McGinnis & Ostrom, 2014) as well as emphasise compact multi-functional mixed-use projects (like metropolicentre) on the surface to reduce demand for residential mobility (Lehmann, 2010; Poddubny, 2008) to physically connect TOD not only with metropolicentre but also with rest of human-oriented transportation system to reduce social isolation (Galychyn et al., 2019; Farreny et al., 2011; Galychyn, 2017b). The physical dimension of organic transportation connects pedestrian-oriented design of TOD along with traffic-calming measures with the sustainable design of the metropolicentre. The surface area of an organic hub is vaguely defined here since both resident-oriented facilities and mean of physical isolation are not specified here (Galychyn et al., 2019); therefore, pedestrian-oriented design and physical design are different units of the system with physical design been a higher hierarchy (Galychyn & Ustundag, 2017; Galychyn et al., 2019).

The sustainable design of metropolicentre has not been yet conceptualized (Galychyn et al., 2019) due to having been a more sophisticated unit than spatial-physical unit (pedestrian-oriented design) (Galychyn & Ustundag, 2017; Galychyn et al., 2019; McGinnis & Ostrom, 2014). However, due to the benefit above, it should be re-conceptualized in terms of neighbourhood. The dimension represents the topological complexity of network structure to capture network growth during lifecycle and reorganization processes within networks (Galychyn et al., 2019; Galychyn, 2017a; John-Paul et al., 2017).

It is crucial to re-conceptualize the neighbourhood as a network to be able to develop the transdisciplinary and multi-dimensional concept of neighbourhood.

5 The Transdisciplinary and Multi-dimensional Healthy Neighbourhood Concept

In this section, the neighbourhood will be re-conceptualized in terms of OT and HOTS framework. According to the most recent definition of the healthy neighbourhood in the heart of the neighbourhood concept located micro-district concept (Galychyn & Ustundag, 2017; Galychyn et al., 2019; Gashenko, 2015; Zosimov, 1976). The recent study has emphasized that this concept is only based on physical and spatial considerations (Galychyn & Ustundag, 2017). As it has been noted before the hierarchical relationship between spatial and physical dimensions exhibits features of degree and eigenvector (Galychyn et al., 2019) give as a hint that neighbourhood structure (Galychyn & Ustundag, 2017; Galychyn et al., 2019) like any other social-ecological-technical system (Galychyn et al., 2019; McGinnis & Ostrom, 2014) emphasizing that traffic flow in the micro-district is in the significant part going through the Palace of Culture located in the geometrical centre of the neighbourhood (Galychyn & Ustundag, 2017; Zosimov, 1976); therefore, both traffic flow pattern and pattern of bicycle and pedestrian path have hub-and-spoke topology (John-Paul et al., 2017). Spatial and physical dimensions were emphasized only as one of the clearest examples of the node's pair. In other words, analyses of links between dimensions in both topologies have shown that both OT and HOTS framework structure through the lens of the neighbourhood are networked. The neighbourhood's structure is a fundamental part of the healthy neighbourhood subsystem (Galychyn et al., 2019; McGinnis & Ostrom, 2014; Metzger, 2000).

During the lifecycle, it subject to change following by both limitations and opportunities as a small change in instances of one of the dimensions result in changes in structural properties and network structure (Galychyn et al., 2019; Galychyn, 2017a) as a whole (Galychyn et al., 2019; McGinnis & Ostrom, 2014). Even though micro-district has means of physical separation (Galychyn & Ustundag, 2017; Farreny et al., 2011; Galychyn, 2017b), or clearly defined physical boundaries that form a Compact Island, the internal structure is complex. While the directness of transit routes is archived, service coverage leaves many opportunities for improvement (Schlossberg, 2004). Also, the disparities of density and service coverage of residential facilities can be observed (Gori et al., 2012). As a result, it leads to disparities in health with Compact Island (micro-district) (Anderson et al., 2004; Galychyn, 2017b). This means that size, (internal and external structure) (Farreny et al., 2011; Galychyn et al., 2019), physical boundaries (Farreny et al., 2011; Galychyn & Ustundag, 2017) and context (Barton et al., 2012; Farreny et al., 2011; Galychyn et al., 2019; Lehmann, 2010) should be the underlying factors for sustainable neighbourhood planning (Farreny et al., 2011; Galychyn et al., 2019). Firstly, we start from micro-level (Bagliani & Dansero, 2011; Browning & Soller, 2014; Galychyn et al., 2019), namely network structure and then step by step the other dimensions will be added in hierarchical order to suggest an updated neighbourhood concept and draw a diagram to show it graphically.

The hub-and-spoke structure has its disadvantages in terms of traffic flow in a similar way as X-shaped form of subway network (Haifa, Israel) (Galychyn, 2011): an increased burden on a central node (Palace of Culture) which have a higher demand for mobility compared to the secondary nodes (Gashenko, 2015; John-Paul et al., 2017; Zosimov, 1976), low density of areas outside of flows (Galychyn, 2011; John-Paul et al., 2017), increased distance and travel time (tiredness of residents and cost of rent for bicycle) (John-Paul et al., 2017; Galychyn, 2011; Vogel, Greiser, & Mattfeld, 2011), a necessity to plan future flow to the remote areas (Galychyn, 2011; John-Paul et al., 2017), can collapse due to the disruption of direct connections following an increase of population in the neighbourhood (Galychyn, 2011; John-Paul et al., 2017).

However, this topology has also advantaged in terms of bicycle service: efficient of distribution of flow at hub and improvement in the quality of bicycle infrastructures (Galychyn, 2011; Galychyn et al., 2019; John-Paul et al., 2017). Therefore, optimal network topology in term of traffic flows and bicycle service should be proposed. Also, duplicate flows should be avoided (Galychyn, 2011; Galychyn et al., 2019; John-Paul et al., 2017). The combined topology seems to be a right solution for traffic flow problem since it provides a best point-to-point connection to the remote areas (Galychyn, 2011; John-Paul et al., 2017) along with an increase in diversity and average coefficient of route network misalignment (directness and service coverage are close to being equal), distance in a radial direction (spokes) significantly reduced (Galychyn, 2011), flow at nodes (multi-purpose facilities such as prestige-oriented and lifestyle) (Lehmann, 2010; Galychyn, 2017b) are equal if the number of uses is equal, cost of rent for bicycle will also decrease (Galychyn et al., 2019; John-Paul et al., 2017; Vogel et al., 2011). However, this topology has gaps in terms of bicycle service: decrease in quality of bicycle infrastructures, increase in bicycle parking at multi-purpose facilities and cost to transport bicycles to the parking lots to ensure that some of the stations will not be abandoned (Galychyn, 2011; Vogel et al., 2011). In other words, high management and maintenance cost that can increase demand for cycling.

To solve this problem size (Davis et al., 2006; Galychyn & Ustundag, 2017; Galychyn et al., 2019), internal and external structure (Galychyn, 2011; Gori et al., 2012), and the physical boundary of the neighbourhood (Davis et al., 2006; Farreny et al., 2011; Galychyn & Ustundag, 2017) should be considered. There more compact and denser neighbourhood describes an amount of waste flows produced (Davis et al., 2006); therefore, neighbourhood size should be minimized to the point where the minimum area of open green spaces is secured and maximize the use of local resources (Davis et al., 2006; Farreny et al., 2011). In other words, the effective use zone of a bicycle (3 km) (Galychyn & Ustundag, 2017), the minimum zone with an availability of 1–3 local resources (sun, water, agricultural areas) to minimize ecological footprint (Farreny et al., 2011) and 'citizen spaces' (Anderson et al., 2004; Browning & Soller, 2014; Galychyn & Ustundag, 2017) and social and physical integration into the human-oriented transportation system (Galychyn & Ustundag, 2017; Galychyn et al., 2019) should be noted to determine the size of the neighbourhood. In order to realize subway-bicycle/walking option, neighbourhood radius should be less than 400 m (1/2 of TOD's radius) from the geometrical centre (space like park zone) (Galychyn et al., 2019).

The physical boundary is a mean of physical isolation like gates and fences in the gated community (Galychyn, 2017b). The protection area (trees and canals) along the road not only transform to the second line of physical isolation but also occupy 'citizen spaces' (Galychyn & Ustundag, 2017; Farreny et al., 2011; Galychyn, 2017b). In other words, the neighbourhood became physically separated from the TOD, organic hub and the rest of the city (Farreny et al., 2011; Galychyn & Ustundag, 2017; Galychyn et al., 2019). Therefore, the physical boundary should be substituted by the spatial boundary (extent of which determined by rules above). This will be discussed in more detailed when the TOD design will be drawn. The calming traffic measures are no longer required for quality of life if the above measures are followed (Farreny et al., 2011; Galychyn & Ustundag, 2017). The complicated security systems utilized in gated communities are not required (Galychyn et al., 2019; Galychyn, 2017b); due to the substitution of a boundary, the establishment of the security control unit of administration against the use of neighbourhood facilities by non-residents, and administration of each healthy neighbourhood to resolve internal conflicts; therefore, the only remaining debate about the contribution of the security cameras to obtain a healthy neighbourhood can be considered here (Galychyn et al., 2019; Galychyn, 2017b). The organic unit specified to resolve socio-environment conflicts within TOD (Galychyn et al., 2019).

Furthermore, the administration of each healthy neighbourhood within TOD has authority aimed to reduce conflicts in smaller scale within the neighbourhood (Galychyn et al., 2019). The issue here is a form of cooperation between the organic unit and administration (Galychyn et al., 2019; Galychyn, 2017b), as well as other actors at the neighbourhood level in terms of OT and HOTS frameworks (Galychyn & Ustundag, 2017; Galychyn et al., 2019). In more detail, it will be discussed during an explanation of TOD design. In gated community also exist administration, but its function limited to the security and maintenance of facilities (Galychyn, 2017b) instead of socio-environmental conflicts (Bagliani & Dansero, 2011; Galychyn et al., 2019).

According to the combined topology (Galychyn, 2011) administration will be divided into three interconnected units placed in the form of a triangle (Galychyn, 2011; Galychyn et al., 2019): security control (Lehmann, 2010; Galychyn et al., 2019; Galychyn & Ustundag, 2017; Galychyn, 2017b), manager of resources (spatial and physical units) (Farreny et al., 2011; Galychyn, 2011; Galychyn et al., 2019) and creativity planning (Galychyn et al., 2019; Cvetinovic, Nedovic-Budic, & Bolay, 2017; Galychyn & Ustundag, 2017; da Costa Lobo & Ustundag, 2007). Security control in terms of the neighbourhood does not mean prevention of trespassing of residents of other neighbourhoods within TOD but rather outside from TOD (Galychyn et al., 2019; Galychyn, 2017b; Browning & Soller, 2014). The activity patterns of residents usually go beyond the spatial boundary of neighbourhood resulted in the formation of ecological communities far beyond spatial boundaries (Browning & Soller, 2014). As a result, residents of one neighbourhood are using services of other community in proximity up to 2600 m from the residence location (Barton et al., 2012). The reasons behind this issue are not the only socio-economic status of residents but also a mix of land uses (lack of strong class-integrated institutions, organizations and amenities within the neighbourhood) (Browning & Soller, 2014) as well as facilities shared by all-income groups within the neighbourhood (school, shopping malls) (Browning & Soller, 2014; Galychyn & Ustundag, 2017; Lehmann, 2010), racially divided common facilities (Farreny et al., 2011) and absence of mixed age and demand related facilities (Lehmann, 2010; Galychyn, 2017b). As a result of this process, 'week' social ties within the community (sharing of routing activity spaces) (Browning & Soller, 2014) which eventually contribute to the low levels of public participation mechanism (Browning & Soller, 2014); therefore, the neighbourhood became socially and physically separated from the immediate surroundings (Browning & Soller, 2014; Farreny et al., 2011). The security control unit of administration is enacted to prohibit the use of facilities within the community for non-residents, which in turn contribute to the integration of neighbourhood structure and social network into an organic hub (Galychyn et al., 2019), including TOD (Galychyn & Ustundag, 2017; Galychyn et al., 2019; Lehmann, 2010).

The resource manager will deal with the management of natural resources (Farreny et al., 2011) and maintenance of spatial units within neighbourhood during lifecycle (Galychyn & Ustundag, 2017; Galychyn et al., 2019; McGinnis & Ostrom, 2014) accompanied by changes in parameters of spatial units (occupied space, density, diversity, design and compactness indicators) (Galychyn et al., 2019) and economic parameters (network effect and path dependence) (Galychyn et al., 2019) of the system as a whole subject to governmental resource policies (Bagliani & Dansero, 2011; Galychyn et al., 2019; Victoria Transport Policy Institute, 2015) and maintenance control performed by organic unit (for example; maintenance check of the biogas plant, rainwater and greywater reuse systems) (Davis et al., 2006; Galychyn et al., 2019; Ustundag, 2012). The triangle of top-tier attributes restricted by historical path dependence (Galychyn et al., 2019; Galychyn, 2017a; John-Paul et al., 2017) can be applied to the network structure of neighbourhood and accompanied by management and maintenance data collected by resource manager can be used to prevent the decline of the neighbourhood due to decrease of demand for community services (Farreny et al., 2011; Galychyn et al., 2019; Metzger, 2000). This unit is collaborating with planners (third party) on issues regarding management of spatial units (Galychyn & Ustundag, 2017; Galychyn et al., 2019; McGinnis & Ostrom, 2014) and with scientists (third party) on an issue regarding the maintenance of physical units (Galychyn & Ustundag, 2017; Galychyn et al., 2019; McGinnis & Ostrom, 2014). Then, information redirected to the creative planning unit (producer of knowledge) (Galychyn et al., 2019; Cvetinovic et al., 2017; McGinnis & Ostrom, 2014; da Costa Lobo & Ustundag, 2007) located with the proximity of residential housing generate decision to socially optimum decision together with residents (consumer) (Galychyn & Ustundag, 2017; Galychyn et al., 2019; McGinnis & Ostrom, 2014) following cultural, social, the standard of living, education, activity, equal opportunity and economic barriers made spatial context (socio-cultural and physical contexts combined) (Cvetinovic et al., 2017; Farreny et al., 2011; Kennedy, Pincetl, & Bunje, 2011; McGinnis & Ostrom, 2014). The group of adults (consumer) are trained by the creative planning unit (producer) through informal sessions to balance the innovation process (Galychyn et al., 2019; Browning & Soller, 2014; da Costa Lobo & Ustundag, 2007).

Later, the creative planning unit sends data to the organic unit to generate socially optimum decision towards safety (Galychyn & Ustundag, 2017; Galychyn et al., 2019; Cvetinovic et al., 2017; McGinnis & Ostrom, 2014; da Costa Lobo & Ustundag, 2007), perform maintenance control for 3–4 healthy neighbourhood and devise regulations regarding management and maintenance of entire TOD (Davis et al., 2006; Galychyn & Ustundag, 2017; Galychyn et al., 2019; Ustundag, 2012).

This structure elaborated not only to integrate neighbourhood structure with the activity-space pattern but because the presence of strong institutions (creativity planning unit) (Galychyn et al., 2019; Browning & Soller, 2014; da Costa Lobo & Ustundag, 2007) and participation in a community-based organization (training of adult groups) will reinforce the social ties even among non-participants (less dense and structurally embedded eco-networks) (Browning & Soller, 2014). The community-based organization in micro-district is the Palace of Culture (Browning & Soller, 2014; Gashenko, 2015; Zosimov, 1976). For the reasons described above its existence is crucial if it divided into few lesser units dispersed through the neighbourhood (Galychyn, 2011; Galychyn et al., 2019; John-Paul et al., 2017), namely one per residential group, following the combined topology of neighbourhood structure (Galychyn, 2011); therefore, it is clusters of activity locations form culturally defined units (activity spaces tied by informal trust and social control, capital-intensive industries based on local resources such as sun, water and energy) (Browning & Soller, 2014; Farreny et al., 2011; Galychyn et al., 2019; Smith, 1980) to cut dependence on external resources (Davis et al., 2006; Farreny et al., 2011; McGinnis & Ostrom, 2014), co-location of education institutions with those industries, mini Palace of Culture (Galychyn et al., 2019) for production of spatial (roof vegetable gardens, communal parking areas) (Farreny et al., 2011; Galychyn & Ustundag, 2017; Galychyn et al., 2019; McGinnis & Ostrom, 2014) and physical units (resident-oriented facilities including ones for sport and recreation of residents within culturally defined unit) (Galychyn & Ustundag, 2017; Galychyn et al., 2019; McGinnis & Ostrom, 2014; Galychyn, 2017b) restricted by cultural, social and economic barriers and local resources policies (Farreny et al., 2011; Galychyn et al., 2019; McGinnis & Ostrom, 2014).

Each of mini Palace of Culture forward data about delivered in collaboration units to the resource management unit to be used for initiation of the process mentioned above (Farreny et al., 2011; Galychyn & Ustundag, 2017; Gashenko, 2015; Zosimov, 1976). Provision of utilities is effectuated by transit company to all healthy neighbourhoods (Galychyn et al., 2019) as well as management of related businesses (retail such as shopping mall Galychyn & Ustundag, 2017; Galychyn et al., 2019), mixed-use and age housing types, prices and ownerships forms (Galychyn & Ustundag, 2017; Lehmann, 2010) located within each neighbourhood to ensure that company can strive to avoid adverse network effects (Galychyn et al., 2019) and to break from path dependence-related restrictions (Galychyn et al., 2019; Galychyn, 2017a). Maintenance of housing or spatial units (Galychyn & Ustundag, 2017; Galychyn et al., 2019; McGinnis & Ostrom, 2014), namely physical units integrated with social transformations (modified perception of the physical unit following affordability, equality security and social design) (Galychyn & Ustundag, 2017; Galychyn et al., 2019; McGinnis & Ostrom, 2014) effectuated by a resource manager (Farreny et al., 2011) with the related cost-shared by community residents (Lehmann, 2010; Galychyn et al., 2019; Galychyn, 2017b).

Although resource manager is a unit of neighbourhood administration (Farreny et al., 2011; Galychyn & Ustundag, 2017; Galychyn et al., 2019; McGinnis & Ostrom, 2014), it is also one of the transit company businesses (Farreny et al., 2011; Galychyn et al., 2019); therefore, the function of resource manager can be referred to as the function of transit company, or a private service provided by the transit company to the residents (Galychyn & Ustundag, 2017; Galychyn et al., 2019; Farreny et al., 2011; Galychyn, 2017b). In order to re-conceptualize a sustainable design of the metropolicentre (Galychyn et al., 2019) in terms of the neighbourhood first, it is essential to look into the diagram in Fig. 2 to understand the composition of the metropolicentre.

Metropolicentre is the multi-functional and mixed-use project that consists of four levels (Galychyn et al., 2019; Poddubny, 2008). First, level-pedestrian crossings, services for residents (retail services, public facilities), private companies (real estate, retail) emerged as a result of diversification of private railway companies into related businesses (Galychyn & Ustundag, 2017; Galychyn et al., 2019; Poddubny, 2008). Underground pedestrian crossings are connected to the lobby of subway station spatially connected to the surface level (TOD). The second level is tunnel aimed at existing automobile users and BRT. The third floor incorporates underground parking and utilities (such as power stations, boiler room and ventilation system) second and third floors can be spatially connected to similar use floors of

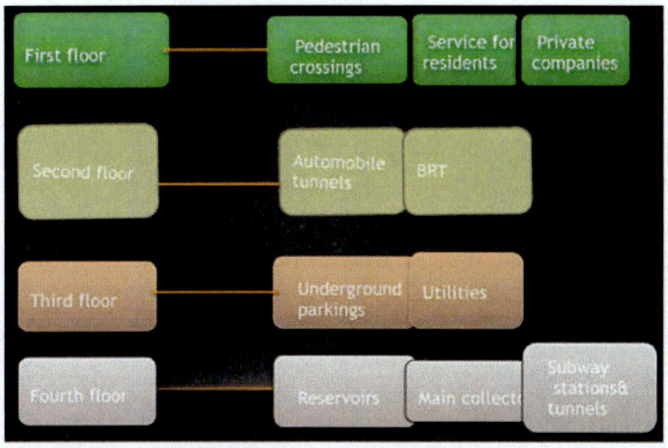

Fig. 2 Metropolicentre (underground level). (*Source* Developed by Galychyn and colleagues, 2019)

other metropolicentres (Galychyn et al., 2019; Poddubny, 2008). The deepest floor incorporates reservoirs, central collector for utilities; deep level placed subway stations and tunnel. Those floors are owned and maintained by other businesses of private railway companies (Galychyn & Ustundag, 2017; Galychyn et al., 2019; Poddubny, 2008), as shown in Fig. 2. The benefits of underground dimension are evident from the image above.

The surface level of the neighbourhood will include the only mixed-use housing with roof vegetable gardens (Galychyn et al., 2019; Lehmann, 2010; Poddubny, 2008), soft transportation network (bicycle and pedestrian paths) (Galychyn & Ustundag, 2017; Galychyn et al., 2019), four culturally defined units (schools, industries and mini Palaces of Culture) (Galychyn & Ustundag, 2017; Smith, 1980; Zosimov, 1976) and a triangle of three administration units discussed above (Cvetinovic et al., 2017; Galychyn et al., 2019). Therefore, space for greening is secured (Farreny et al., 2011), size of the neighbourhood can be reduced by restricted spatial boundary guidelines (effective use zone of the bicycle (Galychyn & Ustundag, 2017), local resources (Farreny et al., 2011; da Costa Lobo & Ustundag, 2007), the green area per person (Farreny et al., 2011; Galychyn & Ustundag, 2017), public-soft transportation integration (Galychyn & Ustundag, 2017; Galychyn et al., 2019). Other benefits can be understood as both reduction of outputs of metabolism process: waste to prevent overcapacity of micro biogas plants and related maintenance cost (Farreny et al., 2011; da Costa Lobo & Ustundag, 2007) and energy to minimize Heat Island Effect (Codoban & Kennedy, 2008). Also, the realization of combined topology with minimum distances/travel time between physical/spatial units (Galychyn & Ustundag, 2017; Galychyn et al., 2019; McGinnis & Ostrom, 2014), where less intermediate nodes will be required (Galychyn, 2011); therefore, it will allow both to reduce traffic flows and increase the quality of management and maintenance of bicycle infrastructure (Galychyn & Ustundag, 2017; Vogel et al., 2011) accompanied by the reduction of costs, and reduction of the number of parking lots required for bicycles (Galychyn & Ustundag, 2017; Vogel et al., 2011).

This will allow utilizing the space in culturally defined units (Galychyn & Ustundag, 2017; Galychyn et al., 2019; Smith, 1980; Zosimov, 1976) for the generation of additional jobs while the part of the service sector will be relocated to the first level of metropolicentre (Galychyn et al., 2019; Poddubny, 2008). More compact mixed-use form compared to the neighbourhood will be archived (Farreny et al., 2011; Galychyn & Ustundag, 2017; Lehmann, 2010); therefore, gain in maintenance cost and less local resources (Farreny et al., 2011; da Costa Lobo & Ustundag, 2007) will be utilized to increase average salary rate for TOD residents (including neighbourhood residents), increase the profit of transit company accompanied by co-location of related businesses (Galychyn et al., 2019) to improve quality of delivery of physical and spatial units (Galychyn et al., 2019; McGinnis & Ostrom, 2014) and, simultaneously, to increase employment rate within the spatial boundary of the neighbourhood (Galychyn et al., 2019; Tokyo Metropolitan Government, 2007; Browning & Soller, 2014; Barton et al., 2012). Last but not least, the increase in overlap of routine activities among residents can be observed that contribute to the intense public participation (Browning & Soller, 2014), and therefore, socially connect a neighbourhood with the other neighbourhoods (social inclusion) (Browning & Soller, 2014; Farreny et al., 2011). In such a neighbourhood, citizens are no longer shared their activities on the street with the vehicles of other residents of the same neighbourhood to shape mini Palaces of Culture (Galychyn & Ustundag, 2017). The only car parking is moved to underground space (Galychyn & Ustundag, 2017; Galychyn et al., 2019; Poddubny, 2008); therefore, this neighbourhood is the first one to prioritize the use of soft transportation modes (Lehmann, 2010).

The spatial-physical units (Galychyn & Ustundag, 2017; Galychyn et al., 2019; McGinnis & Ostrom, 2014) are not observable at the neighbourhood level; therefore, that issue will be discussed in the TOD section.

To sum all up, the improved neighbourhood will be defined as socio-spatial network shaped by bicycle and pedestrian traffic among interconnected units of characterized by the hierarchy of nodes in terms of the units (physical, spatial and spatial-physical), actors and shared activities in spatially delimited settings to maintain its viability. Consequently, the neighbourhood will be classified with prioritization of factors (second-tier attributes) influenced changes in the spatial boundary of the neighbourhood (spatial dimension), built environment dimension (surface level) following an underground dimension and hierarchy of nodes of the neighbourhood as a network in terms of the units, actors, and shared activities (Fig. 3). In this article, it is proposed to exclude the traffic-calming measures from the physical dimension due to the reasons outlined above. Security systems will be substituted with the security cameras. The physical boundary will be excluded from the physical dimension and added to the built environment as the spatial boundary (built environment dimension). The indicators: open green space per person, presence of one or more local resources (sun, water and agricultural areas), preferred transportation options (subway-bicycle/walking, BRT (or bus)-bicycle/walking and options that scored less in terms of the travel distance, pollution, crashes, and the deficit of space) (Galychyn et al., 2019); therefore, it would also be added to the built environment dimension due to integration of soft transformation mode into public transportation network can be archived only by a combination of

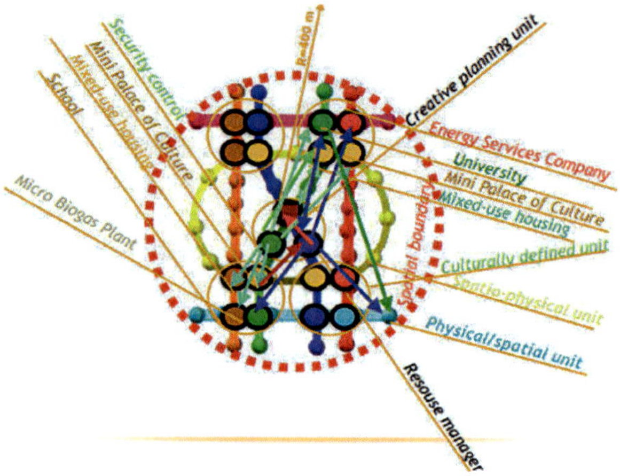

Fig. 3 Conceptual diagram of neighbourhood. (*Source* Developed by Galychyn and colleagues, 2019)

physical and social transformations (spatial transformations that occur at TOD level) modified by the influence of the nearest subway station (biased to the types and number of underground facilities adjusted to the station) (Galychyn & Ustundag, 2017; Galychyn et al., 2019; Poddubny, 2008). The underground dimension emphasizes the importance of the mix of facilities (social and physical infrastructure) (Galychyn et al., 2019; Poddubny, 2008), along with the total catchment area of overlapping services in the underground crossings attached to the subway stations (Galychyn et al., 2019; Poddubny, 2008).

Another second-tier attribute (Galychyn et al., 2019; McGinnis & Ostrom, 2014) of built environment dimension will be the type of community centre by function (school, local administration or cultural centre such as the Palace of Culture) (Browning & Soller, 2014; Galychyn & Ustundag, 2017; Galychyn et al., 2019). In this aspect, the degree (Galychyn et al., 2019; Galychyn, 2017a) of the community centre and the culturally defined unit can be measured (by the number of physical and spatial units included in the community centre) (Galychyn et al., 2019; McGinnis & Ostrom, 2014). The coefficient of route network misalignment (Galychyn, 2011) should be included to determine if directness is balanced with service coverage (travel time is within 15 min or more) (Galychyn, 2011; Gori et al., 2012). The type of network should be defined following a pattern of bicycle/pedestrian paths and the location of the community centre (Galychyn, 2011; Galychyn et al., 2019; John-Paul et al., 2017). The degree of the community centre and a culturally defined unit can also be measured by the number of actors included and the number of essential links presented (user-producer and user services) (Davis et al., 2006; Galychyn et al., 2019; McGinnis & Ostrom, 2014). Also, by measuring several overlapping routine activities at the community centre, it will be possible to determine is network a close to the hub-and-spoke model (John-Paul et al., 2017; Galychyn, 2011; Vogel et al., 2011)-related traffic flow disadvantages or more evolved structure-like combined typology (Galychyn, 2011).

Lastly, it would be essential to emphasize that due 'networked' nature of neighbourhood the structural properties of transportation network also can be applied to the neighbourhood (Galychyn et al., 2019; Galychyn, 2017a; Galychyn, 2011). The network size (Galychyn et al., 2019; Galychyn, 2017a) can help to determine if the diameter of the neighbourhood surpass the spatial boundary. If that the case the size can be reduced till, then it will be able to adjust to the spatial limits. In another case, the neighbourhood consumes more than required for it to maintain viability (Galychyn & Ustundag, 2017; Galychyn et al., 2019; Farreny et al., 2011; Browning & Soller, 2014; da Costa Lobo & Ustundag, 2007). If the medium load of all links is less than capacity, the network effect tends to be negative (Galychyn et al., 2019; Galychyn, 2017a); therefore, the neighbourhood lifecycle represents the decline stage (Metzger, 2000) due to the market failure (Galychyn et al., 2019). The described above indicators of a healthy neighbourhood are summarized in Table 1. For example, in the left column, the network size category indicates that the neighbourhood characterized by a deficit of resources to support its population compared to the right column.

6 Results

In this article, a 'healthy neighbourhood' concept was developed, and the indicator to measure the pattern of the 'healthy neighbourhood' was proposed. This concept was different from the sustainable neighbourhood in terms of the boundary (Davis et al., 2006; Farreny et al., 2011; Galychyn et al., 2019), size (Davis et al., 2006; Galychyn & Ustundag, 2017; Galychyn et al., 2019) and pattern (Galychyn, 2011; Gori et al., 2012) (social, actor and transportation networks). Namely, the physical boundary was substituted by the spatial one to connect the neighbourhood with the rest of the city physically and secure the area for a 'citizen spaces' (Galychyn & Ustundag, 2017; Farreny et al., 2011; Galychyn, 2017b). The size (Galychyn et al., 2019; Poddubny, 2008) is not an overlapping catchment area for facilities of the neighbourhood; therefore, should incorporate the minimum area of open green spaces terms of EU standard (9 m^2 per capita) (Farreny et al., 2011), within the spatial boundary of the neighbourhood, the minimum zone with an availability of 1–3 local resources (sun, water, agricultural areas) (Davis et al., 2006; Farreny et al., 2011) to minimize the ecological footprint and 'citizen spaces' after the physical boundary was substituted by spatial (Farreny et al., 2011; Galychyn & Ustundag, 2017; Galychyn et al., 2019). Besides, the neighbourhood should be situated within an effective use zone of a bicycle ($R = 3$ km^2) as well as TOD ($R = 400$–800 m) in order to capture the benefits of both systems and to realize the subway-bicycle/walking option (Galychyn et al., 2019). The pattern will be optimized by substituting the hub-and-spoke structure of transportation network with the combined one to increased burden on a central node (community centre) (Gashenko, 2015; John-Paul et al., 2017; Zosimov, 1976), low density of areas outside the flows (Galychyn, 2011; John-Paul et al., 2017), accompanied by necessity to plan future flow to those areas (Galychyn, 2011; John-Paul et al., 2017) passenger's disadvantage (tiredness of residents and cost of rent for bicycle) (John-Paul et al., 2017; Galychyn, 2011; Vogel et al., 2011), high vulnerability to disruptions (Galychyn, 2011; John-Paul et al., 2017). However, to improve in quality of bicycle infrastructures, along with the decrease bicycle parking at multi-purpose facilities and remove the requirement to transport bicycles to the parking lots (Galychyn et al., 2019; John-Paul et al., 2017; Vogel et al., 2011) will only work if the combined structure of transportation network is supplied by the size delimited by the rules above (Davis et al., 2006; Galychyn & Ustundag, 2017; Galychyn et al., 2019).

The issues of contemporary neighbourhoods such as out-migration, the deficit of resources and decline of the neighbourhood in economic terms can be combatted by enacting of the specific units within the administration (the central node of actor-network) (Galychyn, 2011): security control (Lehmann, 2010; Galychyn et al., 2019; Galychyn, 2017b), manager of resources (Farreny et al., 2011; Galychyn, 2011; Galychyn et al., 2019) and creative planning (Galychyn et al., 2019; Cvetinovic et al., 2017; da Costa Lobo & Ustundag, 2007). The security unit is aimed to prohibit the use of facilities within the community for non-residents (Galychyn & Ustundag, 2017; Galychyn et al., 2019; Lehmann, 2010). The resource manager unit is aimed to handle the management of natural resources (Farreny et al., 2011) and maintenance of spatial units within the neighbourhood during the lifecycle (Galychyn & Ustundag, 2017; Galychyn et al., 2019; McGinnis & Ostrom, 2014).

The creative planning unit (producer of knowledge) is assigned a function of a mediator between the residents and organic unit1 (Galychyn & Ustundag, 2017; Cvetinovic et al., 2017; McGinnis & Ostrom, 2014; da Costa Lobo & Ustundag, 2007). Also, the Palace of Culture under the combined structure of the transportation network is divided into few lesser units (culturally defined units) (Galychyn, 2011; Galychyn et al., 2019; John-Paul et al., 2017) shaped from the clusters of activity locations dispersed throughout the neighbourhood. Those units incorporate activity spaces tied by informal trust and social control, capital-intensive industries based on local resources (Browning & Soller, 2014; Farreny et al., 2011; Galychyn et al., 2019; Smith,

Table 1 Features of the healthy neighbourhood (*Source* Developed by Galychyn and colleagues, 2019)

Indicators of healthy neighbourhood			
Security systems	Gates and fences, security guards, CCTV	Gates with security guards	CCTV (security cameras)
Physical isolation	Highways and green areas	Green areas (forests, parks) or highways	–
Open green space per person	Less than normative	Near normative	More than normative
Local resources (sun, water, agricultural areas)	1	2	3
Preferred transportation option	Soft transportation modes absent	BRT (or bus)-bicycle/walking	Subway-bicycle/walking
A mix of underground facilities	Social infrastructure	Physical infrastructure	Social and physical infrastructure
The catchment area of underground facilities	The catchment area of facilities outside of neighbourhood zone	Service around half of the neighbourhood area	Service entire territory of neighbourhood
The function of the community centre	School	Local administration	Cultural centre
Degree of a community centre	Physical/or spatial unit	Physical and spatial units	Physical, spatial and spatial-physical units
Coefficient of route network misalignment	Lower than 1	Near 1	More than 1
Network structure	Hub-and-spoke	Transitory	Combined
Degree of the community centre (number of links)	Does not exist	User-producer or user-service	User-producer and user-service
Number of overlapping routine activities at the community centre	Mono-functional	Few uses	Multi-functional
Hierarchy of links (betweenness)	Connected with one crucial node (consist of two or more actors)	Community centre connected with few important nodes	The community centre is connected with all critical nodes within the neighbourhood
Network density (closeness)	Low	Medium	High
Eigenvector (agglomeration)	Mono-link between the community centre and closest complex unit	Few links between the community centre and closest complex unit	All-paths connected between the community centre and closest complex unit
Modes and terminals	Private only	Public–private (integrated)	Public-soft (integrated)
Network size	More than a spatial boundary	Near the spatial boundary	Less the spatial boundary
Type of traffic	Continuous	Divided	Mixed
Volume and direction	Low (mixed traffic and dispersed destinations)	Medium without integration with rapid transit	Highly integrated
Load and capacity	Load less than capacity	Load more than capacity	Load = capacity

1980) to cut dependence on external resources (Davis et al., 2006; Farreny et al., 2011; McGinnis & Ostrom, 2014), co-location of education institutions with those industries, mini Palaces of Culture (Galychyn et al., 2019) for production of spatial (Farreny et al., 2011; Galychyn & Ustundag, 2017; Galychyn et al., 2019; McGinnis & Ostrom, 2014) and physical units (Galychyn & Ustundag, 2017; Galychyn et al., 2019; McGinnis & Ostrom, 2014; Galychyn, 2017b).

The traffic-calming measures were excluded (Farreny et al., 2011; Galychyn & Ustundag, 2017) while security systems were substituted with the security cameras (Galychyn et al., 2019; Galychyn, 2017b) due to the enactment of the spatial boundary. The indicators: open green space per capita (Farreny et al., 2011), presence of one or more local resources (sun, water and agricultural areas) (Davis et al., 2006; Farreny et al., 2011), preferred transportation options (Galychyn et al., 2019) have been introduced in terms of neighbourhood's size (Davis et al., 2006; Galychyn & Ustundag, 2017; Galychyn et al., 2019).

The mix of facilities (physical, social infrastructure), along with the total catchment area of overlapping services in the underground crossings attached to the subway station was added to capture the benefits of underground space in terms of 'healthy neighbourhood' (Galychyn et al., 2019; Poddubny, 2008). The complexity of units, actors and the number of essential links presented (user-producer and user services) (Davis et al., 2006; Galychyn et al., 2019; McGinnis & Ostrom, 2014) were introduced to measure the degree of the community centre (central node of actor, social and transportation networks) (Galychyn et al., 2019; Galychyn, 2017a). Many overlapping routine activities at the community centre were proposed in order to determine the network type (John-Paul et al., 2017; Galychyn, 2011; Vogel et al., 2011). The coefficient of route network misalignment (Galychyn, 2011) should be incorporated to determine if travel time is within 15 min or more (Galychyn, 2011; Gori et al., 2012). Also, the network size (Galychyn et al., 2019; Galychyn, 2017a) was needed to determine if the diameter of the neighbourhood surpass the spatial boundary. Finally, the load of links was presented to determine the stage of the neighbourhood's lifecycle (Galychyn et al., 2019; Metzger, 2000; Galychyn, 2017a). The improved neighbourhood concept introduced in Sect. 4 can transform the neighbourhood into an attraction pole to generate multiply origins and destinations from neighbouring areas to reduce isolation. This aspect is also connected to the neighbourhood's lifecycle (Galychyn et al., 2019; Metzger, 2000; Ustundag, 2012) since it facilitates the transition from decline stagnation to the renewal. From that point, out-migration problems should become a priority of the administration.

After the physical boundary is removed together with the gates and fences, the safe pedestrian-friendly environments can be established with the principles of pedestrian-oriented design implemented. The optimized network in terms of operators benefit (construction and maintenance costs) and passenger's advantage (distance and time) is archived due to the combined structure of transportation network (within the spatial boundary) adjusted to the size of the neighbourhood delimited by the rules mentioned above. In other words, this factor helps to not only to improve the network effect (and to avoid market failure) for producers but also to provide reliable point-to-point connections for users of public transport. To this, the task contributes the security control unit that not only prevents the services (like schools) from been abandoned by users, but also facilitate the maintenance of physical units by the resource manager; therefore, the quality and utilization of services improves gradually.

That means the low levels of public participation are prevented by reinforcing the social ties within the community (sharing of routing activity spaces) (Browning & Soller, 2014); therefore, the out-migration can be prevented within the long-term perspective. The creativity planning units (Galychyn et al., 2019; Browning & Soller, 2014; da Costa Lobo & Ustundag, 2007) and participation in community-based organizations (training of adult groups) will reinforce the social ties even among non-participants (Browning & Soller, 2014) also contributes to the reinforcement of the social ties among the residents by the training of the adults through informal sessions; therefore, the daily routine activities do not go beyond the spatial boundaries. The resource manager unit in cope with the rules regarding the neighbourhood's size will be able to cut the dependence on the external resources that contribute to the reduction of waste to reduce the environmental footprint and the energy to minimize Heat Island Effect (Codoban & Kennedy, 2008) to generate economic benefits (profit from increased closed cycles such as organic waste to electricity).

Finally, the culturally defined unit will provide the opportunity for the reduced the number of actors through spatial co-location of users and producers and to render multi-functional mixed-use projects that not only contribute to the reduction of the traffic demand but also contribute to the effectiveness of management of the spatial unit and the maintenance of the spatial units and also contribute to the physical integration of the neighbourhood to the surrounding context. Lastly, those units also provide a foundation to archive in the long-term an equal redistribution of growth within the spatial boundary of the neighbourhood. In short, the newly introduced network model of healthy neighbourhood contributes to the development of a comprehensive sustainable resource management plans, focusing on human, physical and natural resources of neighbourhoods.

7 Conclusion

In this study, the transdisciplinary and multi-dimensional 'healthy neighbourhood' concept has been developed that was based on the proposal in regards to the boundary (Davis et al., 2006; Farreny et al., 2011; Galychyn & Ustundag, 2017), size (Davis et al., 2006; Galychyn & Ustundag, 2017; Galychyn et al., 2019) and pattern (Galychyn, 2011; Gori et al., 2012) (social, actor and transportation networks). Also, the indicators to measure those second-tier attributes have been proposed.

However, this concept should be applied to the case study to see how it works on the ground, especially in terms of the operation during the lifecycle. Most studies tend to ignore the importance of lifecycle in evaluating the effectiveness of the policies proposed (Davis et al., 2006; Farreny et al., 2011; Metzger, 2000). Besides, the neighbourhood's structure is more complicated than it has been initially thinking (Anderson et al., 2004; Gashenko, 2015; Zosimov, 1976). Firstly, the neighbourhood like HOTS is a social-ecological-technical system characterized by a natural dynamic of ecosystems and dynamic process of complex technological systems (Galychyn & Ustundag, 2017; Galychyn et al., 2019; McGinnis & Ostrom, 2014); therefore, been a multi-layered network (Kivelä et al., 2014), it allows many debates from experts of different fields to analyse its structure and to capture all the processes that occur within them.

The urban planners, civil engineers, architects, physicists, biologists, programmers and geographers might be only the significant specialists involved into the study of the complex networks, even though the full-fledged theory to capture topological complexity of the neighbourhood is still have not been developed (Anderson et al., 2004; Galychyn & Ustundag, 2017; Kivelä et al., 2014; Lehmann, 2010).

Secondly, the neighbourhood is not an isolated system but rather a structural element of the human-oriented transportation system (Galychyn & Ustundag, 2017). This element has many connections with other systems through the social, actor, transportation, Internet, natural and many more (Law, 2009); therefore, this complex task might be left for the future generations since even, at present, the rudimentary theory that takes into account those aspects in terms of the neighbourhood does not exist (Galychyn & Ustundag, 2017). The concept proposed in this study use the technical system to unfold the internal mechanism of a healthy neighbourhood. However, the concept of healthy neighbourhoods involves socio-economic context, the concentration of population, sense of place, community and solidarity. Future studies may accommodate this complexity to develop a universal healthy neighbourhood concept. The proposed multi-criteria and system-based perspective for studying healthy neighbourhoods can assist city managers and policymakers by providing a comprehensive methodology for the development of strategies for sustainable resource management. Therefore, the sustainable management of healthy neighbourhoods is crucial for ensuring the long-term sustainability of society as a whole.

Acknowledgements The authors are wishing to thank the professor Dr. Michelangelo Russo, The University of Naples Federico II, for allowing him to attend the valuable seminar regarding the SES (Socio-Ecological Systems). This seminar has helped an author to develop his main ideas.

References

Anderson, N. B., Bulatao, R. A., & Cohen, B. (2004). *Critical Perspectives on Racial and Ethnic Differences in Health in Late Life. Critical Perspectives on Racial and Ethnic Differences in Health in Late Life*. https://doi.org/10.17226/11086

Bagliani, M., & Dansero, E. (2011). Politics for environment: From nature to the territory. *De Agostini Scuola SpA*, 3–26.

Barton, H., Horswell, M., & Millar, P. (2012). Neighbourhood accessibility and active Travel. In J. Piggin, L. Mansfield & M. Weed (Eds.), *Routledge handbook of physical activity policy and practice*. https://doi.org/10.4324/9781315672779

Browning, C., & Soller, B. (2014). Moving beyond neighborhood: Activity spaces and ecological networks as contexts for youth development. *Cityscape*, 165–196.

Calthorpe, P., Corbett, M., Duany, A., Moule, E., Plater-Zyberk, E., Polyzoides, S., … Weissman, S. (2008). *The Ahwahnee Principles for Resource-Efficient Communities* (pp. 3–26). https://www.lgc.org/wordpress/docs/ahwahnee/ahwahnee_principles.pdf

Codoban, N., & Kennedy, C. A. (2008). Metabolism of neighborhoods. *Journal of Urban Planning and Develoment*, (March), 21–31. https://pdfs.semanticscholar.org/1b07/eda8b95925e0d0e5aafe08281e608d221624.pdf

Cvetinovic, M., Nedovic-Budic, Z., & Bolay, J. C. (2017). Decoding urban development dynamics through actor-network methodological approach. *Geoforum*, *82*(June 2016), 141–157. https://doi.org/10.1016/j.geoforum.2017.03.010

da Costa Lobo, M., & Ustundag, K. (2007). Perception on city planning. *43rd ISOCARP Congress*, 1–7.

Davis, M., Polit, J., & Lamour, M. (2006). Social urban metabolism strategies (SUMS) for Cities. *Procedia Environmental Sciences*, 309–327.

Eremin, A., Ilyin, V., & Lomova, N. (2014). *Actual problems of theory and practice of the Urban governance: What to do with micro-districts?*

Farreny, R., Oliver-solà, J., Marc, M., Escribà, E., Gabarrell, X., & Rieradevall, J. (2011). Transition towards sustainable cities: Opportunities, constraints, and strategies in planning. A neighbourhood ecodesign case study in Barcelona. https://doi.org/10.1068/a43551

Galychyn, O. (2011). Analysis of planning diagrams of underground railroads in large cities of the world. *International Journal of Scientific and Engineering Research.*

Galychyn, O. (2013). Project of subway station "Yuzhnaya" (south station) in Kharkiv. *International Journal of Scientific & Engineering Research*, *4*(5). https://www.ijser.org

Galychyn, O. (2017a). *Embracing the topological complexity of transportation networks*. www.ijsrm.humanjournals.comwww.ijsrm.humanjournals.com

Galychyn, O. (2017b). Physical features of gated communities in Japan and USA : A comparative analysis. *International Journal of Science and Research Methodology* (2). https://ijsrm.humanjournals.com/wp-content/uploads/2018/01/7.Oleksandr-Galychyn.pdf

Galychyn, O., & Ustundag, K. (2017). Organic urbanism: Human-oriented design for metropolises. *Science for Environment Policy, 37*, 396–407. https://doi.org/10.1016/j.proenv.2017.03.005

Galychyn, O., Ahmeti, S., & Ustundag, K. (2019). Organic transportation networks: Human-oriented renewal of modern Megapolises. *The Academic Research Community Publication*, *3*(2), 294. https://doi.org/10.21625/archive.v3i2.617

Gashenko, A. (2015). *Formation of 'Microdistricts ' in Novosibirsk in 1950–2010s*. https://archvuz.ru/2015_2/13

Glazyeva, L., Sukhacheva, V., & Firsova, N. (2014). Actual problems of development and improvement of voronezh residential micro-districts. *Engineering Systems and Structures*, 54–59.

Gori, S., Nigro, M., & Petrelli, M. (2012). The impact of land use characteristics for sustainable mobility: The case study of Rome. *European Transport Research Review, 4*(3), 153–166. https://doi.org/10.1007/s12544-012-0077-6

Hansen, W. G. (1959). How accessibility shapes land use. *Journal of the American Institute of Planners*, 2–5.

John-Paul, R., Claude, C., & Brian, S. (2017). *Geography of transportation networks*. Routledge. https://people.hofstra.edu/geotrans/eng/ch1en/conc1en/ch1c4en.html

Kennedy, C., Pincetl, S., & Bunje, P. (2011). The study of urban metabolism and its applications to urban planning and design. *Environmental Pollution* (Elsevier Ltd.). https://doi.org/10.1016/j.envpol.2010.10.022

Kivelä, M., Arenas, A., Barthelemy, M., Gleeson, J. P., Moreno, Y., & Porter, M. A. (2014). Multilayer networks. *Journal of Complex Networks, 2*(3), 4–17. https://doi.org/10.1093/comnet/cnu016

Law, J. (2009). Actor network theory and material semiotics. *The New Blackwell Companion to Social Theory* (April), 141–159. https://doi.org/10.1002/9781444304992.ch7

Lehmann, S. (2010). Green urbanism: Formulating a series of holistic principles. *OpenEdition* (3). Retrieved from https://journals.openedition.org/sapiens/1057

McGinnis, M. D., & Ostrom, E. (2014). Social-ecological system framework: Initial changes and continuing challenges. *Ecology and Society, 19*(2). https://doi.org/10.5751/ES-06387-190230

Metzger, J. T. (2000). Planned abandonment: The neighborhood life-cycle theory and national urban policy. *Housing Policy Debate, 11*(1), 7–40. https://doi.org/10.1080/10511482.2000.9521359

Paterson, R., Handy, S., Kockelman, K., Bhat, C., Song, J., Rajamani, J., … Waleski, J. (2003). *Techniques for MITIGATING URBAN Sprawl* (3), 1–16.

Poddubny, V. (2008). Justification of engineering solutions for the effective development of underground spaces of largest and large cities. *Ural State Mining University* 62–99.

Schlossberg, M. (2004). From tiger to audit instruments—Measuring neighborhood walkability with street data based on geographic information systems. *Transportation Research Board* 48–56.

Smith, W. D. (1980). Friedrich Ratzel and the origins of Lebensraum. *German Studies Review, 3*(1), 51. https://doi.org/10.2307/1429483

State Building Norms. (1992). *Planning and development of urban and rural settlements* (pp. 2–15).

Tokyo Metropolitan Government. (2007). *Connecting concepts determine tthe exact coverage and influence of the social sphere of sustainability policies for the 10-year project for green*. Regenerating Tokyo's Abundant Greenery.

Ustundag, K. (2012). *An Approach Towards Organic Transportation in medium sized settlements: The Yalova experience*. Mimar Sinan Fine Arts University.

Varlamov. (n.d.). What to do with micro-districts? *Varlamov*. https://varlamov.ru/1142767.html

Victoria Transport Policy Institute. (2015). *Transportation cost and benefits analysis II, roadway land value* (pp. 3–9). https://www.vtpi.org/tca/

Vogel, P., Greiser, T., & Mattfeld, D. C. (2011). Understanding bike-sharing systems using data mining: Exploring activity patterns. *Procedia—Social and Behavioral Sciences, 20*, 514–523. https://doi.org/10.1016/j.sbspro.2011.08.058

Zosimov, G. (1976). *Spatial organisation of the city (Module in the planning structure)* (pp. 35–39). Stroiizdat.

The Impact of the Smart Modern Transportation on Urban Structure of the Cities

Youssef A. Elsayed and Said Hassanien El-Sayed

Abstract

The city urban structure is affected by many factors including the evolution of transport systems, transportation technology, and kinds of transport at the end of the nineteenth century up to the twenty-first century, which led to significant transformation of cities urban structure as a result of its fast development, the technological development in the first two decades of the twenty-first century, and there have been many indications that the new systems of transport will result a fundamental change in the methods and networks of transport, including flying taxis and taxis without a driver systems which will be applied soon in many Arab cities, where some countries are currently support and equipping cities with basic services to implement these systems, which will affect the formation of city urban structures. The aim of the research is to study the impact of smart modern transportation systems on the structure of Arab city, and the changes that will occur in our existing cities as a result of the application of these systems of transport as well as prepare a general perception of the formation of the structures of these cities. The research will follow the inductive approach which starts with the observations, and theories are proposed toward the end of the research process as a result of observations

Keywords

Sustainable development • Transportation technology development • Smart transportation system • City urban structures

1 Sustainable Development (SD)

The (SD) aims to provide the present generation with all basic requirements without affecting the sources of the future generations or damaging the natural environment; the concept of (SD) in its general framework is considered an environmental concept and turned into a comprehensive development concept, considerate main axes as social, economic, and the environmental axis.

1.1 The Concept of Sustainable Cities and SD

With the adoption of (SD) concept and its effects on city urban structure, social and cultural environment, the concept of “sustainable cities” has emerged, aims for creating new form of cities. Economic growth is achieved through an economic base that does not consume or pollute natural resources. The reuse of the product, i.e., recycling as an input in another production process, or the restoration of the energy, invested in this product while relying on sustainable transport (ITS-UK, 2014).

A sustainable city is defined as: the city where the absorptive capacity of local resources and ecosystems is balanced by increasing resource efficiency and minimizing polluting output, characterized as a low or zero-emission city, and thus contributing to reducing Co_2 production, and other organic compounds that exacerbate climate change. Such structural transformations require integrative systems for the management, as recycling of wastes, and also cultural transformations in consumption, recreation, and transition patterns, for example, cities that are relatively sustainable are characterized by cohesive tissue, to reduce transition distances between housing and work or services, and also reduce the use of energy for transport.

Y. A. Elsayed · S. H. El-Sayed (✉)
High Institute of Engineering, Thebes Academy for Sciences, Cairo, Egypt

F. Trapani et al. (eds.), *Advanced Studies in Efficient Environmental Design and City Planning*, Advances in Science, Technology & Innovation, https://doi.org/10.1007/978-3-030-65181-7_48

1.2 Tools for Achieving Sustainable Development (SD)

Global Report on Human Settlements. 2009 entitled "Sustainable Town Planning" includes an assessment of the effectiveness of eco-urban environmental planning systems as a tool to meet the unprecedented challenges facing cities in the twenty-first century (ITS-UK, 2014). To achieve sustainable planning processes, emphasis should be placed on its tools, shown in Fig. 1

2 Benefits of Smart Transportation Systems (STS) (ITS UKnited Kingdom, 2016)

STS has many benefits for the city and the population that help raise the efficiency of the city and society more communicative as follows:

2.1 Health, Safety, and Environment

There are many benefits of STS as creating of traffic-free zones in cities reduce pollution and premature deaths, integration of vehicle systems with mobile communications and advanced mapping technology gives a potential, "smart motorways" improve traffic safety as well as capacity and average speed cameras deliver tangible improvements through casualty reductions, congestion improvements, and improved environmental factors.

2.2 Public Transport

Electric and hybrid public transport vehicles reduce pollution, provision of online information to buses, trains, and their passengers create a better informed traveler and operator and electronic ticketing enables faster, easier travel by public transport and provides management information leading to improvements in the transport system.

2.3 Driver and Traffic Management Benefits

Equipping the vehicle with driver assistance systems to improve the efficiency and safety of road transport, management of vehicle fleets, both private and public transport, via online information and two-way communication between manager and driver, safety and security benefits for drivers and loads, electronic motorway tolling and congestion charging and parking assistance.

2.4 Economic Benefits

Use of "telematics" ITS devices in vehicles leads to a more efficient motor insurance industry and also allows reduced premiums for young drivers, and there are safety benefits by imposing location and timing restrictions on high-risk drivers via insurance products and introduction of road pricing and congestion charging would have major economic as well as environmental and traffic management benefits.

Fig. 1 Tools for achieving sustainable development

3 Types of Transportation

City urban growth is always linked to the accessibility and quality of the transports type used by passengers to reach their areas of residence, especially the business trips, which represent the majority of trips within any city or urban community, and with the technological development in the transport, the development of transport began with the spread of horse-drawn vehicles and the appearance of a horse-drawn bus, then the tram and trolley-bass, which carried more passengers. This development had a strong impact on the city, such as Paris and London, showed the urban strips around the paths of this types and emerged new urban areas around the stops. The block expanded and increased its urban area (Daniels & Warrens, 1980). The need of widening of the streets around the paths of these types make tall buildings and the barricade appeared as radiologically around trains paths and local linear urban transport such as tram and metro, the effects of the use of the regional types of transport as real ways in the growth of urban area around train stations are far from each other, while urban metro with short distances between their stations contributed to make urban strips around the entire path (Bashandy, 1987).

Rubber-coated vehicles including door-to-door car and public transport routes to serve all areas of the city have led to increased growth of urban areas and contributed to filling the spaces of the city that resulted from the presence of foci urban strips formed by the linear transport. New areas, such as suburbs, have grown and have become as important as central areas, especially after urban residents have been attracted to them (Lang et al., 2016).

3.1 Transportation Types

- In the past, the traditional types of transportation were expressed as the main types of access to cities, and it was carried out for people (passengers) and materials and animals (the goods) in all their forms.
- At the end of the twentieth century began the boom in technology and the era of the Internet and telecommunications, which participated traditional types of transportation to carry out some tasks and activities, and it became one of the most important types of communication, but by virtual.
- There are three main types of traditional transport, including road transport, air transport, and water transport. Some vehicle uses two fields such as land–water transport, roads and air transport, or even water, air.
- The road transport was initially based on the progress of horse riding and then horse-drawn vehicles and trains, metro and tramways, motor vehicles that use rubber tires, and some of them became more reliable, enabling them to be without driver. (Daniels & Warrens, 1980)
- The water transport is very large in size, giant ships that can carry more than 10,000 passengers and Massive transport of materials (fish—oil—river buses and this type of transport includes yachts, and floating hotels (Bashandy, 1987).
- The air transport has multiple uses between the transport of passengers and goods in addition to its role in wars, including small and large aircraft, private and mass transit, or air taxi used recently (Bashandy, 1987).

3.2 Spaces and Equipment Needed for Transportation

- **Space**: All Transportations need a space to start from and other to go to, the traditional transportations require many spaces (parking, station, stops) leaner space as road, while the water transport needs to a port and an appropriate waterway. The air transport needs air routes or lands, each place varies according to the type of the transportations and according to destination and origin, and also for service element. Non-conventional kind of transport as virtual communication require only small-sized spaces and sophisticated technology. The Internet starts to grow day by day and increase control. Underlining many of the activities carried out by humans and thus will reduce the need for traditional transport (ITS-UK, 2014).
- **Equipment**: All transport types need a lot of equipment to perform its function as buildings, stations and parking areas, power stations technological devices that control the operation of such devices as computers, satellites, or traditional mechanical equipment such as maintenance areas.
- **The human factor**: Most of the traditional types of transportation needed to be controlled by humans, except for some self-driving, self-guided, remote or satellite-controlled transportations, also flying taxi and on limited routes transportations such as trains and metro.

4 Intelligent Transportation Systems (ITS) (Ezell, 2010)

There are kinds of transportation based on artificial intelligence in management, some of them to be controlled remotely. **(ITS)** deal with vehicles, roads, routes, instructions and systems, some other control movement, and control of traffic and transportation of all types. These types that to have database for transport system are obtained from road sensors, traffic reports, city maps, as well as drivers of vehicles, traffic men, and researchers in this field (Peña, 2015),

ITS Subsystems: works to reduce traffic congestion, road management, security, safety, flight information, and separation between public and private transport, Table 1 showed **(ITS)** and specific.

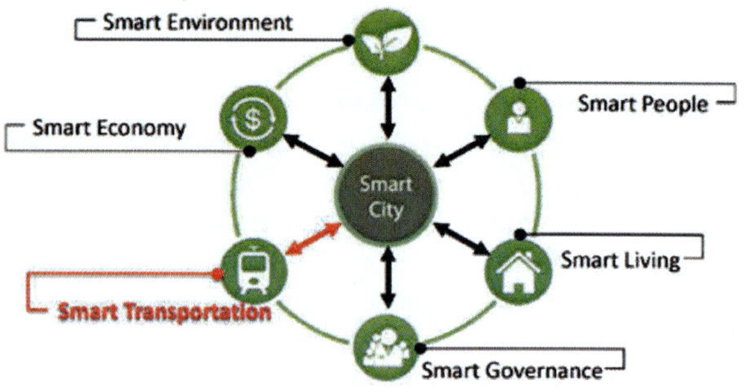

Fig. 2 Smart transportation as apart from smart city

5 Smart Transportation System Requirements (Grob, 2009)

- Achieved smart cities (Fig. 2)
- There is many requirements (STS) as apart smart cities such as an intelligent transportation management (Fig. 3).

6 The Arab Cities (Bashandy, 1987)

Most of Arab cities applied the most recent technology of transportation, but only few of them uses the **(ITS)** as Dubai and Abu Dhabi.

6.1 Kinds of Transportation in the Arab City

- Transport vehicles varied in Arab cities of the countries where they began to develop in a similar way as in European cities, where the horse-drawn vehicles in most Arab countries, Egypt was one of the first countries in the world to use trains in the middle of the nineteenth century, while metro was first operated in Cairo also in 1980 followed by metro of Dubai and the fast train between cities Saudi Arabia. The cars that used rubber tires appeared in Cairo, Gulf cities and were used for private transport, taxi, and public transport. There were also light transportations, such as monorail. The planes appeared at Cairo in the middle of the twentieth century followed by all Arab cities, Hence, most of the traditional Kinds have emerged in most Arab cities, and with the great development as European transport methods, STS have emerged, such as Dubai, where cars and aircraft without driver appeared, and will be followed by Abu Dhabi and Doha to generalize the system in the next few years.
- There are some Arab cities planned to accommodate flying taxy in Gulf countries, while others develop the traditional transport to be supported by smart techniques

Table 1 Showed intelligent transportation systems and specific

Specific	Intelligent transportation systems
Real-time traffic information provision Route guidance/navigation systems Parking information Roadside weather information systems	Advanced traveler information systems (ATIS)
Traffic Operations Centers (TOCs) Adaptive traffic signal control Dynamic message signs (or "variable" message signs) Ramp metering	Advanced transportation Management Systems (ATMS)
Electronic Toll Collection (ETC) Congestion Pricing/Electronic Road Pricing (ERP) Fee-Based Express (HOT) Lanes Vehicle-Miles Traveled (VMT) Usage fees Variable parking fees	ITS-enabled transportation pricing systems
Real-time status information for public transit system (e.g., bus, subway, rail) Automatic Vehicle Location (AVL) Electronic Fare Payment (e.g., smart cards)	Advanced public transportation systems (APTS)
Cooperative Intersection Collision Avoidance System (CICAS) Intelligent Speed Adaptation (ISA)	Vehicle-to-infrastructure integration (VII) and vehicle to-vehicle integration (V2V)

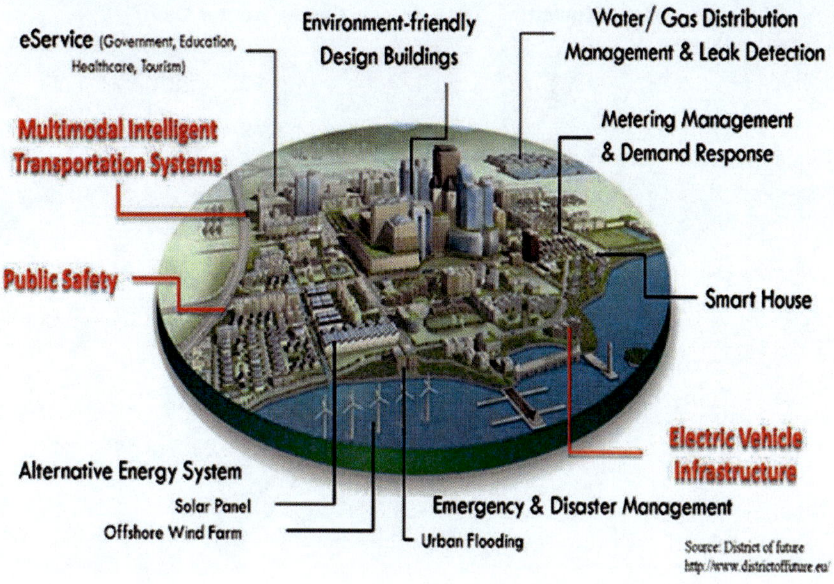

Fig. 3 Element of transportation system to achieved smart city

to enable passengers to use their smart phones via Internet to facilitate ticket reservation find the location of vehicle and stations and also in part at about traffic congestion or any problems may face tripe. The passenger can also work during his journey by providing Internet services systems via satellite.

6.2 The Urban Structure of the Arab City

- Cities began with the beginning of civilizations and many of them emerged in the Arab world as Assyrian and Egyptian cities, which were developed strongly during the nineteenth and twenty century using modern city models as linear and central cities, and the theories of land use distribution, many Arab cities appears such as the circular cities (the city of Mansour Baghdad), or cities with walls as Cairo or coastal cities and ports such as Dubai, Alexandria, and Tripoli in Lebanon. Or follow certain types of planning such as organic planning as in Garden City Cairo, or network mechanic as in the city of Riyadh, or radiation as in Heliopolis in Cairo and Maadi.
- Arab cities are developing strongly as technology evolves, such as Dubai, Doha, and Kuwait. This development is manifested in urban structures such as the city of Kuwait, Cairo, or coastal stripes in Dubai, or in the form of iron grid as in the city of Riyadh and city of Jeddah (KSA).
- The urban structure of the Arab city is affected by site conditions in many cities and also characterized by dependence on the private car as a main transportations of movement, as in most cities of Gulf, Cairo, Damascus, and Baghdad, the paved ways shows interest in the road network in these cities, which indicates a defect between private transport and public transport. The Arab cities also express the development of the city and in conformity with the urban change and show cohesive fabric and narrow streets in the old and historical urban areas, while the spaces and streets are organized and increase its wide in newly planned areas as in new cities around Cairo.
- The structures of the Arab cities are also distinguished by taking the environment into account, and also the social dimension and the religious values, where the elevations are controlled by proportion to the width of the streets, and there is special urban spaces represent the public spaces as of squares on the market or palace of government or important religious buildings, while high-rise buildings are found for multiapartment buildings (Fig. 4).

- There is many indicators that the structures of Arab cities are strongly changing due to global changes in planning patterns and emergence of new types of cities such as environmental or smart cities that use smart transport and Internet cities, these indicators have emerged in Dubai and in many Gulf and Arab cities, investment private cities as Madinaty and Dream Land around Cairo.
- It is expected that smart and modern kinds of transport will be used in some Arab cities, which will have a significant impact on the structures of these cities.

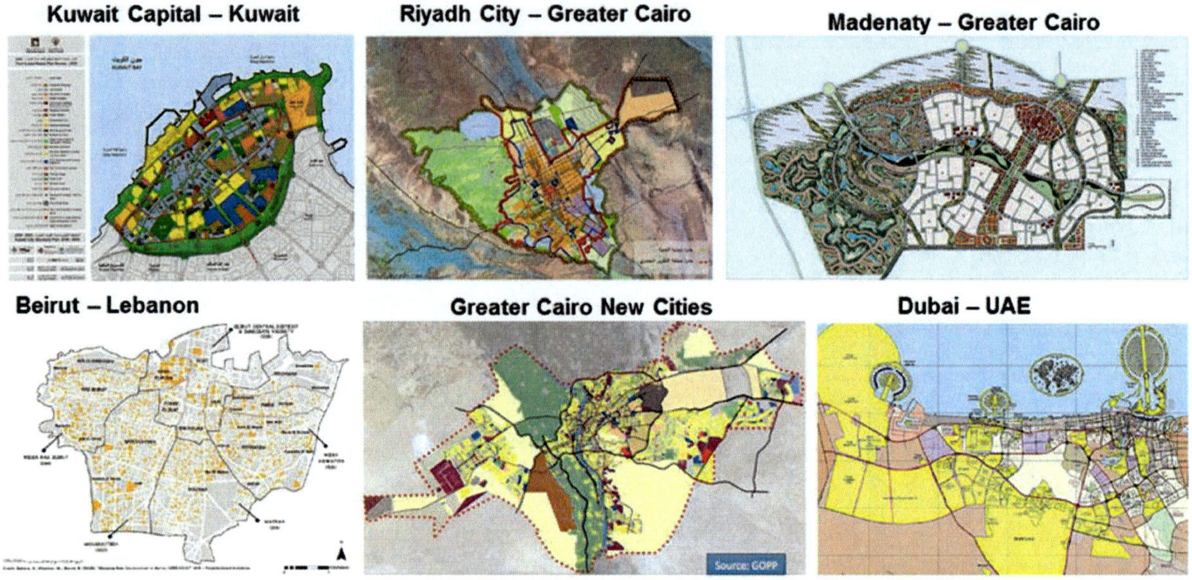

Fig. 4 Arabic city structure in many Arabic countries

7 Analysis of Sustainable Transportation for Some Cities

7.1 City of Vancouver—Canada (Vancouver, 2040–2015)

The vision 2040 of the transportation in the city of Vancouver is more than just mobility. That vision—that falls under the three pillars of sustainability—includes a number of goals. These goals address some challenges that we face and support, among which are the followings:

- Economically sustainable city
- Environmentally sustainable city
- Socially sustainable city.

These goals support each other across issues; striving for an environmental goal, for example, will also move forward social and economic goals and thus will advance overall sustainability. In addition, when we talk, for example, about the City's Greenest City 2020 Action Plan, these goals are aligned with transportation goals. Here, we can mention some relevant Greenest City goals, as mentioned above, the goals of this plan align with the transportation-related goals found in the City's Greenest City 2020 Action Plan (Table 2).

- Pedestrian, bikeways, and transit
- Low fossil fuels
- Healthy lifestyle and clean air.

7.1.1 Greenest City Directions (Vancouver, 2040–2015)

7.1.2 Greenest City Transit

Transit improvements should be supported to increase capacity and ensure service that is comfortable, frequent, fast, fully accessible, and reliable (Table 3).

7.2 Masdar City Model—UAE

The UAE has been implementing road and intelligent transportation systems in many of its cities since the beginning of the twenty-first century to help to ease traffic, reduce accidents, and improve the urban environment. The new cities in the UAE have been planned to reduce the problems stemming from the traffic, especially the environmental ones. The smart transport system in Masdar city and Abu Dhabi city development plan have been implemented as environmental models for smart cities (The masder report on technologies for future smart city transit, 2018) (Fig. 5).

- **Transport and logistics**: UAE has established its strategic position as a global hub for transport and logistics. It has expanded and developed its ports, airports, and road network, as well as measures to increase the capacity of public transport systems to reduce dependence on private vehicles and reduce traffic congestion, and thereby reduce carbon footprint. At 2009, the department of transport in Abu Dhabi launched a comprehensive transport plan in response to Abu Dhabi

Table 2 Showed green city directions in Vancouver

Walking	City for all (public life. healthy and safe walking, comfortable spaces, and enjoyable)	
Land use	Short distance between utilities, more connections to serve cost-effectively with transit Unique and attractive building design and public realm Mixed use and housing types makes it easier to live, work, shop, to achieved high connectivity and accessibility Increase residential rates and employment to reach higher-level density and more service beside cycling and walking distance, and sustainable transportation	
Public spaces	Create a healthy public spaces and streets throughout the city to achieved enabling and encouraging creative uses	
Cycling	Create bikeways characterized comfortable, safe, fun, and suitable for people of all citizens	
Cycling network	Connecting all parts of the city with a highly efficient bicycle network, multi-modal integration, provide many and convenient bicycle parking and facilities, comfortable bikeway's suitable for people of all citizens, regular maintenance of bicycle network and facilities, provide a bicycle network system, combine cycling with other public transportation	
Pedestrian network	City design for pedestrians, free pedestrian and sidewalk on all streets, high accessibility for all citizens, complete all gaps in the pedestrian network, design all street to achieved safety and walking, design network to collect rain water in the streets and spaces, provide attractive pedestrian network	

Source https://vancouver.ca/files/cov/transportation-2040-plan.pdf

Vision 2030, while the Dubai Road and Transport Authority (RTA) operates according to Dubai Strategic Plan 2015.

- **The use of modern public transport**: Dubai Metro contributes to reducing carbon dioxide emissions by more than 645 tons per day due to the decline in the use of private vehicles. Dubai Roads and Transport Authority (RTA) plans to add three new metro lines by 2030 and expand the red line to the Expo 2020. Tram Dubai is the first tram project to be developed outside of Europe and operates a ground-based power supply system. Abu Dhabi has Emara begun the construction of an integrated metro network with a length of 131 km, in addition to the 1200 km railway network which is expected to contribute to the upgrading of the public transport system in the country. There is also the Masdar City project, which aims to build a pedestrian-friendly, vehicle-free city where private parking spaces will be provided only at city entrances and accessible through public transport.
- **Transforming traditional transportation into green and clean solutions**: Following the success of the first "green bus" experience, RTA plans to use biofuels in 118 other buses, with the potential to reduce carbon dioxide emissions by 33% compared to traditional buses.
- **Development of an alternative to mobility and exercise**: Dubai RTA has developed a comprehensive plan for the implementation of 850 km of special routes in the central and new commercial areas in order to stimulate the use of bicycles as alternative solutions for internal mobility. Next bike has developed a European bicycle rental system that includes more than 100 bicycles that can be rented from dedicated stations in both the city center and the Dubai Marina area.
- **Motivating responsible use of private vehicles**: "Salik" has been introduced to address traffic congestion and congestion problems in Dubai. The Roads and Transport Authority (RTA) also legislated the carpooling service in the Emirate with the aim of increasing vehicle occupancy rates and thus traffic management. In turn, the Department

Table 3 Showed green city transit in Vancouver

Accessibility	Support a universally accessible transit system with a goal of equal transit outcomes for people of all ages, incomes, and abilities
Transit network	Development transit reliability and speed using transit-priority measures Support increased water-based transit Combine cycling with other public transportation and trips Support integration with taxis in the city Development rapid transit Improvement local transit Support improved pass transit Provide comfortable connections between parking and network in the city Multi-modal integration Provide a bikeway networks Support a transit system to achieved high accessibility
Transit financing	Support effective prices that encourage transit use Support equitable and stable long-term transit funding sources
Neighborhood impacts	Reduce environmental emissions and noise for transit Highly efficient streets
Road network	Develop a network operation to manage crowding impacts Manage traffic to develop to achieved livability, community, safety for neighborhood Consider impacts to transit, commercial vehicles, and general traffic flow before redesign road space
Local goods and services movement	Support low-impact goods and services movement and delivery Provide for efficient loading and unloading Maintain an efficient network of designated truck routes Support local production and distribution to reduce the need for large-scale transport Manage the impacts of moving goods and services, maintain effective emergency response times, and support economy
Parking	Use off-street parking requirements to support reduced auto ownership and use Support strategies that reduce the need for parking Separate parking and housing costs to increase housing affordability Approach parking as a shared district resource Provide parking to be flexible and adaptable Make it easier for drivers to find available parking spaces Manage parking in neighborhoods Provide accessible parking for persons with disabilities Support cycling, low-carbon vehicles, and car sharing Support efficient loading and servicing
Car sharing	Support increased car sharing
Motor vehicles	Manage crowding, develop road safety, make it easier to drive less, accelerate the shift to low-carbon vehicles, provide charging infrastructure to support electric vehicles
Taxis	Support improved taxi service and safe use of taxis for persons with disabilities
Regional-and-beyond goods and services movement	Support truck movement on key regional routes Protect and develop rail corridors for goods and passenger movement Support Vancouver international airport as universal gateway Support port metro Vancouver efforts to reduce impacts environmental
Emergency response	Accessibility information on traffic calming measures and closures Consider emergency vehicle access in street designs and traffic calming measures
Other demand management tools	Support regional road pricings Improvements income directed toward sustainable transportation

Source https://vancouver.ca/files/cov/transportation-2040-plan.pdf

of Transport (ADTA) has launched "Saf and Tanaqul" service, which aims to facilitate traffic and mobility in the city. This service comes in the context of the Department's efforts to provide convenient, sustainable, and integrated transportation services.

- **Contributing to making the air transport sector greener**: Etihad Airways, in coordination with Boeing, Tackerir, Total and Masdar Institute of Science and Technology, launched the Abu Dhabi Biofuel Initiative Support sustainable biofuel industry in the UAE.

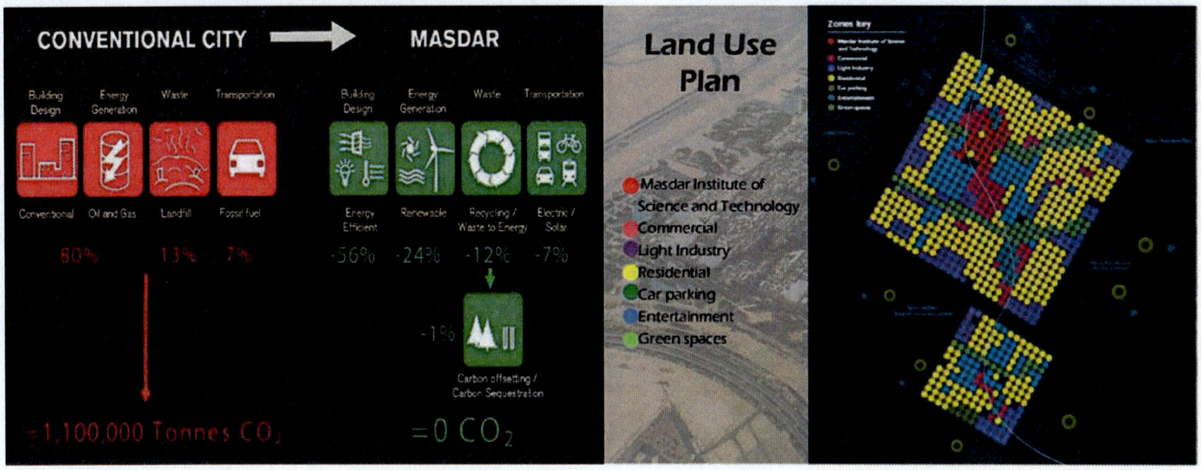

Fig. 5 Masdar city as the first zero-carbon city

8 Comparative Analysis of Sustainable Transportation for Some Cities

See Table 4.

9 Conclusion

- The change in the transportation system and the use of new and modern transportation have strong and effective impact on reshaping road network and its contents such as, parking, smart technology, which simultaneously affect Arab city structure.
- The STS will achieve livability by reducing traffic congestion and trip time, it also increases controls, and improving public safety.
- STS will achieve workability by increasing cities competitive advantage and becoming more attractive.
- ITS will help to apply sustainability by reducing pollution from transportation, improving transportation budgets and achieved smart cities guide lines.
- The modern transportation requirements as (utilities—road network—pedestrian network—cycling network—parking), force changing of city structure to accommodate ITS at all levels of planning for the city.
- The improvement and development of all Arab city roads by redesigning road network to accommodate intelligent transportation, and also by providing footpaths—trail routes—public transport routes.
- The use of transportations and information panels on both sides of the road provide green areas according to global rates, reinforcing pedestrian network and cycling network.

Table 4 showed comparative analysis of sustainable transportation for some cities

The city	Vancouver—Canada	Masdar city—UAE
City goals	Pedestrian, bikeways, and transit Healthy lifestyle and clean air	Zero-carbon city Renewal energy Water, waste
City directions	Walking, mixed housing and land use, public spaces, cycling network, pedestrian network	Sustainable public transportation , communications, technology
City transit	Accessibility, transit network, transit financing, neighborhood impacts, road network, local goods and services movement, parking, car sharing, motor vehicles, goods and services movement, emergency response	Transit (PRT,LRT), parking , pedestrian, accessibility , smart technology, open spaces, optical network, information center, transport and logistics, the use of modern public transport, transforming traditional transportation into green and clean solutions

Source https://vancouver.ca/files/cov/transportation-2040-plan.pdf

References

Bashandy, S. (1987). *The impact of transportation technology on city formation.* Ph.D. thesis, Faculty of Engineering, Cairo University.

Brown, S., Eng, P., Mike Anderson, M. C. I. P., Planner, I. I., Brian Gould, M. C. P., Assistant III, E. P. (2012). Vancouver transportation plan update. In *TAC Annual Meeting (*www.tacatc.ca*).*

Dalton, D. L. (2018). *Expanding access to our communities: A guide to successful mobility management practices in small urban and rural areas* (No. NCHRP Project 20-65, Task 68).

Daniels, P. W., & Warrens, A. M. (1980). *Movements in cities.* University Press.

De La Peña, E. (2015). Smart roads: A vision. *Smart Transportation Alliance.*

Ezell, S. (2010). *Explaining international IT application leadership: Intelligent transportation systems.*

Grob, G. R. (2009, March). Future transportation with smart grids and sustainable energy. In *2009 6th International Multi-Conference on Systems, Signals and Devices* (pp. 1–5). IEEE.

ITS United Kingdom. (2016). *Intelligent transport systems (ITS) and their benefits.*

ITS-UK2014. (2014). *Local authority guide to emerging transport technology.* The Institution of Engineering and Technology.

Lang, N., Rüßmann, M., Mei-Pochtler, A., Dauner, T., Komiya, S., Mosquet, X., & Doubara, X. (2016). Self-driving vehicles, robo-taxis, and the urban mobility revolution. *The Boston Consulting Group, 7,* 2016

Siemens—Mobility Division—Smart Mobility. (2015). *A tool to achieve sustainable cities.* Restricted © Siemens AG.

The Masder Report on Technologies for Future Smart City Transit (2018 January).

Vancouver 2040–2015. (2015). *Moving forward, city of Vancouver.* www.vancouvercity.com, https://vancouver.ca/files/cov/transportation-2040-plan.pdf

A Framework for Branding Rural Towns: A Case Study of Al-Mughayyer, Irbid, Jordan

Anne A. Gharaibeh, Mohammed H. Al-Kaff, Ammar M. Ze'aiter, and Mohammad H. Abuharbid

Abstract

Globalization is increasing competitiveness among cities in drawing attention and attracting visitors as well as stakeholders to them. This is leading to positive impacts on the development of their infrastructure and improving the quality of life for their citizens. In some cases, well-known sites in certain cities draw attention away from other local contexts leading to negative impacts on those sites. In the case of Jordan, famous places such as Petra and the Dead Sea are drawing attention away from other local areas, particularly, rural areas with distinctive characteristics. Here comes the need for branding strategies to highlight the advantages of these areas. However, it is noticeable from previous studies that branding strategy is concentrated on cities rather than rural areas that do not receive sufficient attention from planners and officials. Accordingly, this paper aims to put forth a general framework for branding rural towns. It applies the framework to Al-Mughayyer town to learn how to fit branding to the Jordanian rural context. The study investigates town residents' attitudes toward rural branding. One hundred people answered questions that focused on identifying the important aspects of their town; rating their importance, and suggesting valuable locations, attractions, socio-economic factors in the town and its vicinity. Interviews with residents and local officials clarified many of the in-depth issues related to the town's identity. The researchers employ the geographic information system (GIS) to identify suitable areas that enhance this identity. In the end, this study elucidates that; despite the many attractions that this town possessed, the natural environment was the most important branding factor in this town, where suitable areas appear around agricultural lands and Rahub Spring. The same framework can be used in other towns in order to evaluate and outline the important branding factors which are worth consideration in future development plans.

Keywords

City branding • Rural town branding • Suitability analysis • Green infrastructure

1 Introduction

Competition among countries and cities has recently increased to attract tourists as well as to satisfy residents. Accordingly, to succeed in such competitive conditions, cities and urban areas need to utilize a wide range of strategies and procedures to propose and promote themselves (Rezvanpour & Bayat, 2017). City branding is one of these strategies adopted by many cities to enhance the development of the city by improving its image in the world. If implemented properly, city branding would have a direct impact on the economy and reflect positively on the infrastructure and the quality of life (Gharaibeh, 2019; Prilenska, 2012).

According to Kompaniets and Rauhut (2016), capitals and big cities have paid significant attention in applying city branding strategies because their image is already formed in people's minds, which makes branding strategies easier to develop. In contrast, rural areas may be less accessible, remote, small in size, and more challenging to branding despite its significance. Moreover, they have different resources and population densities compared to large cities. Accordingly, it is hard to use and apply the tools or methods used in big cities in rural areas.

A. A. Gharaibeh · M. H. Al-Kaff (✉) · A. M. Ze'aiter · M. H. Abuharbid
Department of City Planning and Design, College of Architecture and Design, Jordan University of Science and Technology, P.O. Box (3030) Irbid, 22110, Jordan
e-mail: mhalkaff18@ad.just.edu.jo

A. A. Gharaibeh
e-mail: dr-anne@just.edu.jo

F. Trapani et al. (eds.), *Advanced Studies in Efficient Environmental Design and City Planning*, Advances in Science, Technology & Innovation, https://doi.org/10.1007/978-3-030-65181-7_49

1.1 Study Scope

Due to the abundance of branding strategies studies in cities compared to rural areas, this paper aims to address a framework for branding rural towns. Therefore, Al-Mughayyer town has been chosen as a case study for rural towns in Jordan. Al-Mughayyer town in Irbid governorate is located in the northern part of Jordan (Fig. 1). Irbid which was known as Arabella is one of the Decapolis cities in the ancient Hellenistic period (Akasheh et al., 2016). Al-Mughayyer Rahub is 6 km long and 3 km wide. It lies on the eastern side of Esh-Shallaleh Valley and has a lot of natural features, such as Esh-Shallaf and Rahub springs. In addition to archeological sites such as Zeiraqun ruin in Esh-Shallaleh Valley, it is worth mentioning that Zeiraqun goes back to the Bronze Age and contains 31 archeological sites (Kafafi, 2011). It represented one of the urban centers in the northeast of Irbid city during the Bronze Ages which started as a small village that developed into a town then into a city surrounded by city walls (Kafafi, 2011). All these characteristics that have been mentioned above make Al-Mughayyer town an interesting scope to be studied as a case for branding rural towns.

The awareness of the Jordanian community toward Al-Mughayyer town and similar areas is limited, which in turn does not form any pressure on the responsible parties to do necessary measures to preserve and highlight the local heritage. This is visible through the lack of regular maintenance of the existing ponds and springs that have been closed or destroyed as a result of landslides. In addition to the archeological sites in the region that seemed to be abandoned places of neglected stones; moreover, the lack of clear strategies and plans that supported regional and natural tourism did not help to form an effective image of Al-Mughayyer.

2 Literature Review

2.1 Historical Overview

City branding procedure is one of the known strategies and techniques to promote a place for people around the world. According to Rezvanpour and Bayat (2017), city branding started since the emergence of the town by building structures to show power to enemies. In the late 1960s, the idea of city boosterism was established as a management system to advertise and identify a place or a city by its items and products. Later on, in the 1980s, place branding emerged after city boosterism as an advertising process to make a place recognized and give the city an image based on the place (Garcia-Lamarca et al. 2019; Hall, 1993; Harvey, 1989).

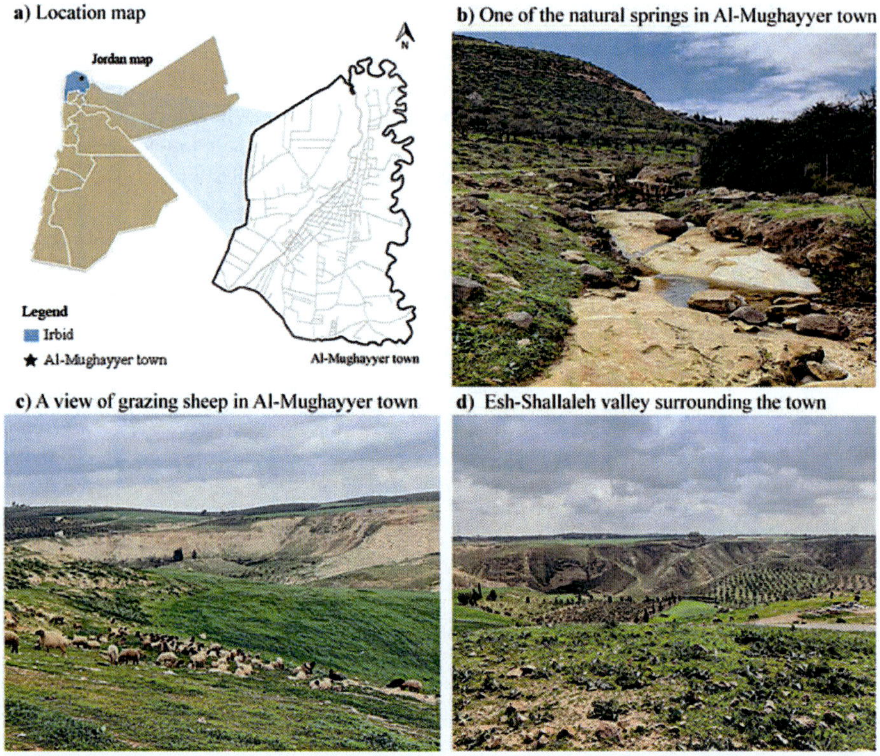

Fig. 1 a) Location map, b) one of the natural springs in Al-Mughayyer town, c) a view of grazing sheep in Al-Mughayyer town, d) a view of the Esh-Shallaleh valley surrounding Al-Mughayyer town (Photos are taken by the authors)

2.2 General Concepts Related to Place Branding

Dinnie (2007) indicated that the idea of place branding was derived from the commercial world and applied on cities. Additionally, he illustrated that place branding may be considered as the management of the place image. Therefore, to understand the city branding strategy some general concepts such as brand image, city image, and place identity must be clarified (Pedeliento & Kavaratzis, 2019; Rodrigues et al. 2019). Rezvanpour and Bayat (2017) explained that the brand has three basic concepts, which are: *Identity*, including the features that the brand owner wants to be perceived; *Image*, which is the real image that remains in the mind of the observer; and *Communication*, which is the way that introduces the brand to the users. According to Abdelaal and Hussein (2012), brand image is the visual image generated by several nonphysical factors to express a certain message which is realized by individuals. However, there are a lot of misunderstandings about the brand, where it is just observed as a name or a logo of a single entity or association. Despite the previous, it is important to understand that brand is beyond a name or a logo; it is a guarantee from an association to give what becomes the principle of that specific brand and its functional, emotional, and social advantage (Purwanti & Genoveva, 2017).

However, Rezvanpour and Bayat (2017) illustrate that the city image is the total impressions of a place or an individual's overall perception. They also illustrate that a city has two integrated images; *the visual image* which is directly perceived when experiencing the city, and *the brand image* which is promoted by various nonphysical mediums to be perceived. However, Prilenska (2012) states that although the image of the city plays an important role in the location choices made by firms, investors, and individuals, and it is essential to make the image well-known both locally and internationally. Under these circumstances, city branding becomes an important tool for image-building and image-communication.

Rezvanpour and Bayat (2017) elucidate that the place identity is categorized into three elements, which are *structural elements* also known as the DNA of place including location and history; *semi-static* that need time to be changed, including physical appearance, land-use planning, the resident's mentality such as the cultural and religious values; and *coloring* including symbols (names, logos), costumes, and folk dances.

2.3 City Branding

City branding is considered as a sub-field of place branding that encourages the marketing of cities (Khirfan & Momani, 2013). It is understood as a set of actions that create a positive image of the city and convey it among different target groups by means of visuals, stories, and events locally and internationally. These actions can be spatial interventions that aim to improve the physical quality of the built environment or non-spatial interventions that concern the social environment, such as event organization (Prilenska, 2012). Sakka (2016) states that city branding considers various aspects as a competitive advantage such as culture, social aspects, architecture, and landscape. On the other hand, Pınar et al. (2017) claim that city branding as a promoting tool concentrates on the nature of place and quality of life in cities and urban areas. However, there are various practical notions of city branding strategy that is worth explaining such as city branding strategies and techniques.

Rezvanpour and Bayat (2017) illustrated that the main goal in city branding is similar to the goals of a product's brand, which is to create loyalty. However, the brand of the place should remain in people's minds, both residents and tourists. They claimed that city planners must take into consideration the essential demands of living and working in urban communities such as affordable and accessible housing, transport, health care, and so on. If these requirements are well established in one city, people will have the opportunity to get a unique index of the city.

According to Prilenska (2012), two city branding strategies have gained exceptional popularity in recent years. The first is a *creative city strategy* that can be applied in cities that are considered as centers of technology where human capital is the driving force of economic progress including educated, skilled, and productive people. The location choice for such cities is influenced by qualitative housing, short travel times, and quality of city spaces such as green and public spaces, and good architecture. The second is the *experience city strategy* that provides opportunities for economic growth for cities with limited high-tech industry options. These cities take advantage of the experiences that they have, e.g., authenticity and nature, including events, activities such as (shopping, walking, sports), services such as (restaurants, art galleries, theaters) and places which are a stage for activities in themselves, such as squares, beaches, and woods.

Similarly, Prilenska (2012) illustrated that there are three techniques for urban branding which have become popular among city authorities: First, related to personality, such as Gaudi architecture in Barcelona. Second, flagship projects which transform the area with its own presence and catalysts for further investment and consumer attraction, such as the Bilbao Museum of Guggenheim. Third, the organization of events, such as the Bayreuth Festival in Germany. However, a branding campaign's success depends on the coherence of the image of the city and the reality.

2.4 City Branding and Green Spaces

Green spaces cannot become a brand of a place unless it is abundant and unique. However, Tobias and Müller Wahl (2013) clarify that the term green space is used for public green spaces in the overbuilding area of an urban settlement, while the term landscape is used for the natural environment and also for agriculture and forestry outside the overbuilding area. Green cities are however more likely to attract talented workers and tourists as the notion of sustainable development tends to be associated with urban greenery (Fok & Law, 2018). Chicago, USA, is a good example of the incorporation of green spaces and trees in successful urban branding. However, Gulsrud et al. (2013) claimed that branding should involve long-term strategy and enhance the main agendas of a city, such as global competitiveness, social cohesion, and health. According to Chan and Marafa (2014), there is a framework for branding green spaces known as Green Brand Hexagon (GBH) consists of six key pillars, which are Green status; including the provision of green resources in cities. Green Space; including the physical aspects or environment. Green Potential; including the willingness to cost more for business investment. Green Pulse; including the utilization of green resources for getting excitement experiences in the whole city. Green Citizenship; including environmentally friendly citizenship and safety in green resources, Green Prerequisites; the overall quality of green spaces and its facilities in addition to the accessibility to it.

2.5 Branding Rural Areas

Kompaniets and Rauhut (2016) illustrated that rural areas are characterized by low population density, weaker activities, and markets. Hence, the transportation network and accessibility to facilities have a lower quality than in urban areas. Rural towns are small or medium sizes, far from the political, cultural, and economic centers, and are marginalized in economic, social, and spatial terms. Kompaniets and Rauhut (2016) also pointed out that there are areas that can be defined as rural in terms of population, activities, and transportation but are located in the vicinity of urban areas. They are affected by the image of the surrounding urban areas. However, Wilson and Muldrow (2010) claimed that branding rural and urban areas share the same goal which is to highlight positive experiences while diminishing the negative experiences without being dishonest.

Wilson and Muldrow (2010) recommended general considerations for branding rural areas. These considerations including understanding the community, through observing the area, speaking with residents, and having a meal in local restaurants, or in other words, immersing in the local culture. *Redefining cultural assets*, which can be by looking for the unique story in the rural community, such as the remarkable history, famous meal, and a common craft. Lastly, determining focal points, which may be based on *character*, such as in Arkansas Delta that is recognized as a culture of agriculture; *place*, such as Atchafalaya, La that is defined by its connection to the river; and *experience*, such as Central Kentucky that has many cultural assets worth preserving in their rural environments and a visitor can absorb the Kentucky story without ever stepping a foot in the big cities. On the other hand, Medica et al. (2010) considered architectural heritage as a tool in branding rural areas, such as The Istrian County which was the first in Croatia that valorized its rural areas using its traditional stone architecture. In a study done in rural Scotland on Loch Fyne Food Fair, the cultural activities rated more on the branding potential than did the environmental aspects of the place (Clark & Rice, 2019). Thus, their concentration was more successfully geared toward branding authentic activities and foods in the fair than place and environment.

3 Problem Statement

Place branding strategy has drawn the attention of many authorities. As a result of the concentration of services and major projects in cities and urban areas, they have become the main places for adopting place branding strategy. This has been reflected in research where they mentioned many cases and examples and illustrated the aspects that were taken into consideration when implementing the place branding strategy. In contrast, rural areas have received less attention in adopting place strategy due to their characteristics—as mentioned earlier—and their weak infrastructure. This has been reflected in research that studied branding rural areas, where there is a lack of information and case studies. On the other hand, Daher (2007) illustrated that the existence of tourist centers that are known or registered internationally by the United Nations Educational, Scientific and Cultural Organization (UNESCO) has drawn attention from other sites with historical and touristic features. For instance, Petra, Jerash, and the Dead Sea have drawn the attention from other regions and more particularly rural areas with distinctive characteristics, whether historical or natural or other characteristics such as Al-Mughayyer town. This lack of attention to other sites has many negative consequences, such as the failure to provide enough financial budget from the concerned authorities to promote archeological and historical features or to improve the infrastructure. This, in turn, led to limit the development of projects, whether urban, architectural, cultural, or touristic that contributed to improving the image of the place and the diversity in the whole image of Jordan. One of the main factors that led to the aggravation of this problem is the lack of public awareness. However, Abu Al Haija (2011) claimed that the perspective of sites differentiates between locals and Jordanian municipality's authorities. For locals, the spirit of the place in all kinds of forms; materials, and spiritual memories are the important aspects, while for local authorities, the financial return is more significant.

Therefore, the problem of the research is to clarify a framework for branding rural towns and furthermore to define a specific framework for Al-Mughayyer town. This framework will enhance Al-Mughayyer image and highlight its identity, which will help to guide the future development plans in Al-Mughayyer and other rural towns in Jordan.

4 Research Questions

The research aims to define a general framework for branding rural towns and apply it on Al-Mughayyer town to elucidate a specific framework to brand it as a mean of developing the region, improving the quality of life for its residents, and attracting visitors. This framework could later be considered in branding other rural towns in Jordan. However, these aims can be achieved by answering the following questions:

- What are the framework aspects for branding rural areas?
- How can this framework be applied in Al-Mughayyer town?
- What identity does the town have from the points of view of locals and officials?
- What are the suitable areas that can adopt development plans to enhance the brand and the identity of Al-Mughayyer town?

5 Methodology

Research questions were answered based on five stages (Fig. 2), as follows:

- The first stage was done by referring to previous studies and case studies in order to clarify a framework for city branding strategy. Then, clarify a specific framework for applying branding strategy in rural towns.
- The second stage was to apply the framework for the rural areas on AL-Mughayyer town, which was done through two phases. The first phase was based on the authors' field observation as well as interviews with 15 local residents and 2 Al-Mughayyer municipality officials to eliminate the features that do not exist and specify a framework for AL-Mughayyer town. The second phase was to evaluate features that are available in the town. The evaluation depended on a questionnaire distributed on 100 people. The questionnaire was based on purposive samples where respondents were selected of different ages, genders, social classes, and even education levels to obtain a comprehensive and clear vision that represents their point of view regarding the town's characteristics. The questionnaire contained two types of questions. The first type was ranking questions using a forced-choice scale that did not allow for a neutral, undecided, or no opinion response to collect more actionable data. The scale allocated number 4 to "very good", 3 to "good ", 2 to "bad" and finally 1 to "very bad" status. On the other hand, the second type of questions was based on open-ended questions to get a clear vision of the town.
- The third stage was to define Al-Mughayyer's identity that may help to form an image of the town. The town's identity was defined from the point of view for each of the local residents and the officials through interviews with them in addition to the answers collected from open-ended questions in the questionnaire.
- The fourth stage was to clarify a specific framework for Al-Mughayyer town based on the results of the second and third stages.
- The fifth stage was to propose the best area that enhances the identity of Al-Mughayyer to guide future plans for developing the town. In this stage, the geographic information system (GIS) was used where the evaluated features have been identified on the map. Then, a selection process was carried out based on several layers extracted from the previous stages. One layer was for the highest-rated features, another layer to represent the sites that form the identity of the town, a third layer for town's slope in addition to a layer represented the infrastructure of the town such as streets and services which would help to determine the best area.

6 Results and Findings

6.1 The First Stage: Clarifying a Framework

Referring to the literature review and based on previous case studies, certain aspects have been specified for branding places. These cases concentrated in cities and urban areas and divided into six aspects which are *historical features*, where it was branded based on its visibility, aesthetic qualities, uniqueness, architectural style, investment opportunities, and accessibility; *natural features*, where it was branded based on its status, space, potential, pulse, citizenship, and prerequisites; *cultural features*, where it was branded based

Fig. 2 Research methodology

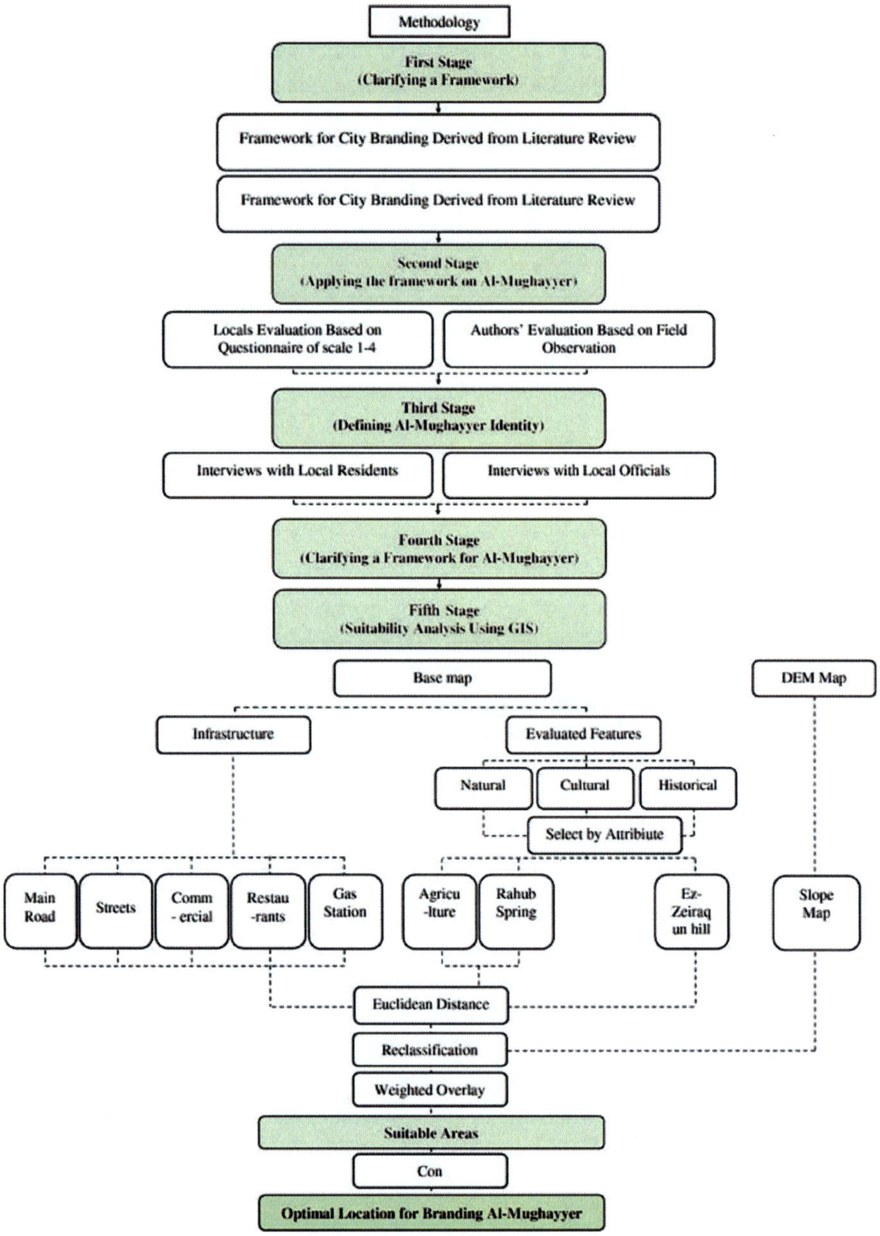

on its common crafts, activities, famous meals, events, rumors and stories; *architectural style*, where it was branded based on its physical appearance and materials; *creative centers*, where it was branded based on its innovations, high-tech industries, and high-quality experiences; and *flagship projects* (Fig. 3a). On the other hand, branding rural towns were based on the three basic assets which were the historical, natural, and cultural assets. In addition to some architectural styles that may be unique in towns such as materials as mentioned earlier regarding Istrian County in Croatia. These four aspects can provide a unique experience in the town (Fig. 3b).

6.2 The Second Stage: Evaluation

Based on field observation and interviews with various residents and officials in the case study of Al-Mughayyer, architectural style criterion was eliminated from the rural branding framework due to the absence of a unique physical appearance or building material in the town. Thus, the criteria for rural branding evaluation have been restricted to the three basic aspects which are historical, natural, and cultural. These aspects included the main features in town based on interviews with Al-Mughayyer municipality's officials as well as local residents.

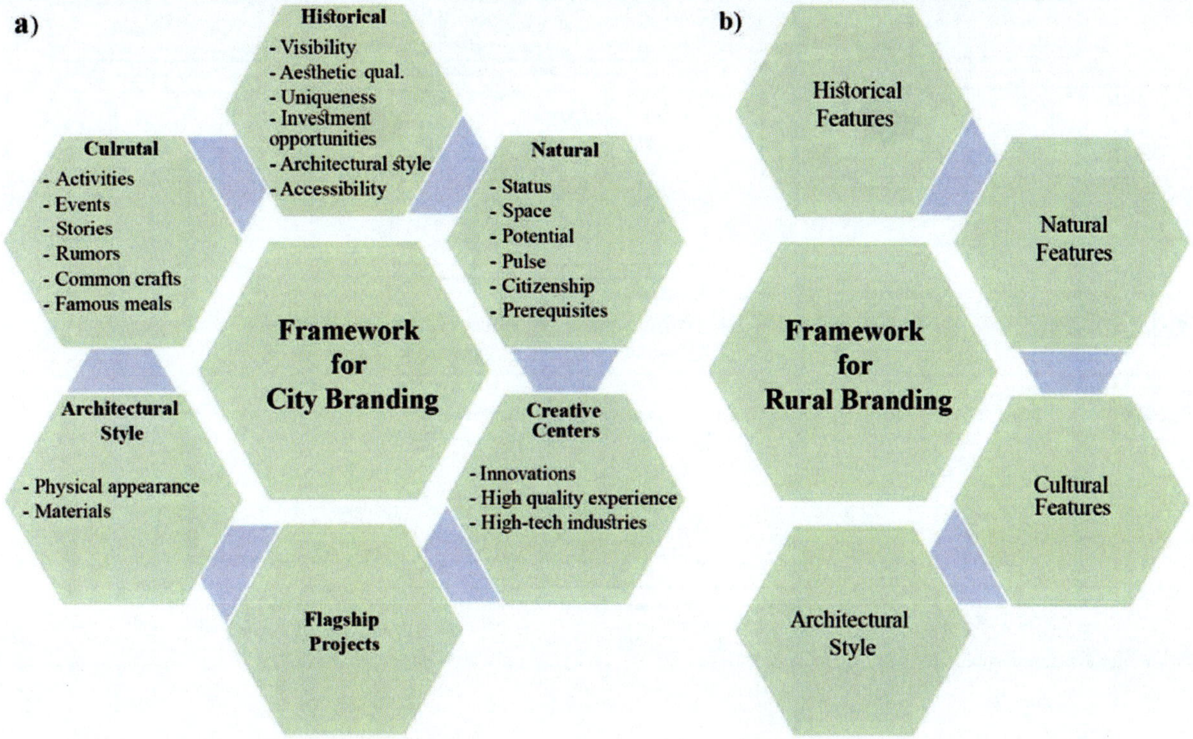

Fig. 3 a) Framework for city branding. b) Framework for rural branding

The historical aspect contained four historical hills which were Al-Fukhar, Um-Qale'tein, Um-Jrain, and Zeiraqun hill. These features have been evaluated based on its symbolic value, spiritual value, aesthetic value, visibility, preservation conditions, and accessibility. The natural aspect contained green spaces which included sports fields, public parks, and green playgrounds, agricultural lands, and springs. These features have been evaluated based on visit frequency, pleasant experience, status, and accessibility. The cultural aspect contained popular activities, meals, crops, and crafts which were evaluated based on open-ended questions (Fig. 4). This stage was divided into two phases, the first was the scale evaluation phase and the second was the open-ended question phase.

6.2.1 First Phase: Residents and Authors Evaluation of Historical and Natural Aspects

Local residents have evaluated agricultural lands and springs with the highest ranks with an average of 3.1 and 2.9 out of 4, respectively, while historical features had the lowest ranks except for Zeiraqun hill which has been evaluated with an average of 2.7 out of 4 (Fig. 5).

On the other hand, authors have evaluated the agriculture and springs with the highest ranks with an average of 3.5 and 2.5 out of 4, respectively, while historical features had the lowest ranks except for Zeiraqun hill receiving an average of 2.6 out of 4 (Fig. 6).

6.2.2 The Second Phase: Open-Ended Questions Regarding the Cultural Aspect

This phase aimed to get a clear vision about the cultural aspect of the town by providing answers with specific names, places, and information. These answers were clarified as a set of charts and diagrams from the respondents' point of view (Fig. 7). For popular crops, the majority of people (43%) have mentioned olive trees as the most popular plant in the town, while the fig, barley, wheat, pomegranate, and others such as almond and field crops had responses of (21%, 12%, 11%, 7%, and 6%), respectively. As for landmarks, the majority of people (63%) have mentioned Rahub spring as the most popular landmark in the town, while Al-Shalalah spring had (16%) of the responses, and other features such as mosques had (21%). For cultural activities, the majority of people (50%) have mentioned religious activities as the most popular activity in the town such as the Prophet Mohammed's birthday celebration. Festivals such as the national day, wedding celebrations, theatrical, and cultural festivals had (43%) of the responses. The other activities such as olive harvesting had (7%). For famous meals,

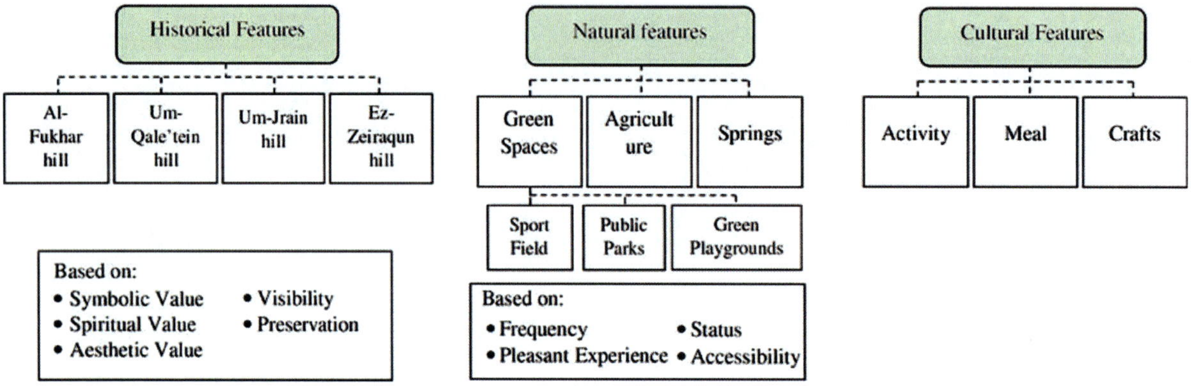

Fig. 4 Criteria to evaluate Al-Mughayyer town based on field observation and interviews

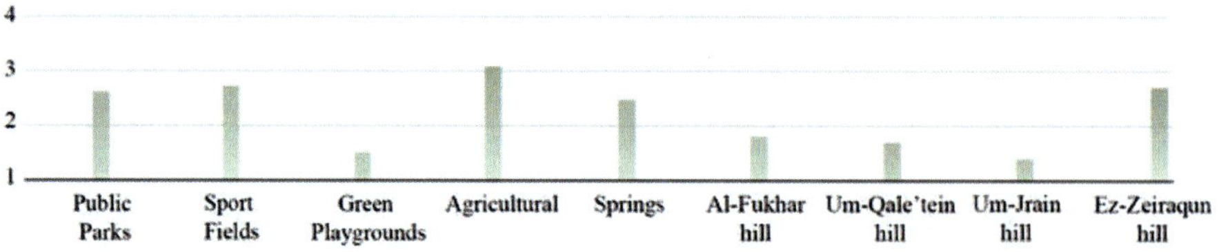

Fig. 5 Residents evaluation

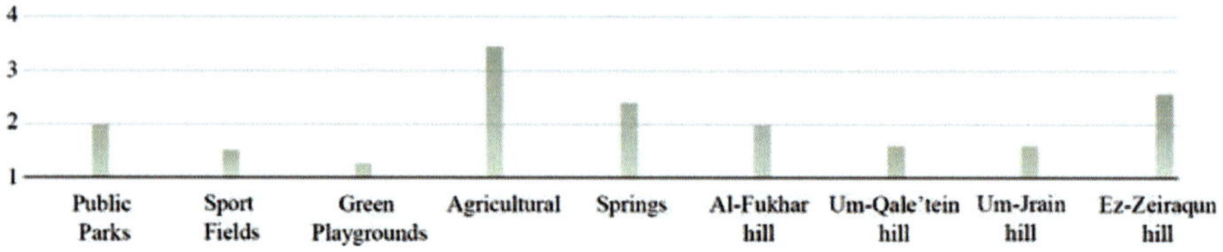

Fig. 6 Authors evaluation

the majority of people (49%) have mentioned Makmourah as the most popular meal in the town, while Mansaf had (40%) of the responses. The other meals such as snacks and Khobaizeh had (11%). For common crafts, the majority of people (73%) have mentioned agriculture as the most popular craft in the town, while handicrafts such as Mosaic had (14%) of the responses and weaving had (11%).

6.3 The Third Stage: Defining Al-Mughayyer Identity

To specify the town identity, various interviews have been done with locals and officials, in addition to the answers of the open-ended questions derived from the questionnaire. Natural aspects were the most mentioned with 70 responses out of 100. Specifically, Rahub spring had the highest weight in local's minds for its representation of the town's identity. Seventeen of the respondents valued historical aspects, while 13 respondents valued the cultural aspects as representatives to the town's identity (Fig. 8).

6.4 The Fourth Stage: Clarifying a Specific Framework for Branding Al-Mughayyer

In order to build a framework for branding rural towns that suits Al-Mughayyer aspects, another modification has been done based on the evaluation and the defining identity stages results. This modification resulted in a new framework specified for Al-Mughayyer town, which focuses mainly on the natural aspect represented in agricultural lands and springs—especially Rahub spring, the historical aspect which has been limited to include Zeiraqun hill, and cultural aspect represented in religious and festivals activities, agricultural craft, olive and fig crops, and Makmourah meal which have a potential for development (Fig. 9).

6.5 The Fifth Stage: Suitability Analysis

According to the highest-rated features and the defined identity, the suitable areas for branding Al-Mughayyer have been selected based on the following phases:

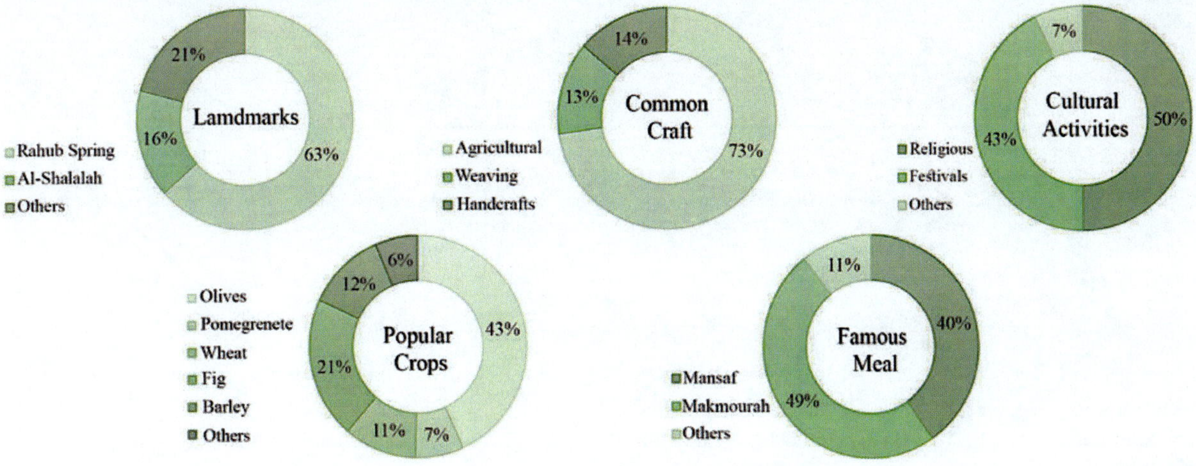

Fig. 7 Open-ended questions results

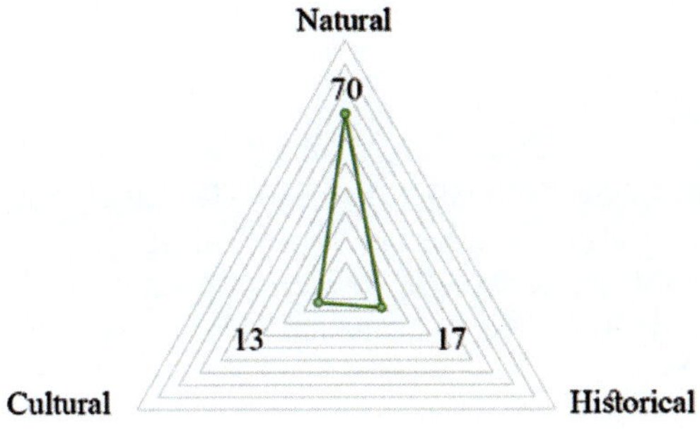

Fig. 8 Town's identity

6.5.1 First Phase: Al-Mughayyer Features

Using the GIS, the town's aspects which divided into natural historical and cultural aspects were digitized in the map to be used in the suitability analysis (Fig. 10).

6.5.2 Second Phase: High Rated Features

After selecting high rated features, only agricultural lands, sports fields, springs, Zeiraqun hill, and the public park have remained. After preparing maps and highlighting the highest-rated features, a comparison between these features was carried out according to the defined identity, which led to eliminating some of the highly-rated features that do not reflect the town identity such as sports fields and public park because these were required services and not a significant feature in the town.

6.5.3 Third Phase: Town's Infrastructure and Slope Map

Al-Mughayyer infrastructure components such as streets and services located on the map to determine suitable areas (Fig. 11a). According to Mulia and Prasetyorini (2013), the maximum slope for open and outdoor areas is (15–25%).

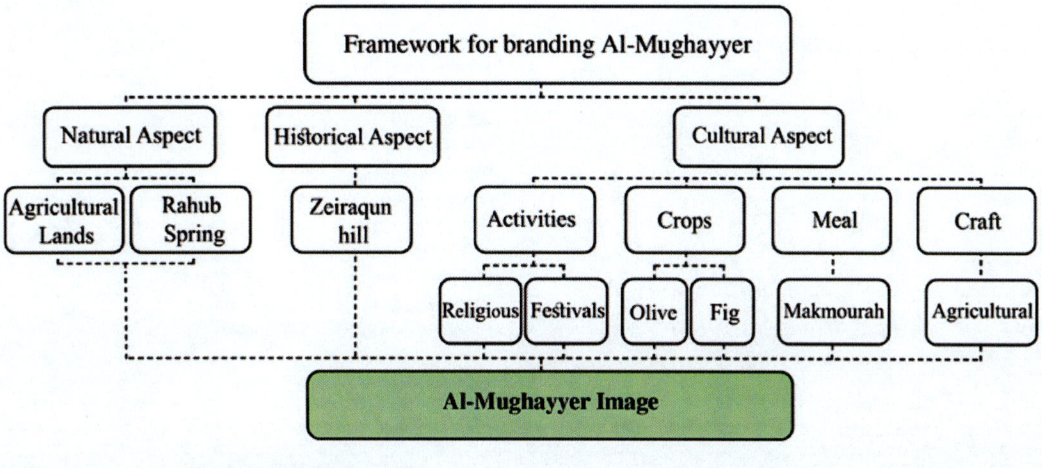

Fig. 9 Specific framework for branding Al-Mughayyer town

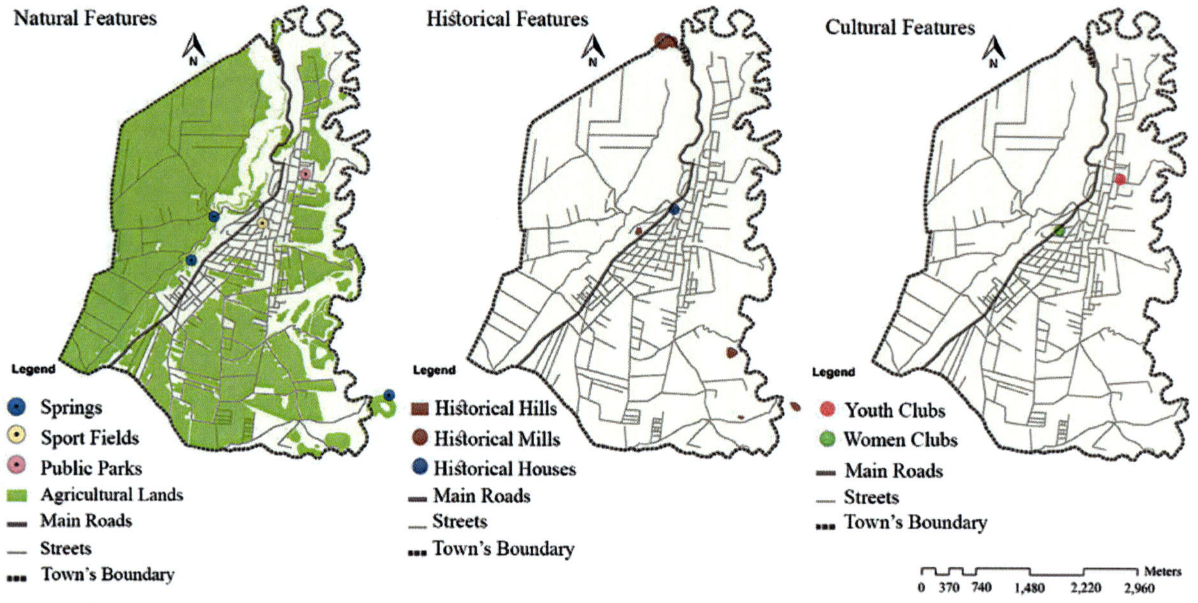

Fig. 10 Al-Mughayyer aspects divided into natural historical and cultural classes

Thus, a slope layer was added to the suitability analysis model to eliminate the areas with steep slopes (Fig. 11b).

6.5.4 Fourth Phase: Distances from Selected Features

Distances from the selected features were calculated to choose the best areas based on their proximity to the desired features (Fig. 12).

Based on the interviews and questionnaire results in addition to discussing these input layers with the local authorities, the researchers distributed the weight as follows. Rahub spring had (35%), the slope had (30%), historical features had (10%), agricultural lands had (15%), restaurants had (2%), the commercial services had (2%), streets had (2%), the main road had (2%), and finally gas station had (2%) (Table 1).

Furthermore, Al-Mughayyer town has been divided into four categories, which are highly suitable, moderately suitable, marginally suitable, and not suitable based on the results of the previous three phases. Highly suitable zones were concentrated around Rahub spring as it had the most influence on the suitability analysis (Fig. 13).

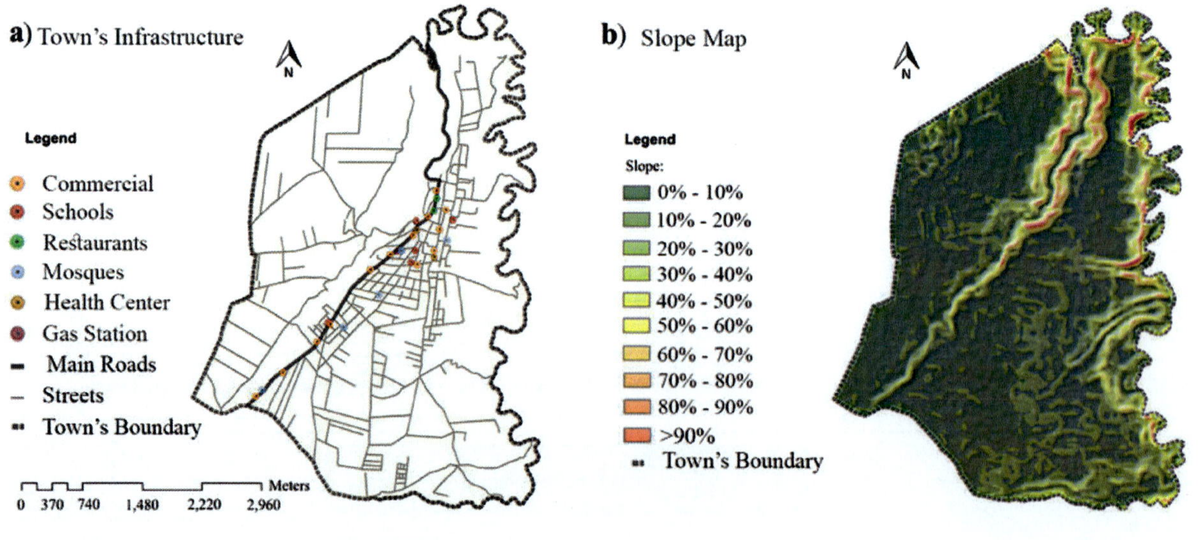

Fig. 11 **a** Town's infrastructure map. **b** Town's slope map

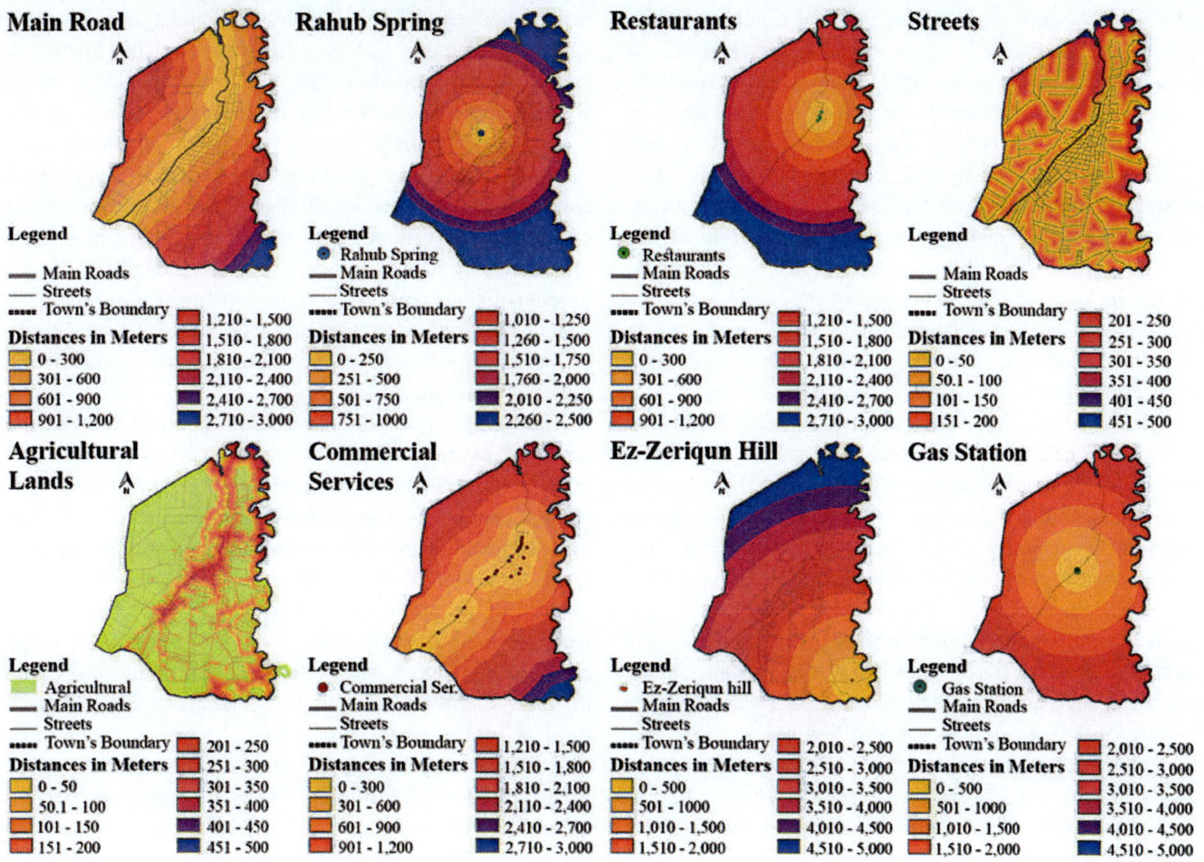

Fig. 12 Maps representing distances from the selected features

Table 1 Influences of input layers

No.	Input layers	Influence (%)
1	Rahub spring	35
2	Slope	30
3	Historical features	10
4	Agricultural lands	15
5	Restaurants	2
6	Commercial services	2
7	Streets	2
8	Main road	2
9	Gas station	2

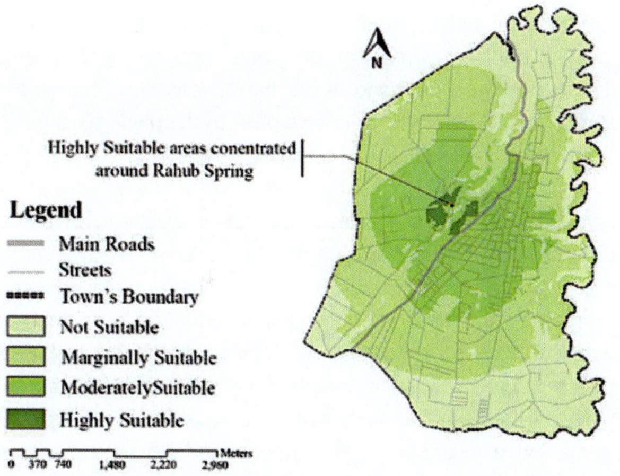

Fig. 13 Map of the suitable zones for branding Al-Mughayyer town

7 Discussion

While there were agreements between the assessments of local residents and the researchers in the evaluation results, there was an obvious difference in some other assessments. These various assessments can be clarified as follows:

Results that were conducted to evaluate the natural, historical and cultural aspects in Al-Mughayyer town showed a consensus among the local residents about the superiority of natural features over the rest of the features. For instance, agricultural lands were evaluated by local residents with an average of 3.1 out of 4 as it was considered places for vacations and family gatherings. However, some resident's opinions were negative due to the lack of children play areas in the town. As for springs, they were evaluated by local residents with an average of 2.9 out of 4, which is also a good assessment. However, the accessibility to these springs and the high visit frequency of the residents to them, as well as their pleasurable locations contributed to raising the assessment of these places. The rest of the features included in the green spaces category such as public parks, sports fields, and green playground areas had varied assessment rates. Public parks obtained an average of 2.6 out of 4 by local residents as it had high frequent visits and it was easy to access. Similarly, sports fields have been evaluated by local residents with an average of 2.7 out of 4, because they were considered as a good forum for young people—which represent the majority of the town population—to practice their various activities. Sports fields had high frequent visits from local residents and easy access. On the other hand, green playgrounds have been evaluated by local residents with an average of 1.5 out of 4. Most of the locals claimed that there were no green playground areas. With regard to the historical features, they obtained lower assessment rates than the natural features. For example, Al-Fukhar, Um-Qale'tein, and Um-Jrain hills got a low evaluation as they had no symbolic or spiritual values for the residents. However, the criterion of accessibility contributed to raising the rate of their evaluation, but they stayed with low historical value. In contrast, Zeiraqun hill had the highest rate with an average of 2.7 out of 4 which is close to a good rate. Its symbolic and spiritual values had a good rate. On the one hand, it had high accessibility which contributed to raising its rate. On the other hand, it had no clear features except for some ruined walls which contributed to decrease its rate.

However, based on field visits and observations, the authors' evaluation was consistent with the residents in evaluating the natural features to some extent with some differences. For instance, agricultural lands had a high evaluation with an average of 3.5 out of 4 which is even more than the residents' evaluation. This high evaluation was due to the diversity of its topography and vegetation which made it a unique area for visitors from urban areas. On the other hand, springs had lower evaluations compared to resident's evaluation with an average of 2.5 out of 4. This was due to its status as it needed maintenance and development plans to become an attractive natural feature for visitors. The rest of the features included in the green spaces category such as public parks and sports fields had also lower evaluations compared to residents' evaluation. Despite the high frequent visits from local residents and easy access to them, sports fields and public parks did not match the basic standards to form an attractive destination. They neither had enough playground equipment nor a unique landscape planning that may help to brand the town. For green playgrounds, they had the lowest rate due to its absence in the town. The historical features were evaluated low by authors as they were not considered reliable elements for branding the town yet. They were neglected and unclear. On the other hand, Zeiraqun hill got an assessment of 2.5 out of 4 because of its importance as a historical hill which was a populated city in the Bronze Age, even though it does not have an obvious physical appearance. However, its historical value in addition to its image in local resident's minds as a historical element may give it a potential development and importance in future.

Open-ended questions were used to study the cultural aspect of the town. Some of the responses were answered by the questionnaire and some by the interviews. For popular crops, olives got the highest percentage. However, figs—although a common product in the North of Jordan—have been considered a significant species particularly favored by people in the area for its distinctive taste. Similarly for famous meals, Mansaf and Makmourah got a high percentage therefore both were chosen for their potential in branding the town. Mansaf is considered as the traditional meal of Jordan while Makmoura is famous in the northern towns. For cultural activities, while religious activities such as the prophet Mohammed's birthday celebration got the highest percentage, other cultural festivals also took a high percentage. Therefore, these festivals were considered as promising features and an opportunity through which they can brand the town. As for the landmarks, it was clear that Rahub spring was the symbol representing the town, the residents explained that it is where the town got its common name known to many as "Al-Mughayyer Rahub" pointing to its natural spring and corridor. As for common crafts, agricultural crafts were the most prevalent in Al-Mughayyer. However, in future, they can brand other crafts such as mosaic craft that was popular in the town and disappeared over time. These answers showed that in branding studies, the qualitative data based on interviews and discussions with locals and authorities are more significant than the quantitative data based on statistics and numbers. Some answers may not get consensus and may have low percentages but they were important for the branding process.

Defining the towns' identity was investigated through interviews with all the respondents to the questionnaire. Discussing open-ended questions with respondents, in addition to discussing what the town represents to them and what distinguishes it, helped to define the towns' identity. Some of the interviews were conducted with residents to clarify the town's identity based on local residents' point of view. The others were interviews with Al-Mughayyer municipality officials such as the head of the Local

Council of Al-Mughayyer municipality to clarify the town's identity based on the official's point of view.

Rahub spring received the highest responses among the local residents as the most famous landmark in the town with 60 responses. The question about the identity of Al-Mughayyer town had a consensus that natural features are the main features to form the towns' identity. This was clearly based on the wide agricultural lands which were famous for olive and fig trees in addition to the variety in the hillsides and slopes which create a stunning view around it. Also, there was a consensus that Rahub spring was considered as the main natural feature in the town which continued to supply it with water.

A new framework for branding Al-Mughayyer town has been clarified by eliminating some criteria and modifying others. Both the natural and historical aspects were retained as the basis of the framework. However, the cultural aspects were opportunities for supporting natural and historical aspects. They may act as triggers for motivating branding such as local foods, agricultural products, crafts, and many seasonal activities.

In order to define the most suitable areas for branding the town's image, the highest-rated natural and historical features such as Rahub spring and Zeiraqun hill were highlighted on the map. These areas have appeared very near to Rahub spring as it represents the town's identity. Conversely, there were not any areas located near to Zeiraqun hill because it does not represent the town identity despite its important historical feature. In these suitable natural areas, one can find an opportunity to brand cultural aspects.

8 Limitations

The study has faced two main limitations. The first was the lack of descriptive analysis of any heritage sites. Due to this limitation, the architectural aspect was eliminated from the study. The second was the lack of studies and research on the town and similar rural towns concerning rural branding.

9 Recommendations

Rural towns have the characteristics and advantages that make them attractive places for branding. Development plans are significant for rural towns, but it is recommended to make plans based on studies that take into account the identity and nature of the town. In particular, development projects should enhance the identity of the town and should be adequate with the nature of its residents, rather than focusing only on the economic benefits. In Al-Mughayyer town, several projects have been established, such as the road network project that covers most areas of the town. However, it is recommended that the natural features of the town, especially the agricultural lands and the water springs, be taken into consideration in future development plans as they represent the town's identity. Moreover, it is recommended to work on projects that increase the productivity of these lands and improve the experience of visitors, whether recreational or investment projects. In future, it is also recommended to have more in-depth interviews and discussions with authorities and locals from various demographic and economic groups in order to have a full understanding of the significant features of rural towns when running such studies.

10 Conclusion

In light of globalization, competition has increased among cities to create an image that defines the place as a distinctive destination that encourages people to visit the city. For that reason, the city branding strategy was adopted by many cities due to its direct impact on the economy, the infrastructure, and the quality of life. The city branding strategy has concentrated particularly in main urban areas and major cities due to the concentration of services, major projects, and well-established infrastructure. However, applying or adopting city branding strategies in rural areas may be more beneficial and more advantageous as it promotes the economy and enhances people's loyalty to their culture. Rural towns are small communities, and therefore, create opportunities for more interaction with locals and understanding of the current significant identity-related features for branding.

In Jordan, most development plans and projects are slightly confined to major cities and areas of historical and tourist attractions. Therefore, Jordan's image could be restricted to these cities and landmarks. However, there are many rural towns in Jordan that—if included in development plans—may increase the diversity of Jordan's image. Moreover, these rural towns may become attractive areas for investors benefitting the economy of Jordan in general and the quality of life for the towns' residents in particular. In this paper, the town of Al-Mughayyer was studied as an example of rural towns in Jordan due to its varied natural, historical, and cultural characteristics. The study relied basically on activating the role of community participation in contributing to decision-making. That was through the evaluation and responses of the town's residents which contributed significantly in the results. However, results have shown a consensus among the residents about the superiority of natural features over the historical and cultural features in Al-Mughayyer. Thus, Rahub spring had the most rated value as it represents the town's identity. The study showed the importance of qualitative data regarding rural branding studies. Interviewing people of different ages, genders, social classes, and other differences help to understand and to identify the identity of the town. This, in turn, leads to development plans that are consistent with the nature and needs of its residents. Moreover, prevent rural towns from becoming places more like cities without a special identity. It prevents losing the characteristics that distinguish them and authentic cultural and natural features. In the long run, the locals as main drivers for future planning can help to render these plans successful. However, future research and studies may take into consideration the results and recommendations of this paper to study suitable projects that may be established in the proposed suitable zones in the town to enhance the identity based on interviews, site visits, discussions, and consultations with locals and authorities in the town.

Acknowledgements This research was conducted as part of the requirements for Planning Workshop 1, a class in the master of Planning and Urban Studies, Department of City Planning and Design, College of Architecture and Design, at Jordan University of Science and Technology. We thank our instructors Dr. Anne Gharaibeh and Dr. Yasmein F. Okour for their help and thank Al-Mughayyer municipality for their cooperation.

References

Abu Al Haija, A. (2011). Jordan: Tourism and conflict with local communities. *Habitat International*, *35*(1): 93–100.https://doi.org/10.1016/j.habitatint.2010.04.002

Abdelaal, M., & Hussein, R. (2012). Branding urban spaces as an approach for city branding-case study: Cairo City, Egypt. *International Journal of Economics and Management Engineering, 6*(11), 3022–3030. https://doi.org/10.5281/zenodo.1074615

Akasheh, T., Haddad, N., Fakhoury, L., Al Ajlouni, F. & Debajah, M. (2016) Jordan Umayyad Route. https://umayyad.eu/sites/default/files/Jordan.pdf

Chan, C., & Marafa, L. (2014). Rebranding Hong Kong "Green": the potential for connecting city branding with green resources. *World Leisure Journal, 56*(1), 62–80. https://doi.org/10.1080/04419057.2013.876587

Clark, J., & Rice, G. (2019). Revitalising rural Scotland: Loch Fyne, branding and belonging. *Journal of Place Management and Development, 13*(1), 18–29. https://doi.org/10.1108/JPMD-06-2019-0044

Daher, R. (2007). *Tourism in the Middle East*. (9th ed.). Channel View Publications.

Dinnie, K. (2007). Competitive identity: the new brand management for nations, cities and regions. *Journal of Brand Management, 14*(6), 474–475. https://doi.org/10.1057/palgrave.bm.2550086

Fok, K., & Law, W. (2018). City re-imagined: Multi-stakeholder study on branding Hong Kong as a city of greenery. *Journal of Environmental Management, 206*, 1039–1051. https://doi.org/10.1016/j.jenvman.2017.11.045

Garcia-Lamarca, M., Anguelovski, I., Cole, H., Connolly, J.J., Argüelles, L., Baró, F., Loveless, S., Frowein, C. & Shokry, G. (2019). Urban green boosterism and city affordability: For whom is the 'branded' green city? *Urban Studies.* Advanced online publication.https://doi.org/10.1177/0042098019885330

Gharaibeh, A.A., Zu'bi, A., Esra'a, M. & Abuhassan, L.B. (2019). Amman (City of Waters); Policy, Land Use, and Character Changes. *Land*, *8*(12): 195-220https://doi.org/10.3390/land8120195

Gulsrud, N., Gooding, S., & Konijnendijk van den Bosch, C. (2013). Green space branding in Denmark in an era of neoliberal governance. *Urban Forestry & Urban Greening*, *12*(3), 330–337. https://doi.org/10.1016/j.ufug.2013.03.001

Hall, P. (1993). Forces shaping urban Europe. *Urban Studies, 30*(6), 883–898. https://doi.org/10.1080/00420989320080831

Harvey, D. (1989). From managerialism to entrepreneurialism: The transformation in urban governance in late capitalism. *Geografiska Annaler, Series B, Human Geography, 71*(1): 3–17. https://doi.org/10.1080/04353684.1989.11879583

Kafafi, Z. (2011). Neither early Bronze Age Cities nor States in the south of the Levant: Another perspective. *Syria, 88*, 47–57. https://doi.org/10.4000/syria.894

Khirfan, L., & Momani, B. (2013). (Re)branding Amman: A 'lived' city's values, image and identity. *Place Branding and Public Diplomacy, 9*(1), 49–65. https://doi.org/10.1057/pb.2013.1

Kompaniets, O. R., & Rauhut, D. (2016). Place marketing of rural areas: Why cannot city marketing strategies be 'copy-pasted'? Paper presented at Regional Studies Association's European Conference, Graz, Austria, 3–6 April 2016

Medica, I., Ruzic, P., & Ruzic, T. (2010). Architecture as a tool for branding in rural Istrian tourism destination. *Turizam, 14*(2), 78–86. https://doi.org/10.5937/Turizam1002078M

Mulia, A., & Prasetyorini, L. (2013). Slope stabilization based on land use methods in Ambang Sub River Basin. *Procedia Environmental Sciences, 17*, 240–247. https://doi.org/10.1016/j.proenv.2013.02.034

Pedeliento, G., & Kavaratzis, M. (2019). Bridging the gap between culture, identity and image: A structurationist conceptualization of place brands and place branding. *Journal of Product & Brand Management, 28*(3), 348–363. https://doi.org/10.1108/JPBM-01-2018-1735

Pınar, I., İğneci, M., & Tütüncüoğlu, M. (2017). Trends and applications in city branding: A case study in Izmir. *Marketing and Branding Research, 4*(1), 25–32. https://doi.org/10.33844/mbr.2017.60417

Prilenska, V. (2012). City branding as a tool for urban regeneration: Towards a theoretical framework. *Architecture and Urban Planning*, *6*. https://doi.org/10.7250/aup.2012.002

Purwanti, A., & Genoveva. (2017). An evaluation of city branding to reinforce the city competitiveness (a case study of Surabaya). *International Journal of Management and Applied Science (IJMAS), 3*(5), 119–124

Rezvanpour, N., & Bayat, A. (2017). Determining effective urban design factors within the branding strategy due to brand city spaces and evaluating city spaces by comparing them to the presented factors. A case study of ChaharBagh Avn, Isfahan, Iran. *Energy Procedia, 115*, 6–17. https://doi.org/10.1016/j.egypro.2017.05.002

Rodrigues, C., Skinner, H., Dennis, C., & Melewar, T.C. (2019). Towards a theoretical framework on sensorial place brand identity. *Journal of Place Management and Development*. Advanced online publication. https://doi.org/10.1108/JPMD-11-2018-0087

Sakka, S. (2016). Sustainability as an effective tool for a place branding: An application on El Gouna City, Egypt. *International Journal of Environmental Science and Development, 7*(11), 861–866. https://doi.org/10.18178/ijesd.2016.7.11.895

Tobias, S., & Müller Wahl, P. (2013). Can place branding support landscape conservation in city-regions? A case study from Switzerland. *Land Use Policy, 30*(1), 266–275. https://doi.org/10.1016/j.landusepol.2012.03.021

Wilson, R., & Muldrow, B. (2010). Telling the untold story: Branding in rural communities. *Forum Journal, 24*(2), 24–30

Urban EcoMobility Hub. A Tool to Design Active Transportation Frameworks

Gerardo M. Arzate Pérez, Miguel Arzate Pérez, and S. A. Gamboa

Abstract

This research used a prospective methodology based on two methods: the logical future and the analysis of forces, to determine with qualitative and quantitative values the forces that intervene in the construction of an urban mobility system. The article shows the key forces obtained as a result of the research, and the summary of their historical characteristics with references in the main cities of the early twentieth century and Mexico City. The second part of the article shows how the research results were transformed into a tool that allows to quickly design a framework to develop new active mobility systems in high-density urban areas, proving a different perspective, that can lead us to more humane design methods and provide us with new mobility alternatives.

Keywords

Urban mobility • Active transportation systems • Ecological transportation

1 Introduction

Mobility has always been part of the human settlement configuration. The approach of human beings toward the use of territory has gone through multiple stages, always characterized by the dominance of human beings over nature, and the use of science and technology to build artificial environments, until society reached the configuration of contemporary mega cities. Today, mobility systems reached every corner of the world, growing around the "anyone, anywhere, anytime slogan of the ubiquitous society that characterizes our times" (De Salvador Agra, 2015), and human beings have become a kind of super nomadic citizens in a surprising way, driving all over the city, from metropolis to suburbs, and traveling at the speed of sound between cities that once were considered far away, causing a strong reconfiguration of urban space, especially during the past five decades. Nevertheless, why when someone compares the cities with the highest growth in the early twentieth century such as London, Paris, Berlin and New York, with the cities that have had the greatest growth in Latin America in the early twentieth-first century, like Mexico City, Bogota and Sao Paulo, all of them had a similar explosive growth of space, large areas of wealth and extreme poverty, but, there are bigger problems in housing and transportation solutions in contemporary cities than in old ones (Hall, 2007). It is also interesting to notice that, in London, during 1905, poor people whose employment was in the city center had problems moving from their home to their working place, so the work of an English man who was not a designer, but a shipowner and sociologist, called Charles Booth[1] (1840–1916), who analyzed poverty from a social point of view, contributed to the design of a better subway system, which began operating in 1907 and currently continues to

G. M. Arzate Pérez (✉) · M. Arzate Pérez
División de Ciencias Y Artes Para El Diseño, Departamento de Medio Ambiente, Universidad Autónoma Metropolitana—Azcapotzalco, Av San Pablo Xalpa 180, Reynosa Tamaulipas, Azcapotzalco, 02200 Ciudad de México, CDMX, México
e-mail: jerry.arzate@gmail.com

S. A. Gamboa
Instituto de Energías Renovables, Universidad Nacional Autónoma de México, Privada Xochicalco S/N., 62580 Temixco, Morelos, México

[1](Liverpool, 1840–Whitwick, 1916). Sociólogo británico, cuyas investigaciones contribuyeron al conocimiento de los problemas sociales de la sociedad inglesa como consecuencia de la Revolución Industrial. Fue consejero privado del gobierno y comisionado encargado de estudiar la pobreza de las clases inglesas más bajas, con el fin de elaborar leyes para mitigar esas condiciones. Su obra principal (Life and Labour of the People in London) se divide en 17 volúmenes, describiendo las condiciones de vida y causas de la pobreza extrema en Londres. Booth fue el primero en utilizar Social Surveys, entrevistando personalmente a 4.076 casos reales, y su metodología sigue utilizándose por la escuela británica y londinenese de ciencias sociales.

F. Trapani et al. (eds.), *Advanced Studies in Efficient Environmental Design and City Planning*, Advances in Science, Technology & Innovation, https://doi.org/10.1007/978-3-030-65181-7_50

efficiently serve the entire city, more than 100 years later. Additionally, the government carried out a project to electrify the streets and extend the tram system, enabling workers to acquire better quality housing at lower prices, improving the quality of life and access to workspaces, and designers of early twentieth century built quality housing based in long-lasting architecture for the working class. In the case of contemporary cities, development has not reached the same accomplishments, Mexican working class, for example, are typically living at informal settlement buildings with a lack of pavement, street lighting, drinking water, drainage, architectural quality, infrastructure and construction specifications. The mobility problems for the working class are worst today than 100 years ago, and it is exacerbated by the size of the city, and millions of people spend more than 4 h every day to commute. The automotive transport system that powers the motion capacity of the city has created big consumption and environmental problems. The 2016 Emission Inventory developed by the Mexican Minister of Environment indicates that urban vehicles consume 60% of all fossil fuel used by the city and the emissions produced by this activity are 16.3 million Tons CO_2e every year (Secretaría de Medio Ambiente de la CDMX, 2016) (Fig. 1).

We can also see this problem in Sao Paulo, which is three times larger than the size of London a century ago; however, Sao Paulo has a much less developed public transport system and the subway has 101 km, while London's tube in 1907 has around 400 km. This can be somewhat perverse, because in relative terms, the lower working classes of twenty-first century cities have a harder time getting to work than workers who lived in London or New York in early 1907 (Hall, 2007). So, it looks like something is going in the wrong direction, when we talk about the impact of mobility on quality of life, energy consumption, environmental impact and land use. It also looks like we could learn something from the old cities to build our future urban mobility, therefore, using prognosis as a methodology to analyze past events, in search of patterns, causes and effects of our actions, is key to building the future.

That is why the methodology was based in two stages. The first stage was called "logical future", and it is an analysis method that look the past events to understand why the present reality is the way it is, and states that if we keep on doing the same, the "logical future" is to obtain the results in the same directions. This method searches for mixed historical evidence (qualitative and quantitative), and synthesizes the data into groups of variables that act under the same conditions. To visualize the evolution of variables, organizing them and understanding their impact on current reality problems is the main goal. The second stage was called "design tool" and is based in a method called force analysis. It consists of analyzing the variables obtained in the logical future, observing them as forces that have an effect The process consists of stage one to create a design tool that, when used, fosters positive strategies in the construction of cities with a future active mobility system.

It is important to notice that an open view of the problem, and the mix between qualitative and quantitative data happens, because contemporary prognosis to design the future is no more a forecast and probability topic just related to numbers and maths. That is why several design teams around the

Fig. 1 Daily rush hour in Pantitlan subway station. Mexico City 2017 (*source* mapa-metro.com)

world are thinking that the design of future mobility has to study different aspects of transportation beside technical and technological issues, because urban mobility is also powered by the land use of the space that we inhabit, and mainly by the dynamics of the people. The actions and desires of each group of inhabitants also shape the transport, not only materially or formally, but from a broader spectrum. The changes that have taken place in the world's mature cities, like Mexico City, go beyond appearance and technological improvements; it is the affluence of society that builds them, the patterns of living that its citizens adopt, and perhaps most critically, the struggle for economic advantage by one city over another (Sudjic, 1991). The energetic and environmental values of urban mobility also are not only related to technology, but people, passengers and users, their perceptions, and their way to choose between different transport systems. When someone decides to use public transportation, private automobile, bicycle, or take a hike, it is increasing the demand of energy, the amount of emissions and the environmental impact of urban mobility.

Therefore, transportation systems are built by the longings and activities of a lot of community members, along with the availability of technological solutions and the configuration of urban space. It is crucial to empower a holistic design vision that includes these values in the prognosis and the design process. A good example of this transportation design vision is explained by Gary Toth, who works at the New Jersey Department of Transportation (NJDOT) and made a review of the New Jersey Future in Transportation program. The goal of this program is to change the decarbonification and car-based design paradigm that almost everybody uses to reduce the vehicle miles traveled (VMT), transportation greenhouse gas emissions, road congestion and traffic patterns. The way to do this, is not by doing quantitative analysis like transport demand calculations, or origin–destination studies, it is just by finding new design values and higher quality, in public spaces, mobility, air and health, so the method he proposes is to apply design strategies focused on this community issues:

- Downsizing new investment in new capacity for state highways.
- Working with communities to create a connected network of local streets.
- Helping communities with land use design.
- Implementing context sensitive street designs (CSDs) on highways.

So, this is one example of many others designers and planners know that the relationship of quality factors can redefine transportation over time. Now more than ever, we need to be more creative in shaping future cities, and we need to move fast. Mexico City, Latin America and Asia will receive the urbanization of the future, and the effects will not stop. The population will grow; demanding more transportation capacity, coverage and technology, and as a result, we will implement new transportation solutions that provoke the territory and population to grow again. It is a cycle in which the city pushes transportation and vice versa (Gabiña, 1998).

All of this gave us the support to propose the sequence of this methodology for the project:

Logical Future

- Critical examination of reality.
- Detailed bibliographic investigation of the history of the problem.
- Fragmentation of the problem into general historical variables—Defining "general historical variable" as the set of events that occurred within the analyzed historical period, which acted as pressures of any kind, or forces that impacted and drove in various ways, the construction of the problem analyzed.
- Fragmentation of general historical variables into particular forces—The list must be obtained through formulations or analytical descriptions, derived from the knowledge of the specialist who coordinates the project, backed by the existing bibliography on the subject.
- Synthesis and logical ordering of historical variables—Synthesize, regroup and order all general and particular variables into categories, according to the relationships and similarities that are found between them. These categories can in turn be separated into internal and external to the problem or topic of study.
- Enumeration of historical variables—Establish the influence of the variables within the construction of the current problem, to from this moment understand the variables as forces (that influence the problem).

Framework Design Tools

- Force-effect relationships. To visualize variables as forces, and to understand the effects that forces can have in the construction of new mobility futures.
- Mobility strategies—Based on the force-effect relationship, generate strategies that counteract or take advantage of the effects of each force to create new mobility futures.
- Framework design tool—To transform the strategies to design tools that can be used as a practical tools to develop new active urban mobility networks.

The final framework design methodology was called EcoHub Mobility. To reach that methodology proposal, the first step was to develop on detail the logical future, we managed to synthesize the variables into five main categories, which were defined as the most influential forces in the space-dweller-transportation dynamic, and the reduction of fuel consumption and emissions. The summary of the investigation of each category with its main characteristics is presented below:

1.1 Energy Demand—Population Grew up from Low Rural to High Urban Consumption

The way of life offered by contemporary cities has become very attractive to many people. "Cities generate more interactions with more people than rural areas because they are central places of trade that benefit those who live there" (Blanchard & Volchenkov, 2009). It means the idea of "modernity" is setting into a city, which is a space considered to—or has the illusion to—has "better quality." This idealization has caused the expansion of urban areas, as well as significant impacts on immigration, disappearance of rural spaces and saturation of urban centers, among others. Thus, in 1950, 30% of the world population lived in urban areas, and today more than half of the world's population already lives in urban areas. Asia hosts 153 "one million cities" and 18 megacities, and in three decades, Latin America will be a highly urbanized region. São Paulo ranks in the top 10 with 18 million inhabitants, Buenos Aires with 12 million inhabitants, Rio de Janeiro with nearly 8 million, Bogota and Santiago with 6.8 million and 5.5 million, respectively (Hall, 2007).

In Mexico, the population has increased sevenfold from 1900 with 15 million to upwards of 110 million inhabitants in 2010. This growth was concentrated in Mexico City, where only 500,000 people were living by 1900, and in 2010, the city became a Metropolis with more than 20 million citizens that produce huge pollution levels, unhealthy spaces, territorial saturation, lack of services and insecurity.

Planners thought that the city would be able to contain the population growth but citizens began to create their own boundaries with informal settlements that had different governors, resources and activities. Today, Mexico City's population is 8.9 million people, with a population density of 6000 citizens/km^2, within 1,485 km^2, and it belongs to the Mexican Valley Metropolitan Zone (ZMVM), where current population exceeds 21 million people in an area of 8000 km^2. This means 2560 citizens/km^2 within 16 municipalities from Mexico City, 59 municipalities from the State of Mexico and one municipality from the Hidalgo State, covering an area of 4700 km^2 (Secretaría de Medio Ambiente del Distrito Federal, 2006). If we translate the mobility of the population to transportation demand, we find these statistics: according to the Origin Destination ZMVM Survey, the ZMVM contains 3.05 million motor vehicles. It is estimated that in one day 6.60 million trips are made by car. The average number of occupants per car is 1.5 people, and the greatest activity of trips is to work and school. 34.6 million trips are made in the ZMVM on a weekday. Of them, 11.15 million are exclusively walking, and 15.57 million in public transportation.

In relation to energy, in 2016, a total of 385.2 Petajoules2 (PJ) were consumed just in Mexico City, of which 356.4 come from the processing of petroleum and 28.8 PJ of primary energies such as natural gas, firewood, and coal. Likewise, an annual per capita consumption of 43.6 gigajoules (GJ) is estimated for the CDMX, which means that each person uses annually, the equivalent of 1300 L of gasoline, and transportation is the sector with the highest demand, with an annual consumption of almost 234 PJ, equivalent to 61% of the total energy (Secretaría de Medio Ambiente de la CDMX, 2016).

So, finding new transportation alternatives in Mexico City is very important and can bring positive impacts on a huge scale. New solutions could also be applied around the world where similar urban problems are appearing. Around the world, the number of urban regions continues to grow. In the year 2030, developing regions of Asia and Africa will have more people living in urban areas than in rural areas; for the first time in history more than 6000 million of inhabitants will live in cities (Bezold et al., 2009). Therefore, the latest report of the UN-Habitat mentioned that the world is becoming an urban environment.

According to the United Nations between 2020 and 2025, the urban population of the world will be doubled, the number of inhabitants in the cities will increase from 47% to more than 61%. Most urban growth will occur in the world's developing cities, and the urban population will double in the next 25 years in Latin America, the Caribbean, Asia and Africa. The cities in these continents will receive the urbanization of the world, and therefore the problems in transportation and technology, greenhouse gas emissions and fuel demand (Unhabitat, 2010).

1.2 Urban Activity—From Agriculture to Network Economies

The transition from rural to urban population was one of the key moments in human history that changed the way we relate to the territory. In 1910, Mexico had 11 million rural habitants and 4 million urban citizens and had 29 million rural inhabitants and 115 million urban citizens by 2010 (INEGI, 2010). Of course, this evolution affected the relationship of human settlements with the primary production sites, such as agriculture, livestock and fisheries. The oldest cities were born on settlements marked by the flow of a river, the goodness of a valley or the benefits of a favorable geographical position for hunting, fishing or planting crops. Ancient Mexico followed the same rules, and the capital of the Aztec empire was established around the big lake of Tenochtitlan, where food was produced in beds lakes known as chinampas. This system covered a surface of more than 12,000 acres of the southern lakes, and produced 10,000,000 kg maize equivalent agricultural surplus, enough to feed 50,000 people for a year (Luna Golya, 2014), in a city that has the same population of 1519 Europe's four largest cities—Paris, Naples, Milan and Venice (Gibson, 1964).

However, life changed toward the end of the nineteenth century—a new formula based on technological development emerged. The industrial revolution would appear, and many things changed. On one hand, the agricultural process changed because of the new industrial conditions of machinery, chemistry and power systems (Bujarin, 1982). On the other hand, the cities began to drive out the primary activities in order to make way for secondary activities: bonded or manufacturing factories that were managed thanks to the invention of the steam engine. Therefore, the value of the land was not linked to the productive capacity of primary goods, but the added value generated by the technological infrastructure to transform natural resources into new products. In Mexico City, this process began with early Spanish colonizers, who changed pre-Hispanic water agricultural-mobility system for a transportation system that has never existed in the ancient city, the one based on the road, horse, wagon and carriage infrastructure that use to work in Europe. Now, only 190 km of canals remain from the original Tenochtitlan. (Luna Golya, 2014). Nevertheless, the real drainage of the southern lakes began in the twentieth century, when government decided to dry the farming city. We changed chinampas and water channels for roads and buildings because we fully changed the vision and activities over the use of the land. We changed agriculture for the activities established by the industrial revolution, and this process reached its climax after WWII, when government supported politics to industrialize the country to cover the demand of industrial products for other countries. Of course, chinampas and agriculture did not work for that modern project. The city was drained, and the politics focused on urbanization, population and industrialization were implemented. Today, Mexico City produces no food; it has to import it all. The city has a huge water pumping system, not used to irrigate the food production, but to supply 2000 gigalitres annually of drinking water and drainage 1600 gigalitres of rain and wastewater (Hernández-Suárez, 2011).

Later on, secondary activities in the city also evolved, and were replaced for tertiary activities. The third sector, focused on trade and services, now lead the activities carried out in the urban centrality, so manufacture and industry were forced to leave the center of the cities, and to be established in the periphery or in less important settlements for the economic interests for the urban life. The result was a new definition of economic development in the urban centrality. During the industrial revolution, the power of the city depended on the core functions of the center. Now centrality is defined more in terms of network. So, we are living in a complex network city, and we experience it every day as a result of multiple complex interactions between networks of transportation, culture, social and economic activities (Blanchard & Volchenkov, 2009).

Today, cities are integrated into networks of cities, connecting urban spaces with other settlements of equal or greater dimensions, allowing them to expand in capacity and infrastructure at a very low cost, and to grow inwards by subdividing into small cities. Therefore, centrality depends on the networks and the participation of the city in those networks, we are talking about an urban centrality related to the cooperation within urban networks. A framework where nodes, places and cities may exercise one or more functions in one or more networks, resulting in different profiles according to the levels of specialization and diversification of the urban economy (Muñoz, 2010). The most representative example of the interest of Mexico City to be part of these networks is the construction of the new city airport, one of the biggest airports in the world. It is the most important infrastructure project for the state. It received around $10,000 million dollars for its construction, because it will receive 120 million of passengers every year and will connect the city with the world.

Contemporary Mexico City changed a lot over the last century. In 1910, the city had 720,000 citizens and a population density of 480 per km^2 (INEGI, 1910). In 2010, the city had almost 9 million inhabitants and a population density of 6000 per km^2 (INEGI, 2010). Each citizen consumes the equivalent to 1300 gasoline liters of energy per year and emits 2.5 t CO_2e every year (Secretaría de Medio Ambiente de la CDMX, 2016). This is the transformation history of Mexico City, it changed from a small unique agricultural and water city into a dry artificial megacity, where conurbations and central areas are joined in a metropolitan area that is trying to participate in a single regional network of metropolis.

1.3 Transportation Technology—From Train to Multimodal Systems

The increasing number of humans in urban areas increases the transportation needs, and this is a driving force to the city, so new technologies applied to transportation are closely related to population, migration and urbanization, and they are key factors in the land expansion of the city (Kates & Parris, 2003).

The first vehicle that changed the mobility of cargo and people between cities was the railway, because it allowed the increase in exchanges between previously remote locations. It also made it possible to generate the first industrial zones outside the city center. In Mexico, the trains had their glorious time between 1880 and 1910, with more than 25,000 km of rail network over the country. The next vehicle was the car, which worked as a detonator of mobility within the city and pushed its boundaries, bringing urban life toward the periphery, an idea that Mexico City brought form the USA and the American way of life after the World War II; the car meant the emergence of new ways of inhabiting the territory, transforming daily mobility and promoting suburban development (Muñoz, 2010).

Later, the aviation industry appeared on the scene, and eventually, the air transportation system matured to become an integral part of the design and growth of the cities. However, poor planning, the focus on economics and the lack of effort to generate projects comprehensively caused air terminals to be dramatically modified over time. The airports were initially built on the periphery, free of urban density; today they are in highly urbanized areas. The natural increase of vehicles, passengers and employees, because of big economical and activity attraction of the airport has a cumulative traffic effect, causing big regional problems. That is why airports are known as "airport cities," where workers are constantly commuting, moving every day to and from the airports (Güller, 2002).

Of course, the direct revenues and the indirect trade from aviation activities create a lot of economic interests, as it happens in other cities around the world. For example, in the London-Heathrow airport as well as in the Paris-Charles de Gaulle and Amsterdam-Schiphol airports, 50% of the airport profits come from commercial activities (Güller, 2002). So, the air transportation system has played a very important role in the planning and socio-economic activity of the city, but also created a big problem around GHG emissions, air quality and mobility. In Mexico City, the environmental impact of the International Airport (only the airships operation) emits 894,360 Tons of CO_2e (Secretaría de Medio Ambiente de la CDMX, 2016), it operates inside 50,000 acres of the Mexican Valley Metropolitan surface, and is trying to coexist with other transportation infrastructures that each day have 5.5 million of particular vehicles, 4.2 million daily subway users, more than 27,000 taxis with 1.1 million passengers and more than 2600 public buses that move 2 million passengers (SEMOVI, 2016). The city also has 46 multimodal transfer stations that help passengers, services and cargo to move through and abroad city centers, and playing a lead role in the construction of contemporary urban mobility.

Therefore, the transportation design system and the concentration of mobility infrastructure such as roads and railway lines shaped Mexico City, and allowed the growth of this urban settlement with many effects, such as the increase of all system variables, the amount of vehicles, number of trips, distance traveled, invested time, economic cost and environmental impact. Even after all these experiences, contemporary Mexican government has always considered transportation design as the consequence of the economic development of the city and is not trying to use transportation design to transform the urban settlement into a better place for living.

Of course, even though it is not a common design action, there are successful examples of the use of transportation design in different modalities as a tool to develop a city. For example, the urban design plan of Amsterdam has established a complete reorientation of the city, due to the enormous success of the airport. The same is occurring in Zurich, where there is a regional strategic plan to make the public transportation more accessible, thinking on the public transportation investment as one of the most important characteristics of that city (Güller, 2002).

1.4 Dwellers Evolution—From Citizen to Territorian

Since the industrial revolution, people have transformed natural into urban areas, leading to expand outwards their own limits, moving from small to big cities, building mega cities and consequently the economic corridors called mega urban regions. This urbanization process has impacted technologically, socioeconomically and culturally, the relationship of the citizens with the territory (Low & Lawrence-Zuniga, 2003).

A contemporary inhabitant moves and occupies space in a multifaceted way and uses the same territory in different styles with different purposes. For example, being a student, worker and tourist within the same city, or living in one place during a week and moving to another one for weekends. Therefore, the traditional metropolis changed into second generation metropolis, where full-time inhabitants are mixed with temporary citizens, who can work in one city but live in other one. After the significant growth of temporary citizens, the third-generation metropolis was born, characterized by the emergence of a third type inhabitant: visitors (Martinotti, 1993). Visitors are temporary residents of urban areas who travel for business or tourism, they nor live neither work in it, but they have been building the third-generation metropolis around the world since last century.

Brazil, one of the best tourism destinations in South America, is a good example of this effect. Since 1990, the number of foreigner visitors grown up very fast. The country received 4.8 million foreigner visitors in 1998, 65% more than in 1997, this required to land almost 870 charter planes in Brazil in the first quarter of 1999, versus 493 in the same period one year before. So, we can easily understand why visitors have a huge impact in the city transport design and the city configuration, forming urban networks to exchange goods and services even they do not dwell space, and just visit it. This huge mobilization, pushed the state to spend in that decade more than US$1.5 billion in transport infrastructure for tourism, investing in a face lift, roads to near-virgin beaches, city's sewage system and the airport expansion (Galanternick, 1999).

This new and contemporary part-time resident is called territoriant, and it belongs to a new city made of fragments of the territories where they live, work, go for shopping or visit at any moment. In Mexico City, the actual transport system design helps 63% of passengers to move outside their municipality, crossing the border between Mexico City and other states, this means a bit more than 4 million passengers every day, number that will raise to 5 million by 2020. The concept of a traditional Mexican citizen has changed from a sedentary to a super dynamic model, where the modern inhabitants uses mobility, transport and infrastructure to optimize the time and space, depending on the requirements and the activities of the day. The territoriant is now moving around the world, that is why the aerial traffic has increased ca. 6% annually, in the cities that have installed new transportation technologies to receive visitors and tourists, and the increase in transportation use around a city is an indicator of economic development of the region (Muñoz, 2010).

1.5 Mobility Evolution—From Centralized to Polycentric

As it has been explained, cities had decentralization at regional level and recentralization at local scale. This process was called concentrated deconcentrating by Dutch planners in the 1960s, concept that explains why several centers were developed in the periphery of the initial city center. This means that the traditional urbanism conception of centrality and periphery has disappeared, and several "centers" have begun to appear within the same city, when this happened, the humanity changed the suburban settlements to polycentric settlements. Now cities are composed by small cities; the centers of production, consumption and services have been cloned at different points of the city; in this way, the life that was exclusive of the center has atomized, generating the diffuse concentration of the edge cities (Garreau, 1992). Traditional approaches say that the change is produced by the concentration of a lot of the household and the economic activity, but recent studies show that traffic congestions and road capacity are the main values that creates a polycentric city (Louf & Barthelemy, 2013), where the outside areas are now peripheral centers or sub-centers, and they host a similar structure and operation than the main city center, but on a smaller scale, encouraging activities of manufacturing, commerce and culture that develops its own social identity, autonomy and less dependence of big downtown. The UN-Habitat 2004/5 report shows that the city has pointed out in complex multi-node regional ways. Urbanization has developed regional systems, where new sub-centers are independent in terms of social economic patterns but they are functionally linked to the big city, and this process has recently been called functional polycentrism, where more than half of the population lives outside the city but is generally part of the city (Unhabitat, 2004). This requires designing urban transport systems with terminals and specialized routes, dedicated to the use of the truck, rail or light rail, to connect multiple centers with multiple peripheries. Meanwhile, the city transport needs to be connected to other polycentric cities, otherwise, the metropolitan area would not be able to offer, exchange or access to other metropolitan networks, and today, contemporary territorial networks are defined by the relationship they establish between different urban spaces to exchange activities, services, goods, tourism, etc. Thus, the definition of future transport systems and technology will depend on the networks involving a city, the specialization in the area and the decided interaction develop, this way the city can interact with other spaces that offer those activities or demand such specialization (Veltz, 1999). It is known that if a city, metropolitan area or region of megacities does not participate in these networks is bound to have problems of development and growth, so it is essential to understand how urban space works, how are being modified and evolving, as well as it is important to study transportation systems to create or destroy different forms of interaction between the different elements of the city.

2 Framework Design Tool

The second step of this research was to carry out the design of the framework, so each of the five variables categories was analyzed, thinking them as forces that influence the city. The variables were translated into forces, which names represent the main effect of the force, and each effect was classified as positive or negative. The result of the process is shown in Table 1.

The second step was to create mobility strategies that stimulate forces with positive effects and discourage forces with negative effects. The conversion can be seen in Table 2.

Table 1 Relationship between variable–force–effect type

Variable category	Force name	Effect type
1. Energy demand	1. High urban energy demand	Negative
2. Urban activity	2. Network economies activity	Positive
3. Transport technology	3. Multimodal transportation systems	Positive
4. Dwellers evolution	4. Territorian lifestyle	Positive
5. Mobility evolution	5. Polycentric cities	Negative

Table 2 Forces converted to mobility strategies

Force/effect type	Mobility strategy
1. High urban energy demand/Negative	Energy economy + active ecological transportation
2. Network economies activity/Positive	Network activities to power economy
3. Multimodal transportation systems/Positive	Connect old transports with new stations
4. Territorian lifestyle/Positive	Flexible, scalable, multimodal transports
5. Polycentric cities/Negative	Polycentric mobility

Table 3 Mobility strategies converted to design tools

Mobility strategy	Framework design tool—final order
Energy economy + active ecological transportation	5. Fuel economy program
Network activities to power economy	4. Active transportation-economy network stations
Connect old transports with new stations	3. Active multimodal central station
Flexible, scalable, multimodal transports	1. Ecological mobility interconnected hubs network
Polycentric mobility	2. Micro-polycentric mobility network

The next step was to explore the possibilities of using, mixing and rearranging strategies as a design tool. The process demands to rename it and set a final order. The result is shown in Table 3.

The final step was to develop a new methodology based on each of the design tools proposed.

El objetivo fue crear una metodología de análisis y diseño, basada en herramientas simples y prácticas que al aplicarlas permitan crear un Mobility Framework de Eco-Hubs, que impulsa estrategias de moviliadad que funcionan como fuerzas con efectos positivos en el espacio urbano.

Since we want people to be the center of the system, the citizens have to be the core of the methodology, so the first tool of the table was the one who gave the name to the framework, and it is an important tool, since it was born from the lifestyle of the present-day citizen, who uses the city in multiple ways, is flexible, and is the contemporary urban inhabitant; The rest of the tools were analyzed to understand their impact in terms of mobility, energy use and environmental impact, and they were ranked based on a logical use sequence of the framework. In this way, the final proposal of the framework presented below was obtained.

EM-Hubs—Design Framework, is a mobility design methodology based in 4 design tools, the designer should follow the sequence to shape a new active transportation design framework that can be used as a reference to develop more detailed systems.

3 EcoMobility Hub—Design tools for a Mobility Framework

3.1 Micro-polycentric Mobility Network

The first tool asks you to select your space of intervention, and to divide the urban area in several action zones called EMhub, to create a polycentric city effect, that allow designers to work in little independent urban spaces that can be interconnected. It is possible to start the system building just one single hub. More hubs can be added through time, so that in the long term, the addition of little mobility systems can work together and cover a big urban zone, so the strategy of building polycentric mobility, as well as flexible, scalable and multimodal transportation, can be reach with this first tool.

Fig. 2 Annually CO_2e tons savings by change to active transportation in an EMhub

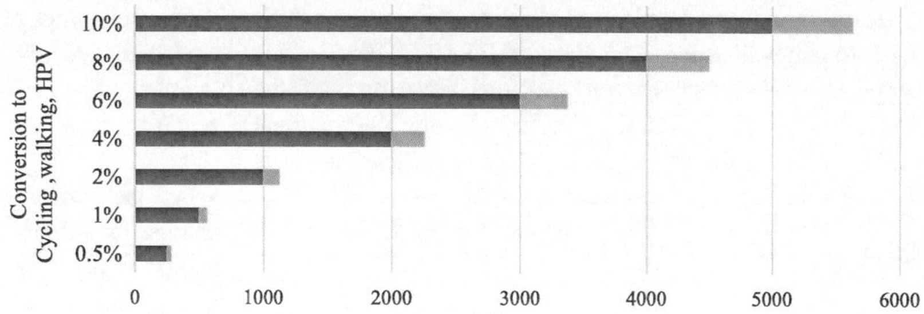

	0.5%	1.0%	2.0%	4.0%	6.0%	8.0%	10.0%
Users walking, cycling, human powered	250	500	1000	2000	3000	4000	5000
CO2e tons saving anually	32	63	126	252	378	504	630

To mark out the surface that will cover each EMhub, it is needed to draw a circle with a radio of 1.5 km approximately (one mile). This strategy allows to design a micro-mobility system that citizens can use to have "last mile" travels inside the EMhub with zero-emission transportation based in walking, cycling, human powered vehicles and electric vehicles.

Even though it seems to be a simple idea, to establish an EMhub in big cities is a powerful tool to combat energy consumption and environmental impact of urban transportation system, it is easy to implement and is scalable. A place like Mexico City allows to work in a 7 km^2 with an average of about 50,000 inhabitants and has an estimated annual CO_2e emission of 75,000 tons specifically from transportation.

So, if we can offer mobility systems that allows citizens to change their transportation mode to active transportation like bicycle, walking or human powered vehicle, every year would be stopped the emission of 1.5 tons of CO_2e per capita.

Figure 2 shows how much CO_2e tons per year would be avoided if 10% of fossil fueled vehicles users would change their transportation mode to bicycle, walking or human powered vehicles.

3.2 Active Multimodal Central Station

Once the area of intervention is defined, it is time to integrate active transportation to the installed transportation infrastructure to connect the new design to the daily citizen mobility. The first step is to select a station from the local transportation system that can become the active multimodal central of the EMhub. Different locations have to be analyzed, using the following impact details, to know the significance of every mode promoted in every possible station, and be able to choose the best option.

Walking

- Speed = 90 m/min.
- Distance traveled every 10 min = 900 m.
- Walkable area around the station = 2.54 km^2.
- Users that can walk to the station = 15,250.

Cycling

- Speed = 12.5 km/h.
- Distance traveled every 10 min = 2 km.
- Cycling area around the station = 12.5 km^2.
- Users that can cycle to the station = 75,000.

3.3 Active Transportation-Economy Network Stations

After selecting the central station, the next step is to integrate active transportation to daily citizen's life, creating stations that offer these mobility modes to the spaces where citizens are living, working, studying, etc. The need to create a system that takes people from and to the areas where commerce, production, business, culture and housing is happening is fundamental so the next step in the design of the network is to locate at least five key stations that covers every one of these categories:

1. Primary station: market, mall, food store, etc.
2. Secondary station: production centers, manufacture centers, etc.
3. Tertiary station: business centers, mall, etc.
4. Cultural station: sports center, school, hospital, etc.
5. Housing station: apartment buildings, housing area, etc.

The same impact details used in the central station can be used to analyze these secondary stations; however, it is better to develop the network stations covering half the impact of the main station, trying to reach the international rule of 400m radial coverage. So, the ideal impact analysis for every station of the network is these:

Walking

- Speed = 90 m/min.
- Distance traveled every 5 min = 450 m.
- Walkable area around the station = 0.65 km^2.
- Users that can walk to the station = 3,900.

Cycling

- Speed = 12.5 km/h.
- Distance traveled every 5 min = 1 km.
- Cycling area around the station = 3.2 km^2.
- Users that can cycle to the station = 19,200.

3.4 Fuel Economy Program

Finally, after visualizing the EMhub coverage, central active station, active stations network and impact areas, the final step is to add fuel economy strategies to the system. To downsize the power and capacity of our vehicles, to choose cars powered with new technologies and to leave the car and use the bus are easy consumption habits that can bring big impact to the reduction of greenhouse gas emissions. The next chart shows the emission reduction in Mexico City by choosing not to use a gas powered SUV and finding another simple option (Table 4).

It is important to work this strategy group with all the users that do not want to or cannot change their mobility to active mode, and the design of the EMhub can add the promotion of simple changes that do not affect their mobility, but reduce their emissions and fossil fuel consumption.

After analyzing the numbers, these are the strategies:

- To promote users to buy smaller vehicles.

When the owner of a SUV changes its vehicle to sedan, its environmental impact immediately reduces 30%.

- To promote users to choose hybrid or electric vehicles.

When the owner of a SUV powered on gasoline decides to change its vehicle to hybrid or electric sedan, its environmental impact immediately reduces 80%. When the owner of a gasoline powered sedan decides to change its vehicle to hybrid or electric power sedan, its environmental impact is reduced by 75%.

- To promote users to choose bus mode.

When the owner of a SUV powered on gasoline decides to leave its vehicle and travel in public bus transportation, its environmental impact immediately reduces 94%. When the owner of a gasoline powered sedan decides to change its vehicle to hybrid or electric power sedan, its environmental impact is reduced by 75%.

4 Conclusion

There is a big necessity to rethink mobility systems and moving toward more sustainable urban life in future cities. We need transportation designs that involve the configuration of very complex systems that look for the interaction of many factors (Morin, 1999). We need to stop designing transportation infrastructure and to start designing transportation systems, looking the vehicle and transport design in relation with the space, dwellers and mobility. It is understood that the production of new transportation systems takes time, usually, long-term periods, which we should plan to analyze the problems, and make decisions around people's quality of life, not around economy.

The history of human being has shown the close relationship between transport, people and land use on the definition of a settlement, so the smart use of energy and low environmental impact of transportation in future depends on the actions of today, but also in the action of reviewing the past to learn about the historical evolution of space, dwellers and transportation.

Table 4 Emissions per km traveled

Vehicle	Power	Emission per vehicle g CO_2/km	Average passengers	Emission per passenger g CO_2/km	Emission reduction changing SUV (%)
SUV	Gasoline	436	1.5	291	0
Sedan	Gasoline	309	1.5	206	29
Sedan	Hybrid	83	1.5	55	73
Sedan	Electric	83	1.5	55	73
Bus	Diesel	1080	40	27	87

EM-hubs is a tool that can create little scalable solutions to big environmental and energy transportation problems, it is compatible with active transportation, cycling projects, walkable cities programs, scootability, and human powered innovations.

Based on the study of the historical relationship between transportation and urban development, we could develop a design tool named EcoMobility Hub (EMhub). This tool helps to design new contemporary urban transportation systems in any urban settlement and helps to reduce the CO_2 emissions and fossil fuel energy consumption around the urban space.

If the implementation of EM-Hubs in Mexico City could avoid only 5% of automobile, SUV, motorcycle, truck and metrobus rides, the city could stop the emission of 82,000 Tons of CO_2e/year that means downsize the gasoline consumption of 245 million liters plus 65 million diesel liters.

Acknowledgements This works was supported by:CONACYT under Scholarship Grant number 40699PRODEP-SEP under Folio:UAM170. Universidad Autónoma Metropolitana, Azcapotzalco – Departamento de Medio Ambiente, Área de Factores del Medio ambiente Natural y Diseño – Project: Estudio taxonómico de vehículos ecológicos urbanos, under folio: N-357.

References

Bezold, C., Craig, B., Juech, C., Michelson, E., Peck, J., & Wilkins, K. (2009). *Foresight for smart globalization: Accelerating & enhancing pro-poor development opportunities*. Institute for alternative futures.

Blanchard, P., & Volchenkov, D. (2009). *Mathematical analysis of urban spatial networks*. Springer-Verlag.

Bujarin. (1982). La agricultura y la industria. La ciudad y el campo. *Investigación Económica, Vol. 41, No. 159/160, Teoría y política de la transformación social*, 423–474.

Culbertson, J. (1969). *Preparing educatinal leaders for the seventies*. University Council for Educational Administration.

De Salvador-Agra, S., & Martínez-Suárez, Y. (2015). Nomadism and intermittent ubiquity in 'off the grid' Shuar people. *Communication & Society, 28*(4), 87–105

Gabiña, J. (1998). *Prospectiva y ordenación del territorio: hacia un proyecto de futuro*. Marcombo.

Galanternick, M. (1999, August). Real Bargain. *Latin Trade, 7*(8).

Garreau, J. (1992). *Edge city, life on the new frontier*. Doubleday.

Gibson, C. (1964). *The Aztecs under Spanish Rule: A history of the Indians of the Valley of Mexico, 1519–1810*. Stanford University Press.

Global Business Network. (2010). The Mont Fleur Scenarios, What will South Africa Will in the year 2002? *Deeper News, 7*(1). Retrieved from Los Escenarios de Mont Fleur: https://www.gbn.org

Godet, M. (2012). *La Prospective*. Retrieved from https://www.laprospective.fr

Gómez, J., & Luján, F. (1990). Papeles de geografía, No. 16. In *Estudio de la red de carreteras en la región de Murcia a través de sus dimensiones básicas* (pp. 125–142).

Güller, G. (2002). *Del aeropuerto a la ciudad aeropuerto*. Gustavo Gili.

Hall, P. (2007). Urban land, housing, and transportation: the global challenge. *Global Urban Development Magazine, 3*(1). Retrieved 2011

Hernández-Suárez, C. (2011). Nueva política del agua y herencias centralizadoras: El consejo de cuenca del valle de México. *Agricultura, sociedad y desarrollo*, 303–327.

INEGI. (1910). *Tercer Censo de Población de los Estados Unidos Mexicanos 1910*. México: INEGI.

INEGI. (2000). *Indicadores de desarrollo Sustentable en México*. INEGI.

INEGI. (2010). *Censo de Población y Vivienda 2010*. INEGI.

Kates, R. W., & Parris, T. T. (2003, Julio 8). Long-Term Trends and a Sustainability Transition. *Proceedings of the National Academy of Sciences of the United States of America, 100*(14).

Labarga, F. (1981). *Fernando Labarga Tejada. Peatones: Circulación y comportamiento, conflictos con el tráfico rodado y sus soluciones*. Ag Grupo, SA.

Louf, R., & Barthelemy, M. (2013). Modeling the polycentric transition of cities. *Physical Review Letters, 111*(19).

Low, S. M., & Lawrence-Zuniga, D. (2003). *The anthropology of space and place: Locating culture*. Blackwell Pub.

Luna Golya, G. G. (2014). *Modeling the aztec agricultural waterscape of lake xochimilco: a gis analysis of lakebed chinampas and settlement*. ProQuest.

Martinotti, G. (1993). *Metropoli. La nuova morfologia sociale della città*. Il Mulino.

Miklos, T. (2007). *Planeación prospectiva, una estrategia para el diseño del futuro*. Limusa.

Molinero, A., & Sánchez, L. (2002). Diseño, Operación y Administración. Transporte Público: Planeación. Universidad Autónoma del Estado de México.

Morin, E. (1999). *Los siete saberes necesarios para la educación del futuro*. UNESCO.

Muñoz, F. (2010). *Urbanalización, paisajes comunes, lugares globales*. Gustavo Gili.

PNUD. (2014). *Índice de desarrollo humano municipal en México: nueva metodología*. Danda Diseño e Impresión.

Secretaría de Medio Ambiente del Distrito Federal. (2006). *Información Climatológica Ambiental del Valle de México*. México: Secretaría de Medio Ambiente del Distrito Federal. Retrieved from Sitio oficial de la Secretaría de Medio Ambiente del Distrito Federal: https://www.sedema.df.gob.mx

SEDESOL - CONAPO. (2012). *Catálogo. Sistema Urbano Nacional 2012*. SEDESOL - CONAPO.

SEMOVI. (2016). *Secretariía de Moviidad - CDMX*. Retrieved Noviembre 12, 2016, from https://data.semovi.cdmx.gob.mx/

Subsecretaría de desarrollo urbano y ordenación del territorio. (2000). *Manual de diseño geométrico de vialidades. Manual normativo – tomo IV*. Dirección general de ordenación del territorio.

Sudjic, D. (1991). *The 100 mile city*. Flamingo.

Unhabitat. (2004). *The State of the World's Cities 2004/2005: Globalization and Urban Culture*. Earthscan.

Unhabitat. (2010). *State of the World's Cities 2010/2011, Bridging The Urban Divide*. Unhabitat.

Veltz, P. (1999). *Mundialización, ciudades y territorios: la economía de archipiélago*. Ariel.

Wardrop, A. (1979). *In search of standars of service for uban public transport*. Department of capital territory.

Young, A., & Cresswell, R. (1982). *The urban transport future*. Construction Press.